Main-Group Elements

			13 IIIA	14 IVA	15 VA	16 VIA	17 VIIA	18 VIIIA
								2 He 4.002602
			5 B 10.811	6 C 12.011	7 N 14.00674	8 O 15.9994	9 F 18.9984032	10 Ne 20.1797
10	11 IB	12 IIB	13 Al 26.981539	14 Si 28.0855	15 P 30.973762	16 S 32.064	17 Cl 35.4527	18 Ar 39.948
28 Ni 58.6934	29 Cu 63.546	30 Zn 65.39	31 Ga 69.723	32 Ge 72.61	33 As 74.92159	34 Se 78.96	35 Br 79.904	36 Kr 83.80
46 Pd 106.42	47 Ag 107.8682	48 Cd 112.411	49 In 114.82	50 Sn 118.710	51 Sb 121.757	52 Te 127.60	53 I 126.90447	54 Xe 131.29
78 Pt 195.08	79 Au 196.96654	80 Hg 200.59	81 Tl 204.3833	82 Pb 207.2	83 Bi 208.98037	84 Po (209)	85 At (210)	86 Rn (222)

Inner-Transition Metals

63 Eu 151.965	64 Gd 157.25	65 Tb 158.92534	66 Dy 162.50	67 Ho 164.93032	68 Er 167.26	69 Tm 168.93421	70 Yb 173.04	71 Lu 174.967
95 Am (243)	96 Cm (247)	97 Bk (247)	98 Cf (251)	99 Es (252)	100 Fm (257)	101 Md (258)	102 No (259)	103 Lr (262)

Chemistry

Chemistry

JAMES P. BIRK ARIZONA STATE UNIVERSITY

HOUGHTON MIFFLIN COMPANY   **BOSTON**   **TORONTO**

GENEVA, ILLINOIS   PALO ALTO   PRINCETON, NEW JERSEY

Sponsoring Editor: Richard Stratton
Development Editor: June Goldstein
Project Editor: Susan Lee-Belhocine
Production/Design Coordinator: Jill Haber
Senior Manufacturing Coordinator: Priscilla Bailey
Marketing Manager: Michael Ginley

To the memory of my father, Albert Birk, who loved books and who would have enjoyed this one.

Warning: This book contains text descriptions of chemical reactions and photographs of experiments that are potentially dangerous and harmful if undertaken without proper supervision, equipment, and safety precautions. DO NOT attempt to perform these experiments relying solely on the information presented in this text.

Credits: A list of credits follows the index.

Cover photograph by Sean Brady.
Cover and interior design by Ron Kosciak, Dragonfly Design.

Values for atomic weights listed in the periodic table on the inside front cover of this book are from the IUPAC report "Atomic Weights of the Elements 1989," *Pure and Applied Chemistry,* Vol. 63, No. 7 (1991), pp. 975–1002 (© 1991 IUPAC).

IBM is a registered trademark of International Business Machines Corporation.
Apple and Macintosh are registered trademarks of Apple Computer, Inc.

Printed in the U.S.A.

Library of Congress Catalog Card Number: 93-78645
Student text ISBN: 0-395-51535-1
Instructor's Annotated Edition ISBN: 0-395-69231-8

123456789-VH-96 95 94 93

Contents in Brief

Contents

18. Electrochemistry   **767**

19. Coordination Chemistry   **810**

20. Nuclear Chemistry   **849**

24. Synthetic and Biological Polymers   **999**

Appendices

Worlds of Chemistry

Preface

The general chemistry course is no stranger to innovation and change. At times, the content of the course has emphasized descriptive chemistry, and at other times chemical principles have dominated. In recent years, instructors have become very concerned about both the content of the general chemistry course and the methods used to teach this content. Increasing attention has been devoted to things that will enhance interest in the course and help students learn and retain more.

This particular project arose out of my sense that there is a need in the general chemistry marketplace for a book that is more student-oriented. I have focused on presenting material in a way that will help students learn and that conveys some of the fascination of chemistry that was transmitted to me by my teachers and by my experiences in the laboratory over the years. Simply put, the goal of this book is to get students *involved* in chemistry.

Length and Flexibility

This text is intended for students majoring in science or engineering. It provides the core material that is usually covered in a two-semester course in general chemistry, as well as additional material that will allow individual instructors a variety of options. There has been a great deal of concern in recent years about the large amount of material that constitutes a course in general chemistry. This book has been designed in a way that addresses this concern. Most chapters contain sections, marked with a box around the section number, that emphasize applications of chemistry. These sections can be omitted, if desired, with little impact on the coverage of topics in later chapters. Because these sections do serve a purpose in the learning of chemistry, however, most instructors will want to include at least some of them. This flexibility gives the instructor a range of choices in course design. In addition, chapters at the end of the book can be omitted without sacrificing comprehensive coverage of general chemistry.

Integration of Descriptive Chemistry

This book integrates descriptive chemistry and the principles of chemistry. Such integration more closely resembles the manner in which we make advances in our knowledge of chemistry. In addition, there is considerable experimental evidence suggesting that students will learn chemistry with a more thorough understanding if it is taught the same way chemical research is done, which, after all, is the way professional chemists learn about new aspects of the behavior of matter. In particular, students seem to understand and retain material more easily if it starts with an investigation of matter that reveals some factual material, then develops models or principles that explain these observations, and finally applies these principles to some new area of chemistry. Of course there is a more pragmatic reason for integrating descriptive chemistry: When this material is presented in large blocks, it is both difficult to teach and difficult to learn. Students can quickly become bored with the fascinating properties of matter if these details are presented in large doses. Thus, most of the descriptive chemistry is presented in bite-sized amounts spread liberally throughout the chapters.

In this book, descriptive chemistry forms a foundation for the development of chemical principles. Concepts are introduced only after a description of an experiment or observation that provides a real situation that the concept can explain. For instance, a prologue called **Encountering Chemistry** opens each chapter, offering a detailed look at a particular chemical phenomenon. The phenomenon is revisited as concepts are developed in the chapter. These points are marked by arrowheads (▶) so that the student can see the connection between the chapter's content and the opening vignette.

Descriptive chemistry is integrated into the chapters in several additional ways:

Marginal notes appear throughout the text to describe the chemicals used in discussions and equations. By this means, students begin to appreciate that they are dealing with real substances.

Color photographs of chemicals and of reactions involving those chemicals appear throughout the book. These photographs give some flavor of the experimental basis of chemistry and show the actual appearance of substances.

Short descriptions of chemicals are sometimes provided in worked examples, practice problems, and end-of-chapter exercises. Again, these descriptions relate applications of concepts to real materials.

Rules of reactivity appear in the margins and show how the patterns of chemical reactions pervade our development of chemistry.

Historical notes, also in the margins set in italics, provide some of the experimental basis for concepts and give insight into some of the people involved in the development of chemistry.

Problem Solving

Students need to use chemical concepts and principles to enhance their understanding in a variety of circumstances. For mathematical models, applications must involve practicing problem-solving methods. Numerous **worked examples** are provided to help students master problem solving using the various concepts. Each example is followed by a practice problem (with the answer provided) for students to work on their own. **End-of-chapter exercises** provide further opportunities to practice problem-solving techniques. **Classified exercises** are divided by key topics, indicating the concept or material being tested. **Additional exercises** provide no such key, thereby challenging the student to apply facts and principles from anywhere in the chapter. More difficult problems are marked with an asterisk (*), and most even-numbered problems (printed in blue) are answered at the back of the book. Starting with Chapter 4, each end-of-chapter exercise set concludes with an activity called **Chemistry in Practice.** Designed to require the application of concepts from several chapters, these problems frequently involve real-world situations. They require the student to select the appropriate data and work through several sets of calculations. Such problems help students gain a sense of connectivity among the myriad topics that are covered and truly test students' understanding of the course material. Chemistry in Practice provides an ideal opportunity for cooperative learning and group problem-solving.

Applications

We provide applications of chemical principles and properties throughout the book. These applications are drawn from a wide range of areas, including industry, the environment, and health and medicine. Each chapter contains a special essay entitled **Worlds of Chemistry,** which applies material in that chapter to some interesting phenomenon that affects our lives. To enhance the cooperative-learning benefits of group discussions, this feature also includes discussion questions about the impact of chemistry on society. **Sections in most chapters (indicated by a blue box around the section number) also apply concepts in that chapter to some area of chemistry, including societal issues and industrial applications.**

Scientific Discovery

Part of the fascination of chemistry comes not only from what is learned but also from the process of discovering how chemists think and operate. Although there is no substitute for doing chemistry, whether performing experiments in the laboratory or working out problems in discussions or on paper, there are ways in which a textbook can actively involve students in thinking and learning about chemistry.

Critical thinking questions, which appear in blue in the margin, guide students through the development of chemical principles and challenge them to think about the implications of applications of these principles. Some questions help students anticipate the material that will come next. This encourages them to ask these questions themselves, thereby becoming actively involved in the learning process. Other questions prompt the reader to think about the influence of chemistry on the world around us. Such questions are ideally suited for group discussion, a form of cooperative learning that has been shown to be an effective way to learn course material.

Chemistry in Practice, the final problem at the end of Chapters 4 through 24, presents a wide range of information that the student (individually or as part of a cooperative-learning group) must use to define the problem and devise an approach to solving it. As such, Chemistry in Practice closely simulates the situations that are encountered in research labs or in other fields of study.

Decision trees, which appear as numbered figures at appropriate points in the text, provide a visual model for the thought processes a chemist goes through in approaching a certain type of problem. The student not only learns a method for solving similar problems but also begins to cultivate an approach to problem solving that can be used when encountering new and unfamiliar questions.

Discussion questions that accompany the material in the boxed essays, *Worlds of Chemistry,* invite students to think critically about the material they have just read and apply what they have learned to situations they encounter in other courses and in everyday life. By interacting with the material, students are encouraged to see that chemistry is a dynamic field with applications to virtually every aspect of today's world.

Organization

The book is organized in a traditional manner. The first five chapters introduce students to matter and energy in a largely macroscopic manner, covering the particulate nature of matter, chemical formulas, chemical reactions, stoichiometry, and thermochemistry. This material provides the background for an extensive laboratory program. Because some instructors prefer to cover thermodynamics at only one point, Chapter 5 (Thermochemistry) can be delayed and coupled with Chapter 12 (Chemical Thermodynamics). A few modifications would be necessary in the intermediate chapters. In particular, the calculation of enthalpies of reaction from average bond energies in Chapter 8 would also have to be postponed.

Chapters 6 through 8 explore electronic structure and bonding, beginning a microscopic view of matter. The next three chapters examine the structure and properties of matter in the three physical states and in solutions. The first of these three chapters, Chapter 9, covers the properties of gases. With the exception of the section on nonideal gases, this chapter could easily be covered earlier in the course, as some instructors prefer.

Chapters 12 and 13 examine the reactivity of matter from the viewpoints of thermodynamics and kinetics, respectively, and the next three chapters cover various aspects of reactions in a state of equilibrium. Redox reactions (Chapter 17) and electrochemistry (Chapter 18) follow. The sequence of chapters on thermodynamics, kinetics, and equilibrium varies considerably in chemistry curricula. The order in this book was carefully chosen to lay a foundation for understanding the dynamic nature and thermodynamic basis of chemical equilibria. Other sequences can accomplish the same goal, and some variation in this order is possible if the prerequisites for topics are satisfied in other ways. These chapters are followed by an introduction to coordination chemistry and nuclear chemistry.

The final four chapters present material that can be considered optional. The first two discuss the chemistry of the main-group elements (Chapter 21) and the transition elements (Chapter 22) from the viewpoint of periodic trends and applications of the principles presented in earlier chapters. A considerable amount of such material was covered in the first eighteen chapters, but these chapters provide an overview of descriptive inorganic chemistry. The last two chapters present a brief introduction to organic chemistry, polymers, and biochemistry.

Ancillaries

A modern course in general chemistry involves many activities outside the purview of the textbook. We have provided a complete package of ancillary materials to assist in the learning and teaching processes.

For the Student

Study Guide, Donald M. Baird, Florida Atlantic University. The study guide provides a review of concepts and key equations, worked examples that build to more complicated examples, and a chapter posttest with answers.

Student's Solutions Manual, James P. Birk, Arizona State University. This manual provides complete worked-out solutions to even-numbered end-of-chapter exercises.

Investigations in Chemistry, James P. Birk and Martha J. Kurtz, Arizona State University. This laboratory manual contains 41 experiments that follow the order of material in the textbook.

For the Instructor

Instructor's Annotated Edition, James P. Birk, Arizona State University, with contributions by Wendy Lou Elcesser, Indiana University of Pennsylvania. The instructor's edition contains annotations concerning applications, the videotapes, the videodisc, the transparencies, transparency masters, commercial software, public domain software on Internet, sources for lecture demonstrations, test bank questions, lab manual experiments that correspond to the chapter, and comments on the chapter.

Instructor's Resource Manual, James P. Birk, Arizona State University. This manual contains curriculum notes on each chapter, suggestions for developing thinking skills, a list of reactivity rules, sources for applications of chemistry, a list of the transparencies derived from the textbook, transparency masters, a list of videotape and videodisc demonstrations that can be used in particular sections, a list of references to demonstrations suitable for specific sections, suggestions for correlations between the textbook and the lab manual, a list of software that can be used with each section of the textbook, and instructions on how to use Internet to obtain public domain software.

Instructor's Resource Manual to accompany **Investigations in Chemistry,** Martha J. Kurtz, Arizona State University. This manual lists correlations of each experiment with appropriate material in the textbook and provides notes on each experiment, including sample data, the time required to carry out each experiment, lists of chemicals and equipment for each experiment, a list of all the chemicals and equipment required in the manual, and instructions for the preparation of all solutions and other materials.

Test Bank, Stan C. Grenda, University of Nevada, Las Vegas. The printed test bank contains approximately 2000 multiple-choice questions. Corresponding text sections and answers are provided for each question. The test items are also available on disk for the IBM PC and the Macintosh. A call-in test service is also available, allowing you to order printed tests by calling Houghton Mifflin's toll-free number.

Solutions Manual, James P. Birk, Arizona State University. The solutions manual contains worked solutions to all within-chapter practice problems and end-of-chapter exercises.

Transparencies. 150 full-color transparencies of figures and tables from the text are included.

Video Lecture Demonstrations. Series A: John Luoma, Cleveland State University; Series B: John J. Fortman and Rubin Batino, Wright State University; and Series C: Patricia L. Samuel, Boston University. More than 80 demonstrations are provided.

Videodisc. The videodisc provides 34 lecture demonstrations plus illustrations and photographs from the text. For more information, contact Houghton Mifflin Company.

Electronic Lecture Manager. This software for Windows™ allows control of the videodisc, line art images, and screens imported from word processors to create customized classroom presentations.

Acknowledgments

Although the author may get credit for writing the book, no project of this magnitude can be completed without help. First and foremost, there is my wife, Kay Gunter, and our daughter, Kara Birk, who continuously provided support by giving that most precious commodity, time, and who encouraged me when one more rewrite seemed beyond reason.

Doug Sawyer, assisted by Bill Sanborn, very ably organized and carried out most of the chemistry that appears in the photographs. Sean Brady and his photographic crew quickly learned the difference between photographing vegetables and photographing chemical reactions in progress.

Colleagues at many colleges and universities helped to guide me through the process of developing the content of this book. Dorothy B. Kurland from the West Virginia Institute of Technology and John M. Goodenow of Lawrence Technological University read the entire manuscript and galleys, respectively, and worked all examples and practice problems to check for accuracy. Charles Marzzacco of Rhode Island College did the same for the page proofs. In addition, John checked all of the answers in the back of the book. Of course, any errors remain my responsibility. The following chemists reviewed all or part of the manuscript in various states of development. I am very grateful to them for their part in improving the manuscript.

John J. Alexander, University of Cincinnati

Anneke S. Allen, Wichita State University

Margaret R. Asirvatham, University of Colorado—Boulder

Peter Baine, California State University, Long Beach

John E. Bauman, University of Missouri—Columbia

Robert H. Bell, University of Richmond

Jesse S. Binford, Jr., University of South Florida

Muriel B. Bishop, Clemson University

William R. Brennen, University of Pennsylvania

Albert W. Burgstahler, University of Kansas

Elliott J. Burrell, Loyola University of Chicago

Kevin D. Cadogan, California State University, Hayward

George Canty, Jr., Fort Valley State College

Loren Carter, Boise State University

James Coke, University of North Carolina at Chapel Hill

Raymond R. Crawford, San Jacinto College—South Campus

D. Scott Davis, Mercer University

James E. Davis, Harvard University

Michael I. Davis, University of Texas at El Paso

Walter Drost-Hansen, University of Miami

Harry A. Eick, Michigan State University

John J. Farrell, Franklin & Marshall College

R. Max Ferguson, Eastern Connecticut State University

John I. Gelder, Oklahoma State University

Donald W. Goebel, Jr., Palm Beach Community College, South Campus

L. Peter Gold, Pennsylvania State University—University Park

Joel M. Goldberg, University of Vermont

Stan C. Grenda, University of Nevada—Las Vegas

David A. Grossie, Wright State University

Harry G. Hajian, Community College of Rhode Island

Michael D. Hampton, University of Central Florida
Colin D. Hubbard, University of New Hampshire
Jerry P. Jasinski, Keene State College
Raymond Kellman, San Jose State University
Richard L. Kiefer, College of William and Mary
Howard C. Knachel, The University of Dayton
Richard W. Kopp, East Tennessee State University
Larry K. Krannich, University of Alabama at Birmingham
Dorothy B. Kurland, West Virginia Institute of Technology
Charles S. Lochner, Pima Community College
James W. Long, University of Oregon
Judith E. Lund, University of Wisconsin—Eau Claire
Vahe M. Marganian, Bridgewater State College
Patricia Milliken Wilde, Triton College
Thomas E. Mines, St. Louis Community College at Florissant Valley
D. J. Morrissey, Michigan State University
David Moseley, Jr., Washington State University
David F. Nachman, Phoenix College
Ronald E. Noftle, Wake Forest University

John A. Paperelli, San Antonio College
Albert L. Payton, Broward Community College
Cortlandt G. Pierpont, University of Colorado—Boulder
Vaughan Pultz, Northeast Missouri State University
David A. Rowley, George Washington University
Dennis J. Sardella, Boston College
Jerry L. Sarquis, Miami University
Barbara A. Sawrey, University of California, San Diego
Paul Schueler, Raritan Valley Community College
Maurice E. Schwartz, University of Notre Dame
Joel I. Shelton, Bakersfield College
Karen A. Singmaster, San Jose State University
Tamar Y. Susskind, Oakland Community College
Wayne Tikkanen, California State University, Los Angeles
Donald D. Titus, Temple University
Frank J. Torre, Springfield College, MA
Vivian Torres, University of Puerto Rico at Mayaguez
William J. Zajdel, University of Redlands
Orville Ziebarth, Mankato State University

Finally, I would be remiss if I did not acknowledge those individuals who most influenced my own education during my journey to becoming a chemist and professor. I am particularly grateful to Sr. Phyllis Plantenberg, now at St. John's University; Dr. Mark Hughes of St. John's University; Dr. James H. Espenson of Iowa State University; and Dr. Jack Halpern of the University of Chicago. Each of these individuals was highly involved in the various stages of my science education, and their influence can be found in this book.

I would welcome any comments about the text and its package. You can contact me at Arizona State University.

James P. Birk
Department of Chemistry
Arizona State University
Tempe, AZ 85287-1604
Internet: birk@asuchm.la.asu.edu

Features

The following pages highlight the features in CHEMISTRY. Pages xvi–xvii of the preface provide more detail on these features and the rest of the text.

Encountering Chemistry

Suppose we want to clean the tarnish from a batch of pennies in a coin collection. We could rub them individually with a polishing agent, but that would be time-consuming and tedious. Another possibility is to soak the pennies in a liquid corrosive enough to remove the tarnish but not so corrosive that the coins would be damaged. Acids are often used to remove tarnish, but there are many different acids, varying in corrosive power. Some acids available to us are vinegar, which contains acetic acid; hydrochloric acid, sold for use in swimming pools; and nitric acid, which is quite powerful. Could we use one of these? Some experiments should tell us if any of them is suitable.

Say we first drop a coin into some acetic acid, the least corrosive of the three acids. Even after an hour, there is no apparent change in the degree of tarnish. This experiment provides no evidence of a chemical change, because no new substances seem to have been produced and no existing substances destroyed.

Next we try the most corrosive of these acids, nitric acid. This experiment produces rather dramatic results. The coin begins to move about as if it were being buffeted by some invisible force. The liquid becomes a greenish-blue color, and a cloud of disagreeable-smelling reddish-brown gas arises from the liquid (Figure 1.1). After getting rid of this obnoxious mixture, we examine the coin again. It is no longer tarnished, but it is also no longer shiny. The surface has become pitted

3

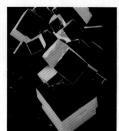

Figure 1.2 Not all matter has the same structure. *Left:* Some solids, such as table salt (sodium chloride), have a highly regular geometric shape; such solids are called crystals. *Right:* Other solids, such as the strands of cotton, have a less regular shape.

composition of that substance is called a **physical change.** For example, you can change water to steam by heating it. If you cool the steam, it changes back to liquid water. This change from a liquid to a gas (or the reverse) is a physical change because both forms involve the same chemical substance, water.

▶ The changes undergone by the pennies described in the chapter opening are not physical changes, however. The tarnishing of copper on long-term exposure to air and the attack on copper by nitric acid to release a reddish-brown gas (nitrogen dioxide) and a greenish-blue substance (copper nitrate) are chemical changes. A **chemical change** converts one substance into one or more other substances that not only have different physical properties but also have different chemical compositions. The **chemical properties** of a substance, then, describe its chemical composition and the changes it can undergo. For example, when electricity is passed through liquid water containing a small amount of sulfuric acid, two gases are given off: hydrogen and oxygen. This process, shown in Figure 1.3, is a chemical

Figure 1.3 The decomposition of liquid water by electricity produces the new chemical substances oxygen and hydrogen in the form of gases. This process involves a chemical change because the new substances (the products) do not have the same composition as the initial substance (the reactant).

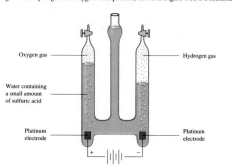

Oxygen gas

Hydrogen gas

Water containing a small amount of sulfuric acid

Platinum electrode

Platinum electrode

Encountering Chemistry

Each chapter-opening vignette, called Encountering Chemistry, introduces an example that is then referred to throughout the chapter. These points are marked by arrowheads. Students see chemical principles as closely linked to real-world objects and phenomena.

Worlds of Chemistry

A boxed essay, called Worlds of Chemistry, appears in every chapter. These essays show students how chemistry is part of our everyday life, including its role in solving a variety of problems from pollution to cancer.

▶ ▶ ▶   **WORLDS OF CHEMISTRY**

Hydrogen: The Fuel of the Future?

The reaction product, water, is not a pollutant, so hydrogen would provide an exceptionally clean fuel.

The price of oil keeps going up. Estimates of the amount of recoverable oil keep going down. Where will we get our fuel in the future? Some scientists think we can get fuel from water in the form of hydrogen, perhaps using solar energy to decompose the water into its component elements.

Hydrogen exists as a gas at room temperature, but it can be converted to a liquid at low temperature and high pressure. Currently, most of the hydrogen that is produced is used to make ammonia and petroleum products. As either a gas or a liquid, however, hydrogen can be used as a fuel. When hydrogen combines with oxygen, it liberates large quantities of heat, which can be used to power vehicles. The space shuttles, for example, use liquid hydrogen and liquid oxygen to fuel the rockets that raise them from the launch pad to the upper atmosphere (Figure 3.22). Electrical power for the space shuttle and other satellites is also provided by the reaction of hydrogen and oxygen in a device called a fuel cell. The reaction product, water, can be collected from the fuel cells for the astronauts to drink.

Closer to earth, hydrogen was once used in airships called dirigibles, not as a fuel, but to provide lift. Hydrogen is highly flammable, reacting explosively with oxygen under some conditions. The disastrous explosions of the airship *Hindenburg* and the space shuttle *Challenger* were due to the explosive combination of hydrogen and oxygen.

Hydrogen is being investigated as a fuel for automobiles and aircraft. Engines can be modified to burn hydrogen. The reaction product, water, is not a pollutant, so hydrogen would provide an exceptionally clean fuel. However, the flammability and explosiveness of hydrogen is a drawback. Hydrogen will have to be stored in some form other than the gaseous or liquid state if it is to be used as a fuel in vehicles operating in populated areas.

One possible storage method is the formation of metal hydrides. Many of the metal hydrides, such as titanium hydride (TiH_3), are readily formed by mixing the metal with hydrogen gas. The hydrogen can be released from the solid metal hydride by adding water or by heating. For example, magnesium hydride, MgH_2, reacts with water to form magnesium hydroxide, $Mg(OH)_2$, and hydrogen gas. Thus, pellets of magnesium hydride could be used as a

Figure 3.22 The space shuttles are launched by liquid-fuel rockets that contain liquid hydrogen and liquid oxygen. This is the space shuttle Columbia.

source of hydrogen gas, which could then be fed to an engine.

Unfortunately, the amount of hydrogen that can be stored by metal hydrides such as this is rather limited. The mass of the metal hydride needed to operate an automobile would be prohibitive. However, not all metal hydrides are simple compounds with a small, whole-number ratio of hydrogen to metal. Some metals, such as platinum and palladium, absorb large quantities of hydrogen gas to form substances called *interstitial hydrides*. Hydrogen molecules apparently fit into small holes in the metal structure. Gentle heating causes the hydrogen gas to be expelled from the metal. These interstitial hydrides thus present considerable potential for transporting hydrogen in a safe manner, should hydrogen ever become a fuel for automobiles.

Questions for Discussion

1. Make a list of possible alternative fuels for the future and discuss their advantages and disadvantages.
2. Would you feel comfortable with a tank of hydrogen in your automobile? Compare your response with how you feel about gasoline, which is also very flammable and explosive under some conditions.
3. If hydrogen is to be used as a fuel, we must get it from some plentiful source, such as water. Where will we get the energy to release the hydrogen from water?

Does the compound contain a metal or an ammonium ion? — **No** → Covalent or Acid

↓ **Yes**

Ionic Compound

Covalent or Acid → Does it contain hydrogen? — **No** → Covalent
Example: carbon disulfide

Does it contain more than one nonmetal? — **No** → Binary Ionic
Example: sodium bromide

↓ **Yes**

Ternary Ionic

Does it contain hydrogen? ↓ **Yes** → Binary Acid or Oxoacid or Covalent

Does it contain oxygen? — **No** → Miscellaneous Ternary Ionic
Example: ammonium cyanide

↓ **Yes**

Ionic Oxoanion

Example: sodium chlorate

Binary Acid or Oxoacid or Covalent → Does it contain oxygen? — **No** → Binary Acid or Covalent

↓ **Yes**

Oxoacid
Example: perchloric acid

Binary Acid or Covalent → Is it aqueous? — **No** → Covalent
Example: hydrogen chloride

↓ **Yes**

Binary Acid
Example: hydrochloric acid

Figure 3.26 A nomenclature decision tree can be used to determine which rules to apply in naming inorganic compounds.

3.6   Some of Our More Important Chemicals

Many consumer products are made up of a number of chemicals. So once you have mastered the naming of inorganic compounds, you are in a position to read the labels of consumer products more intelligently (though you will not be able to interpret the names of all ingredients, because many of them are organic compounds and others are given common names instead of systematic names). For example, read the labels on various brands of toothpaste. You should be able to interpret a number of ingredient names: sodium fluoride (NaF), silicon dioxide (SiO_2, sometimes called silica), and calcium monohydrogen phosphate ($CaHPO_4$). Fluoride compounds are present in toothpaste to prevent dental cavities, and the

Why do you think all these ingredients are found in toothpaste?

Decision trees

We provide figures we call decision trees, a visual model for the thought processes involved in working through problems in chemistry.

Photographs

Photographs and captions work together to illustrate a point in the text. All figures are tied to the text discussion.

(Figure 6.42). Reactivity with oxygen increases as we go down the group. Cesium, for example, reacts so vigorously with oxygen that it ignites spontaneously. Reactions of the elements with oxygen were discussed in Section 4.3.

The alkali metals all react with water to form hydroxides, as illustrated here for rubidium:

$$2Rb(s) + 2H_2O(l) \longrightarrow 2RbOH(aq) + H_2(g)$$

This reaction also becomes more vigorous as we go down the group. Already at potassium, the reaction with water occurs with considerable violence, resulting in the expulsion of glowing bits of metal (Figure 6.43).

Alkaline-Earth Metals

The alkaline-earth metals, Group IIA (2), all have a valence electron configuration of ns^2. The loss of two electrons gives a noble-gas configuration, so these metals form cations with a 2+ charge. These metals tend to be reactive, but not nearly so reactive as the alkali metals. They react with oxygen in air, for example, but when they do, they form an oxide coating that protects them from further reaction. Thus, the alkaline-earth metals do not have to be stored under kerosene to protect them from oxidation (Figure 6.44). The heavier alkaline-earth metals also react with water to form hydroxides, as shown here for calcium:

$$Ca(s) + 2H_2O(l) \longrightarrow Ca(OH)_2(aq) + H_2(g)$$

Again, however, the reactions are not so vigorous as the corresponding reactions of the alkali metals. Compare the reaction of calcium and water (Figure 6.45) with that of potassium and water (Figure 6.43), for instance. As with the alkali metals, the reactivity of the alkaline-earth metals increases as we go down the group.

Transition Metals

The properties of metals in the representative elements change significantly from one group to the next in accord with differences in electronic configuration, ionization energy, electron affinity, and size. In contrast, the properties of the metals in

Figure 6.42 Sodium metal, as well as the other alkali metals, must be stored under a hydrocarbon such as kerosene to prevent reaction with the oxygen in air. This reaction can be quite vigorous for the heavier alkali metals. The metals will be entirely oxidized if left in air long enough.

The alkaline-earth metals are hard, silvery white metals that are not as reactive as the alkali metals. They are used to form alloys with other metals. They have a variety of uses when converted into compounds with nonmetals. Magnesium is the most widely used of these metals; it is primarily alloyed with aluminum to give a strong, lightweight structural metal used in aircraft.

Figure 6.43 Potassium metal reacts explosively with water to form potassium hydroxide.

15.6 I

15.6 Industrial Acids and Bases

▶ The chapter introduction mentioned a numbe[r] as household chemicals. Acids and bases are industrial chemicals. Among the sixteen chem[icals] in excess of 10 billion pounds per year are se[veral] lime (CaO, the basic anhydride of calcium hyd[rox]ide, phosphoric acid, nitric acid, and sodium c[arbonate] produced in high volume, though not quite at t[he same rate.] useful themselves—for example, as cleaning pr[oducts, such as toilet bowl cleaners] (HCl), glass cleaners (NH_3), drain cleaners (NaOH), and rust removers (H_3PO_4)— and are also involved in the production of many other chemicals. This section examines the chemical properties of some important acids and bases.

Why are acids and bases produced in such large quantities?

Sulfuric Acid

Sulfuric acid is manufactured primarily by the contact process, which is described in Sections 4.6 and 14.5. Because it is inexpensive, sulfuric acid is often used when an acid is needed in a process. About two-thirds of the sulfuric acid produced is used to make phosphoric acid, which in turn is used in the production of fertilizers. Sulfuric acid is also important in the production of other fertilizer chemicals.

In concentrated solutions, especially when hot, sulfuric acid is a good oxidizing agent. As such, it can oxidize many nonmetals, such as bromide ion, shown in Figure 15.19.

$$2Br^- + H_2SO_4 \longrightarrow Br_2 + H_2O + SO_3^{2-}$$

Concentrated sulfuric acid has a very strong affinity for water, so it is used as a dehydrating agent. Sulfuric acid is so good at dehydrating that it can remove water from carbohydrates, which do not contain water, but do contain hydrogen and oxygen in a 2:1 ratio. As shown in Figure 15.20, sulfuric acid can convert sugar, $C_{12}H_{22}O_{11}$, to carbon. The dehydrating power of sulfuric acid is used to

Figure 15.19 Hot, concentrated sulfuric acid is a good oxidizing agent. It will oxidize a bromide salt (*Left*) to produce elemental bromine (*Right*).

Sections devoted to descriptive material

Most chapters contain numbered sections that are devoted to descriptive material. These are indicated by a blue box around the section number.

Marginal notes

The marginal notes provide additional descriptive material and historical notes (set in italics), rules of reactivity, and critical thinking questions (in blue).

Relative Lattice Energies   Because the lattice energy provides a major contribution to the strength of an ionic bond, the bond strength should be sensitive to the same factors as lattice energy. Lattice energy is inversely proportional to the size of the ions, and a similar effect is seen in the ionic bond strength. For example, the lattice energy of NaCl is -786 kJ/mol, whereas that of the smaller NaF is -939 kJ/mol. The heat of formation, a measure of the bond strength, is -412 kJ/mol for NaCl and -585 kJ/mol for NaF. Thus, the smaller the ions, the stronger the ionic bond.

We can also examine the effect of ionic charge on bond strength. To isolate charge effects from size effects, we must compare compounds that have ions of about the same size, such as oxide and fluoride ions. The lattice energy for NaF(s) (-939 kJ/mol) is much smaller than the lattice energy for $Na_2O(s)$ (-2600 kJ/mol). Even though it requires more energy to form O^{2-} than F^-, as a comparison of the electron affinities shows, the much greater lattice energy of $Na_2O(s)$ makes it far more stable than NaF(s). Similar calculations with other substances lead to the conclusion that the greater the charges of the ions, the stronger the ionic bond in the solid.

7.3   Covalent Bonds

In contrast to ionic bonds, whose electrostatic forces are spread out over the crystal, covalent bonds involve a much more localized attractive force between atoms, with much weaker forces between discrete molecules. ▶ This contrast is reflected in the different properties of ionic $TiCl_3$ and more covalent $TiCl_4$ described in the chapter introduction. The atoms are held together tightly in covalent molecules, but the molecules usually are not bonded together. As a result, they are often gases or liquids. As solids, they are brittle or soft and have low melting and boiling points and heats of fusion and vaporization because the molecules are not held strongly together. Covalent bonding is typical in the molecular species formed by the nonmetallic elements, such as O_2 and S_8. Some common compounds, including H_2O and CO_2, are also formed as molecules through covalent bonding. In most such compounds, both elements are nonmetals.

The strength of a covalent bond arises because each shared electron interacts simultaneously with two nuclei rather than the one nucleus present in the atomic form. There are repulsions between electrons and between nuclei, but there are also attractions between electrons and nuclei. When the electrons are located between the two nuclei, the attractions are maximized and the repulsions are minimized, resulting in a covalent bond. Depending on the number of electrons shared between the two atoms, the bond is classified as being a single, double, or triple covalent bond.

Single Covalent Bonds

A covalent bond consists of a pair of electrons shared by two atoms. When there is just one such bond between atoms, it is called a **single bond.** Each atom generally has a half-filled orbital in the valence shell, and these orbitals overlap to allow the electron pair to be located around both atoms. For example, in the formation of the

Why are ionic substances usually solids, whereas covalent substances are usually gases or liquids?

The nonmetallic elements that exist as small molecules include H_2, N_2, O_2, F_2, Cl_2, Br_2, I_2, P_4, and S_8. Some common small molecular compounds are H_2O, CO_2, CO, CH_4, CCl_4, NH_3, N_2O_4, P_4O_{10}, SO_2, and SO_3.

How do atoms share electrons to form a bond?

9.9   Chemistry of

When most ammonium salts are heated, they produce ammonia gas. However, when heated carefully, ammonium nitrate produces nitrous oxide, or laughing gas. Caution is required because this salt can explode violently when heated.

Ammonia may also be converted to other ... For example, it may be oxidized at about 500°...

$$4NH_3(g) + 5O_2(g) \longrightarrow 4N...$$
$$2NO(g) + O_2(g) \longrightarrow 2N...$$

catalyst (Figure 9.24). The nitrogen dioxide f...

$$3NO_2(g) + H_2O(l) \longrightarrow 2H...$$

The HNO_3 can be dissolved in water to form ... ammonia to form ammonium nitrate, used in ...

$$NH_3(g) + HNO_3(l) \longrightarrow NH_4NO_3(s)$$

About 20% of the ammonia in the atmosphere comes from industrial processes; the remainder comes from natural sources, especially the hydrolysis of urea from animal urine. Ammonia is colorless, with a characteristic sharp, choking odor. It is very soluble in water (about 700 volumes at STP dissolve in 1 volume of water) and is widely used in cleaning solutions.

Ammonium nitrate is an odorless, transparent, crystalline solid. It is used to make laughing gas, explosives, fireworks, matches, and fertilizers.

Nitrogen Oxides   Other important nitrogen species include the gaseous nitrogen oxides. Dinitrogen oxide, N_2O, is a naturally occurring, relatively unreactive component of the atmosphere. It is also called nitrous oxide. N_2O is formed, along with N_2, by degradation of proteins by soil microorganisms. It can be obtained by the decomposition (at about 250°C) of ammonium nitrate or nitramide:

$$NH_4NO_3(s) \xrightarrow{\Delta} N_2O(g) + 2H_2O(g)$$
$$NH_2NO_2(s) \xrightarrow{\Delta} N_2O(g) + H_2O(g)$$

Mixed with oxygen to prevent asphyxiation, it is used as an anesthetic (laughing gas).

Figure 9.24 A platinum–rhodium gauze catalyst is used to oxidize ammonia in the preparation of nitric acid.

Worked-out examples

The worked-out examples show students the step-by-step solution to a given problem. Each worked example is followed by a practice problem.

For example, when pure water is placed on one side of a semipermeable membrane and a sugar solution on the other side, the volume of the sugar solution increases and the solution becomes more dilute, as shown in Figure 11.20. Pressure would have to be exerted on the solution to prevent osmosis—that is, to prevent water from passing over to the solution side of the membrane. This pressure, called the **osmotic pressure,** is related to the height of solution that will rise up a column separated from pure solvent by a semipermeable membrane (Figure 11.20). Osmotic pressure can be calculated from an equation very similar to the ideal gas law:

$$\pi V = nRT$$

Because the number of moles (n) divided by the volume (V) is the molarity (M) of the solution, this equation can be changed to a more useful form:

$$\pi = MRT$$

Here π is the osmotic pressure (in units of atm), M is the molarity of *particles* in the solution, R is the gas constant (0.08206 L atm/mol K), and T is the absolute temperature (in units of K). Osmotic pressure depends in part on the number of solute particles in the solution, so it is a colligative property. It can be used in the same way as other colligative properties to measure molar masses and to distinguish between electrolytes and nonelectrolytes.

EXAMPLE 11.12

An aqueous solution commonly known as D5W is used for intravenous injection. D5W has the same osmotic pressure as blood fluid. It contains 54.3 g of glucose, $C_6H_{12}O_6$, per liter. The molar mass of glucose is 1.80×10^2 g/mol. Calculate the osmotic pressure in blood cells at 37°C, normal body temperature.

Solution We start with the equation for osmotic pressure:

$$\pi = MRT$$

where M is the molarity of particles in the solution. Because glucose is a covalent compound, we can assume that it is a nonelectrolyte and that the molarity of solute particles is the same as the molarity of glucose.

$$M = \text{moles of solute/liters of solution}$$

$$= \frac{54.3 \text{ g}}{1 \text{ L}} \times \frac{1 \text{ mol}}{180 \text{ g}} = 0.302 \text{ mol/L}$$

The temperature in units of K is

$$T = 37°C + 273° = 310 \text{ K}$$

We can substitute these values into the equation for osmotic pressure.

$$\pi = MRT$$

$$= 0.302 \text{ mol/L} \times 0.08206 \text{ L atm/mol K} \times 310 \text{ K} = 7.68 \text{ atm}$$

Practice Problem 11.12

The total concentration of dissolved particles in a blood cell is about 0.30 M. When a cell is placed in pure water at 5°C, what osmotic pressure is exerted on the cell?

Answer: 6.8 atm

Exercises

Reaction Rates

13.1 Why are reaction rates defined in such a way that they always have positive values?

13.2 Distinguish between initial rates and instantaneous rates.

13.3 Show how stoichiometric coefficients are used to relate the rates of different species involved in a reaction.

13.4 What is a rate law?

13.5 The reaction of ozone with ethylene follows the rate law

$$-\Delta[O_3]/\Delta t = k[O_3][C_2H_4]$$

with a rate constant of $k = 800 \ M^{-1} \ s^{-1}$ at 25°C. What is the rate of reaction if the concentrations are $1.00 \times 10^{-8} \ M \ O_3$ and $4.50 \times 10^{-6} \ M \ C_2H_4$?

13.6 At an initial concentration of N_2O_5 of 1.510 M at 45°C, the concentration decreased by 0.131 M in 2.32 h. Calculate the initial rate of the reaction in units of $M \ s^{-1}$.

$$2N_2O_5 \longrightarrow 4NO_2 + O_2(g)$$

13.7 In the reaction of nitrogen with hydrogen gas (the Haber process),

$$N_2(g) + 3H_2(g) \longrightarrow 2NH_3$$

the rate of reaction was measured to be $-\Delta[H_2]/\Delta t = 2.50 \times 10^{-4} \ M \ s^{-1}$ in a particular experiment. What is the rate expressed in terms of the concentration (a) of nitrogen and (b) of ammonia?

13.8 Express the rate of reaction in terms of each of the species in the reaction

$$3I^- + IO_2^- + 4H^+ \longrightarrow 2I_2 + 2H_2O$$

13.9 At an initial concentration of 0.250 M, the concentration of O_3 decreased by 0.0313 M in 100.0 s. Calculate the initial rate of the reaction

$$2O_3 \longrightarrow 3O_2$$

Reaction Order from Concentration–Time Data

13.10 (a) What is the order with respect to a reactant? (b) How is this related to the overall reaction order?

13.11 How does the rate of a reaction generally change with time?

13.12 What is the relationship between the order with respect to a reactant and the coefficient of that reactant in the balanced equation for the reaction?

13.13 Explain why reaction orders are almost always whole numbers.

13.14 The reaction of hydrocyanic acid with water is described by the equation

$$HCN(aq) + 2H_2O(l) \longrightarrow NH_4^+(aq) + HCO_2^-(aq)$$

The concentrations
Make appropriate p
the order of the rea
determine the valu

Concentration of HCN (M)

Concentration of HCN (M)	
0.100	
0.0984	
0.0961	
0.0923	1.00×10^6
0.0851	2.00×10^6
0.0785	3.00×10^6
0.0724	4.00×10^6
0.0668	5.00×10^6
0.0617	6.00×10^6
0.0569	7.00×10^6
0.0525	8.00×10^6
0.0484	9.00×10^6
0.0447	1.00×10^7
0.0199	2.00×10^7
0.0089	3.00×10^7
0.0040	4.00×10^7

13.15 The decomposition of hypobromite ion follows the equation

$$3BrO^- \longrightarrow BrO_3^- + 2Br^-$$

The rate law is $-\Delta[BrO^-]/\Delta t = k[BrO^-]^n$. According to the following concentration–time data, what are (a) the reaction order and (b) the value of the rate constant?

Concentration of BrO– (M)	Time (s)
0.750	0
0.528	10
0.408	20
0.332	30
0.280	40
0.242	50
0.213	60
0.190	70
0.172	80
0.157	90
0.144	100

13.16 The hydrolysis of methyl chloride in water,

$$CH_3Cl + H_2O \longrightarrow CH_3OH + HCl$$

follows the rate law

$$-\Delta[CH_3Cl]/\Delta t = k[CH_3Cl]^n$$

According to the following concentration–time data, measured at 90°C, what are (a) the reaction order and

End-of-chapter exercises

The exercises include classified problems divided by key topics.

End-of-chapter exercises

Additional exercises challenge the student to apply facts and principles from anywhere in the chapter.

Catalysis

13.70 What is a catalyst?

13.71 How can a catalyst be involved in a reaction but not be used up during the reaction?

13.72 How does a catalyst increase the rate of a reaction?

13.73 The hydrogenation of ethylene,

$$C_2H_4 + H_2 \longrightarrow C_2H_6$$

is slow at room temperature but is accelerated considerably by finely divided nickel metal. Nickel strongly adsorbs hydrogen and weakens the H–H bond. (a) Outline a reaction sequence that explains the catalytic activity of nickel. (b) Why is the metal used in a finely divided state rather than as a foil or bar?

Additional Exercises

13.76 Draw an energy-versus-reaction coordinate diagram for the reaction

$$NO(g) + NO_3(g) \rightleftharpoons 2NO_2(g)$$

The activation energy for the forward reaction is 7.1 kJ/mol and for the reverse reaction is 100.0 kJ/mol. Is the forward reaction endothermic or exothermic? What is the value of ΔH for the forward reaction?

13.77 The reaction of fluorine with chlorine dioxide was studied in the gas state at 250 K.

$$F_2(g) + 2ClO_2(g) \longrightarrow 2FClO_2(g)$$

The following rates were determined for different concentrations of fluorine and chlorine dioxide.

Rate ($M \ s^{-1}$)	**Initial Concentration of**	
	F_2 (M)	ClO_2 (M)
1.2×10^{-3}	0.100	0.0100
4.8×10^{-3}	0.100	0.0400
2.4×10^{-3}	0.200	0.0100

Determine the rate law, including a value of the rate constant.

13.78 Unlike chemical reactions, the rate of radioactive decay is independent of temperature. What must be the activation energy for radioactive decay?

13.79 The reaction

$$5Br^-(aq) + BrO_3^-(aq) + 6H^+(aq) \longrightarrow 3Br_2(aq) + 3H_2O(l)$$

is first-order in Br^- and in BrO_3^- and second-order in H^+. (a) Write a rate law for the disappearance of Br^-. (b) If concentrations are in units of moles per

13.74 The decomposition of acetaldehyde has the following mechanism in the presence of iodine:

$$CH_3CHO + I_2 \longrightarrow CH_3I + CO + HI$$
$$CH_3I + HI \longrightarrow CH_4 + I_2$$

(a) What is the catalyst in this mechanism? (b) Identify a reaction intermediate.

13.75 Identify a catalyst in the following mechanism.

$$Ag^+ + S_2O_8^{2-} \longrightarrow Ag^{3+} + 2SO_4^{2-}$$
$$3Ag^{3+} + 2Cr^{3+} + 7H_2O \longrightarrow 3Ag^+ + Cr_2O_7^{2-} + 14H^+$$

liter, what are the units of the rate constant? (c) Write an equation for the rate of appearance of Br_2. (d) Does the rate constant in this equation have the same units as in part (b)? (e) The same value?

13.80 Hydrogen peroxide solutions are normally reasonably stable, but when metal ions such as Fe^{3+} are added, the hydrogen peroxide decomposes.

$$2H_2O_2(aq) \longrightarrow 2H_2O(l) + O_2(g)$$

You may have noted the evolution of gas bubbles when using H_2O_2 to disinfect a cut. The iron catalyst is in the form of the enzyme peroxidase in this case. In a solution containing 0.025 M $FeCl_3$, the concentration of H_2O_2 varies as follows as a function of time.

Concentration of H_2O_2 (M)	Time (s)
0.80	0
0.72	27
0.64	52
0.56	86
0.48	121
0.40	166
0.32	218
0.24	307
0.00	∞

Determine (a) the order of the reaction with respect to H_2O_2 and (b) the value of the rate constant, assuming the equation

$$-\Delta[H_2O_2]/\Delta t = k[H_2O_2]^n$$

---

* 16.97 A solution that contains 0.0100 M each of Ag^+, Fe^{3+}, and Zn^{2+} is adjusted to a pH of 4.35. The values of K_{sp} are 2.00×10^{-8} for AgOH, 1.10×10^{-36} for $Fe(OH)_3$, and 4.5×10^{-17} for $Zn(OH)_2$.
  a. What precipitates, if any, are formed?
  b. At what pH does each of the metal ions start to precipitate?

c. Is it possible [...] selective pr [...] ing the pH? [...] the concentr [...] still in solut [...] cipitate.

▶ ▶ ▶ Chemistry in Practice

16.98 In the United States, daily direct personal use (not including industrial use) of purified water is about 270 L. For a municipality of 50,000 people, this amounts to 13.5 million liters daily. It is common to add fluoride ion (F^-) in amounts of 2.0 ppm (or mg/L) to municipal water to help prevent tooth decay. Calcium fluoride, CaF_2, with a K_{sp} of 1.70×10^{-10}, is a possible source of F^-. Of concern, however, is the natural hardness of the water, because a high concentration of Ca^{2+} could suppress the solubility of CaF_2 and lower the F^- concentration. The following experiments were done to test the hardness of the water: A standard solution of $CaCl_2$ was prepared by weighing 4.015 g $CaCO_3$ into a volumetric flask, reacting it with a slight excess of $HCl(aq)$, and diluting to 500.0 mL. A 50.00-mL portion of this $CaCl_2$ solution was titrated with an EDTA solution; 39.72 mL of EDTA was required. The EDTA solution was then used to titrate 50.00 mL of the municipal water; 14.92 mL of EDTA was required.
  a. What is the concentration of Ca^{2+} in the municipal water?

b. Is it possible to maintain a concentration of 0.8 ppm of F^- in this water by dissolving CaF_2?

c. If CaF_2 is not sufficiently soluble under these conditions, how much Na_2CO_3 should be added daily to the water supply to lower the concentration of Ca^{2+} enough for it to be possible to maintain this F^- concentration? The solubility product constant of $CaCO_3$ is 9.0×10^{-9}.

d. If CaF_2 is sufficiently soluble in the municipal water, what is the maximum concentration of Ca^{2+} that can exist in this water supply without removing any fluoride ion?

e. How much CaF_2 would this municipality need each year?

f. The municipality wants to manufacture its own CaF_2 by the reaction between $CaCO_3$ and HF. How much of each reactant compound would be needed annually if the reaction gave a 78% yield?

Chemistry in Practice

Most end-of-chapter exercise sets conclude with Chemistry in Practice exercises. These exercises ask students to analyze real-world data prior to solving a problem.

Chemistry

1

► ► ► ►

Chemistry

Tarnish on pennies can be removed with hydrochloric acid.

CHAPTER OUTLINE

Encountering Chemistry

Suppose we want to clean the tarnish from a batch of pennies in a coin collection. We could rub them individually with a polishing agent, but that would be time-consuming and tedious. Another possibility is to soak the pennies in a liquid corrosive enough to remove the tarnish but not so corrosive that the coins would be damaged. Acids are often used to remove tarnish, but there are many different acids, varying in corrosive power. Some acids available to us are vinegar, which contains acetic acid; hydrochloric acid, sold for use in swimming pools; and nitric acid, which is quite powerful. Could we use one of these? Some experiments should tell us if any of them is suitable.

Say we first drop a coin into some acetic acid, the least corrosive of the three acids. Even after an hour, there is no apparent change in the degree of tarnish. This experiment provides no evidence of a chemical change, because no new substances seem to have been produced and no existing substances destroyed.

Next we try the most corrosive of these acids, nitric acid. This experiment produces rather dramatic results. The coin begins to move about as if it were being buffeted by some invisible force. The liquid becomes a greenish-blue color, and a cloud of disagreeable-smelling reddish-brown gas arises from the liquid (Figure 1.1). After getting rid of this obnoxious mixture, we examine the coin again. It is no longer tarnished, but it is also no longer shiny. The surface has become pitted

3

Figure 1.1 Nitric acid and copper are both changed when they are allowed to interact. The copper begins to disappear and go into the liquid, which takes on a greenish-blue color. The nitric acid is converted to a reddish-brown gas called nitrogen dioxide.

and rough, as though some of the metal had been removed. The coin and the acid have certainly undergone some chemical changes, producing some new substances, but they are not the desired changes.

Finally, we try the hydrochloric acid. After a few minutes, the coin begins to lose its dark coating and regain its original shiny appearance. This acid seems to be a good choice for cleaning the coins.

Having found an acid that causes the desired changes, we now add a handful of pennies to the hydrochloric acid. The coins begin to change as expected—that is, all but one do. The one coin is forming a gas bubble on its surface. We notice that this coin, unlike the others, is scratched, as coins are after they have been dropped on a street and driven over by cars. As we continue watching, we see bubbles emerging from the scratches. After several hours, the coin begins to float! What is happening? How do we explain the difference between this coin and all the others, which are behaving as we expected? Are the scratches responsible? Could some other factor be involved?

By the end of this chapter, we will have carried out several investigations that will help us answer these questions. During this process, we will also learn more about what chemistry is, how chemists and other scientists carry out investigations and develop explanations for the behavior of substances, and how numbers and measurements can be used to help us understand this behavior.

1.1   Matter

Chemistry is the study of matter. But what is matter? Copper coins are matter. Nitric acid is matter. The red gas formed by their interaction, known as nitrogen dioxide, is matter. **Matter** is anything that occupies space and is perceptible to the senses, attributes common to all these examples and to all physical substances in the universe. (The term **substance** denotes a kind of matter whose composition does not change.) Chemistry, then, is not just concerned with bubbling, vile mixtures of substances with strange, tongue-twisting names. Chemistry is concerned with all forms of matter, its properties, and the change of one substance into another. This latter concern—chemical change—is the essence of chemistry.

The substances that we have been discussing all fall under the category of chemicals. Although the term *chemical* has taken on the negative connotation of being related to drugs and pollutants, in fact chemicals include any substance that can be consumed or produced by a process involving the interaction of different kinds of matter.

Properties of Matter

Why is an emerald green, but a diamond colorless? Why is foam rubber soft, but a brick hard? Why is air a gas, but copper a solid? Why does table salt occur as tiny cube-shaped crystals, whereas cotton occurs as long strands (Figure 1.2)? Such questions suggest a few of the physical properties of matter. **Physical properties** are characteristics that can be observed without changing the composition of a substance. Besides color, hardness, physical state (solid, liquid, or gas), and shape, physical properties include density, melting and boiling points, solubility (tendency to dissolve), odor, taste, interaction with light, and magnetism. Any process that changes the properties of a substance without changing the chemical

What are some other physical properties?

Figure 1.2 Not all matter has the same structure. *Left:* Some solids, such as table salt (sodium chloride), have a highly regular geometric shape; such solids are called crystals. *Right:* Other solids, such as the strands of cotton, have a less regular shape.

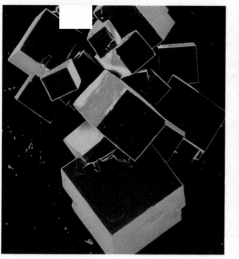
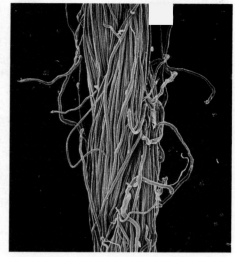

composition of that substance is called a **physical change.** For example, you can change water to steam by heating it. If you cool the steam, it changes back to liquid water. This change from a liquid to a gas (or the reverse) is a physical change because both forms involve the same chemical substance, water.

▶ The changes undergone by the pennies described in the chapter opening are not physical changes, however. The tarnishing of copper on long-term exposure to air and the attack on copper by nitric acid to release a reddish-brown gas (nitrogen dioxide) and a greenish-blue substance (copper nitrate) are chemical changes. A **chemical change** converts one substance into one or more other substances that not only have different physical properties but also have different chemical compositions. The **chemical properties** of a substance, then, describe its chemical composition and the changes it can undergo. For example, when electricity is passed through liquid water containing a small amount of sulfuric acid, two gases are given off: hydrogen and oxygen. This process, shown in Figure 1.3, is a chemical

Figure 1.3 The decomposition of liquid water by electricity produces the new chemical substances oxygen and hydrogen in the form of gases. This process involves a chemical change because the new substances (the products) do not have the same composition as the initial substance (the reactant).

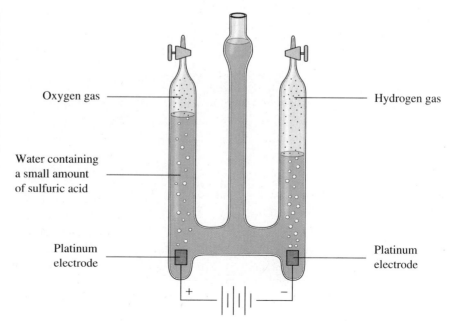

Oxygen gas

Hydrogen gas

Water containing a small amount of sulfuric acid

Platinum electrode

Platinum electrode

$+$ $-$

change that converts water into two substances chemically different from water. During a chemical change, one or more substances (called the **reactants**) are converted into one or more new substances (called the **products**). The process of changing the composition and identity of a substance or substances is a **chemical reaction.**

Composition of Matter

Because matter encompasses the entire physical world, we need to classify it in order to deal with it in a manageable way. Two useful schemes are classification as a pure substance (either element or compound) or a mixture and classification according to state.

Pure Substances and Mixtures   One way to classify matter is based on chemical composition. Some substances, called **pure substances,** have the same chemical composition throughout and from sample to sample, and they cannot be separated into their components by physical means (those not involving chemical reactions). A form of sand called silica, for example, is composed of silicon combined with oxygen in specific proportions. Different grains of silica sand vary in size but have the same composition.

Some examples of matter, such as pencil lead, do not have the same composition in every sample. Pencil lead is made of graphite and clay. The relative amounts of each component may vary in the formulation of pencil leads, depending on the hardness desired. Pencil lead is an example of a **mixture,** a material that consists of two or more pure substances, may vary in composition, and can be separated physically into its component pure substances. For example, you can separate salt from seawater (a mixture composed primarily of salt and water) by evaporating the water (Figure 1.4).

The difference between pure substances and mixtures is exemplified by the difference between acetic acid and vinegar. Acetic acid, a pure substance, is always composed of carbon, hydrogen, and oxygen in the same proportions (40.0% carbon, 6.7% hydrogen, and 53.3% oxygen by mass). White vinegar is a mixture of the pure substances acetic acid and water, but the acetic acid content varies, usually in the range of 4%–6% by mass.

Figure 1.4 Seawater contains large amounts of dissolved solids. The substance present in the largest amount is sodium chloride, or table salt. However, seawater also contains more exotic substances, such as gold; the oceans are estimated to contain about 70 million tons of gold.

Mixtures differ from each other in uniformity of composition. A **homogeneous mixture,** or **solution,** is uniform throughout. A well-mixed sample of vinegar, for example, is uniform in appearance—the components cannot be distinguished visually. Furthermore, any microscopically small portion of the sample has the same composition as any other portion. A mixture that is not uniform throughout—a mixture of salt and pepper, for instance—is called a **heterogeneous mixture.** Different samples of a given heterogeneous mixture may have different compositions.

Elements and Compounds   We have seen that matter may consist of pure substances or of mixtures of substances. Pure substances, in turn, are of two types: elements and compounds. An **element** is a substance that cannot be broken down into simpler substances in a chemical reaction. For example, once you have converted water into hydrogen and oxygen, you cannot break down these products into any simpler substances by the action of heat, light, or electricity or by interaction with other substances. You can convert them into more complex substances, but not into simpler ones. Therefore, oxygen and hydrogen are elements.

The elements are listed on the inside back cover of this book. They are also listed on the inside front cover in the form of a *periodic table,* which contains families of elements that share similar properties. There are presently 109 known elements, of which 83 can be found in natural substances in sufficient quantity to isolate. The 26 elements not isolated from natural sources on earth were synthesized by scientists; some are so unstable that they have only a fleeting existence. Elements are the building blocks of which all matter is composed. Thus, the many examples of matter that we use, see, and read about are all made up of only about 100 different elements in different combinations.

A **compound,** sometimes called a **chemical compound,** is a substance that is composed of two or more elements combined in definite proportions and that has properties different from those of its component elements. Compounds can be chemically decomposed to give back the component elements. For example, mercuric oxide, a red solid, is a *binary compound,* so called because it contains two elements—mercury and oxygen. If this solid is heated sufficiently (to about 500°C), a mixture of oxygen and mercury gases is given off. When the gas is cooled to room temperature, a silvery liquid metal, elemental mercury, is produced, as shown in Figure 1.5.

Elements are generally classified into two main categories: **metals** and **nonmetals.** A few elements do not fit well into either category but have intermediate properties, so they are placed into a third category, **metalloids** or **semimetals.** Generally, metals can be distinguished from nonmetals by their electrical conductivity. Copper, aluminum, iron, and other metals are good conductors of electricity; nonmetal elements, such as carbon (as diamond), chlorine, and sulfur, normally are not. An important exception is carbon in the form of graphite, which is a good electrical conductor. Metalloids, such as silicon, germanium, and arsenic, conduct only a small amount of electricity, a property that makes them important to the semiconductor industry.

We have considered a number of classes and subclasses of matter: mixtures, solutions, heterogeneous mixtures, pure substances, compounds, elements, metals, and nonmetals. A method of classifying matter into these various types is outlined in Figure 1.6.

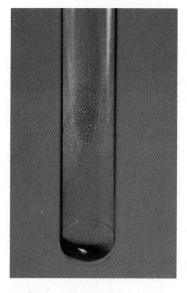

Figure 1.5 Mercury is unique among the metals. It remains liquid below the freezing point of water, so it has been used extensively in thermometers.

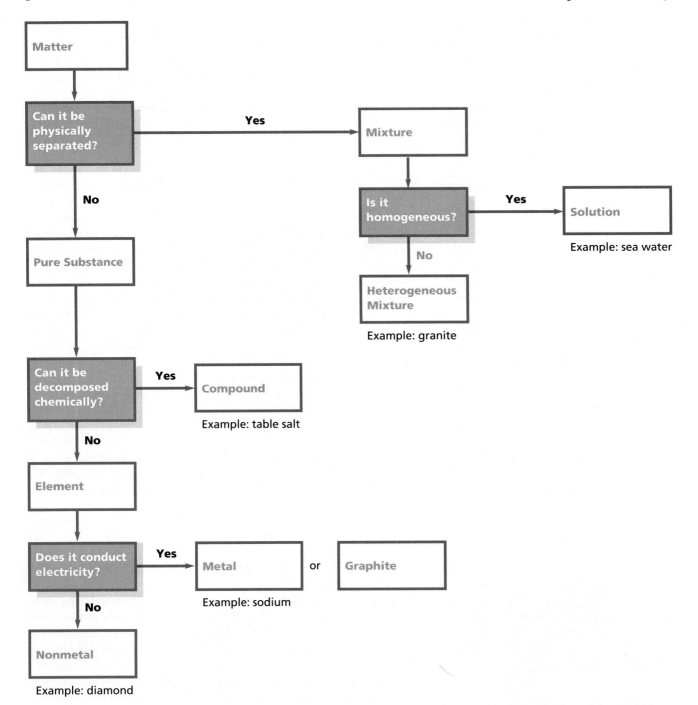

Figure 1.6 We can classify matter by answering the short series of questions in this decision tree.

States of Matter It is also possible to classify matter according to its **physical state**—whether the materials are solids, liquids, or gases. The physical state determines what shape a material assumes and how it occupies a container, as shown in Figure 1.7.

Figure 1.7 A solid, such as a coin, has a characteristic shape that is changed only with difficulty and bears no relation to the shape of the container. A liquid, such as nitric acid, takes the shape of the portion of the container it occupies. A gas, such as nitrogen dioxide, takes the shape of the container and fills it completely.

Draw pictures of the particles of matter as they would appear in each of the physical states— gas, liquid, and solid. Use small circles to represent the particles of matter.

A **solid** has a fixed shape that is characteristic of the sample and has no relation to the shape of the container holding it. (A collection of small pieces may adopt the shape of the container, however.) When solids are placed under pressure, they do not change their volume to any significant extent; for example, a rock does not decrease noticeably in volume when it is squeezed. In other words, solids have low *compressibility,* or tendency to decrease in volume when pressure is applied. This is because the particles of which solids are composed are arranged in a tightly packed, highly ordered structure that does not include much free space into which the particles might be squeezed.

A **liquid** has no fixed shape but takes the shape of the filled portion of its container. Liquids are slightly compressible because their particles have a little free space around them. These particles are not arranged into ordered structures like those in solids, although they are located close to one another.

A **gas** takes no fixed shape or volume but adopts the shape of its container and fills it completely. The particles that make up gases are widely separated and randomly located, with much space around them. As a result, when gases are squeezed, they undergo large changes in volume—they have high compressibility.

Conversion of Matter

Chemistry is concerned with the understanding and control of chemical change. How can a chemist change one substance into another whose properties are more suitable for some specific application, say a nonstick coating on a frying pan? Throughout history, individuals have attempted to make such transformations—to turn lead into gold, for example. This change has been accomplished in relatively recent times, but first much had to be learned about the chemical behavior of matter.

To control chemical change, it is necessary to know how substances behave chemically under different conditions. Some substances are *unstable* and **reactive;** that is, they have a tendency to participate in chemical reactions. Others are **stable** and *inert,* which in a chemical context means not easily decomposed or otherwise chemically changed. A given substance may be stable under some conditions but not under others. ▶ As described in the chapter opening, a penny is stable when placed in acetic acid, but it will react with nitric acid. If we had dropped calcium metal into acetic acid, it would have reacted vigorously because calcium is much more reactive than copper. By selecting the appropriate substances and conditions, then, we can make it possible for chemical changes to occur.

Figure 1.8 Iron, the major component of steel, is used as a structural metal in buildings, automobile bodies, and other products. The rusting of iron requires oxygen from air as well as water, producing the characteristic reddish solid known as rust (or ferric oxide). This solid does not adhere to iron but flakes off, revealing a new iron surface that can be further rusted.

Chemical changes may yield desirable new properties, or they may produce undesirable ones. If a change leads to more desirable properties, chemists want to learn how to bring about that change more efficiently. If a change produces less desirable properties, chemists are interested in preventing it. Corrosion, or rusting, of iron (Figure 1.8) is a major concern in modern industry. About 20% of the iron produced annually in the United States is used to replace rusted objects. Preventing this chemical change, which involves converting a reactive substance into a stable one, is thus of major economic importance. We will examine some methods for preventing corrosion in Chapter 18.

1.2   The Scientific Method

We have seen that chemistry is concerned with matter—its composition, its properties, and its conversion from one form to another. We turn now to the methods chemists use to investigate these phenomena.

Chemistry is one branch of science, a growing and changing body of knowledge obtained by methods based on observation and interpretation and known collectively as the **scientific method.** Although the scientific method is often outlined as consisting of sets of steps and procedures, it is as much a frame of mind as it is anything else. Like other humans, scientists use their intuition and may generalize with insufficient data, but they are distinguished by a willingness to test by experimentation, to organize their facts, and to explain why things happen. Scientists must make use of large bodies of knowledge to gain insight into new observations, perhaps by careful reasoning or insightful analogy but sometimes also by intuition or even luck. Of course, luck by itself is not enough—a receptive mind is needed as well. The scientist must be curious enough about seemingly trivial observations to follow them up and seek to explain them. Louis Pasteur expressed this idea quite succinctly: "Chance favors the prepared mind."

► ► ►   WORLDS OF CHEMISTRY

Plastics and Pollution: A Changing Relationship?

The same properties that make plastics useful also produce ecological problems.

Plastics are a class of man-made materials with properties not found in natural substances. These properties make plastics useful in a wide variety of applications. One of the simplest plastics is polyethylene. To make it, chemists heat the gaseous compound ethylene, which consists of 85.6% carbon and 14.4% hydrogen by mass. The resulting plastic, a tough and flexible solid, is a good electrical insulator and is used to make linings, paper coatings, packaging materials, wire coatings, and other products. It is stable in water and most acids but will react with nitric acid. If we modify the ethylene by replacing the hydrogen with fluorine, we obtain tetrafluoroethylene (24.0% carbon and 76.0% fluorine by mass). Heating this substance yields a plastic with different properties. This plastic, called Teflon, is so stable it does not react with hot nitric acid. But its most outstanding property is that most things do not stick to it, so it is widely used to make nonstick coatings on cookware and to waterproof clothing. Other modifications of the starting material lead to plastics that can be used to make fibers that are corrosion- and mildew-resistant, durable, hard, strong, and flexible.

The same properties that make plastics useful also produce ecological problems. Discarded plastic items normally do not corrode or biodegrade (decompose under the action of bacteria). They have no food value, so animals do not eat them. As a result, they cannot be reabsorbed in any way by the environment. Chemists are developing new plastics that degrade under the action of sunlight after they have been discarded, but this still leaves small pieces of plastic as waste. One recently developed plastic, called polyvinyl alcohol, dissolves in warm or hot water. Hospitals use it to contain infectious clothing so the clothing can be dropped into the laundry with no spread of infection (Figure 1.9). However, the decomposed polyvinyl alcohol is then released into the environment with the laundry wastewater. This plastic is also used to hold toxic substances such as

Figure 1.9 Laundry bags made of polyvinyl alcohol dissolve when placed in water. Such bags are useful for containing infectious materials.

powdered pesticides, which people can mix with water without exposing themselves to the concentrated material.

Questions for Discussion

1. Is the availability of plastic bags that either degrade in sunlight or dissolve in water helpful or harmful to the environment?
2. Is the availability of these bags helpful or harmful to the conservation of natural resources?
3. In the future, landfills may be viewed as national resources because they will be sites in which large concentrations of scarce materials are found. What are the positive and negative implications of this viewpoint?
4. If Teflon does not stick to other things, how can it be used as a coating on frying pans?

There is no one universal definition of the scientific method, but there is agreement on many of its components. These include observations, hypotheses, laws, and theories.

Observations

The scientific method begins with the collection of data through *observation*. A scientist not only observes naturally occurring events but also engages in deliberate experimentation in which the scientist sets conditions, allows events to occur, and observes the result. Experimentation allows scientists to examine events under controlled conditions outside the normal realm of nature and to repeat observations, thus providing a check on accuracy. A common technique involves the isolation and manipulation of one factor at a time in order to determine which of many variables might be affecting the outcome of an experiment.

The chapter opening described a number of observations: ▶ Our pennies underwent no chemical change when placed in acetic acid. The coins reacted with nitric acid to form a blue-green liquid and a red gas. Most of the pennies reacted with hydrochloric acid to remove surface tarnish. One penny, however, reacted with hydrochloric acid to form bubbles. Close examination of this one coin revealed that the bubbles were coming from scratches on its surface. These descriptions are observations made during deliberate and controlled experimentation.

Hypotheses

Which factors would you isolate if you were trying to decide why your desk lamp won't light?

To systematize and correlate many observations and collected data (not necessarily attributed to one scientist alone), scientists propose hypotheses. A **hypothesis** is a tentative explanation for the behavior of matter that accounts for a set of observations and can be tested by further experimentation. In practice, hypotheses are based on small amounts of data, so they start out as intuitive guesses. The systematization of data in a hypothesis allows the scientist to make predictions and devise additional experiments that can be used to test the validity of the hypothesis. If necessary, a hypothesis is modified repeatedly to fit the results of additional experimentation. The building of scientific knowledge by the cyclic interplay of observation and hypothesis formation is shown in Figure 1.10.

Recall the experiments described in the chapter opening. ▶ There we observed that the penny that reacted with hydrochloric acid to form a colorless gas did so along scratches in its surface. This observation might lead us to form a hypothesis: scratched pennies react with hydrochloric acid to form bubbles of a gas. We can test this hypothesis by scratching a number of coins, dropping them

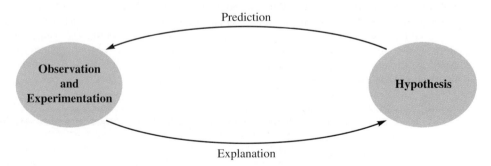

Figure 1.10 Hypotheses are changed and updated by a cyclic process. Explanations of observations are used to develop the hypothesis. The hypothesis is then used to make predictions that can be tested by further experiments.

into hydrochloric acid, and checking for gas bubble formation. Suppose we do this experiment with ten coins and find that six of them form bubbles at the scratches but the other four do not.

Apparently, our hypothesis is only partially correct because it does not explain why some coins do not form bubbles. Some other factor must also be important. Let's consider some other characteristics of pennies. Although all pennies are superficially identical, observation reveals detectable differences. The coins are produced by mints in different cities, usually designated by the first letter of the city name found just below the number indicating the year in which the coin was minted. (Coins minted in Philadelphia have no identifying mark.) We might amend our hypothesis to take the mint into account: pennies react with hydrochloric acid to form bubbles if they are scratched and if they come from a specific mint. However, examination of the reactive and nonreactive coins indicates that members of both groups were produced by a variety of mints. Thus, the amended hypothesis is not correct and must be changed again. Notice that as we move through the process of amending our hypothesis, we are attempting to isolate one variable at a time so that we can determine whether that variable is responsible for the observed results. The variable considered here, the mint, does not affect how the coins behave when mixed with hydrochloric acid.

Another subtle difference in the coins is the year in which they were minted. Examination of the coins indicates that all the reactive coins were minted in 1984 or more recently, and all the nonreactive coins were minted before 1982. This observation leads to another amended hypothesis: scratched coins minted before 1982 do not react with hydrochloric acid to form bubbles, whereas scratched coins minted since 1984 do react. We must test both parts of this hypothesis by scratching more coins and placing them in hydrochloric acid. Additional experiments with many coins support this hypothesis. We now can predict the behavior of copper coins in hydrochloric acid; however, we cannot explain it.

The repeated cycle of making a hypothesis, testing its predictions, and amending it to reflect actual results is central to the scientific method. We must carry this process further, however, if we wish to understand the behavior that the hypothesis predicts.

Laws

Hypotheses that have been supported so extensively that they appear to have universal validity are called **laws.** A scientific law that has been correctly formulated accurately describes the way nature operates under a specified set of conditions. However, not all statements considered to be scientific laws are completely correct. Laws have to be modified as new information comes to light, and some are found to be valid only under certain conditions. For example, observations of the amounts of materials consumed and produced in chemical reactions led to the formulation in the late eighteenth century of the law of conservation of mass: the mass (or quantity of matter) of the products of a chemical reaction equals the mass of the reactants. Within the limits of accuracy of mass measurements on chemical reactions, this law is valid. However, we know now that mass can be converted into energy. Losses of mass accompanying the production of energy are observed in nuclear reactions. Indeed, in atomic power plants, such reactions are carried out specifically to produce large quantities of energy. Chemical reactions involve only small energy changes, so no noticeable mass change occurs.

The conversion of mass to energy was proposed by Albert Einstein 40 years before it was observed experimentally. He developed the now famous equation $E = mc^2$, where E is energy, m is mass, and c is the speed of light. Experimental verification of this equation led to the development of atomic energy and the atomic bomb.

Theories

Laws describe how nature works, but—like hypotheses—they don't explain why it works that way. **Theories** offer explanations of sets of observations, hypotheses, or laws that are applicable in a relatively wide variety of circumstances. For example, atomic theory, which we will examine in Chapter 2, explains many facets of the behavior of matter, including the quantitative aspects of chemical reactions and the law of conservation of mass. Such explanations are usually based on a mathematical or physical model that, if correct, can explain the behavior of matter. Naturally, a theory should fit known facts, so it must be modified if experimentation brings forth incompatible information.

▶ Let's turn our attention again to the behavior of copper coins in hydrochloric acid. We would like to propose a theory that explains the behavior of all the coins. To do so, we might focus on the likelihood that something was changed in the way coins were minted in 1982–1984. A search in the library reveals numerous reports that in 1984 one-cent coins were changed from pure copper to copper-coated disks of zinc. (Actually, the change in coin composition was started in 1982, but it was not until 1984 that all coins were minted in the new way.) If zinc reacts vigorously with hydrochloric acid and if the scratches in our coins expose some of the underlying zinc, we can understand the behavior of the coins. A test with a piece of zinc indicates that this metal does indeed react with hydrochloric acid to release bubbles, as shown in Figure 1.11. We now have an explanation for the behavior of copper coins in hydrochloric acid that is consistent with all the relevant facts. If the zinc is completely removed by reaction with hydrochloric acid, the copper shell that remains may be filled with hydrogen gas and float on the surface of the liquid.

The Scientific Method in Practice

Within the framework of the scientific method, chemists operate with considerable individuality. Some operate by collecting large bodies of facts in the hope of gaining insight into some aspect of the behavior of matter. Such insight may arise from careful reasoning or from analogy with other situations. Some discoveries follow long periods of painstaking work; others arise through serendipity—accidental, fortunate discoveries. However, to make serendipitous discoveries, the chemist must always be prepared to capitalize on unexpected observations.

Consider some historical examples of how several chemists made significant discoveries. In 1879 Thomas Edison invented the light bulb after having collected a massive amount of data on the behavior of electrically heated filaments and engaging in a good deal of trial-and-error experimentation. James Watson and Francis Crick received the 1962 Nobel Prize for their discovery of the double-helix structure of the DNA molecule, the fundamental genetic material. Their proposal resulted from a brilliant insight based largely on evidence that had been collected by other scientists, including Rosalind Franklin and Maurice Wilkins. Watson and Crick certainly did not follow a straightforward path through repeated hypothesis and experimentation to an ultimate explanation. Rather, they made an imaginative intellectual leap in which they used paper models to develop their concept.

The role of serendipity—that is, events happening to individuals capable of capitalizing on them—is not to be underplayed. For example, the discovery of

Figure 1.11 Zinc metal reacts vigorously with hydrochloric acid. The bubbles coming from the metal are hydrogen gas.

Teflon, which now coats much cookware, resulted from an accident and two scientists' receptivity to unexpected observations. In 1938 Roy J. Plunkett and an associate, Jack Rebok, were preparing various materials for use as refrigerants. One of these was a gaseous material composed of carbon and fluorine, with the chemical name *tetrafluoroethylene*. The gas was stored in steel cylinders, and one of them became empty much sooner than expected. Some observers might have dismissed this as the result of a leak. But Plunkett and Rebok were curious. They weighed the cylinder and discovered that it had nearly the same mass as when it was full. Inspecting the cylinder, they found a white, waxy solid coating the inside walls that would not dissolve and that resisted heat and attack by chemicals. It was discovered that the change from gas to solid had been brought about by trace amounts of oxygen that had accidentally leaked into the cylinder. Ultimately, after much further development by many scientists, this discovery led to the applications of Teflon with which we are all so familiar.

1.3 Measurement and the Metric System

To be useful to the experimenter and to others, observation must usually involve measurement. *Measurement* is the determination of the size or magnitude of a particular quantity—the number of nails, the mass of a brick, the length of a wall. Measurements are always defined by both a number and a *unit,* a standard of comparison for expressing the relative magnitude of the quantity being measured. It is important to specify both units and numbers when expressing measurements.

A central issue in measurement is deciding what system of units to use. Most countries use the metric system; the United States still uses primarily the English system. The English system is based on various units that are not related to one another in a consistent manner. Common units of length, for example, are inches, feet (12 inches), yards (3 feet), and miles (1760 yards). The metric system, on the other hand, uses units that are related by some power of ten. Metric units of length are centimeters (10^{-2} meters), decimeters (10 centimeters or 10^{-1} meters), meters, and kilometers (1000 meters). Scientists in all countries use the metric system in their work, but U.S. scientists must still be able to relate the metric system to the English system.

The metric system is convenient to use because of the relationship between its units: converting between related units is simply a matter of shifting the decimal point. The metric system offers an additional convenience. Unlike the English system, it does not use arbitrary names like inch, foot, and yard. Instead, it defines *base units* of measure, and any multiple or fraction of these base units is defined by a special prefix. The most commonly used prefixes are listed in Table 1.1 and illustrated in Figure 1.12.

The numbers used to describe the metric prefixes are expressed in **scientific notation,** a system in which numbers are expressed as products of a number between 1 and 10 multiplied by the appropriate power of ten. For example, the notation 10^6, which describes the prefix *mega-*, means 10 multiplied by itself six times, or 1,000,000; so *mega-* added to a base unit gives a unit a million times larger than the base. The notation 10^{-3} means 1 divided by 10 three times, or 0.001. If you are not familiar with scientific notation, review Appendix A, which explains it in some detail.

Table 1.1 Some Metric Prefixes

Prefix	Factor	Symbol
giga	10^9	G
mega	10^6	M
kilo	10^3	k
deci	10^{-1}	d
centi	10^{-2}	c
milli	10^{-3}	m
micro	10^{-6}	μ
nano	10^{-9}	n
pico	10^{-12}	p
femto	10^{-15}	f
atto	10^{-18}	a

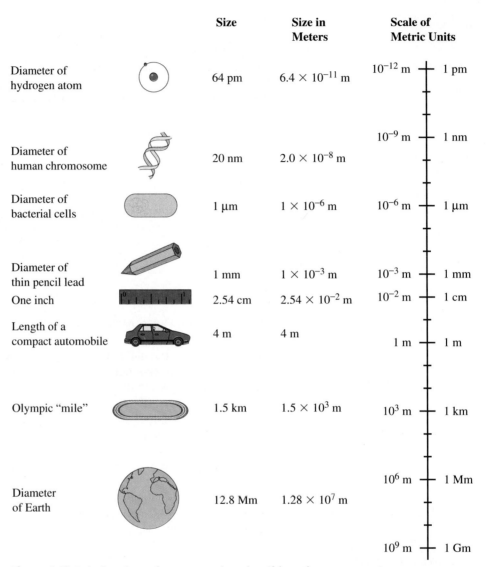

	Size	Size in Meters	Scale of Metric Units
Diameter of hydrogen atom	64 pm	6.4×10^{-11} m	10^{-12} m — 1 pm
Diameter of human chromosome	20 nm	2.0×10^{-8} m	10^{-9} m — 1 nm
Diameter of bacterial cells	1 μm	1×10^{-6} m	10^{-6} m — 1 μm
Diameter of thin pencil lead	1 mm	1×10^{-3} m	10^{-3} m — 1 mm
One inch	2.54 cm	2.54×10^{-2} m	10^{-2} m — 1 cm
Length of a compact automobile	4 m	4 m	1 m — 1 m
Olympic "mile"	1.5 km	1.5×10^{3} m	10^{3} m — 1 km
Diameter of Earth	12.8 Mm	1.28×10^{7} m	10^{6} m — 1 Mm
			10^{9} m — 1 Gm

Figure 1.12 Relative sizes of some metric units of length.

Metric Base Units and Derived Units

Although the metric system of prefixes can be used with any base units, **SI** units (from the French *Système International d'Unités*) are used by most scientists. We will use SI units in this book, with a few exceptions that will be explained as they are introduced. The seven base units in the International System are listed in Table 1.2. All other units are based on these, either by addition of a prefix or by combination.

A number of commonly used units derived from the base units by multiplication and division are given special names in SI. An example is the unit of energy, the *joule*. **Energy,** which is the capacity for doing work, is expressed as force

Table 1.2 Base Units of SI

Unit	Symbol	Quantity
Meter	m	Length
Kilogram	kg	Mass
Second	s	Time
Ampere	A	Electric current
Kelvin	K	Temperature
Mole	mol	Amount of substance
Candela	cd	Luminous intensity

times distance, so one joule (J) is defined as one newton (N), the unit of force, times one meter. Force is mass times acceleration, so the newton itself is a derived unit equal to one kilogram times one meter, divided by one second squared. Thus, in base units, the formula for a joule is $1 \text{ J} = 1 \text{ kg m}^2 \text{ s}^{-2}$. In this formula, s^{-2} means $1/\text{s}^2$. Some of the derived units you will encounter later in this book are given in Table 1.3.

Some quantities are described by units that combine base units or derived units and that have not been given special names. A common example of this is **density,** the ratio of mass to volume, illustrated in Figure 1.13. Density is given in units such as kilograms per cubic decimeter (kg dm^{-3} or kg/dm^3). Although this is the fundamental SI derived unit of density, other units, such as grams per cubic centimeter (g cm^{-3} or g/cm^3) or grams per milliliter (g mL^{-1} or g/mL), are also commonly used.

Table 1.3 Some Derived Units in SI

Symbol	Formula	Unit	Quantity
N	kg m s^{-2}	Newton	Force
Pa	N m^{-2}	Pascal	Pressure
J	N m	Joule	Energy or work or heat
W	J s^{-1}	Watt	Power
V	W A^{-1}	Volt	Electrical potential or electromotive force
Ω	V A^{-1}	Ohm	Electrical resistance
C	A s	Coulomb	Quantity of electricity
Hz	s^{-1}	Hertz	Frequency

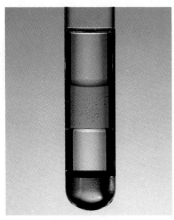

Figure 1.13 Liquids that do not dissolve in one another and that have different densities will float on one another, forming layers. The top layer is gasoline, a mixture of hydrocarbons (containing only chemically combined carbon and hydrogen), with a density of about 0.8 g/mL. Next is water, with a density of 1.0 g/mL; food coloring has been added for visibility. The next layer is carbon tetrachloride, once used in dry cleaning, with a density of 1.59 g/mL. The bottom layer is mercury, the only metal that is liquid at room temperature. Its density is 13.6 g/mL.

Units of Length, Mass, Temperature, and Volume

We will use a variety of units as we proceed through our study of chemistry. A few basic units—those of length, mass, temperature, and volume—are introduced here to set the stage. Some SI units and other units that measure these quantities are compared in Table 1.4. A more comprehensive list can be found in Appendix H.

The base unit of length is the *meter* (m). This unit was formerly defined by the length between two marks on a platinum–iridium bar stored in Paris but is now defined in terms of the speed of light in a vacuum; 1 m is the distance light travels in 1/299,792,458 seconds (abbreviated s). In terms of common English measures of length, 1 m is the same as 39.37 inches (in); 1 mile (mi) is the same as 1.609 kilometers (km).

Mass is the measure of the quantity of matter in a substance. Mass is often confused with weight, and the two terms are often loosely used interchangeably. Technically, however, **weight** is the force exerted on an object by gravity, and so it varies with the strength of gravity, as illustrated in Figure 1.14. Note that for historical reasons the base unit of mass is the *kilogram* (kg), which already contains a prefix meaning 1000. Although 1 kg equals 1000 g, it is the kilogram and not the gram that is the point of reference for mass. Other units of mass, such as the milligram, use a metric prefix preceding gram, not preceding kilogram. A kilogram contains the same amount of matter as 2.205 pounds (lb), a unit in the English system. Some other common units of mass and weight are compared in Table 1.4.

Temperature is a measure of the hotness or coldness of an object. Heat always flows spontaneously from an object of higher temperature to an object of lower temperature, so temperature change is a measure of the direction of heat flow. The base SI unit of temperature is the *kelvin*, a unit on the **Kelvin scale**, which is defined such that the lowest possible temperature is 0 K, or **absolute zero**, theoretically the lowest attainable temperature. Water freezes at 273.15 K and boils at 373.15 K. (Note that we do not use the degree sign, °, with the Kelvin scale.) Also commonly used by scientists is the **Celsius scale**, where 0°C is the temperature at which water freezes and 100°C is the temperature at which water boils. These two scales are related by the following conversion:

$$K = °C + 273.15$$

You may be more familiar with the **Fahrenheit scale**, where the freezing and boiling temperatures of water are 32°F and 212°F, respectively. On this scale freezing and boiling are separated by 180 degrees rather than 100 degrees as on the Celsius and Kelvin scales; therefore, mathematical manipulations of Fahrenheit temperatures are more complex, as for most quantities in nonmetric systems. The following equation can be used to make conversions between the Celsius and Fahrenheit scales:

$$°F = (1.8 × °C) + 32 \qquad \text{or} \qquad °C = \frac{°F - 32}{1.8}$$

All three temperature scales are compared in Figure 1.15.

SI units have not totally displaced other systems. For example, the *atmosphere* and the *torr* are still widely used by scientists to measure pressure. Among other non-SI units still in use by scientists, the unit used to measure **volume,** the

Table 1.4 Some Equivalent Units

Non-Si	SI
Length	
1 in	2.54 cm = 25.4 mm = 0.0254 m
1 ft = 12 in	30.48 cm
1 yd = 3 ft = 36 in	91.44 cm = 0.9144 m
1 mi = 5280 ft = 1760 yd	1609 m = 1.609 km
Mass and weight	
1 oz (avoirdupois)	28.35 g
1 lb (avoirdupois)	453.6 g = 0.4536 kg
1 ton = 2000 lb	0.9072 Mg = 907.2 kg
1 metric ton	1 Mg = 1000 kg
Volume	
1 mL	1 cm^3
1 fl oz (U.S.)	29.57 cm^3
1 qt (U.S. liquid)	946.4 cm^3
1 liter (L)	1 dm^3 = 0.001 m^3 = 1000 cm^3
1 gallon (U.S.)	3785 cm^3
Time	
1 min	60 s
1 h = 60 min	3600 s
1 d = 24 h	86,400 s

space occupied by a substance, is the most common. According to SI, volume (equal to length times width times height) should be given in units of cubic meters (m^3). However, a cubic meter represents a rather large volume, and chemists are

Figure 1.14 Astronaut Linda M. Godwin effortlessly holds up astronaut Jerry L. Ross aboard the orbiting space shuttle Atlantis. Ross's weight has changed, but his mass has not.

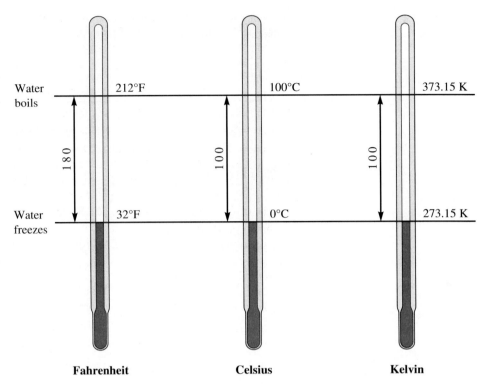

Figure 1.15 The Fahrenheit temperature scale is more finely divided (180 degrees between freezing and boiling water) than are the Celsius or Kelvin scales (100 degrees between freezing and boiling water).

inclined to express volume in the smaller units of liters (L) or metric derivations of liters. A *liter* is the same as a cubic decimeter (dm³) or 1000 cubic centimeters (cm³). Thus, a milliliter (mL), which is 0.001 liter, is the same as a cubic centime-

Figure 1.16 Relative sizes of some volume units.

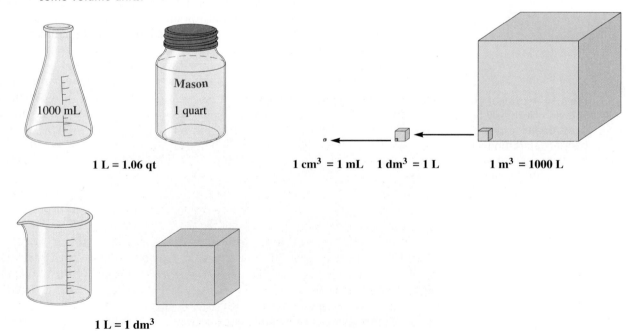

ter. Some other volume conversions are presented in Table 1.4. Figure 1.16 shows the relative sizes of some common volume units.

1.4 Uncertainty in Measurement

▶ Previously, in testing a hypothesis, we dropped ten scratched coins into hydrochloric acid; six of them reacted and four did not. Because the number of coins was fairly small, we could be quite certain that we had counted them correctly. However, if we had used 10,000 coins, it is highly probable that we would have made a counting error. The number of coins used in that experiment and the number that reacted would have had some uncertainty associated with them. Uncertainty of this type is usual with experimental measurements. This section explores the uncertainty associated with numbers, a topic that is important in both planning and interpreting experiments.

Precision and Accuracy

The numbers used in chemistry can be placed into two categories. Some numbers are exact; these are numbers that are established by definition or by counting. Defined numbers are exact because they are assigned specific values: 12 inches in 1 foot, 2.54 centimeters in 1 inch, and 10 millimeters in 1 centimeter, for example. Numbers established by counting, like the ten coins mentioned earlier, are known exactly because they can be counted with no errors. Because defined and counted numbers are known precisely, there is no uncertainty in their values.

Other numbers are not exact; these are numbers obtained by measurement or observation and may also include numbers resulting from a count if the quantity being counted is very large. There is always some uncertainty in the value of such numbers because they depend on how exactly the measuring instrument and the experimenter can measure the values.

Uncertainty in numbers appears in two forms: precision and accuracy. The **precision** of a measured number is the extent of agreement between repeated measurements of its value. **Accuracy** is the difference between the value of a measured number and the expected or correct value. As shown in Figure 1.17, numbers can be precise, they can be accurate, or they can be both. If repeated measurements give values close to one another, the number is precise, whether or not it is accurate. The number is accurate only if it is close to the true value (which might not be known). Although highly precise measurements are often also accurate, it is possible to measure the same incorrect number each time if the same error is made consistently. Thus, precision describes the extent to which measurements are reproducible.

It is customary to report the precision of a number by writing the uncertainty after a plus-or-minus sign (±). For example, the average mass measured for 100 pre-1982 pennies is 3.11 ± 0.03 g. The larger the number after the plus-or-minus sign, the less precisely we know the value of the mass. In this example, we expect most measured values to fall between 3.08 g (3.11 − 0.03) and 3.14 g (3.11 + 0.03). The method for calculating this uncertainty for a series of repeated measure-

Good accuracy,
good precision.

Poor accuracy,
good precision.

Poor accuracy,
poor precision.

Figure 1.17 Accuracy is measured by the extent to which the correct result is obtained. Precision is measured by the reproducibility of a given result.

ments is given in Appendix B. If we do not make a number of repeated measurements, it is not possible to determine the precision. In that case, we must use some other way to designate the approximate precision of the number.

Suppose we measure the pressure in a water pipe over a period of time and record this pressure on a chart as faucets are opened and closed, as shown in Figure 1.18. The height of each peak is a measure of the pressure in the pipe. The precision of the measurement of peak heights depends on the characteristics of the measuring device—in this case, a meterstick. In Figure 1.18B, the meterstick is calibrated with 10-cm divisions. The peak height is somewhere between 10 cm and 20 cm. We can guess that the last digit falls somewhere in the vicinity of 8/10 of the distance between the 10-cm and 20-cm marks, at about 18 cm. This means

Figure 1.18 Measurement of the peak heights on a chart, showing the use of significant figures. See text for details.

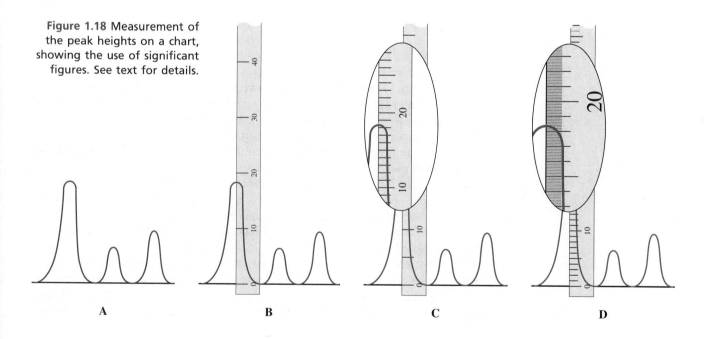

A   B   C   D

that we estimate the height to be between 17 cm and 19 cm. The last digit is uncertain by 1 cm: 18 ± 1 cm. We can estimate the precision of this number by the number of significant figures it contains, two in this case. **Significant figures** in a number are all the digits of which we are absolutely certain, plus one additional digit, which is somewhat uncertain. We are certain of the 1 in 18 cm, but not of the 8. Both these digits are considered significant—that is, they have meaning. Unless we have other information, we assume that there is an uncertainty of at least one unit in the last digit.

Using a meterstick calibrated more finely, say in centimeters, as in Figure 1.18C, we see that the peak height is indeed close to 18 cm but is somewhat greater than this. It appears that the peak reaches about 3/10 of the distance between the 18-cm and 19-cm marks, so we can estimate the height at 18.3 ± 0.1 cm. The number of significant figures is three, two of which are certain (18) and one which is somewhat uncertain (3).

A meterstick calibrated in millimeters, as in Figure 1.18D, shows that the peak height is in fact close to 18.3 cm, falling between the 18.3-cm and 18.4-cm marks. We know this number to three significant figures. We could add another significant figure by estimating the distance between these two marks, but the thickness of the inked lines in the figure makes such an estimate too uncertain to be useful.

Determining the Number of Significant Figures

Each number, then, represents a specific quantity with a certain degree of precision that depends on the manner in which the number was determined. When we work with numbers, we must be able to recognize how many significant figures they contain. How do we do that? First, we must remember that *the nonzero digits are always significant, no matter where they occur*. The only problem in counting significant figures, then, is deciding whether a zero is significant. A guide for making such decisions appears in Figure 1.19, and some examples are given in Table 1.5. The figure and table are based on the following rules:

- A zero alone in front of a decimal point is not significant; it is used simply to make sure we do not overlook the decimal point (**0**.2806, **0**.002806).
- A zero to the right of the decimal point but before the first nonzero digit is simply a placeholder and is not significant (0.**00**2806).
- A zero between significant digits is significant (28**0**6, 28.**0**6, 0.002**80**6).
- A zero at the end of a number and to the right of the decimal point is significant (0.002806**0**, 2806.**0**).
- A zero at the end of a number and to the left of the decimal point (2806**0**) may or may not be significant—we cannot tell by looking at the number. It may be precisely known and thus significant, or it may simply be a placeholder. If we encounter such a number, we have to make our best guess of the meaning intended. To avoid creating an ambiguous number of this kind, we should write the number in scientific notation (see Appendix A) so that the troublesome zero occurs to the right of the decimal point. In that case, it is simple to show whether the zero is significant (2.8060×10^4) or not (2.806×10^4). The power of ten, 10^4, is not included in the count of significant figures, because it simply tells us

There is a story that members of the first surveying party to measure the height of Mount Everest obtained a value of 29,000 ft. They reported the value as 29,002 ft, however, because they were afraid that otherwise people would think that the measurement had been determined to only two significant figures. Was this action ethical? Should they have made this change? Why or why not?

Figure 1.19 The significance of zeros in numbers can be determined by answering this sequence of questions.

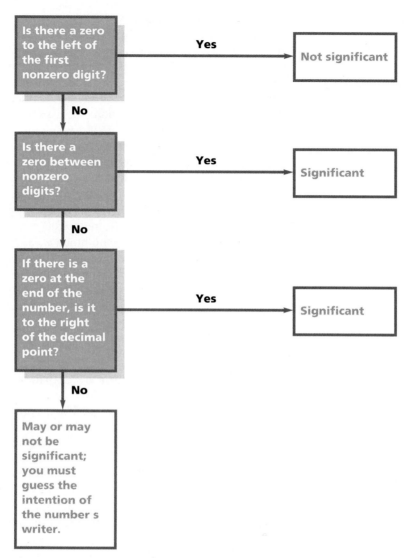

the position of the decimal point.

Although significant figures generally provide an adequate measure of uncertainty, they do not measure uncertainty very exactly. When we use significant figures, we assume that the last significant digit has an uncertainty of ± 1. But what if the uncertainty is actually ± 2 or ± 5? A more exact treatment of uncertainty in measurements is covered in Appendix A.

EXAMPLE 1.1

Determine the number of significant figures in 0.060520.

Solution   Zeros preceding the first nonzero digit are not significant. The zero in the middle of the number is significant. The zero at the end of the number is to the right of the decimal point, so it is significant. Thus, two of the zeros and all three of the nonzero digits are significant, so this number has five significant figures.

Table 1.5 Significant Figures in
Some Numbers

Number	Significant Figures
2.806	Four
28.06	Four
2806	Four
0.2806	Four
0.002806	Four
0.0028060	Five
28060	Four or five, depending on whether the rightmost zero is significant
28060.0	Six
2.80600×10^4	Six

Practice Problem 1.1

Determine the number of significant figures in 5020.01.

Answer: Six significant figures

Significant Figures in Calculations

It is not sufficient to express measured numbers to the proper number of significant figures. We must also be concerned about the proper expression of numbers calculated from measured numbers. (Calculators will not do this for us, so we have to modify their output.) The rules for determining the proper number of significant figures in mathematical manipulations of measured numbers are fairly simple. Note that we consider only measured numbers in deciding on the proper number of significant figures in the answer to a calculation. We need not consider numbers that are known exactly, such as those in conversion equations like 1 ft = 12 in. Exact numbers such as these have no experimental uncertainty, so they can be considered to have an infinite number of significant figures.

Multiplication and Division   In a multiplication or division problem, the product or quotient must have the same number of significant figures as the least precise number in the problem. (The basis for this rule and a more sophisticated way of determining the precision of the result are considered in Appendix A.) Consider the following example:

$$2.5 \times 1.12 = ?$$

The first number has two significant figures, and the second has three. Because the answer must match the number with the fewest significant figures, it should have two:

$$2.5 \times 1.12 = 2.8 \qquad \text{(not 2.80)}$$

Rounding

When a number contains more digits than are allowed by the rules of significant figures, we drop the digits after the last significant figure, using the following procedure known as *rounding:*

- If the first digit being dropped is less than 5, leave the last significant figure unchanged. For example, when rounded to three significant figures, 6.073 becomes 6.07, and 6083 becomes 6.08×10^3.
- If the first digit being dropped is greater than 5 or is 5 followed by digits other than zero, increase the last significant figure by 1. Rounded to three significant figures, 6.077 becomes 6.08, and 60751 becomes 6.08×10^4.
- If the first digit being dropped is 5 followed only by zeros or by no other digits, then increase the last significant figure by 1 if it is odd (6.075 become 6.08) but leave it unchanged if it is even (6.085 becomes 6.08).

EXAMPLE 1.2

Round each of the following numbers to three significant figures: a. 3245; b. 12.263; c. 0.001035; d. 312486; e. 312586.

Solution   a. 3245 becomes 3.24×10^3 because the last digit is 5 and the third digit (4) is even. b. 12.263 becomes 12.3 because the fourth digit (6) is greater than 5. c. 0.001035 becomes 0.00104 because the last digit is 5 and the digit that precedes it (3) is odd. d. 312486 becomes 3.12×10^5 because the fourth digit (4) is less than 5. e. 312586 becomes 3.13×10^5 because the fourth digit is a 5 followed by nonzero digits.

Practice Problem 1.2

Round each of the following numbers to four significant figures: a. 19025; b. 18.533; c. 0.0051235; d. 108492; e. 108592.

Answer: a. 1.902×10^4; b. 18.53; c. 0.005124; d. 1.085×10^5; e. 1.086×10^5

EXAMPLE 1.3

Determine how many significant figures the answers to the following operations should have.

a. 5.27×3.20 b. $(1.5 \times 10^6) \times 317.832$
c. $6.0/2.9783$ d. $(2.01)^3$

Solution

a. $5.27 \times 3.20 = 16.9$
Keep three significant figures because each number being multiplied has three significant figures.
b. $(1.5 \times 10^6) \times 317.832 = 4.8 \times 10^8$
Keep only two significant figures because 1.5×10^6 has two significant figures.
c. $6.0/2.9783 = 2.0$
Keep two significant figures because 6.0 has two significant figures.
d. $(2.01)^3 = 8.12$

Keep three significant figures because 2.01 has three significant figures and this operation is equivalent to $2.01 \times 2.01 \times 2.01$.

Practice Problem 1.3

Determine how many significant figures the answers to the following operations should have.

a. 2.53/0.51   b. 0.00217×5325
c. $(6.05 \times 10^6)/2.00005$ d. $(5.105)^4$

Answer: a. 5.0—two significant figures; b. 11.6—three significant figures; c. 3.02×10^6—three significant figures; d. 679.2—four significant figures

Note in Example 1.3d and Practice Problem 1.3d that powers are just a special case of multiplication. Similarly, roots are a special case of division. Thus, the answer should have the same number of significant figures as the number operated on.

Addition and Subtraction   A sum or difference can only be as precise as the least precise number operated on. Thus, when adding or subtracting we round the sum or difference to the first uncertain digit. If we add 10.1 to 1.91314, for instance, we get 12.0 rather than 12.01314 because there is uncertainty in the tenths position of 10.1, and this uncertainty carries over to the answer. In this case, it is not the number of significant figures that is important but the magnitude of the uncertainty. Consider another example:

$$0.0005032 + 1.0102 = ?$$

The first number has four significant figures, and the second number has five significant figures. However, this information does not determine the number of significant figures in the answer. If we line these numbers up, we can see which digits can be added to give an answer that has significance:

$$
\begin{array}{r}
0.0005032 \\
+ \ 1.0102 \ \ \ \ \\
\hline
1.0107 \ \ \ \ \\
\end{array}
$$

The digits after the 5 in the first number have no corresponding digits in the second number, so it is not possible to add them and obtain digits that are significant in the sum. The answer has five significant figures, dictated by the number with the greatest uncertainty (1.0102).

EXAMPLE 1.4

Carry out the following operations to the correct number of significant figures.

a. $24 + 1.001$ b. $24 + 1.001 + 0.0003$
c. $428 - 0.01$ d. $14.03 - 13.312$

(continued)

Solution

a. $24 + 1.001 = 25$

Because 24 has no significant digits after the decimal point, the answer can have no significant digits after the decimal point either. The number with the greatest uncertainty dictates the precision of the answer.

b. $24 + 1.001 + 0.0003 = 25$

Keep only two significant figures, for the same reason as in the preceding example. Note that 0.0003 has only one significant figure, but it is the magnitude of the uncertainty, not the number of significant figures, that affects the answer.

c. $428 - 0.01 = 428$

Again, the answer must reflect the greatest uncertainty. It makes no difference that 0.01 has only one significant figure.

d. $14.03 - 13.312 = 0.72$

The first number has only two digits after the decimal point, so only two digits should occur after the decimal point in the answer.

Practice Problem 1.4

Carry out the following operations to the correct number of significant figures.

a. $0.2 + 4.009$ b. $2.88 + 0.3452 + 0.0002$
c. $186 - 1.001$ d. $44.18 - 42.814$

Answer: a. 4.2; b. 3.23; c. 185; d. 1.37

1.5   Dimensional Analysis

As we have seen, the scientific method includes making careful observations, frequently involving measurements. However, scientific investigation does not end there. Measurements must be interpreted, often by mathematical manipulation of the data. This book will introduce you to the manipulation of data through problem solving, showing you how to solve problems that require you to apply various principles covered in the text. Learning to solve problems may at first seem a difficult task. This section describes a useful problem-solving approach called *dimensional analysis,* which involves the step-by-step conversion of numbers and their units (or dimensions) into other numbers with different units. An analysis of the units provides clues to the correct solution of many problems.

Equivalence Expressions and Conversion Factors

Consider a problem that can be solved almost automatically: ''Convert 30 minutes into hours.'' You probably gave the answer without thinking: 30 minutes is one half-hour (0.50 h). But how did you know that answer? Let's systematize the problem-solving process. First, even if we don't always consciously think about it, we must decide what the problem is asking for: it requires converting a number of minutes into the corresponding number of hours. Next, we must know the number of minutes in one hour. (This is where information such as that presented in Tables 1.1 and 1.4 can be useful.) From this information, we derive a mathematical expression that shows the equivalence of two quantities having different units:

$$1 \text{ h} = 60 \text{ min}$$

Next we develop a **conversion factor,** a ratio of equivalent quantities that is equal to 1 and that has different units in the numerator and denominator. To develop a conversion factor to convert minutes into hours, we can divide each side of the equivalence expression by 60 min:

$$\frac{1 \text{ h}}{60 \text{ min}} = \frac{60 \text{ min}}{60 \text{ min}} = 1$$

When we multiply a quantity by a factor with a value of 1, we can convert the units of the quantity without changing its value. This means we can multiply the original quantity (30 min) by the conversion factor to get the desired quantity:

Give another example of two numbers that have different units but represent the same quantity.

$$30 \text{ min} \times \frac{1 \text{ h}}{60 \text{ min}} = \frac{30 \times 1 \times \text{min} \times \text{h}}{60 \times \text{min}} = 0.50 \text{ h}$$

Notice how the starting units (minutes) cancel out, leaving the desired units (hours) to accompany the calculated number. In the dimensional-analysis approach, units in the numerator and denominator of a fraction are treated exactly the same as numbers—canceled out, multiplied, divided, squared, or whatever the mathematical operations demand.

Now re-examine the conversion developed above. From the same equivalence expression, 1 h = 60 min, we can derive another conversion factor by dividing through by 1 h instead of by 60 min:

$$\frac{1 \text{ h}}{1 \text{ h}} = \frac{60 \text{ min}}{1 \text{ h}} = 1$$

This conversion factor is also equal to 1. But what happens when the original quantity is multiplied by this conversion factor?

$$30 \text{ min} \times \frac{60 \text{ min}}{1 \text{ h}} = \frac{30 \times 60 \times \text{min} \times \text{min}}{1 \times \text{h}} = \frac{1800 \text{ min}^2}{\text{h}}$$

This answer is clearly wrong.

Every equivalence expression provides two conversion factors. Which one is appropriate to use depends on how the problem is stated. But in every case, using the proper conversion factor cancels out the old units and leaves intact the desired units. Using the improper conversion factor results in nonsensical units (like $1800 \text{ min}^2/\text{h}$) that should signal a clear warning that the conversion factor is upside down or otherwise incorrect.

The incorrect conversion of minutes to hours illustrates one of the most compelling reasons for using dimensional analysis. By arranging the units so that they cancel properly, we can ensure that the numbers that go with the units have also been set down in their proper relationships. If the units in the answer are right, the chances are good that the problem was worked correctly, barring arithmetical errors. If the units are wrong, it is likely that the problem was set up incorrectly. Errors in the setup, rather than in the arithmetic, cause most mistakes in problem solving.

An Approach to Problem Solving

The process just discussed can be applied to many types of problems. This section lists the steps in the procedure and applies them to another sample problem:

A liquid-fertilizer tank has a volume of 6255 ft^3. How many gallons of liquid fertilizer can be placed in this tank?

1. **Decide what the problem is asking for.** First, read the problem carefully. If you are not sure what a term means, look it up. If you need to use an equation to solve the problem, be sure you understand the meaning of each symbol. Look for clues in the problem itself—words or phrases such as *determine, calculate, what mass, how many, what volume,* or *how much.* After deciding what quantity the problem is asking for, write down the units in which this quantity must be stated.

In the fertilizer tank example, we want to find the volume of the tank in units of gallons.

2. **Decide what relationship there is between the information given in the problem and the desired quantity.** If necessary, recall or look up equivalence relationships (such as 1 h = 60 min) that will help you solve the problem. This information may not be given in the problem itself. Often your memory will serve as the source of information necessary to develop the proper relationships; other times you may need to find another source of information.

In the fertilizer tank example, we need to convert volume in cubic feet to volume in gallons, so we need an equivalence relationship between these two quantities. The expression is 1 ft^3 = 7.481 gal.

The additional information that describes the relationship between two quantities won't always be a direct equivalence as in 1 h = 60 min. A series of equivalence expressions and their derived conversion factors may be needed. Sometimes a mathematical equation is required to express the equivalence between two quantities, such as °C = (°F − 32)/1.8. At other times, some chemical principle may have to be applied.

A word of caution: Just as a problem may contain less information than you need to solve it, it may also contain more information than you need. Never assume that you must use all the information contained in the statement of the problem. Examine all information critically and ignore any that is not pertinent.

3. **Set up the problem logically, using the relationships decided on in step 2.** Starting with the relationships you obtained in step 2, develop the conversion factors needed to arrive at the final answer. Be sure to set up the conversion factors so that the old units cancel and the desired units are introduced in the appropriate positions.

In the fertilizer tank example, the conversion factor is 7.481 gal/1 ft^3. The conversion is as follows: ? gal = 6255 $\cancel{\text{ft}^3}$ × $\dfrac{7.481 \text{ gal}}{1 \ \cancel{\text{ft}^3}}$ = 46,790 gal

If you will need a series of conversion factors to find the final answer, it is often helpful to map out the route you will follow to get there. Suppose you need to know how many centimeters are in 1.000 mile, but you can only find a table that gives the number of feet in a mile, the number of inches in a foot, and the number of centimeters in an inch. The "road map" to get from miles to centimeters might look like this:

$$\text{mi} \longrightarrow \text{ft} \longrightarrow \text{in} \longrightarrow \text{cm}$$

The problem setup would follow this progression:

$$? \text{ cm} = 1.000 \text{ mi} \times \frac{5280 \text{ ft}}{1 \text{ mi}} \times \frac{12 \text{ in}}{1 \text{ ft}} \times \frac{2.54 \text{ cm}}{1 \text{ in}} = 1.609 \times 10^5 \text{ cm}$$

4. **Check the answer to make sure it makes sense, both in magnitude and in units.** This step is just as important as the others. You must develop an intuitive feeling for the correct magnitude of physical quantities. Suppose, for example, you were calculating the volume of liquid antacid an ulcer patient needed to neutralize excess stomach acid, and the result of the calculation was 32 liters. Units of liters seems reasonable because the liter is a valid unit of volume, but it should be obvious that something is wrong with the numerical answer. A liter is about the same volume as a quart, which means this patient would need to drink 8 gallons of antacid! The source of the error is probably an arithmetic mistake, or possibly omission of the metric prefix *milli-* (10^{-3}) somewhere, since 32 mL would be an appropriate volume to swallow. In any event, you should recognize a result such as this one as unacceptable and then check the setup and calculations for the error.

In the fertilizer tank example, the units canceled properly, so the answer should be correct if there were no errors in arithmetic. There are about 7.5 gal in 1 ft^3, so the answer should be greater than the volume in cubic feet by something less than a factor of ten. The answer is indeed within this expected range.

Now let's consider a few more examples of dimensional analysis in problem solving. Keep in mind that it is not always possible to carry out a conversion in a single step. Suppose you cannot find an equivalence expression relating the two quantities, for example. If the two quantities can both be related to another quantity, then you can use two equivalence expressions to carry out the conversion, as shown in the next example.

EXAMPLE 1.5

If a laser beam fired from the moon takes 1.30 seconds to reach Earth, what is the distance in meters between the moon and Earth? Light travels in a vacuum at a speed of 3.00×10^{10} cm per second.

Solution

Step 1: The problem asks for distance in units of meters.
Step 2: The speed of light provides a relationship between centimeters and seconds for this particular problem:

$$3.00 \times 10^{10} \text{ cm} = 1 \text{ s}$$

However, we need to relate seconds to meters. A second equivalence expression is needed to convert centimeters to meters:

$$100 \text{ cm} = 1 \text{ m}$$

Step 3: It is useful in a multistep conversion to set up a road map:

$$\text{s} \longrightarrow \text{cm} \longrightarrow \text{m}$$

Then set up the problem with the two conversion factors:

$$? \text{ m} = 1.30 \text{ s} \times \frac{3.00 \times 10^{10} \text{ cm}}{1 \text{ s}} \times \frac{1 \text{ m}}{100 \text{ cm}} = 3.90 \times 10^8 \text{ m}$$

Step 4: Because the units came out right, the setup is probably correct. We know the

(continued)

distance should be quite large, so the magnitude of the answer appears reasonable.

Practice Problem 1.5

The nearest Earth comes to Venus is 4.83×10^{10} m. How long would it take for a laser beam to travel this distance?

Answer: 161 s

Sometimes it is necessary to convert more than one quantity in a single problem. The conversions are simply set up in sequence to convert one quantity at a time.

EXAMPLE 1.6

Gasoline, primarily a mixture of hydrocarbons, has a density of 7.0 lb/gal. What is the density of gasoline in grams per cubic centimeter (g/cm^3)?

Solution

Step 1: The quantity and units required are clearly stated in the problem: density in grams per cubic centimeter.

Step 2: The data are given in units of pounds per gallon, so relationships must be found between the units of mass (pounds and grams) and between the units of volume (gallons and cubic centimeters). Such relationships are given in Table 1.4: 1 lb = 453.6 g and 1 gal = 3785 cm^3.

Step 3: Set up a road map:

$$\text{lb/gal} \longrightarrow \text{g/gal} \longrightarrow \text{g/cm}^3$$

Then set up the problem:

$$\text{Density in g/cm}^3 = 7.0 \, \frac{\text{lb}}{\text{gal}} \times \frac{453.6 \text{ g}}{1 \text{ lb}} \times \frac{1 \text{ gal}}{3785 \text{ cm}^3} = 0.84 \, \frac{\text{g}}{\text{cm}^3}$$

In the answer, note that the number of significant figures (two) is determined by the number of significant figures in the data (7.0 has two) and not by the number of significant figures in the conversion factors, which are known exactly. If we did not have a direct equivalence between gallons and cubic centimeters, we could have taken the less direct path: gal $\rightarrow$ $ft^3 \rightarrow in^3 \rightarrow cm^3$.

Step 4: All the units canceled properly to give the desired units for the answer, so the setup is probably correct. Once you become familiar with the properties of liquids, you will be able to predict that most liquids should have densities not too different from that of water (1.0 g/cm^3). The numerical answer, then, also appears reasonable.

Practice Problem 1.6

Iron has a density of 7.86 g/cm^3. What is the weight in lb of 1.00 ft^3 of iron?

Answer: 4.91×10^2 or 491 lb

Finally, it may be necessary to use a mathematical equation to solve a problem. Dimensional analysis can still be used to convert quantities within the equation to proper units and to check for appropriate setup.

Summary

Chemistry is the study of matter, its properties, and the change of one substance into another. The properties of matter can be categorized as physical or chemical. Matter may consist of pure substances (elements and compounds) or mixtures of substances. Elements can be further classified as metals, nonmetals, or metalloids.

The scientific method is a general investigative approach used by scientists. Observation, or collection of data, is a necessary part of the scientific method. It frequently involves the design of carefully controlled experiments. Hypotheses are proposed to explain the results of experiments and to make predictions that can be tested by further experimentation. Observations often involve measurement. A consistent set of units for making measurements is provided by the Système Internationale (SI), a form of the metric system, which is based on powers of ten. Measurements involve some degree of uncertainty, indicated by the use of the proper number of significant figures. Precision describes the extent to which measurements are reproducible. Accuracy describes the difference between a measurement and the correct or expected value. Interpretation of measurements often involves mathematical manipulations, which can be made easier if the units associated with the numbers are used as clues to the way the manipulations should be carried out.

Key Terms

absolute zero (1.3)	Fahrenheit scale (1.3)	metal (1.1)	reactive substance (1.1)
accuracy (1.4)	gas (1.1)	metalloid	scientific method (1.2)
Celsius scale (1.3)	heterogeneous	(semimetal) (1.1)	scientific notation (1.3)
chemical change (1.1)	mixture (1.1)	mixture (1.1)	SI (1.3)
chemical property (1.1)	homogeneous mixture	nonmetal (1.1)	significant figures (1.4)
chemical reaction (1.1)	(solution) (1.1)	physical change (1.1)	solid (1.1)
compound (chemical	hypothesis (1.2)	physical property (1.1)	stable substance (1.1)
compound) (1.1)	Kelvin scale (1.3)	physical state (1.1)	substance (1.1)
conversion factor (1.5)	law (1.2)	precision (1.4)	temperature (1.3)
density (1.3)	liquid (1.1)	product (1.1)	theory (1.2)
element (1.1)	mass (1.3)	pure substance (1.1)	volume (1.3)
energy (1.3)	matter (1.1)	reactant (1.1)	weight (1.3)

Exercises

Matter

1.1 What is matter?

1.2 Describe various ways to classify matter and give examples for each.

1.3 Compare and contrast physical and chemical properties.

1.4 What are the physical states? In what ways do they differ?

1.5 Classify each of the following as a physical or chemical change.
   a. Ice is formed from water.
   b. A cake is baked.
   c. An apple ripens.
   d. Paper is torn.
   e. Food is digested.
   f. A rubber band is stretched.
   g. Paper is burned.

1.6 Classify each of the following as a physical or chemical property.
   a. decomposition on heating
   b. color
   c. hardness
   d. melting point
   e. ability to burn
   f. brittleness
   g. odor

1.7 Classify each of the following as a pure substance, a homogeneous mixture (solution), or a heterogeneous mixture.
   a. hamburger   b. sugar   c. air
   d. diamond   e. wood   f. milk
   g. table salt

1.8 Classify each of the following as a pure substance, a homogeneous mixture (solution), or a heterogeneous mixture.
   a. ice   b. seawater
   c. soil   d. helium gas
   e. helium gas in a balloon   f. salad dressing
   g. blood

1.9 On heating a red powder, you obtain a colorless gas and a silvery liquid. Classify the red powder as a compound or an element. Can you classify the gas or the liquid?

1.10 When limestone is heated, a gas is released, leaving a white powder called lime. Is limestone a compound or an element? Can you classify lime? Explain.

1.11 Using circles or squares to represent particles, draw a picture showing the difference between pure substances and mixtures.

The Scientific Method

1.12 Explain how a hypothesis is used in scientific research.

1.13 Classify each of the following as a fact, a hypothesis, a law, or a theory.
  a. Bad luck results from walking under a ladder.
  b. Oil floats on water.
  c. Crime rates increase when the moon is full.
  d. Wood burns.
  e. When wood burns, oxygen is consumed.
  f. Heavier-than-air objects always fall toward the center of the earth.
  g. Matter is composed of atoms.

1.14 You observe a piece of balsa wood floating on water and propose the hypothesis that all wood floats on water. Suggest experiments to test this hypothesis.

Measurement and the Metric System

1.15 Why should units be used with numbers expressing measurements?

1.16 How can we express multiples of a given SI base unit?

1.17 Discuss the advantages of the metric system over the English system. Are there any disadvantages?

1.18 Carry out the following operations to the correct number of significant figures.
  a. 4.0 cm + 1.8 cm + 4.2 mm
  b. 2.0 h + 12 min
  c. 16 mm $\times$ 2.0 cm
  d. 6.4 g/2.0 mL
  e. 41 g + 2.00 kg

1.19 The Olympic "mile" is 1.500×10^3 meters. What is this distance in feet?

1.20 Carry out the following unit conversions to the proper number of significant figures.
  a. 3.6 in to mm
  b. 18 s to min
  c. 1.4 L to mL
  d. 18 g to kg
  e. 2.34 m^3 to cm^3
  f. 1.8 ft to m

1.21 Normal body temperature is 98.6°F. What is this temperature in degrees Celsius? in kelvins?

1.22 Calculate the density of the following substances to the proper number of significant figures.
  a. 1.03 cm^3 of uranium with a mass of 6.11 g
  b. 2.65 cm^3 of aluminum with a mass of 7.15 g
  c. 2.50 L of oxygen gas with a mass of 3.58 g
  d. 25 cm^3 of concentrated hydrochloric acid with a mass of 29.8 g
  e. a 10.0-cm cube of gold with a mass of 19.32 kg

1.23 Iron has a density of 7.86 g/cm^3. What is the mass of a strip of iron measuring 2.52 cm $\times$ 11.6 cm $\times$ 1.05 cm?

1.24 What is the volume of 125 g of mercury if its density is 13.6 g/cm^3?

1.25 Platinum has a density of 21.45 g/cm^3. What is the mass of a cube of platinum that measures 1.00 inch on each side?

1.26 The density of water is 1.00 g/cm^3. If 25.0 mL of water has the same mass as 31.7 mL of ethyl alcohol, what is the density of ethyl alcohol?

1.27 At what temperature do the Fahrenheit and Celsius scales have the same value?

1.28 A modern automobile engine operates at $2.50 \times 10^{2\circ}$F. What is this temperature in kelvins?

Significant Figures

1.29 What are the rules for expressing a sum or difference with the correct number of significant figures? How do the rules change for multiplication or division?

1.30 Explain the concepts of accuracy and precision.

1.31 To how many significant figures should conversion factors or physical constants be expressed in a calculation?

1.32 Indicate the number of significant figures in each of the following.
  a. 0.075
  b. 760
  c. 643×10^{12}
  d. 1.002
  e. 0.002
  f. 0.00200
  g. 200.0

1.33 Round each of the following numbers to three significant figures.
  a. 1.742758
  b. 0.013279
  c. 1.005213
  d. 0.005213
  e. 1,279,007
  f. 1.279007×10^6

1.34 Perform the indicated operations and express each answer to the proper number of significant figures.
  a. 4.01×1.5
  b. 2.061/2.00
  c. $2.40 \times (2.01 \times 10^{10})/1.2$
  d. 12 + 1.115

Dimensional Analysis

1.35 Explain how units should be treated in a calculation.

1.36 Two conversion factors can be derived from the same equivalence expression. Explain. How can you decide which is the correct one to use in a given calculation?

1.37 A famous cartoon shows Einstein trying to decide which equation is correct: $E = mc$, $E = mc^2$, $E = mc^3$, $E = mc^4$. Show how this decision could be made with dimensional analysis.

1.38 Assuming 1 year contains 365 days, determine how many minutes there are in an average lifetime of 72 years.

1.39 How many liters are there in a 12.0-gal gas tank?

1.40 What is the mass of a 123-lb woman in kilograms?

1.41 What is the length in meters of a 300.0-ft football field?

1.42 What is the mass in grams of a 1.00-oz letter?

1.43 A standard floor tile is 12-inches square. What is its area in square centimeters?

1.44 Convert a speed limit of 55 mi h^{-1} to km h^{-1}.

1.45 If a runner does the 100-meter dash in 9.7 s, what is the average speed of this runner in mi h^{-1}?

1.46 Light travels at a speed of 3.00×10^8 m s^{-1}. The earth is 9.3×10^7 mi from the sun. How long does it take for sunlight to reach the earth?

each side?

1.48 The density of water is 1.000 g cm^{-3} at 4°C. What is this density in units of lb ft^{-3}?

1.49 If the density of gasoline is 0.70 g cm^{-3}, how much does the gasoline in a full 16.0-gal tank weigh in kilograms? in pounds?

1.50 What is the mass in kilograms of a cubic yard of soil that has a density of 1.28 g cm^{-3}? What is the mass in pounds?

1.51 How many grams of copper are contained in 1.00 lb of an ore that is 16% copper by mass?

1.52 If a laser beam, which travels at a speed of 3.00×10^{10} cm s^{-1}, takes 2.560 s to travel to the moon and back to earth, what is the distance to the moon in meters? in miles?

Additional Exercises

1.53 Perform the indicated operations and express each answer to the proper number of significant figures.
a. $1.20 \times (1.91 \times 10^{15})/2.4$
b. 24.1×2.2
c. $15 + 2.221$
d. $19.52/4.00$

1.54 Classify each of the following as a physical or chemical change.
a. A photograph is developed.
b. Concrete hardens.
c. Butter is cut.
d. Butter is melted.
e. Butter is burned.
f. Iron rusts.
g. Iron is magnetized.

1.55 What is the density of milk in g cm^{-3} if 3.00 L of milk weighs 3.18 kg?

1.56 Carry out the following unit conversions.
a. 36 mm to in
b. 12 h to min
c. 125 mL to L
d. 25 kg to g
e. 154 cm^3 to m^3
f. 2.9 m to ft

1.57 The weather report on television gives the temperature as 92°F. What is the temperature in degrees Celsius? in kelvins?

1.58 Classify each of the following as a pure substance, a homogeneous mixture (solution), or a heterogeneous mixture.
a. a penny
b. a dollar bill
c. gasoline
d. distilled water
e. a martini (no olive)
f. a martini (with olive)
g. popcorn

1.59 What is the volume in milliliters of a 5.00-gal pail? What is the volume in liters?

1.60 Carry out the following operations.
a. 5.0 cm + 3.2 cm + 1.4 mm
b. 2.0 min + 24 sec
c. 25 mm × 4.0 cm^2
d. 2.7 g/3.0 mL
e. 12 mg + 1.98 g

1.61 If a basketball player is 7 ft 2 in tall, what is his height in meters?

1.62 A giant-size box of laundry detergent contains 25 lb. What is this mass in kilograms?

1.63 The distance between two cities is 92 miles. What is this distance in kilometers?

* 1.64 Human blood contains an average of 6.0×10^3 red blood cells in a volume of 0.1 mm × 0.1 mm × 0.1 mm. How many red blood cells are there in each cm^3 of blood? If red blood cells have an average lifetime of about 1 month (30 days) and the adult blood volume is about 5 liters, how many red blood cells are generated every second in the bone marrow?

* 1.65 Battery acid has a density of 1.285 g cm^{-3} and contains 38.0% sulfuric acid by mass. How many grams of pure sulfuric acid are contained in 1.00 L of battery acid?

* 1.66 The recommended daily allowance of vitamin B$_2$ (riboflavin) is 1.7 mg per day for an average adult. If cheese, which contains 5.5×10^{-6} g vitamin B$_2$ per gram of cheese, were a person's only source of vitamin B$_2$, how many pounds of cheese would that person need to consume daily?

* 1.67 The melting point of iodine is 114°C, and its boiling point is 184°C. Suppose we create a new temperature scale in which the melting point of iodine is 0°I and the boiling point of iodine is 100°I. Derive an equation to show the relationship of this temperature scale to the Celsius scale. What is the boiling point of water on the iodine scale? What is absolute zero on the iodine scale?

2

▶ ▶ ▶ ▶

Atoms, Molecules, and Ions

Titanium metal and chlorine gas (in bottle) can combine to form Ti(IV) chloride (in cylinder) or Ti(III) chloride (on filter paper).

Encountering Chemistry

▶ ▶ ▶ ▶

Which properties of these substances are different? Are all their properties different?

In Chapter 1, we noted that all matter is composed of 109 known elements. The chapter-opening photograph shows two of them. One is titanium, a strong, low-density, silver-white metal that is often combined with other metals to add strength and lightness. For example, the metal in airplane wings and spacecraft contains titanium. The other element is chlorine, a nonmetal that normally exists as a pale green gas when it is not combined with other elements. We can obtain the element titanium by breaking down a variety of naturally occurring minerals, such as rutile (which contains titanium combined with oxygen) and ilmenite (which contains iron, titanium, and oxygen). Wherever it comes from, the titanium is the same type of matter, with the same chemical and physical properties. Furthermore, no chemical or common physical processes will allow us to break down titanium into more elementary substances. Titanium can be separated into smaller pieces, but these pieces are still the same element. The same is true for chlorine: we can isolate smaller quantities of chlorine, but these quantities represent the same element.

All chemical substances that are not elements are composed of two or more elements in combination. The opener photograph shows two substances made up of titanium and chlorine: titanium(IV) chloride and titanium(III) chloride (also called titanium tetrachloride and titanium trichloride, respectively). Though both of these substances contain titanium in combination with chlorine, each has properties that are different from those of the other and from those of the original elements. In this chapter, we begin to consider how elements combine to form different chemical substances. We will also see how the composition of these substances is represented and how amounts of the substances are expressed.

2.1   Atomic Theory

What makes titanium and chlorine so different from each other? ▶ And how do they combine to form titanium(IV) chloride and titanium(III) chloride, substances different from either starting material? To begin to answer these questions, we must turn our attention to the atom. **Atoms** are the smallest particles of an element that can still be identified as that element and that retain the element's characteristic chemical properties.

The idea that all matter consists of small, separate particles called atoms probably originated in ancient Greece. After centuries of neglect, the theory was reintroduced in the seventeenth century and was gradually reformulated. A number of investigators contributed to this scientific revolution, but John Dalton (Figure 2.1) is most often associated with the beginnings of modern atomic theory. Dalton's ideas drew on work on the measurement of mass relationships in chemical reactions. Such measurements had led to the formulation of two important laws: the law of conservation of mass and the law of definite proportions.

Figure 2.1 John Dalton proposed his atomic theory in 1804. This theory is a cornerstone of modern chemistry. Dalton began teaching school at the age of twelve and later became a professor of mathematics at Manchester College, Oxford.

Law of Conservation of Mass

In 1785, Antoine Lavoisier (Figure 2.2) clearly showed that there is no measurable change in mass during a chemical reaction—that is, the mass of the substances produced by the reaction (the products) is always equal to the mass of the reacting substances (the reactants). This statement is called the **law of conservation of mass.**

Consider, for example, the reaction between carbon and sulfur. When these two substances are heated, a new substance called carbon disulfide is formed

Figure 2.2 Antoine Lavoisier worked with his wife, Marie, who assisted him in the lab, kept his laboratory notes, and wrote many of his papers. Lavoisier is the person most responsible for demonstrating the importance of careful quantitative measurements. He brought order to the science of chemistry by weighing the substances involved in chemical changes.

Figure 2.3 When solid carbon and sulfur are heated together, they form the liquid carbon disulfide, a vile-smelling, yellow liquid in the usual commercial preparations. If carbon disulfide is carefully purified, it is a colorless, sweet-smelling liquid. Carbon disulfide is used in the preparation of other commercial products, such as rayon, carbon tetrachloride, and soil disinfectants.

What would happen if 5.00 g of sulfur were combined with 1.00 g of carbon?

(Figure 2.3). It is possible to form carbon disulfide by mixing different amounts of carbon and sulfur. For example, if 1.00 g of carbon is combined with 5.34 g of sulfur, 6.34 g of carbon disulfide is formed. If 2.00 g of carbon is combined with 10.68 g of sulfur, 12.68 g of carbon disulfide is formed. In each case, the sum of the masses of the reactants equals the mass of the product.

Both examples just given use the ratio 5.34 g of sulfur to each gram of carbon. It is possible to carry out the reaction with carbon and sulfur in different proportions. If the proportions differ from the ratio of 5.34 to 1.00, some reactant will remain when the reaction is completed, but the sum of the masses of reactants consumed will still equal the mass of carbon disulfide produced. If 6.00 g of sulfur is used, for example, the reaction still produces 6.34 g of carbon disulfide and consumes 5.34 g of sulfur and 1.00 g of carbon, though 0.66 g of sulfur is left over. On the basis of observations such as these, Lavoisier showed that mass is neither lost nor gained during a chemical reaction.

Law of Definite Proportions

The element tin forms a white, powdery substance, called tin(IV) oxide, by reacting with the oxygen in air when heated (Figure 2.4). Table 2.1 gives the results of

Figure 2.4 Tin is a silvery-white metal often used as a protective coating on steel cans. When heated, tin reacts with the oxygen in air to form tin(IV) oxide.

Table 2.1 Composition of
Tin(IV) Oxide

Mass of Tin (g)	Mass of Tin Oxide (g)	Increase in Mass from Oxygen (g)	$\dfrac{\text{Mass Tin}}{\text{Mass Oxygen}}$
5.00	6.35	1.35	3.70
10.0	12.7	2.7	3.7
23.7	30.1	6.4	3.7
73.4	93.2	19.8	3.71

this reaction for tin samples of various size. No matter how much tin is used, the same proportions by mass of tin and oxygen always result. This observation about the composition of substances is summarized in the **law of definite proportions:** all samples of the same pure substance always contain the same proportions by mass of the component elements.

Dalton's Atomic Theory

If matter were continuous rather than composed of particles, could these laws still be observed? Why or why not?

Dalton reasoned that the law of conservation of mass and the law of definite proportions could be explained only if matter was composed of atoms. He proposed as a working hypothesis an atomic theory that formed the basis for all modern developments in chemistry. Dalton's theory can be summarized as follows:

1. All matter is composed of atoms, indivisible particles that are exceedingly small.
2. All atoms of a given element are identical, both in mass and in chemical properties. However, atoms of different elements have different masses and different chemical properties.
3. Atoms are not created or destroyed in chemical reactions.
4. Atoms combine in simple, fixed, whole-number ratios.

A chemical reaction, then, is a way of rearranging atoms into new combinations, resulting in the formation of new chemical compounds, as in the reactions of sulfur and oxygen shown in Figure 2.5. When a sulfur atom and two oxygen atoms combine to form sulfur dioxide, they retain their identities and their characteristic masses, so there can be no change in mass. And because a unit of sulfur dioxide always contains one sulfur atom and two oxygen atoms, it always contains the same proportions by mass of sulfur and oxygen. However, these proportions will be different when sulfur (or sulfur dioxide) reacts with oxygen to form sulfur trioxide.

Modern Atomic Theory

We now know that some of the premises in Dalton's atomic theory are not completely correct. The third and fourth statements are consistent with modern atomic

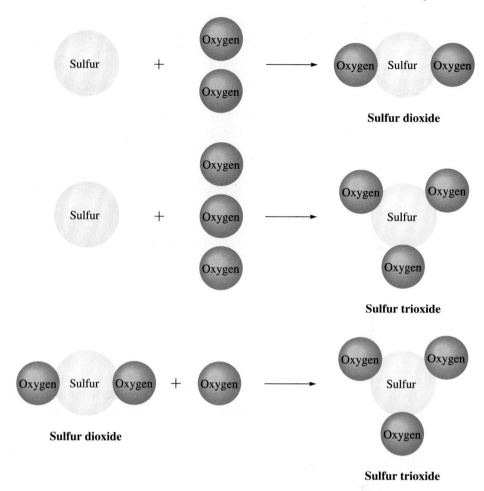

Figure 2.5 Sulfur and oxygen atoms can combine in different ways to form either sulfur dioxide or sulfur trioxide, which have different chemical compositions and different properties.

Sulfur dioxide is a colorless, nonflammable gas with a powerful choking odor. It is used for preserving fruits and vegetables and for bleaching textiles. Sulfur trioxide is a volatile liquid at room temperature that freezes at 16.8°C. It combines with water to form sulfuric acid. These sulfur oxides are formed by the burning of coal or oil that contains sulfur, and they are partially responsible for acid rain.

theory; that is, we cannot create or destroy atoms in chemical reactions, and atoms do combine in fixed proportions. The second statement is almost correct: atoms of a particular element do share identical chemical properties; however, atoms of a given element can actually vary somewhat in mass. It is also not strictly true that, as the first statement says, atoms are indivisible. Atoms can be divided, and they can be combined to form atoms of other elements, but not as the result of ordinary chemical reactions or under ordinary laboratory conditions. These concepts will be explored in the discussion of radioactivity in Chapter 20.

To establish the existence of atoms, Dalton had to rely on the ability of his theory to predict the behavior of matter. However, it is now possible to visualize directly those atoms that for so long were conceptualized on the basis of indirect evidence. Electron microscopes achieve such tremendous magnifications (as great as 10 million times) that they can distinguish between adjacent objects as small as atoms. An example of an electron micrograph is shown in Figure 2.6.

Figure 2.6 A sample of iodine adsorbed on platinum, as seen with a scanning tunneling microscope. The scale shows the size of the atoms (colored pink) in nanometers.

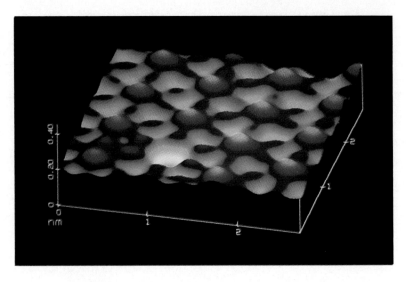

2.2 Structure of the Atom

We now know that atoms are not the small, impenetrable balls Dalton envisaged. They consist of even smaller particles—the fundamental particles that combine to make up all atoms. The three particles of importance to chemistry are electrons, protons, and neutrons. **Electrons** are small, electrically charged particles that are contained within all elements. They are the carriers of electrical current as it passes through wires. The existence of electrons was demonstrated in the early twentieth century in a series of classic experiments with cathode-ray tubes conducted by J. J. Thomson (Figure 2.7). Cathode-ray tubes conduct electricity when subjected to high voltages. By deflecting rays emanating from the cathode (negatively charged electrode) with electrical and magnetic fields, Thomson was able to show that the rays had a negative electrical charge and that they were identical no matter what material the cathode was made from. This result indicated that the negatively charged particles (or electrons) in these rays were common to all matter. Thomson was also able to calculate the ratio of the charge to the mass of the electron from such experiments.

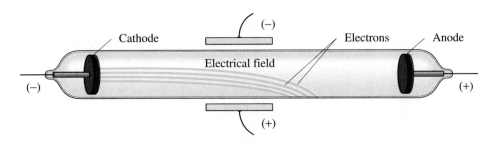

Figure 2.7 Joseph John Thomson, an English physicist at Cambridge University, won the Nobel Prize in physics in 1906 for his work with cathode-ray tubes, the forerunners of modern television tubes.

▶ ▶ ▶ WORLDS OF CHEMISTRY

Scientists Manipulate Single Atoms

Techniques such as this may one day be used to control the reactions of single atoms or molecules and to produce new materials or structures of materials that could not be obtained any other way.

The scanning tunneling microscope (STM), developed in 1986, allows scientists to observe single atoms in a variety of environments. Generally the atoms must be deposited on a clean, flat surface, such as on platinum metal. Unlike ordinary microscopes, which use visible light rays and a series of lenses to make an object visible, STMs use a beam of electrons to make objects as small as atoms "visible." This technique involves processing data with a computer and usually requires image-enhancing technology. A thin, needlelike probe, which emits the electron beam, is moved over the outer layer of the sample being viewed. The STM is generally used to examine surfaces, such as that of silicon used to construct very small microelectronic devices.

Recently scientists have discovered that the STM can also be used to move single atoms on the surface being examined. If the voltage passed through the probe wire is carefully controlled, a single atom or molecule can be picked up, moved to a new location, and deposited there. Figure 2.8 shows the results of a whimsical experiment in which 28 carbon monoxide molecules were moved on a platinum surface to form the shape of a "molecular person."

The ability to move single atoms goes far beyond atomic art. Techniques such as this may one day be used to control the reactions of single atoms or molecules and to produce new materials or structures of materials that could not be obtained any other way. It may be possible, for ex-

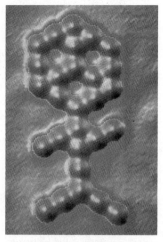

Figure 2.8 By dragging carbon monoxide molecules around on a platinum surface with a scanning tunneling microscope probe, scientists created this piece of atomic art. It is possible to move individual atoms to desired locations in this way.

ample, to assemble extremely small semiconductor devices by relocating the atoms on a silicon surface.

Questions for Discussion

1. If a computer must be used to interpret and enhance an image of an atom, are we really seeing an atom? Is this situation different from using lenses to enhance an image of a plant cell?
2. What advantages might arise from being able to "see" atoms by means of a scanning tunneling microscope?
3. If you had a scanning tunneling microscope, what experiments would you want to perform?

Using these results, Robert Millikan was able to determine the charge on the electron from an experiment with oil droplets (Figure 2.9). The droplets were exposed to radiation that caused them to take on an electrical charge. By measuring the magnitude of the electrical field necessary to cause the droplets to hang suspended in air, Millikan was able to determine the charge on the electron. His experiments provided values of 9.1094×10^{-28} g for the mass and $-1.6022 \times$

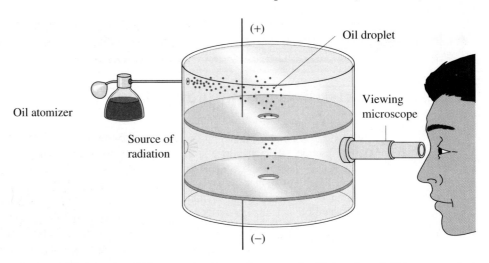

Figure 2.9 Robert A. Millikan, a physics professor at the University of Chicago, carried out a series of experiments in 1909 with oil droplets.

If the electron is so light, what makes up the mass of atoms?

10^{-19} C (coulombs) for the charge of an electron. Because even the lightest atoms weigh more than 10^{-24} g, an electron contributes only a small part of the mass of an atom. In fact, the mass of an electron is smaller than the mass of one hydrogen atom by a factor of 1837.

Atoms are electrically neutral. Therefore, because atoms contain negatively charged electrons, they must also contain positively charged particles. These particles are called **protons,** each of which has a charge equal in magnitude but opposite in sign to the charge on an electron. To maintain electrical neutrality, an atom must have equal numbers of protons and electrons. The mass of a proton, 1.6726×10^{-24} g, is nearly the same as the mass of the hydrogen atom.

The mass of most atoms other than the hydrogen atom is at least twice the sum of the masses of all the protons and electrons the atoms contain. For example, calcium contains 20 protons and 20 electrons, which would have a mass of 3.3471×10^{-23} g. But a calcium atom has a total mass of 6.6359×10^{-23} g. This difference in mass is contributed by an electrically neutral particle called a **neutron.** The mass of the neutron is 1.6749×10^{-24} g, slightly greater than the mass of a proton.

The properties of the fundamental particles of atoms are listed in Table 2.2. To make it easier to deal with electrical charges in matter, we usually express them

Table 2.2 Fundamental Particles of Atoms

Particle	Mass (g)	Charge (C)
Electron, e^-	9.1094×10^{-28}	-1.6022×10^{-19}
Proton, p^+	1.6726×10^{-24}	$+1.6022 \times 10^{-19}$
Neutron, n	1.6749×10^{-24}	0

as a multiple of the charge of an electron or of a proton, instead of in units of coulombs as given in the table. Expressed in this way, the charge of an electron is $1-$, and the charge of a proton is $1+$.

The Nuclear Atom

Early models of atomic structure assumed that the protons and electrons were evenly distributed throughout the atom. Owing in large part to the work of Ernest Rutherford, we now know that most of the atom is empty space. Rutherford discovered a type of particle, called an alpha particle, that has a positive electrical charge $(2+)$. In experiments designed by Rutherford and carried out by his associate, Hans Geiger, and Geiger's student, Ernest Marsden, thin gold foil was bombarded by alpha rays (Figure 2.10). Most of the alpha particles went straight through the foil, but some were deflected by small angles, and a few were actually deflected backwards. This behavior suggested that most of the mass of the atom must be concentrated in a positively charged core, called the **nucleus,** which is surrounded by the electrons. Only if an alpha particle came close enough to the small, heavy nucleus would it be deflected from its original path, or trajectory. Because only a few particles were deflected backwards—a deflection that would

What is in the space not occupied by the nucleus or the electrons?

be caused by a direct hit on the nucleus—the nucleus must be quite small compared with the overall size of the atom. The protons and neutrons are located in the nucleus, so a very small portion of the total volume of the atom contains most of the atomic mass. The electrons move about outside the nucleus, where they determine the outer boundaries of the atom.

The diameter of the nucleus is about 10^{-14} m, and the diameter of the atom is about 10^{-10} m. These relative sizes are comparable to a penny in the center of a baseball field. Although most of the atom's mass is located in the nucleus, it is the electrons and their arrangement around the nucleus that account for the chemical behavior of the elements. In Chapter 6, we will see how this arrangement determines how one element reacts with another.

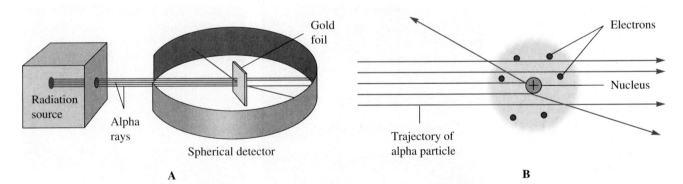

A   **B**

Figure 2.10 Ernest Rutherford, a native of New Zealand, emigrated to England to study with Thomson, became a professor of physics at Manchester and Cambridge universities, and won the Nobel Prize in 1908. (*A*) In his experiment, thin gold foil was bombarded by alpha particles. Most went straight through the foil, but some were deflected by small angles, and a few were deflected backwards. (*B*) The behavior of the alpha particles suggested the nuclear model of the atom.

Atomic Number and Mass Number

▶ As described in the chapter opening, titanium is a silver-white metal and chlorine is a pale green gas. The properties of these elements are different because their atoms contain different numbers of subatomic particles. In fact, we use the number of subatomic particles to distinguish among the elements. The primary differentiation comes from the number of protons. Because atoms are electrically neutral, the number of electrons in an atom is equal to the number of protons. For example, the simplest atom, hydrogen, consists of one proton and one electron (Figure 2.11). As we will see shortly, some forms of hydrogen contain one or two neutrons as well. But it is the presence of one proton that distinguishes all these atoms as hydrogen. All atoms of the next simplest element, helium, contain two protons in the nucleus and two electrons surrounding the nucleus; most helium atoms also contain two neutrons in the nucleus. Lithium contains three protons and three electrons; most lithium atoms also have four neutrons, though some lithium nuclei may contain more or fewer neutrons.

Atoms of an element with the same number of protons but different numbers of neutrons are known as **isotopes.** The isotopes of an element have essentially identical chemical properties, but their physical properties may differ somewhat. An example is hydrogen, which has three isotopes: hydrogen, deuterium, and tritium. Some of the properties of these hydrogen isotopes are listed in Table 2.3. Note the changes in the temperature at which the solid melts (melting point) and in the temperature at which the liquid boils (boiling point).

The number of protons in the nucleus is called the **atomic number (Z).** All atoms of the same element have the same atomic number. Values of atomic numbers of the elements are given in the table on the inside back cover of this book. The number of neutrons in the nucleus is called the **neutron number (N).** The sum of the number of protons (Z) and the number of neutrons (N) is the **mass number (A).** Thus, the isotope of helium with two protons and two neutrons has a mass number of 4 and an atomic number of 2.

Figure 2.11 Hydrogen, a colorless gas, normally exists at room temperature not as single atoms but as particles made up of two associated atoms. Because it is the lightest known gas and has great lifting power, it was used to inflate balloons *(left)* and dirigibles in the early part of this century; but it tends to burn explosively in air *(right)*, so it is no longer used in this way. Hydrogen is now being considered as a potential fuel because it forms only water when it burns, generating no air pollutants. It is also used, along with oxygen, in fuel cells for spacecraft.

Table 2.3 Some Properties of
Hydrogen Isotopes

Name	Hydrogen	Deuterium	Tritium
Symbol	^1_1H	^2_1H	^3_1H
Protons in an atom	1	1	1
Neutrons in an atom	0	1	2
Mass number of an atom	1	2	3
Percentage of all hydrogen in nature	99.9844%	0.0156%	—*
Melting point	13.96 K	18.73 K	20.62 K
Boiling point	20.39 K	23.67 K	25.04 K

*Tritium does not exist in nature, although it can be produced by the radioactive decay of other elements. It is itself radioactive and is transformed into other elements relatively rapidly.

EXAMPLE 2.1

Fluorine, a pale yellow, toxic gas with a sharp, pungent odor, has an atomic number of 9 and a mass number of 19. How many protons, electrons, and neutrons are there in a fluorine atom?

Solution   The atomic number gives the number of protons in the nucleus; thus, fluorine contains 9 protons. The number of electrons in an atom equals the number of protons, so fluorine has 9 electrons. The number of neutrons is obtained from the mass number and the atomic number:

$$N = A - Z$$
$$N = 19 - 9 = 10$$

Fluorine has 10 neutrons.

Practice Problem 2.1

Sodium, a soft, silvery-white metal that reacts vigorously with water, has an atomic number of 11 and a mass number of 23. How many protons, electrons, and neutrons are there in a sodium atom?

Answer: 11 protons, 11 electrons, 12 neutrons

The mass number and atomic number for a given element are often represented by the following notation:

$$\text{Superscript is mass number} \longrightarrow {}^A_Z E \longleftarrow \text{Subscript is atomic number}$$

Here, E represents a one- or two-letter symbol for the element being considered. For example, the notation for the kind of hydrogen atom that contains one proton and no neutrons is ^1_1H, whereas hydrogen with one proton and one neutron (known as deuterium) is represented as ^2_1H. The symbol for helium with two protons and two neutrons is ^4_2He. Another notation for this isotope is He-4.

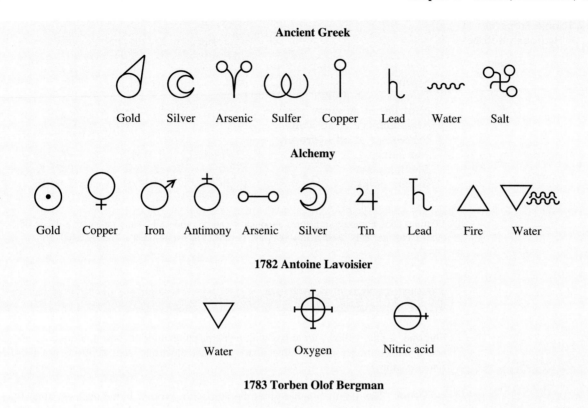

Figure 2.12 Many of the early symbols for the elements, such as those used by the alchemists prior to the seventeenth century, were rather puzzling. The alchemists,

Symbols of the Elements

The symbols used today to designate the elements are based on a system developed by the Swedish chemist Jöns Jakob Berzelius. In this system, the symbol for an element consists of the first one or two letters of its name (H for hydrogen, He for helium, Li for lithium). When the names of two elements start with the same first two letters (magnesium and manganese, for example), the symbol uses the first letter and a later letter (Mg for magnesium, Mn for manganese). For a few elements, the symbols are based on the Latin names or on names from other languages; these are listed in Table 2.4. Some recently discovered elements ($Z >$ 103) have not been given common names. They are named as the Latin word for their atomic number and given a three-letter symbol. For example, element 104, unnilquadium, has the symbol Unq. A list of the modern symbols can be found on the inside front cover of this book. Berzelius's system of symbols is much more convenient to use than earlier systems, whose symbols bear no relation to the names of the elements. Figure 2.12 gives a sampling of the modern and early symbols of the elements.

Pipes used to be made of lead. Why are those who work with pipes called plumbers?

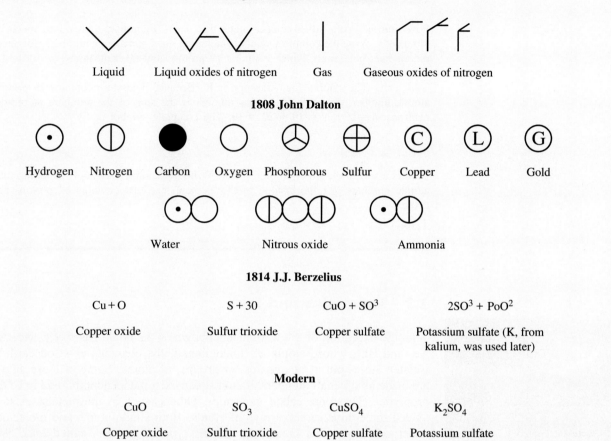

1787 J.H. Hassenfratz and Adet

Liquid   Liquid oxides of nitrogen   Gas   Gaseous oxides of nitrogen

1808 John Dalton

Hydrogen   Nitrogen   Carbon   Oxygen   Phosphorous   Sulfur   Copper   Lead   Gold

Water   Nitrous oxide   Ammonia

1814 J.J. Berzelius

Cu + O   S + 3O   CuO + SO^3 2SO^3 + PoO^2

Copper oxide   Sulfur trioxide   Copper sulfate   Potassium sulfate (K, from kalium, was used later)

Modern

CuO   SO$_3$ CuSO$_4$ K$_2$SO$_4$

Copper oxide   Sulfur trioxide   Copper sulfate   Potassium sulfate

who were interested in transforming base metals like lead into gold, used symbols that could not be easily interpreted by others. Even Dalton's system was difficult to use, for it was an attempt to represent the atom itself within the framework of his atomic model. The system we use today is very close to that proposed by Berzelius. (Reprinted with permission from *Chemical and Engineering News*, April 6, 1976. Copyright © 1976 American Chemical Society.)

Table 2.4 Elements with Symbols Based on Non-English Names

English Name	Symbol	Non-English Name	Language
Antimony	Sb	Stibium	Latin
Copper	Cu	Cuprum	Latin
Gold	Au	Aurum	Latin
Iron	Fe	Ferrum	Latin
Lead	Pb	Plumbum	Latin
Mercury	Hg	Hydrargyrum	Greek
Potassium	K	Kalium	Latin
Silver	Ag	Argentum	Latin
Sodium	Na	Natrium	Latin
Tin	Sn	Stannum	Latin
Tungsten	W	Wolfram	German

EXAMPLE 2.2

Potassium is a soft, silver-colored metal that reacts vigorously with water, much like sodium does. Write the complete symbol, including mass and atomic numbers, for an atom of the element potassium, which contains 19 protons and 20 neutrons.

Solution   The symbol for potassium is K. Because potassium contains 19 protons, the atomic number, Z, is 19. The mass number is the sum of the numbers of protons and neutrons: $A = Z + N = 19 + 20 = 39$. The complete symbol is $^{39}_{19}K$.

Practice Problem 2.2

Lead is a soft, dull, silver-colored metal. Write the complete symbol, including mass and atomic numbers, for the isotope of the element lead that contains 82 protons and 125 neutrons.

Answer: $^{207}_{82}Pb$

2.3   Molecules and Ions

With the exception of a few elements known as the noble gases (He, Ne, Ar, Kr, Xe, and Rn), whose atoms are uncombined, the elements exist on earth not as isolated atoms but as aggregates, or groups, of atoms. Some of these aggregates are made up of small numbers of atoms, can exist independently, and have distinct properties. These are called *molecules*. Others, such as the metals, exist as extended three-dimensional arrays of atoms. Both single atoms and molecules can undergo changes that create yet another type of particle, called *ions*. We will explore molecules and ions in more detail later in this section. First, let's examine a method for representing molecules that will make it easier to discuss them.

Chemical Formulas

Figure 2.13 Water and hydrogen peroxide are both composed of hydrogen and oxygen, but their properties differ greatly. Water is quite stable, whereas hydrogen peroxide readily decomposes into water and oxygen gas. As shown here, addition of manganese dioxide to water causes no change, but hydrogen peroxide decomposes rapidly in the presence of manganese dioxide.

Chemical compounds, like elements, are usually represented by a shorthand notation—the **chemical formula.** We construct chemical formulas by combining the symbols of the elements that form the substance and adding numerical subscripts to indicate either the number of atoms of each element present in a unit of the substance or their relative proportions. There are two kinds of chemical formulas: empirical and molecular. An **empirical formula** contains the simplest ratios of atoms, or the smallest possible whole-number subscripts. A **molecular formula** contains the actual number of atoms in the molecule. The molecular formula is always the same as or some multiple of the empirical formula. Thus, the molecular and the empirical formula for water, which contains two hydrogen atoms and one oxygen atom per unit, is H_2O. (Note that we do not use a subscript of 1.) The molecular formula for hydrogen peroxide, which contains two hydrogen atoms and two oxygen atoms, is H_2O_2, but its empirical formula is HO. The difference between water and hydrogen peroxide is shown in Figure 2.13.

For some substances there is no useful molecular formula because no specific molecule can be identified in the structure. Quartz sand, for example, consists of extended networks of interconnected silicon and oxygen atoms, as shown in Figure 2.14. The atoms occur in a 1 : 2 ratio, and the substance, silicon dioxide, is commonly represented by the empirical formula SiO_2.

When it is meaningful, we will use the molecular formula because it contains more information—not only the ratios of atoms but the actual number of atoms as well. Consider, for example, the substance benzene, a colorless liquid once used extensively as a solvent (dissolving agent) but now known to be carcinogenic, or cancer-causing. Benzene has the empirical formula CH, which tells us that benzene molecules contain one hydrogen atom for each carbon atom. This formula is not specific to benzene, though; a variety of molecules, including acetylene (C_2H_2), have the same empirical formula. In contrast, the *molecular* formula of benzene, C_6H_6, tells us that one molecule of benzene contains six carbon atoms and six hydrogen atoms in chemical combination. Some other examples of empirical and molecular formulas are given in Table 2.5.

Besides telling us what elements a substance contains and in what proportions, formulas also generally tell us something about how the atoms are combined—that is, the substance's structure. Thus, oxygen gas (O_2) has two oxygen atoms connected to each other; phosphorus (P_4) has four connected phosphorus atoms; sulfur (S_8) has eight connected sulfur atoms; and carbon dioxide (CO_2) has two oxygen atoms, each attached to the carbon atom (see Figure 2.15). When we write the formula for a binary compound (containing two elements), such as CO_2, we usually start with the element to which several atoms of the other element are attached. A major exception to this convention is water, with two hydrogen atoms connected to an oxygen atom. The formula for water is usually written H_2O instead of OH_2, which would follow the convention.

Another convention uses the letters *g*, *l*, and *s* in parentheses after a chemical formula to indicate physical state—gas, liquid, or solid. The composition of steam, water, and ice is accurately given by the formula H_2O; we can distinguish among these materials by writing $H_2O(g)$, $H_2O(l)$, and $H_2O(s)$. In addition, the symbol *aq* (for aqueous) is used to show that a substance is dissolved in water to form an aqueous solution. For example, carbon dioxide dissolved in water is represented by $CO_2(aq)$.

Are steam, water, and ice the same substance? If they are, how can they have such different properties?

Figure 2.14 The structure of quartz contains infinite networks of connected silicon and oxygen atoms. This quartz sample, called amethyst, is purple because it contains traces of iron.

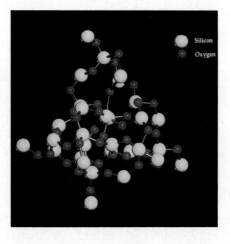

Silicon

Oxygen

Table 2.5 Some Empirical and
Molecular Formulas

Substance	Empirical Formula	Molecular Formula
Water	H_2O	H_2O
Hydrogen peroxide	HO	H_2O_2
Silicon dioxide	SiO_2	$(SiO_2)_x$
Ethylene	CH_2	C_2H_4
Cyclohexane	CH_2	C_6H_{12}
Oxalic acid	HCO_2	$H_2C_2O_4$

Molecules

Most of the substances we have been discussing are made up of small numbers of atoms connected together. These units are **molecules,** the smallest particles of pure substances that retain the composition and properties of those substances and can exist independently. (Some molecules are shown in Figure 2.15.) Molecules, then, are chemical combinations of two or more atoms. Some chemists consider this definition of molecules to include atoms of the noble gases; others assert that uncombined atoms are not molecules. We will generally call the smallest particles of the noble gases atoms, but we will treat them the same as gaseous molecules when we examine the kinetic-molecular theory in Chapter 9. All other elements are found either in molecular form or as extended arrays of atoms under normal conditions. (Under unusual conditions, however, such as at the high temperatures present in a flame, molecules may dissociate, or separate, into atoms.) A number of elements exist in nature primarily as small molecules. For example, several elements, such as the reactive nonmetals known as the halogens, exist commonly as **diatomic molecules,** molecules containing two atoms (see Figure 2.16). These may be gases (H_2, N_2, O_2, F_2, Cl_2), liquids (Br_2), or solids (I_2) under normal conditions. Single atoms of these substances are rarely found because they are extremely reactive and combine with almost any other substance they encounter. **Polyatomic molecules,** which contain two or more atoms, are also known. Ozone (O_3) is an example (see Figure 2.17); P_4 and S_8 are others. Most of the elements that exist uncombined in nature are solids that consist of extremely large numbers of atoms connected together. Examples are carbon (as either graphite or diamond)

Figure 2.15 Structures of atoms of helium and molecules of oxygen, phosphorus, sulfur, and carbon dioxide.

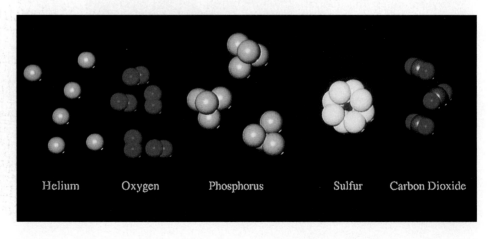

Helium   Oxygen   Phosphorus   Sulfur   Carbon Dioxide

Figure 2.16 Fluorine (yellow) and chlorine (green) are gases, whereas bromine (red) is a liquid and iodine (purple) is a solid at room temperature. These substances belong to a group of elements called the halogens. Fluorine is not shown because it is very reactive and dangerous to handle.

and the metals. The molecules of compounds are, of course, composed of atoms of more than one element. Examples include H_2O (water), SO_2 (sulfur dioxide), CO (carbon monoxide), CO_2 (carbon dioxide), and CH_4 (methane).

Ionic Substances

If an electron is added to an atom, where will it be located in the atom?

In addition to atoms and molecules, a third type of particle occurs in substances. These particles, called **ions,** are atoms or groups of atoms that carry an electrical charge. An ion forms when electrons are removed from or added to an atom or group of atoms (see Figure 2.18). When electrons are removed, the resulting ion has a positive charge and is called a **cation.** When electrons are added, the ion has a negative charge and is called an **anion.** The identity of the element is not altered by these changes in electron number, however; atomic sodium and sodium ion are both the element sodium. They do, however, have very different properties; so-

Figure 2.17 Ozone is a toxic blue gas with a characteristic pungent odor, which can be detected around electric motors and after thunderstorms. In the upper reaches of the atmosphere, ozone is essential because it absorbs ultraviolet radiation, which can cause cancer. Holes, or low concentrations, in the ozone layer of the upper atmosphere, caused by chlorofluorocarbons used in air conditioners, refrigerators, and aerosol cans, have created considerable concern. The decreasing concentration of ozone over the South Pole is shown by the colors in this image. Total ozone is measured in Dobson units, which equal the thickness, in 10^{-2} mm, if the ozone were at 1 atm and O°C. Ozone in the lower atmosphere, produced by automobile exhaust products, causes smog and health problems, so an excess of ozone in the air we breathe is also a problem.

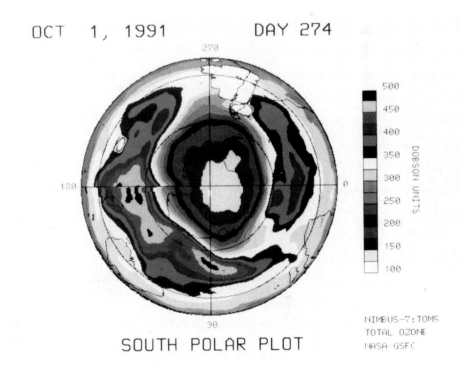

Cl $\xrightarrow{\text{Add 1 electron}}$ Cl$^-$

Atomic chlorine   **Chloride ion**
17 protons   17 protons
17 electrons   18 electrons

Na $\xrightarrow{\text{Remove 1 electron}}$ Na$^+$

Atomic sodium   **Sodium ion**
11 protons   11 protons
11 electrons   10 electrons

O$_2$ $\xrightarrow{\text{Add 2 electrons}}$ O$_2{}^{2-}$

Molecular oxygen   **Peroxide ion**
16 protons   16 protons
16 electrons   18 electrons

Figure 2.18 Ions are formed when electrons are added to or removed from atoms. Ions can also be formed from molecules.

dium is a very reactive metal, whereas sodium ion, associated with a negatively charged ion (as in table salt or seawater, for example), is unreactive.

The ionic charge represents the number of electrons lost (if positive) or gained (if negative). The charge is indicated in the symbol or formula by a superscript number followed by the + or − sign. Removing one electron from the Na atom creates Na$^+$, or sodium ion; Na$^+$ is an example of a **monatomic ion,** an ion formed from one atom. Adding one electron to the Cl atom creates Cl$^-$, chloride ion, another monatomic ion. Adding or removing more than one electron creates ions with a greater charge. Some examples are Ca^{2+} (calcium ion), Al^{3+} (aluminum ion), S^{2-} (sulfide ion), and N^{3-} (nitride ion).

The formation of ions can be represented by a shorthand notation called an equation. In the equations that follow, the arrow simply means a process is occurring, and e$^-$ represents an electron:

$$\text{Na} \longrightarrow \text{Na}^+ + \text{e}^- \qquad \text{Al} \longrightarrow \text{Al}^{3+} + 3\text{e}^-$$
$$\text{Cl} + \text{e}^- \longrightarrow \text{Cl}^- \qquad \text{S} + 2\text{e}^- \longrightarrow \text{S}^{2-}$$

Note that when electrons are being removed to form a cation, they show up on the right-hand side of the equation; electrons added to form an anion are placed on the left-hand side. The numbers in front of the symbol for the electron indicate the number of electrons added or removed. The sum of the charges must be the same on both sides of the arrow.

EXAMPLE 2.3

Chromium (Cr), a hard, blue-white metal used for decorative and corrosion-resistant surfaces, forms an ion with a 3+ charge. Compare the number of electrons in atomic and ionic chromium, and write an equation to represent the formation of the ion.

Solution   Chromium's atomic number is 24, so it contains 24 protons. The neutral atom must also contain 24 electrons. To form the Cr^{3+} ion, the neutral atom must lose three electrons, so the ion must contain 21 electrons. The equation representing this loss is Cr → Cr^{3+} + 3e$^-$.

Practice Problem 2.3

Selenium (Se), a gray solid used as a semiconductor in photocells and xerography, forms an ion with a 2− charge. Compare the number of electrons in atomic and ionic selenium, and write an equation to represent the formation of the ion.

Answer: 34 e$^-$ in Se, 36 e$^-$ in Se^{2-}; Se + 2e$^-$ → Se^{2-}

When electrons are removed from or added to a group of atoms, a **polyatomic ion,** an ion containing two or more atoms, results. An example is the nitrate ion, NO$_3{}^-$. Nitrate ions exist as units, with each oxygen atom attached to the nitrogen atom and the entire group bearing a 1− charge. The most common of the polyatomic anions contain oxygen attached to some other element. These are called **oxoanions** (or sometimes *oxyanions*). Other examples include SO$_4{}^{2-}$ (sulfate), PO$_4{}^{3-}$ (phosphate), ClO$_3{}^-$ (chlorate), and OH$^-$ (hydroxide). It is also possible to form polyatomic cations, such as NH$_4{}^+$ (ammonium) and H$_3$O$^+$ (hydronium).

Ionic compounds are formed when cations combine with anions in proportions that give electrical neutrality—that is, equal positive and negative charge

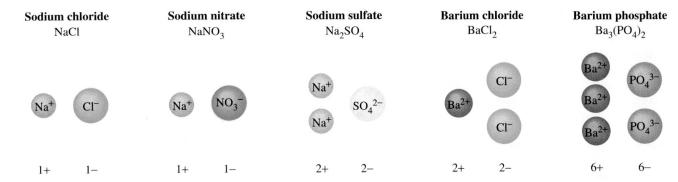

Sodium chloride	Sodium nitrate	Sodium sulfate	Barium chloride	Barium phosphate
$NaCl$	$NaNO_3$	Na_2SO_4	$BaCl_2$	$Ba_3(PO_4)_2$

| 1+ | 1− | 1+ | 1− | 2+ | 2− | 2+ | 2− | 6+ | 6− |

Figure 2.19 Ionic compounds are formed by combinations of cations and anions in proportions that lead to electrical neutrality.

Why do we include the charge in the formula of an ion?

(see Figure 2.19). Compounds formed in this way include $NaCl$ (sodium chloride, or common table salt), $BaCl_2$ (barium chloride), and Na_2SO_4 (sodium sulfate). Note that even though we have to show the charge when we write the formula of an ion, we do not normally show the charges of the ions in an ionic compound.

Let's consider the ionic compounds just mentioned in more detail. The simplest is sodium chloride, $NaCl$, which contains Na^+ and Cl^-. This ionic compound contains one Na^+ ion for every Cl^- ion, so the ionic charges are balanced. Barium chloride, $BaCl_2$, is somewhat more complex. It contains Ba^{2+} and Cl^-. Each barium ion has a 2+ charge, and each chloride ion has a 1− charge. With two chloride ions for each barium ion, the combination contains equal positive and negative charges. Similarly, in sodium sulfate (Na_2SO_4), the sulfate ion, which has a 2− charge, is balanced by two sodium ions, each with a 1+ charge.

It is sometimes necessary to use parentheses when writing formulas for ionic compounds that contain polyatomic ions such as NO_3^-. We use parentheses if the polyatomic ion occurs more than once, to prevent the confusion that would arise from adjacent subscripts, as in aluminum nitrate, $Al(NO_3)_3$.

EXAMPLE 2.4

What is the formula of the compound that could be formed by Cr^{3+} and SO_4^{2-}?

Solution The ions combine in proportions that lead to electrical neutrality. We insert subscripts that equalize the total positive and negative charges. The chromium ion supplies more positive charge than the 2+ required to neutralize the 2− from the sulfate ion, so we cannot simply add a subscript to one of the ions. The simplest approach is to select the largest charge as a reference point. Thus, consider how many sulfate ions would be necessary to neutralize the 3+ charge of chromium. Because each sulfate ion supplies 2−, and we need 3− for neutrality, it will be necessary to have 1.5 sulfate ions, giving $(Cr^{3+})(SO_4^{2-})_{1.5}$. Normally, we want to use whole-number subscripts because it is not possible to have fractions of atoms in compounds. Thus, each subscript must be multiplied by 2, to give $(Cr^{3+})_2(SO_4^{2-})_3$, or $Cr_2(SO_4)_3$. Now there are $(3+) \times 2 = 6+$ charges and $(2-) \times 3 = 6-$ charges, so the compound is electrically neutral.

Practice Problem 2.4

What are the formulas of all the compounds that could be formed by K^+, Fe^{3+}, Br^-, and SO_4^{2-}?

Answer: KBr, K_2SO_4, $FeBr_3$, and $Fe_2(SO_4)_3$

Figure 2.20 The structures of sodium nitrate, NaNO₃, and methane, CH₄, illustrate two distinct ways in which atoms can combine to form chemical compounds. Sodium nitrate contains ions arranged in extensive arrays. No molecule can be distinguished in this structure. Instead, we can recognize a basic formula unit that is repeated indefinitely. In contrast, methane has distinct molecules that can be recognized readily.

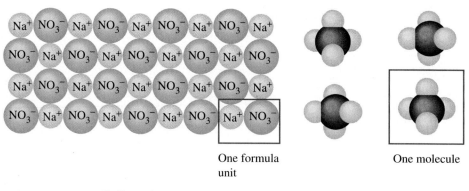

One formula unit

One molecule

Sodium nitrate

Methane

We cannot appropriately describe one unit of an ionic compound as a molecule. Instead, we should use the term **formula unit,** the smallest repeating formula in the structure of the ionic compound. Note the contrast between methane (CH₄) and sodium nitrate (NaNO₃) in Figure 2.20. Methane has easily identifiable molecules and so can be represented by a molecular formula. In ionic compounds, however, the ions combine in regular three-dimensional geometric patterns, so it is not possible to identify one single molecule of, say, NaNO₃. For ionic substances, the chemical formula can only give the ratio of ions of each kind. Thus, the formula NaNO₃ indicates only that there are equal numbers of sodium ions and nitrate ions in the substance; it should not be interpreted to mean that the substance is made up of discrete units of one Na$^+$ ion combined with one NO$_3^-$ ion. The formula for any ionic compound is an empirical formula that represents the formula unit in the structure.

2.4   Atomic and Molecular Weights

Suppose we have a bottle containing a white solid. The bottle is labeled "swimming pool chlorine." We know the contents are not pure chlorine, because that is a pale green gas. By examining the ingredients, we find that the solid contains the elements calcium, oxygen, and chlorine. We want to write a chemical formula for this substance. But how do we proceed? We know chemical formulas represent a count of the relative numbers of the various atoms in a substance. If we could count the atoms of each element in a small sample, we could determine the relative numbers of each and deduce the chemical formula. However, atoms and molecules are too small to see, and even if we could count them, a dust-sized piece of the solid would contain more than 10^{16} of each type of atom. We certainly wouldn't want to count that many atoms individually, even if we could. But fortunately, we don't have to.

The simplest measure of the quantity of atoms or molecules of a substance is the mass of that substance. Indeed, the use of mass as a measure of number is common to a variety of small objects that are considerably larger than atoms or molecules. For example, we purchase nails, potato chips, and grass seed by their mass rather than by the piece. In a similar way, we can use mass to measure out a desired number of atoms or molecules. Once we know the relative masses of the elements in a substance, then, we can determine the chemical formula of that substance. First, though, we need to understand how to measure masses of atoms.

What other ways could we use to determine the numbers of atoms?

Relative Masses of Atoms: Atomic Weights

The term atomic weight *is not really correct. The quantity should be called* atomic mass. *However,* atomic weight *has been used for so long that it has become ingrained in scientific terminology. This text will continue to use the traditional term for the scale of relative masses but will use the term* mass *when referring to the mass of specific atoms or molecules.*

Although single atoms cannot be weighed directly, there are sophisticated techniques by which scientists can indirectly determine individual atomic masses very accurately. For example, one uranium atom weighs 3.9527×10^{-22} g, and one hydrogen atom weighs 1.67380×10^{-24} g. Numbers as small as these are difficult to remember and use. It is more convenient to think of a uranium atom as being about 236 times as heavy as a hydrogen atom. For this reason, scientists have devised a set of relative masses of atoms called **atomic weights.**

Because atomic weights represent relative masses, one element must be selected as the standard against which all others are measured. The modern atomic weight scale uses one isotope of carbon, $^{12}_{6}C$ (or ^{12}C), as the standard. This, the most abundant isotope of carbon, contains 6 protons, 6 neutrons, and 6 electrons. ^{12}C is assigned an atomic weight of exactly 12 units.

The unit of atomic weight is called the **atomic mass unit (amu)** and has the following value:

$$1 \text{ amu} = 1/12 \times \text{mass of } 1 \ ^{12}C \text{ atom} = 1.6606 \times 10^{-24} \text{ g}$$

Clearly, amu measurements are easier to manipulate than are the exceptionally small numbers that represent the absolute masses of individual atoms. Furthermore, comparing numbers of atoms becomes quite simple when atomic mass units are used, as shown in Figure 2.21. If a carbon atom weighs 12 times as much as a hydrogen atom, then 2 carbon atoms weigh 12 times as much as 2 hydrogen atoms, and 100 carbon atoms weigh 12 times as much as 100 hydrogen atoms, and so on. It follows that, given 12 g of carbon and 1 g of hydrogen, the 12 g of carbon contains the same number of atoms as the 1 g of hydrogen. By the same argument, 6 g of carbon contains the same number of atoms as 0.5 g of hydrogen, and 120 g of carbon contains the same number of atoms as 10 g of hydrogen. In each case, the proportion of the mass of carbon atoms to the mass of hydrogen atoms is 12:1, and the proportion of carbon atoms to hydrogen atoms is 1:1. Now we have the ability to weigh a sample of carbon and a sample of hydrogen and determine the relative numbers of atoms of each from their relative masses. This is exactly the information we need to determine a chemical formula.

Would samples of an element collected from different places have exactly the same distribution of isotopes?

One C atom weighs as much as 12 H atoms.

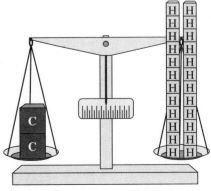

Two C atoms weigh as much as 24 H atoms.

Figure 2.21 Relative atomic masses are useful for counting atoms.

Silver is much used in jewelry because it is easier to work into shapes than any other metal except gold and because it is not chemically attacked by oxygen in air or by water or acids. Silver does react with sulfur and its compounds, however, and such reactions give rise to the black coating (tarnish) often found on silver objects.

However, one complication, the existence of isotopes, remains to be considered. Not all atoms of an element have the same mass, because they can have different neutron numbers and thus different mass numbers. The mass of a collection of atoms, then, is really an average value based on atomic masses of individual isotopes.

Each atomic weight given in tables of the elements (such as the one on the inside back cover of this book) is the average mass of an atom of the element relative to the mass of an atom of ^{12}C, which is set at exactly 12 amu. The atomic weight is based on the average isotopic composition of the element as it is found on earth. For example, natural silver has been found to contain 51.82% ^{107}Ag (106.9051 amu) and 48.18% ^{109}Ag (108.9048 amu). The weighted average of these values, 107.87 amu, is used as the atomic weight of silver.

EXAMPLE 2.5

Calculate the atomic weight of chlorine, which consists of 75.77% ^{35}Cl (34.97 amu) and 24.23% ^{37}Cl (36.95 amu).

Solution   Multiply each isotopic mass by the fraction of atoms existing as that isotope (percent abundance divided by 100). The contribution to the mass from ^{35}Cl is

$$\frac{34.97 \text{ amu}}{^{35}Cl \text{ atom}} \times \frac{75.77 \ ^{35}Cl \text{ atoms}}{100 \text{ Cl atoms}}$$

The contribution to the mass from ^{37}Cl is

$$\frac{36.95 \text{ amu}}{^{37}Cl \text{ atom}} \times \frac{24.23 \ ^{37}Cl \text{ atoms}}{100 \text{ Cl atoms}}$$

The atomic weight of chlorine is the sum of these contributions:

$$\text{Atomic weight} = (34.97 \text{ amu} \times 0.7577) + (36.95 \text{ amu} \times 0.2423)$$
$$= 35.45 \text{ amu}$$

Practice Problem 2.5

Calculate the atomic weight of copper, which consists of 69.09% ^{63}Cu (62.93 amu) and 30.91% ^{65}Cu (64.93 amu).

Answer: 63.55 amu

Molecular and Formula Weights

Recall that most materials exist not as single atoms but rather as some combination of atoms in the form of molecules or ionic substances. It is useful to have a relative scale of masses for these substances as well. The mass of one molecule relative to the mass of one atom of ^{12}C is called the **molecular weight.** The term *molecular weight* is often loosely used to describe masses of ionic compounds as well. However, to be more accurate, the term **formula weight** should be used because ionic substances do not occur as collections of discrete molecules but rather as collections of oppositely charged ions in certain ratios. In this case, the mass used is that of one formula unit of the substance, the formula weight.

To calculate the molecular or formula weight of a substance, we simply add the atomic weights of all the atoms that make up one molecule or one formula unit. For example, the molecule H_2O_2 has two hydrogen atoms and two oxygen atoms. The following equation gives its molecular weight:

Molecular weight H_2O_2 =
atomic weight H + atomic weight H + atomic weight O + atomic weight O

The equation is more efficiently stated this way:

Molecular weight H_2O_2 = (atomic weight H × 2) + (atomic weight O × 2)

We can generalize this equation to find any molecular or formula weight:

Molecular or formula weight = (atomic weight$_x$ × subscript$_x$)
+ (atomic weight$_y$ × subscript$_y$) + . . .

The subscripts come from the molecular or ionic formula. For example, the molecular weight of carbon dioxide, CO_2, is given by the following equation:

Molecular weight = (atomic weight C × 1) + (atomic weight O × 2)

The same procedure is used for ionic compounds—except, of course, that a formula weight is obtained instead of a molecular weight.

EXAMPLE 2.6

Calculate the formula weight of ammonium sulfate, $(NH_4)_2SO_4$, a substance commonly used in chemical fertilizers.

Solution Combine the atomic weights of all the elements occurring in the formula. Ammonium sulfate contains two N atoms, eight H atoms, one S atom, and four O atoms.

Formula weight = (atomic weight N × 2) + (atomic weight H × 8)
+ (atomic weight S) + (atomic weight O × 4)

Now substitute into this formula:

$$
\begin{aligned}
\text{Formula weight} = 14.007 \text{ amu} \times 2 &= 28.014 \text{ amu} \\
1.008 \text{ amu} \times 8 &= 8.064 \text{ amu} \\
32.066 \text{ amu} &= 32.066 \text{ amu} \\
15.9994 \text{ amu} \times 4 &= \underline{63.998 \text{ amu}} \\
& 132.142 \text{ amu}
\end{aligned}
$$

Practice Problem 2.6

Calculate the molecular weight of sulfuric acid, H_2SO_4.

Answer: 98.078 amu

Mass Spectrometry

A mass spectrometer, shown in Figure 2.22, is a variation on the cathode-ray tubes used by J. J. Thomson to measure the charge-to-mass ratio of the electron. This instrument is used to obtain the masses of individual atoms. A substance is vaporized into the mass spectrometer, then subjected to an electric arc or bombarded by

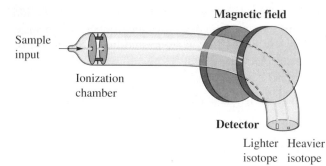

Figure 2.22 The masses of individual atoms are determined with a mass spectrometer. Electrons are removed from atoms (or molecules), and the resultant ions are accelerated through a magnetic field. The amount of bending in the path of the ions is related to the mass and the charge of the ions.

an electron beam to knock electrons from the substance and create cations. Inside the mass spectrometer, the beam of cations is passed through a magnetic field. The smaller the mass of a cation, the more its path is bent, or deflected, by the magnetic field. The variations in deflection allow scientists to distinguish between ions differing by as little as one mass number. The number of ions of each mass number can be recorded electronically.

The distribution of isotopes in an element can be determined by mass spectrometry. A mass spectrum for chlorine is shown in Figure 2.23. After the data are calibrated relative to ^{12}C, the exact mass of each isotope can be determined. The mass of ^{35}Cl is 34.9689 amu and that of ^{37}Cl is 36.9659 amu. The relative intensity of the signals at these mass numbers indicates that the sample of chlorine contains 75.77% ^{35}Cl atoms and 24.23% ^{37}Cl atoms. The atomic weight of chlorine is the weighted average of the masses determined in this way.

Atomic weights are not all given to the same number of significant figures. For example, the atomic weight of chlorine is 35.453 amu and that of fluorine is 18.998403 amu. Atomic weights are determined for a large number of samples of an element collected from different places on earth. The degree of variation in the isotopic distribution in these samples determines the precision of the atomic weight used by chemists. The atomic weight of fluorine is known so precisely because fluorine has only one isotope, so all samples give identical results.

Mass spectrometry can be used to study compounds as well as elements. Suppose we introduce carbon monoxide into a mass spectrometer. We should see peaks corresponding to all isotopic combinations possible for CO^+. These would be the six possible combinations of ^{12}C and ^{13}C with ^{16}O, ^{17}O, and ^{18}O having mass numbers of 28, 29, 29, 30, 30, and 31, so we should see four peaks, two of which arise from different isotopic combinations. Knowing the relative amounts of the isotopes, we can predict the intensities of these peaks, so a mass spectrum can be used as a fingerprint of a compound to identify unknown substances. Larger molecules will often split into fragments, so their fingerprints have many more lines that can be used for identification.

Mass spectrometry is used in space capsules to monitor the air and detect toxic pollutants, such as carbon monoxide. Mass spectrometry can also be used in medicine to analyze the substances in the exhaled breath of patients, assisting in the diagnosis of disease. Similarly, anesthetists use mass spectrometry to monitor the breath of surgery patients.

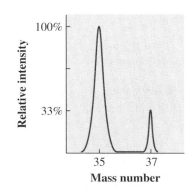

Figure 2.23 The mass spectrum of chlorine atoms contains two peaks, occurring at mass numbers 35 and 37. The exact masses of these isotopes and the relative heights of the two peaks allow the calculation of the atomic weight of chlorine. These peaks also serve as a fingerprint for chlorine in other substances.

The distribution of isotopes of certain elements, such as carbon, changes over long periods of time. Pieces of dead wood, for example, contain amounts of ^{14}C that vary with the age of the wood over periods of thousands of years. A determination of the amount of ^{14}C allows a determination of the age of that wood. We will explore this procedure in more detail in Chapter 20.

2.5 The Mole

Let's consider again our quest to determine the chemical formula of that white solid in the bottle of swimming pool chlorine (Section 2.4). We can now compare the atomic weights of the constituent elements: Ca is 40.08 amu, Cl is 35.453 amu, and O is 15.9994 amu. However, we still have a problem. Although atomic, molecular, and formula weights are very useful for comparing the masses of substances, atomic mass units are not very useful to chemists weighing substances to use in chemical reactions. Common laboratory instruments accurately weigh masses in grams rather than amu. Even if we knew the formula weight of the compound in the bottle, we could not weigh out one or even a few formula units of the substance, because the masses of most atoms and common molecules are less than 10^{-21} g. To obtain a mass of about 1 g to work with, we would have to measure out a number of formula units greater than about 10^{21}. For some idea of the magnitude of this number, consider a teaspoon filled with water. There are as many molecules of water in this teaspoon (about 1.7×10^{23}) as there are teaspoonsful of water in all the oceans on earth. Numbers this large are quite unwieldy, so we would prefer not to work with them. We need some way to measure the amount of a substance other than the number of formula units or molecules. (This is a common practice in other areas as well. We count eggs in units of a dozen; paper in units of a ream, or 500 sheets; paper clips by the gross, or 144 items; and soldiers by the battalion. It is simply more convenient to use a unit that counts by the group rather than by the individual.)

To avoid using extremely large numbers when measuring out chemical substances, we can use a unit called the **mole.** A mole contains 6.022×10^{23} atoms, molecules, or formula units, a figure known as **Avogadro's number.** This figure is the same as the number of atoms in exactly 12 g of ^{12}C. Competence in using the mole is important for chemistry students because the unit is used extensively in chemical calculations. Figure 2.24 shows one mole of various substances.

Molar Mass

The atomic weight, as you have already seen, is the average mass of an atom of an element in atomic mass units. The molecular weight is the mass of one molecule of a substance in atomic mass units. These definitions involve the mass of one atom or molecule (or formula unit), but we can extend the definitions to larger quantities of matter as well.

Recall that the basis of these definitions is that one atom of ^{12}C has a mass of exactly 12 amu. Because 1 amu is 1.6606×10^{-24} g and 1 mole contains 6.022×10^{23} atoms, 1 mole of ^{12}C weighs

$$6.022 \times 10^{23} \text{ atoms} \times \frac{12 \text{ amu}}{\text{atom}} \times \frac{1.6606 \times 10^{-24} \text{ g}}{\text{amu}} = 12.000 \text{ g}$$

Figure 2.24 One mole of atoms or molecules of some common elements and chemical compounds. These substances are Fe_2O_3 (red solid); Hg (silvery liquid); $CuSO_4 \cdot 5H_2O$ (blue solid); NaCl (white solid); S (yellow solid); Fe (nails); Sn (silvery solid), Cu (coins), and H_2O (colorless liquid).

The mass in amu of 1 atom of ^{12}C is the same as the mass of 1 mole of ^{12}C measured in grams. The value of Avogadro's number was selected so these masses would match. The definition of a mole can be restated, then, in terms of the mass of our reference material, ^{12}C: *a mole is the amount of material that contains as many basic particles (atoms, molecules, or formula units) as there are atoms in 12.000 g of ^{12}C*. The term that describes the mass of one mole of a material is **molar mass:** the mass of one mole of basic particles of any substance.

The relationship among all the definitions of mass that we have examined means that the molar mass value for any atom (in units of grams per mole, or g/mol) is numerically the same as the atomic weight value (in amu). Thus, the average mass of one hydrogen atom is 1.008 amu, and the mass of one mole of hydrogen atoms is 1.008 g. The molar mass of a molecular or ionic substance has the same value as the molecular weight or formula weight of that substance. For example, the molecular weight of H_2SO_4 is 98.08 amu; the molar mass of this substance is 98.08 g/mol. Molar masses of molecules (or formula units) are obtained in the same way as molecular weights. The molar mass of a molecule is simply the sum of the molar masses of each of its component atoms.

Moles, Masses, and Particles

Returning again to our white solid swimming pool chlorine, let's suppose we've determined that the amounts of the elements in a sample of the solid are 1.50 g Ca, 2.65 g Cl, and 1.20 g O. We now know the composition of the substance in terms of masses. However, the chemical formula we wish to determine is expressed in terms of relative numbers of *atoms*. Thus, we must somehow convert the masses of Ca, Cl, and O into atoms or moles of atoms. This sort of conversion is often necessary in chemistry. The experimental techniques chemists use most often involve measurement of masses; but counting atoms or moles of atoms makes it easier to discuss molecules, their structure, and changes in molecular composition.

Converting between atoms or molecules, moles, and mass is a straightforward task based on the definitions of mole and molar mass (abbreviated MM):

$$1 \text{ mol} = 6.022 \times 10^{23} \text{ molecules or atoms}$$

$$MM = \text{grams per mole of molecules or atoms}$$

The conversions follow this sequence of operations:

$$\text{Mass} \xleftrightarrow{\text{MM}} \text{moles} \xleftrightarrow{6.022 \times 10^{23}} \text{number of molecules or atoms}$$

The arrows connect the quantities that can be interconverted, and the quantities over the arrows give the conversion factors. Thus, we obtain the number of moles of calcium in our swimming pool chlorine as follows:

$$\text{Moles of Ca} = 1.50 \text{ g Ca} \times \frac{1 \text{ mol Ca}}{40.08 \text{ g Ca}} = 0.0374 \text{ mol Ca}$$

Following a similar procedure, we find that our sample contains 0.0747 mol chlorine and 0.0750 mol oxygen.

EXAMPLE 2.7

How many moles of sugar are there in 1 teaspoon (4.0 g) of sugar, $C_{12}H_{22}O_{11}$?

Solution   Use the molar mass of sugar to convert mass to moles:

$$\text{Mass} \xleftrightarrow{\text{MM}} \text{moles}$$

The molar mass is obtained from the molar masses of the atoms:

$\text{MM} = (\text{MM}_C \times 12) + (\text{MM}_H \times 22) + (\text{MM}_O \times 11)$

$\quad = (12.011 \text{ g/mol C} \times 12 \text{ mol C/mol sugar}) + (1.008 \text{ g/mol H} \times 22 \text{ mol H/mol sugar}) + (15.9994 \text{ g/mol O} \times 11 \text{ mol O/mol sugar}) = 342.3 \text{ g/mol sugar}$

This molar mass provides the following conversion factor:

$$1 \text{ mol sugar} = 342.3 \text{ g sugar}$$

Now use dimensional analysis to cancel out grams and introduce moles:

$$\text{Moles sugar} = 4.0 \text{ g sugar} \times \frac{1 \text{ mol sugar}}{342.3 \text{ g sugar}} = 0.012 \text{ mol sugar}$$

Practice Problem 2.7

Saccharin, $C_7H_5NO_3S$, is an artificial sweetener. How many moles of saccharin are present in a 5.0-g sample of this substance?

Answer: 0.027 mol

To calculate the number of molecules in a sample of a substance, we must know how many moles of that substance are in the sample. We know the number of moles of each element in our sample of swimming pool chlorine, so we can calculate the number of atoms of each element in the sample.

$$\text{Atoms Ca} = 0.0374 \text{ mol} \times \frac{6.022 \times 10^{23} \text{ atoms}}{1 \text{ mol}} = 2.25 \times 10^{22} \text{ atoms}$$

Similar calculations yield 4.50×10^{22} atoms Cl and 4.52×10^{22} atoms O. If the number of moles is not given, then a mass-to-mole conversion must be carried out first.

EXAMPLE 2.8

One 5-grain aspirin tablet contains 0.324 g of acetylsalicylic acid, $C_9H_8O_4$ (MM = 180.2 g/mol). How many molecules of acetylsalicylic acid are in one tablet?

Solution   The necessary conversions are

$$g \longrightarrow moles \longrightarrow molecules$$

The equivalence equations are

$$1 \text{ mol acetylsalicylic acid} = 180.2 \text{ g acetylsalicylic acid}$$
$$1 \text{ mol} = 6.022 \times 10^{23} \text{ molecules}$$

The series of conversions is as follows:

$$\text{Molecules } C_9H_8O_4 = 0.324 \text{ g} \times \frac{1 \text{ mol}}{180.2 \text{ g}} \times \frac{6.022 \times 10^{23} \text{ molecules}}{1 \text{ mol}}$$
$$= 1.08 \times 10^{21} \text{ molecules}$$

Practice Problem 2.8

How many atoms of iron are there in a nail weighing 0.105 g?

Answer: 1.13×10^{21} atoms

Any of these conversions can be carried out in the reverse direction as well. Thus, if we know how many moles are in a sample, we can calculate the mass by using the molar mass as a conversion factor.

2.6   Empirical Formulas and Percent Composition

We are finally ready to determine the empirical formula for the swimming pool chlorine. We have gone through several steps to convert the masses of the constituent elements to number of moles and to number of atoms. Either of these two numbers can be used to obtain an empirical formula; we'll use the number of moles in our sample: 0.0374 mol Ca, 0.0747 mol Cl, and 0.0750 mol O. The empirical formula represents the relative number of moles of atoms of each element. If we select the element present in least amount—Ca—as a reference, we need only divide the moles of Cl and O by the moles of Ca to reduce all these numbers to whole numbers:

$$\frac{0.0747 \text{ mol Cl}}{0.0374 \text{ mol Ca}} = 2.00 \text{ Cl/Ca}$$

$$\frac{0.0750 \text{ mol O}}{0.0374 \text{ mol Ca}} = 2.01 \text{ O/Ca}$$

The value of 2.01 suggests that there may have been a slight error in the experimental determination of the masses of the elements in the sample. But because

Why don't we have to convert moles to numbers of atoms of each element to determine the empirical formula?

2.01 is so close to being a whole-number value, we can safely assume that we have 2 O, as well as 2 Cl, for each Ca, giving the formula $CaCl_2O_2$. Note that this formula can also be written as $Ca(ClO)_2$, which is the formula of calcium hypochlorite.

The remainder of this section gives some other examples of determining empirical formulas from various types of data so you can see the entire process carried out in consecutive steps. You will also be shown how to use moles to determine empirical formulas.

Determining Empirical Formulas

In our swimming pool chlorine calculations, we used the mass of each element in a sample of the substance. Usually we report the composition of a sample in a more general way, by converting the masses to the percent of mass contributed by each element. Such numbers are generally called the *percent composition by mass*, or just *percent composition*. To calculate percent composition, we divide the mass of each element by the total mass and multiply by 100% to convert to a percentage. Our sample contains 1.50 g Ca, 2.65 g Cl, and 1.20 g O, with a total mass of 5.35 g. The percent of Ca in the sample is

$$\% \text{ Ca} = \frac{1.50 \text{ g Ca}}{5.35 \text{ g sample}} \times 100\% = 28.0\%$$

Similar calculations give 49.5% Cl and 22.4% O by mass. We can use this percent composition data to determine the empirical formula.

The series of conversions from percent composition to empirical formula for two elements, A and B, can be represented schematically as follows:

$$\% \text{ A} \xrightarrow{100 \text{ g}} \text{mass A} \xrightarrow{\text{MM of A}} \text{moles A}$$
$$\% \text{ B} \xrightarrow{100 \text{ g}} \text{mass B} \xrightarrow{\text{MM of B}} \text{moles B}$$
$$\longrightarrow \text{moles A/moles B} \longrightarrow \text{formula}$$

The following sections give details of the conversions.

Why is it not necessary to determine the percent composition first?

Empirical Formulas from Combining Masses   Consider first a system in which the masses of the elements combining to form a compound are known. In that case, we need only convert these masses to moles and calculate a mole ratio to deduce the empirical formula.

EXAMPLE 2.9

If titanium metal, Ti, reacts with oxygen gas, O_2, the product is white, powdery titanium oxide, used extensively as a white paint pigment. A sample of 1.44 g Ti reacted completely with O_2, giving 2.39 g of titanium oxide. What is the empirical formula of titanium oxide?

Solution   We already know the combining masses of the elements: 1.44 g of titanium and 0.95 g of oxygen. The amount of oxygen used is obtained from the mass of the product because the reaction must obey the law of conservation of mass:

$$\text{Mass O} = 2.39 \text{ g} - 1.44 \text{ g} = 0.95 \text{ g}$$

(continued)

We can use the molar masses of these elements to calculate the moles of each element:

$$\text{Moles Ti} = 1.44 \text{ g} \times \frac{1 \text{ mol}}{47.9 \text{ g}} = 0.0301 \text{ mol}$$

$$\text{Moles O} = 0.95 \text{ g} \times \frac{1 \text{ mol}}{16.0 \text{ g}} = 0.0594 \text{ mol}$$

To calculate the mole ratio, we divide the larger number (the element present in the greater amount) by the smaller one:

$$\frac{\text{Moles O}}{\text{Moles Ti}} = \frac{0.0594 \text{ mol}}{0.0301 \text{ mol}} = 1.97$$

Because a molecule cannot contain only part of an atom, the mole ratio must be a whole number. Ratios that are close to a whole number (within a few percent) can safely be rounded off to that whole number.

$$\frac{\text{Moles O}}{\text{Moles Ti}} = 1.97 \approx 2$$

There are 2 moles of oxygen for every mole of titanium, so the empirical formula is TiO_2.

Practice Problem 2.9

If 4.32 g Cu combines with 1.09 g S to form 5.41 g of the bluish-black mineral chalcocite, what is the empirical formula of chalcocite?

Answer: Cu_2S

Empirical Formulas from Percent Composition   Example 2.9 considered a system in which the combining masses of two elements in a compound were known. If we know the percent composition instead, we must convert percent composition to relative masses before we calculate moles. The simplest approach is to use a sample weight of exactly 100 g so that the percent composition is numerically equal to the mass of each component.

EXAMPLE 2.10

Determine the empirical formula for the antidecay agent in a certain toothpaste, commonly known as Fluoristan. Fluoristan has the percent composition 75.75% Sn and 24.25% F.

Solution   Assume the sample of Fluoristan has a mass of 100 g so that the masses of tin and fluorine will be numerically equal to the percent composition values:

$$\text{g Sn} = 75.75\% \times 100 \text{ g} = \frac{75.75 \text{ g}}{100 \text{ g}} \times 100 \text{ g} = 75.75 \text{ g}$$

$$\text{g F} = 24.25\% \times 100 \text{ g} = \frac{24.25 \text{ g}}{100 \text{ g}} \times 100 \text{ g} = 24.25 \text{ g}$$

Next, convert the mass of each element into the corresponding number of moles:

$$\text{Moles Sn} = 75.75 \text{ g Sn} \times \frac{1 \text{ mol Sn}}{118.7 \text{ g Sn}} = 0.6382 \text{ mol}$$

$$\text{Moles F} = 24.25 \text{ g F} \times \frac{1 \text{ mol F}}{19.00 \text{ g F}} = 1.276 \text{ mol}$$

Finally, divide by the smaller of the two mole quantities:

$$\frac{\text{Moles F}}{\text{Moles Sn}} = \frac{1.276 \text{ mol}}{0.6382 \text{ mol}} = 2.000$$

Because 2 moles of fluorine are present for each mole of tin, the empirical formula is SnF_2.

Practice Problem 2.10

Determine the empirical formula of aluminum nitride, used in semiconductor electronics. Aluminum nitride has the percent composition 65.82% Al and 34.18% N.

Answer: AlN

Empirical Formulas with More than Two Elements For a compound that contains more than two elements, we follow essentially the same procedure but add a series of conversions for each additional element.

EXAMPLE 2.11

Caffeine occurs in coffee, tea, and some other substances. An analysis of caffeine gives the following percent composition: 49.798% carbon, 28.532% nitrogen, 5.082% hydrogen, and 16.288% oxygen. Calculate the empirical formula of caffeine.

Solution Assume a 100-g sample of caffeine so that the masses of each element will be numerically equal to the percent composition. Then calculate the moles of each element in 100 g of compound:

$$\text{Moles C} = 49.798 \text{ g C} \times \frac{1 \text{ mol C}}{12.011 \text{ g C}} = 4.1460 \text{ mol}$$

$$\text{Moles N} = 28.532 \text{ g N} \times \frac{1 \text{ mol N}}{14.007 \text{ g N}} = 2.0370 \text{ mol}$$

$$\text{Moles H} = 5.082 \text{ g H} \times \frac{1 \text{ mol H}}{1.008 \text{ g H}} = 5.042 \text{ mol}$$

$$\text{Moles O} = 16.288 \text{ g O} \times \frac{1 \text{ mol O}}{15.9994 \text{ g O}} = 1.0180 \text{ mol}$$

To calculate the mole ratio, first find the element present in the smallest amount—oxygen, in this example. Then divide the moles of each of the remaining elements by the moles of oxygen:

$$\frac{\text{Moles C}}{\text{Moles O}} = \frac{4.1460 \text{ mol}}{1.0180 \text{ mol}} = 4.0727 = 4$$

$$\frac{\text{Moles N}}{\text{Moles O}} = \frac{2.0370 \text{ mol}}{1.0180 \text{ mol}} = 2.0010 = 2$$

$$\frac{\text{Moles H}}{\text{Moles O}} = \frac{5.042 \text{ mol}}{1.0180 \text{ mol}} = 4.953 = 5$$

The empirical formula is $C_4N_2H_5O$.

(continued)

Practice Problem 2.11

Lithium oxalate has the following percent composition: 13.62% Li, 23.58% C, and 62.81% O. What is the empirical formula of lithium oxalate?

Answer: $LiCO_2$

Empirical Formulas with Nonintegral Mole Ratios   We have already seen that sometimes the mole ratios calculated in the final step of the process do not come out to be whole numbers. For instance, in Example 2.9 we calculated a mole ratio of 1.97, which we rounded off to a value of 2. The nonintegral values will not always be within a few percent of a whole number, so it will not always be appropriate to round them off. They may, however, have fractional values that correspond to ratios of small whole numbers, such as 1.25 (5/4), 1.33 (4/3), 1.50 (3/2), and 1.67 (5/3). In this case, we multiply each ratio by the smallest whole number that will make all the mole ratios whole numbers.

EXAMPLE 2.12

Sodium thiosulfate is 29.08% sodium, 40.56% sulfur, and 30.36% oxygen by mass. Calculate the empirical formula of sodium thiosulfate.

Solution   If we assume a 100-g sample, the masses of each element will be numerically the same as the percent composition. Calculate the moles of each element in 100 g of compound:

$$\text{Moles Na} = 29.08 \text{ g Na} \times \frac{1 \text{ mol}}{22.99 \text{ g Na}} = 1.265 \text{ mol}$$

$$\text{Moles S} = 40.56 \text{ g S} \times \frac{1 \text{ mol S}}{32.066 \text{ g S}} = 1.265 \text{ mol}$$

$$\text{Moles O} = 30.36 \text{ g O} \times \frac{1 \text{ mol O}}{15.9994 \text{ g O}} = 1.898 \text{ mol}$$

Now calculate the ratio of the moles of each element to the moles of that element present in the smallest amount (sodium is used here):

$$\frac{\text{Moles S}}{\text{Moles Na}} = \frac{1.265 \text{ mol}}{1.265 \text{ mol}} = 1.000$$

$$\frac{\text{Moles O}}{\text{Moles Na}} = \frac{1.898 \text{ mol}}{1.265 \text{ mol}} = 1.500$$

One of the ratios is not close to a whole number, but it corresponds to a ratio of whole numbers (3/2). We can thus multiply this ratio by a factor of 2 to get a whole number. Of course, we must also multiply the other ratio by 2 to maintain the relationship between the ratios:

$$\frac{\text{Moles S}}{2 \text{ moles Na}} = 1.000 \text{ mol} \times 2 = 2.000 = 2$$

$$\frac{\text{Moles O}}{2 \text{ moles Na}} = 1.500 \text{ mol} \times 2 = 3.000 = 3$$

The empirical formula is $Na_2S_2O_3$.

Practice Problem 2.12

Aluminum oxide, commonly used as an abrasive, has the percent composition 52.91% Al and 47.08% O. Calculate the empirical formula of aluminum oxide.

Answer: Al_2O_3

Molecular Formulas from Empirical Formulas   We can derive a molecular formula from an empirical formula if we know the molar mass. The molar mass of a compound can be determined experimentally by any of a variety of methods, which will be discussed in later chapters. Once we know the experimental molar mass, we calculate the molar mass that corresponds to the empirical formula by adding the molar masses of the component elements. If this calculated molar mass is the same as the experimental one, the empirical formula is the same as the molecular formula. If the experimental molar mass is greater than that calculated from the empirical formula, then the molecular formula is some multiple of the empirical formula, as shown in the following example.

EXAMPLE 2.13

An analysis of caffeine was described in Example 2.11. The molar mass of caffeine is 194.2 g/mol. Determine its molecular formula.

Solution   The empirical formula of caffeine was found in Example 2.11 to be $C_4N_2H_5O$. The molar mass is 194.2 g/mol. We calculate the molar mass corresponding to the empirical formula by summing the molar masses of the constituent atoms:

$$MM = (MM_C \times 4) + (MM_N \times 2) + (MM_H \times 5) + (MM_O \times 1)$$
$$= (12.01 \text{ g/mol C} \times 4 \text{ mol C/mol}) + (14.01 \text{ g/mol N} \times 2 \text{ mol N/mol}) + (1.008 \text{ g/mol H} \times 5 \text{ mol H/mol}) + (16.00 \text{ g/mol O} \times 1 \text{ mol O/mol}) = 97.10 \text{ g/mol}$$

The ratio of the experimental molar mass and the calculated molar mass is

$$\text{Experimental MM/empirical MM} = (194.2 \text{ g/mol})/(97.10 \text{ g/mol}) = 2.000$$

Because the true molar mass is two times the molar mass calculated from the empirical formula, the molecular formula must be exactly twice the empirical formula: $C_8N_4H_{10}O_2$.

Practice Problem 2.13

Lithium oxalate has the following percent composition: 13.62% Li, 23.58% C, and 62.81% O. If the molar mass of lithium oxalate is 101.88 g/mol, what is its molecular formula?

Answer: $Li_2C_2O_4$

Determining Percent Composition from Chemical Formulas

We have examined various aspects of determining empirical and molecular formulas from the percent-by-mass composition of pure substances. Sometimes it is

Ammonium nitrate and ammonium sulfate are both odorless white solids used in chemical fertilizers. Ammonium nitrate is also used in matches and explosives and for making laughing gas (nitrous oxide).

necessary to reverse this process and find the percent composition from the chemical formula. For example, if we wanted to determine which substance, NH_4NO_3 (ammonium nitrate) or $(NH_4)_2SO_4$ (ammonium sulfate), was the most efficient source of nitrogen for fertilizing our lawns, we would want to know which of these compounds has the greater percent nitrogen.

To simplify converting a chemical formula to a percent composition, we specify a sample size of one mole of the compound. Then the percent composition of a given element (X) is equal to the mass of that element divided by the molar mass of the compound (which we have set at a mass of one mole) and multiplied by 100% to convert to a percentage:

$$\% \text{ X} = \frac{\text{mass X}}{\text{mass compound}} \times 100$$

$$\% \text{ X} = 100 \times (\text{moles X} \times MM_x / (1 \text{ mol compound} \times MM_{compound})$$

The value of moles X is obtained from the molecular formula. For example, there are 2 moles of nitrogen in 1 mole of NH_4NO_3 (ammonium nitrate), so there are 2 mol $\times$ 14.0 g/mol or 28.0 g of nitrogen in 1 mole of ammonium nitrate. Adding the molar masses of the elements gives 80.0 g/mol for the molar mass of ammonium nitrate. The percent nitrogen in the compound is thus

$$\% \text{ N} = 100 \times 28.0 \text{ g} / 80.0 \text{ g} = 35.0\%$$

Similar calculations give 21.2% N in $(NH_4)_2SO_4$, so NH_4NO_3 will provide more nitrogen for our lawns.

EXAMPLE 2.14

Calculate the percent composition of sulfur and oxygen in sulfur trioxide, SO_3.

Solution   The necessary molar masses are 32.0 g/mol S, 16.0 g/mol O, and 80.0 g/mol SO_3. The percent of S is calculated from the following equation:

$$\% \text{ S} = \frac{(\text{moles S} \times MM_s)}{(1 \text{ mol compound} \times MM_{compound})} \times 100$$

$$\% \text{ S} = \frac{1 \text{ mol S} \times 32.0 \text{ g S/mol S}}{1 \text{ mol compound} \times 80.0 \text{ g/mol compound}} \times 100 = 40.0\%$$

The percent O is calculated in the same way:

$$\% \text{ O} = 100 \times (\text{moles O} \times MM_O) / (1 \text{ mole compound} \times MM_{compound})$$

$$\% \text{ O} = \frac{3 \text{ mol O} \times 16.0 \text{ g O/mol O}}{1 \text{ mol compound} \times 80.0 \text{ g/mol compound}} \times 100 = 60.0\%$$

We could have calculated the percent O from the percent S by subtracting from 100% because the percentages of all the components must, by definition, add up to 100%:

$$\% \text{ O} = 100 - \% \text{ S} = 100 - 40.0 = 60.0\%$$

Practice Problem 2.14

Calculate the percent composition of sulfur and oxygen in sulfur dioxide, SO_2.

Answer: 50.0% S, 50.0% O

Calculations such as these are also used to verify the compositions of newly synthesized compounds. The calculated percent composition, based on the hypothesized formula of the compound, is compared with that obtained by experimental determination of percent composition.

Summary

The elements are composed of atoms, the smallest particles that retain the characteristic chemical properties of the respective elements. Atoms are composed of electrons, protons, and neutrons. The protons and neutrons are found in the nucleus of the atom, a small core that contains the positive charge and most of the mass of the atom. The electrons surround the nucleus and contribute negative charge but very little mass.

The elements can be distinguished from one another by the atomic number, the number of protons in the nucleus. The number of electrons equals the number of protons. Atoms of a given element may differ in the number of neutrons (and hence in mass). Such atoms are called isotopes.

Symbols are used to represent the elements, and chemical formulas are used to represent chemical compounds. Subscripts in the formula indicate how many atoms of each element are present in the compound. An empirical formula provides the simplest whole-number ratios of the elements; a molecular formula gives the actual number of atoms of each element.

The smallest particle of a pure substance that retains the composition and chemical properties of that substance is called a molecule. Some substances contain ions, charged atoms or groups of atoms, rather than molecules. The portion of an ionic substance represented by the empirical formula is called a formula unit.

The average mass of an atom of an element is called the atomic weight, measured in atomic mass units (amu). The amu is defined in terms of an atom of ^{12}C, which weighs exactly 12 amu. The average mass of a molecule—the sum of the component atomic weights—is called the molecular weight. A formula weight is the average mass of one formula unit of an ionic compound.

To work with large numbers of atoms, molecules, or ions, we use another unit, the mole, defined as the amount of a substance that contains Avogadro's number (6.022×10^{23}) of particles. The mass of one mole of a substance is the molar mass of that substance. Molar masses and Avogadro's number provide conversion factors necessary to convert between the mass of a sample of pure matter, the number of moles of the chemical substance, and the number of atoms or molecules in the sample.

Key Terms

anion (2.3)	**chemical formula** (2.3)	**mass** (2.1)	**nucleus** (2.2)
atom (2.1)	**diatomic molecule** (2.3)	**law of definite**	**neutron** (2.2)
atomic mass unit	**electron** (2.2)	**proportions** (2.1)	**neutron number**
(amu) (2.4)	**empirical formula** (2.3)	**mass number (A)** (2.2)	**(N)** (2.2)
atomic number	**formula unit** (2.3)	**molar mass** (2.5)	**oxoanion** (2.3)
(Z) (2.2)	**formula weight** (2.4)	**mole** (2.5)	**polyatomic ion** (2.3)
atomic weight (2.4)	**ion** (2.3)	**molecular formula** (2.3)	**polyatomic**
Avogadro's	**ionic compound** (2.3)	**molecular weight** (2.4)	**molecule** (2.3)
number (2.5)	**isotopes** (2.2)	**molecule** (2.3)	**proton** (2.2)
cation (2.3)	**law of conservation of**	**monatomic ion** (2.3)	

Exercises

Atomic Theory

2.1   State the law of conservation of mass and the law of definite proportions.

2.2   Name and give the charges of the subatomic particles.

2.3   Draw a picture of the atom, showing the location of the subatomic particles.

2.4   What is an element?

2.5   What is meant by the physical properties of the elements? Give some examples.

2.6   What is a chemical reaction?

2.7   When the elements magnesium (a silver metal) and oxygen (a colorless gas) are heated, a new substance (a white solid) is formed. Could this new substance be an element?

2.8   Use the following data to show that the law of conservation of mass holds true. When 3.49 g of vanadium metal is reacted with 12.00 g of fluorine gas, 10.00 g of vanadium pentafluoride (VF_5) is formed and 5.49 g of fluorine gas remains.

2.9   On heating, a white solid forms a silvery solid and a colorless gas. Could this white solid be an element?

2.10 Zinc blende, a naturally occurring form of zinc sulfide, consists of 67.09% zinc and 32.91% sulfur by mass. If 1.34 g of zinc is heated with excess sulfur, 2.00 g of zinc sulfide is formed. Show how these data are consistent with the law of definite proportions.

Atomic Number and Mass Number

2.11 What is meant by mass number and atomic number? Why are they usually not the same?

2.12 What properties of atoms are associated with the nucleus?

2.13 How are isotopes of an element different from one another?

2.14 How many electrons, protons, and neutrons are there in each of the following atoms or ions?
   a. ^9_4Be b. $^{23}_{11}Na$ c. $^{26}_{12}Mg$ d. $^7_3Li^+$ e. $^{19}_9F^-$

2.15 For each of the following nuclear compositions, identify the elements.
   a. $A = 52, Z = 24$ b. $A = 63, Z = 29$
   c. $A = 69, N = 38$ d. $A = 159, N = 94$

2.16 Chlorine has two isotopes, $^{35}_{17}Cl$ and $^{37}_{17}Cl$. How many protons, neutrons, and electrons are there in the atoms of these isotopes?

2.17 Indicate how many electrons, protons, and neutrons there are in each of the following species.
   a. $^{39}_{19}K$ b. $^{238}_{92}U$ c. $^{209}_{83}Bi$
   d. $^{16}_8O^{2-}$ e. $^{52}_{24}Cr^{3+}$

2.18 Give the mass number, atomic number, and neutron number of each of the following atoms.
   a. $^{14}_6C$ b. $^{32}_{15}P$ c. $^{192}_{77}Ir$
   d. $^{27}_{13}Al$ e. $^{209}_{84}Po$

2.19 The element lithium has atoms with a mass number of 6 and atoms with a mass number of 7. How many protons and neutrons are there in each type of atom?

Symbols and Names of Elements

2.20 Discuss the origin of the symbols of the elements.

2.21 Name each of the following elements.
   a. Ti   b. Ta   c. Th   d. Tc   e. Tl

2.22 Give the symbol for each of the following elements.
   a. iron   b. lead   c. silver   d. gold
   e. antimony   f. copper   g. mercury   h. tin
   i. sodium   j. tungsten

2.23 Name each of the following elements.
   a. S   b. Fe   c. Mn   d. Mg
   e. Al   f. Cl   g. Be   h. Rb
   i. Ni   j. Sc   k. Ti   l. Ne

2.24 Give the symbols for each of the following noble-gas elements.
   a. helium   b. neon   c. argon
   d. krypton   e. xenon   f. radon

2.25 Name each of the following elements.
   a. C   b. Ca   c. Cr   d. Co
   e. Cu   f. Cl   g. Cs

2.26 Give the symbols for manganese and magnesium.

2.27 Name each of the following elements.
   a. B   b. Ba   c. Be   d. Br   e. Bi

2.28 List all the elements whose symbols have only one letter.

2.29 Name each of the following elements.
   a. S   b. Si   c. Se   d. Sr   e. Sn

Formulas of Ionic Compounds

2.30 Describe how ions are formed from atoms.

2.31 What are the two major classes of ions? How do they differ?

2.32 What does the formula of a compound tell us?

2.33 What is an ionic compound?

2.34 Write the formula of the compound that would be formed by each of the following sets of ions.
   a. Ba^{2+}, Cl^- b. Fe^{3+}, Br^- c. Ca^{2+}, PO_4^{3-}
   d. Cr^{3+}, SO_4^{2-} e. Na^+, N^{3-} f. Cs^+, ClO_3^-
   g. Ti^{3+}, CO_3^{2-}

2.35 Write the formula for each of the combinations formed by the ions given below.

	Na^+	Ba^{2+}	Al^{3+}
Br^-			
S^{2-}			
SO_4^{2-}			
PO_4^{3-}			

2.36 Iron forms two ions, Fe^{2+} and Fe^{3+}. a. What are the formulas for the compounds the two iron ions can form with the oxide ion, O^{2-}? b. What are the formulas for the compounds the iron ions can form with the chloride ion, Cl^-?

2.37 Sodium ion has a 1+ charge: Na^+. In combination with the sulfate ion, it forms the compound Na_2SO_4. a. What is the charge on the sulfate ion? b. What is the charge on the strontium ion if it forms the compound $SrSO_4$?

Isotopes

2.38 Describe the difference in subatomic particles found in the two isotopes ^{35}Cl and ^{37}Cl.

2.39 The atomic weight of ^{12}C is exactly 12.0000 amu. Account for the fact that the atomic weight of the element carbon is 12.011 amu.

2.40 If the average atomic weight of naturally occurring fluorine is 18.9984 amu and the mass of one atom of $^{19}_9F$ is 18.9984 amu, how many isotopes are present in naturally occurring fluorine?

2.41 The element gallium, a metal that will melt in your hand, has two stable isotopes: $^{69}_{31}Ga$, with a mass of 68.9257 amu, and $^{71}_{31}Ga$, with a mass of 70.9249 amu. If the average mass of gallium is 69.72 amu, what is the relative abundance of these isotopes?

2.42 Magnesium, a low-density metal used in alloys, has three isotopes: $^{24}_{12}Mg$, $^{25}_{12}Mg$, and $^{26}_{12}Mg$. Their relative abundances are 78.99%, 10.00%, and 11.01%, and their atomic masses are 23.985, 24.986, and 25.983 amu, respectively. Calculate the average atomic weight of magnesium.

2.43 Naturally occurring lithium has an average atomic weight of 6.941 amu. If lithium has only the isotopes ^6_3Li of mass 6.0151 amu and ^7_3Li of mass 7.0160 amu, what are the percentages of each isotope in a sample of lithium?

Molecular Weight, Formula Weight, and Molar Mass

2.44 What is the difference between molecular weight and formula weight?

2.45 How are atomic weights used to obtain molecular weights?

2.46 Calculate the molecular weight or formula weight of each of the following compounds.
   a. Hg_2Cl_2 b. K_2SO_4 c. SF_6
   d. $KSbO_3$ e. $NaHSO_4$ f. $C_2H_4Cl_2$
   g. $C_{12}H_{22}O_{11}$ h. Cl_2O_5

2.47 Calculate the molar mass of each of the following substances.
   a. I_2 b. P_4 c. $CrCl_3$
   d. C_4H_8 e. CrO_2Cl_2 f. CaF_2

2.48 If 2.10×10^{23} molecules of a substance weighs 12.0 g, what is the molar mass of this substance?

2.49 If 1.00 mole of LiCl has a mass of 42.394 g, what is the mass of 1 LiCl formula unit in amu? in grams?

Moles, Masses, and Particles

2.50 In general, why must we weigh atoms in order to count them?

2.51 What is a mole? Why do chemists need to use this concept?

2.52 The most sensitive modern chemist's balance can weigh accurately a sample as small as about 10^{-8} g. How many iron atoms are there in a sample this size?

2.53 A sample of ammonia, NH_3, weighs 25.0 g. Calculate the following quantities.
   a. moles of NH_3
   b. number of NH_3 molecules
   c. number of N atoms
   d. moles of H atoms

2.54 A raindrop weighs 0.050 g. How many molecules of water, H_2O, are there in one raindrop?

2.55 Calculate the number of atoms in each of the following quantities.
   a. 36.1 g of argon
   b. 5.00 g of chromium
   c. 1.00 oz of silver
   d. the 44-carat Hope diamond, which is nearly pure carbon (one carat equals 0.200 g)
   e. 2.50 mL of mercury with a density of 13.6 g/mL

2.56 What is the mass of 2.4×10^{22} molecules of SO_2?

2.57 A chemical reaction requires 5.6 moles of zinc acetate, $Zn(CH_3CO_2)_2$. What mass of zinc acetate is needed?

2.58 Which element, Mo, Se, Na, or Br, contains the most atoms in a 1.0-g sample?

2.59 You have a 12.4-g sample of sulfur dioxide, SO_2. Calculate a. the number of molecules of SO_2, b. the moles of SO_2, c. the number of atoms of oxygen, and d. the mass of sulfur in this sample.

2.60 What is the mass of 0.100 mole of $Cu(OH)_2$?

2.61 How many moles, grams, and atoms of vanadium are there in 52.5 g of V_2O_5?

2.62 What mass of iodine contains the same number of atoms as 50.0 g of chlorine?

2.63 What is the mass of one atom of argon in units of amu and in units of grams?

2.64 Calculate the number of moles in 10.0 g of each of the following substances.
  a. $KHCO_3$ b. H_2S c. Se d. $MgSO_4$

2.65 Calculate the mass of sulfur in 100.0 g of each of the following compounds.
  a. SO_2 b. Na_2SO_4 c. $BaSO_4$ d. $KAl(SO_4)_2$

2.66 How many molecules are present in 15.43 g of butyl alcohol, C_4H_9OH?

2.67 What is the mass of Avogadro's number of chromium atoms?

2.68 A beaker contains 100.5 g of water, H_2O. What mass of methanol, CH_3OH, must be added to the beaker for the resulting mixture to contain 2 moles of methanol for every 1 mole of water?

Percent Composition

2.69 You have two colorless gases, each made of sulfur and oxygen. If they have different percent compositions, can they be the same substance?

2.70 Why is it useful to know the percent composition of a substance?

2.71 What is the percent composition of each of the following compounds?
  a. NH_3 b. $FeCl_2$ c. Na_3PO_4 d. KCl

2.72 Calculate the percent nitrogen in each of the following compounds.
  a. $NaNO_3$ b. NH_4Cl c. N_2H_4
  d. N_2O e. N_2O_3

2.73 What is the percent Cl in $NaClO_3$? What mass of Cl is contained in 15.05 g of $NaClO_3$?

2.74 Consider the following minerals as potential sources of copper metal, Cu. Which mineral contains the highest percent copper? Calculate the percent Cu for each mineral.
  a. chalcocite, Cu_2S
  b. malachite, $Cu_2(CO_3)(OH)_2$
  c. cuprite, CuO
  d. azurite, $Cu_3(CO_3)_2(OH)_2$

2.75 Calculate the percent of the first element in each of the following compounds.
  a. $BaCl_2$ b. Na_2CO_3 c. $CuSO_4$ d. $CoCl_2$

2.76 If copper sulfate, $Cu(H_2O)_5SO_4$, is heated strongly enough to drive off all the water, leaving $CuSO_4$, what percent of its mass will remain?

2.77 How much lead can be obtained from 2.00 kg of PbO?

2.78 Calculate the percent oxygen in each of the following iron oxides: FeO, Fe_2O_3, and Fe_3O_4.

Empirical Formulas

2.79 What is the difference between an empirical formula and a molecular formula?

2.80 Why do we normally use an empirical formula instead of percent composition to represent the composition of a substance?

2.81 Eugenol, a chemical substance with the flavor of cloves, consists of 73.19% C, 19.49% O, and 7.37% H. What is the empirical formula of eugenol?

2.82 One of the compounds occurring in cement has the following percent composition: 52.66% Ca, 12.30% Si, and 35.04% O. What is the empirical formula of this compound?

2.83 Para-aminobenzoic acid, often abbreviated PABA, is used in sunscreen formulations to prevent sunburn from ultraviolet radiation. PABA contains 61.30% C, 23.33% O, 10.22% N, and 5.15% H by mass. What is the empirical formula of PABA?

2.84 Coniine is the principal toxic compound found in poison hemlock, used in the self-administered execution of Socrates. Coniine has the percent composition 75.52% C, 13.47% H, and 11.01% N. From these data, calculate the empirical formula of coniine.

2.85 Determine the empirical formulas of the compounds with the following percent compositions.
  a. 72.36% Fe, 27.64% O
  b. 58.53% C, 4.09% H, 11.38% N, 25.99% O
  c. 63.15% C, 5.30% H, 31.55% O
  d. 85.62% C, 14.38% H

2.86 The explosive trinitrotoluene (TNT) has the percent composition 37.01% C, 2.22% H, 18.50% N, and 42.27% O. What is the empirical formula of TNT?

2.87 Strychnine has the percent composition 75.42% C, 6.63% H, 8.38% N, and 9.57% O. What is the empirical formula of strychnine?

2.88 Determine the empirical formulas for the following two gases containing nitrogen and oxygen.
  a. 36.8% N, 63.2% O
  b. 30.4% N, 69.6% O

2.89 What is the empirical formula of the manganese oxide that has the percent composition 69.61% Mn and 30.39% O?

Molecular Formulas

2.90 What information is needed to determine a molecular formula?

2.91 In what way does a molecular formula represent more information than does an empirical formula?

2.92 For each of the following molecular formulas, write the correct empirical formulas.
  a. $C_6H_4Cl_2$ b. C_6H_5Cl c. N_2O_5
  d. N_2O_4 e. $H_2C_2O_4$ f. CH_3CO_2H

2.93 A compound with the empirical formula CH_2O has a molecular weight of 90 amu. What is its molecular formula?

2.94   A gas has the empirical formula NO_2. If its molar mass is 92 g/mol, what is its molecular formula?

2.95   If a compound has a molar mass of 180 g/mol and a percent composition of 40.00% C, 6.72% H, and 53.29% O, what is the molecular formula?

2.96   A compound whose molar mass is 142 g/mol has the percent composition 50.7% C, 9.9% H, and 39.4% N. What is its molecular formula?

Additional Exercises

2.97   A gas has the percent composition 92.3% C and 7.7% H by mass. What is the empirical formula of this gas? If the molar mass of the gas is 78.1 g/mol, what is its molecular formula?

2.98   Chlorine has two isotopes, $^{35}_{17}Cl$ and $^{37}_{17}Cl$. Give the mass number, atomic number, and neutron number for each of these isotopes.

2.99   Tear gas has the percent composition 40.25% C, 6.19% H, 8.94% O, and 44.62% Br. What is its empirical formula?

2.100   A chemical procedure calls for 3.54 moles of dry ice (frozen carbon dioxide, CO_2). What mass of CO_2 should be added?

2.101   How many oxygen atoms are contained in 10.00 g of H_3PO_4?

2.102   Neon, a noble gas used in lights, has three isotopes: $^{20}_{10}Ne$, $^{21}_{10}Ne$, and $^{22}_{10}Ne$. Their relative abundances are 90.92%, 0.257%, and 8.82% and their atomic masses are 19.9924, 20.9838, and 21.9914 amu, respectively. Calculate the average atomic weight of neon.

2.103   Calculate the weight percent of carbon in Na_2CO_3.

2.104   Discuss the validity of the following statement: All pure substances can be decomposed into simpler pure substances by chemical reactions.

2.105   What is the mass of 6.9×10^{24} formula units of $CrPO_4$?

2.106   How many moles of CO_2 are contained in the 960 g of CO_2 exhaled daily by an average person?

2.107   Vanillin, the active component of the flavoring vanilla, has the percent composition 63.15% C, 5.30% H, and 31.55% O. a. What is its empirical formula? b. If the molar mass of vanillin is 152.14 g/mol, what is its molecular formula?

2.108   Indicate how many electrons, protons, and neutrons there are in each of the following species.
    a. ^3_2He b. ^3_1H c. ^2_1H d. $^{244}_{94}Pu$ e. $^{98}_{43}Tc$

2.109   During the preparation of a fungicide, chemists add 1.25 moles of mercuric chloride, $HgCl_2$. What mass of $HgCl_2$ is this?

2.110   Monosodium glutamate, $NaC_5H_8NO_4$, is used extensively as a flavor enhancer. What is the percent composition of this substance?

2.111   Calculate the molar mass of each of the following substances.
    a. $SrCO_3$ b. $MgCl_2$ c. $NiSO_4$ d. NH_4MnO_4
    e. Ar   f. H_2CO_3 g. SiO_2 h. $CaSiF_6$

2.112   If 486 g of magnesium is reacted with 320 g of oxygen, 806 g of magnesium oxide is produced. Which of the laws of composition is demonstrated by these results?

*2.113   The simplest chemical formula for the human body could be written $C_{728}H_{4850}O_{1970}N_{104}Ca_{24}P_{16}K_4S_4Na_3Cl_2Mg$. What is the percent composition of the human body? (Note: This formula accounts for only 99.95% of the body; other elements make up the other 0.05%.)

*2.114   How many grams of oxygen can be obtained from the decomposition of 55.2 g of mercuric oxide, HgO?

*2.115   If 5.00 g of aluminum is burned completely in excess oxygen gas, 9.45 g of an aluminum oxide is formed. What is the empirical formula of this oxide?

*2.116   Dalton's atomic theory is not quite consistent with modern atomic theory. Restate Dalton's theory, taking into account modern discoveries about the atom.

*2.117   One isotope of lithium, ^7Li, is useful in atomic reactors, so it is often separated from natural lithium. The remaining lithium is often converted to chemical compounds, such as LiCl, which are sold to chemists. Can you use the tabulated atomic weight of lithium to calculate the molecular weight of LiCl prepared in this form?

*2.118   Suppose you are a time traveler and you go back to ancient Greece. What evidence could you present to convince Aristotle that matter is made up of atoms?

3

▶ ▶ ▶ ▶

The Elements and Their Compounds

Alka Seltzer contains a carbonate salt and an acid, which react when wet to release carbon dioxide gas.

CHAPTER OUTLINE

Encountering Chemistry

The next time you are in a grocery or drugstore, take a look at a few different brands of antacid. You will see that most of them contain one or two of these ingredients: calcium carbonate, magnesium hydroxide, aluminum hydroxide, and Simethicone, whose chemical formulas are $CaCO_3$, $Mg(OH)_2$, $Al(OH)_3$, and $[(CH_3)_2SiO]_x$ (where x is a large number), respectively. The first three ingredients neutralize stomach acid, and the last relieves the accumulation of gas.

Suppose chemists in a pharmaceutical company want to identify new ingredients suitable for use in antacids. Do they have to test every compound that exists? Fortunately, they do not. Elements and their compounds exhibit patterns and trends in their properties, so the chemists test only compounds of elements that they expect to be similar to calcium, magnesium, and aluminum. There are many metal carbonates (metal ions combined with CO_3^{2-} ions) and metal hydroxides (metal ions combined with OH^- ions) that could neutralize stomach acid. Hydroxides are usually preferable to carbonates because hydroxides don't release carbon dioxide gas in the stomach. Elements expected to have properties similar to those of calcium and magnesium include strontium, barium, and beryllium.

There are, of course, other considerations. Substances used in an antacid must be nontoxic and noncorrosive, stable, readily available, and inexpensive. These criteria eliminate many possible compounds. For example, exposure to small amounts of beryllium compounds can cause illness or death. Strontium does not occur in large quantities in nature, so using it would be expensive. Barium is toxic, and so—at least to some extent—are other metals that would form compounds similar to aluminum hydroxide. The hydroxides of sodium and potassium, while common and inexpensive, are quite corrosive and so could not be swallowed

Why are these elements
expected to have similar
properties?

safely. A consideration of these facts indicates that there is good reason that only a
few ingredients are used in antacid preparations.

There are more than 100 known elements. Taken individually, their physical
and chemical properties, as well as those of their compounds, comprise an unman-
ageably large collection of facts. To avoid hopeless confusion, scientists began
long ago to try to make sense of all this information by classifying the properties of
the elements. These properties, it was found, do not vary randomly from element
to element. Rather, they progress through groups of elements in a generally predictable manner. Further, patterns found in one group of elements are repeated in
other groups. We call this regular patterning *periodicity,* and it serves as the fundamental organizing principle of one of the most useful tools available to the chemist: the periodic table. By arranging the elements in a way that emphasizes both
their differences and their similarities, the periodic table enables chemists to classify and explain a tremendous amount of data.

In this chapter, we examine the periodic table in some detail. We also consider some of the properties of the elements—their ionic charges, compounds,
oxidation numbers, and sources—and systematize them with reference to the periodic table. In addition, we discuss the naming of chemical compounds formed
from the elements.

3.1 The Periodic Table

Consider the elements calcium, strontium, and barium. All of these elements are
silvery-white metals. All react with air to form a coating of tarnish. They are found
naturally in the form of carbonate compounds (containing CO_3^{2-} ions, such as
$CaCO_3$, $SrCO_3$, $BaCO_3$) and sulfate compounds (containing SO_4^{2-} ions, such as
$CaSO_4$, $SrSO_4$, $BaSO_4$), in which the metal occurs as a cation with a 2+ charge
(see Figure 3.1). Generally, their chemical properties are quite similar. Their
physical properties do vary, but in a regular manner—that is, the physical proper-

Figure 3.1 Calcium is a silver-
colored metal that tarnishes
readily in air to form a
protective bluish-gray coating.
It is used in various alloys such
as steels and bearing metals.
Calcium occurs widely in nature in the form of com-
pounds such as calcium
carbonate and calcium sulfate.
Left: Calcium carbonate, a
white solid, is found in limestone and marble. *Right:* Calcium sulfate, also a white
solid, is found in gypsum,
which can be converted to
plaster by heating.

Figure 3.2 Oxygen, sulfur, selenium, and tellurium (counterclockwise from top) show distinct differences in physical state but have similar chemical properties.

Figure 3.3 Dmitri I. Mendeleev (1834–1907) was educated in St. Petersburg, Russia, where he later became a university professor. While writing a chemistry textbook, Mendeleev prepared a card for each known element and listed their properties. When he rearranged the cards, he noticed patterns in the chemical and physical properties of the elements. This discovery led to his publication of a periodic table in 1869.

ties of strontium are very close to the average of the properties of calcium and barium. The average of the atomic weights of calcium (40.08 amu) and barium (137.33 amu) is 88.71 amu, for example, and the atomic weight of strontium is 87.62 amu. The average of the melting points of calcium (839°C) and barium (725°C) is close to the melting point of strontium (769°C). And the average of the densities of calcium (1.54 g mL^{-1}) and barium (3.51 g mL^{-1}) is 2.52 g mL^{-1}, nearly the same as the density of strontium (2.6 g mL^{-1}).

Numerous other groups of three or more elements show the same sort of similarities in behavior. The elements fluorine, chlorine, bromine, and iodine, collectively called the *halogens,* are nonmetals with very similar chemical properties and physical properties that vary in a regular manner, especially for the latter three elements. Another family of elements with related properties contains oxygen, sulfur, selenium, and tellurium (Figure 3.2).

The various families of elements that can be identified as having similar properties can also be arranged so that they show regular trends in their properties. As an example, consider the radius of atoms of a few elements having adjacent atomic numbers. The atomic-radius values for lithium (152 pm), beryllium (111 pm), boron (88 pm), carbon (77 pm), nitrogen (70 pm), oxygen (66 pm), and fluorine (64 pm) decrease as the atomic numbers of the elements increase. Investigation of recurring regularities in the properties of the elements enabled scientists late in the nineteenth century to begin to classify the elements in a meaningful way.

Development of the Periodic Table

Most of the credit for the periodic classification of the elements is given to Dmitri I. Mendeleev (Figure 3.3) and Julius Lothar Meyer (Figure 3.4). Independently and nearly simultaneously, these men arranged the elements in an array

called a **periodic table** and proposed the periodic law to describe the arrangements. The table published by Mendeleev in 1869 is shown in Figure 3.5. Meyer's table is essentially identical to Mendeleev's. However, Mendeleev developed his table largely by examining chemical properties, whereas Meyer based his primarily on physical properties.

The periodic law as formulated by Mendeleev and Meyer describes this periodic nature as follows: the properties of the elements are periodic functions of their atomic weights. Stated more simply, when the elements are arranged in a two-dimensional table in order of increasing atomic weight, elements having similar properties line up in vertical columns, as shown in Figure 3.6.

Notice in Mendeleev's table that the atomic weights of the elements do not progress perfectly from smaller to larger. For example, cobalt and nickel have the same atomic weight (using values of Mendeleev's time), and tellurium appears to be out of sequence. Mendeleev believed that the out-of-order atomic weights were in error. However, we now know that the periodicity of the elements is actually related to the atomic number (a concept unknown at the time) rather than the atomic weight. This relationship was discovered in 1913 by the English physicist H. G. J. Moseley, two years before he was killed in World War I. The **periodic law** is more correctly stated as follows: *the properties of the elements are periodic functions of their atomic numbers.*

$$
\begin{array}{llll}
& & \text{Ti} = 50 & \text{Zr} = 90 & ? = 180 \\
& & \text{V} = 51 & \text{Nb} = 94 & \text{Ta} = 182 \\
& & \text{Cr} = 52 & \text{Mo} = 96 & \text{W} = 186 \\
& & \text{Mn} = 55 & \text{Rh} = 104,4 & \text{Pt} = 197,4 \\
& & \text{Fe} = 56 & \text{Ru} = 104,4 & \text{Ir} = 198 \\
& \text{Ni} = \text{Co} = 59 & \text{Pd} = 106,6 & \text{Os} = 199 \\
\end{array}
$$

H = 1

Be = 9,4   Mg = 24   Cu = 63,4   Ag = 108   Hg = 200

B = 11   Al = 27,4   Zn = 65,2   Cd = 112

C = 12   Si = 28   ? = 68   Ur = 116   Au = 197?

N = 14   P = 31   ? = 70   Sn = 118

O = 16   S = 32   As = 75   Sb = 122   Bi = 210?

F = 19   Cl = 35,5   Se = 79,4   Te = 128?

Li = 7   Na = 23   Br = 80   J = 127

K = 39   Rb = 85,4   Cs = 133   Tl = 204

Ca = 40   Sr = 87,6   Ba = 137   Pb = 207

? = 45   Ce = 92

?Er = 56   La = 94

?Yt = 60   Di = 95

?In = 75,6   Th = 118?

Figure 3.5 Mendeleev arranged the elements to show the periodic nature of their properties and found that this arrangement nearly followed the order of increasing atomic weight. So strong was his belief that periodicity was a fundamental law of nature that he left holes in his table (indicated by question marks) rather than force elements into positions where their properties did not fit. He predicted the existence of germanium, scandium, gallium (which were discovered during the period 1875–1886), rhenium (1925), technetium (1937), and promethium (1945), as well as some of their properties.

6.9 Li	9.0 Be	10.8 B	12.0 C	14.0 N	16.0 O	19.0 F
Soft, silvery, very reactive metal Oxide: Li_2O	Hard, silvery, reactive metal Oxide: BeO	Hard, metalloid Oxide: B_2O_3	Amorphous black or shiny gray solid Hydride: CH_4	Colorless, nonreactive nonmetal; gas Hydride: NH_3	Colorless, reactive nonmetal; gas Hydride: H_2O	Pale yellow, very reactive nonmetal; gas Hydride: HF
23.0 Na	24.3 Mg	27.0 Al	28.1 Si	31.0 P	32.0 S	35.5 Cl
Soft, silvery, very reactive metal Oxide: Na_2O	Hard, silvery, reactive metal Oxide: MgO	Medium-hard, silvery reactive metal Oxide: Al_2O_3	Hard, silvery, metalloid Hydride: SiH_4	Soft, white, red, or black nonmetal; solid Hydride: PH_3	Soft, yellow nonmetal; solid Hydride: H_2S	Pale green, very reactive nonmetal; gas Hydride: HCl
39.1 K						79.9 Br
Soft, silvery, very reactive metal Oxide: K_2O						Red, reactive nonmetal; liquid Hydride: HBr

Figure 3.6 When the elements were arranged by Mendeleev and Meyer in order of increasing atomic weight, similar properties recurred periodically.

The Modern Periodic Table

Mendeleev's periodic table went through several changes to accommodate the discovery of new elements. The modern version of the periodic table is shown in Figure 3.7 (and on the inside front cover).

Examine the classification of the elements shown in Figure 3.7. The elements are divided into **groups** or families (the vertical columns in the table) and **periods** (the horizontal rows in the table). The periods are numbered from 1 through 7 from the top of the table to the bottom. Each group is given a symbol consisting in most cases of a Roman numeral and an uppercase letter A or B. For example, the group to the far left of the periodic table is Group IA. As shown in Figure 3.8, each group can be classified as an A group (the **main-group,** or **representative, elements**) or a B group (the **transition elements,** shown in Figure 3.9). The fourteen members of the **lanthanide** and **actinide** series of elements are usually placed on separate lines at the bottom of the table to conserve space, but they should really be inserted immediately after elements lanthanum (La) and actinum (Ac), respectively. (Their positions are shown by the star and triangle symbols in Figure 3.7.) These groups are referred to as **inner-transition elements.**

Other labeling systems are also used for the periodic groups. The one just described is the Chemical Abstracts Service (CAS) system, widely used in the United States. Europeans generally have used an older system adopted by the International Union of Pure and Applied Chemistry (IUPAC), in which the A designation is used for the first half of the table and the B designation for the second (as shown in Figure 3.7). A recently proposed compromise numbers the

What problems could arise when different labeling systems are used in different countries?

Periodic table of the elements

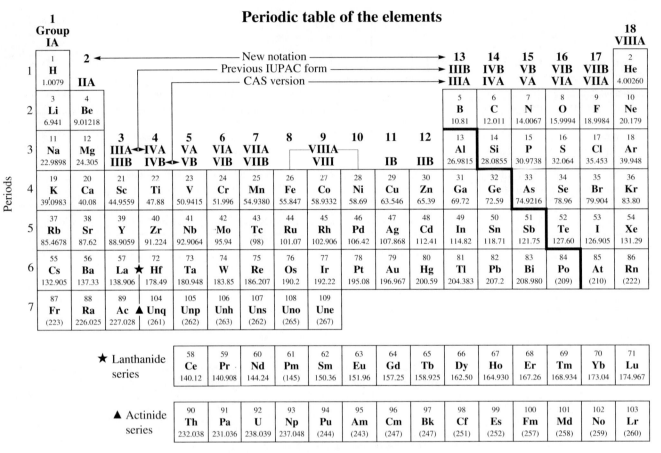

Figure 3.7 A modern periodic table, showing various systems of labeling the columns or families. The Chemical Abstracts Service (CAS) version has been popular in the United States. The International Union of Pure and Applied Chemistry (IUPAC) version has been widely used in Europe. The new notation has been introduced to replace these two conflicting versions. An atomic weight given in parentheses is the mass number of the longest-lived isotope, which is used when the element has no stable isotopes. (Reprinted with permission from *Chemical and Engineering News*, February 4, 1985, p. 27. Copyright 1985 American Chemical Society.)

groups from 1 through 18, using Arabic numbers. This book retains the old CAS labels but also refers to the new notation as needed. For example, the rightmost group will be called Group VIIIA (18).

A convenient comparison within the periodic table involves the metallic character of the elements. We usually refer to an element as a **metal,** a **nonmetal,** or a **metalloid** (also called a **semimetal** or **semiconductor**). Examine the periodic table in Figure 3.10 and note the heavy staircase line running between the elements from boron (B) diagonally down to astatine (At). Elements to the left of this line are metals, whereas elements to the right of the line are nonmetals. Exceptions are the shaded elements along the diagonal line, the metalloids (B, Si, Ge, As, Sb, Te, Po).

Metals are characterized by their luster, their crystalline form as solids, their good electrical and thermal (heat) conductivity, and their tendency to lose electrons to form cations when they form compounds (generally by combining with

Figure 3.8 The elements are grouped in the periodic table as either main-group or transition elements. The transition elements include the inner-transition metals (lanthanides and actinides). Some of the main-group elements are known by group names, such as the alkali metals, alkaline-earth metals, halogens, and noble gases.

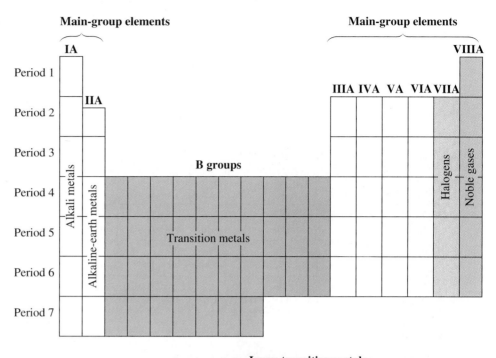

Main-group elements   Main-group elements

What metals and nonmetals are used in your home?

nonmetals). Nonmetals do not conduct electricity well (except for carbon in the form of graphite, which is an electrical conductor). They are also usually heat insulators. They react with one another to form nonionic compounds, and when they combine with metals, they tend to gain electrons to form anions. Metalloids have physical properties more like those of metals, but their chemical reactivity is more like that of nonmetals. Many of their atomic properties are intermediate between those of metals and of nonmetals.

Figure 3.9 The transition elements in Period 4 are all relatively unreactive metals with chemical and physical properties that show many similarities. *Top (from left):* Sc, Ti, V, Cr. *Middle (from left):* Mn, Fe, Co, Ni. *Bottom (from left):* Cu, Zn.

1
Group
IA

1 IA	2 IIA	3 IIIB	4 IVB	5 VB	6 VIB	7 VIIB	8	9 VIII	10	11 IB	12 IIB	13 IIIA	14 IVA	15 VA	16 VIA	17 VIIA	18 VIIIA
1 H																	2 He
3 Li	4 Be											5 B	6 C	7 N	8 O	9 F	10 Ne
11 Na	12 Mg											13 Al	14 Si	15 P	16 S	17 Cl	18 Ar
19 K	20 Ca	21 Sc	22 Ti	23 V	24 Cr	25 Mn	26 Fe	27 Co	28 Ni	29 Cu	30 Zn	31 Ga	32 Ge	33 As	34 Se	35 Br	36 Kr
37 Rb	38 Sr	39 Y	40 Zr	41 Nb	42 Mo	43 Tc	44 Ru	45 Rh	46 Pd	47 Ag	48 Cd	49 In	50 Sn	51 Sb	52 Te	53 I	54 Xe
55 Cs	56 Ba	57 La ★	72 Hf	73 Ta	74 W	75 Re	76 Os	77 Ir	78 Pt	79 Au	80 Hg	81 Tl	82 Pb	83 Bi	84 Po	85 At	86 Rn
87 Fr	88 Ra	89 Ac ▲	104 Unk	105 Unp	106 Unh	107 Uns	108 Uno	109 Une									

Metalloids *Nonmetals* *Metals* **Metals**

★
58 Ce	59 Pr	60 Nd	61 Pm	62 Sm	63 Eu	64 Gd	65 Tb	66 Dy	67 Ho	68 Er	69 Tm	70 Yb	71 Lu
90 Th	91 Pa	92 U	93 Np	94 Pu	95 Am	96 Cm	97 Bk	98 Cf	99 Es	100 Fm	101 Md	102 No	103 Lr

▲

Figure 3.10 The elements can be divided into three classes according to metallic character: metals, nonmetals, and metalloids. Metals are found on the left side of the periodic table, and nonmetals are on the right. Metalloids fall along the diagonal line dividing the metals from the nonmetals.

For several of the groups, descriptive names are frequently used instead of the group labels. All the members of Group IA (1) except hydrogen—Li, Na, K, Rb, Cs, and Fr—are referred to collectively as the **alkali metals.** (Hydrogen is not included because it is not a metal.) All are soft metals with low melting points; cesium has a golden-yellow color and the others are silvery-white. All react readily with oxygen, water, or chlorine. They are called alkali metals because their reaction with water, shown in Figure 3.11, produces strong alkalis, such as sodium

Figure 3.11 Sodium metal, like the other alkali metals, reacts vigorously with water to release hydrogen gas.

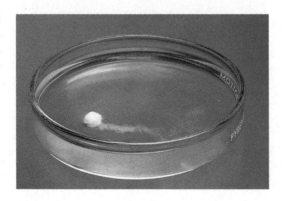

Figure 3.12 The alkaline-earth metals, such as calcium, react with water to release hydrogen gas, but not as vigorously as the alkali metals do. Magnesium reacts only with steam, and beryllium does not react with water.

hydroxide (or lye). Alkalis, also called *bases,* are substances that will neutralize acids.

The Group IIA (2) elements (Be, Mg, Ca, Sr, Ba, Ra) are called the **alkaline-earth metals.** These metals are harder and less reactive than the alkali metals but are nevertheless among the more reactive metals. The name for this group derives from the fact that its members can be found or isolated readily as oxides (compounds containing oxygen, once commonly called *earths*), which are strongly alkaline, or basic. These elements react with water to form alkalis, but more slowly than the alkali metals do (see Figure 3.12).

On the far right are the Group VIIIA (18) elements, called the **noble gases** (Figure 3.13). All these monatomic, colorless gases are relatively unreactive, the property that gives them their group name. Indeed, reactions involving the noble gases are so rare that only a few stable compounds of krypton and xenon have been prepared. A few compounds of argon are known, but they are stable only at low temperatures. The helium compound $He(N_2)_{11}$ exists in an unstable state at very low temperatures. No compounds of neon have yet been prepared.

The Group VIIA (17) elements (fluorine, chlorine, bromine, iodine, and astatine) are called **halogens** and occur as reactive diatomic species. Under normal conditions, F_2 is a light yellow gas, Cl_2 is a pale green gas, Br_2 is a red liquid, and I_2 is a purple solid. At_2 is so unstable and radioactive that it has not been isolated in observable quantities. The name *halogen,* which derives from the Greek words *hals* (''salt'') and *gen* (''to produce''), is appropriate because these elements often form ionic compounds, or salts, called *halides.*

We have seen how the members of several vertical columns (groups) in the periodic table share similarities. The transition elements also exhibit similarities across rows. All the transition elements are less reactive than the alkali and alkaline-earth metals. For example, the alkali and alkaline-earth metals react vigorously with water, but the transition metals react with water only slowly, if at all. Because of this low reactivity, we use the transition metals, such as iron (Fe) and copper (Cu), as structural metals. ▶ These similarities in properties allowed us to decide which compounds might make reasonable antacid ingredients, as discussed in the introduction to this chapter.

Figure 3.13 The noble gases are commonly used to provide colored lights in vacuum tubes similar to the fluorescent lighting tubes used in our homes. Shown here is a street in Taiwan.

3.2 Electrons, Ions, and the Periodic Table

The periodic table is useful to the chemist in predicting properties and explaining trends among the elements. In Chapter 2, we saw that atoms of the elements may form ions by gaining or losing electrons. In this section, we use the periodic table to predict what charges these ions will have. This information, in turn, allows us to predict how atoms or ions will combine to form binary compounds.

Electron Counts of Stable Ions

The noble gases are particularly stable elements. Their stability is associated with the number of electrons that they contain—their electron counts. To achieve a similar stability, atoms of a main-group element tend to gain or lose electrons to form ions with the electron count of the noble gas nearest to that element on the periodic table. The nonmetals usually gain electrons to form anions that have a noble-gas electron count, whereas the metals lose electrons to form cations that have a noble-gas electron count.

Charges of monatomic ions of the main-group elements correlate very nicely with the position of the element in the periodic table, as shown in Figure 3.14. Group IA (1) elements will lose one electron to form a cation with a charge of $1+$; each of these ions has the same number of electrons as the noble gas that precedes it in the periodic table. Group IIA (2) elements will lose two electrons to form cations with a charge of $2+$, again giving the ions the same number of electrons as

1 IA	2 IIA		13 IIIA	14 IVA	15 VA	16 VIA	17 VIIA	18 VIIIA
H 1− 1+								He
Li 1+	Be 2+		B	C 4−	N 3−	O 2−	F 1−	Ne
Na 1+	Mg 2+		Al 3+	Si 4−	P 3−	S 2−	Cl 1−	Ar
K 1+	Ca 2+		Ga 3+ 1+	Ge 4−	As 3−	Se 2−	Br 1−	Kr
Rb 1+	Sr 2+		In 3+ 1+	Sn 4+ 2+	Sb	Te 2−	I 1−	Xe
Cs 1+	Ba 2+		Tl 3+ 1+	Pb 4+ 2+	Bi	Po	At 1−	Rn

Figure 3.14 The negative charge on the monatomic anion of a nonmetal corresponds to its position in the periodic table. The noble gases have a charge of 0; counting backward from there, each successive group through Group IVA (14) accumulates one additional negative charge. A monatomic cation of a representative element has a charge equal to its CAS group number. Some elements have more than one stable ion. When two or more ions are formed by the same element, these ions generally have charges two units apart. Some elements do not form monatomic ions readily, so they have no entries in this table.

the preceding noble gas. Group VIIA (17) elements gain one electron to form anions with a charge of $1-$, whereas Group VIA (16) elements gain two electrons to form anions with a charge of $2-$. These anions also have the same number of electrons as a noble gas. The charges of common ions are summarized in the abbreviated periodic table in Figure 3.14.

With the transition metals, there is usually more than one common ionic charge, and the charges bear no simple relation to position in the periodic table. The charges of the common ions of the transition metals are shown in Figure 3.15. Note that $2+$ and $3+$ are the most common, with a few examples of $1+$ charges.

Charges and Formulas of Oxoanions

The nonmetals (as well as some of the metals) combine with oxygen to form polyatomic ions known as *oxoanions*. Usually there are two common oxoanions for each nonmetal. When this is the case, the two oxoanions differ in the number of oxygen atoms surrounding the nonmetal atom. The name of the oxoanion with the greater number of oxygen atoms combines the beginning of the nonmetal elemental name and the ending *-ate*. The name of the oxoanion with the lesser number of oxygen atoms uses the ending *-ite*. By comparison, the monatomic anion uses the ending *-ide*. For example, a sulfur atom normally adds two electrons to form the sul*fide* ion, S^{2-}. In combination with oxygen, it can form oxoanions such as sul*fate*, SO_4^{2-}, and sul*fite*, SO_3^{2-}. You can remember the formulas and charges of the common oxoanions of the nonmetals by referring to the portion of the periodic table shown in Figure 3.16.

Formulas of Binary Compounds

Elements combine to form binary compounds that can usually be placed into one of two categories. When cations and anions combine and retain their ionic character, as described in Chapter 2, the result is an ionic compound. When atoms combine and do not form ions, the result is a **covalent compound.** Whereas binary ionic compounds generally contain a metal and a nonmetal, binary covalent compounds are usually composed of two nonmetals. The essential difference between these two classes of compounds involves the way in which electrons are distributed between the two atoms, forming ions or molecules. In ionic compounds, electrons

Figure 3.15 Transition-metal cations commonly have $2+$ or $3+$ charges, though a few have $1+$ charges. Elements for which no charges are listed are rarely found as monatomic cations.

3	4	5	6	7	8	9	10	11	12
IIIB	IVB	VB	VIB	VIIB		VIIIB		IB	IIB
Sc	Ti	V	Cr	Mn	Fe	Co	Ni	Cu	Zn
		2+	2+	2+	2+	2+		1+	
3+	3+	3+	3+	3+	3+	3+	2+	2+	2+
Y	Zr	Nb	Mo	Tc	Ru	Rh	Pd	Ag	Cd
3+								1+	2+
La	Hf	Ta	W	Re	Os	Ir	Pt	Au	Hg*
								1+	1+
3+								3+	2+

*Hg with an apparent charge of $1+$ exists as the Hg_2^{2+} ion.

Figure 3.16 The more common oxoanions of the nonmetals. For the elements in the top row and the right column (shaded), one atom combines with three oxygen atoms to form the -ate ion. For the other elements, four oxygen atoms are used to form the -ate ion. The -ite ion, if it exists, has one fewer oxygen atom than the -ate ion. In all cases, the -ate and -ite ions have the same charge. Except for the oxoanions formed by B, C, and N, most oxoanions have the same charge as the monatomic -ide ions of the nonmetals, so their charges can be predicted from position in the periodic table.

B	C	N	O	F
BO_3^{3-} Borate	CO_3^{2-} Carbonate	NO_3^- Nitrate NO_2^- Nitrite	None known	None known
Si SiO_4^{4-} Silicate	P PO_4^{3-} Phosphate PO_3^{3-} Phosphite	S SO_4^{2-} Sulfate SO_3^{2-} Sulfite	Cl ClO_3^- Chlorate ClO_2^- Chlorite	
	As AsO_4^{3-} Arsenate AsO_3^{3-} Arsenite	Se SeO_4^{2-} Selenate SeO_3^{2-} Selenite	Br BrO_3^- Bromate BrO_2^- Bromite	
		Te TeO_4^{2-} Tellurate TeO_3^{2-} Tellurite	I IO_3^- Iodate IO_2^- Iodite	

■ Combines with three O atoms to form -ate ion

□ Combines with four O atoms to form -ate ion

Sodium sulfide, Na₂S, exists as colorless crystals that decompose slowly in the presence of moisture, releasing toxic hydrogen sulfide, H₂S(g). It is used in dehairing hides, processing metal ores, engraving, printing cotton, and manufacturing rubber and dyes.

Is it legitimate to develop a model that is useful even though we know that some of its features are wrong? Explain.

are transferred from one element to the other to form cations and anions; in covalent compounds, electrons are not completely transferred but are shared. In Chapter 7, we will see that some compounds are formed by the partial transfer of electrons between atoms.

Ionic compounds must have no net charge, and the number of cations and anions that combine is determined by this condition of electrical neutrality. We used this principle in Chapter 2 to determine the formulas of binary compounds formed from ions of known charge. Now that we can use the periodic table to predict ionic charges, we can predict the formulas of binary ionic compounds. For example, sodium ion should be Na^+ and sulfide ion should be S^{2-} according to their positions in the periodic table. It requires two sodium ions to provide the 2+ charge needed to balance the 2− charge of the sulfide ion, so these ions should combine to form the electrically neutral ionic compound Na_2S.

Because covalent compounds do not contain ions, we cannot use the same method to predict their composition. In many covalent compounds, the atoms develop partial charges by partial transfer of electrons. However, if *we pretend that electrons are completely transferred to form ions*, we can successfully deduce the composition of one covalent compound formed by two nonmetals. We simply use the condition of electrical neutrality and assume that the element farther to the left in the periodic table will form a cation and that the one farther to the right will form an anion. *The elements will not actually have these charges in the covalent compound they form.* Nevertheless, these imaginary charges, called **oxidation numbers,** are useful for determining the combining ratios of atoms in covalent compounds.

For example, carbon, being in Group IVA (14), would form a 4+ cation to have a noble-gas electron count, and oxygen would form a 2− anion. Thus, carbon has an oxidation number of +4 and oxygen has an oxidation number of −2. Using these numbers, we predict they would combine to form a compound with formula CO_2—carbon dioxide, a compound exhaled in our breath. However, this method does not predict the compound CO, carbon monoxide, which can also be formed between carbon and oxygen. We will consider the composition of covalent compounds in more detail in Section 3.4, where we will examine a more detailed method of keeping track of charges on atoms by assigning oxidation numbers.

3.3 Binary Compounds of Oxygen and Hydrogen

We can understand better the composition of binary compounds by looking at some common examples. Consider compounds formed by oxygen, which combines with nearly all of the elements. The chemical combinations of oxygen are generally quite simple. Binary compounds that contain oxygen, whether ionic or covalent, are called *oxides*. The composition of the oxide that is formed by an element is often determined by its group number. Hydrogen also forms binary compounds, called *hydrides*, with most of the elements. Again, the combining ratios are simple and correlate well with the element's position in the periodic table.

How would you expect the composition of the oxides and hydrides to change across the periodic table?

Oxides

Oxygen is the most abundant element in the earth's crust and is extensively involved in the chemistry of the earth and its atmosphere. As O_2, oxygen makes up about 21% of the volume of the atmosphere. In addition, water consists of 88.9% oxygen by mass. Most minerals contain oxygen in combined form, such as in silicate (Figure 3.17).

Figure 3.18 shows the most common oxides. Most oxides have the composition that would be predicted if oxygen has a charge of $2-$. Ionic oxides contain the O^{2-} ion, and covalent oxides have a composition consistent with this charge. However, there are compounds in which oxygen appears to have other charges. When combined with fluorine, which has such a strong attraction for electrons that

Figure 3.17 One of the anions found in minerals is silicate, SiO_4^{4-}. The silicate salts of many metals are colorful and quite insoluble. The crystal growing packets available in variety stores as "magic rocks" provide the ingredients for growing silicate salts.

Figure 3.18 The elements form many different oxides. This chart shows only the more common ones.

Group 1 IA	2 IIA	3 IIIB	4 IVB	5 VB	6 VIB	7 VIIB	8 (VIII)	9 (VIII)	10 (VIII)	11 IB	12 IIB	13 IIIA	14 IVA	15 VA	16 VIA	17 VIIA	18 VIIIA (Noble gases)
H_2O																	
Li_2O	BeO											B_2O_3	CO_2, CO	N_2O_5, N_2O_4, N_2O_3, NO_2, NO, N_2O		OF_2	
Na_2O	MgO											Al_2O_3	SiO_2, SiO	P_4O_{10}, P_4O_6	SO_3, SO_2	Cl_2O_7, Cl_2O_5, ClO_2, Cl_2O	
K_2O	CaO	Sc_2O_3	TiO_2, Ti_2O_3, TiO	V_2O_5, VO_2, V_2O_3, VO	CrO_3, CrO_2, Cr_2O_3	Mn_2O_7, MnO_2, Mn_2O_3, MnO	Fe_2O_3, Fe_3O_4, FeO	CoO_2, Co_2O_3, CoO	NiO	CuO, Cu_2O	ZnO	Ga_2O_3	GeO_2, GeO	As_2O_5, As_2O_3	SeO_3, SeO_2	Br_2O_5, BrO_2, Br_2O_3, Br_2O	
Rb_2O	SrO	Y_2O_3	ZrO_2	Nb_2O_5	MoO_3, MoO_2	Tc_2O_7, TcO_2	RuO_4, RuO_2	RhO_2, Rh_2O_3, RhO	PdO	Ag_2O_3, AgO, Ag_2O	CdO	In_2O_3	SnO_2, SnO	Sb_2O_5, Sb_2O_4, Sb_4O_6	TeO_3, TeO_2, TeO	I_2O_5, I_4O_9, I_2O_4	XeO_4, XeO_3
Cs_2O	BaO	La_2O_3	HfO_2	Ta_2O_5	WO_3, WO_2	Re_2O_7, ReO_3, Re_2O_5, ReO_2	OsO_4, OsO_2	IrO_2, Ir_2O_3	PtO_2	Au_2O_3	HgO	Tl_2O_3, Tl_2O	PbO_2, Pb_3O_4, PbO	Bi_2O_5, Bi_2O_3			

Figure 3.18 The elements form many different oxides. This chart shows only the more common ones. Note that for most of the elements, the oxide containing the most oxygen is consistent with a formula predicted by the combination of O^{2-} with a cation of the other element, which has a positive charge equal to its CAS group number.

Figure 3.19 Magnesium combines vigorously with oxygen in air at high temperatures to form a white solid, MgO.

it always forms compounds consistent with a $1-$ charge, oxygen appears to have a charge of $2+$, as in F_2O (usually written as OF_2). In another class of oxygen compounds, called the *peroxides,* oxygen has an apparent charge or oxidation number of -1. The peroxides all contain the group O_2^{2-}, either as an anion (Na_2O_2) or in a covalent combination (H_2O_2). The *superoxides,* known only in a limited number of ionic compounds such as KO_2, contain the group O_2^-, with the oxygen having an oxidation number of $-\frac{1}{2}$.

Recall from Figure 3.14 that Group IA (1) elements always take a charge of $1+$. All the oxides of these elements have the formula M_2O (where M is a metal). Similarly, the Group IIA (2) elements, with charges of $2+$, form oxides with the formula MO. Except for water, the oxide of hydrogen, all of the Group IA (1) and IIA (2) oxides, such as MgO (Figure 3.19), are stable white solids that melt at very high temperatures. Because water is a liquid under normal conditions and because it readily dissolves many materials, it is unique among the binary oxides. In Group IIIA (13), oxide formulas are M_2O_3, as would be predicted, but some of these elements also form compounds with the formula M_2O.

Figure 3.20 Sulfur burns in oxygen with a pale blue flame to form gaseous SO_2 and SO_3, which are used to make sulfuric acid. An older name for sulfur is *brimstone,* which means "burning stone."

The oxides of the metals and the nonmetals generally differ significantly. Those of the metals tend to be ionic solids. Nonmetal oxides are mostly covalent compounds, some of which are solids but many of which are liquids or gases. In Group IVA (14) and the groups that follow, most of the oxides are covalent compounds, especially for the lighter elements. The heavier elements in these groups are metals or metalloids and so may form ionic oxides.

The elements in Group IVA (14) and higher usually form more than one stable oxide. We can predict the composition of the oxide that contains the most oxygen (called the highest oxide) by using the following procedure. First, we form the hypothetical cation with the noble-gas electron count the element would have if it were in an ionic compound. For example, carbon would form C^{4+} and nitrogen would form N^{5+} because these elements would need to lose that many electrons to achieve a noble-gas electron count. Next, we combine the hypothetical cation with the oxide ion, O^{2-}. This procedure predicts a formula of CO_2 for carbon and N_2O_5 for nitrogen. The formulas of the lower oxides of these elements generally correspond to the removal of one oxygen atom per atom of the other element. Thus, in Group IVA (14), C forms CO_2, as expected, but it also forms CO, with one oxygen atom removed. In Group VA (15), N forms N_2O_5, N_2O_3, and N_2O (though it also forms some other nitrogen oxides that cannot be predicted by the method just described: NO_2, N_2O_4, and NO). In Group VIA (16), S forms SO_3 and SO_2 (Figure 3.20); in Group VIIA (17), Cl forms Cl_2O_7, Cl_2O_5, Cl_2O_3, and Cl_2O. Fluorine, an unusual element also in Group VIIA (17), forms only F_2O and F_2O_2.

As we move from the left to the right along the top of the periodic table, we no longer find solid oxides as frequently. Most of the oxides formed by the elements in the top right positions of the periodic table exist as gases under normal conditions. Among the gaseous oxides are CO_2, CO, N_2O_4, N_2O_3, NO_2, NO, N_2O, SO_2, F_2O, Cl_2O_5, ClO_2, and Cl_2O. The highest chlorine oxide, Cl_2O_7, is a liquid. The highest nitrogen oxide, N_2O_5, is a solid below 30°C. The highest sulfur oxide, SO_3, exists in the form of a volatile solid or liquid at room temperature. The remainder of the nonmetal oxides, formed with elements in the lower right of the periodic table, are solids.

EXAMPLE 3.1

Predict the formulas of the oxides formed by phosphorus.

Solution   Phosphorus occurs in Group VA (15), so it should have an oxidation number of +5. It would combine with the oxygen anion, whose charge is 2−, in ratios of 2 to 5: P_2O_5. Because phosphorus is a nonmetal and nonmetals usually form more than one oxide, we might expect to find lower oxides as well. We would predict their formulas by removing one oxygen atom per phosphorus atom. In fact, one such oxide, P_2O_3, does exist. The formulas just given are actually empirical formulas. The molecular formulas are P_4O_{10} and P_4O_6. Such information can only be determined from the molecular weights of these oxides.

Practice Problem 3.1

Predict the formulas of the oxides formed by germanium.

Answer: GeO_2 and GeO

Hydrides

Hydrogen, the ninth most abundant element in the earth's crust at 1.4% by mass, is found in a variety of compounds. The most prevalent hydrogen compound is water, but hydrogen is also present in all living matter, as well as in oil and natural gas. Figure 3.21 shows the composition of the common binary hydrogen compounds. We can predict how hydrogen will combine with the main-group elements by assuming that the hydride ion, H^-, combines with metal cations or that the hydrogen ion, H^+, combines with nonmetal anions. Thus, metal hydrides include LiH (assuming Li^+ and H^-) and BeH_2 (assuming Be^{2+} and H^-). Compounds of hydrogen with nonmetals include CH_4 (assuming C^{4-} and H^+) and H_2S (assuming S^{2-} and H^+).

Lithium hydride (LiH) exists as gray crystals formed by direct reaction of the elements. It is used to generate hydrogen by addition of the compound to water. One gram of the solid releases about 2.8 L of hydrogen gas.

Hydrides formed from the very reactive metals, such as the Group IA (1) and IIA (2) elements, are *nonvolatile* solids—that is, they do not evaporate easily. Some of their properties are similar to those of ionic compounds with the more common anions, such as the halides. The hydrogen compounds of the very reactive nonmetals (the halogens) are *volatile* (evaporate easily), as is typical of covalent compounds, but they also show other properties normally associated with ionic compounds. For example, if dissolved in water, they form ions. An example is $HCl(g)$, which forms $H^+(aq)$ and $Cl^-(aq)$ when dissolved in water. Other nonmetal compounds are covalent nonreactive gases, liquids, or solids, depending on the size of the molecules. A major class of hydrogen compounds is the *hydrocarbons,* formed between hydrogen and carbon. The lighter members of this group—such as methane, CH_4 (the major component of natural gas)—are quite volatile

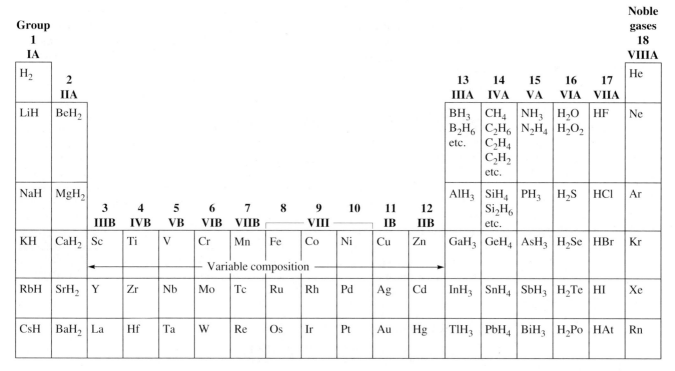

Figure 3.21 The composition of the common binary hydrogen compounds of the elements can be predicted from the element's position in the periodic table.

▶ ▶ ▶ ▕ WORLDS OF CHEMISTRY

Hydrogen: The Fuel of the Future?

The reaction product, water, is not a pollutant, so hydrogen would provide an exceptionally clean fuel.

The price of oil keeps going up. Estimates of the amount of recoverable oil keep going down. Where will we get our fuel in the future? Some scientists think we can get fuel from water in the form of hydrogen, perhaps using solar energy to decompose the water into its component elements.

Hydrogen exists as a gas at room temperature, but it can be converted to a liquid at low temperature and high pressure. Currently, most of the hydrogen that is produced is used to make ammonia and petroleum products. As either a gas or a liquid, however, hydrogen can be used as a fuel. When hydrogen combines with oxygen, it liberates large quantities of heat, which can be used to power vehicles. The space shuttles, for example, use liquid hydrogen and liquid oxygen to fuel the rockets that raise them from the launch pad to the upper atmosphere (Figure 3.22). Electrical power for the space shuttle and other satellites is also provided by the reaction of hydrogen and oxygen in a device called a fuel cell. The reaction product, water, can be collected from the fuel cells for the astronauts to drink.

Closer to earth, hydrogen was once used in airships called dirigibles, not as a fuel, but to provide lift. Hydrogen is highly flammable, reacting explosively with oxygen under some conditions. The disastrous explosions of the airship *Hindenburg* and the space shuttle *Challenger* were due to the explosive combination of hydrogen and oxygen.

Hydrogen is being investigated as a fuel for automobiles and aircraft. Engines can be modified to burn hydrogen. The reaction product, water, is not a pollutant, so hydrogen would provide an exceptionally clean fuel. However, the flammability and explosiveness of hydrogen is a drawback. Hydrogen will have to be stored in some form other than the gaseous or liquid state if it is to be used as a fuel in vehicles operating in populated areas.

One possible storage method is the formation of metal hydrides. Many of the metal hydrides, such as titanium hydride (TiH_3), are readily formed by mixing the metal with hydrogen gas. The hydrogen can be released from the solid metal hydride by adding water or by heating. For example, magnesium hydride, MgH_2, reacts with water to form magnesium hydroxide, $Mg(OH)_2$, and hydrogen gas. Thus, pellets of magnesium hydride could be used as a

Figure 3.22 The space shuttles are launched by liquid-fuel rockets that contain liquid hydrogen and liquid oxygen. This is the space shuttle Columbia.

source of hydrogen gas, which could then be fed to an engine.

Unfortunately, the amount of hydrogen that can be stored by metal hydrides such as this is rather limited. The mass of the metal hydride needed to operate an automobile would be prohibitive. However, not all metal hydrides are simple compounds with a small, whole-number ratio of hydrogen to metal. Some metals, such as platinum and palladium, absorb large quantities of hydrogen gas to form substances called *interstitial hydrides*. Hydrogen molecules apparently fit into small holes in the metal structure. Gentle heating causes the hydrogen gas to be expelled from the metal. These interstitial hydrides thus present considerable potential for transporting hydrogen in a safe manner, should hydrogen ever become a fuel for automobiles.

Questions for Discussion

1. Make a list of possible alternative fuels for the future and discuss their advantages and disadvantages.
2. Would you feel comfortable with a tank of hydrogen in your automobile? Compare your response with how you feel about gasoline, which is also very flammable and explosive under some conditions.
3. If hydrogen is to be used as a fuel, we must get it from some plentiful source, such as water. Where will we get the energy to release the hydrogen from water?

and exist normally as gases. Heavier members of the group—such as octane, C_8H_{18}—are volatile liquids. Gasoline consists of a mixture of hydrocarbons in this general range of composition. Much heavier hydrocarbons, such as $C_{20}H_{42}$, are solids. Hydrocarbons having very high molar masses are found in modern plastics such as polyethylene and polypropylene.

EXAMPLE 3.2

Predict the formula of the compound formed by hydrogen and phosphorus.

Solution   Because phosphorus is a nonmetal, it will combine with hydrogen as though it were forming an anion, even though the compound will be covalent. Phosphorus, in Group VA (15), would tend to adopt a charge (or oxidation number) of 3− as an anion. Hydrogen will form a cation with a charge of 1+. The compound will thus have a formula of H_3P, usually written as PH_3. This is the formula of the very toxic compound phosphine.

Practice Problem 3.2

Predict the formula of the compound formed by hydrogen and calcium.

Answer: CaH_2

3.4   Oxidation Numbers

We have used electron counts and ionic charges (whether real or imagined) to predict how atoms or ions will combine to form compounds. We have also seen that using the periodic table is a highly effective method for predicting both real and hypothetical charges (or oxidation numbers), but it involves some limitations where covalent compounds are concerned. Using the periodic table, we can predict only the largest hypothetical charge for a given element, but compounds exist that are not consistent with this charge. Further, the method is applicable only to binary compounds.

In this section, we consider a more inclusive system for assigning charges. The charges we will assign are called *oxidation numbers*. Although they are assigned somewhat arbitrarily, they allow us to understand and predict the chemical composition of compounds. As noted, the ionic charges, real or hypothetical, that we assigned in the preceding section were actually oxidation numbers.

How would you use oxidation numbers to predict the composition of compounds?

An oxidation number is a number assigned to an atom to indicate how many electrons it has in a compound compared with how many it would have as an isolated atom. The oxidation number is negative if the element in a compound has more electrons than does an uncombined neutral atom of that element, and positive if it has fewer electrons. For example, a cation such as Cu^{2+} has two fewer electrons than atomic Cu, or an electronic charge of 2+, so the oxidation number of copper is +2. An anion such as N^{3-} has three more electrons than atomic N, giving an electronic charge of 3−, so the oxidation number of nitrogen is −3. Table 3.1 lists some simple rules for assigning oxidation numbers. The following points will help you apply the rules.

Table 3.1 Rules for Assigning Oxidation Numbers

1. The oxidation number of the atoms of an uncombined element is 0.
2. The sum of the oxidation numbers of all atoms in a substance must equal the total charge on the species: 0 for molecules; the ionic charge for ions.
3. Fluorine has an oxidation number of -1 in its compounds.
4. Hydrogen has an oxidation number of $+1$ unless it is combined with metals, in which case it has an oxidation number of -1.
5. The position of the element in the periodic table may be useful:
   a. Group IA (1) elements have oxidation numbers of $+1$ in their compounds.
   b. Group IIA (2) elements have oxidation numbers of $+2$ in their compounds.
   c. Group VIIA (17) elements have oxidation numbers of -1 unless combined with oxygen or a halogen closer to the top of Group VIIA (17).
   d. In binary compounds, Group VIA (16) elements have oxidation numbers of -2, unless combined with oxygen or the halogens.
   e. In binary compounds, Group VA (15) elements have oxidation numbers of -3 unless combined with elements to their right in the periodic table.
6. Oxygen usually has an oxidation number of -2 in its compounds. There are some exceptions:
   a. Oxygen has an oxidation number of -1 in peroxides, which contain the O_2^{2-} ion.
   b. Oxygen has an oxidation number of $-\frac{1}{2}$ in superoxides, which contain the O_2^{-} ion.
   c. When combined with fluorine, as in OF_2, oxygen has a positive oxidation number ($+2$ in OF_2).

- Treat the rules as a hierarchy: the first rule that applies takes precedence over any subsequent rules that seem to apply.
- For an isolated atom or a molecule that contains only one element, only rule 1 need be applied. An uncombined element, whether occurring as an atom such as He or a molecule such as H_2, has an oxidation number of 0. Thus, the oxidation number of sulfur in S, S_2, and S_8 is 0.
- A monatomic ion has an oxidation number equal to its ionic charge, according to rule 2. For example, the oxidation numbers of H in H^+, Li in Li^+, and Hg in Hg_2^{2+} are all $+1$. Similarly, the oxidation number is -1 for Cl in Cl^- and for S in S_2^{2-}.
- Rule 2 can be applied whenever oxidation numbers have been assigned to all but one element in a compound.
- Rule 5 uses position in the periodic table to cover situations that are not handled by one of the other rules.

Oxidation numbers are determined fairly simply for binary compounds. One of the elements can be handled by an appropriate rule, and the other by application of rule 2, which is simply a restatement of the condition of electrical neutrality for compounds.

EXAMPLE 3.3

Assign oxidation numbers to the elements in LiH.

Solution   Hydrogen has an oxidation number of -1 when combined with a metal, such as lithium (rule 4). Because the sum of the oxidation numbers must equal zero in a compound, lithium must have an oxidation number of $+1$ (rule 2). (Note that the oxidation number of lithium is also addressed by rule 5a.)

Practice Problem 3.3

Assign oxidation numbers to the elements in NH_3.

Answer: N, -3; H, $+1$

Polyatomic ions that contain two elements can be treated by the same procedure used for binary compounds. Application of rule 2 in this case requires that the sum of the oxidation numbers of the elements must equal the charge on the polyatomic ion.

EXAMPLE 3.4

Assign oxidation numbers to the elements in SO_4^{2-}.

Solution   In SO_4^{2-}, we assign a value of -2 to O (rule 6). Applying rule 2 then gives a value of $+6$ to S. Note that we multiply the oxidation number of each element by the number of atoms of that element to calculate the net charge on the polyatomic ion:

Oxidation number of S + 4 × oxidation number of O = ionic charge

Oxidation number of S + 4 × −2 = −2

Oxidation number of S = −2 + 8 = +6

Note that rule 5d is not used, because the sulfur is combined with oxygen.

Practice Problem 3.4

Assign oxidation numbers to the elements in IO_3^-.

Answer: I, $+5$; O, -2

With compounds that contain three or more elements, it may not be possible to find rules that give each oxidation number directly. But if two of the oxidation numbers can be determined directly, then the oxidation number of the third element can be found by application of rule 2. Assigning oxidation numbers is sometimes difficult with compounds of this type that contain oxoanions. For example, the rules in Table 3.1 indicate that oxygen in $FeSO_4$ has an oxidation number of -2. Because none of the rules apply to iron or sulfur in this compound, it is not possible to assign all the oxidation numbers. However, if the compound contains an oxoanion with a known charge, we have a solution to this problem. The oxoanion can be separated out as a unit, the oxidation numbers assigned to its elements, and the entire compound treated according to rule 2, as in Example 3.5. (The common oxoanions are listed in Figure 3.16.) In the $FeSO_4$ example, we should recognize the sulfate anion, SO_4^{2-}, which was just considered in Example 3.4. Sulfur has an oxidation number of $+6$; using rule 2, we assign an oxidation number of $+2$ to iron:

Oxidation number of Fe + oxidation number of S + 4 × oxidation number of O = 0

Oxidation number of Fe + 6 + 4 × −2 = 0

Oxidation number of Fe = 0 − 6 + 8 = +2

EXAMPLE 3.5

Assign oxidation numbers to the elements in $Mn_3(PO_4)_2$.

Solution This compound contains the phosphate ion, PO_4^{3-}, which can be considered as a unit. The phosphate ion is one of the common oxoanions listed in Figure 3.16. In PO_4^{3-}, oxygen has an oxidation number of -2 (rule 6) and phosphorus can be assigned an oxidation number of $+5$ by application of rule 2, as shown in the following equations:

Oxidation number of P + 4 × oxidation number of O = −3

Oxidation number of P + (4 × −2) = −3

Oxidation number of P = −3 − (4 × −2) = +5

The oxidation number of manganese can then be assigned by rule 2:

3 × oxidation number of Mn + 2 × oxidation number of P + 8 × oxidation number of O = 0

3 × oxidation number of Mn + 2 × 5 + 8 × −2 = 0

3 × oxidation number of Mn = 0 − (2 × 5 + 8 × −2) = +6

Oxidation number of Mn = 6/3 = +2

Practice Problem 3.5

Assign oxidation numbers to the elements in $Cr_2(SO_3)_3$.

Answer: Cr, +3; S, +4; O, −2

A number of the elements, especially the nonmetals and the transition metals, exhibit a wide variety of oxidation numbers in their ions and compounds. A summary of the oxidation numbers of the elements is given in Figure 3.23.

Note the correlation of the oxidation numbers of the main-group elements with their positions in the periodic table. The highest positive oxidation number is the same as the highest cation charge predicted as described in Section 3.3. This oxidation number corresponds to the loss of sufficient electrons to achieve the electron count of the noble gas just previous to the main-group element in the periodic table. Thus, nitrogen with an oxidation number of $+5$ corresponds to the ion N^{5+}, with the electron count of helium. The lowest oxidation number corresponds to the gain of sufficient electrons to achieve the electron count of the next noble gas in the periodic table. For example, nitrogen with an oxidation

Main-group elements (left)   **Main-group elements** (right)

Group 1 IA	2 IIA	3 IIIB	4 IVB	5 VB	6 VIB	7 VIIB	8 VIII	9 VIII	10 VIII	11 IB	12 IIB	13 IIIA	14 IVA	15 VA	16 VIA	17 VIIA	18 VIIIA
H -1, $+1^*$																	He
Li $+1$	Be $+2$											B $+3$	C -4, -2, $+2$, $+4$, Others are possible	N -3, $+3$, $+5^*$ and all others in this range	O -2, -1, $-\frac{1}{2}$	F -1	Ne
Na $+1$	Mg $+2$											Al $+3$	Si -4, $+2$, $+4$	P -3, $+3$, $+5^*$	S -2, $+2$, $+4$, $+6^*$	Cl -1^*, $+1$, $+3$, $+5$, $+7$	Ar
K $+1$	Ca $+2$	Sc $+3$	Ti $+2$, $+3$, $+4$	V $+2$, $+3$, $+4$, $+5$	Cr $+2$, $+3^*$, $+6$	Mn $+2$, $+3^*$, $+4$, $+7$	Fe $+2$, $+3^*$	Co $+2^*$, $+3$	Ni $+2$	Cu $+1$, $+2^*$	Zn $+2$	Ga $+1$, $+3$	Ge -4, $+2$, $+4$	As -3, $+3^*$, $+5$	Se -2, $+4$, $+6$	Br -1^*, $+1$, $+3$, $+5$, $+7$	Kr
Rb $+1$	Sr $+2$	Y $+3$	Zr $+4$	Nb $+5$	Mo $+3$, $+4$, $+6$	Tc $+4$, $+5$, $+7$	Ru $+3$, $+4$	Rh $+3$	Pd $+2$, $+4$	Ag $+1^*$, $+2$	Cd $+1$, $+2^*$	In $+1$, $+3$	Sn -4, $+2^*$, $+4$	Sb -3, $+3^*$, $+5$	Te -2, $+4$, $+6$	I -1^*, $+1$, $+3$, $+5$, $+7$	Xe
Cs $+1$	Ba $+2$	La $+3$	Hf $+4$	Ta $+5$	W $+6$	Re $+4$, $+5$, $+7$	Os $+3$, $+4$, $+8$	Ir $+3$, $+4$	Pt $+2$, $+4$	Au $+1$, $+3$	Hg $+1$, $+2^*$	Tl $+1$, $+3$	Pb $+2$, $+4$	Bi -3, $+3^*$, $+5$	Po	At	Rn

Transition groups span groups 3–12. Noble gases are in group 18 (VIIIA).

Figure 3.23 Some of the common oxidation numbers of the elements can be correlated with position in the periodic table; others, particularly for the transition elements, show no such relationship. Each element also has an oxidation number of 0. Those oxidation numbers marked with an asterisk are the ones that are the most stable, especially for the element as it would exist if its compounds were dissolved in water.

number of -3 corresponds to the ion N^{3-}, with the electron count of neon. The nonmetals in particular have many possible oxidation numbers between these limits. The transition metals do not follow simple periodic trends, so their oxidation numbers cannot usually be predicted from position in the periodic table.

3.5   Nomenclature of Inorganic Compounds

To communicate with each other reliably and efficiently, we must speak the same language. A large part of learning the language of chemistry is mastering the

Compare and contrast the
naming of chemical compounds
and the naming of city streets.

vocabulary associated with the names of substances. Fortunately, the system of names, or the *nomenclature,* chemists use to identify substances is governed by straightforward rules. Once you have learned the rules and practiced applying them, you will be able to name compounds from their formulas and to write formulas based on compound names. This section gives the rules for naming *inorganic compounds,* those compounds that do not contain carbon as the primary element.

Binary Compounds

Some binary compounds are known by their common, or nonsystematic, names, which may bear no direct relation to their composition. For example, dihydrogen oxide, H_2O, is commonly referred to by its nonsystematic name, water, even among chemists. Another example is NH_3, known as ammonia.

How a systematic name is assigned to a compound depends on whether the compound is ionic or covalent. In general, we can assume that a metal and a nonmetal form an ionic compound, whereas two nonmetals form a covalent compound. There are exceptions, but this assumption will work satisfactorily for the purpose of naming inorganic compounds.

Covalent Compounds   Suppose we want to name a covalent compound formed by the two nonmetals carbon and oxygen. Because nonmetals can combine in more than one ratio, we need to identify the appropriate ratio. We'll assume the compound contains two oxygen atoms for each carbon atom, CO_2. To name this compound, we start with the element that is farther toward the left and the bottom of the periodic table—in this case, carbon—simply using its elemental name. To the stem of the second element's name, we add the suffix *-ide:* oxide. We must indicate that the compound contains two oxygen atoms for every carbon atom, and we do this by using the Greek prefix *di-,* which means "two": dioxide. (Other Greek prefixes for numbers are given in Table 3.2.) The name of the compound is carbon dioxide. If there had been more than one carbon atom in the compound, we would have used a Greek prefix with *carbon* as well. The compound Cl_2O_5, for example, is called dichlorine pentoxide, with prefixes before both element names to indicate the relative numbers of atoms.

The prefix *mono-* ("one") is used only with the name of the second element, and then only if the two elements can form more than one compound. Thus, HCl is hydrogen chloride, but CO is carbon monoxide because carbon and oxygen form another compound, CO_2.

Table 3.2 Common Greek
Prefixes That Identify Numbers

Prefix	Number	Prefix	Number
mono-	1	hexa-	6
di-	2	hepta-	7
tri-	3	octa-	8
tetra-	4	nona-	9
penta-	5	deca-	10

EXAMPLE 3.6

Name the compound P_4O_{10}.

Solution   Because this compound is made up of two nonmetals, we can consider it covalent. Phosphorus is named first because it occurs farther toward the left and the bottom of the periodic table, but we must add the prefix *tetra-* to indicate that there are four phosphorus atoms in the molecule. Oxygen is named second, with its ending replaced by *-ide* and *deca-* added at the beginning to indicate that there are ten oxygen atoms in the molecule. The full name is tetraphosphorus decoxide. (Note that the vowel at the end of a prefix is often dropped before an initial vowel of the stem name, as in *decoxide*.)

Practice Problem 3.6

Name the compound As_2O_3.

Answer: Diarsenic trioxide

Knowing the process by which a compound is named provides all the information that we need to write a formula for a covalent compound whose name we know. For example, the name *chlorine dioxide* describes a compound in which one chlorine atom and two oxygen atoms are combined; the formula is thus ClO_2. Similarly, dinitrogen trioxide has two nitrogen atoms and three oxygen atoms; its formula is N_2O_3.

Ionic Compounds   Naming a binary ionic compound is similar to naming a binary covalent compound. We give the name of the metallic element first, simply using its elemental name. To the name of the nonmetallic element we add the suffix *-ide*, just as for covalent compounds. Generally, however, no prefix indicating the atomic ratio is used. Thus, NaCl is sodium chloride, and $ScBr_3$ is scandium bromide. Because the names of these ionic compounds do not indicate the ratio of one component to the other, we cannot use them to write formulas for the compound. Instead, we must use information on the ionic charges expected for the elements, as discussed in Sections 3.2 and 3.4. For example, the compound aluminum oxide contains a metal and a nonmetal, so it should be ionic. Aluminum, in Group IIIA (13), forms a 3+ ion; oxygen, in Group VIA (16), forms a 2− ion. To obtain electrical neutrality, the formula must be Al_2O_3.

Aluminum oxide (Al_2O_3), also known as alumina, is a hard white solid. It occurs in a variety of minerals, including corundum and alundum, which are used as abrasives and polishes. Because of its low electrical conductivity, aluminum oxide is used in a variety of insulating ceramics for electronics parts.

Just as carbon and oxygen can form more than one covalent compound, some metals can combine with a nonmetal to form more than one ionic compound. For example, copper combines with chlorine to form CuCl and $CuCl_2$. In these two compounds, the copper has two different oxidation numbers (+1 and +2). More than one oxidation number can be observed for many metals, including all the transition metals except those in Group IIIB (3) and all the main-group metals in Groups IVA (14), VA (15), and VIA (16).

As previously mentioned, we usually do not use Greek prefixes to distinguish between two ionic compounds with the same elements but different compositions. The preferred method for distinguishing these compounds is called the *Stock system,* after Alfred Stock, who originated it. This system uses Roman numerals in parentheses immediately after the name of the metal to indicate the oxidation number. Thus, CuCl is named copper(I) chloride, and $CuCl_2$ is named copper(II) chloride.

Copper(I) chloride is a white solid that is stable when exposed to air or light under dry conditions. In the presence of moisture, it turns green on exposure to air and blue on exposure to light. Copper(II) chloride is a yellow-brown solid that decomposes to copper(I) chloride on heating. It also exists as a dihydrate, with colors ranging from green to blue. Both compounds are used as pigments and to remove odors and sulfur from petroleum.

Manganese(IV) oxide, MnO_2, is a brownish-black solid that occurs naturally as the mineral pyrolusite. It is used in the manufacture of manganese steels, in alkaline batteries, and as a pigment.

An older, less versatile method for indicating the oxidation number of the metal in an ionic compound is still sometimes used. In this method, the name of the metal is changed by addition of a suffix, either *-ous* or *-ic*. The suffix *-ous* is used to indicate the lower of two oxidation numbers, and the suffix *-ic* to indicate the higher. Of course, this system assumes that there are only two oxidation numbers for the element and that you know what they are. According to this system, the two copper chlorides are called cuprous chloride (CuCl) and cupric chloride ($CuCl_2$). Some additional examples are listed in Table 3.3, along with the corresponding Stock system names.

Constructing formulas from Stock system names is particularly easy because the name gives the oxidation number of the metal. Thus, in manganese(IV) oxide, manganese has an oxidation number of +4. We have already seen that oxygen usually has an oxidation number of −2 in its compounds. To satisfy electrical neutrality, then, there must be two oxygen ions for each manganese ion, giving a formula of MnO_2.

Some ionic compounds, called **hydrates,** contain molecules of water combined in a characteristic proportion in the structure of the solid. Hydrates often have a translucent or glassy appearance but become powdery when the water is driven off by heating. Examples are colorless $MgCl_2 \cdot 6H_2O$ and green $NiBr_2 \cdot 3H_2O$. Here the formula for the compound is followed by a dot and the formula for water. The number preceding the formula for water gives the number of water molecules. The name of a hydrate starts with the name of the compound. This is followed by the word *hydrate* preceded by the Greek prefix indicating the number of water molecules. Thus, $MgCl_2 \cdot 6H_2O$ is named magnesium chloride hexahydrate, and $NiBr_2 \cdot 3H_2O$ is called nickel(II) bromide trihydrate.

Ternary Compounds

Ternary compounds are compounds formed from three elements. Here, we stretch the classification to include a few compounds that contain more than three elements, but the same rules for naming apply.

Ternary Ionic Compounds Some ternary ionic compounds that do not contain oxygen are commonly named according to the rules for binary compounds. If such a compound contains two metals or a metal and hydrogen, both are named, in the

Table 3.3 Names of Some Ionic Compounds with Variable Oxidation Numbers

Formula	Stock System Name	Alternative Name
$FeCl_2$	Iron(II) chloride	Ferrous chloride
$FeCl_3$	Iron(III) chloride	Ferric chloride
CrO	Chromium(II) oxide	Chromous oxide
Cr_2O_3	Chromium(III) oxide	Chromic oxide
CrO_3	Chromium(VI) oxide	
Cu_2S	Copper(I) sulfide	Cuprous sulfide
CuS	Copper(II) sulfide	Cupric sulfide

order given in the formula, followed by the nonmetal name ending in *-ide,* such as sodium aluminum hydride for $NaAlH_4$ or potassium hydrogen sulfide for KHS. Some ternary ionic compounds, however, contain polyatomic ions with special names, and these names are used in the compound names. It is necessary to learn the special ion names because they are not derived systematically. The more important ones are OH^-, hydroxide ion; NH_4^+, ammonium ion; and CN^-, cyanide ion. Thus, NaOH is sodium hydroxide, NH_4Cl is ammonium chloride, and KCN is potassium cyanide.

Sodium hydroxide is usually sold as white pellets or lumps. It absorbs water from air and, in the presence of moisture, will also absorb carbon dioxide from air. It is a very corrosive chemical, particularly to body tissues. It is used to neutralize acids, to make cellophane and rayon from cellulose, to dissolve fabric from rubber tires, and in the production of soaps.

Salts Containing Oxoanions   In Section 3.2, we saw that nonmetals can form oxoanions (polyatomic ions containing oxygen). For some elements, more than one oxoanion can be formed. These oxoanions have the same charge, but they differ in the number of oxygen atoms. Such ions can form ionic salts by combining with cations to achieve electrical neutrality. For example, sulfate ion, SO_4^{2-}, will combine with sodium ions to form Na_2SO_4. The name of such a substance includes the name of the metal ion, as for a binary ionic compound, followed by the name of the oxoanion. Thus, the compound Na_2SO_4 is named sodium sulfate.

As indicated in Figure 3.16, when an element forms more than one oxoanion, the oxoanion with more oxygen ends in *-ate,* and the one with fewer oxygen atoms ends in *-ite.* In these two ions, the nonmetal has a higher oxidation number in the *-ate* ion than in the *-ite* ion. Thus, SO_4^{2-}, in which sulfur has an oxidation number of +6, is sulfate, and SO_3^{2-}, in which sulfur has an oxidation number of +4, is sulfite. The compound Na_2SO_3 is thus named sodium sulfite.

EXAMPLE 3.7

Name the compound $Ca_3(PO_4)_2$.

Solution   The oxoanion is PO_4^{3-}, named phosphate ion because it has more oxygen (and phosphorus has a higher oxidation number) than the other ion, PO_3^{3-}, shown in Figure 3.16. The metal cation is named after the element, calcium. The name of the compound is thus calcium phosphate.

Practice Problem 3.7

Name the compound $FeSO_3$.

Answer: Iron(II) sulfite

If a nonmetal can form more than two oxoanions, with a different oxidation number in each, then we extend the naming system just described by using prefixes. In a series of compounds such as this, the oxoanions in the compound will differ by one oxygen atom, and the oxidation numbers of the nonmetal will differ by two units. For example, the oxoanions of chlorine are ClO_4^- (oxidation number of Cl is +7), ClO_3^- (+5), ClO_2^- (+3), and ClO^- (+1). In such series, one oxoanion is considered to be the most common—ClO_3^-, in this case. This oxoanion is given the *-ate* ending, with no prefix: chlorate ion. The next lower oxidation number is given the *-ite* ending. Thus, ClO_2^- is chlorite ion. If there is an oxoanion with an oxidation number higher than that of the *-ate* ion, its name includes the prefix *per-* along with the suffix *-ate.* We give ClO_4^- the name perchlorate ion. If there is an oxoanion with an oxidation number lower than that of the -ite ion, its name includes the prefix hypo-. The name of ClO^- is hypochlorite ion.

EXAMPLE 3.8

Given that Na_3PO_4 is sodium phosphate, name the compound Na_3PO_2.

Solution Because PO_4^{3-} is phosphate ion, we would predict the formula with one less oxygen atom (and an oxidation number of phosphorus that is lower by two), PO_3^{3-}, to be phosphite ion. This information can also be deduced from Figure 3.16. Removing one more oxygen atom gives PO_2^{3-}, the ion of interest. To indicate the lowest oxidation number of phosphorus, we use the prefix hypo-. This compound is thus sodium hypophosphite.

Practice Problem 3.8

Given that $KBrO_3$ is potassium bromate, name the compound $KBrO_4$.

Answer: Potassium perbromate

Acids

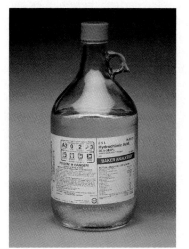

Figure 3.24 Hydrogen chloride is a colorless, nonflammable, corrosive gas with a pungent odor. When dissolved in water, it forms hydrochloric acid, which is sold in a pure state as a solution containing about 38% HCl by mass. In its commercial form, which is often yellow due to iron or chlorine impurities, hydrochloric acid is known as muriatic acid.

Acids are compounds that, when dissolved in water, release hydrogen ions (H^+), giving characteristic properties common to all such solutions. Acids usually contain hydrogen and another nonmetal, but they may also contain oxygen. We generally write the formula with the hydrogen first and write (aq) after the formula to indicate that the compound is an acid and is dissolved in water. For example, $HCl(aq)$ is an acid, but CH_4 is not.

Binary Acids Binary acids in aqueous solution are named with the prefix hydro- followed by the stem of the name of the nonmetal with the suffix -ic and the word acid attached. Thus, $HCl(aq)$ is named hydrochloric acid (Figure 3.24). Some other examples are given in Table 3.4.

EXAMPLE 3.9

Name the compound $H_2Se(aq)$.

Solution Because this compound contains hydrogen and a nonmetal, and because it exists in aqueous solution, it must be an acid. Binary acids have a hydro- prefix and an -ic suffix. The stem of the name derives from the name of the element other than hydrogen. In this case, the element selenium provides the stem selen. Assembling the prefix, the stem, and the suffix, along with the word acid, gives the name hydroselenic acid.

Practice Problem 3.9

Name the compound $HI(aq)$.

Answer: Hydriodic acid

Table 3.4 Names of Some
Binary Acids

Formula	Name	Formula	Name
HF(aq)	Hydrofluoric acid	H_2S(aq)	Hydrosulfuric acid
HCl(aq)	Hydrochloric acid	HCN(aq)	Hydrocyanic acid*
HBr(aq)	Hydrobromic acid		

*Hydrocyanic acid is really a ternary acid—an acid with three components—but is named as if it were a binary acid because of the great stability of the cyanide ion, CN^-.

Oxoacids   **Oxoacids** are acids containing hydrogen, a nonmetal, and oxygen (though a few oxoacids contain a metal instead of the nonmetal). Such species are usually acidic in aqueous solution, releasing hydrogen ions when dissolved in water and giving the characteristic properties of acids. Naming oxoacids is a bit more complex than naming binary acids because the nonmetal often forms a series of oxoacids in which the positive oxidation number of the nonmetal and the number of oxygen atoms vary.

If the nonmetal forms oxoacids with only two different positive oxidation numbers, then the elemental name for the nonmetal is used with the suffix -*ic* and the word *acid* for the higher of the two oxidation numbers. The suffix -*ous* with the word *acid* is used for the lower oxidation number. The names of the oxoanions that are formed by removing the hydrogen ions were discussed earlier; these names end in -*ate* and -*ite*. Thus, H_2SO_4(aq) is sulfuric acid and it contains the sulfate ion, whereas H_2SO_3(aq) is sulfurous acid and it contains the sulfite ion. Oxoacids of nitrogen contain nitrogen with oxidation numbers of $+5$ and $+3$: nitric acid, HNO_3(aq), and nitrous acid, HNO_2(aq). More examples of oxoacids are given in Table 3.5; sulfuric acid and nitric acid are shown in Figure 3.25.

Table 3.5 Names of Some
Oxoacids

Formula	Name
H_3BO_3(aq)	Boric acid
H_2CO_3(aq)	Carbonic acid
H_3AsO_4(aq)	Arsenic acid
HNO_3(aq)	Nitric acid
HNO_2(aq)	Nitrous acid
H_2SO_4(aq)	Sulfuric acid
H_2SO_3(aq)	Sulfurous acid
$HBrO_3$(aq)	Bromic acid
$HBrO_2$(aq)	Bromous acid
Note that some oxoacids containing metals have similar names:	
H_2CrO_4(aq)	Chromic acid
HVO_3(aq)	Vanadic acid
H_2SnO_3(aq)	Stannic acid

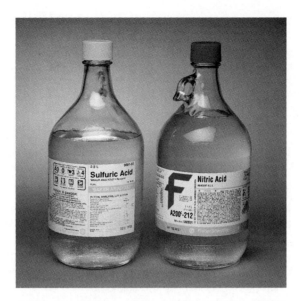

Figure 3.25 Sulfuric acid is a clear, colorless, odorless, oily liquid. The commercial product is 93%–98% H_2SO_4, the remainder being water. Sulfuric acid has a very high affinity for water and will remove moisture from air. Indeed, this affinity is so great that sulfuric acid will char materials like sugar, $C_{12}H_{22}O_{11}$, producing carbon by removing water. Concentrated sulfuric acid is similarly corrosive to all body tissues, so it should be handled with great care.

Nitric acid, HNO_3, is a clear, colorless liquid that fumes in moist air, has a characteristic choking odor, and stains animal tissue bright yellow. It is available commercially as a solution in water that contains 70%–71% HNO_3 by mass. Nitric acid often develops a yellow color from NO_2, which forms by decomposition when nitric acid is exposed to light.

If the nonmetal forms more than two oxoacids, the same prefixes are used as for the oxoanions they contain. For example, $HClO_4(aq)$ is perchloric acid. The naming of the oxoacids and the corresponding oxoanions is summarized in Table 3.6.

Table 3.6 Names of Oxoacids and Corresponding Oxoanions

Oxoacid Name	Oxoanion Name	Example of Acid and Compound Name
hypo- -ous acid	hypo- -ite	Hypochlorous acid [$HClO(aq)$], sodium hypochlorite ($NaClO$)
-ous acid	-ite	Chlorous acid [$HClO_2(aq)$], sodium chlorite ($NaClO_2$)
-ic acid	-ate	Chloric acid [$HClO_3(aq)$], sodium chlorate ($NaClO_3$)
per- -ic acid	per- -ate	Perchloric acid [$HClO_4(aq)$], sodium perchlorate ($NaClO_4$)

Table 3.7 Names, Formulas, and Charges of Some Common Ions

Cations

Formula	Name	Formula	Name
NH_4^+	Ammonium	Li^+	Lithium
Cu^+	Copper(I) or cuprous	K^+	Potassium
H^+	Hydrogen	Ag^+	Silver
H_3O^+	Hydronium	Na^+	Sodium
Ba^{2+}	Barium	Mg^{2+}	Magnesium
Cd^{2+}	Cadmium	Mn^{2+}	Manganese(II) or manganous
Ca^{2+}	Calcium	Hg^{2+}	Mercury(II) or mercuric
Cr^{2+}	Chromium(II) or chromous	Hg_2^{2+}	Mercury(I) or mercurous
Co^{2+}	Cobalt(II) or cobaltous	Sr^{2+}	Strontium
Cu^{2+}	Copper(II) or cupric	Sn^{2+}	Tin(II) or stannous
Fe^{2+}	Iron(II) or ferrous	Zn^{2+}	Zinc
Pb^{2+}	Lead(II) or plumbous		
Al^{3+}	Aluminum	Fe^{3+}	Iron(III) or ferric
Bi^{3+}	Bismuth	Mn^{3+}	Manganese(III) or manganic
Co^{3+}	Cobalt(III) or cobaltic	Sc^{3+}	Scandium
Cr^{3+}	Chromium(III) or chromic		
Pb^{4+}	Lead(IV) or plumbic	Sn^{4+}	Tin(IV) or stannic

EXAMPLE 3.10

Name the compound H_2SeO_4.

Solution   This compound contains hydrogen, oxygen, and a nonmetal, so it is an oxo-acid. Selenium forms two possible oxoacids. In this one, the oxidation number of selenium is +6. As shown in Figure 3.23, +6 is the maximum oxidation number of selenium, so this acid must have the *-ic* suffix. The name is selenic acid.

Practice Problem 3.10

Name the compound H_3AsO_3.

Answer: Arsenous acid

Using a Decision Tree to Name Compounds

The rules we've outlined are adequate to name most of the compounds you will encounter in this book. (Other classes of substances, such as organic compounds,

Table 3.7 Names, Formulas, and Charges of Some Common Ions (cont.)

Anions

Formula	Name	Formula	Name
$C_2H_3O_2^-$	Acetate	H^-	Hydride
HCO_3^-	Hydrogen carbonate	OH^-	Hydroxide
HSO_4^-	Hydrogen sulfate	ClO^-	Hypochlorite
HSO_3^-	Hydrogen sulfite	I^-	Iodide
BrO_3^-	Bromate	NO_3^-	Nitrate
Br^-	Bromide	NO_2^-	Nitrite
ClO_3^-	Chlorate	$C_2O_4^{2-}$	Oxalate
Cl^-	Chloride	ClO_4^-	Perchlorate
ClO_2^-	Chlorite	MnO_4^-	Permanganate
CN^-	Cyanide	SCN^-	Thiocyanate
F^-	Fluoride		
CO_3^{2-}	Carbonate	SO_4^{2-}	Sulfate
CrO_4^{2-}	Chromate	SO_3^{2-}	Sulfite
$Cr_2O_7^{2-}$	Dichromate	S^{2-}	Sulfide
O^{2-}	Oxide	$S_2O_3^{2-}$	Thiosulfate
O_2^{2-}	Peroxide		
AsO_4^{3-}	Arsenate	PO_4^{3-}	Phosphate
AsO_3^{3-}	Arsenite	P^{3-}	Phosphide
N^{3-}	Nitride		
SiO_4^{4-}	Silicate		

will be dealt with as they are introduced.) To help you apply these rules, Table 3.7 presents the names, formulas, and charges of some common ions.

If you don't know which rule to apply to the naming of a particular compound, you might want to use the decision tree shown in Figure 3.26. By answering a few simple questions, you should be able to find the correct rule to apply for most simple compounds.

The questions are designed to distinguish between the various classes of inorganic compounds. Consider an example: $NaHSO_4$. This compound contains a metal, so we follow the "yes" branch in response to the first question in the decision tree. There is more than one nonmetal, so we follow the "yes" branch again in answer to the second question: $NaHSO_4$ must be a ternary ionic compound. The next question asks whether oxygen is present; the answer is yes, so the compound must be an ionic compound containing an oxoanion. The compound should thus be named as the name of the metal (sodium) followed by hydrogen and the anion name (sulfate). The name of the compound is sodium hydrogen sulfate.

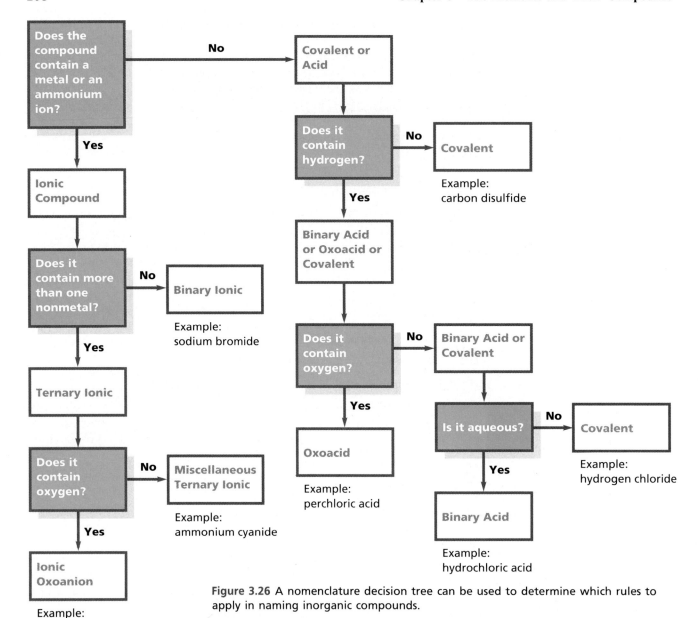

Figure 3.26 A nomenclature decision tree can be used to determine which rules to apply in naming inorganic compounds.

3.6   Some of Our More Important Chemicals

Many consumer products are made up of a number of chemicals. So once you have mastered the naming of inorganic compounds, you are in a position to read the labels of consumer products more intelligently (though you will not be able to interpret the names of all ingredients, because many of them are organic compounds and others are given common names instead of systematic names). For example, read the labels on various brands of toothpaste. You should be able to interpret a number of ingredient names: sodium fluoride (NaF), silicon dioxide (SiO_2, sometimes called silica), and calcium monohydrogen phosphate ($CaHPO_4$). Fluoride compounds are present in toothpaste to prevent dental cavities, and the

Why do you think all these ingredients are found in toothpaste?

phosphate compound and silicon dioxide are abrasives. Toothpaste contains other ingredients to hold water and to stiffen the mixture so the paste will have the proper texture and consistency. ▶ The antacids mentioned in the introduction to this chapter provide more examples of simple chemical ingredients.

Each year, the U.S. chemical industry produces an extremely large number of consumer chemicals—the chemicals, such as sodium fluoride and calcium monohydrogen phosphate, that find their way into consumer products. All these are derived from a much smaller number of industrial chemicals, which in turn come from a few basic raw materials, including a variety of metal ores. For example, calcium monohydrogen phosphate may be prepared from calcium phosphate, which is obtained from phosphate rock. The basic raw materials are listed in Table 3.8.

Included in the raw materials listed in the table are a number of **ores**, naturally occurring materials from which metals can be economically obtained. A few of the less reactive metals occur as the uncombined elements in nature, but most metals are found in the ores as compounds containing relatively few different anions. Table 3.9 lists a number of common ores in which metals are found as carbonate, halide, oxide, phosphate, silicate, sulfate, and sulfide compounds. Because elements that are close to one another in the periodic table have similar properties, it is not surprising that they would often be found together in ores. Figure 3.27 shows the major sources of the elements. Note how substances in the same periodic group are generally found in the same type of ores.

About 50 industrial chemicals are produced in the United States in amounts over 1 billion pounds (4.5×10^8 kg) per year. Table 3.10 lists 16 of these, which are produced in amounts over 10 billion pounds (4.5×10^9 kg) per year. Here, we look briefly at how the first 10 of these chemicals are derived from their sources, the raw materials and ores. Later chapters will describe the processes in more detail.

- Sulfuric acid preparation begins with the burning of sulfur to make sulfur dioxide, SO_2. The SO_2 is combined with oxygen in air to make sulfur trioxide, SO_3, which is then added to water to make sulfuric acid, H_2SO_4. The process is actually more complicated than this because some of the operations require carefully controlled conditions.
- Nitrogen is obtained from air (see Figure 3.28). In order for the nitrogen to be isolated from oxygen and a variety of minor constituents, the air must be liquefied and the components separated by careful evaporation, a technique called

Table 3.8 Basic Industrial Raw Materials

Substance	Formula
Petroleum	$C_nH_{2n+2} + C_nH_n$ + others
Limestone	$CaCO_3$
Salt	$NaCl$
Phosphate rock	$Ca_3(PO_4)_2 + CaF_2 + SiO_2$
Sulfur	S_8
Air	$N_2 + O_2 + CO_2 + H_2 + Ar + He$ + others
Water	H_2O (+ dissolved minerals)
Metal ores	See Table 3.9.

Table 3.9 Sources of Metals

Source	Examples
Pure metal	Ag, As, Au, Bi, Cu, Pd, Pt, Sb
Carbonate ores	$BaCO_3$(witherite), $CaCO_3$ (limestone, calcite), $CuCO_3 \cdot Cu(OH)_2$ (malachite), $FeCO_3$(siderite), $MgCO_3$ (magnesite), $MnCO_3$ (rhodochrosite), $PbCO_3$ (cerussite), $SrCO_3$ (strontianite), $ZnCO_3$ (smithsonite), $CaCO_3 \cdot MgCO_3$ (dolomite)
Halide ores	AgCl (cerargyrite), $MgCl_2$ (in seawater), NaCl (dry halite deposits and in seawater), KCl (sylvite)
Oxide ores	Al_2O_3 (bauxite and many others), Cu_2O (cuprite), Fe_2O_3 (hematite), Fe_3O_4 (magnetite), $FeO \cdot Cr_2O_3$ (chromite), $FeO \cdot WO_3$ (wolframite), MnO_2 (pyrolusite), SnO_2 (cassiterite), TiO_2 (rutile), ZnO (zincite)
Phosphate ores	$Ca_3(PO_4)_2$ (phosphate rock), $CePO_4$, $LaPO_4$, $NdPO_4$, $Th_3(PO_4)_4$
Silicate ores	$Be_3Al_2Si_6O_{18}$ (beryl), $NiSiO_3 \cdot MgSiO_3$ (garnierite), $Sc_2Si_2O_7$ (thortveitite), $ZrSiO_4$ (zircon)
Sulfate ores	$BaSO_4$ (barite), $CaSO_4 \cdot 2H_2O$ (gypsum), $CuSO_4 \cdot 2Cu(OH)_2$ (antlerite), $PbSO_4$ (anglesite), $SrSO_4$ (celestite)
Sulfide ores	FeS_2 (iron pyrite), PbS (galena), ZnS (zinc blende), ZnS (sphalerite), HgS (cinnabar), Cu_2S (chalcocite)

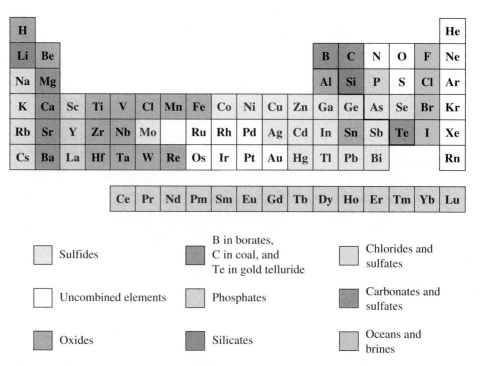

Figure 3.27 The natural sources of the elements include a variety of solid ionic compounds, as well as uncombined elements (metals or components of the atmosphere) and the ocean. Most of the elements can be obtained from more than one source.

Table 3.10 Some Important Industrial Chemicals: Annual U.S. Production in 1991*

Formula	Name	Billions of Pounds	Principal Uses
H_2SO_4	Sulfuric acid	89	Fertilizers, chemicals, oil refining
N_2	Nitrogen	59	Inert atmosphere, low temperatures
O_2	Oxygen	42	Steel making, welding, medical atmospheres, water purification
C_2H_4	Ethylene	40	Manufacture of ethylene glycol, polyethylene, styrene, vinyl chloride
NH_3	Ammonia	36	Fertilizers, nitric acid, cleaning agents, refrigerant, NH_4^+ compounds
CaO	Lime (calcium oxide)	35	Steel making, chemical manufacturing, water treatment, cement
H_3PO_4	Phosphoric acid	25	Fertilizers, detergents
NaOH	Sodium hydroxide	24	Chemicals, aluminum, pulp and paper, soaps, detergents, textile manufacturing
C_3H_6	Propylene	23	Plastics, fibers, solvents
Cl_2	Chlorine	22	Chemical manufacturing, pulp and paper, solvents, plastics, water treatment
Na_2CO_3	Sodium carbonate	21	Glass, chemicals, pulp and paper, detergents
$CO(NH_2)_2$	Urea	17	Fertilizers, plastics, pharmaceuticals
HNO_3	Nitric acid	16	Fertilizers, plastics, explosives, dyes, lacquers
$C_2H_4Cl_2$	Ethylene dichloride	16	Solvent, manufacture of other organic chemicals
NH_4NO_3	Ammonium nitrate	15	Fertilizers, explosives
C_6H_6	Benzene	12	Polystyrene, nylon, other polymers

*Source: From *Chemical and Engineering News,* April 12, 1993, page 11.

Figure 3.28 If air is cooled below about −195°C, it converts to a liquid containing primarily nitrogen and oxygen. By careful warming, these two elements can be separated. Liquid nitrogen, shown here, is commonly used to cool equipment to low temperatures.

distillation. Good separations require repetitions of this process, which also yields oxygen.

- Ethylene comes from petroleum. In a process called *cracking,* large hydrocarbon molecules are broken up by heating, giving C_2H_4 and other smaller molecules. A similar process yields propylene.
- Ammonia is made by the Haber process, in which nitrogen from air is combined with hydrogen (from petroleum or coal and water) at high temperatures and under high pressure.
- Limestone and marble ($CaCO_3$) are the sources for calcium oxide. When the calcium carbonate is heated, gaseous carbon dioxide is released, leaving calcium oxide.
- Phosphoric acid is obtained from phosphate rock by a complex series of steps. Sulfuric acid is used to convert the calcium phosphates to phosphoric acid.
- Sodium hydroxide is derived from sodium chloride, which is obtained either as pure NaCl (halite) or from seawater. The process involves passing electricity through an aqueous solution of sodium chloride, giving sodium hydroxide and chlorine gas. When solid sodium chloride is melted and electricity is passed through this liquid, sodium metal and gaseous chlorine are produced.

These ten chemicals, which come from a fairly small list of raw materials, are used to form a very large number of other chemicals that find numerous applications in our lives.

Summary

Physical and chemical properties of the elements show periodic, or regular, trends as a function of their atomic numbers. The arrangement of the periodic table, which is characterized by groups (columns) and periods (rows), emphasizes the similarities and trends in the properties of the elements. Many of the elements gain or lose electrons to form ions that have the same numbers of electrons as noble gases have. The stability of these ions allows us to use the position of an element in the periodic table to predict the probable charge of its ions and the formulas of binary compounds. Oxoanions of the nonmetals tend to be composed of three or four oxygen atoms surrounding the nonmetal. The charge of such an oxoanion, with a few exceptions, matches that of the anion formed by the nonmetal to achieve a noble-gas electron count.

Compounds can be classified as ionic or covalent. Ionic compounds are generally formed by transferring electrons from a metal to a nonmetal, and covalent compounds are generally formed by sharing electrons between two nonmetals. Properties of simple compounds such as oxides and hydrides vary systematically across the periodic table.

An oxidation number is a number assigned to an atom in a compound to indicate how many electrons it has compared with the number of electrons the uncombined atom would have. Oxidation numbers can be used to predict the composition of covalent compounds as well as ionic compounds. Names can be systematically assigned to species of known formula, or vice versa.

A fairly small number of naturally occurring raw materials are used to prepare a larger number of chemicals important to industry. In turn, these industrial chemicals are used to prepare an extremely great variety of consumer chemicals.

Key Terms

acid (3.5)

actinide (3.1)

alkali metal (3.1)

alkaline-earth

metal (3.1)

covalent

  compound (3.2)

group (3.1)

halogen (3.1)

hydrate (3.5)

inner-transition

  element (3.1)

lanthanide (3.1)

main-group

  (representative)

  element (3.1)

metal (3.1)

metalloid
  (semiconductor,
  semimetal) (3.1)

noble gas (3.1)

nonmetal (3.1)

ore (3.6)

oxidation number (3.2)

oxoacid (3.5)

period (3.1)

periodic law (3.1)

periodic table (3.1)

transition element (3.1)

Exercises

Periodic Table

3.1 What are the vertical columns and the horizontal rows of the periodic table called?

3.2 Identify each of the following elements.
  a. the third noble gas
  b. the second alkali metal
  c. the fifth lanthanide
  d. the fourth transition metal
  e. the second halogen
  f. the fifth alkaline-earth metal
  g. the first member of the third period
  h. the seventh inner-transition metal

3.3 Identify the group and period in which each of the following elements is found.
  a. Zr   b. In   c. As   d. Ne   e. Mg
  f. Bi   g. Fe   h. Rb   i. Br

3.4 Identify the group name of each of the following elements.
  a. I   b. Sr   c. Ar   d. Li   e. Xe   f. F

3.5 Identify each of the following elements as a metal, a nonmetal, or a metalloid.
  a. Ca   b. S   c. B   d. C   e. Se   f. Al

3.6 Identify each of the following elements as a main-group, transition, or inner-transition element.
  a. Ca   b. Cr   c. Pt   d. Eu   e. Fe
  f. Pb   g. Cs   h. U   i. Ar

3.7 Which elements from the following list have a metallic luster?
  a. Si   b. S   c. Sr   d. He   e. H   f. Se

3.8 Indicate which member of each pair has the higher electrical conductivity.
  a. lithium or nitrogen
  b. argon or aluminum
  c. barium or boron
  d. germanium or bromine
  e. graphite or diamond

3.9 In which group(s) of the periodic table would you find ions with a charge of 2+?

3.10 Describe the properties of the alkali metals.

3.11 Describe the properties of the noble gases.

3.12 Describe how metals and nonmetals differ in their physical properties.

3.13 Describe how the metallic nature of the elements varies from one part of the periodic table to another.

3.14 Consider the melting point (mp) and boiling point (bp) of chlorine and iodine:

  Chlorine: mp $= -101°C$ bp $= -35°C$
  Iodine: mp $= 114°C$ bp $= 184°C$

  Predict the values for bromine.

3.15 The density of calcium is 1.55 g/cm^3, while the value for barium is 3.50 g/cm^3. Predict the value for strontium.

Charges of Ions and Formulas of Compounds

3.16 What information is contained in the formula of a compound?

3.17 Distinguish between an ionic compound and a covalent compound.

3.18 Predict the charge of the cation likely to be formed by each of the following elements.
  a. Ca   b. Al   c. Mg   d. Ti   e. Rb   f. Pb

3.19 Predict the charge of the anion likely to be formed by each of the following elements.
  a. S   b. N   c. P   d. Cl   e. O   f. C

3.20 Predict the formula of the ionic compound that would be formed by each of the following pairs of elements.
  a. Li and O   b. Ca and N   c. Pb and O
  d. Rb and Cl   e. Al and C   f. Ba and S

3.21 Write a reasonable formula for a compound formed by each of the following pairs of elements.
  a. N and Cl   b. C and Cl   c. N and H
  d. O and H   e. S and F   f. P and Cl
  g. Si and H   h. S and H   i. P and H

3.22 Write a reasonable formula for a compound formed by each of the following pairs of elements.
  a. Mg and Cl   b. Mg and S   c. Al and N
  d. Na and H   e. Na and S   f. Mg and N
  g. Sn and H   h. Al and O   i. Be and C

3.23 Predict the charge of the oxoanion likely to be formed by each of the following elements. Also give the formula of one or more oxoanions of each element.
  a. S   b. N   c. C   d. Cl   e. B   f. P

3.24 In each of the following anion formulas, replace the X with the symbol of an appropriate element.
   a. XO_3^- b. XO_3^{2-} c. XO_4^-
   d. XO_4^{2-} e. XO_4^{3-} f. XO_4^{4-}

3.25 The following formulas have errors in them. Rewrite each of the formulas correctly.
   a. $LiSO_3$ b. $NaAsO_3$ c. $BaBrO_3$
   d. $HClO_5$ e. $MgNO_3$ f. BO
   g. $CdPO_4$ h. $CsBO_3$ i. SrN

3.26 What are the formulas of the following ions?
   a. arsenite   b. silicate   c. phosphate
   d. borate   e. nitrite   f. chlorite

Simple Compounds

3.27 What is an oxide? a hydride?

3.28 Predict the formula of an oxide of each of the following elements.
   a. Mg   b. Tl   c. S   d. I   e. Cr   f. H

3.29 Identify the physical state of each of the following oxides.
   a. H_2O b. CaO   c. NO_2 d. ClO_2 e. GeO_2

3.30 List some gaseous oxides.

3.31 List some of the chemical compounds in which oxygen occurs in nature.

3.32 Give the formula of a binary compound that could be formed by the elements hydrogen and sulfur.

3.33 If metalloid X forms a hydride with the formula XH_4, what element is it likely to be?

3.34 Which elements are expected to form ionic hydrides?

Oxidation Numbers

3.35 What is an oxidation number?

3.36 Predict a likely oxidation number for an ion of each of the following elements.
   a. Te   b. Rb   c. Mg   d. Sc   e. N   f. I

3.37 Determine the oxidation number of the element indicated.
   a. N in CN^- b. Cl in ClO_3^- c. O in VO_2^+
   d. N in NO_2^- e. Ti in TiO^{2+} f. S in SO_3^{2-}
   g. I in IF_4^+ h. Br in BrO_2^-

3.38 Determine the oxidation number of the element indicated.
   a. O in H_2O_2 b. H in H_2O_2 c. Br in $HBrO_3$
   d. F in NF_3 e. H in CH_4 f. Cr in Na_2CrO_4
   g. Te in TeO_2 h. Be in $BeCl_2$ i. S in Na_2SO_4

Nomenclature

3.39 Why do we need systematic names for ions and compounds?

3.40 Name the following covalent compounds.
   a. PF_5 b. PF_3 c. CO
   d. CO_2 e. SO_2 f. SO_3
   g. N_2O_4 h. NO_2 i. CS_2

3.41 Write formulas for the following covalent compounds.
   a. sulfur tetrafluoride   b. carbon disulfide
   c. tricarbon dioxide   d. dinitrogen pentoxide
   e. chlorine dioxide   f. boron nitride
   g. sulfur trioxide   h. iodine heptafluoride

3.42 Name the following anions.
   a. ClO^- b. HSO_4^- c. BO_3^{3-}
   d. HS^- e. O_2^{2-} f. NO_3^-
   g. BrO_4^- h. CO_3^{2-} i. PO_4^{3-}

3.43 Write formulas for the following anions.
   a. perchlorate   b. hydride   c. sulfite
   d. borate   e. nitrite   f. bromite
   g. hypoiodite   h. hydrogen phosphate

3.44 Name the following ionic compounds.
   a. $(NH_4)_2S$ b. Al_2O_3 c. Ag_2O
   d. $CuBr_2$ e. $Ba(CN)_2$ f. $ZnCl_2$
   g. K_2CO_3 h. KOH   i. MgI_2

3.45 Write formulas for the following ionic compounds.
   a. calcium sulfate   b. potassium hydrogen
   c. barium oxide   sulfide
   e. ammonium sulfate   d. sodium cyanide
   g. barium carbonate   f. cobalt(II) sulfate
   h. ammonium hydroxide

3.46 Name the following binary compounds.
   a. CaO   b. $SbBr_3$ c. PbO_2
   d. ZnI_2 e. NCl_3 f. MgS
   g. $AlCl_3$ h. K_3N i. NaH

3.47 Write formulas for the following binary compounds.
   a. strontium bromide   b. sodium oxide
   c. silver(I) chloride   d. zinc sulfide
   e. dinitrogen trioxide   f. magnesium nitride
   g. carbon disulfide   h. arsenic trihydride

3.48 Name the following transition-metal compounds.
   a. Cu_2O b. $CrCl_2$ c. $FePO_4$
   d. AgCN   e. Au_2O_3 f. $TiCl_3$
   g. CuS   h. $CoBr_2$ i. Mn_2O_3

3.49 Write formulas for the following transition-metal compounds.
- a. copper(II) bromide
- b. cuprous oxide
- c. chromium(III) oxide
- d. gold(I) cyanide
- e. iron(III) nitride
- f. titanium(IV) iodide
- g. ferrous sulfate
- h. chromium(VI) oxide

3.50 Name the following hydrates.
- a. $CuSO_4 \cdot 5H_2O$
- b. $MgSO_4 \cdot 7H_2O$
- c. $CoCl_2 \cdot 6H_2O$
- d. $CuSO_4 \cdot H_2O$

3.51 Write formulas for the following hydrates.
- a. copper(II) nitrate hexahydrate
- b. potassium aluminum sulfate decahydrate
- c. cobalt(II) chloride tetrahydrate
- d. potassium sulfate dihydrate

3.52 Name the following oxoanion compounds.
- a. $NaNO_3$
- b. $Ba(ClO_4)_2$
- c. $CuSO_4$
- d. $CrPO_4$
- e. $NaHSO_4$
- f. $Th(ClO_3)_4$
- g. $CaCO_3$
- h. $NaOBr$
- i. Na_3AsO_4

3.53 Write formulas for the following oxoanion compounds.
- a. barium sulfate
- b. sodium chlorate
- c. copper(II) phosphate
- d. sodium hydrogen carbonate
- e. barium hypobromite
- f. scandium chlorate
- g. strontium arsenite
- h. sodium selenate

3.54 Name the following acids.
- a. $HBr(aq)$
- b. $HBrO_3(aq)$
- c. $HOBr(aq)$
- d. $HBrO_4(aq)$
- e. $HBrO_2(aq)$
- f. $H_3PO_4(aq)$
- g. $H_3BO_3(aq)$
- h. $H_2CO_3(aq)$
- i. $HNO_2(aq)$

3.55 Write formulas for the following acids.
- a. hydrosulfuric acid
- b. perchloric acid
- c. sulfurous acid
- d. hydriodic acid
- e. bromous acid
- f. phosphorous acid
- g. carbonic acid
- h. selenic acid

3.56 Name the following compounds.
- a. $NaKCO_3$
- b. $Cd(OH)Cl$
- c. $KAl(SO_4)_2$
- d. NH_4NO_3
- e. $Ca(HSO_4)_2$
- f. $MgNH_4PO_4$

3.57 Name the following compounds.
- a. $NaSCN$
- b. K_2SO_3
- c. $Pb_3(PO_4)_2$
- d. Cl_2O_7
- e. BN
- f. $(NH_4)_2HPO_4$

3.58 Give the common names for the following compounds.
- a. H_2O
- b. NH_3
- c. NH_4SCN
- d. NH_4OH
- e. $NaCN$

3.59 Write formulas for the following compounds.
- a. lead(II) chloride
- b. europium(III) hydride
- c. magnesium phosphate
- d. barium hydroxide
- e. nitrogen triiodide
- f. cesium aluminum sulfate
- g. ferric oxide
- h. dichlorine pentoxide
- i. calcium nitride
- j. ammonium perchlorate

Classes of Compounds

3.60 Indicate whether each of the following combinations of elements would form covalent or ionic compounds.
- a. sodium and iodine
- b. oxygen and chlorine
- c. selenium and chlorine
- d. chromium and sulfur
- e. lithium and boron

3.61 Classify the following compounds as ionic or covalent.
- a. SiO_2
- b. N_2
- c. NH_4Br
- d. CaS
- e. NO_2
- f. PF_5

3.62 For each of the following, decide whether the compound is an acid. If so, classify it as a binary acid or an oxoacid.
- a. $HCl(aq)$
- b. NO_2
- c. $HClO_3$
- d. Br_2O_5
- e. HPO_3
- f. $Cu(OH)_2$
- g. H_2O
- h. H_2SO_4
- i. NaH_2PO_4

3.63 Which of the following compounds are considered halides?
- a. ClO_2
- b. $CuCl_2$
- c. $NaBrO_3$
- d. $NaBr$
- e. BaO
- f. $CrCl_3$

3.64 Which of the following compounds are considered oxides?
- a. ClO_2
- b. $HClO_3$
- c. O_2
- d. P_2O_5
- e. Na_3PO_4
- f. $NaOH$

3.65 Describe the differences between hydrogen chloride and hydrochloric acid.

Some Important Compounds

3.66 In what kind of ore is copper commonly found?

3.67 Which elements are found in the free form in nature?

3.68 Which elements are present in the earth's atmosphere?

3.69 What raw material would you use to prepare sulfuric acid?

3.70 Which elements occur in the free elemental state as diatomic molecules on earth?

3.71 If you wanted to prepare carbon dioxide industrially, what raw material would you use?

3.72 What important compound can be prepared from coal, water, and air?

3.73 If you wanted to prepare fluorine gas industrially, what raw material would you use?

3.74 What anions are commonly found in metal ores?

Additional Exercises

3.75 Element X forms two solid oxides, X_4O_{10} and X_4O_6. What is the name of this element?

3.76 Classify each of the following compounds as ionic or covalent.

    a. NaH   b. PCl_3 c. CO

    d. $BaSO_4$ e. O_2 f. PCl_5

3.77 Describe the properties of the alkaline-earth metals.

3.78 Write a reasonable formula for a compound formed by each of the following pairs of elements.

    a. Sb and F   b. As and H   c. Ge and H

    d. O and Cl   e. P and F   f. B and Cl

    g. Si and F   h. S and O   i. B and N

3.79 Predict the charge of the ion likely to be formed by each of the following elements.

    a. Cs   b. Br   c. Be

    d. Sc   e. B   f. Se

3.80 The substance XH_3 is a toxic gas. Suggest a possible identity for X.

3.81 Describe the properties of the halogens.

3.82 The following formulas have errors in them. Rewrite each of the formulas correctly.

    a. $Ag(ClO_4)_2$ b. ZnI   c. SiF

    d. CS   e. $ZnNO_3$ f. NaO

    g. BaOH   h. Na_2PO_4 i. CaC

3.83 Determine the oxidation number of the element indicated.

    a. O in O_2 b. F in SF_6 c. Br in BrO_2

    d. S in SF_4 e. C in CH_4 f. Cr in CrO_3

    g. Se in CSe_2 h. Cl in $BeCl_2$ i. S in Na_2S

3.84 Many hydrocarbons can be described by the general formula C_nH_{2n+2}. Describe what happens to the physical state of the hydrocarbons as n is increased.

3.85 Write formulas for the following compounds.

    a. barium chloride   b. palladium(II) hydride

    c. carbon diselenide   d. ammonium hydroxide

    e. chlorine triiodide   f. potassium hydrogen

    g. copper(I) oxide   sulfate

    i. barium phosphide   h. sodium cyanide

                      j. chloric acid

3.86 Identify the group and period in which each of the following elements is found.

    a. Pb   b. Cu   c. Au

    d. Ca   e. P   f. Ga

    g. Pt   h. Li   i. Se

3.87 Name the following compounds.

    a. Cs_2S b. $FeBr_2$ c. NaH_2PO_4

    d. $HClO_2$ e. $(NH_4)_2SO_4$ f. H_2SO_3

    g. $LiNO_2$ h. Cs_3B i. $Cr(OI)_3$

3.88 Describe how metals and nonmetals differ in their chemical properties.

3.89 Indicate which member of each pair has the lower thermal conductivity.

    a. copper or chlorine   b. sulfur or tin

    c. thallium or boron   d. potassium or argon

*3.90 Solve the following puzzle. The code letters A through Z (no J) have been randomly assigned to the 25 elements in a short form of the periodic table that does not include any transition elements. Assign the code letters to the appropriate element positions in the short form of the periodic table.

	I	II	III	IV	V	VI	VII	VIII
1								
2								
3								
4								

The following elements occur in groups: ZRD, PSIF, XEB, TLH, QKA, WVO, YMC, and UNG. The following facts are known:

U has six electrons.

NA_2 is the formula of an oxide.

E is a noble gas.

S is an alkali metal.

D has the lowest atomic weight in its group.

O has an atomic number greater than that of V but smaller than that of W.

C has 15 more electrons than one of the noble gases.

L is an alkaline-earth metal.

O is a halogen.

The atomic weight of T is higher than that of H but lower than that of L.

M has an atomic number smaller by 1 than that of A.

Q has a melting point between those for A and K.

N is in the third period.

I is more metallic than S.

R has the highest atomic weight of its group.

F is a gas.

B contains ten protons.

Q has an atomic weight lower than that of K.

Y is more metallic than either M or C, but not both.

X has an atomic number greater by 1 than that of F.

The atomic weight of P is lower than that of S.

*3.91 If 3.1185 g of silver combines with bromine to form 5.4286 g of a pale yellow solid compound, what is the formula and the name of this compound?

*3.92 Which compound, ammonium chloride or sodium nitrate, contains the greater percentage of nitrogen by mass?

Chemical Equations and Quantitative Relationships

The thermite reaction between iron(III) oxide and aluminum metal.

CHAPTER OUTLINE

Encountering Chemistry

How could we represent all this information in a more compact way?

Two powders—gray aluminum and red iron(III) oxide—are mixed and poured into a ceramic container with a small hole in the bottom. A strip of magnesium ribbon is stuck into the pile of powder and ignited. The powder begins to smoke and soon gives off a magnificent shower of sparks. As this spectacular display continues and then subsides, liquid metal runs out of the hole in the container. The liquid is molten iron. This reaction, called the **thermite reaction,** is sometimes used to weld pieces of iron or steel together, as in reinforcing rod (rebar) in concrete buildings or in the rails for high-speed railroad tracks.

We can use symbols and formulas to describe chemical processes such as the thermite reaction. The symbolic representation of a chemical reaction is called a **chemical equation.** Such equations are veritable storehouses of information concerning chemical reactions.

This chapter begins by describing how to write chemical equations. Next, it gives rules for predicting the products of several kinds of common chemical reactions and illustrates the rules with some chemical reactions of oxygen and oxides. It then considers some quantitative aspects of chemical reactions. Finally, the chapter brings all this material together by discussing some industrial chemistry—the preparation of sulfuric acid and iron (and steel), both extremely important chemical products.

4.1 Writing Chemical Equations

Because writing chemical equations is an important part of chemistry, it is essential that you understand the conventions of the method. You must know what reactants are used and what products are formed in a particular reaction. The physical state—solid (s), liquid (l), or gas (g)—of each reactant and product is also important. It may be necessary to indicate that a reactant or product is in aqueous solution (aq). If we need to indicate that heat must be supplied for the reaction to take place, we place a triangle over the arrow that designates the direction of change in the reaction. Finally, the equation must be balanced. In a **balanced equation,** the number of atoms of each component element is the same on the reactant side as on the product side of the equation.

▶ Let's work through the reaction described in the chapter introduction as an example. By igniting the magnesium ribbon, we produced enough heat to initiate the thermite reaction. In the thermite reaction, aluminum metal reacts with iron(III) oxide to produce molten iron metal and aluminum(III) oxide. Using the proper formulas, we can begin to write the chemical equation that summarizes this experimentally determined information. It is often helpful to write a word equation first, using names instead of formulas, to separate out the critical information about reactant and product identities. From the example description, we can write

$$\text{Aluminum} + \text{iron(III) oxide} \xrightarrow{\Delta} \text{aluminum(III) oxide} + \text{iron}$$

In this word equation, as in all chemical equations, reactants are placed on the left and separated by plus signs, products are placed on the right (also separated by plus signs), and an arrow separates the reactants from the products and indicates the direction of change. Using the rules developed in Chapter 3, we can now substitute formulas for names in this equation. We should also add the physical state of each substance:

Why is this equation not balanced?

$$\text{Al}(s) + \text{Fe}_2\text{O}_3(s) \xrightarrow{\Delta} \text{Al}_2\text{O}_3(s) + \text{Fe}(l)$$

But the equation is not complete as written. According to the law of conservation of mass, matter cannot be created or destroyed by ordinary chemical means. Therefore, the same number of atoms of each element must appear on both sides of the arrow; that is, the equation must be balanced. As written, the equation has the following numbers of atoms:

Element	Atoms in Reactants	Atoms in Products
Al	1	2
O	3	3
Fe	2	1

Oxygen is balanced, but the other two elements are not.

Balancing an equation involves adjusting coefficients in front of the chemical formulas until the reactants and products contain the same number of atoms of each element. The coefficients are generally whole numbers but may sometimes be ratios of whole numbers. The coefficient 1 is never written before a formula. It is important to remember that you cannot change the subscript of an element in a compound to balance an equation. For example, Al_2O_3 cannot be changed to AlO_3—an entirely different (and unknown) compound.

Consider again the equation just given for the thermite reaction. Because the atoms of the reactants and products have not yet been equalized, the equation is referred to as an *unbalanced* (or *skeletal*) equation.

$$Al(s) + Fe_2O_3(s) \xrightarrow{\Delta} Al_2O_3(s) + Fe(l)$$

Note that there are two aluminum atoms in Al_2O_3 but only one in Al. To balance the aluminum atoms, we can place the coefficient 2 in front of Al:

$$2Al(s) + Fe_2O_3(s) \xrightarrow{\Delta} Al_2O_3(s) + Fe(l)$$

To balance iron atoms, we can place the coefficient 2 in front of Fe:

$$2Al(s) + Fe_2O_3(s) \xrightarrow{\Delta} Al_2O_3(s) + 2Fe(l)$$

If we count the number of atoms of each element again, we see that they are now the same on both sides of the equation:

Element	Atoms in Reactants	Atoms in Products
Al	2	2
O	3	3
Fe	2	2

This balanced equation obeys the law of conservation of mass.

The general approach to balancing equations involves changing coefficients one at a time until the number of atoms of each element is the same on both sides of the equation. It is usually easiest to start by adjusting the coefficients for the substances that contain elements that occur least often in the equation. (Because hydrogen and oxygen are common elements, they usually do not make a good starting place.) Then continue with elements that occur in more species. Make a final check by counting the atoms of each element on each side of the equation.

Balancing the thermite reaction equation was quite straightforward. Sometimes, however, changing a coefficient to balance one element unbalances another element. In that case, we must go back and change coefficients to rebalance those elements that have become unbalanced.

EXAMPLE 4.1

Balance the following skeletal equation:

$$KClO_3(s) \xrightarrow{\Delta} KCl(s) + O_2(g)$$

Solution   Potassium and chlorine occur once on each side of the arrow and thus are balanced already, but there are three oxygen atoms on the left and two oxygen atoms on the right. Using the smallest common multiple of 3 and 2—6—we can balance the oxygen atoms by placing a 2 in front of $KClO_3$ and a 3 in front of O_2. Now there are six oxygen atoms on both sides of the equation:

$$2KClO_3(s) \xrightarrow{\Delta} KCl(s) + 3O_2(g)$$

However, now the potassium and chlorine are no longer balanced. The left side of the equation has two atoms of each of these elements, while the right side has only one atom of each. We can rebalance the potassium and chlorine atoms by placing a 2 in front of KCl on the product side of the equation:

$$2KClO_3(s) \xrightarrow{\Delta} 2KCl(s) + 3O_2(g)$$

A final count shows this to be a correct, balanced equation:

Element	Atoms in Reactants	Atoms in Products
K	2	2
Cl	2	2
O	6	6

Practice Problem 4.1

Balance the following skeletal equation:

$$Na(s) + H_2O(l) \longrightarrow NaOH(aq) + H_2(g)$$

Answer: $2Na(s) + 2H_2O(l) \longrightarrow 2NaOH(aq) + H_2(g)$

Coefficients needed to achieve a balanced equation may not always be whole numbers, although whole-number coefficients are preferred. But sometimes a temporary balancing with fractional coefficients facilitates the balancing process, as is the case in the next example.

EXAMPLE 4.2

Balance the following skeletal equation:

$$C_6H_{14}(l) + O_2(g) \longrightarrow CO_2(g) + H_2O(g)$$

Solution   Because oxygen occurs in three substances, we should start by balancing the carbon and hydrogen. There are six carbon atoms on the left, but only one on the right, so we place a 6 in front of CO_2:

$$C_6H_{14}(l) + O_2(g) \longrightarrow 6CO_2(g) + H_2O(g)$$

There are 14 hydrogen atoms on the left and 2 on the right, so we place a 7 in front of H_2O:

$$C_6H_{14}(l) + O_2(g) \longrightarrow 6CO_2(g) + 7H_2O(g)$$

Now all we have left to balance is the oxygen. There are 12 O atoms in six CO_2 molecules and 7 O atoms in seven H_2O molecules, making a total of 19 O atoms in the products. Because we cannot balance the O atoms in the reactants with a whole number unless we unbalance everything else, we use a fractional coefficient, $\frac{19}{2}$, temporarily:

$$C_6H_{14}(l) + \tfrac{19}{2}O_2(g) \longrightarrow 6CO_2(g) + 7H_2O(g)$$

To get whole-number coefficients, we then multiply all coefficients in the equation by 2:

$$2C_6H_{14}(l) + 19O_2(g) \longrightarrow 12CO_2(g) + 14H_2O(g)$$

This is a correct, balanced equation, as the following count confirms:

Element	Atoms in Reactants	Atoms in Products
C	12	12
O	38	38
H	28	28

(continued)

Practice Problem 4.2

Balance the following skeletal equation:

$$C_4H_{10}(g) + O_2(g) \longrightarrow CO_2(g) + H_2O(g)$$

Answer: $2C_4H_{10}(g) + 13O_2(g) \longrightarrow 8CO_2(g) + 10H_2O(g)$

A balanced equation summarizes a great deal of information about a chemical reaction. Consider, for example, the equation that describes the Haber process for preparing ammonia.

$$N_2(g) + 3H_2(g) \longrightarrow 2NH_3(g)$$

Ammonia is prepared by the Haber process. Nitrogen from air is combined with hydrogen from the reaction of carbon monoxide with water. This mixture is heated to about 300°C in an iron–titanium alloy container at a pressure of about 500 atm.

The equation can represent the reaction of a single molecule (of nitrogen, in this case):

$$\text{One } N_2 \text{ molecule} + \text{three } H_2 \text{ molecules} \longrightarrow \text{two } NH_3 \text{ molecules}$$

or of two molecules:

$$\text{Two } N_2 \text{ molecules} + \text{six } H_2 \text{ molecules} \longrightarrow \text{four } NH_3 \text{ molecules}$$

or of any number of molecules, as long as the relative numbers remain the same. The following equation represents the reaction of Avogadro's number of nitrogen molecules:

$$\begin{array}{ccc} 6.022 \times 10^{23} & + \quad 3 \times 6.022 \times 10^{23} & \longrightarrow \quad 2 \times 6.022 \times 10^{23} \\ \text{molecules } N_2 & \text{molecules } H_2 & \text{molecules } NH_3 \end{array}$$

Another way to represent Avogadro's number of molecules is in units of moles:

$$1 \text{ mole } N_2 + 3 \text{ moles } H_2 \longrightarrow 2 \text{ moles } NH_3$$

Using molar masses, we can convert moles to masses; so the equation can also represent the masses of the reactants and products:

$$\begin{array}{ccc} 1 \text{ mol} \times 28.0 \text{ g/mol} + 3 \text{ mol} \times 2.0 \text{ g/mol} & \longrightarrow & 2 \text{ mol} \times 17.0 \text{ g/mol} \\ = 28.0 \text{ g } N_2 \qquad = 6.0 \text{ g } H_2 & & = 34.0 \text{ g } NH_3 \\ \text{Mass of reactants} = 34.0 \text{ g} & & \text{Mass of products} = 34.0 \text{ g} \end{array}$$

Note that the mass of the reactants equals the mass of the products, demonstrating the law of conservation of mass.

Figure 4.1 shows schematically how nitrogen and hydrogen molecules com-

Figure 4.1 A balanced equation follows the law of conservation of mass. The same number of atoms of each element appears on both sides of the equation, though the atoms occur in different chemical forms.

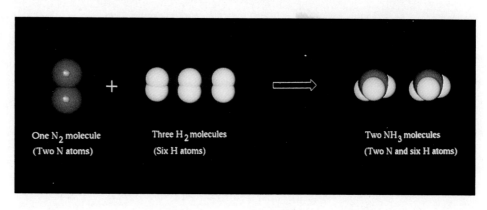

One N₂ molecule
(Two N atoms)

Three H₂ molecules
(Six H atoms)

Two NH₃ molecules
(Two N and six H atoms)

bine to form ammonia. The N_2 and H_2 molecules lose their identities and rearrange to produce NH_3 molecules. Note that during a chemical reaction the number of molecules is not necessarily conserved. It is common, as in this case, for a reaction to yield more or fewer molecules than were initially present, even though the total number of atoms must stay the same.

4.2 Predicting Chemical Reactions

Why would we want to use some approach other than experimentation to predict the products of chemical reactions?

In all the examples of chemical reactions in the preceding section, the identity of the reactants and products was given. But how do we know what products will result from a given set of reactants? One answer is to experiment—carry out the reaction and then isolate and identify the products—or make use of the experimental work of others. We saw in Chapter 3 that the chemical properties of the elements are periodic, meaning that elements in the same group in the periodic table undergo similar reactions. We can use this periodicity to reduce by a considerable extent the difficulty in predicting the products of chemical reactions. We can also use rules to help us predict the products of reactions. Predicting reaction products can be simplified further by classifying chemical reactions into four general categories: direct synthesis, decomposition, single-displacement, and double-displacement reactions (Figure 4.2).

We can also classify reactions according to whether the oxidation number of one or more elements changes. Reactions in which there is a change in oxidation number are called **oxidation–reduction reactions** (or **redox reactions**). One element increases its oxidation number (is **oxidized**) and another decreases its oxidation number (is **reduced**).

Direct Synthesis

In **direct synthesis reactions** (also called **combination reactions**), two elements, an element and a compound, two compounds, or two ions can react to produce a single compound. An important type of direct synthesis reaction involves two elements. *Most metals react with most nonmetals to form ionic compounds.* The products can be predicted from the charges expected for cations of the metal and anions of the nonmetal. (Sometimes covalent compounds are formed, but the products can be predicted in the same way.) The expected charges can be obtained from the periodic table, as explained in Chapter 3. For example, the silvery metal aluminum reacts with any of the halogens (as shown in Figure 4.3 for bromine) to form a solid aluminum halide. The product of the reaction between Al and F_2, for example, can be predicted from the $3+$ charge for aluminum ion and the $1-$ charge for fluoride ion. Because there is a change in the oxidation numbers of the elements, this type of reaction is an oxidation–reduction reaction.

$$2Al(s) + 3F_2(g) \longrightarrow 2AlF_3(s)$$

Similarly, *a nonmetal may react with a more reactive nonmetal to form a covalent compound.* The composition of the product is predicted from the common oxidation numbers of the elements—positive for the less reactive and negative for the more reactive nonmetal. The more reactive nonmetals tend to be located near the upper right side of the periodic table. For example, the yellow solid sulfur reacts with the more active oxygen gas to form colorless gaseous sulfur dioxide,

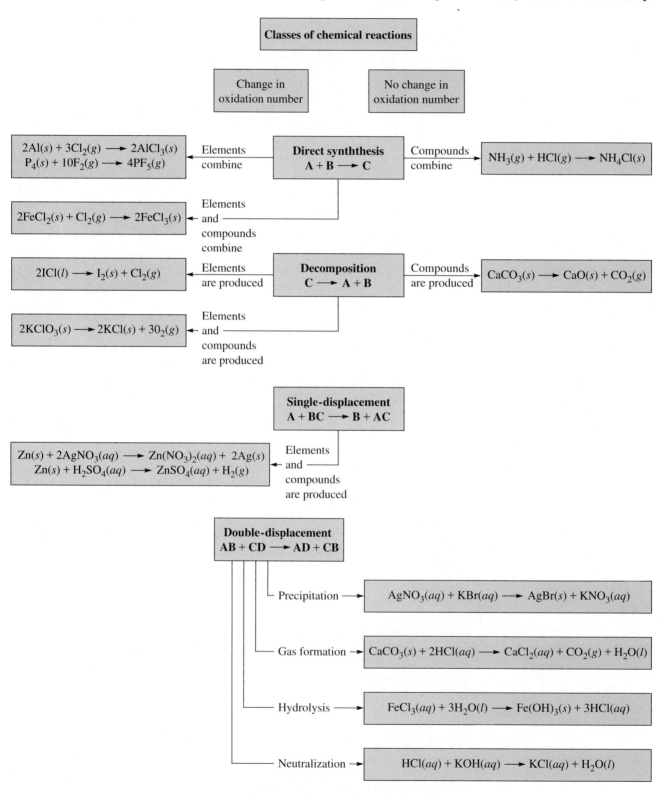

Figure 4.2 Inorganic chemical reactions can be placed into four major classes, some of which involve changes in oxidation number.

Figure 4.3 The reaction between aluminum and bromine is so vigorous that when it is carried out in a darkened room, flashes of light and small flames can be seen on the surface of the liquid bromine. The fumes are vaporized bromine and aluminum(III) bromide.

which has a pungent odor and is one of the substances produced in the preparation of sulfuric acid.

$$S_8(s) + 8O_2(g) \longrightarrow 8SO_2(g)$$

Although two metals usually do not react, they may combine to form a new metal called an **alloy,** as shown in Figure 4.4. Most alloys are simply homogeneous mixtures, not the products of reactions, but some are actually compounds.

A compound and an element may unite to form another compound if in the original compound the element with a positive oxidation number has an accessible higher oxidation number. This type of direct synthesis reaction is often called an **addition reaction.** Carbon monoxide, which is formed by the burning of hydrocarbons or other carbon-containing materials under conditions of oxygen deficiency, is a colorless, odorless, poisonous gas. It reacts with oxygen to form colorless, odorless carbon dioxide, which is a component of the atmosphere and is sold in frozen form as dry ice.

Calcium oxide, commonly known as lime, is a white, powdery solid used in bricks, plaster, mortar, stucco, and other building materials. It is also used in the manufacture of iron and other metals, glass, paper, sodium carbonate, and other industrial chemicals. Dissolved in water, it forms calcium hydroxide, $Ca(OH)_2$, known as slaked lime.

$$2CO(g) + O_2(g) \longrightarrow 2CO_2(g)$$

This reaction is also an oxidation–reduction reaction because the oxidation number of carbon changes from $+2$ to $+4$.

Two compounds may react to form a new compound (another addition reaction). For example, calcium oxide, a white solid also known as lime, reacts with carbon dioxide to form white solid calcium carbonate, which is found naturally as limestone, marble, and seashells.

$$CaO(s) + CO_2(g) \longrightarrow CaCO_3(s)$$

Figure 4.4 A copper coin *(left)* can be coated with zinc to give it a silvery appearance *(center)*. If the coin is heated, the zinc migrates into the copper to give the gold color typical of brass *(right)*, an alloy.

EXAMPLE 4.3

When pure calcium metal is exposed to the oxygen in air, a white coating appears on the surface. Predict the product of the reaction, identify the reaction type, and write a complete, balanced equation for this reaction.

Solution   By consulting Figure 4.2, we can see that the only reaction that will occur between two elements is a direct synthesis reaction. Thus, the reaction of calcium, an alkaline-earth metal, with oxygen should be a direct synthesis reaction, with an ionic compound as the product. The positions of calcium and oxygen in the periodic table indicate they will form an ionic compound containing Ca^{2+} and O^{2-}: CaO. The unbalanced equation for the reaction is

$$Ca(s) + O_2(g) \longrightarrow CaO(s)$$

We balance the equation as follows:

$$2Ca(s) + O_2(g) \longrightarrow 2CaO(s)$$

Practice Problem 4.3

Predict the products and write a balanced equation for the reaction of silicon with oxygen.

Answer: $Si(s) + O_2(g) \longrightarrow SiO_2(s)$

Decomposition Reactions

When a compound undergoes a **decomposition reaction,** it breaks down into simpler compounds or into the elements from which it was made. Rules for predicting the products of decomposition reactions can be formulated for a variety of compounds, primarily ionic compounds. Some of these rules are listed in Table 4.1. As you can see from these rules, the type of decomposition reaction a compound undergoes is determined largely by the type of anion in the compound. Ammonium ion also has characteristic decomposition reactions.

A few binary compounds will decompose to their constituent elements on heating; this is an oxidation–reduction reaction because the elements undergo a change in oxidation number. For example, the oxides and halides of noble metals (primarily gold, platinum, and mercury) decompose when heated. When red solid mercury(II) oxide is heated, it decomposes to liquid metallic mercury and oxygen gas:

$$2HgO(s) \xrightarrow{\Delta} 2Hg(l) + O_2(g)$$

Some nonmetal oxides, such as the halogen oxides, also decompose on heating:

$$2Cl_2O_5(g) \xrightarrow{\Delta} 2Cl_2(g) + 5O_2(g)$$

Other nonmetal oxides, such as dinitrogen pentoxide, decompose to an element and a compound:

$$2N_2O_5(g) \xrightarrow{\Delta} 4NO_2(g) + O_2(g)$$

Many metal salts containing oxoanions will decompose on heating, as shown in Example 4.1. These salts either give off dioxygen gas (usually called oxygen

Oxides of noble metals (Au, Pt, Hg) decompose to the elements.

$$2Au_2O_3(s) \xrightarrow{\Delta} 4Au(s) + 3O_2(g)$$

Halides of noble metals decompose to the elements.

$$PtCl_4(s) \xrightarrow{\Delta} Pt(s) + 2Cl_2(g)$$

Peroxides decompose to oxides and oxygen gas.

$$2H_2O_2(aq) \xrightarrow{\Delta} 2H_2O(l) + O_2(g)$$

Metal carbonates decompose to metal oxides and carbon dioxide gas; however, Group IA(1) metal carbonates do not decompose.

$$NiCO_3(s) \xrightarrow{\Delta} NiO(s) + CO_2(g)$$

Metal nitrates decompose to metal nitrites and oxygen gas if the metal is Al, an alkali metal, or an alkaline-earth metal; other metal nitrates decompose to metal oxides, nitrogen dioxide gas, and oxygen gas.

$$2KNO_3(s) \xrightarrow{\Delta} 2KNO_2(s) + O_2(g)$$

$$2Zn(NO_3)_2(s) \xrightarrow{\Delta} 2ZnO(s) + 4NO_2(g) + O_2(g)$$

Oxoacids decompose to nonmetal oxides and water.

$$H_2SO_3(aq) \xrightarrow{\Delta} SO_2(g) + H_2O(l)$$

Oxoanion salts containing hydrogen ions decompose to oxoanion salts and oxoacids; the oxoacids may decompose further.

$$2NaHCO_3(s) \xrightarrow{\Delta} Na_2CO_3(s) + H_2O(g) + CO_2(g)$$

Metal hydroxides decompose to metal oxides and water.

$$Ca(OH)_2(s) \xrightarrow{\Delta} CaO(s) + H_2O(g)$$

Hydrates lose water to form an anhydrous salt.

$$CuSO_4 \cdot 5H_2O(s) \xrightarrow{\Delta} CuSO_4(s) + 5H_2O(g)$$

Ammonium salts lose ammonia; if the salt contains an anion that is a strong oxidizing agent (such as nitrate, nitrite, or dichromate), oxidation–reduction produces an oxide, water, and nitrogen gas.

$$(NH_4)_2SO_4(s) \xrightarrow{\Delta} 2NH_3(g) + H_2SO_4(l)$$

$$(NH_4)_2Cr_2O_7(s) \xrightarrow{\Delta} Cr_2O_3(s) + 4H_2O(g) + N_2(g)$$

gas or molecular oxygen), forming a metal salt with a different nonmetal anion, or they give off a nonmetal oxide, forming a metal oxide. For example, metal nitrates containing aluminum or Group IA(1) or IIA(2) metals decompose to metal nitrites and oxygen gas:

$$Mg(NO_3)_2(s) \xrightarrow{\Delta} Mg(NO_2)_2(s) + O_2(g)$$

Figure 4.5 Heating copper(II) nitrate *(Left)* yields black copper(II) oxide, red nitrogen dioxide gas, and colorless oxygen gas *(Right)*.

Anhydrous calcium sulfate is found as the mineral anhydrite. Its hydrated forms include plaster of Paris, $CaSO_4 \cdot \frac{1}{2}H_2O$, and gypsum, $CaSO_4 \cdot 2H_2O$, which is used in wallboard and in the manufacture of plaster and portland cement.

All other metal nitrates decompose to the metal oxides, along with nitrogen dioxide and oxygen (Figure 4.5):

$$2Cu(NO_3)_2(s) \xrightarrow{\Delta} 2CuO(s) + 4NO_2(g) + O_2(g)$$

Salts of the oxoanions of the halogens decompose to halides and oxygen on heating:

$$2KBrO_3(s) \xrightarrow{\Delta} 2KBr(s) + 3O_2(g)$$

Carbonates, except for those of the alkali metals, decompose to oxides and carbon dioxide. A common example is calcium carbonate:

$$CaCO_3(s) \xrightarrow{\Delta} CaO(s) + CO_2(g)$$

A number of compounds that contain water or the components of water—hydrates, hydroxides, and oxoacids—lose water when heated. *Hydrates,* or compounds that contain water molecules, lose water to form **anhydrous** compounds, compounds free of molecular water.

$$CaSO_4 \cdot 2H_2O(s) \xrightarrow{\Delta} CaSO_4(s) + 2H_2O(g)$$

Metal hydroxides contain the constituents of water and can be converted to metal oxides by heating:

$$2Fe(OH)_3(s) \xrightarrow{\Delta} Fe_2O_3(s) + 3H_2O(g)$$

Most oxoacids lose water until no hydrogen remains, leaving a nonmetal oxide:

$$H_2SO_4(l) \xrightarrow{\Delta} H_2O(g) + SO_3(g)$$

$$4H_3PO_4(l) \xrightarrow{\Delta} 6H_2O(g) + P_4O_{10}(s)$$

Oxoanion salts that contain hydrogen ions break down into the corresponding oxoanion salts and oxoacids:

$$Ca(HSO_4)_2(s) \xrightarrow{\Delta} CaSO_4(s) + H_2SO_4(l)$$

Finally, some ammonium salts undergo an oxidation–reduction reaction when heated. Common salts of this type are ammonium dichromate, ammonium permanganate, ammonium nitrate, and ammonium nitrite. When these salts decompose, they give off nitrogen gas and water and form an oxide. For example, ammonium dichromate, a yellow solid, decomposes when ignited (Figure 4.6). It produces a grayish green, fluffy solid, chromium(III) oxide, as well as gaseous water and nitrogen:

$$(NH_4)_2Cr_2O_7(s) \xrightarrow{\Delta} Cr_2O_3(s) + 4H_2O(g) + N_2(g)$$

Ammonium nitrate and ammonium nitrite, highly explosive salts, do not form oxides when heated:

$$2NH_4NO_3(s) \xrightarrow{\Delta} 2N_2(g) + 4H_2O(g) + O_2(g)$$

$$NH_4NO_2(s) \xrightarrow{\Delta} N_2(g) + 2H_2O(g)$$

Figure 4.6 *Left:* The decomposition of ammonium dichromate occurs spontaneously once it has been started by heat. *Right:* The voluminous green powder is chromium(III) oxide which is used as an abrasive and as a pigment, especially for coloring glass and in printing fabrics and paper money.

Other ammonium salts will lose ammonia gas on heating:

$$(NH_4)_2SO_4(s) \xrightarrow{\Delta} 2NH_3(g) + H_2SO_4(l)$$

EXAMPLE 4.4

Borax, $Na_2B_4O_7 \cdot 10H_2O$, is used in various laundry and cleaning products. Predict what products result when borax is heated. Write a complete, balanced equation for this reaction.

Solution   Borax is a hydrate, so we predict that it will lose gaseous water to form an anhydrous compound, $Na_2B_4O_7$. The complete, balanced equation is

$$Na_2B_4O_7 \cdot 10H_2O(s) \xrightarrow{\Delta} Na_2B_4O_7(s) + 10H_2O(g)$$

Practice Problem 4.4

Magnesium sulfite trihydrate, $MgSO_3 \cdot 3H_2O$, used in the preparation of paper pulp, decomposes on heating. Predict the products and write a complete, balanced equation for this reaction.

Answer: $MgSO_3 \cdot 3H_2O(s) \xrightarrow{\Delta} MgO(s) + SO_2(g) + 3H_2O(g)$

Single-Displacement Reactions

In a **single-displacement reaction,** a free element displaces another element from a compound to produce a different compound and a different free element. *A more active element displaces a less active element from its compounds.* An example is the reaction between carbon and iron(III) oxide, used in the production of iron metal:

$$3C(s) + Fe_2O_3(s) \xrightarrow{\Delta} 3CO(g) + 2Fe(l)$$

▶ This reaction is similar to the thermite reaction, described in the chapter introduction. Single-displacement reactions are examples of oxidation–reduction reactions because both the displacing element and the displaced element undergo a change in oxidation number. Carbon is oxidized from an oxidation number of 0 to +2; iron is reduced, with its oxidation number changing from +3 to 0.

The element displaced from the compound is always the more metallic element—the one nearer the bottom left of the periodic table. The displaced element need not always be a metal, however. Consider a common type of single-displacement reaction, the displacement of hydrogen from water or from acids by metals.

The very active metals react with water. For example, calcium reacts with water to form calcium hydroxide and hydrogen gas, as shown in Figure 4.7.

$$Ca(s) + 2H_2O(l) \longrightarrow Ca(OH)_2(aq) + H_2(g)$$

Calcium metal has an oxidation number of 0, whereas Ca^{2+} in $Ca(OH)_2$ has an oxidation number of +2, so calcium is oxidized. Hydrogen in water has an oxidation number of +1, whereas in H_2 the oxidation number is 0, so hydrogen is reduced in this reaction.

Some metals, such as magnesium, do not react with cold water but do react slowly with steam:

$$Mg(s) + 2H_2O(g) \longrightarrow Mg(OH)_2(aq) + H_2(g)$$

Other, less active metals do not react with water at all but do react with acids. Iron is an example of such a metal:

$$Fe(s) + 2HCl(aq) \longrightarrow FeCl_2(aq) + H_2(g)$$

Metals that are even less active, such as copper, do not react with acids, unless the anion of the acid (for example, NO_3^-) can do the oxidizing.

How do we know which metals react with water or with acids? We can use an **activity series,** a list of the metals in order of decreasing activity. A short activity series is given in Figure 4.8. The elements at the top of the series, above the line, react with cold water. The elements above hydrogen in the series react with acids, and the elements below hydrogen do not react to release hydrogen gas.

The displacement of hydrogen from water or acids is just one type of single-displacement reaction. Other elements can also be displaced from their compounds. For example, copper metal reduces aqueous solutions of ionic silver compounds, such as silver nitrate, to deposit silver metal (Figure 4.9). The copper is oxidized to form an aqueous solution of an ionic copper(II) compound:

$$Cu(s) + 2AgNO_3(aq) \longrightarrow Cu(NO_3)_2(aq) + 2Ag(s)$$

We can use the activity series of Figure 4.8 to predict which single-displacement reactions of this type will take place. The elemental metal produced is always below the displacing element in the activity series. Thus, iron could be displaced from $FeCl_2$ by zinc metal but not by tin.

In some single-displacement reactions, the displacing element and the displaced element are nonmetals. Because no nonmetals (other than hydrogen) are in the activity series shown in Figure 4.8, we will need another approach to predict which reactions are possible. We can use the periodic table as a general guide. Nonmetals are relatively more active the closer they are to the top right of the periodic table (with the exception of the noble gases). In single-displacement reactions, a more active free nonmetal will displace a less active combined nonmetal from its compound. Chlorine, a pale greenish yellow gas, dissolves in water to

Figure 4.7 Calcium metal reacts with water to produce a solution of calcium hydroxide, along with gaseous hydrogen. This is an example of a single-displacement reaction.

Li
K
Ba   These metals will
Ca   displace hydrogen gas
Na   from water

Mg
Al
Zn
Fe   These metals will
Cd   displace hydrogen gas
Ni   from acids
Sn
Pb

H

Cu   These metals will not
Hg   displace hydrogen gas
Ag   from water or acids
Au

Figure 4.8 An activity series can be used to predict the tendency of metals to displace hydrogen from water or acids.

Figure 4.9 When a copper strip is placed in silver nitrate solution, needles of metallic silver are slowly deposited on the copper. Copper metal dissolves in the solution, imparting the characteristic blue color of Cu^{2+} in solution.

form greenish yellow solutions. If mixed with a colorless aqueous solution of the ionic compound sodium iodide, the solution turns brown (Figure 4.10), a color characteristic of molecular iodine in aqueous solution:

$$Cl_2(aq) + 2NaI(aq) \longrightarrow 2NaCl(aq) + I_2(aq)$$

In this reaction, the chlorine is reduced and the iodine is oxidized.

EXAMPLE 4.5

Iron metal reacts with aqueous solutions of copper sulfate to deposit copper metal, with the iron going into solution as iron(II) ions. Write a complete, balanced equation for this single-displacement reaction and identify the oxidation and reduction processes.

Solution   The reactants are Fe and $CuSO_4(aq)$, and the products are Cu and $FeSO_4(aq)$. Thus, the equation for this reaction is

$$Fe(s) + CuSO_4(aq) \longrightarrow Cu(s) + FeSO_4(aq)$$

The equation is balanced as written.

The oxidation number of copper changes from $+2$ to 0; that of iron changes from 0 to $+2$. Thus, copper is reduced and iron is oxidized.

Practice Problem 4.5

When black copper(II) oxide is heated with gaseous molecular hydrogen, reddish powdered copper metal and gaseous water are formed. Write a complete, balanced equation for this single-displacement reaction and identify the oxidation and reduction processes.

Answer: $CuO(s) + H_2(g) \xrightarrow{\Delta} Cu(s) + H_2O(g)$; copper is reduced and hydrogen is oxidized

Double-Displacement Reactions

When aqueous barium chloride reacts with sulfuric acid, it forms the white solid barium sulfate and hydrochloric acid:

$$BaCl_2(aq) + H_2SO_4(aq) \longrightarrow BaSO_4(s) + 2HCl(aq)$$

Figure 4.10 When an aqueous solution of greenish yellow molecular chlorine is added to an aqueous solution of sodium iodide, yellowish brown molecular iodine is displaced and the chlorine forms colorless aqueous sodium chloride. Molecular iodine is a shiny purple solid, which evaporates easily to form a purple gas and retains its purple color when dissolved in organic liquids. In water, however, it has a brown color.

What do these reactions have
in common?
When potassium hydroxide reacts with nitric acid, it forms water and potassium nitrate:

$$KOH(aq) + HNO_3(aq) \longrightarrow H_2O(l) + KNO_3(aq)$$

When sodium sulfide reacts with hydrochloric acid, it forms sodium chloride and hydrogen sulfide gas:

$$Na_2S(aq) + 2HCl(aq) \longrightarrow 2NaCl(aq) + H_2S(g)$$

These reactions, called **double-displacement reactions,** have two major features in common. First, in these and all double-displacement reactions, two compounds exchange ions or elements to form new compounds. Second, in all such reactions, one of the products is either a compound that will separate from the reaction mixture in some way—commonly as a solid or a gas—or a stable covalent compound, usually water, that doesn't necessarily separate from the reaction mixture. The formation of such compounds provides the driving force for the reactions—the reason why they occur.

Not all combinations of two compounds can undergo a double-displacement reaction. Consider the compounds potassium chloride and sulfuric acid, for example. If these compounds are dissolved in aqueous solution, they form the ions $K^+(aq)$, $Cl^-(aq)$, $H^+(aq)$, and $SO_4^{2-}(aq)$. When these compounds are mixed in aqueous solution, no combination of the component ions can form either a compound that will separate from the solution or a stable covalent molecule; so nothing happens. However, if we eliminate water and drop concentrated sulfuric acid onto solid potassium chloride, gaseous hydrogen chloride and solid potassium sulfate can form, so a double-displacement reaction occurs:

$$2KCl(s) + H_2SO_4(l) \longrightarrow K_2SO_4(s) + 2HCl(g)$$

Double-displacement reactions can be further classified as precipitation, gas-formation, hydrolysis, and acid–base neutralization reactions.

Precipitation Reactions   **Precipitation reactions** are those in which the reactants exchange ions to form an insoluble salt (one that does not dissolve in water). The driving force for reactions of this type is the ability of two of the ions to combine to form an insoluble solid, or **precipitate.** We can predict whether such a compound can be formed by consulting the solubility rules, some of which are listed in Table 4.2. *If a possible product is insoluble, a precipitation reaction should occur.*

Suppose we mix colorless aqueous solutions of barium chloride and sodium sulfate. The following ions would be present: $Ba^{2+}(aq)$, $Cl^-(aq)$, $Na^+(aq)$, and $SO_4^{2-}(aq)$. According to the solubility rules in Table 4.2, most sulfate salts, most sodium salts, and most chloride salts are soluble. However, barium sulfate is insoluble. Because barium ion and sulfate ion could combine to form insoluble barium sulfate, a reaction should occur, as shown in Figure 4.11.

$$BaCl_2(aq) + Na_2SO_4(aq) \longrightarrow BaSO_4(s) + 2NaCl(aq)$$

Gas-Formation Reactions   In another, similar type of reaction, the formation of an insoluble (or only slightly soluble) gas provides a driving force for the reaction. Thus, *a double-displacement reaction should occur if an insoluble gas would be formed.* All gases are soluble in water to some extent, but only a few gases are highly soluble. The common very soluble gases are $HCl(g)$ and $NH_3(g)$. All other

Table 4.2 Some Solubility Rules for Inorganic Salts in Water

Na^+, K^+, NH_4^+	Most salts of sodium, potassium, and ammonium ions are soluble.
NO_3^-	All nitrates are soluble.
SO_4^{2-}	Most sulfates are soluble. Exceptions: $BaSO_4$, $SrSO_4$, $PbSO_4$, $CaSO_4$, Hg_2SO_4, and Ag_2SO_4
Cl^-, Br^-, I^-	Most chlorides, bromides, and iodides are soluble. Exceptions: AgX, Hg_2X_2, PbX_2, and HgI_2
Ag^+	Silver salts, except $AgNO_3$, are insoluble.
O^{2-}, OH^-	Oxides and hydroxides are insoluble. Exceptions: $NaOH$, KOH, NH_4OH, $Ba(OH)_2$, and $Ca(OH)_2$ (somewhat soluble)
S^{2-}	Sulfides are insoluble. Exceptions: salts of Na^+, K^+, NH_4^+ and the alkaline-earth metal ions
CrO_4^{2-}	Most chromates are insoluble. Exceptions: salts of K^+, Na^+, NH_4^+, Mg^{2+}, Ca^{2+}, Al^{3+}, and Ni^{2+}
CO_3^{2-}, PO_4^{3-}, SO_3^{2-}, SiO_3^{2-}	Most carbonates, phosphates, sulfites, and silicates are insoluble. Exceptions: salts of K^+, Na^+, and NH_4^+

gases, generally binary covalent compounds, are sufficiently insoluble to provide a driving force if they are formed as a reaction product. For example, many sulfide salts will react with acids to form gaseous hydrogen sulfide:

$$MnS(s) + 2HCl(aq) \longrightarrow MnCl_2(aq) + H_2S(g)$$

Insoluble gases are often formed by the breakdown of an unstable double-displacement reaction product. For example, carbonates react with acids to form carbonic acid, H_2CO_3, an unstable substance that readily decomposes into water

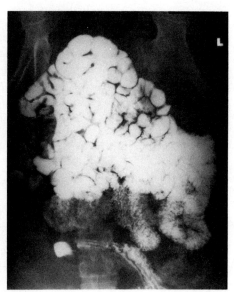

Figure 4.11 *Left:* Soluble barium chloride reacts with soluble sodium sulfate to form insoluble barium sulfate and soluble sodium chloride. *Right:* Barium compounds are extremely toxic, but barium sulfate is so insoluble that people preparing to have their digestive systems X-rayed swallow barium sulfate blended with water to make their digestive system visible to the X rays.

and carbon dioxide. When calcium carbonate reacts with hydrochloric acid, it forms calcium chloride and carbonic acid:

$$CaCO_3(s) + 2HCl(aq) \longrightarrow CaCl_2(aq) + H_2CO_3(aq)$$

The carbonic acid decomposes into water and carbon dioxide:

$$H_2CO_3(aq) \longrightarrow H_2O(l) + CO_2(g)$$

The net reaction is

$$CaCO_3(s) + 2HCl(aq) \longrightarrow CaCl_2(aq) + H_2O(l) + CO_2(g)$$

Sulfites react with acids in a similar manner to release sulfur dioxide.

Hydrolysis Reactions   **Hydrolysis** is the reaction of water with a substance to form either hydroxide (OH^-) or hydrogen (H^+) ions (or both). Often, when water reacts with a binary compound, the more metallic element forms a compound containing the hydroxide ion, and the less metallic element forms a compound containing the hydrogen ion. The resulting compounds may be covalent and not contain ions, but they will still contain the OH or H groups.

Many metal ions, especially those with a charge of 3+ or greater, undergo hydrolysis in aqueous solution. When aluminum fluoride, a white solid, is dissolved in water, a gelatinous white precipitate of aluminum hydroxide is formed:

$$AlF_3(s) + 3H_2O(l) \longrightarrow Al(OH)_3(s) + 3HF(aq)$$

Water splits into hydroxide and hydrogen ions. The hydroxide ions combine with aluminum ions to form an insoluble precipitate. Hydrogen ions remain in solution with the nonmetal anion, F^-.

Similar behavior occurs when water reacts with covalent compounds containing two nonmetals. *A hydrolysis reaction occurs if the nonmetal with a positive oxidation number can form a stable oxoacid.* When phosphorus trichloride, a clear, colorless liquid, is added to liquid water, a hydrolysis reaction produces an aqueous solution of phosphorous acid and hydrochloric acid:

$$PCl_3(l) + 3H_2O(l) \longrightarrow H_3PO_3(aq) + 3HCl(aq)$$

Acid–Base Neutralization Reactions   A **neutralization reaction** is a double-displacement reaction of an acid and a base. Recall that *acids* are compounds that will release hydrogen ions; **bases** are compounds that can neutralize acids by reacting with the hydrogen ions released by the acids. The most common bases are hydroxide and oxide compounds of the metals. Normally, *an acid reacts with a base to form a salt and water.* Neutralization reactions occur because of the formation of the very stable covalent water molecule, H_2O, from hydrogen and hydroxide ions. The most common laboratory example of an acid–base neutralization is the reaction between hydrochloric acid and sodium hydroxide:

$$HCl(aq) + NaOH(aq) \longrightarrow NaCl(aq) + H_2O(l)$$

EXAMPLE 4.6

Bauxite ore, the principal source of aluminum, contains the mineral alumina, Al_2O_3. To retrieve the alumina from the ore, the alumina is converted to a soluble salt by treatment with sulfuric acid, giving aluminum sulfate and water. Write a complete, balanced equation for this reaction and identify the reaction type.

Solution The reactants are Al_2O_3 and $H_2SO_4(aq)$. The products are $Al_2(SO_4)_3$ and H_2O. The reactants in this double-displacement reaction are a base and an acid, so it is also a neutralization reaction. Writing the skeletal equation gives

$$Al_2O_3(s) + H_2SO_4(aq) \longrightarrow Al_2(SO_4)_3(aq) + H_2O(l)$$

The complete, balanced equation is

$$Al_2O_3(s) + 3H_2SO_4(aq) \longrightarrow Al_2(SO_4)_3(aq) + 3H_2O(l)$$

Practice Problem 4.6

Calcium oxide, a white powder, is a base. When added to sulfuric acid, it forms calcium sulfate and water. Write a complete, balanced equation for this reaction and identify the reaction type.

Answer: $CaO(s) + H_2SO_4(aq) \longrightarrow CaSO_4(s) + H_2O(l)$; acid–base neutralization

---

4.3 Chemistry of Oxygen and Its Compounds

We have already seen that oxygen is the most abundant terrestrial element and is extensively involved in the chemistry of the earth and its atmosphere. We often think of oxygen in terms of the air we breathe. Molecular oxygen, O_2, makes up about 21% of the molecules in air. Oxygen also makes up 49.5% by mass of the earth's crust, primarily as silicon dioxide (quartz and other minerals) and other oxides and as various oxoanion minerals, including silicates, carbonates, sulfates, and nitrates. Oxygen is present in the water molecule in the oceans, lakes, streams, and other bodies of water. The chemistry of oxygen and some of its compounds provides a number of excellent examples of the various types of chemical reactions.

Direct Synthesis of Oxides

RULES OF REACTIVITY

Most metals react with molecular oxygen to form oxides.

Almost all the elements react directly with molecular oxygen, O_2, and all these reactions are direct synthesis reactions. The products of such reactions are summarized in Figure 4.12. Most of the metals—lithium from Group IA(1), all of Group IIA(2) except barium, and most other metals—normally form oxides, containing O^{2-}, when reacting with O_2 (see Figure 4.13):

$$4Li(s) + O_2(g) \longrightarrow 2Li_2O(s)$$
$$4Fe(s) + 3O_2(g) \longrightarrow 2Fe_2O_3(s)$$

RULES OF REACTIVITY

Most nonmetals react directly with molecular oxygen to form oxides.

All the nonmetals except the noble-gas elements and the halogens react directly with oxygen to form oxides. (Oxides of the halogens are known but cannot be made by direct synthesis.) Recall from Chapter 3 that the nonmetals usually have more than one oxidation number, separated by two units. Similarly, they usually have more than one oxide, different by one oxygen atom per nonmetal

Group 1 IA	2 IIA	3 IIIB	4 IVB	5 VB	6 VIB	7 VIIB	8	9	10 VIIIB	11 IB	12 IIB	13 IIIA	14 IVA	15 VA	16 VIA	17 VIIA	18 VIIIA
H_2O																	
Li_2O	BeO											B_2O_3	CO_2 CO	NO_2 NO		OF_2	
Na_2O_2	MgO											Al_2O_3	SiO_2	P_4O_{10} P_4O_6	SO_3 SO_2		
K_2O_2 KO_2	CaO	Sc_2O_3	TiO_2	V_2O_5	Cr_2O_3	MnO_2	Fe_2O_3	CoO	NiO	CuO	ZnO	Ga_2O_3	GeO_2	As_2O_3	SeO_2		
RbO_2 Rb_2O_2	SrO	Y_2O_3	ZrO_2	Nb_2O_5	MoO_3	TcO_2	RuO_4	RhO_2	PdO		CdO	In_2O_3	SnO_2		TeO_2		
CsO_2 Cs_2O_2 Cs_2O	BaO_2 BaO	La_2O_3	HfO_2	Ta_2O_5	WO_3	ReO_2	OsO_4	IrO_2			HgO	Tl_2O	PbO	Bi_2O_3	PoO_2		

Figure 4.12 Products of direct synthesis reactions of elements with oxygen. Note the general consistency of the formulas within each periodic group. Other oxides are also known but are not formed by direct synthesis reactions between the elements.

Carbon monoxide is a toxic, colorless gas, a product of the combustion of carbon-containing materials under conditions of insufficient oxygen.

How can we predict the products of the reaction of compounds with oxygen?

atom. The particular oxide that results from a direct synthesis reaction depends on the amount of oxygen present. With carbon, for example, either carbon monoxide or carbon dioxide can be formed, the latter requiring more oxygen:

$$2C(s) + O_2(g) \longrightarrow 2CO(g)$$
$$C(s) + O_2(g) \longrightarrow CO_2(g)$$

Phosphorus can also form two oxides by direct synthesis reactions:

$$P_4(s) + 3O_2(g) \longrightarrow P_4O_6(s)$$
$$P_4(s) + 5O_2(g) \longrightarrow P_4O_{10}(s)$$

We can predict the products of the reaction of oxygen with compounds by considering the reactions of oxygen with the component elements. Usually, the products will be the same. For example, silane, SiH_4, reacts with oxygen to form the same products silicon and hydrogen would form when reacting with oxygen:

$$Si(s) + O_2(g) \longrightarrow SiO_2(s)$$
$$2H_2(g) + O_2(g) \longrightarrow 2H_2O(l)$$
$$SiH_4(g) + 2O_2(g) \longrightarrow SiO_2(s) + 2H_2O(l)$$

RULES OF REACTIVITY

Compounds react with molecular oxygen to form the same products their component elements would form when reacting with oxygen.

Peroxides and Superoxides

Although most of the metals form oxides when they react with oxygen, some of the more reactive metals form peroxides or superoxides instead. Peroxides contain oxygen in the form of the O_2^{2-} ion; superoxides, in the form of the O_2^- ion. The oxidation number of oxygen in peroxides is -1, whereas its oxidation number in

Figure 4.13 Iron exposed to air and moisture rusts, forming Fe_2O_3, which flakes off and exposes more metal. Aluminum is widely used as a structural metal where strength is not required because it does not corrode like iron. It forms an oxide coating, but the coating does not flake off, and so no new metal is exposed to agents of corrosion. However, if the aluminum is coated with mercury, the oxide coating no longer sticks to the metal, and the metal deteriorates rapidly in air. The white powder on and near the aluminum foil is aluminum oxide.

superoxides is $-\frac{1}{2}$. The most active metals, rubidium and cesium, normally form superoxides:

$$Rb(s) + O_2(g) \longrightarrow RbO_2(s)$$

The somewhat less active metals, sodium and barium, form peroxides:

$$2Na(s) + O_2(g) \longrightarrow Na_2O_2(s)$$

Potassium is intermediate between these elements in activity and may form either the peroxide or the superoxide, depending on the amount of oxygen with which it is reacted. With sufficient oxygen, the superoxide is formed:

$$K(s) + O_2(g) \longrightarrow KO_2(s)$$

The peroxide is formed if less oxygen is present:

$$2K(s) + O_2(g) \longrightarrow K_2O_2(s)$$

Hydration Reactions

In Figure 4.2, we saw that direct synthesis reactions could involve elements or compounds. An example of a direct synthesis involving compounds is hydration, which involves the oxide water. A **hydration reaction** involves the addition of water molecules to a compound, giving a new form of the compound called a *hydrate*. An example of a hydrate is the blue compound copper(II) sulfate pentahydrate (Figure 4.14). This compound can be formed by the addition of water to anhydrous copper(II) sulfate, a white solid:

$$CuSO_4(s) + 5H_2O(l) \longrightarrow CuSO_4 \cdot 5H_2O(s)$$

Compounds that absorb water to form hydrates are said to be **hygroscopic.** On exposure to water vapor, a hygroscopic compound spontaneously takes on water to form a hydrate. A useful drying agent in the laboratory is anhydrous calcium chloride, $CaCl_2(s)$, shown in Figure 4.15. This compound absorbs two moles of water per mole of $CaCl_2$ to give the hydrated compound $CaCl_2 \cdot 2H_2O(s)$.

$$CaCl_2(s) + 2H_2O(l) \longrightarrow CaCl_2 \cdot 2H_2O(s)$$

Figure 4.14 Hydrated copper sulfate is blue, but anhydrous copper sulfate is white. When the anhydrous form is exposed to a liquid, such as alcohol, it turns blue if any water is present, as shown in the center of the white $CuSO_4$.

Figure 4.15 Some substances, such as calcium chloride, have such an affinity for water that they remove water from moist air. Calcium chloride *(left)* first forms a hydrate *(center)*, then continues absorbing water until it forms a solution *(right)*.

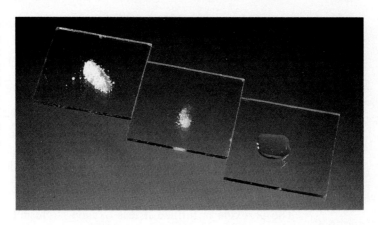

Acidic and Basic Anhydrides

Water and metal oxides react to form bases. The simplest of these, as we saw earlier, are substances containing the hydroxide ion. For this reason, metal oxides are called **basic anhydrides.**

$$\underset{\text{basic anhydride}}{Na_2O(s)} + H_2O(l) \longrightarrow \underset{\text{base}}{2NaOH(aq)}$$

RULES OF REACTIVITY

Metal oxides react with water to form hydroxide bases.

Water and nonmetal oxides react to form acids, so nonmetal oxides are called **acidic anhydrides.**

$$\underset{\text{acidic anhydride}}{SO_2(g)} + H_2O(l) \longrightarrow \underset{\text{acid}}{H_2SO_3(aq)}$$

RULES OF REACTIVITY

Nonmetal oxides react with water to form oxoacids.

Some of the oxides show both acidic and basic behavior: the oxides can release either H^+ or OH^- when water is added, depending on what else is present for these ions to react with. Such oxides are said to be **amphoteric.** Figure 4.16 shows the location of acidic and basic anhydrides and amphoteric oxides in the periodic table.

Acid–Base Reactions of Oxides

Oxides are the anhydrides of acids and bases, and they exhibit the chemical behavior of acids and bases. Metal oxides, like the hydroxides they form when reacting with water, will react with acids in a neutralization reaction to give a salt and water:

RULES OF REACTIVITY

Acids neutralize bases to form salts and water. Metal oxides neutralize nonmetal oxides to form salts.

$$Fe_2O_3(s) + 6HNO_3(aq) \longrightarrow 2Fe(NO_3)_3(aq) + 3H_2O(l)$$
$$Fe(OH)_3(s) + 3HNO_3(aq) \longrightarrow Fe(NO_3)_3(aq) + 3H_2O(l)$$

Conversely, nonmetal oxides, like the acids they form when reacting with water, will react with bases to give a salt and water:

$$SO_3(g) + 2NaOH(aq) \longrightarrow Na_2SO_4(aq) + H_2O(l)$$
$$H_2SO_3(aq) + 2KOH(aq) \longrightarrow K_2SO_3(aq) + 2H_2O(l)$$

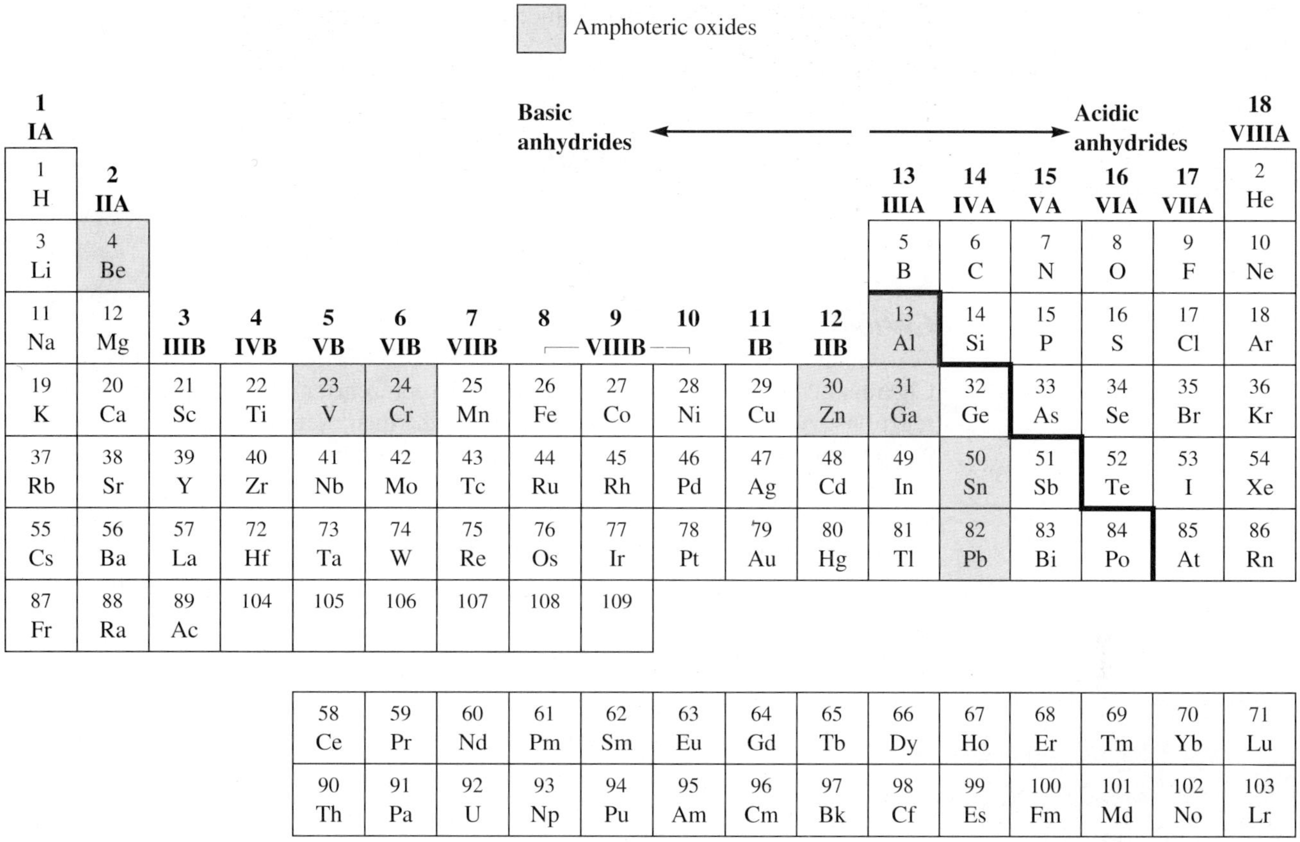

Figure 4.16 Elements whose oxides are acidic anhydrides are found on the right side of the periodic table. These are mostly nonmetals. Elements on the left side of the periodic table—the metals—mostly form basic anhydrides. A few of the metals, especially those near the metalloids, form amphoteric oxides.

It is also possible to carry out an acid–base reaction between two oxides; metal oxides will react with nonmetal oxides to form ionic compounds:

$$Na_2O(s) + SO_2(g) \longrightarrow Na_2SO_3(s)$$

If you examine these reactions carefully, you can see that the reactions of acids, bases, and their anhydrides produce the same products, differing only in the amount of water that forms:

$$Ca(OH)_2(aq) + H_2CO_3(aq) \longrightarrow CaCO_3(s) + 2H_2O(l)$$
$$CaO(s) + H_2CO_3(aq) \longrightarrow CaCO_3(s) + H_2O(l)$$
$$Ca(OH)_2(aq) + CO_2(g) \longrightarrow CaCO_3(s) + H_2O(l)$$
$$CaO(s) + CO_2(g) \longrightarrow CaCO_3(s)$$

The oxides of some of the elements—those identified in Figure 4.16 as being amphoteric oxides—can behave as either acidic anhydrides or basic anhydrides, depending on the other reactant. One of these amphoteric oxides is aluminum oxide, which reacts with hydrochloric acid, acting as a base:

$$Al_2O_3(s) + 6HCl(aq) \longrightarrow 2AlCl_3(aq) + 3H_2O(l)$$

In a reaction with sodium hydroxide, aluminum oxide acts as an acid:

$$Al_2O_3(s) + 2NaOH(aq) + 3H_2O(l) \longrightarrow 2NaAl(OH)_4(aq)$$

4.4 Quantitative Relationships in Chemical Reactions

In what ways can chemical equations be put to practical use?

Chemical equations are important tools for representing and controlling chemical reactions. Also important is a knowledge of the quantitative relationships in chemical reactions—the amounts of reactants and products involved. For example, chemists prepare large quantities of phosphoric acid for conversion to phosphate salts. These compounds are found in fertilizers and detergents, and they give many soft drinks their acidic taste. Phosphoric acid may be prepared by reaction of phosphate rock (Figure 4.17), which is largely calcium phosphate, with sulfuric acid:

$$Ca_3(PO_4)_2(s) + 3H_2SO_4(aq) \longrightarrow 3CaSO_4(s) + 2H_3PO_4(aq)$$

Using more of the reactants than necessary to produce the desired amount of phosphoric acid would be wasteful. Using excess sulfuric acid could also lead to a contaminated product. Given the amount of phosphoric acid needed, chemists need to know exactly how much phosphate rock and sulfuric acid to use in the reaction.

The study of quantitative relationships in chemical reactions is called **stoichiometry.** It deals with what amounts of reactants must be combined to obtain the desired amount of product, what amount of product results from given amounts of reactants, and related calculations.

Usually we want to measure out the amounts of reactants or products by their mass, or possibly their volume, if they are liquids. We might, for example, want to know how much sulfuric acid and phosphate rock we need to prepare 1000 kg of phosphoric acid. We know that the total mass of the reactants must equal the total mass of the products, but how do we apportion the mass among these species? Chemical equations provide us with information on the relative amounts of reactants and products in units of moles, but we cannot measure anything directly in moles. To obtain information on the relative masses of each reactant and product in a chemical equation, we will have to perform a number of numerical conversions between mass and moles.

Figure 4.17 Some phosphoric acid is made by reacting elemental phosphorus with oxygen gas and then with water. However, most phosphorus is recovered from phosphate rock and other phosphate minerals. Phosphate rock was deposited in the ancient oceans and originated from urine, animal skeletons, and guano from fish-eating birds and bats. This is a somewhat bizarre link in the food chain.

Mass–Mole Conversions

Recall that to obtain the amount of material in units of moles from the mass in units of grams (or vice versa), we must use the molar mass of the material as a conversion factor. The molar mass is obtained as the sum of the molar masses of the component elements, each first multiplied by the number of atoms of that element in the compound. For example, suppose we want to know the number of moles of

Molding Our Environment with Chemical Reactions

In the creation of buildings, as in many of the other ways in which we alter our environment, we use a variety of chemical reactions.

Earthquakes cause the ground to shift, buildings to collapse, and freeway overpasses to break apart. But why do these buildings and freeways stand up in the first place? In the creation of buildings, as in many of the other ways in which we alter our environment, we use a variety of chemical reactions.

A gray powder containing calcium silicates, calcium aluminates, and calcium ferrates is mixed first with water, then with large particles of silicon dioxide. This mixture is poured into forms, where it slowly hardens into any desired shape—a dam, a sidewalk, or a skyscraper. The gray powder is cement and the final hardened mixture is concrete. The changes described involve chemical reactions like those we have discussed in this chapter.

Cement is made from a mixture of limestone and clay. Limestone is calcium carbonate, and clay contains a mixture of aluminum silicates and other minerals. These materials are crushed and heated in a kiln. The limestone decomposes to form quicklime, or calcium oxide:

$$CaCO_3(s) \xrightarrow{\Delta} CaO(s) + CO_2(g)$$

The calcium oxide, a basic anhydride, reacts with aluminum and silicon oxides, which are acidic anhydrides. These acid–base reactions of oxides begin once the mixture is heated to about 1500°C, at which point it begins to melt and form ionic salts:

$$CaO(l) + Al_2O_3(l) \xrightarrow{\Delta} CaAl_2O_4(l)$$

$$2CaO(l) + SiO_2(l) \xrightarrow{\Delta} Ca_2SiO_4(l)$$

These salts can undergo further acid–base reactions:

$$2CaO(l) + CaAl_2O_4(l) \xrightarrow{\Delta} Ca_3Al_2O_6(l)$$

$$CaO(l) + Ca_2SiO_4(l) \xrightarrow{\Delta} Ca_3SiO_5(l)$$

Any iron oxides present in the clay undergo similar reactions. The resulting mixture of calcium silicates, aluminates, and ferrates is cooled and comes out of the kiln as lumps called clinker. The clinker is ground up and mixed with small amounts of gypsum (calcium sulfate). This mixture, called portland cement, contains various proportions

Figure 4.18 At high magnification, it is possible to see the fibers growing from particles of portland cement. These fibers strengthen the concrete.

of four primary compounds: Ca_3SiO_5, Ca_2SiO_4, $Ca_3Al_2O_6$, and $Ca_4Al_2Fe_2O_{10}$.

The setting of concrete consists of two steps. The first step, occurring within a few hours, is solidification, in which silicate gels form around the sand and rocks in the mixture. The tricalcium silicate undergoes hydration and hydrolysis to form a gelatinous hydrated silicate on the surface of the sand and rock particles:

$$2Ca_3SiO_5(s) + 6H_2O(l) \longrightarrow$$
$$Ca_3Si_2O_7 \cdot 3H_2O(s) + 3Ca(OH)_2(s)$$

These gels form a protective coating around the cement and prevent complete hydration of the cement particles. Dicalcium silicate, Ca_2SiO_4, hydrates and hydrolyzes in a similar manner. Tricalcium aluminate also hydrates, forming $Ca_3Al_2O_6 \cdot 6H_2O$, but it does not hydrolyze, so no $Ca(OH)_2$ is formed in this reaction.

The next step is hardening, which occurs over long time periods, even years, as dry cement particles remove water from the hydrated silicate gels. In addition, the calcium hydroxide formed in the solidification process absorbs carbon dioxide from air and converts some of the calcium back to calcium carbonate. The process of hardening is not fully understood, but as water is removed from the silicate gels, the particles develop fibers that interlock, as shown in Figure 4.18, and bond the particles together.

(continued)

Questions for Discussion

1. What other chemical reactions might be important when constructing buildings or bridges?
2. Why does cement not always have the same composition?
3. Why are iron bars, called rebar, sometimes embedded in cement?

4. What happens to the carbon dioxide produced when limestone is converted to quicklime? Should it be released into the atmosphere or should it be captured? Find some newspaper or magazine articles about the greenhouse effect and discuss this issue with your classmates.

calcium phosphate, $Ca_3(PO_4)_2$, in a 126-g sample of the compound. We need the molar mass:

Molar mass of $Ca_3(PO_4)_2$ = (3 × molar mass of Ca)
+ (2 × molar mass of P) + (8 × molar mass of O) = 310.18 g/mol

We can use this molar mass to convert the mass of calcium phosphate into the corresponding number of moles:

$$\text{Moles } Ca_3(PO_4)_2 = 126\text{ g} \times 1\text{ mol}/310.18\text{ g} = 0.406\text{ mol}$$

Conversely, if the number of moles is known, then multiplying by the molar mass will convert this quantity to the corresponding mass. Conversions of this sort, introduced in Chapter 2, are readily carried out by use of dimension analysis.

EXAMPLE 4.7

A sample of iron ore consisting of Fe_2O_3 weighs 542 g. How many moles of Fe_2O_3 are in this sample?

Solution   The molar mass of iron(III) oxide is the sum of the molar masses of the component elements, each multiplied by the subscript in the formula. The value is 159.7 g/mol. The conversion from mass to moles is carried out by use of this molar mass:

$$\text{mol } Fe_2O_3 = 542\text{ g } Fe_2O_3 \times \frac{1\text{ mol } Fe_2O_3}{159.7\text{ g } Fe_2O_3} = 3.39\text{ mol}$$

Practice Problem 4.7

What mass of iron(III) oxide is needed to provide 2.35 moles of iron(III) oxide?

Answer: 375 g

Mole–Mole Conversions

Once we know the number of moles of a substance involved in a reaction, the balanced equation for the reaction provides the information needed to convert the number of moles of that substance to the corresponding number of moles of any

Why must we use a balanced equation in stoichiometric calculations?

other substance involved in that reaction. The coefficients in the balanced equation give the relative number of moles (or molecules) of each substance. For example, consider the reaction for the preparation of phosphoric acid:

$$Ca_3(PO_4)_2(s) + 3H_2SO_4(aq) \longrightarrow 3CaSO_4(s) + 2H_3PO_4(aq)$$

For every mole of calcium phosphate that reacts, 3 mol of sulfuric acid must react, and 3 mol of calcium sulfate and 2 mol of phosphoric acid will be produced. Thus, if 1.5 mol of calcium phosphate reacts, then 4.5 mol of sulfuric acid reacts with it to produce 4.5 mol of calcium sulfate and 3.0 mol of phosphoric acid. Again, dimensional analysis allows us to carry out conversions of this sort with ease.

EXAMPLE 4.8

Consider the reaction of iron(III) oxide with phosphoric acid:

$$Fe_2O_3(s) + 2H_3PO_4(aq) \longrightarrow 2FePO_4(aq) + 3H_2O(l)$$

If we wish to produce 1.40 mol of iron(III) phosphate, how many moles of iron(III) oxide do we need?

Solution   The coefficients in the equation indicate that two mol of $FePO_4$ are produced from each mole of Fe_2O_3: 2 mol $FePO_4$/1 mol Fe_2O_3. This mole ratio can be used to convert the number of moles of one of these substances into the corresponding number of moles of the other.

$$\text{Moles } Fe_2O_3 = 1.40 \text{ mol } FePO_4 \times \frac{1 \text{ mol } Fe_2O_3}{2 \text{ mol } FePO_4} = 0.700 \text{ mol}$$

Practice Problem 4.8

In the reaction of iron(III) oxide with phosphoric acid, how many moles of water are produced from the reaction of 1.58 mol of phosphoric acid?

Answer: 2.37 mol

Mass–Mass Conversions

When working with chemical reactions, we often want to measure out a known mass of the reactants and calculate the mass of products that will be produced. We might also want to produce a specific mass of product and need to calculate the appropriate mass of reactants. If we know the mass of one substance (A) involved in a chemical reaction, we can calculate the mass of any other substance (B) by following this sequence of conversions:

$$\text{Mass of A} \longrightarrow \text{moles of A} \longrightarrow \text{moles of B} \longrightarrow \text{mass of B}$$

As we have just seen, we can convert mass to or from moles by using the molar mass as the conversion factor, and we can convert moles of one substance to moles of another substance by using the coefficients in the balanced equation as a conversion factor. Normally, the amount of material is measured in mass rather than in moles, so this sequence of conversions is common in stoichiometric calculations. The calculation usually involves three steps:

1. Convert the mass of one substance into the number of moles of that substance, using the molar mass for the conversion.
2. Convert the moles of one substance into moles of a second, related substance, using coefficients in the balanced equation for the conversion.
3. Convert moles of the second substance into its mass, using its molar mass for the conversion.

Mass Conversions in a Single Reaction   Once again consider the following reaction:

$$Ca_3(PO_4)_2(s) + 3H_2SO_4(aq) \longrightarrow 3CaSO_4(s) + 2H_3PO_4(aq)$$

If 46.0 g of calcium phosphate reacts, how much phosphoric acid will be produced? We solve this problem by following the three steps just listed.

Step 1. Convert mass of calcium phosphate to moles of calcium phosphate, using the molar mass of calcium phosphate, 310.18 g/mol:

$$mol\ Ca_3(PO_4)_2 = 46.0\ \cancel{g\ Ca_3(PO_4)_2} \times 1\ mol\ Ca_3(PO_4)_2/310.18\ \cancel{g\ Ca_3(PO_4)_2}$$
$$= 0.148\ mol$$

Step 2. Using the coefficients from the balanced equation, convert the number of moles of calcium phosphate (1 mol) to the corresponding number of moles of phosphoric acid (2 mol):

$$mol\ H_3PO_4 = 0.148\ \cancel{mol\ Ca_3(PO_4)_2} \times 2\ mol\ H_3PO_4/1\ \cancel{mol\ Ca_3(PO_4)_2} = 0.296\ mol$$

Step 3. Convert moles of phosphoric acid to mass of phosphoric acid, using the molar mass of 98.00 g/mol:

$$Mass\ H_3PO_4 = 0.296\ \cancel{mol\ H_3PO_4} \times 98.00\ g\ H_3PO_4/1\ \cancel{mol\ H_3PO_4} = 29.0\ g$$

This same sequence of conversions (or variations on it) is used in all stoichiometric calculations.

EXAMPLE 4.9

One step in the commercial production of iron is the reaction of hematite, Fe_2O_3, with carbon monoxide:

$$Fe_2O_3(s) + 3CO(g) \xrightarrow{\Delta} 2Fe(l) + 3CO_2(g)$$

How much iron (in grams) could be obtained from 50.0 g of hematite in this process?

Solution   Before we begin, we should note that the equation is balanced—always check this first to be sure your solution will be right. We also need to know the relevant molar masses: 55.85 g/mol of Fe and 159.7 g/mol of Fe_2O_3. Now we must carry out the following sequence of conversions, with the steps identified in brackets:

$$Mass\ of\ Fe_2O_3 \xrightarrow{[1]} moles\ of\ Fe_2O_3 \xrightarrow{[2]} moles\ of\ Fe \xrightarrow{[3]} mass\ of\ Fe$$

This series of conversions can be set up as a single equation, just as it appears above. However, in this first example, we will develop the answer by introducing each step in the conversion one at a time. In later examples, we will proceed right to the final setup.

Step 1: Mass of Fe_2O_3 is converted to moles of Fe_2O_3, with molar mass as the conversion factor:

$$\text{mol } Fe_2O_3 = 50.0 \text{ g } Fe_2O_3 \times \frac{1 \text{ mol } Fe_2O_3}{159.7 \text{ g } Fe_2O_3}$$

Step 2: Moles of Fe_2O_3 are converted to moles of Fe, with the coefficients in the balanced equation providing the conversion factor:

$$\text{mol Fe} = 50.0 \text{ g } Fe_2O_3 \times \frac{1 \text{ mol } Fe_2O_3}{159.7 \text{ g } Fe_2O_3} \times \frac{2 \text{ mol Fe}}{1 \text{ mol } Fe_2O_3}$$

Step 3: Moles of Fe are converted to mass of Fe:

$$\text{g Fe} = 50.0 \text{ g } Fe_2O_3 \times \frac{1 \text{ mol } Fe_2O_3}{159.7 \text{ g } Fe_2O_3} \times \frac{2 \text{ mol Fe}}{1 \text{ mol } Fe_2O_3} \times \frac{55.85 \text{ g Fe}}{1 \text{ mol Fe}} = 35.0 \text{ g}$$

Practice Problem 4.9

Copper(II) oxide can be converted to copper metal by reaction with heated gaseous hydrogen:

$$CuO(s) + H_2(g) \longrightarrow Cu(s) + H_2O(g)$$

What mass of copper would be produced if 22.0 g of copper(II) oxide reacted with hydrogen gas?

Answer: 17.6 g

In addition to determining how much product is formed from a given amount of reactant, we can calculate how much of a second reactant is needed. The calculation follows exactly the same procedure as that in Example 4.9. In the same way, we can calculate the amount of a reactant required to produce a specific amount of product.

Mass Conversions in Sequences of Reactions The same types of conversions can be carried out in processes that involve two or more reactions. Coefficients from more than one equation must be used to obtain the mole ratio of two substances. But as long as we use the coefficients from one equation at a time, the series of conversions should not become confusing. We simply introduce another mole–mole conversion for each additional reaction in the sequence. For example, if we make phosphoric acid by our familiar process,

$$Ca_3(PO_4)_2(s) + 3H_2SO_4(aq) \longrightarrow 3CaSO_4(s) + 2H_3PO_4(aq)$$

and then convert the phosphoric acid to sodium phosphate,

$$H_3PO_4(aq) + NaOH(aq) \longrightarrow Na_3PO_4(aq) + 3H_2O(l)$$

we can calculate how much sodium phosphate is produced from a given amount of calcium phosphate by this series of conversions:

Mass of calcium phosphate $\xrightarrow{[1]}$ moles of calcium phosphate $\xrightarrow{[2a]}$ moles of phosphoric acid $\xrightarrow{[2b]}$ moles of sodium phosphate $\xrightarrow{[3]}$ mass of sodium phosphate

EXAMPLE 4.10

The carbon monoxide used in the reaction with hematite is obtained in part by the burning of coke (a form of carbon):

$$2C(s) + O_2(g) \longrightarrow 2CO(g)$$

If 25.0 g C is converted to CO, which is then reacted with an excess of Fe_2O_3, how much iron is produced?

Solution   We must use the equations for formation of CO and for reaction of CO with hematite:

$$2C(s) + O_2(g) \longrightarrow 2CO(g) \qquad \text{(used in steps 1 and 2a below)}$$
$$Fe_2O_3(s) + 3CO(g) \longrightarrow 2Fe(l) + 3CO_2(g) \qquad \text{(used in steps 2b and 3 below)}$$

The conversion sequence is

Mass of C $\xrightarrow{[1]}$ moles of C $\xrightarrow{[2a]}$ moles of CO $\xrightarrow{[2b]}$ moles of Fe $\xrightarrow{[3]}$ mass of Fe

The molar mass of carbon is used as the conversion factor in step 1. The coefficients from the first equation are used for the conversion factor in step 2a; coefficients from the second equation are used in step 2b. Finally, the molar mass of iron is used to convert moles of iron to mass of iron in step 3.

$$\text{g Fe} = 25.0 \text{ g C} \times \frac{1 \text{ mol C}}{12.01 \text{ g C}} \times \frac{2 \text{ mol CO}}{2 \text{ mol C}} \times \frac{2 \text{ mol Fe}}{3 \text{ mol CO}} \times \frac{55.85 \text{ g Fe}}{1 \text{ mol Fe}} = 77.5 \text{ g}$$

Practice Problem 4.10

Hydrogen is produced by the reaction of carbon monoxide with steam:

$$CO(g) + H_2O(g) \longrightarrow CO_2(g) + H_2(g)$$

If 155 g of carbon monoxide is reacted with water to form hydrogen gas and if the hydrogen is completely reacted with copper(II) oxide, what mass of copper metal will be formed?

Answer: 352 g

Limiting Reactants

You will often find that not all the substances present in a chemical reaction mixture react completely. Usually, at least one substance is present in excess of the amount required, and some of this substance is left over after the reaction is completed. In such a case, the reactant that is not present in excess and that does react completely is called the **limiting reactant,** or **limiting reagent.** The concept can be illustrated by an analogy. Suppose we are filling egg cartons with eggs. If each carton holds 12 eggs, and we have eight cartons and 60 eggs, we will run out of eggs before all the cartons are filled. The 60 eggs will fill five cartons, leaving three empty. The eggs are the limiting reactant in this analogy.

The limiting reactant, then, is the reactant that determines how much product will be formed. The amount of product that, in principle, can be formed from a complete reaction of a given amount of the limiting reactant is called the **theoretical yield.** To predict the amount of product that will result from a reaction, we need to use the mass of the limiting reactant in our calculation of the theoretical yield, which follows these steps:

Mass of limiting reactant $\longrightarrow$ moles of limiting reactant $\longrightarrow$
                           moles of product $\longrightarrow$ mass of product (theoretical yield)

To calculate the theoretical yield, we must know which reactant is limiting. To find this out, we calculate the number of moles of each reactant present and compare the actual mole ratio with that required in the reaction, as in the following example.

EXAMPLE 4.11

The reaction of hematite with carbon monoxide was introduced in Example 4.9. When this reaction involves 40.0 g Fe_2O_3 and 15.0 g CO, which species is the limiting reactant? How much Fe will be produced? How much of the reactant present in excess will remain?

Solution   Recall that the equation is

$$Fe_2O_3(s) + 3CO(g) \longrightarrow 2Fe(l) + 3CO_2(g)$$

First, we must calculate the moles of each reactant required according to the balanced equation:

$$\text{Moles } Fe_2O_3 = 40.0 \text{ g } Fe_2O_3 \times \frac{1 \text{ mol } Fe_2O_3}{159.7 \text{ g } Fe_2O_3} = 0.250 \text{ mol}$$

$$\text{Moles CO} = 15.0 \text{ g CO} \times \frac{1 \text{ mol CO}}{28.0 \text{ g CO}} = 0.536 \text{ mol}$$

Next, we calculate the mole ratio of CO to Fe_2O_3:

$$\text{Required mole ratio} = \frac{3 \text{ mol CO}}{1 \text{ mol } Fe_2O_3} = 3 \text{ mol CO/mol } Fe_2O_3$$

Then we calculate the actual mole ratio of CO to Fe_2O_3:

$$\text{Actual mole ratio} = \frac{0.536 \text{ mol CO}}{0.250 \text{ mol } Fe_2O_3} = 2.14 \text{ mol CO/mol } Fe_2O_3$$

Comparing the ratios, we can see that the actual mole ratio of CO to Fe_2O_3 (2.14) is less than the required mole ratio (3). Thus, there will not be enough CO to react with all the Fe_2O_3. The CO is the limiting reactant, and some Fe_2O_3 will remain when the reaction is complete. Note that we could just as well have compared the ratio of Fe_2O_3 to CO. In that case, the actual mole ratio would have been 0.466 and the desired mole ratio would have been 0.333. Comparison of the two would have shown that Fe_2O_3 was present in excess.

    Now we can calculate the theoretical yield of iron based on the amount of CO present, following the same procedure as in Example 4.9:

Mass of CO $\longrightarrow$ moles of CO $\longrightarrow$ moles of Fe $\longrightarrow$ mass of Fe

$$\text{g Fe} = 0.536 \text{ mol CO} \times \frac{2 \text{ mol Fe}}{3 \text{ mol CO}} \times \frac{55.85 \text{ g Fe}}{1 \text{ mol Fe}} = 20.0 \text{ g}$$

Once we know how much CO reacts, we can follow a similar procedure to determine the amount of Fe_2O_3 that reacted:

Mass of CO $\longrightarrow$ moles of CO $\longrightarrow$ moles of Fe_2O_3 $\longrightarrow$ mass of Fe_2O_3

$$\text{g } Fe_2O_3 = 0.536 \text{ mol CO} \times \frac{1 \text{ mol } Fe_2O_3}{3 \text{ mol CO}} \times \frac{159.7 \text{ g } Fe_2O_3}{1 \text{ mol } Fe_2O_3} = 28.5 \text{ g}$$

(continued)

The mass of Fe_2O_3 remaining is the original mass of Fe_2O_3 less the mass that reacted:

$$\text{g } Fe_2O_3 = 40.0 \text{ g} - 28.5 \text{ g} = 11.5 \text{ g}$$

Practice Problem 4.11

If 132.0 g CuO is reacted with 5.000 g H_2, what mass of Cu would be prepared by the following reaction?

$$CuO(s) + H_2(g) \longrightarrow Cu(s) + H_2O(g)$$

Answer: 105.4 g

Why would the amount of product obtained from a chemical reaction be less than the theoretical yield?

The result of an actual chemical reaction is rarely the theoretical yield. Because of inefficiency in handling materials, losses from side reactions (unintended reactions that give other products), solubility of the product, and sometimes other factors, the amount of product obtained is often less than expected. In other words, the **actual yield** is usually less than the theoretical yield. This relationship is often expressed in terms of the **percent yield,** which gives the actual yield as a percentage of the theoretical yield:

$$\text{Percent yield} = \frac{\text{actual yield}}{\text{theoretical yield}} \times 100\%$$

Consider again the reaction of calcium phosphate with sulfuric acid. Starting with 46.0 g of calcium phosphate, we calculated a theoretical yield of 29.0 g. Suppose we actually carried out this process and recovered only 24.5 g of phosphoric acid. The percent yield would be

$$\text{Percent yield} = \frac{24.5 \text{ g}}{29.0 \text{ g}} \times 100\% = 84.5\%$$

EXAMPLE 4.12

Sugar cane produces sucrose ($C_{12}H_{22}O_{11}$) from carbon dioxide and water through photosynthesis:

$$12CO_2(g) + 11H_2O(l) \longrightarrow C_{12}H_{22}O_{11}(s) + 12O_2(g)$$

If the reaction consumes 30.0 g of carbon dioxide and produces 6.84 g of sucrose, what is the percent yield of sucrose?

Solution   First, calculate the theoretical yield, following the procedure described previously:

Mass of CO_2 $\longrightarrow$ moles of CO_2 $\longrightarrow$ moles of sucrose $\longrightarrow$ mass of sucrose

$$\text{g sucrose} = 30.0 \text{ g } CO_2 \times \frac{1 \text{ mol } CO_2}{44.0 \text{ g } CO_2} \times \frac{1 \text{ mol sucrose}}{12 \text{ mol } CO_2} \times \frac{342 \text{ g sucrose}}{1 \text{ mol sucrose}} = 19.4 \text{ g}$$

Because the actual yield is 6.84 g of sucrose, the percent yield is

$$\text{Percent yield} = \frac{6.84 \text{ g sucrose}}{19.4 \text{ g sucrose}} \times 100\% = 35.3\%$$

Practice Problem 4.12

If the photosynthesis reaction consumes 20.0 g of water and produces 2.45 g of sucrose, what is the percent yield of sucrose?

Answer: 7.10%

4.5   Molarity and Solution Stoichiometry

Because chemical reactions between solids are normally slow, we usually carry out such reactions by dissolving the materials in a liquid. Most chemical reactions carried out in the laboratory or used in industrial processes occur in liquid solution.

Why is it appropriate to call air a solution?

A *solution* is any homogeneous mixture dispersed on a molecular or ionic scale. (Recall from Chapter 1 that a mixture consists of two or more substances and may vary in composition from sample to sample. Within a given sample of a homogeneous mixture, however, the composition is uniform.) The substance being dissolved is called the *solute* (usually present in lesser amount), and the substance doing the dissolving is called the *solvent* (usually present in greater amount). Salt water, for example, is a liquid solution in which sodium chloride, the solute, is dissolved in water, the solvent.

Molarity

There are a variety of ways to express the **concentration** of a solution—that is, the relative amounts of solute and solvent. One of the most common uses the concentration unit of **molarity (M).** The molarity of a solution is the number of moles of solute dissolved in one liter of solution:

$$\text{Molarity} = \text{moles of solute/liters of solution}$$

Suggest some other ways for expressing the concentration of a solution.

A solution having a volume of exactly 1 L and containing 58.44 g (1.000 mol) of sodium chloride is a 1.000 M sodium chloride solution. If the liter of solution had half this much sodium chloride, 29.22 g, the solution would have a concentration of 0.500 M. Solutions of known molarity are usually prepared in volumetric flasks, as shown in Figure 4.19.

It is important to remember that molarity involves liters of solution, not liters of solvent. Depending on the concentration of the solution, this difference may or may not be significant. **Dilute solutions** contain only a small amount of solute. The volume of such solutions is essentially the same as the volume of solvent. However, the difference can be significant for **concentrated solutions,** those containing a large amount of solute. The distinction between dilute and concentrated is somewhat nebulous, meaning different things for different substances. However, as a general rule, we can take *dilute* to mean concentrations of less than 1 M and *concentrated* to mean higher concentrations.

To calculate molarity, we first calculate the number of moles of solute contained in the solution, then divide by the volume in liters of that solution. We can calculate moles of solute from the mass of solute in the usual manner, using the molar mass of solute as a conversion factor.

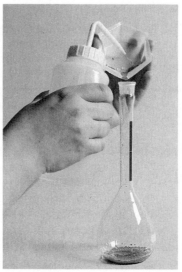
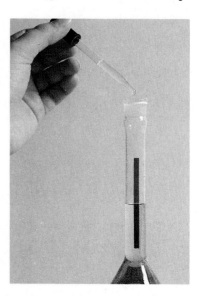

Figure 4.19 Solutions of known molarity are usually prepared in volumetric flasks, which are calibrated to hold a specific volume. *Left:* To prepare 250 mL of a 0.100 *M* solution of $CuSO_4 \cdot 5H_2O$, we would weigh 6.24 g (0.0250 mol) of the solute and add it to the flask. *Center:* We would then add some solvent (water) and dissolve the solute by swirling the flask. *Right:* We would continue to add solvent with occasional swirling until the bottom of the liquid surface just matches the marked line on the neck of the flask, giving a total volume of solution of exactly 250 mL. Finally, we would insert the stopper and invert the flask several times to mix the contents completely.

EXAMPLE 4.13

A solution is prepared from 17.0 g NaCl dissolved in sufficient water to give 150.0 mL of solution. What is the molarity of this solution?

Solution We know the volume of solution, but we must find the moles of solute before we can calculate the molarity. The moles of sodium chloride can be calculated from its mass by use of the molar mass, 58.44 g/mol. We then divide by the volume in liters to get molarity.

$$\text{Mass NaCl} \longrightarrow \text{moles NaCl} \longrightarrow \text{molarity}$$

$$\text{Molarity} = 17.0 \text{ g NaCl} \times \frac{1 \text{ mol NaCl}}{58.4 \text{ g NaCl}} \times \frac{1}{150.0 \text{ mL}} \times \frac{1000 \text{ mL}}{1 \text{ L}} = 1.94 \ M$$

Practice Problem 4.13

A solution is prepared from 22.5 g KNO_3 dissolved in sufficient water to give 250.0 mL of solution. What is the molarity of this solution?

Answer: 0.890 *M*

If we know the volume and the molarity of a solution, we can calculate how many moles of solute it contains. Thus, we can also calculate the mass of solute in that volume of solution. This is the reverse of the process just outlined.

EXAMPLE 4.14

Seawater is about 0.86 M in NaCl. What mass of NaCl (in grams) is contained in 100.0 mL of seawater having this concentration?

Solution To determine mass of solute, first determine moles of solute. Multiply the molarity by the volume of solution to obtain the moles of solute. Then convert moles to mass, using the molar mass as a conversion factor.

$$\text{Molarity} \longrightarrow \text{moles NaCl} \longrightarrow \text{mass NaCl}$$

$$\text{g NaCl} = 100.0 \; \cancel{\text{mL}} \times \frac{1 \; \cancel{\text{L}}}{1000 \; \cancel{\text{mL}}} \times \frac{0.86 \; \cancel{\text{mol NaCl}}}{1 \; \cancel{\text{L solution}}} \times \frac{58.4 \; \text{g NaCl}}{1 \; \cancel{\text{mol NaCl}}} = 5.0 \; \text{g}$$

Practice Problem 4.14

Seawater contains 0.0524 M $MgCl_2$. What mass of $MgCl_2$ is contained in 500.0 mL of seawater?

Answer: 2.49 g

Finally, we can calculate the volume of solution required to make a solution of a specific concentration from a given amount of solute. Again, we need use only conversions arising from the definition of molarity and the gram-to-mole conversion using molar mass.

EXAMPLE 4.15

Suppose you wish to prepare a 0.150 M solution of NaCl, starting with 40.0 g NaCl. What will be the volume of the solution?

Solution Convert the mass of solute to moles of solute, using the molar mass. Then, using the molarity of the solution as a conversion factor, convert moles of solute to volume of solution.

$$\text{Mass NaCl} \longrightarrow \text{moles NaCl} \longrightarrow \text{volume solution}$$

$$\text{Volume solution} = 40.0 \; \cancel{\text{g NaCl}} \times \frac{1 \; \cancel{\text{mol NaCl}}}{58.4 \; \cancel{\text{g NaCl}}} \times \frac{1 \; \text{L solution}}{0.150 \; \cancel{\text{mol NaCl}}} = 4.57 \; \text{L}$$

Practice Problem 4.15

Suppose you wish to prepare a 0.225 M solution of $CaCl_2$, starting with 1.89 g $CaCl_2$. What will be the volume of the solution?

Answer: 0.0757 L or 75.7 mL

Stoichiometric Relationships in Solutions

Using the types of calculation just described, we can apply stoichiometry to reactions that occur in solution. The only difference from the systems we have already examined is the addition of the volume term.

To see how the calculations are different, we first consider an example in which a reactant in solution reacts with a solid. We use molar masses to convert between moles and mass of the solid. Coefficients from the balanced equation allow us to convert from moles of one reactant to moles of the other. Finally, we use the molarity of the solution to convert between moles of dissolved reactant and the volume of solution needed to carry out the reaction.

EXAMPLE 4.16

What volume of 0.1248 M HCl should be added to a 5.000-g sample of calcium carbonate (limestone) for the calcium carbonate to react completely according to the following equation?

$$CaCO_3(s) + 2HCl(aq) \longrightarrow CaCl_2(aq) + H_2O(l) + CO_2(g)$$

Solution   Use the molar mass of calcium carbonate to convert mass of calcium carbonate to moles. Calculate moles of hydrochloric acid needed from moles of calcium carbonate according to the coefficients in the balanced equation. Use the molarity of the hydrochloric acid solution to convert from moles of hydrochloric acid to the volume of the hydrochloric acid solution required.

Mass $CaCO_3 \longrightarrow$ moles $CaCO_3 \longrightarrow$ moles HCl $\longrightarrow$ volume HCl solution

Volume HCl solution

$$= 5.000 \text{ g } CaCO_3 \times \frac{1 \text{ mol } CaCO_3}{100.1 \text{ g } CaCO_3} \times \frac{2 \text{ mol HCl}}{1 \text{ mol } CaCO_3} \times \frac{1 \text{ L HCl}}{0.1248 \text{ mol HCl}}$$

$$= 0.8005 \text{ L (or } 800.5 \text{ mL)}$$

Practice Problem 4.16

What mass of sodium carbonate will completely neutralize 50.0 mL of a 0.238 M HCl solution? The reaction is

$$2HCl(aq) + Na_2CO_3(s) \longrightarrow 2NaCl(aq) + CO_2(g) + H_2O(l)$$

Answer: 0.631 g

Now consider a reaction in which both reactants are in solution. We can determine how much product results from such a reaction by combining calculations involving stoichiometry and molarity. We start by using the volumes and molarities of the reactant solutions to calculate the moles of the reactants. Then we can derive moles of product from moles of the limiting reactant, using the coefficients in the balanced equation. To convert moles of product to mass of product, we use the molar mass if the product is a solid. If the product remains in solution, we convert moles of product to concentration using the combined volumes of reactant solutions as the volume of the product solution.

EXAMPLE 4.17

If 25 mL of a 0.45 M $AgNO_3$ solution is added to 55 mL of a 1.2 M NaCl solution, what mass of AgCl will be precipitated?

Solution This is a double-displacement reaction:

$$AgNO_3(aq) + NaCl(aq) \longrightarrow AgCl(s) + NaNO_3(aq)$$

Because we have two volumes of reactants to consider, we must first decide which is the limiting reactant by calculating the moles of each. We do this by multiplying the molarity by the volume in liters.

$$\text{Moles } AgNO_3 = \frac{0.45 \text{ mol } AgNO_3}{1 \text{ L solution}} \times 25 \text{ mL solution} \times \frac{1 \text{ L}}{1000 \text{ mL}} = 0.011 \text{ mol}$$

$$\text{Moles NaCl} = \frac{1.2 \text{ mol NaCl}}{1 \text{ L solution}} \times 55 \text{ mL solution} \times \frac{1 \text{ L}}{1000 \text{ mL}} = 0.066 \text{ mol}$$

Mole ratio = 0.011 mol $AgNO_3$/0.066 mol NaCl = 0.17

The desired mole ratio, determined from the equation, is 1, whereas the actual mole ratio is 0.17, so the limiting reactant is $AgNO_3$. Now we can calculate the amount of AgCl. We convert the moles of limiting reactant to moles of AgCl, using coefficients from the equation. Then we convert this quantity to mass of AgCl, using its molar mass.

$$\text{Moles } AgNO_3 \longrightarrow \text{moles AgCl} \longrightarrow \text{mass AgCl}$$

$$\text{g AgCl} = 0.011 \text{ mol } AgNO_3 \times \frac{1 \text{ mol AgCl}}{1 \text{ mol } AgNO_3} \times \frac{143 \text{ g AgCl}}{1 \text{ mol AgCl}} = 1.6 \text{ g}$$

Practice Problem 4.17

If 25.0 mL of a 0.380 M $BaCl_2$ solution is added to 75.0 mL of a 0.152 M H_2SO_4 solution, what mass of $BaSO_4$ will be precipitated?

Answer: 2.22 g

Titration and Dilution

If we are to use solutions to carry out chemical reactions, we must be able to determine and adjust their concentrations. Two techniques are commonly used for these purposes: titration and dilution, respectively.

Titration is the process of determining the concentration of one substance in solution by reacting it with a solution of another substance whose concentration is known. To carry out the process, we add one solution to a known volume of the other solution, as shown in Figure 4.20, until the reaction between the two substances is just complete—that is, until chemically equivalent amounts of the two reactants are present. We can often tell when we have reached this point, called the *equivalence point,* by a change in the color of an **indicator,** a chemical reactant that has been added to the system for this purpose. The point at which the indicator changes color is actually the *endpoint,* which may be slightly different from the equivalence point but is usually a good approximation. The volume of the reactant solution of known concentration is used to calculate the moles of that reactant. The

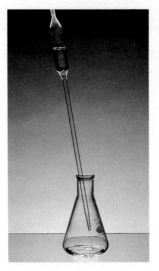

Figure 4.20 Titration. *Left:* A device called a pipet is used to transfer 25.00 mL of 0.1000 *M* H_2SO_4 into a flask. *Center:* A solution of NaOH whose concentration is unknown is then carefully added to the flask with a measuring device called a buret until chemically equivalent amounts of the dissolved substances are present. *Right:* This equivalence point is detected when the solution changes from colorless to faint pink because of a reaction between NaOH and an indicator called phenolphthalein. The equivalence point is reached when we have added 25.05 mL of NaOH solution. The concentration of the NaOH solution can be calculated from these data; see Example 4.18.

moles of the other reactant are obtained from the coefficients in the balanced reaction equation, and this value is used with the volume of the solution of unknown concentration to calculate its molarity.

Titration techniques are commonly applied to reactions of acids and bases to determine how much base is chemically equivalent to a given quantity of acid, a condition attained at the equivalence point of the reaction. The calculations follow this sequence of conversions:

$$\text{Volume of A} \longrightarrow \text{moles of A} \longrightarrow \text{moles of B} \longrightarrow \text{volume of B}$$

EXAMPLE 4.18

It requires 25.05 mL of NaOH solution of unknown concentration to titrate 25.00 mL of 0.1000 *M* H_2SO_4 solution. The chemical equation that summarizes the reaction is

$$H_2SO_4(aq) + 2NaOH(aq) \longrightarrow 2H_2O(l) + Na_2SO_4(aq)$$

What is the concentration of the NaOH solution?

Solution   We have two known volumes, one known concentration, and one unknown concentration. The series of conversions is

$$\text{Molarity } H_2SO_4 \longrightarrow \text{moles } H_2SO_4 \longrightarrow \text{moles NaOH} \longrightarrow \text{molarity NaOH}$$

We convert the molarity of H_2SO_4 to moles of H_2SO_4, using the volume of H_2SO_4 solution in liters; and we convert moles of H_2SO_4 to moles of NaOH, using the coefficients in the balanced equation. Finally, we convert moles of NaOH to molarity of NaOH, using the volume of its solution in liters.

$$M \text{ of NaOH} = \frac{0.1000 \text{ mol } H_2SO_4}{1 \text{ L solution}} \times 25.00 \text{ mL } H_2SO_4 \text{ solution} \times \frac{1 \text{ L}}{1000 \text{ mL}} \times$$

$$\frac{2 \text{ mol NaOH}}{1 \text{ mol } H_2SO_4} \times \frac{1}{25.05 \text{ mL NaOH solution}} \times \frac{1000 \text{ mL}}{1 \text{ L}} = 0.1996 \; M$$

Practice Problem 4.18

We need 28.18 mL of 0.2437 M HCl solution to completely titrate 25.00 mL of $Ba(OH)_2$ solution of unknown concentration. What is the molarity of the barium hydroxide solution?

Answer: 0.1373 M

Figure 4.21 Dilution. *Left:* A pipet is used to measure 25.00 mL of a 0.0200 M solution of $KMnO_4$. *Left center:* The solution is delivered into a volumetric flask. *Right center:* Water is added until the solution fills the flask to the marked 250.0-mL level. *Right:* Because the diluted solution has a known total volume, the new concentration can be readily calculated to be 0.00200 M.

To lower the concentration of a solution, we use the process of **dilution,** which consists of adding solvent to a known amount of solution of known concentration, as described in Figure 4.21. This procedure produces a solution of known total volume, so its molarity can be calculated easily. First we calculate the number of moles of solute contained in the volume of the more concentrated solution (V_{con}) by multiplying the volume in liters by the molarity (M_{con}):

$$\text{Moles}_{con} = V_{con} \times M_{con}$$

Because we are adding only solvent, the moles of solute do not change on dilution:

$$\text{Moles}_{dil} = \text{moles}_{con}$$

The moles of solute in the diluted solution also equal the product of molarity (M_{dil}) and volume (V_{dil}) in liters:

$$\text{Moles}_{dil} = V_{dil} \times M_{dil}$$

Because the number of moles of solute is unchanged, we get the equation

$$M_{dil}V_{dil} = M_{con}V_{con}$$

We can rearrange this equation to solve for the molarity of the diluted solution:

$$M_{dil} = M_{con}V_{con}/V_{dil}$$

Thus, if 25.00 mL of a 0.1000 M sodium hydroxide solution is diluted to 125.0 mL, the concentration of the diluted solution is

$$M_{dil} = 0.1000 \text{ mol/L} \times 25.00 \text{ mL}/125.0 \text{ mL} = 0.02000 \text{ mol/L}$$

Variations on this calculation can be carried out as long as any three of the four variables in the equation are known.

EXAMPLE 4.19

If 85.2 mL of 6.25 M HCl solution is diluted to a final volume of 250.0 mL, what is the molarity of the diluted solution of HCl?

Solution   The numbers of moles of hydrochloric acid are the same before and after dilution, so we can use the equation for dilution:

$$M_{dil} = M_{con}V_{con}/V_{dil}$$

The values of the quantities can be inserted and the equation can be solved for the molarity of the diluted solution:

$$M_{HCl} = 6.25 \text{ mol/L} \times 85.2 \text{ mL}/250.0 \text{ mL} = 2.13 \text{ mol/L}$$

Practice Problem 4.19

If 42.8 mL of 3.02 M H_2SO_4 solution is diluted to a final volume of 500.0 mL, what is the molarity of the diluted solution of H_2SO_4?

Answer: 0.259 mol/L

4.6   Some Important Industrial Processes

As pointed out in Chapter 3, relatively few raw materials form the basis for many chemical substances used by industry. (The more important industrial chemicals are listed in Table 3.10.) In this section, we consider methods for converting raw materials into some economically valuble and industrially beneficial substances—namely, sulfuric acid, iron, and steel.

Production of Sulfuric Acid

Sulfuric acid, H_2SO_4, is produced in greater amounts than any other industrial chemical in the United States. Although it has some consumer uses as an acid (for example, in lead storage batteries), it is used primarily to manufacture fertilizers and other chemicals, to recover metals, and to refine petroleum.

If sulfuric acid were expensive to produce, do you think it would be used in such large quantities? How do the principles of stoichiometry bear on the economic aspects of chemical production?

RULES OF REACTIVITY

A nonmetal will usually combine with oxygen to form an oxide in which the nonmetal has its lowest common positive oxidation number. If sufficient oxygen is present and the nonmetal has a higher accessible oxidation number, another oxide may be formed in which the nonmetal has a higher oxidation number.

Sulfuric acid can be produced fairly inexpensively. Some comes from sulfur dioxide (SO_2) emitted as a gas when sulfide ores such as iron pyrites, FeS_2, are processed. Most sulfuric acid, however, is made from sulfur (S_8). Vast deposits of sulfur occur under the limestone rock that caps subterranean salt domes along the Gulf of Mexico. Because these areas are covered by wet, marshy land, conventional mining techniques do not work. In the 1890s, the petroleum scientist Herman Frasch developed a method for melting the sulfur and forcing it to the surface through pipes, as shown in Figure 4.22.

The manufacture of sulfuric acid from sulfur involves several steps. First, sulfur is easily converted to sulfur dioxide by burning it in air at about 1000°C, a direct synthesis reaction:

$$S_8(s) + 8O_2(g) \xrightarrow{1000°C} 8SO_2(g)$$

Once started, this reaction gives off enough heat to maintain the high temperature required for the reaction to occur rapidly. The next step is the conversion of sulfur dioxide to sulfur trioxide, another direct synthesis reaction. This conversion is commonly achieved through the *contact process,* in which solid vanadium(V) oxide (V_2O_5) is used as a catalyst. A **catalyst** is a substance that takes part in an

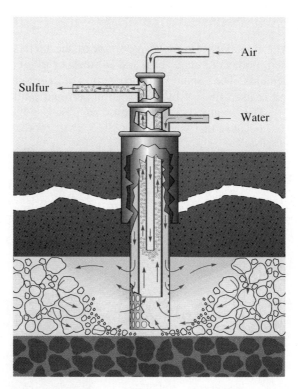

Figure 4.22 The Frasch process for recovering sulfur from underground deposits. Water is heated under pressure to about 160°C and is pumped down the outer pipe. It flows out the holes and melts the sulfur, which has a melting point of 119°C. The sulfur is more dense than water, so it collects at the bottom of the well. The pressurized water forces the molten sulfur partway up the middle pipe. Compressed air forced down the small center pipe fills the sulfur with bubbles, causing the sulfur to flow the rest of the way up the middle pipe to the surface, where it is collected and cooled back to the solid state.

overall chemical process and increases the rate of that process but is ultimately recovered in an unchanged form.

$$2SO_2(g) + O_2(g) \xrightarrow[450°C]{V_2O_5} 2SO_3(g)$$

The final step in the process is the hydration of sulfur trioxide, another direct synthesis reaction:

$$SO_3(g) + H_2O(l) \longrightarrow H_2SO_4(aq)$$

However, this step is complicated by the fact that SO_3 does not dissolve readily in H_2O, so this reaction is carried out indirectly. The SO_3 is dissolved in concentrated sulfuric acid, giving pyrosulfuric acid (also called oleum):

$$SO_3(g) + H_2SO_4(l) \longrightarrow H_2S_2O_7(l)$$

Pyrosulfuric acid reacts with water in a hydrolysis reaction to give sulfuric acid:

$$H_2S_2O_7(l) + H_2O(l) \longrightarrow 2H_2SO_4(aq)$$

This process is outlined in Figure 4.23.

Production of Iron and Steel

Iron is perhaps the most important of the metals. It is certainly the most widely used, both in its pure form and as steel, or iron containing small amounts of other elements. The annual U.S. production of iron and steel exceeds 280 billion pounds and is used mostly by the construction and automobile industries.

The commercial preparation of iron is not aimed at producing highly pure iron, because that is very difficult. Rather, production methods control the amount

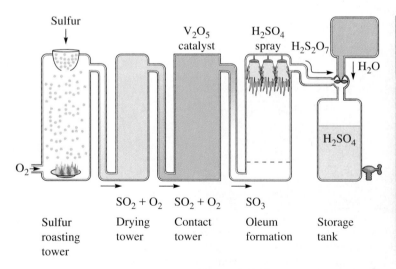

Figure 4.23 The contact process. *(Left)* Sulfur is burned in a roasting tower to form a mixture of sulfur dioxide and oxygen gases. This mixture is dried, then passed through the V_2O_5 catalyst in the contact tower. The resulting sulfur trioxide is passed through a spray of sulfuric acid in a scrubbing tower and mixed with water to form a sulfuric acid solution. *(Right)* The various towers in a sulfuric acid production plant.

of impurities or additives in the iron, thus allowing steel to be produced. Primary sources of iron are the ores hematite (Fe_2O_3) and magnetite (Fe_3O_4). ▶ These ores are reduced to the metal by carbon monoxide in a blast furnace (Figure 4.24), using a reaction similar to the thermite reaction described in the chapter introduction. Iron ore, coke (coal heated to remove the volatile components), and limestone ($CaCO_3$) are loaded into the top of the furnace; then a blast of hot air enters at the bottom. The temperature in the furnace varies from 2000°C near the bottom to about 200°C at the top. The coke reacts with oxygen in the air to form carbon monoxide:

$$2C(s) + O_2(g) \longrightarrow 2CO(g)$$

RULES OF REACTIVITY

Metal carbonates, other than those of the alkali metals, decompose to metal oxides and carbon dioxide gas when heated.

The limestone is decomposed by heat:

$$CaCO_3(s) \xrightarrow{\Delta} CaO(s) + CO_2(g)$$

One of the products of this reaction, carbon dioxide, also reacts with the coke to form carbon monoxide:

$$CO_2(g) + C(s) \longrightarrow 2CO(g)$$

Carbon monoxide reacts with the iron ore in a stepwise fashion as it passes down the furnace and increases in temperature:

$$3Fe_2O_3(s) + CO(g) \longrightarrow 2Fe_3O_4(s) + CO_2(g)$$
$$Fe_3O_4(s) + CO(g) \longrightarrow 3FeO(s) + CO_2(g)$$
$$FeO(s) + CO(g) \longrightarrow Fe(l) + CO_2(g)$$

The iron in the oxide is reduced successively to lower oxidation numbers until it is finally reduced to metallic iron in the liquid state. The molten iron collects at the bottom of the furnace, where it can be drained off.

Figure 4.24 Iron ore is reduced to liquid iron in a blast furnace.

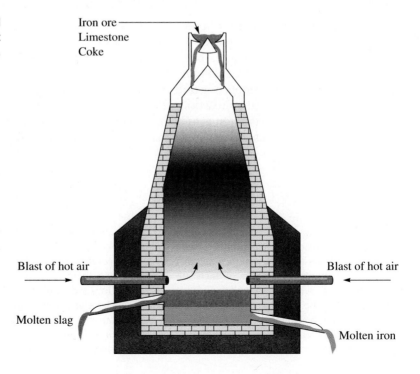

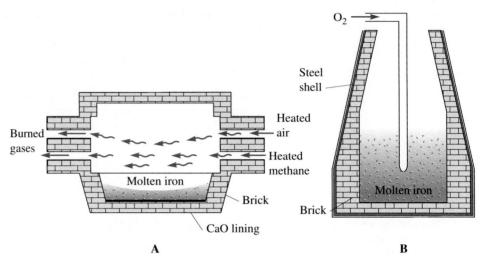

Figure 4.25 Iron is purified by the burning off of carbon that has dissolved in it during reduction. (*A*) In an open-hearth furnace, heated air is mixed with a fuel such as methane, which is used to burn off the carbon. (*B*) In a basic oxygen furnace, oxygen is injected into the molten iron, where it combines with dissolved carbon to form carbon oxides.

RULES OF REACTIVITY

Metal oxides react with nonmetal oxides to form metal oxoanion salts. Similar reactions can occur between two metal oxides if one of them shows some amphoteric behavior. The oxides of aluminum and zinc are common amphoteric oxides.

The calcium oxide formed by the decomposition of the limestone forms a molten slag (refuse that separates from metal ores) that floats on the iron. This slag contains any silica or alumina present as impurities in the ore:

$$CaO(l) + SiO_2(l) \longrightarrow CaSiO_3(l)$$
$$CaO(l) + Al_2O_3(l) \longrightarrow CaAl_2O_4(l)$$

The slag may be drained off and then used in the manufacture of cement.

The iron recovered from a blast furnace, called *pig iron*, contains 3%–4% carbon. This carbon must either be removed to obtain pure iron or must be present in controlled amounts to produce steel. Removal of the carbon is accomplished in either an open-hearth furnace or a basic oxygen furnace (Figure 4.25). Carbon is removed by reaction with air in the first type and with pure oxygen in the second. When the iron is to be made into steel, measured amounts of carbon and of metals such as chromium or manganese are added to the molten iron.

Summary

Chemical equations represent chemical reactions, with formulas of the reactants and products separated by an arrow. Each formula includes a coefficient to indicate the relative number of moles of each substance. A chemical equation in which the number of atoms of each component element is the same on the reactant side as on the product side is said to be balanced. The products of chemical reactions must be identified in the laboratory. However, it is possible to make reasonable predictions of the products by recognizing that reactions can be placed into four classes: direct synthesis, decomposition, single-displacement, and double-displacement reactions. Reactions in which there is a change in oxidation number are called oxidation–reduction reactions.

The chemistry of oxygen provides many examples of reaction types. Most elements react directly with molecular oxygen to form oxides. Most nonmetals and transition metals form more than one oxide. The more active alkali metals and alkaline-earth metals form peroxides or superoxides instead of oxides in their direct synthesis reactions with molecular oxygen. Some ionic compounds will undergo hydration reactions to form hydrated salts. Metal oxides react with water to form hydroxide bases, whereas nonmetal oxides react with water to form oxoacids.

The quantitative calculations of stoichiometry are used to determine the masses of reactants or products involved in chemical reactions. In reactions with more than one reactant, one of them, called the limiting reactant, may be consumed before the other is completely reacted.

The concentration of a solute in a solution is commonly expressed in terms of molarity: the moles of solute per liter of solution. Through the use of appropriate conversions and stoichiometric calculations, it is possible to calculate the amounts of reactants and products for reactions occurring in solution.

Key Terms

acidic anhydride (4.3)
activity series (4.2)
actual yield (4.4)
addition reaction (4.2)
alloy (4.2)
amphoteric (4.3)
anhydrous (4.2)
balanced equation (4.1)
base (4.2)
basic anhydride (4.3)
catalyst (4.6)
chemical equation
  (p. 118)

concentrated
  solution (4.5)
concentration (4.5)
decomposition
  reaction (4.2)
dilute solution (4.5)
dilution (4.5)
direct synthesis reaction
  (combination
  reaction) (4.2)
double-displacement
  reaction (4.2)
hydration reaction (4.3)

hydrolysis (4.2)
hygroscopic (4.3)
indicator (4.5)
limiting reactant
  (limiting
  reagent) (4.4)
molarity (M) (4.5)
neutralization
  reaction (4.2)
oxidation–reduction
  reaction (redox
  reaction) (4.2)
oxidize (4.2)

percent yield (4.4)
precipitate (4.2)
precipitation
  reaction (4.2)
reduce (4.2)
single-displacement
  reaction (4.2)
solute (4.5)
solvent (4.5)
stoichiometry (4.4)
theoretical yield (4.4)
thermite reaction (p.
  118)

Exercises

Chemical Equations

4.1 What is a chemical equation?

4.2 Why must a chemical equation be balanced?

4.3 Why can't subscripts be changed to balance a chemical equation?

4.4 What is the difference between a chemical equation and a chemical reaction?

4.5 Describe the identifying characteristics of each of the following classes of reactions: direct synthesis, decomposition, single-displacement, and double-displacement.

4.6 Distinguish between oxidation and reduction.

4.7 Give examples of a hydrate, an anhydrous compound, and a hygroscopic compound.

4.8 What is a precipitate?

4.9 Distinguish between an acid and a base. Give examples of each and of acidic and basic anhydrides.

4.10 What are the distinguishing characteristics of an amphoteric oxide?

4.11 Complete and balance each of the following direct synthesis reactions.

a. $Ca(s) + N_2(g) \xrightarrow{\Delta}$
b. $SO_3(g) + H_2O(l) \longrightarrow$
c. $Li_2O(s) + H_2O(l) \longrightarrow$
d. $SO_2(g) + O_2(g) \longrightarrow$
e. $Al(s) + O_2(g) \longrightarrow$
f. $CaO(s) + H_2O(l) \longrightarrow$

4.12 Complete and balance each of the following decomposition reactions.

a. $NaHCO_3(s) \xrightarrow{\Delta}$
b. $Na_2B_4O_7 \cdot 10H_2O(s) \xrightarrow{\Delta}$
c. $Cu(NO_3)_2(s) \xrightarrow{\Delta}$
d. $HgO(s) \xrightarrow{\Delta}$
e. $(NH_4)_2Cr_2O_7(s) \xrightarrow{\Delta} Cr_2O_3(s) + N_2(g) + ?$
f. $HBrO(aq) \longrightarrow HBrO_3(aq) + HBr(aq)$

4.13   Complete and balance each of the following single-displacement reactions.
   a. $Cl_2(g) + NaBr(aq) \longrightarrow$
   b. $Na(s) + H_2S(g) \longrightarrow$
   c. $Al(s) + H_2SO_4(aq) \longrightarrow$
   d. $K_2S(aq) + I_2(aq) \longrightarrow$
   e. $Cs(s) + H_2O(l) \longrightarrow$

4.14   Complete and balance each of the following double-displacement reactions.
   a. $CaCO_3(s) + H_2SO_4(aq) \longrightarrow$
   b. $SbCl_3(aq) + H_2S(aq) \longrightarrow$
   c. $NaOH(aq) + H_3PO_4(aq) \longrightarrow$
   d. $H_2SO_4(aq) + KOH(aq) \longrightarrow$

4.15   Write a balanced equation to describe any precipitation reaction that might occur when each of the following pairs of substances are mixed.
   a. $K_2CO_3(aq) + BaCl_2(aq)$
   b. $CaS(aq) + Hg(NO_3)_2(aq)$
   c. $MgSO_4(aq) + BaCl_2(aq)$
   d. $K_2SO_4(aq) + MgCl_2(aq)$
   e. $MgCl_2(aq) + Pb(NO_3)_2(aq)$

4.16   Balance the following chemical equations. Classify each as a direct synthesis, decomposition, single-displacement, or double-displacement reaction.
   a. $CaCl_2(aq) + Na_2SO_4(aq) \longrightarrow$
      $CaSO_4(s) + NaCl(aq)$
   b. $Ba(s) + HCl(aq) \longrightarrow BaCl_2(aq) + H_2(g)$
   c. $Al(s) + O_2(g) \xrightarrow{\Delta} Al_2O_3(s)$
   d. $FeO(s) + CO(g) \xrightarrow{\Delta} Fe(s) + CO_2(g)$
   e. $CaO(s) + H_2O(l) \longrightarrow Ca(OH)_2(aq)$
   f. $Na_2CrO_4(aq) + Pb(NO_3)_2(aq) \longrightarrow$
      $PbCrO_4(s) + NaNO_3(aq)$
   g. $KI(aq) + Cl_2(g) \longrightarrow KCl(aq) + I_2(aq)$
   h. $NaHCO_3(s) \xrightarrow{\Delta}$
      $Na_2CO_3(s) + CO_2(g) + H_2O(g)$
   i. $CaO(s) + SO_3(g) \xrightarrow{\Delta} CaSO_4(s)$
   j. $PCl_5(g) \xrightarrow{\Delta} PCl_3(g) + Cl_2(g)$
   k. $Ca_3N_2(s) + H_2O(l) \xrightarrow{\Delta}$
      $Ca(OH)_2(aq) + NH_3(g)$

4.17   Write a complete, balanced equation for each of the following reactions.
   a. When solid sodium hydride is added to water, hydrogen gas is released and aqueous sodium hydroxide is formed.
   b. Solid copper(II) nitrate is heated to produce solid copper(II) oxide, gaseous nitrogen dioxide, and oxygen gas.
   c. Solid magnesium reacts with hydrochloric acid to form gaseous hydrogen and aqueous magnesium chloride.

   d. Solid calcium phosphate reacts with a mixture of solid silicon dioxide and solid carbon to form solid phosphorus, solid calcium silicate, and gaseous carbon monoxide.

4.18   Balance the following equations and identify each as a decomposition, direct synthesis, single-displacement, or double-displacement reaction.
   a. $GaH_3 + N(CH_3)_3 \longrightarrow (CH_3)_3NGaH_3$
   b. $Ga + GaCl_3 \longrightarrow Ga[GaCl_4]$
   c. $GeCl_2 + Cl_2 \longrightarrow GeCl_4$
   d. $Si_2Cl_6 \longrightarrow Si_6Cl_{14} + SiCl_4$
   e. $N_2(g) + CaC_2(s) \longrightarrow C(s) + CaNCN(s)$
   f. $N_2(g) + Mg(s) \longrightarrow Mg_3N_2(s)$
   g. $NH_4Cl(s) \longrightarrow NH_3(g) + HCl(g)$

Stoichiometry

4.19   Define *limiting reactant*.

4.20   Distinguish between theoretical yield, actual yield, and percent yield.

4.21   How is a chemical equation used as a source of conversion factors for stoichiometric calculations?

4.22   Describe how stoichiometric calculations make use of the law of conservation of mass.

4.23   A small tank of propane, used for home soldering jobs, contains 405 g of propane, C_3H_8. What mass of CO_2 and what mass of H_2O are obtained when a tankful of propane is burned?

$$C_3H_8(g) + 5O_2(g) \longrightarrow 3CO_2(g) + 4H_2O(g)$$

4.24   Blue vitriol, or copper(II) sulfate pentahydrate, is often added to swimming pools to kill algae. It is prepared by the reaction between copper metal and hot sulfuric acid to give copper(II) sulfate, $CuSO_4(aq)$, sulfur dioxide, $SO_2(g)$, and water. Write a balanced equation for the reaction used to prepare copper(II) sulfate. Assume that 1.00 mole of copper is reacted with 1.00 mole of sulfuric acid.
   a. How many moles of copper(II) sulfate will be formed?
   b. What mass of copper(II) sulfate will be formed?
   c. How many moles of sulfur dioxide will be formed?
   d. How many molecules of sulfur dioxide will be formed?

4.25   The explosive TNT (trinitrotoluene) is prepared by the reaction of toluene with nitric acid:

$$C_7H_8(l) + 3HNO_3(aq) \longrightarrow$$
$$C_7H_5N_3O_6(s) + 3H_2O(l)$$

How much toluene and how much nitric acid (in kg) would be required to prepare 1.00 metric ton (1.00×10^3 kg) of TNT?

4.26 Diiodine pentoxide is used in respirators to remove carbon monoxide from air:

$$I_2O_5(s) + 5CO(g) \longrightarrow I_2(s) + 5CO_2(g)$$

What mass of carbon monoxide could be removed from air by a respirator that contains 50.0 g of diiodine pentoxide?

4.27 Solid fats for cooking (shortening) are made by hydrogenation of vegetable oils. What mass of H_2 is needed to hydrogenate 454 g (1 lb) of an oil with the average composition $C_3H_5(C_{17}H_{31}CO_2)_3$?

$$C_3H_5(C_{17}H_{31}CO_2)_3 + 6H_2 \longrightarrow$$
$$C_3H_5(C_{17}H_{35}CO_2)_3$$

4.28 Aspirin ($C_9H_8O_4$) is produced by the reaction of salicylic acid ($C_7H_6O_3$) with acetic anhydride:

$$C_7H_6O_3 + O(OCCH_3)_2 \longrightarrow$$
$$C_9H_8O_4 + CH_3CO_2H$$

a. How much salicylic acid is required to produce the 2.00×10^8 g of aspirin used daily in the United States?

b. How much acetic acid (CH_3CO_2H) is obtained as well?

4.29 Because magnesium generates intense light when it burns, it is used in flashbulbs and flares. How much magnesium oxide will be produced if 0.486 g Mg is reacted with 0.160 g O_2?

$$2Mg(s) + O_2(g) \longrightarrow 2MgO(s)$$

4.30 Sugar cane and other plants produce sucrose, a sugar, by photosynthesis:

$$12CO_2 + 11H_2O \longrightarrow C_{12}H_{22}O_{11} + 12O_2$$

Ethyl alcohol, used in alcoholic beverages, can be produced by fermentation of this sugar:

$$C_{12}H_{22}O_{11} + H_2O \longrightarrow 4C_2H_5OH + 4CO_2$$

If 25.0 g CO_2 is used up in the photosynthesis step, how much C_2H_5OH will be obtained?

4.31 Phenol (C_6H_5OH), often used as a disinfectant in stables, cesspools, drains, and so on, is a common water pollutant. It can be converted to less harmful oxalic acid ($H_2C_2O_4$) by reaction with ozone (O_3):

$$C_6H_5OH + 11O_3 \longrightarrow 3H_2C_2O_4 + 11O_2$$

a. What mass of ozone would be required to react with 125.0 g of phenol?

b. What mass of oxalic acid would be produced?

4.32 Phosphorus sulfide, P_4S_3, is used in the heads of wooden matches. This material is manufactured by heating a mixture of red phosphorus and sulfur:

$$8P_4 + 3S_8 \longrightarrow 8P_4S_3$$

If 10.0 g of phosphorus and 17.0 g of sulfur are reacted, how much phosphorus sulfide will be produced?

4.33 Benzene, C_6H_6, burns in air according to the following equation:

$$2C_6H_6 + 15O_2 \longrightarrow 12CO_2 + 6H_2O$$

How many moles of CO_2 and H_2O will be produced in the complete combustion of 200.0 g of benzene?

4.34 What mass of the first product will be obtained in the following decomposition reactions, assuming each starts with 175.0 g of reactant?

a. $2H_2O_2(aq) \longrightarrow 2H_2O(l) + O_2(g)$

b. $2HgO(s) \longrightarrow 2Hg(l) + O_2(g)$

c. $2KClO_3(s) \longrightarrow 2KCl(s) + 3O_2(g)$

d. $2PbO_2(s) \longrightarrow 2PbO(s) + O_2(g)$

4.35 In order to purify zirconium, impure zirconium is heated with iodine according to the following equation:

$$Zr(s) + 2I_2(g) \longrightarrow ZrI_4(s)$$

The zirconium iodide is then decomposed to produce pure zirconium metal by the Van Arkel DeBoar process.

$$ZrI_4(s) \longrightarrow Zr(s) + 2I_2(g)$$

If one ton of impure zirconium yields 1075 lb of pure zirconium, what percent of the impure sample was zirconium?

4.36 In the natural nitrogen cycle, nitrates are converted to dinitrogen oxide when glucose and potassium nitrate react:

$$C_6H_{12}O_6 + KNO_3 \longrightarrow$$
$$CO_2 + H_2O + KOH + N_2O$$

a. Balance this equation.

b. What mass of CO_2 will be obtained from 25.0 g of glucose?

4.37 Metal chlorides such as praseodymium chloride, $PrCl_3$, can be prepared by heating praseodymium oxide, Pr_2O_3, with ammonium chloride to yield the chloride, $PrCl_3$, plus water and ammonia.

a. Write a balanced equation for this reaction.

b. If 50.0 g Pr_2O_3 is used, what mass of $PrCl_3$ will be produced?

4.38 Insoluble oxides (acid anhydrides) of some of the less metallic (or nonmetallic) elements will generally dissolve in a base, as in the following examples:

$$N_2O_5(s) + H_2O(l) \longrightarrow 2H^+(aq) + 2NO_3^-(aq)$$
$$Sb_2O_5(s) + 2OH^-(aq) + 5H_2O(l) \longrightarrow$$
$$2Sb(OH)_6^-(aq)$$

a. What mass of N_2O_5 is required to prepare 50.0 millimoles of NO_3^- ion?

b. How many millimoles of $Sb(OH)_6^-$ would be obtained from 28.5 g Sb_2O_5?

4.39 A 1.050-g sample containing carbon, hydrogen, and oxygen was burned in oxygen, and 2.006 g of carbon dioxide and 1.232 g of water were obtained.

a. How many moles of carbon and of hydrogen did the sample contain?

b. What is the empirical formula of the compound?

c. The molar mass of the compound was determined to be 46.07 g/mol. What is its molecular formula?

4.40 The mineral stibnite, Sb_2S_3, can be heated with oxygen to form antimony(III) oxide, which is used as a paint pigment and for flameproofing canvas.

$$2Sb_2S_3(s) + 9O_2(g) \longrightarrow 2Sb_2O_3(s) + 6SO_2(g)$$

What mass of antimony(III) oxide could be prepared from 125 kg of stibnite?

4.41 When dissolved in water or dilute acids, ionic peroxides give hydrogen peroxide, H_2O_2, and all are powerful oxidizers. Carbon dioxide reacts with peroxides to give CO_3^{2-}:

$$2CO_2(g) + 2Na_2O_2(s) \longrightarrow 2Na_2CO_3(s) + O_2(g)$$

What mass of sodium carbonate is formed from 195 kg Na_2O_2?

4.42 What mass of copper(II) oxide must be reacted with hydrogen gas to produce 7.50 g of copper metal? The reaction equation is

$$CuO(s) + H_2(g) \longrightarrow Cu(s) + H_2O(g)$$

4.43 Chromium can be prepared by the Goldschmidt process:

$$Cr_2O_3(s) + 2Al(s) \longrightarrow 2Cr(l) + Al_2O_3(s)$$

If 125 g of chromium(III) oxide is heated with 55.0 g of aluminum, how much chromium metal would be formed?

4.44 The interaction of ozone, O_3, with potassium hydroxide gives ozonides according to the following equation:

$$3KOH(s) + 2O_3(g) \longrightarrow$$
$$2KO_3(s) + KOH \cdot H_2O(s) + \tfrac{1}{2}O_2(g)$$

The ozonide KO_3 slowly decomposes to KO_2 and oxygen:

$$2KO_3(s) \longrightarrow 2KO_2(s) + O_2(g)$$

What mass of KO_2 will be produced by the reaction of 75.0 g KOH?

4.45 The hydrolysis of xenon hexafluoride, XeF_6, gives xenon oxofluoride, $XeOF_4$:

$$XeF_6 + H_2O \longrightarrow XeOF_4 + 2HF$$

What mass of $XeOF_4$ can be prepared from 78.5 g XeF_6?

4.46 Hydrazine, N_2H_4, is used as a rocket fuel. What mass of nitrogen dioxide would be produced by burning 745 g of hydrazine?

$$N_2H_4 + 3O_2 \longrightarrow 2NO_2 + 2H_2O$$

4.47 The principal ingredient of the self-defense agent MACE is chloroacetophenone, $C_6H_5COCH_2Cl$. This compound is prepared by chlorination of acetophenone, $C_6H_5COCH_3$.

$$C_6H_5COCH_3 + Cl_2 \longrightarrow C_6H_5COCH_2Cl + HCl$$

What mass of chloroacetophenone would be produced by the reaction of 4.00 g of acetophenone with 10.4 g of chlorine?

4.48 Some household drain cleaners consist of a mixture of granules of sodium hydroxide and powdered aluminum. When dissolved in water, the sodium hydroxide reacts with the aluminum to produce hydrogen gas.

$$2Al(s) + 2NaOH(aq) + 6H_2O(l) \longrightarrow$$
$$2NaAl(OH)_4(aq) + 3H_2(g)$$

The sodium hydroxide helps dissolve grease, and the hydrogen gas provides mixing and scrubbing action. What mass of hydrogen gas would be formed from a mixture of 2.48 g Al and 4.76 g NaOH?

4.49 A common abrasive is SiC, which is sold commercially as Carborundum. It is prepared by the combination of SiO_2 and carbon:

$$SiO_2 + 3C \longrightarrow SiC + 2CO$$

What mass of SiC is formed from 10.0 g SiO_2 and 5.00 g C?

Solutions

4.50 What is a solution? Give five examples of solutions that can be found in your home.

4.51 What is the difference between a dilute and a concentrated solution?

4.52 Distinguish between a solute and a solvent.

4.53 Define *molarity*.

4.54 Calculate the molarity of each of the following solutions.

a. 122 g of acetic acid, CH_3CO_2H, in 1.00 L of solution

b. 185 g of sucrose, $C_{12}H_{22}O_{11}$, in 1.00 L of solution

c. 70.0 g of hydrogen chloride, HCl, in 0.600 L of solution

4.55 Calculate the molarity of each of the following solutions.

a. 6.30 g HNO_3 dissolved in 255 mL of solution

b. 49.0 g H_2SO_4 dissolved in 125 mL of solution

c. 2.80 g KOH dissolved in 525 mL of solution

d. 7.40 g $Ca(OH)_2$ dissolved in 200.0 mL of solution

4.56 Calculate what volume of the following solutions is required to obtain 0.250 moles of each solute.

a. 0.250 M $AlCl_3$ b. 3.00 M HCl

c. 1.50 M H_2SO_4 d. 0.750 M NaCl

4.57 Calculate the mass of solute in each of the following solutions.

a. 250.0 mL of 1.50 M KCl

b. 250.0 mL of 2.05 M Na_2SO_4

c. 150.0 mL of 0.245 M $CaCl_2$

d. 1450 mL of 0.00187 M H_2SO_4

4.58 How many mL of 0.200 M HCl must be added to completely neutralize 30.0 mL of 0.155 M NaOH?

4.59 How much water must be added to 935.0 mL of 0.1074 M HCl to obtain a solution that is exactly 0.1000 M?

4.60 If 26.35 mL of 0.2473 M HCl is diluted to 250.0 mL, what is the concentration of the diluted solution?

4.61 If 25.00 mL of 0.0485 M HCl is titrated with a solution of KOH, it requires 27.63 mL of KOH solution to reach the endpoint. What is the molarity of the KOH solution?

4.62 What is the concentration of a solution prepared by diluting 35.0 mL of 0.150 M KBr to 250.0 mL?

4.63 If you wish to prepare a 0.055 M solution of $NaNO_3$, to what volume would you have to dilute 25 mL of 3.0 M $NaNO_3$?

4.64 What volume of 12.0 M HCl must be diluted with water to a total volume of 750 mL to give a 0.250 M solution?

4.65 Calculate the molarity of each of the following solutions.

a. 50 mL of 0.50 M H_2SO_4 diluted to 125 mL

b. 10 mL of 1.0 M HCl diluted to 5000 mL

c. 80 mL of 0.125 M NaOH diluted to 200 mL

4.66 Calculate the molarity of an NaOH solution if 25.00 mL of this solution requires 36.30 mL of 0.1250 M HCl for titration.

4.67 Cadmium hydroxide, used in storage battery electrodes, is prepared by precipitation from a solution containing cadmium chloride and potassium hydroxide:

$$CdCl_2(aq) + 2KOH(aq) \longrightarrow$$
$$Cd(OH)_2(s) + 2KCl(aq)$$

What mass of cadmium hydroxide could be prepared from 125 mL of 0.250 M $CdCl_2$ mixed with 125 mL of 0.450 M KOH?

4.68 An excess of silver nitrate solution is added to 50.0 mL of hydrochloric acid of unknown concentration. The following reaction occurs:

$$AgNO_3(aq) + HCl(aq) \longrightarrow AgCl(s) + HNO_3(aq)$$

The precipitate, silver chloride, is isolated, dried, and found to weigh 0.658 g. What is the molarity of the hydrochloric acid solution?

4.69 Basic copper carbonate, $CuCO_3 \cdot Cu(OH)_2$, occurs in nature as the mineral malachite. It is used as a fungicide for seed treatment, in pyrotechnics, as a paint pigment, and as an additive in animal and poultry feeds to prevent copper deficiency. It is prepared by the reaction of aqueous solutions of copper sulfate and sodium carbonate:

$$2CuSO_4(aq) + 2Na_2CO_3(aq) + H_2O(l) \longrightarrow$$
$$CuCO_3 \cdot Cu(OH)_2(s) + 2Na_2SO_4(aq) + CO_2(g)$$

What mass of basic copper carbonate will be produced if 5.00 L of 0.250 M copper sulfate is mixed with 2.50 L of 0.525 M sodium carbonate?

Oxygen Chemistry

4.70 In what forms can the element oxygen be found in nature?

4.71 a. What is an oxide? a peroxide? a superoxide?

4.72 a. What is an acidic anhydride? a basic anhydride? Give examples of each.

4.73 Write a formula for the product of the reaction of each of the following metals with molecular oxygen.

a. Cs   b. Pb   c. Al   d. Mg   e. Fe   f. Li

4.74 Write a formula for the product of the reaction of each of the following nonmetals with molecular oxygen.

a. H_2 b. C   c. S_8 d. P_4 e. Se   f. B

4.75 Write a formula for the product of the reaction of each of the following compounds with molecular oxygen.

a. CH_4 b. CO

c. CuS   d. Li_3P

e. CS_2 f. SCl_4

4.76 Write a balanced equation for the reaction between each of the following pairs of substances.

a. $Na_2O(s) + H_2O(l)$ b. $CaO(s) + H_2SO_4(aq)$

c. $NaOH(aq) + Cl_2O(g)$ d. $Ba(OH)_2(aq) +$

e. $Al_2O_3(s) + HCl(aq)$ $HNO_3(aq)$

g. $BaO(s) + CO_2(g)$ f. $Al_2O_3(s) + NaOH(aq)$

i. $NaOH(aq) + H_2SO_3(aq)$ h. $NaOH(aq) + SO_2(g)$

j. $Na_2O(s) + SO_2(g)$

4.77 Write the formula for the acid formed by the reaction of each of the following compounds with water.

a. SO_3 b. SO_2 c. N_2O_5 d. P_4O_{10}

e. Cl_2O_5 f. Cl_2O_7 g. B_2O_3 h. Br_2O

Industrial Chemistry

4.78 List some major uses of sulfuric acid.

4.79 Outline the industrial procedure for preparing sulfuric acid.

4.80 How does steel differ from iron?

4.81 Outline the industrial process for preparing iron.

4.82 Write an equation describing the production of sulfur dioxide from sulfur and from iron pyrites.

4.83 Describe the Frasch process for mining sulfur from salt domes.

4.84 Explain the purpose of adding limestone to a blast furnace in the preparation of iron.

4.85 Write formulas for the major components of the slag formed in a blast furnace during the preparation of iron.

4.86 Vanadium(V) oxide, used as a catalyst in the preparation of sulfuric acid, is produced by the decomposition of ammonium vanadate:

$$2NH_4VO_3 \xrightarrow{\Delta} V_2O_5 + 2NH_3 + H_2O$$

If 75.2 g of vanadium(V) oxide is produced by this reaction, how much ammonia is also produced?

4.87 The low-grade iron ore taconite, which contains Fe_3O_4, is concentrated and made into pellets for processing. If one ton of taconite pellets yields 545 lb of iron, what percentage of the pellets is Fe_3O_4?

Additional Exercises

4.88 Sodium tripolyphosphate, $Na_5P_3O_{10}$, is used as a water-softening agent in detergents. It is prepared by the dehydration of a mixture of sodium dihydrogen phosphate and sodium hydrogen phosphate:

$$NaH_2PO_4 + 2Na_2HPO_4 \xrightarrow{\Delta} Na_5P_3O_{10} + 2H_2O$$

What mass of sodium tripolyphosphate would be produced by the reaction of 15.0 g NaH_2PO_4 and 32.0 g Na_2HPO_4?

4.89 Bismuth selenide, Bi_2Se_3, is used in semiconductor research. It can be prepared directly from the elements:

$$2Bi + 3Se \longrightarrow Bi_2Se_3$$

What mass of bismuth selenide could be produced from 15.2 g of bismuth and 12.8 g of selenium?

4.90 When 25.00 mL of 0.1035 M H_2SO_4 is titrated with a solution of NaOH, it requires 45.37 mL of NaOH solution to reach the endpoint. What is the molarity of the NaOH solution?

4.91 In the following reaction, identify the substance that is reduced and the substance that is oxidized:

$$FeCl_3(aq) + CrCl_2(aq) \longrightarrow$$
$$FeCl_2(aq) + CrCl_3(aq)$$

4.92 Write a balanced equation for the reaction between each of the following pairs of substances:
  a. $BaO(s) + H_2CO_3(aq)$
  b. $LiOH(aq) + SO_2(g)$
  c. $P_4O_{10}(s) + NaOH(aq)$
  d. $Ca(OH)_2(aq) + H_2SO_4(aq)$
  e. $Cr_2O_3(s) + HCl(aq)$
  f. $Fe_2O_3(s) + H_2SO_4(aq)$

4.93 Calculate the molar concentration of chloride ion in each of the following solutions.
  a. 5.00 mL of 0.0100 M LiCl diluted to 50.0 mL
  b. 5.00 mL of 0.0652 M NaCl diluted to 500.0 mL
  c. 100.0 mL of 0.0100 M $BaCl_2$ diluted to 500.0 mL
  d. 100.0 mL of 0.250 M $CaCl_2$ diluted to 2000.0 mL

4.94 Heating causes the copper mineral malachite, $CuCO_3 \cdot Cu(OH)_2$, to decompose:

$$CuCO_3 \cdot Cu(OH)_2(s) \xrightarrow{\Delta}$$
$$2CuO(s) + CO_2(g) + H_2O(g)$$

The CuO is recovered, washed, and dried. If 122 g CuO is obtained from 205 g of malachite, what is the percent yield of this reaction?

4.95 Write a balanced equation to describe any precipitation reaction that might occur when each of the following pairs of substances are mixed.
  a. $ZnSO_4(aq) + Ba(NO_3)_2(aq)$
  b. $Ca(NO_3)_2(aq) + K_3PO_4(aq)$
  c. $ZnSO_4(aq) + BaCl_2(aq)$
  d. $KOH(aq) + MgCl_2(aq)$
  e. $CuSO_4(aq) + BaS(aq)$

4.96 Ethyl alcohol, used in alcoholic beverages and gasohol, can be produced by fermentation of sugar:

$$C_{12}H_{22}O_{11}(aq) + H_2O(l) \longrightarrow$$
$$4C_2H_5OH(aq) + 4CO_2(g)$$

How much ethyl alcohol could be obtained from a 5.00-lb bag of sugar?

4.97 Chloroform, a substance that appears frequently in spy stories, has a fast and powerful anesthetic action. It can react with oxygen to form poisonous phosgene gas, which has been used in gas warfare:

$$2CHCl_3 + O_2 \longrightarrow 2COCl_2 + 2HCl$$

What mass of phosgene would be formed from 125 g of chloroform?

4.98 When 25.00 mL of 0.0245 M H_2SO_4 is titrated with a solution of $Ba(OH)_2$, it requires 11.29 mL of $Ba(OH)_2$ solution to reach the equivalence point. What is the molarity of the $Ba(OH)_2$ solution?

*4.99 Nitric acid is manufactured by the following sequence of reactions, known as the Ostwald process:

$$N_2 + 3H_2 \longrightarrow 2NH_3$$
$$4NH_3 + 5O_2 \longrightarrow 4NO + 6H_2O$$
$$2NO + O_2 \longrightarrow 2NO_2$$
$$4NO_2 + 2H_2O + O_2 \longrightarrow 4HNO_3$$

If the percent yield is 48% for the first step, 95% for the second step, 98% for the third step, and 76% for the fourth step, how much nitric acid (in grams) could be obtained from 75.0 g N_2?

*4.100 Oxygen difluoride can be prepared by bubbling gaseous fluorine into a 0.5 M solution of NaOH:

$$2F_2(g) + 2NaOH(aq) \longrightarrow$$
$$OF_2(g) + 2NaF(aq) + H_2O(l)$$

Oxygen difluoride can be used to prepare compounds such as O_2AsF_6, containing the dioxygen cation, O_2^+, by the following reaction:

$$4OF_2 + 2AsF_5 \longrightarrow 2O_2AsF_6 + 3F_2$$

If 14.0 g F_2 is bubbled through 650 mL of 0.500 M NaOH to prepare OF_2 with a 78.0% yield, how much O_2AsF_6 can be prepared?

*4.101 To obtain pure chromium, chemists first treat the ore chromite with molten alkali and oxygen to convert chromium(III) to chromium(VI), which is then dissolved in water and eventually precipitated as sodium dichromate.

$$4FeCr_2O_4 + 8NaOH + 7O_2 \longrightarrow$$
$$4Na_2Cr_2O_7 + 2Fe_2O_3 + 4H_2O$$

The sodium dichromate is then reduced with carbon to chromium(III) oxide.

$$Na_2Cr_2O_7 + 2C \longrightarrow Cr_2O_3 + Na_2CO_3 + CO$$

This oxide is reduced with aluminum.

$$Cr_2O_3 + 2Al \longrightarrow Al_2O_3 + 2Cr$$

What mass of chromium is obtained when 485 g of an ore containing 45% chromite is treated as outlined?

▶ ▶ ▶ Chemistry in Practice

4.102 A compound of vanadium contains 29.40% V, 61.37% Cl, and 9.23% O by mass. If 2.21135 g of this compound is reacted with excess $AgNO_3$, 5.48593 g AgCl is formed. The vanadium product contains 56.02% V and 43.98% O by mass.

   a. Write formulas for the vanadium compounds.

   b. What are the names of the silver compounds?

   c. What are the oxidation numbers of the elements in the four compounds?

   d. Write a balanced equation for the reaction between the vanadium compound and $AgNO_3$.

   e. Use this information to calculate the atomic weight of vanadium.

5

Thermochemistry

▶ ▶ ▶ ▶

The reaction between zinc and iodine generates so much heat that some iodine is vaporized.

CHAPTER OUTLINE

Encountering Chemistry

Zinc reacts with iodine in a direct synthesis reaction to make zinc iodide. When the two solids are mixed, very little happens. However, once some water is added, the reaction occurs vigorously:

$$Zn(s) + I_2(aq) \longrightarrow ZnI_2(aq)$$

The reaction mixture becomes so hot that excess iodine is vaporized, producing a purple cloud above the reaction mixture. In this reaction, energy is given off and heats the reaction components. Other reactions use up energy. For example, when solid barium hydroxide octahydrate reacts with solid ammonium chloride, energy is removed from the container, which becomes so cold that it will freeze water, as shown in Figure 5.1:

$$Ba(OH)_2 \cdot 8H_2O(s) + 2NH_4Cl(s) \longrightarrow BaCl_2 \cdot 2H_2O(s) + 2NH_3(aq) + 8H_2O(l)$$

Figure 5.1 When solid barium hydroxide octahydrate reacts with solid ammonium chloride, the reaction uses energy, causing the container to become so cold that it freezes water.

Chemical reactions, then, involve not only a change in the chemical composition of the components but also a change in energy manifested in a loss or gain of heat.

The investigation of the heat changes that accompany chemical reactions is the study of thermochemistry, which is the focus of this chapter. The chapter begins by defining the various forms of energy and explaining how they are interconverted. It turns next to the measurement of the heat associated with chemical reactions. It explains the heat changes that accompany changes of state and changes of chemical composition. Then it describes how numerical values are assigned to the heat changes that accompany certain reactions and how these values are used to predict the values for other reactions. Finally, the chapter examines the chemistry involved in the use of coal as a fuel and as a raw material for the production of other chemicals.

5.1   Energy

A change in a physical object is accompanied by an energy change. A basketball, if nudged over the edge of a step, will roll down a stairway, releasing energy in the process. This release or production of energy is related to the spontaneous process of rolling down the stairs. The basketball will not roll back to the top of the stairs by itself. It will do so only if energy is used to roll it there.

Similarly, chemical changes are usually accompanied by energy changes. Some chemical reactions produce energy, whereas others use energy. ▶ We saw in the chapter introduction that the reaction of zinc and iodine produces energy. Another example is the reaction of sodium and chlorine to form sodium chloride, shown in Figure 5.2:

$$2Na(s) + Cl_2(g) \longrightarrow 2NaCl(s)$$

In this reaction, enough energy is released to cause sparks to fly from the sodium metal. This reaction occurs spontaneously, without having to be forced. But sodium chloride does not spontaneously decompose to give back sodium metal and

Figure 5.2 Sodium, a reactive alkali metal, and chlorine, a reactive halogen, react vigorously in a direct synthesis reaction to form sodium chloride, commonly known as salt.

chlorine gas. It can be decomposed, but energy must be used to melt the solid, and more energy must be used to pass electricity through the liquid to obtain the elements, a process called **electrolysis**:

$$2\text{NaCl}(l) \xrightarrow{\text{electrolysis}} 2\text{Na}(l) + \text{Cl}_2(g)$$

RULES OF REACTIVITY

Only a few chlorides, those of unreactive metals like gold and platinum, can be decomposed to the elements by heating. Other metal chlorides can be decomposed to the elements only by electrolysis of the molten salts.

Only when energy is added to the sodium chloride will it once again form the elements from which it was produced.

But what is energy? **Energy** is defined as the capacity to do work or to transfer heat. **Work,** usually taken to mean mechanical work, is the result of a force acting over a distance. Thus, work is done when a person pushes a basketball up a stairway or when compressed gases, resulting from the combustion of a fuel, push the piston in the cylinder of an automobile engine. Not all reactions can be made to do work directly. However, when a reaction produces heat, the heat can be used to do work indirectly—by transfer to a water boiler in a turbine-operated generator, for example.

Forms of Energy

Energy takes many different forms, and it can be converted from one form to another. The primary types of energy are *kinetic energy* and *potential energy*. **Kinetic energy** is the energy of motion. The basketball rolling downstairs possesses kinetic energy. An object of mass m moving at a constant velocity v has a kinetic energy (E) given by the equation

$$E = \tfrac{1}{2}mv^2$$

What types of energy transformations are involved in everyday activities such as getting from your home to your chemistry class?

Potential energy is energy possessed by an object because of its position. Thus, the basketball resting at the top of the stairs has potential energy. If the ball did not contain this stored energy, it could not release energy when it rolled down the stairs. Any object in a position to be rolled, dropped, thrown, or otherwise induced into motion has potential energy that will be converted to kinetic energy once the motion is initiated.

An object may have kinetic energy, potential energy, or both, but it also has *internal energy*. The particles that make up the object—the atoms, molecules, or ions—have kinetic energy from their motion and potential energy from their positions with respect to one another. The sum of the kinetic and potential energies of these fundamental particles is the **internal energy** of the object. The total energy of a basketball, then, is the sum of its kinetic energy (if it is in motion), its potential energy, and its internal energy.

Other types of energy—chemical, mechanical, electrical, and thermal energy, for example—are really just forms of kinetic or potential energy. For example, chemical compounds can release **chemical energy,** energy associated with a chemical reaction. Chemical energy is potential energy arising from the positions of the atoms, ions, and electrons in the compounds. A compound can release its potential energy by undergoing a chemical reaction or change in composition to substances that store less potential energy.

Chemical compounds can also have kinetic energy. The molecules in gases move around much faster than do the molecules in liquids or solids, so gases have more kinetic energy. Thus, a change of physical state involves a change in kinetic energy. Molecules also move faster as the temperature is raised. The motion of molecules or atoms is associated with *thermal energy,* or heat—kinetic energy that changes with temperature.

Electrical energy is associated with the passage of electricity—the motion of electrons, generally in metals. Electrical energy added to molten sodium chloride converts the compound to new substances that contain more potential energy. Electrical current passed through the filament of a light bulb causes the metal to glow red as the electrons pass through it, increasing the motion of the atoms; so electrical energy shows up as kinetic energy in this case. The light given off by a light bulb is also a form of energy, called *radiant energy* or light energy.

Mechanical energy is energy associated with work, the exertion of a force through a distance. Because motion is involved, this is a form of kinetic energy.

Nuclear energy is released when nuclei of one element are converted to nuclei of another element. Examples of such conversions are *fission,* a process in which a nucleus breaks up into lighter nuclei, and *fusion,* in which two small nuclei combine to form a heavier nucleus.

All these forms of energy can be converted into one another. For example, we convert mechanical energy to potential energy if we carry a brick from the street to the top of a building. If we attach the brick to a windlass and drop it, the potential energy is converted back to mechanical energy. If the windlass is attached to a generator, the mechanical energy is converted to electrical energy. If the electrical energy is passed through molten sodium chloride, chemical energy results. If sodium and chlorine then react, the chemical energy is converted to heat. The conversions we can carry out are endless.

Energy can be equated with the motion of the fundamental particles of matter, such as atoms, molecules, ions, electrons, and nuclei. Can this definition of energy be reconciled with the description of different forms of energy?

Law of Conservation of Energy

Energy can be converted from one form to another, but the amount of energy remains the same—that is, energy cannot be created or destroyed. This principle is known as the **law of conservation of energy.** Thus, when we use energy, we are not "using it up." However, we may be converting it into forms that are not useful. When gasoline is burned in an engine, the chemical energy in the gasoline is converted into heat energy and mechanical energy. The mechanical energy does useful work, but the heat energy is generally lost to the atmosphere.

Conversions of energy, then, are usually not completely efficient, as shown in Figure 5.3. The efficiency of the conversion is the percentage of energy that ends up in the desired form. Consider the series of conversions from the chemical energy stored in coal to the radiant energy we see when we turn on an incandescent lamp. When coal is burned in a steam boiler, about 88% of its chemical energy is converted to thermal energy, which can be used to heat water and form steam. A steam turbine can convert about 46% of the thermal energy in the steam to mechanical energy, and an electrical generator can convert about 98% of this mechanical energy to electrical energy. Finally, when the electrical energy is used to operate a light bulb, only about 4% is converted to light, while the rest is lost to the surroundings as heat. In the overall process of burning coal to operate a light bulb, only 1.4% of the chemical energy ends up as light energy; the rest has been lost along the way because of inefficient conversions. Most of the "lost" energy ends up as heat, which is dissipated to the air or to the bodies of water used to cool power plants.

Figure 5.3 The conversion of one kind of energy to another is usually not completely efficient, because some energy is converted into undesired forms. Some typical conversion efficiencies are shown here.

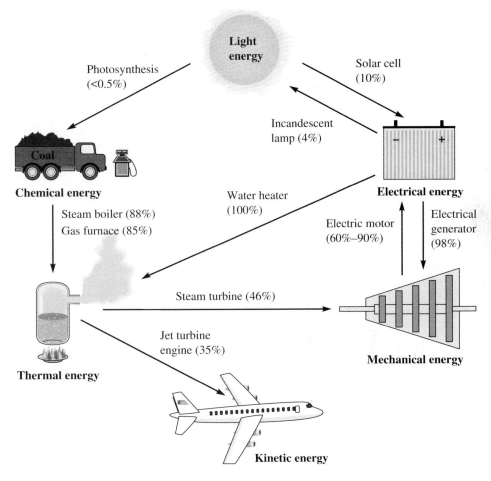

Units of Energy

Energy is measured by chemists in units of *joules* or *calories*. One **calorie (cal)** is the amount of heat energy needed to raise the temperature of 1 g of water by 1°C (from 14.5°C to 15.5°C). A **joule (J),** the SI unit of energy, is smaller than a calorie: 1 cal = 4.18400 J. The calorie used by chemists should not be confused with the **Calorie** used by nutritionists, which is actually a kilocalorie (kcal), or 1000 cal.

5.2 Calorimetry and Measurement of Heat

As mentioned in the chapter introduction, thermochemistry is concerned with the heat produced or used in a chemical reaction. In this context, **heat** means energy that is transferred between two objects because of a difference in their temperatures. If we place a beaker of water on a heated plate, energy is transferred. The plate loses heat and gets cooler, and the beaker and water gain heat and get hotter. We say that heat is transferred to the *system* from its *surroundings*. The **system** is that part of the universe in which we are interested. When we heat a beaker of water, the beaker and water constitute the system. The **surroundings** are the rest of the universe, the part in which we are not currently interested.

In mathematical manipulations, heat gain or loss is usually represented by the symbol q. When we add heat to a system, q has a positive value ($q > 0$). Conversely, negative values of q ($q < 0$) mean that the system gives off heat to the surroundings. A physical or chemical change that causes heat to be given off by a system is called an **exothermic process.** A physical or chemical change that absorbs heat is called an **endothermic process.** For example, the reaction of gaseous acetylene, C_2H_2, with chlorine gas gives off heat (Figure 5.4), so it is an exothermic reaction:

$$C_2H_2(g) + Cl_2(g) \longrightarrow 2C(s) + 2HCl(g) + \text{heat}$$

In contrast, a mixture of carbon and silicon dioxide must be heated continuously for a reaction to occur, so the reaction is endothermic:

$$\text{Heat} + 2C(s) + SiO_2(s) \longrightarrow 2CO(g) + Si(s)$$

The reaction of carbon with silicon dioxide is used in the semiconductor industry to produce silicon. This reaction is an example of a large group of reactions in which a metal or metalloid oxide reacts with carbon at high temperatures to produce the element and carbon monoxide.

Heat Capacity

When heat is added to a substance, an increase in temperature results (unless the heat is immediately converted into another form of energy). The amount of heat that must be added to a substance to raise its temperature by 1°C is a unique property of each substance, called its **heat capacity.** Heat capacity varies with the amount of substance and can be measured according to either the number of moles or the mass of the substance. These two kinds of heat capacity measurements are called the *molar heat capacity* and the *specific heat*.

The **molar heat capacity** is the amount of heat required to raise the temperature of 1 mol of a substance by 1°C. We can use the molar heat capacity (C_m) of

Figure 5.4 Acetylene, C_2H_2, can be produced by the reaction of calcium carbide, CaC_2, with water: $CaC_2(s) + 2H_2O(l) \longrightarrow Ca(OH)_2(s) + C_2H_2(g)$. Chlorine gas is produced by the reaction $NaOCl(aq) + 2HCl(aq) \longrightarrow NaCl(aq) + Cl_2(g) + H_2O(l)$. Acetylene reacts with chlorine and produces so much heat that the acetylene bursts into flame: $C_2H_2(g) + Cl_2(g) \longrightarrow 2C(s) + 2HCl(g)$.

► ► ►   WORLDS OF CHEMISTRY

Energy: A Dwindling Resource

Some scientists believe that the increased use of fossil fuels will lead to a gradual warming of the atmosphere, a phenomenon called the greenhouse effect.

From the energy expended by a hurricane or an earthquake to the energy required to make an automobile run, energy affects our lives in ways impossible to ignore. Often, when we talk about energy, we are referring to the energy people have channeled to provide heat, light, and power for homes and industries. As the sources we have relied on for this energy are depleted or are found to pose threats to our environment, efforts to develop practical new sources are intensified.

Consider, for example, electrical energy. The chief U.S. sources of this energy in recent years are oil, natural gas, coal, hydroelectric energy, and nuclear fission reactors (see Figure 5.5). Of these five major sources, four are not renewable within short time periods. In addition, all may damage the environment.

Coal, oil, and natural gas—the fossil fuels—were formed from plant and animal material over millions of years. Once they have been used up, they cannot be replaced within our lifetime. Fossil fuels are converted to electricity by combustion. The heat generated by burning fuel is used to boil water. The resulting steam is passed through huge turbines, which generate electricity (Figure 5.6). The steam is cooled somewhat by this process but must be cooled further to prevent heat pollution of the surrounding areas. The steam is passed through cooling towers, familiar sights at power plants, and the condensed water is returned to the boilers.

A typical combustion reaction is that of natural gas, which is largely composed of methane, CH_4:

$$CH_4(g) + 2O_2(g) \longrightarrow CO_2(g) + 2H_2O(g)$$

A problem with this and similar reactions is the carbon dioxide they create. Over the last hundred years or so, because of the widespread burning of fossil fuels, the carbon dioxide concentration in the earth's atmosphere has been increasing. Carbon dioxide in the atmosphere absorbs heat radiation given off by earth and reflects it back. Some scientists believe that the increased use of fossil fuels will lead to a gradual warming of the atmosphere, a phenomenon called the greenhouse effect.

Nuclear fission reactors currently supply about 18% of the electrical energy needs of the United States. Nuclear

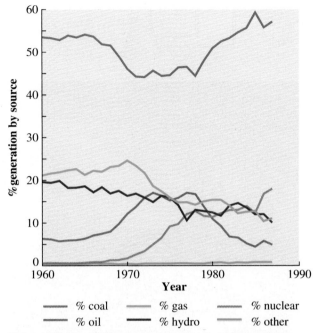

Figure 5.5 Electrical energy in the United States is produced primarily from nonrenewable fuels. Data from *Chemical and Engineering News*, Sept. 20, 1982; *The World Almanac, 1989*; and the U.S. Department of Energy.

fission is a process in which atoms are split to form lighter atoms. This process produces heat, which is used to boil water and turn a turbine. Nuclear fission relies on a limited resource, a particular isotope of uranium. Concerns about safety and the disposal of highly radioactive and toxic wastes also limit the desirability of nuclear fission. In the future, energy may be available from the nuclear fusion process, in which lighter atoms are combined to form heavier atoms. This is the process by which the sun generates its radiant energy. However, the technology for controlling fusion is not yet available.

The fifth major source of U.S. electricity, hydroelectric power, is a renewable resource because water is continually recycled through the atmosphere by evaporation and precipitation as rain and snow. Water from a reservoir behind a dam is allowed to fall at a controlled rate, turning turbines that operate electrical generators. But the percentage of U.S. electrical energy needs met by hydroelectric power is decreasing. The number of rivers suitable for the

(continued)

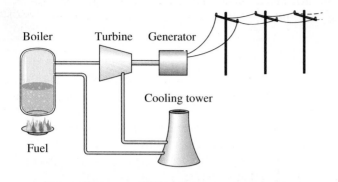

Boiler   Turbine   Generator

Cooling tower

Fuel

Figure 5.6 This is a coal-fired plant at West Jefferson, AL. Water is heated in boilers and passed through turbines in the building with the tall smokestack. Water from condensed steam is cooled in the large cooling towers.

construction of dams is limited, dams are very expensive to build and maintain, and the construction of a dam changes the ecology of the river. Some rivers, such as the Salt River in Arizona, have so many dams along their course that they are normally dry below the dams.

What about the "other" sources of energy referred to in Figure 5.5? These include geothermal energy (heat from natural underground sources), wind, tides, and solar power. We can see that, in total, their use has increased over the past three decades, but they still contribute only about 0.5% of U.S. electrical energy needs. It is difficult to predict whether we can rely on these sources to produce much more energy in the future.

Questions for Discussion

1. Discuss the advantages and disadvantages of using coal or oil as a fuel for producing electricity.
2. Within a generation or two, it is likely that there will be no inexpensive sources of oil or gas to use as fuels. Which substitute sources would be the most desirable? What criteria should be considered when making a choice?
3. The Geysers steam field, which lies north of San Francisco, has been used as an economical geothermal source of electricity since 1960. Why has geothermal energy not been used extensively in other locations?
4. Why has solar energy not yet been used extensively as a source of electricity?
5. More lives are shortened by cigarette smoking than would be shortened by a probable accident in a nuclear power plant. Does this mean we should not be so cautious about nuclear safety? Other than an accident in a power plant, what risks might be associated with nuclear energy?

a substance, along with the moles (n) of the substance and the temperature change (ΔT), to calculate the change in heat (q) associated with a temperature change.

$$q = n \times C_m \times \Delta T$$

Here, C_m is stated in units of joules per mole per degree Celsius (J/mol °C) or calories per mole per degree Celsius (cal/mol °C), and ΔT is the difference between the final temperature and the initial temperature. When the temperature increases, q has a positive value; heat has been added to the system. When the temperature decreases, q is negative; heat has been removed from the system and added to the surroundings.

Specific heat is the amount of heat required to raise the temperature of 1 g of a substance by 1°C. Thus, whereas molar heat capacity is the heat capacity per mole of substance, specific heat is the heat capacity per gram of substance. We can

Heat, a form of energy, flows from a hotter object to a colder object in contact with it. Describe the changes in these objects in terms of their atomic motion.

also calculate the change in heat (q) associated with a temperature change in a substance by multiplying the mass in grams (m) of that substance by its specific heat and by the temperature change (ΔT).

$$q = m \times \text{specific heat} \times \Delta T$$

Specific heat is given in units of joules per gram per degree Celsius (J/g °C or J g^{-1} °C^{-1} or J g^{-1} K^{-1}) or calories per gram per degree Celsius (cal/g °C). Because the calorie can be defined as the amount of energy required to raise the temperature of 1 g of water by 1°C, the specific heat of water is 1.000 cal/g °C or 4.184 J/g °C; the molar heat capacity of water is 18.02 cal/mol °C or 75.37 J/mol °C. Some representative values of specific heat and molar heat capacity for various substances are presented in Table 5.1.

Using the equations just given, we can calculate the heat changes involved in changing the temperature of a substance. Each equation involves four quantities. If any three of them are known, the fourth can be calculated readily.

Table 5.1 Some Specific Heats and Molar Heat Capacities

Substance	Specific Heat J/g °C	Molar Heat Capacity J/mol °C
Aluminum	0.895	24.3
Barium	0.192	26.4
Calcium	0.656	26.3
Carbon (diamond)	0.508	6.10
Carbon (graphite)	0.708	8.50
Chromium	0.450	23.4
Copper	0.377	23.9
Gold	0.129	25.4
Iodine	0.214	27.2
Iron	0.448	25.1
Lead	0.129	26.8
Lithium	3.54	24.6
Silicon	0.712	20.0
Silver	0.234	25.5
Sulfur (rhombic)	0.705	22.6
Tin	0.222	26.4
Uranium	0.117	27.8
Water (liquid)	4.184	75.37
Water (ice)	2.03	36.5
Water (steam)	2.02	36.3
Zinc	0.377	24.6

EXAMPLE 5.1

A simple solar energy heater consists of a glass-topped box containing rocks. Sunlight heats the rocks during the day, and air is blown over them at night to heat a house. Suppose the box contains 7.5×10^4 g of rocks, which have a specific heat of 0.49 J/g °C. How much heat is stored by the rocks if they are at a room temperature of 18°C (64°F) during the night and are heated to 43°C (109°F) during the day?

Solution   The heat is given by the following equation:

$$q = m \times \text{specific heat} \times \Delta T$$

Substituting values, we get

$$q = 7.5 \times 10^4 \text{ g} \times 0.49 \text{ J/g °C} \times (43°C - 18°C)$$
$$= 7.5 \times 10^4 \times 0.49 \times 25 \text{ J}$$
$$= 9.2 \times 10^5 \text{ J or } 9.2 \times 10^2 \text{ kJ}$$

Note that q is positive, which means that heat has been absorbed by the system (the rocks) from the surroundings.

Practice Problem 5.1

The specific heat of aluminum is 0.895 J/g °C. If 156 g of aluminum at 75.0°C is cooled to 25.5°C, how much heat is transferred?

Answer: −6.91 kJ

The change in temperature of a substance of known specific heat provides a means to measure the quantity of heat transferred in a process. Using the same equations, we can calculate how much the temperature of a system will change when a known amount of heat is added.

EXAMPLE 5.2

If 7.84×10^3 J of heat is added to 25.0 g of water at 25.0°C, what will be the final temperature of the water? The specific heat of water is 4.184 J/g °C.

Solution   We use the equation for specific heat:

$$q = m \times \text{specific heat} \times \Delta T$$

Substituting values gives

$$7.84 \times 10^3 \text{ J} = 25.0 \text{ g} \times 4.184 \text{ J/g °C} \times \Delta T$$

Next we solve for the temperature change:

$$\Delta T = 7.84 \times 10^3 \text{ J}/(25.0 \text{ g} \times 4.184 \text{ J/g °C}) = 75.0°C$$

We can obtain the final temperature (T_f) from the initial temperature (T_i) and the temperature change because

$$\Delta T = T_f - T_i$$

On rearranging and substituting values

$$T_f = \Delta T + T_i = 75.0°C + 25.0°C = 100.0°C$$

Practice Problem 5.2

If 1.52×10^3 J of heat is added to 72.5 g of aluminum at 37.2°C, what will be the final temperature of the aluminum? The specific heat of aluminum is 0.895 J/g °C.

Answer: 60.6°C

A rock can store a certain amount of heat, with the amount depending on the size, the heat capacity, and the temperature of the rock. But how do we determine the heat capacity of the rock? We can design a heat-transfer process that will allow us to determine the amount of heat transferred to the rock. In doing so, we will apply the law of conservation of energy, which states that energy cannot be lost in a process but only changed from one form to another.

The movement of heat from a hot object to a colder object results in a temperature change for both objects. As the hotter object cools off, the temperatures of the objects eventually become the same. We can express this situation mathematically as follows:

$$\text{Heat lost} + \text{heat gained} = 0 \quad \text{or} \quad \text{Heat lost} = -\text{heat gained}$$

The hotter object loses as much heat during the cooling process as the cooler object gains.

We can use this relationship to determine the heat capacity of a substance (as long as no heat is transferred to the surroundings). If we know the heat capacity of the cooler object, we can determine the amount of heat it gained by measuring the temperature change. We commonly measure specific heats by dropping a hot object of known mass into a volume of water in an insulated container (to prevent heat from being transferred to the surroundings). From the temperature change of the water and its specific heat, we can calculate how much heat the water has gained. The heat lost by the object of interest is simply the negative of this value. From the heat transferred, the mass of the object, and its temperature change, we can calculate the object's specific heat.

EXAMPLE 5.3

A 100.0-g rock is heated to 78.24°C. The rock is then placed in 100.0 g of water at 25.00°C. The final temperature of the mixture is 33.43°C. The specific heat of water is 4.184 J/g °C. What is the specific heat of the rock?

Solution   We start with the following equation:

$$q = m \times \text{specific heat} \times \Delta T$$

We use this equation to calculate the heat gained by the water

$$q_{\text{water}} = 100.0 \text{ g} \times 4.184 \text{ J/g °C} \times (33.43°C - 25.00°C) = 3.53 \times 10^3 \text{ J}$$

and the heat lost by the rock

$$q_{\text{rock}} = 100.0 \text{ g} \times \text{specific heat} \times (33.43°C - 78.24°C)$$
$$= -4.481 \times 10^3 \text{ g °C} \times \text{specific heat}$$

(continued)

Substituting these two heat-change equations into the equation for the conservation of energy allows us to solve for the remaining unknown quantity, the specific heat of the rock:

$$q_{water} = -q_{rock}$$
$$3.53 \times 10^3 \text{ J} = -(-4.481 \times 10^3 \text{ g } °C) \times \text{specific heat}$$
$$\text{Specific heat} = 3.53 \times 10^3 \text{ J}/4.481 \times 10^3 \text{ g } °C = 0.788 \text{ J/g } °C$$

Practice Problem 5.3

A 150.0-g sample of copper is heated to 89.3°C. The copper is then placed in 125.0 g of water at 22.5°C. The final temperature of the mixture is 29.0°C. What is the specific heat of the copper?

Answer: 0.38 J/g °C

Calorimetric Measurements

The temperature changes that accompany a chemical reaction are generally measured in a **calorimeter,** an apparatus that isolates a system from its surroundings. A calorimeter consists of an insulated vessel and a temperature-measuring device. One type of calorimeter, called a *bomb calorimeter,* is shown in Figure 5.7. Bomb calorimeters are generally used to determine the heat changes involved in combustion reactions, such as the reaction of octane (C_8H_{18}, a component of gasoline) with oxygen:

$$2C_8H_{18}(l) + 25O_2(g) \longrightarrow 16CO_2(g) + 18H_2O(g)$$

RULES OF REACTIVITY

The products of combustion reactions involving the reaction of a compound with oxygen are the same as those the component elements form in a direct synthesis reaction with oxygen.

The octane and oxygen are placed in the bomb, a steel container built to withstand the high pressures of excess oxygen introduced to make the reaction proceed rapidly. The octane is ignited by an electrical spark and burns inside the bomb. The heat generated by this combustion reaction is transferred to the bomb and the water in which it is immersed. The insulation prevents heat from being transferred to the surroundings. We know that the heat gained by the water and the calorimeter must equal the heat produced by the reaction. Therefore, we can use the temperature change of the water and the bomb to calculate the amount of heat generated by the combustion reaction.

Suppose we burn 0.600 g of octane in a bomb calorimeter containing 2.500×10^3 g of water, with a bomb that has a heat capacity of 1.09×10^4 J/°C.

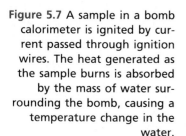

Figure 5.7 A sample in a bomb calorimeter is ignited by current passed through ignition wires. The heat generated as the sample burns is absorbed by the mass of water surrounding the bomb, causing a temperature change in the water.

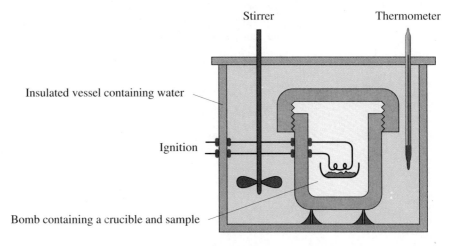

The temperature of the water and the bomb changes from 24.984°C to 26.328°C. To determine how much heat the reaction produces, we first calculate the heat gained by the bomb, which is a function of its heat capacity and the temperature change:

$$q_{bomb} = C_{bomb} \times \Delta T = 1.09 \times 10^4 \text{ J/°C} \times (26.328°C - 24.984°C) = 1.46 \times 10^4 \text{ J}$$

Next, we calculate the heat gained by the water, determined from its specific heat and the temperature change:

$$q_{water} = m_{water} \times \text{specific heat}_{water} \times \Delta T$$
$$= 2.500 \times 10^3 \text{ g} \times 4.184 \text{ J/g °C} \times (26.328°C - 24.984°C) = 1.406 \times 10^4 \text{ J}$$

The heat produced by the combustion reaction can now be obtained from the equation for conservation of energy, which in this situation includes three terms:

$$q_{combustion} = -(q_{water} + q_{bomb}) = -(1.46 \times 10^4 \text{ J} + 1.406 \times 10^4 \text{ J}) = -2.87 \times 10^4 \text{ J}$$

Because we know how much octane was burned (0.600 g), we could also calculate the heat of combustion per gram or per mole.

A simpler calorimeter, which consists only of an insulated container, is often used to measure the heat generated by reactions in solution. For example, the reaction of aqueous hydrochloric acid with aqueous sodium hydroxide can be carried out directly in the insulated container:

$$HCl(aq) + NaOH(aq) \longrightarrow NaCl(aq) + H_2O(l)$$

RULES OF REACTIVITY

Acids react with bases to give a salt and water.

The heat produced by the reaction is transferred to the solvent in the solution and to the calorimeter. Assuming no heat loss to the surroundings, we again have three terms in the equation for conservation of energy:

$$q_{reaction} = -(q_{solvent} + q_{calorimeter})$$

To calculate how much heat the reaction has produced, we use a procedure identical in principle to the one described for the bomb calorimeter.

EXAMPLE 5.4

The chemical reaction between hydrochloric acid and sodium hydroxide is carried out by mixing 50.0 g of dilute HCl(aq) solution with 50.0 g of dilute NaOH(aq) solution in a calorimeter. Because the solution is dilute, it consists mostly of water, and its specific heat can be considered identical to that of water, 4.184 J/g °C. The heat capacity of the calorimeter has been determined to be 24.0 J/°C. If the temperature of the system increases by 6.00°C, what is the heat generated by the reaction?

Solution From the increase in temperature and the heat capacity, we can calculate the heat gained by the calorimeter:

$$q_{calorimeter} = C_{calorimeter} \times \Delta T = 24.0 \text{ J/°C} \times 6.00°C = 144 \text{ J}$$

To calculate the heat gained by the solvent, we multiply its mass times its specific heat times the temperature change:

$$q_{solvent} = m_{solvent} \times \text{specific heat}_{solvent} \times \Delta T$$
$$= 100.0 \text{ g} \times 4.184 \text{ J/g °C} \times 6.00°C = 2.51 \times 10^3 \text{ J}$$

(continued)

The heat produced by the reaction is obtained from the equation for conservation of energy:

$$q_{reaction} = -(q_{solvent} + q_{calorimeter}) = -(144 \text{ J} + 2.51 \times 10^3 \text{ J}) = -2.65 \times 10^3 \text{ J}$$

Practice Problem 5.4

The reaction between 62.5 g of dilute hydrochloric acid and 62.5 g of dilute sodium hydroxide is carried out in a calorimeter. The specific heat of the solution is 4.184 J/g °C. The heat capacity of the calorimeter has been determined to be 92.0 J/°C. If the temperature of the system increases by 7.24°C, what is the heat generated by the reaction?

Answer: -4.45×10^3 J

5.3   Enthalpy Change and Changes of Physical State

Even though chemists still use the term *calorimeter*, they have switched from units of calories to units of joules. Discuss the merits of extending this change in units to nonscientific terminology. For example, should we change soda labels to read "low-Joule cola," as some countries have done?

A hot rock contains energy. If we drop the rock into water, heat is transferred to the water, causing an increase in the water's temperature. The rock contains some specific amount of energy that can be transferred as heat, but we do not know how much. All we can measure is the amount of heat transferred to the water. In all our discussions so far in this chapter, we have been dealing with heat changes.

The amount of heat transferred depends somewhat on how the process is carried out, at constant pressure or at constant volume. (*Pressure* is the amount of force exerted on a substance per unit area.) The difference between these conditions is primarily important for processes that involve gases. Because most chemical processes, whether in the laboratory, the ocean, the atmosphere, or our bodies, are carried out at constant pressure, we will focus on the heat at constant pressure, symbolized q_p. We give such a heat change a special name, the change in **enthalpy, *H*:**

$$q_p = \Delta H$$

where the symbol delta, Δ, is used to denote a change in a property.

Enthalpy, then, is related to the amount of energy that can be transferred as heat by a substance at constant pressure. We don't know how much energy is represented by the enthalpy, because we can only measure changes in the heat content of the substance. The enthalpy change, which can be measured, is the difference in enthalpy between an initial and a final state, neither of which can be measured:

$$\Delta H = H_f - H_i$$

In discussing changes in enthalpy (ΔH), we use the same sign conventions as for heat (q):

$\Delta H > 0$ Heat added to the system (endothermic process)

$\Delta H < 0$ Heat removed from the system (exothermic process)

An enthalpy change is associated with any transition, or phase change, of a pure substance from one physical state—solid, liquid, or gas—to another. These transitions and their names are listed in Table 5.2, and the relationship between

Table 5.2 Changes of State

Change	Name
Solid $\longrightarrow$ liquid	Fusion (melting)
Liquid $\longrightarrow$ solid	Crystallization (freezing)
Liquid $\longrightarrow$ gas	Vaporization
Gas $\longrightarrow$ liquid	Condensation
Solid $\longrightarrow$ gas	Sublimation
Gas $\longrightarrow$ solid	Deposition

them is shown in Figure 5.8. In general, the solid form of a substance contains less energy than the liquid form, which contains less energy than the gaseous form. The change in enthalpy for each of these transitions is called the enthalpy change of the process or, more commonly, the heat of the process. For example, ΔH_{fusion} is called the enthalpy change of fusion, or heat of fusion.

The enthalpy change for a given change of state is the negative of the value for the reverse process:

$$\Delta H_{fusion} = -\Delta H_{crystallization}$$
$$\Delta H_{vaporization} = -\Delta H_{condensation}$$
$$\Delta H_{sublimation} = -\Delta H_{deposition}$$

These equations must be true because the initial and final states are just the opposite for the reverse process. Thus, the enthalpy change, or heat of fusion, is given by

$$\Delta H_{fusion} = H_{liquid} - H_{solid}$$

whereas the heat of crystallization is

$$\Delta H_{crystallization} = H_{solid} - H_{liquid}$$

Each type of change of state takes place at one constant temperature. Suppose we have a piece of ice whose temperature is $-40.0°C$. If we heat the ice at a constant pressure of 1 atm, it will go through the temperature and phase changes shown in Figure 5.9. The temperature will rise until the ice reaches $0.0°C$, at which point it will begin to melt. The temperature will stay constant until the ice has melted completely because the added energy is being used to overcome the forces that hold the water molecules together in ice. When the ice has melted, the

Describe properties of molecules in the three states of matter that are consistent with the energy content in these states.

Figure 5.8 Changes in physical state.

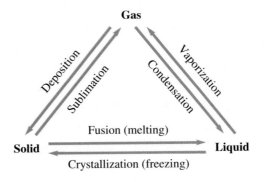

Figure 5.9 Changes of state for H$_2$O at 1 atm pressure as the temperature changes over a period of time because of a constant input of heat. When the H$_2$O undergoes a change of physical state, the temperature does not change until the change of state is completed.

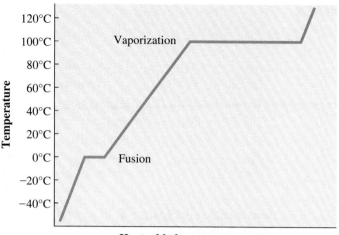

Heat added at a constant rate

temperature will start to rise again at a rate determined by the specific heat of water. When the water reaches 100.0°C, it will begin to boil; the water will stay at 100.0°C until it has completely vaporized. Continued heating will then increase the temperature of the steam to, say, 120.0°C.

Why is water used to extinguish fires?

It is possible to calculate the heat required to carry out the entire process by which ice is transformed to steam, or any other process involving temperature changes and changes in physical state. We can represent the process as follows:

Ice at $-40.0°C \xrightarrow{[1]}$ ice at $0.0°C \xrightarrow{[2]}$ water at $0.0°C \xrightarrow{[3]}$

water at $100.0°C \xrightarrow{[4]}$ steam at $100.0°C \xrightarrow{[5]}$ steam at $120.0°C$

The enthalpy change for the entire process is simply the sum of the enthalpy changes for the individual steps in the process:

$$\Delta H = \Delta H_{\text{heat ice}} + \Delta H_{\text{fusion}} + \Delta H_{\text{heat water}} + \Delta H_{\text{vaporization}} + \Delta H_{\text{heat steam}}$$

Values of the enthalpy changes that accompany changes of state can be measured for many pure substances, such as water. The enthalpy changes for the other steps are calculated from the number of moles or the mass of the substance and its molar heat capacity or specific heat.

Suppose we start with 1.000 mol of ice at $-40.0°C$. To calculate how much heat is required to raise the temperature of ice to $0.0°C$, the point at which it will begin to melt, we use the following equation:

$$\Delta H = n\, C_{\text{m}}\, \Delta T$$
$$\Delta H_1 = 1.000\ \text{mol} \times 36.5\ \text{J/mol·°C} \times [0.0°C - (-40.0°C)] = 1.46 \times 10^3\ \text{J}$$

The molar heat of fusion for the melting of ice is 6009.5 J/mol. To obtain the enthalpy change, we multiply the heat of fusion by the number of moles:

$$\Delta H_2 = n\, \Delta H_{\text{fusion}} = 1.000\ \text{mol} \times 6009.5\ \text{J/mol} = 6.010 \times 10^3\ \text{J}$$

The temperature of the water, now liquid, is then raised by 100.0°C. The heat required for this part of the process is calculated from the molar heat capacity of liquid water:

$$\Delta H_3 = n\, C_{\text{m}}\, \Delta T = 1.000\ \text{mol} \times 75.37\ \text{J/mol·°C} \times (100.0°C - 0.0°C) = 7.537 \times 10^3\ \text{J}$$

The evaporation of water, which takes place as the liquid boils, requires 4.07×10^4 J/mol, the molar heat of vaporization:

$$\Delta H_4 = n \; \Delta H_{\text{vaporization}} = 1.000 \; \cancel{\text{mol}} \times 4.07 \times 10^4 \; \text{J/}\cancel{\text{mol}} = 4.07 \times 10^4 \; \text{J}$$

The steam is then heated further to 120.0°C, a temperature change of 20.0°C. The enthalpy change is determined from the molar heat capacity of steam:

$$\Delta H_5 = n \; C_m \; \Delta T = 1.000 \; \cancel{\text{mol}} \times 36.3 \; \text{J/}\cancel{\text{mol}}\cdot\cancel{°C} \times (120.0\cancel{°C} - 100.0\cancel{°C}) = 726 \; \text{J}$$

The total heat required is the sum of the enthalpy changes for each step of the process:

$$\Delta H_{\text{process}} = 1.46 \times 10^3 \; \text{J} + 6.010 \times 10^3 \; \text{J} + 7.537 \times 10^3 \; \text{J} + 4.07 \times 10^4 \; \text{J} + 726 \; \text{J}$$
$$= 5.64 \times 10^4 \; \text{J}$$

In these calculations, we used moles of the substance to determine the enthalpy change for each step. However, it is more common to measure the amount of a substance by its mass. We must then convert the mass of the substance into the corresponding number of moles to calculate the enthalpy change.

EXAMPLE 5.5

Calculate the enthalpy change associated with melting 1.00 g of ice at 0°C. The molar heat of fusion of ice is 6009.5 J/mol.

Solution   The enthalpy change is given by the product of the number of moles and the molar heat of fusion:

$$\Delta H = n \; \Delta H_{\text{fusion}}$$

Because the amount of ice is given in grams, we must first convert to moles. The calculation follows this series of conversions:

$$\text{g } H_2O \longrightarrow \text{mol } H_2O \longrightarrow \text{J}$$

The molar mass of water is used to convert grams to moles, and the molar heat of fusion is used to convert moles to joules:

$$\Delta H = 1.00 \; \cancel{\text{g } H_2O} \times \frac{1 \; \cancel{\text{mol } H_2O}}{18.0 \; \cancel{\text{g } H_2O}} \times \frac{6009.5 \; \text{J}}{1 \; \cancel{\text{mol } H_2O}} = 334 \; \text{J}$$

Practice Problem 5.5

Calculate the enthalpy change for boiling 125 g of water at 100.0°C. The molar heat of vaporization of water is 4.07×10^4 J/mol.

Answer: 2.83×10^5 J or 283 kJ

5.4   Enthalpy Change and Chemical Reactions

Just as enthalpy changes can be calculated for processes like changes of state and changes in temperature, they can be calculated for chemical reactions. At the beginning of this chapter, thermochemistry was defined in a general way as the investigation of the heat changes that accompany chemical reactions. More specifically, **thermochemistry** involves the measurement and manipulation of ΔH values associated with chemical processes. The heat accompanying a chemical reaction is often called the **heat of reaction** (or the **enthalpy of reaction**). A change in enthalpy is associated with the atomic rearrangements resulting from a chemical reaction:

Why do changes in enthalpy accompany chemical reactions?

$$\Delta H = H_{\text{products}} - H_{\text{reactants}}$$

Values of ΔH are usually reported along with the chemical equation to which they pertain in one of two ways. The value of the enthalpy change can be written alongside the chemical equation, as for the following reaction between hydrogen and iodine:

$$H_2(g) + I_2(s) \longrightarrow 2HI(g) \qquad \Delta H = 53.00 \text{ kJ}$$

Hydrogen is a colorless, odorless, diatomic gas that is flammable or explosive when mixed with air or oxygen. Iodine exists as diatomic molecules that form a purple-black solid with a metallic luster and a characteristic odor. Iodine vaporizes as a purple diatomic gas. Hydrogen iodide is a colorless, nonflammable gas that has an acrid odor and fumes when exposed to moist air.

Such a representation is called a **thermochemical equation.** Alternatively, the enthalpy change can be included in the equation along with the reactants or products, indicating whether heat must be added to the chemical system when the reaction occurs or whether the reaction produces heat. For example, in the equation just given, the enthalpy change is positive, so the products must have a greater enthalpy than the reactants. Therefore, heat must be added to the reactants in this reaction, as represented by the following equation:

$$H_2(g) + I_2(s) + 53.00 \text{ kJ} \longrightarrow 2HI(g)$$

In either type of representation, several conventions apply. First, we should specify the temperature because values of ΔH are slightly sensitive to temperature. If no temperature is specified, a temperature of 25°C is assumed. A pressure of 1 atm is always assumed.

The value of ΔH is given for the number of moles of reactants and products indicated by the coefficients in the balanced equation. The value of ΔH per mole may differ depending on the chosen substance. In the hydrogen–iodine equation, for example, the change in enthalpy of 53.00 kJ pertains to 1 mol of H_2 or 1 mol of I_2 or 2 mol of HI. The enthalpy change per mole for HI, then, is simply 53.00 divided by 2:

$$\Delta H = 53.00 \text{ kJ/mol } H_2$$
$$\Delta H = 53.00 \text{ kJ/mol } I_2$$
$$\Delta H = 53.00 \text{ kJ/2 mol HI or } 26.50 \text{ kJ/mol HI}$$

Thus, ΔH is proportional to the amount of substance reacting or being produced in a reaction. ▶ Consider the equation for the reaction between zinc and iodine, discussed in the chapter introduction. Although the reaction did not occur in the solid state, it would be exothermic if it did:

$$Zn(s) + I_2(s) \longrightarrow ZnI_2(s) \qquad \Delta H = -209.1 \text{ kJ}$$

Reaction of 1 mol of $Zn(s)$ evolves 209.1 kJ of heat. Reaction of 2 mol, then, evolves 2×209.1 kJ, or 418.2 kJ. If other amounts are involved, we calculate the change in enthalpy by multiplying the number of moles by the enthalpy change per mole.

EXAMPLE 5.6

Calculate the enthalpy change involved in the combustion of 3.20 g of methane, $CH_4(g)$. The equation is

$$CH_4(g) + 2O_2(g) \longrightarrow CO_2(g) + 2H_2O(l) \qquad \Delta H = -890.4 \text{ kJ}$$

Solution Here we must combine the thermochemical manipulations with stoichiometric calculations according to the following progression:

$$g\ CH_4 \longrightarrow mol\ CH_4 \longrightarrow kJ$$

First, we convert grams of methane to moles:

$$\text{Moles } CH_4 = 3.20\ \text{g } CH_4 \times \frac{1\ \text{mol } CH_4}{16.0\ \text{g } CH_4} = 0.200\ \text{mol}$$

Next, we calculate the enthalpy change:

$$\Delta H = 0.200\ \text{mol } CH_4 \times \frac{-890.4\ \text{kJ}}{1\ \text{mol } CH_4} = -178\ \text{kJ}$$

This set of calculations can also be carried out in one continuous set of conversions, as described in Chapter 4 for stoichiometric calculations:

$$\Delta H = 3.20\ \text{g } CH_4 \times \frac{1\ \text{mol } CH_4}{16.0\ \text{g } CH_4} \times \frac{-890.4\ \text{kJ}}{1\ \text{mol } CH_4} = -178\ \text{kJ}$$

Practice Problem 5.6

Calculate the enthalpy change involved in the combustion of 45.2 g of ethane, $C_2H_6(g)$. The equation is

$$2C_2H_6(g) + 7O_2(g) \longrightarrow 4CO_2(g) + 6H_2O(g) \qquad \Delta H = -2.86 \times 10^3 \text{ kJ}$$

Answer: -2.15×10^3 kJ

In thermochemical equations, we must also specify the physical state of the reactants and products because the value of ΔH generally depends on physical state. For convenience, scientists have selected a set of standard conditions, called the **standard state,** under which values of ΔH are commonly given. The standard state is chosen to be the physical state in which a substance is stable at 1 atm pressure and a reference temperature, usually 25°C. An enthalpy change measured under these conditions is called a *standard enthalpy change* and is symbolized $\Delta H°$. An enthalpy change measured under any other conditions has a different value.

Consider the effect on the value of ΔH if we started our reaction with $I_2(g)$ instead of $I_2(s)$. We would require less heat to convert reactants to products; in fact, the reaction would become exothermic:

$$H_2(g) + I_2(g) \longrightarrow 2HI(g) \qquad \Delta H = -9.44 \text{ kJ}$$

This difference is due to the heat required to vaporize solid iodine:

$$I_2(s) \longrightarrow I_2(g) \qquad \Delta H = 62.44 \text{ kJ}$$

The energy relationships here are represented in the enthalpy diagram in Figure 5.10. This figure illustrates that the conversion of $H_2(g)$ and $I_2(s)$ into $HI(g)$ involves the same change in enthalpy overall, regardless of whether the $H_2(g)$ and $I_2(s)$ react directly or the iodine is vaporized before the reaction occurs.

Enthalpy is one of several energy-related properties of matter that are called *state functions*. A **state function** is a quantity whose value depends only on the initial and final states of a system and not on the path taken between them. Consider as an analogy the profit made by buying and selling shares of stock. The profit is the difference between the value when the stock is sold and the value when it is bought. It doesn't matter whether the stock changed value in a single step and then remained constant in value or whether the value fluctuated wildly. The profit depends only on the initial and final values.

Similarly, ΔH must be the same regardless of the number of steps involved in getting from reactants to products. Thus, ΔH is independent of both the number and the nature of the intermediate steps in an overall process. This is essentially a statement of a principle called **Hess's law:** if a reaction can be regarded as the sum of two or more other reactions, ΔH for the overall reaction equals the sum of the enthalpy changes for the component reactions.

Consider again the example outlined in Figure 5.10:

$$I_2(s) \longrightarrow I_2(g) \qquad \Delta H = 62.44 \text{ kJ}$$
$$H_2(g) + I_2(g) \longrightarrow 2HI(g) \qquad \Delta H = -9.44 \text{ kJ}$$
$$H_2(g) + I_2(s) \longrightarrow 2HI(g) \qquad \Delta H = 53.00 \text{ kJ}$$

We can react $I_2(s)$ with $H_2(g)$ in two ways, either in one step (the third reaction) or by first vaporizing solid iodine to give $I_2(g)$ (the first reaction) and then reacting $I_2(g)$ with $H_2(g)$ (the second reaction). The first two reactions added together give the third reaction:

$$I_2(s) + H_2(g) + \cancel{I_2(g)} \longrightarrow \cancel{I_2(g)} + 2HI(g)$$

Because $I_2(g)$ appears on both sides of the equation, it can be canceled out:

$$I_2(s) + H_2(g) \longrightarrow 2HI(g)$$

Figure 5.10 Enthalpy changes in the hydrogen–iodine system. The horizontal lines indicate the enthalpy for each state, but we do not know the absolute value of the enthalpy. We can measure only the differences between the various states.

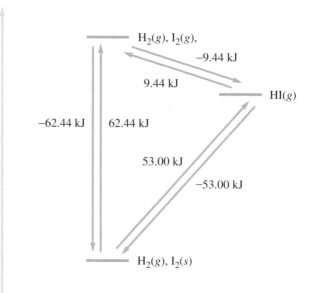

Similarly, adding together the first two values of ΔH gives the third value:

$$\Delta H = 62.44 \text{ kJ} - 9.44 \text{ kJ} = 53.00 \text{ kJ}$$

We need note only one additional characteristic of enthalpy changes to be able to manipulate thermochemical equations correctly. Because ΔH is a state function, the value of ΔH for a reaction is equal in magnitude to, but opposite in sign from, ΔH for the reverse reaction. Thus, for the $H_2(g) + I_2(g)$ system, we have

$$H_2(g) + I_2(g) \longrightarrow 2HI(g) \qquad \Delta H = -9.44 \text{ kJ}$$
$$2HI(g) \longrightarrow H_2(g) + I_2(g) \qquad \Delta H = +9.44 \text{ kJ}$$

Enthalpy goes up or down depending on the direction of the chemical change, but the enthalpy change has the same magnitude.

Hess's law provides a method for calculating enthalpy changes for processes that might be difficult to measure directly. The method involves manipulating chemical equations with known values of ΔH to give the desired chemical equation, then manipulating the values of ΔH in the same manner. If the coefficients in the equation are doubled, for example, the value of ΔH is doubled. If the equation is reversed, the sign of the value of ΔH is reversed. If two equations are added, the associated values of ΔH are also added.

An alternative to the mining and burning of coal as a fuel is the gasification of coal underground to produce a mixture of gases known as water gas:

$$C(s) + H_2O(g) \longrightarrow CO(g) + H_2(g) \qquad \Delta H = +131.4 \text{ kJ}$$

This process requires heat, which is a disadvantage. But compared to coal, the gases are easier to remove from underground and to transport to the site where fuel is needed. The water gas can be used as a fuel (though not in homes, because carbon monoxide is poisonous). When it is burned, the carbon monoxide is converted to carbon dioxide, the hydrogen is converted to water, and 525.1 kJ of heat per mole is produced:

$$CO(g) + H_2(g) + O_2(g) \longrightarrow CO_2(g) + H_2O(g) \qquad \Delta H = -525.1 \text{ kJ}$$

Determine the net enthalpy change that results from the gasification of coal, which uses heat, and the combustion of water gas, which produces heat.

First, we add the two equations:

$$CO(g) + H_2(g) + O_2(g) + C(s) + H_2O(g) \longrightarrow CO_2(g) + H_2O(g) + CO(g) + H_2(g)$$

Next, we cancel the substances appearing on both sides of the equation:

$$C(s) + O_2(g) \longrightarrow CO_2(g)$$

Because we added the two equations, we also add the two values of ΔH:

$$\Delta H = +131.4 \text{ kJ} - 525.1 \text{ kJ} = -393.7 \text{ kJ}$$

The net thermochemical equation is

$$C(s) + O_2(g) \longrightarrow CO_2(g) \qquad \Delta H = -393.7 \text{ kJ}$$

This value of ΔH is exactly the same as the value we would obtain by burning coal directly.

We must sometimes use more than two thermochemical equations to obtain the desired result. The procedure is identical to that already used, complicated only by the need to manipulate more equations.

EXAMPLE 5.7

Consider the reaction of ethane with hydrogen to give methane:

$$C_2H_6(g) + H_2(g) \longrightarrow 2CH_4(g)$$

Calculate the value of ΔH for this reaction, given the following thermochemical equations:

$$C(s) + 2H_2(g) \longrightarrow CH_4(g) \qquad \Delta H = -74.9 \text{ kJ}$$
$$2C(s) + 3H_2(g) \longrightarrow C_2H_6(g) \qquad \Delta H = -84.5 \text{ kJ}$$

Solution   We must manipulate these equations so they will add together to give the desired equation. We multiply the first equation by 2 because we need 2 mol of CH_4 in the overall equation; therefore, we also multiply ΔH by 2:

$$2C(s) + 4H_2(g) \longrightarrow 2CH_4(g) \qquad \Delta H = -149.8 \text{ kJ}$$

Next, we reverse the second equation so ethane will be a reactant:

$$C_2H_6(g) \longrightarrow 2C(s) + 3H_2(g) \qquad \Delta H = +84.5 \text{ kJ}$$

Note that reversing the equation changes the sign of ΔH. Now we add the two equations:

$$C_2H_6(g) + 2C(s) + 4H_2(g) \longrightarrow 2C(s) + 3H_2(g) + 2CH_4(g)$$

We can cancel any substances that appear on both sides of the equation:

$$C_2H_6(g) + H_2(g) \longrightarrow 2CH_4(g)$$

Because this is the correct overall equation, we add the values of ΔH of the component equations to get the overall value of ΔH:

$$\Delta H = -149.8 \text{ kJ} + 84.5 \text{ kJ} = -65.3 \text{ kJ}$$

Thus, the final thermochemical equation is

$$C_2H_6(g) + H_2(g) \longrightarrow 2CH_4(g) \qquad \Delta H = -65.3 \text{ kJ}$$

Practice Problem 5.7

Determine the enthalpy change that would result from the incomplete combustion (combustion with insufficient oxygen) of coal to give carbon monoxide:

$$2C(s) + O_2(g) \longrightarrow 2CO(g)$$

The following thermochemical equations are known:

$$C(s) + O_2(g) \longrightarrow CO_2(g) \qquad \Delta H = -393.7 \text{ kJ}$$
$$2CO(g) + O_2(g) \longrightarrow 2CO_2(g) \qquad \Delta H = -566.0 \text{ kJ}$$

Answer: -221.4 kJ

5.5 Heats of Formation

We have seen that we can calculate the enthalpy change for a reaction from thermo-chemical equations for other reactions that add up to the reaction of interest. But how do we obtain values of ΔH for the many millions of chemical reactions that are possible? The task is simplified by use of the **heat of formation** (or **enthalpy of formation**), ΔH_f, the enthalpy change associated with the formation of 1 mol of a compound from its component elements in their stable form:

$$\text{Elements} \longrightarrow 1 \text{ mol of compound} \qquad \Delta H = \Delta H_f$$

Knowing the values of ΔH_f for the substances involved in a process allows us to carry out thermochemical calculations even if we do not know the thermochemical equation and associated enthalpy change for the process. As a simple example, suppose we wanted to determine the amount of heat needed to vaporize water at 100.0°C:

$$H_2O(l) \longrightarrow H_2O(g) \qquad \Delta H = ?$$

The heats of formation of liquid and gaseous water at 100.0°C and 1.00 atm provide us with the necessary information.

$$H_2(g) + \tfrac{1}{2}O_2(g) \longrightarrow H_2O(l) \qquad \Delta H_f = -283.1 \text{ kJ}$$
$$H_2(g) + \tfrac{1}{2}O_2(g) \longrightarrow H_2O(g) \qquad \Delta H_f = -242.4 \text{ kJ}$$

Note that the heat of formation depends on the physical state of the compound. Notice too that when we write equations for ΔH_f, we write them so that only 1 mol of product is formed, even if this means using some fractional coefficients for the elements.

Why would the heats of formation of gaseous and liquid water be different?

We can manipulate heats of formation in the same way we manipulate thermochemical equations. We rearrange the formation equations—by changing their direction or multiplying by an integer—so they will add up to give the equation of interest.

To find the heat needed to vaporize water, we reverse the equation for formation of liquid water to get liquid water on the reactant side. The enthalpy change for this reaction, which we will label ΔH_1, is the negative of the heat of formation of liquid water because we have simply reversed the formation equation:

$$H_2O(l) \longrightarrow H_2(g) + \tfrac{1}{2}O_2(g) \qquad \Delta H_1 = -\Delta H_f \text{ of } H_2O(l) = +283.1 \text{ kJ}$$

We will label the enthalpy change for the formation of gaseous water as ΔH_2.

$$H_2(g) + \tfrac{1}{2}O_2(g) \longrightarrow H_2O(g) \qquad \Delta H_2 = \Delta H_f \text{ of } H_2O(g) = -242.4 \text{ kJ}$$

Now we add these two equations and cancel substances appearing on both sides:

$$\cancel{H_2(g)} + \cancel{\tfrac{1}{2}O_2(g)} + H_2O(l) \longrightarrow H_2O(g) + \cancel{H_2(g)} + \cancel{\tfrac{1}{2}O_2(g)}$$

Because we added the two equations, we also add their enthalpy changes:

$$H_2O(l) \longrightarrow H_2O(g) \qquad \Delta H = \Delta H_1 + \Delta H_2$$
$$= -\Delta H_f \text{ of } H_2O(l) + \Delta H_f \text{ of } H_2O(g)$$
$$= 283.1 \text{ kJ} - 242.4 \text{ kJ} = 40.7 \text{ kJ}$$

Notice that ΔH for the process equals ΔH_f of the product (gaseous water) minus ΔH_f of the reactant (liquid water). This observation is universally true: the enthalpy change of a reaction equals the sum of the heats of formation of the products minus the sum of the heats of formation of the reactants:

$$\Delta H = \Sigma \, n \, \Delta H_f(\text{products}) - \Sigma \, m \, \Delta H_f(\text{reactants})$$

where the symbol Σ (sigma) means "sum" and n and m are the stoichiometric coefficients of the reactants and products.

We have already seen that values of ΔH depend on temperature, pressure, and physical state. When specifying values of heats of formation, it is common to give them for the substances in the standard state. The **standard enthalpy of formation, ΔH_f°,** is the enthalpy change accompanying the formation of 1 mol of a substance from its elements in their stable form at 25°C and 1 atm. Values of the standard enthalpy of formation for a number of substances are listed in Table 5.3 (see also Appendix C for a more complete list). Note that the standard enthalpy of formation of an element in its most stable form is assigned a value of 0.00 kJ. Thus, when an element occurs in a reaction, we use a value of 0 for its ΔH_f° because the final and initial states are identical.

We can use values of ΔH_f° to calculate the standard enthalpy change of a reaction by subtracting the sum of the standard enthalpies of formation of the reactants from the sum of the standard enthalpies of formation of the products. Consider the preparation of sulfuric acid, which is described in Chapter 4. One step in this process is the reaction of sulfur trioxide with water. (Although this reaction is carried out indirectly, Hess's law states that the enthalpy change should be the same.) To carry out such a reaction, we need to know whether the reaction is exothermic or endothermic. If it is exothermic, we need to know how much heat it will produce because we will need to design our synthesis so the heat will be carried away. We may need a cooling tower to keep the reaction under control and avoid explosions or other hazards. If the reaction is endothermic, we need to know how much heat has to be supplied and how much fuel we will need to maintain the reaction.

To find out what we need to know, we must calculate the standard enthalpy change involved in the formation of sulfuric acid by reaction of sulfur trioxide with liquid water. First, we must write a balanced equation:

$$SO_3(g) + H_2O(l) \longrightarrow H_2SO_4(l) \qquad \Delta H^\circ = \, ?$$

The general equation to calculate the enthalpy change of a reaction is

$$\Delta H = \Sigma \, n \, \Delta H_f(\text{products}) - \Sigma \, m \, \Delta H_f(\text{reactants})$$

We substitute the information from the balanced equation into the general equation:

A large amount of heat is also generated by burning sulfur to form sulfur dioxide during the preparation of sulfuric acid. This heat is used to make steam to generate electricity. As a result, it is possible to sell sulfuric acid for less than the cost of the raw materials.

$$\Delta H^\circ = \Delta H_f^\circ \text{ of } H_2SO_4(l) - \Delta H_f^\circ \text{ of } H_2O(l) - \Delta H_f^\circ \text{ of } SO_3(g)$$

Now we can use standard enthalpies of formation obtained from Table 5.3 to solve the equation. We must use the values appropriate to the physical states in which the reactants and products occur.

$$\Delta H^\circ = -813.989 \text{ kJ} - (-285.83 \text{ kJ}) - (-395.7 \text{ kJ}) = -132.5 \text{ kJ}$$

Thus, the reaction is exothermic, and we must devise some means of removing the heat produced.

Table 5.3 Standard Enthalpies of Formation at 25.0°C

Substance	ΔH_f° (kJ/mol)	Substance	ΔH_f° (kJ/mol)
$Al_2O_3(s)$	−1676	$HCl(g)$	−92.307
$Br_2(l)$	0.0	$HBr(g)$	−36.4
$Br_2(g)$	+30.91	$HI(g)$	+26.5
$C(s, \text{ diamond})$	+1.887	$KCl(s)$	−435.868
$CO(g)$	−110.52	$LiCl(s)$	−408.8
$CO_2(g)$	−393.51	$MgCl_2(s)$	−641.8
$CH_4(g)$	−74.81	$MgCl_2 \cdot 2H_2O(s)$	−1280
$C_2H_6(g)$	−84.68	$Mg(OH)_2(s)$	−924.7
$C_2H_4(g)$	+52.26	$NH_3(g)$	−46.11
$C_2H_2(g)$	+226.7	$N_2O(g)$	+82.05
$C_3H_8(g)$	−103.85	$NO(g)$	+90.25
$C_6H_6(l)$	+49.028	$NO_2(g)$	+33.2
$CH_3OH(l)$	−238.7	$NaF(s)$	−571
$HCO_2H(g)$	−363	$NaCl(s)$	−411.00
$CS_2(l)$	+89.70	$NaBr(s)$	−360
$CS_2(g)$	+117.4	$NaI(s)$	−288
$CCl_4(l)$	−135.4	$Na_2O_2(s)$	−504.6
$C_2H_5OH(l)$	−277.7	$NaOH(s)$	−426.8
$CH_3CHO(g)$	−167	$O_3(g)$	+143
$CH_3CO_2H(l)$	−484.5	$PbO_2(s)$	−277
$CaO(s)$	−635.5	$PbSO_4(s)$	−920.1
$Ca(OH)_2(s)$	−986.59	$SO_2(g)$	−296.83
$CaSO_4(s)$	−1432.7	$SO_3(g)$	−395.7
$CuO(s)$	−157	$H_2SO_4(l)$	−813.989
$Fe_2O_3(s)$	−824.2	$SiO_2(s)$	−910.94
$H_2O(l)$	−285.83	$SiH_4(g)$	+34
$H_2O(g)$	−241.82	$ZnO(s)$	−348.3
$HF(g)$	−271	$Zn(OH)_2(s)$	−642.2

Many chemical equations do not have stoichiometric coefficients of 1, as those in the preceding example did. In such cases, we must multiply by the stoichiometric coefficients when summing the enthalpies of formation.

EXAMPLE 5.8

Calculate the standard enthalpy change for the combustion of methane:

$$CH_4(g) + 2O_2(g) \longrightarrow CO_2(g) + 2H_2O(l) \qquad \Delta H^\circ = ?$$

(continued)

Solution   The simplest way to carry out this calculation is to use standard enthalpies of formation in the general equation:

$$\Delta H = \Sigma\, n\, \Delta H_f(\text{products}) - \Sigma\, m\, \Delta H_f(\text{reactants})$$

For this reaction, the equation becomes

$$\Delta H^\circ = 2\Delta H_f^\circ \text{ of } H_2O(l) + \Delta H_f^\circ \text{ of } CO_2(g) - \Delta H_f^\circ \text{ of } CH_4(g) - 2\Delta H_f^\circ \text{ of } O_2(g)$$

Substituting values from Table 5.3 gives

$$\Delta H^\circ = 2 \times (-285.83 \text{ kJ}) + (-393.51 \text{ kJ}) - (-74.81 \text{ kJ}) - 2 \times 0 \text{ kJ} = -890.36 \text{ kJ}$$

Practice Problem 5.8

Calculate the standard enthalpy change for the reaction

$$N_2O(g) + 4H_2(g) \longrightarrow 2NH_3(g) + H_2O(g) \qquad \Delta H^\circ = ?$$

Answer: -416.09 kJ

5.6   Coal

The principles of thermochemistry can be used to evaluate chemical reactions as sources of heat. Combustion reactions are a recurrent theme in discussions of energy sources: they provide about 90% of the energy in the United States. Coal was once the major source of energy-producing combustion reactions in all industrial countries, but its use decreased as petroleum and natural gas became readily available. Now, as the supplies of these more convenient fuels are being depleted, coal is again playing a major role as an energy source.

What are some advantages and disadvantages of using coal as a fuel?

Many coal deposits began forming about 300 million years ago. Plants growing in swamps died and gradually built a thick layer of decaying matter on the swamp floor, forming peat. The peat was buried under layers of sediment and gradually subjected to increased pressure and heat over many years, causing the peat to convert to coal. The nature of the coal depends on the intensity of the heat and pressure to which it was subjected and the length of time it has been buried beneath the earth's surface.

The major classes of coal are listed in Table 5.4. The carbon content and the heat energy that can be obtained by combustion decrease as we proceed down the classes listed in the table.

The hardest coal, with the highest carbon content and highest heat content, is anthracite. Anthracite is also the oldest and least plentiful coal. The high heat content of anthracite makes it good for heating homes and boilers in electricity-generating plants and in industrial heating operations.

Bituminous coals are the most plentiful and the most widely used. These coals are used as fuel to generate steam in electrical power plants and are also used to make coke for steel manufacture. The lower classes of coal are also used for generating electricity.

Hydrocarbons from Coal

Coal can be converted to liquid or gaseous hydrocarbons for use as fuels or as chemical *feedstock,* starting materials to produce other materials. The lower grades

Table 5.4 Classes of Coal

	Carbon Content*			Heating Value	
Class	**% Fixed Carbon**	**% Volatile Hydrocarbons**	**% C**	**kJ/g**	**Btu/lb†**
Anthracite	91	9	93	30–35	13,000–14,000
Bituminous	65	35	79	23–35	10,000–14,000
Subbituminous	54	46	70	18–23	8000–10,000
Lignite	50	50	58	14–16	6000–7000
Peat	30	70	26	2–9	1000–4000

*Fixed carbon (elemental carbon) and volatile hydrocarbons (compounds of carbon and hydrogen) are usually distinguished in coal analyses. The percent carbon equals the percent fixed carbon plus the percent carbon found in the volatile hydrocarbons, assuming that there are no water or minerals in the coal. The numbers presented here are averages for a large number of different coals; the values for a given sample may vary considerably from these averages.
†The Btu (British thermal unit) is the amount of heat required to raise the temperature of 1 lb of water by 1°F; 1 Btu = 1.055 kJ.

are more easily converted than anthracite. The simplest method of producing these materials is by heating coal to form coke. Table 5.5 lists some of the products obtained from the volatile substances driven off in this process. They include coal tars, gases, and ammonia. The coal gas burns like natural gas and can be used as a fuel; however, it has a low heating value and produces soot.

The conversion of coal to coke by heating is not a very flexible or efficient process for producing hydrocarbons, so newer, more sophisticated processes are being developed. In the Bergius process, a mixture of coal and coal tar is reacted with hydrogen gas (or *hydrogenated*) at high temperature (475°C) and pressure (200 atm) with a catalyst. The Bergius process was the first of several hydrogenation processes developed in the early twentieth century to produce petroleum fuels

Table 5.5 Some Products from Coking of Coal

Product	**Pounds/ Ton of Coke**	**Product**	**Pounds/ Ton of Coke**
Benzene	11.80	Methane	132.0
Naphthalene	6.48	Carbon monoxide	43.2
Toluene	2.72	Hydrogen	30.4
Phenanthrene	2.26	Ethylene	19.6
Xylenes	1.33	Hydrogen sulfide	6.7
Phenol	0.95	Propylene	3.4
Anthracene	0.64	Hydrogen cyanide	1.7
Pyridine	0.16	Ammonium sulfate	20.0
Quinoline	0.13		

such as fuel oil, gasoline, and diesel oil from coal. Interest in such processes has now increased with the growing awareness that oil and natural gas are in limited supply. The basic reaction in the Bergius process is

$$nC + (n + 1)H_2 \longrightarrow C_nH_{2n+2}$$

where n is an integer. The compounds that can be produced by this reaction include CH_4, C_2H_6, and C_3H_8. The mixture of liquid hydrocarbons that results from the Bergius process can be distilled to produce gasoline, kerosene, and other products.

Gasification of Coal

Several processes react coal with steam to yield synthesis gas (also called water gas), a mixture of carbon monoxide and hydrogen, plus a certain amount of methane, CH_4, which is distilled from the coal during heating. The production of synthesis gas occurs by the following reaction:

$$C(s) + H_2O(g) \longrightarrow CO(g) + H_2(g)$$

One of these processes is the Lurgi process, developed in Germany around 1930 and currently in use in South Africa (see Figure 5.11). The resulting mixture of gases is sometimes used directly as a fuel, called intermediate Btu gas, with about 40% of the energy content of natural gas. A high Btu gas can be produced by conversion of the carbon monoxide to methane.

Synthesis gas can be converted to a wide variety of products, as shown in Figure 5.12. Conversions to methane, gasoline, and diesel fuel are important uses of synthesis gas because these substances are widely used as fuels. Synthesis gas is converted to methane in a two-step process. First, synthesis gas is heated with steam in the presence of suitable catalysts to enrich the mixture in hydrogen. Some of the carbon monoxide reacts with steam to form carbon dioxide and hydrogen in a reaction known as the water–gas shift reaction:

$$CO(g) + H_2O(g) \longrightarrow CO_2(g) + H_2(g)$$

The carbon dioxide can be removed by reaction with calcium hydroxide:

$$Ca(OH)_2(aq) + CO_2(g) \longrightarrow CaCO_3(s) + H_2O(l)$$

The reaction mixture still contains some carbon monoxide, but the mixture is enriched in hydrogen. This mixture can undergo a methanation reaction in the presence of the proper catalysts:

$$CO(g) + 3H_2(g) \longrightarrow CH_4(g) + H_2O(g)$$

Other hydrocarbons can be formed by similar processes.

Liquefaction of Coal

Coal liquefaction involves conversion of coal to synthetic crude oil through the use of processes such as the Fischer–Tropsch synthesis, which starts with synthesis gas:

$$nCO + (2n + 1)H_2 \longrightarrow C_nH_{2n+2} + nH_2O$$

Figure 5.11 The Lurgi process is a pressurized process in which coal descends into a gasifier, where it is dried and then reacted with oxygen and steam. Carbon in the bottom of the gasifier is burned with oxygen to supply sufficient heat for the reactions occurring in the upper part of the gasifier. Steam is injected into the reactor to react with the red-hot carbon, forming synthesis gas. (From Dan Todd in "The Gasification of Coal," by Harry Parry. Copyright © 1974 by Scientific American, Inc. All rights reserved.)

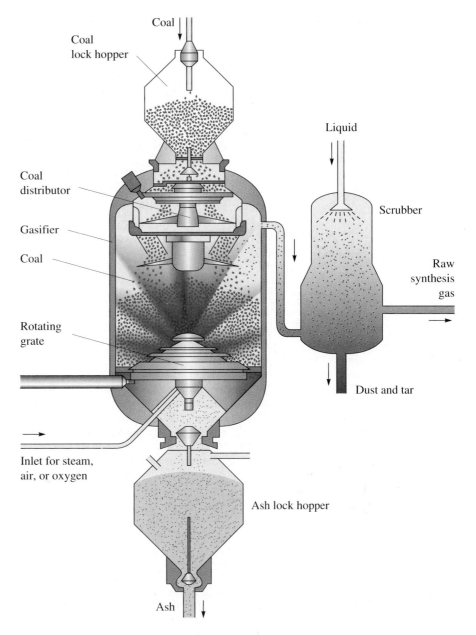

Methanol, also known as methyl alcohol or wood alcohol, was originally obtained by distillation from wood. This colorless, poisonous, flammable liquid dissolves readily in water and is used as a solvent, a fuel, and an antifreeze in air brakes. Acetic acid (called glacial acetic acid when it is not dissolved in water) is a colorless liquid with a pungent odor used in a wide variety of manufacturing processes. Vinegar derives its odor from the 5% of acetic acid (and other acids) that it contains.

This process occurs at about 200°C and high pressures in the presence of a nickel or iron catalyst. The process can also be used to produce unsaturated hydrocarbons, which contain less hydrogen than the hydrocarbons shown in the preceding equation:

$$n\text{CO} + 2n\text{H}_2 \longrightarrow \text{C}_n\text{H}_{2n} + n\text{H}_2\text{O}$$

This process could produce the volatile liquid hydrocarbon hexane, C_6H_{14}, the unsaturated hydrocarbon hexene, C_6H_{12}, or others. Depending on the reaction conditions, the ratio of reactant amounts, and the type of catalyst, mixtures of different products are obtained. These include not only hydrocarbons but also oxyhydrocarbons (compounds containing carbon, hydrogen, and oxygen) such as methanol (CH_3OH) and acetic acid ($\text{CH}_3\text{CO}_2\text{H}$).

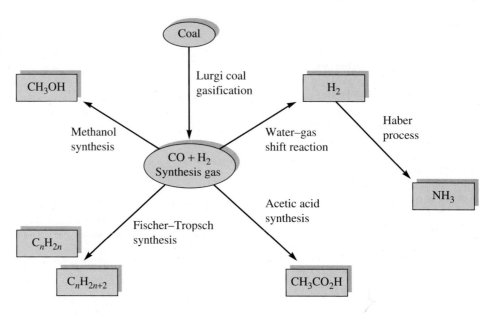

Figure 5.12 Synthesis gas formed by the gasification of coal can be converted to a variety of other products.

Summary

Energy changes accompany physical and chemical changes. Energy occurs in a variety of forms that can be converted to one another, but it cannot be created or destroyed. However, conversions of energy to new forms are rarely completely efficient. Some of the energy ends up in undesired forms, such as heat, which is readily lost to the surroundings.

Chemical reactions that produce heat are exothermic; those that consume heat are endothermic. The amount of energy that can be transferred to a substance by an increase in temperature is measured by the heat capacity. Enthalpy is the heat content of a substance at constant pressure and a given temperature. The enthalpy change occurring when an object changes temperature is the sum of the enthalpy changes for any changes of state and those due to a change in energy contained by the object as heat, as measured by the heat capacity.

Enthalpy changes for chemical reactions are often represented in the form of thermochemical equations and can be manipulated in the same way as the chemical equations they accompany. Enthalpy is a state function; an enthalpy change depends only on the initial and final states of a system and not on the path taken between them. Thus, we can calculate the enthalpy change accompanying the reaction of any amount of reactant. Enthalpy changes can be calculated from the difference in heats of formation of the products and reactants.

Coal is a major energy source, as well as a source of carbon for chemical syntheses. Coal can be burned directly or converted to other gaseous or liquid substances that can be burned to provide heat.

Key Terms

calorie (cal) (5.1)
Calorie (5.1)
calorimeter (5.2)
chemical energy (5.1)
electrolysis (5.1)
endothermic
  process (5.2)
energy (5.1)
enthalpy (H) (5.3)
exothermic process (5.2)

heat (5.2)
heat capacity (5.2)
heat of formation
  (enthalpy of
  formation; ΔH_f) (5.5)
heat of reaction
  (enthalpy of
  reaction) (5.4)
Hess's law (5.4)
internal energy (5.1)

joule (J) (5.1)
kinetic energy (5.1)
law of conservation of
  energy (5.1)
molar heat
  capacity (5.2)
potential energy (5.1)
specific heat (5.2)
standard enthalpy of
  formation (ΔH_f°) (5.5)

standard state (5.4)
state function (5.4)
surroundings (5.2)
system (5.2)
thermochemical
  equation (5.4)
thermochemistry (5.4)
work (5.1)

Exercises

Energy

5.1   What is energy?

5.2   Describe several different types of energy.

5.3   Describe how some common types of energy can be transformed from one type to another.

5.4   If energy cannot be created, what is the source of the energy that is released when gasoline is burned?

5.5   Convert an energy of 526 cal to units of joules.

5.6   Convert an energy of 145 kJ to units of calories.

5.7   Convert an energy of 876 J to units of Calories.

5.8   Sodium cyclamate, $NaC_6H_{12}NSO_3$, was a popular nonsugar sweetener until it was banned by the U.S. Food and Drug Administration. Its sweetness is about 30 times that of sugar (or sucrose, $C_{12}H_{22}O_{11}$). The heats of combustion are 3226 kJ/mol for cyclamate and 5646 kJ/mol for sucrose. What is the Calorie savings in using 1.00 g of cyclamate in place of 30.0 g of sucrose?

5.9   Calculate the kinetic energy of an automobile moving at a speed of 24 m/s and having a mass of 2.6×10^3 kg.

Specific Heat and Heat Capacity

5.10   Define *specific heat*, *heat capacity*, and *molar heat capacity*.

5.11   What is the specific heat of lead in joules if the molar heat capacity of lead is 26.8 J/mol °C?

5.12   How much heat must be added to 528 g of copper at 22.3°C to raise the temperature of the copper to 49.8°C?

5.13   Calculate the temperature change that results when 1.05 kJ is supplied to 30.0 g of aluminum.

5.14   a. If the specific heat of sulfur is 0.705 J/g °C, what is its molar heat capacity?
b. What is the heat capacity of a 28.5-g sample of sulfur?

5.15   If 20.0 g of chromium cools from 50.0°C to 25.5°C and loses 220.6 J, what is the specific heat of chromium in joules per gram per degree Celsius?

5.16   What is the specific heat of lead if 658 J of heat increases the temperature of 1.00 kg of lead by 5.10°C?

5.17   If 20.0 g of iron at 24.5°C is added to 105 g of water at 55.0°C, what will be the final temperature of the mixture?

5.18   What mass of zinc heated to 105.2°C must be added to 125 g of water at 24.8°C so the final temperature of the mixture will be 42.5°C?

5.19   Calculate the heat energy required to convert 10.0 g of ice at −10.0°C to steam at 100.0°C.

5.20   Calculate the heat energy required to convert 248 g of ice at −42.0°C to steam at 125.0°C. Consult Section 5.3 and Table 5.1 for values of needed quantities.

5.21   What is the temperature of a 1.25-mole sample of tin initially at 26.2°C if 155 J of heat is added to the tin?

5.22   What is the enthalpy change when 1.25 g of water vapor (steam) at 185.3°C is cooled to 102.1°C?

5.23   What is the final temperature of a system prepared by mixing 100.0 g of hot water (at 98.6°C) with 50.0 g of cold water (at 0.0°C)?

Calorimetry

5.24   What information is provided by a thermochemical equation?

5.25   Draw a diagram showing the essential parts of a calorimeter. Describe the function of each part.

5.26   Distinguish between a system and the surroundings.

5.27   A 6.00-g sample of coal produced enough heat when burned to raise the temperature of 2010 g of water from 24.0°C to 41.5°C.
a. Calculate the heat of combustion of coal in units of joules per gram.
b. If the coal were pure carbon, what would be the heat of combustion in units of joules per mole?

5.28   The enthalpy change of the following reaction at 25°C is known:

$$MgO(s) + H_2O(l) \longrightarrow Mg(OH)_2(s)$$
$$\Delta H = -37.7 \text{ kJ}$$

Suppose this reaction is carried out in a calorimeter containing 1250 g of water initially at 25.0°C. The calorimeter has a heat capacity of 110.5 J/°C. If 80.0 g of magnesium oxide is added to the water and reacts completely, what is the final temperature?

5.29   A 1.10-g sample of naphthalene ($C_{10}H_8$), a substance used in mothballs, was burned in a bomb calorimeter:

$$C_{10}H_8(s) + 12O_2(g) \longrightarrow 10CO_2(g) + 4H_2O(l)$$

The temperature changed from 24.36°C to 28.45°C.
a. Assuming the heat capacity of the calorimeter was 1.07 kJ/°C, calculate the amount of heat produced by the combustion of the naphthalene sample.
b. Calculate the heat of combustion of naphthalene in units of kilojoules per mole.

5.30   When 5.25 g of ice at 0.0°C was added to 125 g of water at 60.3°C in a calorimeter, what was the final temperature if the calorimeter had a heat capacity of 15.0 J/°C and all the ice melted? The enthalpy of fusion of ice at 0.0°C is 6009.5 J/mol.

Enthalpy and Change of State

5.31   Describe how standard heats of formation can be used to calculate the enthalpy change accompanying the melting of a solid.

5.32 Draw a graph of temperature versus time to plot what happens when a sample of bromine is heated at a constant rate from $-40°C$ to $100°C$. The melting and boiling points of bromine are $-7.3°C$ and $58.8°C$.

5.33 Calculate the enthalpy change for the evaporation of 105.0 g of water at $100.0°C$.

5.34 What is the enthalpy change when 155 g of liquid water is frozen to ice at $0.0°C$?

5.35 Calculate the enthalpy change when 2.50 moles of steam is condensed to liquid water at $100.0°C$.

5.36 Calculate the enthalpy change when 1.85 moles of ice is melted to liquid water at $0.0°C$.

5.37 If 22.0 g of water vapor at $125.0°C$ is condensed to the liquid at $15.3°C$, what is the enthalpy change?

5.38 Calculate the enthalpy change when 15.0 g of water at $52.5°C$ is converted to steam at $238.2°C$.

5.39 What is the enthalpy change when 105 g of ice at $-15.2°C$ is converted to liquid water at $35.6°C$?

5.40 Calculate the enthalpy change when 72.5 g of liquid water at $75.3°C$ is cooled to ice at $-102.2°C$.

5.41 Calculate the enthalpy change when 55.0 g of steam at $145.3°C$ is converted to ice at $-54.6°C$.

5.42 A 1.50-kg sample of ice increases in temperature from $-32.3°C$ to $124.8°C$. How much heat is gained by the ice?

5.43 Using standard heats of formation at $25°C$, calculate the standard heat of vaporization of water at $25°C$.

$$H_2O(l) \longrightarrow H_2O(g)$$

Enthalpy and Stoichiometry

5.44 Distinguish between exothermic and endothermic reactions.

5.45 Draw a diagram that shows the relative enthalpies of the substances involved in the reaction of graphite with oxygen gas to form carbon dioxide.

5.46 Graphite is a more stable form of carbon than is diamond. Predict the relative values of their heats of formation.

5.47 For the following thermochemical equation, $\Delta H = -2.15 \times 10^3$ kJ:

$$2C_2H_6(g) + 7O_2(g) \longrightarrow 4CO_2(g) + 6H_2O(l)$$

If enough ethane (C_2H_6) is burned to liberate 10.0 kJ of heat, what mass of liquid water will be produced at the same time?

5.48 Iron ore is reduced to iron commercially by use of carbon monoxide. Calculate the standard enthalpy change for the reduction of 2.55×10^3 kg of iron ore, Fe_3O_4, according to the equation

$$Fe_3O_4(s) + 4CO(g) \longrightarrow 3Fe(s) + 4CO_2(g)$$

5.49 A common method for detecting the presence of oxygen in a mixture of inert gases is to insert a broken light bulb connected to an electrical current. If the exposed tungsten filament reacts with oxygen gas to form tungsten(VI) oxide, the filament burns out. Calculate the enthalpy change accompanying the reaction of 0.100 g of tungsten with oxygen:

$$2W(s) + 3O_2(g) \longrightarrow 2WO_3(s)$$

5.50 Sulfur is burned (reacted with O_2) to form sulfur dioxide in the first step of the production of sulfuric acid:

$$S_8(s) + 8O_2(g) \longrightarrow 8SO_2(g)$$

Calculate the amount of heat released when 245 g of sulfur is burned in air.

5.51 Given the following balanced thermochemical equation at $25°C$, calculate the amount of heat produced when 4.45 g of propane, C_3H_8, is burned completely.

$$C_3H_8(g) + 5O_2(g) \longrightarrow 3CO_2(g) + 4H_2O(g)$$
$$\Delta H = -2025 \text{ kJ}$$

5.52 Consider the hydrogen–iodine reaction:

$$H_2(g) + I_2(s) \longrightarrow 2HI(g) \qquad \Delta H = 51.88 \text{ kJ}$$

What is the enthalpy change if 155 g of $HI(g)$ is produced with appropriate amounts of the reactants?

5.53 Methane, the principal component of natural gas, burns according to the equation

$$CH_4(g) + 2O_2(g) \longrightarrow CO_2(g) + 2H_2O(g)$$
$$\Delta H = -890.4 \text{ kJ}$$

How many grams of methane must be burned to supply 2.50×10^3 kJ of heat?

5.54 Given the following thermochemical equation for the burning of butane, a fuel often used in cigarette lighters, calculate the amount of butane necessary to produce 1.25×10^3 kJ of heat.

$$2C_4H_{10}(g) + 13O_2(g) \longrightarrow 8CO_2(g) + 10H_2O(l)$$
$$\Delta H = -5754 \text{ kJ}$$

5.55 The following thermochemical equation describes the burning of methanol:

$$2CH_3OH(g) + 3O_2(g) \longrightarrow 2CO_2(g) + 4H_2O(l)$$
$$\Delta H = -1453 \text{ kJ}$$

What mass of CO_2 would be produced by burning enough methanol to produce 857 kJ of heat?

5.56 If enough ethane is burned according to the following equation to produce 245 g of water, how much heat will be liberated?

$$2C_2H_6(g) + 7O_2(g) \longrightarrow 4CO_2(g) + 6H_2O(l)$$
$$\Delta H = -2150 \text{ kJ}$$

Thermochemical Equations and Hess's Law

5.57 Define Hess's law. Explain how it can be used to carry out thermochemical calculations.

5.58 Select some chemical system and draw a diagram to illustrate Hess's law.

5.59 Calculate the enthalpy change for the reaction

$$2Cu(s) + S(s) \longrightarrow Cu_2S(s)$$

Use the following enthalpy data:

$$S(s) + O_2(g) \longrightarrow SO_2(g) \qquad \Delta H = -296.8 \text{ kJ}$$

$$Cu_2S(s) + O_2(g) \longrightarrow 2Cu(s) + SO_2(g)$$
$$\Delta H = -217.3 \text{ kJ}$$

5.60 Calculate the enthalpy change for the reaction

$$4HCl(g) + O_2(g) \longrightarrow 2Cl_2(g) + 2H_2O(l)$$

Use the following enthalpy data:

$$H_2(g) + Cl_2(g) \longrightarrow 2HCl(g) \qquad \Delta H = -184.6 \text{ kJ}$$

$$2H_2(g) + O_2(g) \longrightarrow 2H_2O(l) \qquad \Delta H = -571.7 \text{ kJ}$$

5.61 Calculate the enthalpy change for the reaction

$$2H_2(g) + O_2(g) \longrightarrow 2H_2O(l)$$

Use the following enthalpy data:

$$2H_2O_2(l) \longrightarrow 2H_2O(l) + O_2(g) \; \Delta H = -196.6 \text{ kJ}$$

$$H_2O_2(l) \longrightarrow H_2(g) + O_2(g) \qquad \Delta H = +187.6 \text{ kJ}$$

5.62 Calculate the enthalpy change for the reaction

$$HgCl_2(s) + Hg(l) \longrightarrow Hg_2Cl_2(s)$$

Use the following enthalpy data:

$$2Hg(l) + Cl_2(g) \longrightarrow Hg_2Cl_2(s) \quad \Delta H = -224 \text{ kJ}$$

$$Hg(l) + Cl_2(g) \longrightarrow HgCl_2(s) \quad \Delta H = -265.2 \text{ kJ}$$

5.63 Calculate the heat of the reaction to convert graphite to diamond:

$$C(graphite) \longrightarrow C(diamond)$$

Use the following thermochemical equations:

$$C(graphite) + O_2(g) \longrightarrow CO_2(g)$$
$$\Delta H = -393.51 \text{ kJ}$$

$$2C(diamond) + O_2(g) \longrightarrow 2CO(g)$$
$$\Delta H = -224.84 \text{ kJ}$$

$$2CO(g) + O_2(g) \longrightarrow 2CO_2(g) \; \Delta H = -565.98 \text{ kJ}$$

5.64 Calculate the enthalpy change for the reaction

$$2S(s) + 3O_2(g) \longrightarrow 2SO_3(g)$$

Use the following thermochemical equations:

$$S(s) + O_2(g) \longrightarrow SO_2(g) \qquad \Delta H = -296.8 \text{ kJ}$$

$$2SO_2(g) + O_2(g) \longrightarrow 2SO_3(g) \quad \Delta H = -197.0 \text{ kJ}$$

5.65 Consider the reaction of ethylene with hydrogen to give methane:

$$C_2H_4(g) + 2H_2(g) \longrightarrow 2CH_4(g)$$

What is the value of ΔH for this reaction, given the following thermochemical equations?

$$C(s) + 2H_2(g) \longrightarrow CH_4(g) \qquad \Delta H = -74.9 \text{ kJ}$$

$$2C(s) + 2H_2(g) \longrightarrow C_2H_4(g) \qquad \Delta H = 52.2 \text{ kJ}$$

5.66 Calculate the enthalpy change for the reaction

$$4Al(s) + 3O_2(g) \longrightarrow 2Al_2O_3(s)$$

Use the following thermochemical equations:

$$2Zn(s) + O_2(g) \longrightarrow 2ZnO(s) \quad \Delta H = -696.0 \text{ kJ}$$

$$2Al(s) + 3ZnO(s) \longrightarrow Al_2O_3(s) + 3Zn(s)$$
$$\Delta H = -631.5 \text{ kJ}$$

Heats of Formation

5.67 Why are the standard heats of formation of zinc metal and chlorine gas both equal to zero?

5.68 Carbon in the form of diamond has a standard heat of formation at 25°C of 1.897 J/mol. Given that diamond consists of elemental carbon, explain why this value is not zero.

5.69 Determine the enthalpy of formation of $Ca(OH)_2(s)$, given the following data:

$$H_2(g) + \tfrac{1}{2}O_2(g) \longrightarrow H_2O(l) \qquad \Delta H = -285.8 \text{ kJ}$$

$$Ca(s) + \tfrac{1}{2}O_2(g) \longrightarrow CaO(s) \qquad \Delta H = -635.5 \text{ kJ}$$

$$CaO(s) + H_2O(l) \longrightarrow Ca(OH)_2(s)$$
$$\Delta H = -65.3 \text{ kJ}$$

5.70 Very pure silicon used in semiconductors can be prepared by the reduction of silicon tetrachloride with sodium metal:

$$SiCl_4(l) + 4Na(s) \longrightarrow 4NaCl(s) + Si(s)$$

Calculate the standard enthalpy change for this reaction at 25°C.

5.71 Calculate the standard enthalpy change involved in the hydrolysis of zinc oxide:

$$ZnO(s) + H_2O(l) \longrightarrow Zn(OH)_2(s)$$

5.72 Metallic copper can be produced from copper(II) oxide by reaction with hydrogen:

$$CuO(s) + H_2(g) \longrightarrow Cu(s) + H_2O(l)$$

The standard enthalpy change for this reaction is -128.8 kJ. If the standard enthalpy of formation of liquid water is -285.83 kJ/mol, what is the standard enthalpy of formation of copper(II) oxide?

5.73 Use standard enthalpies of formation to find the enthalpy change that accompanies the following reaction:

$$Fe_2O_3(s) + 3H_2(g) \longrightarrow 2Fe(s) + 3H_2O(g)$$

5.74 What is the change in enthalpy for the following reaction?

$$2C_2H_6(g) + 7O_2(g) \longrightarrow 4CO_2(g) + 6H_2O(g)$$

5.75 Calculate the standard molar enthalpy of formation of $CO_2(g)$ from the following standard enthalpy changes:

$$2C(s) + 2H_2(g) \longrightarrow C_2H_4(g) \quad \Delta H° = +52.3 \text{ kJ}$$
$$C_2H_4(g) + H_2(g) \longrightarrow C_2H_6(g) \quad \Delta H° = -137.0 \text{ kJ}$$
$$2C_2H_6(g) + 7O_2(g) \longrightarrow 4CO_2(g) + 6H_2O(l)$$
$$\Delta H° = -3119.4 \text{ kJ}$$
$$2H_2(g) + O_2(g) \longrightarrow 2H_2O(l) \quad \Delta H° = -571.6 \text{ kJ}$$

5.76   Mercuric sulfide, HgS, exists in two forms, one red and the other black. Using heats of formation, determine $\Delta H°$ at 25°C for the red-to-black transition.

5.77   Calculate the standard enthalpy of formation of $SO_3(g)$ at 25°C, given the following equations:

$$S_8(s) + 8O_2(g) \longrightarrow 8SO_2(g) \quad \Delta H = -2375 \text{ kJ}$$
$$2SO_2(g) + O_2(g) \longrightarrow 2SO_3(g) \quad \Delta H = -198 \text{ kJ}$$

5.78   Calculate the enthalpy change for the following reaction at 25°C:

$$BaO(s) + SO_3(g) \longrightarrow BaSO_4(s)$$

5.79   The standard enthalpy of the following reaction at 25°C is 286 kJ:

$$3O_2(g) \longrightarrow 2O_3(g)$$

What is the standard molar enthalpy of formation of ozone, O_3?

5.80   How much heat will be liberated when 22.8 g of sodium metal reacts with excess chlorine gas to form sodium chloride? The equation is

$$2Na(s) + Cl_2(g) \longrightarrow 2NaCl(s)$$

5.81   Calculate the heat of reaction (at 25°C) for the thermal decomposition of calcium carbonate:

$$CaCO_3(s) \longrightarrow CaO(s) + CO_2(g)$$

5.82   Use standard enthalpies of formation to find the enthalpy change, $\Delta H°$, that accompanies the following reaction:

$$P_4(g) + 6H_2(g) \longrightarrow 4PH_3(g)$$

Additional Exercises

5.83   The thermite reaction for the formation of chromium from chromium(III) oxide and aluminum involves the reaction

$$Cr_2O_3(s) + 2Al(s) \longrightarrow 2Cr(s) + Al_2O_3(s)$$

What is $\Delta H°$ at 25°C for this reaction?

5.84   Calculate the enthalpy change for the oxidation of ethane to acetic acid:

$$2C_2H_6(g) + 3O_2(g) \longrightarrow$$
$$2CH_3CO_2H(l) + 2H_2O(l)$$

The following thermochemical equations are known:

$$2C(s) + 3H_2(g) \longrightarrow C_2H_6(g) \quad \Delta H = -84.68 \text{ kJ}$$
$$2C(s) + O_2(g) + 2H_2(g) \longrightarrow CH_3CO_2H(l)$$
$$\Delta H = -484.5 \text{ kJ}$$
$$H_2(g) + \tfrac{1}{2}O_2(g) \longrightarrow H_2O(l) \quad \Delta H = -285.8 \text{ kJ}$$

5.85   What is the final temperature of the mixture formed by the addition of 195 g of silicon at 86.1°C to 345 g of water at 12.8°C?

5.86   Given the following data, calculate the heat of formation of $H_2O_2(g)$:
Enthalpy of vaporization of H_2O_2 = 51.6 kJ/mole
Enthalpy of decomposition of liquid hydrogen peroxide into liquid water and oxygen gas = −98.0 kJ/mole
$\Delta H_f°$ of $H_2O(l)$ = −285.8 kJ/mole

5.87   What mass of silver heated to 102.1°C must be added to 145 g of water at 32.1°C so that the final temperature of the mixture will be 56.4°C?

5.88   Calculate the enthalpy change for the reaction

$$3Fe(s) + 2O_2(g) \longrightarrow Fe_3O_4(s)$$

Use the following enthalpy data:

$$8Al(s) + 3Fe_3O_4(s) \longrightarrow 4Al_2O_3(s) + 9Fe(s)$$
$$\Delta H = -3450 \text{ kJ}$$
$$4Al(s) + 3O_2(g) \longrightarrow 2Al_2O_3(s) \quad \Delta H = -3352 \text{ kJ}$$

5.89   Consider the hydrogen–oxygen reaction:

$$2H_2(g) + O_2(g) \longrightarrow 2H_2O(g) \quad \Delta H = -484 \text{ kJ}$$

What is the enthalpy change if 238 g of water is produced with appropriate amounts of the reactants?

5.90   If 250.0 mL of 1.00 M $HNO_3(aq)$ at 25.0°C is mixed with 250.0 mL of 1.00 M K_2SO_4 at 25.0°C, the mixture has a temperature of 22.4°C. What is the enthalpy change for the following reaction?

$$H^+(aq) + SO_4^{2-}(aq) \longrightarrow HSO_4^-(aq)$$

The calorimeter has a heat capacity of 1.20 J/°C. The specific heat of the mixture is 4.10 J/g°C, and its density is 1.00 g/mL. Calculate the enthalpy change for the amounts used; then determine the enthalpy of reaction in units of joules per mole.

5.91   Given the following data at 25°C, determine the standard molar enthalpy of formation of carbon dioxide.

$$CO(g) + \tfrac{1}{2}O_2(g) \longrightarrow CO_2(g) \quad \Delta H° = -283.0 \text{ kJ}$$
$$C(graphite) + \tfrac{1}{2}O_2(g) \longrightarrow CO(g)$$
$$\Delta H° = -110.5 \text{ kJ}$$

5.92 If 20.0 g of aluminum at 22.5°C is added to 115 g of water at 57.3°C, what will be the final temperature of the mixture?

5.93 Calculate the enthalpies of reaction with oxygen of the following substances and indicate which you would choose as the best fuel for burning in air: AsH_3, B_2H_6, CH_4, and C_2H_6.

5.94 Prior to smelting a metal sulfide ore such as copper(II) sulfide, the ore is roasted, or heated with oxygen in air:

$$2CuS(s) + 3O_2(g) \longrightarrow 2CuO(s) + 2SO_2(g)$$

Calculate the standard enthalpy change for this reaction at 25°C.

5.95 When 150.0 g of water at 70.3°C was mixed with 95.0 g of water in a calorimeter at 21.9°C, the final temperature was 50.1°C. What is the heat capacity of the calorimeter?

5.96 Calculate the enthalpy change for the combustion of methane, given the following thermochemical equations:

$$C(s) + 2H_2(g) \longrightarrow CH_4(g) \qquad \Delta H = -74.81 \text{ kJ}$$
$$C(s) + O_2(g) \longrightarrow CO_2(g) \qquad \Delta H = -393.51 \text{ kJ}$$

$$H_2(g) + \tfrac{1}{2}O_2(g) \longrightarrow H_2O(l) \qquad \Delta H = -285.8 \text{ kJ}$$

The balanced equation for the combustion of methane is

$$CH_4(g) + 2O_2(g) \longrightarrow CO_2(g) + 2H_2O(l)$$

*5.97 Consider the equation for the reaction of sodium hydroxide with hydrogen peroxide to form sodium peroxide:

$$2NaOH(s) + H_2O_2(l) \longrightarrow Na_2O_2(s) + 2H_2O(l)$$

The standard enthalpy change for this reaction is -34.86 kJ. Calculate the enthalpy change that would occur if 25.0 g NaOH were allowed to react with 37.5 g H_2O_2.

*5.98 According to the law of Dulong and Petit, a graph of the specific heat versus the inverse of atomic weight should be a straight line for metallic elements. The slope of this straight line is about 24. What physical quantity is represented by the slope of this line?

*5.99 The heat of combustion of methane gas (CH_4) is -802.34 kJ/mol. If 85% of the heat from this reaction is transferred to water at 100°C, what mass of water can be converted to steam by burning 1.25 kg of methane?

▶ ▶ ▶ Chemistry in Practice

5.100 Dinitrogen tetroxide reacts with carbon monoxide to form carbon dioxide and a gaseous oxide of nitrogen, which contains 63.65% N and 36.35% O by mass. Known standard heats of formation are:

Compound	ΔH_f°, kJ/mol
$N_2O_4(g)$	9.7
$CO(g)$	-110
CO_2	-393

Under standard conditions, reaction of a mixture of 40.35 g N_2O_4 and 51.16 g CO in a calorimeter raised the temperature of 3255.0 g H_2O by 25.09°C.

a. Determine the formula of the oxide of nitrogen and name it.

b. Write a balanced equation for the reaction.

c. Is this reaction an oxidation–reduction reaction?

d. Determine how much of the oxide of nitrogen is formed.

e. How much heat is produced by the reaction?

f. Calculate the standard heat of formation of the oxide of nitrogen.

g. What is the standard heat of reaction per mole of N_2O_4?

6

▶ ▶ ▶ ▶

Structure of the Atom

Metals or metal compounds can color flames. Clockwise from lower left: LiCl, $SrCl_2$, NaCl, and $CuCl_2$. Center: KCl.

CHAPTER OUTLINE

Encountering Chemistry

▶ ▶ ▶ ▶

When methanol is ignited, it undergoes a combustion reaction and gives off a bluish light. If we dissolve some sodium chloride in the methanol, the flame turns yellow. Solutions containing salts of other metals give different colors: calcium, orange; barium, green; strontium and lithium, red; potassium, purple. When electricity is passed through an iron coil on a kitchen stove, the iron becomes hot and then glows, giving off red light. When electricity is passed through an incandescent light bulb, the tungsten filament heats up and gives off white light. Electricity passed through a small amount of neon gas in a sealed tube causes red light to be emitted.

What do these observations have to do with chemistry? Combustion reactions, such as that of methanol, are exothermic:

$$2CH_3OH(l) + 3O_2(g) \longrightarrow 2CO_2(g) + 4H_2O(g) + \text{heat}$$

However, the conversion of chemical to thermal energy is not completely efficient; some chemical energy is converted into energy in the form of light, giving the blue glow. Similarly, heated metals lose some of their thermal energy as light.

The explanation of how metal ions such as those in sodium chloride change the color of a flame is not as simple. The colors are produced by changes in the energy of the electrons in the metal ions. The way electrons behave when subjected to thermal energy is one of the basic experimental observations that led to the development of the modern model of the atom, which is described in this chapter and summarized in Figure 6.1.

Figure 6.1 The historical development of the modern model of the atom. (Adapted from ''Presenting the Bohr Atom,'' by Blanca L. Haendler, *J. Chem. Educ.*, Vol. 59, p. 372 (May 1982). Used by permission.)

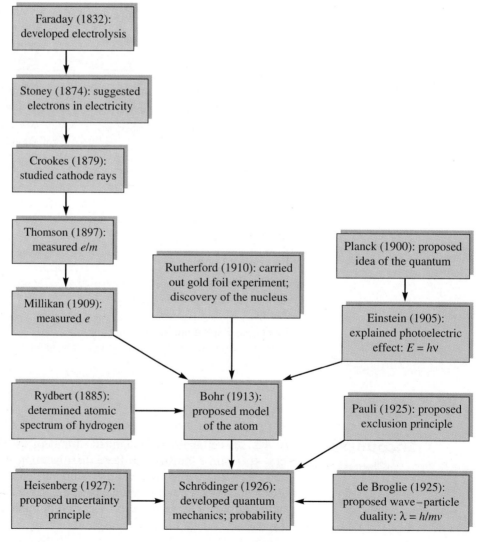

6.1 Electromagnetic Radiation

▶As described in the chapter introduction, when atoms are heated or subjected to electrical discharges, they emit light. The characteristics of the light provide clues to the electronic structure of the atoms. To understand atomic structure, then, we need to know something about light.

What characteristics do heat and light have in common?

Visible light is just one form of **electromagnetic radiation,** which also includes radio waves, microwaves, X rays, infrared light, ultraviolet light, and gamma rays. Electromagnetic radiation is a form of energy usually described as being propagated through space by waves having electrical and magnetic components. The electrical and magnetic components lie perpendicular to one another, as shown in Figure 6.2. All electromagnetic radiation can be characterized by a few properties: velocity, wavelength, amplitude, and frequency.

Velocity, Wavelength, Amplitude, and Frequency of Light

The *velocity,* or speed, of electromagnetic radiation, no matter what type, is the same: 2.997925×10^8 m s^{-1} in a vacuum. Both radio waves and light from the sun, for example, travel at this speed. This velocity is the horizontal distance that a wave crest, the highest point of the wave (or a wave trough, the lowest point), travels during a given time.

The movement of electromagnetic waves through space is associated with fluctuations or oscillations in the electrical and magnetic fields. At a given point in space, the magnitude of the fields periodically decreases to a minimum value and then increases to a maximum value as the wave crest moves forward. Two important wave characteristics, *wavelength* and *amplitude* (height) of the wave, are shown in Figure 6.3. The **wavelength** is the horizontal distance between two corresponding points on a wave. As shown in Figure 6.3, the corresponding points may be two wave crests (maxima), or two wave troughs (minima), or any other two points that recur along the wave. The wavelength is a measure of the energy of the radiation. The **amplitude** is the vertical height of the wave at its crest or depth of the wave at its trough.

Related to a wave's velocity and wavelength is its **frequency,** the number of wave crests (or other recurring points) that pass a stationary point in a second. The wavelength, frequency, and velocity of radiation are interrelated by the equation

$$\lambda \nu = c$$

where the Greek letter λ (lambda) is the wavelength in units of meters (or meters per cycle); the Greek letter ν (nu) is the frequency in units of cycles per second (s^{-1}, also called **hertz, Hz**); and c is the speed of light (2.998×10^8 m s^{-1}). Any compatible units can be used. For example, if the wavelength is given in centimeters, then the speed of light must be given in units of centimeters per second (2.998×10^{10} cm s^{-1}). Note that the frequency and wavelength are inversely related. The longer the wavelength, the lower the frequency. All electromagnetic radiation travels at the same velocity, but the wavelength (and thus the frequency) changes for different types of radiation.

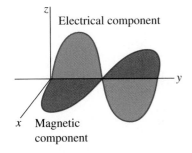

Figure 6.2 Electromagnetic radiation propagates itself in the form of waves that follow a sine curve. The electrical field and the magnetic field lie in perpendicular planes (*yz* and *xy,* respectively).

Figure 6.3 Waves are characterized by their wavelength and their amplitude. The frequency of a wave is the number of wave crests passing a given point during a unit of time.

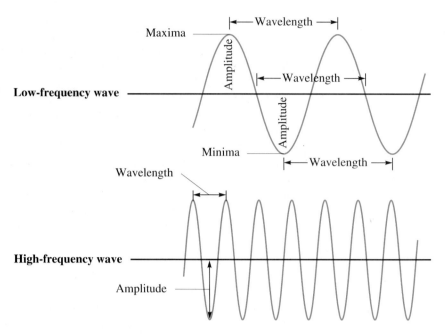

EXAMPLE 6.1

Calculate the wavelength in meters and in nanometers of the radiowaves produced by an FM station broadcasting at a frequency of 101.5 MHz.

Solution   First we convert megahertz (MHz) to s^{-1} so that the frequency will be in appropriate units to convert to the corresponding wavelength:

$$\nu = 101.5 \text{ MHz} \times (10^6 \text{ Hz/MHz}) \times (1 \text{ s}^{-1}/\text{Hz}) = 1.015 \times 10^8 \text{ s}^{-1}$$

We find the corresponding wavelength by rearranging the equation relating wavelength, frequency, and velocity as follows:

$$\lambda = c/\nu$$

Entering appropriate values yields

$$\lambda = \frac{2.998 \times 10^8 \text{ m s}^{-1}}{1.015 \times 10^8 \text{ s}^{-1}} = 2.954 \text{ m}$$

To find the wavelength in nanometers, we must know the relation between meters and nanometers: 1 nm = 10^{-9} m.

$$\lambda = 2.954 \text{ m} \times (1 \text{ nm}/10^{-9} \text{ m}) = 2.954 \times 10^9 \text{ nm}$$

Practice Problem 6.1

Calculate the wavelength in meters and in nanometers of the light generated by a heat lamp operating at a frequency of $2.5 \times 10^{13} \text{ s}^{-1}$.

Answer: 1.2×10^{-5} m; 1.2×10^4 nm

Types of Electromagnetic Radiation

Although visible light is the only form of electromagnetic radiation our eyes can detect, we all are familiar with other types as well. For example, microwaves are used for cooking, infrared light is used in heat lamps, ultraviolet light causes sunburn, radio waves are used for communications, and X rays are used for medical diagnosis. These types of radiation differ in wavelength and frequency, as shown in Figure 6.4. Note that each type is characterized not by a single wavelength but by a range of wavelengths. Thus, the identification of radiation as a certain type is somewhat ambiguous near the boundaries. Nevertheless, the best description of radiation is given by its characteristic wavelength. Note also that within the range of visible light, the wavelength ranges can be further broken down into ranges corresponding to particular colors.

Radiation is useful in the investigation of matter because matter responds in various ways to being irradiated. For example, atoms vibrate in molecules with frequencies equivalent to the frequency of infrared light, so the interaction between molecules and infrared light can provide information on the vibrations of atoms and can be used to identify the types of connections between atoms in molecules (see Figure 6.5). The energy changes that electrons undergo in atoms correspond to light in the visible and ultraviolet ranges, so such light can be used to study the energies of electrons in atoms, as described in the next section.

Figure 6.4 Types of electromagnetic radiation and their characteristic wavelengths.

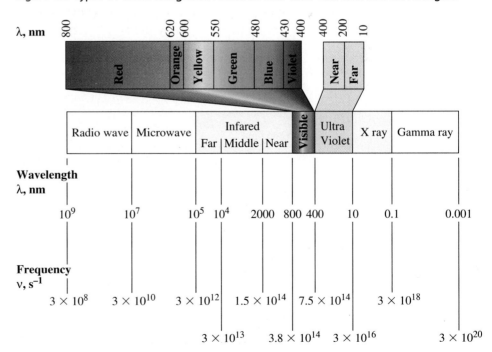

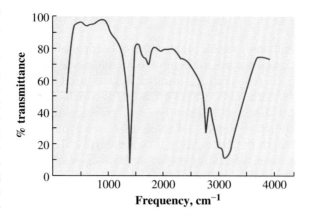

Figure 6.5 When a thin film of a substance is subjected to infrared light, certain frequencies are absorbed, while other light at other frequencies passes through the film unchanged. The frequencies that are absorbed make up a molecular "fingerprint" and can be used to identify connections between atoms in molecules and ionic compounds, as in ammonium chloride shown here. Vibration of the nitrogen and hydrogen atoms relative to one another results in absorption of light at about 3100 cm^{-1} and 1395 cm^{-1}. Any substance containing the ammonium ion absorbs light at these frequencies.

6.2   Atomic Spectra

We have seen that electromagnetic radiation has an extremely wide range of wavelengths. Devices such as prisms can be used to separate this radiation into its component wavelengths. For example, when white light—which contains all the components of visible radiation—is passed through a prism (Figure 6.6), the various wavelengths are bent to different degrees and thus separated. As a result, we see a rainbow of colors. This sequence of colors is known as a **spectrum,** the resolution, or separation, of radiation into its component wavelengths. The spectrum shown in Figure 6.6 is a **continuous spectrum** because all the component wavelengths of light are present.

How many radiation-emitting devices do you use regularly? In which part of the spectrum does their radiation occur?

Not all spectra are continuous. Consider some lamps available for lighting. Although a tungsten lamp emits light that is nearly white, a sodium lamp gives off

Figure 6.6 A continuous spectrum can be generated from white light.

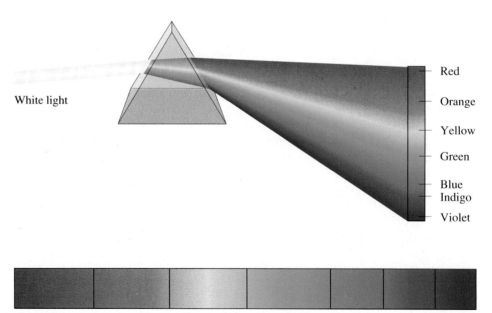

Figure 6.7 A line spectrum is generated when light from a gas-discharge tube passes through a prism. The slit helps to focus the light on the prism.

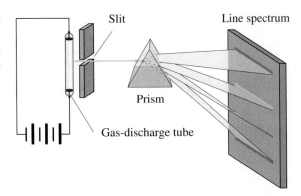

light with a yellowish color and a neon lamp emits orange-red light. The sodium and neon lamps produce light of different colors because they do not generate light of all wavelengths. When the light produced by such lamps is separated into its component wavelengths with a prism, the result is not a continuous spectrum but a **line spectrum,** which consists of a number of lines of various colors corresponding to various wavelengths (see Figure 6.7).

Line spectra from various lamps are different because of the elements from which the light is issued. Many lamps contain a gas or a mixture of gases at low pressure. When an electrical discharge is passed through such a lamp—the "neon" lamp in Figure 6.8, for example—the gas glows and emits a characteristic color. Each element generates its own unique line spectrum, ▶ as demonstrated by the different colors generated by various metal salts described in the chapter introduction.

Line spectra can be generated in two ways. When a gaseous element is heated or subjected to electrical discharges, it will emit light of specific wavelengths. Such a spectrum is called an **emission spectrum.** Alternatively, when white light

Figure 6.8 When an electrical current is passed through a gas-discharge tube containing neon at low pressure, an orange-red color is generated. This color may be altered by addition of other substances, such as mercury or argon, to the tube. Such tubes are widely used in "neon" signs.

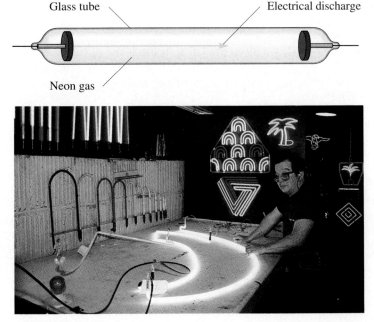

Figure 6.9 An absorption spectrum results when white light is passed through a gas. All wavelengths pass through except those absorbed by the gas. Hydrogen gas gives a spectrum with four lines in the wavelengths of visible light.

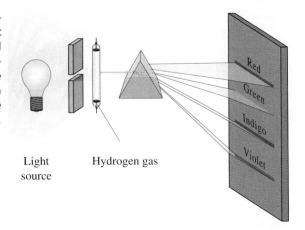

Light source   Hydrogen gas

Johann J. Balmer, a Swiss schoolteacher and physicist, derived the equation that describes the visible hydrogen spectrum in 1885. The lines observed by Balmer occur at wavelengths of 656.3 nm (red), 486.1 nm (green), 434.1 nm (blue), and 410.2 nm (violet). Other lines can be observed in the violet region as well.

is passed through a sample of the gaseous substance, an **absorption spectrum** is created (Figure 6.9). At most wavelengths, light comes out of the sample at the same intensity with which it entered; but at some wavelengths, light exits with diminished intensity because it has been absorbed by the sample. As shown in Figure 6.10, the wavelengths emitted in an emission spectrum are exactly the same as the wavelengths absorbed to produce an absorption spectrum.

The simplest line spectrum is associated with hydrogen gas. The lines that occur in the visible region of the spectrum for hydrogen are shown in Figure 6.10. This region of the hydrogen spectrum is now known as the Balmer series of lines. The positions of the lines can be described by an equation:

$$\nu = 3.2894 \times 10^{15} \text{ s}^{-1}\left(\frac{1}{2^2} - \frac{1}{n^2}\right)$$

where n can be any positive integer greater than 2. The wavelengths (λ) of the lines can be calculated from the frequencies (ν) by use of the equation introduced earlier:

$$\lambda\nu = c$$

Figure 6.10 Absorption and emission spectra have lines at the same wavelengths. In an absorption spectrum, the lines are wavelengths missing from a continuous spectrum. In an emission spectrum, the lines are the only wavelengths found. These spectra are for hydrogen gas.

If we insert the value of the speed of light and combine these two equations, we get the Balmer equation for the wavelengths, in nanometers, of the lines observed in the visible region of the hydrogen spectrum:

$$\lambda = 364.56\frac{n^2}{n^2 - 4}$$

Absorption spectrum

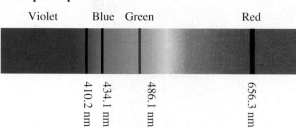

Violet   Blue Green   Red

410.2 nm   434.1 nm   486.1 nm   656.3 nm

Emission spectrum

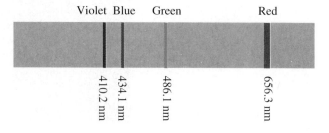

Violet Blue   Green   Red

410.2 nm 434.1 nm   486.1 nm   656.3 nm

Because n can be any integer greater than 2, we can substitute $n = 3, 4, 5, 6, \ldots$ into this equation to calculate the observed wavelengths for hydrogen.

EXAMPLE 6.2

Using Balmer's equation, calculate the wavelength of the first line in the visible spectrum of hydrogen.

Solution   The first line in the spectrum corresponds to $n = 3$ in the equation

$$\lambda = 364.56 \frac{n^2}{n^2 - 4}$$

Substitution gives

$$\lambda = 364.56 \ 3^2/(3^2 - 4) = 364.56 \times 9/(9 - 4) = 364.56 \times 9/5 = 656.2 \text{ nm}$$

The observed position of this line is at a wavelength of 656.3 nm. The difference is within the rounding error that is expected for derivation of the constant in the equation.

Practice Problem 6.2

Calculate the wavelength of the fourth line in the Balmer series for hydrogen.

Answer: 410.1 nm (observed line is 410.2 nm)

Other series of lines exist in the infrared and the ultraviolet regions of the hydrogen spectrum and can be described by equations similar to that for the visible region. The various series of lines for hydrogen are summarized in Table 6.1. All of them are accommodated by a single equation developed in 1885 by the Swedish scientist J. R. Rydberg:

$$\frac{1}{\lambda} = R\left(\frac{1}{n_1^{\ 2}} - \frac{1}{n_2^{\ 2}}\right)$$

where n_2 is greater than n_1 and both are positive integers. In this equation, R has the value $109737.31 \text{ cm}^{-1}$ for hydrogen. The Rydberg constant, R_h, listed inside the back cover, is related to this constant by $R_h = hcR$, where c is the speed of light and h is Planck's constant, discussed in the next section.

Table 6.1 Series of Lines Observed for Atomic Hydrogen

Series	Region of Spectrum	Rydberg Equation Parameters n_1	n_2
Lyman series	Ultraviolet	1	2, 3, 4, . . .
Balmer series	Visible	2	3, 4, 5, . . .
Paschen series	Near infrared	3	4, 5, 6, . . .
Brackett series	Infrared	4	5, 6, 7, . . .
Pfund series	Far infrared	5	6, 7, 8, . . .

6.3   Quantum Theory

Investigate the lives of the
scientists whose research is
described in this section. Many
of them received the Nobel
Prize in physics or chemistry
for their work.

We have seen that atoms emit radiation when they are heated or subjected to electrical discharges and that atoms of different elements emit radiation of different wavelengths. A number of discoveries made in the early twentieth century fundamentally changed scientists' ideas about how energy and matter interact in such processes.

Planck's Equation

At the end of the nineteenth century, certain assumptions about matter and energy were well accepted: Matter is made up of particles, and energy is made up of waves. Energy in the form of heat, when added to a solid, causes the atoms to vibrate, or oscillate. This oscillation transfers the energy the atoms have absorbed back out as waves of radiation. The transfer of energy was thought to be continuous—any amount could be absorbed, and a corresponding amount would be radiated. However, this explanation failed to account for a number of experimental results. The problem centered on the radiation emitted by heated black solids, called *blackbody radiation* (Figure 6.11).

In 1900, the German physicist Max Planck derived an equation that would explain blackbody radiation, but to do so he had to assume that the energy of the vibrating atoms could have not any value, as had been assumed, but only multiples of small packets, or *quanta,* of energy. This concept is called **quantum theory.** Planck proposed that the frequency of blackbody radiation is the same as the frequency of the vibrating atoms, which is related to the energy (E) transmitted by the following equation:

$$E = nh\nu = nhc/\lambda$$

where n is an integer and h is a constant, known as **Planck's constant,** with the value $h = 6.626 \times 10^{-34}$ J s. The numbers n are called **quantum numbers,** and the fact that they are integers means that the energy transmitted to and from the solid can have only certain values—$h\nu$, $2h\nu$, $3h\nu$, and so on. The $h\nu$ units, then, are the quanta of energy. Quanta for low-frequency radiation contain less energy than those for high-frequency radiation.

Figure 6.11 When a black solid is heated, it first gives off a red glow (it is called red-hot), then yellow and blue light as the temperature increases. The mixture of wavelengths gives white light (the solid is white-hot). The spectra emitted at the lower and higher temperatures are actually quite similar, as shown here, although the frequency of maximum intensity changes. Wave theory predicted that there should be a continuous increase in intensity as the frequency increases, contrary to the observed behavior. The inability to explain this phenomenon was one of the limitations of wave theory.

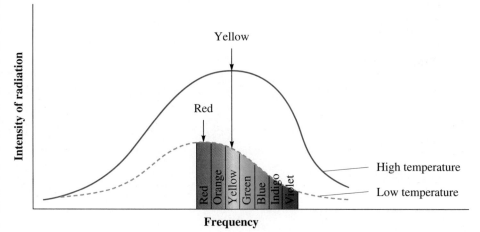

Einstein and Photons

In 1905, Albert Einstein extended Planck's discovery in an explanation of the *photoelectric effect*, the emitting of electrons from an active metal when a light of sufficiently high frequency shines on it (see Figure 6.12). Einstein proposed that electromagnetic radiation has some of the properties of particles and that the energy of these "particles," called **photons,** is related to the frequency of the radiation by Planck's equation ($E = h\nu$). Essentially, Einstein was arguing that energy is quantized not just in its interactions with matter, as Planck had demonstrated, but all the time, by its very nature.

EXAMPLE 6.3

Calculate the energy of a photon of light having a wavelength of 485 nm (green light).

Solution   The equation to be used is Planck's equation: $E = h\nu = hc/\lambda$. Consistent units are required for wavelength and speed of light:

$$c = 2.998 \times 10^8 \text{ m s}^{-1}$$
$$\lambda = 485 \text{ nm} \times (10^{-9} \text{ m/1 nm}) = 4.85 \times 10^{-7} \text{ m}$$

Using a value of 6.626×10^{-34} J s for Planck's constant, we obtain

$$E = (6.626 \times 10^{-34} \text{ J s})(2.998 \times 10^8 \text{ m s}^{-1})/(4.85 \times 10^{-7} \text{ m}) = 4.10 \times 10^{-19} \text{ J}$$

This is a rather small amount of energy, but if enough light were used to generate a mole of photons, the energy would be substantial:

$$E = 4.10 \times 10^{-19} \text{ J/photon} \times 6.022 \times 10^{23} \text{ photons/mol} = 2.47 \times 10^5 \text{ J/mol}$$

Practice Problem 6.3

Calculate the energy of a photon in a radar beam having a wavelength of 1.00 m.

Answer: 1.99×10^{-25} J

Figure 6.12 The photoelectric effect. Electrons (e^-) are emitted from an active metal as soon as light of sufficiently high frequency shines on the metal. The flow of electrons is detected by the electroscope. Wave theory was not able to explain this effect. Einstein proposed that photons striking the electrons dislodge them from their atoms if the photons have sufficient energy. The number of electrons ejected depends on the number of photons, which in turn depends on the brightness, or intensity, of the light. The energy of the ejected electrons depends on the energy of the photons, which depends on the frequency of the light, as described by Planck's equation.

The de Broglie Relationship

Einstein had shown that light waves may behave like particles. In 1925, the French physicist Louis de Broglie proposed that matter might also show wavelike properties. De Broglie concluded that the electron, thought to be a particle because it has

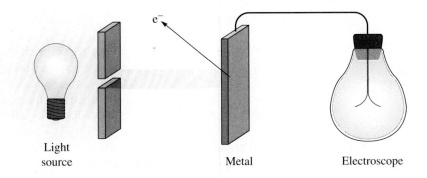

Light source   Metal   Electroscope

mass, is somehow connected with waves. Experiments carried out in 1927 by Clinton Davisson and Lester Germer at the Bell Research Labs in New York showed that a beam of electrons could indeed be made to undergo diffraction (change in direction or intensity), a phenomenon normally associated with waves, as described in Figure 6.13.

The wave–particle conflict worried physicists early in the century but is generally not considered a problem today. We have come to accept the fact that atomic particles have dual natures: in some experiments, their wave properties are most obvious; in others, they behave more like particles.

De Broglie developed an equation to describe the wave properties of particles. He reasoned that if moving particles have the properties of waves, then they should have a characteristic wavelength (λ) depending on their mass (m) and velocity (v):

$$\lambda = h/mv$$

where h is Planck's constant. The greater the mass and velocity of a particle, the shorter its wavelength, so very large particles will have wavelengths so short they cannot be observed. The de Broglie relationship is important only in the behavior of very small particles—atoms and parts of atoms.

EXAMPLE 6.4

Calculate the wavelength associated with an electron whose mass is 9.1×10^{-28} g and whose velocity is 1.2×10^7 cm s^{-1}.

Solution   An additional piece of information needed is the value of Planck's constant, $h = 6.626 \times 10^{-34}$ J s. A joule is equivalent to a kg m^2 s^{-2}, so let's convert our data to corresponding units:

$$m = 9.1 \times 10^{-28} \text{ g} \times 1 \text{ kg}/1000 \text{ g} = 9.1 \times 10^{-31} \text{ kg}$$
$$v = 1.2 \times 10^7 \text{ cm s}^{-1} \times 1 \text{ m}/100 \text{ cm} = 1.2 \times 10^5 \text{ m s}^{-1}$$

Then we can substitute into the de Broglie relationship:

$$\lambda = h/mv$$
$$= \frac{6.626 \times 10^{-34} \text{ kg m}^2 \text{ s}^{-1}}{9.1 \times 10^{-31} \text{ kg} \times 1.2 \times 10^5 \text{ m s}^{-1}} = 6.1 \times 10^{-9} \text{ m or 6.1 nm}$$

Practice Problem 6.4

Calculate the wavelength associated with a helium atom whose mass is 6.6×10^{-24} g and whose velocity is 1.4×10^5 cm s^{-1}.

Answer: 7.2×10^{-11} m or 0.072 nm

Bohr's Model

Physicists around the turn of the century had also been working to develop a satisfactory model of the atom. Ernest Rutherford, some of whose work was described in Chapter 2, had shown that the atom must be a positively charged nucleus surrounded by negatively charged electrons. But what are the electrons doing? If they were standing still, they should be attracted by the nucleus, and the atom should collapse. Only by moving at tremendous speed could the electrons keep from falling into the nucleus. However, if electron motion were responsible for the emission of light observed in atomic line spectra, then atoms with constantly mov-

Figure 6.13 When a beam of X rays (*A*) or a beam of electrons (*B*) is passed through a crystal of any metal, it is diffracted, or broken up, into areas of greater and lesser intensity. Because diffraction is a phenomenon normally associated with waves, the diffraction of electrons supported de Broglie's prediction that particles would exhibit some properties of waves.

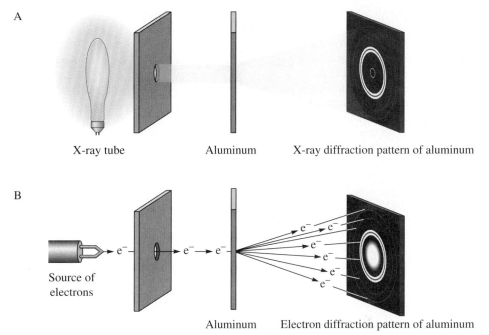

A

X-ray tube   Aluminum   X-ray diffraction pattern of aluminum

B

Source of electrons

Aluminum   Electron diffraction pattern of aluminum

ing electrons should radiate light continuously. The electrons would lose energy and the atom would eventually collapse. Clearly, this is not the case.

Niels Bohr, a Danish scientist, used quantum theory to develop a model of the atom that solved these problems. In Bohr's model, proposed in 1913, the single electron in the hydrogen atom revolved about the nucleus in one of several fixed orbits, each with a specific radius (Figure 6.14). While the electron was in an orbit, it had a fixed amount of energy. If it jumped from one orbit to another closer to the nucleus, then energy in the form of light would be emitted; if it jumped to an orbit farther away, light energy would be absorbed. On the basis of this model, Bohr derived an equation that related the energy of the electron to the principal quantum number, *n*, the number associated with the orbits shown in Figure 6.14. Because the energy can have values related only to the integral values of the principal quantum number, it is quantized. Energy can be related to the frequency or wavelength of the emitted or absorbed light by Planck's quantum equation, and Bohr used this relationship to calculate correctly the wavelengths of light in the line spectrum of the hydrogen atom.

Bohr's model worked very well for the hydrogen atom, allowing a large body of data to be explained and correlated. Unfortunately, attempts to extend the theory to atoms containing more than one electron didn't work. Even the hydrogen spectrum was later found to be more complex than early measurements had indicated, and thus more complex than could be explained with Bohr's model.

The Heisenberg Uncertainty Principle

An inherent conceptual problem in the Bohr model emerged at around the same time de Broglie was formulating his equation. The Bohr model had placed electrons in specific and well-defined orbits. But according to a young German physicist, Werner Heisenberg, it is impossible to delineate atomic structure so precisely.

Figure 6.14 In the Bohr model of the hydrogen atom, the nucleus is at the center of the atom and the electron is located in one of the circular orbits, each of which is characterized by a unique quantum number (*n*) and radius. The Lyman series of spectral lines is associated with transitions from orbits of higher principal quantum number to the *n* = 1 orbit, whereas the Balmer series of spectral lines is associated with transitions into the *n* = 2 orbit. The wavelengths of these lines calculated by Bohr exactly matched the observed values.

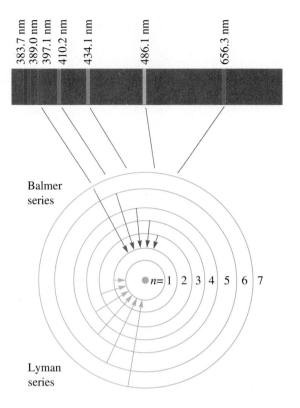

Specifically, according to the **Heisenberg uncertainty principle,** it is not possible to know both the velocity and the position of an atomic particle with a high level of certainty.

Small particles like electrons in atoms can be ''seen'' only by means of their interaction with radiation or with particles of appropriate wavelength, such as a beam of electrons. But the radiation or other particles interact so strongly with the particle that they change the particle's position. An experimenter using radiation to locate an electron is somewhat like a person trying to determine the location of a pool ball on a table in a dark room. This person could roll other pool balls across the table until an impact was heard. But then the ball would no longer be in its original position; the impact would have moved it.

We are always dealing with considerable uncertainty when trying to determine and describe the behavior of electrons. The best that can be determined is the *probability* of finding them at a given point in space.

6.4   The Quantum Mechanical Model

The ideas of de Broglie and Heisenberg are brought together in a highly mathematical model of the atom, the **quantum mechanical model,** which treats electrons not as particles but as wave patterns. This model is still in use today.

At the heart of the quantum mechanical model is the **wave mechanics** of the Austrian scientist Erwin Schrödinger. Schrödinger applied de Broglie's wave relationship to the behavior of electrons in atoms. The equation that describes the energy of the electrons is called the Schrödinger wave equation. The mathematics involved is beyond the scope of this book, so the equation will not be presented here. We should note, however, that when the wave equation is applied to a real system like the hydrogen atom, it cannot be solved unless the energy has certain discrete allowed values, which are related to one another by three variables that can have only integral values. Thus, this mathematical approach leads automatically to the concept of quantized energy and quantum numbers.

Orbitals

A key part of the wave equation is the wave function, an equation that describes the wavelike properties of the electron. A wave function by itself has no physical meaning; it is simply a mathematical expression that must be deduced to solve the wave equation. However, the square of a wave function does have physical meaning. It allows us to calculate the probability of finding an electron at any position in space. Because it deals with probability, it is consistent with the Heisenberg uncertainty principle. This model does not define exact electron paths, as the Bohr model did. Instead, the location of an electron is described by a new concept, that of the **orbital.** An orbital is a three-dimensional region in space where it is possible to find the electron most of the time. The orbital, then, can be considered a diffuse volume of electronic charge about a nucleus. Orbitals are pictured generally as having distinct shapes and volumes. The volume is determined by setting some arbitrary limit, such as 90%, on the fraction of time the electron will spend within that volume. The orbital is described by the wave function, which in turn is defined by three quantum numbers, n, l, and m_l.

The Principal Quantum Number, n

How does the idea of orbitals explain the observed spectra of hydrogen (and other) atoms?

The energy of an electron is characterized in the wave equation by the same **principal quantum number,** n, that we have already encountered. It may be any positive integer:

$$n = 1, 2, 3, \ldots$$

As in Bohr's model, the principal quantum number correlates with the distance of the electron from the nucleus; but in the quantum mechanical model, it is an average distance rather than a fixed distance. The greater the value of the principal quantum number, the farther from the nucleus will the electron most probably be found and the higher will be the energy of the electron. All electrons having the same value of n are said to be in the same electronic **shell** or principal energy level.

Recall that orbitals are the regions in space occupied by electrons. An orbital that is not occupied by an electron is really just an imaginary construct. Although it is sometimes useful to speak of the energy of an orbital, we actually mean the energy an electron would have if it occupied that orbital. All orbitals with the same value of n are in the same principal energy level or shell. The size of the orbitals is also related to n; the larger the value of n, the larger the orbital.

The Azimuthal Quantum Number, l

The **azimuthal quantum number,** l, describes the orbital's shape. Although others are possible, we commonly deal with orbitals having four different shapes,

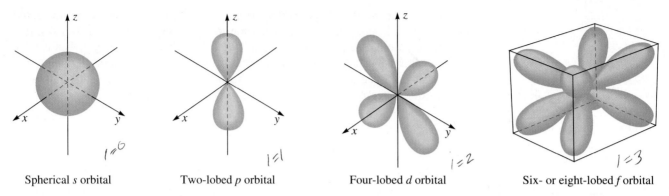

Spherical s orbital   Two-lobed p orbital   Four-lobed d orbital   Six- or eight-lobed f orbital

Figure 6.15 The four common orbital shapes.

labeled s, p, d, and f, as shown in Figure 6.15. The existence of different types of orbitals was verified experimentally by the discovery that each line in the hydrogen atom spectrum was actually composed of several closely spaced lines, as shown in Figure 6.16. Within an energy level or shell, l can be any integer from 0 to $n - 1$, where n is the principal quantum number:

$$0 \leq l \leq n - 1$$

Table 6.2 lists the value of the azimuthal quantum number for each type of orbital.

Each shell or principal energy level contains as many different types of orbitals as the value of the principal quantum number, n. Thus, for $n = 1$, there is only one, the s orbital; for $n = 2$, there are both s and p orbitals; for $n = 3$, there are s, p, and d orbitals; for $n = 4$ (and higher), there are s, p, d, and f orbitals. We generally use a shorthand designation for the energy level and type of orbital consisting of the principal quantum number followed by the letter corresponding to

The letters used to label the different types of orbitals come from old spectroscopic descriptions of the lines corresponding to these orbitals: sharp, principal, diffuse, and fundamental.

Figure 6.16 The hydrogen atom spectrum consists of more lines than early measurements had indicated. Each previously known line actually consists of several closely spaced lines of very nearly the same wavelength and energy, a phenomenon called the fine structure of the spectrum. In the presence of a magnetic field, these lines are split into even more lines. These lines can be classified as four distinct types, which can be related to the four different kinds of orbitals in which the electron might be found.

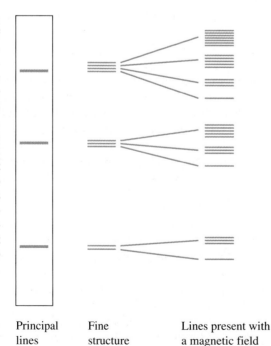

Principal lines   Fine structure   Lines present with a magnetic field

Table 6.2 Orbitals and the Azimuthal Quantum Number, *l*

Orbital Designation	Azimuthal Quantum Number	Shape
s	0	Spherical
p	1	Two-lobed (dumbbell)
d	2	Four-lobed
f	3	Six- or eight-lobed

Figure 6.17 The *p* orbitals are oriented along the three principal axes labeled *x*, *y*, and *z*. The subscripts *x*, *y*, and *z* are also used as labels to distinguish the three orbitals.

the azimuthal quantum number. Thus, an electron in the first energy level ($n = 1$) and in the orbital for which $l = 0$ is designated $1s$, whereas an electron in the orbital with $n = 3$ and $l = 2$ is designated $3d$. A set of orbitals having both the same principal quantum number, n, and the same azimuthal quantum number, l, is called a **subshell.**

The Magnetic Quantum Number, m_l

Subshells become distinguishable only when an atom is placed in a magnetic field. Figure 6.16 shows that the lines in the atomic hydrogen spectrum split when measured in a magnetic field; the resulting lines are associated with the presence of subshells. The magnetic effect occurs because the orbitals in a subshell are pointing in different directions and so interact with a magnetic field in different ways.

The orbitals in a subshell are distinguished from one another by a third quantum number, the **magnetic quantum number, m_l,** which specifies the orientation of an orbital in space and can take any integral value from $-l$ to $+l$, including 0. The possible values of m_l are 0, ± 1, ± 2, . . . , $\pm l$. The number of orbitals in a subshell, which is the number of values possible for m_l, is calculated as $2l + 1$. The spherical s orbital ($l = 0$) has only one orbital in a subshell because m_l can have only the value 0. The two-lobed p orbitals ($l = 1$) occur in sets of three because m_l can have values of -1, 0, and $+1$. These three values correspond to the two-lobed shapes pointing along the x, y, and z axes in Figure 6.17. We often use axes as subscripts to distinguish the orbitals: p_x, p_y, and p_z. The d orbitals ($l = 2$) occur in sets of five, with m_l having values of -2, -1, 0, $+1$, and $+2$. The orientations of these four-lobed orbitals are shown in Figure 6.18. The f orbitals occur in sets of seven ($m_l = -3$, -2, -1, 0, $+1$, $+2$, $+3$). They have either six or eight lobes, but we need not be concerned here with their orientation.

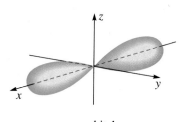

p_x orbital

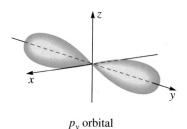

p_y orbital

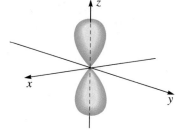

p_z orbital

The Spin Quantum Number, m_s

When atoms having an odd number of electrons are passed through a nonuniform magnetic field, the beam of atoms is split into two beams. This behavior suggests that in these atoms there are two kinds of electrons with different magnetic properties. The magnetic properties of the two kinds of electrons must cancel one another because the effect is not observed for atoms with an even number of electrons. To accommodate this behavior, it was necessary to add a fourth quantum number, the **spin quantum number** (m_s). This quantum number is described as the spin of the

Figure 6.18 The *d* orbitals are four-lobed shapes oriented differently around the nucleus. The subscripts are used as labels to distinguish among the five orbitals. The labels *xy, xz,* and *yz* indicate the planes in which the corresponding orbitals lie; $x^2 - y^2$ and z^2 indicate the axes along which those two orbitals lie. Note that one of the orbitals, d_{z^2}, appears to be a two-lobed shape inserted into a ring. This shape relates to the mathematical treatment of wave functions, which actually gives rise to an extra orbital; the averaging of two of them to give the required five orbitals results in this shape.

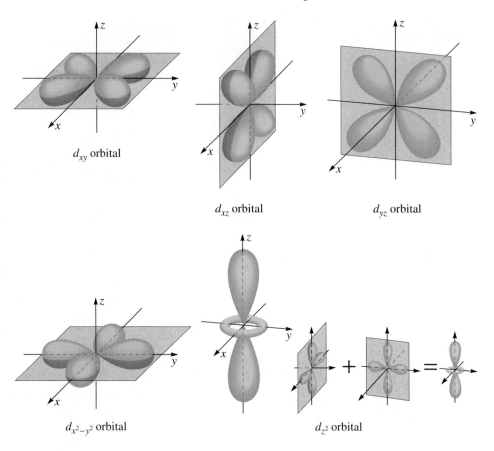

d_{xy} orbital

d_{xz} orbital d_{yz} orbital

$d_{x^2-y^2}$ orbital d_{z^2} orbital

electron about its own axis, which can occur in either a clockwise or a counterclockwise direction. The spin quantum number can have a value of either $+\frac{1}{2}$ or $-\frac{1}{2}$.

In summary, the quantum mechanical model of the atom requires four quantum numbers to describe the behavior of electrons in atoms. Each quantum number has a physical significance that helps to develop a picture of the atom as well as a mathematical model. The values of the quantum numbers and their physical significance are summarized in Table 6.3.

Table 6.3 Allowed Values of the Quantum Numbers

Quantum Number	Values	Number of Values	Significance
Principal, n	1, 2, 3, . . .	—	Distance from nucleus
Azimuthal, l	0, 1, 2, . . . , $n-1$	n	Shape of orbital
Magnetic, m_l	0, ± 1, ± 2, . . . $\pm l$	$2l + 1$	Orientation of orbital
Spin, m_s	$-\frac{1}{2}$, $+\frac{1}{2}$	2	Direction of electron spin

Fireworks: Entertaining Electronic Transitions

Fireworks have been around for a long time, dating back to the discovery of black powder by the Chinese before A.D. 1000.

Many major holidays are celebrated with fireworks displays (Figure 6.19). The bright flashes of white and colored lights provide excitement and a holiday feeling. Unfortunately, reports of the magnificent displays often include accounts of injuries suffered by individuals who misused fireworks. What causes the colored displays and what causes the injuries?

Fireworks have been around for a long time, dating back to the discovery of black powder by the Chinese before A.D. 1000. However, early fireworks were mostly noisy explosives and rockets, unlike the colorful visual effects now available. Both early and modern fireworks use black powder—a mixture of potassium nitrate (also known as saltpeter), charcoal, and sulfur—as a propellant. Small quantities of this mixture will burn, but bulk quantities or smaller quantities that are enclosed will explode when ignited. It is this explosiveness that carries the risk of injury.

The brilliant colors of today's fireworks are produced by adding finely divided metals or metal salts and an oxidizing agent in the form of pellets to the black powder. When these pellets are ignited and ejected from the rocket, they produce the familiar streaks of color, which are due to the energy changes accompanying electronic transitions in the heated metal ions. Common oxidizing agents are chlorate, perchlorate, and nitrate salts. The oxidizing agent is present to help control the temperature of the explosion so that the electronic transitions will emit light in the visible region of the spectrum. The addition of magnesium, aluminum, or titanium metals produces brilliant white light along with booming noises. Red flames are produced by adding strontium salts, such as the carbonate or nitrate. (Strontium is also used to produce the red flame in highway flares.) Green flames result from the addition of barium salts, whereas blue results from copper salts and yellow from sodium salts.

Although only a small number of chemicals are used to create fireworks, the chemistry involved is complex. It is not possible to obtain the desired visual and auditory effects by simply mixing the chemicals. The materials must have the proper particle sizes and be in a proper state of hydration to work. None of these chemical operations can be carried

Figure 6.19 The brilliant colors of fireworks result from various metal atoms emitting light when heated to high temperatures.

out safely by an amateur. They require considerable training and appropriate working conditions to prevent unwanted explosions.

Questions for Discussion

1. What is the environmental impact of dispersing metal salts into the atmosphere? Are any of them harmful if ingested?
2. Ammonium nitrate is used in some fertilizers as a source of nitrogen. Why is this salt used in preference to other ammonium salts, such as ammonium chloride? Given that this salt is highly explosive and has been known to explode in large quantities with great loss of property and life, do you think it should be used as a fertilizer?
3. Compare the relative temperatures to which the salts of various metals—strontium, barium, copper, and sodium—must be heated to give the characteristic colors of the electronic transitions.
4. People are injured frequently by fireworks. In your opinion, should chemical companies continue to produce materials that can cause injuries if misused?

6.5 Electronic Configurations of Multielectron Atoms

How do we know how many orbitals are needed to hold all the electrons in an atom? Which orbitals are used to hold the electrons?

Many properties of atoms can be understood in terms of the number and relative energies of their electrons. To examine these effects, we must be able to predict the **electronic configurations** of the elements, or the distribution of electrons among orbitals. To predict the configurations, we use an approach called the **Aufbau principle** (*Aufbau* is a German term for "building up"). We start by assuming that an atom consists of an atomic nucleus of appropriate charge surrounded by empty atomic orbitals. Then we place electrons into these orbitals, successively filling the orbitals in order of increasing energy.

The Pauli Exclusion Principle

We have mentioned filling the orbitals with electrons. But how many electrons can an orbital hold? As described by the **Pauli exclusion principle,** originated by the Austrian physicist Wolfgang Pauli in 1925, no two electrons in an atom may have the same set of four quantum numbers. The important implication of this principle is that an orbital may hold no more than two electrons because the spin quantum number has only two different values.

Energies of Orbitals

Section 6.4 stated that the energies of electrons are described by the principal quantum number, n. However, in multielectron atoms, the azimuthal quantum number, l, also affects the energy levels. In the hydrogen atom, with only one electron, all the orbitals in a given principal energy level or shell have the same energy; for this reason, they are called **degenerate.** The degeneracy is shown for a few energy levels in Figure 6.20. However, the energy of an electron in a multielectron atom also increases with the value of the azimuthal quantum number; that

Figure 6.20 Orbital energies in the hydrogen atom and in single-electron ions, such as He$^+$, are determined only by the principal quantum number, n. All orbitals in a given principal energy level or shell have the same energy—we say they are degenerate.

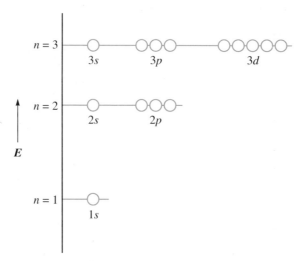

is, the energy depends on the type of orbital the electron is occupying. An electron in a 2*p* orbital, for example, is higher in energy than one in a 2*s* orbital, and an electron in a 3*d* orbital is higher in energy than one in a 3*p* orbital.

Electrons are assumed to go into the available orbitals of lowest energy first. An atom with its electrons in the lowest-energy orbitals is said to be in its **ground state,** the state of lowest energy. If electrons occupy any other orbitals, the atom is in an **excited state.** ▶ The transitions between excited states and ground states give rise to the colors of heated salts, as described in the chapter introduction.

The increases in energy with increasing value of the principal quantum number (*n*) and, within a given energy level, with increasing value of the azimuthal quantum number (*l*) are illustrated in Figure 6.21. Note that there is some overlap in energy as a result of the two trends. For example, the 4*s* orbital is actually lower in energy than the 3*d* orbitals even though the 4*s* orbital has a higher principal quantum number and would be expected to have a higher energy.

The relative energies of the various orbitals can be summarized by the following scheme, in order of increasing principal quantum number (*n*):

$$ns < (n - 2)f < (n - 1)d < np$$

In using the scheme, we skip any orbitals that cannot exist. The entire ordering of the energies of the orbitals, summarized in Figure 6.22, is as follows:

$$1s < 2s < 2p < 3s < 3p < 4s < 3d < 4p < 5s < 4d$$
$$< 5p < 6s < 4f < 5d < 6p < 7s < 5f < 6d < 7p$$

Hund's Rule

We have seen that in filling orbitals with electrons, we place each succeeding electron into the unfilled orbital having the lowest energy. However, suppose

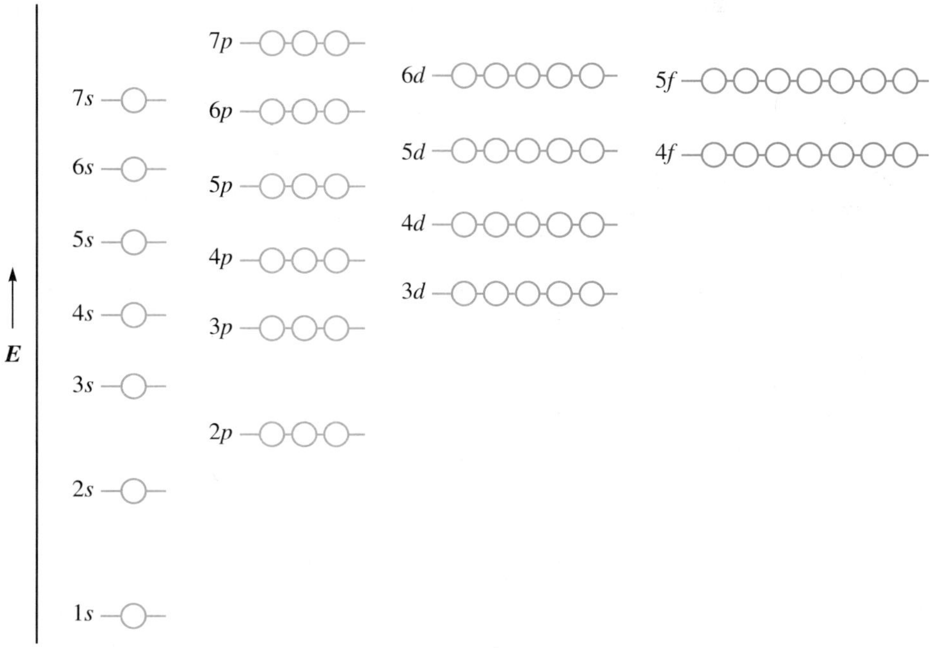

Figure 6.21 In multielectron atoms and ions, all of the orbitals in the same shell or principal energy level no longer have the same energy—all are no longer degenerate. The energy of the orbitals in a shell increases with the azimuthal quantum number. This increase causes some overlap in energy between different shells. For example, the 4*s* orbital has lower energy than the 3*d* orbitals, and the 4*f* orbitals have lower energy than the 5*d* orbitals.

Figure 6.22 The order for filling the atomic orbitals usually follows the sequence outlined here.

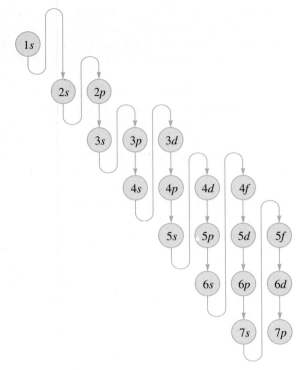

Why would electrons occupy different orbitals when the orbitals have the same energy, rather than pair up in a single orbital?

orbitals occur in subshells differing only in spatial orientation—that is, they have the same values of n and l but different values of m_l. All the orbitals in a subshell have the same energy. There are essentially two ways to fill these orbitals: each separate orbital could be filled with two electrons before the next is filled; or each could get one electron until all have one, and then each could get a second electron. The latter approach is proposed by **Hund's rule:** electrons are distributed in subshells of orbitals of identical energy in such a way as to give the maximum number of unpaired electrons.

Hund's rule is a consequence of the fact that electrons are negatively charged and thus prefer to stay as far from one another as they can. Because it requires some energy to overcome the interelectronic repulsions and place a second electron into the same orbital, the next electron will go into an empty orbital of the same energy, if possible. When a second electron is placed in these orbitals, some energy must be expended to overcome the interelectronic repulsions, but this amount of energy is usually less than that required to start filling other orbitals of higher energy.

Electronic Configurations of Some Atoms

Let's assign electrons to orbitals for the ground state of the first few elements in the periodic table. We will also develop a shorthand method for representing the electronic configuration of an atom by using a variation on the designations developed earlier for orbitals. For example, if four electrons are present in the $3p$ orbitals, we write $3p^4$. The electronic configurations of the first ten elements are summarized in Figure 6.23.

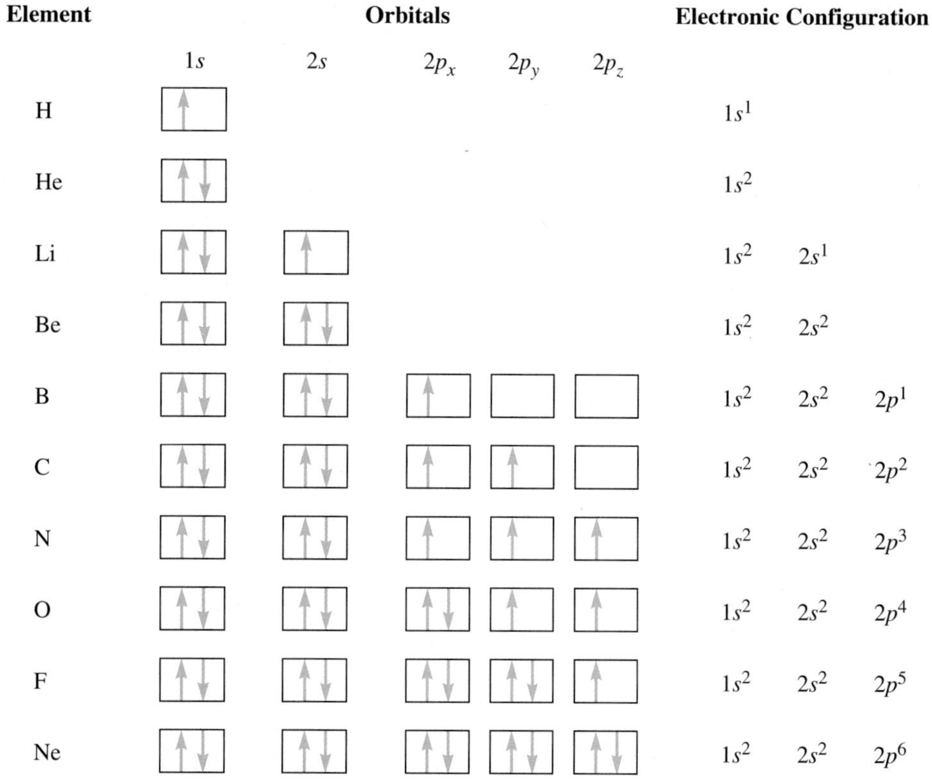

Figure 6.23 The electronic configurations of the ground state of the first ten elements.

Element	Orbitals					Electronic Configuration

The first element in the periodic table, hydrogen, has only one electron. The lowest-energy orbital is the $1s$ orbital, so the electron occupies that orbital, represented here by a box:

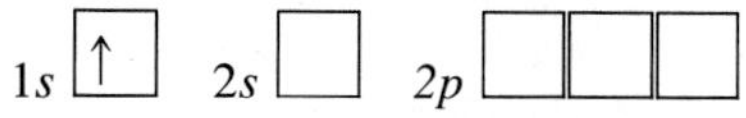

$$1s\ \boxed{\uparrow}\quad 2s\ \boxed{}\quad 2p\ \boxed{}\boxed{}\boxed{}$$

The arrow represents the electron, and the direction of the arrow distinguishes between the two possible spin orientations of the electron, the values of m_s. We write this configuration as $1s^1$.

Helium has two electrons. Both occupy the $1s$ orbital because that orbital is much lower in energy than the next highest, the $2s$ orbital:

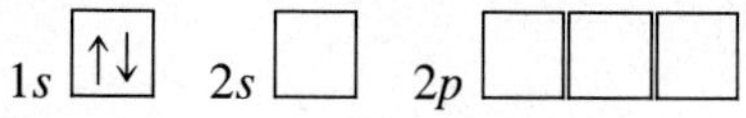

$$1s\ \boxed{\uparrow\downarrow}\quad 2s\ \boxed{}\quad 2p\ \boxed{}\boxed{}\boxed{}$$

This configuration is $1s^2$.

In a similar way, lithium ($1s^2 2s^1$) and beryllium ($1s^2 2s^2$) have electrons in the $1s$ and $2s$ orbitals, as shown in Figure 6.23.

The next elements from boron through neon add electrons to the set of three $2p$ orbitals. Boron, with five electrons, begins to occupy the $2p$ orbitals:

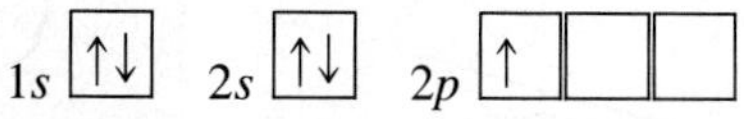

$$1s\ \boxed{\uparrow\downarrow}\quad 2s\ \boxed{\uparrow\downarrow}\quad 2p\ \boxed{\uparrow}\boxed{}\boxed{}$$

This electronic configuration is $1s^2 2s^2 2p^1$. Carbon, with six electrons, adds another electron to the $2p$ orbitals:

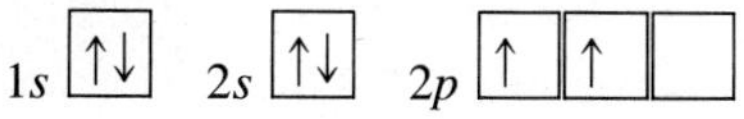

$$1s\ \boxed{\uparrow\downarrow}\quad 2s\ \boxed{\uparrow\downarrow}\quad 2p\ \boxed{\uparrow}\boxed{\uparrow}\boxed{}$$

Its configuration is $1s^2 2s^2 2p^2$. Nitrogen has seven electrons. Three occupy the $2p$ subshell:

$$1s \boxed{\uparrow\downarrow} \quad 2s \boxed{\uparrow\downarrow} \quad 2p \boxed{\uparrow}\boxed{\uparrow}\boxed{\uparrow}$$

The electronic configuration is $1s^2 2s^2 2p^3$. We can further differentiate the p orbitals by using subscripts. The configuration for nitrogen, for example, can be written $1s^2 2s^2 2p_x^1 2p_y^1 2p_z^1$. The elements oxygen through neon successively fill each p orbital. This series of electronic configurations is shown in Figure 6.23.

The next eight elements follow exactly the same pattern as the first ten. That is, they have the configuration of the first ten ($1s^2 2s^2 2p^6$) and then successively fill the $3s$ and then the $3p$ orbitals, ending up with a configuration of $1s^2 2s^2 2p^6 3s^2 3p^6$ at argon. The pattern of completing an outermost shell at a noble-gas element continues through the list of elements.

Scandium is a relatively soft, silvery white metal that has a slight yellowish color after exposure to the oxygen in air. It is primarily recovered with other metals where it is present in low concentrations. But because no uses have been found for the metal that cannot be satisfied by less expensive alternatives, only a few kilograms of scandium are produced each year.

Starting with element 19, a somewhat different pattern emerges. Potassium and calcium add electrons to the $4s$ orbital. Then scandium adds its last electron to a $3d$ orbital:

$$\text{Sc} \qquad 1s^2 2s^2 2p^6 3s^2 3p^6 4s^2 3d^1$$

The next two elements, titanium and vanadium, fill normally, with each electron going into another $3d$ orbital:

$$\text{Ti} \qquad 1s^2 2s^2 2p^6 3s^2 3p^6 4s^2 3d^2$$
$$\text{V} \qquad 1s^2 2s^2 2p^6 3s^2 3p^6 4s^2 3d^3$$

However, at chromium (Figure 6.24), we see a deviation from the expected four electrons in the $3d$ orbitals:

$$\text{Cr} \qquad 1s^2 2s^2 2p^6 3s^2 3p^6 4s^1 3d^5$$

One of the $4s$ electrons is removed and is added to the last of the $3d$ orbitals, giving half-filled $4s$ and $3d$ subshells. Most deviations from the regular filling of the

Figure 6.24 Chromium *(Left)*, a hard metal, is a component of stainless steel and is used to plate objects like automobile bumpers *(Right)* because it takes a high shine and is corrosion-resistant.

orbitals are similar to this one. Such deviations occur because half-filled and completely filled d subshells (and also f subshells) are particularly stable; in several elements, lower-energy orbitals lose electrons to achieve these stable configurations. The next such exception occurs at copper, where the $3d$ orbitals are filled prematurely, at the expense of the $4s$ orbital:

$$\text{Cu} \qquad 1s^2 2s^2 2p^6 3s^2 3p^6 4s^1 3d^{10}$$

Then zinc follows the normal pattern:

$$\text{Zn} \qquad 1s^2 2s^2 2p^6 3s^2 3p^6 4s^2 3d^{10}$$

And gallium (Figure 6.25) begins to fill the $4p$ orbitals:

$$\text{Ga} \qquad 1s^2 2s^2 2p^6 3s^2 3p^6 4s^2 3d^{10} 4p^1$$

Table 6.4 shows the progression of filling the orbitals for all the elements. The table also suggests that notations for electronic configurations can become rather long. Often we shorten them by using the symbol of a noble gas, enclosed in square brackets, to represent the notation that corresponds to the noble-gas configuration. For example, the electronic configuration of argon is $1s^2 2s^2 2p^6 3s^2 3p^6$ and that of neon is $1s^2 2s^2 2p^6$; thus, the configuration of argon can be shortened to $[\text{Ne}]3s^2 3p^6$. The complete and shorthand notation for each of the noble gases is given by the following:

$[\text{He}] = 1s^2$

$[\text{Ne}] = 1s^2 2s^2 2p^6 = [\text{He}]2s^2 2p^6$

$[\text{Ar}] = 1s^2 2s^2 2p^6 3s^2 3p^6 = [\text{Ne}]3s^2 3p^6$

$[\text{Kr}] = 1s^2 2s^2 2p^6 3s^2 3p^6 4s^2 3d^{10} 4p^6 = [\text{Ar}]4s^2 3d^{10} 4p^6$

$[\text{Xe}] = 1s^2 2s^2 2p^6 3s^2 3p^6 4s^2 3d^{10} 4p^6 5s^2 4d^{10} 5p^6 = [\text{Kr}]5s^2 4d^{10} 5p^6$

$[\text{Rn}] = 1s^2 2s^2 2p^6 3s^2 3p^6 4s^2 3d^{10} 4p^6 5s^2 4d^{10} 5p^6 6s^2 4f^{14} 5d^{10} 6p^6 = [\text{Xe}]6s^2 4f^{14} 5d^{10} 6p^6$

The shortened notation gives another benefit: it focuses attention on the electrons that are important in chemical processes and in the determination of the chemical behavior of the elements. These electrons are the **valence electrons,** those in the highest or outermost principal energy level or shell, called the **valence shell.**

Figure 6.25 *Left:* Gallium is a silvery metal that fractures in conchoidal patterns, much like glass. *Right:* It has a low melting point, 29.78°C, so it will melt in the palm of one's hand over a period of time. Gallium is used in semiconductors; gallium arsenide, GaAs, is being investigated as a semiconductor material.

Table 6.4 Electron Distributions in the Elements

		1s	2s	2p	3s	3p	3d	4s	4p	4d	4f	5s	5p	5d	5f	6s	6p	6d	7s
H	1	1																	
He	2	2																	
Li	3	2	1																
Be	4	2	2																
B	5	2	2	1															
C	6	2	2	2															
N	7	2	2	3															
O	8	2	2	4															
F	9	2	2	5															
Ne	10	2	2	6															
Na	11	2	2	6	1														
Mg	12	2	2	6	2														
Al	13	2	2	6	2	1													
Si	14	2	2	6	2	2													
P	15	2	2	6	2	3													
S	16	2	2	6	2	4													
Cl	17	2	2	6	2	5													
Ar	18	2	2	6	2	6													
K	19	2	2	6	2	6		1											
Ca	20	2	2	6	2	6		2											
Sc	21	2	2	6	2	6	1	2											
Ti	22	2	2	6	2	6	2	2											
V	23	2	2	6	2	6	3	2											
Cr	24	2	2	6	2	6	5	1											
Mn	25	2	2	6	2	6	5	2											
Fe	26	2	2	6	2	6	6	2											
Co	27	2	2	6	2	6	7	2											
Ni	28	2	2	6	2	6	8	2											
Cu	29	2	2	6	2	6	10	1											
Zn	30	2	2	6	2	6	10	2											
Ga	31	2	2	6	2	6	10	2	1										
Ge	32	2	2	6	2	6	10	2	2										
As	33	2	2	6	2	6	10	2	3										
Se	34	2	2	6	2	6	10	2	4										
Br	35	2	2	6	2	6	10	2	5										
Kr	36	2	2	6	2	6	10	2	6										
Rb	37	2	2	6	2	6	10	2	6			1							
Sr	38	2	2	6	2	6	10	2	6			2							

Table 6.4 Electron Distributions (cont.)

		1s	2s	2p	3s	3p	3d	4s	4p	4d	4f	5s	5p	5d	5f	6s	6p	6d	7s
Y	39	2	2	6	2	6	10	2	6	1		2							
Zr	40	2	2	6	2	6	10	2	6	2		2							
Nb	41	2	2	6	2	6	10	2	6	4		1							
Mo	42	2	2	6	2	6	10	2	6	5		1							
Tc	43	2	2	6	2	6	10	2	6	5		2							
Ru	44	2	2	6	2	6	10	2	6	7		1							
Rh	45	2	2	6	2	6	10	2	6	8		1							
Pd	46	2	2	6	2	6	10	2	6	10									
Ag	47	2	2	6	2	6	10	2	6	10		1							
Cd	48	2	2	6	2	6	10	2	6	10		2							
In	49	2	2	6	2	6	10	2	6	10		2	1						
Sn	50	2	2	6	2	6	10	2	6	10		2	2						
Sb	51	2	2	6	2	6	10	2	6	10		2	3						
Te	52	2	2	6	2	6	10	2	6	10		2	4						
I	53	2	2	6	2	6	10	2	6	10		2	5						
Xe	54	2	2	6	2	6	10	2	6	10		2	6						
Cs	55	2	2	6	2	6	10	2	6	10		2	6			1			
Ba	56	2	2	6	2	6	10	2	6	10		2	6			2			
La	57	2	2	6	2	6	10	2	6	10		2	6	1		2			
Ce	58	2	2	6	2	6	10	2	6	10	2	2	6			2			
Pr	59	2	2	6	2	6	10	2	6	10	3	2	6			2			
Nd	60	2	2	6	2	6	10	2	6	10	4	2	6			2			
Pm	61	2	2	6	2	6	10	2	6	10	5	2	6			2			
Sm	62	2	2	6	2	6	10	2	6	10	6	2	6			2			
Eu	63	2	2	6	2	6	10	2	6	10	7	2	6			2			
Gd	64	2	2	6	2	6	10	2	6	10	7	2	6	1		2			
Tb	65	2	2	6	2	6	10	2	6	10	9	2	6			2			
Dy	66	2	2	6	2	6	10	2	6	10	10	2	6			2			
Ho	67	2	2	6	2	6	10	2	6	10	11	2	6			2			
Er	68	2	2	6	2	6	10	2	6	10	12	2	6			2			
Tm	69	2	2	6	2	6	10	2	6	10	13	2	6			2			
Yb	70	2	2	6	2	6	10	2	6	10	14	2	6			2			
Lu	71	2	2	6	2	6	10	2	6	10	14	2	6	1		2			
Hf	72	2	2	6	2	6	10	2	6	10	14	2	6	2		2			
Ta	73	2	2	6	2	6	10	2	6	10	14	2	6	3		2			
W	74	2	2	6	2	6	10	2	6	10	14	2	6	4		2			
Re	75	2	2	6	2	6	10	2	6	10	14	2	6	5		2			

(continued)

Table 6.4 (cont.)

		1s	2s	2p	3s	3p	3d	4s	4p	4d	4f	5s	5p	5d	5f	6s	6p	6d	7s
Os	76	2	2	6	2	6	10	2	6	10	14	2	6	6		2			
Ir	77	2	2	6	2	6	10	2	6	10	14	2	6	7		2			
Pt	78	2	2	6	2	6	10	2	6	10	14	2	6	9		1			
Au	79	2	2	6	2	6	10	2	6	10	14	2	6	10		1			
Hg	80	2	2	6	2	6	10	2	6	10	14	2	6	10		2			
Tl	81	2	2	6	2	6	10	2	6	10	14	2	6	10		2	1		
Pb	82	2	2	6	2	6	10	2	6	10	14	2	6	10		2	2		
Bi	83	2	2	6	2	6	10	2	6	10	14	2	6	10		2	3		
Po	84	2	2	6	2	6	10	2	6	10	14	2	6	10		2	4		
At	85	2	2	6	2	6	10	2	6	10	14	2	6	10		2	5		
Rn	86	2	2	6	2	6	10	2	6	10	14	2	6	10		2	6		
Fr	87	2	2	6	2	6	10	2	6	10	14	2	6	10		2	6		1
Ra	88	2	2	6	2	6	10	2	6	10	14	2	6	10		2	6		2
Ac	89	2	2	6	2	6	10	2	6	10	14	2	6	10		2	6	1	2
Th	90	2	2	6	2	6	10	2	6	10	14	2	6	10		2	6	2	2
Pa	91	2	2	6	2	6	10	2	6	10	14	2	6	10	2	2	6	1	2
U	92	2	2	6	2	6	10	2	6	10	14	2	6	10	3	2	6	1	2
Np	93	2	2	6	2	6	10	2	6	10	14	2	6	10	4	2	6	1	2
Pu	94	2	2	6	2	6	10	2	6	10	14	2	6	10	6	2	6		2
Am	95	2	2	6	2	6	10	2	6	10	14	2	6	10	7	2	6		2
Cm	96	2	2	6	2	6	10	2	6	10	14	2	6	10	7	2	6	1	2
Bk	97	2	2	6	2	6	10	2	6	10	14	2	6	10	8	2	6	1	2
Cf	98	2	2	6	2	6	10	2	6	10	14	2	6	10	10	2	6		2
Es	99	2	2	6	2	6	10	2	6	10	14	2	6	10	11	2	6		2
Fm	100	2	2	6	2	6	10	2	6	10	14	2	6	10	12	2	6		2
Md	101	2	2	6	2	6	10	2	6	10	14	2	6	10	13	2	6		2
No	102	2	2	6	2	6	10	2	6	10	14	2	6	10	14	2	6		2
Lr	103	2	2	6	2	6	10	2	6	10	14	2	6	10	14	2	6	1	2

EXAMPLE 6.5

Write the electronic configuration of the phosphorus atom.

Solution   Phosphorus is element 15, so it has 15 electrons. The first 2 go into the $1s$ orbital, the next 2 into the $2s$ orbital, and the next 6 into the $2p$ orbitals. At this point, 10 electrons have been added, leaving 5. Of these, 2 go into the $3s$ orbital, leaving 3 for the $3p$ orbitals, each going into a different p orbital. The electronic configuration is thus

$$P \qquad 1s^2 2s^2 2p^6 3s^2 3p^3 \qquad \text{or} \qquad [\text{Ne}]3s^2 3p^3$$

Practice Problem 6.5

Write the electronic configuration of the selenium atom.

Answer: Se $1s^22s^22p^63s^23p^64s^23d^{10}4p^4$ or $[Ar]4s^23d^{10}4p^4$

Electronic Configurations of Monatomic Ions

We can describe the electronic configurations of monatomic ions by adding or subtracting electrons from those of the parent atoms. For an anion, which has more electrons than the atom of the corresponding element, the additional electrons are simply added according to the Aufbau principle. For example, to form the oxide ion, O^{2-}, from the oxygen atom, we add two electrons:

$$O \qquad 1s^22s^22p^4$$
$$O^{2-} \qquad 1s^22s^22p^6$$

EXAMPLE 6.6

Write the electronic configuration of the P^{3-} ion.

Solution   The phosphorus atom has a total of 15 electrons. The phosphide ion has 3 more, for a total of 18. Its electronic configuration is the same as that of element 18, argon. The first 10 electrons go into the 1s, 2s, and 2p orbitals. The last 8 fill the 3s and 3p orbitals:

$$P^{3-} \qquad 1s^22s^22p^63s^23p^6 \quad \text{or} \quad [Ne]3s^23p^6 \quad \text{or} \quad [Ar]$$

Practice Problem 6.6

Write the electronic configuration of the bromide ion, Br^-.

Answer: Br^- $1s^22s^22p^63s^23p^64s^23d^{10}4p^6$ or $[Ar]4s^23d^{10}4p^6$

The formation of cations requires the removal of electrons from the parent atom. The electrons are not necessarily removed simply in the reverse of the order suggested by the Aufbau principle, however. The Aufbau principle involves a successive increase in the nuclear charge by $1+$ at the same time as an electron is added outside the nucleus. The formation of a cation involves the removal of an electron with no change in the nuclear charge. In general, when electrons are removed from an atom, those electrons in the outermost shell (highest n) are removed first. Thus, a 4s electron should be removed before a 3d electron. Consider iron as an example:

$$Fe \qquad [Ar]4s^23d^6$$

When the Fe^{2+} ion is formed, two electrons must be removed. These electrons come from the 4s orbital:

$$Fe^{2+} \qquad [Ar]4s^03d^6$$

The removal of one more electron will form Fe^{3+}. That electron will come from

the $3d$ orbital that contains two electrons, because the $3d$ subshell is now outermost:

$$\text{Fe}^{3+} \qquad [\text{Ar}]4s^03d^5$$

EXAMPLE 6.7

Write the electronic configuration of the V^{3+} ion.

Solution   The electronic configuration of vanadium, element 23, is:

$$\text{V} \qquad 1s^22s^22p^63s^23p^64s^23d^3$$

Three electrons must be removed to form the vanadium(III) ion. The first two come from
the $4s$ orbital because it has the highest principal quantum number. The last electron will be
removed from a $3d$ orbital:

$$\text{V}^{3+} \qquad 1s^22s^22p^63s^23p^64s^03d^2$$

Practice Problem 6.7

Write the electronic configuration of the chromium(III) ion, Cr^{3+}.

Answer: $\text{Cr}^{3+} \qquad 1s^22s^22p^63s^23p^64s^03d^3$

6.6 Electronic Configuration and the Periodic Table

How are the electronic configurations of the elements related to the structure of the periodic table?

In Chapter 3, we saw that when the elements are arranged in order of increasing
atomic number, their physical and chemical properties recur periodically. For example, magnesium, calcium, and strontium are all hard, very reactive, silvery
metals, whereas neon, argon, and krypton are all unreactive colorless gases. This
periodicity stems from the electronic configurations of the elements—in particular, the valence electron configurations.

Consider the horizontal rows, or periods, in the table. Each starts with an
element that has one valence electron in an s orbital, and the length of the periods
is explicable in terms of the orbitals available in each shell, as shown in Figure
6.26. The first period, with elements having $n = 1$ as the valence shell, has only
one available orbital, the $1s$ orbital. This period contains only two elements because no more than two electrons can occupy the $1s$ orbital. The second period,
containing Li through Ne, has eight elements because the valence shell of these
elements includes the $2s$ and the three $2p$ orbitals, which can hold a total of eight
electrons. The third shell includes nine orbitals in the $3s$, $3p$, and $3d$ subshells.
However, the third period has only eight elements (Na through Ar) because the $3d$
orbitals are not occupied by electrons until after the fourth shell has begun to fill.
(Recall that the energy of the $4s$ orbital is less than that of the $3d$ orbitals.) Thus, in
the third period, it is only the $3s$ and $3p$ orbitals that hold valence electrons. The
fourth-period elements use the $4s$, $3d$, and $4p$ orbitals to hold the valence electrons. Thus, this period has nine orbitals available and includes 18 elements (K
through Kr). The fifth period also has 18 elements (Rb through Xe), using the $5s$,

Figure 6.26 The periodicity of chemical and physical properties stems from the periodicity of electronic configurations. Within each group, the same type of orbital is being filled. Adjacent groups are filling the same type of orbital as long as there is still room in that type of orbital.

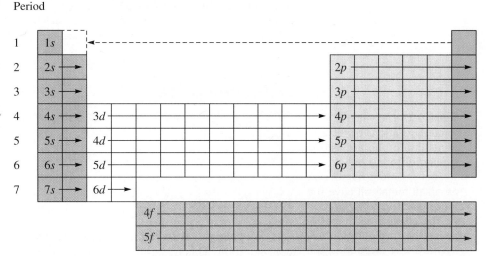

4*d*, and 5*p* orbitals. The sixth period expands to 32 elements (Cs through Rn, including Ce through Lu). At this point, the 4*f* orbitals become available. These elements use the 6*s*, 4*f*, 5*d*, and 6*p* orbitals for their valence electrons, so 16 orbitals are available. The seventh period is similar to the sixth but has less than 32 elements simply because that many elements have not yet been discovered or synthesized.

How are the electronic configurations of elements within a group of the periodic table similar to one another?

Note in Figure 6.26 that the periodic table can be broken up into blocks that conform not only to the orbitals being filled but also to the classifications introduced in Chapter 3. The *s*-block and *p*-block elements are the representative, or main-group, elements. The *d*-block elements are the transition elements, and the *f*-block elements are the inner-transition elements.

In the periodic table, elements that line up in vertical columns, or groups, have similar properties. For example, the elements in Group IA (1) (except hydrogen because it is not a metal) are soft, silvery white metals with low melting points. The similarity in the valence electron configurations of these elements—all are ns^1—causes all the other properties to be similar. The valence electron configurations of the elements are shown in the periodic table in Figure 6.27.

Earlier, we wrote electronic configurations by using the Aufbau principle along with a knowledge of the order in which orbitals are filled. A simpler approach uses the periodic table as a guide. As we just saw, the periodic table has a recurring pattern in the configuration of the valence electrons. The structure of the table exactly corresponds to the relative order of orbital energies that we used earlier:

$$ns < (n-2)f < (n-1)d < np$$

We can use the location of the element in the periodic table to tell us the electronic configuration by starting with hydrogen and moving through the orbitals being filled until we get to the element of interest.

EXAMPLE 6.8

Write the electronic configuration of tellurium, element 52.

(continued)

Solution   The noble gas in the preceding period is krypton. Tellurium is the fourth element in the p block of the fifth period. The $5s$ and $4d$ orbitals should be filled with electrons, and the $5p$ orbitals should contain four electrons. Thus, the configuration is

$$\text{Te} \qquad [\text{Kr}]5s^2 4d^{10} 5p^4$$

Practice Problem 6.8

Write the electronic configuration of gallium, element 31.

Answer: Ga $[\text{Ar}]4s^2 3d^{10} 4p^1$

Figure 6.27 The valence electron configurations of the elements in a group are usually similar, differing only in the principal quantum number, n. The alkali metals all have the valence electron configuration ns^1. The alkaline-earth metals all have the configuration ns^2. With a few exceptions, the next ten groups (the transition elements) have the configuration $ns^2(n-1)d^x$, where x changes from 1 to 10 across the set of groups. The next six groups are all filling the np orbitals and have a configuration of ns^2np^x ($x = 1$ to 6) or $ns^2(n-1)d^{10}np^x$.

Ionic Charges

We have just seen that the noble gases have filled valence shells, a very stable configuration. This means that the noble gases have a very low reactivity and form only a few compounds. Other stable configurations correspond to completely filled or half-filled subshells. We have seen this in the electronic configuration of chromium and copper. The valence shell of chromium contains a half-filled $4s$ subshell and a half-filled $3d$ subshell, whereas that of copper contains a half-filled $4s$ subshell and a completely filled $3d$ subshell. Stability of electronic configuration has a number of important effects. One of these involves the formation of ions. Chapter

Main-group elements

Main-group elements
np^x

ns^1																	2 **He** $1s^2$
1 **H** $1s^1$	ns^2																
3 **Li** $2s^1$	4 **Be** $2s^2$											5 **B** $2p^1$	6 **C** $2p^2$	7 **N** $2p^3$	8 **O** $2p^4$	9 **F** $2p^5$	10 **Ne** $2p^6$
11 **Na** $3s^1$	12 **Mg** $3s^2$											13 **Al** $3p^1$	14 **Si** $3p^2$	15 **P** $3p^3$	16 **S** $3p^4$	17 **Cl** $3p^5$	18 **Ar** $3p^6$
19 **K** $4s^1$	20 **Ca** $4s^2$	21 **Sc** $3d^1$	22 **Ti** $3d^2$	23 **V** $3d^3$	24 **Cr** $4s^1 3d^5$	25 **Mn** $3d^5$	26 **Fe** $3d^6$	27 **Co** $3d^7$	28 **Ni** $3d^8$	29 **Cu** $4s^1 3d^{10}$	30 **Zn** $3d^{10}$	31 **Ga** $4p^1$	32 **Ge** $4p^2$	33 **As** $4p^3$	34 **Se** $4p^4$	35 **Br** $4p^5$	36 **Kr** $4p^6$
37 **Rb** $5s^1$	38 **Sr** $5s^2$	39 **Y** $4d^1$	40 **Zr** $4d^2$	41 **Nb** $5s^1 4d^4$	42 **Mo** $5s^1 4d^5$	43 **Tc** $4d^5$	44 **Ru** $5s^1 4d^7$	45 **Rh** $5s^1 4d^8$	46 **Pd** $5s^0 4d^{10}$	47 **Ag** $5s^1 4d^{10}$	48 **Cd** $4d^{10}$	49 **In** $5p^1$	50 **Sn** $5p^2$	51 **Sb** $5p^3$	52 **Te** $5p^4$	53 **I** $5p^5$	54 **Xe** $5p^6$
55 **Cs** $6s^1$	56 **Ba** $6s^2$	57 **La** $5d^1$	72 **Hf** $5d^2$	73 **Ta** $5d^3$	74 **W** $5d^4$	75 **Re** $5d^5$	76 **Os** $5d^6$	77 **Ir** $5d^7$	78 **Pt** $6s^1 5d^9$	79 **Au** $6s^1 5d^{10}$	80 **Hg** $5d^{10}$	81 **Tl** $6p^1$	82 **Pb** $6p^2$	83 **Bi** $6p^3$	84 **Po** $6p^4$	85 **At** $6p^5$	86 **Rn** $6p^6$
87 **Fr** $7s^1$	88 **Ra** $7s^2$	89 **Ac** $6d^1$	104 **Unq** $6d^2$	105 **Unp** $6d^3$	106 **Unh** $6d^4$	107 **Uns**	108 **Uno**	109 **Une**									

Transition elements
$(n-1)d^x$

$(n-2)f^x$

	58 **Ce** $4f^2$	59 **Pr** $4f^3$	60 **Nd** $4f^4$	61 **Pm** $4f^5$	62 **Sm** $4f^6$	63 **Eu** $4f^7$	64 **Gd** $4f^7 5d^1$	65 **Tb** $4f^9$	66 **Dy** $4f^{10}$	67 **Ho** $4f^{11}$	68 **Er** $4f^{12}$	69 **Tm** $4f^{13}$	70 **Yb** $4f^{14}$	71 **Lu** $4f^{14} 5d^1$
†	90 **Th** $6d^2$	91 **Pa** $5f^2 6d^1$	92 **U** $5f^3 6d^1$	93 **Np** $5f^4 6d^1$	94 **Pu** $5f^6$	95 **Am** $5f^7$	96 **Cm** $5f^7 6d^1$	97 **Bk** $5f^9$	98 **Cf** $5f^{10}$	99 **Es** $5f^{11}$	100 **Fm** $5f^{12}$	101 **Md** $5f^{13}$	102 **No** $5f^{14}$	103 **Lr** $5f^{14} 6d^1$

Inner-transition elements

How do the common ionic charges of the elements correlate with their electronic configurations?

3 listed the most common charges for the monatomic ions and stated that the common charges, except for the transition elements, could be predicted from position in the periodic table. Atoms of an element tend to lose or gain electrons to form ions that contain the same number of electrons as the noble gas nearest to that element in the periodic table.

The formation of monatomic ions can be explained in terms of a tendency to achieve a stable electronic configuration. Thus, in forming Ca^{2+}, calcium loses two electrons to achieve the electronic configuration of argon:

$$Ca \qquad 1s^2 2s^2 2p^6 3s^2 3p^6 4s^2$$
$$Ca^{2+} \qquad 1s^2 2s^2 2p^6 3s^2 3p^6 4s^0 = [Ar]$$

Like calcium, the other main-group metals form cations with noble-gas electronic configurations. Thus, Na, Mg, and Al form the ions Na^+, Mg^{2+}, and Al^{3+}, which all have the electronic configuration of the preceding noble gas, Ne: $1s^2 2s^2 2p^6$. In general, any horizontally adjacent main-group metals form ions with the same electronic configuration. They are said to be an **isoelectronic series.**

The transition metals do not behave so simply. They tend to form 2+ and 3+ ions and usually form more than one ion.

The nonmetals add electrons to form anions with the configuration of the noble gas in the same period. For example, selenium gains two electrons to form Se^{2-}, which has the same configuration as the noble gas in the same period, Kr:

$$Se \qquad 1s^2 2s^2 2p^6 3s^2 3p^6 4s^2 3d^{10} 4p^4$$
$$Se^{2-} \qquad 1s^2 2s^2 2p^6 3s^2 3p^6 4s^2 3d^{10} 4p^6 = [Kr]$$

Like the main-group metals, adjacent nonmetals form anions with the same electronic configuration. Thus, N^{3-}, O^{2-}, F^-, and Ne are isoelectronic and all have the same electronic configuration: $1s^2 2s^2 2p^6$.

6.7 Noble-Gas Chemistry

Early versions of the periodic table made no allowance for the existence of the noble gases, Group VIIIA (18). Their discovery around the turn of the twentieth century profoundly influenced the development of theories of atomic structure and bonding.

The first hint of the existence of the noble gases came in 1868, when the characteristic spectral lines of helium were observed in the solar spectrum during an eclipse. It was not until 1895 that helium was isolated on earth, however. The Scottish chemist Sir William Ramsay isolated helium as a decay product of radioactive uranium.

Ramsay, in collaboration with the English physicist Lord Rayleigh (John William Strutt), had already discovered argon in 1894. In measuring the densities of simple gases, these investigators found that N_2 from the atmosphere was 0.5% heavier than N_2 prepared by the decomposition of NH_4NO_2. On reacting nitrogen from air with magnesium to form magnesium nitride,

$$3Mg(s) + N_2(g) \xrightarrow{\Delta} Mg_3N_2(s)$$

they found that a small amount of gas did not react. This gas was denser than N_2 and had a new set of spectral lines. It was named argon ("the lazy one") because it seemed chemically unreactive.

Table 6.5 Physical States of the Noble Gases at 1 atm Pressure

Temperatures (°C) for Conversion Between Physical States		
	Gas ⟷ Liquid	⟷ Solid
Helium	−268.9	−272.2 (26 atm)
Neon	−245.9	−248.7
Argon	−185.7	−189.2
Krypton	−152.3	−156.6
Xenon	−108.1	−111.9
Radon	−62	−71

Figure 6.28 Crystals of xenon tetrafluoride. The crystals are about 0.1 mm–0.2 mm in length.

By 1898, Ramsay and his assistant Morris William Travers had isolated three other noble gases—Kr, Ne, and Xe—by careful separation from liquid air. Radon was discovered in radium emanations in 1900 and was isolated by Ramsay in 1908.

The sources of the noble gases have not changed much since Ramsay's day. Neon, argon, krypton, and xenon are isolated by careful separation from liquid air. Natural gas wells in Kansas, Oklahoma, and Texas are the primary sources of helium. Radon is formed by the radioactive decay of radium, thorium, uranium, and actinium. The most stable isotope of radon is very short-lived: half disappears by radioactive decay in 3.8 days. Because of the intense radioactivity, it is impossible to accumulate much radon (most studies have used less than 10^{12} atoms of Rn), so little is known about its behavior and chemical properties.

We do know that the noble gases are colorless, odorless, generally unreactive substances in all physical states. As shown in Table 6.5, they condense to liquids and the liquids solidify at extremely low temperatures. In fact, helium cannot be obtained as a solid at 1 atm pressure even near absolute zero; it must be subjected to 26 atm pressure to solidify.

Because their filled outermost electronic shells are very stable, the noble gases tend to be chemically inert. But they are not completely unreactive. Until 1962, there were no known compounds of the noble gases. In that year, however, Neil Bartlett, a chemist at the University of British Columbia, prepared a compound containing xenon. He had prepared the ionic compound $O_2^+[PtF_6]^-$ by reaction of O_2 with PtF_6. Noting from spectroscopic measurements that it required about the same amount of energy to remove an electron from xenon as from molecular oxygen, he reasoned that if oxygen could form a cation, xenon should be able to do so as well. When he tried to react Xe with PtF_6, the result was a substance that appeared to be $Xe^+[PtF_6]^-$, although it is now thought to be $[XeF]^+[Pt_2F_{11}]^-$. A tremendous flurry of activity followed, and soon a number of other noble-gas compounds had been produced, including KrF_2, $KrF_2 \cdot 2SbF_5$, XeF_2, XeF_4, XeF_6, $XeOF_4$, XeO_2F_2, XeO_3, XeO_4, Na_4XeO_6, and RnF_x (x unknown). Crystals of one of the compounds, XeF_4, are shown in Figure 6.28.

The noble gases are put to a variety of uses. Helium is nonflammable and much less dense than air, so it is used extensively for balloons and dirigibles (Figure 6.29). Deep-sea divers breathe helium–oxygen mixtures instead of the nitrogen–oxygen mixtures that make up air (Figure 6.30). Divers who breathe air may, on rising to the surface, suffer a painful and sometimes dangerous condition

Figure 6.29 Helium is used in balloons and dirigibles.

called the bends, which results from the formation of nitrogen bubbles in body tissues. Because of its smaller atomic size, helium escapes from body tissues much faster than nitrogen does, so helium is much less likely to cause this condition.

Helium and argon are used as an inert gas shield for arc welding. They prevent the oxygen in air from attacking the metal and producing a weak weld. Helium, neon, and argon serve a similar purpose in the production of ultrapure silicon and germanium crystals for the semiconductor industry. Helium has also been used extensively for pressurizing liquid-fuel rockets; the *Saturn* booster rocket used on the Apollo moon shots required 370 million liters of helium gas for firing. Liquid helium, the coldest of liquid refrigerants, provides a coolant for studies of the behavior of materials at temperatures of 0–5 K.

Neon at low pressure in a gas-discharge tube gives off a bright orange-red glow. This is the most intense discharge produced by a noble gas at ordinary voltages and currents, so neon is widely used in advertising signs. Some discharge tubes use mixtures of neon and argon or other gases.

Argon, or sometimes krypton, fills incandescent light bulbs and inhibits the evaporation of the tungsten filament, thereby prolonging the life of the bulb. Other noble gases could be used instead, but argon is the cheapest and most plentiful. Argon also forms a protective atmosphere for the production of reactive elements such as titanium.

Figure 6.30 Deep-sea divers breathe helium–oxygen mixtures.

Xenon produces a beautiful blue glow when subjected to an electrical discharge in a vacuum tube. Xenon is used in electron tubes, stroboscopes, bactericidal lamps, lamps used to excite ruby lasers, and other instruments in which a high-intensity light source is needed.

6.8   Electronic Configuration and Periodic Properties of Atoms

In comparing the observed properties of the elements, we generally focus on three factors related to electronic configuration: the increase in nuclear charge with increasing atomic number, the pairing energy (electron repulsions), and the shielding effect of inner electrons on nuclear charge. This section briefly explains these factors before going on to discuss several properties they affect: atomic size, ionization energy, and electron affinity.

Nuclear Charge

The nuclear charge increases from one atom to the next across the periodic table, along with the number of electrons. The greater the nuclear charge, the greater the attraction between the positively charged nucleus and the negatively charged electrons. (This attraction is proportional to the square of the nuclear charge.) Thus, it becomes more difficult to remove electrons from an atom as the nuclear charge increases because the electrons are more tightly held in the atom. In a series of adjacent elements where electrons are being added to the same set of orbitals, the electrons are held increasingly more tightly as the orbitals are filled because the nuclear charge is increasing. For example, the elements carbon, nitrogen, oxygen, and fluorine all have their outermost electrons in the $2p$ orbitals. As the attraction between nucleus and electrons increases in this series of elements, the average distance between the electrons and the nucleus decreases.

Pairing Energy

The effect of increasing nuclear charge is sometimes offset by the **pairing energy,** the energy required to place two electrons into the same orbital. Two electrons confined to the same volume of space will repel each other because they both have a negative charge.

Shielding Effect

The effect of nuclear charge may be further offset by a phenomenon related to energy levels. On the average, electrons in the outermost energy level (the valence shell) spend their time farther away from the nucleus than do the inner electrons; thus, the outermost electrons receive not the entire nuclear charge but only the nuclear charge as shielded or screened by the inner electrons. This nuclear charge cancellation is known as the **shielding effect.** The charge felt by the outermost electrons is called the **effective nuclear charge** and is approximately equal to the total nuclear charge minus the number of electrons that are located closer to the nucleus. Consider the element lithium, for example, which has three electrons. The first two electrons are in the $1s$ orbital, where they feel the effect of the entire 3+ charge of the nucleus. The third electron is in the $2s$ orbital, which is farther from the nucleus. There, it feels a charge closer to 1+ than to 3+ because the inner two electrons effectively screen out about two units of charge.

Atomic Size

How do you expect the atomic size of the elements to vary throughout the periodic table?

We turn now to how these three factors—nuclear charge, pairing energy, and shielding effect—can be used to predict and explain the variation of properties of the elements in the periodic table. We first consider the variation in the atomic size of the elements. The electron cloud in an atom is not sharply defined, so the edge of an atom cannot be discerned exactly. Thus, different ways of determining atomic size may well result in different values.

Metallic and Covalent Radii   A common measure of the size of an atom is the distance between the centers of two adjacent atoms in a solid substance, where it is assumed that the two atoms will be in contact with each other, as shown in Figure 6.31. (This distance can be measured with X-ray diffraction, explained in some detail in Chapter 10.) The **metallic radius** is one-half the distance between the centers of two adjacent atoms in a solid metal; the **covalent radius** is one-half the distance between the centers of two atoms in a gaseous diatomic molecule, such as N_2 or F_2 (Figure 6.31).

Metallic and covalent radii for many elements are shown in Figure 6.32. Some of these values are plotted against atomic number in Figure 6.33. In a given group, as the figures show, the atomic radius increases as the atomic number increases. As we go down a group in the periodic table, we see an increase in the principal quantum number, n. Thus, from the top of the group to the bottom, the outermost electrons of the elements are located in orbitals ever farther from the nuclei. The effective nuclear charge, however, remains fairly constant.

Figure 6.31 Metallic and covalent radii are found from the distance between the centers of two adjacent atoms, either in metals or in covalent molecules. The radius is half the interatomic distance.

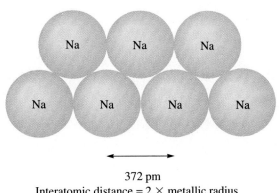

372 pm
Interatomic distance = 2 × metallic radius
Metallic radius = 186 pm

128 pm
Interatomic distance = 2 × covalent radius
Covalent radius = 64 pm

Metallic radii (pm)

H																	**He**
Li 123	**Be** 89											**B** 80	**C**	**N**	**O**	**F**	**Ne**
Na 158	**Mg** 136											**Al** 125	**Si** 117	**P**	**S**	**Cl**	**Ar**
K 203	**Ca** 174	**Sc** 144	**Ti** 132	**V** 122	**Cr** 119	**Mn** 118	**Fe** 117	**Co** 116	**Ni** 115	**Cu** 118	**Zn** 121	**Ga** 125	**Ge** 124	**As** 121	**Se** 117	**Br**	**Kr**
Rb 216	**Sr** 191	**Y** 162	**Zr** 145	**Nb** 134	**Mo** 130	**Tc** 127	**Ru** 125	**Rh** 125	**Pd** 128	**Ag** 134	**Cd** 138	**In** 142	**Sn** 142	**Sb** 137	**Te** 139	**I**	**Xe**
Cs 235	**Ba** 198	**La** 169	**Hf** 144	**Ta** 134	**W** 130	**Re** 128	**Os** 126	**Ir** 127	**Pt** 130	**Au** 134	**Hg** 139	**Tl** 144	**Pb** 150	**Bi** 151	**Po**	**At**	**Rn**

Covalent radii (pm)

					He
B 89	**C** 77	**N** 70	**O** 66	**F** 64	**Ne**
Al	**Si** 117	**P** 110	**S** 104	**Cl** 99	**Ar**
Ga	**Ge** 122	**As** 121	**Se** 117	**Br** 114	**Kr**
In	**Sn** 140	**Sb** 141	**Te** 137	**I** 133	**Xe**
Tl	**Pb** 146	**Bi** 151	**Po**	**At**	**Rn**

Figure 6.32 Metallic and covalent radii for most of the elements.

Figure 6.33 A plot of metallic radius against atomic number reveals periodic trends.

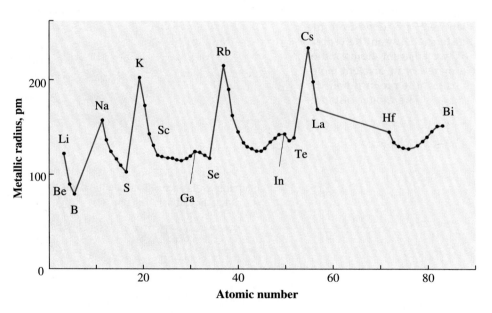

As atomic number increases within a given period, the opposite effect on atomic radius is observed. In this case, the principal quantum number, n, remains the same, so the orbitals in the valence shell are at roughly the same distance from the nucleus. However, the nuclear charge increases, which draws the valence electrons closer to the nucleus, making the atoms somewhat smaller. Consider atomic radii of the second-period elements with $n = 2$ to see this effect: Li = 123 pm, Be = 89 pm, B = 80 pm, C = 77 pm, N = 70 pm, O = 66 pm, F = 64 pm. The effective nuclear charge increases from about 1+ in lithium to about 7+ in fluorine and is not offset by a changing shielding effect. The general trends in atomic radii within periods or groups of the periodic table are summarized in Figure 6.34.

Ionic Radius   The size of ions formed from atoms can be estimated from the distance between centers of cations and anions in solid materials, or the **ionic radius.** X-ray diffraction can be used to determine the distance between the centers of adjacent ions, but it is not always easy to determine where one ion ends and the other begins, so the values are somewhat uncertain. Some values are shown in Figure 6.35, and the relative sizes of various ions are shown in Figure 6.36.

When we want to predict or explain the relative sizes of ions, it is not reasonable to compare just any ions, because too many factors may be involved. To isolate some of the factors, we can consider an isoelectronic series of ions. In general, the ionic radius decreases as the atomic number increases in an isoelectric series, as in the following ions:

Ion	Radius	Electronic Configuration	Nuclear Charge
S^{2-}	184 pm	$1s^2 2s^2 2p^6 3s^2 3p^6$	16+
Cl^-	181 pm	$1s^2 2s^2 2p^6 3s^2 3p^6$	17+
K^+	133 pm	$1s^2 2s^2 2p^6 3s^2 3p^6$	19+
Ca^{2+}	99 pm	$1s^2 2s^2 2p^6 3s^2 3p^6$	20+
Sc^{3+}	81 pm	$1s^2 2s^2 2p^6 3s^2 3p^6$	21+

Note that all have the argon configuration. In this series, because the number of electrons is constant, the decrease in ionic radius is due to a shrinking of the electron cloud as the nuclear charge increases with atomic number.

We can also isolate some factors affecting ionic size by comparing the elements in a group, whose ions all have the same charge. Ionic size increases in a given group as atomic number increases. Consider the alkali metals, where we see a regular increase in radius as we go down the group:

$$Li^+ = 60 \text{ pm}, Na^+ = 95 \text{ pm}, K^+ = 133 \text{ pm}, Rb^+ = 148 \text{ pm}, Cs^+ = 169 \text{ pm}$$

Figure 6.34 Trends in atomic radii within the periodic table.

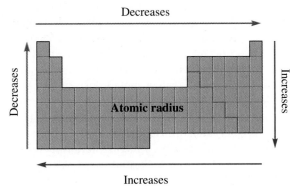

Ionic radii (pm)

	1+											3+	4+	5+	6+	7+	
H																	**He**
Li 60	**Be** 31 (2+)											**B** 20	**C**	**N**	**O**	**F**	**Ne**
Na 95	**Mg** 65	3+				2+						**Al** 50	**Si**	**P**	**S**	**Cl**	**Ar**
K 133	**Ca** 99	**Sc** 81	**Ti** 90	**V** 88	**Cr** 84	**Mn** 80	**Fe** 76	**Co** 74	**Ni** 72	**Cu** 70	**Zn** 74	**Ga** 62	**Ge**	**As**	**Se**	**Br**	**Kr**
Rb 148	**Sr** 113	**Y** 93	**Zr**	**Nb**	**Mo**	**Tc**	**Ru**	**Rh**	**Pd** 86	**Ag**	**Cd** 97	**In** 81	**Sn** 71	**Sb**	**Te**	**I**	**Xe**
Cs 169	**Ba** 135	**La** 115	**Hf**	**Ta**	**W**	**Re**	**Os**	**Ir**	**Pt**	**Au**	**Hg** 110	**Tl** 95	**Pb** 84	**Bi** 74	**Po**	**At**	**Rn**
Fr 176	**Ra** 140	**Ac** 118															

3+

Ce 111	**Pr** 109	**Nd** 108	**Pm** 106	**Sm** 104	**Eu** 103	**Gd** 102	**Tb** 100	**Dy** 99	**Ho** 97	**Er** 96	**Tm** 95	**Yb** 94	**Lu** 93

3+

Th 114	**Pa** 112	**U** 111	**Np** 109	**Pu** 107	**Am** 106	**Cm**	**Bk**	**Cf**	**Es**	**Fm**	**Md**	**No**	**Lw**

3+

Ti 76	**V** 74	**Cr** 69	**Mn** 66	**Fe** 64	**Co** 63	**Ni** 62	**Cu**	**Zn**
Zr	**Nb**	**Mo**	**Tc**	**Ru**	**Rh**	**Pd**	**Ag**	**Cd**
Hf	**Ta**	**W**	**Re**	**Os**	**Ir**	**Pt**	**Au**	**Hg**

1+	2+
Ga 113	**Ge** 93
In 132	**Sn** 112
Tl 140	**Pb** 120

4−	3−	2−	1−	
			H 208	**He**
B	**C** 260	**N** 171	**O** 140	**F** 136	**Ne**
Al	**Si** 271	**P** 212	**S** 184	**Cl** 181	**Ar**
Ga	**Ge** 272	**As** 222	**Se** 198	**Br** 195	**Kr**
In	**Sn** 294	**Sb** 245	**Te** 221	**I** 216	**Xe**
Tl	**Pb**	**Bi**	**Po**	**At**	**Rn**

Figure 6.35 Values of ionic radii for most of the elements.

We see a similar trend among the nonmetals—in the halogens, for example:

$$F^- = 136 \text{ pm}, \quad Cl^- = 181 \text{ pm}, \quad Br^- = 195 \text{ pm}, \quad I^- = 216 \text{ pm}$$

Atomic number increases as we go down a group, with valence electrons located in orbitals ever farther from the nucleus. Nuclear charge also increases with atomic number. However, the increased nuclear charge is shielded from the valence electrons by the inner electrons, so the effective nuclear charge is approximately constant. The net effect is increasing size from the top of a group to the bottom.

We can also compare the size of ions with the size of the atoms from which they were formed. When metals lose one or more electrons to form cations, the nuclear charge remains the same, while the number of electrons decreases. The remaining electrons are pulled more tightly to the nucleus, so the ions are considerably smaller than the parent atoms. Just the opposite effect is observed in the nonmetals. Anions are formed from atoms by addition of one or more electrons. These extra electrons are not attracted as strongly by the nucleus, which retains its original charge; so anions are much larger than their parent atoms.

Figure 6.36 The sizes of monatomic ions show regular trends across the periodic table. The sizes are radii in pm.

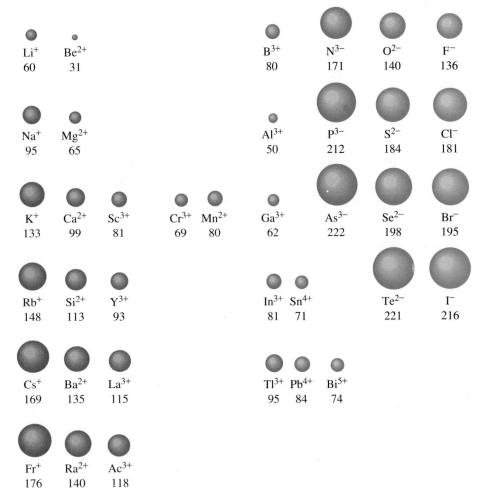

EXAMPLE 6.9

Arrange the following isoelectronic ions in order of increasing radius: Al^{3+}, F^-, Mg^{2+}, N^{3-}, Na^+, O^{2-}.

Solution The size of isoelectronic ions increases with decreasing nuclear charge. Thus, N^{3-}, with a nuclear charge of 7+, should be the largest ion; and Al^{3+}, with a nuclear charge of 13+, should be the smallest ion. The order is

$$Al^{3+} < Mg^{2+} < Na^+ < F^- < O^{2-} < N^{3-}$$

Practice Problem 6.9

Arrange the following isoelectronic ions in order of decreasing radius: Br^-, Se^{2-}, Rb^+, Sr^{2+}, Y^{3+}.

Answer: $Se^{2-} > Br^- > Rb^+ > Sr^{2+} > Y^{3+}$

Ionization Energy

What effect do you expect changes in nuclear charge, pairing energy, and nuclear shielding to have on the energy required to remove electrons from an atom?

Another property of atoms that shows periodic variations is ionization energy. **Ionization energy** (or **ionization potential**) is the minimum energy required to remove the outermost electron from a gaseous atom to form a gaseous ion. We assume that in the gas phase the atom and the ion are isolated from all external forces, so the ionization energy is simply the energy with which the atom binds the electron. The magnitude of the ionization energy thus gives a quantitative measure of the stability of the electronic structure of the isolated atom. Atoms with a low ionization energy do not bind their electrons very tightly, so such atoms are very reactive. The first ionization energies of the elements are given in Figure 6.37.

Except for hydrogen, which has only one electron, the elements have more than one ionization energy. There is one value for each electron in an atom. The first ionization energy (IE_1) of lithium, for example, corresponds to the removal of the first electron:

$$\text{Li}(g) \longrightarrow \text{Li}^+(g) + e^- \qquad IE_1 = 520 \text{ kJ/mol}$$

The second ionization energy corresponds to the removal of one electron from the lithium(I) ion:

$$\text{Li}^+(g) \longrightarrow \text{Li}^{2+}(g) + e^- \qquad IE_2 = 7298 \text{ kJ/mol}$$

The third ionization energy for lithium measures the energy needed to remove the last electron:

$$\text{Li}^{2+}(g) \longrightarrow \text{Li}^{3+}(g) + e^- \qquad IE_3 = 11,815 \text{ kJ/mol}$$

The trend observed here for lithium is common to all the elements, as you can see (for a few elements) in Table 6.6. That is, the ionization energy becomes larger with the removal of each electron. As electrons are removed, the nuclear charge is felt more strongly by the remaining electrons, so each is more difficult to remove than the last.

Figure 6.37 First ionization energies of the elements.

First ionization energies (kJ/mol)

H 1311.7																	He 2371.8
Li 520.0	Be 899.2											B 800.4	C 1086.1	N 1401.8	O 1313.5	F 1680.6	Ne 2080.1
Na 495.7	Mg 737.5											Al 577.4	Si 786.2	P 1011.5	S 999.3	Cl 1255.6	Ar 1520.1
K 418.6	Ca 589.6	Sc 630.85	Ti 657.8	V 650.6	Cr 652.6	Mn 717.1	Fe 759.1	Co 759.8	Ni 736.5	Cu 745.2	Zn 906.1	Ga 578.6	Ge 766.2	As 946.8	Se 940.7	Br 1142.7	Kr 1350.4
Rb 402.9	Sr 549.2	Y 615.3	Zr 659.5	Nb 663.9	Mo 684.7	Tc 702.0	Ru 710.5	Rh 719.9	Pd 804.1	Ag 730.8	Cd 867.5	In 558.2	Sn 708.4	Sb 833.5	Te 869.0	I 1008.6	Xe 1170.1
Cs 375.6	Ba 502.7	La 541.7	Hf 675.7	Ta 760.6	W 770.1	Re 759.8	Os 842.5	Ir 880	Pt 864.5	Au 889.9	Hg 1006.8	Tl 589.1	Pb 715.4	Bi 703.1	Po 813.0	At	Rn 1036.8
Fr	Ra 509.1	Ac															

Ce 540	Pr 529	Nd 532	Pm	Sm 540	Eu 547	Gd 594	Tb 577	Dy	Ho	Er 587	Tm 561	Yb 600	Lu 593
Th 671	Pa	U 587	Np	Pu	Am 580	Cm	Bk	Cf	Es	Fm	Md	No	Lw

Element	1st	2nd	3rd	4th	5th	6th	7th	8th	9th	10th
H	1312									
He	2372	5250								
Li	520	7298	11815							
Be	899	1757	14849	21006			Inner electrons			
B	801	2427	3660	25026	32827					
C	1086	2353	4620	6222	37830	47277				
N	1402	2856	4582	7475	9445	53266	64360			
O	1314	3388	5300	7469	10989	13326	71334	84078		
F	1681	3374	6050	8408	11023	15164	17868	92038	106434	
Ne	2081	3952	6122	9370	12178	15238	19999	23069	115379	131431

Valence electrons

Table 6.6 Successive Ionization Energies for Some Elements (kJ/mol)

The first ionization energies of the elements are plotted against atomic numbers in Figure 6.38, revealing the periodicity mentioned earlier. The noble gases all fall at the maximum values for each period, and the alkali metals fall at the minimum values. Because ionization energy is a measure of the difficulty of removing an electron to form a cation, it is reasonable that ionization energy should increase on going from the metals to the nonmetals across a period, consistent with the tendency of the metals to form cations and the nonmetals to form anions. This variation reflects the increase in nuclear charge across a period. The valence electrons are all at approximately the same distance from the nucleus, so the electrons are held more tightly from left to right across the periodic table.

Ionization energy decreases from the top to the bottom of a group, as you can see by examining the maximum values, those for the noble gases, in Figure 6.37. The valence electrons are located farther from the nucleus in the members of a group lower in the periodic table. For example, the valence electrons in Xe are farther from the nucleus than the valence electrons in Ne. The nuclear charge increases down a group (10 in Ne and 54 in Xe), but the inner electrons shield some of the nuclear charge from the valence electrons, so all members of a group experience approximately the same effective nuclear charge. Electrons farther from the nucleus, then, are easier to remove. The general tendency is for ionization energy to increase from bottom to top and from left to right across the periodic table, as summarized in Figure 6.39.

Superimposed on the general trend for ionization energy is a fine structure, additional maxima and minima, which can be seen in Figure 6.38. This fine structure gives rise to exceptions to the expected trends and can be explained by examining the electronic configuration of the atoms. Consider the trend shown by Li, Be, and B. The ionization energy of beryllium is greater than that of lithium,

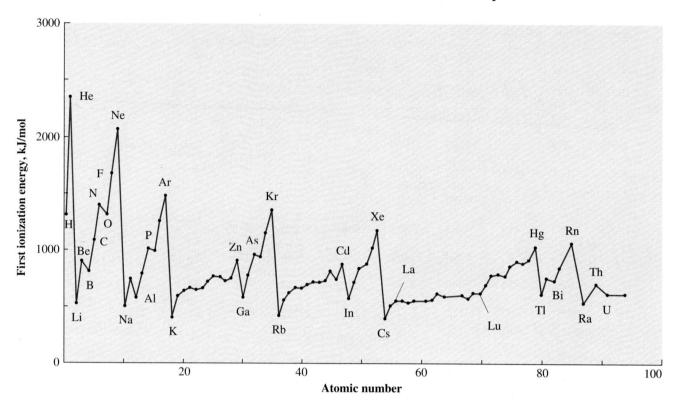

Figure 6.38 The first ionization energy varies in a regular and periodic manner with atomic number.

consistent with the increase in nuclear charge. But when the first electron is added to a *p* orbital, as in boron, the ionization energy drops somewhat, instead of increasing as expected. Because on average the 2*s* orbital is located closer to the nucleus than is the 2*p* orbital, electrons in the 2*p* orbital will experience a smaller effective nuclear charge because of some shielding by the 2*s* electrons. As the *p* orbitals are being filled, ionization energy increases with the increase in nuclear charge. However, addition of the fourth electron to the *p* orbitals is accompanied by a slight decrease in ionization energy, so the ionization energy of oxygen is less than that of nitrogen, opposite to the general trend. This effect is due to pairing energy. At the fourth *p* electron, it is necessary for two electrons to begin occupying the same *p* orbital, so electron repulsions offset the increase in nuclear charge. Nuclear charge increases further as the *p* orbitals continue to fill, and so ionization energy continues to increase.

Figure 6.39 Trends in ionization energy within the periodic table.

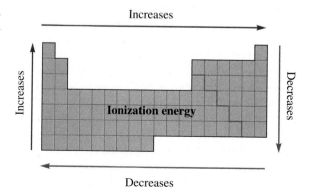

Although the transition metals and the inner-transition metals show an increase in ionization energy as the atomic number increases across a period, as Figure 6.37 shows, the increase is relatively small. Consider the first transition series (Sc through Zn) as an example. The last electron in each element is being added to a 3d orbital, and the electron being ionized is in the 4s orbital. The 4s orbital is located farther from the nucleus than is the 3d orbital. Increases in nuclear charge accompanied by increases in the number of 3d electrons have only a small effect on the difficulty of removing a 4s electron. The 3d electrons are shielding the 4s electron from the increased nuclear charge to a large extent, so the ionization energy increases only slowly. The lanthanide and actinide series show a similar pattern because of the shielding effects of the inner 4f or 5f electrons.

EXAMPLE 6.10

Which element, carbon or fluorine, should have the largest first ionization energy?

Solution   The ionization energy increases from left to right across a period, with exceptions for the first and fourth electron in the p orbitals. Because neither of these elements is one of the exceptions to the general trend, fluorine should have a larger ionization energy than carbon.

Practice Problem 6.10

Which element, beryllium or boron, will have the largest first ionization energy?

Answer: Beryllium (because boron is one of the exceptions to the general trend)

Electron Affinity

Another measure of the stability of various electronic configurations is the **electron affinity.** This quantity is defined as the enthalpy change that occurs when a gaseous atom (A) adds an electron to form a gaseous anion:

$$A(g) + e^- \longrightarrow A^-(g)$$

A negative value for electron affinity means energy has been released, whereas a positive value indicates that energy has been consumed. Electron affinities are not easy to measure, so they are not available for all the elements, and some of the values that are available were determined indirectly or were calculated rather than measured. Some known values are shown in the periodic table in Figure 6.40. A plot of electron affinity against atomic number, shown in Figure 6.41, demonstrates the periodic nature of this energy quantity.

The regular variation in the values of electron affinity can be explained in much the same way as the values of ionization potential. Electron affinity increases (becomes more negative) from left to right across the periodic table. It also increases from bottom to top of the table, but this trend has many exceptions and is not very pronounced. The increase across a period is consistent with the decrease in size and increase in nuclear charge, both of which lead to a greater attraction of the nucleus for an additional electron. A major deviation from the general trend across a period is found at the alkaline-earth metals. Each of these elements has a filled ns subshell (ns^2), so the electron added to form an anion will go into the np

Electron affinity (kJ/mol)

Group 1 IA	Group 2 IIA											Group 13 IIIA	Group 14 IVA	Group 15 VA	Group 16 VIA	Group 17 VIIA	Noble gases
H −73																	**He**
Li −60	**Be**											**B** −27	**C** −122	**N** 0.0	**O** −141	**F** −328	**Ne**
Na −53	**Mg**											**Al** −42	**Si** −134	**P** −72	**S** −200	**Cl** −349	**Ar**
K −48	**Ca**	**Sc** −18	**Ti** −8	**V** −57	**Cr** −64	**Mn**	**Fe** −16	**Co** −64	**Ni** −112	**Cu** −118	**Zn**	**Ga** −29	**Ge** −130	**As** −78	**Se** −195	**Br** −325	**Kr**
Rb −47	**Sr**	**Y** −30	**Zr** −41	**Nb** −86	**Mo** −72	**Tc** −53	**Ru** −101	**Rh** −109	**Pd** −54	**Ag** −126	**Cd**	**In** −29	**Sn** −116	**Sb** −103	**Te** −190	**I** −295	**Xe**
Cs −46	**Ba** −48	**La**	**Hf**	**Ta** −31	**W** −79	**Re** −14	**Os** −106	**Ir** −151	**Pt** −205	**Au** −223	**Hg**	**Tl** −19	**Pb** −35	**Bi** −91	**Po** −183	**At** −270	**Rn**
Fr −45	**Ra**																

Figure 6.40 Values of the electron affinity for some elements. Missing values are estimated to be either very close to zero or possibly positive.

orbital, located farther from the nucleus. Because the added electron is not attracted as expected by the nucleus, the value of electron affinity is smaller (more positive) than expected. We see a similar effect at Group VA (15): N, P, As, Sb, Bi. Here, addition of another electron requires the pairing of electrons in one of the

Figure 6.41 Electron affinity varies regularly with atomic number.

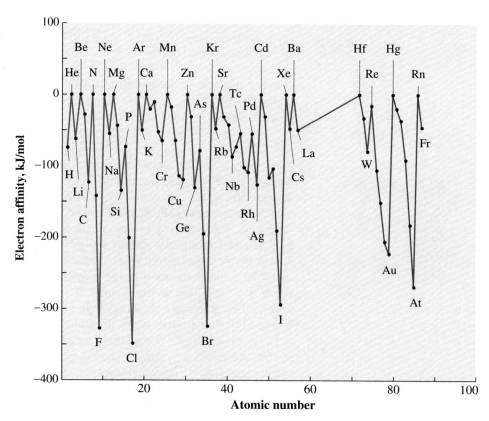

np orbitals, so more energy than expected is required. The greatest (most negative) values of electron affinity belong to the halogens, where addition of one electron results in the very stable, filled valence-shell configuration of the noble gases.

6.9 Periodicity and Chemical Reactivity

Chemical reactions of the elements involve changes in the electronic configurations—that is, addition, removal, or sharing of electrons—to give more stable configurations. For example, when a copper penny reacts with nitric acid, the copper loses electrons to form Cu^{2+}, and nitrogen in the nitrate ion gains electrons to form $NO_2(g)$. The aqueous copper(II) and nitrate ions remain in solution:

$$Cu(s) + 4HNO_3(aq) \longrightarrow Cu(NO_3)_2(aq) + 2NO_2(g) + 2H_2O(l)$$

Which elements would you expect to be reactive or unreactive, based on their electronic configurations?

Because the ease of loss or gain of electrons correlates with position in the periodic table, it should be no surprise that the chemical reactivity of the elements correlates with position in the periodic table.

Noble Gases

The noble gases were formerly called the inert gases because of their lack of reactivity. We have seen that these elements have completely filled outer orbitals, a highly stable electronic configuration. From values of the electron affinity and the ionization energy, we can surmise that it is very difficult either to add an electron to a noble gas or to remove an electron from one. Thus the noble gases undergo little chemical change; some exceptions were discussed in Section 6.7.

Alkali Metals

The alkali metals are all so reactive that they do not occur as free metals in nature. They are found only in the form of compounds, such as sodium chloride and potassium chloride, which can be found in seawater and in solid deposits.

At the opposite side of the periodic table are the alkali metals, Group IA (1), which are extremely reactive. These elements all have a valence electron configuration of ns^1. Because their ionization energies are very low, it is very easy to remove one electron to form cations having noble-gas electronic configurations. The ionization energy decreases as we move from the top of the alkali-metal group to the bottom because the effective nuclear charge remains about the same while the atomic size increases. Thus, the heavier members of the group are rather more reactive than the lighter members.

Lithium, the first member of the group and a very reactive metal, is not so reactive as the other alkali metals. It is the only such metal that can be melted and poured in dry air without losing its metallic luster because of the formation of an oxide coating. Lithium does not react with dry oxygen below 100°C. In contrast, the heavier alkali metals react quickly with the oxygen in air at room temperature, as shown below for sodium:

$$4Na(s) + O_2(g) \longrightarrow 2Na_2O(s)$$
$$2Na(s) + O_2(g) \longrightarrow Na_2O_2(s)$$

All the alkali metals must be stored under a hydrocarbon such as kerosene to prevent their eventual reaction with oxygen in air to form oxides or peroxides

Figure 6.42 Sodium metal, as well as the other alkali metals, must be stored under a hydrocarbon such as kerosene to prevent reaction with the oxygen in air. This reaction can be quite vigorous for the heavier alkali metals. The metals will be entirely oxidized if left in air long enough.

The alkaline-earth metals are hard, silvery white metals that are not as reactive as the alkali metals. They are used to form alloys with other metals. They have a variety of uses when converted into compounds with nonmetals. Magnesium is the most widely used of these metals; it is primarily alloyed with aluminum to give a strong, lightweight structural metal used in aircraft.

(Figure 6.42). Reactivity with oxygen increases as we go down the group. Cesium, for example, reacts so vigorously with oxygen that it ignites spontaneously. Reactions of the elements with oxygen were discussed in Section 4.3.

The alkali metals all react with water to form hydroxides, as illustrated here for rubidium:

$$2Rb(s) + 2H_2O(l) \longrightarrow 2RbOH(aq) + H_2(g)$$

This reaction also becomes more vigorous as we go down the group. Already at potassium, the reaction with water occurs with considerable violence, resulting in the expulsion of glowing bits of metal (Figure 6.43).

Alkaline-Earth Metals

The alkaline-earth metals, Group IIA (2), all have a valence electron configuration of ns^2. The loss of two electrons gives a noble-gas configuration, so these metals form cations with a 2+ charge. These metals tend to be reactive, but not nearly so reactive as the alkali metals. They react with oxygen in air, for example, but when they do, they form an oxide coating that protects them from further reaction. Thus, the alkaline-earth metals do not have to be stored under kerosene to protect them from oxidation (Figure 6.44). The heavier alkaline-earth metals also react with water to form hydroxides, as shown here for calcium:

$$Ca(s) + 2H_2O(l) \longrightarrow Ca(OH)_2(aq) + H_2(g)$$

Again, however, the reactions are not so vigorous as the corresponding reactions of the alkali metals. Compare the reaction of calcium and water (Figure 6.45) with that of potassium and water (Figure 6.43), for instance. As with the alkali metals, the reactivity of the alkaline-earth metals increases as we go down the group.

Transition Metals

The properties of metals in the representative elements change significantly from one group to the next in accord with differences in electronic configuration, ionization energy, electron affinity, and size. In contrast, the properties of the metals in

Figure 6.43 Potassium metal reacts explosively with water to form potassium hydroxide.

Figure 6.44 The alkaline-earth metals are (from the left) beryllium, magnesium, calcium, strontium, and barium. Though reactive, these metals do not have to be handled in the absence of air to maintain their elemental integrity.

The transition metals are widely used as structural metals. The primary structural metal is iron, usually converted to steel by alloying with carbon or other transition metals, such as vanadium, manganese, or chromium. These metals impart hardness and corrosion resistance to the iron. Titanium is used as a hard, chemically resistant, light structural metal in applications such as aircraft. Chromium is highly resistant to corrosion and takes a high shine, so it is used to plate other metals.

the transition groups tend not to vary much from one group to the next. These metals all have valence electrons in the ns orbital, as well as electrons in the inner $(n - 1)d$ orbitals. For most of the transition metals, the valence electron configuration is ns^2, while a few have the configuration ns^1. On average, the electrons in the d orbitals are found farther from the nucleus than those in s or p orbitals, so they do not shield the valence electrons as well, and the valence electrons experience greater effective nuclear charge. For this reason, electrons in the s orbitals in the transition metals are held much more tightly than the electrons in s orbitals in the alkali metals or alkaline-earth metals. As a result, the transition metals are less reactive.

The lower reactivity of the transition metals is reflected in the way we use these metals and in their occurrence in nature. They are used in coins and in structural applications (iron in building materials, for example) because they do not corrode rapidly when exposed to water or air. (Iron, as we know, rusts more rapidly than is desirable for its common uses.) Some of them can be found in nature as the free metal rather than as metal compounds, again because of their low reactivity with water and oxygen.

Halogens

The halogens all exist at room temperature as diatomic species: F_2 is a green gas, Cl_2 is a yellow-green gas, Br_2 is a red liquid, and I_2 is a violet solid.

Because they have high ionization energies and high electron affinities, the nonmetals tend to react by adding electrons (or by sharing electrons) rather than giving up electrons. An example is the halogens, Group VIIA (17), all of which have the valence electron configuration ns^2np^5, which is one electron short of a noble-gas

Figure 6.45 Calcium metal reacts slowly with water to form calcium hydroxide and hydrogen gas.

configuration. The reactivity of these elements decreases as we move from the top of the group to the bottom, with fluorine being the most reactive. The atoms of the heavier elements are larger, so the addition of an electron occurs farther from the nucleus, while the effective nuclear charge remains about the same. Thus, the addition of an electron releases more energy for fluorine than for iodine, making fluorine more reactive. At sufficiently high temperature, fluorine reacts with all metals to form fluoride salts. Many of the reactions with fluorine, like the reaction with uranium shown below, are vigorous enough to ignite the metals and cause vaporization of the metal halide:

$$U(s) + 2F_2(g) \longrightarrow UF_4(g)$$

Iodine, near the bottom of the group, is also reactive, but it is usually less violent in its reactions than are the other members of the group. The other nonmetals show similar trends, with reactivity decreasing from the top of a group to the bottom.

Summary

Much of what we know about the behavior of electrons in matter comes from observing the interaction of matter and electromagnetic radiation. When an electrical discharge is passed through a gas, the gas emits light at only certain wavelengths, which can be predicted accurately. For the wavelengths emitted by hydrogen gas, we use the Rydberg equation, which contains a variable that can have only integral values. Similarly, the energy of radiation arising from the heating of solids is described by the Planck equation, which also contains a variable, called a quantum number, that can have only integral values. These equations indicate that energy is absorbed and radiated by matter in small bits, or quanta. Bohr applied the idea of quantization of energy to electrons in atoms. His orbital model of the hydrogen atom provided an explanation for the line spectrum of hydrogen by indicating that radiation would be emitted or absorbed when the electron jumped from one orbit to another. However, this model could not explain the spectra of other elements.

A more successful model of the atom is provided by quantum mechanics, which assumes that the energies of electrons in atoms can be described in much the same way as are the energies of waves. Each electron occurs in a shell characterized by a principal quantum number, n. Within each shell are orbitals, the volumes of space in which the electrons can be found most of the time. An orbital is characterized by an azimuthal quantum number, l, which denotes its shape (labeled s, p, d, or f). Orbitals occur in sets or subshells (one s, three p, five d, and seven f), with each orbital in a subshell pointing in a different direction characterized by a different value of the magnetic quantum number, m_l. Each orbital can be occupied by two electrons, which differ by the value of the spin quantum number, m_s.

In multielectron atoms, the electronic configuration of an atom—a listing of the orbitals occupied by electrons—can be determined by the Aufbau principle. Electrons are placed into orbitals in order of increasing energy: $ns < (n - 2)f < (n - 1)d < np$. The electrons are placed in different orbitals of a subshell when possible. The electronic configurations of the elements correlate with their positions in the periodic table. All members of a group have equivalent valence electron configurations.

Properties of the elements, such as atomic and ionic radii, ionization energy, electron affinity, and chemical reactivity, show periodic variations with atomic number. These trends can all be explained by considering three factors: increases in nuclear charge with increasing atomic number, the pairing energy that exists when two electrons occupy the same orbital, and the shielding effect of inner electrons.

Key Terms

absorption
   spectrum (6.2)
amplitude (6.1)
Aufbau principle (6.5)
azimuthal quantum

number (6.4)
continuous
   spectrum (6.2)
covalent radius (6.8)
degenerate (6.5)

effective nuclear
   charge (6.8)
electromagnetic
   radiation (6.1)
electron affinity (6.8)

electronic
   configuration (6.5)
emission spectrum (6.2)
excited state (6.5)
frequency (6.1)

ground state (6.5)

Heisenberg uncertainty
   principle (6.3)

hertz (Hz) (6.1)

Hund's rule (6.5)

ionic radius (6.8)

ionization energy
   (ionization
   potential) (6.8)

isoelectronic series (6.6)

line spectrum (6.2)

magnetic quantum
   number (6.4)

metallic radius (6.8)

orbital (6.4)

pairing energy (6.8)

Pauli exclusion
   principle (6.5)

photon (6.3)

Planck's constant (6.3)

principal quantum
   number (6.4)

quantum mechanical
   model (6.4)

quantum number (6.3)

quantum theory (6.3)

shell (6.4)

shielding effect (6.8)

spectrum (6.2)

spin quantum
   number (6.4)

subshell (6.4)

valence electrons (6.5)

valence shell (6.5)

wavelength (6.1)

wave mechanics (6.4)

Exercises

Electromagnetic Radiation

6.1 Draw a picture that shows the wavelength, frequency, and amplitude of electromagnetic radiation.

6.2 List the various types of radiation in order of increasing energy.

6.3 Why would you expect gamma rays to be more hazardous to humans than ultraviolet radiation would be?

6.4 Which has the higher energy—visible or ultraviolet light?

6.5 Calculate the energy of a photon having a wavelength of 501 nm.

6.6 What is the energy of light with a frequency of 5.00×10^{14} s^{-1}?

6.7 Calculate the energy of an X ray with a wavelength of 0.500 nm.

6.8 A photon of light has an energy of 95.0 kJ/mol.
   a. What is the wavelength of this light?
   b. What is its frequency?

6.9 What is the wavelength of a photon having a frequency of 2.00×10^{15} s^{-1}?

6.10 a. What is the frequency of gamma rays having a wavelength of 2.00×10^{-5} nm?
   b. What is the energy of this radiation?

6.11 What is the wavelength of light having a frequency of 1.8×10^{15} Hz?

6.12 What is the frequency of visible light having a wavelength of 472 nm?

6.13 What is the wavelength of an AM radio station broadcasting at a frequency of 100.5 kHz?

Quantum Theory

6.14 Describe how Bohr applied the ideas of Planck and Einstein to develop a model of the atom.

6.15 What is a quantum number?

6.16 a. What contribution did de Broglie make to the development of the modern model of the atom?
   b. What experiment supported his description of the wave–particle duality?

6.17 Calculate the wavelength of a proton having a velocity of 1.80×10^8 cm s^{-1}. The mass of a proton is 1.67×10^{-24} g.

6.18 A neutron of mass 1.67×10^{-24} g exhibits a wavelength of 5.02×10^{-10} cm. What is the velocity of this neutron?

6.19 What is the velocity of an electron having a de Broglie wavelength of 0.100 nm? The mass of the electron is 9.11×10^{-28} g.

6.20 A light meter on a camera works because of the photoelectric effect. Explain.

6.21 Calculate the wavelength of an electron with a mass of 9.11×10^{-28} g that is traveling at the speed of light.

Quantum Numbers

6.22 What are the four quantum numbers?

6.23 Explain the physical significance of each of the four quantum numbers.

6.24 List the possible values of each of the four quantum numbers.

6.25 What does the Pauli exclusion principle say about the possible values of the four quantum numbers?

6.26 How many electrons can go into each orbital?

6.27 List the quantum numbers for each electron in the sodium atom.

6.28 How many values of the azimuthal quantum number, l, are possible in the $n = 5$ main energy level?

6.29 List the quantum numbers for each electron in the H$^-$ ion.

6.30 What is the value of the principal quantum number, n, for an electron in a p orbital in the valence shell of sulfur?

6.31 What is the value of the azimuthal quantum number, l, for an electron in a p orbital in the valence shell of bromine?

6.32 For krypton in its ground state, indicate how many electrons have each of the following quantum-number values.
   a. $n = 3$ b. $l = 1$ c. $m_l = -1$ d. $l = 2$
   e. $n = 4$ f. $m_s = +\frac{1}{2}$

6.33 What are possible values of m_l for a d orbital?

6.34 What are possible values of m_s for an electron in a p orbital?

6.35 List the possible values of m_l when $l = 1$.

6.36 List the quantum numbers for each electron in the following atoms.
     a. Be   b. F   c. N

Electronic Configurations

6.37 What shorthand notation is used to represent electronic configurations?

6.38 Explain the application of Hund's rule to writing electronic configurations.

6.39 How can we use the noble-gas electronic configuration to shorten the electronic configuration of other elements?

6.40 How can the periodic table be used to help write electronic configurations?

6.41 Sketch the shape of a typical s, p, d, and f atomic orbital.

6.42 Write the electronic configurations of each of the following.
     a. Sb   b. Sc   c. Br   d. N   e. Ca   f. Mn

6.43 Write the electronic configurations of each of the following.
     a. K^+ b. Ti^{3+} c. Be^{2+} d. Ni^{2+} e. Cr^{2+}
     f. Mn^{3+}

6.44 Write the electronic configurations of each of the following.
     a. S^{2-} b. Se^{2-} c. Br^- d. N^{3-} e. C^{4-}
     f. H^-

6.45 Indicate the number of unpaired electrons in each of the following atoms.
     a. Ba   b. Bi   c. Ne   d. Si   e. Gd

6.46 Identify the atom represented by each of the following electronic configurations.
     a. $[Ar]4s^2 3d^1$ d. $[Kr]5s^2 4d^{10} 5p^5$
     b. $[Ar]4s^2 3d^{10} 4p^4$ e. $[Kr]5s^2 4d^{10} 5p^2$
     c. $[Ar]4s^1 3d^5$

6.47 a. How many unpaired electrons are there in the oxygen atom?
     b. How many unpaired electrons are there in the oxide ion, O^{2-}?

6.48 What ions could have the electronic configuration $1s^2 2s^2 2p^6 3s^2 3p^6$?

6.49 What ion (or ions) could have the electronic configuration $[Ar]4s^0 3d^2$?

Atomic and Ionic Radii

6.50 Describe different measures of atomic radii.

6.51 Compare atomic radii for metals and nonmetals.

6.52 Compare the relative sizes of atoms and cations or anions formed from them.

6.53 Which of the following has the largest atom: B, C, N, or F?

6.54 Arrange the following atoms in order of increasing size: Ar, Ca, K, Sc.

6.55 Which of the following has the largest atom: C, N, O, or S?

6.56 Arrange the following ions in order of increasing size: Sc^{3+}, Y^{3+}, La^{3+}.

6.57 Arrange the following ions in order of decreasing size: Br^-, Cl^-, F^-, I^-.

6.58 Which of the following has the smallest atom: P, S, O, or F? Which has the largest atom?

6.59 Would you expect Ba or Ba^{2+} to be larger? Explain your answer.

6.60 Arrange the following ions in order of increasing size: Br^-, Cl^-, N^{3-}, P^{3-}, S^{2-}.

6.61 Which ion would you expect to be larger, Fe^{2+} or Fe^{3+}? Explain your answer.

6.62 Which member of each pair has the smaller radius?
     a. K or Cs   b. Rb or Sr   c. B or C   d. P or O

6.63 Would you expect S or S^{2-} to be larger? Explain your answer.

Ionization Energy

6.64 Compare and contrast ionization energy and electron affinity.

6.65 Sketch the variation in ionization energy for the first 20 elements. Explain the features of your sketch.

6.66 Arrange sulfur, silicon, phosphorus, and aluminum in order of increasing first ionization energy.

6.67 Compare the expected values of the ionization energy for metals and nonmetals.

6.68 Arrange neon, sodium, magnesium, and aluminum in order of increasing first ionization energy.

6.69 Which member of each pair should have the larger first ionization energy?
     a. S or Cl   d. K or Rb
     b. Ne or O   e. Si or Ge
     c. Sr or Ca

6.70 For any element, the second ionization energy is always larger than the first. Explain this observation.

6.71 Arrange magnesium, silicon, sulfur, and argon in order of decreasing first ionization energy.

6.72 Which member of each pair should have the smaller first ionization energy?
     a. N or O   d. Cs or Rb
     b. S or O   e. Si or N
     c. Li or Be

Electron Affinity

6.73 Compare the values of electron affinity that would be expected for a metal and a nonmetal. Explain the difference.

6.74 Arrange the following elements in order of increasing (more negative) electron affinity: O, S, Si, C, N.

6.75 Which of the following elements has the greatest (most negative) electron affinity: C, Si, O, S, or Se?

6.76 The second electron affinity (for addition of an electron to a singly charged anion) is always positive and is always more positive than the first electron affinity. Explain this observation.

6.77 Indicate which of the following elements has the most negative electron-affinity value: O, S, Se, or Cl. Explain your answer.

Chemical Reactivity

6.78 Name an element that can be classified as a member of each of the following groups:
  a. transition metals   f. noble gases
  b. inner-transition metals g. nonmetals
  c. alkali metals   h. main-group elements
  d. alkaline-earth metals   i. metalloids
  e. halogens

6.79 Explain why the noble gases are not very reactive.

6.80 Why do elements in the same group have similar chemical properties?

6.81 Write the formulas for four ions that are isoelectronic with neon.

6.82 Which element should be more reactive, O or S? Explain your answer.

6.83 Indicate which of the following elements should have chemical properties similar to those of oxygen: N, F, Cl, S, C.

6.84 Indicate the charge on the ions that would most likely be formed by each of the Period 3 elements (Na through Ar).

6.85 Write the formula of the compound that should be formed by each of the following pairs of elements:
  a. K and O   d. Sc and Cl
  b. Ba and Cl   e. Mg and N
  c. Sr and O   f. Na and S

Additional Exercises

6.86 Which member of each pair should have the larger first ionization energy?
  a. O or S   b. Kr or Br   c. Mg or Ca
  d. Na or Li   e. Al or Ga

6.87 Calculate the wavelength of an electron that has a mass of 9.11×10^{-28} g and is traveling around an atom with a velocity of 5.05×10^6 m s^{-1}.

6.88 Arrange the following types of radiation in order of increasing wavelength: infrared, visible, ultraviolet, gamma rays, radio waves, microwaves, X rays.

6.89 Why is the aluminum ion smaller than the aluminum atom?

6.90 Explain why the first ionization energy is larger for nitrogen than for oxygen.

6.91 List the quantum numbers for each electron in the following atoms.
  a. B   b. O   c. He   d. Li

6.92 Identify the atom represented by each of the following electronic configurations.
  a. $[Ar]4s^2 3d^3$ d. $[Kr]5s^0 4d^{10}$
  b. $[Ar]4s^2 3d^{10} 4p^2$ e. $[Kr]5s^2 4d^{10} 5p^5$
  c. $[Ar]4s^1 3d^{10}$

6.93 Although numerous xenon compounds have been synthesized, very few other noble-gas compounds have been prepared. Why is this so?

*6.94 Discuss any correlation between the discontinuities in a graph of ionization energy against atomic number and the commonly used generalization that particularly stable electronic configurations have filled shells, filled subshells, or half-filled subshells.

*6.95 Predict the shape you would expect for a graph of the ratio of the first to the second ionization energy (IE_1/IE_2) against atomic number. Make such a graph for the first 10 elements and explain its predominant features.

*6.96 Explain why the variation in density of the transition elements within a given period is less than the variation in density of the main-group elements.

*6.97 a. Use the Rydberg equation and the Planck equation to calculate the energy of the electronic transition in the hydrogen atom in which $n_1 = 1$ and $n_2 = \infty$.
  b. What property of the atom is obtained by this calculation?
  c. How does your calculated value compare to the tabulated value found in this chapter?

▶ ▶ ▶ Chemistry in Practice

6.98 An element is a solid at room temperature and has a metallic luster. When the solid is vaporized, radiation with a wavelength of 165 nm causes the gaseous atoms to eject an electron. No reaction occurs within an hour when this element is placed in water, but it does react with steam to produce hydrogen gas. A similar reaction occurs with aqueous acids. When the element is heated in oxygen, it forms a white solid that contains 39.68% O by mass. What is the element?

7

Bonding in Molecules and Compounds

▶ ▶ ▶ ▶ ▶

Ti(IV) chloride fumes to form TiO₂ when exposed to moist air. Ti(III) chloride is more stable in moist air.

CHAPTER OUTLINE

Encountering Chemistry

itanium(III) chloride is a purple solid that can be exposed to air with no change (under some conditions); in water, it forms a purple solution. In contrast, titanium(IV) chloride, or titanium tetrachloride, is a volatile, colorless liquid that forms a white solid, TiO_2, when exposed to moist air:

$$TiCl_4(g) + 2H_2O(g) \longrightarrow TiO_2(s) + 4HCl(g)$$

This reaction, used to make military smoke screens, also occurs when $TiCl_4$ is mixed with liquid water.

Consider two other volatile, colorless liquids: carbon tetrachloride and silicon tetrachloride. Carbon tetrachloride neither dissolves in water nor reacts with water at ordinary temperatures, and it does not fume when exposed to moist air. However, silicon tetrachloride, like titanium tetrachloride, fumes in moist air and reacts with liquid water to form silicic acid, a hydrated form of silicon dioxide:

$$SiCl_4(l) + 3H_2O(l) \longrightarrow H_2SiO_3(s) + 4HCl(g)$$

Because carbon and silicon are next to one another in the same group of the periodic table, we might have expected them to behave similarly. Although some similarities are observed, there are distinct differences in the chemical behavior of these two elements and their compounds, just as there are distinct differences between the titanium(III) and titanium(IV) chlorides.

What causes such differences? Much of the answer has to do with **chemical bonds,** the forces that hold atoms together in molecules or ionic compounds. A chemical reaction is a process that changes one arrangement of atoms, with its particular chemical bonds, into another arrangement. Consequently, understanding chemical bonding helps us understand the chemical and physical properties of the elements and their compounds.

This chapter examines various types of bonds and some of their properties. Among these properties is bond strength, which can be related to the enthalpy changes in chemical reactions. The chapter also examines a way of representing bonds using the electron-dot formula, which allows us to predict the bonding in molecules.

7.1   Chemical Bonds and Electronic Configurations

Consider the physical and chemical properties, listed in Table 7.1, of sodium, a metal; chlorine, a nonmetal; and sodium chloride, a compound formed by a reaction between the two. The three substances are shown in Figure 7.1. Clearly, they differ significantly in their properties. In fact, these substances are typical of substances containing the three major types of chemical bonds: metallic, covalent, and ionic. Table 7.2 lists some properties of these types of substances.

What is it about these three types of bonds that results in such different properties? The way elements interact to form bonds can be understood in terms of their electronic configurations. In bonding, they tend to lose, gain, or share electrons to achieve a more stable configuration, especially a noble-gas configuration. **Ionic bonding** involves a complete transfer of one or more electrons from a metal

Table 7.1 Properties of Sodium, Chlorine, and Sodium Chloride

	Sodium	**Chlorine**	**Sodium Chloride**
Formula	Na	Cl_2	NaCl
Physical state	Silvery Metallic Solid	Yellow-green Gas	Colorless Crystalline Solid
Molar mass (g/mol)	22.99	70.91	58.44
Melting point (°C)	98	-101	801
Boiling point (°C)	883	-35	1413
Electrical conductivity:			
Solid	Very high	Very low	Very low
Liquid	Very high	Very low	High
Dissolves in:	Other metals	CCl_4, less in H_2O	H_2O, Not in CCl_4
In water, forms:	$NaOH + H_2$	$HCl + HOCl$ (slowly forms O_2)	$Na^+ + Cl^-$ (No chemical change)

Figure 7.1 Sodium and chlorine *(left)* react to form sodium chloride *(right).*

to a nonmetal, forming a metal cation and a nonmetal anion held together by *electrostatic forces,* or attractions between oppositely charged ions offset by repulsions between ions of like charge. In **covalent bonding,** atoms share electron pairs to form molecules. The focus of this chapter is on the nature of ionic and covalent bonds.

Metallic bonding involves the sharing of free and mobile valence electrons by the nuclei of the metal atoms. The metal cations are fixed into position in a solid network, which is immersed in a "sea" of freely moving electrons (see Figure 7.2). The interaction between the positively charged metal cations and the negatively charged electrons binds the metal atoms together. This model accounts for the high electrical conductivity of metals: the electrons are free to move about and conduct an electrical current. The electrons, being mobile, are removed relatively

Table 7.2 General Properties of Substances with Metallic, Covalent, and Ionic Bonds

Metallic	Covalent	Ionic
Lustrous solids	Gases, liquids, or solids	Crystalline solids
Malleable and ductile	Brittle and weak or soft and waxy as solids	Hard and brittle
Usually high melting point	Low melting point	Very high melting point
High boiling point	Low boiling point	Very high boiling point
High heat of vaporization	Low heat of vaporization and fusion	High heat of vaporization and fusion
High density	Low density	High density
Good electrical and heat conductor	Poor electrical and heat conductor	Good electrical conductor when molten
Soluble in other metals	Generally soluble in hydrocarbon solvents	Often soluble in water

Figure 7.2 Metallic bonding involves the interaction between fixed metallic cations and a "sea" of mobile and free electrons.

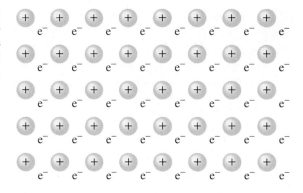

easily in oxidation–reduction reactions, so metals typically donate electrons to nonmetals when they react. Other properties of metals are related to their structures, which will be considered in detail in Chapter 10.

How do metals and nonmetals differ in the way they bond to other elements?

We can see the connection between bonding and electronic configurations of the elements especially well by examining the elements immediately preceding or following the noble gases in the periodic table. Each element immediately following a noble gas is a metal and is strongly *electropositive*. That is, it has a strong tendency to give up electrons, thereby achieving a noble-gas electronic configuration, which is unusually stable because of its filled outermost shell. By giving up electrons, the metal achieves a positive electrical charge, forming a cation. Each element immediately preceding a noble gas in the periodic table is a nonmetal and is highly *electronegative,* which means it has a strong tendency to add electrons and achieve a noble-gas configuration. In so doing, it achieves a negative electrical charge, forming an anion.

When a metal and a nonmetal bond, the metal loses electrons to achieve a more stable electronic configuration, while the nonmetal gains electrons. Thus, an ionic bond, with transfer of electrons from metal to nonmetal, is formed. When two nonmetals bond, both would have to gain electrons to attain noble-gas configurations. Because this is not possible, the nonmetals share electrons, forming a covalent bond. This compromise allows both atoms to have an electronic configuration that is more stable. This electronic configuration also approximates a noble-gas configuration, with the shared electrons being a part of the electronic configuration of both atoms.

The Octet Rule

How many bonds will an atom form when combining with another atom?

Often, the number of bonds an element forms is related to the element's tendency to donate, accept, or share a specific number of electrons in order to achieve a noble-gas configuration. Thus, fluorine, with a configuration of $1s^2 2s^2 2p^5$, normally forms one bond because it requires one electron to complete its valence shell and attain a noble-gas configuration. Sodium ([Ne]$3s^1$) must lose one electron to reach a noble-gas configuration, so it also usually forms one bond. When these two elements combine, they do so in a 1:1 ratio to form the ionic compound NaF. Fluorine forms a covalent compound with hydrogen by sharing electrons. Both fluorine and hydrogen tend to form one bond because they each need one electron to complete their valence shell, so they combine in a 1:1 ratio to form HF. In either

Sodium fluoride (NaF), a white, crystalline solid, is prepared by melting cryolite (Na_3AlF_6) with sodium hydroxide (NaOH) or by adding NaOH or Na_2CO_3 to a 40%-by-mass solution of HF. Among other applications, NaF is used as a pesticide, for fluoridation of drinking water, and for removal of HF from exhaust gases to reduce air pollution (NaF reacts with HF to form solid $NaHF_2$).

case, whether forming a covalent or an ionic bond, fluorine is adding an electron to achieve a configuration of $1s^2 2s^2 2p^6$, the stable electronic configuration of neon.

Oxygen, with a configuration of $1s^2 2s^2 2p^4$, must add two electrons to attain the configuration of neon, so it often forms bonds to two atoms when those atoms tend to form only one bond. For example, oxygen combines with two atoms of hydrogen or sodium, as in H_2O or Na_2O. Carbon usually forms four bonds because its configuration is $1s^2 2s^2 2p^2$, and it needs to add four electrons to achieve the same configuration as neon, as in CH_4 or CCl_4. In CO_2, carbon shares its four valence electrons with two oxygen atoms, each of which tends to form two bonds.

Elements in the first two periods of the periodic table have only *s* and *p* orbitals filled or partially filled in their valence shells. These orbitals can hold a maximum of eight electrons. For these elements, filled valence shells, like those of the noble gases (except helium), contain a set of eight electrons, or an *octet*. The tendency to achieve an electronic configuration with eight valence electrons is known as the **octet rule.** Hydrogen, with only a $1s$ orbital in its valence shell, tends to have two electrons rather than eight. Many elements in Periods 3 through 7 also tend to follow the octet rule, although with the availability of *d* orbitals, they can have more than eight valence electrons. Thus, sulfur forms H_2S, which obeys the octet rule, but it can also form SCl_4 and SCl_6, in which sulfur has more than eight valence electrons by making use of its unfilled $3d$ orbitals. We will explore the octet rule in more detail in Sections 7.3–7.6.

Lewis Symbols

A convenient way to represent valence electrons consists of dots surrounding the symbol of an element. These representations, devised by G. N. Lewis, are known as **Lewis symbols** or **electron-dot symbols** (see Figure 7.3). The dots representing valence electrons are placed singly on the four sides of the elemental symbol, in any order, and then paired as necessary. An element with a filled valence shell is surrounded by four pairs of dots—an octet. For elements without a filled valence shell, the Lewis symbol shows the number of unpaired electrons remaining when all the valence electrons have been placed, which is the number of bonds that can be formed.

Figure 7.3 Gilbert Newton Lewis, a chemistry professor at the University of California, Berkeley, proposed the concept of covalent bonding in 1916. Lewis devised a shorthand method for representing electronic configurations that originally consisted of a series of concentric cubes with the nucleus at the center and electrons at the corners. More modern representations consist of dots surrounding the symbol of the element. (Reprinted from *Journal of American Chemical Society.* Published 1916 by the American Chemical Society.)

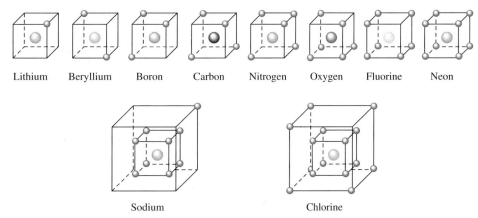

Lithium   Beryllium   Boron   Carbon   Nitrogen   Oxygen   Fluorine   Neon

Sodium   Chlorine

In the first period (an exception to the octet rule), the valence shell holds only two electrons, so only one side of the elemental symbol is used. Hydrogen, the first element, has one valence electron, so it is represented as H ·, which corresponds to the electronic configuration $1s^1$. Helium is He :, with two valence electrons, representing a filled shell, $1s^2$.

The second-period elements follow a progression beginning with one valence electron for Li and ending with eight for Ne:

$$\text{Li}\cdot \quad \cdot\text{Be}\cdot \quad \cdot\dot{\text{B}}\cdot \quad \cdot\dot{\text{C}}\cdot \quad :\dot{\text{N}}\cdot \quad :\dot{\text{O}}: \quad :\dot{\text{F}}: \quad :\ddot{\text{Ne}}:$$

Successive periods are represented similarly, with the exception that the transition elements have more electrons to be represented. This means all the noble gases (except He) have eight valence electrons occurring in four pairs because they come at the ends of periods.

EXAMPLE 7.1

Write the Lewis symbol for the arsenic atom.

Solution   Arsenic has five valence electrons, so it has one electron pair and three unpaired electrons:

$$:\dot{\text{As}}\cdot$$

Practice Problem 7.1

Write the Lewis symbol for the aluminum atom.

Answer: $\cdot\dot{\text{Al}}\cdot$

7.2   Ionic Bonds

An ionic bond results from the complete transfer of one or more electrons from a metal to a nonmetal. The loss and gain of electrons involved in this transfer occurs simultaneously because the electrons lost by the metal must have someplace to go. In other words, a metal loses electrons to form a cation only when a nonmetal is available to gain the electrons and form an anion. The following equations use Lewis symbols to illustrate this process for the formation of sodium chloride:

$$\text{Na}\cdot \longrightarrow \text{Na}^+ + e^-$$
$$\cdot\ddot{\underset{..}{\text{Cl}}}: + e^- \longrightarrow :\ddot{\underset{..}{\text{Cl}}}:^-$$
$$\text{Na}^+ + :\ddot{\underset{..}{\text{Cl}}}:^- \longrightarrow (\text{Na}^+)(:\ddot{\underset{..}{\text{Cl}}}:^-) \text{ or NaCl}$$

The sodium atom loses one electron to form the sodium ion, which has a noble-gas electronic configuration. That electron is transferred to the chlorine atom, which becomes the chloride ion; this ion also has a noble-gas configuration. The ions combine to form sodium chloride, a compound held together by the electrostatic forces between ions.

Because sodium needs to lose one electron and chlorine needs to gain one, these atoms combine in a 1:1 ratio, which also leads to electrical neutrality. Other

Magnesium chloride is a white salt that normally occurs as a hydrate, $MgCl_2 \cdot 6H_2O$. When heated, the hydrate undergoes hydrolysis to form MgO and HCl. The anhydrous salt can be formed by heating the hydrate in an atmosphere of HCl. Magnesium chloride is found in nature in solid deposits and in seawater and salt springs. It is used primarily as a source of metallic magnesium.

ions may combine in different ratios. Consider, for example, the formation of magnesium chloride:

$$Mg: \longrightarrow Mg^{2+} + 2e^-$$
$$2 \cdot \ddot{C}l: + 2e^- \longrightarrow 2 : \ddot{C}l:^-$$
$$Mg^{2+} + 2 : \ddot{C}l:^- \longrightarrow (Mg^{2+})(:\ddot{C}l:^-)_2 \text{ or } MgCl_2$$

Magnesium loses two electrons to achieve a noble-gas configuration. Because it takes only one electron to fill the chlorine atom's octet, two chlorine atoms are required to use the two electrons given up by the magnesium atom. We now have a magnesium ion and two chloride ions, which combine in an electrically neutral 1:2 ratio to form ionic magnesium chloride.

The chemical formulas of ionic compounds, then, represent the ratios of ions necessary to maintain electrical neutrality. These are not the formulas of molecules, because there are no single units in ionic crystals that can be identified as molecules. Rather, the formulas represent the simplest ratios of ions.

Structures of Ionic Crystals

The forces or ionic bonds holding ions together in ionic compounds are electrostatic, consisting of attractions between oppositely charged ions offset by repulsions between ions of like charges. How do these ions fit together to make a stable ionic compound? For convenience, we can consider ions to be spherical particles with the charge concentrated at the center of the sphere. The effects of the charge are felt equally in all directions. Each ion is surrounded by ions of opposite charge in a pattern, called a **crystal lattice,** that extends in all directions. The result is an **ionic crystal,** a solid structure in which ions are arranged in some regular geometric pattern that maximizes the attractive forces between cations and anions.

The characteristic patterns of ionic crystals are largely determined by the charges and sizes of the ions. For example, the structure of many 1:1 salts is the same as that of sodium chloride (Figure 7.4A and B), with each ion surrounded by six ions of opposite charge, or cesium chloride (Figure 7.4C), with each ion surrounded by eight ions of opposite charge. Larger cations tend to adopt the eight-ion cesium chloride structure; smaller cations, the six-ion sodium chloride structure. Very small cations may have other structures; these will be covered in more detail in Chapter 10. For 1:2 salts (or 2:1 salts), the structures are usually the same as that of fluorite, shown in Figure 7.4D.

Why does the ionic structure change as the relative sizes of the cations and anions change?

The number of ions surrounding an ion of opposite charge in a compound is called the **coordination number.** In NaCl, for example, the coordination number of both Na^+ and Cl^- is 6; in CsCl, the coordination number of both ions is 8. When ions quite different in size combine, the coordination number is usually 6. When the ions are similar in size, the coordination number is more likely to be 8. Thus, periodic trends in ionic sizes show up in the structures of ionic compounds. Salts formed from the smaller metals near the top of the periodic table are more likely to have a sodium chloride structure, whereas those formed from larger metals near the bottom of the table tend to have a cesium chloride structure.

The properties of ionic compounds are related to their structures and to the strength of the interactions between ions in the structures. The strength of an ionic bond is usually high, so separating the ions from the solid structure requires a lot of energy. Because the processes of melting and boiling involve the separation of

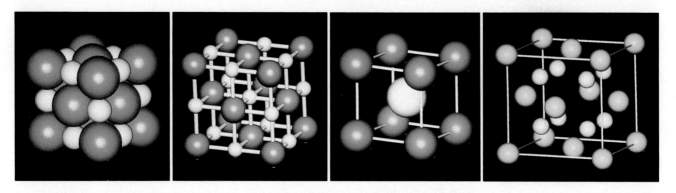

Figure 7.4 Some common ionic structures. The sodium chloride structure is represented in two ways. The first *(A)*, a more accurate representation, shows the ions touching. The second *(B)* is an expanded version of this structure, in which it is easier to see the positions of all the ions. Cesium chloride *(C)* and fluorite *(D)* represent some other structures found among ionic compounds.

ions or groups of ions from one another, ionic compounds have high melting points and boiling points, as well as high **heats of vaporization** (the energy required to vaporize 1 mole of a liquid) and **heats of fusion** (the energy required to melt 1 mole of a solid).

The high attractive forces also make ionic crystals hard and brittle. If struck hard enough, ionic crystals will shatter. A small amount of movement along a layer in the crystal results in ions of like charge being adjacent, as shown in Figure 7.5. Along this layer, there are no longer strong attractive forces between ions of opposite charge, but rather strong repulsive forces between ions of like charge. Thus, the crystal breaks apart along this layer.

Solid ionic salts are poor electrical conductors because the ions (and their electrons) are not mobile but are held in place in the crystal lattice by the strong ionic forces. However, in a molten ionic salt, the rigid crystal structure is disrupted and the ions in the liquid state are free to move. Thus, melting greatly increases the electrical conductivity of ionic salts; see Figure 7.6.

Figure 7.5 When a crystal is struck, layers of ions move relative to one another. The strong forces of attraction between ions of opposite charge are replaced by strong repulsions. The layers fly apart, causing the crystal to break between the layers.

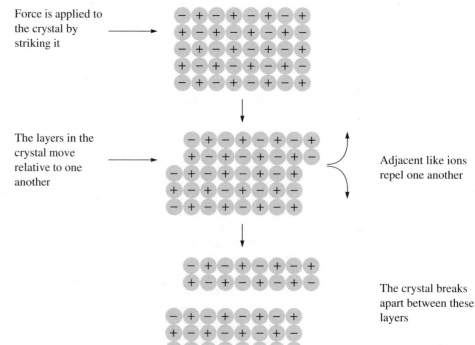

The Born–Haber cycle was proposed by Max Born, Kazimir Fajans, and Fritz Haber in 1919. Haber received a Nobel Prize for developing a process for producing ammonia from hydrogen and nitrogen. Born is best known for his theoretical work in quantum mechanics, for which he too received a Nobel Prize. (He is also the grandfather of the singer Olivia Newton John.) Fajans, a professor of chemistry at the University of Michigan, worked on radioactive transformations and thermochemistry.

The Born–Haber Cycle

Ionic salts are quite stable, largely because of the strong attractive forces between ions of opposite charge. This stability is reflected in the energy change that accompanies the reaction of a metal and a nonmetal to form an ionic salt—the heat of formation, ΔH_f. When this energy change is negative, we know the salt is more stable than the elements from which it was formed.

The heat of formation is often difficult to measure directly. However, we can examine the energy change involved in forming the ionic compound by using the **Born–Haber cycle,** shown in Figure 7.7. The principle behind the Born–Haber cycle is Hess's law. Recall from Chapter 5 that this law states that the energy change accompanying a chemical reaction is the same no matter what pathway is taken to bring about the change. The Born–Haber cycle describes a pathway in which the elements in their natural state are converted first to gaseous atoms, then to gaseous ions, and finally to the crystalline solid. The energy changes in these processes are either known or can be calculated, and their sum must equal the heat of formation. Besides allowing us to determine the heat of formation of ionic compounds, the Born–Haber cycle lets us examine the various factors that affect the heat of formation. As an example, let's look at the cycle for the reaction of sodium and chlorine:

$$Na(s) + \tfrac{1}{2}Cl_2(g) \longrightarrow NaCl(s)$$

Heat of Atomization The first step is the conversion of the elements in their natural state to gaseous atoms. Sodium, a solid metal, must be vaporized. The energy accompanying this process is the *heat of atomization:*

$$Na(s) \longrightarrow Na(g) \qquad \Delta H_{atom} = 109 \text{ kJ/mol}$$

The positive value of the enthalpy change indicates that this process requires energy; this is generally true for the atomization of metals. Values of the heats of atomization of the elements are found in Figure 7.8.

Chlorine is a diatomic gas in its natural state, so the diatomic molecule must be broken into gaseous atoms. The energy for this process is called the **bond energy,** the energy required to break a bond between two gaseous atoms. For chlorine, the bond energy is the amount of energy required to form 2 mol of chlorine atoms from 1 mol of diatomic chlorine molecules. Because the reaction uses only 1 mol of atoms, we divide the bond energy by 2 to obtain the heat of atomization:

$$\tfrac{1}{2}Cl_2(g) \longrightarrow Cl(g) \qquad \Delta H_{atom} = \tfrac{1}{2}\Delta H_{bond} = 120 \text{ kJ/mol}$$

Because this heat of atomization is positive, we can see that it requires energy to atomize nonmetals.

Ionization Energy and Electron Affinity The next step is the ionization of the gaseous atoms to form a gaseous metal cation and a gaseous nonmetal anion. These conversions are just those discussed in Chapter 6 for the ionization energy and the electron affinity. The ionization energy (*IE*) is the energy required to remove an electron from a gaseous atom:

$$Na(g) \longrightarrow Na^+(g) + e^- \qquad IE = 494 \text{ kJ/mol}$$

Figure 7.6 When the wires from a conductivity apparatus are placed into solid sodium acetate *(left),* nothing happens. However, if we melt the sodium acetate first *(right),* the bulb glows, indicating that the molten salt conducts electricity. The ions are free to move and carry electrical charge in the molten state, but not in the solid state.

Figure 7.7 The Born–Haber cycle provides a means of examining the various energy factors involved in the formation of an ionic salt (MX) from a metal (M) and a nonmetal, shown here as a diatomic gas (X$_2$). The sum of the energy changes involved in atomizing the reactants, converting the gaseous atoms to ions, and crystallizing the gaseous ions to give the solid product equals the energy change for the one-step process that directly converts reactants into products.

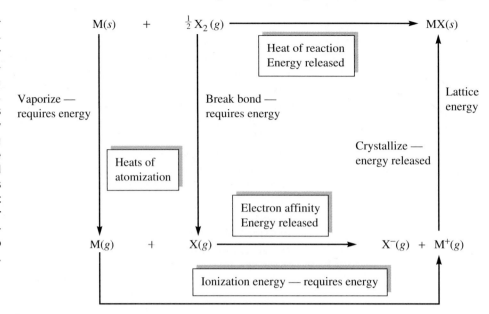

The electron affinity (*EA*) is the energy released when an electron is added to a gaseous atom:

$$Cl(g) + e^- \longrightarrow Cl^-(g) \qquad EA = -349 \text{ kJ/mol}$$

The ionization energy of all substances is positive (requires energy) and increases as additional electrons are removed. The electron affinity is negative in this case, releasing some energy; but the overall energy balance in the cycle to this point is still on the positive side:

Sum of energies = 109 kJ/mol + 120 kJ/mol + 494 kJ/mol − 349 kJ/mol = 374 kJ/mol

Lattice Energy The final step in the cycle is the combination of the gaseous ions to form the solid crystalline salt. The energy involved here is the **lattice energy,**

Figure 7.8 Heats of atomization of the elements at 25°C (kJ/mol).

H 218																	He 0
Li 161	Be 326											B 560	C 717	N 473	O 249	F 79	Ne 0
Na 109	Mg 146											Al 330	Si 439	P 332	S 277	Cl 121	Ar 0
K 90	Ca 178	Sc 375	Ti 469	V 515	Cr 397	Mn 285	Fe 416	Co 428	Ni 430	Cu 339	Zn 131	Ga 276	Ge 377	As 301	Se 227	Br 112	Kr 0
Rb 82	Sr 163	Y	Zr	Nb	Mo	Tc	Ru	Rh	Pd	Ag 285	Cd 112	In 243	Sn 301	Sb 264	Te 197	I 107	Xe 0
Cs 78	Ba 176	La	Hf	Ta	W	Re	Os	Ir	Pt	Au	Hg 61	Tl 182	Pb 195	Bi 210	Po	At	Rn 0
Fr	Ra	Ac															

which cannot be measured, because the process of bringing together gaseous ions to form a crystal is largely hypothetical. Instead, we can calculate the lattice energy (U) with an equation that relates it to the product of the ionic charges (Z^+ and Z^-), the distance between the centers of the ions (sum of ionic radii, $d_\pm$), and a proportionality constant (A, the *Madelung* constant) that depends only on the structure of the crystal lattice:

$$U = - A \, Z^+ \, Z^-/d_\pm$$

This equation gives a lattice energy for sodium chloride of -786 kJ/mol. The lattice energy makes a large negative contribution to the overall energy of the process, consistent with the fact that gaseous ions are considerably less stable than ionic salts.

The energy for formation of sodium chloride from the elements can now be calculated by applying Hess's law:

$$\Delta H_f = \Delta H_{atom} \text{ of Na} + \Delta H_{atom} \text{ of Cl}_2 + IE + EA + U$$

Inserting the values we have accumulated, we obtain

$$\Delta H_f = 109 \text{ kJ/mol} + 120 \text{ kJ/mol} + 494 \text{ kJ/mol} - 349 \text{ kJ/mol} - 786 \text{ kJ/mol}$$
$$= -412 \text{ kJ/mol}$$

The sodium chloride salt is 412 kJ/mol more stable than metallic sodium and gaseous chlorine, largely because of the stability of the crystal lattice. If it were not for the large negative value of the lattice energy, the heat of formation would be positive (endothermic) and it is likely that the process would not occur, because reactions tend to go in a direction that leads to the release of energy. The lattice energy, then, can be taken as one measure of the strength of an ionic bond.

EXAMPLE 7.2

Calculate the heat of formation of sodium fluoride, which crystallizes in the sodium chloride lattice. The heat of atomization of Na(s) is 109 kJ/mol, half the bond energy of $F_2(g)$ is 79 kJ/mol, the ionization energy of sodium atoms is 494 kJ/mol, the electron affinity of fluorine atoms is -328 kJ/mol, and the lattice energy is -939 kJ/mol.

Solution Values are given for all the quantities in the Born–Haber cycle, so we can apply Hess's law:

$$\Delta H_f = \Delta H_{atom} \text{ of Na} + \Delta H_{atom} \text{ of F}_2 + IE + EA + U$$
$$= 109 \text{ kJ/mol} + 79 \text{ kJ/mol} + 494 \text{ kJ/mol} - 328 \text{ kJ/mol} - 939 \text{ kJ/mol}$$
$$= -585 \text{ kJ/mol}$$

Practice Problem 7.2

Calculate the heat of formation of lithium fluoride, which crystallizes in the sodium chloride lattice. The heat of atomization of Li(s) is 161 kJ/mol, half the bond energy of $F_2(g)$ is 79 kJ/mol, the ionization energy of lithium atoms is 520 kJ/mol, the electron affinity of fluorine atoms is -328 kJ/mol, and the lattice energy is -1107 kJ/mol.

Answer: -675 kJ/mol

Relative Lattice Energies   Because the lattice energy provides a major contribution to the strength of an ionic bond, the bond strength should be sensitive to the same factors as lattice energy. Lattice energy is inversely proportional to the size of the ions, and a similar effect is seen in the ionic bond strength. For example, the lattice energy of NaCl is -786 kJ/mol, whereas that of the smaller NaF is -939 kJ/mol. The heat of formation, a measure of the bond strength, is -412 kJ/mol for NaCl and -585 kJ/mol for NaF. Thus, the smaller the ions, the stronger the ionic bond.

We can also examine the effect of ionic charge on bond strength. To isolate charge effects from size effects, we must compare compounds that have ions of about the same size, such as oxide and fluoride ions. The lattice energy for NaF(s) (-939 kJ/mol) is much smaller than the lattice energy for Na_2O(s) (-2600 kJ/mol). Even though it requires more energy to form O^{2-} than F^-, as a comparison of the electron affinities shows, the much greater lattice energy of Na_2O(s) makes it far more stable than NaF(s). Similar calculations with other substances lead to the conclusion that the greater the charges of the ions, the stronger the ionic bond in the solid.

7.3   Covalent Bonds

In contrast to ionic bonds, whose electrostatic forces are spread out over the crystal, covalent bonds involve a much more localized attractive force between atoms, with much weaker forces between discrete molecules. ▶ This contrast is reflected in the different properties of ionic $TiCl_3$ and more covalent $TiCl_4$ described in the chapter introduction. The atoms are held together tightly in covalent molecules, but the molecules usually are not bonded together. As a result, they are often gases or liquids. As solids, they are brittle or soft and have low melting and boiling points and heats of fusion and vaporization because the molecules are not held strongly together. Covalent bonding is typical in the molecular species formed by the nonmetallic elements, such as O_2 and S_8. Some common compounds, including H_2O and CO_2, are also formed as molecules through covalent bonding. In most such compounds, both elements are nonmetals.

The strength of a covalent bond arises because each shared electron interacts simultaneously with two nuclei rather than the one nucleus present in the atomic form. There are repulsions between electrons and between nuclei, but there are also attractions between electrons and nuclei. When the electrons are located between the two nuclei, the attractions are maximized and the repulsions are minimized, resulting in a covalent bond. Depending on the number of electrons shared between the two atoms, the bond is classified as being a single, double, or triple covalent bond.

Why are ionic substances usually solids, whereas covalent substances are usually gases or liquids?

The nonmetallic elements that exist as small molecules include H_2, N_2, O_2, F_2, Cl_2, Br_2, I_2, P_4, and S_8. Some common small molecular compounds are H_2O, CO_2, CO, CH_4, CCl_4, NH_3, N_2O_4, P_4O_{10}, SO_2, and SO_3.

How do atoms share electrons to form a bond?

Single Covalent Bonds

A covalent bond consists of a pair of electrons shared by two atoms. When there is just one such bond between atoms, it is called a **single bond.** Each atom generally has a half-filled orbital in the valence shell, and these orbitals overlap to allow the electron pair to be located around both atoms. For example, in the formation of the

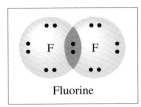

Fluorine

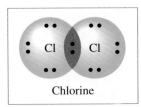

Chlorine

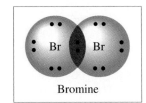

Bromine

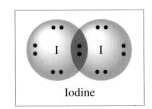

Iodine

Figure 7.9 Covalent bonding gives each atom in a halogen molecule an octet of electrons, as shown by the circles.

hydrogen molecule from hydrogen atoms, each hydrogen atom has an electron in the $1s$ orbital. When the atoms come together, the orbitals can overlap, and the atoms can share an electron pair:

$$H \cdot + \cdot H \longrightarrow H : H$$

Here we have shown the bond in the hydrogen molecule by combining Lewis symbols into a **Lewis formula** (or **electron-dot formula**) in which all the atoms are shown separately and in which the electrons are represented by dots. Lewis formulas are sometimes simplified by showing the bonds not as a pair of dots but as a line, as in H—H.

By sharing electrons, both hydrogen atoms have achieved a share in a noble-gas electronic configuration, that of helium. Many of the other elements form covalent bonds that will allow them to share in a total of eight valence electrons, thus obeying the octet rule.

Consider the halogens, for example, whose distribution of electrons is shown in Figure 7.9. Each atom in these structures can be assigned eight valence electrons—an octet—provided the electrons shared between the two atoms are counted for each atom. The halogens also react with other elements so as to form octets of electrons—as in HF and CCl_4, shown in Figure 7.10. The fluorine and chlorine atoms have octets of electrons, as does the carbon atom, whereas hydrogen has its usual two electrons. Each of the covalent bonds in these molecules is a single bond because it arises from the sharing of a single electron pair.

Figure 7.10 Covalent compounds containing halogens have an octet of electrons (shown by the circles) around each halogen atom.

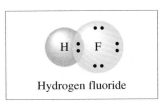

Hydrogen fluoride

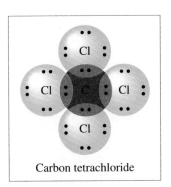

Carbon tetrachloride

Multiple Bonds

Some combinations of atoms do not have enough electrons to satisfy the octet rule by forming a single bond. To achieve an octet, the atoms must share more than one pair of electrons. In O_2, for example, the oxygen atoms must share two pairs of electrons to achieve an octet; the nitrogen atoms in N_2 share three pairs (see Figure 7.11). As you might expect, this bonding pattern reflects the number of bonds normally formed by these elements. Oxygen usually forms two bonds, and nitrogen normally forms three bonds.

The sharing of two pairs of electrons is called a **double bond,** and the sharing of three pairs of electrons is called a **triple bond.** In Lewis formulas, double bonds are represented by two pairs of dots or two lines; triple bonds are represented by three pairs of dots or three lines.

The characteristics of single, double, and triple bonds differ. Normally, breaking a double bond requires more energy than breaking a single bond; in other words, the bond energy is greater for double bonds. Furthermore, the **bond length** (the distance between the centers of the bonded atoms) is greater in single than in double bonds and greater in double than in triple bonds. The bond energies and bond lengths for some bonds involving oxygen, nitrogen, and carbon atoms are

Figure 7.11 The two atoms in the oxygen molecule *(left)* must share two pairs of electrons to achieve an octet. In the nitrogen molecule *(right),* the atoms must share three pairs of electrons.

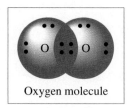

Oxygen molecule

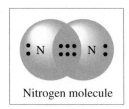

Nitrogen molecule

listed in Table 7.3. These data follow the general trends just outlined, which may be summarized as follows: the more electrons shared between two atoms, the more tightly those two atoms are held together.

Valence Electrons and Number of Bonds

Elements combine in a number of different ways to achieve octets. Oxygen typically forms two bonds, but it can do so in different ways. When oxygen atoms combine to form molecular oxygen, O_2, they form a double bond:

$$:\ddot{O}\cdot + \cdot\ddot{O}: \longrightarrow :\ddot{O}::\ddot{O}:$$

But in combining with hydrogen atoms, oxygen may form two single bonds, one with each of two hydrogens:

$$H\cdot + \cdot\ddot{O}\cdot + \cdot H \longrightarrow H:\overset{\overset{\textstyle H}{\cdot\cdot}}{\underset{\cdot\cdot}{O}}:$$

Table 7.3 Characteristics of Some Single, Double, and Triple Bonds

Molecule	Bond	Bond Length (pm)	Bond Energy (kJ/mol)
H_2	H—H	74	432
O_2	O=O	121	494
H_2O_2	O—O	148	142
N_2	N≡N	110	942
N_2H_2	N=N	125	418
N_2H_4	N—N	145	167
C_2H_2	C≡C	121	835
C_2H_4	C=C	133	602
C_2H_6	C—C	154	346
CH_3CN	C≡N	116	887
—	C=N	128	615
H_3C–NH_2	C—N	147	305
CO	C≡O	113	1072
H_2CO	C=O	120	799
H_3C—OH	C—O	143	358

Hydrogen peroxide, H_2O_2, is a colorless liquid that is caustic to the skin. Highly concentrated solutions are used as a rocket propellant; very dilute solutions are available for use as a disinfectant or bleach. The fizzing that is experienced when hydrogen peroxide solution is poured on cut tissue results from the decomposition into oxygen and water, catalyzed by an enzyme called peroxidase.

Hydrazine, N_2H_4, is a colorless, oily liquid that fumes in air and has an odor much like that of ammonia. Hydrazine explodes if heated in the presence of air. It is used as a reducing agent and a rocket fuel.

And two oxygen atoms may be joined by a single bond with two hydrogen atoms supplying the additional electrons to form hydrogen peroxide:

$$\text{H} \cdot + \cdot \overset{..}{\text{O}} \cdot + \cdot \overset{..}{\text{O}} \cdot + \cdot \text{H} \longrightarrow \overset{\text{H H}}{\underset{}{:\overset{..}{\text{O}}:\overset{..}{\text{O}}:}}$$

Similarly, a nitrogen atom, with five valence electrons, shares three electron pairs with another nitrogen atom to form molecular nitrogen:

$$:\overset{\cdot}{\text{N}} \cdot + \cdot \overset{\cdot}{\text{N}}: \longrightarrow :\text{N}:::\text{N}:$$

Nitrogen forms single bonds with three hydrogen atoms in ammonia:

$$\text{H} \cdot + \text{H} \cdot + \text{H} \cdot + \cdot \overset{\cdot}{\text{N}}: \longrightarrow \underset{\text{H}}{\overset{\text{H}}{:\overset{}{\text{N}}:\text{H}}}$$

And like oxygen, nitrogen can bond to itself with a single bond if another element is also present, as in hydrazine:

$$\text{H} \cdot + \text{H} \cdot + \cdot \overset{\cdot}{\text{N}}: + \cdot \overset{\cdot}{\text{N}}: + \text{H} \cdot + \text{H} \cdot \longrightarrow \text{H}:\overset{\text{H}}{\underset{..}{\text{N}}}:\overset{\text{H}}{\underset{..}{\text{N}}}:\text{H}$$

EXAMPLE 7.3

Use electron-dot structures to describe the bonding between sulfur and chlorine atoms. Give the formula of the compound they would form.

Solution Sulfur has six valence electrons and so can fill its octet by sharing two electrons. Thus, a sulfur atom tends to form two single bonds or one double bond. Chlorine has seven valence electrons and needs to share one by forming one covalent bond. Thus, we would expect that two chlorine atoms would combine with one sulfur atom to form SCl_2, with single bonds between the atoms:

$$\begin{array}{c} :\overset{..}{\text{Cl}}: \\ :\overset{..}{\text{Cl}}:\overset{..}{\text{S}}: \end{array}$$

Practice Problem 7.3

Use electron-dot structures to describe the bonding between nitrogen and fluorine atoms. Give the formula of the compound they would form.

Answer: NF_3, with a single bond between the nitrogen atom and each fluorine atom:

$$\begin{array}{c} :\overset{..}{\text{F}}: \\ :\text{N}:\overset{..}{\text{F}}: \\ :\overset{..}{\text{F}}: \end{array}$$

Structures of Covalent Molecules

We saw earlier that the normal structures for ionic compounds are extended arrays of ions called crystals. The structures of covalent substances are quite different.

Table 7.4 Molecules Formed by the Nonmetals

			H_2	He
C_n	N_2	O_2	F_2	Ne
Si_n	P_4	S_8	Cl_2	Ar
	As_4	Se_8	Br_2	Kr
	Sb_4	Te_8	I_2	Xe

Consider the structures of the nonmetallic elements. The nonmetals typically exist as diatomic or polyatomic molecules, with the exception of the noble gases, which are monatomic. Formulas for the common forms of these elements are shown in Table 7.4. Hydrogen, oxygen, nitrogen, fluorine, and chlorine exist as diatomic gases; bromine is a diatomic liquid; and iodine is a diatomic solid. The structure of these molecules is shown in Figure 7.12A. Sulfur (Figure 7.12B), selenium, and tellurium exist as solids containing eight-membered rings of covalently bonded atoms. Phosphorus (Figure 7.12C), arsenic, and antimony are solids containing

Figure 7.12 The structures of covalent elements vary from diatomic gases *(A)*, to small molecules *(B, C)*, to extended networks of connected atoms *(D)*.

A. Chlorine, oxygen, nitrogen, and other diatomic gases

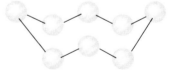

B. Sulfur, selenium, and tellurium

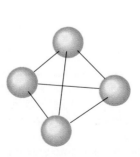

C. Phosphorus, arsenic, and antimony

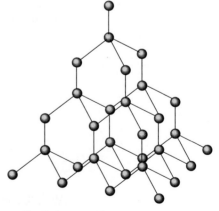

D. Carbon (as diamond) and silicon

four-atom molecules. Two elements, carbon (in the form of diamond) and silicon, do not form small molecules but exist as the very large molecules C_n and Si_n (with *n* having very large values). These are essentially extended networks of atoms, each bonded to four other atoms in what is called a *tetrahedral* structure, as shown in Figure 7.12D.

These structures are typical of those in covalent compounds formed between the nonmetals as well. Generally, these substances are small molecules with simple structures. For example, hydrogen combines with the halogens to form diatomic hydrogen halides. The tetrahedral arrangement of atoms around a given carbon atom in the diamond structure is typical of carbon compounds, such as CH_4 and CCl_4, in which carbon is surrounded by four atoms of the other element in a tetrahedral arrangement.

7.4   Writing Lewis Formulas

Let's look more closely at how to develop Lewis formulas for common molecules and ions. Although there are exceptions to the octet rule, we will continue to use it here as we examine a number of helpful rules for writing Lewis formulas and apply them to the covalent compound ammonia, NH_3. The rules are also presented as a flowchart in Figure 7.13.

Ammonia is a colorless, corrosive gas with a very pungent odor. It is produced by the reaction between nitrogen and hydrogen gases in the presence of iron. Ammonia is used in fertilizers, explosives, synthetic fibers, and in the manufacture of nitric acid.

Rule 1: Write an *atomic skeleton,* a formula with the symbols of the elements placed in their correct locations with respect to one another.

- To write correct formulas, you must know which atoms are connected to which. The arrangement of atoms is usually symmetrical. When two elements are bonded together, the one present in greater number generally surrounds the one present in lesser number. The central atom tends to be the one that is the more metallic. Hydrogen is rarely the central atom, because it usually forms only one bond. Where both hydrogen and oxygen are present, the hydrogen is usually bonded to the oxygen. If both oxygen and carbon are present, hydrogen may also be bonded to carbon.

The atomic skeleton of ammonia has hydrogen atoms surrounding the nitrogen atom:

$$H$$
$$H\ N\ H$$

Rule 2: Count the total number of valence electrons. The periodic table can be used as a guide.

- Be sure to take into account any ionic charge, subtracting for positive charge and adding for negative charge.

The number of valence electrons in ammonia can be obtained from the sum for each atom: five from nitrogen and one from each hydrogen, for a total of eight. If we were considering the ammonium ion, NH_4^+, which has nitrogen covalently bonded to hydrogen, we would also count eight valence electrons. This ion has five electrons from nitrogen and four from the four hydrogens, but we must remove one to give the $1+$ charge on the ion.

Figure 7.13 Procedure for writing Lewis formulas.

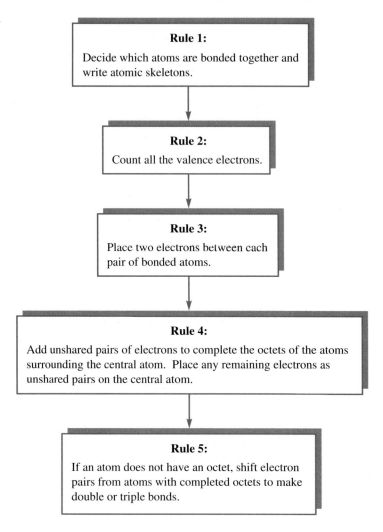

Rule 1:

Decide which atoms are bonded together and write atomic skeletons.

Rule 2:

Count all the valence electrons.

Rule 3:

Place two electrons between each pair of bonded atoms.

Rule 4:

Add unshared pairs of electrons to complete the octets of the atoms surrounding the central atom. Place any remaining electrons as unshared pairs on the central atom.

Rule 5:

If an atom does not have an octet, shift electron pairs from atoms with completed octets to make double or triple bonds.

Rule 3: Place a single bond—two electrons—between each pair of bonded atoms.

- Note carefully which atoms are connected to which. Place a bond between only those atoms that are connected.

In ammonia, we place a bond between the nitrogen atom and each hydrogen atom, but not between the hydrogen atoms:

$$\begin{array}{c} \text{H} \\ \text{\.{.}} \\ \text{H} : \text{N} : \text{H} \end{array}$$

Rule 4: If you have not placed all the valence electrons in the formula, add any remaining electrons as unshared electron pairs. Remember that the total number of electrons for each atom is very often an octet.

- When placing electrons around the atoms, recall that Period 1 elements have two electrons; Period 2 elements have eight; and Period 3–7 elements usually have eight electrons, but ten or twelve are also possible.

- Add pairs of electrons to complete the octet of atoms surrounding the central atom first. Then add any remaining electrons in pairs to the central atom.
- If the total number of electrons is odd, one atom must have a deficiency. Generally, this atom will be the one with the most metallic character.
- If more electrons are present than are needed to give each atom an octet, add the extra electrons to the central atom.

Ammonia has eight valence electrons, and we used six of these to form bonds (Rule 3). The remaining two electrons should be placed on the nitrogen atom because the hydrogen atoms already have the two electrons they need:

$$\text{H} \\ \overset{..}{\underset{..}{\text{H} : \text{N} : \text{H}}}$$

Rule 5: If necessary to satisfy the octet rule, shift electrons from nonbonded (unshared) positions on atoms with completed octets to positions between atoms to make multiple bonds.

- Sometimes there is more than one choice for the position of a multiple bond. In such cases, do not shift electrons away from the more nonmetallic to the more metallic element. Note, however, that the choice for position of the multiple bond is not always clear-cut.

Because the Lewis formula for ammonia satisfies the octet rule as we have written it, we need not apply Rule 5. Work through Example 7.4 to see how the rules apply to a molecule that has a double bond.

EXAMPLE 7.4

Write a Lewis formula for sulfur dioxide, SO_2.

Solution

1. The oxygens are symmetrically attached to the sulfur atom:

$$\text{O} \quad \text{S} \quad \text{O}$$

2. The sulfur atom contributes six valence electrons, as does each oxygen atom: $6 + 2(6) = 18$ valence electrons.
3. Two electrons must be placed between each pair of bonded atoms, for a total of four electrons:

$$\text{O} : \text{S} : \text{O}$$

4. Fourteen electrons remain. Each oxygen atom needs six more electrons to fill its octet:

$$\overset{..}{\underset{..}{: \text{O}}} : \text{S} : \overset{..}{\underset{..}{\text{O} :}}$$

The remaining two electrons should be added to the sulfur atom:

$$\overset{..}{\underset{..}{: \text{O}}} : \overset{..}{\text{S}} : \overset{..}{\underset{..}{\text{O} :}}$$

5. The sulfur atom has only six electrons, so a pair must be shifted from one of the oxygen atoms, resulting in a double bond with that oxygen.

$$\overset{..}{\underset{..}{: \text{O}}} :: \overset{..}{\text{S}} : \overset{..}{\underset{..}{\text{O} :}}$$

(continued)

But because the oxygen atoms are located symmetrically, it could have been the other oxygen atom that donated the electron pair to the sulfur:

$$:\overset{..}{\underset{..}{O}}:S::\overset{..}{\underset{..}{O}}:$$

These two Lewis formulas are equivalent.

Practice Problem 7.4

Write a Lewis formula for the nitrite ion, NO_2^-.

Answer: $:\overset{..}{\underset{..}{O}}::N:\overset{..}{\underset{..}{O}}:^-$ or $^-:\overset{..}{\underset{..}{O}}:N::\overset{..}{\underset{..}{O}}:$

Hydrogen cyanide, called hydrocyanic acid when dissolved in water, is a colorless gas with the characteristic odor of bitter almonds. It is prepared by the oxidation of mixtures of methane and ammonia. HCN is highly poisonous and has been used to execute prisoners in the gas chamber. It is primarily used to exterminate rodents and insects in ships.

If the central atom lacks two electrons, as in Example 7.4, it must form one double bond with a surrounding atom. If it lacks four electrons, a triple bond may be required. For example, the Lewis formula for hydrogen cyanide, $H:C:::N:$, includes a triple bond between carbon and nitrogen. Of the atoms surrounding the central atom in hydrogen cyanide, only nitrogen could shift an unshared electron pair. The result is a triple bond. Under other circumstances, a central atom deficient by four electrons may instead form two double bonds, as illustrated in Example 7.5.

EXAMPLE 7.5

Write a Lewis formula for carbon dioxide, CO_2.

Solution

1. The oxygens are symmetrically attached to the carbon:

    O   C   O

2. The carbon supplies four valence electrons, and each oxygen supplies six, for a total of sixteen valence electrons.

3. Two electrons must be placed between each pair of bonded atoms, for a total of four electrons:

    O:C:O

4. Twelve electrons remain. Each oxygen atom needs six more electrons to fill its octet. This distribution uses all remaining electrons:

    $:\overset{..}{\underset{..}{O}}:C:\overset{..}{\underset{..}{O}}:$

5. The carbon atom has only four electrons, so two pairs must be shifted. Because both oxygens have unshared pairs, one pair of electrons can be shifted from each, forming two double bonds:

    $:\overset{..}{O}::C::\overset{..}{O}:$

    It would have been possible to shift both pairs from a single oxygen, forming a triple bond:

    $:\overset{..}{\underset{..}{O}}:C:::O:$

    However, it is usually preferable to maintain a symmetrical arrangement of electrons.

Practice Problem 7.5

Write a Lewis formula for nitrous oxide, N_2O, which has the arrangement N N O.

Answer: $:\ddot{N}::N::\ddot{O}:$

Formal Charge

How do we know whether to use two double bonds or a single and triple bond?

We have already seen that we can write more than one Lewis formula for some molecules and ions. Example 7.5 gave two quite different Lewis formulas for carbon dioxide, $:\ddot{O}::C::\ddot{O}:$ and $:\ddot{O}:C:::O:$. The concept of formal charge can sometimes help us decide which member of a set of possible electron-dot formulas is the most likely. **Formal charges** are charges assigned to atoms in molecules or ions according to a particular set of rules for counting electrons. Unshared electrons are assigned to the atom on which they are found, and shared electrons are divided equally between the sharing atoms. The formal charge on an atom is then calculated as the difference between the number of valence electrons in the isolated atom and the number of electrons assigned to the atom in the molecule:

Formal charge = number of valence electrons − number of assigned electrons

The sum of the formal charges in a molecule or ion must equal the total charge on that species.

Consider the first formula of carbon dioxide. Here, the shared electrons are divided equally between carbon and oxygen:

Each oxygen is assigned six electrons—four unshared electrons and two of the shared electrons—and the carbon atom is assigned four of the shared electrons. Because oxygen has six valence electrons in the isolated atom and six assigned electrons, its formal charge is zero. Similarly, the formal charge of carbon must be zero because it is assigned four electrons and also has four valence electrons as an isolated atom. All the formal charges are zero, so they add up to zero, the charge on the CO_2 molecule.

Now consider the second Lewis formula of carbon dioxide, in which the arrangement of electrons is not symmetrical:

The oxygen on the left is assigned seven electrons, and the oxygen on the right is assigned five. The carbon is assigned four electrons. The formal charges are as follows:

Formal charge of left oxygen = 6 valence electrons − 7 assigned electrons = 1−
Formal charge of right oxygen = 6 valence electrons − 5 assigned electrons = 1+
Formal charge of carbon = 4 valence electrons − 4 assigned electrons = 0

Again, the sum of the formal charges is zero.

In general, the most important Lewis formulas for a molecule or ion are the formulas in which the atoms have formal charges closest to zero. On this basis, the first formula of carbon dioxide is the preferable form. If it is not possible to achieve a formal charge of zero on all atoms, the negative formal charge should be on the atom that is the least metallic (most electronegative).

EXAMPLE 7.6

Determine the formal charges on the atoms in sulfur dioxide:

$$: \ddot{O} : \ddot{S} :: \ddot{O} :$$

Solution　First, calculate the number of electrons assigned to each atom:

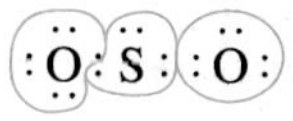

$$: \ddot{O} : \ddot{S} : \ddot{O} :$$

The oxygen on the left is assigned seven electrons, the sulfur is assigned five, and the oxygen on the right is assigned six. Then determine the formal charges by subtracting the assigned electrons from the number of valence electrons in the isolated atoms:

Formal charge of left oxygen = 6 valence electrons − 7 assigned electrons = 1−
Formal charge of sulfur = 6 valence electrons − 5 assigned electrons = 1+
Formal charge of right oxygen = 6 valence electrons − 6 assigned electrons = 0

Practice Problem 7.6

Determine the formal charges on the atoms in ammonia, NH_3, and the ammonium ion, NH_4^+.

Answer: H is zero in both; N is zero in NH_3, but 1+ in NH_4^+.

Although formal charges are useful in determining the plausibility of a given Lewis electron-dot formula, they should not be taken too literally. Formal charges do not represent the actual distribution of charge within molecules.

7.5　Resonance

Why do we sometimes draw more than one Lewis formula for a molecule? Which one should be used to represent the bonding in the molecule?

Lewis formulas do not always give accurate pictures of the chemical bonding as it is known to exist in nature. Consider the ozone molecule, for example. The Lewis formula of this molecule can be drawn in two ways, each having one double bond and one single bond:

$$: \ddot{O} :: \ddot{O} : \ddot{O} : \qquad : \ddot{O} : \ddot{O} :: \ddot{O} :$$

But neither of these Lewis formulas accurately describes the bonding. Measurements show that the length of both bonds in the ozone molecule is the same: 127.8 pm. By comparison, measurements of other molecules show that the normal length of an oxygen–oxygen single bond is 148 pm, and that of an oxygen–oxygen double bond is 121 pm. Furthermore, the properties of the bonds in ozone are not characteristic of oxygen–oxygen single or double bonds but fall somewhere between these types. The bonds in ozone are, then, unlike any bonds we have discussed so far.

The concept of *resonance* allows us to represent the sort of bonding that occurs in the ozone molecule while retaining the usefulness and simplicity of Lewis formulas. According to this concept, the electronic structure of molecules like ozone is represented nót by a single Lewis formula but by two or more Lewis formulas, each illustrating a different aspect of the true electronic structure. The actual molecule is a composite of the formulas drawn and is called a **resonance hybrid.**

The term *resonance* may seem to suggest a resonation, or oscillation, between the various structures, but such is not the case. The true structure is not one of the Lewis formulas some of the time and the other the rest of the time. Rather, the actual structure always has some of the characteristics of each contributing formula. It is a kind of ''average'' structure.

Using Lewis formulas to represent the bonding in molecules like sulfur trioxide is similar to using a camera or a photocopy machine to represent the structure of a person's hand. Figure 7.14 shows two aspects of a hand. The hand is not represented sometimes by the front view and sometimes by the rear view, but is best thought of as a composite of the two.

To represent a resonance hybrid, we draw the contributing structures in the usual way and connect them with double-headed arrows. The two structures for ozone, for instance, are drawn as follows:

$$:\overset{..}{O}::\overset{..}{O}:\overset{..}{O}: \quad\longleftrightarrow\quad :\overset{..}{O}:\overset{..}{O}::\overset{..}{O}:$$

The composite structure is sometimes depicted as a single formula with a dashed or dotted line to show the sharing of the double-bond character over two sets of bonded atoms:

$$:\overset{..}{O}\cdots\overset{..}{O}\cdots\overset{..}{O}:$$

It is important to note that resonance forms differ only in the locations of the electrons, not in the relative positions of the atoms. Thus, all resonance forms of ozone have an atomic arrangement in which one oxygen atom is between the other two.

The azide ion, N_3^-, occurs in metal salts and in hydrazoic acid or hydrogen azide, HN_3. Hydrazoic acid has an intolerable pungent odor that causes an intense feeling of tightening in the throat. Sodium azide is a colorless salt that decomposes rapidly on heating or when mixed with salts of copper or other metals. The rapid production of nitrogen gas under these conditions is used to inflate automobile air bags. Azide salts of the heavier metals are very unstable and explode spontaneously.

Figure 7.14 Two views of a hand, neither of which accurately represents the hand's true structure, which is a composite of the two.

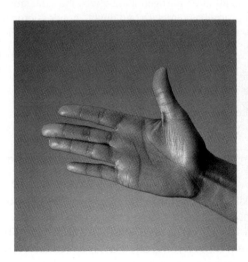
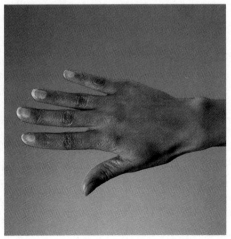

The most common resonance forms have a central atom bonded to two (or more) other atoms, which are identical to each other. One bond is double and the other (or others) is (are) single. The double bond can be exchanged with any single bond to give a valid resonance form. Similarly, single and triple bonds can be exchanged with two double bonds. The resonance forms of the azide ion are developed in the following example to illustrate these points.

EXAMPLE 7.7

Write a Lewis structure for the azide ion, N_3^-.

Solution

1. The nitrogens are attached in a linear manner:

$$N \quad N \quad N^-$$

2. Each nitrogen contributes five valence electrons, and the negative ionic charge accounts for one: $3(5) + 1 = 16$ valence electrons.

3. Two electrons must be placed between each pair of bonded atoms, for a total of four electrons:

$$N : N : N^-$$

4. Twelve electrons remain. Each surrounding nitrogen atom needs six more electrons to fill its octet, which uses all the remaining electrons.

$$: \overset{..}{\underset{..}{N}} : N : \overset{..}{\underset{..}{N}} : \; ^-$$

5. The central atom has only four electrons, so two pairs of electrons must be shifted. Because each surrounding nitrogen has unshared pairs, one pair of electrons can be shifted from each, forming two double bonds:

$$: \overset{..}{N} :: N :: \overset{..}{N} : \; ^-$$

It would have been possible to shift both pairs from a single outer nitrogen, forming a triple bond:

$$: \overset{..}{\underset{..}{N}} : N ::: N : \; ^- \quad \text{or} \quad : N ::: N : \overset{..}{\underset{..}{N}} : \; ^-$$

Because all three nitrogen atoms have the same attraction for electrons, the more symmetrical arrangement with two double bonds is probably best, but the other two arrangements may also contribute to the true picture of the bonding in the azide ion:

$$: \overset{..}{N} :: N :: \overset{..}{\underset{..}{N}} : \; ^- \longleftrightarrow : \overset{..}{\underset{..}{N}} : N ::: N : \; ^- \longleftrightarrow : N ::: N : \overset{..}{\underset{..}{N}} : \; ^-$$

Practice Problem 7.7

Write a Lewis structure for the thiocyanate ion, NCS^-.

Answer: $: \overset{..}{\underset{..}{N}} :: C :: \overset{..}{\underset{..}{S}} : \; ^- \longleftrightarrow : N ::: C : \overset{..}{\underset{..}{S}} : \; ^- \longleftrightarrow : \overset{..}{\underset{..}{N}} : C ::: S : \; ^-$

Most of the common molecules and ions whose bonding must be represented by resonance hybrids contain oxygen. The Lewis formulas of some nonmetal oxides and oxoanions are shown in Figure 7.15. The nature of the bonds between oxygen and other nonmetals makes the resulting compounds generally quite stable with respect to decomposition, though not necessarily with respect to other reac-

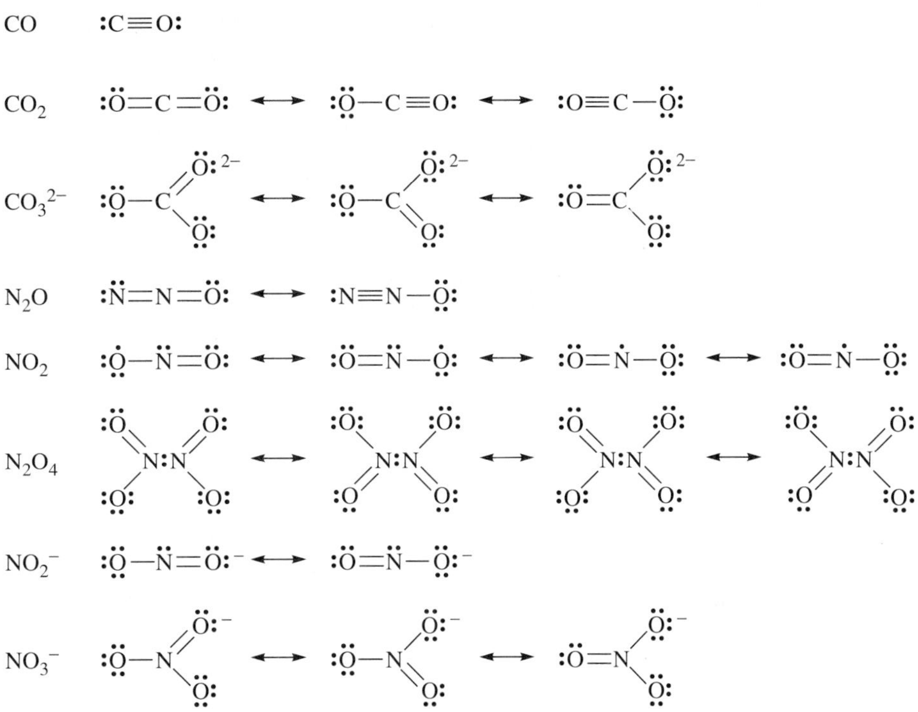

Figure 7.15 Lewis formulas of some nonmetal oxides and oxoanions.

CO $:C{\equiv}O:$

CO_2 $:\ddot{O}{=}C{=}\ddot{O}:$ $\longleftrightarrow$ $:\ddot{O}{-}C{\equiv}O:$ $\longleftrightarrow$ $:O{\equiv}C{-}\ddot{O}:$

CO_3^{2-} (resonance structures shown)

N_2O $:\ddot{N}{=}N{=}\ddot{O}:$ $\longleftrightarrow$ $:N{\equiv}N{-}\ddot{O}:$

NO_2 $:\ddot{O}{-}\dot{N}{=}\ddot{O}:$ $\longleftrightarrow$ $:\ddot{O}{=}\dot{N}{-}\ddot{O}:$ $\longleftrightarrow$ $:\ddot{O}{=}N{-}\ddot{O}:$ $\longleftrightarrow$ $:\ddot{O}{=}N{-}\ddot{O}:$

N_2O_4 (resonance structures shown)

NO_2^- $:\ddot{O}{-}\ddot{N}{=}\ddot{O}:^-$ $\longleftrightarrow$ $:\ddot{O}{=}\ddot{N}{-}\ddot{O}:^-$

NO_3^- (resonance structures shown)

tions. An example of the relation between stability and Lewis formulas can be found in nitrogen chemistry. Oxygen combines with nitrogen to form a number of species. All contain nitrogen–oxygen double or triple bonds, so they tend to be stable. An unusual case is nitrogen dioxide (Figure 7.16), a reddish brown compound that coexists with a colorless substance, dinitrogen tetroxide:

$$2NO_2 \rightleftharpoons N_2O_4$$

Figure 7.16 Nitrogen dioxide is a deadly poisonous, reddish brown gas. It is formed by the air oxidation of nitric oxide (NO), a pollutant from high-temperature combustion processes, such as in automobile engines. Nitrogen dioxide dissolves in water to form nitric oxide and nitric acid and is used in the manufacture of nitric and sulfuric acids. It also forms colorless dinitrogen tetroxide, N_2O_4, in amounts that depend on temperature, as shown here. The higher the temperature *(from left to right),* the greater the proportion of NO_2 to N_2O_4.

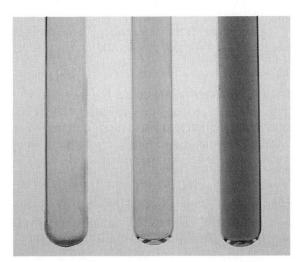

The resonance forms of NO_2 (shown in Figure 7.15) all have an unpaired electron; there are not enough electrons to satisfy the octet rule. The most important of the resonance forms has an unpaired electron on the nitrogen atom, which gives the nitrogen a formal charge of $1+$. The oxygen atom that is singly bonded has a formal charge of $1-$. When two nitrogen dioxide molecules combine to form dinitrogen tetroxide, they do so by forming a single bond between the two nitrogen atoms. This substance satisfies the octet rule, unlike nitrogen dioxide; but it is not very stable. Both nitrogen atoms have a formal charge of $1+$. Even though formal charges are not accurate representations of electron distribution, when adjacent atoms have formal charges of the same sign, the bond between them is usually weak, probably because of electrostatic repulsions. The weak N—N bond causes dinitrogen tetroxide to dissociate readily to give nitrogen dioxide.

7.6   Exceptions to the Octet Rule

How can molecules such as SF_6 have more than four bonds to an atom if bonding is governed by the octet rule?

We have seen that some molecules cannot satisfy the octet rule, because they have either too few or too many electrons. Furthermore, measurements of bond lengths and bond energies suggest that elements in Periods 3–7 sometimes form double bonds even when they can satisfy the octet rule without doing so. What causes these variations?

Exceptions to the octet rule can generally be explained in terms of the very principle on which the rule is based. The origin of the octet rule lies in the presence of four valence orbitals, the ns and np orbitals, in the outer shell of the elements. Elements in the second period have only these valence orbitals. Thus, they can form no more than four bonds, because each atom must contribute one electron to a bond and the electrons must be located in an orbital.

We find elements in the third and higher periods, such as sulfur, in compounds with ten or even twelve electrons surrounding them. These compounds are consistent with the principle behind the octet rule. The number of bonds formed is equal to the number of half-filled valence orbitals available to the atom, which can include some of the nd orbitals. This more general method of counting valence electrons explains the formulas of a much larger number of compounds than does the octet rule.

Exceptions to the octet rule generally fall into three categories: odd-electron molecules, incomplete octets, and expanded valence shells.

Odd-Electron Molecules

Nitric oxide, NO, is a colorless gas that condenses to a pale blue liquid when cooled sufficiently. We can prepare it by passing mixtures of nitrogen and oxygen gases through an electric arc. NO also forms in mixtures of these two gases, such as that found in air, when heated sufficiently. This reaction occurs readily in automobile engines, so NO is a common air pollutant. It readily combines with oxygen to form nitrogen dioxide and is used primarily in the manufacture of nitric acid.

In a molecule that contains an odd number of valence electrons, one electron is unpaired, so one of the atoms does not have an octet of electrons. Such molecules are fairly reactive. An example is nitric oxide, NO:

$$\cdot \ddot{N} :: \ddot{O} :$$

The presence of the unpaired electron in nitric oxide means it will react with any species that can readily supply an electron. It may, for example, react with chlorine to form nitrosyl chloride:

$$: \ddot{Cl} : \ddot{N} :: \ddot{O} :$$

► ► ► WORLDS OF CHEMISTRY

Is Levitation in Our Future?

Superconductivity would have applications in electrical utilities, industry, electronics, transportation, and medicine.

Can a train move at 300 mph? Some scientists think that superconductors will make such rapid travel possible. But what is a superconductor?

Metals conduct electricity relatively well. However, even in metals, resistance slows the current and heats the wire. If a metal wire is cooled to temperatures close to absolute zero (0 K), electrical resistance disappears and the metal becomes superconducting.

Superconductivity would have applications in electrical utilities, industry, electronics, transportation, and medicine. Power companies could use underground superconductive transmission lines to carry electricity with no losses or heating, so electrical generation could be carried out farther from population centers. Superconducting coils could be used to store power at night, so it could be released during daytime peak-usage periods. Superconducting microchips could lead to smaller, faster computers.

One of the properties of a superconductor is its ability to levitate above a magnet. A magnetic field passed through a superconductor causes it to generate an opposing magnetic field and repel the magnet (Figure 7.17). A potential application of this phenomenon is 300-mph trains that would travel levitated above the tracks, thereby reducing friction. Prototypes of such a train have already been tested in Japan.

Nuclear magnetic resonance scanners are used in medicine to make images of tissues inside the body. These scanners are very expensive to operate because the magnets must be operated at liquid-helium temperature ($-268.9°C$). With high-temperature superconducting magnets, the expense of running the scanners would decrease and the scanners could be made more powerful.

Every superconductor has a temperature above which it loses its superconducting properties. Most materials must be cooled by expensive liquid helium to become superconducting, so they cannot be used in practical applications, where temperatures above 77 K are typical. Recently, chemists have been investigating ceramics as possible high-temperature superconductors. In 1986, K. Alex Muller and

Figure 7.17 A magnet hovers above a chilled superconductive ceramic disk because the magnetic field causes the superconductor to generate an opposing magnetic field.

J. Georg Bednorz at IBM's Zurich Research Laboratory discovered a barium–lanthanum–copper oxide that becomes superconducting at 35 K. This led to a flurry of activity in other laboratories, and in 1987, Paul C. W. Chu of the University of Houston synthesized a metal oxide ceramic, $YBa_2Cu_3O_7$, that is superconducting at 95 K. Since that time, similar ceramic materials have been prepared that become superconductors at ever higher temperatures, making a room-temperature superconductor a more likely possibility.

Questions for Discussion

1. How might room-temperature superconductors be used?
2. What are some advantages of locating electrical generators farther from population centers?
3. Should the government or private industry provide the funding to try to develop a room-temperature superconductor?
4. Assume that the ultimate superconductor will be a material like $YBa_2Cu_3O_7$. Will we find ample supplies of the component elements in the United States or will it be necessary to import them?

Nitric oxide is readily oxidized or reduced by other substances to give nitrosonium ion, NO^+, or nitrosyl ion, NO^-. These species are even-electron ions with complete octets:

$$:N:::O:^+$$
$$:\overset{..}{N}::\overset{..}{O}:^-$$

Color is characteristic of odd-electron molecules. Nitric oxide is deep blue in the liquid and solid states, although most of the nitrogen oxides are colorless. Another nitrogen oxide with an odd number of electrons is the reddish brown compound nitrogen dioxide.

Incomplete Octets and Coordinate Covalent Bonding

Boron trifluoride is a colorless gas with a pungent, suffocating odor. It fumes in moist air, forming a dense white cloud of HBF_4 and H_3BO_3. It is used in neutron-radiation detectors for monitoring radiation in the earth's atmosphere and in space, to catalyze organic synthesis reactions, as a fumigant, and to prepare KBF_4, which is used in the electroplating of nickel, tin, and lead, as well as in the preparation of elemental boron.

Some atoms, notably boron, participate in covalent bonding but do not have enough valence electrons to form an octet. Boron has only three valence electrons, which can be used to form three covalent bonds. Even when boron bonds with atoms that have additional electrons, like fluorine or chlorine, the molecule contains an incomplete octet:

$$
\begin{array}{c}
\overset{..}{:F:} \\
\overset{..}{:F}:\overset{..}{B}:\overset{..}{F:}
\end{array}
$$

Boron is electropositive and somewhat metallic in nature, whereas the atoms to which it will bond are very electronegative and nonmetallic. Thus, the surrounding atoms will not donate any of their unshared electron pairs to the boron atom, because they strongly attract additional electrons rather than give up a share in those they already have. As a result, boron has an incomplete octet.

The existence of an incomplete octet often leads to the formation of molecules in which an electron pair is donated from one atom to another. Usually, both atoms contribute one electron to a single covalent bond. However, it is possible for one atom to contribute both electrons. An example is the covalent bond formed between ammonia and boron trifluoride:

$$H_3N: + BF_3 \longrightarrow H_3N:BF_3$$

This type of bond is called a **coordinate covalent bond.** Once formed, such bonds are indistinguishable from other covalent bonds; only the source of the electrons differs. In general, boron completes its octet in simple covalent molecules only when it can form a coordinate covalent bond. This happens when a molecule in which boron has one pair of electrons missing from its valence octet reacts with another molecule that contains an unshared pair of electrons, such as ammonia.

Expanded Valence Shells

Elements in Periods 3–7 have empty valence d orbitals available to them. These orbitals can hold electrons that are in the valence s and p orbitals when the atom is in the ground state. Suppose a valence p orbital holds two electrons and one of them enters an empty valence d orbital. The result is two unpaired electrons, which can be used to form covalent bonds. As a result of their accessible valence d

orbitals, the elements in Periods 3–7 may form more than four bonds and the central atom may have more than eight electrons around it. Although it requires energy to move an electron into an empty valence d orbital, this energy is more than repaid by the formation of a covalent bond. Phosphorus, sulfur, and the heavier halogens and noble gases are notable for their ability to expand their octets.

Phosphorus forms two different molecules with chlorine, PCl_3 and PCl_5. Phosphorus has a valence electron configuration of $3s^2 3p^3$, so we would predict the existence of PCl_3 based on the octet rule. If one of the electrons in the $3s$ orbital is moved to another orbital, such as a $3d$ orbital, then phosphorus has five unpaired electrons: $3s^1 3p^3 3d^1$. The formation of five bonds with five chlorine atoms to pair up these five unpaired electrons does not follow the octet rule, but it follows the principle on which the octet rule is based—the pairing up of electrons during bond formation.

Species such as PF_5, SF_4, SbF_5, $IO_3F_2^-$, IF_4^+, ClF_3, BrF_3, I_3^-, and XeF_2 are molecules or ions with five electron pairs around the central atom. Species with six electron pairs around the central atom include SF_6, PF_6^-, IF_5, BrF_5, $XeOF_4$, XeF_4, and BrF_4^-.

7.7 Bond Polarity and Electronegativity

How do we decide whether elements will tend to transfer or share electrons when they form bonds?

The use of Lewis formulas to represent covalent bonding and of Lewis symbols to represent ionic bonding results in a deceptively simple picture. In reality, a covalent bond formed between two different nonmetals usually does not result in an equal sharing of electrons. The bonding in most of these molecules involves partial transfer of electrons from one atom to the other and unequal sharing of the electron pair. The resulting bond has some characteristics of an ionic bond and some characteristics of a covalent bond.

A whole range of bonding types can be thought of as having this mixed character. For convenience, the bonds in a compound are usually classified as the dominant type. Thus, the bonding in NaCl is considered ionic, and the bonding in CO is called covalent, even though each has some of the other character. But we can classify bonds further to describe their nature more accurately.

Figure 7.18 In HCl, the electrons in the bond lie, on average, closer to the chlorine atom than to the hydrogen atom. This unequal sharing leads to a partial negative charge ($\delta-$) on chlorine and a partial positive charge ($\delta+$) on hydrogen.

$$\overset{\delta+}{H} \;\; \overset{\delta-}{:\!\ddot{C}\!l\!:}$$

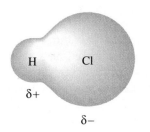

Polar and Nonpolar Covalent Bonds

An example of a compound with unequal sharing of electrons in a covalent bond is hydrogen chloride, HCl. Although the aqueous solution of this substance, known as hydrochloric acid, is more common, hydrogen chloride also exists as a colorless gaseous molecule. In this molecule, on the average, the electron pair is found closer to the chlorine atom than to the hydrogen atom, as pictured in Figure 7.18. Although primarily covalent, HCl does have some ionic character because the electron from the hydrogen atom has been partially transferred to the chlorine atom. The amount of ionic character in a principally covalent bond can be judged from the **polarity** of the bond, or the degree of separation of electronic charge. The greater the transfer of electrons, the greater the polarity. The range of possible bonding patterns is shown in Figure 7.19.

Figure 7.19 Types of bonds.

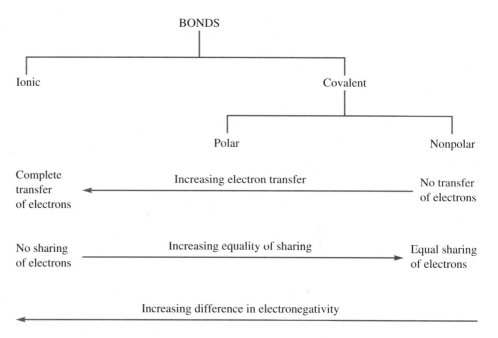

Covalent bonds can be subdivided on the basis of polarity into two general types. **Nonpolar covalent bonds** are those in which electrons are equally shared and therefore charge is evenly distributed between the atoms. In **polar covalent bonds,** the electrons are located on the average closer to one of the atoms than to the other. This unequal sharing leads to the development of partial charges on the two atoms, as shown in Figure 7.18 for HCl. One of the atoms has more electron density than it had as a free atom, so it has a partial negative ($\delta-$) charge. The other atom has lost electron density, so it develops a partial positive ($\delta+$) charge.

Electronegativity

Linus Pauling, a chemist formerly at the California Institute of Technology, received the Nobel Prize in chemistry in 1954 for his work on the chemical bond. He also received the Nobel Peace Prize in 1963 for his activities in opposition to nuclear weapons testing.

The concept of a separation of charge in a bond and the simultaneous presence of some ionic and some covalent character was developed by Linus Pauling. Pauling examined bond energies and dissimilarities in the covalent bonds formed between like elements (as in H_2 and F_2) and those formed between different elements (as in HF). The bonds between different elements appeared to be stronger. Pauling proposed that a partial transfer of electrons, resulting in partial ionic character, would explain the extra bond stability. This partial transfer occurs because one atom attracts the shared electrons more strongly than the other. Pauling called this ability to attract bonding electrons the **electronegativity** of the atom. If the bonding electrons are equally shared, the two atoms must have the same electronegativity, and the bond is nonpolar covalent. The greater the difference in electron-attracting ability, or electronegativity, the more ionic character the bond has. The more ionic character, the greater the partial ionic charges developed on the atoms.

Pauling set up a scale of relative electronegativity values based on a value of 2.1 for hydrogen. Fluorine (4.0) has the highest electronegativity on this scale, and cesium and francium (0.7) the lowest. Values for most of the elements are given in Figure 7.20.

H 2.1																	He
Li 1.0	Be 1.5											B 2.0	C 2.5	N 3.0	O 3.5	F 4.0	Ne
Na 0.9	Mg 1.2											Al 1.5	Si 1.8	P 2.1	S 2.5	Cl 3.0	Ar
K 0.8	Ca 1.0	Sc 1.3	Ti 1.5	V 1.6	Cr 1.6	Mn 1.5	Fe 1.8	Co 1.8	Ni 1.8	Cu 1.9	Zn 1.6	Ga 1.6	Ge 1.8	As 2.0	Se 2.4	Br 2.8	Kr
Rb 0.8	Sr 1.0	Y 1.2	Zr 1.4	Nb 1.6	Mo 1.8	Tc 1.9	Ru 2.2	Rh 2.2	Pd 2.2	Ag 1.9	Cd 1.7	In 1.7	Sn 1.8	Sb 1.9	Te 2.1	I 2.5	Xe
Cs 0.7	Ba 0.9	La 1.1	Hf 1.3	Ta 1.5	W 1.7	Re 1.9	Os 2.2	Ir 2.2	Pt 2.2	Au 2.4	Hg 1.9	Tl 1.8	Pb 1.8	Bi 1.9	Po 2.0	At 2.2	Rn
Fr 0.7	Ra 0.9	Ac 1.1															

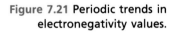

Figure 7.20 Electronegativity values for the elements.

Electronegativity values show periodic trends, as shown in Figure 7.21. The nonmetals in general have higher values than the metals, in accord with the tendency of metals to give up electrons and nonmetals to gain electrons in the formation of chemical bonds. Indeed, the metals are often said to be electropositive, emphasizing their difference from the electronegative nonmetallic elements. Electronegativity values tend to increase from bottom to top of a group and from left to right across a period. This trend is shown in Figure 7.22. The values within the transition-metal series and the inner-transition-metal series are fairly constant within a period, as can be seen in Figures 7.20 and 7.21.

Figure 7.21 Periodic trends in electronegativity values.

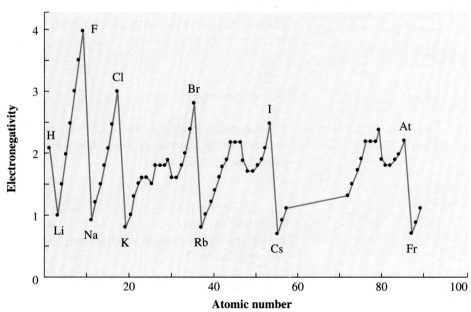

Figure 7.22 Trends in electronegativity within the periodic table.

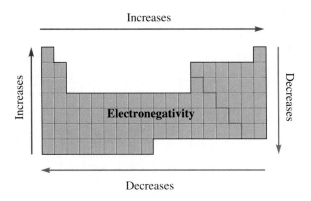

7.8   Strength of Covalent Bonds

▶ This chapter began by stating that chemical reactions, such as that between silicon tetrachloride and water, involve the making and breaking of chemical bonds:

$$SiCl_4(l) + 3H_2O(l) \longrightarrow H_2SiO_3(s) + 4HCl(g)$$

Silicon tetrachloride, SiCl₄, is a colorless liquid with a suffocating odor. It fumes in the presence of water or moist air and is decomposed by water to form silicic acid and hydrochloric acid. It is prepared by direct reaction of the elements and is used to prepare pure silicon and to produce smoke screens in warfare.

Here we break Si—Cl and H—O bonds and make Si—O, H—O, and H—Cl bonds. Because different bonds have different strengths, the energy change accompanying a reaction depends on the types of bonds in the reactants and products. If we could associate an energy with the strengths of bonds, we could use this information to calculate the energy changes of reactions. In Section 7.3, we saw that a triple bond is generally stronger than a double bond, which is generally stronger than a single bond. However, not all bonds of a given type are equally strong. For example, it is easier to break the single bond between hydrogen and chlorine in HCl than that between hydrogen and fluorine in HF. To compare the strengths of bonds, we must look more closely at their bond energies.

Bond Energy

Why can we use bond energies to calculate energy changes only for reactions involving gaseous reactants and products? What other energy quantities would we need if some of the substances are liquids or solids?

For a diatomic molecule, the *bond energy* (or **bond dissociation energy,** D) is the energy change that occurs in the reaction in which the gaseous molecule is separated into gaseous atoms:

$$H_2(g) \longrightarrow 2H(g) \qquad D_{H-H} = 432 \text{ kJ/mol}$$

The subscript in the notation for the bond energy indicates the atoms whose bond is being broken. The bond energy for H_2 is equal in magnitude but opposite in sign to the energy (ΔH) released when gaseous hydrogen atoms combine to form a gaseous molecule:

$$2H(g) \longrightarrow H_2(g) \qquad \Delta H = -432 \text{ kJ/mol}$$

In some groups of diatomic molecules, bond energies display an obvious trend. For example, consider the alkali metals, which can form diatomic molecules in the gaseous state if heated sufficiently:

Molecules: Li_2 Na_2 K_2 Rb_2 Cs_2
Bond energy (kJ/mol):   105   72   49   45   44

All these bond energies are quite small, and they decrease as the atomic number increases. Among the hydrogen halides, bond energies are comparatively large; they too decrease as the atomic number increases:

Molecules: HF   HCl   HBr   HI
Bond energy (kJ/mol):   565   428   362   295

These chemically related molecules, all of which contain single bonds, have bond energies that are similar for adjacent elements in the group and that vary smoothly.

In contrast, bond energies for molecules of elements in the same period but different groups of the periodic table may vary widely, if the molecules have different bond types. For example, the bond energy of O_2 is 494 kJ/mol, only slightly more than half of that of its neighbor N_2, which is 942 kJ/mol. Moreover, the bond energy of O_2 is more than three times as large as that of F_2, 155 kJ/mol. These differences result from the nature of the covalent bond in these species—a single bond in F_2, a double bond in O_2, and a triple bond in N_2. However, even here, there are some regularities in the bond energies. Examine the following examples of bond energies (in kJ/mol) for different types of bonds:

$$C—C \quad 346 \qquad N—N \quad 167 \qquad O—O \quad 142 \qquad F—F \quad 155$$
$$C{=}C \quad 602 \qquad N{=}N \quad 418 \qquad O{=}O \quad 494$$
$$C{\equiv}C \quad 835 \qquad N{\equiv}N \quad 942$$

Note the general similarity of most single bonds (or double or triple bonds) across a period and the increase in bond strength from a single to a double to a triple bond.

Average Bond Energy

The same two atoms may bond in a variety of different compounds. Do bond energies vary in different environments? Consider these bonds between hydrogen and oxygen:

$$H_2O(g) \longrightarrow H(g) + OH(g) \qquad D_{H-OH} = 502 \text{ kJ/mol}$$
$$OH(g) \longrightarrow H(g) + O(g) \qquad D_{O-H} = 427 \text{ kJ/mol}$$
$$HOOH(g) \longrightarrow H(g) + OOH(g) \qquad D_{H-OOH} = 431 \text{ kJ/mol}$$

Although the energy of the O—H bond is sensitive to its environment, the variation is not very large. Variations of a similar magnitude occur in other bonds as well.

The approximate constancy of energies for the bonds between two given atoms is significant, for it suggests that the principal factors that determine the energy of a particular bond are the intrinsic properties of the two bonded atoms. Thus, it is possible to characterize any bond, such as the O—H or C—H bond, by an **average bond energy** ($\bar{D}$), the approximate energy necessary to break that bond in any compound in which it occurs. This is different from the bond energy, which describes a particular bond in a particular molecule. Some average bond energies are listed in Table 7.5.

Table 7.5 Average Bond
Energies (in kJ/mol)

	C	N	O	F	Si	P	S	Cl	Br	I
H—	411	386	459	565	318	322	363	428	362	294
C—	346	305	358	485	318	264	272	327	285	213
C=	602	615	799				573			
C≡	835	887	1072							
N—	305	167	201	283				313	243	
N=	615	418	607							
N≡	887	942								
O—	358	201	142	190	452	335		218	201	201
O=	799	607	494			544	323			
S—	272			284	293		268	255	217	
F—	485	283	190	155	565	490	284	249	249	278
Cl—	327	313	218	249	381	326	265	240	216	208
Br—	285	243	201	249	310	264	217	216	190	175
I—	213		201	278	234	184		208	175	149

Average Bond Energies and Heats of Formation

By using average bond energies, we can estimate the energy released when a gaseous molecule is formed from gaseous atoms. Because they are defined as the energy required to break bonds, bond energies have positive values. When bonds are formed, energy is released, so the energy change for such a reaction must be the negative of the sum of the bond energies. For example, let's calculate the energy change that occurs when chloroform, CH_3Cl, is formed from its component atoms:

$$3H(g) + C(g) + Cl(g) \longrightarrow CH_3Cl(g)$$

The energy change is related to the energies of the bonds in chloroform by the following equation:

$$\Delta H = -[3\bar{D}_{C-H} + \bar{D}_{C-Cl}]$$

Substituting values of the average bond energies from Table 7.5 yields

$$\Delta H = -[3(411 \text{ kJ/mol}) + 327 \text{ kJ/mol}] = -1560 \text{ kJ/mol}$$

We do not find every element in the gaseous state, and it is useful to calculate the energy change for the formation of a compound from the elements as they occur naturally. The component elements of chloroform exist naturally as hydrogen and chlorine in the form of gaseous diatomic molecules and carbon in the form of solid graphite. The reaction of interest, then, is

$$\tfrac{3}{2}H_2(g) + C(graphite) + \tfrac{1}{2}Cl_2(g) \longrightarrow CH_3Cl(g)$$

The energy change for this reaction is the heat of formation or enthalpy of formation, introduced in Chapter 5: the amount of energy released when 1 mol of a compound is formed from its elements in their most stable form. Let's assume the reaction can be broken into two steps, one in which the elements are converted to gaseous atoms and a second in which the gaseous atoms combine to form chloroform:

$$\text{Step 1: } \tfrac{3}{2}H_2(g) + C(\text{graphite}) + \tfrac{1}{2}Cl_2(g) \longrightarrow 3H(g) + C(g) + Cl(g)$$
$$\text{Step 2: } 3H(g) + C(g) + Cl(g) \longrightarrow CH_3Cl(g)$$

We have already calculated ΔH for the second process:

$$\text{Step 2: } \Delta H = -1560 \text{ kJ/mol}$$

For the first process, ΔH is calculated as follows:

$$\text{Step 1: } \Delta H = \tfrac{3}{2}\bar{D}_{H-H} + \tfrac{1}{2}\bar{D}_{Cl-Cl} + \Delta H_{\text{atomization for graphite}}$$
$$= \tfrac{3}{2}(432 \text{ kJ/mol}) + \tfrac{1}{2}(240 \text{ kJ/mol}) + 715 \text{ kJ/mol}$$
$$= 1483 \text{ kJ/mol}$$

Using Hess's law, we combine the two results to get the enthalpy change for the overall reaction:

$$\Delta H = 1483 \text{ kJ/mol} - 1560 \text{ kJ/mol} = -77 \text{ kJ/mol}$$

If we had measured ΔH for this reaction using a calorimeter, we would have obtained -82 kJ/mol. The discrepancy between calculated and measured values is often this large because we are using average bond energies rather than exact bond energies. Nevertheless, such calculations are useful to indicate the approximate strengths of various bonds, as well as to provide estimates in cases where direct measurements are difficult to obtain.

Average Bond Energies and Heats of Reaction

In general, the heat of reaction for any gaseous chemical reaction can be calculated from average bond energies by use of the following equation:

$$\Delta H = \Sigma \bar{D}_{\text{reactant bonds}} - \Sigma \bar{D}_{\text{product bonds}}$$

Consider the Haber process for preparing ammonia:

$$N_2(g) + 3H_2(g) \longrightarrow 2NH_3(g)$$

The reactants contain one $N\equiv N$ and three $H-H$ bonds, and the product contains six $N-H$ bonds. Thus, the enthalpy of reaction is given by the following equation:

$$\Delta H = [\bar{D}_{N\equiv N} + 3\bar{D}_{H-H}] - [6\bar{D}_{N-H}]$$

Substituting values from Tables 7.3 and 7.5 yields

$$\Delta H = [942 \text{ kJ/mol} + 3(432 \text{ kJ/mol})] - [6(386 \text{ kJ/mol})]$$
$$= -78 \text{ kJ/mol} \qquad (\text{Measured value} = -46 \text{ kJ/mol})$$

Calculating enthalpies of reaction from average bond energies can help us decide whether reactions are feasible, as in the following example.

EXAMPLE 7.8

Methane and bromine react to form bromomethane, CH_3Br. Two possible reactions might give this product:

$$CH_4 + Br_2 \longrightarrow CH_3Br + HBr$$

$$2CH_4 + Br_2 \longrightarrow 2CH_3Br + H_2$$

Which reaction is more probable?

Solution   Use average bond energies to calculate the heat of reaction for both possible reactions. For the first reaction, the equation is

$$\Delta H_1 = \bar{D}_{Br-Br} + 4\bar{D}_{C-H} - \bar{D}_{H-Br} - \bar{D}_{C-Br} - 3\bar{D}_{C-H}$$

Because some average bond energies appear more than once, we can combine terms to give the simpler equation:

$$\Delta H_1 = \bar{D}_{Br-Br} + \bar{D}_{C-H} - \bar{D}_{H-Br} - \bar{D}_{C-Br}$$

Next, we insert values from Table 7.5:

$$\Delta H_1 = 190 \text{ kJ/mol} + 411 \text{ kJ/mol} - 362 \text{ kJ/mol} - 285 \text{ kJ/mol} = -46 \text{ kJ/mol}$$

We use the same procedure for the second reaction:

$$\Delta H_2 = \bar{D}_{Br-Br} + 8\bar{D}_{C-H} - \bar{D}_{H-H} - 2\bar{D}_{C-Br} - 6\bar{D}_{C-H}$$
$$= \bar{D}_{Br-Br} + 2\bar{D}_{C-H} - \bar{D}_{H-H} - 2\bar{D}_{C-Br}$$
$$= 190 \text{ kJ/mol} + 2(411 \text{ kJ/mol}) - 432 \text{ kJ/mol} - 2(285 \text{ kJ/mol}) = 10 \text{ kJ/mol}$$

The first reaction is exothermic; the second is endothermic. We predict that the first reaction will be the one that actually occurs because reactions tend to proceed in a direction that releases energy.

Practice Problem 7.8

Hydrogen and oxygen can combine to form two different products, water or hydrogen peroxide:

$$2H_2 + O_2 \longrightarrow 2H_2O$$

$$H_2 + O_2 \longrightarrow H_2O_2$$

Which reaction is favored?

Answer: $\Delta H_1 = -478$ kJ/mol, and $\Delta H_2 = -134$ kJ/mol; so the first reaction is favored.

Summary

Different chemical bonds give substances different properties. In bonding, elements tend to lose, gain, or share electrons to achieve a more stable configuration. According to the octet rule, the tendency is to fill the valence orbitals with eight electrons (hydrogen usually has two).

Ionic bonding involves the complete transfer of one or more electrons from a metal to a nonmetal, often giving both a noble-gas configuration. In ionic compounds, the ions are held together in crystal lattices by electrostatic forces. Covalent bonds result from the sharing of a pair of electrons between two atoms. In some molecules, two atoms share two electron pairs to form a double bond or three electron pairs to form a triple bond.

Lewis formulas can be used to represent the bonding in molecules and ions. If more than one Lewis formula is possible, the formal charges of the atoms can be used to examine their relative importance. When a single Lewis formula does not adequately represent the arrangement of electrons in a covalent molecule or ion, the actual bonding is a composite of all the resonance hybrids.

Some elements have a stronger tendency to attract electrons (measured by their electronegativity values), and the electrons in a covalent bond are not shared equally. Such covalent bonds are said to be polar.

The strength of bonds depends on the elements that are bonded and on the number of electron pairs being shared. Bond strength can be measured by the bond energy, the energy that must be supplied to break the bond.

Key Terms

average bond energy (7.8)
bond energy (bond dissociation energy) (7.2)
bond length (7.3)
Born–Haber cycle (7.2)
chemical bond (p. 260)
coordinate covalent bond (7.6)

coordination number (7.2)
covalent bonding (7.1)
crystal lattice (7.2)
double bond (7.3)
electronegativity (7.7)
formal charge (7.4)
heat of fusion (7.2)
heat of vaporization (7.2)

ionic bonding (7.1)
ionic crystal (7.2)
lattice energy (7.2)
Lewis formula (electron-dot formula) (7.3)
Lewis symbol (electron-dot symbol) (7.1)
metallic bonding (7.1)
nonpolar covalent bond (7.7)

octet rule (7.1)
polar covalent bond (7.7)
polarity (7.7)
resonance hybrid (7.5)
single bond (7.3)
triple bond (7.3)

Exercises

Types of Bonding

7.1 What is a chemical bond?

7.2 Describe the differences between ionic, covalent, and metallic bonding.

7.3 Which type of elements are most likely to form substances using ionic, covalent, or metallic bonding?

7.4 Which of the following substances are likely to be ionically bonded?
a. HF b. LiF c. PCl_3 d. $CaBr_2$ e. CCl_4

7.5 Identify the bonding in each of the following substances as ionic, covalent, or metallic.
a. Na b. NaCl c. F_2 d. Fe e. Si
f. CO g. $FeCl_3$ h. SiO_2

7.6 Which of the following substances are ionic?
a. CH_4 b. Ca_3N_2 c. $BeCl_2(g)$ d. SF_4
e. KBr f. HCl(g)

7.7 Which of the following substances are covalent?
a. BrCl b. $CaCO_3$ c. SO_2 d. CaO
e. H_2O f. NO

7.8 Which of the following chlorides are ionic?
a. NaCl b. $SiCl_4$ c. CCl_4 d. ICl
e. $BaCl_2$

7.9 Classify each of the following species according to the primary type of bonding.
a. $NH_4{}^+$ b. $SbCl_3$ c. CsCl d. CO_2
e. CaO f. O_2

7.10 Classify each of the following species according to the primary type of bonding.
a. P_4 b. $CaCl_2$ c. $SiCl_4$ d. BN
e. CaC_2 f. PCl_5

Lewis Symbols

7.11 What is a Lewis symbol?

7.12 Why would we represent valence electrons by means of a Lewis symbol instead of designating the valence electron configuration?

7.13 Draw Lewis symbols showing the valence electrons of each of the following atoms.
a. As b. I c. Se d. Sr e. Cs f. Ar

7.14 Draw Lewis symbols showing the valence electrons of each of the following atoms.
a. Kr b. Sb c. F d. In e. Ba f. Si

7.15 Draw Lewis symbols showing the valence electrons of each of the following ions.
a. Cl^- b. Pb^{2+} c. S^{2-} d. Ba^{2+}
e. B^{3+} f. Al^{3+}

Ionic Bonding

7.16 a. What is a crystal lattice?
b. Why do ions stay together in this structure?

7.17 What is a coordination number?

7.18 a. Describe the sodium chloride structure.
   b. What are the coordination numbers of each ion in this structure?

7.19 What is the Born–Haber cycle?

7.20 Write a Lewis formula for each of the following ionic salts.
   a. LiCl   b. $BaCl_2$ c. Na_2O d. LiOH
   e. CsF   f. BaS

7.21 K^+ is known in nature, but not K^{2+}. Why?

7.22 Give the formulas for the oxides, sulfides, chlorides, phosphates, and nitrates of lithium and barium.

7.23 Calculate the heat of formation of lithium chloride assuming that it crystallizes in the sodium chloride lattice and that the following energy quantities apply: the heat of vaporization of Li(s) is 161 kJ/mol; half the bond energy of $Cl_2(g)$ is 120 kJ/mol; the ionization energy of lithium atoms is 520 kJ/mol; the electron affinity of chlorine atoms is −349 kJ/mol; and the lattice energy is −901 kJ/mol.

7.24 Calculate the lattice energy of lithium fluoride assuming that it crystallizes in the sodium chloride lattice and that the following energy quantities apply: the heat of vaporization of Li(s) is 161 kJ/mol; half the bond energy of $F_2(g)$ is 79 kJ/mol; the ionization energy of lithium atoms is 520 kJ/mol; the electron affinity of fluorine atoms is −328 kJ/mol; and the heat of formation is −669 kJ/mol.

Covalent Bonding and Lewis Formulas

7.25 Why does hydrogen exist as a diatomic molecule?

7.26 Distinguish between single, double, and triple bonds.

7.27 What is a coordinate covalent bond?

7.28 How do we decide whether a Lewis formula should have double or triple bonds?

7.29 Write a Lewis formula for each of the following.
   a. NO_3^- b. SO_4^{2-} c. SO_3^{2-} d. NO_2^-
   e. NO^+ f. H_3O^+

7.30 Write a Lewis formula for each of the following.
   a. HNO_2 (HONO)   b. NH_4^+ c. H_2CO
   d. $(CH_3)_2CO$ e. OH^- f. CH_3NO_2

7.31 Write a Lewis formula for each of the following.
   a. OF_2 b. $SiCl_4$ c. PBr_3 d. CSe_2 e. NCl_3

7.32 Two chlorine atoms combine to form Cl_2. Is the bond normal covalent or coordinate covalent?

7.33 A chlorine atom combines with a hydrogen atom to form HCl.
   a. Is the bond normal covalent or coordinate covalent?
   b. What if H^+ combines with Cl^- to form HCl?

7.34 Write an electron-dot formula for each of the following.
   a. HCN   b. H_3CCN c. H_3CBr d. C_2H_2
   e. C_2H_4 f. C_2H_6

7.35 Write an electron-dot formula for each of the following.
   a. NH_2OH b. CCl_4 c. C_2H_3Cl d. C_2Br_2
   e. $N_2H_5^+$ f. HOCl

7.36 Write an electron-dot formula for each of the following.
   a. $HClO_2$ b. $HClO_3$ c. $HClO_4$ d. H_2SO_4
   e. BrF_3 f. ClO_3^-

Resonance

7.37 Define *resonance*.

7.38 Why is it necessary to introduce the concept of resonance into the theory of covalent bonding?

7.39 Indicate whether or not each of the following species exhibits resonance.
   a. O_2 b. H_2O c. SO_2 d. ClO_2^-
   e. NO_2 f. CO_2 g. SO_3^{2-} h. SO_3

7.40 Write a Lewis formula, including the resonance forms, for each of the following molecules.
   a. SO_2 b. SO_3 c. CO_2 d. CO_3^{2-}
   e. NO_3^- f. NO_2

7.41 Write a Lewis formula, including the resonance forms, for each of the following molecules or ions.
   a. CS_2 b. NCO^- c. NO_2^- d. SO_4^{2-}
   e. C_6H_6 (in which the carbon atoms are bonded in a six-membered ring)   f. SO_3^{2-}

7.42 Determine the formal charge of each indicated atom.
   a. N in NO_2 b. C in CO   c. O in CO
   d. N in NCS^- e. C in NCS^- f. S in NCS^-

7.43 Draw the best Lewis formula for each of the following species. Indicate the formal charge of each atom. Draw any necessary resonance forms.
   a. CO_3^{2-} b. HN_3 c. $SiCl_4$ d. NH_4^+
   e. KrF_4 f. PF_5

The Octet Rule

7.44 What is the octet rule?

7.45 In SF_6, the sulfur atom has a share in 12 electrons. Explain why this observation is consistent with the principle behind the octet rule.

7.46 Decide whether the indicated atom obeys the octet rule. If it doesn't, indicate how the octet rule is broken.
   a. O in H_2O b. Ne atom   c. F atom
   d. S in SF_4 e. F in SF_4 f. H in H_2O

7.47 Gaseous aluminum chloride exists as Al_2Cl_6, with chlorines bridging between the two aluminum atoms: $Cl_2AlCl_2AlCl_2$. Discuss the bonding in aluminum trichloride, $AlCl_3$, and the reason for the tendency to form Al_2Cl_6.

7.48 Decide whether the indicated atom or ion obeys the octet rule. If it doesn't, indicate how the octet rule is broken.
  a. Mg^{2+} ion   b. Cl^- ion   c. Na atom
  d. O^{2+} ion   e. F in F_2 f. F in HF

7.49 An atom in each of the following molecules does not obey the octet rule. Decide which atom violates the rule and explain the violation.
  a. NO_2 b. XeF_2 c. XeF_6 d. ICl_3
  e. SF_4 f. BH_3 g. $BeCl_2$ h. ClO_2

7.50 Describe the bonding in S_2Cl_2. Why would this molecule not exist as SCl?

Bond Polarity and Electronegativity

7.51 a. What is electronegativity?
  b. Compare the electronegativity of electropositive and electronegative elements.

7.52 Describe the origin of polarity in bonds.

7.53 Arrange each series of atoms in order of increasing electronegativity.
  a. Br, Cl, F, N, O   c. B, C, H, Se, Si
  b. C, F, H, N, O   d. C, Ca, Cl, Cs, Cu

7.54 Decide which molecule in each pair is polar. Explain why that molecule is polar and the other is not.
  a. HF and H_2 b. ICl and I_2 c. H_2 and HI

7.55 Arrange the following series of bonds in order of increasing polarity.
  a. O—H, C—H, H—H, F—H
  b. CaO, BeO, SrO, MgO
  c. H—F, F—F, H—H, H—I
  d. B—F, O—F, C—F, H—F

Bond Strength

7.56 What is bond energy?

7.57 How can average bond energies be used to calculate the enthalpy change for a chemical reaction?

7.58 How would you expect the bond strength to vary for a series of closely related molecules as the bond length increases?

7.59 Which of the following molecules would you expect to have the highest bond energy: N_2, P_2, or As_2?

7.60 Use bond energies to calculate the heats of reaction for each of the following gas-phase reactions.
  a. $CH_2Cl_2 + Cl_2 \longrightarrow CCl_4 + H_2$
  b. $2HCl \longrightarrow Cl_2 + H_2$
  c. $2CH_3CH_3 + 7O_2 \longrightarrow 4CO_2 + 6H_2O$
  d. $N_2 + 3Cl_2 \longrightarrow 2NCl_3$
  e. $2CO + O_2 \longrightarrow 2CO_2$

7.61 Use bond energies to calculate the heats of reaction for each of the following gas-phase reactions.
  a. $CH_4 + 2O_2 \longrightarrow CO_2 + 2H_2O$
  b. $N_2 + O_2 \longrightarrow 2NO$
  c. $H_2C{=}CH_2 + H_2 \longrightarrow H_3C{-}CH_3$
  d. $H_2C{=}CH_2 + Cl_2 \longrightarrow ClH_2C{-}CH_2Cl$
  e. $CH_4 + CCl_4 \longrightarrow 2CH_2Cl_2$

7.62 Calculate the S—S bond energy, given that the enthalpy of reaction for the following reaction is 928 kJ: $S_8 + 8O_2 \longrightarrow 8SO_2$.

7.63 Calculate the N—O bond energy, given that the enthalpy of reaction for the following reaction is +92 kJ: $2NO + O_2 \longrightarrow 2NO_2$.

7.64 Which of the following molecules would you expect to have the lowest bond energy: HF, HCl, or HBr?

7.65 Which of the following molecules would you expect to have the strongest bond: C—F, C—Cl, or C—Br?

7.66 Calculate the enthalpy of reaction in kilojoules for each of the following reactions.
  a. $4Cl_2 + 2CH_4 + O_2 \longrightarrow 2CO + 8HCl$
  b. $4NH_3 + 5O_2 \longrightarrow 4NO + 6H_2O$
  c. $Cl_2 + H_2O \longrightarrow HCl + HOCl$
  d. $2C_6H_{14} + 19O_2 \longrightarrow 12CO_2 + 14H_2O$
  e. $4NH_3 + 3O_2 \longrightarrow 2N_2 + 6H_2O$

7.67 Calculate the P—Cl bond energy, given that the enthalpy of reaction for the following reaction is −412 kJ: $PCl_3 + Cl_2 \longrightarrow PCl_5$.

Additional Exercises

7.68 Draw Lewis symbols showing the valence electrons of each of the following atoms.
  a. Br   b. Pb   c. S   d. Ca   e. Be   f. Xe

7.69 Arrange the following atoms in order of decreasing electronegativity: Br, Cl, F, I.

7.70 Which of the following substances are likely to be ionically bonded?
  a. H_2 b. Li_2O C. BCl_3 d. ClBr   e. SiO_2

7.71 Chlorine atoms combine with a boron atom to form BCl_3.
  a. Is the bond normal covalent or coordinate covalent?
  b. What if B^{3+} combines with Cl^- to form BCl_3?

7.72 Decide which molecule in each pair is polar. Explain why that molecule is polar and the other is not.
  a. F_2 and HF   b. FCl and Cl_2 c. O_2 and BO

7.73 Write a Lewis formula, including the resonance forms, for each of the following molecules and ions.
   a. OCN^- b. N_3^- c. ClO_2^- d. PO_4^{3-}
   e. SeO_2 f. H_2CO_3

7.74 Draw Lewis symbols showing the valence electrons of each of the following ions.
   a. P^{3-} b. In^{3+} c. Se^{2-} d. Be^{2+} e. C^{4-}

7.75 Use bond energies to calculate the heats of reaction for each of the following gas-phase reactions.
   a. $CH_4 + Br_2 \longrightarrow CH_3Br + HBr$
   b. $2H_2 + O_2 \longrightarrow 2H_2O$
   c. $H_2 + O_2 \longrightarrow H_2O_2$
   d. $H_2C{=}CH_2 + H_2O \longrightarrow H_3C{-}CH_2OH$
   e. $C + H_2O \longrightarrow CO + H_2$

7.76 Draw the best Lewis formula for each of the following species. Indicate the formal charge of each atom. Draw any necessary resonance forms.
   a. ClO_4^- b. NO_2^- c. XeF_4 d. NCO^-
   e. HCO_2^- f. BF_3

7.77 Calculate the heat of formation of cesium chloride assuming that it crystallizes in the cesium chloride lattice and that the following energy quantities apply: the heat of vaporization of $Cs(s)$ is 78 kJ/mol; half the bond energy of $Cl_2(g)$ is 120 kJ/mol; the ionization energy of cesium atoms is 376 kJ/mol; the electron affinity of chlorine atoms is -349 kJ/mol; and the lattice energy is -620 kJ/mol.

7.78 Write a Lewis formula for each of the following.
   a. $NaBr$ b. CaF_2 c. Na_2S d. K_2O
   e. $MgBr_2$ f. MgO

7.79 Decide whether each indicated atom or ion obeys the octet rule. If it doesn't, indicate how the octet rule is broken.
   a. Pb^{2+} ion   b. S^{2-} ion   c. C atom
   d. N^{3-} ion   e. N in N_2 f. N in NH_3

7.80 Classify each of the following species according to the primary type of bonding.
   a. HCN b. $AgCl$ c. S_8 d. CH_4
   e. $NaCl$ f. $CoCl_2$

7.81 Write a Lewis formula for each of the following molecules.
   a. N_2H_2 b. CS_2 c. AsF_3 d. CO_2
   e. CO f. SCl_2

7.82 Write a Lewis formula for each of the following.
   a. HNO_3 b. SO_3 c. SF_4 d. NO_2
   e. N_2H_4 f. PH_3

7.83 Calculate the $C{\equiv}O$ bond energy, given that the enthalpy of reaction for the following reaction is -558 kJ: $2CO + O_2 \longrightarrow 2CO_2$.

*7.84 Use the Born–Haber cycle to explain why magnesium and fluorine form MgF_2 rather than MgF. The first ionization energy of magnesium is 738 kJ/mol, and the second ionization energy is 1451 kJ/mol. The lattice energy is -1080 kJ/mol for MgF and -2930 kJ/mol for MgF_2.

*7.85 Would MgF_3 be more stable than MgF_2? The lattice energy is -4500 kJ/mol for MgF_3 and -2930 kJ/mol for MgF_2. The ionization energies for the successive removal of electrons from Mg are 738, 1451, and 7733 kJ/mol. Explain your answer.

*7.86 Robert S. Mullikan proposed that electronegativity could be defined as the average of the first ionization energy and the electron affinity for an element: $(IE - EA)/2$, where the minus sign arises because of the unusual sign convention for electron affinity.
   a. Calculate the electronegativities of the halogens by this definition.
   b. Compare these values with the Pauling values. What relationship, if any, exists between the two systems?

*7.87 An ion containing one carbon, one nitrogen, and one oxygen atom with a charge of $1-$ has three isomers, which differ in the arrangement of the atoms. Write Lewis formulas for each isomer, showing all resonance structures. Use formal charges to decide whether any of these isomers is likely to be less stable than the others. Explain your reasoning.

▶ ▶ ▶ Chemistry in Practice

7.88 A gaseous covalent compound contains 29.67% sulfur and 70.33% fluorine by mass.
   a. Write a Lewis formula for this compound.
   b. Use bond energies to estimate the heat of formation of this compound (assuming gaseous sulfur and fluorine as reactants).

   c. The compound reacts with water vapor to form sulfur dioxide and hydrogen fluoride. How much heat is released by the reaction of 54.0 g of this compound with excess water?
   d. How much sulfur dioxide is produced by this reaction?

Covalent Bonds and Properties of Molecules

Various odors can be detected with the sense of smell, which has a chemical basis.

CHAPTER OUTLINE

Encountering Chemistry

▶ ▶ ▶ ▶

The functioning of the senses in the human body has long intrigued chemists. The senses must have a chemical basis, but it is not a simple matter to determine how this chemistry operates. We know that there are receptor sites for taste and odor—places where molecules can fit. These receptors are fashioned from the protein that makes up tissue. Proteins are interlinked in various ways, and through these interlinkings, cavities in their structure may be created. The cavities can act as receptors for other molecules if they are the right size and shape to fit the cavity and if they interact properly with the protein atoms, resulting in a biological effect.

It has been difficult to relate the structures of molecules to the tastes they have. The common tastes are salty, sour, bitter, and sweet. Other tastes, such as astringent, soapy, and metallic, are also recognized. Sweet-tasting substances are highly attractive to humans, and for good reason. Such substances are rarely toxic, whereas the other tastes often warn of undesirable properties. For example, a sweet substance, lactose, is found in mothers' milk, where it undoubtedly provides encouragement for babies to consume more. Other sweet substances include glucose and sucrose, which occur naturally in plants. A common feature of these substances is the size of a particular portion of the molecule—presumably the part that fits into the receptor site. Sweet molecules commonly have an —H atom (or —OH group) separated from an —O atom by about 0.30 nm, as illustrated in Figure 8.1. Chemists working to develop artificial sweeteners would like to be able to take advantage of this information. But because there is no generally ac-

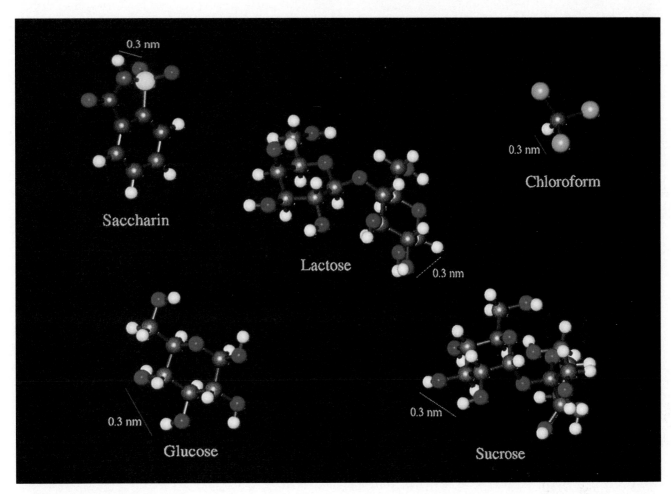

Saccharin

Lactose

Chloroform

Glucose

Sucrose

Figure 8.1 Some sweet molecules containing —H and —OH groups that probably fit into a taste receptor site.

cepted theory of exactly how taste works, the search for artificial sweeteners has been largely a matter of trial and error.

Perhaps as many as 30 different odors can be distinguished by humans. A popular theory proposes that odors can be explained by the existence of seven kinds of receptor cavities in the nose. When a volatile molecule enters the nose and interacts with a particular cavity, an electrical signal is generated and sent to the brain. A molecule that can fit into more than one cavity generates a composite odor. The proposed receptor cavities and some molecules that give each of the seven primary odors are shown in Figure 8.2. Whether or not this theory will ultimately be found to be correct, it is likely that theories of odor will always attempt to correlate the odor with the structure and shape of the molecule that causes the odor.

Taste and smell are representative of innumerable interactions between molecules that depend on molecular shape, which in turn depends on the bonding in the molecule. In this chapter, we begin with a discussion of the shapes of molecules. We then turn to an examination of some properties of molecules and to two theories of covalent bonding, valence bond theory and molecular orbital theory, that explain some properties of molecules that the model we have already examined cannot.

Odor	Chemical and common examples	Molecular shape		Receptor shape
Camphoraceous	Hexachloroethane Camphor Mothballs	Spherical		
Musky	Xylene musk Pentadecanolactone Musk perfumes	Disk		
Floral	α-amyl pyridine Phenylethylmethylethyl carbinol Roses	Disk and tail		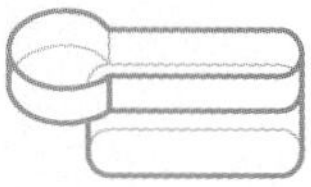
Pepperminty	Menthol Mint candy	Wedge		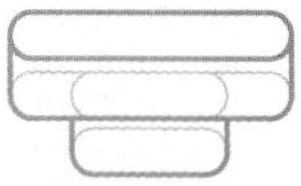
Ethereal	Diethyl ether Dry-cleaning fluids	Rod		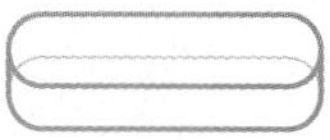
Pungent	Formic acid Vinegar	Simple (electron attracting)		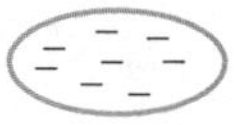
Putrid	Hydrogen sulfide Ethyl sulfide Rotten eggs	Simple (affinity for + centers)		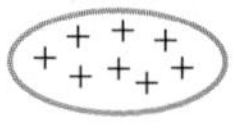

Figure 8.2 Primary odors, the corresponding receptors, and approximate molecular shapes that fit the receptors. Two of the receptors do not involve shape but use simple electrostatic attractions.

8.1   The Valence-Shell Electron-Pair Repulsion Model

Why do molecules have different shapes? How can we predict these shapes?

▶ In the chapter introduction, we saw that the shapes of molecules seem to determine their tastes and odors. Other properties of molecules also correlate with their shapes. The shapes of molecules can be largely explained by the assumption that electrons, being negatively charged, tend to avoid one another. This tendency has relatively little effect on the electrons within a pair because the distance between them is limited by the orbital they occupy. However, electron pairs in different orbitals stay as far apart as possible. The relative locations of the electron pairs determine what shape and thus what properties a molecule has. The tendency of electron pairs to adjust the orientation of their orbitals to maximize the distance between pairs is the basis of the **valence-shell electron-pair repulsion (VSEPR) theory.**

   To predict the geometric shape of a molecule (or ion) using the VSEPR model, we first need to know the molecule's electronic structure. Specifically, we need to know the number of groups of electrons around the central atom—information available from the Lewis formula. A group of electrons can be considered to be either a single unpaired electron or an unshared electron pair on the central atom, or the electrons involved in a bond between two atoms. Because they occupy the same general area in space between the bonded atoms, the electrons in a double or triple bond are considered to be one group, equivalent to the electron pair in a single bond. The number of bonding electron groups is thus equal to the number of atoms bonded to the central atom. We arrange the electron groups, both bonding and unshared, around the central atom as far apart as possible. The result is a shape characterized by certain angles between the central atom and the atoms bonded to it, called **bond angles.** These shapes are derived from the geometric arrangement of the electron groups around the central atom. Figure 8.3 shows the geometries that achieve the maximum distance between two to six electron pairs, along with the corresponding angles between them.

The geometries shown in this figure are adopted by any set of objects that are attached at a common point and that try to stay as far apart as possible. For

Figure 8.3 Structures predicted by the maximum separation of electron pairs around an atom (A).

Number of Electron Groups	Geometric Shape and Angles	Geometric Arrangement of Electron Groups
Two	Linear, 180°	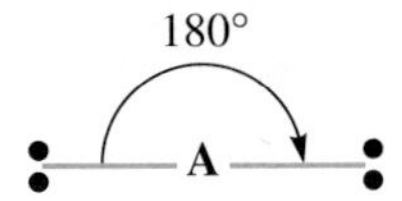
Three	Trigonal planar, 120°	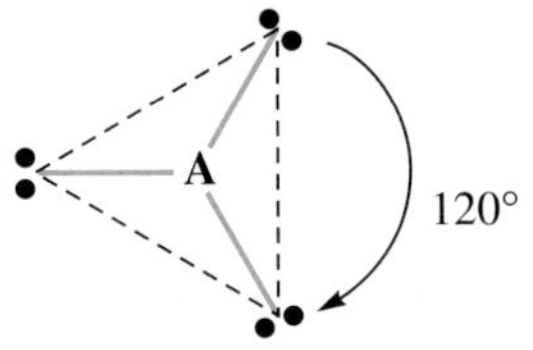
Four	Tetrahedral, 109.5°	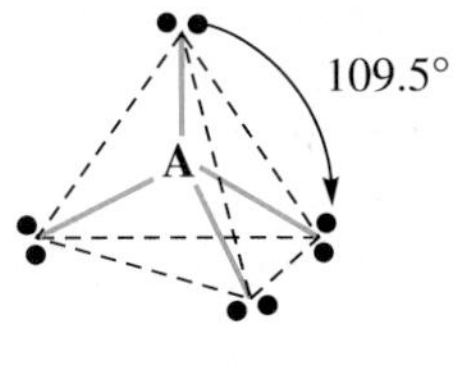
Five	Trigonal bipyramidal, 90° or 120°	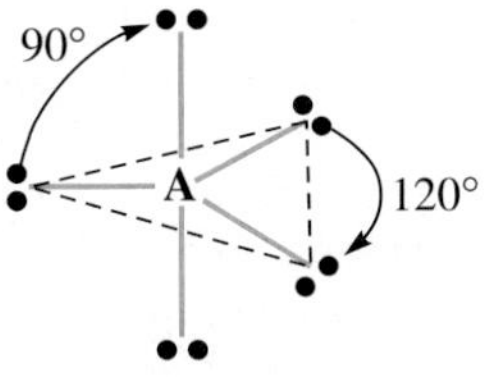
Six	Octahedral, 90°	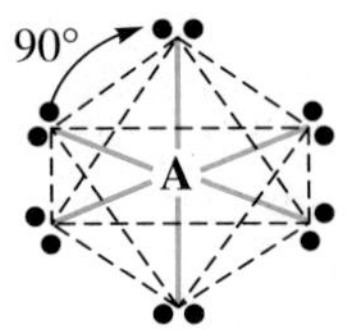

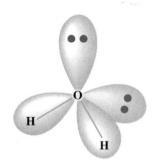

Figure 8.5 The four electron pairs in the water molecule arrange themselves around the oxygen atom in orbitals that point to the corners of a tetrahedron. Two of these positions are occupied by hydrogen atoms; the other two are occupied by unshared pairs of electrons that form part of the oxygen atom. The shape of this molecule is called bent, not tetrahedral, because the molecular shape depends on the locations of the atoms, not of all the electron pairs.

example, the plastic eggs shown in Figure 8.4 adopt the same geometries as pairs of electrons around a nucleus. These structures, which we will call *parent structures,* form the basis for identifying a molecule's shape. If the central atom contains any unshared pairs of electrons, the molecular shape is not that of the parent structure but is derived from it. The shape we assign a molecule includes only the bonding electron groups within the parent structure; the shape is a description of the locations of the atoms in the molecule.

Consider the water molecule, which has two bonding pairs and two unshared pairs of electrons. The four electron pairs in H_2O arrange themselves tetrahedrally around the oxygen atom, as described in Figure 8.3 and shown in Figure 8.5. But because we consider only the two bonding pairs when we describe the molecule's shape, we do not call the water molecule tetrahedral; rather, we describe it as bent. All possible shapes for molecules that have two to six electron groups are summarized in Figure 8.6.

VSEPR theory provides a basis for understanding the differences in the structures of related molecules. For example, CO_2 is linear, whereas SO_2, with the same number of atoms, is bent. The electronic configurations and shapes of these molecules are shown in Figure 8.7. Carbon in CO_2 has two bonding groups and no unshared electron pairs, so it has a linear structure. Sulfur in SO_2 has two bonding electron groups and one unshared electron pair, so the three electron groups lie at the corners of a planar triangle to maximize their separation. Two of the three corners of the triangle are occupied by oxygen atoms, so the shape of the molecule is properly described as bent. The unshared pair of electrons on sulfur makes the difference in the shapes—and thus, in part, the properties—of these two molecules.

Figure 8.6 Arrangement of electron pairs and molecular shapes.

Formula	Number of Electron Groups	Number of Bonding Groups	Number of Unshared Groups	Molecular Shape		Examples
AB_2	2	2	0	Linear 		$HgCl_2$, $BeCl_2(g)$, CO_2, HCN
AB_3	3	3	0	Trigonal planar 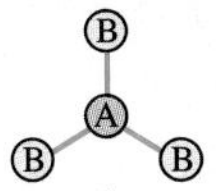		BF_3, BCl_3, BH_3, SO_3
AB_2	3	2	1	Bent 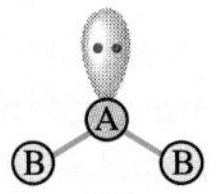		$SnCl_2$, SO_2, NO_2^-
AB_4	4	4	0	Tetrahedral 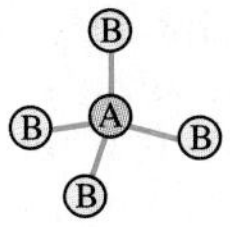		CH_4, $SiCl_4$, $POCl_3$
AB_3	4	3	1	Trigonal pyramidal 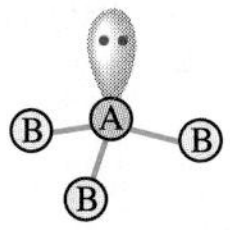		NH_3, PF_3
AB_2	4	2	2	Bent 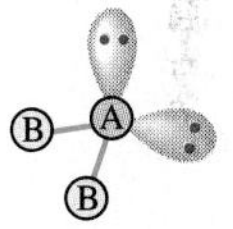		H_2O, ICl_2^+, F_2O, BrO_2, SCl_2
AB_5	5	5	0	Trigonal bipyramidal 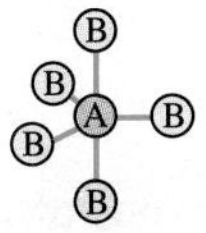		PF_5, PCl_5, SbF_5, $IO_3F_2^-$
AB_4	5	4	1	Distorted tetrahedral 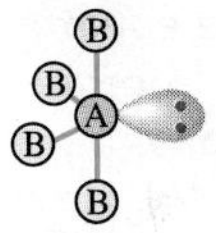		SF_4, IF_4^+
AB_3	5	3	2	T-shape 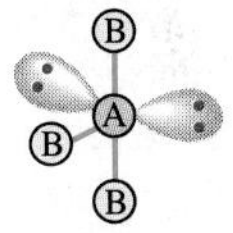		ClF_3, BrF_3
AB_2	5	2	3	Linear 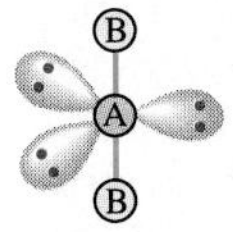		I_3^-, ICl_2^-, XeF_2
AB_6	6	6	0	Octahedral 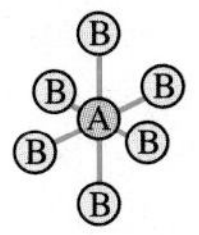		SF_6, PF_6^-
AB_5	6	5	1	Square pyramidal 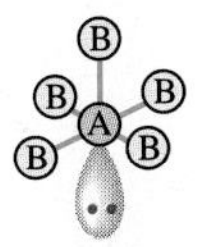		IF_5, BrF_5, $XeOF_4$
AB_4	6	4	2	Square planar 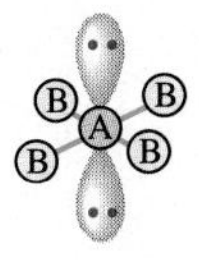		XeF_4, BrF_4^-

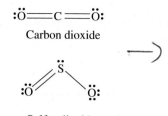

Carbon dioxide

Sulfur dioxide

Figure 8.7 Carbon dioxide is linear, and sulfur dioxide is bent. The difference arises from the unshared pair of electrons on the sulfur in sulfur dioxide, a feature not found on carbon in carbon dioxide.

Of all the parent structures, only the trigonal bipyramid contains nonequivalent positions; that is, it is the only one in which not all the bond angles are the same. Two positions, called **axial,** are opposite one another and 90° from the triangular plane outlined by the other three positions, which are called **equatorial,** as shown in Figure 8.8. The equatorial positions are 120° from one another. For some molecules, we may need to decide which type of position an unshared pair will occupy. Because an unshared electron pair is under the influence of only one atomic nucleus, whereas bonded electrons are under the influence of two nuclei, an unshared pair occupies more space than a shared pair. For this reason, it will occupy a position—one in the equatorial plane—that minimizes the number of 90° interactions with other electron pairs.

In real molecules, unshared pairs of electrons occupy more space than bonded pairs, so they repel bonded pairs somewhat, resulting in bond angles that are slightly smaller than VSEPR theory would predict. The deviation is about 2° per unshared pair. Thus, ammonia, with one unshared pair of electrons, has a bond angle of 107° rather than 109.5°. Water, with two unshared pairs of electrons, has a bond angle of about 105°.

To predict the shape of a molecule according to VSEPR theory, we can follow four steps:

1. Draw any valid Lewis electron-dot formula (as described in Chapter 7).
2. Count bonding electron groups (equal to the number of atoms bonded to the central atom), and count unshared electron pairs.
3. Add the numbers of bonding electron groups and unshared electron pairs. The total gives the parent structure.

Figure 8.8 Two types of positions—axial and equatorial—are available to orbitals arranged in a trigonal bipyramidal shape *(A)*. Unshared electron pairs, which need more space, are located in the equatorial orbitals, resulting in the distorted tetrahedral *(B)*, T-shape *(C)*, and linear *(D)* shapes that are derived from the trigonal bipyramidal parent structure.

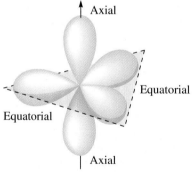

A. Trigonal bypyramid

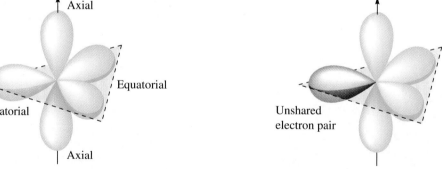

B. Distorted tetrahedron

C. T-shape

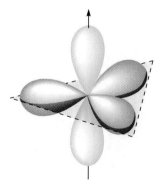

D. Linear

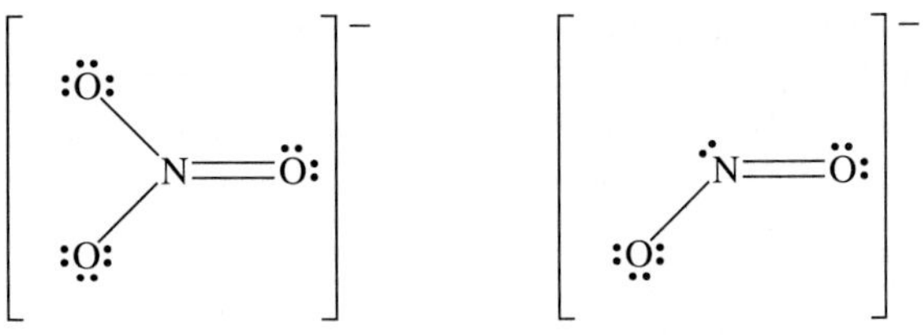

$$\left[\begin{array}{c} \ddot{\text{O}} \\ \text{N} = \ddot{\text{O}} \\ \ddot{\text{O}} \end{array} \right]^{-} \qquad \left[\begin{array}{c} \dot{\text{N}} = \ddot{\text{O}} \\ \ddot{\text{O}} \end{array} \right]^{-}$$

A. Nitrate ion   **B. Nitrite ion**

Figure 8.9 (A) Because the nitrate ion has three groups of bonding electrons and no unshared pairs, the ion has a trigonal planar shape. (B) The nitrite ion has two groups of bonding electrons and one unshared pair, so it is bent.

4. The molecular shape is derived from the parent structure by considering the number of positions in this structure that are occupied by bonded atoms.

Let's predict the shape of the nitrate ion, NO_3^-, by this procedure.

1. First, we write a valid Lewis formula:

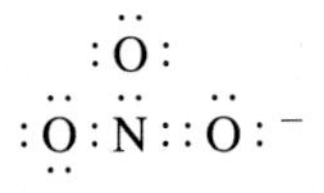

$$\ddot{\text{O}} \\ :\ddot{\text{O}}:\text{N}::\ddot{\text{O}}:^{-}$$

There are two other resonance forms, which have the double bond in other positions, but we need obtain only one Lewis formula.

2. Next, we count the bonding electron groups and unshared electron pairs. The nitrogen atom has three groups of bonding electrons and no unshared pairs of electrons around it.
3. The total number of electron groups is three, so the parent structure is trigonal planar.
4. Because there are no unshared electron pairs around nitrogen, the shape of this ion is the same as the parent structure, as shown in Figure 8.9A.

Now consider the nitrite ion, NO_2^-.

1. A valid Lewis formula is

$$:\ddot{\text{O}}::\text{N}:\ddot{\text{O}}:^{-}$$

2. The nitrogen atom is bonded to two oxygen atoms, so it has two bonding electron groups. It also has one unshared pair of electrons.
3. Because there are three groups of electrons around the nitrogen, the parent structure is trigonal planar, as in NO_3^-.
4. But one position is occupied by an unshared pair of electrons, so only two corners of the trigonal plane are occupied by atoms. The shape of the ion is bent (Figure 8.9B).

Nitrates occur naturally in deposits of Chile saltpeter, $NaNO_3$. The nitrate ion is colorless, so its salts are colorless unless the metal ion in the salt has a color. It is used in the manufacture of nitric acid and sulfuric acid, in pottery enamels, in matches, and in fertilizer. Because the nitrate salts of most metals are soluble in water, many ionic compounds are prepared in the form of nitrates.

EXAMPLE 8.1

Predict the shape of SF_4.

Solution

1. A valid Lewis formula is

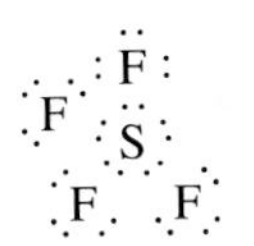

(continued)

Nitrite, like nitrate, is a colorless ion. Sodium nitrite is a white solid that slowly oxidizes to sodium nitrate when exposed to air. It is used in preparing dyes and in curing meat. There are limits on the amount of nitrite that is allowed in food because it can deactivate red blood cells in young children.

2. The sulfur atom is bonded to four fluorine atoms and has one unshared pair of electrons.
3. The parent structure for five electron groups is a trigonal bipyramid.
4. The unshared electron pair occupies one position in the triangular plane, so the fluorine atoms occupy two equatorial positions and both axial positions. This shape is called a distorted tetrahedron.

Practice Problem 8.1

Predict the shape of BrF_3.

Answer: T-shape

8.2   Dipole Moments

What conditions are necessary for a molecule to be polar?

In Chapter 7, we saw that bonds could be polar if the bonded atoms shared electrons unequally, resulting in the formation of partial charges on the atoms. Molecules can also be polar if the electrons are distributed unequally and one end of the molecule has a partial charge relative to the other end.

Before looking more closely at how molecular shape can affect polarity, let's consider how polarity is measured. Polarity of a bond shows up as a separation of charge within the bond, the more electropositive atom being partially positive and the more electronegative atom partially negative. The degree of charge separation in such bonds is measured experimentally as the **dipole moment,** symbolized by the Greek letter mu, μ. The dipole moment is the product of the charge magnitude (q) and the distance between the charges, the bond length (d):

$$\mu = q \times d$$

The dipole moment is usually measured in units of debye (D), named after the chemist Peter Debye. The debye is related to the charge in coulombs (C) and the distance in meters by the equation

$$1 \text{ D} = 3.34 \times 10^{-30} \text{ C m}$$

Experimental values of the dipole moments of the hydrogen halides are listed in Table 8.1.

The dipole moment is related, though not simply and only in part, to the amount of ionic character in the bond of a diatomic molecule. Unshared electron pairs apparently also contribute to the dipole moment. If a bond were completely

Table 8.1 Dipole Moments of the Hydrogen Halides

Hydrogen Halide	Dipole Moment (D)	Bond Length (pm)	Electronegativity Difference
HF	1.91	92	1.9
HCl	1.03	127	0.9
HBr	0.79	141	0.7
HI	0.38	161	0.4

ionic, there would be a complete separation of charges to give ions. On the other hand, a nonpolar covalent bond would have no charge separation and the dipole moment would be zero. Most polar bonds lie between these two extremes and have some covalent character and some ionic character.

Electronegativity and Bond Polarity

Chapter 7 indicated that the greater the difference in electronegativity between two bonded atoms, the more polar the bond formed between them. This can be seen for the hydrogen halides in Table 8.1. The greater the electronegativity difference, the greater the dipole moment and the more polar the bond. A general rule suggests that a bond that is 50% ionic correlates with an electronegativity difference of 1.7. A bond between atoms whose electronegativity difference equals at least 1.7 is considered predominantly ionic, whereas a difference of less than 1.7 corresponds to a predominantly covalent bond. Of course, in either case the bond generally has characteristics of both types of bonding.

Polar and Nonpolar Polyatomic Molecules

Will a molecule with polar bonds necessarily also be polar?

In a diatomic molecule, the dipole moment lies along the direction of the bond. Because there are only two atoms, one atom must be partially positive ($\delta+$) and the other must be partially negative ($\delta-$). This can be represented by an arrow, with the $+$ sign at the position of partial positive charge and the arrowhead at the position of partial negative charge:

$$\overset{\delta+ \quad \delta-}{\text{H---Cl}}$$
$$\underset{}{+\!\longrightarrow}$$

Dipole moments, then, have not only quantity but also direction. In the diatomic molecules we have been considering so far, the dipole moment of the bond and the dipole moment of the molecule are identical because there is only one bond.

The BH_3 molecule is not stable and cannot be isolated, but it has been detected and is thought to exist fleetingly during transformations of other boron–hydrogen compounds. The simplest stable boron hydride is diborane, B_2H_6.

In polyatomic molecules, however, the dipole moments of the bonds and the dipole moment of the molecule may be different. A polyatomic molecule that has nonpolar bonds cannot be polar. For example, in BH_3, boron and hydrogen have nearly the same electronegativity, so each bond is nonpolar and the molecule must be nonpolar. Carbon–hydrogen bonds are also almost nonpolar, so hydrocarbons are generally considered to be essentially nonpolar. But a polyatomic molecule with polar bonds may be polar or nonpolar, depending on its geometry. The dipole moments of some molecules are listed in Table 8.2.

Beryllium chloride, $BeCl_2$, is a white to faintly yellow solid. It is very soluble in water, where it hydrolyzes to $Be(OH)_2$ and HCl. In a vacuum, $BeCl_2$ vaporizes at 300°C to form gaseous covalent molecules. Along with beryllium metal and most of its compounds, $BeCl_2$ is quite toxic. It is used mostly to prepare beryllium metal for controlling neutrons in nuclear reactors.

Consider the gaseous molecule beryllium chloride ($BeCl_2$). Although this molecule has two polar Be—Cl bonds, it is nonpolar. The bonds in $BeCl_2$ are 180° apart, and the dipoles point in opposite directions. Thus, the charge separation places negative charge at the two chlorine positions and positive charge at the beryllium position:

$$\overset{\delta- \quad \delta+ \quad \delta-}{\text{Cl ---Be ---Cl}}$$
$$\longleftarrow\!+\!+\!\longrightarrow$$

No net dipole

Table 8.2 Values of the Dipole Moment for Various Molecules

Compound	Dipole Moment (D)
HCN	2.93
HCl	1.03
HBr	0.79
HI	0.38
H_2O	1.85
H_2S	0.95
NH_3	1.49
SO_3	0.00
SO_2	1.61
CO_2	0.00
CO	0.12
NO	0.16
CH_3F	1.81
CH_3Cl	1.87
CH_3Br	1.80
CH_3I	1.64
C_2H_5Cl	2.05
CHF_3	1.61
CH_2Cl_2	1.58
CH_3OH	1.69
C_2H_5OH	1.69

Draw a picture showing how the electrons would be distributed in the Be—Cl bonds in the gaseous $BeCl_2$ molecule.

Sulfur hexafluoride is a very stable, dense, colorless, odorless gas. It is used as an insulating medium in electrical circuit breakers.

The dipoles are exactly equal in magnitude and opposite in direction, so they cancel out. This leaves the molecule as a whole with a dipole moment of zero, so the molecule is nonpolar.

Any molecule in which the relationship of the bonding pairs is completely symmetrical, such as in one of the parent geometries identified in Figure 8.3 (linear, trigonal planar, tetrahedral, trigonal bipyramidal, or octahedral), experiences this sort of cancellation and so is nonpolar. Thus, cancellation of dipoles also occurs in planar triangular molecules such as BF_3, tetrahedral molecules such as CH_4 and CF_4, trigonal bipyramidal molecules such as PF_5, and octahedral molecules such as SF_6. In all these molecules, the central atom is surrounded by identical atoms. We can think of all these shapes as symmetrical.

However, if the atoms surrounding the central atom are not identical, the molecule's shape is not symmetrical. Such molecules are polar if the bonds are polar. ▶ Chloroform, $CHCl_3$, one of the sweet molecules mentioned in the chapter introduction, is polar, as are CH_3Cl and CH_2Cl_2. In these molecules, the atoms surrounding the carbon are not identical, so the polarity of the bonds does not cancel, resulting in a net polarity for the molecule.

WORLDS OF CHEMISTRY

Adhesives: Holding Things Together with Chemical Bonds

We have a wide array of modern adhesives that can hold almost anything together, from airplane parts to human skin, even though no general theory of adhesion can explain exactly how they work.

Oil spills, such as the massive spill off the coast of Alaska caused by an accidental grounding of the *Exxon Valdez*, are ecological disasters. Beaches become coated with gooey, thick substances. Fish die. Birds become coated with oil, as shown in Figure 8.10, and they can no longer fly. The oil can be cleaned from the birds and other animals and from the beaches, but only with great effort. But why does the oil stick to sand and animals?

Sticking two surfaces together can be accomplished with a group of substances called *adhesives* that include glues, pastes, gums, and numerous synthetic materials. The adhesion was once thought to take place by filling in minute holes and crevices in the surfaces. Although this may play a small role, adhesion is now thought to arise primarily from chemical forces that hold atoms and molecules of the materials together. The strength of these forces depends on factors related to the structure of the adhesive molecules, such as polarity. These same adhesive forces cause oil to stick to bird feathers.

Natural adhesives are made from animals or plants. For many years, glue has been made of gelatin obtained by boiling down animal bones, sinews, and skins. Such glue is sold as a solid and must be melted before use. Vegetable products include mucilage, made of starch and used to make light adhesives such as wallpaper paste and office paste. The adhesive on envelopes and stamps is gum arabic, made from the sap of the acacia tree. Natural rubber, dissolved in an appropriate solvent, is used in the rubber cement that sticks paper, rubber, and leather together. Clear adhesives, such as household cement and modeling glue, are made from cellulose, found in all plants.

A variety of synthetic adhesives provide strong bonds between materials that natural adhesives do not bond well, such as metals. These adhesives fall into two classes: thermosetting adhesives that become hard when heated and thermoplastic adhesives that melt when heated. Among the thermosetting adhesives are the phenolics, used extensively to make plywood and particleboard, and epoxy resins, which set when a hardening agent is added. The thermo-

Figure 8.10 Oil sticks to bird feathers because of attractions between the molecules that make up the oil and the molecules that make up the feathers.

plastic adhesives include the vinyl resins, such as those used to adhere layers of glass in automobile windshields.

Another class of adhesives includes the modern superglues, which are cyanoacrylate resins that rapidly bond to almost any material. Although most adhesives must be applied to a clean, dry surface, the cyanoacrylates can actually be sprayed on human skin to suture an incision after surgery. The bleeding stops and the incision is closed.

We have a wide array of modern adhesives that can hold almost anything together, from airplane parts to human skin, even though no general theory of adhesion can explain exactly how they work. The nature of the interactions between the adhesives and the surfaces to which they stick is not yet thoroughly understood.

Questions for Discussion

1. Lubricants are used to prevent the adhesion of two metal surfaces. Suggest an explanation of how lubricants might work.

(continued)

2. Waterproofing fabrics involves exposing the fabric to some chemical substance such as $(CH_3)_3SiCl$. The molecules in the fabric bond to the silicon. Which part of the molecule, the CH_3— or the Cl end, must form the new outer coating of the fabric to repel the water?

3. Discuss at a molecular level why a grease spot might prevent a glue from working.

4. Discuss whether economic factors or ecological factors are more important in determining whether oil should be transported by sea or by pipeline. Should offshore drilling for oil be allowed?

EXAMPLE 8.2

Predict whether H_2S is polar or nonpolar.

Solution   The molecule H_2S has polar bonds because there is a difference of 0.4 in electronegativity between the sulfur and the hydrogen. There are four electron pairs around the sulfur, only two of which are used in bonds, so the molecule is bent. Because the bonds are polar and the molecule is not perfectly symmetrical, the molecule is polar.

Practice Problem 8.2

Predict whether C_2H_6, NO_2, CO_2, SO_2, and SO_3 are polar or nonpolar.

Answer: nonpolar, polar, nonpolar, polar, nonpolar

Polarity is important in determining many properties related to molecular interactions. For example, polar molecules condense from the gaseous state to the liquid state at higher temperatures than do nonpolar molecules. Polar liquids freeze to solids at higher temperatures than do nonpolar liquids. Ionic salts and polar liquids dissolve better in polar liquids than in nonpolar liquids. Conversely, nonpolar liquids dissolve better in other nonpolar liquids than in polar liquids.

8.3   Valence Bond Theory

How can we explain why atoms bond together to form molecules with their specific geometries?

Chapter 7 described the formation of a covalent bond in terms of the donation of an unpaired electron by each of two atoms to form a shared electron pair and represented this phenomenon by Lewis electron-dot formulas. While this approach explains the formation of bonds, it tells us little about the shapes of molecules. On the other hand, VSEPR theory predicts and explains the shapes of molecules but does not address the formation of bonds. We turn now to valence bond theory, which is concerned with both the formation of bonds and the shapes of molecules.

▶ Understanding the biological functions of molecules, such as those involved in the senses of taste and smell, requires that we consider both bond formation and shape.

Figure 8.11 The formation of a bond between two hydrogen atoms by overlap of their 1*s* orbitals.

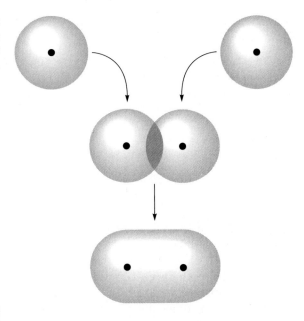

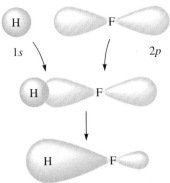

Figure 8.12 Formation of a bond between hydrogen and fluorine atoms by overlap of the hydrogen 1*s* orbital with the fluorine 2*p* orbital.

Orbital Overlap

Valence bond theory is an explanation of covalent bond formation based on quantum theory. Its central premise is that covalent bonds are formed when half-filled orbitals in the outer, or valence, shells of two atoms overlap so that a pair of electrons, one electron from each atom, occupies the overlapped orbital. The two electrons must be of opposite spin. The formation of such a bond between two hydrogen atoms is pictured in Figure 8.11. Each atom has one electron in a 1*s* orbital. When the atoms approach closely enough, the two orbitals overlap and a bond is formed, partly because of the attraction between the nuclei and the electrons. The probability of finding the electrons in the regions of overlap between the two nuclei is greater than the probability of finding them outside these regions, which helps cancel repulsions between nuclei and maximizes attractions between the electrons and both nuclei.

The description of bonds forming by simple orbital overlap works well for any diatomic molecule. For example, the formation of hydrogen fluoride can be visualized in terms of the overlap of the half-filled 1*s* orbital on hydrogen with the half-filled 2*p* orbital on fluorine (Figure 8.12). The bond in the fluorine molecule, F_2, is formed by the overlap of a half-filled 2*p* orbital on each fluorine atom (Figure 8.13).

Each of the bonds in Figures 8.11, 8.12, and 8.13 is a **sigma (σ) bond,** a covalent bond in which the electron density is greatest along the bond axis, a line between the two nuclei. If viewed down the bond axis, combinations involving *s* and *p* orbitals, like those in Figures 8.11, 8.12, and 8.13, look circular, just as a cross section of an atomic *s* orbital looks circular.

Covalent bonds can also form by a side-to-side overlap of *p* orbitals, as shown in Figure 8.14. The result is a **pi (π) bond,** in which the greatest electron density lies above and below the internuclear axis. Consider the bonding between nitrogen atoms having the electronic configuration $1s^2 2s^2 2p_x{}^1 2p_y{}^1 2p_z{}^1$. The three unpaired electrons on each atom are located in perpendicular *p* orbitals, which are oriented

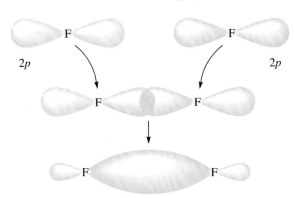

Figure 8.13 Formation of a bond between two fluorine atoms by overlap of their $2p$ orbitals.

so that if one end-to-end p orbital overlap occurs (resulting in a sigma bond, as in Figure 8.13), the other two p orbitals cannot overlap in the same fashion. Rather, they are aligned parallel to the corresponding orbital on the other atom. In N_2, the three bonds between the two nitrogen atoms consist of one sigma bond and two pi bonds, as shown in Figure 8.15. In general, a single bond is a sigma bond, a double bond is a combination of a sigma bond and a pi bond, and a triple bond is a sigma bond and two pi bonds.

What bond angles would be expected for polyatomic molecules formed by the overlap of s and p orbitals?

The simple picture of orbitals overlapping to form covalent bonds does not work so well for polyatomic molecules and ions. One difficulty is illustrated by the bonding in the largely covalent gaseous molecule $BeCl_2$. Beryllium has an electronic configuration of $1s^2 2s^2$, with no unpaired electrons. In order for it to form covalent bonds, it must have unpaired electrons. At least conceptually, an electron could be moved to a $2p$ orbital, giving the configuration $1s^2 2s^1 2p^1$. The two unpaired electrons in this configuration could be shared by overlap of the $2s$ beryllium orbital with a $3p$ orbital on one chlorine atom and of the $2p$ beryllium orbital with a $3p$ orbital on the other chlorine atom. The $3p$ chlorine orbital is the one containing chlorine's unpaired electron. Because the energies of a $2s$ orbital and a $2p$ orbital are different, it might be expected that the two bonds formed would have different bond strengths and different bond lengths. Such is not the case, however. The two bonds in $BeCl_2$ are identical in properties.

Another difficulty involves bond angles. Oxygen, for example, has a valence electronic configuration of $2s^2 2p^4$, whereas that of nitrogen is $2s^2 2p^3$. Bond formation of oxygen with hydrogen to make water or of nitrogen with hydrogen to make ammonia could be visualized in terms of the $1s$ orbital of hydrogen overlapping the half-filled $2p$ orbital of oxygen or nitrogen, as shown in Figure 8.16. This picture may look reasonable, but the bond angles present a problem. The bond angles shown for each molecule are 90°, as would be expected from the fact that the p orbitals are perpendicular. But as actually measured, the H—O—H angle in water is 104.5°, and the H—N—H angle in ammonia is 107°.

Valence bond theory must be modified to explain both the equivalence of bonds apparently formed by electrons in different types of orbitals and the discrepancy between the bond angles predicted by the theory and those actually observed. The concept of hybridization is useful in this regard.

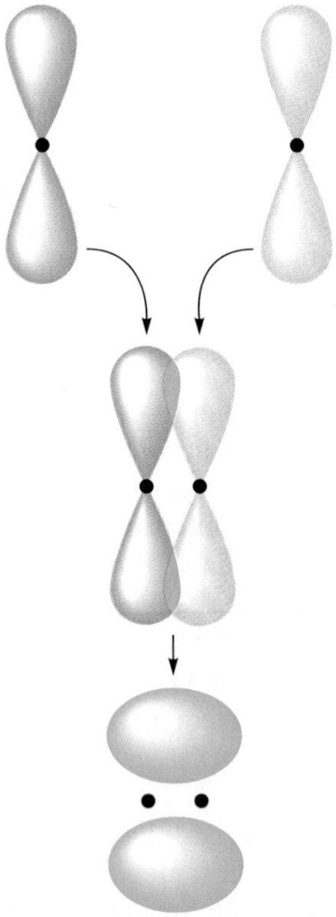

Hybridization

Figure 8.14 The sideways overlap of two atomic p orbitals to give a pi bond.

The problems just described arise because of an invalid assumption that electrons in the presence of two or more nuclei in a molecule behave the same as they would

Figure 8.15 A valence bond picture of the bonding in the dinitrogen molecule. One *p* orbital from each nitrogen atom overlaps in an end-to-end fashion to give a sigma bond. The other two *p* orbitals on each atom overlap in a side-to-side fashion to form two pi bonds.

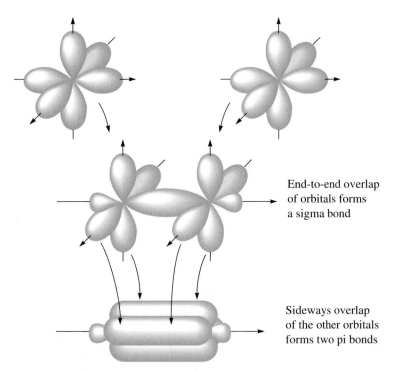

End-to-end overlap of orbitals forms a sigma bond

Sideways overlap of the other orbitals forms two pi bonds

in the presence of one nucleus in an atom. In fact, the presence of another nucleus changes the way an electron travels in the space near the nuclei. The process of **hybridization** adjusts for this change by combining two or more atomic orbitals from the same atom, as shown in Figure 8.17.

We can examine the energy changes involved in hybridization by dividing the process into two hypothetical steps. We begin by assuming that all the electrons are in ground-state orbitals (the orbitals of lowest energy). These electrons are energized to unpair as many electrons as needed to form the proper number of bonds, giving an excited-state (higher-energy) electronic configuration. Usually, valence *s* and *p* electrons are unpaired to the maximum extent possible in these excited-state orbitals. This unpairing consumes energy, but the formation of additional bonds will repay this energy. To equalize the energy of the orbitals and to

Figure 8.16 A valence bond picture of the bonding in water and ammonia assuming that oxygen and nitrogen 2*p* orbitals are overlapped with hydrogen 1*s* orbitals. The predicted bond angles of 90° are not correct.

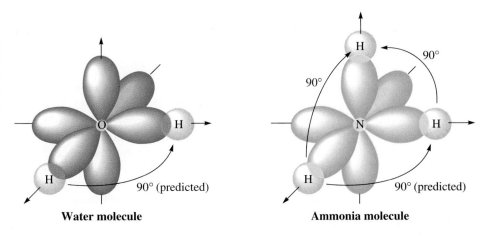

Water molecule   **Ammonia molecule**

Figure 8.17 Formation of hybrid orbitals from atomic orbitals. Here, one *s* and three *p* orbitals are mixed to produce four hybrid orbitals (called *sp*³).

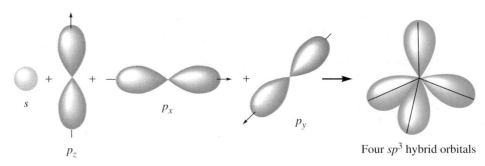

Four *sp*³ hybrid orbitals

achieve proper bond angles, the orbitals that contain electrons are then mixed together to give new orbitals, each containing one or two electrons. The new orbitals, called **hybrid orbitals,** are the same in number as the atomic orbitals that combined to make them; but unlike the original orbitals, they are all identical. Thus, they give bonds with identical bond lengths and bond energies. The angles between the hybrid orbitals are as large as possible, allowing the electrons to avoid one another as much as possible, resulting in the same bond angles that are predicted by VSEPR theory. The process can be summarized as follows:

$$\text{Ground-state atomic orbitals} \xrightarrow[\text{electrons}]{\text{unpair}} \text{Excited-state atomic orbitals} \xrightarrow{\text{mix}} \text{Hybrid orbitals}$$

Methane, CH₄, is a colorless, odorless, flammable gas. It is the principal component (about 85%) of natural gas. It is also known as marsh gas because it results from the decay of plant matter in the absence of oxygen, an environment common to marshes. Methane is the major constituent of the atmospheres of several planets: Jupiter, Saturn, Uranus, and Neptune.

Consider methane as an example of this process. Experiments show that this molecule has four equivalent C—H bonds pointing to the corners of a tetrahedron, with bond angles of 109.5°. Atomic carbon has a valence electronic configuration of $2s^2 2p^2$. The two electrons in the 2*s* orbital must be unpaired if they are to be shared with an electron in the hydrogen 1*s* orbital, allowing the formation of four bonds. This gives an excited-state electronic configuration:

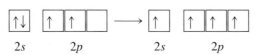

An equal number of hybrid orbitals is formed from the excited-state orbitals that contain electrons:

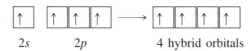

Now four equivalent C—H bonds pointing in the correct directions can be formed by the overlapping of the hydrogen 1*s* orbitals with the carbon hybrid orbitals.

The hybrid orbitals are named according to the number and kind of orbitals used to make them. Thus, the mixing of an *s* orbital with a *p* orbital gives two *sp* hybrid orbitals, and the mixing of an *s* orbital with three *p* orbitals gives four *sp*³ hybrid orbitals (the superscript indicates how many *p* orbitals were used).

Some common hybrid orbitals are given in Table 8.3. As you can see, these orbitals are associated with characteristic bond angles and geometries. Let's look more closely at how hybridization leads to these molecular shapes.

Table 8.3 Properties of Some Common Hybrid Orbitals

Hybrid Orbital	Component Orbitals	Angle Between Orbitals	Geometry of Molecule
sp (2)	1 s, 1 p	180°	Linear
sp^2 (3)	1 s, 2 p	120°	Trigonal planar
sp^3 (4)	1 s, 3 p	109.5°	Tetrahedral
dsp^2 (4)	1 d, 1 s, 2 p	90°	Square planar
dsp^3 (5)	1 d, 1 s, 3 p	120°, 90°	Trigonal bipyramidal
d^2sp^3 (6)	2 d, 1 s, 3 p	90°	Octahedral

Hybridization and Bond Angles

Boron has a valence electronic configuration of $2s^2 2p^1$. To form three bonds, the electrons must be unpaired and hybrid orbitals must be formed:

$$\text{B } (2s^2 2p^1) \longrightarrow \text{B } (2s^1 2p^1 2p^1) \longrightarrow \text{B } (\{sp^2\}^1 \{sp^2\}^1 \{sp^2\}^1)$$

The result of hybridization is the formation of three sp^2 hybrid orbitals that point to the corners of an equilateral triangle and so are 120° apart. Each orbital holds one electron, which it can share by overlapping with a half-filled orbital on another atom, such as chlorine:

$$\text{B } (\{sp^2\}^1 \{sp^2\}^1 \{sp^2\}^1) + 3\text{Cl } (3s^2 3p^5) \longrightarrow \text{BCl}_3$$

The bonding in triangular planar boron trichloride is shown in Figure 8.18.

A very common bonding pattern involves sp^3 hybrid orbitals. Such hybridization is used by carbon, nitrogen, oxygen, and many of their family members in a large variety of compounds, including those found in living organisms. Consider the hybridization of carbon:

How is the bonding in water and ammonia related to that in methane?

$$\text{C } (2s^2 2p^2) \longrightarrow \text{C } (2s^1 2p^1 2p^1 2p^1) \longrightarrow \text{C } (\{sp^3\}^1 \{sp^3\}^1 \{sp^3\}^1 \{sp^3\}^1)$$

Figure 8.18 The bonding in boron trichloride. According to valence bond theory, this molecule is formed by overlap of chlorine 3p orbitals with boron sp^2 hybrid orbitals.

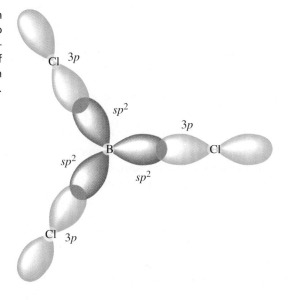

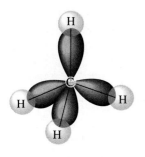

A. Methane, CH$_4$

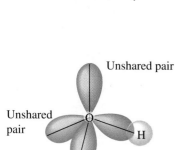

Unshared pair

Unshared pair

H

B. Water, H$_2$O

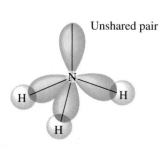

Unshared pair

H   H

H

C. Ammonia, NH$_3$

Figure 8.19 The bonding in methane, water, and ammonia. According to valence bond theory, these molecules are formed by the overlap of hydrogen 1s orbitals with sp^3 hybrid orbitals on the other atom.

With an unpaired electron in four hybrid orbitals, carbon can make four equivalent bonds by overlap with half-filled orbitals of other atoms—for example, with hydrogen:

$$\text{C } (\{sp^3\}^1\{sp^3\}^1\{sp^3\}^1\{sp^3\}^1) + 4\text{H } (1s^1) \longrightarrow \text{CH}_4$$

The bonding in CH$_4$ is shown in Figure 8.19A.

Some atoms in which sp^3 hybridization seems to occur cannot make four bonds because some of the hybrid orbitals are filled. Examples are nitrogen and oxygen:

$$\text{N } (\{sp^3\}^2\{sp^3\}^1\{sp^3\}^1\{sp^3\}^1)$$
$$\text{O } (\{sp^3\}^2\{sp^3\}^2\{sp^3\}^1\{sp^3\}^1)$$

How do we know these atoms use sp^3 hybridization if we cannot tell from the number of bonds they form? It is suggested by the shapes of the molecules they form, such as H$_2$O and NH$_3$. For example, water (Figure 8.19B) has two O—H bonds with a bond angle of 104.5°. Although sp hybrids could form the necessary two bonds, the resulting bond angle would be 180°. Similarly, ammonia (Figure 8.19C) has bond angles of 107° rather than the 120° that would be expected if it used sp^2 hybridization.

It appears, then, that in forming a molecule an atom uses hybridization appropriate to the total number of electron pairs around that atom in the molecule. In water, there are two bonding pairs and two unshared pairs of electrons around the oxygen atom. Because four pairs of electrons must be accommodated, sp^3 hybridization is used.

Hybridization and Multiple Bonds

Earlier, this section described multiple bonds for diatomic molecules in terms of the overlap of atomic p orbitals, either end-to-end to form sigma bonds or side-to-side to form pi bonds. Although we did not need to invoke hybridization to explain the bonding in diatomic molecules, in polyatomic molecules we generally must assume that an atom hybridizes its orbitals to form equivalent bonds rather than simply overlapping ground-state atomic orbitals. The central atom in polyatomic molecules that contain double or triple bonds undergoes that hybridization which is needed to form the sigma bonds. The pi bonds are formed by side-to-side overlap of unhybridized p orbitals. Thus, in a molecule that contains a double bond, one of the p orbitals is not used in the formation of hybrid orbitals, so the hybrids formed are sp^2 hybrids:

$$2s^22p^2 \longrightarrow 2s^12p^12p^12p^1 \longrightarrow \{sp^2\}^1\{sp^2\}^1\{sp^2\}^12p^1$$

Consider ethylene, C$_2$H$_4$, which contains a C=C double bond. The H—C—H bond angle is about 120°, which is characteristic of sp^2 hybridization. The formation of sigma and pi bonds in ethylene is shown in Figure 8.20A.

In a molecule that contains a triple bond, such as acetylene, C$_2$H$_2$, two of the p orbitals are not used in hybridization, so sp hybrids are formed:

$$2s^22p^2 \longrightarrow 2s^12p^12p^12p^1 \longrightarrow \{sp\}^1\{sp\}^12p^12p^1$$

The unhybridized p orbitals are used in pi-bond formation, and the sigma bonds are formed with sp hybrid orbitals. The bonding in acetylene is illustrated in Figure 8.20B.

Figure 8.20 Hybrid orbitals in double and triple bonds. *(A)* Ethylene (C_2H_4) uses sp^2 hybridization for the sigma bonds and unhybridized *p* orbitals for pi-bond formation. *(B)* Acetylene (C_2H_2) uses *sp* hybridization for the two sigma bonds and two unhybridized *p* orbitals on each carbon for pi-bond formation.

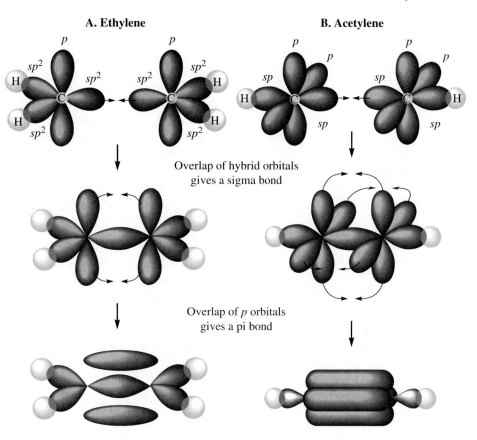

A. Ethylene

B. Acetylene

Overlap of hybrid orbitals gives a sigma bond

Overlap of *p* orbitals gives a pi bond

If an atom forms two double bonds to other atoms, the pi bonds of these double bonds must be formed by side-to-side overlap of two *p* orbitals on the central atom with a *p* orbital on each of the double-bonded atoms. The two pi bonds are perpendicular to one another, as shown in Figure 8.21 for carbon dioxide.

EXAMPLE 8.3

Describe the bonding in carbon dioxide in terms of the carbon orbitals used. The carbon–oxygen bonds are both double bonds, O=C=O.

Solution The carbon must make two sigma bonds to the oxygen atoms, so *sp* hybridization must be used. The double bond to each carbon is the result of the overlap of a *p* orbital on carbon with a *p* orbital on oxygen. The carbon atom must make two such bonds, so it uses two perpendicular *p* orbitals.

Practice Problem 8.3

Describe the bonding in carbon monoxide in terms of the carbon orbitals used. The carbon–oxygen bond is a triple bond, C≡O.

Answer: *sp* orbitals for the sigma bond and the unshared pair on carbon; overlap of *p* orbitals for the two pi bonds

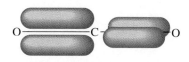

Figure 8.21 The pi bonds in carbon dioxide are formed by the overlap of unhybridized *p* orbitals on the carbon atom with *p* orbitals on the two oxygen atoms. Because two carbon *p* orbitals must be used, they will be perpendicular, and the resulting pi bonds are also perpendicular.

Hybridization and Delocalized Orbitals

Unhybridized p orbitals for pi bonds along with hybridized orbitals for sigma bonds allow for the spreading out, or delocalization, of electrons that was represented by resonance forms in Lewis formulas. When two or more atoms surrounding the central atom in a molecule with a double bond are identical, each has an identical unhybridized p orbital, which can overlap with a p orbital on the central atom to form a pi bond. The electron pair used to form the pi bond is delocalized over all the resulting pi orbitals.

This phenomenon can be illustrated by the hybridization in sulfur dioxide. The Lewis formula is

$$\ddot{:}\ddot{O}::\ddot{S}:\ddot{O}: \quad \longleftrightarrow \quad :\ddot{O}:\ddot{S}::\ddot{O}:$$

Figure 8.22 The structure of the sulfur dioxide molecule showing the sigma-bond framework and the delocalization of the pi bond: *(A)* sigma framework using sp² hybrids; *(B)* resonance structures with a localized pi bond; *(C)* delocalized pi bond.

The double bond is delocalized between the two sulfur–oxygen bonds. Figure 8.22 shows the orbitals used to bond these atoms together. The sigma bonds and unshared pairs on each atom use sp^2 hybridization. The remaining two electron pairs, which can be considered a pi bond and an unshared pair, use the remaining unhybridized p orbital on each atom. Because these p orbitals are all aligned, each oxygen orbital can overlap the sulfur orbital. This leads to a pi bond that is spread out over the entire molecule.

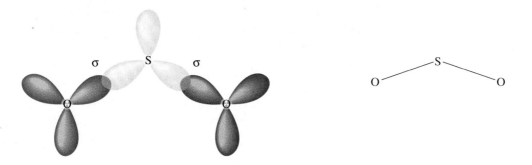

A. Sigma framework using sp^2 hybrids

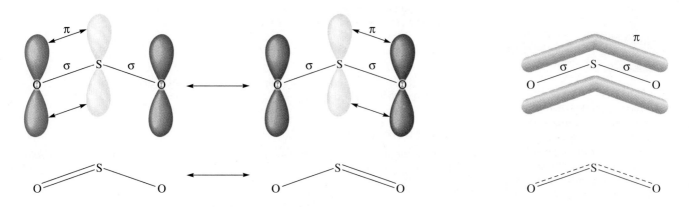

B. Resonance structures with a localized pi bond   **C. Delocalized pi bond**

8.4   Structure of Carbon and Its Compounds

So far in our discussion of structure and bonding, we have usually looked to small molecules for examples. ▶ However, the same principles hold for larger molecules, such as the sweet molecules described in the chapter introduction. Such molecules can be considered to have many central atoms, with the structure around each being the same as if there were only one. Carbon and its compounds provide good examples.

How are the structures of small molecules and large molecules related?

Structural Forms of Carbon

Two important structural forms of pure carbon are diamond and graphite (Figure 8.23). Diamond has a structure consisting of infinite networks of carbon atoms bonded to one another in a tetrahedral arrangement with each C atom bonded to four other C atoms (Figure 8.24A). The strength of these bonds makes diamond hard. The C—C—C bond angles are all 109.5° in this structure, and each carbon uses sp^3 hybrid orbitals for bond formation. Thus, each carbon atom has the same structure as the carbon in methane, CH_4.

Graphite consists of layers of planar, interconnected six-membered rings (Figure 8.24B). The layers are held together by rather weak intermolecular forces, which gives graphite its slippery characteristics. All the C—C—C bond angles are 120°, and each carbon uses sp^2 hybrid orbitals for sigma-bond formation with three other carbon atoms. The *p* orbital perpendicular to the plane remains unhybridized on each carbon atom and can overlap with the corresponding *p* orbital on each adjacent carbon atom to form delocalized pi bonds, similar to what we saw earlier for the SO_2 molecule (Figure 8.22). The bonding can be represented as alternating single and double bonds, but the double-bond character is actually delocalized over the entire sigma-bond framework.

Hydrocarbons

Carbon bonds with hydrogen to form a wide array of compounds called **hydrocarbons** (Figure 8.25) that form the subject of an entire branch of chemistry known as

Figure 8.23 Diamond *(right)* is extremely hard, colorless when pure, and crystalline. It is quite rare; about 90% of it is found in Africa. Since 1955, however, industry has produced it synthetically as small crystals by subjecting carbon to high temperatures and pressures. Graphite *(left and in pencil lead)* is a soft, slippery solid used as a lubricant. It is also used to make electrodes for electric arc furnaces (which are used in the production of steel) and for electrolysis processes. Although graphite is found worldwide, much is prepared synthetically from amorphous carbon.

Figure 8.24 *(A)* Diamond consists of a network of sp^3 hybridized carbon atoms connected by single bonds. *(B)* Graphite has a layered structure in which sp^2 hybridized carbon atoms are connected by alternating single and double bonds. The layers are slightly offset from one another. Carbon atoms are not shown in the graphite structure but are found at each intersection of bonds.

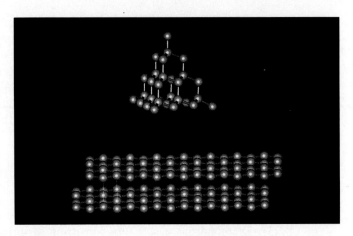

Ethane, C_2H_6, is a colorless, odorless, flammable gas that constitutes about 9% of natural gas. It is used in the manufacture of chlorofluorocarbons, as a refrigerant, and as a fuel gas. Bottled gas or suburban propane contains about 5% ethane.

Heptane, C_7H_{16}, is a colorless, volatile, flammable liquid used as a standard in testing the "knocking" of gasoline engines.

organic chemistry. The structures of hydrocarbon molecules are mostly based on those found in diamond and graphite.

Methane, the simplest hydrocarbon, has one carbon atom bonded to four hydrogens in a tetrahedral arrangement. The tetrahedral carbon repeating unit in diamond, using sp^3 hybridization, provides the carbon framework for methane (Figure 8.26). The next larger hydrocarbon is ethane, C_2H_6. If we could remove two adjacent carbons from the diamond structure and add hydrogens to each of the bonds that had to be broken, the result would be the structure of ethane (Figure 8.26). In ethane, each carbon uses sp^3 hybridization, with the atoms bonded to it arranged at the corners of a tetrahedron. This interconnected tetrahedral arrangement is continued even in long chains of carbon atoms. Suppose we could remove a chain of seven carbon atoms from the diamond structure (Figure 8.26). If we gave each carbon atom enough hydrogens to form four bonds, we would obtain C_7H_{16}, or heptane. Hydrocarbons can also be found in branched arrangements, where some carbon atoms may be attached to three or four other carbon atoms.

Figure 8.25 Types of hydrocarbons.

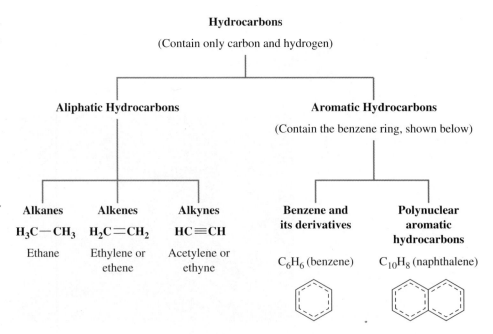

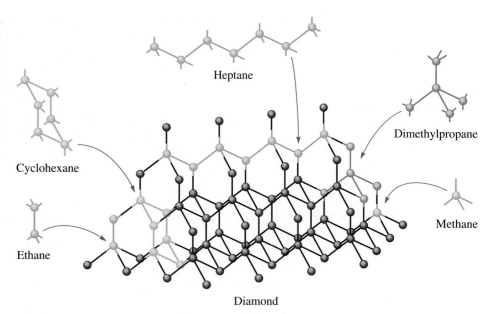

Figure 8.26 Diamond is made up of carbon atoms that are bonded together using sp^3 hybridization. The carbon framework of hydrocarbons with single bonds can be identified as fragments of this structure. Hydrogen atoms would be found at the end of each bond not connected to another carbon.

Ethene, C_2H_4, is a colorless, flammable gas that burns with a luminous flame. It is used in welding and cutting torches, in the manufacture of alcohol and other simple organic substances, and in the manufacture of polyethylene plastics, used extensively in plastic bags. It is also used commercially to accelerate the ripening of fruit.

Such structures are related to the diamond structure as well. All of these hydrocarbons, which contain only C—C bonds, are called **alkanes.**

Hydrocarbons that contain a double bond **(alkenes)** are related conceptually to the graphite structure. The simplest such compound is ethene (or ethylene), C_2H_4, which has a double bond between the carbon atoms: $H_2C = CH_2$. If the graphite structure is pictured as consisting of alternating single and double bonds instead of a delocalized pi-bond arrangement, two adjacent carbons with a double bond are not difficult to find (Figure 8.27). Other arrangements are also possible. Some carbon chains contain one double bond but more than two carbon atoms, such as propylene, CH_3—$CH = CH_2$. These molecules all contain sp^2 hybridization on those carbons involved in the double bond and sp^3 hybridization on single-bonded carbons. It is also possible to find chains containing more than one double

Figure 8.27 Graphite is made of carbon atoms bonded together using sp^2 hybridization. The carbon framework of hydrocarbons with double bonds can be identified as fragments of this structure. Hydrogen atoms are located at the end of any bond not connected to another carbon. Carbon atoms are found at any point in the structure where two or more bonds come together.

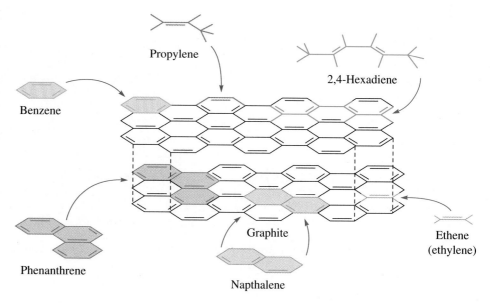

Acetylene, C_2H_2, is a colorless gas that burns with a very sooty flame. It is explosive under some conditions when mixed with air, and it forms explosive compounds with copper and silver. It is used as a fuel in welding torches.

bond, such as 2,4-hexadiene, CH_3—CH=CH—CH=CH—CH_3. Finally, many hydrocarbon structures are based on a six-membered ring with alternating single and double bonds (or, more accurately, delocalized bonds). Examples are benzene (C_6H_6), naphthalene ($C_{10}H_8$), and phenanthrene ($C_{14}H_{10}$). All these molecules are shown in Figure 8.27.

The one class of hydrocarbons shown in Figure 8.25 whose structures cannot be derived from those of diamond or graphite are the **alkynes.** These compounds contain a triple bond formed from sp hybrids on each carbon and the overlap of two p orbitals on each carbon to form two pi bonds. This bonding leads to a linear

Table 8.4 Functional Groups for Some Classes of Organic Compounds

Class	Functional Group	Example	Formula
Alcohol	—OH	Methanol	CH_3—OH
		Ethanol	C_2H_5—OH
Ether	—O—	Dimethyl ether	CH_3—O—CH_3
		Diethyl ether	C_2H_5—O—C_2H_5
Aldehyde		Formaldehyde	
		Acetaldehyde	
Ketone		Acetone	
Carboxylic acid		Acetic acid	
		Propionic acid	
Ester		Methyl acetate	
		Ethyl acetate	
Amine		Trimethylamine	

Ethanol, CH₃CH₂OH, is a clear, colorless liquid with a pleasant odor and a burning taste. It is manufactured by the fermentation of carbohydrates, such as sugar and starch, or from ethylene. It is used in various dilutions in alcoholic beverages and as a solvent in industry and in pharmaceutical solutions.

Phenol, C₆H₅OH, also known as carbolic acid, is a white solid obtained from coal tar. It is used as a disinfectant in toilets, drains, and stables and in the manufacture of resins, medical compounds, and dyes.

arrangement of the triple-bonded carbon atoms and the adjacent atoms, such as in the linear molecule acetylene: H—C≡C—H.

Functional Groups

Other carbon compounds that fall within the realm of organic chemistry are derivatives of the hydrocarbons in which other groups are substituted for one or more hydrogens on the hydrocarbon framework. The properties of these derivatives depend on the atoms involved in the substitution. The group that is substituted is called a **functional group** and is the part of the molecule that is unique to the class of compounds and gives it its characteristic properties. For example, if a hydrogen is replaced by a hydroxyl group (—OH), the compound is an **alcohol.** Methyl alcohol (or methanol) is CH_3OH, derived from methane. Substituting a hydroxyl group for a hydrogen on ethane gives ethyl alcohol (or ethanol), CH_3CH_2OH. Alcohols derived from benzene rings include phenol, C_6H_5OH. The most common functional groups and examples of organic molecules falling into these classes of substances are given in Table 8.4.

8.5 Molecular Orbitals

The valence bond model explains many features of the structure and bonding of molecules and polyatomic ions. However, there are some experimental facts this theory cannot accommodate. For example, when oxygen gas, O_2, is liquefied, it has a pale blue color. Color in small molecules generally indicates that the substance is **paramagnetic**—that is, unpaired electrons are present and so the molecules are attracted to a magnetic field (Figure 8.28). (A molecule of a **diamagnetic** substance has no unpaired electrons and is weakly repelled by a magnetic field.) Results of various measurements of the interaction of oxygen with magnetic fields support this indication: O_2 is paramagnetic, with two unpaired electrons. Yet neither the Lewis formula nor the valence bond picture of O_2 includes unpaired electrons. The molecule obeys the octet rule perfectly, with a double bond between the oxygen atoms. Similarly, although valence bond theory can be made to accommodate the bonding in unusual species such as the noble-gas compound XeF_4 by the use of expanded octets, the theory failed to predict the existence of such species before they were first prepared experimentally. Finally, valence bond theory cannot be used to calculate bond energies (and thus molecular stabilities) except on the basis of experimental results.

If a theory or model of bonding cannot accommodate all known facts about the structure and stability of molecules, should we completely abandon it, or does it still have some utility?

An alternate theory, *molecular orbital theory,* has been developed to correct these shortcomings. This model is too complex to apply to most molecules without the use of advanced mathematics. Here, we examine its application to simple molecules to demonstrate the basic approach.

Molecular orbital theory uses wave functions to describe how electrons are distributed in molecules, in the same way that quantum mechanics uses wave functions to describe how electrons are distributed in atoms. The areas in which the probability of finding an electron is high are called **molecular orbitals.** Molecular orbitals can be used quantitatively to calculate the geometry, energy levels, and other properties of molecules, but these are rather elegant and difficult calculations. From qualitative characteristics of molecular orbitals, however, we can deduce some important characteristics of molecules.

Figure 8.28 When liquid O_2 is poured between the poles of a magnet, it is held in place by the magnetic field, a phenomenon observed only for paramagnetic substances.

Combining Atomic Orbitals

One method for describing molecular orbitals involves the use of **linear combinations of atomic orbitals,** or LCAO. This method uses a framework provided by the atomic nuclei, constructs the molecular orbitals that the atomic orbitals can form, then fills the molecular orbitals with electrons in order of increasing energy. Only two electrons may occupy a molecular orbital—and then only if they have opposite spins. (This is a statement of the Pauli exclusion principle, which applies to molecular orbitals as well as to atomic orbitals.) Thus, two molecular orbitals must be formed from two atomic orbitals. The molecular orbitals are obtained simply from linear combinations of atomic orbitals achieved through either addition or subtraction. Mathematically, this means either adding or subtracting the wave functions of the atomic orbitals. In our more pictorial approach, it means overlapping the orbitals and either adding or subtracting electron density in the areas of overlap. When we add the overlap, we build up electron density between the nuclei, leading to a stable bonding condition and a molecular orbital lower in energy than the separate atomic orbitals. This molecular orbital is called a **bonding molecular orbital.** When we subtract the overlap, we decrease electron density between the nuclei. Because this leads to an unstable condition in which the nuclei will repel one another, the resulting molecular orbital is higher in energy than the separate atomic orbitals. This molecular orbital is called an **antibonding molecular orbital.** These combinations are shown for two *s* orbitals in Figure 8.29.

A third type of molecular orbital is the **nonbonding molecular orbital,** which is essentially an atomic orbital in the molecule that does not overlap with another. These orbitals retain their atomic characteristics. Generally, inner-shell orbitals do not overlap, so they are usually nonbonding orbitals. In some cases, the orientation of valence-shell orbitals prevents them from overlapping. Thus, as shown in Figure 8.30, an *s* orbital cannot overlap effectively with a *p* orbital in a sideways fashion, so a combination of these orbitals could give only nonbonding orbitals.

Figure 8.29 Two *s* orbitals overlap to form molecular orbitals. If the regions of overlap are added, a bonding molecular orbital with enhanced electron density between the nuclei results. If the regions of overlap are subtracted, electron density is removed from between the nuclei, and an antibonding molecular orbital results. Thus, combination of two atomic orbitals results in two molecular orbitals, a lower-energy bonding molecular orbital and a higher-energy antibonding molecular orbital.

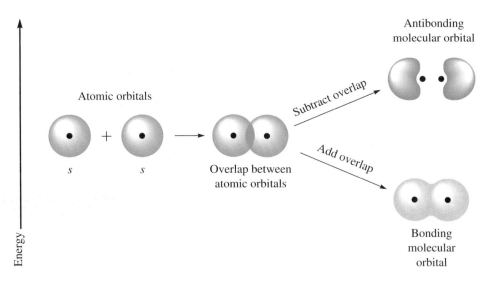

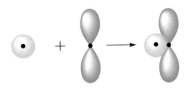

Figure 8.30 The sideways overlap of a *p* orbital with an *s* orbital does not lead to sufficient sharing of electron density to give rise to bonding or nonbonding molecular orbitals. These orbitals retain their atomic characteristics and are considered nonbonding orbitals.

Combinations of *s* and *p* Atomic Orbitals

The molecular orbitals that result from the allowed combinations of *s* and *p* atomic orbitals are shown in Figure 8.31. These can be classified into four groups. One distinction is based on shape. Depending on the cross section of the orbital, as in valence bond theory, the molecular orbital can be either a sigma (σ) or a pi (π) orbital. The sigma orbital has a circular cross section, whereas the pi orbital has a node in the middle, with electron density on opposite sides of the bond axis. The other distinction is the classification as bonding or antibonding.

To identify molecular orbitals according to these classifications, we use the appropriate Greek letter with a subscript to indicate the atomic orbitals from which they originated and an asterisk if the orbital is antibonding. Some possible designations for molecular orbitals are as follows:

$$\sigma_{1s} \quad \sigma_{1s}^* \quad \sigma_{2s} \quad \sigma_{2s}^* \quad \sigma_{2p_x} \quad \sigma_{2p_x}^* \quad \sigma_{1s2p_x} \quad \sigma_{1s2p_x}^* \quad \pi_{2p_y} \quad \pi_{2p_y}^*$$

8.6   Molecular Orbital Diagrams for Diatomic Molecules

How can we explain the bonding in simple molecules using molecular orbital theory?

We have seen that molecular orbitals can be constructed from linear combinations of atomic orbitals and that each molecular orbital can hold two electrons. How do we determine which orbitals will contain electrons? We must know the order of increasing energies of the orbitals because electrons fill the orbitals in this order, just as in atoms.

Let's consider a simple example, the bonding of two hydrogen atoms in the hydrogen molecule. We can see from Figure 8.31 that the combination of the 1*s* orbitals on two hydrogen atoms should give a sigma bonding molecular orbital and a sigma antibonding molecular orbital. The average energy of the two new orbitals must equal the average energy of the atomic *s* orbitals from which they were formed; but whereas the *s* orbitals were equal in energy, the antibonding molecular orbital is higher in energy than the bonding orbital. (More sophisticated calculations indicate that the antibonding orbital has a higher energy than is suggested by this simple approach.) These relationships are shown in Figure 8.32, the *energy-level diagram* for the H_2 molecule. Just as in filling atomic orbitals, electrons go first into the lowest-energy molecular orbital—here, the bonding orbital. Because H_2 contains only two electrons, one from each hydrogen atom, all the electrons in the molecule can fit in the bonding orbital.

The electronic configuration of the H_2 molecule as set out in the molecular orbital energy-level diagram can be represented as follows:

$$(\sigma_{1s})^2(\sigma_{1s}^*)^0$$

This notation uses the symbols for all the molecular orbitals, with superscripts to indicate how many electrons each holds.

When, as in the H_2 molecule, the number of electrons in bonding orbitals exceeds the number in antibonding orbitals, the energy level of the molecule is lower than the energy levels of the separate atoms. Because matter tends to change to a state of minimum energy, the molecule is stable and is formed spontaneously by combination of atoms.

Figure 8.31 Molecular orbitals that result from the allowed combinations (addition and subtraction) of *s* and *p* atomic orbitals. Each set of two orbitals forms one bonding and one antibonding molecular orbital.

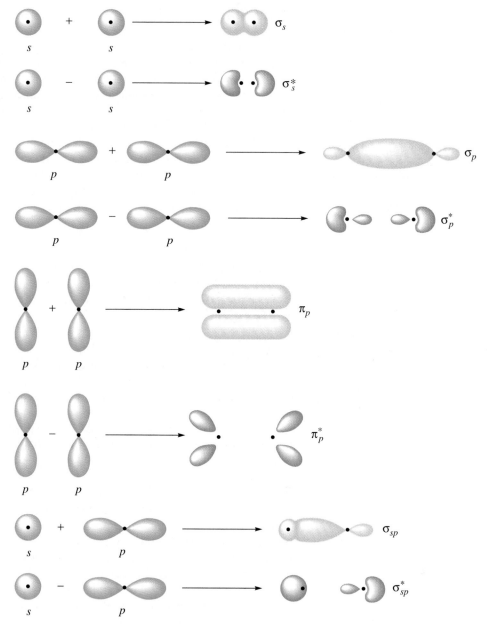

In contrast, if the number of electrons in antibonding orbitals equals or exceeds the number in bonding orbitals, the species is unstable and should not exist. Consider the helium molecule, He_2, a theoretical entity. No such molecule has ever been detected, and the molecular orbital energy-level diagram shown in Figure 8.33 suggests that it should not exist. The molecule would contain four electrons. The first two would go into the sigma bonding orbital, and the next two would have to go into the sigma antibonding orbital. The net result would be an increase in energy, with no bond being formed. The energy-level diagram, then, not only gives the electronic configuration of a molecule or polyatomic ion but also predicts whether the molecule or ion could be formed.

Figure 8.32 The hydrogen molecule, H_2, has two electrons in a sigma bonding molecular orbital, leading to a single bond between the atoms.

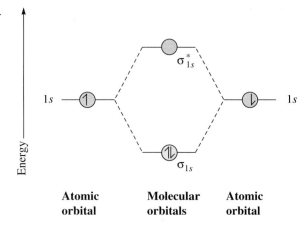

Another property that can be examined in terms of molecular orbital energy-level diagrams is the **bond order,** a measure of the number of covalent bonds between two atoms. We can estimate it by assuming that two bonding electrons are equal to one covalent bond, whereas antibonding electrons subtract from the bonding, as represented in the following equation:

Bond order $= \frac{1}{2}$(number of bonding electrons − number of antibonding electrons)

The bond order gives us an idea of the bond strength between atoms in various molecules and polyatomic ions. As noted in the discussion of valence bond theory, the greater the number of covalent bonds between two atoms—and thus, the higher the bond order—the stronger and the shorter the bond. For example, let's look again at the H_2 molecule. It has two electrons in a bonding orbital and no electrons in the antibonding orbital, so the bond order is 1:

Bond order of $H_2 = \frac{1}{2}(2 − 0) = 1$

The bond energy of H_2 is 430 kJ/mol, and the bond length is 74 pm.

The molecular orbital energy-level diagram also predicts the magnetic properties of the molecule or ion. In the H_2 molecule (Figure 8.32), there are no unpaired electrons, so the molecule is diamagnetic.

Figure 8.33 The hypothetical helium molecule, He_2, has two electrons in a sigma bonding molecular orbital and two electrons in a sigma antibonding orbital. The energies of the bonding and antibonding orbitals offset one another, so no bond will form.

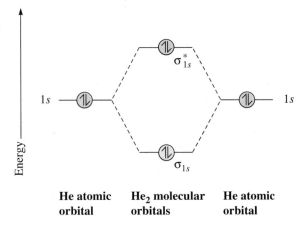

Compare the results for the H_2 molecule with those for the molecular ion H_2^+, which results from the removal of one electron from the H_2 molecule. This ion has an electronic configuration of $(\sigma_{1s})^1(\sigma_{1s}^*)^0$, giving a bond order of $\frac{1}{2}(1 - 0) = \frac{1}{2}$. The bond is weaker than a single bond. The bond energy is lower (242 kJ/mol) and the bond longer (106 pm) than in the H_2 molecule.

The bond orders, bond energies, bond lengths, and magnetic properties of some diatomic molecules and ions are given in Table 8.5. We turn now to a more detailed discussion of a few of these molecules and ions.

The relative energies of all the molecular orbitals used in molecules and ions formed from the second-period elements are shown in Figure 8.34. The exact energies of molecular orbitals are determined by experimental approaches, such as

Table 8.5 Molecular Orbital Descriptions and Properties of Diatomic Molecules

Molecule	Bond Order	Bond Energy (kJ/mol)	Bond Length (pm)	Paramagnetic (P) or Diamagnetic (D)
H_2^+	$\frac{1}{2}$	242	106	P
H_2	1	432	74	D
He_2^+	$\frac{1}{2}$	240	108	P
He_2	0	—	—	D
Li_2	1	105	267	D
Be_2	0	—	—	D
B_2	1	293	159	P
C_2	2	602	134	D
BN	2	385	128	D
N_2^+	$2\frac{1}{2}$	840	112	P
N_2	3	942	110	D
C_2^{2-}	3	835	120	D
CO	3	1070	128	D
CN^-	3	887	114	D
NO^+	3	—	106	D
O_2^+	$2\frac{1}{2}$	625	112	P
NO	$2\frac{1}{2}$	631	115	P
NO^-	2	607	121	P
O_2	2	494	121	P
O_2^-	$1\frac{1}{2}$	395	135	P
O_2^{2-}	1	142	149	D
F_2	1	155	142	D
Ne_2	0	—	—	D

molecular spectroscopy, which monitors the light emitted or absorbed by electrons as they are transferred from one orbital to another. This approach shows that there are actually two different orderings of the energy levels, differing in the energies of the σ_{2p} and the two π_{2p} orbitals. The order presented in Figure 8.34 is adequate for the present discussion, however.

The molecular orbital energies parallel those of the atomic orbitals from which they were formed. We feed the appropriate number of valence electrons into the energy-level pattern in Figure 8.34 to determine the electronic configuration of a molecule or ion. Again, electrons fill the lowest-energy orbital first. When two orbitals are of the same energy, such as the π_{2p_y} and the π_{2p_z} orbitals, one electron goes into each before any electrons are paired, consistent with Hund's rule.

We begin with the boron molecule, B_2, which is known to exist as a stable species and to have two unpaired electrons. These are the same characteristics we would predict after looking at the molecular orbital energy-level diagram in Figure 8.35. The valence electrons fill the σ_{2s} and σ_{2s}^* orbitals, leaving two electrons to go into the next level, the π_{2p_y} and π_{2p_z} orbitals. Because these two orbitals have the same energy, one electron enters each, giving the following valence electronic configuration:

$$(\sigma_{2s})^2(\sigma_{2s}^*)^2(\pi_{2p_y})^1(\pi_{2p_z})^1$$

There are four valence electrons in bonding orbitals and two in antibonding orbitals, so the bond order predicted by this diagram is $\frac{1}{2}(4-2)$, or 1. Note that the

The B_2 molecule is observed only in the gaseous state at low pressures and high temperatures. Normally, boron is a black solid metalloid. Although boron constitutes only 0.001% of the earth's crust, it occurs in highly concentrated deposits as ores such as borax, $Na_2B_4O_7 \cdot 10H_2O$, and kernite, $Na_2B_4O_7 \cdot 4H_2O$, which are found in the Mojave Desert in California.

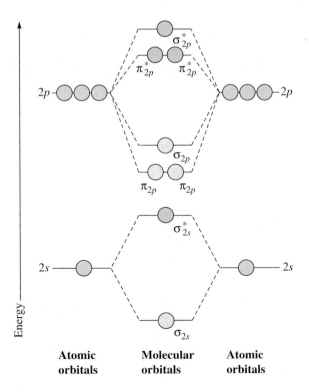

Energy

Atomic orbitals   Molecular orbitals   Atomic orbitals

Figure 8.34 General molecular orbital energy-level diagram for second-period diatomic molecules. The 1s orbitals are not included, because they are not valence orbitals and thus are nonbonding.

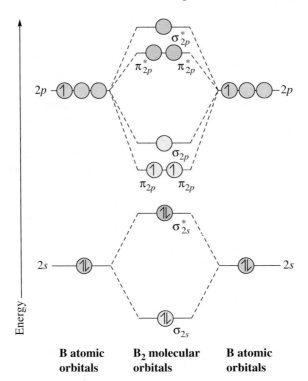

Figure 8.35 Molecular orbital energy-level diagram for the boron molecule, B_2.

bond order of 1 in this molecule arises from two unpaired electrons in two different π molecular orbitals. In valence bond theory, by contrast, a single bond is a sigma bond. It is possible to analyze ions based on the B_2 molecule using this same molecular orbital diagram.

EXAMPLE 8.4

What is the bond order for the molecular ion B_2^+?

Solution   The molecular orbital energy-level diagram is similar to Figure 8.35, but with one of the highest-energy electrons removed:

$$(\sigma_{2s})^2(\sigma_{2s}^*)^2(\pi_{2p_y})^1(\pi_{2p_z})^0$$

The bond order is $\frac{1}{2}(3 - 2)$, or $\frac{1}{2}$.

Practice Problem 8.4

What is the bond order for the molecular ion B_2^-?

Answer: $1\frac{1}{2}$

The nitrogen molecule, N_2, is diamagnetic and very stable, with a very high bond energy. Its molecular orbital energy-level diagram (Figure 8.36) corresponds to the following electronic configuration:

$$(\sigma_{2s})^2(\sigma_{2s}^*)^2(\pi_{2p_y})^2(\pi_{2p_z})^2(\sigma_{2p_x})^2$$

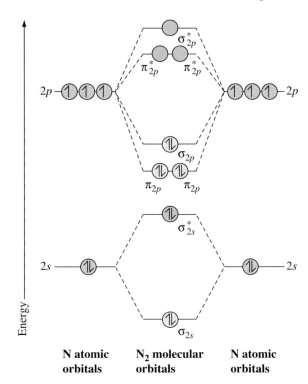

Figure 8.36 Molecular orbital energy-level diagram for the nitrogen molecule, N_2.

N atomic
orbitals

N_2 molecular
orbitals

N atomic
orbitals

Energy

The bond order is $\frac{1}{2}(8 - 2)$, or 3. This triple bond consists of two pi bonds and a sigma bond, as would be predicted by valence bond theory. The high bond order is consistent with the very high bond energy.

The oxygen molecule, O_2, is paramagnetic, a fact that could not be predicted by valence bond theory. This paramagnetism, as well as the bond order of 2, consistent with the bond energy of 494 kJ/mol, is predicted by the molecular orbital energy-level diagram shown in Figure 8.37. The electronic configuration is

$$(\sigma_{2s})^2(\sigma_{2s}^*)^2(\pi_{2p_y})^2(\pi_{2p_z})^2(\sigma_{2p_x})^2(\pi_{2p_y}^*)^1(\pi_{2p_z}^*)^1$$

There are eight electrons in bonding orbitals and four in antibonding orbitals, for a bond order of $\frac{1}{2}(8 - 4)$, or 2. The two unpaired electrons in the antibonding pi orbitals give rise to the paramagnetism of the O_2 molecule.

Molecular orbital theory can also be applied to *heteronuclear* diatomic molecules, or molecules made up of atoms of different elements. In general, an atom bonds to an atom of a different element in much the same way it bonds to an atom identical to itself, provided the number of electrons is the same. Thus, it is to be expected that, for example, N_2, CO, CN^-, and NO^+ have similar bonding and properties because they all have ten valence electrons. Indeed, these species all have high bond energies corresponding to a bond order of 3, and the bonding in all can be described in terms of the orbital occupancy described in Figure 8.36 for N_2. The molecular orbital energy-level diagram is somewhat in error for the heteronuclear molecules, however, in that it assumes that the atomic orbital energies are the same in both atoms. In fact, the atomic orbital energies of the more electronegative element are somewhat lower than those of the other element. This discrepancy generally does not cause any fundamental differences, but it does give rise to different bond energies. A look back at Table 8.5, for example, shows that

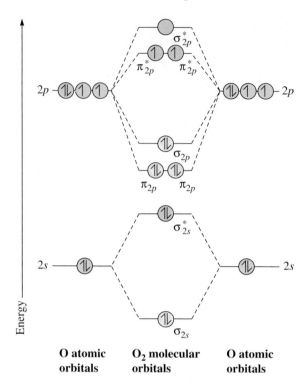

Figure 8.37 Molecular orbital energy-level diagram for the oxygen molecule, O_2. (This diagram gives the correct magnetic properties of O_2, even though spectroscopic experiments show that the energies of the π_{2p} and σ_{2p} orbitals should be reversed.)

whereas the bond energy of N_2 is 942 kJ/mol, that of CO is 1070 and that of CN^- is 887.

8.7 Chemical Reactivity of Oxygen Species

There are four known dioxygen species, the O_2 molecule and three ions: O_2^{2-}, the peroxide ion; O_2^-, the superoxide ion; and O_2^+, the dioxygenyl cation. We can understand some of the properties of these substances in terms of their bonding as described by molecular orbital theory.

The molecular orbital energy-level diagram for O_2 is shown in Figure 8.37, and its valence electronic configuration and bonding properties were described in the previous section. Molecular oxygen has a strong tendency to oxidize other substances; its bond length is 121 pm.

Addition of one electron to O_2 forms the superoxide ion, O_2^-:

$$(\sigma_{2s})^2(\sigma_{2s}^*)^2(\pi_{2p_y})^2(\pi_{2p_z})^2(\sigma_{2p_x})^2(\pi_{2p_y}^*)^2(\pi_{2p_z}^*)^1(\sigma_{2p_x}^*)^0$$

The superoxide ion has a bond order of $1\frac{1}{2}$, less than that in O_2 because the extra electron enters a π_{2p}^* orbital. The bond length is 135 pm, and the bond strength is 395 kJ/mol. These values differ from the values for O_2 because of the decreased bond order. Metal superoxides are generally a yellow-orange color, consistent with the presence of an unpaired electron.

Addition of two electrons to molecular oxygen results in the peroxide ion, O_2^{2-}:

$$(\sigma_{2s})^2(\sigma_{2s}^*)^2(\pi_{2p_y})^2(\pi_{2p_z})^2(\sigma_{2p_x})^2(\pi_{2p_y}^*)^2(\pi_{2p_z}^*)^2(\sigma_{2p_x}^*)^0$$

The peroxide ion has eight valence bonding electrons and six antibonding electrons, for a net bond order of 1. The bond is correspondingly weaker (bond energy of 140 kJ/mol) and longer (149 pm) than the bond in O_2.

The dioxygenyl cation is formed by the oxidation of O_2 with substances that have unusually strong oxidizing power. An example is PtF_6, which reacts with O_2 to give an orange solid containing the O_2^+ ion:

$$O_2 + PtF_6 \longrightarrow O_2^+PtF_6^-$$

This oxidation process removes one electron, resulting in the following valence electronic configuration:

$$(\sigma_{2s})^2(\sigma_{2s}^*)^2(\pi_{2p_y})^2(\pi_{2p_z})^2(\sigma_{2p_x})^2(\pi_{2p_y}^*)^1(\pi_{2p_z}^*)^0(\sigma_{2p_x}^*)^0$$

The bond order is $\frac{1}{2}(6 - 1)$, or $2\frac{1}{2}$, consistent with the bond strength of 625 kJ/mol and the bond length of 112 pm. As expected, O_2^+ has a single unpaired electron.

When four electrons are added to O_2, the result is the oxide ion, O^{2-}. The hypothetical O_2^{4-} would have a bond order of 0, so it cannot be formed:

$$(\sigma_{2s})^2(\sigma_{2s}^*)^2(\pi_{2p_y})^2(\pi_{2p_z})^2(\sigma_{2p_x})^2(\pi_{2p_y}^*)^2(\pi_{2p_z}^*)^2(\sigma_{2p_x}^*)^2$$

We can examine the relative stabilities of these species by looking at reactions of O_2 with metals (such reactions were first presented in Chapter 4). Electrons added to O_2 enter antibonding orbitals, with more energy required for addition to the sigma antibonding orbital than to the pi antibonding orbitals (see Figure 8.37). Thus, the ability to form a particular ion from O_2 depends on the amount of energy that can be supplied by the metal.

The most reactive metals, such as Cs, Rb, and K, normally form superoxides:

$$Cs(s) + O_2(g) \longrightarrow CsO_2(s)$$

These metals, as well as Na and Ba, also may form peroxides:

$$2Na(s) + O_2(g) \longrightarrow Na_2O_2(s)$$

Lithium, all alkaline-earth metals except Ba, and all other metals normally form oxides:

$$4Li(s) + O_2(g) \longrightarrow 2Li_2O(s)$$

These reactions can be understood in terms of the energy required to add to O_2. Each succeeding electron must enter an antibonding orbital of higher energy. The most reactive metals will react with oxygen to form superoxides at relatively low temperatures, where the energy is supplied primarily by the metal being oxidized. At higher temperatures, these reactive metals will instead form peroxides or oxides, with the higher temperatures providing the additional energy necessary to transfer electrons into the higher-energy antibonding orbitals. The lattice energy of the metal oxide, peroxide, or superoxide is also an important factor in determining the reaction product (see Chapter 7). For example, a more stable lattice is formed between Cs^+ and O_2^- than between $2Cs^+$ and O^{2-}.

Of the four species just discussed, O_2 and O^{2-} are the most stable. O_2^- and O_2^{2-} are higher in energy, and therefore less stable, because electrons occupy high-energy antibonding orbitals in these ions. This instability is reflected in the tendency of superoxide and peroxide ions to undergo internal oxidation–reduction to produce oxygen species that are more stable because they use lower-energy orbitals. The reactions are as follows:

$$2O_2^{2-} \longrightarrow 2O^{2-} + O_2$$
$$2O_2^- \longrightarrow O_2^{2-} + O_2$$

In these reactions, the superoxide or peroxide ion is being both oxidized and reduced.

Superoxides have considerable oxidizing power. For example, they react vigorously with water:

$$2O_2^- + 2H_2O \longrightarrow O_2 + H_2O_2 + 2OH^-$$

Superoxides are not very stable substances. When heated, they will decompose into the corresponding peroxides:

$$2KO_2 \longrightarrow K_2O_2 + O_2$$

At high pressures, oxygen is toxic to many organisms. This toxicity may be attributable to the superoxide ion, although the precise mechanism is still a matter of controversy. One proposal suggests that certain species in living cells reduce O_2 to O_2^-. The resulting superoxide would be toxic to the cells. However, organisms also contain proteins, called superoxide dismutases, that promote the decomposition of superoxide and thus protect the organism from this toxicity.

Summary

The shapes of molecules can be predicted by use of VSEPR theory, which assumes that electron pairs try to stay as far apart as possible. The total number of electron groups determines the parent structure: linear, trigonal planar, tetrahedral, trigonal bipyramidal, or octahedral. The molecular shape is determined by the number of unshared electron pairs occupying positions in the parent structure.

A polar molecule has centers of positive and negative charge at different points within the molecule. A molecule is polar only if the bonds are polar and the molecule is unsymmetrical.

According to valence bond theory, covalent bonds are formed by the overlap of half-filled atomic orbitals, which allows the atoms to share electron pairs. Hybridization of orbitals is necessary if the theory is to predict and explain bond angles and bond strengths correctly, as well as account for the delocalization of electrons. The structures and bonding in hydrocarbons provide extensive examples of the application of valence bond theory.

According to the molecular orbital theory of bonding, atomic orbitals are combined mathematically to describe molecular orbitals, regions of space in molecules and polyatomic ions where electrons are most likely to be found. Molecular orbitals are classified as bonding, antibonding, or nonbonding, and they can be arranged into a sequence of increasing energy. Placing the appropriate number of electrons into orbitals of increasing energy gives the electronic structure of molecules, just as for atoms. Molecular orbital energy-level diagrams provide information about bond orders, bond strengths, bond lengths, and magnetic properties of diatomic molecules and ions.

Key Terms

alcohol (8.4)
alkanes (8.4)
alkenes (8.4)
alkynes (8.4)
antibonding molecular
   orbital (8.5)
axial position (8.1)
bond angle (8.1)
bonding molecular

orbital (8.5)
bond order (8.6)
diamagnetic
   substance (8.5)
dipole moment (8.2)
equatorial position (8.1)
functional group (8.4)
hybridization (8.3)
hybrid orbitals (8.3)

hydrocarbon (8.4)
linear combination of
   atomic orbitals (8.5)
molecular orbital (8.5)
molecular orbital
   theory (8.5)
nonbonding molecular
   orbital (8.5)
paramagnetic

substance (8.5)
pi (π) bond (8.3)
sigma (σ) bond (8.3)
valence bond
   theory (8.3)
valence-shell electron-
   pair repulsion
   (VSEPR) theory (8.1)

Exercises

VSEPR and Molecular Structure

8.1 How can VSEPR theory be used to predict molecular structures?

8.2 Why are unshared pairs of electrons on a central atom not considered to be part of the molecular shape?

8.3 Predict the shapes and give approximate bond angles for each of the following molecules.
- a. $BeCl_2$
- b. PH_3
- c. SCl_2
- d. SO_2
- e. H_2Te
- f. SiH_4
- g. BBr_3
- h. H_2O

8.4 For the following species, identify what main-group element (X) could form each compound. Unshared electron pairs, denoted as e, are identified for the central atom.
- a. XF_4e
- b. XCl_2
- c. XBr_4
- d. XBr_4e
- e. XF_2e_3
- f. XF_2e_2
- g. XCl_3e
- h. XCl_2e_2

8.5 Predict the shapes and give approximate bond angles for each of the following molecules or ions.
- a. OCS
- b. FNO
- c. FCN
- d. HN_3
- e. NCS^-
- f. NO_2^-
- g. XeF_2
- h. NH_2^-

8.6 Describe the molecular geometry of each of the following.
- a. PF_3
- b. PF_5
- c. SF_2
- d. SF_4
- e. SF_6
- f. ICl_2^-
- g. BrF_4^-
- h. IO_3^-

8.7 For each of the following, write a formula for a species that has the specified structure and general formula. The letter e stands for an unshared electron pair.
- a. trigonal bipyramidal, XY_5
- b. linear, XYe_3
- c. linear, XY
- d. bent, XY_2e
- e. bent, XY_2e_2
- f. distorted tetrahedral, XY_4e

8.8 Draw each of the following geometric arrangements.
- a. octahedral
- b. trigonal bipyramidal
- c. tetrahedral
- d. trigonal planar
- e. bent

8.9 In which of the following molecular shapes would you expect to find one or more unshared pairs of electrons on the central atom?
- a. octahedral
- b. linear
- c. tetrahedral
- d. distorted tetrahedral
- e. trigonal planar
- f. bent

8.10 Predict the bond angles in each of the following gaseous molecules.
- a. NH_3
- b. H_2O
- c. HCl
- d. HCN
- e. BF_3
- f. H_2CO
- g. PCl_3
- h. PCl_5

8.11 Predict the shape of each of the following molecules.
- a. NF_3
- b. SO_2Cl_2
- c. CBr_4
- d. F_2CO
- e. PH_3
- f. HOCl

Molecular Polarity

8.12 Distinguish between bond polarity and molecular polarity.

8.13 Why does molecular polarity depend not only on bond polarity but also on the geometry of the molecule?

8.14 Explain how carbon tetrachloride can have polar bonds but still be a nonpolar molecule.

8.15 Compare the electronegativity differences between two atoms that form a nonpolar covalent bond, a polar covalent bond, or an ionic bond.

8.16 Which member of each pair is the most polar?
- a. CCl_4 or CH_2Cl_2
- b. CH_3F or CH_3Br
- c. HI or HCl
- d. NF_3 or NH_3
- e. OF_2 or H_2O

8.17 Decide which member of each pair is polar, and explain why that molecule is polar and the other is not.
- a. SO_2 and CO_2
- b. SO_2 and SO_3
- c. $SeCl_2$ and $BeCl_2$
- d. CH_4 and CH_3I

8.18 Explain why the first molecule of each pair is polar and the second is not.
- a. CH_2Cl_2, CCl_4
- b. PF_3, BF_3
- c. BF_2Cl, BF_3
- d. SCl_4, CCl_4

8.19 Decide whether each of the following molecules has a dipole moment.
- a. HI
- b. CHF_3
- c. SO_2Cl_2
- d. PF_5

8.20 Hydrocarbons are all essentially nonpolar substances. Explain this observation.

Hybridization

8.21 What is hybridization? What are hybrid orbitals?

8.22 Why is it necessary for valence bond theory to include the concept of hybrid orbitals?

8.23 What is a sigma bond? What is a pi bond?

8.24 Why does carbon not use d orbitals for hybridization?

8.25 Compare the hybrid orbitals that must be used to form single, double, or triple bonds between carbon atoms.

8.26 What hybrid orbitals must be used to obtain each of the parent structures predicted by VSEPR theory?

8.27 What is the hybridization of the central atom in each of the following molecules and ions?
- a. ICl_4^-
- b. $BeCl_2(g)$
- c. $TeCl_4(g)$
- d. PO_4^{3-}
- e. $ICl_3(g)$
- f. $SF_6(g)$

8.28 Draw a picture and describe in words the shape of each of the following sets of hybrid orbitals.
- a. sp
- b. sp^3
- c. d^2sp^3
- d. sp^2
- e. dsp^3

8.29 What is the hybridization of the central atom in each of the following molecules and ions?
- a. SO_3^{2-}
- b. $NF_3(g)$
- c. $SCl_4(g)$
- d. CO_3^{2-}
- e. $ICl_5(g)$
- f. $SO_2(g)$

8.30 What is the hybridization of the central atom in each of the following molecules and ions?
- a. $XeF_6(g)$
- b. $BCl_3(g)$
- c. $PCl_3(g)$
- d. SO_4^{2-}
- e. $PCl_5(g)$
- f. $CO_2(g)$

8.31 Describe the orbitals used in the bonding for each of the following molecules. Describe the type of bond between each pair of bonded atoms.
   a. CH_3CO_2H b. C_2H_2 c. O_2 d. CH_2Cl_2
   e. IF_3 f. ClF_5

8.32 Describe the orbitals used in the bonding for each of the following molecules. Describe the type of bond between each pair of bonded atoms.
   a. $HCCl_3$ b. C_2H_4 c. N_2 d. H_2CO_3
   e. XeF_4 f. SF_6

8.33 Which of the following molecules or ions contain triple bonds?
   a. N_2 b. CO_2 c. NO^+ d. CO e. O_2
   f. S_8 g. CO_3^{2-} h. BF_3

8.34 Acetylene has the formula C_2H_2. Indicate the number of sigma and pi bonds in this molecule.

Structure of Organic Molecules

8.35 What are the different classes of hydrocarbons?
8.36 What is a functional group?
8.37 Describe the structures of diamond and graphite.
8.38 Diamond is one of the hardest substances known, whereas graphite is soft enough to use as a lubricant. Both substances are composed of pure carbon. Discuss the difference in hardness of these two forms of carbon in terms of their structures.
8.39 Explain how the structures of diamond and graphite are related to the structures of hydrocarbons.
8.40 Describe the structure of benzene, C_6H_6 (a cyclic compound), in terms of the structures of diamond and/or graphite.
8.41 Describe the structure of hexane, C_6H_{14} (a long chain of carbon atoms), in terms of the structures of diamond and/or graphite.
8.42 Describe the structure of ethylene, C_2H_4, in terms of the structures of diamond and/or graphite.
8.43 Identify the functional group and the class of organic substance for each of the following molecules.
   a. CH_3OH f. CH_3NH_2
   b. C_2H_6 g. H_3COCH_3
   c. C_2H_2 h. CH_3CO_2H
   d. C_6H_6 i. CH_3CHO
   e. $(CH_3)_2CO$ j. $CH_3CH{=}CH_2$

Types of Molecular Orbitals

8.44 What is meant by linear combination of atomic orbitals (LCAO)?
8.45 What is a molecular orbital? How is a molecular orbital similar to or different from an atomic orbital?
8.46 How many electrons can occupy a molecular orbital?
8.47 Describe a sigma and a pi molecular orbital.
8.48 Distinguish between bonding, antibonding, and nonbonding molecular orbitals.
8.49 Sketch the shape of the σ_{2p}^* molecular orbital.
8.50 Sketch the shape of the σ_{2p} molecular orbital.
8.51 Sketch the shape of a π_{2p} molecular orbital.

8.52 Sketch the shape of the σ_{2s} molecular orbital.
8.53 Describe the ways in which a pi molecular orbital can be formed by linear combinations of atomic orbitals.
8.54 What molecular orbitals can be formed by the linear combination of atomic p orbitals?
8.55 What molecular orbitals can be formed by the linear combination of atomic s and p orbitals?
8.56 Why are the energies of bonding molecular orbitals lower than those of the atomic orbitals from which they were formed?

Molecular Orbital Diagrams

8.57 A certain diatomic molecule has its molecular orbitals arranged in the following order:

$$(\sigma_{1s}) (\sigma_{1s}^*) (\sigma_{2s}) (\sigma_{2s}^*) [(\pi_{2p_y}) (\pi_{2p_z})] (\sigma_{2p_x})$$
$$[(\pi_{2p_y}^*) (\pi_{2p_z}^*)] (\sigma_{2p_x}^*)$$

where orbitals contained in square brackets have the same energy.
   a. If the molecule contains 19 electrons, what is the highest-energy molecular orbital occupied by an electron or electrons?
   b. What is the bond order of this molecule?
   c. Is this molecule paramagnetic?
   d. Which should be most stable: the neutral molecule, the cation formed from this molecule, or the anion formed from this molecule? Explain your answer.

8.58 Arrange the following diatomic species in order of increasing bond order: CN, CN^-, CN^+.
8.59 Describe the bonding in BO in terms of its molecular orbital energy-level diagram.
8.60 Arrange the following diatomic species in order of decreasing bond order: NO, NO^-, NO^+.
8.61 What is the bond order in each of the following species?
   a. HHe b. H_2^+ c. H_2^- d. He_2

8.62 Arrange the following diatomic species in order of increasing bond order: BN, BN^-, BN^+.
8.63 What is the number of unpaired electrons in each of the following species?
   a. CN b. BO c. NO d. C_2 e. Be_2

8.64 Indicate whether each of the following substances is diamagnetic or paramagnetic.
   a. BO b. B_2 c. BN d. C_2 e. CN

8.65 Write out the molecular orbital electronic configuration of each of the following molecules or ions.
   a. O_2 b. O_2^+ c. O_2^- d. O_2^{2-}

8.66 Arrange the following species in order of increasing bond strength: O_2, O_2^{2+}, O_2^+, O_2^{2-}, O_2^-.
8.67 What is the bond order in a molecule that has the following molecular orbital electronic configuration?

$$(\sigma_{2s})^2(\sigma_{2s}^*)^2(\sigma_{2p_x})^2[(\pi_{2p_y})^1(\pi_{2p_z})^1]$$

8.68 Describe the bonding in F_2^+ in terms of its molecular orbital energy-level diagram.

Additional Exercises

8.69 What is the number of unpaired electrons in each of the following species?
 a. O_2 b. O_2^+ c. O_2^- d. N_2 e. B_2

8.70 Indicate which of the following molecules would be expected to have the highest bond energy: N_2, O_2, or F_2.

8.71 Explain how carbon tetrafluoride can have polar bonds but still be a nonpolar molecule.

8.72 Write out the molecular orbital electronic configuration of each of the following molecules or ions.
 a. BN b. BO c. CN d. CN^-

8.73 Compare the explanations of covalent bonding offered by valence bond theory and molecular orbital theory.

8.74 Describe the molecular geometry of each of the following.
 a. SCl_4 b. NCl_2^+ c. SF_5^- d. PCl_4^+
 e. $SiBr_6^{2-}$ f. GeH_4 g. $SOCl_2$

8.75 What is the hybridization of the central atom in each of the following molecules and ions?
 a. $SO_2(g)$ b. $BeCl_2(g)$ c. $BCl_3(g)$ d. SO_4^{2-}
 e. $ICl_5(g)$ f. $SF_4(g)$

8.76 Describe the molecular geometry of each of the following halogen species.
 a. BrO_3^- b. BrF_2^+ c. ICl_3 d. IF_5 e. Br_2O
 f. ClO_2^- g. OF_2 h. IO_6^{5-}

8.77 Describe the structure and bonding in sulfuric acid, H_2SO_4, and in its two ions, HSO_4^- and SO_4^{2-}.

8.78 Describe the orbitals used in the bonding of the following molecules. Also describe the type of bond between each pair of bonded atoms.
 a. $HClO_2$ b. CO_2 c. Br_2 d. $TeCl_4$
 e. BrF_5 f. HCN

8.79 Describe the bonding in Ne_2^+ in terms of its molecular orbital energy-level diagram.

8.80 Arrange the following diatomic species in order of decreasing bond order: O_2, O_2^{2-}, O_2^-, O_2^+.

8.81 What is the bond order of each of the following species?
 a. HF b. NO^- c. BO d. NO e. C_2^{2-}

8.82 Indicate whether each of the following substances is diamagnetic or paramagnetic.
 a. O_2 b. F_2 c. HF d. NO^- e. NO

8.83 Arrange the following diatomic species in order of increasing bond order: CO, CO^-, CO^+.

8.84 Explain why carbon tetrachloride is nonpolar and why sulfur tetrachloride is polar.

8.85 Decide which member of each pair of gaseous molecules is polar, and explain why that molecule is polar while the other is not.
 a. $BeCl_2$ and OCl_2 d. BCl_3 and $AsCl_3$
 b. PH_3 and BH_3 e. SiH_4 and NH_3
 c. SCl_4 and SCl_6

8.86 A certain diatomic molecule has its molecular orbitals arranged in the following order:

$$(\sigma_{1s})\,(\sigma_{1s}^*)\,(\sigma_{2s})\,(\sigma_{2s}^*)\,[(\pi_{2p_y})\,(\pi_{2p_z})]\,(\sigma_{2p_x})$$
$$[(\pi_{2p_y}^*)\,(\pi_{2p_z}^*)]\,(\sigma_{2p_x}^*)$$

where orbitals contained in square brackets have the same energy.
 a. If the molecule contains 12 electrons, what is the highest-energy molecular orbital occupied by an electron or electrons?
 b. What is the bond order of this molecule?
 c. Is this molecule paramagnetic?
 d. Which should be most stable: the neutral molecule, the cation formed from this molecule, or the anion formed from this molecule? Explain your answer.

*8.87 Benzene (C_6H_6) and cyclohexane (C_6H_{12}) both have a six-membered ring of carbon atoms. Compare the Lewis structures and valence bond descriptions of the bonding in these molecules. How are their shapes different? Predict the relative C—C bond lengths and C—C—C bond angles in these compounds.

*8.88 It is possible to measure first ionization energies for molecules, as we do for atoms. Use molecular orbital theory to predict the relative values of the first ionization energy for C_2, N_2, O_2, and F_2.

▶ ▶ ▶ Chemistry in Practice

8.89 A fluoride of xenon was prepared by reacting 0.2045 g of Xe with excess F_2 to form 0.3229 g of product.
 a. Describe the bonding in this xenon fluoride in terms of its Lewis formula and a valence bond structure.
 b. What is the shape of the molecule?
 c. Is the molecule polar?
 d. If the standard enthalpy of formation of this compound is −215 kJ/mol and the standard enthalpy of atomization of fluorine is 155 kJ/mol, what is the average Xe—F bond energy?

9

▶ ▶ ▶ ▶ ▶

Gases

Earth's atmosphere, extending to beyond 50 km, is visible only where condensed water vapor forms clouds.

CHAPTER OUTLINE

Encountering Chemistry

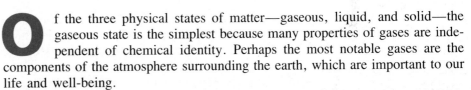

Of the three physical states of matter—gaseous, liquid, and solid—the gaseous state is the simplest because many properties of gases are independent of chemical identity. Perhaps the most notable gases are the components of the atmosphere surrounding the earth, which are important to our life and well-being.

The atmosphere is a layer of gas attracted to earth by gravity. This gas layer is estimated to have a total mass of about 5.2×10^{18} kg, only about 0.03% of the mass of earth. Different levels of the atmosphere have different characteristics. Extending to about 11 km from the earth's surface is the part of the atmosphere called the *troposphere,* which contains most of the atmosphere's gas molecules, primarily N_2, O_2, CO_2, H_2O, and Ar. Beyond this, from 11 km to about 50 km, is the *stratosphere*. This area contains some ozone, O_3, in addition to other gases. Ozone is toxic to humans and animals when they breathe it, and it is responsible in part for photochemical smog at low altitudes. In the stratosphere, however, ozone absorbs harmful ultraviolet radiation before it reaches the earth. The discovery of a "hole" in the ozone layer—actually, a depletion in the concentration of stratospheric ozone—above Antarctica has caused alarm because of the potentially devastating effects, such as skin cancer, that increased ultraviolet radiation could have

Table 9.1 Volume Percent of Various Gases in Clean, Dry Air*

Gas	Volume Percent	Gas	Volume Percent	Gas	Volume Percent
N_2	78.09	CH_4	0.00015	O_3	0.000002
O_2	20.94	Kr	0.0001	NH_3	0.000001
Ar	0.93	H_2	0.00005	NO_2	0.0000001
CO_2	0.032	N_2O	0.000025	SO_2	0.00000002
Ne	0.0018	CO	0.00001		
He	0.00052	Xe	0.000008		

*The amounts of minor components depend on the location and the extent of air pollution.

on humans. Beyond the stratosphere is the *ionosphere,* which contains small amounts of ions, such as NO^+, O^+, $O_2{}^+$, and electrons.

The composition of the atmosphere at ground level is given in Table 9.1. The principal components are nitrogen and oxygen, which make up 99% of the total volume of air. The composition may vary from that given in the table as a result of air pollution or humidity. Air pollution adds a number of components, especially the oxides of carbon, nitrogen, and sulfur.

Carbon dioxide, CO_2, is a colorless, odorless gas with a slight acid or sour taste when dissolved in water. Carbon dioxide gives the "bite" to carbonated beverages.

Both oxygen and carbon dioxide gas are vital to the maintenance of life. Oxygen is produced by plants and consumed by the respiratory systems of animals. Carbon dioxide plays a complementary role; it is exhaled by animals and used as a primary nutrient by plants. But carbon dioxide is not just a waste product for animals. The production of some carbon dioxide during the breakdown of nutrients in the body helps to maintain proper blood acidity essential for good health. Carbon dioxide also helps regulate breathing. A common malfunction of the breathing mechanism is hiccups. One cure involves breathing into a paper bag, which builds up the level of CO_2 in the lungs. The CO_2 can begin to trigger breathing at the appropriate times, and the hiccups disappear. High levels of carbon dioxide in the body are not beneficial, however. Breathing air that contains more than 10% CO_2 causes humans to lose consciousness, and continued exposure to high CO_2 levels can lead to respiratory failure and death.

This chapter examines the physical properties of gases, especially the behavior of gases when subjected to pressure or temperature changes. This behavior is described by several laws that allow us to predict how gases respond to changes in their environment. To explain the gas laws, the chapter develops a model for the behavior of gases at a molecular level. Deviations from this model are also discussed, and the chapter ends with an examination of the chemical properties of some small gaseous molecules.

9.1　Properties of Gases, Liquids, and Solids

▶ The gases found in the atmosphere, whether prominent like O_2 and N_2 or minor like CO and Ar, share some common properties. They all have a relatively low density, which is partly responsible for the fact that they float above the earth. They all can be compressed into a container, such as an aerosol can, if we exert a force on them; and they all escape if we open the container. They all expand if heated, a property that causes winds.

We have seen that many of the properties of a substance depend on its chemical identity. A number of properties, however, are typical of the physical state of the substance regardless of chemical identity. These properties are related largely to differences in the distance between atoms or molecules in gases, liquids, and solids. In this section, we briefly consider some properties of all three states. Later sections of this chapter examine gas properties in more detail. The properties of liquids and solids are described in Chapter 10.

Gases consist of rapidly moving molecules (or atoms for the noble gases) that are relatively far apart. Gas particles do not have much effect on one another unless they collide. They have essentially no attraction for one another, so they move about freely, taking the volume and shape of their containers. The large spaces between particles explain why gases have low density and high compressibility.

The particles in liquids, most of which are molecules, are fairly close together and are moving about, but not as rapidly as the particles in gases. Molecules in liquids have some attraction for one another, causing them to stay closer together than gas molecules do and to move about in a way that reflects the movement of other molecules close by. Liquids have a characteristic volume because the molecules are held close together by attractions between them. However, because the molecules are moving, liquids take the shape of their containers. Because the molecules are closer together than gas molecules are, liquids have a higher density and lower compressibility than gases do.

The particles in solids—atoms, molecules, or ions—are generally held as close together as possible by very strong attractive forces. Thus, solids have characteristic shapes and volumes. The absence of space between particles explains the relatively high density and low compressibility typical of this state.

Solids are characterized by a high degree of order in the arrangement of their particles, whereas liquids are less ordered and gaseous particles are completely disordered (see Figure 9.1). To convert a solid to a liquid, energy must be added to overcome the forces holding the solid particles in place. The lesser forces between liquid particles must be overcome by addition of still more energy to convert the

Which properties of gases are common to all gases and which are unique to a given gas?

Reconcile the idea that solids take a characteristic shape with the observation that table salt takes the shape of the salt shaker.

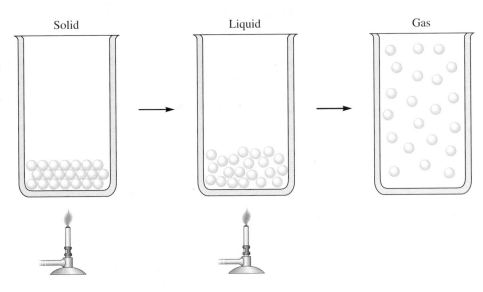

Figure 9.1 As matter is heated, it is converted from a solid to a liquid to a gas. During these conversions, the amount of disorder and the distance between particles (atoms or molecules) increases.

Solid   Liquid   Gas

substance to the gaseous state, in which there are essentially no forces between particles. A common method of adding energy is by heating. Thus, the conversion of a solid to a liquid to a gas generally occurs as the temperature increases, as illustrated in Figure 9.1. The higher energy state of gases causes gas particles to move about rapidly, thereby giving gases their distinct properties.

9.2   Pressure

Do gases occupy the same volume under all conditions?

If we fill a teakettle with ice and heat it, at a certain point the ice melts to form liquid water. The volume of the water is nearly the same as the volume of the ice. On continued heating, the water evaporates, with steam (gaseous water) completely filling the kettle. The whistle we hear when the water begins to boil is caused by the gaseous water trying to expand its volume and escape through the small hole in the spout. This greatly expanded volume accompanied by greatly lowered density is the most noticeable difference between gases and liquids or solids.

Because a gas completely fills its container, its volume as measured by the volume of the container is a typical measure of the amount being handled. However, volume is highly dependent on the conditions to which the gas is subjected. In order to use volume as a measure of amount, it is necessary to know both the temperature (discussed in Chapter 1) and the pressure of the gas.

Pressure *(P)* is the amount of force applied per unit area:

$$\text{Pressure } (P) = \text{force/area}$$

Force equals mass multiplied by acceleration. Generally, for our purposes force is equivalent to weight. To get an idea of the relative magnitudes of force and pressure, imagine a 115-lb person stepping on an upright nail. If the nail point has an area of 0.0100 in^2, the pressure generated is

$$P = \text{force/area} = 115 \text{ lb}/0.0100 \text{ in}^2 = 11,500 \text{ lb/in}^2$$

This pressure is enough to cause the nail to penetrate the person's foot.

A tremendous pressure can be exerted by a moderate force, then, if the force is applied to a small area. The pressure exerted by the air molecules in the atmosphere is considerably smaller than the example just considered, only 14.7 lb/in^2. However, the pressure of the atmosphere is quite significant. For example, it is large enough to crush a metal can that has been evacuated, as shown in Figure 9.2.

The pressure of the atmosphere or any other pressure can be expressed in many different units. The official SI unit of force is the newton (N), which is defined in terms of a mass in kilograms (kg) multiplied by an acceleration in meters per second squared (m/s^2):

$$\text{Unit of force} = 1 \text{ N} = 1 \text{ kg m/s}^2$$

The SI unit of area is the square meter (m^2). Because pressure is defined as force divided by area, we can derive its SI unit as follows:

$$\text{Unit of pressure} = \text{N/m}^2 = (\text{kg m})/(\text{s}^2 \text{ m}^2) = \text{kg/(m s}^2)$$

This SI pressure unit is called a *pascal* (Pa).

A kilogram is a fairly large mass and a square meter is a large area, so the pascal is not especially useful for measuring the pressures of gases. A common

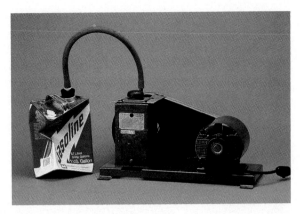

Figure 9.2 When the air inside a metal can is evacuated with a vacuum pump, the pressure of the outside air crushes the can.

method of measuring the pressure of gases, especially those in the atmosphere, uses a **barometer,** which is a tube closed at one end, filled with liquid mercury, and then inverted into a container of liquid mercury. The height of the mercury column adjusts to equal the atmospheric pressure exerted on the pool of mercury, as shown in Figure 9.3. Thus, changes in atmospheric pressure cause changes in the height of the mercury column. Units based on barometric measurements, such as inches of mercury (in Hg) or centimeters of mercury (cm Hg) or millimeters of mercury (mm Hg), are commonly used to express atmospheric pressures. Weather forecasters use such units, for example.

The pressure of the atmosphere at sea level and 0°C under normal weather conditions corresponds to 76.0 cm Hg or 29.9 in Hg. This pressure is the basis for the standard **atmosphere (atm),** a unit related to the barometric units as follows:

$$1 \text{ atm} = 76.0 \text{ cm Hg} = 760 \text{ mm Hg}$$
$$1 \text{ atm} = 29.9 \text{ in Hg}$$

The pressures of gases in chemical systems are often fractions of 1 atm, so they are commonly measured in units of mm Hg, which has been given the name **torr.** Because 1 torr is the same as 1 mm Hg, the torr is related to the unit of atmospheres as follows:

$$1 \text{ atm} = 760 \text{ torr}$$

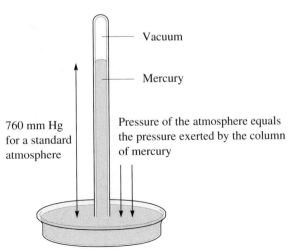

Vacuum

Mercury

760 mm Hg for a standard atmosphere

Pressure of the atmosphere equals the pressure exerted by the column of mercury

Figure 9.3 In a barometer, the pressure exerted by a column of mercury in a closed tube equals the pressure of the atmosphere.

The common barometer is not used to measure the pressure of gases in chemical systems. Chemists sometimes use a related apparatus, essentially a U-tube containing mercury, as pictured in Figure 9.4. The difference in heights of the mercury columns is used as a measure of gas pressure. Because atmospheric pressure is pushing down on the open tube and the pressure of the gas is pushing down on the closed tube, the pressure of the gas must equal atmospheric pressure plus or minus the difference in mercury levels. If the open tube has the higher mercury level, for example, then the pressure of the gas in the bulb must be greater than atmospheric pressure, and the difference is added to atmospheric pressure to determine the gas pressure.

EXAMPLE 9.1

A sample of methane gas is placed in the U-tube apparatus of Figure 9.4. The mercury column reaches 235 mm in the closed tube and 428 mm in the open tube. If atmospheric pressure is 735 torr, what is the pressure of the methane in the bulb?

Solution   The pressure of the gas is the atmospheric pressure corrected for the difference in the mercury levels, or 193 mm Hg:

$$\text{Pressure difference} = 428 \text{ mm Hg} - 235 \text{ mm Hg} = 193 \text{ mm Hg}$$

Because 1 mm Hg = 1 torr, the pressure difference is 193 torr. The mercury level is lower in the tube connected to the gas bulb, so the gas must have a higher pressure than the atmosphere. Thus, the pressure difference must be added to the atmospheric pressure to give the gas pressure:

$$\text{Gas pressure} = \text{atmospheric pressure} + \text{pressure difference}$$
$$= 735 \text{ torr} + 193 \text{ torr} = 928 \text{ torr}$$

Practice Problem 9.1

A sample of nitrogen gas is placed in the U-tube apparatus of Figure 9.4. Atmospheric pressure is 722 torr. The mercury column is at 435 mm in the closed tube and 128 mm in the open tube. What is the pressure of the nitrogen in the bulb?

Answer: 415 torr

Figure 9.4 A U-tube is used to measure the pressure of a gas in a closed container.

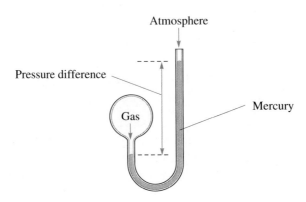

In this book, we will generally use units of either atm or torr because these units are most useful for the kinds of measurements chemists make. However, it may sometimes be necessary to convert to other units. The SI unit of pressure, the pascal, is related to the standard atmospheric pressure as follows:

$$1 \text{ atm} = 101{,}325 \text{ Pa}$$

Units of pounds per square inch (lb/in^2) are usually used to measure gas pressures in containers of compressed gases, such as gas cylinders or automobile tires. This unit can also be related to atmospheres:

$$1 \text{ atm} = 14.7 \text{ lb}/in^2$$

Conversions between these units are most easily carried out by use of their common relationships to the standard atmosphere.

EXAMPLE 9.2

Express the pressure 735 torr in units of atmospheres and of pascals.

Solution The relationship between torr and atom is

$$1 \text{ atm} = 760 \text{ torr}$$

This relationship can be used as the source of a conversion factor:

$$P \text{ (atm)} = 735 \text{ torr} \times \frac{1 \text{ atm}}{760 \text{ torr}} = 0.967 \text{ atm}$$

We can convert to pascals by use of the following equivalence:

$$1 \text{ atm} = 101{,}325 \text{ Pa}$$

$$P \text{ (Pa)} = 0.967 \text{ atm} \times \frac{101{,}325 \text{ Pa}}{1 \text{ atm}} = 9.80 \times 10^4 \text{ Pa}$$

Practice Problem 9.2

The pressure of a gas is 8.25×10^4 Pa. What is this pressure expressed in units of atm and of torr?

Answer: 0.814 atm; 619 torr

9.3 The Gas Laws

▶ Scientists have extensively investigated the various properties of gases, such as those described in the chapter introduction, and have found that under normal atmospheric conditions, all gases exhibit essentially the same physical properties. Such behavior is called *ideal gas behavior*. While they are exhibiting ideal behavior, gases are called **ideal gases.** Because of this consistency, it is possible to describe certain aspects of the behavior of gases in terms of general gas laws. These laws define various relationships of volume, pressure, temperature, and amount.

List the ways in which we can change the volume of air in a balloon.

Boyle's Law

The effect of pressure on the volume occupied by a gas is expressed by **Boyle's law:** in a given mass of gas, at constant temperature, volume varies inversely with pressure. A qualitative view of this relationship is shown in Figure 9.5. A sample of gas is enclosed in a tube with a piston. Placing additional pressure on the piston causes it to move deeper into the tube, compressing the gas to a smaller volume. If the piston is not moving, the pressure on both sides must be equal, so the pressure of the gas has increased. The qualitative result should be intuitively obvious if you consider your own experiences with gases. For example, what happens when you enclose a balloon in your hands and squeeze? Its volume decreases. And what happens to a small gas bubble released at the bottom of a large fish tank? As the bubble rises, the pressure on it decreases because there is not as much water pressing down on it. The bubble grows larger as it rises because of the decreasing pressure. These effects can be illustrated with a plot of volume versus pressure (Figure 9.6). As pressure increases, volume decreases. Doubling the pressure halves the volume, tripling the pressure cuts the volume to one-third the original value, and so on.

To understand this relationship, consider how gas molecules in a container behave. They are constantly moving around, eventually bouncing off the container walls. This impact transmits force and thus pressure to the walls. When a gas is compressed to a smaller volume, the molecules do not have to travel as far to hit the walls. The more frequently the molecules collide with the walls, the more force is transmitted to the walls and the higher the resulting pressure.

Draw pictures of gas molecules showing why pressure increases when volume decreases.

The volume occupied by a gas, then, is inversely proportional to its pressure:

$$V \propto 1/P \qquad \text{or} \qquad V = k_B/P$$

Here, V is volume, P is pressure, and k_B is a proportionality constant. This equation can be rearranged to the following form:

$$PV = k_B$$

The constant should have the same value for all values of pressure and volume,

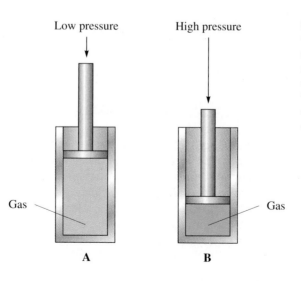

Low pressure   High pressure

Gas   Gas

A   **B**

Figure 9.5 A gas at constant temperature before compression *(A)* and after compression *(B)*. When the pressure is increased, the volume decreases. The product of pressure and volume is a constant, in accord with Boyle's law.

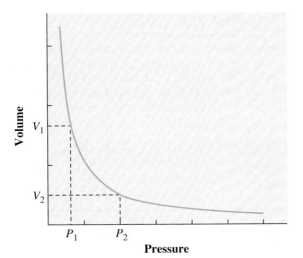

Figure 9.6 As the pressure on a gas increases at constant temperature, the volume occupied by the gas decreases.

provided the amount (in terms of mass or moles) of gas and the temperature remain the same. Thus, the same equation should be observed for different sets of P and V values:

$$P_1V_1 = k_B \qquad P_2V_2 = k_B$$

The subscripts could refer to any two points on the curve in Figure 9.6. If both P_1V_1 and P_2V_2 equal k_B, they must also equal one another:

$$P_1V_1 = P_2V_2$$

This equation is a mathematical expression of Boyle's law and holds for all ideal gases as long as the temperature and mass (or moles) remain constant.

Boyle's law can be used for many pressure-volume calculations. For example, we can calculate what pressure is needed to change the volume occupied by a gas or what pressure will result if the volume is changed. We can also calculate the volume that results from a change in the pressure exerted on a gas, as in the following example.

EXAMPLE 9.3

A balloon contains 512 mL of helium when filled at 1.00 atm. What would be the volume of the balloon if it were subjected to 2.50 atm pressure?

Solution We use Boyle's law and make P_1 the original pressure, 1.00 atm; P_2 the new pressure, 2.50 atm; and V_1 the original volume, 512 mL. V_2 is the new, unknown volume. Boyle's law can be rearranged to solve for V_2:

$$P_1V_1 = P_2V_2$$

$$V_2 = V_1 \times \frac{P_1}{P_2}$$

$$V_2 = 512 \text{ mL} \times \frac{1.00 \text{ atm}}{2.50 \text{ atm}} = 205 \text{ mL}$$

(continued)

A common error in calculations such as this is inverting the two values of pressure. Always check your calculation to see whether it makes sense. In this case, pressure is being increased, so volume must be decreased. If we are to get a smaller volume when we multiply the original volume by the ratio of pressures, that ratio must be less than 1. Therefore, the smaller of the two pressures (P_1 here) must be the numerator of the ratio. The final volume must be smaller than the original volume, as our calculation indicates.

Practice Problem 9.3

What pressure is needed to compress 455 mL of oxygen gas at 2.50 atm to a volume of 282 mL?

Answer: 4.03 atm

Charles's Law

When a balloon is inflated and placed outside in hot sunlight, it gets larger. It may even be stretched beyond its limits and burst. When a balloon is inflated and placed in a bath of very cold liquid nitrogen, it deflates to a very small volume, but it increases to its original volume if allowed to warm up to room temperature again, as shown in Figure 9.7.

Jacques Charles was a French physicist and a pioneer in hot-air and hydrogen balloons. Joseph Louis Gay-Lussac, a French professor of chemistry at the École Polytechnique and the Sorbonne, was also an early balloonist. He ascended to an altitude of 23,000 feet in 1804 to study the atmosphere, a record that held for several decades.

The effect of temperature on volume was investigated in 1787 by Jacques Charles and in 1802 by Joseph Louis Gay-Lussac. These investigators found that a sample of gas of a fixed mass and at a fixed pressure increases in volume linearly with temperature. That is, a plot of volume versus temperature under these conditions is a straight line, as shown in Figure 9.8. This relationship is *Charles's law,* (sometimes known as the *Charles–Gay-Lussac law*).

In their calculations, Charles and Gay-Lussac used the volume occupied by a gas at a temperature of 0°C as a reference point. They found that the gas's volume increased by $\frac{1}{273}$ of the reference volume for each 1°C rise in temperature. Similarly, the volume decreased by $\frac{1}{273}$ of the reference volume for every 1°C decrease in temperature. If we use the best-known value of 273.15, this relationship of volume to temperature can be expressed as

$$V = V°(1 + t/273.15)$$

Figure 9.7 *Left:* When an air-filled balloon is cooled in liquid nitrogen to 77 K, the volume of the air is reduced drastically. *Right:* When the balloon is warmed back to room temperature, the gas regains its original volume.

Figure 9.8 Effect of temperature on the volume of a gas. Plots of volume against temperature for 1.0-g samples of oxygen gas at different pressures can be extrapolated, or extended, in a straight line from the experimental data. All extrapolate to the same temperature, −273.15°C or 0 K, which is called absolute zero.

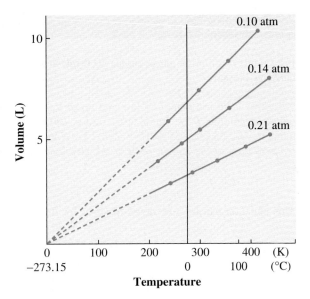

Can we decrease the volume occupied by a gas to zero?

In this statement of Charles's law, V is volume, $V°$ is volume at 0°C, and t is the temperature in degrees Celsius. This relationship implies that a gas would have no volume at all if it were cooled to −273.15°C. We can see this in Figure 9.8, where all the volume-temperature lines intercept at −273.15°C and zero volume. In fact, however, all real gases become liquids before they reach this temperature.

The absolute temperature scale, known as the Kelvin scale, arose from these observations. The zero value on the Kelvin scale (0 K) occurs at −273.15°C. Using the absolute temperature scale yields a simpler version of **Charles's law:** the volume occupied by a gas changes in direct proportion to its absolute temperature (T) at constant pressure.

$$V \propto T$$

The absolute temperature (T) is related to the temperature (t) on the Celsius scale as follows:

$$T = t + 273.15$$

We can thus rearrange the first Charles equation to give

$$V = V°(273.15 + t)/273.15 = (V°/273.15)T$$

Then we can replace $V°/273.15$ with a constant, k_C, and rearrange the resulting equation to give a ratio of volume to absolute temperature:

$$V = k_C T \quad \text{or} \quad V/T = k_C$$

As with Boyle's law, any two sets of the ratio can be equated because both equal the constant:

$$\frac{V_1}{T_1} = \frac{V_2}{T_2}$$

This equation can be used to calculate the change in the volume occupied by a gas resulting from a change in temperature at constant pressure or to calculate the temperature change required to achieve a specific change in volume.

EXAMPLE 9.4

If a sample of chlorine gas occupies 50.0 mL at 100.0°C, what is the volume at 25.0°C at constant pressure?

Solution   We use Charles's law and make V_1 the initial volume, 50.0 mL; V_2 the new, unknown volume; T_1 the initial temperature, 100.0°C + 273.15 = 373.2 K; and T_2 the new temperature, 25.0°C + 273.15 = 298.2 K. *Note that it is absolutely essential to convert temperatures to the Kelvin scale if the simpler form of the equation is to be used:*

$$\frac{V_1}{T_1} = \frac{V_2}{T_2}$$

This equation can be rearranged to solve for the unknown volume:

$$V_2 = V_1 \times \frac{T_2}{T_1}$$

Alternatively, it can be noted that the ratio of temperatures is a correction term to convert a volume to its new value. If temperature is increasing, the volume should become larger, so the ratio should have the larger absolute temperature divided by the smaller. If the temperature is decreasing, as in this case, the opposite is true.

$$V_2 = 50.0 \text{ mL} \times \frac{298.2 \text{ K}}{373.2 \text{ K}} = 40.0 \text{ mL}$$

The volume is indeed smaller, as expected for a decrease in temperature.

Practice Problem 9.4

A sample of carbon monoxide gas occupies 150.0 mL at 25.0°C. It is then cooled at constant pressure until it occupies 100.0 mL. What is the new temperature?

Answer: 198.8 K or −74.4°C

Combined Gas Law

Other investigations of the interdependence of volume, pressure, and temperature can be summarized by the **combined gas law,** which states that for a constant amount of gas, the volume is proportional to the absolute temperature divided by the pressure:

$$V \propto T/P$$

This proportionality can be expressed as the following equation, with k being a constant:

$$V = kT/P \qquad \text{or} \qquad PV/T = k$$

As usual, for any two sets of conditions, the proportionality constant can be deleted:

$$\frac{P_1 V_1}{T_1} = \frac{P_2 V_2}{T_2}$$

This equation is the form we would expect to result from the merging of Boyle's law and Charles's law. If temperature is held constant, the temperature terms cancel out, giving Boyle's law:

$$\frac{P_1 V_1}{\cancel{T_1}} = \frac{P_2 V_2}{\cancel{T_2}}$$

$$P_1 V_1 = P_2 V_2$$

If pressure is held constant, the pressure terms cancel, giving Charles's law:

$$\frac{\cancel{P_1} V_1}{T_1} = \frac{\cancel{P_2} V_2}{T_2}$$

$$\frac{V_1}{T_1} = \frac{V_2}{T_2}$$

The combined gas equation can be used to predict and calculate what happens to one of the gas properties—P, V, or T—when one or both of the other properties are changed. The equation can be rearranged to any form necessary to solve for a given unknown variable, provided the other five variables are known.

EXAMPLE 9.5

If a sample of argon gas occupies 2.50 L at 100.0°C and 5.00 atm, what volume will it occupy at 0.0°C and 1.00 atm?

Solution   We use the combined gas law rearranged to solve for a new volume. V_1 is 2.50 L, V_2 is unknown, P_1 is 5.00 atm, P_2 is 1.00 atm, T_1 is 100.0°C + 273.15 = 373.2 K, and T_2 is 0.0°C + 273.15 = 273.2 K:

$$V_2 = V_1 \times \frac{P_1}{P_2} \times \frac{T_2}{T_1}$$

$$= 2.50 \text{ L} \times \frac{5.00 \text{ atm}}{1.00 \text{ atm}} \times \frac{273.2 \text{ K}}{373.2 \text{ K}} = 9.15 \text{ L}$$

Note that the new volume is simply the old volume multiplied by two factors, a pressure adjustment and a temperature adjustment. The same commonsense approach used in earlier examples to decide how to set up the pressure or temperature ratio can be used here. In this case, the decrease in pressure causes an increase in volume, which is offset somewhat by a decrease in temperature.

Practice Problem 9.5

A sample of hydrogen gas occupies 1.25 L at 80.0°C and 2.75 atm. What volume will it occupy at 185.0°C and 5.00 atm?

Answer: 0.892 L

Gay-Lussac's Law of Combining Volumes

During his investigations of gases and balloons, Gay-Lussac also examined gaseous chemical reactions. In 1808, he measured the volumes occupied by gases

When Gay-Lussac proposed his law of combining volumes, he was working with Claude Berthollet, a French chemist who did not accept Dalton's concept of atoms, which had been proposed in 1804. Gay-Lussac agreed with Berthollet, so he did not recognize the implications of his law.

before and after reaction and found that, under conditions of constant pressure and temperature, the volumes occupied by reacting gases and their gaseous products were always in the ratio of small whole numbers. For example, as shown in Figure 9.9, 2.5 L of hydrogen gas reacts with 2.5 L of chlorine gas to produce 5.0 L of hydrogen chloride gas. In other words, one volume of hydrogen gas reacts with one volume of chlorine gas to form two volumes of hydrogen chloride gas, a ratio of 1:1:2. **Gay-Lussac's law of combining volumes** states that gases combine in simple whole-number volume proportions.

Although Gay-Lussac did not understand the underlying reasons for this behavior, we can now see that the proportions of reacting gases are the same as the proportions of reacting molecules. That hydrogen reacts with an equal volume of chlorine to form twice that volume of hydrogen chloride is explained by the numbers of molecules involved in the reaction:

$$H_2(g) + Cl_2(g) \longrightarrow 2HCl(g)$$

Avogadro's Hypothesis

In 1811, Amadeo Avogadro, an Italian physicist, recognized the implications of Gay-Lussac's law of combining volumes in terms of atomic theory. He hypothesized that the volume occupied by a gas at a given temperature and pressure is proportional to the number of gas molecules and thus to the moles of a gas (n):

$$V \propto n \qquad \text{or} \qquad V = k_A n$$

Draw a picture of gas molecules that illustrates why the volume or pressure should increase as the number of gas molecules increases.

where k_A is a proportionality constant. You can see that this makes sense by thinking again about blowing up a balloon. When you blow in one breath, the balloon expands to a small volume. A second breath expands the balloon to a greater volume. Eventually, the pressure starts to change as well, but the volume depends on the amount of air blown into the balloon.

Avogadro's hypothesis states that at a given pressure and temperature, equal volumes of all gases contain equal numbers of moles (or molecules). Experimentation shows that 1 mol of any ideal gas occupies a volume of 22.414 L at 0°C and 1 atm. This volume is called the **molar volume,** and these conditions are called **standard temperature and pressure (STP).** We can calculate the volume occupied by a gas under any other conditions of temperature and pressure from the volume at STP by using the combined gas law. The volume at STP can be calculated from the number of moles in the sample and the molar volume, 22.414 L.

Figure 9.9 One volume of hydrogen gas reacts with one volume of chlorine gas to produce two volumes of hydrogen chloride gas.

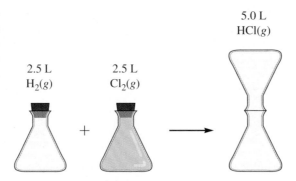

EXAMPLE 9.6

What volume is occupied by 3.00 mol of argon gas at 25.0°C and 0.750 atm?

Solution   First, we calculate the volume at STP, which we will call V_1, by multiplying the number of moles of argon gas by the molar volume:

$$V_1 = 3.00 \text{ mol} \times 22.414 \text{ L/mol} = 67.2 \text{ L}$$

Then we use the combined gas law to calculate the volume at the new conditions. We use the following values: $V_1 = 67.2$ L, $P_1 = 1.00$ atm, $P_2 = 0.750$ atm, $T_1 = 0.0°C + 273.15 = 273.2$ K, and $T_2 = 25.0°C + 273.15 = 298.2$ K. V_2 is the new, unknown volume:

$$V_2 = V_1 \times \frac{P_1}{P_2} \times \frac{T_2}{T_1}$$

$$= 67.2 \text{ L} \times \frac{1.00 \text{ atm}}{0.750 \text{ atm}} \times \frac{298.2 \text{ K}}{273.2 \text{ K}} = 97.8 \text{ L}$$

Because the pressure is less than standard pressure and the temperature is greater than standard temperature, the ratio in both correction terms should be greater than 1, and the final volume should be larger than the initial volume.

Practice Problem 9.6

What volume will 7.50 mol of chlorine gas occupy at 75.0°C and 2.35 atm?

Answer: 91.1 L

Ideal Gas Law

We have seen that the volume (V) occupied by a gas can be related to three other properties, the pressure (P), the absolute temperature (T), and the number of moles (n). These relationships are expressed by the following laws:

Boyle's law: $V = k_B/P$ at constant n, T

Charles's law: $V = k_C T$ at constant n, P

Avogadro's hypothesis: $V = k_A n$ at constant P, T

These three equations can be summarized in a single relationship, for if volume is proportional to each of the three factors, it must be simultaneously proportional to all of them:

$$V \propto nT/P$$

Using a proportionality constant, R, we obtain the following equation:

$$V = RnT/P$$

Rearranging yields an equation known as the **ideal gas law**:

$$PV = nRT$$

Table 9.2 Some Values of the Universal Gas Constant (*R*)

Pressure Unit	Volume Unit	Value of *R*
Atmospheres	Liters	0.08206 L atm/mol K
Atmospheres	Milliliters	82.06 mL atm/mol K
Torr	Liters	62.4 L torr/mol K
Torr	Milliliters	6.24×10^4 mL torr/mol K
Pascals	Cubic decimeters	8.3145 kPa dm^3/mol K

Why is the relationship PV = nRT called the ideal gas law?

The constant *R* is called the *universal gas constant*. Using the observation that 1 mole of a gas at STP occupies 22.414 L, we can calculate the value of *R* as follows:

$$R = PV/nT$$
$$= 1 \text{ atm} \times 22.414 \text{ L}/(1 \text{ mol} \times 273.15 \text{ K}) = 0.08206 \text{ L atm/mol K}$$

This value is valid only if we measure volume in liters and pressure in atmospheres. If we use any other units, we must use other values of *R*, some of which are listed in Table 9.2. In this chapter, we will always use units of liters and atmospheres, converting measurements to these units if necessary, to avoid having to recall more than one value of *R*.

The ideal gas law includes all the information summarized by Boyle's law, Charles's law, Avogadro's hypothesis, and the combined gas law. Like these, the ideal gas law is valid only for an ideal gas. It indicates that the volume occupied by an ideal gas is directly proportional to the absolute temperature and to the number of moles of gas molecules and inversely proportional to the pressure. Having one equation that relates all these variables gives us a useful tool.

9.4 Calculations with the Ideal Gas Law

When we are dealing with conversions between different values of volume, pressure, and temperature, we generally use the combined gas law. The real utility of the ideal gas law arises when we are concerned with the mass or the number of moles of the gas. Let's look at some calculations in which the amount of gas is important.

Moles of a Gas

How many quantities must be known to use the ideal gas law to determine an unknown quantity?

If we are given the *P*, *V*, *T* conditions for a sample of a gas, we have enough information to calculate the number of moles of the gas by using the ideal gas law. We need only rearrange the equation $PV = nRT$ to solve for the number of moles.

Remember that all the other quantities must be in units that match the value used for the universal gas constant.

EXAMPLE 9.7

The volume of a propane cylinder used to operate a torch for home plumbing repairs is 0.960 L. The surrounding atmospheric conditions are 25.0°C and 745.0 torr. How many moles of propane gas remain in the cylinder when it is "empty"? The cylinder is empty when all the liquid propane has vaporized and the pressure and temperature inside the cylinder are the same as atmospheric conditions.

Solution   First, we must convert to the desired units of T and P:

$$T = 25.0°C + 273.15° = 298.2 \text{ K}$$
$$P = 745.0 \text{ torr} \times 1 \text{ atm}/760 \text{ torr} = 0.9803 \text{ atm}$$

Next, we rearrange the ideal gas law and substitute in the appropriate values to calculate the number of moles:

$$PV = nRT$$
$$n = PV/RT$$
$$n = \frac{0.9803 \text{ atm} \times 0.960 \text{ L}}{0.08206 \text{ L atm/mol K} \times 298.2 \text{ K}} = 0.0385 \text{ mol}$$

Practice Problem 9.7

The volume of an oxygen cylinder used as a portable breathing supply is 2.025 L. When the cylinder is emptied at 29.2°C, it has a pressure of 723 torr. How many moles of oxygen gas remain in the cylinder?

Answer: 0.0776 mol

Mass of a Gas

When you know the number of moles of a gas, it is a straightforward matter to convert to mass, provided the molar mass is known. Again, you solve for moles using the ideal gas law.

EXAMPLE 9.8

What mass of propane is left in the cylinder described in Example 9.7?

Solution   The molar mass (*MM*) of propane is obtained from its formula, C_3H_8:

$$MM = (3 \text{ mol C} \times 12.011 \text{ g/mol C}) + (8 \text{ mol H} \times 1.008 \text{ g/mol H}) = 44.10 \text{ g/mol}$$

The mass (*m*) is the molar mass multiplied by the number of moles:

$$m = n \times MM$$
$$= 0.0385 \text{ mol} \times 44.10 \text{ g/mol} = 1.70 \text{ g}$$

(continued)

Practice Problem 9.8

The volume of an oxygen cylinder used as a portable breathing supply is 1.85 L. What mass of oxygen gas remains in the cylinder when it is "empty" if the pressure is 755 torr and the temperature is 18.1°C?

Answer: 2.46 g

Gas Density

▶ The density of gases is an important property that affects many aspects of the earth's atmosphere, described in the chapter introduction, and many of the interactions between humans and the atmosphere. Sea breezes and land breezes, for example, exist because the density of air changes as the temperature changes. Air molecules move from cooler, more dense regions to warmer, less dense regions in an attempt to equalize the densities. We experience this movement as a wind or breeze. The ubiquitous Goodyear blimp can hover above sporting events because the density of helium is lower than the density of air under identical conditions. We can calculate the density of a gas by using the ideal gas law because density is related to mass.

How much helium is required to lift a blimp? What information do you need to answer this question?

First, we note that the number of moles of a substance can be related to its molar mass:

$$n = m/MM$$

Substituting this equivalence into the ideal gas law, we obtain

$$PV = (m/MM)RT$$

Recalling that density is mass divided by volume, we can rearrange this equation to obtain an expression for the density of an ideal gas:

$$d = m/V = P \times MM/RT$$

This expression can be used to calculate the density of a gas under given conditions of pressure and temperature. Note that the identity of the gas must be known because the density depends in part on the molar mass of the substance.

EXAMPLE 9.9

Assuming that gasoline has the formula C_8H_{18}, calculate the density of gasoline vapor at 125.0°C and 720.0 torr.

Solution   We must know the molar mass of C_8H_{18}:

$MM = (8 \text{ mol C} \times 12.011 \text{ g/mol C}) + (18 \text{ mol H} \times 1.008 \text{ g/mol H}) = 114.2 \text{ g/mol}$

We must also convert T and P to desired units:

$$T = 125.0°C + 273.15° = 398.2 \text{ K}$$
$$P = 720.0 \text{ torr} \times 1 \text{ atm}/760 \text{ torr} = 0.9474 \text{ atm}$$

Now we can use the ideal gas law, rearranged to calculate density:

$$d = m/V = P \times MM/RT$$

$$d = \frac{0.9474 \text{ atm} \times 114.2 \text{ g/mol}}{0.08206 \text{ L atm/mol K} \times 398.2 \text{ K}} = 3.311 \text{ g/L}$$

Practice Problem 9.9

Bromine gas has the formula Br_2. Calculate the density of bromine gas at 50.0°C and 785.0 torr.

Answer: 6.22 g/L

Molar Mass of a Gas

Knowing the molar mass of an unknown substance can enable us to identify it. If the substance can be obtained in the form of a gas, its molar mass is particularly easy to determine.

The number of moles of a substance is related to its mass and its molar mass, as we have already seen:

$$n = m/MM$$

Substituting this relationship into the ideal gas law yields

$$PV = (m/MM)RT$$

We have rearranged this equation to solve for density. We can also rearrange it to solve for molar mass:

$$MM = mRT/PV$$

Because of this relationship, it is possible to measure the molar mass of a gas by determining the volume occupied by a known mass of gas under specified conditions.

EXAMPLE 9.10

Freon-11 is one of a group of compounds known as chlorofluorocarbons, with a general empirical formula CF_nCl_{4-n}. A sample of the refrigerant Freon-11 is vaporized. If 0.2173 g of Freon occupies 43.6 mL at 50.0°C and 735 torr, what is its molar mass?

Solution First, we must convert to appropriate units:

$$P = 735 \text{ torr} \times 1 \text{ atm}/760 \text{ torr} = 0.967 \text{ atm}$$

$$V = 43.6 \text{ mL} \times 1 \text{ L}/1000 \text{ mL} = 0.0436 \text{ L}$$

$$T = 50.0°C + 273.15 = 323.2 \text{ K}$$

Now we substitute values of the variables into the equation $MM = mRT/PV$:

$$MM = \frac{0.2173 \text{ g} \times 0.08206 \text{ L atm/mol K} \times 323.2 \text{ K}}{0.967 \text{ atm} \times 0.0436 \text{ L}} = 137 \text{ g/mol}$$

(continued)

Of the possible identities for Freon-11, we can select $CFCl_3$ because its molar mass calculated from the sum of the molar masses of the component atoms is 137.4 g/mol. None of the other possible formulas would give a molar mass close to the measured value.

Practice Problem 9.10

If 1.48 g of an unknown gas occupies 132 mL at 25.0°C and 722 torr, what is its molar mass?

Answer: 289 g/mol

Stoichiometric Calculations

Stoichiometric calculations for reactions of gases may involve not only moles or mass of reactants and products but also volumes of gaseous species. Recall that Gay-Lussac found that volumes of gases in reactions occur in small whole-number ratios. These ratios correspond to the stoichiometric coefficients in the balanced equation for the reaction. Consider the reaction of silane with chlorine to form silicon tetrachloride:

$$SiH_4(g) + 4Cl_2(g) \longrightarrow SiCl_4(g) + 4HCl(g)$$

Silane, SiH_4, is a colorless gas with a repulsive odor. It decomposes slowly in the presence of water and ignites in air at elevated temperatures. Silane is made by the reaction of aluminum silicide or magnesium silicide with hydrochloric acid: $Mg_2Si(s) + 4HCl(aq) \rightarrow 2MgCl_2(aq) + SiH_4(g)$. It is used as a source of very pure silicon for semiconductors.

If 1.19 L of $SiH_4(g)$ at 25°C and 720 torr is allowed to react with 4.76 L of $Cl_2(g)$ under the same conditions, they will form exactly 1.19 L of $SiCl_4(g)$ and 4.76 L of $HCl(g)$. The ratio of these volumes is 1 SiH_4:4 Cl_2:1 $SiCl_4$:4 HCl.

Simple conversions of this sort work properly only if all the substances are gases at the same pressure and temperature. A more general approach can be developed by use of the ideal gas law. This approach combines the calculation of moles of a gas from volume, or vice versa, with stoichiometric calculations to find volumes occupied by a gas or moles of a gas in a chemical reaction.

Product Volume from Reactant Volume   We can calculate the volume of one gaseous substance from the volume of another gaseous substance involved in a chemical reaction. If temperature and pressure are constant, the volume-to-volume relationship is given by Gay-Lussac's law of combining volumes. If these conditions vary, the mole-to-volume relationship is given by the ideal gas law and the mole-to-mole relationship is provided by the balanced chemical equation.

EXAMPLE 9.11

A tank of hydrogen gas has a volume of 8.00 L and an internal pressure of 50.0 atm at a temperature of 25.0°C. What volume of gaseous water is produced by the following reaction at 100.0°C and 0.947 atm if all the hydrogen gas reacts with copper oxide?

$$CuO(s) + H_2(g) \longrightarrow Cu(s) + H_2O(g)$$

Solution   First, we calculate the moles of hydrogen gas:

$$\text{mol } H_2 = \frac{50.0 \text{ atm} \times 8.00 \text{ L}}{0.08206 \text{ L atm/mol K} \times 298.2 \text{ K}} = 16.3 \text{ mol}$$

Next, we use the coefficients from the balanced equations to calculate the moles of water:

$$\text{mol } H_2O = 16.3 \; \cancel{\text{mol } H_2} \times \frac{1 \text{ mol } H_2O}{1 \; \cancel{\text{mol } H_2}} = 16.3 \text{ mol } H_2O$$

Finally, we calculate the volume of water vapor, using the ideal gas law solved for V:

$$V = nRT/P$$
$$= 16.3 \; \cancel{\text{mol}} \times 0.08206 \text{ L } \cancel{\text{atm}}/\cancel{\text{mol}} \; \cancel{K} \times 373.2 \; \cancel{K}/0.947 \; \cancel{\text{atm}} = 527 \text{ L}$$

Practice Problem 9.11

A tank of hydrogen gas has a volume of 7.49 L and an internal pressure of 22.0 atm at a temperature of 32.0°C. What volume of gaseous water is produced by the following reaction at 100.0°C and 0.975 atm if all the hydrogen gas reacts with iron(III) oxide?

$$Fe_2O_3(s) + 3H_2(g) \longrightarrow 2Fe(s) + 3H_2O(g)$$

Answer: 207 L

Moles or Mass from Volume   It is often necessary to calculate the number of moles of a product from a known volume of a gaseous reactant. For example, we might want to know how much chromium(III) chloride, shown in Figure 9.10, could be prepared from the reaction of a given volume of chlorine gas with chromium metal, as in the following equation:

$$2Cr(s) + 3Cl_2(g) \longrightarrow 2CrCl_3(s)$$

In this procedure, the volume of the reactant chlorine is used to calculate the number of moles of the reactant. This result is used with the coefficients from the balanced chemical equation to calculate the moles of product $CrCl_3$. The mass of this product can be obtained from the moles of product and its molar mass.

EXAMPLE 9.12

If 3.45 L of $CO_2(g)$, measured at 45.0°C and 1.37 atm, is passed through a bed of lime (CaO), how many moles of product calcium carbonate are formed? How many grams? The balanced equation is as follows:

$$CaO(s) + CO_2(g) \longrightarrow CaCO_3(s)$$

(continued)

Figure 9.10 Chromium(III) chloride, $CrCl_3$, is a lustrous, violet solid that feels greasy. It also exists as a variety of hydrates, with colors ranging from green to purple, all having the formula $CrCl_3 \cdot 6H_2O$. Chromium(III) chloride is used in tanning leather, to inhibit corrosion, as a waterproofing agent, and in dyeing textiles.

Solution   First, we use the ideal gas law solved for n to calculate the moles of CO_2:

$$n = PV/RT$$

$$= \frac{1.37 \text{ atm} \times 3.45 \text{ L}}{0.08206 \text{ L atm/mol K} \times (45.0° + 273.15)\text{K}} = 0.181 \text{ mol}$$

We calculate the number of moles of calcium carbonate from the number of moles of carbon dioxide by means of the coefficients in the balanced equation:

$$\text{Moles } CaCO_3 = 0.181 \text{ mol } CO_2 \times \frac{1 \text{ mol } CaCO_3}{1 \text{ mol } CO_2} = 0.181 \text{ mol } CaCO_3$$

Then we obtain the mass of calcium carbonate from its molar mass, 100.1 g/mol:

$$\text{g } CaCO_3 = 0.181 \text{ mol } CaCO_3 \times \frac{100.1 \text{ g } CaCO_3}{1 \text{ mol } CaCO_3} = 18.1 \text{ g}$$

Practice Problem 9.12

If 2.38 L of SO_3 gas, measured at 65.0°C and 1.05 atm, is passed through liquid water, how many moles of the product sulfuric acid are formed? How many grams? The balanced equation is

$$SO_3(g) + H_2O(l) \longrightarrow H_2SO_4(l)$$

Answer: 0.0900 mol; 8.83 g

Calculating the volume of a product from the mass of a reactant in a chemical reaction proceeds similarly. The steps in this procedure are simply carried out in the reverse order of those just described.

9.5   Dalton's Law

So far we have been talking about the properties of pure gases. ▶ Mixtures of gases, such as the earth's atmosphere (described in the chapter introduction), have the same properties, provided all gases in the mixtures are ideal gases. We often deal with mixtures of gases when we conduct chemical reactions. Even when a reaction produces only one gaseous product, the procedure used to collect this product may introduce another gas. For example, a technique often used to collect gases from a chemical reaction is the displacement of water from an inverted bottle, as shown in Figure 9.11. This technique involves bubbling the gas through a tube into an inverted bottle filled with water until the bottle is full. For example, oxygen gas can be produced by heating solid potassium bromate just past its melting point:

How can we determine the properties of a gas in a mixture?

$$2KBrO_3(l) \xrightarrow{\Delta} 2KBr(s) + 3O_2(g)$$

RULES OF REACTIVITY

When heated, some metal oxoanion salts, particularly those of the halogens, decompose to release oxygen gas.

Because oxygen gas is not very soluble in water and does not react with it, this technique should collect the gas in a pure state, uncontaminated by the components of air; but this is not the case. When the gas is bubbled through the water, some of the water evaporates, so the collected oxygen gas consists of a mixture of the oxygen and some water vapor. The amount of water contained in the gas is most readily measured by the pressure it exerts at that temperature, called the *vapor pressure*. Vapor pressures for water are listed in Table 9.3.

Figure 9.11 Collecting a gaseous reaction product by water displacement.

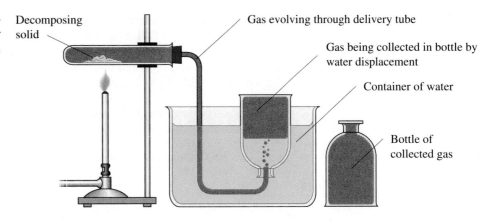

Decomposing solid

Gas evolving through delivery tube

Gas being collected in bottle by water displacement

Container of water

Bottle of collected gas

To determine the pressure exerted by the gas, we need only to subtract the vapor pressure of water from the total pressure of the mixture. This calculation is an application of **Dalton's law of partial pressures,** which states that gases in a mixture behave independently and exert the same pressure they would exert if they were in the container alone. Thus, the total pressure exerted by a mixture of gases is the sum of the "partial" pressures of the component gases:

$$P_{\text{total}} = P_A + P_B + P_C + \cdots$$

For our example involving water vapor, the equation can be stated as follows:

$$P_{\text{total}} = P_{\text{oxygen}} + P_{\text{water}}$$

Table 9.3 Vapor Pressure of Water at Various Temperatures

Temperature (°C)	Vapor Pressure (torr)	Temperature (°C)	Vapor Pressure (torr)
0	4.6	28	28.3
5	6.5	29	30.0
10	9.2	30	31.8
15	12.8	35	42.2
16	13.6	40	55.3
17	14.5	45	71.9
18	15.5	50	92.5
19	16.5	60	149.4
20	17.5	70	233.7
21	18.6	80	355.1
22	19.8	90	525.8
23	21.1	100	760.0
24	22.4	110	1074.6
25	23.8	150	3570.5
26	25.2	200	11659.2
27	26.7	300	64432.8

The total pressure in this case is atmospheric pressure because the pressure inside the bottle is the same as that outside the bottle. The vapor pressure of water, found in Table 9.3, is subtracted from the total pressure to give the partial pressure of the gas. This pressure can be combined with the volume of gas and the temperature to calculate the moles of gas collected.

EXAMPLE 9.13

Suppose we produce gaseous oxygen by the following reaction:

$$2KClO_3(s) \xrightarrow{\Delta} 2KCl(s) + 3O_2(g)$$

We collect 1.500 L of O_2 over water at 27°C and 758 torr. What volume does the oxygen gas alone occupy at STP? The vapor pressure of water at 27°C is 26.7 torr.

Solution First, we calculate the pressure of the oxygen gas using a rearrangement of Dalton's law:

$$\begin{aligned} P_{\text{oxygen}} &= P_{\text{total}} - P_{\text{water}} \\ &= 758 \text{ torr} - 26.7 \text{ torr} = 731 \text{ torr} \end{aligned}$$

We must convert the pressure in torr to units of atm:

$$P_{\text{oxygen}} = 731 \text{ \sout{torr}} \times 1 \text{ atm}/760 \text{ \sout{torr}} = 0.962 \text{ atm}$$

Then we can calculate the number of moles of oxygen using the ideal gas equation solved for n:

$$\begin{aligned} n &= PV/RT \\ &= \frac{0.962 \text{ \sout{atm}} \times 1.500 \text{ \sout{L}}}{0.08206 \text{ \sout{L atm}/mol \sout{K}} \times 300 \text{ \sout{K}}} = 0.0586 \text{ mol} \end{aligned}$$

Now we can calculate the volume at STP using the same equation solved for V:

$$\begin{aligned} V &= nRT/P \\ &= 0.0586 \text{ \sout{mol}} \times 0.08206 \text{ L \sout{atm}/\sout{mol} \sout{K}} \times 273 \text{ \sout{K}}/1.00 \text{ \sout{atm}} = 1.31 \text{ L} \end{aligned}$$

Practice Problem 9.13

Suppose the following reaction is used to produce gaseous oxygen:

$$2KClO_3(s) \xrightarrow{\Delta} 2KCl(s) + 3O_2(g)$$

We collect 2.25 L of O_2 over water at 18.0°C and 722.8 torr. What volume would the oxygen gas alone occupy at STP?

Answer: 1.96 L

9.6 Graham's Law of Effusion and Diffusion

Gas particles are continuously moving around, a feature associated with the lack of order and the large distance between particles that characterize gases. You may have noticed that dust particles suspended in air seem to bounce around. This motion, called *Brownian motion,* is caused by gas particles colliding with the dust particles.

How do we know gas particles are moving?

There are essentially two ways to detect the motion of gas particles: through **effusion,** the passage of a gas through a small opening (a pinhole) or porous barrier, and through **diffusion,** the movement of gas molecules through space. If you have ever had a balloon filled with helium, you have observed effusion. The balloon loses its buoyancy after about a day and eventually shrinks to a fraction of its original size. These effects are caused by the escape of helium atoms through the natural holes in the structure of the rubber, or effusion. Anyone with a normal sense of smell is quite familiar with diffusion. Substances such as perfume or dead fish can be smelled from fairly long distances because the gaseous molecules that cause the odors diffuse rapidly through the atmosphere.

Rates of Effusion and Diffusion

Because gas particles occasionally collide with one another, their courses are not completely straight. A gas particle will travel in a straight line until it hits another particle, and then both will change their direction of travel. Thus, the rate at which a gas diffuses across a room is not a straightforward measure of the rate at which each gas particle is moving. There should be some connection, though. We would expect that the faster that gas molecules moved, the faster they would diffuse. Similarly, the rate of effusion should be closely related to the rate of motion of the gas particles.

Molecules of different gases move at different rates. Lighter molecules move faster than heavier molecules. And we do find a relationship between these rates and the rates of both effusion and diffusion. We can observe this relationship in the effusion of gases out of balloons, for example. Hydrogen gas escapes faster than helium gas, which escapes faster than nitrogen gas.

Graham's law states that the rate of effusion of a gas is inversely proportional to the square root of the gas's molar mass. According to Graham's law, the rates of effusion (r) of two gases at the same temperature and pressure are related as follows:

$$\frac{r_1}{r_2} = \frac{\sqrt{MM_2}}{\sqrt{MM_1}}$$

For example, the rate of effusion of oxygen gas ($MM = 32$ g/mol) is about 25% that of hydrogen ($MM = 2$ g/mol). The ratio of these molar masses is 16 and the square root of this ratio is 4, so hydrogen should effuse four times as fast as oxygen:

$$\frac{r_{H_2}}{r_{O_2}} = \frac{\sqrt{MM_{O_2}}}{\sqrt{MM_{H_2}}} = \frac{\sqrt{32}}{\sqrt{2}} = \sqrt{16} = 4$$

$$r_{H_2} = 4 \times r_{O_2}$$

Simple methods for studying the rates of effusion and diffusion of gases are shown in Figures 9.12 and 9.13. Experiments with rates of diffusion show that they too follow Graham's law. The rates of diffusion and effusion are not identical, but rates of diffusion are related in the same way as rates of effusion.

We saw earlier that the density of a gas is proportional to its molar mass ($d = MM \times P/RT$), so Graham's law can also be expressed in terms of relative densities:

$$\frac{r_1}{r_2} = \frac{\sqrt{d_2}}{\sqrt{d_1}}$$

Figure 9.12 We can measure
the rate of effusion of a gas
by inverting a tube of the gas
into a cylinder of water. The
top of the tube is open to the
atmosphere through a short
section of capillary tube. The
times required for a selected
volume of two different gases
to escape from the tube provide relative rates. For example, 10.0 mL of CH₄ takes 94 s
to effuse through a particular
capillary, whereas the same
volume of O₂ takes 133 s to
effuse. The rates are
0.106 mL/s for CH₄ and
0.0752 mL/s for O₂. The ratio
of rates is 1.41, the same as
the square root of the inverse
ratio of molar masses.

Figure 9.12 We can measure the rate of effusion of a gas by inverting a tube of the gas into a cylinder of water. The top of the tube is open to the atmosphere through a short section of capillary tube. The times required for a selected volume of two different gases to escape from the tube provide relative rates. For example, 10.0 mL of CH_4 takes 94 s to effuse through a particular capillary, whereas the same volume of O_2 takes 133 s to effuse. The rates are 0.106 mL/s for CH_4 and 0.0752 mL/s for O_2. The ratio of rates is 1.41, the same as the square root of the inverse ratio of molar masses.

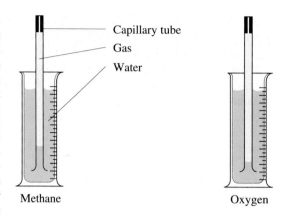

Capillary tube
Gas
Water

Methane

Oxygen

Thus, a measurement of the relative rates of effusion or diffusion of two gases can be used to determine their relative molar masses or their relative densities.

EXAMPLE 9.14

A sample of HI(g) diffuses through air at the rate of 0.0962 cm/s, and the diffusion rate of butylamine gas is 0.126 cm/s. Given that the molar mass of HI is 127.9 g/mol, determine the molar mass of butylamine.

Solution   We can apply Graham's law to this problem:

$$\frac{r_1}{r_2} = \frac{\sqrt{MM_2}}{\sqrt{MM_1}}$$

First, we calculate the relative rates of diffusion, making r_1 the diffusion rate of HI and r_2 the diffusion rate of butylamine:

$$r_1/r_2 = r_{HI}/r_{butylamine}$$
$$= (0.0962 \text{ cm/s})/(0.126 \text{ cm/s}) = 0.763$$

Now we can substitute this value and the value for MM_1, the molar mass of HI, into the equation. MM_2, the molar mass of butylamine, is the unknown:

$$0.763 = \sqrt{MM_{butylamine}}/\sqrt{127.9 \text{ g/mol}}$$

We then square each side of the equation and solve:

$$MM_{butylamine}/127.9 \text{ g/mol} = (0.763)^2$$
$$MM_{butylamine} = 127.9 \text{ g/mol} \times 0.582 = 74.4 \text{ g/mol}$$

For comparison, we can calculate the molar mass of butylamine from the molar masses of the component elements, given the formula: $C_4H_9NH_2$. This theoretical molar mass is 73.1 g/mol.

Practice Problem 9.14

A sample of tear gas is released in the last row of seats in a lecture hall. At the same time, a sample of laughing gas is released in the first row of seats. After a time, the last 10 rows of students are crying and the first 20 rows of students are laughing. If laughing gas is N_2O, with a molar mass of 44.0 g/mol, what is the molar mass of tear gas?

Answer: 176 g/mol

HCl(g)   NH₄Cl(s)   NH₃(g)

Figure 9.13 We can measure the relative rates of diffusion of two gases by introducing a gas at each end of a horizontal tube. The gases must react in such a way that they can be detected when they meet. When reaction begins, we measure the distance each has traveled. These distances are related to the rates of diffusion. Two gases commonly used in this experiment are $NH_3(g)$ and $HCl(g)$; they form a white powder, $NH_4Cl(s)$, when they react.

Figure 9.14 The gaseous diffusion plant, now at Paducah, Kentucky, used to separate uranium isotopes.

Separation of Isotopes

The fact that different gases diffuse at different rates can be used to separate gases of different mass. The first large-scale application of gaseous diffusion was the separation of the uranium isotopes needed for the development of the atomic bomb. This separation was carried out at Oak Ridge National Laboratories in Tennessee as part of the Manhattan Project of World War II. Gaseous diffusion is still used to isolate the lighter isotope of uranium (Figure 9.14).

Uranium consists primarily of two isotopes, U-235 and U-238. Only U-235 is useful for atomic fission reactions, so this isotope must be separated from uranium samples and concentrated. The U-235 in naturally occurring uranium is present in relatively low abundance (0.7%). To achieve a separation, the uranium is reacted with fluorine, giving UF_6. This substance is a solid at room temperature, but it vaporizes at 56.5°C, so it is readily converted to a gas. This gas is passed through a porous barrier. The two uranium isotopes differ only in mass: $^{235}UF_6$ has a molar mass of 349.03 g/mol; $^{238}UF_6$ has a molar mass of 352.04 g/mol. The lighter U-235 isotope passes through the barrier slightly faster, so the gas becomes enriched in the lighter isotope. To achieve the needed degree of enrichment, the gas must be passed through thousands of barriers.

Uranium hexafluoride, UF_6, is a volatile, white, crystalline solid that melts at 64.8°C and sublimes at 56.5°C. It is formed by the reaction between uranium metal and fluorine gas.

Fluorine has only one isotope, ^{19}F. Explain why this is relevant to the selection of UF_6 for separating uranium isotopes.

EXAMPLE 9.15

Calculate the relative rates of effusion of $^{235}UF_6$ (MM = 349.03 g/mol) and $^{238}UF_6$ (MM = 352.04 g/mol).

Solution Even though the two gases are chemically the same, Graham's law still pertains:

$$\frac{r_1}{r_2} = \frac{\sqrt{MM_2}}{\sqrt{MM_1}}$$

We can substitute appropriate values into the equation and solve it:

$$\frac{r_{235}}{r_{238}} = \frac{\sqrt{352.04 \text{ g/mol}}}{\sqrt{349.03 \text{ g/mol}}} = 1.0043$$

(continued)

This result means that the lighter isotope passes through the barrier 1.0043 times as fast as the heavier isotope, leading to a gaseous mixture that is slightly enriched in the lighter isotope.

Practice Problem 9.15

Calculate the relative rates of effusion of HCl(g) ($MM = 36.46$ g/mol) and DCl(g) ($MM = 37.46$ g/mol). D is deuterium, an isotope of hydrogen with a molar mass of 2.01 g/mol.

Answer: 1.014

9.7   Kinetic-Molecular Theory of Gases

What properties of a gas are necessary for it to exhibit ideal behavior?

In discussing the behavior of gases, this chapter has mentioned several times that what is being described is ideal gas behavior. What does this statement mean? The simplest response is that the behavior strictly obeys the ideal gas law and other simple equations. But here, as elsewhere, it is helpful to have a theoretical model that will allow us to think more systematically about the behavior of interest. The **kinetic-molecular theory of gases** is such a model. This theory puts forth five postulates that allow us to describe and predict ideal gas behavior:

1. **Gases are composed of small and widely separated particles (molecules or atoms).** Because the particles are widely separated, the actual volume of the particles is very small compared with the total volume occupied by the gas. The total volume occupied by a gas is mostly empty space. This postulate correctly predicts that the volume occupied by a gas is much larger than that of a liquid or a solid, where the particles are much closer together. Thus, gases have relatively low densities. They are also highly compressible because the particles can easily be squeezed much closer together.

Some size comparisons are useful. The atomic radius of xenon is 130 pm. Using geometric formulas, we can calculate the volume of a xenon atom to be 9.20×10^{-30} m^3, whereas the actual volume of 1 mol of xenon atoms is 5.54×10^{-6} m^3, or 5.54×10^{-3} L. However, the total volume occupied by 1 mol of gaseous xenon at STP is 22.4 L. Thus, only 0.025% of the volume occupied by the gas is actually taken up by the atoms.

2. **Molecules of a gas behave independently of one another.** Because gas particles are far apart, they do not influence one another; there are essentially no intermolecular forces of attraction or repulsion. This postulate explains the existence of Dalton's law of partial pressures. If gas particles are independent of one another, the presence of more than one kind of particle is irrelevant to the total pressure exerted by the gas. Only the total number of particles is important in determining the pressure.

3. **Each molecule in a gas is in rapid, straight-line motion** until it collides with another molecule or with its container. Collisions of gas particles are perfectly *elastic*—that is, energy may be transferred from one particle to another, but there is no net loss of energy. This postulate explains the diffusion of gases through space and the expansion of gases to fill their containers.

4. **The pressure of a gas arises from the sum of the collisions of the particles with the walls of the container.** This postulate explains Boyle's law, which states

that pressure is inversely proportional to volume. The smaller the volume of the container, the more collisions per unit area and therefore the greater the pressure. This postulate also predicts that pressure should be proportional to the number of moles of gas particles. The more gas particles, the greater the number of collisions with the walls, so the greater the pressure.

5. **The *average* kinetic energy of gas particles depends only on the absolute temperature.** This postulate provides insight into the motion of particles in gases, which affects pressure as well as effusion and diffusion. The postulate also leads us to an understanding of how gases condense into liquids, discussed in the next section. Let's look at it more closely.

First of all, what is average kinetic energy? Not all gas particles move at the same speed, or velocity. In a sample of a gas, there is a distribution of velocities. Because the velocity of a gas particle depends in part on its kinetic energy, there is also a distribution of kinetic energies. A typical distribution of velocities is shown in Figure 9.15 for two different temperatures. At low temperatures, the range of velocities (and kinetic energies) for most of the particles is fairly narrow. The average (or mean) velocity is the value at the maximum in the curve. This average velocity, which is observed for a relatively small number of molecules, is the value used to calculate the average kinetic energy.

The rest of the molecules have velocities either lower or higher than the average value. Thus, more particles have the average kinetic energy than any other energy, but significant numbers of particles have kinetic energies either higher or lower than the average. As the temperature is increased, the average velocity (and kinetic energy) of the collection of particles increases. The number actually having the average value decreases, however, and more particles have velocities and kinetic energies higher and lower than the average. The behavior of individual gas molecules is thus rather complex. But the *average* behavior of a collection of gas molecules can be described by two equations resulting from postulate 5.

The first equation states that the average kinetic energy due to molecular motion is directly proportional to absolute temperature:

$$KE_{av} = \tfrac{3}{2}kT$$

Figure 9.15 Distribution of velocities of gaseous particles.

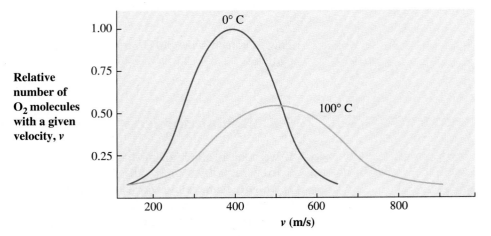

In this equation (which is the same for all gas molecules), k, the Boltzmann constant, is equal to the universal gas constant divided by Avogadro's number:

$$k = R/N$$

Classical mechanics uses another expression for kinetic energy, one that relates it to m, the mass of a gas particle, and v, its velocity:

$$KE = \tfrac{1}{2}mv^2$$

For a collection of particles, not all of which have identical velocities, we adjust this equation to calculate an average kinetic energy by using an average velocity of the particles:

$$KE_{av} = \tfrac{1}{2}m(v_{av})^2$$

These relationships explain the direct proportionality between pressure and absolute temperature. A higher temperature means a higher average kinetic energy, which in turn means a greater average velocity. If the particles are moving faster, they strike the walls more frequently and at greater energy, so the pressure increases as temperature increases.

The two equations for average kinetic energy also allow us to explain the behavior described by Graham's law of diffusion. Two different gases at the same temperature have the same average kinetic energy:

$$KE_{1,av} = \tfrac{3}{2}kT \text{ and } KE_{2,av} = \tfrac{3}{2}kT$$

These two expressions may be equated:

$$KE_{1,av} = KE_{2,av}$$

Because KE_{av} equals not only $\tfrac{3}{2}kT$ but also $\tfrac{1}{2}m(v_{av})^2$, we can substitute as follows:

$$\tfrac{1}{2}m_1(v_{1,av})^2 = \tfrac{1}{2}m_2(v_{2,av})^2$$

Thus, the following must be true:

$$m_1/m_2 = (v_{2,av}/v_{1,av})^2$$

This expression can be rearranged into the same form as Graham's law:

$$v_{2,av}/v_{1,av} = (m_1/m_2)^{\frac{1}{2}}$$

Thus, Graham's law arises from the expression for the average velocity of gas particles.

9.8   Deviations from Ideal Behavior

Why do gases liquefy under some conditions?

If we cool a sample of methane, it liquefies when the temperature drops below $-161°C$. But if we apply sufficient pressure to the methane gas, it can be liquefied at a higher temperature. For example, at $-82°C$, pressures of about 45 atm cause liquefaction. This behavior is not predicted by the gas laws, which apply perfectly and under all conditions only to ideal gases. All real gases deviate to some extent from ideal behavior. Why do we see these deviations?

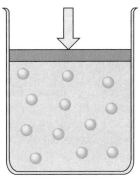

A Low pressure

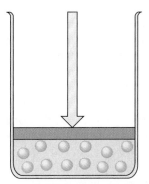

B High pressure

Figure 9.16 At low pressure (A), gas molecules are far apart, so their volume is only a small fraction of the total volume occupied by the gas. At high pressure (B), the volume of the gas molecules accounts for a higher proportion of the total gas volume.

Two critical assumptions are made in the model for ideal gas behavior, the kinetic-molecular theory:

1. Gas molecules occupy no volume.
2. There are no attractive forces between gas molecules.

As long as the pressure does not get much higher than atmospheric pressure and the temperature does not fall much below 0°C, molecules of a real gas are so far apart and occupy so large a total volume that these assumptions are reasonably valid. However, at high pressures, where molecules are close together, and at low temperatures, where they are moving relatively slowly, these two assumptions are no longer valid for real gases.

The pressure of a gas, as we have seen, can be affected by the amount of gas, the volume, and the temperature. The pressure will increase if the volume is decreased, for example. When the pressure is very high, the volume is very low. As a result, the fraction of the volume actually occupied by gas molecules may become significant, as shown in Figure 9.16. When pressures are extremely high, the gas can no longer be compressed by increases in pressure, because the molecules cannot get any closer together. The measured volume of a real gas is greater than the predicted volume of an ideal gas under these conditions, so the pressure-volume relationship is no longer valid.

Under conditions where molecules are close together (high pressure) or where they are moving slowly (low temperature), interactions between molecules may become important. Attractive forces between the molecules result in fewer collisions with the walls than is expected for an ideal gas, leading to a pressure that is lower than expected.

But why do the gas molecules attract one another? Actually, gas molecules that are relatively small, with their electrons held tightly by the nucleus, do not have much attraction for one another. Such molecules (or atoms) are said to have a low **polarizability**—that is, they are not deformed much by electrostatic forces in their vicinity. Larger molecules, with their electrons held more loosely, are more polarizable. They can be deformed by charges in their vicinity, giving them temporary polarity, which causes them to be attracted to one another.

The attractive forces that arise from polarization of adjacent nonpolar molecules or atoms are called **London dispersion forces.** The origin of these forces is shown in Figure 9.17. If the electrons in one atom are temporarily located on one side of that atom, it becomes a temporary dipole, which attracts the electrons in an adjacent atom, converting that atom into a temporary dipole. The temporary dipoles weakly attract one another. These attractive forces are relatively small compared with forces between permanent dipoles but are significant in causing deviations from ideal behavior in gases.

Intermolecular forces of the type illustrated in Figure 9.17 become more important at low temperatures. Only adjacent molecules or atoms can attract one another in this manner. The higher the temperature, the faster the molecules are moving and the less time they spend adjacent to one another.

Figure 9.17 Because of electron movements, adjacent molecules form temporary dipoles that interact with one another, resulting in weak attractive forces.

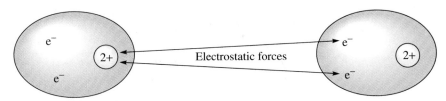

Electrostatic forces

A simple way to examine deviations from ideality is to look at the **compressibility factor,** z, a measure of the change in volume with change in pressure described by the following equation:

$$z = PV/nRT$$

If a gas is ideal, it obeys the ideal gas law ($PV = nRT$) and the compressibility factor has a value of 1. However, for most gases, the value of z departs from a value of 1 as the pressure exerted on the gas increases. The most common behavior involves an initial negative deviation followed by positive deviations, as shown in Figure 9.18. Consider oxygen gas at 0°C as an example. At near atmospheric pressure, z is equal to 1. At increased but still low pressures, the value of z is less than 1. At about 350 atm, the value of z increases back to 1, and it continues to increase as the pressure increases beyond 400 atm.

Under conditions where z is less than 1, the gas pressure is lower than that of an ideal gas because of attractive forces between gas molecules. When z begins to increase, we are seeing the effect of the molecules occupying a growing proportion of the total volume of the gas. Thus, we see two opposing effects: negative deviations caused by intermolecular attractions and positive deviations caused by the volume of the gas molecules.

These factors may not be of equal importance under all conditions. For example, consider oxygen gas above 100°C, shown in Figure 9.18. Here, the kinetic energy of the molecules, which increases with temperature, is great enough to overcome the intermolecular forces no matter what the pressure, so only the positive deviations caused by the volume of the molecules are observed.

The relative importance of the two factors also differs for various gases at the same temperature, as shown in Figure 9.19. Here, the difference depends on the size of the molecules. The negative deviations shown in Figure 9.19 increase as the molecular size increases. Larger molecules have a greater polarizability and stronger molecular interactions. The greatest deviation in the figure is observed for ammonia gas, which has polar molecules. Here, in addition to London dispersion forces, there are interactions between permanent dipoles. These interactions create forces much stronger than the forces in the other molecules, which are nonpolar.

Figure 9.18 Compressibility factor of oxygen gas as a function of pressure.

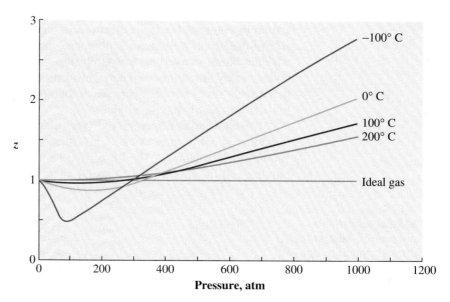

Figure 9.19 Compressibility factor of various gases at 0°C as a function of pressure. In each case, the deviation from ideal gas behavior is proportional to the polarizability of the electron cloud on the molecule. The data for CO_2 are at 40°C because it liquefies at 0°C under high pressure.

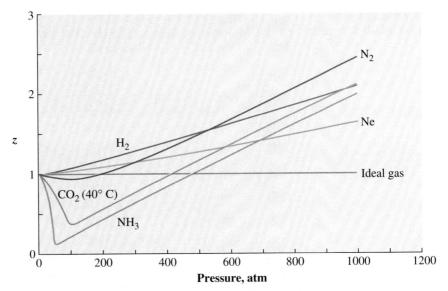

van der Waals Equation

The two factors leading to deviations from ideal behavior, molecular volume and molecular attractions, are taken into account in a variety of modifications of the ideal gas law. The simplest of these and the one that most directly corresponds to the factors as described here is known as the **van der Waals equation:**

$$(P + an^2/V^2)(V - nb) = nRT$$

Values of the constants a and b are shown in Table 9.4 for several gases.

The first constant, a, is a correction for intermolecular forces. The constant is added to the real pressure because real pressure is less than (or equal to) ideal pressure:

$$P_{ideal} = P_{real} + an^2/V^2$$

The expression n/V is simply the concentration of molecules, and the number of interactions between molecules is proportional to this concentration squared. As the pressure increases, the volume decreases, so the correction term increases as the square of the pressure increases. At moderate pressures, the correction term becomes negligible and drops out of the equation.

The second constant, b, is a measure of molecular volume. Recall that the real volume of a gas is greater than the ideal volume because—contrary to the assumption of the kinetic-molecular theory—gas molecules occupy volume. Thus, the molecular volume term has to be subtracted from the real volume:

$$V_{ideal} = V_{real} - nb$$

What would the compressibility factor be if it were derived from the van der Waals equation? What would Figure 9.18 look like if this compressibility factor were used?

At moderate temperatures and pressures, the actual volume of the gas molecules is small compared with the total volume of the gas. Under these conditions, the volume correction drops out of the equation. Thus, under normal atmospheric conditions, the van der Waals equation is reduced to the ideal gas law equation.

The size of gas molecules affects the values of a and b. We would expect larger molecules to occupy greater volumes and so have greater values of b than

Table 9.4 Some Values of the van der Waals Constants

Gas Molecule or Atom	a (L^2 atm/mol^2)	b (L/mol)
He	0.03412	0.02370
Ne	0.211	0.0171
H_2	0.2444	0.02661
H_2O	5.464	0.0305
O_2	1.360	0.03183
Ar	1.345	0.03219
NH_3	4.170	0.03707
N_2	1.390	0.03913
Kr	2.32	0.0398
HCl	3.667	0.04081
CO_2	3.592	0.04267
CH_4	2.253	0.04278
N_2O	3.782	0.04415
Xe	4.19	0.0510
Cl_2	6.49	0.0562
SO_2	6.714	0.05636
CCl_4	20.39	0.1383

smaller molecules have. The value of a is also generally larger for larger molecules because larger molecules are more polarizable and hence exhibit greater intermolecular attractive forces. Note in Table 9.4 that molecules such as HCl, NH_3, and H_2O, which are permanent dipoles, have unusually high values of a.

The characteristics we have just discussed enable real gases to be condensed to liquids. Cooling a gas causes intermolecular forces of attraction to increase as the kinetic energy of the molecules decreases. When the gas molecules no longer have enough velocity to overcome the intermolecular forces, the gas becomes a liquid.

How much a particular gas must be cooled to form a liquid depends on the magnitude of the intermolecular forces and the velocity of the gas molecules. Generally, larger, heavier molecules have greater intermolecular forces and lower velocities than lighter molecules do, so they do not have to be cooled as much for liquefaction.

Recall that because pressure pushes the gas molecules closer to one another, it increases the interactions between them. Under some conditions, gases can be liquefied by application of pressure alone. This only works at temperatures below a particular temperature for each gas, called the **critical temperature.**

9.9 Chemistry of Small Gaseous Molecules

Most of this chapter has focused on the physical properties of gases in general, which are usually the same for all gases. But what chemical substances are normally found as gases? Where do they come from? And, besides for blowing up

▶ ▶ ▶   **WORLDS OF CHEMISTRY**

Refrigeration: A Necessity or a Destructive Convenience?

Cycles of liquefaction and evaporation form the basis for refrigeration and air conditioning.

In 1987, environmental scientists began to find evidence for a loss of ozone from the stratosphere above Antarctica. This ozone protects us from the harmful effects of ultraviolet radiation. The disappearance of ozone is now being observed in more northern regions as well. What is causing this modification of our atmosphere? The blame has been placed on chemicals used in refrigeration and air conditioning.

Refrigeration is an important part of food storage. At one time, coldness to keep food fresh was supplied by a block of ice in an icebox. Now we use refrigerators powered by electricity and based on the condensation of gases, which is also used in air conditioners. But how is gas condensation used in refrigerators?

When a gas is liquefied, energy in the form of heat is given off because the internal energy of a gas is greater than that of a liquid. If the heat is carried away in some fashion, the liquid will be cooler than the gas was and the surroundings will be hotter than they were before. Conversely, if a liquid is evaporated to a gas, heat is absorbed from the surroundings. The gas is warmer than the liquid was, and the surroundings become cooler.

Cycles of liquefaction and evaporation form the basis for refrigeration and air conditioning. If a gas is compressed and caused to liquefy and the evolved heat is removed, the liquid will have a lower kinetic energy than the gas from which it was formed. If the liquid is then allowed to evaporate in an insulated container, the resulting gas will be cooler than before the cycle began. Repetition of such cycles results in the removal of heat from the gas until the temperature is low enough for the liquid to exist at lower pressures.

Refrigeration works in just this way. As a refrigerant is alternately vaporized to absorb heat and liquefied to give it up, following the path shown in Figure 9.20, heat is transported from one place to another. Fans assist in the transfer of heat between the refrigerant and the surroundings. In refrigerators, heat is removed from the inside of the refrigerator and pumped into the room outside. With air conditioning, the heat is removed from the inside of a room and transferred to the outside of the building.

A number of substances are suitable as refrigerants, including common small molecules such as SO_2 and NH_3.

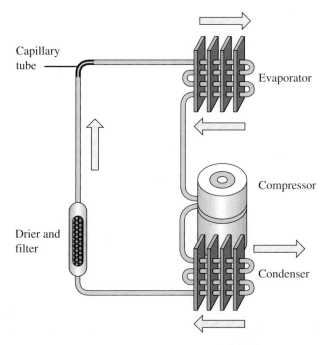

Figure 9.20 As a refrigerant is alternatively vaporized to absorb heat and liquefied to produce heat, it follows the path shown. The liquid is pushed through the capillary tube by the compressor. The liquid then evaporates, taking up heat. The warmed gas is then pumped through the condenser, where it loses heat and changes back to a liquid.

However, these substances are corrosive to metals and would be toxic to humans if they leaked from the refrigeration system. The most popular refrigerants are chlorofluorocarbons, such as Freon-11 ($CFCl_3$), Freon-12 (CF_2Cl_2), and Freon-113 ($CFCl_2CF_2Cl$). These compounds destroy stratospheric ozone if they escape into the atmosphere. The chemical industry has been trying to develop new refrigerants that have the desirable properties of the chlorofluorocarbons but do not harm the environment.

Questions for Discussion

1. List ways you use refrigeration and air conditioning.
2. In what ways do others use refrigeration and air conditioning that impact your life?

(continued)

3. Discuss the desirability of ending various uses of refrigeration and air conditioning to protect the environment.

4. What are some of the compounds that are being investigated to replace chlorofluorocarbons as refrigerants?

How do gaseous molecules differ?

balloons and tires and for breathing, how do we use them? This section explores these questions.

Common Gaseous Molecules

The common gaseous molecules—substances that exist in a stable state as gases under atmospheric conditions—are all formed from elements that fall toward the right side of the periodic table (see Figure 9.21). ▶ Most of the common gases exist as components of air, as enumerated in Table 9.1 at the beginning of this chapter. These gases are the products of naturally occurring processes, such as the cycles shown in Figure 9.22. The figure depicts the nitrogen cycle, which controls levels of N_2, N_2O, NO_2, NO, and NH_3; the sulfur cycle, which controls H_2S, SO_2, and SO_3; and the carbon cycle, which controls CO_2, CH_4, and CO.

Why are the common gases either monatomic or small molecules?

You can see from the figure that many of the gases normally considered pollutants arise in part from natural processes. The nitrogen oxides, for example, come primarily from human pollution but are also formed naturally by lightning and solar radiation. Similarly, carbon dioxide is a product of the combustion of fossil fuels but also participates in photosynthesis in plants and is a respiration product of animals. Gases such as the nitrogen oxides and carbon dioxide become problems only when people produce them in amounts too large to be accommodated by the natural cycles. For carbon dioxide, this has occurred over the last century, and especially over the last several decades, because of the burning of fossil fuels. Along with water vapor, CO_2 is responsible for the absorption of infrared radiation from the sun and for the reradiation of this energy back to the earth's surface. Some scientists believe the heightened CO_2 levels are causing a warming of the earth, the so-called greenhouse effect. The primary human pollutants in the air are carbon monoxide, nitrogen oxides, hydrocarbons, sulfur oxides, and particulates (solids). The primary contributors to this pollution are transportation, stationary fuel combustion (such as power plants), industrial processes, and solid waste disposal.

Some of the gaseous molecules shown in Figure 9.21 do not exist in appreciable amounts in nature, but only in the chemical lab. Some of these substances are so reactive that they quickly combine with other components of the atmosphere and so lose their identities. Examples are F_2 and Cl_2. Others, such as HF, HCl, HBr, HI, NH_3, and SO_3, are very soluble in water and so are carried out of the atmosphere in appreciable amounts.

			H	He
			H_2	
C	N	O	F	Ne
CO	N_2	O_2	F_2	
CO_2	N_2O	O_3	HF	
CH_4	NO			
	NO_2			
	NH_3			
		S	Cl	Ar
		H_2S	Cl_2	
		SO_2	HCl	
		SO_3		
			Br	Kr
			HBr	
			I	Xe
			HI	

Figure 9.21 Some common gaseous molecules.

Gaseous Oxygen Species

Molecular Oxygen   Oxygen exists in the atmosphere as a colorless, odorless diatomic molecule, O_2. Most O_2 is obtained commercially by distillation from air,

Figure 9.22 Biogeochemical
cycles.

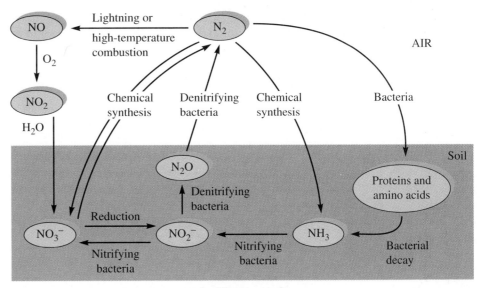

A. Nitrogen cycle

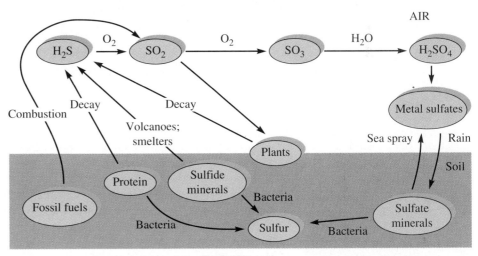

B. Sulfur cycle

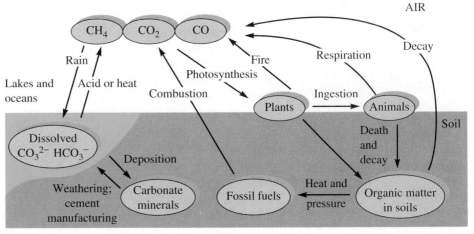

C. Carbon cycle

but a small amount is also obtained in a very pure state by the electrolysis of aqueous solutions or by the thermal decomposition of oxoanion, peroxide, or oxide salts (see Figure 9.23):

$$2KBrO_3(s) \xrightarrow{\Delta} 2KBr(s) + 3O_2(g)$$

$$2BaO_2(s) \xrightarrow{\Delta} 2BaO(s) + O_2(g)$$

$$2HgO(s) \xrightarrow{\Delta} 2Hg(l) + O_2(g)$$

RULES OF REACTIVITY

Oxides of noble metals such as mercury and gold can be decomposed by heating.

Oxygen is widely used as an industrial oxidizing agent in applications such as the conversion of coke to CO for the reduction of iron oxides, the oxyacetylene torch for high-temperature welding, and hydrogen–oxygen fuel cells.

Ozone   Ozone (O_3) is an **allotrope** of oxygen (O_2)—a different form of the same element in the same physical state. Ozone has a characteristic pungent odor associated with rainstorms. It is produced from O_2 by electric arc discharges, by lightning, and by ultraviolet light. Ozone is a very strong and fast-acting oxidizing agent, second only to fluorine (F_2) among simple substances. The product of reactions with ozone is usually O_2:

$$2NO_2(g) + O_3(g) \longrightarrow N_2O_5(g) + O_2(g)$$
$$2I^-(aq) + O_3(g) + 2H^+(aq) \longrightarrow I_2(aq) + O_2(g) + H_2O(l)$$

Ozone is used as a germicide and as a bleaching agent for varnishes, waxes, and fats. ▶ As discussed in the chapter introduction, although ozone is toxic to humans, it is a necessary component of the stratosphere.

Gaseous Nitrogen Species

Molecular Nitrogen   Nitrogen exists in the atmosphere as a colorless, odorless diatomic molecule. Nitrogen gas is isolated by distillation from liquid air.

The nitrogen molecule contains a nitrogen–nitrogen triple bond with a very high bond energy of 942 kJ/mol, so N_2 is very unreactive—nearly as inert as a noble gas. Thus, the conversion of atmospheric N_2 to certain other nitrogen species, called *nitrogen fixation,* is not a simple process. It is an important one, however. Plants need nitrogen to grow, but the N_2 gas in the air is useless to them.

Several natural processes fix nitrogen in the soil in forms usable by plants. The most important natural fixation processes involve certain bacteria that live in soil and on alfalfa and other legumes. Smaller amounts of nitrogen are added to soil by conversion of N_2 (and O_2) to NO by lightning, oxidation of NO to NO_2 by O_2, reaction of NO_2 with H_2O to form nitric acid, and by eventual deposition on the soil of nitric acid, ammonium nitrate, and metal nitrates formed by reaction with particulates. Much nitrogen is also fixed by industry in the synthesis of fertilizers and then added to the soil.

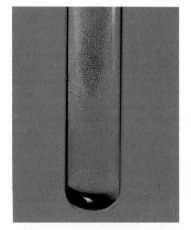

Figure 9.23 When mercury(II) oxide is heated, oxygen gas and mercury metal are produced.

Ammonia   One of the most widely used sources of nitrogen-containing fertilizers is synthetic ammonia. This compound is formed by the Haber process, discussed briefly in Section 3.6:

$$N_2(g) + 3H_2(g) \longrightarrow 2NH_3(g)$$

RULES OF REACTIVITY

When most ammonium salts are heated, they produce ammonia gas. However, when heated carefully, ammonium nitrate produces nitrous oxide, or laughing gas. Caution is required because this salt can explode violently when heated.

Ammonia may also be converted to other nitrogen-containing compounds. For example, it may be oxidized at about 500°C over a platinum–rhodium gauze

$$4NH_3(g) + 5O_2(g) \longrightarrow 4NO(g) + 6H_2O(g)$$
$$2NO(g) + O_2(g) \longrightarrow 2NO_2(g)$$

catalyst (Figure 9.24). The nitrogen dioxide formed is reacted with water:

$$3NO_2(g) + H_2O(l) \longrightarrow 2HNO_3(l) + NO(g)$$

The HNO_3 can be dissolved in water to form aqueous nitric acid or reacted with ammonia to form ammonium nitrate, used in fertilizers:

$$NH_3(g) + HNO_3(l) \longrightarrow NH_4NO_3(s)$$

Ammonium nitrate is an odorless, transparent, crystalline solid. It is used to make laughing gas, explosives, fireworks, matches, and fertilizers.

About 20% of the ammonia in the atmosphere comes from industrial processes; the remainder comes from natural sources, especially the hydrolysis of urea from animal urine. Ammonia is colorless, with a characteristic sharp, choking odor. It is very soluble in water (about 700 volumes at STP dissolve in 1 volume of water) and is widely used in cleaning solutions.

Nitrogen Oxides   Other important nitrogen species include the gaseous nitrogen oxides. Dinitrogen oxide, N_2O, is a naturally occurring, relatively unreactive component of the atmosphere. It is also called nitrous oxide. N_2O is formed, along with N_2, by degradation of proteins by soil microorganisms. It can also be obtained by the decomposition (at about 250°C) of ammonium nitrate or nitramide:

$$NH_4NO_3(s) \xrightarrow{\Delta} N_2O(g) + 2H_2O(g)$$

$$NH_2NO_2(s) \xrightarrow{\Delta} N_2O(g) + H_2O(g)$$

Mixed with oxygen to prevent asphyxiation, it is used as an anesthetic (laughing gas).

Figure 9.24 A platinum–rhodium gauze catalyst is used to oxidize ammonia in the preparation of nitric acid.

Nitrogen(II) oxide, also known as nitric oxide (NO), can be formed by reaction of copper metal with aqueous nitric acid:

$$3Cu(s) + 8H^+(aq) + 2NO_3^-(aq) \longrightarrow 2NO(g) + 4H_2O(l) + 3Cu^{2+}(aq)$$

It is also formed in high-temperature ($> 1200°C$) combustion processes involving air, such as in automobile engines:

$$N_2(g) + O_2(g) \xrightarrow{\Delta} 2NO(g)$$

This colorless, odorless gas reacts rapidly with O_2 to form NO_2.

Nitrogen dioxide is a reddish brown gas with a pungent, choking odor. It exists also as colorless N_2O_4, with the proportion of NO_2 to N_2O_4 increasing with temperature. The gas at 1 atm pressure condenses to a liquid at temperatures below 21.15°C, where it is nearly all colorless N_2O_4. In the atmosphere, the substance exists almost completely as NO_2, which has been implicated in the formation of smog.

Hydrogen

Molecular hydrogen is a colorless, odorless, flammable gas. Because the H_2 molecule is so light, it can easily escape the earth's gravitational field, so very little is found in the atmosphere.

Hydrogen is very widely used. It is obtained commercially primarily from coal and hydrocarbons. Coal is heated to drive off volatile hydrocarbons, yielding coke, which is reacted with steam in the so-called water–gas reaction:

$$C(s) + H_2O(g) \longrightarrow CO(g) + H_2(g)$$

Steam reforming, the reaction of hydrocarbons with steam in the presence of a nickel catalyst, yields the same products in different proportions:

$$CH_4(g) + H_2O(g) \longrightarrow CO(g) + 3H_2(g)$$

The relative proportions of carbon monoxide and hydrogen gases can be adjusted by a "shift" reaction, which is carried out at about 350°C with an iron–chromium catalyst:

$$CO(g) + H_2O(g) \longrightarrow CO_2(g) + H_2(g)$$

In the laboratory, hydrogen can be produced on a small scale by the reaction of an acid with an active metal:

$$Zn(s) + H_2SO_4(aq) \longrightarrow ZnSO_4(aq) + H_2(g)$$

Aluminum will release hydrogen when placed in a basic solution (Figure 9.25):

$$2Al(s) + 2NaOH(aq) + 6H_2O(l) \longrightarrow 2Na^+(aq) + 2Al(OH)_4^-(aq) + 3H_2(g)$$

Because of its fairly high bond energy (432 kJ/mol), hydrogen is relatively inert at room temperature, but it becomes reactive at high temperatures or pressures. It is used as a gaseous fuel at high temperatures in various industrial operations and in some rockets. It is also used in the hydrogen–oxygen fuel cell, used in space voyages.

Figure 9.25 The reaction between aluminum metal and alkali metal hydroxides is involved in some solid drain cleaners. Sodium hydroxide chemically breaks down fats, while the hydrogen produced by the reaction between aluminum and hydroxide ion causes a churning action that helps to break up the clog in the drain.

Gaseous Carbon Species

Carbon Monoxide Carbon forms two stable gaseous oxides, CO and CO_2. Carbon monoxide is colorless, odorless, and tasteless. It is also toxic because it displaces oxygen from hemoglobin (Hb) in the blood:

$$HbO_2 + CO \longrightarrow HbCO + O_2$$

Carbon monoxide results from incomplete combustion of carbon-based fuels. Approximately 80% of the CO in the atmosphere comes from automobiles. It is formed commercially by the water–gas or steam-reforming reactions discussed previously. Carbon monoxide can also be formed by the high-temperature reaction of carbon dioxide and carbon-containing materials such as coke:

$$CO_2(g) + C(s) \longrightarrow 2CO(g)$$

This reaction provides a method for gasifying coal, making mining operations easier. Although CO can be used as a fuel, its toxicity makes it unsuitable for use in homes. CO is used primarily as a fuel in industrial plants and in synthetic processes. It is also used as a reducing agent in recovering iron from its ores:

$$Fe_2O_3(s) + 3CO(g) \longrightarrow 2Fe(l) + 3CO_2(g)$$

This process was discussed in detail in Section 4.6.

Figure 9.26 Dry ice is solid carbon dioxide at a temperature of −78.5°C. The smoke is water vapor condensing from air.

Carbon Dioxide Carbon dioxide exists as a gas above −78.5°C at 1 atm pressure. At temperatures below this, at 1 atm pressure, CO_2 exists as a solid rather than a liquid. The solid, called dry ice (Figure 9.26), can be liquefied at −56°C if sufficient pressure is applied. Carbon dioxide results from the complete combustion of coal, oil, natural gas, or any other carbon-based fuel:

$$C(s) + O_2(g) \longrightarrow CO_2(g)$$
$$CH_4(g) + 2O_2(g) \longrightarrow CO_2(g) + 2H_2O(g)$$

Figure 9.27 Limestone, or calcium carbonate, is a very common metal carbonate that is used to produce carbon dioxide for many uses and lime, CaO, for use in cement. When calcium carbonate is heated, it forms calcium oxide and carbon dioxide gas. A white precipitate of barium carbonate forms when barium hydroxide solution is exposed to carbon dioxide.

It is also produced by the decomposition of carbonates (Figure 9.27) such as limestone:

$$CaCO_3(s) \xrightarrow{\Delta} CaO(s) + CO_2(g)$$

RULES OF REACTIVITY

All metal carbonates, except those of the alkali metals, decompose when heated to form metal oxides and carbon dioxide gas.

Carbon dioxide provides the carbonation in soft drinks. It is also used in fire extinguishers and, to a small extent, in the manufacture of sodium carbonate.

Gaseous Sulfur Species

Hydrogen Sulfide   Sulfur combines with hydrogen to form H_2S, hydrogen sulfide, an extremely poisonous gas that smells like rotten eggs. Hydrogen sulfide is the largest source of sulfur in the atmosphere. It comes primarily from the decay of organic matter, from biological reduction of sulfate, from volcanoes, and from weathering of minerals (Figure 9.28).

Figure 9.28 Hydrogen sulfide is released into the air by weathering of sulfide minerals, such as iron pyrites, which is also called fool's gold because of its golden metallic sheen.

Figure 9.29 Acid rain resulting from sulfur oxides in the air causes damage to limestone and marble statues. The calcium carbonate in the statues reacts with acid to form soluble salts and carbon dioxide.

Sulfur Oxides Sulfur forms two important gaseous oxides, SO_2 and SO_3. Much SO_2 is formed by the oxidation of hydrogen sulfide in the atmosphere:

$$2H_2S(g) + 3O_2(g) \longrightarrow 2SO_2(g) + 2H_2O(g)$$

Approximately one-third of the SO_2 in the atmosphere arises from human sources, primarily from the burning of fossil fuels and the processing of sulfide ores. Under the action of sunlight or of metal oxide catalysts present in particulates, SO_2 can be oxidized further to sulfur trioxide:

$$2SO_2(g) + O_2(g) \longrightarrow 2SO_3(g)$$

In the presence of water vapor, SO_3 forms sulfuric acid:

$$SO_3(g) + H_2O(g) \longrightarrow H_2SO_4(l)$$

RULES OF REACTIVITY

Metal oxides will combine with nonmetal oxides to form metal oxoanion salts. For example, calcium oxide will combine with sulfur trioxide to form calcium sulfate.

Metal oxides in the atmosphere combine with sulfur trioxide to form metal sulfates. The sulfuric acid is returned to earth in rainfall, resulting in the acid rain that has caused widespread damage to forests and human-made structures alike, as shown in Figure 9.29.

The primary use of sulfur oxides is in the synthesis of sulfuric acid (described in Section 4.6). Sulfur dioxide is also used as a reducing agent in processes such as the bleaching of pulpwood.

Summary

The properties of gases are described by several gas laws. Boyle's law states that volume is inversely proportional to pressure. According to Charles's law, volume is directly proportional to absolute temperature. Both effects are considered in the combined gas law. Gay-Lussac's law of combining volumes indicates that the volumes of gases used or produced in chemical reactions are related by small whole-number ratios given by the coefficients in the balanced reaction equation. According to Avogadro's hypothesis, the volume of a gas is proportional to the number of gas molecules and thus to the moles of the gas. The dependence of volume on pressure, temperature, and number of moles is summarized by the ideal gas law: $PV = nRT$. Dalton's law of partial pressures states that the total pressure of gases in a mixture is the sum of the partial pressures the gases would have if they were alone under the same conditions. Graham's law of effusion and diffusion describes the relation between the rates of movement of gases and their molar masses or densities.

The kinetic-molecular theory of gases provides a model for an ideal gas, one that obeys the gas laws. This theory assumes that gas molecules occupy no volume and do not attract one another; but when gases are at high pressures or low temperatures, these assumptions fail and gases do not follow the ideal gas law. The van der Waals equation is a gas law that takes account of molecular attractions and molecular volume. The conditions under which gases depart from ideal behavior are exactly the conditions necessary to liquefy gases.

Common gaseous molecules are formed primarily by the nonmetallic elements. Most of these molecules are present in the atmosphere. Several of them play critical roles in the interactions of the atmosphere with solid and liquid components of earth. The common gaseous molecules are put to many industrial uses.

Key Terms

allotropes (9.9)	factor (9.8)	volumes (9.3)	polarizability (9.8)
atmosphere (atm) (9.2)	critical	Graham's law (9.6)	pressure (9.2)
Avogadro's	temperature (9.8)	ideal gas (9.3)	standard temperature
hypothesis (9.3)	Dalton's law of partial	ideal gas law (9.3)	and pressure
barometer (9.2)	pressures (9.5)	kinetic-molecular theory	(STP) (9.3)
Boyle's law (9.3)	diffusion (9.6)	of gases (9.7)	torr (9.2)
Charles's law (9.3)	effusion (9.6)	London dispersion	van der Waals
combined gas law (9.3)	Gay-Lussac's law of	forces (9.8)	equation (9.8)
compressibility	combining	molar volume (9.3)	

Exercises

Volume-Pressure Relations

9.1   What is pressure? What is atmospheric pressure?

9.2   How is pressure measured?

9.3   What are the common units of pressure? How are they related to one another?

9.4   What is Boyle's law? State the mathematical expression for this law.

9.5   Given a fixed quantity of a gas at constant temperature, calculate the new volume the gas would occupy if the pressure were changed as shown in the following table:

	Initial Volume	Initial Pressure	Final Pressure
a.	3.00 L	2.00 atm	5.00 atm
b.	30.0 mL	60.0 torr	90.0 torr
c.	2.50 mL	20.0 torr	255 torr
d.	1.50 L	1.50 atm	725 torr
e.	3.25 L	825 torr	456 torr
f.	350.0 mL	50.0 torr	20.0 torr

9.6   Given a fixed quantity of a gas at constant temperature, calculate the new pressure the gas would exert if the volume were changed as shown in the following table:

	Initial Pressure	Initial Volume	Final Volume
a.	602 torr	205 mL	1512 mL
b.	0.00100 torr	250.0 mL	2.50 mL
c.	20.0 torr	905 L	27.5 L
d.	745 torr	155 mL	1.55 L
e.	5.30 atm	1.85 L	4.50 L

9.7   How many cubic meters of helium gas at 1.00 atm are required to inflate a balloon of 6.00×10^2 m^3 to a pressure of 1.30 atm?

9.8   What pressure is required to compress 925 L of N_2 at 1.25 atm into a container whose volume is 6.35 L?

9.9   How much will the volume of 186 mL of Cl_2 change if the pressure is increased from 0.945 atm to 1.76 atm?

9.10   What volume of H_2 must be collected at 725 torr to have 0.450 L of the gas at 1.00 atm?

9.11   A 2.50-L flask is filled with air at 3.00 atm. What size flask is needed to hold this much air at 735 torr?

9.12 Boyle used a U-tube to investigate gas properties. A gas was trapped in the closed arm of the U-tube at 29.9 in Hg and the pressure was varied by adding mercury to the open arm. The following data were observed by Boyle:

Length of Gas Column (in)	Difference Between Mercury Levels (in)
48	0.0
44	2.8
40	6.2
36	10.1
32	15.1
28	21.2
24	29.7
22	35.0
20	41.6
18	48.8
16	58.2
14	71.3
12	88.5

Graph these data to show the relationship between volume and pressure.

Pressure-Temperature Relations

9.13 What is absolute temperature? How is it related to gas properties?

9.14 Assume there is a fixed amount of gas in a rigid container and the volume of the gas does not change. Calculate the pressure the gas would exert if the temperature were changed as shown in the following table:

	Initial Pressure	Initial Temperature	Final Temperature
a.	302 torr	0.0°C	105.0°C
b.	735 torr	25.0°C	0.0°C
c.	155 torr	273 K	373 K
d.	3.25 atm	25°C	206 K
e.	2.50 atm	500 K	298 K

9.15 A steel tank contains acetylene gas at a pressure of 7.25 atm at 18.5°C. What is the pressure at 37.2°C?

9.16 Assume there is a fixed amount of gas in a rigid container and the volume of the gas does not change. Calculate the temperature to which the gas would have to be changed to achieve the change in pressure shown in the following table:

	Initial Temperature	Initial Pressure	Final Pressure
a.	30.0°C	1525 torr	1015 torr
b.	250.0°C	0.500 atm	1042 torr
c.	255 K	500.0 torr	1000.0 torr
d.	205 K	2.35 atm	1.20 atm
e.	25.0°C	243 torr	735 torr

9.17 Use the following data for 1.00 L of helium gas to determine the value of absolute zero in °C:

Temperature, °C	Pressure, atm
100	1.366
50	1.183
0	1.000
−50	0.817
−100	0.634
−150	0.451
−183	0.331
−208	0.239

Volume-Temperature Relations

9.18 How is the volume occupied by a gas related to its temperature? What mathematical expression describes this relationship?

9.19 What is Charles's law?

9.20 What is the expected volume of a gas at absolute zero? Why is this result not actually observed?

9.21 For a fixed amount of gas held at constant pressure, calculate the new volume the gas would occupy if the temperature were changed as shown in the following table:

	Initial Volume	Initial Temperature	Final Temperature
a.	5.00 L	30°C	0°C
b.	212 mL	−60.0°C	501.0°C
c.	22.4 L	0.0°C	100.0°C
d.	152 mL	45 K	450 K
e.	156 mL	45°C	450°C

9.22 A bubble of air has a volume of 0.350 mL at 24.2°C. If the pressure remains constant, what is the volume of the bubble at 72.5°C?

9.23 If 79.0 L of helium at 27.0°C is compressed to 32.0 L at constant pressure, what is the new temperature?

9.24 For a fixed amount of gas held at constant pressure, calculate the temperature to which the gas would have to be changed to achieve the change in volume shown in the following table:

	Initial Temperature	Initial Volume	Final Volume
a.	0.0°C	60.0 mL	120.0 mL
b.	−37°C	255 mL	75 mL
c.	145 K	87.5 L	135 L
d.	100.0°C	250.0 L	100.0 mL
e.	27.5°C	125 mL	148 mL

Volume-Pressure-Temperature Relations

9.25 How is the product of pressure and volume of a gas related to the absolute temperature?

9.26 For a gas under a given initial set of conditions, calculate the final value for the following indicated variable if the other two variables change as described.

Initial

	Volume	Pressure	Temperature
a.	2.50 L	0.500 atm	20.0°C
b.	1.12 L	760.0 torr	0.0°C
c.	455 mL	200.0 torr	27.0°C
d.	601 mL	900.0 torr	−10.0°C
e.	237 mL	75.0 atm	147 K
f.	125 L	0.250 atm	25.0°C

Final

	Volume	Pressure	Temperature
a.	?	760.0 torr	0.0°C
b.	?	700.0 torr	25.0°C
c.	20.0 mL	?	327°C
d.	1200.0 mL	?	0.0°C
e.	474 mL	150.0 atm	?
f.	62.0 L	100.0 torr	?

9.27 What volume of O_2 at STP can be pumped into a 0.500-L tank at 24.5°C to give a pressure of 3.50 atm?

9.28 If 22.0 L of N_2 at 25.0°C and 725 torr is heated to 134°C and compressed to 4.50 L, what is the new pressure?

Mole-Volume Relations

9.29 What is standard temperature and pressure (STP)?

9.30 What is the volume occupied by one mole of a gas at STP?

9.31 How is the volume occupied by a gas related to the number of moles of that gas? What other variable must be held constant to show experimentally that this relationship is valid?

9.32 Given the following amounts of gases, calculate the number of moles of each gas and the volume that amount of gas would occupy at STP.
a. 6.8 g NH_3 d. 5.6 g N_2
b. 48 g O_2 e. 8.8 g He
c. 2.2 g CO_2

9.33 Given the following volumes of gases at STP, calculate the number of moles of each gas and the mass of the gas.
a. 7.62 L CH_4 d. 135 mL H_2
b. 8.96 L N_2 e. 150.0 mL Xe
c. 38.1 L CO

9.34 Given the following amounts of gases, calculate the number of moles of each gas and the volume that amount of gas would occupy at 100.0°C and 15.0 atm.
a. 6.8 g NH_3 d. 5.6 g N_2
b. 48 g O_2 e. 8.8 g He
c. 2.2 g CO_2

9.35 Given the following volumes of gases at 87.5°C and 722 torr, calculate the number of moles of each gas and the mass of the gas.
a. 7.62 L CH_4 d. 135 mL H_2
b. 8.96 L N_2 e. 150.0 mL Xe
c. 38.1 L CO

Ideal Gas Law

9.36 What is an ideal gas?

9.37 What is the ideal gas law? What is it used for?

9.38 How can the ideal gas law be used to calculate the density of a gas?

9.39 Calculate the density in g/L of each of the following gases at STP.
a. NH_3 b. N_2 c. N_2O d. NO   e. NO_2

9.40 Calculate the density in g/L of each of the following gases at 25.0°C and 735 torr.
a. NH_3 b. N_2 c. N_2O d. NO   e. NO_2

9.41 Given the following densities of gases at STP, calculate the molar mass of each gas.
a. 1.785 g/L   d. 0.905 g/L
b. 1.340 g/L   e. 0.714 g/L
c. 2.052 g/L

9.42 Given the following densities of gases at 27.0°C and 745 torr, calculate the molar mass of each gas.
a. 2.436 g/L   d. 3.450 g/L
b. 0.842 g/L   e. 1.706 g/L
c. 1.325 g/L

9.43 If a 5.00-L container is filled to a pressure of 840 torr at 50.0°C with each of the following gases, calculate the mass of each gas in its container.
a. H_2 b. O_2 c. CH_4 d. CO_2 e. SO_2

9.44 A sample of an unknown liquid weighing 0.495 g is collected as vapor in a 127-mL flask at 98°C. The pressure of the vapor in the flask is then measured and found to be 691 torr. What is the molar mass of the liquid?

Graham's Law

9.45 What is the relationship between rate of diffusion of a gas and its molar mass?

9.46 Write a mathematical expression for Graham's law.

9.47 A sample of sulfur dioxide gas has a density of 2.25 g/L; a sample of neon under the same conditions has a density of 0.160 g/L. What are the relative rates of diffusion of these two gases?

9.48 What is the molar mass of a gas that diffuses 0.250 times as fast as hydrogen gas under the same conditions?

9.49 If an unknown gas diffuses 1.12 times as fast as oxygen gas, what is its molar mass?

9.50 If equal amounts of neon and argon are placed in a porous container, which gas will escape faster? How much faster?

9.51 What would be the velocity of a methane molecule (CH_4) under conditions where a nitrogen molecule has a velocity of 3.35×10^4 cm/s?

9.52 If an unknown gas effuses 1.12 times as fast as C_2H_2, what is its molar mass?

9.53 A sample of O_2 having a density of 1.76 g/L effuses at a rate of 2.25 mL/min. Another sample of O_2 effuses at a rate of 4.36 mL/min. What is the density of the second gas sample?

Kinetic-Molecular Theory

9.54 State the postulates of the kinetic-molecular theory.

9.55 Explain in molecular terms why the pressure of a gas increases with increasing temperature when the volume is held constant.

9.56 Explain in molecular terms why, at constant pressure, the volume occupied by a gas decreases as the temperature is decreased.

9.57 At constant temperature, the pressure of a gas is inversely proportional to its volume. Explain this relationship in molecular terms.

9.58 Explain why the pressure of a gas sample is directly proportional to the moles of gas molecules in the sample.

Dalton's Law

9.59 What is Dalton's law?

9.60 How is Dalton's law used to determine the volume a moist gas would have if the moisture were removed?

9.61 A sample of wet oxygen gas saturated with water vapor at 22°C exerts a total pressure of 728 torr. What is the partial pressure of O_2 in the flask? The vapor pressure of water at 22°C is 20.0 torr.

9.62 A total of 0.400 L of hydrogen gas is collected over water at 18°C. The total pressure is 742 torr. If the vapor pressure of water at 18°C is 16 torr, what is the partial pressure of hydrogen?

9.63 A tank contains 78.0 g of N_2 and 42.0 g of Ne at a total pressure of 3.75 atm and a temperature of 50.0°C. Calculate the following:
  a. moles of N_2 c. partial pressure of N_2
  b. moles of Ne   d. partial pressure of Ne

Gases in Chemical Reactions

9.64 What is Gay-Lussac's law?

9.65 What is the relationship between the volume of a gaseous reactant and the volume of a gaseous product?

9.66 The average annual production of elemental sulfur is about 1.1×10^7 kg. Most of this sulfur is burned to give sulfur dioxide for use in the production of sulfuric acid:

$$S(s) + O_2(g) \longrightarrow SO_2(g)$$

What volume of oxygen gas measured at 27°C and 737 torr would be required to burn all this sulfur?

9.67 The explosion of nitroglycerin can be described by the following equation:

$$4C_3H_5(ONO_2)_3(l) \longrightarrow$$
$$12CO_2(g) + 10H_2O(g) + 6N_2(g) + O_2(g)$$

What is the total volume of gases produced at 2.00 atm and 275°C from 1.00 g of nitroglycerin?

9.68 Hexane burns according to the following equation:

$$2C_6H_{14}(g) + 19O_2(g) \longrightarrow 12CO_2(g) + 14H_2O(g)$$

What volume of CO_2 will be formed by burning 9.00 L of hexane, with the two volumes being measured under the same conditions? What volume of oxygen will be needed?

9.69 When nitric acid is synthesized from ammonia, the first step in the process is

$$4NH_3(g) + 5O_2(g) \longrightarrow 4NO(g) + 6H_2O(g)$$

What volume of NO at STP can be formed from 1250 L of NH_3 at 325°C and 4.25 atm?

9.70 Nitrous oxide can be formed by thermal decomposition of ammonium nitrate:

$$NH_4NO_3(s) \xrightarrow{\Delta} N_2O(g) + 2H_2O(g)$$

What mass of ammonium nitrate would be required to produce 145 L of N_2O at 2850 torr and 42°C?

9.71 Nitric oxide is produced in the reaction between copper metal and nitric acid:

$$3Cu(s) + 8HNO_3(aq) \longrightarrow$$
$$3Cu(NO_3)_2(aq) + 4H_2O(l) + 2NO(g)$$

What mass of copper is required to produce 15.0 L of NO at 725 torr and 20.0°C?

9.72 What volume of hydrogen is required to react with 12 L of oxygen? Each volume is measured under the same conditions.

$$2H_2(g) + O_2(g) \longrightarrow 2H_2O(g)$$

9.73 Arsenic compounds are detected by means of Marsh's test. Some metallic zinc is added to an acid solution of the material to be tested; the mixture is then heated. The arsenic is converted to arsine, AsH_3, which is decomposed by heating to form an "arsenic mirror." The reaction is

$$4Zn(s) + H_3AsO_4(aq) + 8HCl(aq) \longrightarrow$$
$$4ZnCl_2(aq) + AsH_3(g) + 4H_2O(l)$$

What volume of AsH_3 at 0.955 atm and 25.0°C is evolved by 7.0×10^{-7} g of H_3AsO_4, the smallest amount that can be detected?

9.74 The Mond process for purification of nickel involves passing CO gas over the hot impure metal, producing nickel carbonyl gas, which is swept away by the gas stream:

$$Ni(s) + 4CO(g) \longrightarrow Ni(CO)_4(g)$$

What volume of CO at 3.05 atm and 25.0°C is needed to react with 1.00 kg of nickel?

Gaseous Molecules

9.75   What are some common gaseous molecules?

9.76   What are the common constituents of the atmosphere?

9.77   Write balanced equations for a synthesis of each of the following molecules.
   a. H_2 d. HNO_3
   b. CO   e. NH_3
   c. CO_2 f. H_2SO_4

9.78   Give the color of each of the following compounds.
   a. NO   b. N_2O c. NO_2 d. N_2O_4 e. N_2

Additional Exercises

9.79   What will be the volume at STP of 3.00 mol oxygen gas that occupies 12.0 L at 25.0°C?

9.80   Oxygen can be produced by thermal decomposition of mercuric oxide:

$$2HgO(s) \xrightarrow{\Delta} 2Hg(l) + O_2(g)$$

What volume of O_2 is produced at 50.0°C and 0.947 atm by decomposition of 27.0 g of HgO?

9.81   What pressure is required to compress 275 L of nitrogen gas at 1.00 atm into a container whose volume is 12.0 L?

9.82   What volume of SO_3 will be formed from 15 L of O_2, when each gas is under the same conditions, by the following reaction?

$$2SO_2(g) + O_2(g) \longrightarrow 2SO_3(g)$$

9.83   A sample of argon measuring 3.00 L at 27.0°C and 765 torr was bubbled through and collected over water. The pressure of the new system is 778 torr at 15.0°C. What is the volume occupied by the gas? The vapor pressure of water at 15.0°C is 13 torr.

9.84   Two identical balloons are separately filled with hydrogen and helium gas. When half of the hydrogen has been lost by effusion, how much of the helium will have been lost?

9.85   An oxygen cylinder having a capacity of 4.00 L has an internal pressure of 152 atm at 45.0°C. What temperature is required to give a pressure of 995 atm?

9.86   If 22.0 L of nitrogen gas at STP are heated to 167°C and compressed to a volume of 7.00 L, what will be the final pressure?

9.87   Consider the combustion of butene:

$$C_4H_8(g) + 6O_2(g) \longrightarrow 4CO_2(g) + 4H_2O(g)$$

What volume of butene at 188°C and 2.50 atm can be burned with 12.0 L of O_2 at 745 torr and 25.0°C?

9.88   A 2.00-L tank contains a mixture of 72.0 g N_2 and 66.0 g Ar. Calculate the total pressure of gas in the tank at 45.0°C and the partial pressure of each gas.

9.89   If 88 L of helium at 27°C is compressed to a volume of 29 L at constant pressure, what will be the new temperature?

9.90   Copper oxide can be reduced to copper metal by heating in a stream of hydrogen gas:

$$CuO(s) + H_2(g) \xrightarrow{\Delta} Cu(s) + H_2O(g)$$

What volume of hydrogen at 27.0°C and 722 torr would be required to react with 95.0 g of CuO?

9.91   If xenon and argon are placed in a porous container, which gas will escape faster? How much faster?

9.92   What would be the velocity of a CO_2 molecule under conditions where an O_2 molecule has a velocity of 2.50×10^3 cm/s?

9.93   What volume of hydrogen gas must be collected at 675 torr pressure in order to have 375 mL containing the same amount of gas at 1.00 atm and constant temperature?

9.94   A cement company uses 1.00×10^5 kg of limestone daily. The limestone decomposes on heating, forming lime and carbon dioxide:

$$CaCO_3(s) \xrightarrow{\Delta} CaO(s) + CO_2(g)$$

What volume of CO_2 at 735 torr and 25.0°C is released into the atmosphere daily by this company?

9.95   If all volumes of gases are measured under the same conditions, what volume of the product will be formed from reaction of 10.0 L of H_2 in each of the following reactions?
   a. $S(s) + H_2(g) \longrightarrow H_2S(g)$
   b. $N_2(g) + 3H_2(g) \longrightarrow 2NH_3(g)$
   c. $C_6H_6(g) + 3H_2(g) \longrightarrow C_6H_{12}(g)$

9.96   Oxygen gas can be generated by heating potassium chlorate:

$$2KClO_3(s) \longrightarrow 2KCl(s) + 3O_2(g)$$

What volume of oxygen gas, collected by displacement of water and measured at 735.0 torr and 70.0°C, will be formed by the decomposition of 13.5 g of potassium chlorate?

*9.97   Calcium carbide, CaC_2, is made by heating lime, CaO, with carbon. The lime is made by heating limestone, $CaCO_3$. Acetylene can be made from the calcium carbide by reacting it with water:

$$CaC_2(s) + 2H_2O(l) \longrightarrow Ca(OH)_2(aq) + C_2H_2(g)$$

If 12.5 L of acetylene at 0.750 atm can be made from 5.00 g of limestone, what is the temperature?

*9.98 Using Avogadro's hypothesis, Stanislao Cannizarro demonstrated in 1858 that the hydrogen molecule contained an even number of hydrogen atoms. The data available to him were measured gas densities and the percent composition by mass of several gases:

Gas at 100°C and 1 atm	Gas Density in g/L	% H
Hydrogen	0.0654	100.0
Methane	0.520	25.1
Ammonia	0.552	17.7
Water	0.584	11.2
Hydrogen sulfide	1.106	5.92
Hydrogen chloride	1.183	2.76

Use these data to demonstrate that hydrogen gas contains two atoms (or a multiple of two) of hydrogen per molecule.

*9.99 The density of air at 760.0 torr and 25.0°C is 1.186 g/L.
   a. Calculate the average molar mass of air.
   b. From this value, and assuming that air contains only molecular nitrogen and molecular oxygen gases, calculate the mass % of N_2 and of O_2 in air.

*9.100 A sample of Freon 11 is vaporized at 50.0°C and 760.0 torr in a flask with only a small opening, then cooled to 0°C to condense any Freon vapor (with a boiling point of 23.7°C) remaining in the flask. The flask is then filled with water. The following data were obtained:

Mass of flask before Freon was added = 92.3162 g

Mass of flask and condensed Freon = 93.5335 g

Mass of flask filled with water at 25.0°C = 328.0 g

   a. Calculate the density of Freon vapor at 50.0°C and 760.0 torr.
   b. Calculate the molar mass of Freon-11.
   c. If the composition by mass of Freon-11 is 8.74% C, 77.43% Cl, and 13.83% F, what is the molecular formula?

▶ ▶ ▶ Chemistry in Practice

9.101 A gaseous compound was prepared by the reaction of 10.0 g of molecular fluorine with 15.0 g of sulfur trioxide gas, yielding 19.8 g of the compound. The compound was composed of 27.0% S, 32.0% F, and 41.0% O by mass. A 75.0-mL sample of the gas, measured at 728 torr and 25.0°C, weighed 0.346 g. Spectroscopic measurements indicated that one of the fluorine atoms was bonded to an oxygen atom.

   a. What is the formula of the molecule?
   b. What is the molar mass?
   c. Write a balanced equation for the reaction.
   d. What is the theoretical yield and the percent yield of the reaction?
   e. Write a valid Lewis structure for the molecule.
   f. Predict the shape of the molecule.

10

▶ ▶ ▶ ▶

Liquids, Solids, and Changes of State

The three states of water are present in the solid iceberg, the liquid ocean, and as water vapor in the air.

CHAPTER OUTLINE

Encountering Chemistry

▶ ▶ ▶ ▶

We have seen that Earth's atmosphere is a mixture of gases that, through biogeochemical cycles, interact with the planet's surface. Let's look for a moment at that surface—a floating solid mass called the crust, most of it covered by bodies of water.

Water is the most plentiful and important liquid on our planet, although it is not very abundant in the universe as a whole. And though its liquid state is probably most familiar to us, it is the only common substance that exists in all three physical states on Earth under atmospheric conditions. The primary source of water is the oceans, and vast quantities also exist in glaciers and clouds. Water is essential for all known life forms; it acts as a solvent for chemical reactions, as a chemical reactant, and as a transport medium for chemical substances and heat. In addition, water is necessary in many types of industrial applications. The most common heat transfer agent, it is used as a coolant in condensers and cooling towers. It is also used for steam generation in power and heating plants and as a solvent. Industries use vast quantities of water. The production of 1 ton of steel requires about 25,000 gal of water; petroleum refineries use about 20 gal of water in the production of 1 gal of petroleum product. Paper mills require about 180,000 gal of water to produce 1 ton of paper.

Although the oceans cover most of our planet's surface, it is the solid material of the crust with which we associate the word *earth*. We drive our cars over it, build our houses on it, and use it to grow our food. The crust is about 35 km thick under the continents but only 6 km thick under the oceans. It is a complex mixture of solid materials, many of which are used as raw materials in chemical processes or as structural materials. For example, phosphate rock, primarily $Ca_3(PO_4)_2$, is used to prepare elemental phosphorus and phosphoric acid, and cement is a complex mixture of calcium silicates and aluminates.

The crust is the source of many useful materials, but it is only a small part of our planet. The crust floats on a solid layer, called the mantle, about 2900 km thick. Inside the mantle is a liquid metallic outer core 2000 km thick that consists primarily of iron and nickel. Finally, at the earth's center, there is a solid iron–nickel ball, the inner core, that is 2740 km in diameter. These layers are too far from the surface of the earth to provide any raw materials.

Phosphate rock is only one of the solids useful to us, and water but one of the liquids. This chapter examines the properties and structures of the solid and liquid states. In addition, it explores the transitions among the three states of matter.

10.1   Intermolecular Forces

In Chapter 9, we saw that one reason why real gases sometimes deviate from ideal-gas behavior is the existence of intermolecular forces of attraction between the particles. These forces are also involved in the liquefaction of gases. Indeed, if it were not for intermolecular forces, the liquid state could not exist, and many solid substances would exist as gases instead.

Why do polar liquids boil at higher temperatures than similar nonpolar liquids?

What are the forces that attract molecules to one another? In Chapter 9 we briefly discussed one such force, the London dispersion force, which is one of a general class of interactions between dipoles called van der Waals forces. These and other types of intermolecular forces are listed in order of increasing strength in Figure 10.1 and are discussed in more detail in the following paragraphs. All intermolecular forces are much weaker than bonding forces. The strengths of ionic and covalent bonds are included in the figure for comparison.

London Dispersion Forces (Instantaneous-Dipole–Induced-Dipole Forces)

Electrons moving around an atom or nonpolar molecule are at times not distributed symmetrically. As we noted in Section 9.8, this situation leads to the formation of a temporary dipole, or **instantaneous dipole,** which—like any dipole—contains partial charges. The positive end of the dipole exerts an attractive force on nearby electrons, causing an adjacent atom to develop into another temporary dipole that we call an **induced dipole.** This effect is translated to more atoms, resulting in a

Figure 10.1 Intermolecular and Bonding Forces.

Type of Force	Type of Interaction	Energy of Force (kJ/mol)
London dispersion force (instantaneous-dipole–induced-dipole force	A temporary dipole induces formation of another dipole to which it is attracted. 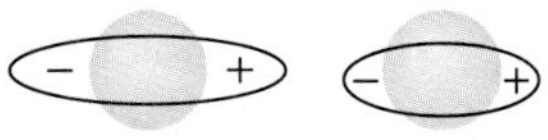	0.05–2
Dipole–induced-dipole force	A permanent dipole induces formation, in a nonpolar molecule, of a temporary dipole to which the permanent dipole is attracted. 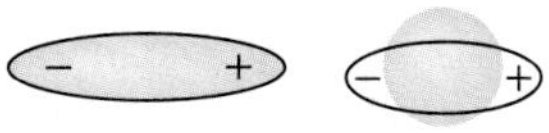	0.05–2
Ion–induced-dipole force	An ion induces formation, in a nonpolar molecule, of a temporary dipole to which the ion is attracted. 	1–3
Dipole–dipole force	Polar molecules attract one another.	1–5
Ion–dipole force	An ion is attracted to a polar molecule. 	10–20
Hydrogen bond	Two dipoles, one of them containing hydrogen bonded to an electronegative element and the other containing an electronegative element, attract one another. 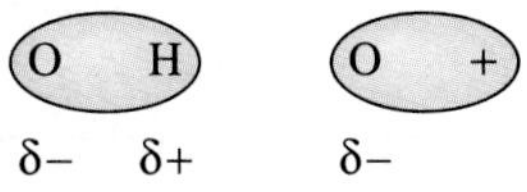	20–40
Covalent bond	Nuclei of two atoms attract electrons shared between them. 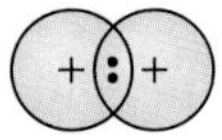	200–400
Ionic bond	Cations and anions attract one another. 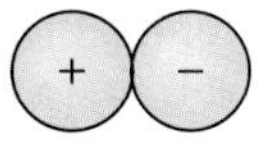	300–600

kind of electronic choreography in which the motion of the electrons in one atom is correlated with the motion of electrons in nearby atoms. The attractions between these temporary dipoles are the London dispersion forces mentioned in Chapter 9. These are the only intermolecular forces found in nonpolar substances. Although they are quite weak, they are sufficient to cause substances normally found as gases, such as neon and methane, to liquefy at high pressure or low temperature.

How do intermolecular forces vary with molecular size?

London dispersion forces tend to be stronger the larger the atom or molecule. Thus I_2 molecules interact with one another more strongly than Br_2 molecules, which interact more strongly than Cl_2 molecules. It is for this reason that at room temperature Cl_2 is a gas, Br_2 a liquid, and I_2 a solid.

Dipole (or Ion)–Induced-Dipole Forces

Like an instantaneous dipole, a permanent dipole can induce adjacent nonpolar molecules to become temporary dipoles by polarizing their electron clouds. Ions can have a similar effect. The induced dipoles are then attracted to the permanent dipoles. Because these forces require both a polar molecule (or ion) and a nonpolar molecule, they occur only between two different substances. The forces between induced dipoles and permanent dipoles or ions are somewhat stronger than London dispersion forces. The solubility of nonpolar gases such as oxygen and nitrogen in polar liquids such as water arises from attractive forces of this type.

Dipole–Dipole Forces

Among permanent dipoles, the partially positive end of one molecule attracts the partially negative ends of other molecules. Attractions between dipoles are somewhat stronger than forces involving induced dipoles. The relative strength of the dipole–dipole force depends on the dipole moments. The greater the dipole moment of the molecule, the greater the force. The solubility of one polar liquid in another polar liquid arises because of dipole–dipole forces between unlike molecules. A polar liquid, such as CH_2Cl_2, dissolves to some extent in the polar liquid water (1 mL dissolves in 50 mL of water), whereas a nonpolar liquid, such as CCl_4, is nearly insoluble (1 mL dissolves in 2000 mL of water).

Ion–Dipole Forces

Nickel sulfate hexahydrate, $NiSO_4 \cdot 6H_2O$, is a blue-green, crystalline solid. When heated to 100°C, it loses five of the water molecules. Complete dehydration does not occur unless the salt is heated to 280°C. The resulting anhydrous salt, $NiSO_4$, is greenish yellow. The hydrated salt is very soluble in water, somewhat soluble in methanol (CH_3OH), and less soluble in ethanol (C_2H_5OH).

Ion–dipole forces also result from attractions between species containing polarized electrical charges. Ions have full electron charges rather than partial charges, however, so ion–dipole forces tend to be somewhat stronger than dipole–dipole forces. Ion–dipole forces are involved in the hydration of metal ions, either in solution or in solid salts. In Chapter 3 we noted that some ionic salts occur as hydrates. In most such hydrates—$NiSO_4 \cdot 6H_2O$, for example—the negative ends of water molecules surround the cation. When this salt is dissolved in water, the cation remains in a hydrated form, $Ni(OH_2)_6^{2+}$. The sulfate ion in solution is also surrounded by water molecules, the positive ends of which point toward the anion. The solubility of ionic salts in polar solvents such as water is due to ion–dipole forces such as these.

Hydrogen Bonds

Especially strong dipole–dipole forces exist between polar molecules that contain hydrogen attached to a small, highly electronegative element (nitrogen, oxygen, or fluorine). These forces are called **hydrogen bonds,** even though there is no bond between molecules in the normal sense. The strength of most hydrogen bonds is in the range of 20–40 kJ/mol, although a few examples have energies as high as 150 kJ/mol. Hydrogen bonds are highly directional: a hydrogen atom is covalently bonded to an electronegative atom in one molecule, and this bond axis points at an electronegative atom in an adjacent molecule. Some examples are shown in Figure 10.2. The hydrogen is closer to the atom to which it is covalently bonded than to the other molecule. Though hydrogen bonds are longer and weaker than covalent bonds, they are strong enough to result in some unusual molecular properties.

Hydrogen bonding is also an important force in living systems, because it maintains much of the structure of proteins and nucleic acids. The hydrogen bonds stabilize their molecular shapes, thereby protecting their biological functioning.

Consider, for example, the boiling points shown in Figure 10.3. In general, boiling point increases with mass within a series of related elements. Earlier discussions of periodicity pointed out similar relationships for many properties. Boiling point increases within each of the three series shown in Figure 10.3A as the atoms or molecules get larger, because larger atoms are more readily polarized, and London dispersion forces become greater. This same effect is also seen in the four series of polar molecules shown in Figure 10.3B—the increases parallel those of the nonpolar Group IVA (14) hydrides. But the general trend does not always hold. NH_3, H_2O, and HF (Figure 10.3B) have much higher boiling points than we would expect, and many of their other properties are unusual as well. The anomalous behavior of these compounds results from the strength of the hydrogen bonds they contain. ▶ Many of the uses of water described in the chapter introduction are related to the unusual properties of water caused by hydrogen bonding.

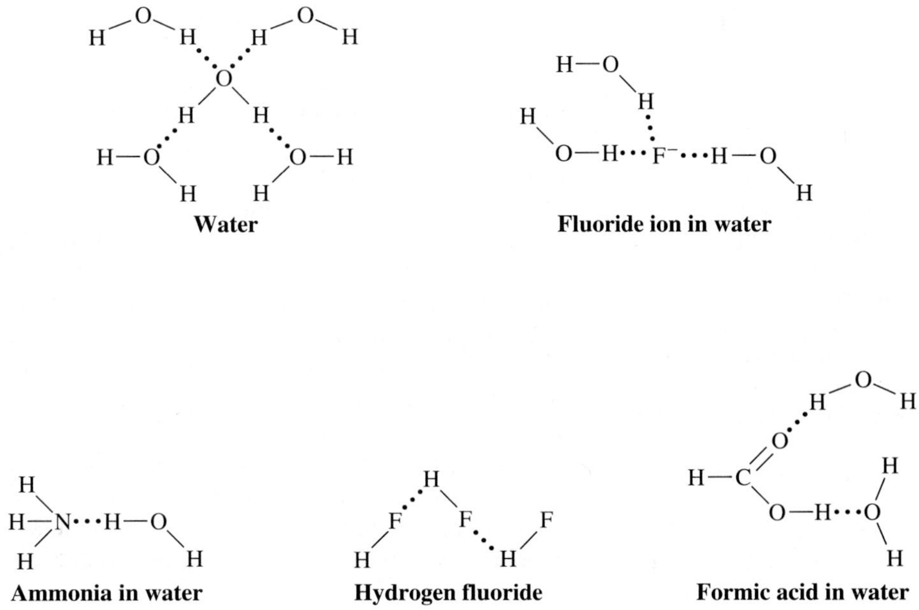

Figure 10.2 Hydrogen bonds form between a hydrogen atom on one molecule aligned with a center of electron density on a highly electronegative atom in the other molecule or ion. It is customary to represent a hydrogen bond with three dots between the hydrogen and the electronegative atom in the other molecule.

Water

Fluoride ion in water

Ammonia in water

Hydrogen fluoride

Formic acid in water

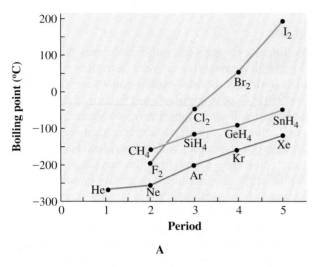

A

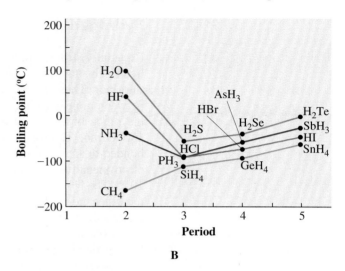

B

Figure 10.3 *(A)* The boiling points of these nonpolar substances vary smoothly, increasing with size within each series. *(B)* The boiling points of the nonmetal hydrides vary in a systematic and periodic manner. The unusually high values for water, hydrogen fluoride, and ammonia are due to hydrogen bonding.

Trends in Intermolecular Forces

The strengths of intermolecular forces vary over a wide range. As indicated in Figure 10.1, strength increases over this range in the following order:

London dispersion < dipole (or ion)–induced-dipole <

dipole–dipole < ion–dipole < hydrogen bonds

As we have seen, we can predict trends in various properties of liquids, such as solubility and boiling points, if we know the relative strengths of their intermolecular forces. How do we know what intermolecular forces to expect in a given substance? Three of the forces, dipole–induced-dipole, ion–induced-dipole, and ion–dipole forces, can be observed only in mixtures of substances, because they involve different types of molecules (or ions). The other forces can be observed in mixtures or in pure substances.

A procedure for deciding which intermolecular force is likely to be most important in a given substance is presented as a decision tree in Figure 10.4. Let's consider the pure substances methyl alcohol (CH_3OH) and methyl mercaptan (CH_3SH). What would we predict for their relative boiling points and solubilities in water? We must decide on the intermolecular forces present in each substance to predict their relative strengths. Both are pure substances, so it is necessary first to identify each substance as nonpolar or polar. Each has a methyl group (CH_3—) and a hydrogen atom bonded to an atom (O or S) that contains two unshared electron pairs, so each is bent and is polar. CH_3OH has the hydrogen atom bonded to an oxygen atom, whereas CH_3SH does not have a hydrogen atom bonded to any of F, O, or N. Thus CH_3OH experiences hydrogen bonding forces, and CH_3SH experiences dipole–dipole forces. Because hydrogen bonding forces are stronger, we expect CH_3OH to have the higher boiling point and the greater solubility in water. And sure enough, measurements indicate boiling points of 64.7°C for CH_3OH and 5.95°C for CH_3SH. Methanol forms solutions with water in any proportions, whereas only 23.3 g of CH_3SH dissolves in 1 L of water.

Methyl mercaptan, CH_3SH, occurs in "sour" natural gas in west Texas, as well as in coal tar and petroleum. It is also found in the urine of individuals who have recently eaten asparagus. Commercially it is prepared from methanol and hydrogen sulfide in the presence of a catalyst. Methyl mercaptan is a flammable gas with the odor of rotten cabbage. It is used in the synthesis of jet fuels, pesticides, fungicides, and plastics.

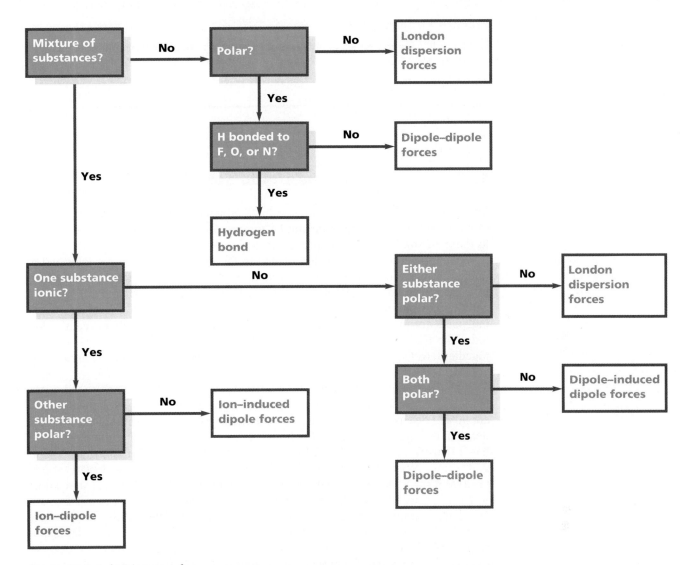

Figure 10.4 A decision tree for determining the type of intermolecular forces between molecules, atoms, or ions.

EXAMPLE 10.1

List the following molecules in order of increasing strength of their intermolecular forces: F_2, CO, HF, Kr.

Solution F_2 and Kr are nonpolar, so they can involve only London dispersion forces. Because the krypton atom contains many more electrons than the fluorine molecule, it is much more readily polarized and so should have stronger London forces. Both CO and HF are polar molecules, so both undergo dipole–dipole interactions. However, HF contains a hydrogen atom bonded to a very electronegative element, so it involves hydrogen bonding. The order of increasing strength in intermolecular forces is thus

$$F_2 < Kr < CO < HF$$

(continued)

Practice Problem 10.1

List the following molecules in order of increasing strength of their intermolecular forces: CO_2, NH_3, N_2, SO_2.

Answer: $N_2 < CO_2 < SO_2 < NH_3$

10.2   Properties of the Liquid State

Intermolecular forces cause liquids to have properties intermediate between those of gases and solids, some more similar to those of gases, and others more similar to those of solids. A few properties are unique to the liquid state.

Which properties of liquids can be explained by the distance between particles and the magnitude of intermolecular forces?

The properties of liquids are related to the distance between particles and to intermolecular forces. Particles in a liquid are much closer together than particles in a gas. Even so, liquid particles are not fixed in position but move about and collide with one another, though much more slowly and with considerably less independence than gas particles. Because liquid particles move about readily, liquids take the shape of their container. However, because the particles are held together fairly closely by intermolecular forces, liquids have definite characteristic volumes.

The densities of the states of matter are related to how close together the particles are. Generally, liquids are denser than gases by a factor of about 1000 and are 90%–95% as dense as solids. As you might expect, most substances are more dense in the solid state than in the liquid. For example, liquid aluminum at 659°C has a density of 2.38 g/mL, whereas the density of solid aluminum at 25°C is 2.70 g/mL. Water is an exception. Liquid water has its greatest density at 4°C (1.0000 g/mL). At 0°C, solid water has a density of 0.9167 g/mL, and liquid water a density of 0.9998 g/mL. In other words, water expands when it freezes. One happy result of this property is that ice floats on water, allowing plant and animal life to survive in lakes and streams. The expansion of freezing water also is responsible for much weathering action. When water freezes in cracks in rock, for example, the rock may be broken up, a process that helps create soil. (We can see a similar effect in frozen roadbeds, engine blocks, and glass containers, so this action is not always beneficial.)

How would life be affected if ice were more dense than liquid water?

Viscosity

Like gases, liquids are **fluids**—that is, they can flow. Most liquids flow rather readily, though not so readily as gases. Solids, in contrast, are generally not considered to flow at all over normal time periods, such as days or weeks. The resistance of a substance to flow is its **viscosity.** Gases have very low viscosity and solids have very high viscosity. In general, the viscosity of liquids is low. Water has a viscosity about average for liquids. At room temperature, water is about 50 times as viscous as dry air, but its viscosity is only about 10^{-16} that of glacier ice.

Liquids vary considerably in viscosity, depending on the magnitude of intermolecular forces and on molecular size. At the low end of the range is liquid helium (helium is a liquid below a temperature of 4.2 K). Liquid helium placed in

Figure 10.5 Molasses has such high viscosity that it pours only very slowly.

an open vessel flows up the sides and out of the container, as a gas would. Highly viscous liquids include oils (castor oil is about 1000 times as viscous as water), honey, syrup, molasses (Figure 10.5), and hot tar.

Some liquids are so viscous that they appear to be solids. Because they lack the highly ordered molecular structure characteristic of solids, however, they are often classified as liquids. They include glasses, waxes, and many plastics. In spite of their high viscosity, these substances—even glass—may begin to flow eventually.

Surface Tension

Why do liquids tend to assume spherical shapes?

A property of liquids that has no counterpart in solids or gases is surface tension. Small droplets of liquids tend to have spherical shapes; water on a waxed surface beads up; soap bubbles and raindrops are spherical. All these phenomena result from surface tension, which causes liquids to minimize their surface areas by assuming shapes as nearly spherical as possible.

Molecules within the liquid are subject to forces of attraction about equally in all directions, as shown in Figure 10.6. At the surface of the liquid, however, the forces are unbalanced, because there are fewer forces above the liquid surface than below it. The net effect is that, at the surface, forces of attraction tend to pull molecules toward the interior of the liquid, giving rise to near-spherical shapes.

In order for the surface of a liquid to increase—for the surface to be deformed from its preferred spherical shape—molecules must move from the interior of the liquid to the surface. But to do so, they must overcome intermolecular forces. This takes energy, and that energy is the **surface tension**—the amount of work necessary to increase the surface area of a liquid by a unit amount. In general, the greater the intermolecular forces in a liquid, the greater the surface tension. Surface tension decreases somewhat with increasing temperature. Water at 80°C, for example, has only 83% of the surface tension of water at 0°C. This happens because intermolecular forces are overcome to some extent by the greater kinetic energy at higher temperatures.

We can visualize surface tension by thinking about soap bubbles. It is possible to blow quite large bubbles (Figure 10.7) as long as the surface tension of the liquid is sufficiently high. The higher the surface tension, the stronger the film that is the bubble's surface, and the larger the bubble can be before it bursts. The

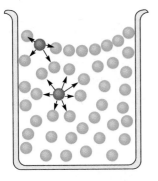

Figure 10.6 Molecules at the surface of a liquid are subject to unbalanced intermolecular interactions, because fewer attractive forces are exerted above the liquid surface than below it. Molecules under the surface are subject to forces of attraction that are about equal in all directions.

Figure 10.7 A bubble mixture with high surface tension allows very large soap bubbles to form. When the bubble becomes too large for the surface tension to support, the bubble bursts.

surface of other liquids can also be considered as a thin film. The surface tension of water is great enough to float a steel needle. However, if a detergent is added to the water, the surface tension is lowered, the surface film is weakened, and the needle sinks. (Laundry detergents work in part by reducing surface tension so the water can wet the clothing better, allowing the detergent to surround the dirt particles.)

Another manifestation of surface tension is observed when a capillary tube is placed in water. As shown in Figure 10.8, water rises up a glass capillary tube, a phenomenon called *capillary action*. Water–glass interactions are stronger than the intermolecular forces in water, so water rises in a capillary. How high it rises depends on the size of the capillary; water rises until the increased magnitude of the water–glass interactions is just offset by the pull of gravity on the resulting water column. Capillary action is partly responsible for the rise of sap in trees.

A related phenomenon can be seen in the curved surface of a liquid in containers. This curved surface, called a **meniscus,** may be either concave or convex. As Figure 10.8 shows, the surface is concave for water because the water–glass interactions are stronger than the water–water interactions. In contrast, the intermolecular forces between the mercury atoms are much stronger than the attraction between the mercury and the glass container, so mercury tends to withdraw into itself, forming a convex surface.

Figure 10.8 Meniscus and capillary action in water and mercury.

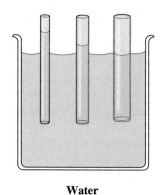

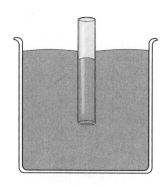

Water   **Mercury**

10.3 Properties of Water and Ice

In what ways is water different from other liquids? Why is it different?

▶ The chapter introduction mentioned that water is a compound that is not only essential to life but is also useful in many less vital ways. The usefulness of water arises from its unique properties, which in turn arise from the structure, bonding, and intermolecular interactions in water. Water is a nonlinear molecule with an H—O—H bond angle of 104.4° and an O—H bond length of 96 pm. The oxygen atom in water uses sp^3 hybridization to accommodate the two shared electron pairs and the two unshared electron pairs. Repulsion between the two unshared pairs and the shared pairs is expected to close down the bond angle by about 4° from the 109.5° tetrahedral bond angle predicted by VSEPR theory, and this is approximately what is observed. Oxygen has a considerably higher electronegativity (3.5) than hydrogen (2.1), so the O—H bond is polar, electron density being shifted toward the oxygen end of the bond. Water is quite polar, with a dipole moment of 1.86 debyes.

But the high polarity of water is not enough to explain its properties. The most important factor is hydrogen bonding, which gives unusually high levels of intermolecular attraction between the molecules. Hydrogen bonding effects are especially pronounced in water because water has two hydrogen atoms and two unshared electron pairs, giving each molecule the opportunity to participate in four hydrogen bonds. These intermolecular forces account for the high viscosity of water, because they keep molecules from moving past one another easily. The strength of the hydrogen bonds also makes it unusually difficult to vaporize water. A fairly large amount of energy is required to overcome the hydrogen bonds and increase the kinetic energy and the velocity of the water molecules enough so that they can escape to the gas phase. Thus water's heat capacity and heat of vaporization are high.

The high heat capacity of water is an important feature for living organisms. Most organisms can survive over a fairly narrow internal-temperature range—much smaller than the normal environmental changes they experience. Water within the cell structure of an organism can absorb a lot of heat while undergoing only small changes in temperature, so it helps to regulate the internal temperature of the organism.

Through evaporation and condensation, water also helps to regulate the temperature of the environment. Evaporation requires large amounts of heat energy, whereas condensation releases that energy. Thus heat is removed from hot climates by evaporation of bodies of water and of water in plant cells. Heat is released by condensation of water vapor in the atmosphere as rain or snow. Even though snow is cold, its formation raises the surrounding temperature by evolving heat. Citrus growers use the same principle to protect their crops from freezing during cold nights. They spray water on the trees, and the freezing of this water releases enough heat to keep the water inside the plant cells from freezing.

In liquid water, hydrogen bonding causes the molecules to form clusters with solidlike structures. The arrangement of the molecules in the clusters is tetrahedral, as shown in Figure 10.9. These clusters are constantly forming and disappearing, because the kinetic energy in the liquid causes the molecules to move about. Upon freezing, however, the molecules are fixed into position in a tetrahedral arrangement somewhat like that of diamond, with water molecules occupying the lattice positions and with hydrogen bonds holding molecules together. Ice, in

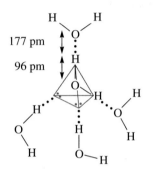

Figure 10.9 The tetrahedral arrangement of water molecules in the liquid and solid states is due to strong hydrogen bonding forces.

177 pm

96 pm

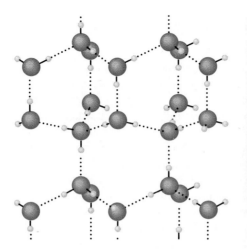

its stable low-pressure form, has a fairly open structure, with open channels passing through the crystals (Figure 10.10 Left). The tetrahedral arrangement of molecules in ice, with the resulting staggered hexagonal symmetry, leads to the hexagonal shapes of snow crystals, which show a wide variety of patterns, all with six
sides (Figure 10.10 Right). And it is this open structure that causes ice to have a
lower density at 0°C than liquid water.

10.4   The Solid State

The most obvious property of solids is their definite and rigid shapes and sizes.
Solids are not very compressible, and they usually do not flow. Attractive forces
hold the particles in solids together in a rigid structural array.

Some solids, such as tar, flow.
What is different about these
solids?

   There are two classes of solids: amorphous and crystalline. If the temperature
of a liquid is lowered very rapidly, there is a chance that the particles will solidify
in a partially chaotic state, particularly if the particles are long chains. In this case,
the particles are arranged somewhat randomly and the resulting solid is said to be
an **amorphous solid**—a solid "without form." Glass, rubber, and most plastics belong in this class. Structurally, these substances are usually classified as
liquids, because the relative disorder characteristic of a liquid is frozen into their
structures. In other respects, these substances behave like solids.

Why do crystalline solids
typically have symmetrical
planar faces?

   If, in contrast, a liquid solidifies slowly, allowing the array of particles to
become well ordered, the result is a **crystalline solid.** Most solids are of this type.
The quartz crystals shown in Figure 10.11 illustrate the well-ordered and symmetrical arrangement of planar faces typical of a crystalline solid. This section describes the structures of crystalline solids.

Crystals and Crystal Lattices

Crystalline solids are composed of **crystals**—orderly, repeating, three-dimensional assemblages of fundamental particles (atoms, molecules, or ions). The pattern formed by these arrays of particles is called the **crystal lattice** or crystal
structure. One layer in the crystal structure of a typical metal is shown in Figure

Figure 10.11 Quartz is silicon dioxide, SiO$_2$, in the form of transparent crystals. When melted, it forms a glass. It is nearly insoluble in water and acids but is attacked by hydrofluoric acid, forming gaseous silicon tetrafluoride. The presence of other elements as impurities creates a variety of quartz minerals, such as agate, amethyst, and smoky quartz, shown here.

10.12. The orderliness of the arrangement of atoms is quite apparent. Small, three-dimensional, repeating units called unit cells are responsible for this orderliness. The **unit cell** in a crystal lattice is the smallest portion of the lattice that makes up a repeating unit. Note that the contents of each of the "boxes" in this figure are identical; we could reconstruct the figure by lining up boxes. In a real crystal, which is three-dimensional, more than one layer of atoms must be considered, but the development of a unit cell is just the same. Constructing a crystal structure by assembling unit cells is like laying a floor by aligning identical tiles or building a wall by stacking up identical bricks. A three-dimensional analogy is constructing a square-based pyramid by stacking identical blocks. The dimensions of a block determine the shape and dimensions of the pyramid.

Seven different systems of units cells can be identified, which differ in the relative lengths of the sides and in the angles between them. Here we will consider only the most common, the cubic crystal system, which has all edges the same length and 90° angles between adjacent sides.

The cubic crystal system has three variants (Figure 10.13), which differ in the arrangement and number of particles.

1. **Simple cubic.** Particles are located at the corners of the cube. Because each of the eight corners is shared with seven other cubes, each contains one-eighth of a particle, for a net one particle per unit cell ($8 \times \frac{1}{8} = 1$).

Figure 10.12 A crystal lattice, shown in only two dimensions here, is an orderly arrangement of particles built up from identical unit cells. The unit cell contains the smallest repeating unit in the lattice.

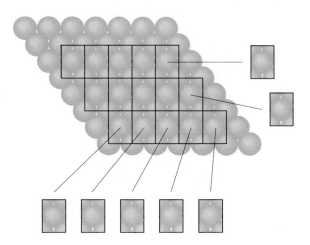

Figure 10.13 Unit cells in the cubic crystal system. Determining the fraction of each particle that actually resides in the cube enables us to calculate the number of atoms per unit cell in the crystal.

Simple cubic
$8 \times \frac{1}{8} = 1$

Body-centered cubic
$8 \times \frac{1}{8} + 1 = 2$

Face-centered cubic
$8 \times \frac{1}{8} + 6 \times \frac{1}{2} = 4$

2. **Body-centered cubic.** Particles are located at the corners of the cube, as in the simple cubic arrangement, but an additional, unshared particle is located at the center of the cube. Thus there are two particles per unit cell ($8 \times \frac{1}{8} + 1 = 2$).

3. **Face-centered cubic.** Particles are located at the corners and in the center of each of the six faces of the cube. Half of each face-centered particle is considered to reside in the cube, so the net number of particles per unit cell is four ($8 \times \frac{1}{8} + 6 \times \frac{1}{2} = 4$).

Knowing how atoms are arranged in a unit cell of an atomic crystal enables us to predict either the size of the cell or the atomic radius, given the other quantity. To do this, sometimes we must apply the Pythagorean theorem: the square of the hypotenuse of a right triangle equals the sum of the squares of the legs. Planes containing adjacent atoms in the three cubic systems are shown in Figure 10.14. In the simple cubic unit cell, shown in Figure 10.14A, each edge of each face (a) is twice the radius (r) of an atom:

$$a = 2r$$

In a face-centered cubic unit cell, the face contains adjacent atoms along a diagonal that forms the hypotenuse of a right triangle, as shown in Figure 10.14C. In

Figure 10.14 Calculating the relationship between atomic size and unit cell size. *(A)* One face of a simple cubic unit cell is outlined by the square. *(B)* A diagonal through four corners and the body center of a body-centered cubic unit cell. *(C)* One face of a face-centered cubic unit cell is outlined by the square.

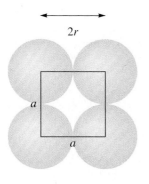
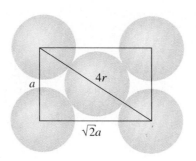
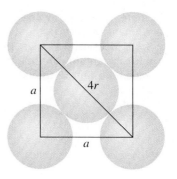

A   **B**   **C**

this case, the relationship is

$$(4r)^2 = a^2 + a^2 = 2a^2 \quad \text{or} \quad a = 2\sqrt{2}r$$

In a body-centered cubic unit cell (Figure 10.14B), the adjacent atoms occur along a diagonal plane through two edges and the body center. In this case the length of the diagonal, or hypotenuse, is $\sqrt{2}a$, so the relationship is

$$(4r)^2 = 2a^2 + a^2 = 3a^2 \quad \text{or} \quad a = 4r/\sqrt{3}$$

EXAMPLE 10.2

The radius of aluminum atoms is 125 pm. If aluminum crystallizes in a face-centered cubic lattice, what is the size of the unit cell?

Solution In a face-centered cube, as shown in Figure 10.13C, atomic radius (r) and unit cell size (a) are related as follows:

$$(4r)^2 = a^2 + a^2$$

Inserting the value for the atomic radius gives the following value for the unit cell size:

$$(4 \times 125)^2 = 2a^2$$
$$a^2 = 125000$$
$$a = 354 \text{ pm}$$

Practice Problem 10.2

The radius of copper atoms is 118 pm. If copper crystallizes in a face-centered cubic lattice, what is the size of the unit cell?

Answer: 334 pm

The planes shown in Figure 10.14 are portions of **lattice planes,** which connect equivalent points in unit cells throughout a crystal lattice. Some representative planes are depicted in Figure 10.15. Note that some lattice planes correspond with faces of unit cells, whereas others do not. Because each face of a crystal must fall along one of the planes within the crystal lattice, these planes are responsible for the characteristic angles between the faces of a crystal, such as the quartz crystal shown in Figure 10.11. Lattice planes are also responsible for the characteristic breaking patterns of crystalline solids, because crystals break along one or more of the planes. In contrast, amorphous solids (such as glass) break in random patterns.

Figure 10.15 Some lattice planes in a cubic crystal system.

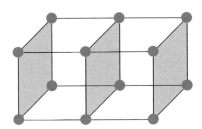 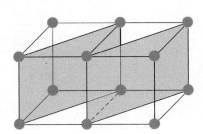 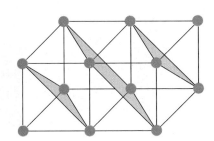

Figure 10.16 A beam of X rays is diffracted by lattice planes when the angle and wavelength have specific values related to the distance between planes. The pattern that results on the photographic plate is characteristic of the crystal and can be used to locate atoms in the unit cell.

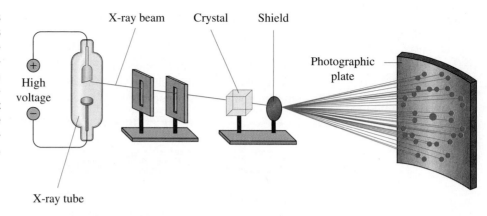

Why would we want to know the crystal structure of a solid?

Lattice planes are also important in determining the crystal structure of solids through X-ray diffraction (or X-ray crystallography), the most widely used method of determining structure. A beam of X rays is directed onto a small crystal, as shown in Figure 10.16. If the combination of the wavelength of the X rays, the distance between two parallel planes, and the angle between the plane and the X-ray beam is just right, the beam is bent, or diffracted, back out of the crystal. To get diffraction from the many lattice planes in a crystal, the crystal is slowly rotated in the X-ray beam. A typical diffraction pattern, recorded photographically, is shown in Figure 10.17. Note the symmetry apparent in this pattern, which looks much like the pattern of light spots reflected from a light ball. By careful mathematical analysis of such data, experimenters can determine the size of a unit cell and the exact positions of atoms within it. Such analyses are now routinely carried out with sophisticated computer programs.

Metal Structures and Closest Packing

The crystal structures of most metallic solids can be understood on the basis of their tendency to pack together efficiently. We can think of the metal atoms as identical spheres. If we pack the spheres in a layer as close together as possible, the result is a planar layer in which each sphere is touching six other spheres

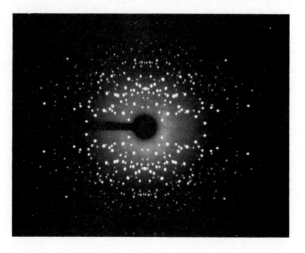

Figure 10.17 This X-ray diffraction pattern was generated by a crystal of a sulfur ylide, $C_9H_4O_2S(CH_3)_2$.

Figure 10.18 Arrangement *(A)* uses space more efficiently than arrangement *(B)*, both for the packing of spheres in a planar layer and for packing layer on layer. 90.7% of the area is occupied by the spheres in *(A)* and 78.5% of the area is occupied by the spheres in *(B)*.

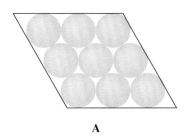

A   B

Penguins in Antarctica keep warm during very cold times by huddling together in a nearly close-packed arrangement. What drives penguins and atoms in metals to behave so similarly?

(Figure 10.18A). Such an arrangement is much more efficient than one in which rows of spheres are lined up in a square pattern (Figure 10.18B).

Of course, the crystal structure of the metal contains many layers. Here again, the efficient packing exhibits the pattern shown in Figure 10.18A. The second layer is set down into depressions in the first layer, rather than on top of the spheres in the first layer. As indicated in Figure 10.19, there are two possible positions in which a sphere can be set into a depression. We'll label with A the positions of spheres in the first layer. Spheres in the second layer can be placed at either positions B or positions C. Let's assume that the second layer is placed in the B depressions, giving the two layers depicted in Figure 10.20.

In placing the third layer, we again have a choice between two sets of depressions, C and A. If the third layer occupies the A positions, the result is a **hexagonal closest-packed structure.** If the third layer occupies the C positions, the structure is a **cubic closest-packed structure.** These two arrangements are shown in Figure 10.21. In both, the packing pattern for successive layers follows the same progression as for the first layers.

Hexagonal closest-packed: ABABABAB . . .

Cubic closest-packed: ABCABCABCABC . . .

Figure 10.22A presents a side view of the cubic closest-packed structure, along with a view of four layers (CABC) that have been expanded for clarity. When a unit made up of these layers is rotated, we can recognize the resulting structure as a face-centered cubic unit cell. A similar view of hexagonal closest-packed layers is shown in Figure 10.22B.

These two closest-packed arrangements are the most efficient methods of locating identical spheres. Each uses 74% of the available volume. In each, a given sphere is surrounded by 12 other spheres (6 in the same layer, 3 in the layer above, and 3 in the layer below). This number (12) is called the **coordination number.**

Figure 10.19 One closest-packed layer of identical spheres, showing the positions of atoms and of depressions between atoms.

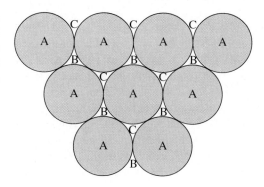

Figure 10.20 Two closest-packed layers of identical spheres, with the first layer in the A positions and the second layer in the B positions or the C positions.

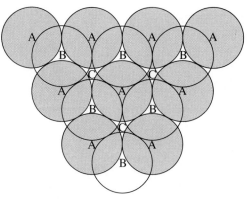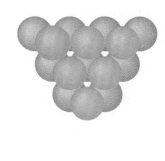

The second layer occupies the B positions.

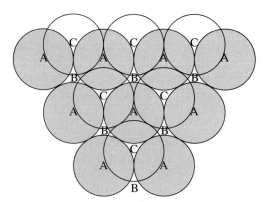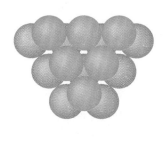

The second layer occupies the C positions.

As indicated in Figure 10.23, more than two-thirds of the metals use one of these closest-packed arrangements. Most of the rest exhibit a body-centered cubic structure, which uses 68% of the available volume and has a coordination number of 8. Some of the metals toward the right side of the periodic table exhibit the diamond structure, which uses only 34% of the available volume and has a coordination number of 4. This inefficient packing arrangement arises because of strong tetrahedral bonds between the atoms in these elements.

Structures of Ionic Crystals

How are ionic crystals related to metallic crystals? How are they different?

Many ionic crystal structures are related to closest-packed arrangements. Because closest-packed structures use only 74% of the available volume, there are holes left within them. These holes are of two types. The B positions between A and B layers are called **tetrahedral holes** because they are surrounded by *four* spheres. The C positions between A and B layers are called **octahedral holes** because they are surrounded by *six* spheres. (Although an octahedron has eight faces, these faces connect only six points.) Two tetrahedral holes and one octahedral hole are associated with each sphere. A closest-packed structure has four spheres per unit cell, so each unit cell contains eight tetrahedral holes and four octahedral holes. The location of the holes is shown in Figure 10.24.

Figure 10.21 With the second layer in the B positions, there are two possible positions for the third layer. *(A)* The third layer occupies the C positions, resulting in a cubic closest-packed structure. *(B)* The third layer occupies the A positions, resulting in a hexagonal closest-packed structure.

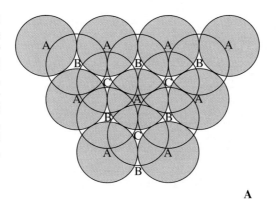

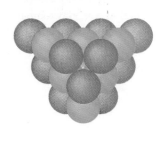

A

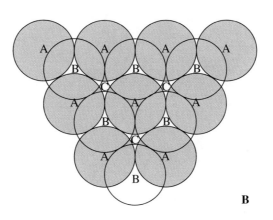

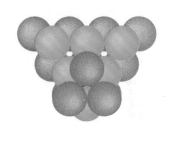

B

Figure 10.22 Layers in cubic and hexagonal closest-packed structures. *(A)* Cubic closest-packed structures. When a four-layer unit is rotated, it is seen to be a face-centered cubic arrangement. *(B)* Hexagonal closest-packed structure.

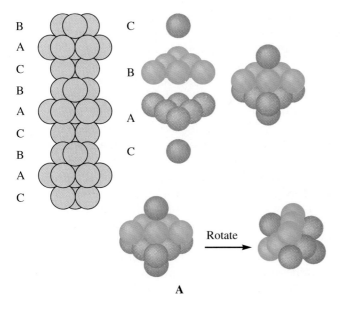

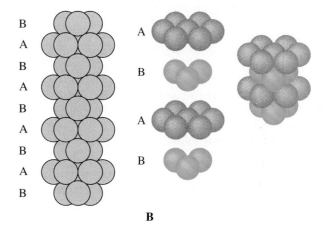

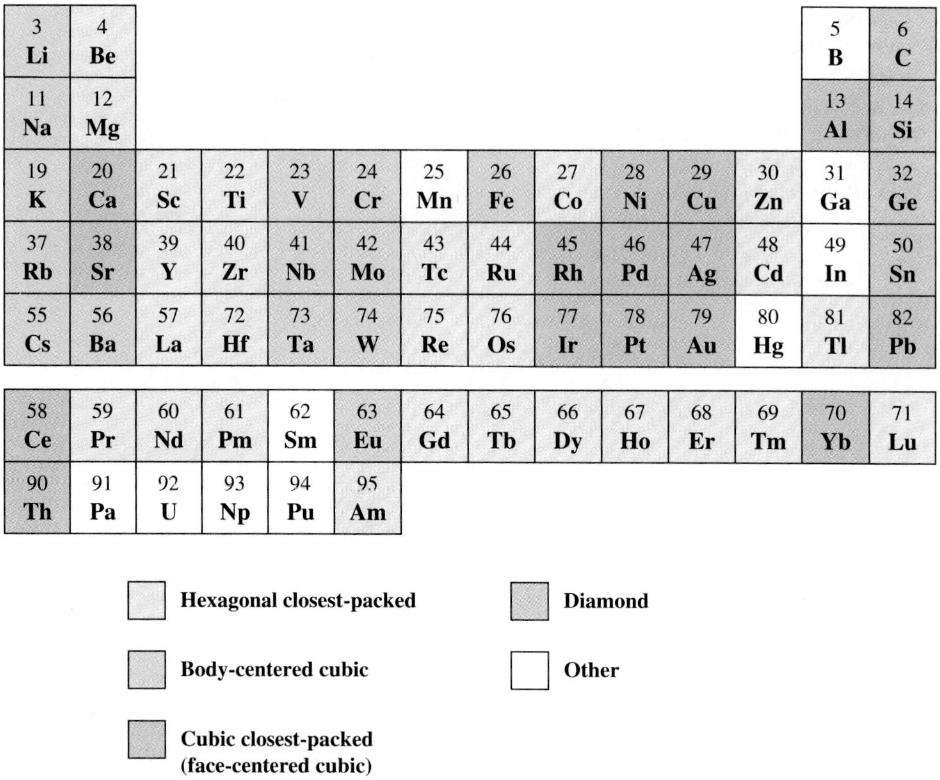

Figure 10.23 Structures of the metals at room temperature.

3 Li	4 Be												5 B	6 C
11 Na	12 Mg												13 Al	14 Si
19 K	20 Ca	21 Sc	22 Ti	23 V	24 Cr	25 Mn	26 Fe	27 Co	28 Ni	29 Cu	30 Zn	31 Ga	32 Ge
37 Rb	38 Sr	39 Y	40 Zr	41 Nb	42 Mo	43 Tc	44 Ru	45 Rh	46 Pd	47 Ag	48 Cd	49 In	50 Sn
55 Cs	56 Ba	57 La	72 Hf	73 Ta	74 W	75 Re	76 Os	77 Ir	78 Pt	79 Au	80 Hg	81 Tl	82 Pb

58 Ce	59 Pr	60 Nd	61 Pm	62 Sm	63 Eu	64 Gd	65 Tb	66 Dy	67 Ho	68 Er	69 Tm	70 Yb	71 Lu
90 Th	91 Pa	92 U	93 Np	94 Pu	95 Am								

- Hexagonal closest-packed
- Body-centered cubic
- Cubic closest-packed (face-centered cubic)
- Diamond
- Other

Use geometry to show that these radius ratios correspond to an exact fit of an atom in the hole in the host structure.

Sodium chloride is produced commercially by evaporation of sea water, salt lake water, or brine from underground salt deposits. It exists as white, cubic crystals. It is used as a source of chlorine gas, sodium metal, and their compounds. Table salt contains small amounts of magnesium chloride, which are added to keep the sodium chloride crystals dry enough to pour freely.

Cesium chloride exists as a white, crystalline solid. It is used to make fluorescent X-ray screens and to prepare cesium metal.

In ionic crystals, the larger ion (usually the anion) occupies closest-packed sites, and the other ion fits into the appropriate type of hole for its size. The size of the holes depends on the size of the larger ions that make up the lattice. A tetrahedral hole provides an exact fit for a smaller ion whose radius is 0.225 times the radius of the closest-packed larger ions. For an octahedral hole, the ratio is 0.414. Especially large ions may pack in a simple cubic arrangement with a hole in the body center that is 0.732 times as large as the ions.

We rarely find structures in which the smaller ions can "rattle around" in the holes, nor do we find many structures in which the smaller ions push the larger ones farther apart to make larger holes, unless no other structure is available that can provide larger holes. Thus there is a range of ion sizes that is typical of various structures for ionic salts. The structures of some common ionic salts are shown in Figure 10.25. See the structure of sodium chloride. In this arrangement, cubic closest-packed (face-centered cubic) chloride ions surround sodium ions, which occupy all the octahedral holes. The sodium chloride structure is adopted by many 1:1 salts, including all alkali metal halides except cesium halides. For most of these salts, the fit of cations into the octahedral holes gives good cation–anion contacts and strong ionic bonding. The lithium salts LiCl, LiBr, and LiI are exceptions. Because the lithium ion is small, it does not fit the holes very well, so anion–anion contact is not offset by cation–anion contact. As a result, these salts are not so stable as other alkali metal halides. They have unusually low lattice energies and low melting points.

Because cesium ions are too large to fit into the octahedral holes, cesium halides have the cesium chloride structure, in which the chloride ions adopt a simple cubic structure and the cesium ion occupies the hole in the body center.

Figure 10.24 Holes in closest-packed structures. *(A)* tetrahedral hole *(B)* octahedral hole *(C)* Location of holes in a cubic closest-packed structure. In an actual structure, the spheres touch.

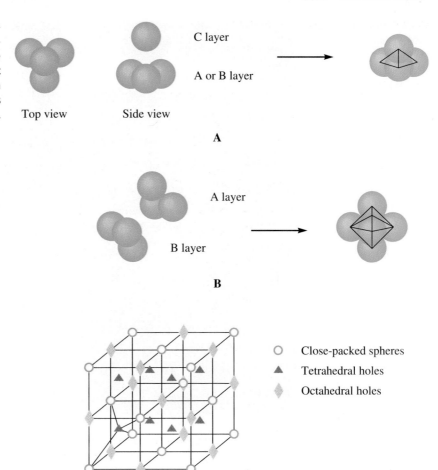

Top view   Side view

C layer

A or B layer

A

A layer

B layer

B

○ Close-packed spheres
▲ Tetrahedral holes
◆ Octahedral holes

C

Zinc sulfide exists in two forms: sphalerite (or zinc blende) and wurtzite. The wurtzite form is prepared commercially by precipitation from solution, followed by heating to 725°C in the absence of air (to prevent oxidation to the sulfate). The white or grayish white powder is used as a pigment in paint, linoleum, leather, and rubber. Zinc sulfide is also used in X-ray and television screens.

Calcium fluoride exists in nature as the mineral fluorite or fluorospar. This solid occurs as a white powder or as colorless, clear, cubic crystals. It is a primary source of fluorine and its compounds and is also used to make optical windows that transmit ultraviolet light.

Where the smaller ions are too small to fit into octahedral holes effectively, many 1:1 salts adopt a cubic closest-packed structure in which the smaller ions fill half the tetrahedral holes. The typical salt here is ZnS, in a crystal modification known as zinc blende. It is also possible to accommodate 2:1 and 1:2 salts by using all the tetrahedral holes in a cubic closest-packed structure. The typical salt of this type is fluorite, CaF_2. Though other structures are also known, a remarkably large number of ionic structures can be understood on the basis of the tendency to maximize packing efficiency and cation–anion interaction.

EXAMPLE 10.3

Examine the structure of sodium chloride and determine the number of sodium and chloride ions in a unit cell.

Solution   The sodium chloride structure (see Figure 10.25) consists of a cubic closest-packed (or face-centered cubic) arrangement of chloride ions with sodium ions in all the octahedral holes. There is a chloride ion at each of the 8 corners, but because each is shared among 8 unit cells, the corner ions contribute a total of 1 to the unit cell. In addition, there is a chloride ion in the center of each of the 6 faces. Because each of these ions is shared

(continued)

Figure 10.25 Some common structures of ionic crystals. The ions are shown pulled apart here so they will all be visible. In the actual structures, the ions touch.

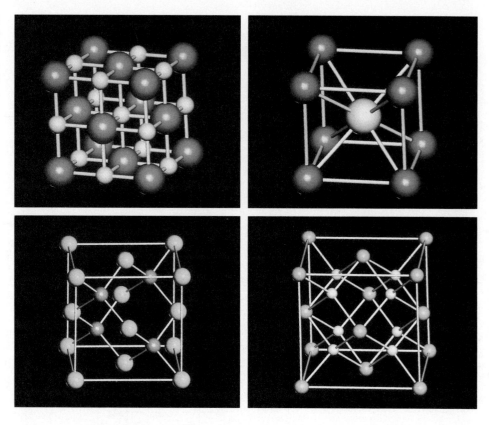

between 2 unit cells, the chloride ions contribute 3 ions to the unit cell. Thus the unit cell contains a total of 4 chloride ions. We can also count the sodium ions. There is 1 sodium ion at the center of the unit cell, and there are 12 along its 12 edges. Each of these 12 is shared with 4 other unit cells, so they contribute 3 to the unit cell, giving a total of 4 sodium ions.

Practice Problem 10.3

Examine the structure of cesium chloride and determine the number of cesium and chloride ions in a unit cell.

Answer: one of each

10.5   Types of Crystalline Solids

Solids show a wide range of properties. For most solids, these properties can be correlated with and attributed to the types of forces holding the fundamental particles together. A useful classification system, presented in Table 10.1, arranges solids into four groups: molecular, ionic, metallic, and network covalent. Next, we consider each of these classes in turn.

Why is it useful to classify solids?

Type of Solid	Fundamental Particles	Attractive Forces	Properties	Examples
Molecular	Nonpolar molecules	London dispersion forces	Low melting point, soft, poor heat conductors, electrical insulators	Noble gases, CH_4, CCl_4, CO_2, C_6H_6, I_2, H_2, O_2
Molecular	Polar molecules	Dipole–dipole forces	Low to moderate melting point, variable hardness, may be brittle, nonconductors	H_2O, NH_3, SO_2, $C_{12}H_{22}O_{11}$ (sugar)
Ionic	Cations and anions	Ionic bonds	High melting point, hard, brittle, nonconductors when solid, electrical conductors when molten	NaCl, CaF_2, KNO_3, $CaCO_3$, $CuSO_4$
Metallic	Atoms	Attractions between delocalized electrons and fixed cations; partially covalent	Low melting point and soft (even liquid) to high melting point and hard; good conductors; malleable and ductile	Hg, Na, Fe, Au, Ag, Cu, Cr
Network covalent	Atoms	Covalent bonds	Very high melting point, very hard, somewhat brittle, nonconductors or semiconductors	C (diamond), SiC (carborundum), SiO_2 (silica), BN (boron nitride), Al_2O_3 (alundum), Cr_2O_3 (corundum)

Table 10.1 Classification of Solids

Molecular Solids

Iodine is an example of a molecular solid. It is a fairly soft solid that normally occurs as bluish black plates with a metallic luster. It can be vaporized readily to form a violet gas and has an unpleasant odor even as a solid. Iodine is used in germicidal and antiseptic preparations. As iodide salts, it is used as a trace dietary additive to prevent malfunction of the thyroid gland.

Molecular solids are mostly associated with the elements on the right side of the periodic table, as shown in Figure 10.26. The particles in such solids are either atoms, as in the case of the noble gases, or molecules (either nonpolar or polar).

In nonpolar molecular solids, the forces holding molecules together are very weak London dispersion forces. These solids thus tend to form soft crystals that have rather low melting points and are poor heat conductors and good electrical insulators. Included in this group are not only the noble gases and the diatomic elements but also small polyatomic molecules, such as carbon dioxide and sulfur hexafluoride, that are nonpolar because of their molecular symmetry.

Polar molecular solids consist of molecules held together by the attractive forces between permanent dipoles, which are stronger than London dispersion forces. As a result, polar molecular solids are often somewhat harder than nonpolar molecular solids and have low to moderate melting points. Because they contain no ions, they are electrical insulators. A familiar example is pure water.

1 Group IA	2 IIA	3 IIIB	4 IVB	5 VB	6 VIB	7 VIIB	8	9 VIII	10	11 IB	12 IIB	13 IIIA	14 IVA	15 VA	16 VIA	17 VIIA	18 VIIIA
1 H																	2 He
3 Li	4 Be											5 B	6 C	7 N	8 O	9 F	10 Ne
11 Na	12 Mg											13 Al	14 Si	15 P	16 S	17 Cl	18 Ar
19 K	20 Ca	21 Sc	22 Ti	23 V	24 Cr	25 Mn	26 Fe	27 Co	28 Ni	29 Cu	30 Zn	31 Ga	32 Ge	33 As	34 Se	35 Br	36 Kr
37 Rb	38 Sr	39 Y	40 Zr	41 Nb	42 Mo	43 Tc	44 Ru	45 Rh	46 Pd	47 Ag	48 Cd	49 In	50 Sn	51 Sb	52 Te	53 I	54 Xe
55 Cs	56 Ba	57 La ★	72 Hf	73 Ta	74 W	75 Re	76 Os	77 Ir	78 Pt	79 Au	80 Hg	81 Tl	82 Pb	83 Bi	84 Po	85 At	86 Rn
87 Fr	88 Ra	89 Ac ▲	104	105	106												

★	58 Ce	59 Pr	60 Nd	61 Pm	62 Sm	63 Eu	64 Gd	65 Tb	66 Dy	67 Ho	68 Er	69 Tm	70 Yb	71 Lu
▲	90 Th	91 Pa	92 U	93 Np	94 Pu	95 Am	96 Cm	97 Bk	98 Cf	99 Es	100 Fm	101 Md	102 No	103 Lr

Figure 10.26 Molecular solids. The elements that form molecular solids are shaded. They form either monatomic crystals or molecular crystals.

Ionic Solids

Ionic solids contain cations and anions arranged in crystalline lattices. These solids thus tend to consist of species formed from the reaction of metals with nonmetals. The forces holding the particles together are cation–anion attractive electrostatic forces, offset somewhat by cation–cation and anion–anion repulsive electrostatic forces. Because these electrostatic forces are rather strong, ionic crystals have high melting points. It is difficult to displace ions, so the crystals tend to be hard. They are also brittle. Though displacement is difficult, only a small amount of displacement is necessary to break a crystal (review Figure 7.5). Ionic solids are electrical nonconductors because the ions are fixed in position. However, melted or dissolved solids, in which the ions are free to move and carry electrical charge, are good conductors.

Metallic Solids

As shown in Figure 10.27, metallic bonding is found in the elements on the left side and in the center of the periodic table. In metals, the electrons are at least partially delocalized. There are no localized covalent bonds between metals atoms as there are in molecular and network covalent solids, though there may be some covalent contribution to the bonding.

The properties of metals can be changed by the addition of impurities. Hard, strong steel is formed from soft iron by the addition of a small amount of carbon or of some other metal such as chromium, vanadium, nickel, molybdenum, or tung-

How many of the metals we use in our daily lives are actually pure metallic elements?

1 Group IA																	18 VIIIA
1 H	2 IIA											13 IIIA	14 IVA	15 VA	16 VIA	17 VIIA	2 He
3 Li	4 Be											5 B	6 C	7 N	8 O	9 F	10 Ne
11 Na	12 Mg	3 IIIB	4 IVB	5 VB	6 VIB	7 VIIB	8	9 VIII	10	11 IB	12 IIB	13 Al	14 Si	15 P	16 S	17 Cl	18 Ar
19 K	20 Ca	21 Sc	22 Ti	23 V	24 Cr	25 Mn	26 Fe	27 Co	28 Ni	29 Cu	30 Zn	31 Ga	32 Ge	33 As	34 Se	35 Br	36 Kr
37 Rb	38 Sr	39 Y	40 Zr	41 Nb	42 Mo	43 Tc	44 Ru	45 Rh	46 Pd	47 Ag	48 Cd	49 In	50 Sn	51 Sb	52 Te	53 I	54 Xe
55 Cs	56 Ba	57 La ★	72 Hf	73 Ta	74 W	75 Re	76 Os	77 Ir	78 Pt	79 Au	80 Hg	81 Tl	82 Pb	83 Bi	84 Po	85 At	86 Rn
87 Fr	88 Ra	89 Ac ▲	104	105	106												

★	58 Ce	59 Pr	60 Nd	61 Pm	62 Sm	63 Eu	64 Gd	65 Tb	66 Dy	67 Ho	68 Er	69 Tm	70 Yb	71 Lu
▲	90 Th	91 Pa	92 U	93 Np	94 Pu	95 Am	96 Cm	97 Bk	98 Cf	99 Es	100 Fm	101 Md	102 No	103 Lr

Figure 10.27 Metallic solids. The shaded elements exist as metals.

sten. Metals formed by the mixing of two or more elements, called *alloys*, have properties different from those of their parent elements (Figure 10.28). Some typical alloys are described in Table 10.2.

Network Covalent Solids

Most network covalent solids are found among the semiconducting or metalloid elements in the transition region between metals and nonmetals in the periodic table (Figure 10.29). These substances consist of giant molecules encompassing

Figure 10.28 Copper is a reddish, lustrous metal that has high electrical conductivity. It is readily available and is widely used in coins. We can plate a copper coin with a thin layer of zinc by placing the coin in an alkaline solution with some zinc metal and heating it. If the coin is then dried and heated in a flame, the zinc migrates into the copper crystal lattice and forms the alloy brass. The copper coin on the left is unadulterated, the one in the center has a coating of zinc, and the one on the right is coated with brass.

Alloy	Metals	Properties	Uses
Chrome steel	96%–98% Fe, 2%–4% Cr	Very hard, shock-resistant	Files, safes, ball bearings
Dental amalgam	70% Ag, 25% Pb, 3% Cu, 2% Hg	Easily worked	Dental fillings
Duralumin	90% Al, 5.5% Mn, 4% Cu, 0.5% Mg	Low density, high tensile strength	Airplane and automobile parts
14-karat gold	58% Au, Ag, Cu, others	Harder than pure gold	Jewelry
Monel metal	69% Ni, 33% Cu, 7% Fe	Resistant to corrosion, bright finish	Kitchen fixtures
Nichrome	60% Ni, 40% Cr	High melting point, low electrical conductivity	Heat coils in stoves, toasters, ovens
Pewter	85% Sn, 6.8% Cu, 6% Bi, 1.7% Sb		Utensils
Plumber's solder	67% Pb, 33% Sn	Low melting point, soft	Soldering joints
Stainless steel	80.6% Fe, 0.4% C, 18% Cr, 1% Ni	Resistant to corrosion	Tableware
Sterling silver	92.5% Ag, 7.5% Cu	Bright finish	Tableware
Tungsten steel	80%–90% Fe, 10%–20% W	Hard at high temperatures	Cutting tools, high-speed drills
Yellow brass	67% Cu, 33% Zn	Ductile, polishes	Hardware
Wood's metal	50% Bi, 25% Pb, 2.5% Sn, 12.5% Cd	Low melting point	Fuse plugs, automatic sprinklers

Table 10.2 Examples of Alloys

Like carbon, boron nitride (BN) has both a diamond structure and a graphite-type structure; the two forms have properties similar to those of their carbon analogs. Boron nitride is prepared by igniting compounds of boron with nitrogen compounds. The most common form is the graphite-like modification, which is used in lubricants. The cubic modification with the diamond structure is among the hardest substances known.

the entire crystal. The atoms in network covalent solids are connected by strong covalent bonds. The electrons in most such solids are completely localized, so they are usually poor electrical conductors. As we will see later, some of these substances are semiconductors. Because the bonds tend to be strong, the solids tend to have high melting points and to be very hard—among them are the hardest substances known.

The crystals of network covalent solids have very stable three-dimensional structures because of the specific directional nature of the covalent bonds. There are planes in crystals, however, and movement along these can cause fractures. Thus even a diamond can be "cut" if the crystal is struck in the correct direction, though force applied in the wrong direction may lead to the reduction of a diamond crystal to powder.

The structure of diamond is shown in Figure 10.30. Many network covalent solids exhibit the diamond structure or some modification of it. A typical modification appears in the silicon dioxide crystal, in which oxygen atoms form bridges between the silicon atoms.

A quite different sort of structure is found in the graphite form of carbon, which consists of a two-dimensional network structure with London dispersion forces rather than covalent bonds holding the layers together. Layers of atoms in

1
Group

IA																	VIIIA
1 **H**	**2** **IIA**											**13** **IIIA**	**14** **IVA**	**15** **VA**	**16** **VIA**	**17** **VIIA**	**2** **He**
3 **Li**	4 **Be**											5 **B**	6 **C**	7 **N**	8 **O**	9 **F**	10 **Ne**
11 **Na**	12 **Mg**	**3** **IIIB**	**4** **IVB**	**5** **VB**	**6** **VIB**	**7** **VIIB**	**8**	**9** VIII	**10**	**11** **IB**	**12** **IIB**	13 **Al**	14 **Si**	15 **P**	16 **S**	17 **Cl**	18 **Ar**
19 **K**	20 **Ca**	21 **Sc**	22 **Ti**	23 **V**	24 **Cr**	25 **Mn**	26 **Fe**	27 **Co**	28 **Ni**	29 **Cu**	30 **Zn**	31 **Ga**	32 **Ge**	33 **As**	34 **Se**	35 **Br**	36 **Kr**
37 **Rb**	38 **Sr**	39 **Y**	40 **Zr**	41 **Nb**	42 **Mo**	43 **Tc**	44 **Ru**	45 **Rh**	46 **Pd**	47 **Ag**	48 **Cd**	49 **In**	50 **Sn**	51 **Sb**	52 **Te**	53 **I**	54 **Xe**
55 **Cs**	56 **Ba**	57 **La** ★	72 **Hf**	73 **Ta**	74 **W**	75 **Re**	76 **Os**	77 **Ir**	78 **Pt**	79 **Au**	80 **Hg**	81 **Tl**	82 **Pb**	83 **Bi**	84 **Po**	85 **At**	86 **Rn**
87 **Fr**	88 **Ra**	89 **Ac** ▲	104	105	106												

★	58 **Ce**	59 **Pr**	60 **Nd**	61 **Pm**	62 **Sm**	63 **Eu**	64 **Gd**	65 **Tb**	66 **Dy**	67 **Ho**	68 **Er**	69 **Tm**	70 **Yb**	71 **Lu**
▲	90 **Th**	91 **Pa**	92 **U**	93 **Np**	94 **Pu**	95 **Am**	96 **Cm**	97 **Bk**	98 **Cf**	99 **Es**	100 **Fm**	101 **Md**	102 **No**	103 **Lr**

Figure 10.29 Network covalent solids. Network covalent solids are formed primarily by the elements shaded here: metalloids and elements close to them in the periodic table.

this structure can slide easily over one another, so graphite is a good lubricant. Graphite also has delocalized electrons, so it is a good electrical conductor in the direction of its layers, in contrast to most network covalent solids.

Band Theory: Metals and Semiconductors

In the simplest picture of metallic bonding, positive ions (consisting of the metal nuclei and all the electrons except the valence electrons) are located at the lattice points, and the valence electrons move relatively freely within the crystal.

The valence electrons of metals are delocalized, in part because they are so loosely bound (that is, they have low ionization potentials). In contrast to the other electrons, the valence electrons no longer reside in orbitals localized on single atoms. Rather, they reside in sets of orbitals that span the entire crystal—that is, in molecular orbitals. Recall from Chapter 8 that one molecular orbital is formed for each atomic orbital contributed by the atoms. The number of molecular orbitals in a metallic crystal would be very large even if each atom in a crystal contributed only one atomic orbital to the formation of the molecular orbitals.

The molecular orbitals in metallic crystals are all very close to one another in energy, so (rather than having discrete energy levels, like molecular substances) the crystals have a continuous range of allowable energies available to their electrons. This continuous range is called an *energy band*. The development of an energy band from atomic orbitals is depicted in Figure 10.31. We can also construct energy bands for metalloids and nonmetals.

Figure 10.30 Structures of some network covalent solids.

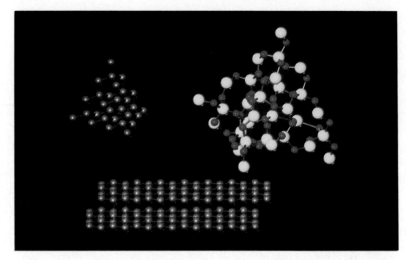

There are two types of energy bands: the **valence band** and the **conduction band.** Electrons paired in localized atomic orbitals or used in covalent bond formation are found in the lower-energy valence band. Delocalized electrons, such as the valence electrons of metals, are found in the higher-energy conduction band.

The nature of various elements can be explained in terms of these energy bands. The critical factor is the existence of an energy gap between them (Figure 10.32). The size of the energy gap helps to determine whether an element is an electrical conductor or an insulator. In metals, electrons either reside permanently in a partially filled conduction band or reside in a valence band so close in energy to a conduction band that some of the electrons can be promoted to the conduction band with very little energy. Electrons in the conduction band are responsible for conductivity. Hence metals are electrical conductors.

In contrast, nonmetals are electrical insulators because no electrons reside permanently in the conduction band, and the gap between energy bands is too large to be easily bridged by electrons in the valence band.

Most network covalent solids, because they are connected by strong covalent bonds, contain no delocalized electrons and so do not conduct electricity. Diamond, for example, is a good electrical insulator. However, silicon and some other

Figure 10.31 Interaction of atomic orbitals to form molecular orbitals. The greater the number of interacting atomic orbitals, the closer together the energies of the molecular orbitals. With a large enough number of interacting atomic orbitals, the molecular orbital energies spread out into a band that is almost continuous.

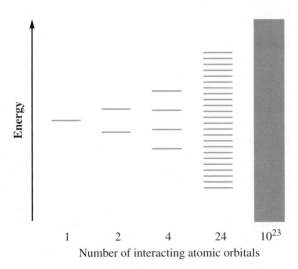

Energy

1   2   4   24 10^{23}

Number of interacting atomic orbitals

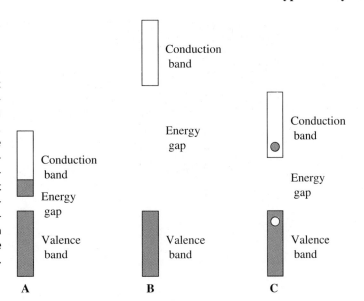

Figure 10.32 Energy gaps between energy bands in a metal, a nonmetal, and a semiconductor. *(A)* In sodium, the valence band is full, but electrons also occupy the conduction band, so electrical conductance can occur readily. *(B)* In diamond, the valence band is full, and the conduction band is empty, so electrical conductance is not possible. *(C)* In silicon, electrons promoted from the valence band to the conduction band leave vacancies in the valence band.

substances, even though they have the same crystal structure as diamond, are *intrinsic semiconductors:* they have a low electrical conductance in their natural pure state. What makes these substances different?

Electrical conductivity results generally from the presence of electrons in the conduction band (shown in Figure 10.32). In semiconductors, a few electrons cross the energy gap to enter the conduction band, resulting in a low electron flow and a certain level of electrical conductivity. Silicon and like substances are semiconductors because the covalent bonds are weaker than those of carbon, resulting in a smaller energy gap. Intrinsic semiconductors have a few electrons in the conduction band and a corresponding number of vacant orbitals (''holes'') in the valence band.

Semiconductor behavior can be enhanced by the **doping** (or blending) of a pure crystal with other materials. The resulting *extrinsic semiconductors* are essentially alloys in which one element is substituted for another. Elements that enhance semiconductor behavior have either one more or one fewer valence electron than the element being doped.

Most commercial semiconductors are made from pure crystals of silicon or germanium, each having four valence electrons. If these substances are doped with an element that has five valence electrons (such as arsenic, phosphorus, or antimony), an *n*-type semiconductor (*n* for negative) results. This procedure requires only a small amount of dopant—for example, 1 atom of arsenic per 10^7 atoms of silicon is typically sufficient to give appropriate levels of conductivity. The arsenic atom replaces a silicon atom in the crystal lattice, but because it has an extra electron, that electron ends up in the conduction band.

An *n*-type semiconductor, then, has extra electrons. It should not be surprising that there are also *p*-type semiconductors (*p* for positive), which are deficient in electrons. Doping a Group IVA (14) element with a Group IIIA (13) element, such as boron or aluminum, yields a *p*-type semiconductor. Because these dopants have only three valence electrons, ''holes'' are created in the valence band. Electrical conduction results from migration of these ''holes''—in other words, from electron movement—through the valence band.

Among applications of semiconductors are transistors, widely used in many electronic devices such as computers. Some solar cells use silicon semiconductors,

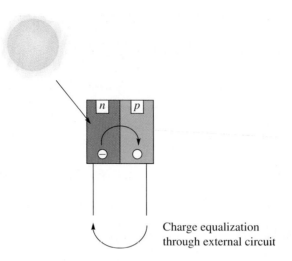

Charge equalization
through external circuit

as shown in Figure 10.33. Here, energy from sunlight causes the ejection of electrons from an *n*-type semiconductor. The electrons pass into the ''holes'' in a *p*-type semiconductor, resulting in charge separation between *n* and *p* parts of the semiconductor arrangement. The only way for electrons to return to the *n*-type side to balance the charge is through an external circuit, which thus carries a flow of electricity.

10.6   Phase Changes

We have now examined the structure and properties of each of the physical states of matter. But many phenomena involve changes between these states or the coexistence of states. ▶ As mentioned in the chapter introduction, water is the only common substance that exists in all three physical states under atmospheric conditions. An iceberg floats in water and gradually melts to produce more liquid water. Liquid water in underground pools is expelled from a geyser as steam. When two or more different forms of a substance exist as a heterogeneous system, each form is called a **phase.** When an iceberg floats in warm water, water vapor is present in the air, so three phases of water—the gaseous, liquid, and solid phases—are present in the system. Transitions between these phases are called *phase changes*. This section examines phase changes and the conditions that affect them.

Under what conditions can different states of matter coexist?

Liquid–Gas Conversions

Why do liquids become cooler when they evaporate?

Like molecules in a gas, molecules in a liquid at any given temperature have a range of kinetic energies, and these energies follow the same kind of curve as that shown in Figure 9.15. If a molecule with a high kinetic energy happens to be close to the liquid's surface, and if it is moving in the right direction, it may be able to overcome the intermolecular forces of attraction and escape into the gas phase. This process is called **evaporation.**

   Because the molecules of highest kinetic energy are the ones most likely to escape from the body of the liquid, the average kinetic energy of the remaining

WORLDS OF CHEMISTRY

Liquid Crystals: Sending Messages with Molecules

Increasingly, machines are being designed to communicate with their human users via molecules.

What do wristwatches, some computer screens, calculators, and postcard thermometers have in common? Liquid-crystal displays (LCDs). Increasingly, machines are being designed to communicate with their human users via molecules. But how can molecules be used as messengers?

Certain substances, called liquid crystals, can flow like liquids but have the structural order characteristic of solids. The liquid crystalline state occurs for select substances during melting as a transition between the solid state and the true liquid state. Most liquid crystals change color with a change of structural order, which can be controlled by temperature, mechanical stress, magnetic fields, or electrical fields.

Liquid crystalline substances all consist of long, narrow, rigid organic molecules. Most of them contain substituted benzene rings, C_6H_6, introduced in Chapter 8. One example is the molecule 4-methoxybenzylidene-4'-n-butylaniline, which behaves as a liquid crystal over the temperature range 21°C–47°C:

$$CH_3O \underset{}{\overset{}{\bigcirc}} \overset{H}{\underset{}{C}} = N \underset{}{\overset{}{\bigcirc}} C_4H_9$$

Generally, the group connecting the two benzene rings contains a double or triple bond, which gives the molecule linear rigidity.

Liquid crystals can be classified into three structural types. Nematic liquid crystals have parallel molecules that can slip past one another (see Figure 10.34). Smectic liquid crystals also contain parallel molecules, but here the molecules are arranged in layers that can slip past one another (Figure 10.35). The displays on watches and other devices use smectic or nematic liquid crystals. When an electrical field is applied to a film of the liquid crystal, the molecular alignment changes, and parts of the film turn black, forming the shapes we see on the watch face or screen.

Cholesteric liquid crystals are nematic crystals in which each layer is rotated with respect to adjacent layers. The successive layers form a helical arrangement. Cholesteric liquid crystals reflect different colors of light under different conditions, such as changing temperatures. Because of this property, they are used in mood rings, which

Nematic structure

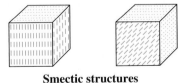

Smectic structures

Cholesteric structure

Figure 10.34 Molecular arrangements in liquid crystals.

Figure 10.35 A smectic liquid crystal magnified 320 times, showing layers of parallel molecules.

(continued)

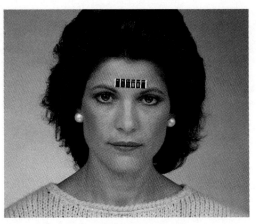

Figure 10.36 Liquid-crystal thermometers contain a thin film of a cholesteric liquid crystalline material. The color of this film changes with temperature.

change color with the temperature of the skin, and in liquid-crystal thermometers (Figure 10.36).

Questions for Discussion

1. Draw the structure of 4-methoxybenzylidene-4'-n-butylaniline, showing why it should be a liquid crystal.
2. Brainstorm with your classmates some possible uses of liquid crystals.
3. Displays on LCD computer monitors, such as those on laptop computers, cannot be changed as fast as those on tube-type monitors. Suggest a reason for the slowness of LCD monitors.
4. Why do LCD computer monitors change color when you touch them?

liquid molecules decreases as evaporation proceeds. Therefore, unless energy is supplied, the temperature of the liquid decreases as a result of evaporation. To evaporate a liquid at constant temperature, we must supply an amount of heat called the **heat of vaporization** or **enthalpy of vaporization, ΔH_{vap},** which is the amount of heat needed to evaporate 1 mol of a liquid at constant temperature. Some representative values of heats of vaporization are given in Table 10.3.

Evaporating a liquid requires overcoming the intermolecular forces between liquid molecules. Accordingly, the highest values for heat of vaporization in Table 10.4 belong to molecules that involve hydrogen bonding, which is much stronger than other intermolecular forces. The molecules with lower values include both polar and nonpolar molecules. The values for molecules with large electron clouds, such as C_6H_6 and CCl_4, tend to be at the higher end of the group of nonpolar molecules because of their larger London dispersion forces.

EXAMPLE 10.4

Predict which molecule, NH_3 or PH_3, should have the higher heat of vaporization.

Solution   Ammonia (NH_3) should have a higher heat of vaporization than phosphine because of hydrogen bonding, even though phosphine, with its greater size and number of electrons, would be expected to have greater London dispersion forces.

Practice Problem 10.4

Predict which molecule, H_2S or H_2O, would have the higher heat of vaporization.

Answer: H_2O

The cooling effect associated with evaporation is significant in many processes. Some examples include cooling by perspiration, evaporative cooling

Table 10.3 Heats of Vaporization of Some Substances at 25°C

Substance	ΔH_{vap} (kJ/mol)	Nonpolar	Polar	Hydrogen-Bonded
Ar	6.3	×		
CH_4	9.2	×		
C_2H_5Cl	24.3		×	
CS_2	27.7	×		
CH_2Cl_2	29.0		×	
$SiCl_4$	30.0	×		
Br_2	30.9	×		
$CHCl_3$	31.4		×	
CCl_4	32.5	×		
C_6H_6	33.9	×		
CH_3OH	38.0		×	×
C_2H_5OH	42.6		×	×
H_2O	44.0		×	×
N_2H_4	44.8		×	×
H_2O_2	51.5		×	×

(swamp cooling) in homes, and the use of canvas water bags for desert travel. The wind chill factor also is based on evaporation. The cooling effect of cold air on skin is enhanced if it is windy, because moving air evaporates water from the skin faster. The increased cooling is equivalent to that produced by a lower air temperature, which is the temperature given in weather reports in cold climates as the wind chill factor.

The extent of evaporation differs depending on the liquid's container. In an open container, a liquid evaporates completely. The more energetic molecules overcome the intermolecular forces and escape into the gas phase. The container is open, so these gaseous molecules can diffuse away into the surrounding atmosphere. This process continues until all the liquid has evaporated. If the container is insulated, the liquid cools, and because the average energy of the molecules decreases, the process goes more slowly. If the container is not insulated, the average energy of the liquid remains constant, because the liquid absorbs heat from its surroundings and remains at constant temperature. Some of the evaporated molecules may return to the liquid state if they happen to diffuse toward the liquid rather than away from it, but most of the evaporated molecules stay in the gaseous state.

In a closed container, gas-phase molecules are more likely to re-enter the liquid phase. Because the gaseous molecules cannot diffuse completely away from the liquid, some of them diffuse toward it. When they hit the liquid, intermolecular forces prevent them from returning directly to the gas phase. The return of gas-phase molecules to the liquid phase is called **condensation.**

Starting with a pure liquid in a closed container, only evaporation occurs at first, because there are no gas-phase molecules. As evaporation continues, molecules begin to accumulate in the gas phase. Then some of these molecules undergo condensation. Evaporation and condensation continue at rates that depend on the

Figure 10.37 Establishment of an equilibrium between a liquid and its vapor. The amount of substance in the vapor state increases until the rates of evaporation and condensation (represented by the arrows) are just equal. After that, the amount of vapor is constant even though evaporation and condensation continue. The partial pressure of the vapor increases until it reaches the value of the vapor pressure.

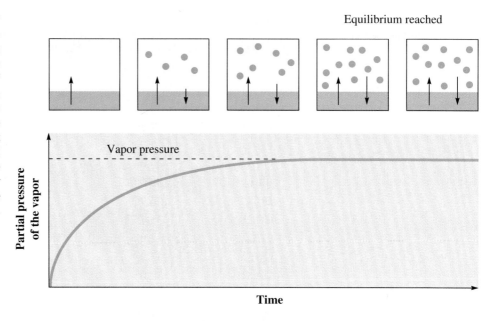

temperature and the number of molecules eligible to undergo each process. The changes that occur during evaporation in a closed container are depicted in Figure 10.37. Ultimately, the rate of condensation equals the rate of evaporation, and the number of gas-phase molecules becomes constant. This situation, in which opposing processes occur at equal rates, is called a state of **equilibrium.** It can be represented by an equation such as

$$\text{Liquid} \underset{\text{condensation}}{\overset{\text{evaporation}}{\rightleftharpoons}} \text{gas}$$

The double arrows stand for a state of equilibrium involving two reversible processes.

In a state of equilibrium, the rate of evaporation equals the rate of condensation, so the amount of substance in the gas state becomes constant, even though evaporation and condensation continue to occur. The partial pressure of the gas molecules above the liquid at equilibrium is called the **vapor pressure** of the liquid.

The vapor pressure of a given liquid depends largely on two factors: the magnitude of the intermolecular forces and the temperature. The vapor pressures at 25°C of several liquids, as well as the primary intermolecular forces at work in each, are given in Table 10.4. Note that the stronger intermolecular forces present in polar substances or in metals or large molecules are associated with lower vapor pressures. The stronger the forces, the greater the energy needed to overcome them and effect evaporation. Thus, at a given temperature, strongly interacting substances evaporate less readily than weakly interacting substances and so have lower vapor pressures.

Why do different liquids have different vapor pressures?

EXAMPLE 10.5

Explain why the vapor pressure of water is about 40 torr at 35°C, whereas that of diethyl ether, $(C_2H_5)_2O$ is about 760 torr.

Table 10.4 Equilibrium Vapor Pressure of Some Liquids at 25°C

Liquid	Vapor Pressure (torr)	Primary Intermolecular Forces
BrF$_5$	400	London dispersion forces
CCl$_4$	100	London dispersion forces
H$_2$O	24	Dipole–dipole forces, hydrogen bonding
Hg	1.8×10^{-3}	Metallic bonding
Silicone oils	$\approx 10^{-8}$	Dipole-dipole forces, also large London dispersion forces associated with very large molecules having many intermolecular interactions

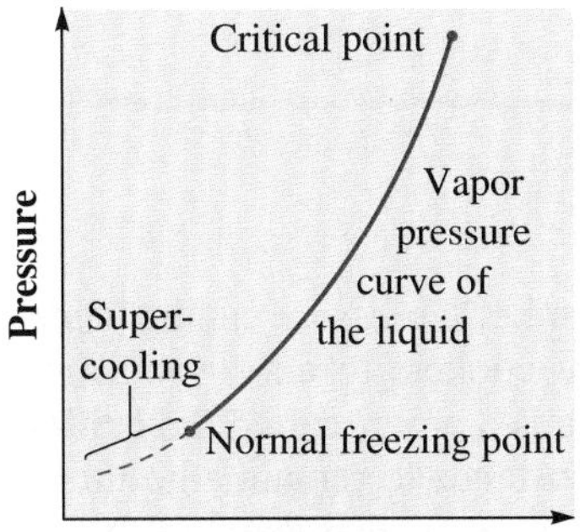

Figure 10.38 The vapor pressure of a liquid increases with temperature. The vapor-pressure curve terminates at the upper end at the critical point and at the lower end at the normal freezing point. A liquid sometimes can be cooled below its freezing point, a phenomenon called supercooling.

Solution Both molecules are polar and so undergo dipole–dipole interactions. However, diethyl ether, unlike water, does not involve hydrogen bonding. Because hydrogen bonding is much stronger than dipole–dipole forces, the intermolecular forces are much stronger in water than in diethyl ether. Thus, at a given temperature, more diethyl ether evaporates, giving a higher vapor pressure.

Practice Problem 10.5

Which substance, Br$_2$(l) or Hg(l), would be expected to have the higher vapor pressure at a given temperature?

Answer: Br$_2$

The vapor pressure of a given liquid also depends on the temperature. As the temperature of a liquid increases, the average kinetic energy of the liquid molecules increases, the rate of evaporation increases, and the vapor pressure exerted by the liquid increases, as shown in Figure 10.38. The data for water given in Table 9.3 are typical. When the vapor pressure equals the external pressure, the liquid begins to boil.

The temperature of a pure, boiling liquid remains constant until all the liquid has vaporized. Adding more heat to the liquid increases the rate of evaporation rather than the temperature. The temperature at which boiling occurs—that at which the vapor pressure equals the external pressure of the atmosphere—is called the **boiling point.** The boiling point for a given liquid depends on the prevailing atmospheric pressure, as shown in Figure 10.39. The temperature at which boiling occurs when the atmospheric pressure is exactly 1 atm (standard pressure) is called the **normal boiling point.** An atmospheric pressure of 1 atm is normal only at sea level, however. At higher altitudes, atmospheric pressure is less than 1 atm and liquids boil at lower temperatures. The normal boiling point of water, for example, is 100.0°C. At 9000 ft above sea level, the atmospheric pressure is about 550 torr, and water boils at 91°C. It takes nearly twice as long to cook an egg in boiling water at this elevation than at sea level, because the egg must cook at a lower temperature.

Figure 10.39 Water was boiled in the flask long enough to drive all air out of the space above the water. The flask was then stoppered and inverted. Upon cooling of the flask with ice, the pressure of the water vapor decreases sufficiently (because of condensation) that the temperature of the water is above the boiling point and the water boils.

EXAMPLE 10.6

Arrange the following liquids in order of increasing boiling point: CH_4, GeH_4, SiH_4.

Solution   None of these molecules is polar, so we must consider only the magnitude of the London dispersion forces in each liquid. These forces increase as the molecule becomes larger and thereby has a more readily polarized electron cloud. As intermolecular forces increase, the energy required to vaporize the liquids increases, and the temperature needed to supply the energy increases. Thus we predict that the order of boiling points should be

$$CH_4\ (-164°C) < SiH_4\ (-112°C) < GeH_4\ (-89°C)$$

Practice Problem 10.6

Arrange the following liquids in order of increasing boiling point: $CHCl_3$, $CHBr_3$, CHI_3.

Answer: $CHCl_3 < CHBr_3 < CHI_3$

A liquid held under a pressure higher than atmospheric pressure can be heated above its normal boiling point. The use of pressure cookers to cook food more rapidly is based on this phenomenon. However, there is a point, called the **critical point,** beyond which it is no longer possible for a substance to remain a liquid, no matter how much pressure is applied. This critical point is characterized by the minimum pressure (**critical pressure**) at the maximum temperature (**critical temperature**) at which a substance can still exist as a liquid. At any temperature higher than the critical temperature, no amount of additional pressure can prevent the liquid from evaporating completely. Beyond the critical temperature, the kinetic energy becomes so great that no matter how close together the molecules are forced, the intermolecular forces cannot overcome the kinetic energy, and the substance cannot liquefy.

Liquid–Solid Conversions

As liquids are cooled, the average kinetic energy decreases. When the intermolecular forces completely overcome the kinetic energy of the liquid molecules, the molecules become fixed rigidly into position in the solid state, and the liquid freezes. The freezing of a liquid into the solid state occurs at a characteristic temperature called the **freezing point,** which is the temperature at which the solid and liquid coexist in a state of equilibrium. When this state of equilibrium is reached under a pressure of exactly 1 atm, the temperature is called the **normal freezing point.** The temperature remains constant from the time a liquid starts freezing until all the liquid has solidified. Only after that does removal of more heat cause a further decrease in temperature of the solid.

Why does pressure cause the freezing point of some liquids to increase and that of others to decrease?

Pressure usually has a small effect on the freezing point of a liquid. For example, the freezing point of water is 0.0098°C at 5 torr, 0.0023°C at 1 atm, −5.00°C at 590 atm, and −10.00°C at 1090 atm. Pressure may either raise or lower the freezing point, depending on whether the solid or the liquid is more

Figure 10.40 *(A)* Most liquids are less dense than their solid forms. In these liquids, an increase in pressure causes the freezing point to increase—the solid melts or the liquid freezes at a higher temperature. *(B)* In water, however, an increase in pressure results in melting at a lower temperature, because the density of solid water is less than that of liquid water.

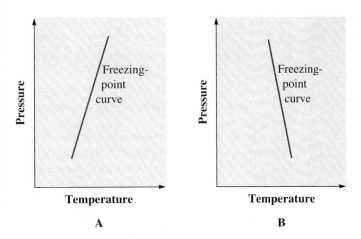

A

B

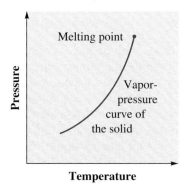

Figure 10.41 The vapor pressure of a solid increases as the temperature increases. The vapor-pressure curve terminates at its upper end when the solid melts to become a liquid at the melting point. This temperature is the same as the freezing point of the liquid.

dense. If the solid is more dense than the liquid, as in most substances, a decrease in pressure causes the solid to melt. This results in a decrease in the freezing point (a positive slope in the plot of pressure and temperature shown in Figure 10.40A).

Water is an example of a substance whose density is lower in the solid state than in the liquid state, resulting in the negative slope in Figure 10.40B. This unusual property allows us to perform such experiments as passing a string directly through an ice cube without dividing the ice. Pressure applied to the ice cube with the string causes the cube to melt in the area of increased pressure, and the string sinks into the ice cube. Liquid water passes above the string, where the pressure is lower, and then freezes again.

Melting is just the reverse of freezing; it involves the same equilibrium process, but with the direction reversed. Thus the **melting point** of a solid and the freezing point of its liquid state are the same temperature.

$$\text{Solid} \; \underset{\text{freezing}}{\overset{\text{melting}}{\rightleftharpoons}} \; \text{liquid}$$

Solid–Gas Conversions

It is possible for a substance to change directly from the solid state to the gaseous state without going through the liquid state. Thus solids, like liquids, have a characteristic vapor pressure. The vapor pressures of most solids are rather low, much lower than those of their liquids. However, a few have fairly high vapor pressures—mothballs, iodine, and dry ice, $CO_2(s)$, are examples—and can change directly from the solid to the gas phase. Like that of a liquid, the vapor pressure of a solid increases as the temperature increases (Figure 10.41).

The evaporation of a solid is commonly, though somewhat inexactly, called **sublimation.** In the strictest sense, sublimation refers to the process

$$\text{Solid} \; \underset{\text{deposition}}{\overset{\text{sublimation}}{\rightleftharpoons}} \; \text{gas}$$

of purifying a solid by first evaporating it to a gas and then resolidifying the gas, as shown for iodine in Figure 10.42.

Figure 10.42 When iodine is heated, it evaporates or sublimes to give a purple gas. The gas resolidifies on the colder surfaces of the test tube filled with ice.

Phase Equilibria

So far we have been concerned with equilibria between the phases of matter. Substances also exist under conditions other than these equilibrium situations, especially when they are undergoing changes in conditions. However, substances that are not at equilibrium change in such a way as to move toward a state of equilibrium.

The state of equilibrium can be reached only where two phases coexist. Consider a gas whose pressure is less than the equilibrium vapor pressure. If we add a liquid to the container, it evaporates, increasing the pressure of the gas by increasing the number of molecules in the gas phase. The liquid evaporates either completely or until the equilibrium vapor pressure is reached. In the latter case, some liquid is still present, and the two phases are in a state of equilibrium.

Heating and Cooling Curves

We can reach a state of equilibrium in another way. If we cool the gas at constant pressure, its volume decreases, but it remains as a gas until the temperature has been lowered enough so that the pressure of the gas just matches the vapor pressure of the liquid at that temperature. Any further removal of heat causes condensation of the gas to a liquid, in proportion to the amount of heat removed. When both gas and liquid are present in a state of equilibrium, removal of heat results not in a decreased temperature but only in further condensation of gas to liquid. This shows up as the first plateau in the cooling curve in Figure 10.43. The heat removed, on a per mole basis, is called the heat of condensation, ΔH_{cond}, which is equal in magnitude but opposite in sign to the heat of vaporization, ΔH_{vap} (described in Section 5.3).

When all the gas has been condensed to a liquid, the system is no longer at equilibrium, and removal of heat once again results in decreased temperature. The temperature drop halts again at the freezing point of the liquid, where we have another state of equilibrium; it is shown as the second plateau in Figure 10.43. The heat removed now corresponds to the heat of crystallization, ΔH_{cryst}, which is equal in magnitude but opposite in sign to the heat of fusion, ΔH_{fus} (see Section 5.3).

Once all the liquid has crystallized to the solid, further removal of heat lowers the temperature below the freezing point of the liquid. All the changes that occur during the cooling of a substance are illustrated in the cooling curve shown in Figure 10.43. The reverse process simply proceeds backwards along this same curve, which then becomes a heating curve. The same two plateaus are observed during phase changes.

In cooling curves for some substances, we may observe a liquid becoming colder than the equilibrium freezing point. This phenomenon is called **supercooling.** Solidification requires some matrix around which a solid can form—perhaps imperfections in a container wall, pieces of dust, or other solid particles. Thus in a very clean system, solidification may be delayed. Usually, after supercooling a degree or two, the substance begins to crystallize, and the temperature returns to the freezing point. Supercooling is common in such substances as glass, tars, rubbers, and plastics.

Figure 10.43 Cooling curve for a pure substance at constant pressure.

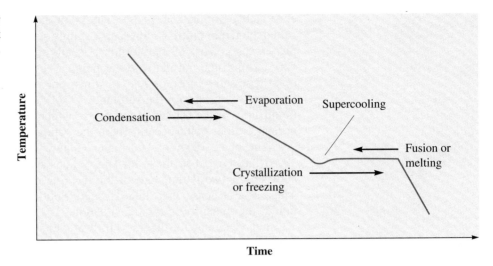

Phase Diagrams

We have described a variety of phenomena related to the different phases of a substance and the conditions under which they coexist in equilibrium or nonequilibrium states. The phases of a substance and their existence at various pressures and temperatures can be represented in a **phase diagram.** This diagram contains all the information we need to construct cooling curves, vapor-pressure curves, and melting-point curves, as well as to determine the phase or phases at which a substance is stable under specific conditions. The features of a typical phase diagram are shown in Figure 10.44. Note that each of the lines in the diagram represents an equilibrium condition.

We have already discussed most of the features of phase diagrams. The solid lines represent pressure–temperature conditions under which two phases of a substance can coexist in a state of equilibrium. In any other area, the substance can

Figure 10.44 Typical phase diagram for a pure substance.

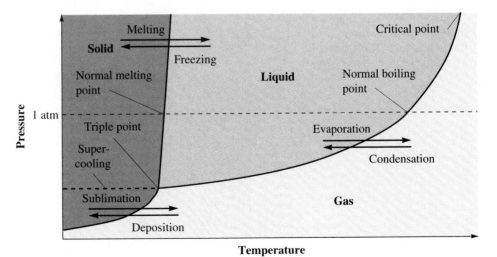

exist stably in only one phase. The point at which the three solid lines meet—called the **triple point**—represents the only set of conditions under which all three phases can coexist in equilibrium. Usually the temperature at the triple point is very close to the normal freezing point. The dashed line to the left of the triple point represents supercooling, in which only the liquid state exists, even though conditions are right for the substance to solidify. The vapor-pressure curve of the liquid stops at the critical point, above which the liquid can no longer exist. The normal boiling point occurs where the vapor pressure is exactly 1 atm.

The phase diagrams for water (Figure 10.45) and carbon dioxide (Figure 10.46) are typical of those for simple substances. The major distinction in form between the two is the slope of the freezing-point curve. As we have seen, solid water is less dense than liquid water. Solid carbon dioxide is denser than the liquid, the much more typical situation.

Let's examine the diagrams for water and carbon dioxide to see what information can be obtained from them. Since we normally operate at an external pressure of 1 atm, consider the phase diagrams at this pressure. For water (Figure 10.45), starting at a constant pressure of 1 atm, raising the temperature results ultimately in the transformation of solid ice to liquid water at a temperature of 0°C. The melting process continues at 0°C until no solid remains. The temperature can then be raised further by addition of heat until it reaches 100°C, where water boils. This temperature remains constant until all the liquid has been converted to the gas. Further addition of heat raises the temperature of the resulting steam (water vapor). By comparison, the temperature of carbon dioxide (Figure 10.46) at 1 atm increases until it reaches −78.6°C and then remains constant until the solid has completely evaporated. The solid does not melt first, because the vapor pressure of the solid reaches 1 atm at a temperature below the melting point, which is −56.6°C. To melt solid carbon dioxide, it is necessary to raise the temperature above −56.6°C while the pressure is above 5.11 atm. In contrast, solid water must be heated at a pressure greater than only 4.59 torr (0.00604 atm) to be converted to the liquid. At any pressure below the triple-point pressure, it is not possible to melt a solid, but only to evaporate it. Solid water can sublime as well, at tempera-

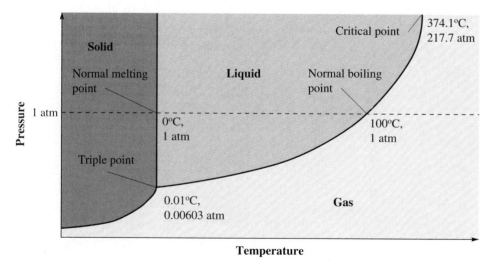

Figure 10.45 Phase diagram for water. The curves are not drawn to scale; this allows all features to be seen more easily.

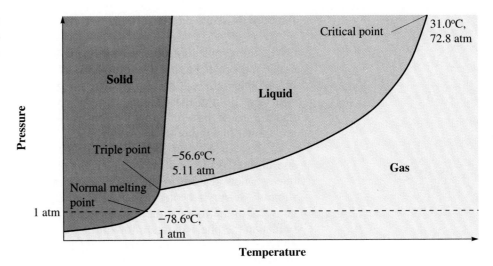

tures below 0°C. At these temperatures, solid water evaporates because it is in equilibrium with the gaseous state, and no liquid state can exist. Wet clothing hung outside in very cold weather slowly dries because of this sublimation process.

EXAMPLE 10.7

Describe the changes that occur in a sample of water at -0.01°C when the external pressure is raised from 0.001 atm to 10 atm.

Solution   Consult Figure 10.45. You may find it useful to draw a vertical pencil line at -0.01°C, which is just below the normal freezing point of water and is below the triple point as well. At this temperature, the sample is a gas at pressures lower than the vapor-pressure curve of solid water. Increasing the pressure past this curve causes the gas to solidify. Increasing the pressure much past 1 atm, however, causes the solid to begin to melt. At higher pressures, the water is in the liquid state at this temperature.

Practice Problem 10.7

Describe the changes in a sample of carbon dioxide at -56.0°C when the external pressure is raised from 0.001 atm to very high pressures.

Answer: gas → liquid → solid

10.7 Forms of Elemental Sulfur

The element sulfur is an interesting study in the equilibria between phases, because it exists in a number of allotropic solid forms as well as in the liquid and gaseous phases.

Rhombic and Monoclinic Sulfur

Of the numerous allotropes of sulfur, the most prevalent are two modifications of cyclic molecules having the formula S_8. These modifications, rhombic sulfur and monoclinic sulfur, differ in the way the S_8 molecules pack together in the solid structure. Both of these forms of sulfur consist of odorless, bright yellow crystals that are nearly insoluble in water but very soluble in carbon disulfide, CS_2. They are shown in Figure 10.47.

The existence of more than one solid phase, which is exhibited by many substances besides sulfur, is called *polymorphism*. The extra solid phase introduces a new feature to the phase diagram of sulfur, shown in Figure 10.48. If rhombic sulfur is heated to 95.31°C at a pressure of 5.1×10^{-6} atm, it slowly converts to the monoclinic form. Note that this temperature is one of three triple points on the phase diagram for sulfur. At 95.31°C and 5.1×10^{-6} atm, the three phases in equilibrium are rhombic, monoclinic, and gaseous sulfur. The temperature required for conversion from rhombic to monoclinic sulfur at pressures up to 1 atm are not much above this triple-point temperature; the transition temperature is 95.39°C at 1 atm.

Liquid Sulfur

When sulfur is heated to about 115°C at 1 atm, it melts to give a yellow, transparent, mobile liquid. Further heating produces little change up to about 159°C. Above this temperature the liquid rapidly turns brown, and it becomes increasingly viscous as the temperature is raised to about 200°C. The change in viscosity is quite phenomenal; it increases by a factor of about 10,000 between 160°C and 187°C. This phenomenon is caused by the formation of long, intertwined chains

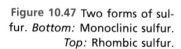

Figure 10.47 Two forms of sulfur. *Bottom:* Monoclinic sulfur. *Top:* Rhombic sulfur.

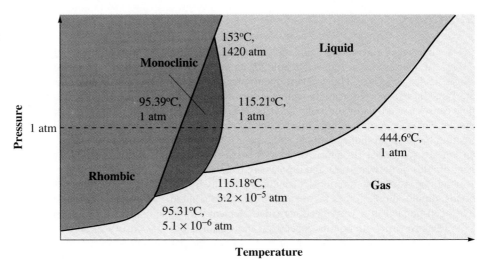

Figure 10.48 Phase diagram for sulfur. The curves are not drawn to scale.

that restrict the flow of molecules past one another. Above 200°C, the chains begin to break up and the viscosity decreases, until at the normal boiling point of 444.6°C, the liquid is once again mobile and yellow.

Gaseous Sulfur

Sulfur vapor contains various species, such as S_8, S_6, S_4, and S_2, in amounts that depend on the temperature. At very high temperatures, significant proportions of the molecules dissociate into sulfur atoms.

Plastic Sulfur

Figure 10.49 Plastic sulfur.

If we heat sulfur to about 200°C and then cool it very rapidly—for example, by pouring it into cold water—we produce a soft, rubbery mass called plastic sulfur (Figure 10.49). Plastic sulfur consists of long, coiled chains that can uncoil or recoil under stress, giving the rubberlike properties. If it is left standing, plastic sulfur slowly crystallizes back to rhombic sulfur.

Uses of Sulfur

About 85% of the sulfur consumed annually in the United States is used in the manufacture of sulfuric acid. Smaller amounts go into carbon disulfide, fungicides, medicines, and germicides. Sulfur dioxide, produced by burning sulfur, is used in the pulp and paper industry as a bleaching agent. Sulfur is bonded with rubber in a process called vulcanization. Rubber consists of long-chain hydrocarbons that are rather sticky. Vulcanizing gets rid of the sticky properties and hardens the rubber. When sulfur is bonded between the hydrocarbon chains, they become more rigid.

Summary

Substances exist in the liquid and solid states because of intermolecular forces that hold the atoms or molecules together. The weakest intermolecular forces are London dispersion forces, in which an instantaneous dipole causes a nearby atom or molecule to become an induced dipole. These temporary dipoles attract one another. The attraction between an induced dipole and a polar molecule or an ion are somewhat stronger. Attractions between permanent dipoles or between ions and dipoles are even stronger. Among the strongest intermolecular forces are hydrogen bonds: dipole–dipole forces between molecules that contain hydrogen bonded to a very electronegative element.

The properties of liquids depend on the distance between molecules and on the strength of intermolecular forces. Viscosity, or resistance to flow, is one such property. A unique property of liquids is surface tension. Many properties of liquid water and ice can be explained by the existence of strong hydrogen bonding forces. The hydrogen bonds cause water molecules to arrange themselves in tetrahedral patterns, which are rapidly fluctuating in the liquid but are frozen into the solid.

Most solids exist in a crystalline form. Crystals contain well-ordered and symmetrical arrangements of planar faces, which are reflected in a similarly ordered arrangement of atoms, molecules, or ions in a crystal lattice. The lattice is built from unit cells. The most efficient packing of spheres in a cube is achieved in the cubic closest-packed or face-centered cubic structure or in the hexagonal closest-packed structure. Most metals adopt one of these closest-packed structures. Many ionic salts adopt structures based on the closest-packed structures. Crystalline solids can be classified as molecular solids, ionic solids, metallic solids, and network covalent solids. The properties of metals can be explained by a form of molecular orbital theory called the band theory of solids, which also explains the behavior of semiconductors.

The transitions among the three states of matter are called phase changes. Such changes occur with changes in pressure and temperature. Both solids and liquids exhibit vapor pressure. Changes in physical state that occur with temperature changes at constant pressure can be described by means of cooling or heating curves. All the information related to the phases of a substance and the conditions under which they exist can be summarized in a phase diagram.

Key Terms

amorphous solid (10.4)
boiling point (10.6)
condensation (10.6)
conduction band (10.5)
**coordination
    number** (10.4)
critical point (10.6)
critical pressure (10.6)
**critical
    temperature** (10.6)
crystal (10.4)
crystal lattice (10.4)

crystalline solid (10.4)
**cubic closest-packed
    structure** (10.4)
doping (10.5)
equilibrium (10.6)
evaporation (10.6)
fluid (10.2)
freezing point (10.6)
**heat (enthalpy) of
    vaporization** (10.6)
**hexagonal closest-
    packed**

structure (10.4)
hydrogen bond (10.1)
induced dipole (10.1)
**instantaneous
    dipole** (10.1)
lattice plane (10.4)
melting point (10.6)
meniscus (10.2)
**normal boiling
    point** (10.6)
**normal freezing
    point** (10.6)

octahedral holes (10.4)
phase (10.6)
phase diagram (10.6)
sublimation (10.6)
supercooling (10.6)
surface tension (10.2)
tetrahedral holes (10.4)
triple point (10.6)
unit cell (10.4)
valence band (10.5)
vapor pressure (10.6)
viscosity (10.2)

Exercises

Intermolecular Forces
10.1 What is meant by the term *intermolecular force*?
10.2 List various types of attractive forces between atoms and molecules.

10.3 Should a hydrogen bond be called a bond? How is it similar to or different from a covalent bond?
10.4 Why are intermolecular forces greater for polar molecules than for nonpolar molecules?

10.5 Why are intermolecular forces greater for large molecules than for small molecules?

10.6 Indicate the primary type of intermolecular force operating in each of the following pure substances.
  a. C_6H_6 e. CH_3NH_2
  b. NaCl   f. SO_2
  c. CS_2 g. CO_2
  d. $CHCl_3$ h. SiO_2

10.7 Indicate the nature of the interactions that must be overcome to melt or boil each of the following substances.
  a. krypton   f. lithium bromide
  b. silicon dioxide   g. diamond
  c. carbon dioxide   h. sulfur dioxide
  d. methane   i. carbon tetrachloride
  e. ammonia

10.8 What are the major types of intermolecular attractions between members of the following pairs?
  a. Cl_2 and H_2O d. NH_3 and H_2O
  b. H_2 and O_2 e. HBr and CCl_4
  c. CCl_4 and CH_4 f. CH_3OH and C_2H_5OH

10.9 Which of the following substances can participate in hydrogen bonding?
  a. H_2O d. PH_3
  b. NH_3 e. HCO_2H
  c. H_2Se

10.10 Which of the following substances can form hydrogen bonds with water molecules?
  a. SO_2 b. HF   c. BeH_2 d. CH_4
  e. CH_3—O—CH_3

10.11 The boiling point of helium is 4 K, and that of H_2 is 20 K. Explain why hydrogen boils at a higher temperature than helium.

10.12 List the following substances in order of increasing strength of intermolecular forces: $CaBr_2$, H_2O, He, I_2, N_2.

10.13 Which substance in each of the following groups is most likely to be a gas at room temperature and atmospheric pressure?
  a. Ar, Cr, Hg, I_2 b. Cl_2, Br_2, F_2, I_2

Properties of Liquids and Solids

10.14 Which properties of liquids are different from those of solids?

10.15 Which properties of liquids are different from those of gases?

10.16 What is viscosity?

10.17 What is surface tension?

10.18 Arrange the following substances in order of increasing boiling point: CH_3OH, C_2H_5OH, C_3H_7OH.

10.19 Arrange the following substances in order of increasing boiling point: AsH_3, NH_3, PH_3.

10.20 Arrange the following substances in order of increasing melting point: CH_4, GeH_4, SiH_4, SnH_4.

10.21 Explain how an increase in the strength of intermolecular forces affects each of the following properties of a substance.
  a. vapor pressure of a solid
  b. hardness of a molecular solid
  c. boiling point of a liquid
  d. viscosity of a liquid
  e. melting point of a molecular solid

10.22 Why does the viscosity of a liquid decrease as the temperature increases?

Water and Ice

10.23 Why is water an important liquid?

10.24 How many hydrogen atoms surround each oxygen atom in ice?

10.25 What molecular properties of ammonia and water cause the physical properties of these substances to show similarities?

10.26 Discuss the features of the water molecule that endow water with unique properties.

10.27 Ice is lower in density than liquid water for reasons that lie in the structure arising from hydrogen bonding between water molecules. Both HF and NH_3 undergo strong hydrogen bonding. Why are they more dense as solids than as liquids?

10.28 Explain why the heat of vaporization of water (44.0 kJ/mol) is greater than the heat of fusion of ice (6.01 kJ/mol).

Phase Changes and Phase Diagrams

10.29 List possible phase changes.

10.30 What is a phase diagram?

10.31 What is the role of the vapor pressure of a liquid in determining its boiling point?

10.32 Skin swabbed with alcohol is cooled below room temperature, even though the alcohol is at room temperature. Explain.

10.33 Why do liquids boil at lower temperatures in the mountains?

10.34 Provide a molecular explanation for the increase in vapor pressure with increasing temperature.

10.35 Explain the process of evaporation in terms of molecular behavior in the liquid state.

10.36 Explain the process of sublimation in terms of molecular behavior in the solid state.

10.37 Which substance in each pair has the higher heat of vaporization?
  a. Cl_2 or I_2
  b. CH_3OH or CH_3SH
  c. PF_3 or NF_3

10.38   Which substance in each pair has the higher vapor pressure at a given temperature?
   a. C_2H_5OH or CH_3OH
   b. CH_3OH or H_2O
   c. NH_3 or H_2O
   d. Cl_2 or Br_2
   e. PH_3 or AsH_3

10.39   Explain why $(CH_3)_2C{=}O$ boils at 56.2°C, whereas H_2O boils at 100°C.

10.40   What changes occur when the temperature is raised at constant pressure in a sample of water that was initially at -173°C and 100 torr?

10.41   What are the allowable temperatures of the following systems?
   a. $H_2O(l)$ in equilibrium with $H_2O(g)$ at 1 atm pressure
   b. $H_2O(l)$ in equilibrium with $H_2O(s)$ at 1 atm pressure
   c. $H_2O(l)$, $H_2O(g)$, and $H_2O(s)$ in equilibrium
   d. $H_2O(l)$ at 1 atm pressure

10.42   What changes occur when the external pressure is raised at constant temperature in a sample of water that was initially at 527°C and 1 torr?

10.43   How many phases are represented a. off a line in a phase diagram? b. on a line in a phase diagram? c. where three lines meet in a phase diagram?

10.44   Sketch a phase diagram for carbon dioxide from the following data. Label all regions, lines, and points on your diagram.
   Vapor pressure of solid: 0.89 atm at -80°C, 1.97 atm at -70°C, 4.07 atm at -60°C
   Vapor pressure of liquid: 6.78 atm at -50°C, 9.93 atm at -40°C
   Melting point: -56.5°C at 9.93 atm, -56.4°C at 12.4 atm
   Triple point: -56.6°C at 5.11 atm
   Critical point: 31.0°C at 72.8 atm

10.45   What happens when a line on a phase diagram is crossed because of a change in pressure or temperature?

10.46   What changes occur when the temperature is raised at constant pressure in a sample of carbon dioxide that was initially at -173°C and 6 atm?

10.47   Indicate whether each of the following processes is exothermic or endothermic.
   a. freezing of water
   b. evaporation of water
   c. melting of ice
   d. condensation of steam
   e. sublimation of ice

10.48   Explain why the vapor pressure of H_2O is 72 torr at 45°C, whereas that of CS_2 is 760 torr.

10.49   What changes occur when the external pressure is raised at constant temperature in a sample of carbon dioxide that was initially at -73°C and 1 torr?

Crystal Structures of Metals

10.50   What are the common structures of metals?

10.51   What is a unit cell?

10.52   Describe the relationship between the cubic closest-packed structure and the face-centered cubic unit cell.

10.53   Describe the arrangement of layers in the cubic closest-packed and hexagonal closest-packed structures.

10.54   Calcium has a face-centered cubic structure. How many atoms of calcium are in a unit cell?

10.55   Chromium has a body-centered cubic structure. What is the coordination number of chromium?

10.56   The radius of iridium atoms is 127 pm. If iridium crystallizes in a face-centered cubic lattice, what is the size of an edge of the unit cell?

10.57   The radius of potassium atoms is 203 pm. If potassium crystallizes in a body-centered cubic lattice, what is the size of an edge of the unit cell?

10.58   Strontium crystallizes in a face-centered cubic lattice. If the edge of a unit cell is 540 pm, what is the radius of a strontium atom?

10.59   Tungsten crystallizes in a body-centered cubic lattice. If the edge of a unit cell is 300 pm, what is the radius of a tungsten atom?

10.60   Silver metal crystallizes in a face-centered cubic lattice whose unit cell has sides 409 pm long. Use this information and the atomic weight of silver to calculate the density of silver metal.

Crystal Structures of Ionic Salts

10.61   Which ionic salts represent common structures found in a variety of ionic salts?

10.62   What is the relationship between the structures of metals and the structures of ionic salts?

10.63   What is the coordination number of sodium ions in NaCl?

10.64   Examine the zinc blende structure and determine the number of zinc and sulfide ions in a unit cell.

10.65   Consulting a diagram of the unit cell of sodium chloride, show that the stoichiometry of this compound must be 1:1—that is, that it must have the formula NaCl.

10.66   An ionic crystal has a cubic lattice. The unit cell contains an anion (A) at each corner of the cube and a cation (C) in the center of each face of the cube. What is the formula of this compound?

10.67   Sodium bromide crystallizes in the sodium chloride lattice. The edge of a unit cell is 580 pm. If the radius of the bromide ion is 195 pm, what is the radius of the sodium ion?

Types of Solids

10.68   Distinguish between amorphous and crystalline solids.

10.69 What are the four major types of crystalline solids?

10.70 Distinguish between the structures of molecular and ionic solids. Give examples of each and contrast their properties.

10.71 Distinguish between the structures of metallic and ionic solids. Give examples of each, and contrast their properties.

10.72 Explain why sodium is malleable but sodium chloride is brittle.

10.73 What type of crystalline solid has a high melting point, conducts electricity molten but not as a solid, and is brittle?

10.74 What type of crystalline solid conducts electricity as a solid and when molten, is ductile, and has a lustrous appearance?

10.75 What forces hold atoms or molecules together in each of the following substances?
   a. diamond   c. $H_2O(s)$
   b. $CO_2(s)$ d. $SiO_2(s)$

10.76 Predict what type of solid each of the following compounds should form.
   a. $CaCl_2$ b. N_2 c. Ti   d. BN   e. SO_3

10.77 Which of the following solids is a molecular solid?
   a. glass (SiO_2)   d. Fe
   b. ice (H_2O)   e. dry ice (CO_2)
   c. $CaCl_2$

10.78 Which of the following solids is a covalent network solid?
   a. glass (SiO_2)   d. Fe
   b. ice (H_2O)   e. dry ice (CO_2)
   c. $CaCl_2$

10.79 Which of the following solids is an ionic solid?
   a. glass (SiO_2)   d. Fe
   b. ice (H_2O)   e. dry ice (CO_2)
   c. $CaCl_2$

10.80 Compare NaCl, SiC, $SiCl_4$, and Fe with respect to electrical conductivity, hardness, melting point, and vapor pressure at room temperature.

Semiconductors

10.81 What is a semiconductor?

10.82 Compare the difference in energy between the valence band and the conduction band in conductors, insulators, and semiconductors.

10.83 In each case, is the result an *n*-type or a *p*-type semiconductor?
   a. silicon doped with Al
   b. silicon doped with As
   c. silicon doped with Ga
   d. silicon doped with Sb
   e. silicon doped with In

Additional Exercises

10.84 If graphite can be transformed into diamond at high pressures, which of these forms of carbon has the higher density?

10.85 Examine the fluorite structure and determine the number of calcium and fluoride ions in a unit cell.

10.86 Explain why the heat of fusion of sodium chloride is greater than that of water.

10.87 Explain why wet laundry hung out in freezing weather eventually dries.

10.88 Arrange Br_2, Cl_2, and I_2 in order of increasing boiling point.

10.89 Match the compounds in the first list with the boiling points in the second list. Then explain the trend you have described.

CH_4	0°C
C_2H_6	−42°C
C_3H_8	−88°C
C_4H_{10}	−162°C

10.90 When a vacuum pump is connected to a container of water, the water begins to boil. Explain this phenomenon.

10.91 Lithium has a body-centered cubic structure. How many atoms of lithium are in a unit cell?

10.92 The radius of a gold atom is 134 pm. If gold crystallizes in a face-centered cubic lattice, what is the length of an edge of the unit cell?

10.93 What changes occur when the external pressure is raised at constant temperature in a sample of carbon dioxide that was initially at 27°C and 1 atm?

10.94 Explain why diamond is a good insulator, whereas graphite is a good conductor of electricity.

10.95 Which characteristic of molecules changes in magnitude from one state to another and largely determines whether a substance is a gas, a liquid, or a solid?

*10.96 A metal crystallizes in both a body-centered cubic and a face-centered cubic lattice. If the body-centered cubic form has a density of 15.0 g/cm^3, what is the density of the face-centered cubic form?

*10.97 Tungsten metal crystallizes in a body-centered cubic structure with a unit-cell edge of 316 pm. The density of tungsten is 19.3 g/cm^3, and its molar mass is 183.9 g/mol. Calculate Avogadro's number from this information.

*10.98 Use a Born–Haber cycle to calculate the lattice energy for crystalline rubidium chloride. The following data may be useful: heat of sublimation of Rb = 86 kJ/mol, ionization potential of Rb = 398 kJ/mol, heat of formation of RbCl = −431 kJ/mol, bond energy of Cl$_2$ = 226 kJ/mol, electron affinity of Cl = −349 kJ/mol.

*10.99 Use the ratio of ionic radii (cation/anion) to account for the different structures of NaCl and CsCl.

▶ ▶ ▶ Chemistry in Practice

10.100 By means of X-ray diffraction, the structure of an intermetallic compound of copper and gold is found to be cubic. The edge of the unit cell is 384 pm long. The compound contains approximately equal masses of the two elements. A 253.24-g sample of the compound was placed in a 250.00-mL volumetric flask that weighed 102.33 g when empty. The flask was then filled with water at 25°C (density = 0.9971 g/mL), to give a total mass of 582.64 g.

    a. What is the volume of the unit cell?

    b. What is the volume of a mole of unit cells?

    c. What is the mass of a mole of unit cells?

    d. What are possible formula weights of the compound?

    e. How many atoms of copper and of gold are in one formula unit? That is, what is the empirical formula?

    f. How many formula units are in one unit cell?

    g. Is the unit cell face-centered cubic or body-centered cubic?

    h. What is the theoretical percent composition of the compound?

    i. How much gold is needed to prepare 1.500 kg of the compound?

Solutions

▶ ▶ ▶ ▶

Iron(III) nitrate and potassium ferrocyanide do not react in the solid state but react rapidly to form Prussian blue when dissolved in water.

CHAPTER OUTLINE

Encountering Chemistry

▶ ▶ ▶ ▶

A white solid, $Fe_2(SO_4)_3 \cdot 9H_2O$, is mixed with a pale yellow solid, $K_4Fe(CN)_6 \cdot 3H_2O$. Even if these solids are allowed to stay in this mixture for a long time, they do not react with one another. But if water is poured onto this solid mixture, the solids dissolve and immediately react to form a new deep blue solid, iron(III) ferrocyanide, or Prussian blue.

$$2Fe_2(SO_4)_3(aq) + 3K_4Fe(CN)_6(aq) \longrightarrow Fe_4[Fe(CN)_6]_3(s) + 6K_2SO_4(aq)$$

This experiment illustrates a major use of solutions—as a medium for carrying out chemical reactions between compounds.

Water has a wide variety of uses, as we noted in Chapter 10, but it is rarely used in its pure form. Even the much-sought-after "pure" mountain spring water is not pure. Pure H_2O is a colorless, odorless, tasteless liquid. As water falls through the air as rain, it dissolves gases. As it percolates up through the ground and runs down to stream beds, it dissolves a variety of minerals and gases, which give it a characteristic taste. As water flows in streams and rivers on its way to the oceans, it gathers more minerals. Finally, it becomes "salty" ocean water. All of these water samples are in fact solutions.

We often think of solutions as liquids, but there are gaseous and solid solutions as well. Air, described in Chapter 9, is a mixture of many gases. When the mixture is homogeneous, air is a solution. And among the alloys, we find solid solutions. Gold used in jewelry is hardened by the addition of other metals, such as copper. Liquid mercury can also dissolve in metals to form solids, called amalgams. Dental amalgam is a mixture of silver, mercury, and a few other metals. This amalgam is solid like silver, hard enough to be used for the reconstruction of teeth, and bright and shiny.

This chapter examines solutions, explaining how to express their composition, how the solution process works, what reactions take place in solution, and how dissolved substances affect the vapor pressure and other properties of liquids. Finally, the chapter describes colloids, mixtures that resemble solutions in some ways.

11.1 Concentration

Before we can begin to discuss the properties of solutions, we must consider their concentrations, because this property affects all others. By **concentration,** we mean the relative amounts of the substances that make up the solution. The terminology associated with solutions includes both qualitative and quantitative expressions for concentration. We examine both here.

Chapter 4 introduced some terms necessary for discussing solutions. A solution is any homogeneous mixture; that means it contains two or more substances, looks the same throughout, and may vary in composition from sample to sample. The substances that make up a solution may be labeled the solvent and solute. Usually, the solvent is the substance present in the greater amount, so it tends to be the component that resembles the solution most in physical properties. For example, as Table 11.1 shows, the physical state of the solution is generally the same as the physical state of the solvent. The solute is the substance that is dissolved or dispersed in the solvent. There is only one solvent in a solution, but a solution can contain more than one solute. Sometimes it is difficult to distinguish between solvent and solute. For example, ethanol (C_2H_5OH) and water can dissolve in one another in all possible proportions, so the relative amounts cannot be used to identify which liquid is the solvent and which the solute. Water is conventionally considered to be the solvent in all such mixtures, even when ethanol is present in greater quantity.

Draw a picture of a solution and of a heterogeneous mixture that shows the difference between them at a molecular level.

Table 11.1 Examples of Solutions

Physical State of			Examples
Solution	Solute	Solvent	
Gas	Gas	Gas	Air, freons, auto exhaust fumes
Liquid	Gas	Liquid	Oxygen in water
Liquid	Liquid	Liquid	Ethanol in water, bromine in water
Liquid	Solid	Liquid	Sodium chloride in water
Solid	Gas	Solid	Hydrogen in palladium
Solid	Liquid	Solid	Mercury in silver
Solid	Solid	Solid	Alloys, such as copper in gold

Solubility

The composition of a solution can be described in terms of **solubility,** a ratio that identifies the maximum amount of a solute that will dissolve in a particular solvent to form a stable solution under specified conditions. Solubility, then, is a measure of concentration. Solubility is often expressed as grams of solute per 100 g of solvent, especially for solids dissolved in liquids. Other concentration units can be used as well; we will discuss them shortly.

If excess solute is present when the maximum concentration is reached, it remains undissolved, and the system is in a state of equilibrium between the dissolved solute and the undissolved solute. For example, if 1 g of lead iodide is added to 100 mL of water, only about 0.1 g PbI_2 dissolves (Figure 11.1). An equilibrium exists between lead iodide in solution and in the solid phase. Such a solution is said to be a **saturated solution.** ▶ The mixture of Prussian blue and water described in the chapter introduction is another example of a saturated solution. The establishment of equilibrium to give a saturated solution is depicted in Figure 11.2. Once the solution is saturated, undissolved solute continues to dissolve, but at the same time, previously dissolved solute is deposited from solution. The two processes occur at exactly the same rate, so the amount of dissolved solute stays the same. This is a dynamic equilibrium, like the evaporation of a liquid discussed in Chapter 10.

A solution that contains less than the maximum amount of solute is called **unsaturated.** One that contains more than the expected maximum is called **supersaturated.** As you might expect, supersaturation is generally a very unstable situation that quickly reverts to the expected saturation level when the solution is disturbed by agitation or when a small ''seed'' crystal of solute is added, as shown in Figure 11.3. A supersaturated solution is usually obtained by preparing a saturated solution at a high temperature and then carefully cooling the hot solution.

Another set of terms used to describe solids reflects their degree of solubility in a given solvent. A solid may be called *insoluble, slightly soluble, soluble,* or *very soluble.* All these terms are relative. Strictly speaking, there is no insoluble substance—even glass dissolves in water to a slight extent. However, for practical purposes, solids with solubility less than about 0.1 g per 100 g of solvent are considered insoluble, whereas solutes with solubility greater than about 2 g per 100 g of solvent are considered very soluble.

Other terms sometimes used as a qualitative measure of the amount of solute dissolved in a solvent are *dilute* and *concentrated.* The distinction between these

Figure 11.1 Lead iodide, PbI_2, is a heavy, odorless, bright yellow, crystalline powder. It is not very soluble; 1 g requires 1350 mL of water. It is used to make a gold color in applications such as printing, photography, and bronzing.

Figure 11.2 Dissolution in saturated and unsaturated solutions. *(A)* In unsaturated solutions, dissolution occurs more rapidly than deposition, so concentration increases. *(B)* In saturated solutions, the rate of dissolution equals the rate of deposition, so there is no change in concentration.

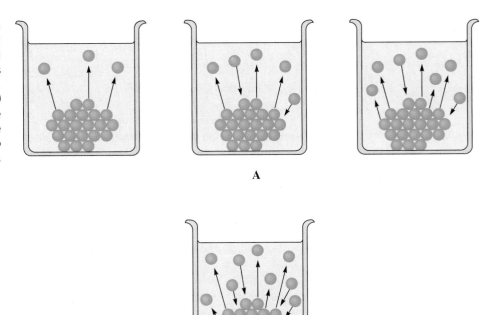

A

B

terms depends on the particular solutions being compared. It is most meaningful to use these terms as relative descriptors, as when we describe one solution as more concentrated or more dilute than another.

When both the solute and the solvent are liquids, yet another system of terminology is used. Liquids such as ethanol and water are called completely **miscible liquids,** which means that they can form solutions in all proportions. *Partially miscible* liquids form solutions only over a limited range of compositions. For example, it is possible to form a solution by adding butanol to water up to a concentration of about 20% butanol by volume. No higher level of concentration

Figure 11.3 Sodium thiosulfate, $Na_2S_2O_3$, is an ionic salt that readily forms supersaturated solutions. If a small crystal is added to the solution, or if the container wall is scratched, crystallization begins, and the supersaturated solution reverts to a saturated solution in equilibrium with the solid. Sodium thiosulfate, also known as hypo, is used to develop black and white photographs.

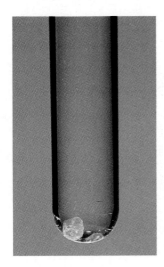

can be achieved. Liquids such as cyclohexane (C_6H_{12}) and water are essentially insoluble in one another and are called *immiscible*.

Measuring Concentrations of Solutions

Why are there so many different ways of measuring concentration?

Suppose we need to describe the concentration level of a substance in exact terms. We can use one of several concentration units. The particular unit we choose depends on the purpose for which the solution is being used and on the method by which it is prepared. All the units express concentration as a relative quantity—the amount of solute in a given amount of solvent or in a given amount of solution.

$$\text{Concentration} = \frac{\text{amount of solute}}{\text{amount of solvent or solution}}$$

Some commonly used units are summarized in Table 11.2 and will be discussed here.

Solubility Units   As mentioned earlier, the solubility of solids in liquids is often expressed in units of mass of solute per unit mass of solvent, usually grams of solute per 100 g of solvent (or g solute/g solvent × 100, as expressed in Table 11.2). This unit is used mostly to describe saturated solutions. If we know the mass of solute and the mass of solvent, we can calculate the solubility simply by substituting these values into the equation. If we know the mass of solute and the mass of solution, we can determine the mass of solvent by subtracting the mass of solute from the mass of solution. Then we can carry out the rest of the calculation.

Table 11.2 Common Concentration Units

Unit	Definition
Solubility* (g solute/100 g solvent)	$\dfrac{\text{grams of solute}}{\text{grams of solvent}} \times 100$
Percent by mass	$\dfrac{\text{grams of solute}}{\text{grams of solution}} \times 100$
Percent by volume	$\dfrac{\text{volume of solute}}{\text{volume of solution}} \times 100$
Mass/volume	$\dfrac{\text{grams of solute}}{\text{volume of solution}}$
Mole fraction (χ)	$\dfrac{\text{moles of one component}}{\text{total moles of all components}}$
Molality (m)	$\dfrac{\text{moles of solute}}{\text{kilograms of solvent}}$
Molarity (M)	$\dfrac{\text{moles of solute}}{\text{liters of solution}}$
Normality (N)	$\dfrac{\text{equivalents of solute}}{\text{liters of solution}}$

*This unit is generally used to describe the concentration of saturated solutions.

EXAMPLE 11.1

Calculate the solubility of KCl (in g/100 g H_2O) if 10.0 g of H_2O will dissolve 3.55 g of KCl at 25°C.

Solution   The necessary quantities, mass of solute (3.55 g KCl) and mass of solvent (10.0 g H_2O), are given in the problem. It is only necessary to substitute these quantities into the equation for solubility.

$$\text{Solubility} = \frac{\text{g solute}}{\text{g solvent}} \times 100$$

$$= \frac{3.55 \text{ g KCl}}{10.0 \text{ g } H_2O} \times 100$$

$$= 35.5 \text{ g KCl/100 g } H_2O$$

Practice Problem 11.1

Calculate the solubility of NH_4Cl (in g/100 g H_2O) if 22.0 g H_2O dissolves 8.18 g NH_4Cl at 20°C.

Answer: 37.2 g NH_4Cl/100 g H_2O

An important application of solubility values involves the recovery of dissolved substances from solution. Because solubility varies with temperature, it is often possible to recover a dissolved solid by changing the temperature, most often to a lower value. It is also possible to recover some dissolved solid by evaporating some solvent. In either case, the point at which saturation occurs is lowered by the change in temperature, and enough solute is deposited to bring the solution back to equilibrium. If we determine the masses of solute needed to saturate the solution under both sets of conditions, and find the difference between the two, we have identified the amount of solute recovered.

EXAMPLE 11.2

Salts such as sylvine, KCl, can be mined by dissolving them in hot water and then precipitating them by cooling the solution. A solution saturated at 80°C contains KCl, the solute, and 250 g H_2O, the solvent. The solution is cooled to 20°C. How much KCl is recovered? The solubility of KCl is 51.1 g KCl/100 g H_2O at 80°C and 34.0 g KCl/100 g H_2O at 20°C.

Solution   We must calculate the mass of KCl that can be dissolved in 250 g H_2O at each temperature, using the following solubilities:

$$\text{Mass dissolved at 80°C} = \frac{51.1 \text{ g KCl}}{100 \text{ g } H_2O} \times 250 \text{ g } H_2O = 127.8 \text{ g KCl}$$

$$\text{Mass dissolved at 20°C} = \frac{34.0 \text{ g KCl}}{100 \text{ g } H_2O} \times 250 \text{ g } H_2O = 85.0 \text{ g KCl}$$

The amount of KCl that is recovered is the difference between the masses that are dissolved at the two temperatures.

$$\text{Mass precipitated at 20°C} = 127.8 \text{ g KCl} - 85.0 \text{ g KCl} = 42.8 \text{ g KCl}$$

(continued)

Practice Problem 11.2

A solution saturated at 90°C contains KCl, the solute, and 185 g of H_2O, the solvent. The solution is cooled to 40°C. How much KCl is recovered? The solubility of KCl is 54.0 g KCl/100 g H_2O at 90°C and 40.0 g KCl/100 g H_2O at 40°C.

Answer: 25.9 g KCl

Percent by Mass   **Percent by mass** differs from solubility only in the denominator term, which gives mass of solution rather than mass of solvent.

$$\text{Percent by mass} = \frac{\text{grams of solute}}{\text{grams of solution}} \times 100$$

If necessary, we can determine the mass of solution by adding the mass of solute to the mass of solvent. We can also determine the mass of solution from the volume of solution, provided we know the density of the solution as well.

$$\text{Mass of solution} = \text{density of solution} \times \text{volume of solution}$$

EXAMPLE 11.3

What is the percent-by-mass concentration of acetic acid, CH_3CO_2H, in a vinegar solution that contains 54.5 g of acetic acid per liter of solution and has a density of 1.005 g/mL?

Solution   The mass of solute is given, 54.5 g acetic acid. The mass of solution must be calculated from the volume and density.

$$\text{Mass of solution} = \text{density} \times \text{volume of solution}$$
$$= 1.005 \text{ g/mL} \times 1 \text{ L} \times 1000 \text{ mL/L} = 1005 \text{ g}$$

The percent by mass can then be calculated.

$$\text{Percent by mass} = \frac{\text{grams of solute}}{\text{grams of solution}} \times 100$$
$$= \frac{54.5 \text{ g}}{1005 \text{ g}} \times 100 = 5.42\%$$

Practice Problem 11.3

What is the percent-by-mass concentration of acetic acid, CH_3CO_2H, in a vinegar solution that contains 2.65 g of acetic acid per 125.0 mL of solution and has a density of 1.003 g/mL?

Answer: 2.11% by mass

Percent by Volume   Calculating **percent by volume** is similar to calculating percent by mass. We simply substitute volumes for masses in the equation.

$$\text{Percent by volume} = \frac{\text{volume of solute}}{\text{volume of solution}} \times 100$$

There is an important difference beyond this simple substitution, however. Masses are additive; that is, the mass of solution is equal to the sum of the masses of solute and solvent. Volumes are not always additive. It is possible for the volume of a solution to be different from the sum of the volumes of the solute and solvent. Depending on the attractive forces between solute and solvent, the solution may have a greater or a lesser volume than the separated components. Percent by volume is most often used when liquids are dissolved in liquids, because it is easy to measure amounts of liquids by their volumes.

EXAMPLE 11.4

To prepare a solution, we dissolve 12.5 mL of ethanol in sufficient water to give a total volume of 85.4 mL. What is the percent-by-volume concentration of ethanol?

Solution The two required quantities, volume of solute and volume of solution, are given in the problem. We need only substitute these quantities into the expression for volume percent.

$$\text{Percent by volume} = \frac{\text{volume of solute}}{\text{volume of solution}} \times 100$$

$$= \frac{12.5 \text{ mL ethanol}}{85.4 \text{ mL solution}} \times 100$$

$$= 14.6\%$$

Practice Problem 11.4

To prepare a solution, we dissolve 205 mL of ethanol in sufficient water to give a total volume of 235 mL. What is the percent-by-volume concentration of ethanol?

Answer: 87.2% by volume

Mole Fraction The **mole fraction** is a concentration unit that expresses the number of moles of a given component (χ_A) as a proportion of total moles.

$$\chi_A = \frac{\text{moles A}}{\text{total moles of all components}}$$

By definition, the sum of the mole fractions of all components in a solution must add up to exactly 1.

EXAMPLE 11.5

A gaseous solution consists of 0.450 mol of nitrogen gas and 0.225 mol of oxygen gas. What is the mole fraction of each component?

Solution We can determine the mole fraction of either component by using the equation. Let's start with N_2.

$$\chi_{N_2} = \frac{\text{moles nitrogen}}{\text{moles nitrogen} + \text{moles oxygen}}$$

$$= \frac{0.450 \text{ mol}}{0.450 \text{ mol} + 0.225 \text{ mol}} = 0.667$$

(continued)

We can use the result to find the mole fraction of oxygen. We simply subtract it from 1.

$$\chi_{O_2} = 1 - \chi_{N_2}$$
$$= 1 - 0.667 = 0.333$$

Practice Problem 11.5

A gaseous solution consists of 1.64 mol of hydrogen gas and 4.87 mol of helium gas. What is the mole fraction of each component?

Answer: $\chi_{H_2} = 0.252$, $\chi_{He} = 0.748$

Molality   **Molality** (m) is concentration expressed as moles of solute per kilogram of solvent.

$$\text{Molality} = \frac{\text{moles solute}}{\text{kilograms solvent}}$$

Molality is useful for examining temperature-dependent properties, because it involves only masses and does not vary with temperature. It is also useful for examining properties that depend on the number of solute particles in solution, as we will see later. To calculate molality, we must determine the number of moles of solute and the mass of solvent. Note that it is the mass of the solvent, not the mass of the solution, that we need.

EXAMPLE 11.6

A solution contains 22.5 g of methanol, CH_3OH, dissolved in sufficient water to give a total mass of 105.3 g. The molar mass of CH_3OH is 32.04 g/mol. What is the molality of the aqueous methanol solution?

Solution   We can calculate the number of moles of solute from the mass of solute and its molar mass.

$$\text{Moles } CH_3OH = 22.5 \text{ g} \times 1 \text{ mol}/32.04 \text{ g} = 0.702 \text{ mol}$$

Subtracting the mass of solute from the mass of solution yields the mass of solvent.

$$\text{Mass } H_2O = 105.3 \text{ g solvent} - 22.5 \text{ g } CH_3OH = 82.8 \text{ g}$$

This mass must be converted to units of kilograms.

$$\text{Mass } H_2O = 82.8 \text{ g} \times 1 \text{ kg}/1000 \text{ g} = 0.0828 \text{ kg}$$

Now we have all the information we need to calculate the molality.

$$\text{Molality} = \frac{\text{moles solute}}{\text{kilograms of solvent}}$$
$$= \frac{0.702 \text{ mol } CH_3OH}{0.0828 \text{ kg } H_2O} = 8.48 \ m$$

Practice Problem 11.6

A solution contains 78.2 g of potassium hydroxide, KOH, dissolved in sufficient water to give a total mass of 245.8 g. The molar mass of KOH is 56.11 g/mol. What is the molality of the aqueous potassium hydroxide solution?

Answer: 8.29 *m*

Molarity   Molarity (*M*) is one of the most commonly used expressions of concentration. You may remember from Chapter 4 that it is the moles of solute divided by the volume of solution in liters.

$$\text{Molarity} = \frac{\text{moles of solute}}{\text{liters of solution}}$$

Note that here we compare the moles of solute to the volume of solution, not to the volume of solvent. Calculating molarity is relatively straightforward, given the moles or mass of solute and the volume of solution. We will consider some manipulations of molarity later in the chapter.

Normality   **Normality** (*N*) is similar to molarity, but instead of measuring the number of moles of solute, it measures the number of *equivalents* of solute.

$$\text{Normality} = \frac{\text{equivalents of solute}}{\text{liters of solution}}$$

This concentration unit is most useful for acids and bases that supply more than one hydrogen ion or hydroxide ion per formula unit. For these species, the number of equivalents equals the number of moles of hydrogen ions or hydroxide ions that are supplied by 1 mol of the acid or base. If an acid releases two hydrogen ions per acid molecule, as in H_2SO_4, 1 mol of the acid supplies 2 mol of hydrogen ions. Thus, 1 mol of H_2SO_4 is the same as 2 equivalents. A 1 *M* H_2SO_2 solution has a normality of 2 *N*.

Conversions of Concentration Units   It is possible to convert any of the concentration units to one of the others. Here, we examine the conversion of percent by mass to molarity, one of the conversions we most often need to perform. A typical situation involves an aqueous acid, such as hydrochloric acid or nitric acid. These acids are sold as aqueous solutions whose concentration is expressed as percent by mass. When we use them in chemical reactions, however, we want to know their molarity so that we can measure out a volume of solution containing a stoichiometric number of moles. To get the volume of solution, we must know the density of the solution as well as its mass.

EXAMPLE 11.7

Concentrated hydrochloric acid, an aqueous solution of HCl, contains 37.2% HCl by mass and has a density of 1.19 g/mL. Calculate the molarity of HCl in this solution.

(continued)

Solution   We start by calculating the mass of HCl in some amount of solution—exactly 100 g for convenience. Because percent by mass is the mass of solute per 100 g of solution, the mass of HCl in 100 g of solution is 37.2 g. Next we calculate the moles of solute, using the molar mass as a conversion factor.

$$\text{Moles of HCl} = 37.2 \text{ g} \times \frac{1 \text{ mol}}{36.5 \text{ g}} = 1.02 \text{ mol}$$

Now we need to calculate the volume of 100 g of solution. To do this, we use the density.

$$\text{Volume of solution} = 100 \text{ g} \times \frac{1 \text{ mL}}{1.19 \text{ g}} \times \frac{1 \text{ L}}{1000 \text{ mL}} = 0.0840 \text{ L}$$

Now we can calculate molarity.

$$\text{Molarity} = \frac{1.02 \text{ mol HCl}}{0.0840 \text{ L solution}} = 12.1 \text{ mol/L} = 12.1 \; M$$

Practice Problem 11.7

Concentrated potassium hydroxide, an aqueous solution of KOH, contains 45.0% KOH by mass and has a density of 1.46 g/mL. Calculate the molarity of KOH in this solution.

Answer: 11.7 *M*

11.2   Factors That Affect Solubility

Commercially important salts are found in the form of deposits in dry lake beds such as Searles Lake in California, in Great Salt Lake in Utah, and in the Dead Sea in Israel, as well as in underground salt domes and deposits in numerous locations. The recovery and purification of these salts depend in large part on a knowledge of the factors affecting their solubility in various solvents.

Substances differ widely in their solubilities or miscibilities. Oil and water don't mix; ethanol and water mix in all proportions. A grease stain is more easily removed with a commercial chlorinated hydrocarbon cleaner than with water. Why do some substances dissolve in one solvent but not in another? What makes dissolution occur at all?

Why does an ionic salt dissolve in a polar solvent but not in a nonpolar solvent?

A number of factors are important in determining the solubility of a material in a given solvent. A general rule is "like dissolves like": Polar covalent solvents or ionic solvents dissolve polar covalent or ionic solutes, for example, whereas nonpolar covalent solvents dissolve nonpolar covalent solutes. This section examines several factors that affect solubility, but first it gives a general description of the solution process.

The Solution Process

A solution contains solute particles uniformly distributed within the volume occupied by the solvent particles, so its formation must involve the destruction and formation of intermolecular forces. Both solute and solvent, assuming that they are liquids or solids, are held together by intermolecular forces. If the particles of the solute are dispersed throughout the solvent, they must be separated from one an-

Figure 11.4 Interactions between water molecules and cations and anions provide the energy necessary to overcome both the intermolecular forces between water molecules and the ionic bond in a crystal such as solid sodium chloride.

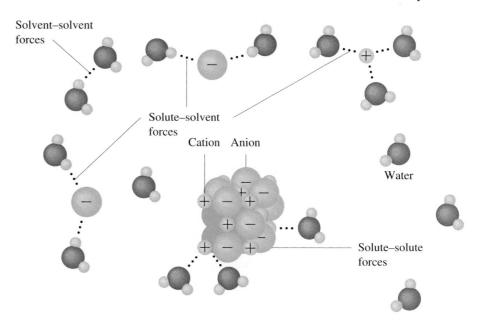

other—their intermolecular forces must be overcome. Once that has occurred, the solvent particles can no longer interact in the same way as before, so some intermolecular forces must be overcome in the solvent as well.

The energy required to break the existing intermolecular forces can be supplied, at least in part, by new intermolecular forces formed between solute and solvent particles. Figure 11.4 depicts such a process for the formation of a solution between an ionic solute, such as sodium chloride, and water. (This solution process is representative of the interactions involved in other types of solutions as well.) Here the existing forces are ionic bonds, which hold the cations and anions together, and the dipole–dipole and hydrogen bonding interactions among water molecules. The new forces created in the solution are strong dipole–ion forces, which will stabilize the ions in their new environment. This process is called *hydration* in water and *solvation* in other solvents. The hydrated ions are surrounded by a sheath of water molecules oriented such that the partially negative polar ends of the water molecules are next to the cations and the partially positive polar ends are next to the anions, as shown in Figure 11.5.

Figure 11.5 Hydrated cations and anions interact with the appropriate polar end of the surrounding water molecules— the negatively charged end around cations and the positively charged end around anions.

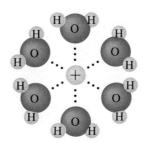

A. Hydrated cation

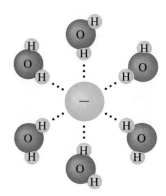

B. Hydrated anion

The energy changes involved in the solution process are summarized in Figure 11.6 for the dissolution of a solid in a liquid. Similar energy changes occur in other types of solutions. As we have seen, energy is required to separate solute particles from one another and to separate solvent particles from one another. New interactions between solute and solvent particles release energy called the *heat of solvation.* The difference between the energy required for separation and the heat of solvation is called the **heat of solution.** When the separation processes require more energy than the solvation process releases, the heat of solution is positive, and the overall dissolution process is endothermic. When the solvation process provides more energy than is needed to separate the pure solute and solvent particles, the heat of solution is negative, and the overall dissolution process is exothermic. Even in the latter case, the solute–solvent interactions must be strong for the solution process to occur at all. If these interactions are not at least almost as strong as the solute–solute and solvent–solvent interactions, no solution forms.

One other factor is important in the solution process. This factor, called **entropy,** is a measure of the tendency of matter to become disordered or random in its distribution. Entropy will be discussed in more detail in Chapter 12. For now, it

List some examples from your everyday life of the tendency to become disordered.

Figure 11.6 Energy is required to separate solute particles from one another and to separate solvent particles from one another. New interactions between solute and solvent particles release energy. The relative amounts of energy involved in these processes determine whether the dissolution process is exothermic or endothermic.

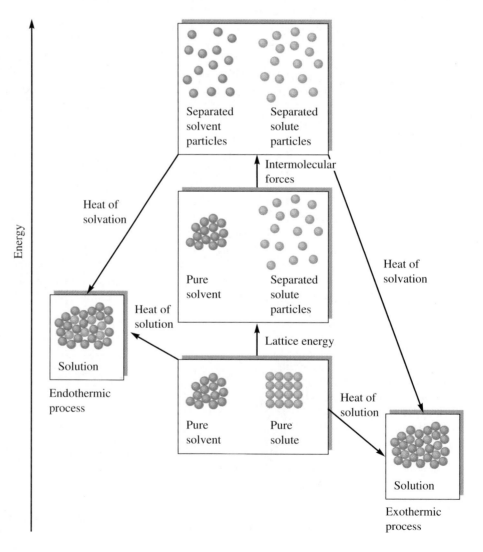

is only necessary to recognize that matter spontaneously changes from a state of order to a state of randomness, unless energy changes oppose this tendency. Examining your desk after an evening of studying should allow you to recognize the natural tendency toward disorder that pervades our universe.

Next we examine the various factors that affect solubility. We concentrate on solutions in which the solvent is a liquid, because such solutions are the most common.

Effect of Molecular Structure on Solubility

Among the factors that affect solubility is molecular structure. We noted earlier that "like dissolves like." This rule is invoked primarily to explain the solubility of liquids in liquids, where the molecular structures of solvent and solute are important. The forces in ionic salts are all much alike, and forces between gaseous molecules are very weak. Covalent molecular liquids involve a variety of forces, which were described in Chapter 10. These forces affect solubility, as well as a variety of other properties, such as melting point and boiling point.

The solubility (or miscibility) of liquids in other liquids depends largely on the polarity of the molecules in the solute and the solvent. Polar liquids dissolve in other polar liquids, because the intermolecular forces between solute and solvent molecules are strong enough to overcome the strong solute–solute and solvent–solvent interactions. Nonpolar liquids do not dissolve in polar liquids—at least, not to any great extent—because the intermolecular forces between solvent and solute are not strong enough to overcome the solute–solute and solvent–solvent forces. Nonpolar liquids dissolve in one another, because the forces between the molecules are weak and also because matter has a tendency to become more random. A variety of forces between molecules play a role in the solution process. These intermolecular forces and their relative strengths were discussed in Chapter 10. Ultimately, it is the relative strengths of these forces that determines the solubility of one liquid in another.

Table 11.3 shows the miscibility of several liquids. Let's consider some of these data in terms of the polarity of the molecules. Benzene and carbon tetrachloride are completely nonpolar. Water is very polar and has strong hydrogen bonding interactions. Ethylene glycol and glycerol have polar hydroxyl groups, but they are attached to nonpolar hydrocarbons. Pyridine and ethyl alcohol have nonpolar and polar ends.

Glycerol, $HOCH_2CHOHCH_2OH$, is a syrupy liquid with a sweet taste; it is toxic in large doses. It freezes at 17.8°C, but an aqueous solution containing 67% glycerol by mass freezes at −46°C, so it is often used in shock absorber fluids and windshield washer fluids. It is also used in the manufacture of dynamite, cosmetics, liquid soaps, candy, inks, lubricants, and glues.

Ethylene glycol is a slightly viscous liquid that also has a sweet taste. This liquid is extremely toxic, however. It is used extensively in antifreeze fluids, because an aqueous solution that is 58% by mass ethylene glycol has a freezing point of −48°C. It is also used as a solvent in paints, plastics, printer's inks, and ballpoint pen inks.

pyridine   ethyl alcohol

Acetone and diethyl ether are somewhat polar but have nonpolar hydrocarbon groups on the outside of the molecules.

acetone   diethyl ether

Table 11.3 Miscibility of Some Liquids

	Acetone	Benzene	Carbon Tetrachloride	Diethyl Ether	Ethyl Alcohol	Ethylene Glycol	Glycerol	Pyridine	Water
Acetone $CH_3(CO)CH_3$	—	M*	M	M	M	M	I	M	M
Benzene C_6H_6	M	—	M	M	M	I	I	M	I
Carbon Tetrachloride CCl_4	M	M	—	M	M	I	I	M	I
Diethyl Ether $C_2H_5—O—C_2H_5$	M	M	M	—	M	I	I	M	I
Ethyl Alcohol CH_3CH_2OH	M	M	M	M	—	M	M	M	M
Ethylene Glycol $HO—CH_2CH_2—OH$	M	I	I	I	M	—	M	M	M
Glycerol $HOCH_2CH(OH)CH_2OH$	I	I	I	I	M	M	—	M	M
Pyridine C_5H_5N	M	M	M	M	M	M	M	—	M
Water H_2O	M	I	I	I	M	M	M	M	—

*M = miscible; I = immiscible

Water is miscible with the more polar of the other molecules, because the hydrogen bonds in water must be disrupted in order for a solution to be formed. All the organic liquids that dissolve in water have oxygen or nitrogen or oxygen–hydrogen groups that can hydrogen-bond to water. Benzene is immiscible with other liquids that are quite polar, because the intermolecular forces between benzene molecules and solvent molecules are weak. A similar pattern is observed for carbon tetrachloride. Ethyl alcohol, with distinct nonpolar and polar ends, is the only liquid that is miscible with all the other liquids.

Essentially the same factors related to molecular structure are important in solutions of solids dissolved in liquids as in solutions of liquids dissolved in liquids. In order for a solid to be soluble in a given solvent, the energy given up by the solvation process must compensate for all (or at least most) of the energy required to break the forces within the solid (the lattice energy) and within the pure solvent (intermolecular forces). Thus ionic solids are generally soluble in polar solvents, where very strong interactions can occur between the dipoles of the solvent and the dissolved ions. A nonpolar solvent cannot interact with the dissolved ions strongly enough to overcome the ionic forces within the crystal.

Effect of Temperature on Solubility

Why does solubility vary with temperature?

As shown in Figure 11.6, energy is released (in an exothermic process) or absorbed (in an endothermic process) as a solution is formed. For an endothermic process, raising the temperature of the solution supplies more energy for overcoming the lattice energy or intermolecular forces of the solute and the intermolecular forces in the solvent. Thus solubility increases as temperature increases. This is the case for most ionic solids when water is the solvent, and for most liquid–liquid solutions. For example, the solubility of potassium chloride in water increases as temperature increases. In contrast, for an exothermic process, an increase in temperature leads to a decrease in solubility. In this case, the added energy overcomes the stronger solute–solvent forces and solubility decreases. An example is calcium hydroxide in water.

Calcium hydroxide, or slaked lime, is prepared commercially by the hydration of lime, CaO. Aqueous solutions of Ca(OH)$_2$ are known as limewater. The solutions and the solid readily absorb carbon dioxide from air to form calcium carbonate, or limestone. Calcium hydroxide is used in building materials such as mortar, plaster, and cement; in the preparation of paper pulp; and in water treatment.

Solubility data for some typical salts are given in Table 11.4. This table also lists values for the heats of solution of these solids. Note that for the solids with a positive heat of solution, solubility increases with increasing temperature. For the solids with a negative heat of solution, solubility decreases with increasing temperature. However, there are a number of exceptions to this simplified description of the effects of temperature on solubility. In practice, it is necessary to determine solubility experimentally at various temperatures.

Some solubility data are displayed graphically in Figure 11.7. In the figure, a positive slope corresponds to an endothermic solution process, and a negative

Table 11.4 Effect of Temperature on the Solubility of Some Salts in Water

| Temperature (°C) | Solubility (g/100 g water) | | | | |
	KCl	NaCl	NH$_4$Cl	Li$_2$SO$_4$	Ca(OH)$_2$
0	27.6	35.7	29.4	35.3	0.185
10	31.0	35.8	33.3	35.0	0.176
20	34.0	36.0	37.2	34.2	0.165
30	37.0	36.3	41.1	33.5	0.153
40	40.0	36.6	45.8	32.7	0.141
50	42.6	37.0	50.4	32.5	0.128
60	45.5	37.3	55.2	31.9	0.116
70	48.3	37.8	60.2	—	0.106
80	51.1	38.4	65.6	30.7	0.094
90	54.0	39.0	71.3	—	0.085
100	56.7	39.8	77.3	29.9	0.077
Heat of Solution at 25°C (J/mol)					
	984	222	844	−1720	−927

Figure 11.7 Solubility of some ionic salts in water as a function of temperature.

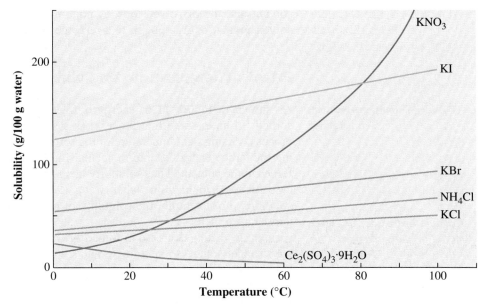

slope corresponds to an exothermic process. We can use such curves to predict the conditions necessary to dissolve a salt or to recover a salt from solution. For example, consider the solubility curve for potassium nitrate. A solution saturated with KNO_3 at 80°C contains 1.70×10^2 g of KNO_3 per 100 g of water. When this solution is cooled to 20°C, the solubility becomes 32 g/100 g water. The amount of solid KNO_3 that can be recovered when this solution is cooled from 80°C to 20°C is given by the difference in solubilities.

$$\text{Mass } KNO_3 \text{ recovered} = 1.70 \times 10^2 \text{ g/100 g water} - 32 \text{ g/100 g water}$$
$$= 138 \text{ g/100 g water}$$

We can also employ the curves in deciding how much water we must use to dissolve a given amount of solid at a particular temperature.

Although the solubility of a liquid dissolved in a liquid is also dependent on temperature, practical applications of this effect are not very common, so we will not consider it further.

Most gases (such as N_2, O_2, Ar, and CO_2) are soluble in water only to a small extent, as Table 11.5 shows. However, some gases, such as HCl and NH_3, are very soluble because they interact strongly with water (Figure 11.8). At 0°C, hydrogen chloride has a solubility of 82.3 g/100 g water, whereas the solubility of ammonia in water is 89.9 g/100 g water. No matter how weak or strong the interactions between gas molecules and the solvent, gases grow less soluble in liquids as the temperature is raised. That means gases can be removed from solution by boiling. ▶ The flat taste you may have noticed in boiled (pure) water—mentioned in the chapter introduction—is due in part to the removal of dissolved gases, including CO_2, the most soluble of the normal atmospheric gases. The effect of temperature on solubility in water is shown for some typical gases, the major constituents of air, in Table 11.5. An important consequence of this effect is the thermal pollution of water, such as occurs around power plants cooled by a river or stream. When the water temperature of the river or stream is increased, the

Could fish live in water that had been boiled and then cooled?

Table 11.5 Solubility of Some Gases in Water at 1 atm Pressure

Temperature (°C)	Solubility (g/100 g H$_2$O)			
	N$_2$	O$_2$	Ar	CO$_2$
0	0.00294	0.00695	0.00025	0.335
10	0.00231	0.00537	—	0.232
20	0.00190	0.00434	—	0.169
30	0.00162	0.00359	—	0.126
40	0.00139	0.00308	—	0.0973
50	0.00122	0.00266	0.0000995	0.0761
60	0.00105	0.00227	—	0.0576
70	0.000851	0.00186	—	—
80	0.000660	0.00138	—	—
90	0.000380	0.000790	—	—
100	0	0	0	0

solubility of oxygen in the water decreases, making it more difficult for fish to survive.

Effect of Pressure on Solubility

The solubility of gases, but not of liquids or solids, in liquids is strongly affected by pressure. The solubility of a gas in a liquid is directly proportional to the pressure of the gas above the liquid. This makes sense in terms of dynamic equilibrium. Increasing the number of gas particles above the solution should increase the number of particles that dissolve and so should increase solubility, as depicted in Figure 11.9.

Figure 11.8 Ammonia is so soluble in water that a flask of ammonia gas will draw up water, forming a fountain. All the ammonia will dissolve, and the flask will fill with water. Phenolphthalein added to the water turns pink as the solution becomes basic.

Figure 11.9 The amount of dissolved gas in a solution is proportional to the partial pressure of that gas above the solution. The more gas particles above the solution, the more gas particles in solution.

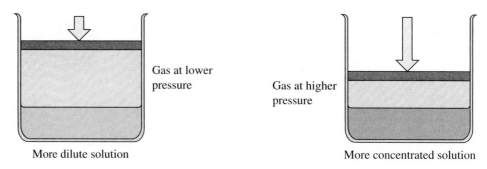

Gas at lower pressure

Gas at higher pressure

More dilute solution

More concentrated solution

The relationship between solubility and pressure, called **Henry's law,** is summarized by the following equation:

$$\text{Solubility of gas} = k \times \text{pressure of gas}$$
$$C = k \times P$$

where k is a constant at a given temperature for a given gas and solvent. When the solubility is given in grams per 100 g of water and the pressure is given in torr, the Henry's law constant has units of g/100 g H_2O/torr. When other concentration and pressure units are used, of course, the Henry's law constant has other units. The Henry's law relationship is demonstrated by the solubility of O_2 in H_2O, given in Table 11.6. Note the constant value of k obtained when solubility is divided by pressure. If we know the Henry's law constant for a gas, then we can calculate the solubility of that gas from the pressure of the gas above the solution.

EXAMPLE 11.8

The Henry's law constant for oxygen gas in water is 1.38×10^{-3} M/atm at 20°C. What is the solubility of oxygen in water at 20°C if the oxygen pressure is 0.21 atm, as it is in dry air at sea level?

Solution   In this case, we will calculate the solubility in units of molarity, because these are the units given for the Henry's law constant in the problem. The pressure must be in atmospheres for the same reason. The equation is

$$C = k \times P$$

Substituting values gives

$$C = 1.38 \times 10^{-3} \ M/\text{atm} \times 0.21 \ \text{atm} = 2.9 \times 10^{-4} \ M$$

Practice Problem 11.8

The Henry's law constant for nitrogen gas in water is 9.07×10^{-7} M/torr at 20°C. What is the solubility of nitrogen in water at 20°C if the nitrogen pressure is 0.78 atm?

Answer: $5.4 \times 10^{-4} \ M$

You observe the effect of pressure on solubility every time you open a bottle or can of a carbonated beverage. These beverages are bottled under a pressure of

Table 11.6 Solubility of
Oxygen in Water at 25°C

Pressure of Oxygen (torr)	Solubility (g/100 g H_2O)	Henry's Law Constant, k (g/100 g H_2O/torr)
175	0.00095	5.4×10^{-6}
300	0.00160	5.33×10^{-6}
414	0.00220	5.31×10^{-6}
610	0.00325	5.33×10^{-6}
760	0.00408	5.37×10^{-6}

CO_2 greater than atmospheric pressure. When you remove the bottle cap, the pressure of CO_2 is decreased (you can hear the excess gas escape). Some of the CO_2 that had been dissolved in the solution escapes and forms gas bubbles.

Pressure also affects the solubility of gases in the bloodstream. Hospital patients who have difficulty breathing are placed in oxygen tents, where the oxygen pressure is higher than it is in air. With a higher oxygen pressure, more oxygen dissolves in the patient's bloodstream with each breath, so the patient does not have to breathe so often to maintain an adequate oxygen supply.

As described in Chapter 9, when divers breathe a mixture containing nitrogen gas, they may develop the bends. Nitrogen dissolves in the bloodstream at the high pressure of great depths. If a diver surfaces too rapidly, causing a rapid decrease in pressure, the nitrogen comes out as bubbles in the bloodstream, rather than being passed back out through the lungs. Gas mixtures containing helium—called helium cocktails—instead of nitrogen do not cause the bends, because helium is not so soluble as nitrogen and the lighter helium atoms can diffuse more rapidly out of the bloodstream.

11.3 Electrolyte Solutions

What happens to solute particles when they are dissolved in a solution?

We can analyze the factors that determine whether a given solute will dissolve in a solvent in terms of the changes in intermolecular forces that accompany the conversion between the pure solute and solvent and the solution. However, such an analysis does not generally enable us to predict the condition of the solute once it has dissolved. Most ionic solids, for example, dissociate into ions when they dissolve, but covalent solids may or may not ionize. It would be useful to have some general rules to help us predict whether a solute will dissociate or ionize to give ions, at least for the common solvent water.

Chapter 10 introduced a classification scheme for solids. Depending on its class, a solid may or may not be soluble in water. If it is soluble, it may or may not give ions, regardless of whether it is classified as an ionic solid. We can decide whether a compound has dissociated into ions in solution by determining whether the solution conducts electricity, a property associated with the presence of dissolved ions. (The ability of a solution to conduct electricity can be determined with a fairly simple apparatus, such as the one shown in Figure 11.10.)

 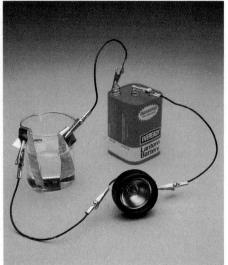

A solute that dissociates into ions in aqueous solution, such as NaCl, is an **electrolyte.**

$$NaCl(s) \xrightarrow{H_2O(l)} Na^+(aq) + Cl^-(aq)$$

In contrast, a solute that retains its molecular identity in solution and yields no ions, such as CH_3OH, is a **nonelectrolyte.**

$$CH_3OH(l) \xrightarrow{H_2O(l)} CH_3OH(aq)$$

An electrolyte can be further described as weak, if it dissociates only partially into its ions, or as strong, if it dissociates completely. As you might expect, a solution of a strong electrolyte passes more current than a solution of a weak electrolyte of comparable concentration. In the simple apparatus shown in Figure 11.10, a strong electrolyte gives an intense light, whereas a weak electrolyte causes the bulb to glow faintly. A nonelectrolyte gives no light at all.

We should note that although electrically conductive solutions usually arise from the dissolution of an ionic salt, they can also come from the dissolution of covalent substances that interact strongly with water to produce ions through hydrolysis (discussed in Chapter 4). The most common examples of this phenomenon are hydrogen chloride and ammonia, which dissolve in water to give hydrochloric acid and aqueous ammonia.

$$HCl(g) + H_2O(l) \longrightarrow H_3O^+(aq) + Cl^-(aq)$$
$$NH_3(g) + H_2O(l) \rightleftharpoons NH_4^+(aq) + OH^-(aq)$$

On the basis of electrical conductivity experiments, some general rules have been formulated to predict which species will be strong electrolytes. These rules are given in Table 11.7. The degree of dissociation of a substance can be represented by the **van't Hoff factor** (i), which is the average number of particles produced by one solute formula unit when it dissolves. As noted earlier, when NaCl dissolves, it forms two ions, $Na^+(aq)$ and $Cl^-(aq)$, so it has a value of i equal to 2. As shown by the following list, the van't Hoff factor has a lower limit of 1 and an upper limit equal to the number of ions per formula unit.

Table 11.7 Rules for Determining the Strength of Electrolytes

Rule 1: Most ionic salts are strong electrolytes in aqueous solution. Exceptions are some compounds that might be predicted to be ionic salts but are physically soft, have low melting points, and are molecular compounds; these include mercury(II) chloride, $HgCl_2$, and lead(II) acetate, $Pb(CH_3CO_2)_2$. Such compounds retain their molecular identity when dissolved in water and are weak electrolytes or nonelectrolytes.

Rule 2: Most acids are weak electrolytes in aqueous solution. Exceptions are the strong acids, which are also strong electrolytes: $HCl(aq)$, or hydrochloric acid; $HBr(aq)$, or hydrobromic acid; $HI(aq)$, or hydroiodic acid; $HClO_4(aq)$, or perchloric acid; $HNO_3(aq)$, or nitric acid; and $H_2SO_4(aq)$, or sulfuric acid.

Rule 3: Most bases are weak electrolytes. However, most hydroxides, if they are soluble, are strong electrolytes. Thus NaOH, RbOH, KOH, $Ca(OH)_2$, and the like are strong electrolytes. Hydroxides of metal ions with a charge greater than 2+ are usually quite insoluble. $NH_3(aq)$, sometimes written NH_4OH, is a weak electrolyte.

Rule 4: A species not accounted for in the first three rules is not likely to be dissociated into ions in solution.

Nonelectrolyte	$i = 1$
Strong electrolyte	i = number of ions in empirical formula
Weak electrolyte	$1 < i <$ maximum number of ions

Table 11.8 gives the value of the van't Hoff factor for several strong electrolytes, nonelectrolytes, and weak electrolytes.

Table 11.8 Degrees of Dissociation (van't Hoff Factors) of Some Substances

Substance	Expected van't Hoff Factor
Nonelectrolytes	
$C_2H_4(OH)_2$ (ethylene glycol)	$i = 1$
CH_3OH (methanol)	$i = 1$
$C_{12}H_{22}O_{11}$ (sucrose)	$i = 1$
C_2H_5OH (ethanol)	$i = 1$
Strong Electrolytes	
NaCl	$i = 2$
HCl	$i = 2$
$CaCl_2$	$i = 3$
K_2SO_4	$i = 3$
AlF_3	$i = 4$
$FeCl_3$	$i = 4$
K_3PO_4	$i = 4$
Weak Electrolytes	
CH_3CO_2H	$1 < i < 2$, depending on concentration
$NH_3(aq)$ (sometimes written NH_4OH)	$1 < i < 2$, depending on concentration

11.4   Reactions in Aqueous Solutions

How do we write equations for reactions involving ions?

A chemical equation, as you know, is a shorthand method of representing the chemical changes that occur in a chemical reaction. Chapter 4 explained how to balance chemical equations. That discussion included a number of examples of reactions in solution, but it left out one important factor. The formulas for reactions occurring in solution should represent the species as they occur in the solution.

For example, consider the reaction of aqueous solutions of lead nitrate and potassium chromate, both ionic salts and strong electrolytes. As shown in Figure 11.11, this reaction produces a yellow solid (lead chromate) and an aqueous solution of potassium nitrate. In Chapter 4 we might have written the following equation for this reaction:

$$Pb(NO_3)_2(aq) + K_2CrO_4(aq) \longrightarrow PbCrO_4(s) + 2KNO_3(aq)$$

Lead chromate, $PbCrO_4$, is a yellow or orange-yellow powder that occurs in nature as the minerals crocoite *and* phoenicochroite. *It is also known as Chrome yellow, Cologne yellow, King's yellow, Leipzig yellow, and Paris yellow. As these names suggest, lead chromate is used primarily as a pigment—in oil and water colors, inks for printing on fabrics, and porcelain glazes. It is a very insoluble substance; only 0.2 mg dissolves in 1 L of water.*

This is often called a **molecular equation,** because it represents the substances as though they existed as molecules (or formula units) rather than dissociating into ions. The following **ionic equation** represents the substances more appropriately, because both reactants are strong electrolytes that dissociate into ions in solution.

$$Pb^{2+}(aq) + 2NO_3^-(aq) + 2K^+(aq) + CrO_4^{2-}(aq) \longrightarrow$$
$$PbCrO_4(s) + 2K^+(aq) + 2NO_3^-(aq)$$

Note that potassium ion and nitrate ion occur on both sides of the equation and do not participate in the reaction. Such ions, called *spectator ions,* need not be written in the final form of the chemical equation. Thus the reaction can be represented as follows:

$$Pb^{2+}(aq) + CrO_4^{2-}(aq) \longrightarrow PbCrO_4(s)$$

This is the **net ionic equation.**

Let's reexamine some of the reactions described in Chapter 4 to see how much better net ionic equations represent the chemical changes that occur. ▶ We will consider two types of reactions: precipitation reactions, such as that described in the chapter introduction, and acid–base neutralization reactions. Both types are

Figure 11.11 When a solution of potassium chromate is poured into a solution of lead nitrate, lead chromate precipitates from solution. The resulting solution contains potassium and nitrate ions.

examples of double-displacement reactions, in which two compounds exchange elements or groups of elements to form new compounds.

An ionic precipitation reaction is one in which the cation from a solution of one salt combines with the anion from a solution of another salt to form an insoluble solid substance, a *precipitate*. The reaction between lead nitrate and potassium chromate is one such reaction. For a precipitation reaction to occur, then, the reactants must be soluble, but at least one cation–anion combination in the mixture of reactants must be insoluble or no precipitate will form. The solubilities of ionic substances can be used to predict whether a precipitation reaction will occur, as well as to identify the insoluble product in any mixture of aqueous solutions of ionic salts.

Silver nitrate, $AgNO_3$, is a colorless, odorless, crystalline solid. It decomposes when heated to 440°C, giving Ag, N_2, O_2, and nitrogen oxides. It is quite soluble in water; 1 g requires only 0.4 mL H_2O. It is used in photography, the manufacture of mirrors, silver plating, indelible inks, porcelain glazes, and chemical analyses for chloride.

Consider the reaction between colorless aqueous solutions of $AgNO_3$ and KCl, which gives a white precipitate and a colorless solution (see Figure 11.12). At the point when they are mixed, both compounds are dissociated because they are strong electrolytes, so the ions present before any reaction occurs are $Ag^+(aq)$, $NO_3^-(aq)$, $K^+(aq)$, and $Cl^-(aq)$. Possible formulas for the precipitate, then, are $AgNO_3$, KCl, AgCl, and KNO_3. The first two choices can be eliminated. We started with solutions of $AgNO_3$ and KCl; if they could have precipitated as solids, they would have done so before being mixed. That leaves AgCl and KNO_3. Chapter 4 presented some generalized rules about solubility, including two that are pertinent to choosing between these possibilities. Most salts of K^+ are soluble, and most salts of NO_3^- are soluble. On the basis of these rules, we can conclude that the precipitate is unlikely to be KNO_3. Another solubility rule is also pertinent: With only a few exceptions, Ag^+ salts are insoluble. Thus it is logical to identify the precipitate as AgCl.

Appendix B presents a more extensive set of rules that can serve as a guide for making predictions about the solubility of a wide variety of ionic salts in water. It should be noted again that no solid, including the so-called insoluble salts, is totally insoluble. Also note that, although the salts of some oxoanions are insoluble, if these oxoanions contain hydrogen atoms attached to oxygen atoms, they form soluble salts. Thus $Ca_3(PO_4)_2$ is insoluble, but $Ca(H_2PO_4)_2$ is soluble.

In general, for ionic precipitation reactions, it is reasonable to assume that if a reaction occurs, it corresponds to an exchange of ionic partners. For the silver

Figure 11.12 When aqueous solutions of silver nitrate and potassium chloride are mixed, a white precipitate of silver chloride is formed.

chloride example, we have

silver nitrate + potassium chloride $\longrightarrow$ silver chloride + potassium nitrate

Molecular equation:

$$AgNO_3(aq) + KCl(aq) \longrightarrow AgCl(s) + KNO_3(aq)$$

Ionic equation:

$$Ag^+(aq) + NO_3^-(aq) + K^+(aq) + Cl^-(aq) \longrightarrow AgCl(s) + K^+(aq) + NO_3^-(aq)$$

Net ionic equation:

$$Ag^+(aq) + Cl^-(aq) \longrightarrow AgCl(s)$$

The net ionic equation provides the best representation of this reaction, because it focuses on the chemical changes that are occurring.

EXAMPLE 11.9

Write a balanced net ionic equation for the reaction between copper sulfate and barium chloride.

Solution   According to the solubility rules in Appendix B, both of these ionic salts should be soluble. They are also strong electrolytes, according to the rules in Table 11.7. Thus the solution should contain the ions $Cu^{2+}(aq)$, $SO_4^{2-}(aq)$, $Ba^{2+}(aq)$, and $Cl^-(aq)$. Insoluble salts that might form include both compounds that would result from an exchange of partners: $CuCl_2$ and $BaSO_4$. Again, we can consult the solubility rules in Appendix B to check on the solubility of these ionic salts. The rules indicate that copper forms a soluble chloride salt but that barium sulfate should be insoluble. Thus an ionic precipitation reaction should produce barium sulfate by the following molecular equation:

$$CuSO_4(aq) + BaCl_2(aq) \longrightarrow CuCl_2(aq) + BaSO_4(s)$$

The ionic equation is

$$Cu^{2+}(aq) + SO_4^{2-}(aq) + Ba^{2+}(aq) + 2Cl^-(aq) \longrightarrow$$
$$Cu^{2+}(aq) + 2Cl^-(aq) + BaSO_4(s)$$

Copper(II) and chloride ions occur on both sides of the equation, so they are spectator ions. Eliminating them yields the net ionic equation

$$SO_4^{2-}(aq) + Ba^{2+}(aq) \longrightarrow BaSO_4(s)$$

Practice Problem 11.9

Write a balanced net ionic equation for the reaction between ammonium carbonate and chromium(III) chloride.

Answer: $3CO_3^{2-}(aq) + 2Cr^{3+}(aq) \longrightarrow Cr_2(CO_3)_3(s)$

Net ionic equations also allow us to focus on the important part of acid–base neutralization reactions, in which an acid reacts with a base, normally forming a salt and water. The formation of the water molecule, H_2O, from hydrogen and hydroxide ions is common to all aqueous neutralization reactions. The acid–base reaction involves the donation of a hydrogen ion by an acid and the acceptance of that hydrogen ion by a base. A common example is the reaction between hydrochloric acid and sodium hydroxide, both of which are strong electrolytes.

Hydrochloric acid + sodium hydroxide $\longrightarrow$ sodium chloride + water

Molecular equation:

$$HCl(aq) + NaOH(aq) \longrightarrow NaCl(aq) + H_2O(l)$$

Ionic equation:

$$H^+(aq) + Cl^-(aq) + Na^+(aq) + OH^-(aq) \longrightarrow Na^+(aq) + Cl^-(aq) + H_2O(l)$$

Net ionic equation:

$$H^+(aq) + OH^-(aq) \longrightarrow H_2O(l)$$

This net ionic equation is the fundamental neutralization equation that is common to all strong-acid–strong-base reactions.

When the acid is a weak electrolyte, the net ionic equation is different, however. Consider the neutralization of acetic acid by potassium hydroxide.

Acetic acid + potassium hydroxide $\longrightarrow$ potassium acetate + water

Molecular equation:

$$CH_3CO_2H(aq) + KOH(aq) \longrightarrow KCH_3CO_2(aq) + H_2O(l)$$

Ionic equation:

$$CH_3CO_2H(aq) + K^+(aq) + OH^-(aq) \longrightarrow K^+(aq) + CH_3CO_2^-(aq) + H_2O(l)$$

Net ionic equation:

$$CH_3CO_2H(aq) + OH^-(aq) \longrightarrow CH_3CO_2^-(aq) + H_2O(l)$$

In this case potassium ion is a spectator ion. However, because acetic acid does not ionize appreciably (it is a weak electrolyte), acetate ion is not a spectator ion.

11.5 The Chemistry of Water Treatment

Water is one of the most abundant and important substances found on Earth. In liquid form, it occurs not only in streams, lakes, and oceans, which cover over 70% of Earth's surface, but also underground in the water table. It exists in solid form as snow and ice at the polar ice caps and in other cold regions. It is found in the atmosphere as gaseous water vapor and clouds.

Impurities in Water

▶ This chapter began by noting how seldom we encounter pure water. Because it has excellent solvent properties, water contains various amounts of substances dissolved from the materials with which it comes in contact. Many atmospheric gases as well as solids and liquids are significantly soluble in water.

Among the gases, carbon dioxide is particularly soluble, and it has a dramatic effect on the solubility of various minerals in water. Dissolved carbon dioxide has acidic properties and can be considered to form carbonic acid, H_2CO_3.

$$CO_2(aq) + H_2O(l) \longrightarrow H_2CO_3(aq)$$

The presence of dissolved CO_2 or H_2CO_3 contributes to the dissolution of normally insoluble carbonate, phosphate, and sulfite minerals. For example, limestone can be dissolved upon reaction with carbonic acid (Figure 11.13).

$$CaCO_3(s) + H_2CO_3(aq) \longrightarrow Ca^{2+}(aq) + 2HCO_3^-(aq)$$

A significant amount of the solids found in water results from natural processes such as erosion of stream beds and surface erosion by rain water. Some of these substances dissolve and form aqueous ions, such as calcium, magnesium, iron, aluminum, sulfate, carbonate, phosphate, and chloride ions. Others are particles suspended in water but not in solution with it. For example, clay particles may be present because of erosion and make the water cloudy. Particles may also be formed when metal ions are exposed to bases, causing precipitation of metal hydroxides:

$$Fe^{3+}(aq) + 3OH^-(aq) \longrightarrow Fe(OH)_3(s)$$

Iron(III) hydroxide, Fe(OH)₃, is a reddish brown powder. It occurs in nature as various minerals. Upon heating, it loses water to form iron(III) oxide, Fe₂O₃. Iron(III) hydroxide is used as a pigment and a catalyst.

With so many substances dissolved in water, where do we get water that is safe to drink?

Water supplies may contain not only naturally occurring substances but also municipal, industrial, and agricultural wastes. Bacteria grow in some such wastes and can cause a number of diseases if the water is used for drinking. In addition, the contamination of water with phosphates from detergents has caused rapid growth of algae, which remove dissolved oxygen, resulting in the death of aquatic life. The absence of dissolved oxygen also results in the conversion of rotting plants to bad-smelling hydrogen sulfide and methane instead of to carbon dioxide and sulfate.

Drinking water is regulated to a certain extent by the U.S. Environmental Protection Agency (EPA). However, EPA standards cover only the traditionally known contaminants and make no mention of some harmful substances whose toxic effects are only now being discovered. Furthermore, not all drinking water supplies meet these standards.

Figure 11.13 A basic solution containing calcium(II) ions deposits calcium carbonate when carbon dioxide is bubbled through it. As more carbon dioxide dissolves, however, the calcium carbonate redissolves as calcium bicarbonate (or calcium hydrogen carbonate).

Purification of Municipal Water Supplies

In most municipalities, drinking water sources are sufficiently contaminated that the water must be treated to meet EPA standards. A typical treatment procedure is shown in Figure 11.14.

In nature, standing water is partially purified by the settling of suspended solids. Municipal plants that process water to make it suitable for drinking also use a settling procedure, and they often assist it chemically because it is rather slow. A mixture of aluminum sulfate, or alum, and lime, or some other basic material, is added to the water. These chemicals react to form a very gelatinous precipitate of $Al(OH)_3$.

$$Al_2(SO_4)_3(aq) + 3CaO(s) + 3H_2O(l) \longrightarrow 2Al(OH)_3(s) + 3CaSO_4(s)$$

This precipitate carries suspended material down with it when it settles. The water is then filtered through sand to remove the remainder of the suspended material.

The next step is aeration. The water is sprayed through air to increase the amount of dissolved oxygen, which converts dissolved organic materials to carbon dioxide and water.

$$CH_4(aq) + 2O_2(aq) \longrightarrow CO_2(aq) + 2H_2O(l)$$

Finally, if water is to be used for drinking or cooking, it must be freed of harmful bacteria. Bacteria can be killed by chemical treatment with an oxidizing substance such as permanganate (MnO_4^-), hydrogen peroxide (H_2O_2), ozone (O_3), or chlorine (Cl_2). Chlorine or a chlorinating agent is normally used for domestic water supplies, because these substances are relatively inexpensive and readily available. When chlorine is dissolved in water, it reacts with water to form hydrochloric acid and hypochlorous acid.

$$Cl_2(g) + H_2O(l) \longrightarrow H^+(aq) + Cl^-(aq) + HOCl(aq)$$

Other chlorinating agents, such as sodium hypochlorite ($NaOCl$) and calcium hypochlorite ($Ca(OCl)_2$), are also used. These chemicals are the usual choice for treating swimming pool water.

Figure 11.14 Water is pumped from a well or reservoir into settling tanks. After the larger suspended particles settle out, an aluminum salt and a base, such as lime, are added and form aluminum hydroxide, which drags down smaller suspended particles. The water is filtered through sand to remove any remaining solids and then sprayed through air to begin the oxidizing of any dissolved organic materials. After this process is completed, bacteria are killed by chlorination. The water is then stored until needed for distribution through the city water mains.

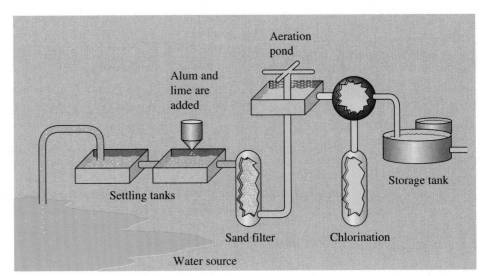

Figure 11.15 Boiler scale, caused by precipitation of calcium and magnesium carbonates, can completely clog pipes.

How do we remove the dissolved substances in hard water to make soft water?

Hard Water and Water Softening

Water containing the cations Ca^{2+}, Mg^{2+}, and sometimes Fe^{3+}, along with anions such as Cl^-, SO_4^{2-}, and HCO_3^-, is called **hard water,** because a hard deposit of the metal carbonates ($CaCO_3$ and $MgCO_3$) forms in pipes and pots when the water is heated or evaporated. When hard water is boiled, hydrogen carbonate ions decompose to form gaseous carbon dioxide and carbonate ions.

$$2HCO_3^-(aq) \xrightarrow{\Delta} CO_2(g) + H_2O(l) + CO_3^{2-}(aq)$$

The carbonate ions form insoluble precipitates with calcium and magnesium ions, forming a scale of $CaCO_3$ and $MgCO_3$, in a process much like the deposition of limestone in caves. This scale can cause clogging of pipes and reduces heat transport in boiler systems (Figure 11.15).

Calcium, magnesium, and iron(III) ions also cause an undesirable precipitate to form when soap is added.

$$2C_{17}H_{35}CO_2^-(aq) + Ca^{2+} \longrightarrow Ca(C_{17}H_{35}CO_2)_2(s)$$

Soap contains anions of fatty acids. The salts formed between these cations and the fatty acids in soaps are insoluble and form a scum that floats to the surface. This scum is sticky and gives rise to bathtub ring and other unpleasant effects. Hard water requires more soap before suds begin to form, because the metal ions precipitate the soap until they are removed from solution.

Hardness is usually reported as though it were due entirely to dissolved $CaCO_3$, given in milligrams per liter (which is the same as parts per million, ppm). A hardness of 1 ppm corresponds to a concentration of Ca^{2+} of $1 \times 10^{-5}\,M$. Water would generally be described as very soft if it had a hardness of 30–50 ppm; 50–100 ppm is moderately hard, 100–300 ppm is hard, and greater than 300 ppm is very hard.

The unit ppm is frequently used in reporting the concentration of pollutants in water and air. This unit simply means that in a mixture, there is 1 part by mass of the subject material in 1 million parts of the mixture. Thus in water, 1 ppm corresponds to 1 g of pollutant per 1 million g of water (or, more accurately, of solution). By shifting the decimal point, you can see that this is equal to 0.001 g per 1000 g, or 1 mg/kg. Because 1 L of water weighs 1 kg, 1 ppm also corresponds to 1 mg/L.

Hardness is classified as permanent (Cl^- or SO_4^{2-} salts) or temporary (HCO_3^- salts), depending on whether it can be removed by boiling. Boiling removes temporary hardness by converting soluble bicarbonates into insoluble carbonates, thereby removing the hardness from water.

$$Ca^{2+}(aq) + 2HCO_3^-(aq) \xrightarrow{\Delta} CaCO_3(s) + CO_2(g) + H_2O(l)$$

The hardness remaining after boiling is permanent hardness.

Hardness can also be removed chemically—a process known as softening. Somewhat surprisingly, hardness can be removed by the addition of more calcium salts in the form of calcium oxide (lime) or calcium hydroxide (slaked lime).

$$Ca(HCO_3)_2(aq) + Ca(OH)_2(s) \longrightarrow 2CaCO_3(s) + 2H_2O(l)$$

This method, called the lime-soda process, is often used in municipal water treatment plants like the one in Figure 11.14.

On a smaller scale—in detergents, for example—water is softened by addition of other chemicals, called sequestering agents, such as sodium tripolyphosphate, $Na_5P_3O_{10}$. These substances tie up Ca^{2+} and Mg^{2+} ions. Although some such materials work as water softeners by precipitating the calcium out as an insoluble salt, the more effective sequestering agents make complex ions with the metal ions, keeping them in solution but making them unavailable to form a precipitate with soap.

$$Ca^{2+}(aq) + P_3O_{10}^{5-}(aq) \longrightarrow CaP_3O_{10}^{3-}(aq)$$

Various sequestering agents have been used in detergents as water softening agents. These include washing soda (Na_2CO_3), borax ($Na_2B_4O_7 \cdot 10H_2O$), sodium phosphate (Na_3PO_4), and various polyphosphates ($Na_{n+2}P_nO_{3n+1}$). Hardness can be removed from water by ion exchange processes—the processes used in most home water softeners. Ion exchangers use large insoluble molecules that contain replaceable sodium ions to exchange Ca^{2+}, Mg^{2+}, and other cations with Na^+.

$$Ca^{2+}(aq) + Na_2Al_2Si_4O_{12}(s) \longrightarrow 2Na^+(aq) + CaAl_2Si_4O_{12}(s)$$

The water then contains Na^+ instead of Ca^{2+} or Mg^{2+}. Boiler scale no longer forms if the water contains carbonate or hydrogen carbonate ions, because sodium carbonate is soluble.

11.6   Colligative Properties

We can think of certain properties of solutions as properties of the pure solvent modified by the presence of solute particles. Generally, such properties vary only with the number of solute particles (molecules or ions) present in a specific quantity of solvent, not with the identity of the solute. These properties, which include vapor pressure, freezing point, boiling point, and osmotic pressure, are called **colligative properties.**

Vapor Pressure

Consider a solution of table sugar, or sucrose, in water. If a tube containing the sugar solution and another tube containing pure water are placed into a sealed container, over a period of time the volume of the sugar solution increases, and the volume of the pure water decreases (see Figure 11.16). Both liquids have a vapor

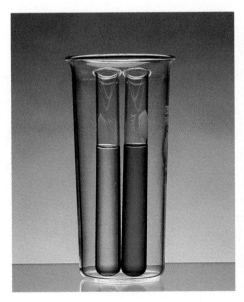

Figure 11.16 Water evaporates from the tube of pure water (blue) and condenses into the tube of sugar solution (colorless). A small amount of blue dye was added to the water to make it easier to distinguish between the two tubes.

pressure, but the pure water has a vapor pressure greater than that of the sugar solution. Because the system is sealed, water that evaporates cannot escape from the system. Rather, water evaporates from the liquid that has the greater vapor pressure and condenses in the liquid that has the lower vapor pressure. The presence of solute molecules, then, has lowered the vapor pressure of the solvent.

Why does the presence of a solute lower the vapor pressure of the solution?

We can explain this phenomenon in simple terms by assuming the solute is nonvolatile and considering the vapor pressure to depend on what proportion of liquid-phase molecules at the surface of the liquid can escape into the gas phase (Figure 11.17). If some of the solvent molecules at the surface are replaced by nonvolatile solute molecules, then fewer solvent molecules can escape into the gas phase, and the vapor pressure is lowered. The vapor pressure of the solution is lower than the vapor pressure of the pure solvent at all temperatures at which the solution is liquid, as shown in Figure 11.18.

Decreases in vapor pressure for *ideal solutions,* in which solute–solvent interactions are equal to solute–solute plus solvent–solvent interactions, follow **Raoult's law,** which states that the partial pressure of a component of a solution equals the vapor pressure of the pure component multiplied by its mole fraction in the solution. For a solution of B (the solute) in A (the solvent), this is represented as follows:

$$P_A = \chi_A P_A^\circ \qquad P_B = \chi_B P_B^\circ$$

Here P is the observed partial pressure of the substance in the solution, χ is the mole fraction of the substance in the solution, and P° is the vapor pressure of the pure substance.

In a solution, the mole fraction of the solvent is always less than 1, because the sum of the mole fractions of solute and solvent must equal 1, by definition. Thus Raoult's law asserts that the vapor pressure of the solvent in a solution should be less than that of the pure solvent. We will consider only nonvolatile solutes, for which $P_B^\circ = 0$, to simplify the mathematical application of Raoult's law.

We can use Raoult's law to show that the lowering of vapor pressure (ΔP) is a colligative property, proportional to the concentration of the solution. The vapor-pressure lowering is the difference between the vapor pressures of the solvent and the solution.

$$\Delta P = P_A^\circ - P_A$$

Substituting for P_A from Raoult's law yields

$$\Delta P = P_A^\circ - \chi_A P_A^\circ = (1 - \chi_A)P_A^\circ$$

Figure 11.17 The vapor pressure of a solution is lower than that of the pure solvent, because the presence of nonvolatile solute particles on the surface of the solution reduces the number of solvent molecules that can evaporate.

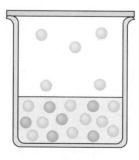

Solution

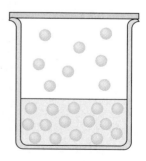

Pure liquid

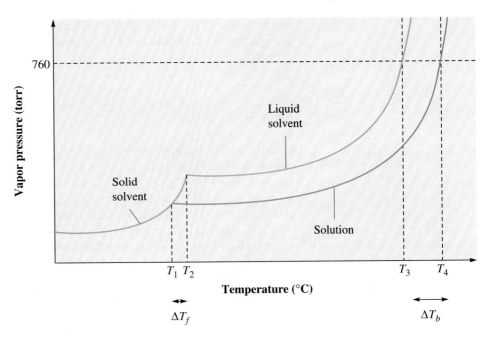

Figure 11.18 The addition of nonvolatile solute lowers the vapor pressure of a solvent, causing an elevation of the boiling point and a depression of the freezing point. T_1 = the freezing point of the solution; T_2 = the freezing point of the pure solvent; T_3 = the boiling point of the pure solvent; T_4 = the boiling point of the solution; ΔT_f = the freezing-point depression; ΔT_b = the boiling-point elevation.

Because the sum of the mole fractions must equal 1 ($\chi_A + \chi_B = 1$), we get the following equation:

$$\Delta P = P_A^\circ \chi_B$$

The vapor-pressure lowering is directly proportional to the mole fraction of the solute. For dilute solutions, χ_B is just the ratio of moles of B to moles of A, so the lowering of vapor pressure is a colligative property—that is, it is proportional to the amount of B in the solution.

Consider the solution of sucrose again. Suppose that its temperature is 25°C and that the mole fraction of sucrose is 0.25. We can determine the vapor-pressure lowering from the vapor pressure of water at this temperature, 23.8 torr. The vapor-pressure lowering is given by the equation

$$\Delta P = P_{\text{water}}^\circ \chi_{\text{sucrose}} = 23.8 \text{ torr} \times 0.25 = 6.0 \text{ torr}$$

The vapor pressure of the sucrose solution is the difference between the vapor pressure of pure water and the vapor-pressure lowering, 23.8 torr − 6.0 torr, or 17.8 torr.

Boiling Point

A solute lowers the vapor pressure of the solution. Predict what effect this should have on the boiling point of the solution.

We have already seen that the vapor pressure of liquids increases with temperature. In Figure 11.18, you can see that when the temperature is sufficiently high that the vapor pressure of the liquid equals the surrounding atmospheric pressure, the liquid reaches its boiling point. The addition of a solute affects the boiling point because it affects the vapor pressure. The presence of solute means that the temperature must be raised further to raise the vapor pressure to atmospheric pressure, so the boiling point is raised as well. The boiling-point elevation is shown in Figure 11.18 as ΔT_b. Because the lowering of vapor pressure is proportional to the concentration of the solute, the elevation of the boiling point is also proportional to the concentration of the solute. Thus boiling-point elevation is also a colligative property.

Freezing Point

Recall that the freezing point of a liquid is the temperature at which the vapor pressures of the liquid and its solid form are equal. For pure solvent, this is the point labeled T_2 in Figure 11.18. The lowering of vapor pressure with addition of a solute lowers the freezing point to T_1. Again, the freezing-point depression is proportional to the concentration of solute.

Why is calcium chloride added to cement that is to be poured during cold weather?

This colligative property finds practical application in the use of ethylene glycol (HO—C_2H_4—OH) as an antifreeze in automobile radiators. Ethylene glycol lowers the freezing point of radiator coolant (and incidentally raises the boiling point, preventing boil-over during the summer). Another widespread application in cold climates is the use of sodium chloride or calcium chloride to melt ice on roads and sidewalks. These compounds work by lowering the freezing point of water.

Colligative Properties and Molar Masses

Chemists often use colligative properties to determine the molar mass of a solute. Changes in vapor pressure are difficult to measure accurately without special equipment, but changes in freezing point and boiling point are not difficult to measure. Here we examine how they can be used to determine molar masses.

Freezing-Point Depression   Freezing-point depression is related to the molality of a solution as follows:

$$\Delta T_f = K_f m$$

The term ΔT_f is the difference in freezing point between pure solvent and solution. In terms of the labeled points on Figure 11.18, this difference is given by

$$\Delta T_f = T_2 - T_1$$

The term K_f is a constant that is characteristic of the solvent, the **freezing-point constant.** Values of K_f in units of degrees Celsius per mol of solute per kilogram of solvent (°C/mol solute/kg solvent) are included in Table 11.9 for some typical solvents. The term m is the molal concentration of the solution.

$$m = \frac{\text{moles of solute}}{\text{kilograms of solvent}} = \frac{(\text{mass of solute/molar mass of solute})}{\text{mass of solvent in kilograms}}$$

To determine the molar mass of a solute from the change in freezing point, we start by using the freezing-point depression to calculate the molality of the solution. If we know the mass of solute and the mass of solvent in the solution, then we can use the value of the molality to determine the molar mass of the solute. This method is frequently used to determine the molar mass of an unknown substance; it provides information that can help identify that substance if the empirical formula is also known.

Table 11.9 Freezing-Point
Constants and Boiling-Point
Constants for Some Solvents

Solvent	K_f $\left(\dfrac{°C/\text{mol solute}}{\text{kg solvent}}\right)$	T_f (°C)	K_b $\left(\dfrac{°C/\text{mol solute}}{\text{kg solvent}}\right)$	T_b (°C)
Water (H_2O)	1.86	0.00	0.52	100.0
Benzene (C_6H_6)	5.12	5.53	2.53	80.1
Cyclohexane (C_6H_{12})	20.0	6.47	2.79	80.7
Naphthalene ($C_{10}H_{18}$)	6.9	80.2	5.65	218
Ethanol (C_2H_5OH)	1.99	−115	1.22	78.4
Tertiary butanol (C_4H_9OH)	8.3	25.6	—	82.4
Carbon tetrachloride (CCl_4)	31.8	−22.96	5.03	76.5
Methanol (CH_3OH)	—	−97.8	0.80	64.7
Acetic acid (CH_3CO_2H)	3.9	16.7	3.07	118

EXAMPLE 11.10

Naphthalene normally freezes at 80.2°C and has a K_f value of 6.9°C/mol/kg. A solution containing 1.19 g of sulfur dissolved in 10.20 g of naphthalene freezes at 77.1°C. Calculate the molality of the solution and the molar mass of the solute, sulfur.

Solution   Adding sulfur lowers the freezing point by the following amount:

$$\Delta T_f = 80.2°C − 77.1°C = 3.1°C$$

We can use this value in determining the molality of the solution.

$$\Delta T_f = K_f m$$
$$3.1°C = 6.9°C/\text{mol sulfur/kg naphthalene} \times m$$
$$m = 3.1°C/6.9°C/\text{mol sulfur/kg naphthalene}$$
$$= 0.45 \text{ mol sulfur/kg naphthalene}$$

We can use the molality, combined with the mass of naphthalene, to calculate the number of moles of sulfur in the solution.

$$\text{Moles sulfur} = 0.45 \text{ mol/kg} \times 1 \text{ kg}/1000 \text{ g} \times 10.20 \text{ g} = 0.0046 \text{ mol}$$

Finally, we can use the moles of sulfur and the mass of sulfur to calculate the molar mass.

$$\text{Molar mass} = 1.19 \text{ g}/0.0046 \text{ mol} = 260 \text{ g/mol}$$

Because the molar mass of sulfur is greater than the molar mass of atomic sulfur, 32 g/mol, we can see that the sulfur molecule is polyatomic. In fact, it contains eight atoms per molecule.

Practice Problem 11.10

When 235 g of glucose is added to 2.35×10^3 g of water, the freezing point of the water is lowered by 1.03°C. The freezing-point constant of water is 1.86°C/mol/kg. What is the molar mass of glucose?

Answer: 181 g/mol

Boiling-Point Elevation   Like freezing-point depression, boiling-point elevation ($T_4 - T_3$ in Figure 11.18) is related to the molality of the solute. The same equation applies, with ΔT_b substituted for ΔT_f and a **boiling point constant,** K_b, substituted for K_f:

$$\Delta T_b = K_b m$$

Some values of K_b are included in Table 11.9. The equation can be used in the same way as that for freezing-point depression to measure molar masses of solutes.

Colligative Properties and the van't Hoff Factor   We have used a measured value of freezing-point depression (or boiling-point elevation) to calculate the molality of a solution and thus the molar mass of a solute. If we know the molality of a solution, we can reverse this process and calculate the freezing-point depression or boiling-point elevation expected for that solution. However, the colligative properties measured for many solutions are larger than would be predicted from such calculations. This is due to the fact that many solutes dissociate into ions when they dissolve, and colligative properties are proportional to the number of particles. Thus, when the measured value of the freezing-point depression or boiling-point elevation is greater than the calculated value, we can assume the solute is an electrolyte and has dissociated into ions.

The number of particles can be taken into account by use of the van't Hoff factor, i. As we have seen, i is the number of particles formed from each solute molecule (or formula unit) when the solute dissolves.

The equations for freezing-point depression and boiling-point elevation can be modified by inclusion of the van't Hoff factor:

$$\Delta T_f = iK_f m$$
$$\Delta T_b = iK_b m$$

When these equations are used, calculated and measured values of the freezing-point depression and the boiling-point elevation should agree. We can also use these equations to determine the van't Hoff factor, which gives us a way to distinguish between an electrolyte ($i > 1$) and a nonelectrolyte ($i = 1$).

EXAMPLE 11.11

The freezing point of pure naphthalene is 80.21°C. A solution of 0.6432 g of iodoform, CHI_3, in 4.950 g of naphthalene freezes at 77.93°C. Calculate the van't Hoff factor for CHI_3, and decide whether it is an electrolyte or a nonelectrolyte.

Solution   We will use the expression for freezing-point lowering to solve this problem:

$$\Delta T_f = iK_f m$$

We have enough information to evaluate all variables in this equation except the van't Hoff factor. The freezing-point depression is given by

$$\Delta T_f = 80.21°C - 77.93°C = 2.28°C$$

The molality can be calculated from the masses of CHI_3 and naphthalene and the molar mass of CHI_3, 393.8 g/mol.

$$\text{Molality} = \frac{\text{moles of } CHI_3}{\text{kilograms of naphthalene}}$$

$$= \frac{0.6432 \text{ g CHI}_3/393.8 \text{ g/mol}}{0.004950 \text{ kg naphthalene}} = 0.3300 \ m$$

The freezing-point constant of naphthalene is 6.9°C/mol/kg. We can solve the freezing-point lowering expression for the van't Hoff factor:

$$i = \Delta T_f / K_f m$$

Substituting values, we obtain

$$i = \frac{2.28°C}{6.9°C/mol/kg \times 0.3300 \text{ mol/kg}} = 1.0$$

Because the van't Hoff factor equals 1, the solute, CHI_3, must be a nonelectrolyte.

Practice Problem 11.11

When 11.3 g of sodium chloride is dissolved in 325 g of water, the freezing point is lowered by 2.22°C. Calculate the van't Hoff factor for sodium chloride.

Answer: i = 2

Osmosis

Why does limp lettuce become crisp again when it is soaked in water? Why do prunes lose their wrinkles when soaked in water?

Osmosis is a process in which solvent molecules (most important among them, water molecules) diffuse through a barrier that does not allow the passage of solute particles. Osmosis makes it possible for water to pass through plant and animal membranes such as the cell walls of all living organisms, so the process is intimately involved in life processes. Membranes that allow the passage of some substances but not others are called **semipermeable membranes.** In crossing semipermeable membranes, water tends to travel from the side that is more dilute in solute to the side that is more concentrated, as shown in Figure 11.19. The water actually flows in both directions, but it flows more rapidly in the direction that leads to equalization of the solute concentration on both sides of the membrane.

Figure 11.19 Osmosis occurs because only solvent molecules can pass through a semipermeable membrane. The solvent travels in both directions, but more solvent travels in the direction that leads to dilution of the more concentrated solution.

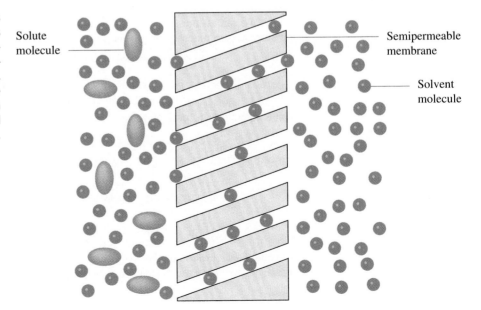

Solute molecule

Semipermeable membrane

Solvent molecule

For example, when pure water is placed on one side of a semipermeable membrane and a sugar solution on the other side, the volume of the sugar solution increases and the solution becomes more dilute, as shown in Figure 11.20. Pressure would have to be exerted on the solution to prevent osmosis—that is, to prevent water from passing over to the solution side of the membrane. This pressure, called the **osmotic pressure,** is related to the height of solution that will rise up a column separated from pure solvent by a semipermeable membrane (Figure 11.20). Osmotic pressure can be calculated from an equation very similar to the ideal gas law:

$$\pi V = nRT$$

Because the number of moles (n) divided by the volume (V) is the molarity (M) of the solution, this equation can be changed to a more useful form:

$$\pi = MRT$$

Here π is the osmotic pressure (in units of atm), M is the molarity of *particles* in the solution, R is the gas constant (0.08206 L atm/mol K), and T is the absolute temperature (in units of K). Osmotic pressure depends in part on the number of solute particles in the solution, so it is a colligative property. It can be used in the same way as other colligative properties to measure molar masses and to distinguish between electrolytes and nonelectrolytes.

EXAMPLE 11.12

An aqueous solution commonly known as D5W is used for intravenous injection. D5W has the same osmotic pressure as blood fluid. It contains 54.3 g of glucose, $C_6H_{12}O_6$, per liter. The molar mass of glucose is 1.80×10^2 g/mol. Calculate the osmotic pressure in blood cells at 37°C, normal body temperature.

Solution   We start with the equation for osmotic pressure:

$$\pi = MRT$$

where M is the molarity of particles in the solution. Because glucose is a covalent compound, we can assume that it is a nonelectrolyte and that the molarity of solute particles is the same as the molarity of glucose.

$$M = \text{moles of solute/liters of solution}$$

$$= \frac{54.3 \text{ g}}{1 \text{ L}} \times \frac{1 \text{ mol}}{180 \text{ g}} = 0.302 \text{ mol/L}$$

The temperature in units of K is

$$T = 37°C + 273° = 310 \text{ K}$$

We can substitute these values into the equation for osmotic pressure.

$$\pi = MRT$$

$$= 0.302 \text{ mol/L} \times 0.08206 \text{ L atm/mol K} \times 310 \text{ K} = 7.68 \text{ atm}$$

Practice Problem 11.12

The total concentration of dissolved particles in a blood cell is about 0.30 M. When a cell is placed in pure water at 5°C, what osmotic pressure is exerted on the cell?

Answer: 6.8 atm

Figure 11.20 A tube with a semipermeable membrane on the end can be used to measure osmotic pressure. When solvent and solution are separated by a semipermeable membrane, the solvent moves through the membrane to dilute the solution, resulting in an increase in the volume of solution. The solution rises in the tube until the pressure caused by the column of solution just equals the osmotic pressure.

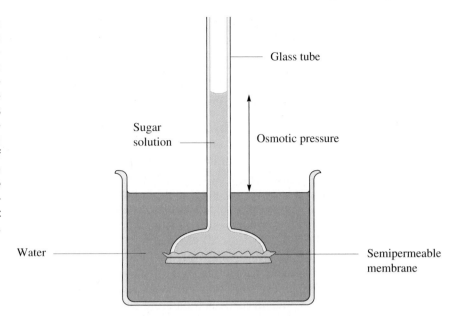

Glass tube

Sugar solution

Osmotic pressure

Water

Semipermeable membrane

Measurements of osmotic pressure are frequently used to determine the molar mass of large molecules such as proteins or polymers. Often, because they are not highly soluble, such molecules form only dilute solutions; it is easy to measure the resulting small values of osmotic pressure but not easy to measure small changes in freezing point or boiling point. The equation for osmotic pressure is used to calculate the molarity of the solution, which in turn makes it possible to calculate the molar mass.

EXAMPLE 11.13

A solution of polystyrene plastic contains 12.3 g of polystyrene dissolved in sufficient toluene to give 1.00×10^2 mL of solution. The osmotic pressure of this solution is 0.0320 atm at 20°C. What is the molar mass of this sample of polystyrene?

Solution Calculate the molarity of the solution from the osmotic pressure equation.

$$\pi = MRT$$
$$M = \pi/RT$$

$$= \frac{0.0320 \text{ atm}}{0.08206 \text{ L atm/mol K} \times 293 \text{ K}} = 0.00133 \text{ mol/L}$$

The molar mass can now be calculated from the definition of molarity.

$$MM = \frac{12.3 \text{ g}}{100 \text{ mL}} \times \frac{1000 \text{ mL}}{1 \text{ L}} \times \frac{1 \text{ L}}{0.00133 \text{ mol}} = 9.25 \times 10^4 \text{ g/mol}$$

Practice Problem 11.13

A solution of a polymer contains 25.9 g of the polymer dissolved in sufficient benzene to give 2.50×10^2 mL of solution. The osmotic pressure of this solution is 0.128 atm at 30°C. What is the molar mass of this sample of polymer?

Answer: 2.01×10^4 g/mol

11.7   Desalination of Sea Water

▶More than 97% of Earth's water is contained in the oceans, which are salty because of the dissolution of minerals described in the chapter introduction. Unfortunately, there are few uses to which people can put this abundant resource. Salt water cannot serve household purposes or be used for irrigation. Plants can be killed by salty irrigation water, because water from cell fluids flows toward the more highly concentrated external fluid, leaving the cells to collapse. Water used for industrial cooling must also be relatively salt-free, because water containing salt is quite corrosive. Because the ability of freshwater sources to supply all our demands for pure water is becoming strained, much interest has focused on the purification of sea water via *desalination methods* that are largely based on colligative properties of solutions or on the electrolytic nature of the dissolved salts.

Propose one or more methods for recovering pure water from sea water. What chemical principles are involved in your method(s)?

At least 70 elements, in ionic or molecular forms, are found in sea water. Table 11.10 shows the major dissolved constituents. Some of these dissolved materials are recovered for their own sake from sea water. For example, sodium chloride is extracted by evaporation for use as table salt and for the preparation of sodium metal and chlorine gas. Magnesium metal is recovered from sea water to the extent of over 0.5 million tons per year in the United States. The process involves reacting lime, which can be made from limestone or sea shells, with sea water to produce insoluble magnesium hydroxide.

$$CaCO_3(s) \xrightarrow{\Delta} CaO(s) + CO_2(g)$$
$$CaO(s) + Mg^{2+}(aq) + H_2O(l) \longrightarrow Mg(OH)_2(s) + Ca^{2+}(aq)$$

Magnesium hydroxide is an amorphous white powder that is nearly insoluble in water. It is used in antacid and laxative preparations. As a suspension in water, it is known as milk of magnesia.

The magnesium hydroxide is separated out and purified by conversion to $MgCl_2$. This salt is melted, and electrical current is passed through the molten salt to produce magnesium metal and chlorine gas.

$$MgCl_2(l) \xrightarrow{electricity} Mg(s) + Cl_2(g)$$

Other substances are also removed from sea water, one of the more important being the water itself.

Table 11.10 Major Dissolved Constituents of Sea Water

Constituent	Concentration (g/kg H_2O)
Cl^-	19.35
Na^+	10.77
SO_4^{2-}	2.71
Mg^{2+}	1.29
Ca^{2+}	0.41
K^+	0.40
HCO_3^-	0.145
Br^-	0.067
H_3BO_3	0.027
Sr^{2+}	0.008
F^-	0.001

► ► ► **WORLDS OF CHEMISTRY**

The Low-Salt Diet: A Prescription for Good Health

Intravenous medication and nourishment administered to hospital patients must have the same concentration of dissolved particles as body fluids.

Have you ever noticed that your hands feel swollen after you have eaten a lot of potato chips or other salty foods? The swelling is extra water, which your body has "retained" to keep the salt concentration in your body fluids at the normal level. How is this concentration controlled? By osmosis. But why is the concentration important?

The fluid inside body cells contains primarily potassium, phosphate, carbonate, and sulfate ions, along with dissolved protein, whereas the fluid outside the cells contains primarily sodium and chloride ions. If the extra sodium chloride from the salty food were not diluted, the fluid outside the cells would have a greater concentration of dissolved particles than the fluid inside. Water would flow across the semipermeable cell membrane toward the more highly concentrated solution, and the cell would begin to collapse.

Human body fluids normally contain about 0.88% by mass of salts, corresponding approximately to 0.15 M NaCl. Because the ion concentration is approximately 0.15 M Na$^+$ plus 0.15 M Cl$^-$, the concentration of particles in fluids injected intravenously must be 0.30 M to match the ion concentration in blood cells. One such solution is D5W, which is used to maintain fluid levels in patients being transported in ambulances. This solution consists of 0.30 M glucose. Glucose is a nonelectrolyte, so D5W has the same concentration of dissolved particles as 0.15 M NaCl. Solutions with the same concentration of dissolved particles as

body fluids are called *isotonic solutions*. Solutions more concentrated than isotonic solutions are called *hypertonic solutions*, and less concentrated solutions are called *hypotonic solutions*. Figure 11.21 shows blood cells in all three kinds of solutions. A normal cell in a hypotonic solution swells because of the effects of osmosis, and it may even burst (called hemolysis). As we have already suggested, a cell in a hypertonic solution shrinks (called crenation).

You can see why intravenous medication and nourishment administered to hospital patients must have the same concentration of dissolved particles as body fluids and why foods with high salt content can cause health problems. Similarly, care must be taken in the storage of body organs for transplants. To remain viable, they must be immersed in an isotonic solution.

Questions for Discussion

1. The concentration of ions in cellular fluids is nearly the same as the concentration of ions in sea water. Does this coincidence have any significance?
2. A traditional method of preserving meat from the harmful effects of bacteria involves packing the meat in salt. Why does this keep the meat from spoiling?

(continued)

Figure 11.21 Red blood cells change their shape and size, depending on the ionic environment in which they are placed. *Left:* Blood cells in an isotonic solution. *Center:* Blood cells in a hypotonic solution. *Right:* Blood cells in a hypertonic solution.

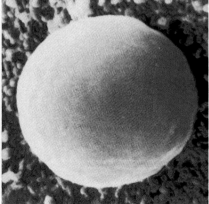
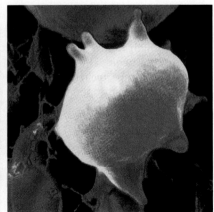

3. Explain why shipwreck survivors can die from lack of water, in spite of being surrounded by vast quantities of water.

4. Nicotine patches are used to help smokers overcome their addiction. Other types of transdermal patches are used to slowly deliver measured quantities of medications, such as scopolamine, which is used to combat motion sickness. How do the drugs get into the body? Cite some possible advantages and disadvantages of this method of drug delivery.

Distillation

The primary desalination process in operation around the world today is distillation. As shown in the schematic diagram in Figure 11.22, the method is fairly simple in theory. Actual distillation systems are somewhat more complex. To conserve energy, they may use heat removed from the vapor as it condenses to heat up the brine. Units such as that shown in Figure 11.22 may be connected to one another so that the vapor and brine from one unit are fed to an adjacent unit. This more efficient design interconnects evaporators, giving a series of condenser–evaporator combinations. The design may also make use of lower pressure to reduce the temperature to which the brine must be heated and thus speed up the evaporation process. Such an arrangement, shown in Figure 11.23, is called flash distillation. This method is usually carried out in a series of distillation chambers of successively lower pressure—and hence successively lower temperature requirements.

Freezing

When a solution of a salt in water freezes, the ice formed is essentially pure water, whereas the liquid that remains is more concentrated in salt than the original

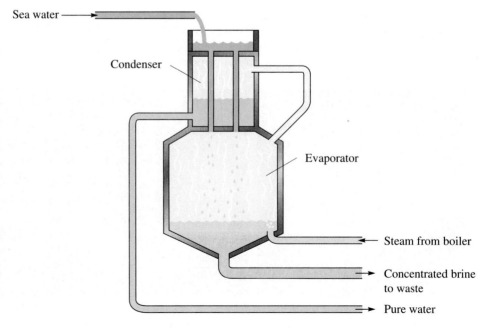

Figure 11.22 Desalination by distillation. Water is heated and evaporated, using steam as the heat source; the vapor is condensed; the pure water is removed; and the concentrated brine is removed when the salt concentration becomes so high that crystallization could begin. The vapor and brine can be recirculated for more efficient operation.

Sea water

Condenser

Evaporator

Steam from boiler

Concentrated brine to waste

Pure water

Figure 11.23 Multistage flash distillation. *Top:* A schematic drawing of the process. *Bottom:* One of the world's largest multistage distillation plants is located in Qatar. The plant produces 5 million gallons of pure water daily.

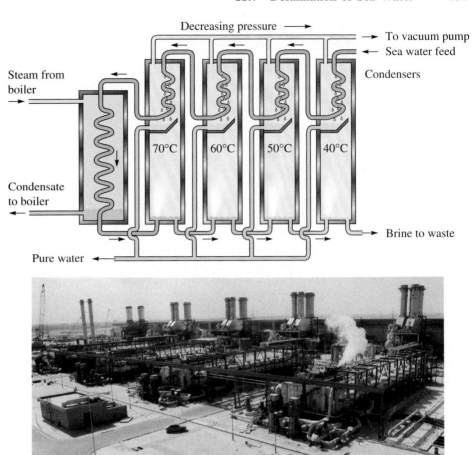

solution because some water has been removed. Fresh water can therefore be obtained by a process that involves partially freezing sea water, separating and washing the ice, and then melting the ice. To freeze the sea water, this process (called *flash evaporation*) evaporates it in a vacuum chamber, as shown in Figure 11.24. Evaporation of water uses heat, which is removed from the sea water. The

Figure 11.24 Pure water can be removed from sea water by freezing. Here, heat is removed by flash evaporation in a vacuum chamber.

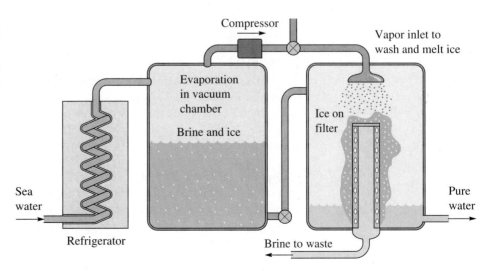

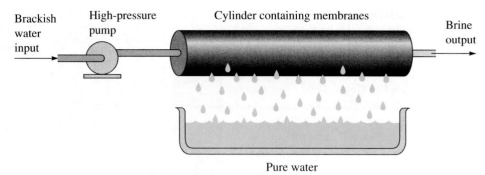

Figure 11.25 Brackish water can be purified by reverse osmosis.

Figure 11.25 Brackish water can be purified by reverse osmosis.

sea water cools enough to freeze partially, creating an ice–brine mixture. This mixture is pumped through a filter, and the ice crystals are washed with fresh water to remove any remaining brine. The water vapor formed by evaporation of sea water in the flash evaporator is pressurized and condensed onto the washed ice. The heat released by condensation helps to melt the ice.

Reverse Osmosis

Desalination can also be carried out by **reverse osmosis** (Figure 11.25). In this process, sea water is forced through a tube containing a semipermeable membrane. The pressure exerted on the sea water, about 25 atm, is greater than the osmotic pressure of sea water. As a result, water flows from the side of the membrane that is more concentrated in dissolved salts to the side with a lower concentration (the reverse of osmosis). At present this method is used only to purify *brackish water*, salty water from streams or marshes that has only about one-fifth of the salt content of sea water.

11.8   Colloids

This chapter has been concerned with the behavior of solutions. But not every mixture is a solution. When one substance is dispersed in another, the resulting mixture may show the properties of a solution, a colloid, or a suspension. A detailed comparison of the properties of these three types of systems appears in Table 11.11.

In a solution, a substance dissolves in a solvent in the form of individual ions or molecules 0.2 nm–1 nm in size. A solution is stable—that is, it does not spontaneously separate into its components—and is transparent. As we have seen, solutions exhibit colligative properties that depend only on the concentration of solute particles.

When the particles dispersed through a substance are greater than about 200 nm in diameter, a **suspension** results. Because of their size, the particles settle out of the solvent under the force of gravity; they may also be filtered out. Suspensions are neither homogeneous nor stable. Suspended particles are visible with the naked eye or with a light microscope. Suspensions are usually opaque (muddy water is an example) and exhibit no colligative properties.

Table 11.11 Comparison of Solutions, Colloids, and Suspensions

	Solution	Colloid	Suspension
Size of particles	< 1 nm	2 nm–200 nm	> 200 nm
Effect of gravity	None	None	Settle out
Effect of centrifuge	None	Settle out	Settle out
Effect of ordinary filters	Pass through	Pass through	Retained
Effect of ultrafilters	Pass through	Retained	Retained
Colligative properties	Large	Small	None
Visibility with light microscope	Invisible	Barely visible	Visible
Visibility with electron microscope	Invisible	Visible	Visible
Light-scattering properties	None	Particles scatter Tyndall beam	Particles scatter all light
Brownian movement	Not visible	Visible	None
Electrical charge	Maybe	Many	Negligible

Why do colloids not exhibit colligative properties?

If the dispersed particles are large enough to be seen with an electron microscope (2 nm) but too small to be seen with a light microscope (200 nm), the system is a **colloid.** A colloid, like a solution, is a stable system, but it is nonhomogeneous and is subject to **coagulation,** or aggregation into larger particles. A colloid can pass through filter paper but cannot diffuse through a semipermeable membrane. Colloidal particles, unlike dissolved particles, do not significantly change the colligative properties of a liquid. Some colloids are opaque, others transparent or translucent. In either case, unlike solutions, they scatter light.

This light-scattering effect, called the **Tyndall effect** (Figure 11.26), is the simplest way to distinguish between solutions and colloids. When a beam of light is passed through a solution, the path of the beam is not visible because the particles are not large enough to scatter the beam; thus solutions are transparent. With a colloidal system, however, the larger particles scatter the light beam, making it clearly visible when viewed from the side. We observe this effect when we see

Figure 11.26 The Tyndall effect is observed for a colloid *(Right),* but not for a solution *(Left).*

Figure 11.27 An opal is an example of a colloidal solid.

Arsenic(III) sulfide, As$_2$S$_3$, is a yellow or orange powder in the solid phase. It is used in the manufacture of infrared-transmitting glass, linoleum, oilcloth, and photoconductors. It is also used as a pigment and in removing hair from hides.

dust particles floating in a beam of sunlight. The effect is also responsible for the blue color of the sky.

Colloidal systems are very important to life processes, geological processes, and industrial processes. Body fluids such as the blood, lymphatic fluid, and bile are colloidal systems, as are muddy rivers containing clay and silt particles. Some gem stones, such as opals (Figure 11.27), are colloidal systems. Fog, smoke, and smog are also colloidal systems, so a knowledge of colloidal properties is important in the fight to reduce air pollution in large cities.

Colloidal systems are formed when dispersed particles of the proper size are insoluble in the dispersing medium. The systems are stabilized in two different ways. Some colloids, called *lyophilic* (solvent-loving) or *hydrophilic* (water-loving) are stabilized by the strong attraction between dispersed particles and the solvent. These are typically large covalent substances, such as starch and gelatin, dispersed in water. Other colloids, called *lyophobic* (solvent-hating) or *hydrophobic* (water-hating) arc stabilized by the formation of charges on the surface of the suspended particles, as shown in Figure 11.28. The particles attain a net charge by absorbing excess cations or anions from the solution in which they are formed, and the charged particles repel one another, which prevents coagulation.

It is possible to reverse the charge effect in lyophobic colloids and thus bring about coagulation by adding to the system salts containing oppositely charged ions. The newly added ions are strongly attracted to the original, absorbed ions, so the particles get together and form larger particles. This is what causes suspended clay and silt particles in river water to coagulate and form river deltas as the water flows into the ocean. The salt in ocean water neutralizes the colloidal charges, and the particles settle out.

Different ions have different neutralizing effects. A negatively charged colloid, such as As$_2$S$_3$ in water, is coagulated more effectively by CaCl$_2$ than by NaCl, because the higher charge on the calcium ions neutralizes the surface charge more effectively. Conversely, a positively charged colloid, such as Fe(OH)$_3$ in water, is coagulated better by Na$_3$PO$_4$ than by Na$_2$SO$_4$, which in turn works better than NaCl.

Figure 11.28 Lyophobic colloids are stabilized by the formation of surface charges. These charges prevent particles from coming together to form larger particles, which would settle out.

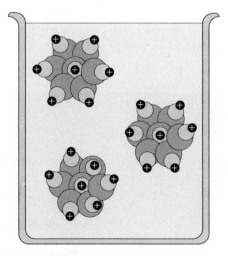

Iron(III) oxide solution

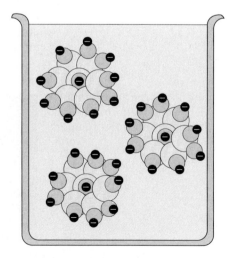

Arsenic(III) sulfide solution

We can classify colloidal systems according to the physical state of the dispersed particles and the dispersing medium, as outlined in Table 11.12. **Sols,** solids dispersed in liquids, may be formed by dispersion (breaking down larger particles) or condensation (building up from smaller particles, usually by a chemical reaction). For example, aqueous sodium thiosulfate, $Na_2S_2O_3$, mixed with sulfuric acid generates a sol containing colloidal sulfur particles. Other examples include the formation of colloidal arsenic(III) sulfide by reaction of arsenic(III) oxide with hydrosulfuric acid and the hydrolysis of iron(III) chloride:

$$As_2O_3(s) + 3H_2S(aq) \longrightarrow As_2S_3(s) + 3H_2O(l)$$
$$FeCl_3(aq) + 3H_2O(l) \longrightarrow Fe(OH)_3(s) + 3HCl(aq)$$

A **gel** is a colloidal system in which the dispersed phase is a semirigid mass enclosing all the dispersing medium. In JELLO®, the gelatin first dissolves in water and then sets into a solid mass that encloses the water. The camping fuel known as canned heat or STERNO® is a mixture of calcium acetate solution and ethyl alcohol that sets into a flammable gel.

A **smoke** is a colloidal system consisting of solid particles dispersed in a gaseous medium. Such particles, like some colloidal particles in liquids, may have an electrical charge. The Cottrell precipitator shown in Figure 11.29 uses electricity to remove dust and smoke from air, principally in smokestacks. An electrical discharge neutralizes the charge on the particles, so they can coagulate and precipitate.

An **emulsion** is a colloidal dispersion in which one liquid is dispersed as small droplets throughout another liquid with which it is immiscible. Because the

Table 11.12 Types of Colloidal Systems

Colloidal System	Dispersed Particles	Dispersing Medium	Examples
Solid sol	Solid	Solid	Smoky quartz, ruby glass, opal
Sol (fluid)	Solid	Liquid	Starch in water, milk, many paints
Gel (nonfluid)	Solid	Liquid	Silica gel, gelatin, jelly
Smoke	Solid	Gas	NH_4Cl in air, smoke, dust in air
Solid emulsion	Liquid	Solid	Water in butter
Emulsion	Liquid	Liquid	Butterfat in milk, salad dressing, mayonnaise, cold creams
Fog	Liquid	Gas	Water droplets in air, aerosols
Solid foam	Gas	Solid	Ivory soap, angel food cake, marshmallows
Foam	Gas	Liquid	Soap suds, whipped cream, shaving cream
—	Gas	Gas	Form solutions only

Figure 11.29 Cottrell precipitators remove colloidal particles from air, principally in smokestacks.

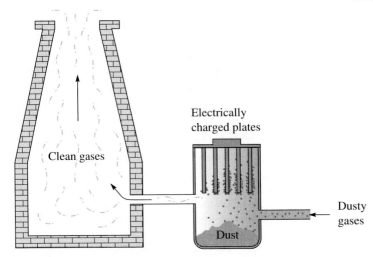

Clean gases

Electrically charged plates

Dusty gases

Dust

two liquids have no attraction for one another, emulsions typically are relatively unstable, and the two liquids separate into layers after standing for a long time. The common salad dressing made from vegetable oil and vinegar is an example of such an emulsion. To stabilize an emulsion, we can add an **emulsifying agent.** This substance is generally made up of particles with polar ends and nonpolar ends. One end is attracted to the surface of the dispersed droplets, giving the droplets a coating that is soluble in the dispersing medium (Figure 11.30). Egg white can be used as an emulsifying agent in salad dressing, for example.

A **foam** is a colloidal dispersion of a gas in a liquid. Some common examples are soap suds, whipped cream, and shaving cream. Expanded styrofoam is an example of a solid foam, which is formed when the liquid phase subsequently solidifies by *polymerization*—the formation of large molecules by combinations of small molecules.

Figure 11.30 An emulsified drop of oil in water.

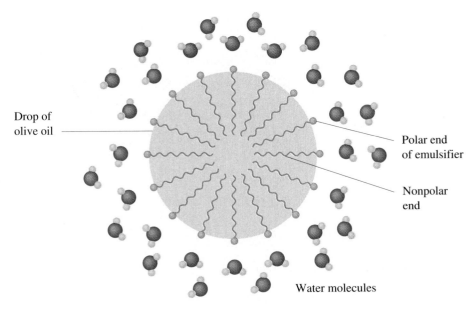

Drop of olive oil

Polar end of emulsifier

Nonpolar end

Water molecules

Summary

Solutions can be formed from matter in any combination of physical states. The units percent by mass, percent by volume, mole fraction, molality, molarity, and normality are used to express the concentration of solutions. The factors that affect solubility can be understood in terms of a model of the solution process that examines the forces that must be overcome (intermolecular forces and lattice energies) and those that are established when solute and solvent combine (solvation energies) to form a solution. Entropy is another important factor that provides a driving force for the solution process. Polar liquids generally dissolve other polar liquids and ionic solids. Nonpolar liquids dissolve other nonpolar liquids and covalent solids. Most solution processes of solids in water are endothermic, and the solubility increases with temperature. Gases become less soluble in liquids with increasing temperature, but increasing pressure makes them more soluble.

Electrolytes dissociate into ions when dissolved in solution; nonelectrolytes do not. The degree of dissociation of a substance in solution can be represented by the van't Hoff factor, the average number of particles produced by one solute formula unit when it dissolves.

Chemical equations for aqueous reactions can be written as molecular equations, ionic equations, or net ionic equations. Net ionic equations, which omit spectator ions, are especially useful for precipitation and acid–base neutralization reactions.

Water contains many impurities, some of which may make it unfit to drink. Municipal water supplies must be treated to remove suspended particles, some metals, dissolved organic substances, and bacteria. Hard water contains calcium and magnesium ions. Water may be softened by boiling, chemical precipitation or sequestering, or ion exchange procedures.

Colligative properties of solutions depend only on the concentration of solute particles, not on their identity. These properties include vapor pressure, boiling point, freezing point, and osmotic pressure.

Colloids are stable dispersions of one substance in another. The particles do not settle out and cannot be filtered out with normal filters, and they scatter light. Colloidal systems can be formed between substances in any combination of physical states, except between two gases, which can form only solutions.

Key Terms

boiling-point constant
(K_b) (11.6)
coagulation (11.8)
colligative
   properties (11.6)
colloid (11.8)
concentration (11.1)
electrolyte (11.3)
emulsifying agent (11.8)
emulsion (11.8)
entropy (11.2)
foam (11.8)

freezing-point constant
(K_f) (11.6)
gel (11.8)
hard water (11.5)
heat of solution (11.2)
Henry's law (11.2)
ionic equation (11.4)
miscible liquids (11.1)
molality (11.1)
molecular
   equation (11.4)
mole fraction (11.1)

net ionic equation (11.4)
nonelectrolyte (11.3)
normality (11.1)
osmosis (11.6)
osmotic pressure (11.6)
percent by mass (11.1)
percent by
   volume (11.1)
Raoult's law (11.6)
reverse osmosis (11.7)
saturated solution (11.1)
semipermeable

membrane (11.6)
smoke (11.8)
sol (11.8)
solubility (11.1)
supersaturated
   solution (11.1)
suspension (11.8)
Tyndall effect (11.8)
unsaturated
   solution (11.1)
van't Hoff factor (11.3)

Exercises

Units of Concentration

11.1   What is a solution?

11.2   How do you distinguish between the solute and the solvent in a solution?

11.3   What is a saturated solution?

11.4   What feature do all concentration units have in common?

11.5   A 3.830 M sucrose ($C_{12}H_{22}O_{11}$) solution has a density of 1.473 g/mL. Calculate the (a) percent by mass, (b) mole fraction, and (c) molality of this solution.

11.6   A concentrated solution of formic acid (HCOOH) contains 36.62 g HCOOH dissolved in 73.63 g H_2O. This solution has a density of 1.081 g/mL. Calculate the (a) percent by mass, (b) mole fraction, (c) molality, and (d) molarity of HCOOH in the solution.

11.7 If a solution has a mole fraction of 0.50 in ethanol, C_2H_5OH, and 0.50 in water, what is the molality of C_2H_5OH in the solution?

11.8 What is the molality of a 0.700 *M* sucrose ($C_{12}H_{22}O_{11}$) solution if the density of the solution is 1.06 g/mL?

11.9 To what volume of water must 5.0 g NaOH be added to make a 0.100 *M* solution?

11.10 A solution of sulfuric acid contains 24.0% by mass of H_2SO_4 and has a density of 1.17 g/mL. What is the molarity of H_2SO_4 in this solution?

11.11 What is the percent by mass of H_3PO_4 in a 9.0 *M* solution with a density of 1.40 g/mL?

11.12 What is the molarity of a concentrated hydrobromic acid solution if the solution is 48% HBr by mass and has a density of 1.50 g/mL?

11.13 Hydrobromic acid, used primarily for chemical analyses, is sold as a concentrated solution that contains 48% HBr by mass and has a density of 1.50 g/mL. What is the molality of concentrated hydrobromic acid?

11.14 Calculate the molarity of each of the following solutions.
a. 12.4 g KCl dissolved in water to give a volume of 900.0 mL
b. 14.2 g Na_2SO_4 dissolved in water to give a volume of 400.0 mL
c. 45.4 g Li_3PO_4 dissolved in water to give a volume of 300.0 mL

11.15 What volume of each of the following solutions would contain 0.25 mol of the solute?
a. 3.0 *M* HCl   c. 1.5 *M* H_2SO_4
b. 0.25 *M* $Al(OH)_3$ d. 0.75 *M* NaCl

11.16 Calculate the mass of solute in each of the following solutions.
a. 250 mL 1.5 *M* KOH
b. 150 mL 3.0 *M* H_2SO_4
c. 500 mL 0.50 *M* $Ca(OH)_2$

11.17 A concentrated solution of NaBr is prepared from 40.0 g NaBr dissolved in 60.0 g H_2O. If the density of this solution is 1.416 g/mL, what is the molarity of NaBr in the solution?

Solubilities of Substances

11.18 Distinguish between the terms *soluble* and *insoluble*.

11.19 List some substances that are soluble in water and some that are insoluble.

11.20 Describe a model for the energy changes involved in the dissolution process.

11.21 Explain why some substances are soluble and others insoluble in a given solvent.

11.22 Describe the effect of pressure on the solubility of gases in liquids.

11.23 How do you prepare a supersaturated solution?

11.24 Which member of each pair would you expect to be more soluble in benzene, C_6H_6? Explain your answer.
a. $CH_3CH_2NH_2$, $CH_3CH_2CH_3$
b. $CH_3CH_2CH_2CH_2OH$, CH_3OH
c. CH_4, CH_3Cl
d. $CH_3CH_2CH_3$, $CH_3—O—CH_3$
e. CH_3CO_2H, $CH_3CH_2CH_2CO_2H$
f. H_2O, H_2S

11.25 Which member of each pair would you expect to be more soluble in water? Explain your answer.
a. $CH_3CH_2NH_2$, $CH_3CH_2CH_3$
b. $CH_3CH_2CH_2CH_2NH_2$, CH_3NH_2
c. CH_4, CH_3Cl
d. $CH_3CH_2CH_3$, $CH_3—O—CH_3$
e. CH_3CO_2H, $CH_3CH_2CH_2CO_2H$
f. $HOCH_2CH_2OH$, $HOCH_2CH_2CH_2CH_2OH$
g. $CH_3CH_2CH_2OH$, CH_3OH

11.26 Lithium sulfate is a colorless, crystalline solid used as an antidepressant. Calculate the molarity of a solution of lithium sulfate, assuming that 200.0 mL of this solution yields 11.0 g Li_2SO_4 when evaporated to dryness.

11.27 Sodium nitrate, which is mined as the mineral Chile saltpeter, is purified by recrystallization. What mass of sodium nitrate should be recovered when 450.0 mL of 0.750 *M* $NaNO_3$ solution is evaporated to dryness?

11.28 Potassium chloride, found in nature as the mineral sylvite, is used to replenish electrolytes in the body, and in photography. A solution of KCl in 100 g H_2O is saturated at 20°C. How much additional KCl must be added to saturate this solution at 80°C? The solubility of KCl is 34.0 g/100 g H_2O at 20°C and 51.1 g/100 g H_2O at 80°C.

11.29 If 400.0 g of a solution saturated with NH_4Cl at 20°C is heated to completely evaporate the water, how much solid NH_4Cl should be recovered? The solubility of NH_4Cl at 20°C is 37.2 g/100 g H_2O.

11.30 How much NH_4Cl must be dissolved in 75 g of water to give a saturated solution at 70°C if the solubility is 60.2 g/100 g H_2O?

Electrolytes

11.31 Distinguish between an electrolyte and a nonelectrolyte.

11.32 Describe an experimental method for determining whether a solid is an electrolyte or a nonelectrolyte.

11.33 What is the difference between a strong electrolyte and a weak electrolyte?

11.34 Indicate whether each of the following substances is an electrolyte or a nonelectrolyte.
a. NaOH(*aq*)   d. $C_{12}H_{22}O_{11}$(*aq*)
b. HCl(*aq*)   (sucrose solution)
c. NaCl(*aq*)   e. CH_3CH_2OH(*aq*)
                              f. H_2O

11.35 Indicate whether solutions of the following substances should be good conductors, poor conductors, or nonconductors of electricity.
   a. sugar ($C_{12}H_{22}O_{11}$)   e. KI
   b. SO_3 f. CH_3OH
   c. NaCl   g. NaOH
   d. I_2

Reactions in Aqueous Solutions

11.36 What is a precipitation reaction?

11.37 Distinguish among a molecular, an ionic, and a net ionic equation.

11.38 Why is it necessary to identify a substance as an electrolyte or a nonelectrolyte when writing a net ionic equation?

11.39 What are spectator ions?

11.40 Consulting the rules for electrolytes and for solubility of salts, predict whether reactions should occur between solutions of the following salts. If so, write balanced molecular and net ionic equations for the reactions.
   a. $Sr(NO_3)_2 + H_2SO_4$
   b. $Zn(NO_3)_2 + Na_2SO_4$
   c. $CuSO_4 + BaS$
   d. $(NH_4)_2CO_3 + CrCl_3$
   e. $Al(OH)_3 + HCl$

11.41 Write and balance net ionic equations for the following reactions.
   a. $NaCl + Ag_2SO_4 \longrightarrow Na_2SO_4 + AgCl$
   b. $BaCO_3 + H_2SO_4 \longrightarrow BaSO_4 + H_2O + CO_2$
   c. $BaCl_2 + Ag_2SO_4 \longrightarrow BaSO_4 + AgCl$
   d. $Cr_2(SO_4)_3 + KOH \longrightarrow Cr(OH)_3 + K_2SO_4$
   e. $K_2CrO_4 + PbCl_2 \longrightarrow KCl + PbCrO_4$

Water Chemistry

11.42 What is hard water?

11.43 Distinguish between temporary and permanent hardness.

11.44 Hard water containing 120 ppm of calcium ions is passed through an ion exchanger. What is the concentration of sodium ions in the resulting solution?

11.45 Indicate whether water containing each of the following substances is permanently hard or temporarily hard.
   a. $CaSO_4$ d. $Ca(HCO_3)_2$
   b. $MgCl_2$ e. $Mg(HCO_3)_2$
   c. $Fe(NO_3)_3$

11.46 Write equations for removal of hardness by the following processes.
   a. ion exchange   c. addition of Na_3PO_4
   b. boiling   d. addition of CaO

Colligative Properties

11.47 What is a colligative property?

11.48 List some colligative properties.

11.49 Describe the connection between vapor pressure and boiling point.

11.50 Explain why the boiling point of a solvent increases when a solute is dissolved in the solvent.

11.51 Explain how the lower vapor pressure of a solution affects the freezing point of the solution relative to that of the solvent.

11.52 Explain why calcium chloride is spread on icy highways.

11.53 Why must the concentration of a solution injected into the bloodstream be carefully controlled?

11.54 What is the boiling point of a 1.50 m solution of sucrose, $C_{12}H_{22}O_{11}$, in water?

11.55 The boiling point of an aqueous solution is 100.56°C. What is the freezing point of this solution?

11.56 When 2.00 g of a solid is added to 50.0 g of water, the freezing point of the solution is −0.42°C. What is the molar mass of the solid?

11.57 Methanol was one of the first substances used as an automobile antifreeze. What mass of CH_3OH must be added to 2.00 kg of water to prevent the water from freezing at −40.0°C?

11.58 A 2.35-g sample of a nonionizing solid is dissolved in 1.000 kg of water. The freezing point of the solution is −0.125°C. What is the molar mass of the solid?

11.59 What is the freezing point of a 0.500 m solution of sucrose, $C_{12}H_{22}O_{11}$, in water?

11.60 The osmotic pressure of a 0.100 M sugar solution is 0.225 atm. Calculate the osmotic pressure of a 0.300 M sugar solution.

11.61 In what mass of water must we dissolve 5.00 g of ethylene glycol ($C_2H_6O_2$) antifreeze to obtain a solution that will freeze at −8.5°C?

11.62 When 3.72 g of dichloroethane is dissolved in 25.0 g of carbon tetrachloride, the boiling point is raised from 76.5°C to 84.1°C. The value of K_b for CCl_4 is 5.03°C/mol/kg. Calculate the molar mass of dichloroethane.

11.63 What is the osmotic pressure inside a cell that contains physiological saline solution (0.30 M total ions) when the cell is placed in pure water at body temperature (37°C)?

11.64 At 35.0°C, the vapor pressure of pure ethanol is 30.0 torr, and the vapor pressure of pure chloroform is 10.4 torr. What vapor pressure is predicted by Raoult's law for each component in a mixture of 1.00 mol of ethanol with 4.00 mol of chloroform?

11.65 A solution of 5.00 g of polyisobutylene in 1.00 L of benzene has an osmotic pressure of 5.00×10^{-5} atm at 25°C. What is the molar mass of this polymer?

11.66 A solution of 60.0 g of a protein in 1.00 L of aqueous solution has an osmotic pressure of 2.20×10^{-4} atm at 25°C. What is the molar mass of this protein?

Colloids

11.67 Distinguish among a solution, a suspension, and a colloid.

11.68 List some properties that could be used to identify a colloid.

11.69 Describe the various types of colloids.

11.70 Indicate whether each of the following substances is a solution, a colloid, or a suspension.
   a. sea water   f. fog
   b. air   g. muddy water
   c. milk   h. grape jelly
   d. a gold ring   i. waffle syrup
   e. smoke

Additional Exercises

11.71 A concentrated solution of nitric acid (HNO_3) is made from 36.14 g HNO_3 dissolved in 47.06 g H_2O. This solution has a density of 1.275 g/mL. Calculate the percent by mass, mole fraction, molality, and molarity of HNO_3 in the solution.

11.72 A skyscraper in Pittsburgh is supported by hollow columns filled with water to prevent buckling in case of fire. Potassium carbonate was added to the water to keep it from freezing during cold weather. If the solution is 38% K_2CO_3 by mass, what is the freezing point of this solution in degrees Celsius and degrees Fahrenheit?

11.73 When pear and apple slices are canned, they may retain their shape or they may shrink. Some other fruits, such as plums, swell and may even rupture in the canning process. Propose an explanation for these observations.

11.74 A simple application of the basic principles of colligative properties can be used to entertain friends. Suppose you are sitting at the dinner table. The table has been cleared, but salt, pepper, sugar, and glasses containing ice remain. Ask a friend to remove a piece of ice from a glass using only what is on the table and a piece of string. No knife, fork, spoon, or fingers are allowed to touch the ice, and no loop is allowed in the string. How would you do this?

11.75 Consulting the rules for electrolytes and for solubility of ionic salts as needed, write balanced molecular and net ionic equations for the following reactions in aqueous solution.
   a. copper + silver perchlorate → cupric perchlorate + silver
   b. manganese dioxide + hydrochloric acid → manganous chloride + chlorine + water
   c. calcium hydroxide + ammonium chloride → calcium chloride + water + ammonia
   d. barium chloride + sulfuric acid → barium sulfate + hydrochloric acid

11.76 How many grams of ethylene glycol ($C_2H_6O_2$) antifreeze must be added to 2.00 L of water ($d = 1.00$ g/mL) to lower the freezing point of the water to $-10.0°C$?

11.77 From the form of the solubility curve for KCl (Figure 11.7), decide whether the dissolution of KCl in water

is an exothermic or an endothermic process. What does your answer indicate about the relative magnitude of the lattice energy and the solvation energy for KCl?

11.78 A solution is made from 0.50 g of glucose, $C_6H_{12}O_6$, dissolved in an unknown amount of water. If the density of the solution is 1.0 g/mL and the osmotic pressure of the solution is 720 torr at 25°C, how much water was used to make the solution?

11.79 Nitrogen gas has a solubility of 6.9×10^{-4} M in water at 20°C and 1 atm pressure. What is the solubility of N_2 at 20°C and 230 torr?

11.80 A concentrated solution of NaOH is prepared from 30.0 g NaOH dissolved in 70.0 g H_2O. If the density of this solution is 1.330 g/mL, what is the molarity of NaOH in the solution?

11.81 How much Li_2SO_4 should crystallize out of solution when 200 g of H_2O is saturated with Li_2SO_4 at 0°C and then heated to 80°C? The solubility of Li_2SO_4 is 35.3 g/100 g H_2O at 0°C and 30.7 g/100 g H_2O at 80°C.

11.82 The sap of the maple sugar tree is a very dilute aqueous solution of sucrose ($C_{12}H_{22}O_{11}$). Maple syrup is made by boiling the sap until the boiling point has risen by 4°C. Calculate the mass of sucrose per liter of the final solution.

11.83 How should the concentration of oxygen in the water of an unfrozen lake during the winter compare with that during the summer? Plot the solubility of O_2 as a function of temperature (see Table 11.5), and compare the solubilities at typical winter and summer temperatures.

11.84 An 8.090 M sulfuric acid solution has a density of 1.445 g/mL. Calculate (a) the percent by mass, (b) the mole fraction of sulfuric acid, and (c) the molality of this solution.

11.85 At 20.0°C, the vapor pressure of pure ethanol is 4.50 torr, and the vapor pressure of pure methanol is 9.00 torr. What vapor pressure does Raoult's law predict for each component in a mixture of 1.00 mol of ethanol with 1.43 mol of chloroform?

11.86 A solution of 2.95 g of polyvinyl chloride in 1.00 L of benzene has an osmotic pressure of 7.50×10^{-5} atm at 25°C. What is the molar mass of this polymer?

11.87 Calculate the molar concentration of chloride ion in the following solutions.

   a. 0.450 g LiCl dissolved in 1.00 L H_2O
   b. 3.85 g NaCl dissolved in 500.0 mL H_2O
   c. 4.55 g $BaCl_2$ dissolved in 1.00 L H_2O
   d. 1.85 g $MgCl_2$ dissolved in 160.0 mL H_2O

11.88 Plot the solubility of O_2 in H_2O as a function of pressure (see Table 11.6). From this graph, calculate the value of the Henry's law constant (k).

11.89 When fresh strawberries are coated with sugar, they shrivel and a pool of juice appears. Propose an explanation.

*11.90 The temperatures at various times for distilled water and for a solution of 5.80 g of urea in 50.0 g of water are given below. Construct cooling curves to determine the freezing point of each liquid, and use these data to calculate the molar mass of urea.

Distilled Water		**Urea Solution**	
Time(s)	Temperature (°C)	Time(s)	Temperature (°C)
0	3.15	0	1.20
15	1.62	15	−0.80
30	0.20	30	−2.78
45	0.02	45	−3.60
60	0.02	60	−3.70
75	0.02	75	−3.80

*11.91 A phosphorus iodide compound contains 10.86% phosphorus and 89.14% iodine. If 0.385 g of the substance is dissolved in 5.025 g of the solvent carbon disulfide, CS_2, the boiling point increases by 0.322°C. The boiling point constant for CS_2 is 2.34°C/m. What is the molecular formula of the phosphorus iodide?

*11.92 Aluminum bronze is an alloy containing Al and Cu and is considered to be a solid solution of Al in Cu. A piece of this alloy is reacted with excess sodium hydroxide, which removes the aluminum. The hydrogen gas produced occupies 1.416 L at 27.0°C and 735.0 torr. The copper remaining is filtered off and dried and is found to weigh 9.001 g. The alloy has a density of 7.60 g/mL. Calculate (a) the mole fraction, (b) the molality, and (c) the molarity of Al in this alloy.

*11.93 Hydrogen peroxide solutions decompose to form water and oxygen gas:

$$2H_2O_2(aq) \longrightarrow 2H_2O(l) + O_2(g)$$

What volume of $O_2(g)$, measured at STP, is produced by the decomposition of 245 mL of a 0.136 M H_2O_2 solution?

▶ ▶ ▶ Chemistry in Practice

11.94 An isotonic solution is needed for an experiment with some blood cells. A bottle containing a white powder is found in the chemical stockroom, but only a small part of the label is still on the bottle. The only recognizable information on the label are the letters *ose,* which suggest a sugar. An inventory indicates that the only sugars in the stockroom are glucose $(C_6H_{12}O_6)$, raffinose $(C_{18}H_{32}O_{16})$, ribose $(C_5H_{10}O_5)$, and sucrose $(C_{12}H_{22}O_{11})$. A 0.1023-g sample of the powder is burned in an oxygen atmosphere. The product gases are passed through two tubes. The first tube contains anhydrous calcium chloride to absorb water vapor. The mass of this tube changes from 102.3254 g to 102.3868 g. The second tube contains inert pellets coated with sodium hydroxide to absorb carbon dioxide. The mass of this tube changes from 124.1168 g to 124.2667 g. A solution prepared by dissolving 4.275 g of the powder in 25.0 g of water has a freezing point of −1.77°C.

   a. What is the white powder?
   b. Is the powder an electrolyte or a nonelectrolyte?
   c. Write balanced chemical equations to represent the reactions described.
   d. How much of this white powder would have to be dissolved to make 0.500 L of an aqueous solution that would be an isotonic solution with an osmotic pressure of 7.68 atm at 37°C?

12

▶ ▶ ▶ ▶ ▶

Chemical Thermodynamics

Chromium(VI) oxide and ethanol react spontaneously when mixed.

CHAPTER OUTLINE

Encountering Chemistry

When we add chromium(VI) oxide to liquid ethanol, an oxidation–reduction reaction occurs, producing chromium(III) oxide and acetaldehyde:

$$2CrO_3(s) + 3CH_3CH_2OH(l) \longrightarrow Cr_2O_3(s) + 3CH_3CHO(g) + 3H_2O(g)$$

This exothermic reaction generates so much heat that the ethanol bursts into flame. No external energy is needed to cause the reaction to go to completion. Rather, the energy needed for the reaction is supplied by the internal energy of the components. This reaction is **spontaneous**—that is, it proceeds without restraint or effort.

Not all spontaneous reactions are exothermic. When the reddish pink solid $CoSO_4 \cdot 7H_2O$ is added to colorless liquid thionyl chloride, $SOCl_2$, the solid turns blue as a vigorous reaction occurs, producing gaseous products:

$$CoSO_4 \cdot 7H_2O(s) + 7SOCl_2(l) \longrightarrow CoSO_4(s) + 7SO_2(g) + 14HCl(g)$$

This reaction is shown in Figure 12.1. The mixture of solid and liquid reactants and products becomes cold, because this is an endothermic reaction. But even though the reaction has to use heat to go to completion, it too is spontaneous. This reaction is spontaneous not because of internal energy, but because of entropy, the tendency toward randomness.

Other reactions, called *nonspontaneous reactions,* cannot occur unless external energy is continuously added. The decomposition of potassium thiosulfate is nonspontaneous, but can be made to occur by heating:

$$K_2S_2O_3(s) \xrightarrow{\Delta} K_2SO_3(s) + S(s)$$

493

Figure 12.1 Cobalt(II) sulfate heptahydrate is dehydrated by thionyl chloride *(left)*, which disappears as it is converted into gaseous products *(center)*. As can be seen from the temperature decrease *(right)*, this spontaneous reaction is endothermic.

The spontaneous and nonspontaneous reactions just described represent extremes in a continuum. There are also reactions that begin spontaneously but do not completely convert the reactants into products. For example, when hydrogen gas is mixed with nitrogen gas, some ammonia gas is formed, but the reaction does not go to completion. When all apparent chemical reaction stops, both reactants and products are still present:

$$3H_2(g) + N_2(g) \rightleftharpoons 2NH_3(g)$$

This reaction is reversible, as the parallel arrows indicate. But whether it starts with a mixture of hydrogen and nitrogen or with pure ammonia, the same proportions of all the species are present when the composition of the reaction mixture becomes constant. This final composition is in a state of chemical equilibrium analogous to the physical equilibria discussed previously: vaporization (Chapter 9) and dissolution (Chapter 10). Like those physical processes, chemical equilibrium is dynamic. Even though no net change in composition occurs, the reaction is proceeding simultaneously and at the same rate in the forward and the reverse directions.

These three types of processes—spontaneous, nonspontaneous, and equilibrium reactions—can be better understood via **thermodynamics,** the study of energy changes in physical and chemical systems. Chapter 5 considered one aspect of thermodynamics—thermochemistry, the changes in enthalpy accompanying chemical changes. This chapter takes a closer look at the relationship of energy to chemical reactions, paying special attention to the factors that affect spontaneity, the laws governing these factors, and the use of these laws to determine whether a reaction should occur spontaneously or what the extent of an equilibrium reaction should be. The chapter concludes with an application of thermodynamic principles to pollution.

12.1 Spontaneous Processes and Chemical Reactions

▶ A spontaneous process, such as any of the reactions described in the chapter introduction, occurs by itself without external intervention. A hot object cools down, but it does not heat up unless energy from an external source is added. Cooling is a spontaneous process; heating above the surrounding temperature is a nonspontaneous process. Similarly, gases expand of their own accord, but they never compress themselves. Round objects roll downhill but never uphill (at least not when they start from a state of rest). Spontaneous changes are important because they can be made to do work. For example, the expanding gases produced by a spontaneous combustion reaction in an automobile engine do work on the pistons of the engine.

What makes a process spontaneous?

Two factors are important in determining the spontaneity of a process. *Enthalpy (H)* is the first of these factors. Many spontaneous processes are exothermic; that is, they occur with the evolution of heat, resulting in a decrease in the enthalpy of the system. As we noted in Chapter 5, systems tend to change from a state of higher potential energy to a state of lower potential energy—to minimize their potential energy. This is why exothermic reactions, in which the reactants have a higher enthalpy than the products, tend to be spontaneous. The decrease in enthalpy results in a more stable chemical system. Exothermic reactions are characterized by an enthalpy change that has a negative value: $\Delta H < 0$. Conversely, an endothermic reaction has $\Delta H > 0$.

However, because not all spontaneous processes evolve heat or other forms of energy, some other factor must also be important in determining the spontaneity of a process. Nature exhibits a tendency to go from a more ordered to a more random arrangement of particles—that is, from a state of order to a state of disorder. Randomness, then, as measured by *entropy (S)*, is the second driving force for spontaneity. A gas expands because its particles get farther apart and thus are more randomly distributed. A solid melts and a liquid evaporates in spite of the requirement for energy input because these changes give rise to states of higher disorder.

List, from your own experience, some examples of the tendency to go to a state of disorder.

Changes in the randomness of a system are accompanied by changes in entropy, ΔS. An increase in randomness or entropy corresponds to a positive value of ΔS, and a decrease corresponds to a negative value:

Increase in randomness $\Delta S > 0$
Decrease in randomness $\Delta S < 0$

The melting of a solid and the evaporation of a liquid (Figure 12.2), as well as the expansion of a gas, involve an increase in randomness and a positive change in entropy.

Figure 12.2 Processes that result in greater disorder are accompanied by an increase in entropy.

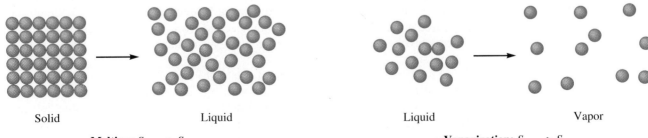

Solid   Liquid   Liquid   Vapor

Melting: $S_{liquid} > S_{solid}$ **Vaporization:** $S_{vapor} > S_{liquid}$

12.2   Enthalpy and the First Law of Thermodynamics

As we have seen, one of the factors involved in determining whether a process is spontaneous is internal energy, which is related to enthalpy. Internal energy and enthalpy were first introduced in Chapter 5, which also discussed various aspects of the *first law of thermodynamics*. We begin by reviewing these concepts.

The first law of thermodynamics is essentially the same as the law of conservation of energy, which states that energy is neither created nor destroyed in a process but is merely converted from one form to another. The forms of energy that are of most interest in chemical systems are internal energy, heat, and work.

Internal Energy, Heat, and Work

We have previously seen energy in a variety of forms. Which of these forms correspond to internal energy?

Internal energy (E) is the amount of energy stored within a system (that part of the universe in which we are currently interested). Processes in which the internal energy of the system decreases are often spontaneous processes, because matter tends toward a state of lower internal energy. Energy can be transferred between the system and its surroundings (the rest of the universe) by two mechanisms: heat (q) and work (w). In either case, energy must be conserved, so there must be a corresponding change in the internal energy of the system.

▶ Let's consider an example from the chapter introduction. The reaction between solid cobalt(II) sulfate heptahydrate and liquid thionyl chloride absorbs heat from the surroundings and generates gaseous products:

$$CoSO_4 \cdot 7H_2O(s) + 7SOCl_2(l) \longrightarrow CoSO_4(s) + 7SO_2(g) + 14HCl(g)$$

Cobalt(II) sulfate heptahydrate, $CoSO_4 \cdot 7H_2O$, exists as crystals ranging in color from pink to red. When heated, it dehydrates to blue $CoSO_4$. This anhydrous salt reverts to the hydrated red form when exposed to moisture, so it is used as a humidity indicator. The salt is also used in cobalt electroplating baths, as a drier for inks and varnishes, and as a pigment in ceramics, enamels, and glazes.

The expansion from the small volume of the solid and liquid to the much greater volume of the gases represents work done by the system—the chemical substances. The energy required to do the work involved in expanding the gases must decrease the internal energy of the system. Conversely, the energy gained as heat from the surroundings in this example must increase the internal energy of the system.

We can state these observations more quantitatively. The system absorbs heat, so the value of q is positive. The system also does work; thus the value of w is negative, because work done by the system represents energy it loses to the surroundings. The change in the internal energy of the system (ΔE) is simply the sum of these effects.

$$\Delta E = q + w$$

This equation is a mathematical statement of the first law of thermodynamics. It provides a means of determining the change in internal energy, which cannot be measured directly. Only by combining heat and work can we determine whether a process involves a decrease in internal energy—a change to a more stable, lower-energy state.

The way a process is carried out (the path) can alter the relative amounts of heat and work used or produced, but the change in internal energy resulting from the process is independent of the path. Recall that a property whose value depends only on the starting point and the ending point of a process, not on the pathway between, is called a *state function*. Internal energy, then, is a state function.

EXAMPLE 12.1

Much of the early work on heat and its relationship to work and energy was carried out through studies of the drilling of cannon barrels by Sir Benjamin Thompson, an American who moved to London in 1776. Among his many accomplishments, he claimed to have created a meringue-topped ice cream—a forerunner of baked Alaska—as a byproduct of investigations into the resistance of stiffly beaten egg whites to heat transfer.

During the reaction of 1 mol of ethylene with 1 mol of chlorine, the system loses 218.4 kJ of heat, while the surroundings do 5.0 kJ of work on the system.

$$C_2H_4(g) + Cl_2(g) \longrightarrow C_2H_4Cl_2(l)$$

What change in internal energy of the system results from this process?

Solution   Because the system loses heat, the value of the heat term is negative: $q = -218.4$ kJ. Work done by the surroundings on the system has a positive value: $w = 5.0$ kJ. The change in internal energy is the sum of the heat and work terms.

$$\Delta E = q + w = -218.4 \text{ kJ} + 5.0 \text{ kJ} = -213.4 \text{ kJ}$$

The internal energy decreases by 213.4 kJ during this process.

Practice Problem 12.1

During the decomposition of 1 mol of ammonium chloride, the system gains 176.9 kJ of heat, while the system does 5.0 kJ of work.

$$NH_4Cl(s) \longrightarrow NH_3(g) + HCl(g)$$

What change in internal energy of the system results from this process?

Answer: $+171.9$ kJ

Pressure–Volume Work

We have seen that chemical energy can be converted to other types of energy, including work. Most chemical reactions are carried out under one of two conditions: in a container open to the atmosphere or in a closed container. These conditions correspond to constant pressure or constant volume. If a reaction involves gases, it is a simple matter to obtain work from the reaction. One common way to convert chemical energy to work, then, involves pressure–volume work, or *PV* work. *PV* work is done when a substance changes its volume in opposition to a restraining pressure. Piston engines such as the one shown in Figure 12.3 involve

Figure 12.3 In a steam engine, steam expands inside a cylinder, driving a piston connected to wheels that do work. The steam engine in the photo is Puffing Billy, an Australian railroad engine.

this sort of work. Because a change in volume is required, *PV* work is important for gases, rather than liquids or solids; the volumes of both of the latter cannot change much, and their volume changes can be ignored with negligible error.

PV work can be done in chemical systems by the expansion or contraction of a gas, as in the piston engine, and by the production or consumption of a gaseous substance. When a gas expands or is produced by a chemical reaction, the system is doing work. When a gas contracts or is consumed by a chemical reaction, the surroundings do work on the system. Work done by the system, then, involves an increase in volume: $\Delta V > 0$.

When chemical reactions are used to do *PV* work, the most common experimental conditions involve volume changes at constant pressure, as, for example, when a reaction is carried out at atmospheric pressure. Consider the decomposition of hydrogen peroxide:

$$2H_2O_2(aq) \longrightarrow 2H_2O(l) + O_2(g)$$

This reaction produces a gas, which expands and can be made to lift a weight, thus doing work, as shown in Figure 12.4. Here the system is under pressure from the atmosphere and from the weight. Both pressures remain constant throughout the process, so the process is carried out at constant pressure. The positive change in volume represented by the production of O_2 gas corresponds to work done by the system and thus to a decrease in the internal energy of the system. Because work done by the system decreases its internal energy, it has a negative value, as indicated by the equation

$$w = -P\Delta V$$

The change in volume, ΔV, equals the difference between the final volume and the initial volume, $V_f - V_i$.

We now have a way to calculate the amount of work involved in a volume change of a gas at constant pressure. For chemical reactions involving only *PV* work on an ideal gas at constant temperature, we can convert this equation to a new equation that is simpler to use. This conversion uses the ideal gas law:

$$PV = nRT$$

Figure 12.4 When hydrogen peroxide decomposes, it produces oxygen gas. The reaction is accelerated by the addition of a potassium iodide catalyst. As the gas is produced, it expands and can do work. Here it lifts a weight.

When we carry out a reaction at constant pressure and temperature, common experimental conditions, the only variables in the equation are volume and number of moles. Thus a change in gas volume occurs in a reaction at constant pressure and temperature only when the number of moles of gas changes, as a result of the production or consumption of gas molecules by the reaction. From the ideal gas equation, we get the relationship

$$P\Delta V = \Delta nRT$$

and the equation for work becomes

$$w = -P\Delta V = -\Delta nRT$$

We now have a means for determining the pressure–volume work associated with a reaction at constant temperature.

EXAMPLE 12.2

Calculate the *PV* work that occurs in the following reaction of 1 mol O_2 with 2 mol H_2 at constant pressure and a temperature of 25°C.

$$2H_2(g) + O_2(g) \longrightarrow 2H_2O(g)$$

Solution   The work is given by $w = -\Delta nRT$, because the reaction occurs at constant pressure and temperature. The number of moles of gaseous products is 2 mol; that of reactants is 3 mol (2 mol H_2 + 1 mol O_2). The change in moles of gas molecules is thus

$$\Delta n = 2 \text{ mol} - 3 \text{ mol} = -1 \text{ mol}$$

The temperature of 25°C corresponds to an absolute temperature of

$$T = 25°C + 273 = 298 \text{ K}$$

In order to have work in units of energy (J), we must use the value 8.314 J/K mol for *R*, the gas constant. Now we can find a value for *w*, work.

$$w = -\Delta nRT = -(-1 \text{ mol})(8.314 \text{ J/K mol})(298 \text{ K}) = +2.48 \times 10^3 \text{ J, or } +2.48 \text{ kJ}$$

Practice Problem 12.2

Calculate the *PV* work that occurs in the following decomposition reaction of 0.500 mol NI_3 at constant pressure and a temperature of 25°C.

$$2NI_3(s) \longrightarrow N_2(g) + 3I_2(g)$$

Answer: -2.48 kJ

Enthalpy

In Chapter 5 we saw that heat changes for processes at constant pressure are called enthalpy changes (ΔH). Because we have limited ourselves to calculations of work under the common conditions of constant pressure and temperature, we can do the same for heat. Substituting the enthalpy change for heat in the equation for internal energy at constant pressure, we get a new statement of the first law of thermodynamics under these conditions:

$$\Delta E = \Delta H - P\Delta V$$

Here not only ΔE but also ΔH and ΔV are state functions; they depend only on the initial and final states, not on the path between them. Thus, at constant temperature and pressure, the contributions to internal energy from heat and work are fixed.

How can we determine the enthalpy change of a reaction?

We do not have any method for directly measuring the change in internal energy of a system, but we can use the foregoing equation, together with measurements of enthalpy change and PV work, to determine changes in internal energy. In fact, the enthalpy change alone is usually a good approximation to the change in internal energy for reactions carried out at constant pressure. This is because, for most reactions, $P\Delta V$ is a fairly small term compared with ΔE or ΔH. For example, in the decomposition of hydrogen peroxide, the enthalpy change at 25°C is -189.5 kJ/mol O_2, whereas the PV work equals -2.48 kJ/mol O_2. Because decreases in internal energy are usually associated with spontaneous processes, enthalpy changes can provide an indication of the spontaneity of chemical reactions and other processes. Chapter 5 described a number of ways to determine the enthalpy change for a reaction, including the use of Hess's law, enthalpies of formation, and calorimetric measurements.

12.3   Entropy and the Second Law of Thermodynamics

Matter tends to change spontaneously to a state of lower internal energy and higher disorder. We have just seen that, under some conditions, enthalpy changes can be used to determine whether a change involves a decrease or an increase in internal energy. Entropy changes are used to determine whether a system undergoes an increase or a decrease in randomness, which also contributes to the possible spontaneity of a process.

How can we predict whether entropy will increase or decrease?

Increases in entropy are caused by increases in randomness of arrangement and by increased freedom of motion. For example, a mixture of N_2 and O_2 has a higher entropy than the two gases separately under the same conditions, because a greater variety of arrangements is possible for the molecules. Gases have greater disorder than liquids, which have greater disorder than solids, because of the increased freedom of particles to move around. The following list gives some examples of situations in which we expect entropy to increase and cites reasons for the increases.

- The temperature being increased. Molecular motion increases with temperature in all states of matter.
- Solids melting to form liquids. Particles have greater freedom of motion and more possible arrangements.

$$NaCl(s) \longrightarrow Na^+(l) + Cl^-(l)$$
$$H_2O(s) \longrightarrow H_2O(l)$$

- Solids dissolving in liquids. Again, freedom of motion increases.

$$NaCl(s) \longrightarrow Na^+(aq) + Cl^-(aq)$$

- Gases being formed from liquids or solids. Freedom of motion increases.

$$H_2O(l) \longrightarrow H_2O(g)$$
$$I_2(s) \longrightarrow I_2(g)$$

▶ ▶ ▶   WORLDS OF CHEMISTRY

Explosives: The Winds of Change

The use of explosives in the construction industry is just one of the ways we apply chemical reactions to change our environment.

An electrical circuit is closed, causing a blasting cap to detonate. This detonation causes tubes of dynamite to explode, which in turn causes bricks and dust to fly off and a building to collapse (Figure 12.5). The use of explosives in the construction industry is just one of the ways we apply chemical reactions to change our environment. Unfortunately, humanity has also seen fit to use explosives for military purposes, in an attempt to effect political and social changes. But what are explosives and how do they work?

Gunpowder may have set off fireworks in China as early as the ninth century. Although we still use gunpowder today, other substances, such as trinitrotoluene (TNT) and nitroglycerin, are more common. Dynamite is nitroglycerin stabilized by absorption into a solid support such as sawdust.

All these explosives have certain features in common. They are solids or liquids that explode when detonated by being hit. Detonation is usually brought about with a blasting cap containing lead azide, $Pb(N_3)_2$, or mercury fulminate, $Hg(NCO)_2$. An electrical spark inside the blasting cap detonates it to produce a shock that sets off decomposition of the primary explosive. The following equation describes the decomposition of nitroglycerin:

$$4C_3H_5(NO_3)_3(l) \longrightarrow 12CO_2(g) + 10H_2O(g)$$
$$+ 6N_2(g) + O_2(g) \qquad \Delta H = -6600 \text{ kJ}$$

and the equation for decomposition of TNT is

$$2H_3CC_6H_2(NO_2)_3(s) \longrightarrow 3N_2(g) + 5H_2O(g) + 7CO(g)$$
$$+ 7C(s) \qquad \Delta H = -1894 \text{ kJ}$$

These decomposition reactions occur rapidly and produce shock waves and gaseous molecules that create tremendous pressures, as high as 8×10^4 atm. The pressures do work on the surroundings and can considerably damage them, as shown in Figure 12.5.

Nitroglycerin contains sufficient oxygen atoms to convert the entire molecule to colorless gases, so it produces a smokeless explosion. TNT, on the other hand, contains fewer oxygen atoms and gives a smoky explosion because it is not completely converted to gases. For this reason, TNT is often mixed with ammonium nitrate, NH_4NO_3, to increase the number of oxygen atoms and so make the explosion less smoky.

Figure 12.5 Dynamite is used to demolish old buildings. The direction of the explosion must be carefully controlled so that the building collapses on its foundations instead of on the surrounding areas.

The reactions are highly exothermic—an advantage for explosives, because the heat released by the reaction helps to speed up the reaction and increases the pressure of the gaseous products. The production and expansion of the gases lead to a large increase in entropy as well. The combination of large decreases in enthalpy and large increases in entropy results in highly spontaneous reactions. A good explosive, of course, must be stable under normal conditions, during which the decomposition of the explosive should be very slow. Detonation adds enough energy to increase the speed of the reaction. Once started, the reaction proceeds spontaneously and rapidly.

Questions for Discussion

1. Alfred Nobel made a fortune by manufacturing dynamite. The Nobel Peace Prize was established with proceeds from his estate. Is it appropriate for profits from explosives to be used in this way?

(continued)

2. Explosives provide just one example of materials produced by chemists that can be used for either peaceful or military purposes. Should chemists avoid producing such materials? Who is ethically responsible for the harm that may result?

3. Discuss the idea that reactions can be highly spontaneous and yet not occur under normal conditions.
4. Why must an explosive be contained (in a drill hole, for example) to be most effective?

- Gas molecules being produced in a chemical reaction. Freedom of motion and possible arrangements increase.

$$N_2O_4(g) \longrightarrow 2NO_2(g)$$

$$CaCO_3(s) \xrightarrow{\Delta} CaO(s) + CO_2(g)$$

- Elements reacting to form compounds, all in the same physical state. More arrangements become possible.

$$H_2(g) + Cl_2(g) \longrightarrow 2HCl(g)$$

In this example, unlike the previous two examples of chemical reactions, the number of particles is not greater in the products than in the reactants. But this reaction does give a more random arrangement of atoms of each element. Entropy changes for such processes are generally fairly small.

Entropy Change of the Universe

We have two tendencies, then—toward lower enthalpy and toward higher entropy—that affect the spontaneity of a process. But how do we decide which tendency predominates? We know that a positive change in entropy, corresponding to an increase in randomness, favors a spontaneous process. Yet some processes, such as the freezing of water at temperatures below 0°C, proceed spontaneously even though they appear to involve a decrease in randomness. In this case, the change to a state of lower energy apparently predominates.

When water freezes, heat is transferred from the water (the system) to the surroundings. This leads not only to a lower energy in the system but also to increased motion of the molecules in the surroundings. The entropy of the system decreases, but the randomness and entropy of the surroundings increase as a result of the added heat. We can examine both the total energy change and the entropy change of the system by considering the *total* entropy change, or the entropy change of the universe (system plus surroundings), which determines the spontaneity of a process. This fact is embodied in the **second law of thermodynamics:** The entropy of the universe tends toward a maximum.

The entropy change of the universe for a process can be represented as follows:

$$\Delta S_{universe} = \Delta S_{system} + \Delta S_{surroundings}$$

$\Delta S_{universe}$ must be positive for the process to occur spontaneously. We can see how this principle applies to physical processes by considering again the freezing of water. At temperatures below 0°C, the freezing process is spontaneous because the entropy of the surroundings increases more (from the transfer of heat from the water to the surroundings) than the entropy of the system (the water) decreases, leading to a net increase in the entropy of the universe. At temperatures above 0°C, the entropy of the system would still decrease from freezing, and the entropy of the surroundings would increase. However, the positive value of $\Delta S_{surroundings}$ would be less than the negative value of ΔS_{system}, so the entropy change of the universe would be negative. That is why water does not spontaneously freeze at temperatures above 0°C.

Chemical reactions present a similar picture. Consider, for example, the spontaneous reaction between hydrogen and oxygen to produce liquid water:

$$2H_2(g) + O_2(g) \longrightarrow 2H_2O(l)$$

In this reaction, three molecules are converted to two molecules. Further, the product molecules are in the liquid state and the reactant molecules are in the gaseous state. The decreasing number of molecules and the conversion of gas to liquid both contribute to a decrease in entropy of the system. This reaction is highly exothermic. Enough heat is lost to the surroundings to cause the surroundings to increase in entropy more than the system decreases in entropy. The entropy change of the universe, the sum of the smaller negative entropy change of the system and the larger positive entropy change of the surroundings, is positive in this reaction, so the reaction is spontaneous.

12.4 Absolute Entropy and the Third Law of Thermodynamics

You may recall from Chapter 5 that it is possible to measure only enthalpy changes of a substance; we cannot assign an absolute value for enthalpy. However, it is possible to determine the **absolute entropy** of a substance for any physical state and temperature. This value is based on a reference point of absolute zero temperature (0 K), as stated in the **third law of thermodynamics:** The absolute entropy of a perfect crystal of a pure substance is zero at 0 K. A perfect crystal at 0 K, then, represents a condition of no disorder (Figure 12.6).

The absolute entropy of a substance at a specified temperature can be calculated from the amount of heat required to raise the temperature of the substance from 0 K to that temperature, including the heat required for any phase changes the substance undergoes as the temperature rises. Consider, for example, the absolute entropy of tetraphosphorus decoxide. The entropy of P_4O_{10}, given in units of J/mol K, increases with temperature, as shown in Figure 12.7. Until the substance undergoes a phase change, the entropy increases slowly from 0 J/mol K at 0 K to 177

A temperature of absolute zero has never actually been obtained experimentally. How, then, is it possible to measure absolute entropies?

Figure 12.6 In a perfect crystal, the particles are all oriented in the same way. Such a crystal has an entropy of zero at 0 K, where all molecular motion and vibrations have stopped. In an imperfect crystal, some of the particles are oriented differently. Such a crystal has an entropy greater than zero, even at 0 K.

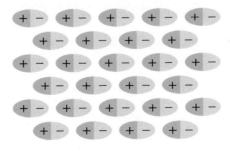

Perfect crystal

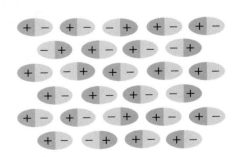

Imperfect crystal

Figure 12.7 The standard entropy of tetraphosphorus decoxide, P_4O_{10}, at different temperatures. The entropy increases gradually with temperature, except at the temperature where the solid undergoes sublimation, where the entropy jumps considerably.

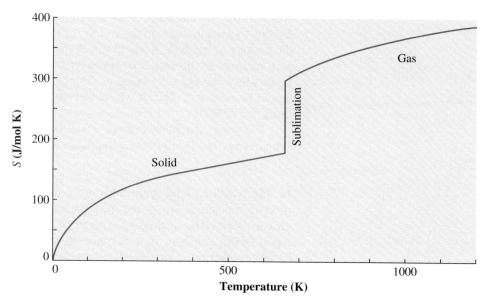

J/mol K at 632 K. At this temperature P_4O_{10} sublimes, converting the solid to the gaseous state. As is to be expected, the entropy of the gas is much greater than that of the solid, because the amount of disorder is greater. The entropy of gaseous P_4O_{10} at 632 K is 293 J/mol K. As the temperature increases further, the entropy of the gas continues to increase slowly; it reaches a value of 363 J/mol K at 1000 K.

The change in absolute entropy with temperature can be calculated from heat measurements, but the procedure is somewhat complicated, so we usually simply rely on values listed in tables. The entropy change associated with a phase change is not so complex, however. It is related to the enthalpy change and the temperature of the phase change by the following equation, which is valid when a system is at equilibrium.

$$\Delta S = \Delta H / T$$

For example, for the melting (fusion) of a solid, the entropy change is

$$\Delta S_{fus} = \Delta H_{fus} / T_{fus}$$

The enthalpy change for the sublimation of P_4O_{10} at 632 K is 73.6 kJ/mol. Substituting these values into the equation gives a value of 116 J/mol K for the entropy change.

EXAMPLE 12.3

Calculate the entropy change for the vaporization of water at 100.0°C. The enthalpy of vaporization is 40.5 kJ/mol.

Solution   The entropy change is calculated from the equation

$$\Delta S_{vap} = \Delta H_{vap} / T_{vap}$$

After converting 100.0°C to 373.2 K, we substitute values into this equation to obtain the entropy change.

$$\Delta S_{vap} = (40.5 \text{ kJ/mol})(1000 \text{ J/kJ}) / 373.2 \text{ K} = 109 \text{ J/mol K}$$

Practice Problem 12.3

Calculate the entropy change for the melting of ice at 0.00°C. The enthalpy of fusion is 5.98 kJ/mol.

Answer: 21.9 J/mol K

Standard Entropies

Extensive tables of absolute entropies are available for the elements and their compounds. It is convenient to report values of entropies under *standard-state* conditions—the pure substance for a liquid or solid, 1 atm pressure for a gas, and 1 *M* concentration for a dissolved substance. When we measure absolute entropies under standard-state conditions, we call them **standard entropy** values, designated $S°$. Selected values at 25°C are given in Table 12.1. Additional values can be found in Appendix C.

Because entropy is a state function, standard entropy values can be used to calculate the standard entropy changes for phase changes or for chemical reactions at 25°C. To carry out such calculations, we simply add the standard entropies of the products and subtract the sum of the standard entropies of the reactants, taking into account the number of moles of each substance, as indicated in the equation

$$\Delta S° = \Sigma n S°_{\text{products}} - \Sigma n S°_{\text{reactants}}$$

where n equals the appropriate stoichiometric coefficient. For example, the decomposition of potassium chlorate is represented by the equation

Potassium chlorate is a white powder that decomposes when heated, and it does so explosively if contaminated with organic material. It is used in explosives, fireworks, and matches.

$$2\text{KClO}_3(s) \xrightarrow{\Delta} 2\text{KCl}(s) + 3\text{O}_2(g)$$

The entropy change for this reaction is given by the equation

$$\Delta S° = 3S°_{\text{O}_2(g)} + 2S°_{\text{KCl}(s)} - 2S°_{\text{KClO}_3(s)}$$
$$= 3(205.0) + 2(82.7) - 2(143.0) = 494.4 \text{ J/mol K}$$

EXAMPLE 12.4

Calculate the standard entropy change at 25°C for the reaction

$$\text{CaO}(s) + \text{SO}_3(g) \longrightarrow \text{CaSO}_4(s)$$

which is used to remove sulfur oxides from smokestack gases.

Solution   The standard entropy change is given by the standard entropies of the product less those of the reactants:

$$\Delta S° = S°_{\text{CaSO}_4} - (S°_{\text{SO}_3} + S°_{\text{CaO}})$$

Values of the standard entropies can be obtained from Table 12.1 and substituted into this equation.

$$\Delta S° = 106.7 \text{ J/mol K} - (256.6 \text{ J/mol K} + 39.7 \text{ J/mol K})$$
$$= -189.6 \text{ J/mol K}$$

(continued)

Table 12.1 Standard Entropies of Some Substances in J/mol K at 25°C

Substance	Entropy ($S°$)	Substance	Entropy ($S°$)
$Al(s)$	28.3	$HCl(g)$	186.80
$Al_2O_3(s)$	50.92	$HBr(g)$	198.59
$Br_2(l)$	152.23	$HI(g)$	206.48
$Br_2(g)$	245.35	$I_2(s)$	116.14
C(graphite)	5.740	$K(s)$	63.6
C(diamond)	2.38	$KCl(s)$	82.68
$CO(g)$	197.56	$Mg(s)$	32.5
$CO_2(g)$	213.6	$MgCl_2(s)$	89.5
$CH_4(g)$	186.15	$MgO(s)$	27.
$C_2H_6(g)$	229.5	$Mg(OH)_2(s)$	63.14
$C_2H_4(g)$	219.5	$N_2(g)$	191.5
$C_2H_2(g)$	200.8	$NH_3(g)$	192.3
$C_3H_8(g)$	269.9	$N_2O(g)$	219.7
$C_6H_6(l)$	172.8	$NO(g)$	210.65
$C_6H_6(g)$	269.2	$NO_2(g)$	239.9
$CH_3OH(l)$	126.8	$Na(s)$	51.0
$CH_3OH(g)$	239.7	$NaF(s)$	58.6
$CS_2(l)$	151.3	$NaCl(s)$	72.38
$CS_2(g)$	237.7	$NaBr(s)$	83.7
$CCl_4(l)$	216.4	$NaI(s)$	91.2
$CCl_4(g)$	309.7	$Na_2O_2(s)$	95.
$C_2H_5OH(l)$	160.7	$NaOH(s)$	64.
$C_2H_5OH(g)$	282.6	$O_2(g)$	205.03
$CH_3COOH(l)$	159.8	$Pb(s)$	64.81
$CH_3COOH(g)$	282.4	$PbO_2(s)$	68.7
$Ca(s)$	41.6	$PbSO_4(s)$	147.
$CaO(s)$	39.7	$S(s)$	31.8
$Ca(OH)_2(s)$	76.1	$SO_2(g)$	248.1
$CaSO_4(s)$	106.7	$SO_3(g)$	256.6
$Cl_2(g)$	222.96	$H_2S(g)$	205.7
$Cu(s)$	33.15	$H_2SO_4(l)$	156.90
$CuO(s)$	42.63	$Si(s)$	18.8
$F_2(g)$	202.7	$SiO_2(s)$	41.84
$Fe(s)$	27.3	$SiH_4(g)$	204.5
$Fe_2O_3(s)$	87.40	$SiF_4(g)$	282.4
$H_2(g)$	130.57	$SiCl_4(l)$	239.7
$H_2O(l)$	69.91	$SiCl_4(g)$	330.6
$H_2O(g)$	188.71	$Zn(s)$	41.6
$H_2O_2(l)$	109.6	$ZnO(s)$	43.64
$HF(g)$	173.67	$ZnSO_4(s)$	120.

Practice Problem 12.4

Calculate the standard entropy change at 25°C for the reaction

$$CaO(s) + H_2O(l) \longrightarrow Ca(OH)_2(s)$$

This reaction is the conversion of lime to slaked lime, one of the reactions that occurs in the preparation of concrete from cement.

Answer: -33.5 J/mol K

To calculate entropy changes for phase changes, such as melting or vaporization, we can subtract the absolute entropy of the initial state of the substance from that of its final state. For example, the entropy change for sublimation of germanium is 25°C, shown in the chemical equation

$$Ge(s) \longrightarrow Ge(g)$$

can be obtained from the equation

Germanium is a grayish white, brittle, lustrous metalloid. The crystalline element has the diamond structure. Germanium melts at 937°C and boils at 2700°C. It is relatively stable in air, although it oxidizes slowly in air above 600°C. It is used in electronic semiconductor devices, in dental alloys, and in glass that can transmit infrared light.

$$\Delta S_{subl} = S^{\circ}_{Ge(g)} - S^{\circ}_{Ge(s)} = 167.8 - 42.4 = 125.4 \text{ J/mol K}$$

EXAMPLE 12.5

Calculate the standard entropy change at 25°C for the phase change of liquid water to gaseous water:

$$H_2O(l) \longrightarrow H_2O(g)$$

Solution   The entropy change is given by

$$\Delta S^{\circ} = S^{\circ}_{H_2O(g)} - S^{\circ}_{H_2O(l)}$$

Inserting values from Table 12.1 gives

$$\Delta S^{\circ} = 188.71 \text{ J/mol K} - 69.91 \text{ J/mol K} = 118.80 \text{ J/mol K}$$

This value is different from that calculated in Example 12.3 because of the difference in temperature.

Practice Problem 12.5

Calculate the standard entropy change at 25°C for the phase change of gaseous methanol to liquid methanol:

$$CH_3OH(g) \longrightarrow CH_3OH(l)$$

Answer: -112.9 J/mol K

12.5   Change in Gibbs Free Energy

How can we predict spontaneity by considering both the tendency toward lower energy and the tendency toward greater randomness?

This section continues our quest to develop a means of deciding whether a reaction or process is expected to be spontaneous. We saw earlier that spontaneous reactions can be used to do work on the surroundings. But how much work? This can be decided by considering the changes in both the enthalpy and the entropy.

Gibbs free energy was named after the thermodynamicist J. Willard Gibbs (1839–1903). His work led to the application of thermodynamics to chemical systems. He was the first American scientist to receive a Ph.D. degree from an American university (Yale, in 1863).

The maximum amount of energy available to do useful work on the surroundings is called the change in **Gibbs free energy** or simply the change in **free energy,** symbolized ΔG. Of the energy available in a system, some must be used to do work within the system, such as rearranging atoms to convert reactants to products or changing the degree of randomness. The remainder of the available energy can be transferred to the surroundings as heat that can be used to do work on the surroundings. The change in free energy is the enthalpy change of the system minus the energy needed to do work within the system, which is given by the entropy change of the system multiplied by the absolute temperature ($T\Delta S$). The equation for the change in free energy is thus

$$\Delta G = \Delta H - T\Delta S$$

Because a spontaneous process transfers energy from the system to the surroundings, the free energy of the system must decrease. Thus ΔG must be negative for a spontaneous process. We now have a mathematical criterion for spontaneity. This equation tells us whether a process should occur spontaneously at constant temperature and pressure.

▶ Consider again the reaction between chromium(VI) oxide and ethanol:

$$2CrO_3(s) + 3CH_3CH_2OH(l) \longrightarrow Cr_2O_3(s) + 3CH_3CHO(g) + 3H_2O(g)$$

This is an exothermic process—so exothermic that the ethanol bursts into flame—so the value of ΔH is negative. Because gaseous molecules are created from a solid and a liquid, the entropy increases in this process. Absolute temperature is always positive, so the term $-T\Delta S$ is negative. Taking the sum of two negative values results in a value of ΔG that is negative. Thus we predict correctly that this reaction is spontaneous at all temperatures.

If the free-energy change accompanying a process is negative, then the process occurs spontaneously. If the free-energy change is positive, the process does not occur spontaneously but the reverse process is spontaneous. (When we reverse the process, we change the sign of the free-energy change.) If the free-energy value does not change, no net spontaneous change can occur in either the forward or the reverse direction. This is a system in a state of equilibrium, in which the forward and reverse processes balance one another.

Spontaneity can be measured by the change in free energy, ΔG, which is related to enthalpy and entropy as well as to temperature. Let's look more closely at the relationship of temperature to spontaneity.

The following list gives the conditions for spontaneity at various temperatures.

Spontaneous processes ($\Delta G < 0$):

$\Delta H < 0$	$\Delta S > 0$	spontaneous at all temperatures
$\Delta H < 0$	$\Delta S < 0$	spontaneous at low temperatures ($T < \Delta H/\Delta S$)
$\Delta H > 0$	$\Delta S > 0$	spontaneous at high temperatures ($T > \Delta H/\Delta S$)

Nonspontaneous processes ($\Delta G > 0$):

$\Delta H > 0$	$\Delta S < 0$	not spontaneous at any temperature
		(spontaneous in the reverse direction at all temperatures)

As you can see, when the enthalpy and entropy terms both favor spontaneity, the process occurs spontaneously at any temperature; and when both oppose spontaneity, the process cannot occur at all. When the two factors work against one another, the temperature determines whether the process occurs spontaneously. Using these principles, we can decide whether a reaction should occur spontaneously at a given temperature. (This treatment assumes the values of ΔH and ΔS are independent of temperature, which is not strictly true.) The temperature limit for spontaneity is determined from the equation for free-energy change by setting the free-energy change equal to zero, inserting values of ΔH and ΔS, and solving for the temperature.

EXAMPLE 12.6

Sodium bicarbonate decomposes according to the reaction

$$2NaHCO_3(s) \longrightarrow Na_2CO_3(s) + H_2O(g) + CO_2(g)$$

For this decomposition reaction, $\Delta H = 64.5$ kJ/mol and $\Delta S = 167$ J/mol K. Is the reaction spontaneous at 355 K?

Solution   The enthalpy change and entropy change are both greater than zero, so the reaction should be spontaneous at some temperatures but not at others. To decide whether it is spontaneous at 355 K, we must calculate the value of ΔG to see whether it is negative at this temperature. We calculate the change in free energy from the equation

$$\Delta G = \Delta H - T\Delta S$$

Inserting values for each known term, we obtain

$$\Delta G = 64.5 \text{ kJ/mol} - (355 \text{ K} \times 167 \text{ J/mol K} \times 1 \text{ kJ/1000 J})$$
$$= 64.5 \text{ kJ/mol} - 59.3 \text{ kJ/mol} = 5.2 \text{ kJ/mol}$$

Because the change in free energy is positive, the reaction is not spontaneous at this temperature.

Practice Problem 12.6

The hydration of sulfur trioxide occurs by the reaction

$$SO_3(g) + H_2O(l) \longrightarrow H_2SO_4(aq)$$

Here $\Delta H = -227$ kJ/mol and $\Delta S = -309$ J/mol K. Is this reaction spontaneous at 1450 K?

Answer: No, because $\Delta G = +221$ kJ/mol.

12.6 Standard Free-Energy Changes

How can we determine the value of the free-energy change for a process?

The change in free energy gives us a useful criterion for deciding on the spontaneity of a reaction or process. A negative value of ΔG indicates the reaction is spontaneous. Thus it would be helpful to have values of ΔG for reactions.

We have seen that we can calculate the free-energy change for a reaction if we know the enthalpy change, the entropy change, and the temperature by using the equation

$$\Delta G = \Delta H - T\Delta S$$

Values of ΔH and ΔS for reactions are readily available if the reactants and products are in their standard states. Substituting standard-state values for enthalpy and entropy changes into this equation yields the **standard free-energy change, $\Delta G°$**:

$$\Delta G° = \Delta H° - T\Delta S°$$

The value of $\Delta G°$ tells us whether a reaction or phase change is spontaneous when all substances are in their standard states.

EXAMPLE 12.7

Calculate the standard free-energy change at 25.0°C for the chemical reaction

$$CaO(s) + SO_3(g) \longrightarrow CaSO_4(s)$$

Also decide whether the process is spontaneous under standard-state conditions at 25.0°C.

Solution   We can use the equation for standard free-energy changes:

$$\Delta G° = \Delta H° - T\Delta S°$$

We must, of course, use values of the standard enthalpy and entropy changes. As described in Chapter 5, the standard enthalpy change can be obtained from standard heats of formation given in Appendix C. We subtract the standard heats of formation of the reactants from the standard heats of formation of the products.

$$\begin{aligned}\Delta H° &= \Delta H°_{f,CaSO_4(s)} - \Delta H°_{f,CaO(s)} - \Delta H°_{f,SO_3(g)} \\ &= -1432.7 \text{ kJ/mol} + 635.5 \text{ kJ/mol} + 395.7 \text{ kJ/mol} \\ &= -401.5 \text{ kJ/mol}\end{aligned}$$

The standard entropy change can be obtained from standard absolute entropies in Table 12.1 or Appendix C.

$$\begin{aligned}\Delta S° &= S°_{CaSO_4(s)} - S°_{CaO(s)} - S°_{SO_3(g)} \\ &= 106.7 \text{ J/mol K} - 39.7 \text{ J/mol K} - 256.6 \text{ J/mol K} \\ &= -189.6 \text{ J/mol K}\end{aligned}$$

After converting 25.0°C to 298.2 K and -401.5 kJ/mol to -4.015×10^5 J/mol, we substitute in the values.

$$\begin{aligned}\Delta G° &= -4.015 \times 10^5 \text{ J/mol} - (298.2 \text{ K} \times -189.6 \text{ J/mol K}) \\ &= -4.015 \times 10^5 \text{ J/mol} + 5.654 \times 10^4 \text{ J/mol} = -3.450 \times 10^5 \text{ J/mol} \\ &= -345.0 \text{ kJ/mol}\end{aligned}$$

This reaction is spontaneous at 25.0°C under standard-state conditions, because the free-energy change is negative.

Practice Problem 12.7

Calculate the standard free-energy change at 25.0°C for the reaction

$$C(s,\text{graphite}) + O_2(g) \longrightarrow CO_2(g)$$

for which the value of $\Delta H°$ is -393.51 kJ/mol and the value of $\Delta S°$ is 2.83 J/mol K.

Answer: -394.35 kJ/mol

Standard Free Energy of Formation

We can calculate the standard free-energy change from the standard enthalpy change and the standard entropy change at a given temperature, as in the preceding example. Another alternative is also possible. We can derive the standard free-energy change from values of the **standard free energy of formation** (ΔG_f°), the change in free energy that occurs when 1 mol of a compound in its standard state is formed from its elements in their standard states. Values of the standard free energy of formation have been determined for a wide variety of substances. Table 12.2 gives some representative values, and a longer list appears in Appendix C. The value of the standard free energy of formation of the stable form of an element is zero.

Why can we apply Hess's law to free energy?

In Chapter 5 we used Hess's law to calculate the standard enthalpy change for a process from the standard enthalpies of formation of the substances involved in the process. Hess's law also applies to free energy, so we can calculate the standard free-energy change of a process by subtracting the sum of the free energies of formation of the reactants from the sum of the free energies of formation of the products:

$$\Delta G^\circ = \Sigma n \Delta G_f^\circ(\text{products}) - \Sigma n \Delta G_f^\circ(\text{reactants})$$

where n is the stoichiometric coefficient for each substance. For example, the hydrolysis of calcium phosphide is described by the equation

$$Ca_3P_2(s) + 6H_2O(l) \longrightarrow 3Ca(OH)_2(s) + 2PH_3(g)$$

Calcium phosphide, Ca_3P_2, is a red-brown powder that is decomposed by moist air or water, forming phosphine gas (PH_3), which is spontaneously flammable and very toxic. Calcium phosphide is used for signal fires and in rodent poisons.

The standard free-energy change for this reaction can be calculated from standard free energies of formation of the reactants and products, according to the equation

$$\Delta G^\circ = 3\Delta G_{f,Ca(OH)_2(s)}^\circ + 2\Delta G_{f,PH_3(g)}^\circ - \Delta G_{f,Ca_3P_2(s)}^\circ - 6\Delta G_{f,H_2O(l)}^\circ$$

EXAMPLE 12.8

Calculate the standard free-energy change at 25.0°C for the decomposition reaction

$$2SO_3(g) \longrightarrow 2SO_2(g) + O_2(g)$$

Solution   We use the equation for ΔG° and substitute values for the standard free energies of formation from Table 12.2.

$$\Delta G^\circ = \Sigma n \Delta G_f^\circ(\text{products}) - \Sigma n \Delta G_f^\circ(\text{reactants})$$
$$= 2\Delta G_{f,SO_2(g)}^\circ + \Delta G_{f,O_2(g)}^\circ - 2\Delta G_{f,SO_3(g)}^\circ$$
$$= (2 \text{ mol} \times -300.2 \text{ kJ/mol}) + (1 \text{ mol} \times 0.0 \text{ kJ/mol}) - (2 \text{ mol} \times -371.1 \text{ kJ/mol})$$
$$= -600.4 \text{ kJ} + 0.0 \text{ kJ} + 742.2 \text{ kJ}$$
$$= 141.8 \text{ kJ}$$

Remember that this is the free-energy change for the decomposition of 2 mol SO_3. If we wanted to know the free-energy change for the decomposition of 1 mol SO_3, we would have to divide 141.8 kJ by the number of moles:

$$\Delta G^\circ = 141.8 \text{ kJ/2mol} = 70.9 \text{ kJ/mol } SO_3$$

This reaction is not spontaneous.

(continued)

Table 12.2 Standard Free Energies of Formation in kJ/mol at 25°C

Substance	ΔG_f°	Substance	ΔG_f°
Al(s)	0.0	HCl(g)	−95.299
Al$_2$O$_3$(s)	−1582.	HBr(g)	−53.43
Br$_2$(l)	0.0	HI(g)	1.7
Br$_2$(g)	3.142	I$_2$(s)	0.0
C(graphite)	0.0	K(s)	0.0
C(diamond)	2.900	KCl(s)	−408.32
CO(g)	−137.15	Mg(s)	0.0
CO$_2$(g)	−394.36	MgCl$_2$(s)	−592.33
CH$_4$(g)	−50.75	MgO(s)	−569.57
C$_2$H$_6$(g)	−32.9	Mg(OH)$_2$(s)	−833.75
C$_2$H$_4$(g)	68.12	N$_2$(g)	0.0
C$_2$H$_2$(g)	209.2	NH$_3$(g)	−16.5
C$_3$H$_8$(g)	−23.49	N$_2$O(g)	104.2
C$_6$H$_6$(l)	124.50	NO(g)	86.57
C$_6$H$_6$(g)	129.66	NO$_2$(g)	51.30
CH$_3$OH(l)	−166.4	Na(s)	0.0
CH$_3$OH(g)	−162.0	NaF(s)	−541.0
CS$_2$(l)	65.27	NaCl(s)	−384.03
CS$_2$(g)	67.15	NaBr(s)	−347.
CCl$_4$(l)	−65.27	NaI(s)	−282.
CCl$_4$(g)	−60.63	Na$_2$O$_2$(s)	−451.
C$_2$H$_5$OH(l)	−174.9	NaOH(s)	−381.
C$_2$H$_5$OH(g)	−168.6	O$_2$(g)	0.0
CH$_3$COOH(l)	−390.0	Pb(s)	0.0
CH$_3$COOH(g)	−374.0	PbO$_2$(s)	−217.4
Ca(s)	0.0	PbSO$_4$(s)	−811.2
CaO(s)	−604.2	S(s)	0.0
Ca(OH)$_2$(s)	−896.76	SO$_2$(g)	−300.19
CaSO$_4$(s)	−1320.3	SO$_3$(g)	−371.1
Cl$_2$(g)	0.0	H$_2$S(g)	−33.6
Cu(s)	0.0	H$_2$SO$_4$(l)	−690.101
CuO(s)	−130.	Si(s)	0.0
F$_2$(g)	0.0	SiO$_2$(s)	−856.67
Fe(s)	0.0	SiH$_4$(g)	56.9
Fe$_2$O$_3$(s)	−742.2	SiF$_4$(g)	−1572.7
H$_2$(g)	0.0	SiCl$_4$(l)	−619.90
H$_2$O(l)	−237.18	SiCl$_4$(g)	−617.01
H$_2$O(g)	−228.59	Zn(s)	0.0
H$_2$O$_2$(l)	−120.4	ZnO(s)	−318.3
HF(g)	−273.	ZnSO$_4$(s)	−874.5

Practice Problem 12.8

Calculate the standard free-energy change at 25.0°C for the reaction

$$CO_2(g) + 4HCl(g) \longrightarrow CCl_4(l) + 2H_2O(l)$$

Answer: +235.9 kJ

12.7   Free-Energy Changes and Equilibrium

We have used both ΔG and $\Delta G°$ as criteria for spontaneity in chemical and physical processes. Negative values indicate that a process is spontaneous; positive values indicate the opposite. But what if the value for ΔG is zero? As we noted earlier, a process for which ΔG equals zero is in a state of equilibrium. Equilibrium is a very important nonstandard condition in which the amounts of reactants and products remain constant even though reaction is still occurring in the forward and reverse directions. This section explores the relationship between a condition of equilibrium and the thermodynamic quantities for that reaction or process.

How can amounts of reactants and products remain constant when the reactions are still occurring?

For example, consider the standard free-energy change at 25.0°C for the condensation of gaseous water:

$$H_2O(g) \longrightarrow H_2O(l)$$

From Table 12.2, we calculate $\Delta G°$ to be -8.6 kJ/mol. This result indicates that under standard-state conditions, gaseous water spontaneously condenses to liquid water at 25°C. However, this result seems contrary to experience. Normally, water vapor does not condense to the liquid at room temperature (unless the air is very humid). On the contrary, liquid water at 25°C normally evaporates. The problem is that we rarely find this process occurring under standard-state conditions in our daily lives. What we do find in nature is liquid and gaseous water coexisting in a state of equilibrium in which the liquid is in the standard state (the pure liquid) but the gas is not—it has a pressure less than 1 atm. Because we cannot use the standard free-energy change for equilibrium reactions or for other nonstandard systems, we need a method for determining ΔG from $\Delta G°$.

Calculation of Nonstandard Free-Energy Changes

To examine nonstandard conditions, we need to have a quantity that describes these conditions. Such a quantity is the **reaction quotient** (Q), which relates the pressures or concentrations of the reaction products to those of the reactants:

$$Q = \Pi P_{prod}{}^{a}/\Pi P_{react}{}^{b}$$

where the exponents a and b correspond to the coefficients in the balanced equation for the reaction, and Π indicates that terms must be multiplied together. In aqueous solution, concentrations are used instead of pressures. Pressure and concentration are not appropriate measures of the amount of a pure solid or liquid, so these substances do not appear in the reaction quotient. ▶ For example, the chapter introduction discussed the Haber process, by which ammonia is prepared from nitrogen and hydrogen:

$$N_2(g) + 3H_2(g) \rightleftharpoons 2NH_3(g)$$

The reaction quotient is based on the pressures (P) of the reactants and products, which are all gases:

$$Q = P_{NH_3}^2/P_{N_2}P_{H_2}^3$$

In the reaction quotient, each pressure (or concentration) is raised to a power equal to the stoichiometric coefficient in the balanced equation.

A somewhat different example is the following reaction used in the production of zirconium metal.

$$ZrO_2(s) + 2C(s) + 2Cl_2(g) \rightleftharpoons ZrCl_4(s) + 2CO(g)$$

Zirconium is a grayish white, lustrous metal that appears bluish black when powdered. It is used as a structural material in nuclear reactors because it does not react readily with neutrons and because it resists corrosion and heat. Hydrated zirconium oxide is used in poison ivy creams and deodorants because it can absorb many organic and inorganic compounds. Zirconium oxide is used to produce highly reflective glazes on ceramics.

Here some of the substances are not in the gaseous state. The reaction quotient is based only on the pressures of the gaseous species:

$$Q = P_{CO}^2/P_{Cl_2}^2$$

The reaction quotient can be used to find free-energy changes under nonstandard conditions. We can calculate the value of ΔG from the tabulated or calculated values of $\Delta G°$ and the reaction conditions by using the following equation:

$$\Delta G = \Delta G° + RT \ln Q$$

Here the gas constant R is equal to 8.314 J/mol K, ln is the symbol for the natural logarithm, and Q is the reaction quotient. The reaction quotient is important because it is a measure of the extent of reaction—the relative amounts of reactants and products at a given time. In the standard state, all gaseous species are present with a partial pressure of 1 atm, and Q has a value of 1; under this circumstance, ln $Q = 0$ and $\Delta G = \Delta G°$. When no products are present, Q has a value of zero, and it has an infinitely large value when no reactants are present. Under conditions between these two extremes, adjusting $\Delta G°$ by taking the extent of reaction into account enables us to calculate the actual value of ΔG.

EXAMPLE 12.9

The following reaction has a standard free-energy change of 3.4 kJ at 25°C:

$$H_2(g) + I_2(g) \longrightarrow 2HI(g)$$

If we increase the pressure of H_2 from 1.00 atm to 10.0 atm while leaving the other pressures at 1.00 atm, what is the value of ΔG?

Solution   To apply the equation relating ΔG to $\Delta G°$, we must calculate the value of the reaction quotient.

$$Q = P_{HI}^2/P_{H_2}P_{I_2}$$
$$= (1.00 \text{ atm})^2/(1.00 \text{ atm})(10.0 \text{ atm}) = 0.100 \text{ atm}$$

Now we use this value in the following equation:

$$\Delta G = \Delta G° + RT \ln Q$$
$$= 3.4 \times 10^3 \text{ J} + (8.314 \text{ J/mol K} \times 298 \text{ K} \times \ln 0.100)$$
$$= 3.4 \times 10^3 \text{ J} + (8.314 \text{ J/mol K} \times 298 \text{ K} \times -2.303)$$
$$= 3.4 \times 10^3 \text{ J} - 5.7 \times 10^3 \text{ J} = -2.3 \times 10^3 \text{ J}$$
$$= -2.3 \text{ kJ}$$

Practice Problem 12.9

The following reaction has a standard free-energy change of 3.4 kJ at 25°C:

$$H_2(g) + I_2(g) \longrightarrow 2HI(g)$$

If we decrease the pressure of HI to 0.10 atm while leaving the other pressures at 1.00 atm, what is the value of ΔG?

Answer: $\Delta G = -8.0$ kJ

Spontaneity and Equilibrium

We have seen that when the free-energy change is negative, the reaction proceeds spontaneously in the forward direction. This means that the free energy must go from higher to lower values as a spontaneous reaction converts reactants into products. Consider the water gas shift reaction of carbon monoxide with steam at 100°C, used to produce hydrogen gas:

$$CO(g) + H_2O(g) \rightleftharpoons CO_2(g) + H_2(g)$$

The value of ΔG changes with the extent of reaction as shown in Figure 12.8. As the reaction proceeds, the free energy decreases until it reaches its minimum value, corresponding to $\Delta G = 0$. The reactant and product pressures have their equilibrium values at this point and do not change spontaneously. Reactions always proceed spontaneously toward a state of equilibrium.

As we saw for the water gas shift reaction, the magnitude and sign of the free-energy change depend on the extent of reaction. Let's examine the course of this reaction, as shown in Figure 12.8, in more detail. A spontaneous reaction such as this has a negative value of ΔG that gets closer to zero as the reaction proceeds. As reactants are converted to products, the denominator terms of the reaction quotient decrease, and the numerator terms increase. As a result, the value of Q increases as reactants are spontaneously converted to products. Ultimately, Q increases enough to cause ΔG to equal zero. At this point, the forward reaction no

Figure 12.8 The free energy of the water gas shift reaction varies as the reaction proceeds from reactants to products. The minimum free energy ($\Delta G = 0$) occurs at the point at which the reaction reaches a state of equilibrium. From any other point on the curve, the system changes spontaneously to the equilibrium composition.

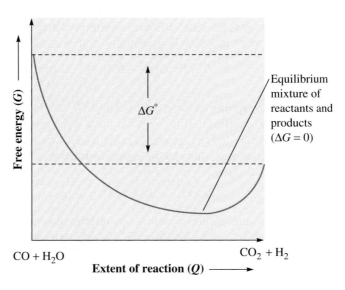

longer occurs spontaneously, and the system ceases to undergo any net change; it is in a state of equilibrium.

Reactions that reach a state of equilibrium are special—they can be carried out in either the forward or the reverse direction. Suppose we mix $CO_2(g)$ and $H_2(g)$ to carry out the water gas shift reaction in the reverse direction:

$$CO_2(g) + H_2(g) \rightleftharpoons CO(g) + H_2O(g)$$

The initial value of ΔG for this reaction is negative because the extent of reaction is less than the equilibrium value. Thus this reaction proceeds spontaneously. As these reactants are converted to products, the value of Q increases until ultimately ΔG equals zero. At this point, the reaction no longer occurs spontaneously and the system ceases to undergo any net change.

In either case, when net change ceases, the system has come to a state of equilibrium. This equilibrium arises because the reaction is reversible—it can proceed in either direction depending on the reaction conditions. Such a situation exists only when both reactants and products are present. In principle, all chemical reactions reach a state of equilibrium, but the equilibrium position of some reactions lies so far toward products that they essentially go to completion.

Under conditions of equilibrium, with $\Delta G = 0$, the equation for free-energy change becomes

$$\Delta G = 0 = \Delta G° + RT \ln Q_{equil}$$

We call Q_{equil} the **equilibrium constant** and give it the symbol K. The equilibrium constant depends on reactant and product pressures or concentrations just as the reaction quotient does, but the partial pressures or concentrations in the function must be the equilibrium values.

The magnitude of an equilibrium constant can indicate not only whether a reaction proceeds spontaneously but also how far it proceeds. The relationship among extent of reaction at equilibrium, magnitude of the equilibrium constant, and sign of the standard free-energy change is illustrated in Figure 12.9. Three ranges of values can be used to give a qualitative estimate of how far a reaction will proceed before it reaches a state of equilibrium.

Dichlorine monoxide, Cl_2O, is a yellowish brown gas with a disagreeable odor. It explodes on contact with organic matter or in the presence of a spark or heat. It decomposes at a moderate rate at room temperature. In water it forms HOCl.

- The equilibrium constant is small ($K \ll 1$). This situation is associated with positive values of $\Delta G°$. The equilibrium lies very far toward the reactants, so not much reaction will occur. In the equilibrium state, mostly reactants are present. The reverse reaction occurs spontaneously and nearly completely, whereas the forward reaction is essentially nonspontaneous.

  Example: $2Cl_2(g) + O_2(g) \rightleftharpoons 2Cl_2O(g)$
  $$K = 3.81 \times 10^{-17} \qquad \Delta G° = 93.7 \text{ kJ/mol}$$

- The equilibrium constant is large ($K \gg 1$). Here the value of $\Delta G°$ is negative. The equilibrium lies very far toward the products. The forward reaction is spontaneous and occurs nearly completely, whereas the reverse reaction is essentially nonspontaneous. In the equilibrium state, primarily products are present.

  Example: $H_2(g) + Br_2(l) \rightleftharpoons 2HBr(g)$
  $$K = 2.12 \times 10^9 \qquad \Delta G° = -53.2 \text{ kJ/mol}$$

- The equilibrium constant has intermediate values ($K \approx 1$). The value of $\Delta G°$ is near zero. The reaction will go approximately halfway before it reaches a state of equilibrium, and similar amounts of reactants and products are present in the

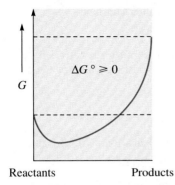 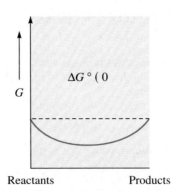 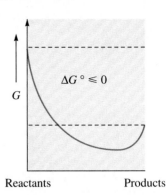

Figure 12.9 The position of equilibrium depends on the magnitude of $\Delta G°$ and on the equilibrium constant. Positive $\Delta G°$ ($K \ll 1$) gives an equilibrium mixture favoring reactants. A $\Delta G°$ of approximately zero ($K \approx 1$) gives an equilibrium mixture intermediate between reactants and products. Negative $\Delta G°$ ($K \gg 1$) gives an equilibrium mixture favoring products.

Nitrogen dioxide, NO_2, is a poisonous, reddish brown gas that liquefies below 21.1°C. It combines with itself in an equilibrium reaction to form some dinitrogen tetroxide, N_2O_4, which is a colorless gas.

equilibrium state. The reaction will occur spontaneously in either direction, starting with pure reactants or with pure products. However, the reaction will not go to completion in either direction.

$$\text{Example:} \quad N_2O_4(g) \rightleftharpoons 2NO_2(g)$$
$$K = 0.653 \qquad \Delta G° = +1.05 \text{ kJ/mol}$$

Values of equilibrium constants can be calculated from standard free-energy changes by the use of the equation

$$\Delta G° = -RT \ln K$$

Because standard free-energy changes can be calculated from the standard free energies of formation of the reactants and products, it is possible to determine the spontaneity and extent of reaction without measuring values of equilibrium constants for all possible reactions.

EXAMPLE 12.10

The following reaction has a standard free-energy change of 78.7 kJ at 25°C.

$$2CH_4(g) \longrightarrow C_2H_6(g) + H_2(g)$$

Calculate the equilibrium constant for this reaction, and decide whether the position of equilibrium will be closer to reactants or products.

Solution   We know the standard free-energy change and the temperature, so we can use the equation

$$\Delta G° = -RT \ln K$$

After converting 25°C to 298 K and 78.7 kJ to 7.87×10^4 J, we insert the values.

$$7.87 \times 10^4 \text{ J} = -8.314 \text{ J/mol K} \times 298 \text{ K} \times \ln K$$

Dividing through gives

$$\ln K = -31.8$$

(continued)

Taking the antilog gives us the value of the equilibrium constant:

$$K = 2 \times 10^{-14}$$

Because the equilibrium constant is very small, the reaction will not proceed very far before reaching equilibrium. This is in agreement with the positive value of the standard free-energy change.

Practice Problem 12.10

The following reaction has a standard free-energy change of -70.6 kJ at 25°C.

$$2NO(g) + O_2(g) \longrightarrow 2NO_2(g)$$

Calculate the equilibrium constant for this reaction, and decide whether the position of equilibrium will be closer to reactants or products.

Answer: K $= 2 \times 10^{12}$; the equilibrium lies far toward products.

12.8   Thermodynamics and Pollution

What effect does pollution have on your life? What could be done to reduce the amount of pollution?

Pollution is one of the major problems facing the world today. Pollutants, many resulting from chemical processes, are increasingly found distributed throughout our environment—in our air and water and on our land. What does pollution have to do with thermodynamics? One obvious relationship is in the area of thermal pollution. Another involves the connection among entropy, pollution, and recycling.

Thermal Pollution

We have already seen that many manufacturing processes and electric power plants use water as a coolant, often drawing it from a nearby waterway. If heated water is returned to the waterway, the resulting increase in temperature lowers the solubility of oxygen in the water, reducing the oxygen available to aquatic life. Many fish kills have occurred downstream from industrial locations because of this thermal pollution.

Let's look at an example of how this sort of pollution can be avoided. Power stations operate by burning coal or other fuel (or by carrying out nuclear fission reactions). The heat from this process is used to convert water to steam. The steam is forced through a turbine, which drives the electrical generators. Cooling the water that condenses from the steam before returning it to the environment prevents thermal pollution. Modern power plants use cooling towers, such as that shown in Figure 12.10. Cooling towers work much like automobile radiators, except that most of them do not have a fan. The shape of the tower (Figure 12.11) causes air to come in through the bottom and to rise. Meanwhile, hot water enters a series of vanes in the bottom of the tower. As the water passes down through these vanes, the air passes up, evaporating some of the water and cooling the rest in the process. The humid warm air rises and eventually leaves the tower. Outside it mixes with cooler air, causing the fog we usually see around cooling towers. Thus,

Figure 12.10 In a coal-fired
power plant, water is heated
in large boilers and the steam
is used to drive turbines.
Water from the condensed
steam is cooled in the concrete cooling towers.

Figure 12.10 In a coal-fired power plant, water is heated in large boilers and the steam is used to drive turbines. Water from the condensed steam is cooled in the concrete cooling towers.

although this sort of process prevents thermal pollution of waterways, the thermal effect ends up in the air instead.

Effects of Entropy

All the processes involved in the operation of a power plant inexorably increase the entropy of the universe. We may cool the water, but some of the water is vaporized, and heat is transferred to the atmosphere; both result in increased entropy.

Figure 12.11 Cold air enters the bottom of a cooling tower, passes through hot water, cooling it by evaporation, and then exits through the top of the tower. The hot, humid air causes fog to form as it contacts the cooler outside air.

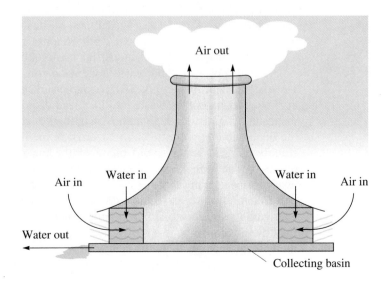

Burning fuel such as coal or petroleum to produce carbon dioxide and water vapor converts large molecules in the solid or liquid state into many small molecules in the gaseous state that are passed out the smokestack and dispersed into the atmosphere. Both the solid-to-gas conversion and the dispersing of material lead to increased entropy.

Entropy is an important consideration not only in the processes just described but also in our aggregate use of natural resources, from the digging of ores through the manufacture of consumer products to the disposal or re-use of the products. Ores usually consist of a small amount of the material we want, such as Al_2O_3, mixed in a larger amount of another material, the *matrix*. Generally, the concentration of the desired material is very low indeed, giving evidence of past effects of entropy. In order to reverse these effects, we must expend much energy in collecting and concentrating ores, in recovering the desired element from the ore, and in purifying the element. All these activities, though designed to reduce entropy with respect to a particular element, transfer heat to the surroundings and so cause the entropy of the universe to increase.

Once the element is put to use, entropy still remains a consideration. Consider what happens when Al_2O_3 is used to prepare aluminum sheet for beverage cans or household foil, for example. Consumer products such as aluminum beverage cans contain several elements. In addition to aluminum, for example, the beverage can contains a plastic liner and is covered with paint. Because all these materials start with pure substances, randomness in the distribution of the elements involved— and thus entropy—is increased during construction of the product.

The use of consumer products often results in entropy increases as well. Aluminum cans are made in only a few locations, but once they have been filled with soft drinks, dispersed to many retail stores, purchased, and emptied, there has been a considerable increase in randomness. Recycling programs can reverse some of this increase, but only at the expense of energy for the human activity involved and for transportation.

Aluminum cans are far from the only consumer products people discard. Every day a wide variety of castoffs (including old newspapers, plastic and metal containers, garbage, construction wastes, and old cars) are thrown away. Along with solid sludge derived from sewage and other waste waters, they are classified as solid waste, which ranks with water and air pollution as one of our major environmental concerns. The United States produces over 4 billion tons of solid waste annually. Animal and agricultural wastes make up about 2.5 billion tons; mining residues contribute about 1 billion tons; municipal and commercial wastes about 300 million tons; and industrial waste about 150 million tons.

The disposal of solid waste involves a considerable cost in energy and also has a significant impact on the environment. Entropy must be decreased and energy used to collect the waste into localized disposal facilities. Most solid waste is disposed of on the land, in dumps, where piles of refuse are left to decompose in the open and can present health hazards. Some is also buried in sanitary landfills (Figure 12.12), which avoid many of the health hazards associated with open dumps but require large amounts of open land, a commodity that is no longer available near many large cities, especially in the eastern United States. A small percentage of waste is incinerated, giving ash, carbon dioxide, and water. However, unless the incinerator has appropriate air pollution controls, exhaust gases may merely convert the solid waste to unacceptable air pollution with a resulting increase in entropy.

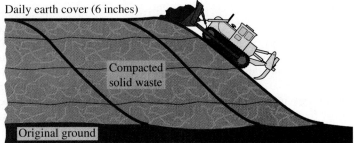

Figure 12.12 *Left:* Some solid waste is buried in sanitary landfills. *Right:* Layers of refuse are daily compacted and covered with a thin layer of earth.

Only a very small amount of solid waste is subjected to recycling. Yet resource recovery is potentially the most effective solution to our solid-waste problem. As natural resources become more scarce, resource recovery from solid wastes will no doubt become a feasible way to solve problems of scarcity as well. Expending the energy required to overcome the high entropy of wastes will become economical once wastes have higher concentrations of scarce resources than do natural sources. Even now, it is often cheaper to recover an element from a disposed product than to extract it from an ore. For example, it is estimated that the energy required to recycle aluminum cans is only about 5% of that required to recover aluminum from bauxite ores.

Summary

Spontaneous processes occur with a decrease in energy, but they may be exothermic or endothermic, so an additional factor is necessary to explain why a process is spontaneous. This factor is the randomness or disorder of the system, measured by the change in entropy, ΔS.

Internal energy, the energy stored within a system, can be transferred to the surroundings as heat or as work. The most common type of work in chemical systems is pressure–volume work, in which a substance, usually a gas, changes its volume. Heat, in a process at constant pressure, is the enthalpy change of the system, ΔH. Usually, pressure–volume work is small compared with the enthalpy change, so the enthalpy change is a good approximation to the change in internal energy.

A positive change in entropy may arise from increases in randomness of the arrangement of particles or from increased freedom of motion. The second law of thermodynamics states that the entropy of the universe always increases in any process. A transfer of heat to the surroundings leads to increased molecular motion and to an increase in the entropy of the surroundings. The third law of thermodynamics states that the absolute entropy of a perfect crystal of a pure substance at 0 K is zero.

The difference between the enthalpy change of the system and the amount of energy required to change the entropy of the system is called the free-energy change, ΔG. It is the amount of energy available to do useful work on the surroundings. A negative value of ΔG corresponds to a spontaneous process.

A reaction proceeds in the direction that leads to a free-energy change of zero. The composition of the reaction mixture at this point is the equilibrium composition, and the system is said to be in a state of equilibrium. When ΔG has a value of zero, the composition of the reaction mixture is the equilibrium composition, which is summarized by the equilibrium constant, K. The value of the equilibrium constant indicates how far a reaction proceeds before it reaches a state of equilibrium, where net changes in composition stop.

Key Terms

absolute entropy
  (S) (12.4)
equilibrium constant
  (K) (12.7)
free energy (G) (12.5)
Gibbs free energy

(G) (12.5)
reaction quotient
  (Q) (12.7)
second law of
  thermodynamics (12.3)
spontaneous

reaction (p. 493)
standard entropy (S°)
  (12.4)
standard free-energy
  change (ΔG°) (12.6)
standard free energy of

formation
  (ΔG°$_f$) (12.6)
thermodynamics (p. 494)
third law of
  thermodynamics (12.4)

Exercises

Internal Energy and Enthalpy Changes

12.1 Distinguish between internal energy and enthalpy.

12.2 What mathematical relationship exists between internal energy and enthalpy?

12.3 Why can enthalpy be used as a good approximation of internal energy in many chemical systems?

12.4 Give an example of a reaction in which internal-energy changes and enthalpy changes have the same value.

12.5 During a certain process, the system gains 872 kJ of heat while doing 268 kJ of work on the surroundings. What is the change in internal energy of the system arising from this process?

12.6 Calculate the PV work that occurs in the following reaction of 0.750 mol PCl_3 with 0.750 mol Cl_2 at constant pressure and a temperature of 25°C.

$$PCl_3(g) + Cl_2(g) \longrightarrow PCl_5(g)$$

12.7 Calculate the PV work that occurs in the following reaction of 1.50 mol O_2 with 3.00 mol H_2 at constant pressure and a temperature of 25°C.

$$2H_2(g) + O_2(g) \longrightarrow 2H_2O(g)$$

12.8 Propane is widely used as a fuel for camp stoves. The standard enthalpy change for the combustion of propane is −2043.96 kJ/mol propane:

$$C_3H_8(g) + 5O_2(g) \longrightarrow 3CO_2(g) + 4H_2O(g)$$

What is the change in the internal energy under standard conditions for the combustion of 1 mol of propane?

Entropy Changes

12.9 Define the term *entropy*.

12.10 Describe what happens when entropy increases.

12.11 Explain how a decrease in entropy of the system leads to an increase in entropy of the surroundings.

12.12 What are the second and third laws of thermodynamics?

12.13 Lithium chloride is a colorless salt used in pyrotechnics. Does the entropy increase, decrease, or remain the same when crystalline lithium chloride melts? Explain your answer.

12.14 Mercury is the only metal that exists as a liquid below 25°C. Arrange the following substances in order of increasing entropy: $Hg(g)$, $Hg(l)$, $Hg(s)$.

12.15 Under appropriate conditions, bromine and hydrogen gases react to form hydrogen bromide gas. Indicate which of the following has the largest entropy: $Br_2(g)$, $H_2(g)$, or $HBr(g)$.

12.16 Does the entropy increase, decrease, or remain the same when the following processes occur?
  a. $2H_2O(g) \longrightarrow 2H_2(g) + O_2(g)$
  b. $2NO(g) + O_2(g) \longrightarrow 2NO_2(g)$
  c. $PCl_5(g) \longrightarrow PCl_3(g) + Cl_2(g)$
  d. $CuSO_4(s) + 5H_2O(l) \longrightarrow CuSO_4 \cdot 5H_2O(s)$
  e. $2SO_2(g) + O_2(g) \longrightarrow 2SO_3(g)$
  f. $CaCO_3(s) \longrightarrow CaO(s) + CO_2(g)$

12.17 In each case tell which reaction, the forward or the reverse, is favored by entropy changes.
  a. $4NH_3(g) + 3O_2(g) \longrightarrow 2N_2(g) + 6H_2O(g)$
  b. $4NO(g) + 6H_2O(g) \longrightarrow 4NH_3(g) + 5O_2(g)$
  c. $C(s) + CO_2(g) \longrightarrow 2CO(g)$
  d. $2O_3(g) \longrightarrow 3O_2(g)$
  e. $CH_3OH(l) \longrightarrow CH_3OH(g)$

12.18 When placed on the skin, ethanol evaporates and has a cooling effect. Calculate the entropy change for vaporization of ethanol at 25°C if the enthalpy of vaporization is 42.6 kJ/mol.

12.19 Liquid ammonia is used as a solvent in chemical research and as a fertilizer. The enthalpy of vaporization of liquid ammonia is 23.3 kJ/mol, whereas its entropy of vaporization is 97.2 J/mol K. What is the normal boiling point of liquid ammonia?

12.20 Calculate standard entropy changes for the following reactions at 25°C.
  a. $2H_2O(g) \longrightarrow 2H_2(g) + O_2(g)$
  b. $2NO(g) + O_2(g) \longrightarrow 2NO_2(g)$
  c. $PbSO_4(s) \longrightarrow PbO_2(s) + SO_2(g)$
  d. $2SO_2(g) + O_2(g) \longrightarrow 2SO_3(g)$
  e. $Ca(OH)_2(s) \longrightarrow CaO(s) + H_2O(g)$

Free-Energy Changes

12.21 What is free energy?

12.22 How is a change in free energy related to changes in enthalpy and entropy?

12.23 Why is it necessary to define a standard state when discussing free energies?

12.24 Distinguish between free-energy changes and standard free-energy changes.

12.25 Methanol is a flammable, poisonous liquid used as an industrial solvent. The standard free energy of formation of $CH_3OH(l)$ is -166.4 kJ/mol. Write the chemical reaction that corresponds to this value.

12.26 The standard free energies of formation of $Hg(g)$ and $Hg(l)$ are $+31.8$ kJ/mol and 0.00 kJ/mol, respectively. Comment on the significance of these values.

12.27 Carbon tetrachloride is a colorless, nonflammable, dense liquid that was widely used as a dry-cleaning solvent until it was discovered to be a probable carcinogen. The standard free energies of formation of $CCl_4(l)$ and $CCl_4(g)$ are -68.6 kJ/mol and -64.0 kJ/mol, respectively. Explain these relative values.

12.28 Anhydrous calcium chloride can be used as a drying agent. Could it be prepared from its component elements? (a) Calculate the change in free energy for the following reaction at 25°C:

$$Ca(s) + Cl_2(g) \longrightarrow CaCl_2(s)$$
$$\Delta H = -795 \text{ kJ}, \ \Delta S = -150 \text{ J/K}$$

(b) Is this reaction exothermic or endothermic? (c) Is it spontaneous as written?

12.29 Calculate the change in free energy at 25°C for the following reactions, using standard free energies of formation.
   a. $ZnO(s) + SO_3(g) \longrightarrow ZnSO_4(s)$
   b. $2C(graphite) + O_2(g) \longrightarrow 2CO(g)$
   c. $2C(diamond) + O_2(g) \longrightarrow 2CO(g)$
   d. $CS_2(g) + 3O_2(g) \longrightarrow CO_2(g) + 2SO_2(g)$
   e. $4Fe(s) + 3O_2(g) \longrightarrow 2Fe_2O_3(s)$

12.30 Lead(II) chloride is a white, crystalline solid used to manufacture pigments. (a) Calculate the change in free energy for the following reaction at 25°C:

$$PbCl_2(s) \longrightarrow Pb(s) + Cl_2(g)$$
$$\Delta H = 359 \text{ kJ}, \ \Delta S = 153 \text{ J/K}$$

(b) Is this reaction exothermic or endothermic? (c) Is it spontaneous as written?

12.31 The water gas reaction can be used to synthesize hydrogen gas. (a) Calculate the value of $\Delta G°$ at 25°C for the reaction

$$C(graphite) + H_2O(g) \rightleftharpoons CO(g) + H_2(g)$$

(b) At what temperature does the value of $\Delta G°$ equal zero?

12.32 Calculate $\Delta G°$, $\Delta H°$, and $\Delta S°$ for the following reactions at 25°C.

   a. $Fe_2O_3(s) + 3CO(g) \longrightarrow 2Fe(s) + 3CO_2(g)$
   b. $2SO_2(g) + O_2(g) \longrightarrow 2SO_3(g)$
   c. $SO_3(g) + H_2O(l) \longrightarrow H_2SO_4(l)$
   d. $2C(graphite) + 2H_2O(l) \longrightarrow CH_3COOH(l)$
   e. $2H_2S(g) + 3O_2(g) \longrightarrow 2H_2O(g) + 2SO_2(g)$

Spontaneity of Processes

12.33 What is a spontaneous process?

12.34 What factors must we take into account to predict the spontaneity of a process?

12.35 Sodium hydroxide solutions are used in chemical laboratories to determine the concentrations of solutions of acids. When sodium hydroxide dissolves in water, heat is given off. Which factor, enthalpy or entropy, is responsible for the spontaneity of this process, or are both responsible?

12.36 The combustion of methane, widely used as a source of heat, is a spontaneous reaction:

$$CH_4(g) + 2O_2(g) \longrightarrow CO_2(g) + 2H_2O(g)$$

Which factor, enthalpy or entropy, is responsible for the spontaneity of this process, or are both responsible?

12.37 Under what temperature conditions should reactions that have the following characteristics be spontaneous?
   a. exothermic, positive entropy change
   b. endothermic, positive entropy change
   c. exothermic, negative entropy change
   d. endothermic, negative entropy change

12.38 Would you expect the following reactions to be spontaneous at 25°C under standard-state conditions?
   a. $N_2(g) + 3H_2(g) \longrightarrow 2NH_3(g)$
   b. $N_2(g) + O_2(g) \longrightarrow 2NO(g)$
   c. $2N_2(g) + O_2(g) \longrightarrow 2N_2O(g)$
   d. $C(graphite) + 2H_2(g) \longrightarrow CH_4(g)$
   d. $2PbO_2(s) + 2SO_3(g) \longrightarrow 2PbSO_4(s) + O_2(g)$

12.39 Calcium carbide, a grayish black solid used as a source of acetylene gas, is prepared from lime and carbon. In the following chemical reaction, $\Delta H = 462$ kJ/mol and $\Delta S = 222$ J/mol K.

$$CaO(s) + 3C(s) \longrightarrow CaC_2(s) + CO(g)$$

Is this reaction spontaneous at 425 K?

12.40 Ethylene gas, used to form polyethylene plastics and to accelerate the ripening of fruit, can be prepared from acetylene gas. In the following chemical reaction, $\Delta H = -174$ kJ/mol and $\Delta S = -112$ J/mol K.

$$C_2H_2(g) + H_2(g) \longrightarrow C_2H_4(g)$$

In what temperature range is this reaction spontaneous?

Reaction Quotients

12.41 What is a reaction quotient?

12.42 Write the reaction quotient for each of the following reactions.
   a. $2CO(g) + O_2(g) \longrightarrow 2CO_2(g)$
   b. $C_2H_2(g) + 2H_2(g) \longrightarrow C_2H_6(g)$
   c. $C_2H_4(g) + H_2(g) \longrightarrow C_2H_6(g)$
   d. $SiO_2(s) + 4HF(g) \longrightarrow SiF_4(g) + 2H_2O(g)$
   e. $2H_2O_2(l) \longrightarrow 2H_2O(l) + O_2(g)$

Free-Energy Changes for Nonstandard Conditions

12.43 What are standard conditions?
12.44 How is the reaction quotient used to calculate the free-energy change for nonstandard conditions?
12.45 The synthesis of methanol is represented as follows:

$$CO(g) + 2H_2(g) \rightleftharpoons CH_3OH(g)$$

Calculate the change in free energy at 25°C for each of the following conditions.
   a. 1.00 atm CH_3OH, 1.00 atm CO, 1.00 atm H_2
   b. 1.00 atm CH_3OH, 1.00 atm CO, 0.100 atm H_2
   c. 10.0 atm CH_3OH, 10.0 atm CO, 1.00 atm H_2
   d. 0.100 atm CH_3OH, 1.00 atm CO, 0.100 atm H_2
   e. 1.00 atm CH_3OH, 0.100 atm CO, 0.100 atm H_2

12.46 The water gas shift reaction, shown here, is used to increase the yield of hydrogen gas produced by the reaction of carbon with steam (the water gas reaction).

$$CO(g) + H_2O(g) \rightleftharpoons CO_2(g) + H_2(g)$$

What is the change in free energy at 25°C for the water gas shift reaction if the partial pressure of each substance is 0.0100 atm?

Free-Energy Change and Equilibrium

12.47 What is a state of equilibrium?
12.48 What is an equilibrium constant?
12.49 How is an equilibrium constant used to describe the relative amounts of reactants and products for a reaction in a state of equilibrium?

12.50 How are an equilibrium constant and a standard free-energy change related?
12.51 What is the value of the free-energy change for a reaction in a state of equilibrium?
12.52 Use standard free-energy values to calculate the equilibrium constant for the vaporization of water at 25°C.
12.53 Dinitrogen tetroxide is a colorless gas, whereas nitrogen dioxide, because of its unpaired valence electron, is reddish brown. The equilibrium constant for the following reaction has a value of 15 at a given temperature.

$$N_2O_4(g) \rightleftharpoons 2NO_2(g)$$

What would you expect to be the sign of the value of the change in standard free energy?

12.54 One step in the synthesis of sulfuric acid is the oxidation of sulfur dioxide to sulfur trioxide. The change in standard free energy for this reaction is negative.

$$2SO_2(g) + O_2(g) \rightleftharpoons 2SO_3(g)$$

Would you expect the equilibrium constant to be greater than or less than 1? Explain your answer.

12.55 Nitrosyl bromide is a brown liquid that boils at −2°C. The equilibrium constant for the following reaction at 25°C is 1.00×10^{-2}.

$$2NOBr(g) \rightleftharpoons 2NO(g) + Br_2(g)$$

What is the value of $\Delta G°$ at 25°C for this reaction?

12.56 Calculate the standard free-energy change and the value of the equilibrium constant for the following reactions at 25°C.
   a. $I_2(g) + 2HBr(g) \rightleftharpoons Br_2(g) + 2HI(g)$
   b. $4HCl(g) + O_2(g) \rightleftharpoons 2H_2O(g) + 2Cl_2(g)$
   c. $2H_2S(g) + 3O_2(g) \rightleftharpoons 2H_2O(g) + 2SO_2(g)$
   d. $CH_4(g) + CO_2(g) \rightleftharpoons CH_3OH(g) + CO(g)$
   e. $CO(g) + NO_2(g) \rightleftharpoons CO_2(g) + NO(g)$

Additional Exercises

12.57 Potassium chloride is a colorless salt used in electrolyte-replenishing solutions. (a) Calculate the change in free energy for the following reaction at 25°C.

$$2KCl(s) \longrightarrow 2K(s) + Cl_2(g)$$
$$\Delta H° = 870 \text{ kJ}, \ \Delta S° = 185 \text{ J/K}$$

(b) Is this reaction exothermic or endothermic? (c) Is it spontaneous as written?

12.58 Does the entropy increase, decrease, or remain the same when the following processes occur?
   a. $HCl(g) + NH_3(g) \longrightarrow NH_4Cl(s)$
   b. $H_2O(g) \longrightarrow H_2O(s)$
   c. $P_4(s) + 5O_2(g) \longrightarrow P_4O_{10}(s)$
   d. $CaO(s) + SO_3(g) \longrightarrow CaSO_4(s)$
   e. $CaSO_4(s) + 2H_2O(l) \longrightarrow CaSO_4 \cdot 2H_2O(s)$

12.59 Indicate which of the following has the largest entropy: $Br_2(l)$, $Cl_2(g)$, or $BrCl(g)$.

12.60 Calculate the value of $\Delta G°$ at 25°C for the reaction

$$CH_4(g) + 3CO_2(g) \rightleftharpoons 4CO(g) + 2H_2O(g)$$

At what temperature does the value of $\Delta G°$ equal zero?

12.61 Acetic acid, CH_3CO_2H, is a liquid in the pure state. It has a pungent odor and produces skin burns. Calculate the entropy change for vaporization of acetic acid at 25°C if the enthalpy of vaporization is 52.3 kJ/mol.

12.62 Calculate the change in free energy at 25°C for each of the following reactions.

    a. $2CO(g) + O_2(g) \longrightarrow 2CO_2(g)$
    b. $C_2H_2(g) + 2H_2(g) \longrightarrow C_2H_6(g)$
    c. $C_2H_2(g) + H_2(g) \longrightarrow C_2H_4(g)$
    d. $MgO(s) + H_2O(l) \longrightarrow Mg(OH)_2(s)$
    e. $SiO_2(s) + 4HF(g) \longrightarrow SiF_4(g) + 2H_2O(g)$
    f. $2H_2O_2(l) \longrightarrow 2H_2O(l) + O_2(g)$
    g. $C_2H_5OH(l) + O_2(g) \longrightarrow$
                  $CH_3COOH(l) + H_2O(l)$

12.63 Calcium hydroxide is prepared by adding water to lime and is used in plaster and cement. When calcium hydroxide dissolves in water, the solution becomes cold. Which factor, enthalpy or entropy, is responsible for the spontaneity of this process, or are both responsible?

12.64 Tell whether you would expect each of the following reactions to be spontaneous at 25°C under standard-state conditions.

    a. $C(graphite) \longrightarrow C(diamond)$
    b. $Si(s) + 2Cl_2(g) \longrightarrow SiCl_4(l)$
    c. $SiCl_4(g) + 2F_2(g) \longrightarrow SiF_4(g) + 2Cl_2(g)$
    d. $3Zn(s) + Al_2O_3(s) \longrightarrow 3ZnO(s) + 2Al(s)$
    e. $Cu(s) + H_2O_2(l) \longrightarrow CuO(s) + H_2O(l)$

12.65 Indicate what change, positive or negative, occurs in the enthalpy and the entropy for each of the following reactions.

    a. $CO_2(s) \longrightarrow CO_2(g)$
    b. $CaCO_3(s) + heat \longrightarrow CaO(s) + CO_2(g)$
    c. $O_2(l) \longrightarrow O_2(s)$
    d. $NaCl(s) \longrightarrow Na^+(l) + Cl^-(l)$
    e. $NH_3(l) \longrightarrow NH_3(g)$

12.66 The following chemical reaction has $\Delta H° = 572$ kJ and $\Delta S° = 326$ J/K.

$$2H_2O(l) \longrightarrow 2H_2(g) + O_2(g)$$

In what temperature range is this reaction spontaneous?

12.67 Calculate standard entropy changes for the following reactions at 25°C.

    a. $6CO_2(g) + 15H_2(g) \longrightarrow C_6H_6(l) + 12H_2O(l)$
    b. $4NH_3(g) + 7O_2(g) \longrightarrow 4NO_2(g) + 6H_2O(g)$
    c. $Si(s) + PbO_2(s) \longrightarrow SiO_2(s) + Pb(s)$
    d. $SO_3(g) + CO(g) \longrightarrow SO_2(g) + CO_2(g)$
    e. $2H_2O_2(l) \longrightarrow O_2(g) + 2H_2O(l)$

12.68 Calculate the value of the equilibrium constant at 25°C for the reaction

$$CH_4(g) + 2O_2(g) \rightleftharpoons CO_2(g) + 2H_2O(g)$$

12.69 During a certain process, the system loses 534 kJ of heat, and the surroundings do 237 kJ of work on the

system. What change in the internal energy of the system arises from this process?

12.70 Arrange the following substances in order of increasing entropy: $Cu(g)$, $Cu(l)$, $Cu(s)$.

12.71 The equilibrium constant for the following reaction at 45°C is 0.65.

$$N_2O_4(g) \rightleftharpoons 2NO_2(g)$$

What is the value of $\Delta G°$ at 45°C for this reaction?

12.72 Write the reaction quotient for each of the following reactions.

    a. $2HF(g) \rightleftharpoons H_2(g) + F_2(g)$
    b. $2NO(g) + O_2(g) \rightleftharpoons 2NO_2(g)$
    c. $Ca(s) + Cl_2(g) \rightleftharpoons CaCl_2(s)$
    d. $SiF_4(g) + 2H_2O(l) \rightleftharpoons SiO_2(s) + 4HF(g)$
    e. $2HgO(s) \rightleftharpoons 2Hg(l) + O_2(g)$
    f. $NH_4Cl(s) \rightleftharpoons NH_3(g) + HCl(g)$

12.73 (a) Calculate the change in free energy for the following reaction at 25°C.

$$Pb(s) + Cl_2(g) \longrightarrow PbCl_2(s)$$
$$\Delta H° = -359 \text{ kJ}, \Delta S° = -153 \text{ J/K}$$

(b) Is this reaction exothermic or endothermic? (c) Is it spontaneous as written?

12.74 Explain the relative values of the free energies of formation of $I_2(g)$ (19.4 kJ/mol) and $I_2(s)$ (0.00 kJ/mol).

12.75 The equilibrium constant at 440°C for the following reaction is 0.020.

$$2HI(g) \rightleftharpoons H_2(g) + I_2(g)$$

What would you expect to be the sign of the value of the change in standard free energy?

12.76 Calculate $\Delta G°$, $\Delta H°$, and $\Delta S°$ for the following reactions at 25°C.

    a. $CH_4(g) + 2O_2(g) \longrightarrow CO_2(g) + 2H_2O(g)$
    b. $2C_2H_5OH(l) + O_2(g) \longrightarrow 2CH_3COOH(l) + 2H_2(g)$
    c. $SO_2(g) + 3H_2(g) \longrightarrow H_2S(g) + 2H_2O(l)$
    d. $2HCl(g) + I_2(s) \longrightarrow 2HI(g) + Cl_2(g)$
    e. $2H_2O_2(l) \longrightarrow 2H_2O(l) + O_2(g)$

12.77 Does the entropy increase, decrease, or remain the same when crystalline calcium chloride melts? Explain your answer.

12.78 In a chemical reaction, $\Delta H = -142$ kJ/mol and $\Delta S = -64.0$ J/mol K. Is this reaction spontaneous at 1125 K?

12.79 Calculate the standard free-energy change and the value of the equilibrium constant for the following reactions at 25°C.

    a. $CS_2(g) + 4H_2(g) \rightleftharpoons CH_4(g) + 2H_2S(g)$
    b. $4H_2(g) + CO_2(g) \rightleftharpoons 2H_2O(g) + CH_4(g)$
    c. $MgO(s) + H_2O(l) \rightleftharpoons Mg(OH)_2(s)$
    d. $2N_2(g) + 6H_2O(g) \rightleftharpoons 4NH_3(g) + 3O_2(g)$
    e. $SiO_2(s) + 4H_2(g) \rightleftharpoons SiH_4(g) + 2H_2O(g)$

* 12.80 When a pesticide is applied to crops, it is sprayed over large areas. Plants absorb the pesticide. Animals later eat the plants. Rain washes the pesticide into lakes. The pesticide concentrates in fatty tissues in fish. (a) Discuss the entropy changes involved in these processes. (b) What entropy changes are necessary to rid the environment of this pesticide?

* 12.81 Would you expect the mixture of $N_2(g)$ and $O_2(g)$ present in air to form significant amounts of $N_2O(g)$, $NO(g)$, or $NO_2(g)$? Provide evidence to support your answers.

* 12.82 Given the following changes in standard free energy:

$$TiO_2(s) + 4HCl(g) \longrightarrow$$
$$TiCl_4(l) + 2H_2O(g) \qquad \Delta G° = 76.3 \text{ kJ}$$

$$2TiCl_4(l) + H_2(g) \longrightarrow$$
$$2TiCl_3(s) + 2HCl(g) \qquad \Delta G° = 53.8 \text{ kJ}$$

Calculate the standard free energy of the following reaction and decide whether it is spontaneous.

$$2TiCl_3(s) + 4H_2O(g) \longrightarrow$$
$$2TiO_2(s) + 6HCl(g) + H_2(g)$$

* 12.83 What conditions are necessary for $Ag_2O(s)$ to decompose spontaneously into its elements at 25°C?

▶ ▶ ▶ Chemistry in Practice

12.84 You are given a sample of a colorless gas, told that it is a compound that contains 29.67% sulfur and 70.33% fluorine, and asked to synthesize 1.50 kg of this gas. A 125.0-mL portion of the gas weighs 0.5340 g at 735.4 torr and 25.0°C. As starting materials for your synthesis, you have solid sulfur (S), fluorine gas (F_2), sulfur oxide gases (SO_2 or SO_3), and hydrogen fluoride gas (HF).

  a. What is the gas you need to synthesize?
  b. Propose some possible reactions that could be used to synthesize the gas.
  c. Which of these reactions is spontaneous?
  d. Which reaction would you choose for the synthesis? Explain your choice.
  e. Would you have to supply heat to carry out the synthesis, or would heat be evolved?
  f. What mass of sulfur-containing starting material (S, SO_2, or SO_3,) would be needed to prepare 1.50 kg of the gas?
  g. What volume of fluorine-containing starting material (F_2 or HF), measured at 953.5 torr and 22.8°C, would be needed to prepare 1.50 kg of the gas?

Chemical Kinetics

▶ ▶ ▶ ▶

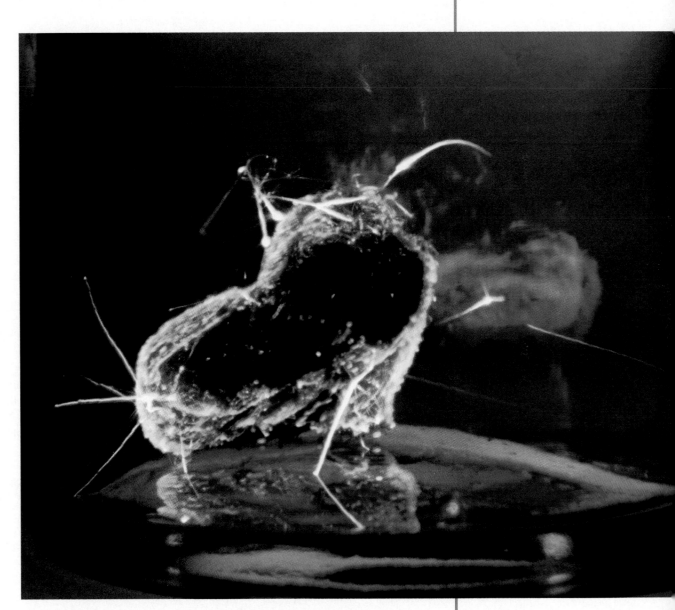

Steel wool burns when heated and placed in an oxygen atmosphere.

CHAPTER OUTLINE

Encountering Chemistry

Why does the reaction occur faster when the iron is more finely divided? Why does it occur faster when the oxygen concentration is increased?

A piece of steel wool is placed in a flame. As it is heated, it begins to glow pale red. The steel wool is removed from the flame and the color fades. No chemical reaction is apparent. The steel wool is heated again and then placed in a bottle of oxygen gas. There, it gives off flashes of light as it burns. The iron reacts with the pure oxygen in a direct synthesis reaction, forming iron(III) oxide, even though it did not react with the oxygen in air.

Iron filings sprinkled into a flame glow brightly and burn up, as shown in Figure 13.1. Yet iron in the form of steel wool did not burn—it merely became hot. Apparently the size of the iron particles affects the behavior of iron when it is heated in air.

A mixture of hydrogen and oxygen gases kept at room temperature shows no reaction even over a long period of time, though the formation of water, as represented by the following equation, is highly favored thermodynamically ($\Delta G° = -237$ kJ/mol).

$$2H_2(g) + O_2(g) \longrightarrow 2H_2O(l)$$

However, when this mixture of gases is exposed to an electrical spark or a flame, the reaction occurs explosively, as shown in Figure 13.2.

Figure 13.1 Although bulky pieces of iron do not burn in air when heated, iron filings do.

A solution of N_2O_5 in carbon tetrachloride is stable for long times when kept cold. When warmed up, however, it decomposes by the reaction

$$2N_2O_5 \longrightarrow 4NO_2 + O_2(g)$$

How can we explain all these observations? We need to understand how fast a reaction occurs and what causes it to occur at this rate. These considerations are the subject matter of chemical kinetics. **Kinetics** is the study of the rates of chemical reactions and other processes and the factors that affect these rates: concentrations, temperature, pressure, catalysis, solvent, ionic medium, and so on. This chapter describes how reaction rates are measured, how they depend on concentration and temperature, how analysis of the stepwise manner in which reactions occur explains the dependence of rate on concentration, and how rates can be changed by the addition of catalysts.

Figure 13.2 A mixture of hydrogen and oxygen gases subjected to a spark or a flame burns explosively.

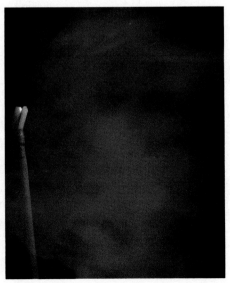

13.1 Reaction Rates

Chemical reactions occur at widely varying rates. Some occur almost instantaneously, and some take place so slowly that for practical purposes, they do not happen at all. Most have rates between these extremes. The usefulness of a chemical reaction in the synthesis of a desired compound or as a source of energy depends on its rate, which, in turn, depends on a number of factors. To examine the various factors that affect how fast a reaction proceeds, we must be able to determine rates.

What is a reaction rate? How do we measure it?

The **rate of reaction** is defined as the change in concentration of one of the reactants or products per unit of time. The determination of rates, then, involves measurements of both concentration and time.

Calculating Reaction Rates

Before we measure a rate of reaction directly or calculate it from concentrations measured at different times, we must be careful to specify a reference point for that rate. Depending on the reaction stoichiometry and on which concentration is measured—that of one of the reactants or that of one of the products—the rate may be different.

For example, consider the decomposition reaction of the hypochlorite ion in basic aqueous solution:

$$3ClO^-(aq) \longrightarrow 2Cl^-(aq) + ClO_3^-(aq)$$

The hypochlorite ion, ClO^-, is found in salts such as NaOCl and $Ca(OCl)_2$. Anhydrous sodium hypochlorite is an unstable, explosive white salt. When dissolved in water, however, it is quite stable. Dilute aqueous solutions, marketed under names such as Clorox™, are used as bleaches and disinfectants. Calcium hypochlorite is a stable white salt that is used as an algicide, bactericide, deodorant, and disinfectant. It is one of the common forms of swimming pool "chlorine" and is used as a bleach in sugar refining.

Because the stoichiometric coefficients are not the same, the concentrations of the species do not change at the same rate. However, the rate of appearance of products must be the same as the rate of disappearance of reactants; otherwise, the law of conservation of matter would not be obeyed. We can reconcile these two observations by making allowance for the stoichiometric ratios of species. For example, because three ClO^- ions react to form one ClO_3^- ion, ClO^- disappears three times as fast as ClO_3^- appears. Similarly, Cl^- must be formed twice as fast as ClO_3^- is formed, but only $\frac{2}{3}$ as fast as ClO^- disappears. These relationships can be summarized as follows:

$$\begin{aligned}
\text{Rate of reaction} &= \text{rate of appearance of } ClO_3^- \\
&= \tfrac{1}{3} \text{ rate of disappearance of } ClO^- \\
&= \tfrac{1}{2} \text{ rate of appearance of } Cl^-
\end{aligned}$$

If we express the rate of a reaction as the rate of change of a given reactant or product divided by the stoichiometric coefficient, the rate of reaction is the same no matter which species is used as a reference point. The important thing to remember is that no matter which reference point we use, we must know exactly how the rate of reaction is being defined in order to interpret it properly.

As we have said, the measurement of rate involves measurements of concentration taken at various times. However, the rates of most reactions change as the reaction proceeds, so we cannot simply measure the concentration at any two different times to determine the rate of reaction. ▶ Let's consider the decomposition of dinitrogen pentoxide dissolved in carbon tetrachloride, mentioned in the chapter introduction, to see how we convert concentration–time data to rates.

$$2N_2O_5 \longrightarrow 4NO_2 + O_2(g)$$

Table 13.1 Concentration of N_2O_5 as a Function of Time at 45°C

Time (s)	Δt (s)	$[N_2O_5]$ (M)	$-\Delta[N_2O_5]$ (M)	Rate, $-\frac{1}{2}\Delta[N_2O_5]/\Delta t$ (M s⁻¹)
0		2.330		
1.11×10^4	1.11×10^4	2.080	0.250	1.13×10^{-5}
1.92×10^4	8.10×10^3	1.910	0.170	1.05×10^{-5}
3.16×10^4	1.24×10^4	1.670	0.240	9.68×10^{-6}
5.24×10^4	2.08×10^4	1.350	0.320	7.69×10^{-6}
7.20×10^4	1.96×10^4	1.110	0.240	6.12×10^{-6}
1.127×10^5	4.07×10^4	0.720	0.390	4.79×10^{-6}
1.389×10^5	2.62×10^4	0.550	0.170	3.24×10^{-6}
1.886×10^5	4.97×10^4	0.340	0.220	2.21×10^{-6}

RULES OF REACTIVITY

Like other nonmetal oxides that are acid anhydrides, N_2O_5 adds water to form an oxoacid—nitric acid in this case.

$$N_2O_5(s) + H_2O(l) \longrightarrow 2HNO_3(aq)$$

When in the gas phase or dissolved in solvents such as carbon tetrachloride, dinitrogen pentoxide decomposes to give red nitrogen dioxide (which is soluble in carbon tetrachloride) and oxygen gas (which is not soluble).

The concentrations of N_2O_5 measured at several times after N_2O_5 is dissolved in carbon tetrachloride at 45°C are given in Table 13.1 and graphed in Figure 13.3. The notation $[N_2O_5]$ represents the concentration of N_2O_5 in moles per liter. Because rate is a change in concentration per unit of time, we can select two points, designated by the subscripts 1 and 2, to calculate the rate over this period of time.

$$\text{Rate} = -\tfrac{1}{2}\frac{[N_2O_5]_2 - [N_2O_5]_1}{t_2 - t_1}$$

We include $\frac{1}{2}$ in the equation because the stoichiometric coefficient of N_2O_5 is 2. The concentration of a reactant decreases over time, so we include a minus sign in

Figure 13.3 The data from Table 13.1 are plotted with concentration on the y-axis and time on the x-axis. The concentration of a reactant decreases with increasing time.

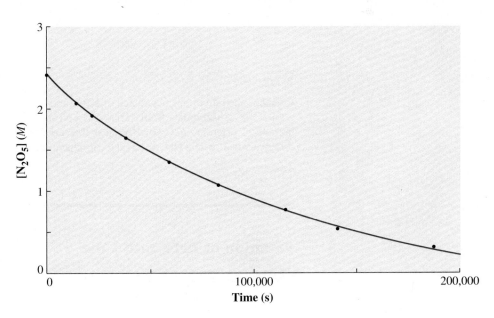

order to make the rate a positive number. We can also express the rate of disappearance of N_2O_5 as follows, where $\Delta[N_2O_5]$ represents the change in concentration of N_2O_5, and Δt represents the change in time:

$$\text{Rate} = -\tfrac{1}{2}\Delta[N_2O_5]/\Delta t$$

Using this equation as a reference point for defining the rate of reaction, we represent the rate of appearance of the products as follows. Remember that the stoichiometry must be taken into account.

$$\text{Rate} = \Delta[O_2]/\Delta t = \tfrac{1}{4}\Delta[NO_2]/\Delta t$$

EXAMPLE 13.1

Calculate the rate of disappearance of N_2O_5 and the rate of appearance of NO_2 at a time of 4.20×10^4 s, using data from Table 13.1 and starting with 2.330 M N_2O_5. Also calculate the rate of reaction, and show that it is the same whether it is calculated from the rate of disappearance of reactant or the rate of appearance of product.

Solution   In Table 13.1, the time 4.20×10^4 s is found in the middle of the time range from 3.16×10^4 s to 5.24×10^4 s. The concentration of N_2O_5 has the values 1.670 M and 1.350 M at these times. The rate of disappearance of N_2O_5 is given by

$$-\Delta[N_2O_5]/\Delta t = -\frac{[N_2O_5]_2 - [N_2O_5]_1}{t_2 - t_1}$$

$$= -\frac{1.350 - 1.670}{5.24 \times 10^4 - 3.16 \times 10^4}$$

$$= -\frac{(-0.320\ M)}{2.08 \times 10^4\ \text{s}} = 1.54 \times 10^{-5}\ M\ \text{s}^{-1}$$

The rate of appearance of NO_2 is twice the rate of disappearance of N_2O_5, because 4 mol NO_2 are produced for every 2 mol N_2O_5 that react.

$$\Delta[NO_2]/\Delta t = 2 \times -\Delta[N_2O_5]/\Delta t = 2 \times 1.54 \times 10^{-5}\ M\ \text{s}^{-1} = 3.08 \times 10^{-5}\ M\ \text{s}^{-1}$$

The rate of reaction is given by the following equations:

$$\text{Rate} = -\tfrac{1}{2}\Delta[N_2O_5]/\Delta t = \tfrac{1}{2}(1.54 \times 10^{-5}\ M\ \text{s}^{-1}) = 7.70 \times 10^{-6}\ M\ \text{s}^{-1}$$

or $$\text{Rate} = \tfrac{1}{4}\Delta[NO_2]/\Delta t = \tfrac{1}{4}(3.08 \times 10^{-5}\ M\ \text{s}^{-1}) = 7.70 \times 10^{-6}\ M\ \text{s}^{-1}$$

Practice Problem 13.1

Calculate the rate of disappearance of N_2O_5 and the rate of appearance of NO_2 at a time of 2.54×10^4 s, using data from Table 13.1 and starting with 2.330 M N_2O_5. Also calculate the rate of reaction, and show that it is the same whether it is calculated from the rate of disappearance of reactant or the rate of appearance of product.

Answer: $-\Delta[N_2O_5]/\Delta t = 1.94 \times 10^{-5}\ M\ \text{s}^{-1}$, $\Delta[NO_2]/\Delta t = 3.88 \times 10^{-5}\ M\ \text{s}^{-1}$, rate = $9.70 \times 10^{-6}\ M\ \text{s}^{-1}$

Variation of Rate with Time

For the equation to yield an accurate value for the rate, the time interval (Δt) between the two points should be as small as possible, because the rate of most

Figure 13.4 An average rate is the slope of a line drawn between two points on a curve that are equidistant in time from the time of interest. An instantaneous rate is the slope of a line drawn tangent, or exactly parallel, to the curve at the time of interest.

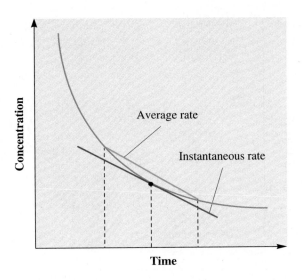

reactions decreases as the reaction proceeds. The decrease in rate is shown by the curvature in the plot of $[N_2O_5]$ against time in Figure 13.3. The most accurate rate is the one for a particular "point" in time, called the **instantaneous rate.** Rates calculated for longer time intervals are **average rates.** The difference between the two types is shown in Figure 13.4. The average rate comes closer to the instantaneous rate as the time interval gets smaller.

Measurement of the Initial Rate

How can we measure instantaneous rates?

Examine Figure 13.3 carefully and note that the first part of this plot of concentration against time looks more like a straight line than later parts. This observation holds for most chemical reactions. We can determine instantaneous rates with more accuracy here than in later, more highly curved parts of the graph. We also know the concentrations better early in the reaction, because they are close to the initial concentrations that we selected for the experiment.

To determine instantaneous rates in this way, we draw a straight line through the initial data, as shown in Figure 13.5, ignoring any data that begin to curve upward from this straight line. We calculate the slope of the straight line from Δconcentration/Δtime. The negative value of the slope of the initial part of the concentration–time curve is called the **initial rate,** the rate at the very beginning of the reaction. We can measure this slope reasonably accurately, especially for slow reactions, by drawing a tangent to the concentration–time curve as shown in Figure 13.5. Calculations of initial rates are essentially the same as the rate calculations we have done so far.

Consider the experiment in which the initial concentration of N_2O_5 is 2.330 *M* at 45°C. To determine the initial rate, we draw a line tangent to the initial part of the curve, as shown in Figure 13.5. We calculate the slope of this line from two points—say, at times of 0 s and 5.00×10^4 s. The concentrations on the line at these times are 2.330 *M* and 1.172 *M*, respectively. The concentration changes by -1.158 *M* in 5.00×10^4 s. The slope of the line is given by

$$\text{Slope} = \Delta\text{concentration}/\Delta\text{time} = -1.158 \ M/5.00 \times 10^4 \ \text{s} = -2.32 \times 10^{-5} \ M \ \text{s}^{-1}$$

Figure 13.5 We can determine the initial rate of the reaction by drawing a straight line through the initial portion of the concentration–time curve. We obtain the initial rate by taking the negative value of the slope of this straight line.

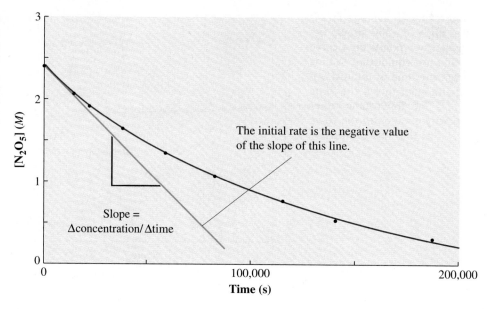

The initial rate is the negative value of the slope of this line.

Slope = Δconcentration/Δtime

The rate of change in N_2O_5 concentration is the negative of the slope.

$$-\Delta[N_2O_5]/\Delta t = -\text{slope} = 2.32 \times 10^{-5} \ M \ s^{-1}$$

The initial rate of the reaction is obtained from the equation

$$\text{Rate} = -\tfrac{1}{2}\Delta[N_2O_5]/\Delta t = \tfrac{1}{2}(2.32 \times 10^{-5} \ M \ s^{-1}) = 1.16 \times 10^{-5} \ M \ s^{-1}$$

Note that this value is slightly larger than the first value of the rate of reaction given in Table 13.1. This is because it is the rate for a time of 0 s, whereas that in the table is the average for the first 1.11×10^4 s.

13.2 Effect of Concentration on Rates

▶ In the chapter introduction, we saw that the rate of oxidation of iron was dependent on the oxygen concentration. Many reaction rates are concentration-dependent. For example, consider the reaction between sodium carbonate and hydrochloric acid.

$$Na_2CO_3(s) + 2HCl(aq) \longrightarrow 2NaCl(aq) + H_2O(l) + CO_2(g)$$

Sodium carbonate, Na_2CO_3, is a white, odorless, hygroscopic salt. It occurs more commonly as a hydrate, such as the decahydrate, $Na_2CO_3 \cdot 10H_2O$, which is known as washing soda. Na_2CO_3 is used in the manufacture of other sodium salts, glass, and soap, for washing wool and various textiles, as a general cleanser, and for softening water.

As shown in Figure 13.6, the CO_2 gas can be collected, and the rate of reaction can be measured from the volume of gas produced per unit of time. You can see that the solutions in the figure produce different amounts of CO_2. That is because all the solutions have different concentrations of $HCl(aq)$, and the rate of production of gas increases as the concentration of hydrochloric acid increases. This experiment illustrates one of the factors that affect the rate of reaction—concentration.

Our examination of the rate of decomposition of N_2O_5 demonstrated that the rate of that reaction decreases as the reaction proceeds. This is true of most reactions. Why? Because the concentration of reactants decreases during the course of the reaction. Molecules must come together to react, and the frequency with which molecules collide is proportional to their concentrations, so we might expect the

RULES OF REACTIVITY

Like other carbonate salts, sodium carbonate reacts with acids to generate carbon dioxide gas.

Figure 13.6 The reaction of sodium carbonate with hydrochloric acid produces carbon dioxide gas. The three flasks contain different concentrations of hydrochloric acid, *from left to right:* 1.0 *M*; 3.0 *M*; 6.0 *M*. In each case, sodium carbonate is the limiting reagent. The different volumes of gas in the balloons indicate that the higher the concentration of hydrochloric acid, the faster carbon dioxide is produced.

rate of a reaction to depend on the concentration of reactants in some way. Although the rate usually does depend on the concentration of reactants, the concentration of products or catalysts may also have an effect. (Recall from Chapter 4 that a catalyst is a substance that changes the rate of reaction but can be recovered unchanged from the reaction mixture.)

The Rate Law

The dependence of the rate of reaction on the concentration of each species is represented by the **rate law,** or **rate equation.** For the ozone–nitric oxide reaction

$$O_3(g) + NO(g) \longrightarrow O_2(g) + NO_2(g)$$

the rate law is given by the equation

$$\text{Rate} = k[O_3][NO]$$

The proportionality constant k is called the **rate constant.**

We can give the rate law a general form. Suppose substances A and B react in the presence of catalyst D to give product C, as described by the equation

$$a\text{A} + b\text{B} \xrightarrow{\text{D}} c\text{C}$$

The rate law usually is some variation on the following:

$$\text{Rate} = k[\text{A}]^Z[\text{B}]^Y[\text{C}]^X[\text{D}]^W$$

The reaction product (C) does not always appear in the rate law (the exponent of its concentration is often zero). If the reaction involves a catalyst (species D), its concentration usually appears in the rate law, even though the catalyst does not appear in the balanced reaction equation.

The exponent for each concentration is called the *order* of that species and is usually an integer. The values assigned to the order of each species must be determined experimentally. If no exponent is given, the order is understood to be 1. The sum of the exponents is the **reaction order.** For an nth-order reaction in solution, the units of the rate constant are $M^{-(n-1)}\text{ s}^{-1}$. The order of each species tells us how the rate depends on the concentration of that species. For the ozone–nitric

Figure 13.7 *Left:* Vanadium(III) exists as a blue-gray hydrated ion, $V^{3+}(aq)$, in aqueous solution. It can be found in green to black salts such as VCl_3, VF_3, and V_2O_3. The V^{3+} ion is slowly oxidized in air to sky-blue VO^{2+}. *Right:* Vanadium(IV) exists in aqueous solution as the sky-blue hydrated oxoion, $VO^{2+}(aq)$. it occurs in the solid state in salts such as vanadium(IV) sulfate (also known as vanadyl sulfate), $VOSO_4$. This salt is used in dyes, colored glass, and blue and green pottery glazes.

What information can we obtain from a rate law?

oxide reaction, the reaction is first-order in ozone, first-order in nitric oxide, and zero-order in the products, molecular oxygen and nitrogen dioxide. The reaction is second-order overall. The rate constant has units of M^{-1} s^{-1}.

Knowing the orders of the species and a value of the rate constant enables us to calculate the rate of reaction under the range of conditions for which the rate law is valid. For example, vanadium(III) in aqueous solution is slowly oxidized by dissolved oxygen (Figure 13.7):

$$4V^{3+}(aq) + O_2(aq) + 2H_2O(l) \longrightarrow 4VO^{2+}(aq) + 4H^+(aq)$$

The *experimental* rate law for this reaction is given by the equation

$$\text{Rate} = k[V^{3+}][O_2]/[H^+]$$

The rate constant was determined experimentally to be 8.3×10^{-6} s^{-1} at 25°C. The reaction was determined to be first-order in V^{3+}, first-order in O_2, and inverse-first-order (order of -1) in H^+. The reaction is first-order overall; the sum of the orders is $1 + 1 - 1 = 1$.

Let's see how we can use these results to calculate the rate of reaction. Suppose we have a solution containing 0.100 M V^{3+} and 1.00 M H^+. Saturated with oxygen gas, the solution contains 0.0124 M O_2. We can calculate the rate of reaction from the rate law by using these concentrations and the value of the rate constant.

$$\text{Rate} = k[V^{3+}][O_2]/[H^+] = 8.3 \times 10^{-6}\ s^{-1} \times 0.100\ M \times 0.0124\ M/1.00\ M =$$
$$1.03 \times 10^{-8}\ M\ s^{-1}$$

This result is consistent with the slowness of the reaction, which we can observe; only 1.03×10^{-8} M of the 1.24×10^{-2} M of O_2 in the solution reacts during 1 s. Calculating the rate for other concentrations involves the same procedure, so we can summarize a lot of information about rates with the rate law and a value of the rate constant.

Reaction Order and Stoichiometry

Can the rate law be predicted from the reaction stoichiometry?

As we have noted, the order for each species must be determined experimentally. *The rate law has no necessary relationship to the stoichiometry of the reaction.* In some cases, there appears to be a connection between reaction order and reaction stoichiometry. Consider the reaction between nitric oxide and oxygen:

$$2NO(g) + O_2(g) \longrightarrow 2NO_2(g)$$

This reaction happens to have a rate law in which the reaction orders match the stoichiometric coefficients of the reactants:

$$\text{Rate} = k[NO]^2[O_2]$$

This reaction is second-order in NO, first-order in O_2, and third-order overall. However, reaction orders and stoichiometric coefficients do not necessarily match. Consider the reaction between bromide and bromate ions in acidic aqueous solution.

$$5Br^-(aq) + BrO_3^-(aq) + 6H^+(aq) \longrightarrow 3Br_2(aq) + 3H_2O(l)$$

The rate law for this reaction is

$$\text{Rate} = k[BrO_3^-][Br^-][H^+]^2$$

This reaction is first-order in BrO_3^- and in Br^-, second-order in H^+, and fourth-order overall.

Initial Rates and the Determination of Rate Laws

How can we determine the form of the rate law and the value of the rate constant?

To obtain a rate law, we must *experimentally* determine the order of reaction with respect to each species and the value of the rate constant. One way to determine reaction orders experimentally involves measuring initial rates for different starting concentrations. ▶ Consider again the decomposition of dinitrogen pentoxide dissolved in carbon tetrachloride, described in the chapter introduction.

$$2N_2O_5 \longrightarrow 4NO_2 + O_2(g)$$

The rate of reaction is

$$\text{Rate} = -\tfrac{1}{2}\Delta[N_2O_5]/\Delta t$$

Using initial rates, we can determine the dependence of the rate of the concentration of each species by inspection. Table 13.2 shows initial rates measured in a series of experiments with different initial concentrations of reactant and products. Examining the first three experiments indicates that the rate does not change when we change the concentration of either NO_2 or O_2. This means that the reaction is zero-order with respect to both products.

We note from the table, however, that the rate does change when we change the initial concentration of N_2O_5. The rate law must then have the following form:

$$\text{Rate} = k[N_2O_5]^n$$

This reaction is *n*th-order in $[N_2O_5]$ and *n*th-order overall. In general, *n* is most likely to equal 0, 1, or 2. We consider other integers, positive or negative, if these most common possibilities do not work. Only in rare cases do reaction orders have nonintegral values. The rate varies with concentration for the three common reaction orders $n = 0$, 1, and 2 as follows:

- Zero-order reaction: rate is independent of concentration (because a number raised to the zero power equals 1); rate = k

Table 13.2 Dependence of Initial Rate of Decomposition of N_2O_5 on Concentrations at 45°C

$[N_2O_5]$ (M)	$[NO_2]$ (M)	$[O_2]$ (M)	Rate $(M\ s^{-1})$	Rate/$[N_2O_5]$ (s^{-1})
2.21	2.00	1.00	1.14×10^{-5}	5.16×10^{-6}
2.21	1.00	2.00	1.12×10^{-5}	5.07×10^{-6}
2.21	0.00	0.00	1.13×10^{-5}	5.11×10^{-6}
2.00	0.00	0.00	1.05×10^{-5}	5.25×10^{-6}
1.79	0.00	0.00	9.67×10^{-6}	5.40×10^{-6}
1.51	0.00	0.00	7.83×10^{-6}	5.19×10^{-6}
1.23	0.00	0.00	6.31×10^{-6}	5.13×10^{-6}
0.92	0.00	0.00	4.81×10^{-6}	5.23×10^{-6}

- First-order reaction: rate is proportional to the concentration; rate $= k[N_2O_5]$
- Second-order reaction: rate is proportional to the square of the concentration; rate $= k[N_2O_5]^2$

Examination of the rates in Table 13.2 clearly shows that the rates vary with the concentration of N_2O_5, so the reaction cannot be zero-order. We can determine whether either of the other orders (1 or 2) is the correct one by comparing rates. A simple approach is to find two experiments in which the concentration varies by a factor of 2. The rate should then vary by 2 raised to the power of the order, as shown by the following ratio of two rates.

$$\text{Rate}_2/\text{rate}_1 = (k[N_2O_5]_2{}^n)/(k[N_2O_5]_1{}^n) = 2^n$$

If the reaction is first-order ($n = 1$), the rate varies by a factor of 2, whereas if it is second-order ($n = 2$), the rate varies by a factor of 4. The fifth and the last experiments in Table 13.2 have N_2O_5 concentrations that vary by a factor of about 2, so we can use these data.

$$\text{Rate}_2/\text{rate}_1 = 9.67 \times 10^{-6}/4.81 \times 10^{-6} = 2.01 \approx 2$$

The rate varies by a factor of about 2, so we can conclude that the reaction is first-order with respect to N_2O_5. This reaction is thus first-order overall and is said to follow *first-order kinetics*.

We can verify this conclusion, and complete the determination of the rate law, by rearranging the rate law:

$$\text{Rate} = k[N_2O_5]$$

Solving for the rate constant, we get the equation

$$k = \text{rate}/[N_2O_5]$$

If we have correctly evaluated the order, the rate constant should be the same, within experimental error, for all concentrations. The last column in Table 13.2 lists rate constants calculated in this way. These values are indeed essentially constant, so our conclusion that the reaction is first-order in N_2O_5 and first-order overall is correct. The rate constant is usually given as the average of all the values determined in this way: $k = (5.19 \pm 0.07) \times 10^{-6}\ s^{-1}$ for the decomposition of N_2O_5 at 45°C.

EXAMPLE 13.2

Consider the decomposition of nitramide, described by the equation

$$NH_2NO_2(aq) \longrightarrow N_2O(g) + H_2O(l)$$

Some experimental data for this reaction at 25°C follow.

Initial Concentration of NH_2NO_2 (M)	Initial Rate (M s^{-1})
0.400	1.88×10^{-5}
0.200	9.32×10^{-6}
0.100	4.76×10^{-6}

Determine the rate law, including the value of the rate constant, assuming that the rate depends only on the concentration of NH_2NO_2, not on the concentrations of the products.

Solution   The rate is defined as

$$Rate = k[NH_2NO_2]^n$$

To determine the reaction order, we assume that n can have a value of 0, 1, or 2. If n is 0, the rate should be independent of the concentration of NH_2NO_2. We can see from the table that this is not the case, so the reaction cannot be zero-order. Next we examine ratios of concentrations and rates to determine whether the reaction is first- or second-order. The ratio of concentrations in the first two experiments is given by

$$[NH_2NO_2]_1/[NH_2NO_2]_2 = 0.400 \; M/0.200 \; M = 2$$

If the reaction is first-order, the ratio of rates should equal 2; if it is second-order, the ratio should equal 4. The measured ratio of rates is

$$Rate_1/rate_2 = 1.88 \times 10^{-5} \; M \; s^{-1}/9.32 \times 10^{-6} \; M \; s^{-1} = 2.02$$

This ratio is very close to 2, so the reaction follows first-order kinetics. We would reach the same conclusion by comparing other rate ratios. Thus the rate law is

$$Rate = k[NH_2NO_2]$$

To obtain the rate constant, we rearrange the rate law:

$$k = rate/[NH_2NO_2]$$

Then we divide the rate by $[NH_2NO_2]$ for each experiment.

Experiment 1: $k = 1.88 \times 10^{-5} \; M \; s^{-1}/0.400 \; M = 4.70 \times 10^{-5} \; s^{-1}$
Experiment 2: $k = 9.32 \times 10^{-6} \; M \; s^{-1}/0.200 \; M = 4.66 \times 10^{-5} \; s^{-1}$
Experiment 3: $k = 4.76 \times 10^{-6} \; M \; s^{-1}/0.100 \; M = 4.76 \times 10^{-5} \; s^{-1}$

The average value of the rate constant for this reaction is $(4.71 \pm 0.04) \times 10^{-5} \; s^{-1}$.

Practice Problem 13.2

Consider the decomposition of ozone in the gas phase:

$$2O_3(g) \longrightarrow 3O_2(g)$$

Some experimental data for this reaction at 80°C follow.

Initial Concentration of O_3 (M)	Initial Rate (M s^{-1})
6.00×10^{-3}	5.03×10^{-7}
3.00×10^{-3}	1.28×10^{-7}
1.50×10^{-3}	3.08×10^{-8}

(continued)

Determine the rate law, including the value of the rate constant, assuming that the rate depends only on the concentration of ozone.

Answer: Rate $= k[O_3]^2$, $k = 0.0140 \pm 0.0002$ M^{-1} s^{-1}

Graphical Determination of the Rate Law

How can we determine the rate law if we are unable to measure initial rates?

The use of initial rates to determine the rate law relies on the assumption that the rate is constant initially and that accurate concentration measurements can be made in the initial part of the reaction. These assumptions are usually valid only for very slow reactions. Furthermore, the accuracy of the measured rates depends on the accuracy of two concentration measurements. If one concentration measurement is inaccurate, the initial rate is also inaccurate. Finally, a number of experiments must be carried out just to determine the reaction order.

An alternative approach involves the measurement of concentration–time data in a single experiment throughout the course of the reaction, not just during the initial part. These data are then manipulated graphically to determine the rate law.

Rate laws, of course, are expressed in terms of rate–concentration relationships. To convert them to concentration–time relationships, it is necessary to use integration procedures from calculus. The result is an *integrated rate law,* a form of the rate equation that expresses concentration as a function of time. Here we are concerned not with the process of integration but only with the resulting integrated rate laws.

Consider the general form of a rate law in which the rate depends on a single reactant, substance A:

$$-\Delta[A]/\Delta t = k[A]^n$$

The rate of reaction, measured in terms of the disappearance of substance A, depends on the concentration of A, with an order of n. We will discuss integrated rate laws for the three most common rate laws, zero-order, first-order, and second-order.

Zero-Order Rate Laws   A simple zero-order rate law has the form

$$-\Delta[A]/\Delta t = k[A]^0 = k$$

Using calculus, we can convert this equation to a relation between concentration and time.

$$[A] = [A]_0 - kt$$

$[A]$ is the concentration of A measured at the time of interest (t), and $[A]_0$ is the initial concentration of A (at $t = 0$). If a reaction follows a zero-order rate law, then a plot of $[A]$ against t should be linear, because this equation is just the equation for a straight line with an intercept of $[A]_0$ and a slope of $-k$. If the reaction is not zero-order, then this equation is not valid, and a plot of $[A]$ against time will not be linear.

The most common examples of zero-order reactions are the decompositions of some gaseous molecules occurring on the surfaces of metals, such as the decomposition of nitrous oxide on gold.

$$2N_2O(g) \xrightarrow{\text{Au}} 2N_2(g) + O_2(g)$$

Such reactions are zero-order in the gaseous molecule, because if the sites on the metal surface available for reaction all become occupied by an adsorbed molecule, the rate of reaction becomes constant and is not dependent on the amount of gaseous reactant. Another example is the production of hydrogen gas by reaction of a metal, such as magnesium, in acid, as shown in Figure 13.8.

$$Mg(s) + 2HCl(aq) \longrightarrow MgCl_2(aq) + H_2(g)$$

RULES OF REACTIVITY

Active metals, such as magnesium, react with acids to form metal salts and hydrogen gas.

We cannot measure the concentration of a solid, so we must express the rate law in terms of a substance other than $Mg(s)$. Experimentally, it is simplest to determine the amount of hydrogen gas produced. Because they have the same stoichiometric coefficients, the rates can be related as follows:

$$\Delta[H_2]/\Delta t = -\Delta[Mg]/\Delta t$$

The same change in sign shows up in the integrated rate law for the formation of a product, which has the following form:

$$[H_2] = [H_2]_0 + kt$$

A plot of $[H_2]$ against time, made by using data collected when a small piece of magnesium metal was reacted with 0.500 M HCl in a closed container, is indeed linear, as shown in Figure 13.9.

First-Order Rate Laws   A first-order law for the disappearance of reactant A has the form

$$-\Delta[A]/\Delta t = k[A]$$

The corresponding integrated rate law is

$$\ln [A] = \ln [A]_0 - kt$$

If we plot $\ln [A]$ against t, we should get a straight line, with the intercept $\ln [A]_0$ and the slope $-k$. If the reaction is not first-order, this equation is not valid, and a

Figure 13.8 Magnesium reacts with hydrochloric acid, producing bubbles of hydrogen gas.

Figure 13.9 A zero-order plot of the concentration of hydrogen gas against time for the reaction of magnesium with hydrochloric acid. The intercept is zero because no hydrogen was present initially. The slope equals the zero-order rate constant.

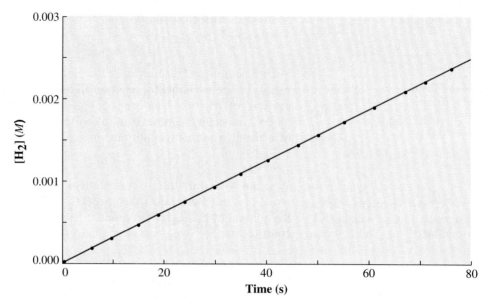

plot of $\ln [A]$ against time will not be linear. The rate law can be reorganized by taking the antilogarithm of both sides, which yields the equation

$$[A] = [A]_0 e^{-kt}$$

Consider the concentration–time data collected for the decomposition of N_2O_5 and listed in Table 13.1. Earlier we used data from several initial-rate experiments to establish that this is a first-order reaction. We can also use the data from a single experiment to determine the rate law by plotting $\ln [N_2O_5]$ against time, as shown in Figure 13.10. These data conform to a straight line, indicating that the reaction is indeed first-order. The rate law, then, takes the first-order form (the term $\frac{1}{2}$ arises from the stoichiometric coefficient for N_2O_5).

$$-\tfrac{1}{2}\Delta[N_2O_5]/\Delta t = k[N_2O_5] \quad \text{or} \quad -\Delta[N_2O_5]/\Delta t = 2k[N_2O_5]$$

Figure 13.10 A plot of $\ln [N_2O_5]$ against time for data from Table 13.1. The nearly perfect straight line indicates that the reaction follows first-order kinetics.

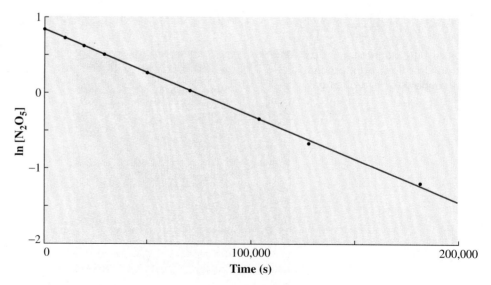

How do we know this reaction is not zero-order? Suggest a method to establish that it is not.

The slope of the straight line can be calculated from any two points on the line, say at times of 0 and 1.00×10^5 s.

$$\text{Slope} = (\ln [N_2O_5]_1 - \ln[N_2O_5]_2)/(t_1 - t_2)$$
$$= (0.85 - (-0.20)/(0 \text{ s} - 1.00 \times 10^5 \text{ s})$$
$$= (1.05)/(-1.00 \times 10^5 \text{ s}) = -1.05 \times 10^{-5} \text{ s}^{-1}$$

The rate constant is obtained from the slope via the relation

$$\text{Slope} = -2k$$

Thus $2k = 1.05 \times 10^{-5} \text{ s}^{-1}$ and $k = 5.25 \times 10^{-6} \text{ s}^{-1}$ for this reaction.

Second-Order Rate Laws A second-order rate law for the disappearance of a substance A has the form

$$-\Delta[A]/\Delta t = k[A]^2$$

Using calculus, we can transform this equation into the integrated rate law

$$\frac{1}{[A]} = \frac{1}{[A]_0} + kt \quad \text{or} \quad [A]^{-1} = [A]_0^{-1} + kt$$

If a reaction is second-order, a plot of $1/[A]$ against time should give a straight line. The intercept of the straight line should be $1/[A]_0$, and the slope should be the rate constant, k. If the plot is not linear, then the reaction is not second-order.

An example of a second-order reaction is the decomposition of uranium(V) into uranium(IV) and uranium(VI) in acidic aqueous solution, which is described by the equation

$$2UO_2^+(aq) + 4H^+(aq) \longrightarrow U^{4+}(aq) + UO_2^{2+}(aq) + 2H_2O(l)$$

Uranium, when dissolved in acidic aqueous solutions, can form several ions. Uranium(VI) exists as a stable oxoion, UO_2^{2+}, which has a bright yellow color. Uranium(IV) forms the apple-green U^{4+}, which undergoes hydrolysis to form hydroxide species unless it is kept in strongly acidic solution. This ion can be oxidized to uranium(VI) by the oxygen in air. Uranium(V) exists as a very unstable ion with the formula UO_2^+.

If the reaction is second-order with respect to UO_2^+, the following equation describes the rate law:

$$-\tfrac{1}{2}\Delta[UO_2^+]/\Delta t = k[UO_2^+]^2 \quad \text{or} \quad -\Delta[UO_2^+]/\Delta t = 2k[UO_2^+]^2$$

The integrated rate law is

$$[UO_2^+]^{-1} = [UO_2^+]_0^{-1} + 2kt$$

Data collected for a $1.00 \times 10^{-3} M$ solution of UO_2^+ in $0.100 M$ acid were used to construct a plot of $[UO_2^+]^{-1}$ against time, as shown in Figure 13.11. Because this plot is linear, the reaction is indeed second-order with respect to UO_2^+. The rate constant is obtained from the slope of the straight line and has a value of $2k = 13.0$ $M^{-1} \text{ s}^{-1}$, or $k = 6.50 M^{-1} \text{ s}^{-1}$. Plots of $[UO_2^+]$ against time and of $\ln [UO_2^+]$ against time are not linear, indicating that the reaction is neither zero-order nor first-order with respect to UO_2^+.

Half-Lives and Reaction Rates

We can use the rate law, with the value of the rate constant, to calculate the concentration at any desired time (by inserting the time into the integrated rate law) or to calculate the time that it would take to reduce the concentration to any desired value (by inserting the target concentration into the integrated rate law).

Figure 13.11 A linear plot of $[UO_2^+]^{-1}$ against time indicates that the decomposition of uranium(V) is second-order. The slope of the line equals the second-order rate constant, 13.0 $M^{-1} s^{-1}$.

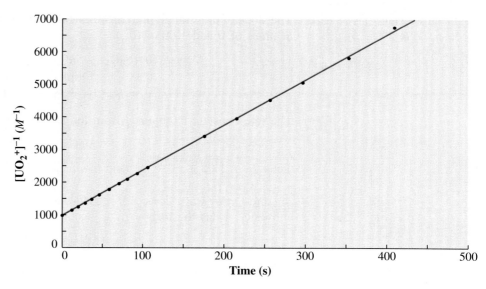

EXAMPLE 13.3

The integrated rate law for the decomposition of dinitrogen pentoxide at 45°C in carbon tetrachloride solution is given by the equation

$$\ln [N_2O_5] = \ln [N_2O_5]_0 - 2kt$$

The rate constant, k, has the value $5.25 \times 10^{-6} s^{-1}$. Calculate the time required for the concentration of N_2O_5 to be reduced to one-fifth of its original value.

Solution   We need to find the time required for $\frac{4}{5}$ of the original N_2O_5 to react. We will let the original concentration of N_2O_5 be $[N_2O_5]_0$. Then the concentration at the desired time, t, will be $[N_2O_5]_0/5$. We can use the integrated form of the rate law to carry out this calculation.

$$\ln [N_2O_5] = \ln [N_2O_5]_0 - (1.05 \times 10^{-5} s^{-1})t$$

We insert the value of the concentration.

$$\ln([N_2O_5]_0/5) = \ln [N_2O_5]_0 - (1.05 \times 10^{-5} s^{-1})t$$

Then we rearrange terms to collect all terms that involve the initial concentration.

$$\ln [N_2O_5]_0 - \ln(0.200 [N_2O_5]_0) = \ln([N_2O_5]_0/0.200 [N_2O_5]_0) = (1.05 \times 10^{-5} s^{-1})t$$

The initial concentration can be eliminated from the equation.

$$\ln(1/0.200) = \ln(5.00) = 1.05 \times 10^{-5} t$$

Evaluating $\ln(5.00)$ by taking the antilogarithm and solving for t gives

$$t = 1.61/1.05 \times 10^{-5} = 1.53 \times 10^5 s$$

Practice Problem 13.3

Calculate how much time it would take for the concentration of N_2O_5 to be reduced to three-fourths of its original value, given the integrated rate law

$$\ln[N_2O_5] = \ln[N_2O_5]_0 - (1.05 \times 10^{-5} \text{ s}^{-1})t$$

Answer: 2.74×10^4 s

One application of calculations such as those in Example 13.3 is the determination of half-life. The **half-life** of a reaction is the time required for a given concentration of reactant to be reduced to half its initial value. Depending on the reaction order, the half-life may or may not depend on the concentration. We just used a rate law to determine the time it takes for the initial concentration of a reactant to be reduced to some lesser concentration. The rate laws can be used similarly to determine the time it takes to reduce the initial concentration to half its value.

Consider the integrated rate law for a first-order reaction of a substance A:

$$\ln([A]/[A]_0) = -kt$$

At a time equal to one half-life, symbolized $t_{\frac{1}{2}}$, $[A]$ will equal $\frac{1}{2}[A]_0$.

$$\ln(\tfrac{1}{2}[A]_0/[A]_0) = -kt_{\frac{1}{2}}$$

Solving for $t_{\frac{1}{2}}$ gives

$$t_{\frac{1}{2}} = -\ln(\tfrac{1}{2})/k$$

The natural logarithm of $\frac{1}{2}$ is -0.693, so we get the equation

$$t_{\frac{1}{2}} = 0.693/k$$

For first-order reactions, then, the half-life depends only on the rate constant. It is independent of the initial concentration. This means that all successive half-lives have the same value. In other words, it takes as long to convert $[A]_0$ to $[A]_0/2$ as it does to convert $[A]_0/2$ to $[A]_0/4$ or to convert $[A]_0/4$ to $[A]_0/8$. This is illustrated in Figure 13.12, where successive half-lives are marked on a plot of concentration against time for the first-order decomposition of N_2O_5.

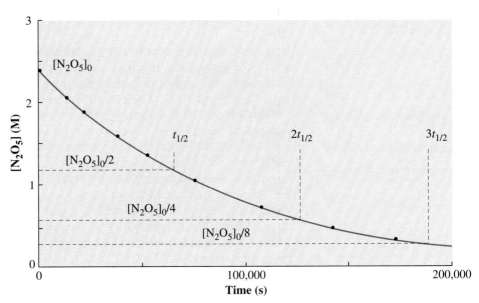

Figure 13.12 Successive half-lives for a first-order reaction, the decomposition of dinitrogen pentoxide.

EXAMPLE 13.4

The rate constant for the N_2O_5 decomposition reaction is 1.05×10^{-5} s^{-1} if the rate law is expressed as

$$-\Delta[N_2O_5]/\Delta t = k[N_2O_5]$$

What is the half-life of this reaction?

Solution This is a first-order reaction, so the half-life is related to the rate constant by the equation

$$t_{\frac{1}{2}} = 0.693/k$$

Inserting the value of the rate constant, we obtain

$$t_{\frac{1}{2}} = 0.693/1.05 \times 10^{-5} \text{ s}^{-1} = 6.60 \times 10^4 \text{ s}$$

Practice Problem 13.4

The half-life for a first-order reaction is $t_{\frac{1}{2}} = 4.62$ s. What is the rate constant for this reaction?

Answer: 0.150 s^{-1}

Following a similar procedure, we obtain the following equation for zero-order reactions:

$$t_{\frac{1}{2}} = [A]_0/2k$$

Here, the time required to convert an initial concentration of reactant to half its value depends both on the initial concentration and on the rate constant. The smaller the rate constant (the slower the reaction), the longer the half-life. This result also indicates that for a zero-order reaction, the more reactant we start with, the longer it takes for half of it to react.

For a second-order reaction, the relationship between half-life and rate constant is given by the equation

$$t_{\frac{1}{2}} = 1/k[A]_0$$

Here, the half-life depends both on the initial concentration and on the rate constant. The greater the initial concentration of reactant, the shorter the half-life. This means that, unlike first-order reactions, second-order reactions have increasingly longer half-lives. These results are summarized in Table 13.3.

Half-lives are commonly given as a characteristic property of radioactive isotopes of the elements. Because the decay of radioactivity is a first-order process, the half-life of a radioactive isotope describes the length of time required for the isotope to reduce its radioactivity to half the initial value. Each successive half-life is identical to the first. By analyzing half-lives, we can determine when the radioactivity of a sample will be reduced enough to make it safe to handle. For example, cobalt-60, used as a radiation source to treat cancer, has a half-life of 5.2 years. After 10 half-lives, 99.9% of the radioactivity will have disappeared. If we assume the concentration of radioactivity is 100% initially, dividing by 2 ten times gives a concentration of 0.09%, which is the amount of radioactivity left after 10 half-lives. Thus it takes about 52 years for Co-60 radiation to be reduced to 0.1% of its initial value.

Table 13.3 Rate Laws and Half-Lives for Reactions of Various Orders

Order	Rate Law	Integrated Rate Law	Half-Life
0	$-\Delta[A]/\Delta t = k$	$[A] = [A]_0 - kt$	$t_{\frac{1}{2}} = [A]_0/2k$
1	$-\Delta[A]/\Delta t = k[A]$	$\ln[A] = \ln[A]_0 - kt$	$t_{\frac{1}{2}} = 0.693/k$
2	$-\Delta[A]/\Delta t = k[A]^2$	$[A]^{-1} = [A]_0^{-1} + kt$	$t_{\frac{1}{2}} = 1/k[A]_0$

Rate Laws Involving Concentrations of More Than One Species

The rate laws we have considered so far have depended on the concentration of only one species. However, most rate laws depend on the concentrations of two or more species. These rate laws can also be solved to give integrated rate laws, but the procedures can be quite complex.

If we can measure initial rates, we can use a method similar to that used earlier, but we must carry out at least two series of experiments. In each series, only one substance's concentration varies; those of the others are held constant.

Why must we set up the experiment such that only one concentration changes?

Consider the reaction between arsenite ion and cerium(IV) ion in acidic aqueous solution, for example.

$$AsO_3^{3-}(aq) + 2Ce^{4+}(aq) + H_2O(l) \longrightarrow AsO_4^{3-}(aq) + 2Ce^{3+}(aq) + 2H^+(aq)$$

One series of experiments is summarized in the following list.

Cerium is a ductile, malleable, iron-gray metal used in alloys with iron to make spark metal, used in the flints of gas lighters. It reacts slowly with cold water and rapidly with hot water or acids to release hydrogen gas and form aqueous cerium(III) ions. These ions can be oxidized to cerium(IV) ions. Cerium(IV) can be isolated as the sulfate salt, which usually occurs as a hydrate, Ce(SO₄)₂·4H₂O. Cerium(IV) is yellow; cerium(III) is colorless. Cerium(IV) oxide, a white or pale yellow powder, is used to decolorize glass.

Initial Molar Concentration of		**Initial Rate, M s^{-1}**
AsO_3^{3-}	Ce^{4+}	
0.0100	0.120	6.34×10^{-5}
0.0200	0.120	1.27×10^{-4}

The experiments vary the concentration of arsenite ion but keep the cerium(IV) concentration constant. We use these experiments to determine the order with respect to arsenite ion. Doubling the concentration of arsenite ion doubles the rate, so the reaction is first-order in arsenite ion. A second series of experiments (including one in common with the first series) varies the cerium(IV) concentration but keeps the arsenite concentration constant.

Initial Molar Concentration of		**Initial Rate, M s^{-1}**
AsO_3^{3-}	Ce^{4+}	
0.0100	0.120	6.34×10^{-5}
0.0100	0.0600	1.58×10^{-5}

We use these experiments to determine the order with respect of cerium(IV). Here, doubling the Ce^{4+} concentration quadruples the rate ($6.34 \times 10^{-5}/1.58 \times 10^{-5} = 4.01$), so the reaction is second-order in cerium(IV).

The rate law for the reaction is given by the following equation, which combines the information determined from the experiments.

$$\text{Rate} = k[AsO_3^{3-}][Ce^{4+}]^2$$

We can evaluate the rate constant by rearranging the rate law:

$$k = \text{rate}/[\text{AsO}_3{}^{3-}][\text{Ce}^{4+}]^2$$

The values of k obtained from the three experiments are 0.440, 0.441, and 0.439 M^{-2} s^{-1}.

13.3   Effect of Temperature on Rates

Nitrogen and oxygen gases coexist in air. However, when we put air in the cylinder of an automobile engine to burn gasoline, nitrogen and oxygen react to form nitric oxide (NO). Why do nitrogen and oxygen react under these conditions? Because the temperature in the engine is very high. Similarly, when the space shuttle re-enters the atmosphere, it becomes extremely hot, causing oxygen and nitrogen molecules in air to dissociate into atoms.

$$\text{N}_2(g) \longrightarrow 2\text{N}(g)$$

$$\text{O}_2(g) \longrightarrow 2\text{O}(g)$$

These atoms can recombine to form nitric oxide.

$$\text{N}(g) + \text{O}(g) \longrightarrow \text{NO}(g)$$

Many industrial chemical processes are carried out at high temperature to speed up the reactions. An example is the Haber process for synthesizing ammonia from nitrogen and hydrogen gases:

$$\text{N}_2(g) + 3\text{H}_2(g) \rightleftharpoons 2\text{NH}_3(g)$$

This process is usually carried out at about 500°C because it is very slow at lower temperatures.

The rates of these reactions become higher as the temperature increases.

Indeed, the rates of almost all reactions increase with increasing temperature. This effect is illustrated in Figure 13.13 for the reaction between chromium(III) and manganese(III) in acidic aqueous solution.

$$2\text{Cr}^{3+}(aq) + 6\text{Mn}^{3+}(aq) + 7\text{H}_2\text{O}(l) \longrightarrow \text{Cr}_2\text{O}_7{}^{2-}(aq) + 6\text{Mn}^{2+}(aq) + 14\text{H}^+(aq)$$

Generally, we find that when the reaction rate increases with temperature, the rate constant also increases with temperature. Consider the decomposition of nitrogen dioxide, for example.

$$2\text{NO}_2(g) \longrightarrow 2\text{NO}(g) + \text{O}_2(g)$$

Figure 13.13 *Left:* Blue-gray chromium(III) is mixed with red manganese(III). *Center:* When the solutions are cold, they react only slowly. *Right:* But they rapidly form the yellow-orange dichromate ion when hot.

The dichromate ion is found in salts such as orange-red potassium dichromate, $K_2Cr_2O_7$. Dichromate salts are extensively used as oxidizing agents and for tanning hides. They are also used in dyes, paints, porcelain glazes, batteries, and safety matches. In basic aqueous solutions, the dichromate ion is converted to the yellow chromate ion, CrO_4^{2-}.

Why are reactions generally faster at higher temperatures?

This reaction follows a second-order rate law:

$$-\Delta[NO_2]/\Delta t = k[NO_2]^2$$

Figure 13.14 shows a plot of some values of the second-order rate constant against the absolute temperature. This plot is typical of that observed for most rate constants. Generally, the rate constant increases as an exponential function of temperature.

Several theories have been proposed to explain the effect of temperature on the rate constant. We will examine the simplest of these: collision theory and transition state theory. Other, more complex theories are similar but provide a more detailed thermodynamic analysis of the energy changes involved.

Collision Theory

According to **collision theory,** three conditions must be met before a reaction between two molecules can take place. The molecules must collide with one another, they must be oriented in such a way that the proper bonds can be broken and made, and they must have sufficient energy to ensure that these changes can occur.

1. The molecules must collide in order to react. Two facts about collisions are relevant here. First, *collision frequencies*—rates of collisions between molecules—can be calculated, and they turn out to be higher than the rates of most reactions. Thus, the orientation and energy factors must serve to slow down reactions. Second, the effect of temperature on reaction rates cannot be due solely to the collision frequency. A change in temperature of 10°C usually changes a reaction rate by 100% to 300% but increases the collision frequency by only about 2%.

2. Molecules do not always react when they collide. (This helps explain why the collision rate is higher than the reaction rate.) As shown in Figure 13.15 for the reaction

$$O_3(g) + NO(g) \longrightarrow O_2(g) + NO_2(g)$$

Figure 13.14 The rate constant for the decomposition of nitrogen dioxide increases with temperature, as most rate constants do. The relationship is exponential, giving a curve such as that shown here.

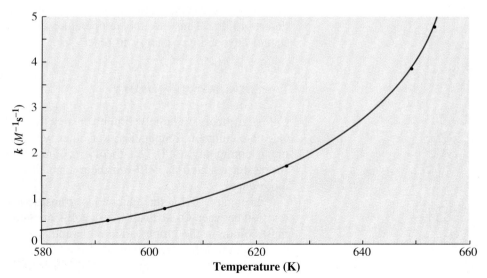

Figure 13.15 The orientation
of two molecules must be
appropriate, or no chemical
reaction can occur when they
collide.

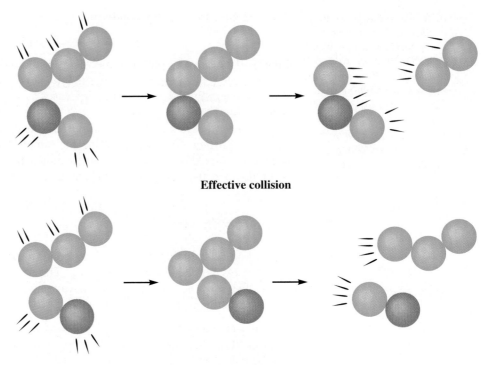

Effective collision

Ineffective collision

in order for the collision to result in the formation of products, the molecules must collide in such a way that new bonds can be made. If the molecules are improperly oriented, no chemical reaction can occur, and the molecules simply rebound from one another unchanged. The orientation of molecules does not depend on temperature.

3. Even when properly oriented molecules collide, they may not react, because they may not have enough energy to break bonds. It is this energy factor that explains most of the effect of temperature on the reaction rate. As the temperature increases, the average kinetic energy of the molecules increases, as shown in Figure 13.16. Thus, as the temperature increases, more and more molecules should have enough energy to react.

Transition-State Theory

Collision theory is largely limited to reactions of gases, where it is possible to calculate collision frequencies. For other types of reactions, **transition-state theory** is commonly used. This theory recognizes the same basic requirements for a successful reaction as collision theory, but it explains the energy requirement in more detail.

In Figure 13.17, the elements of transition-state theory are illustrated for the reaction between O_3 and NO. As in collision theory, in order to react, the reactants must collide in the proper orientation and with sufficient energy. Reactant molecules that meet these requirements form an **activated complex** or **transition state,** a short-lived, unstable, high-energy species that must be formed for reactants to be

Figure 13.16 The average kinetic energy of molecules increases as the temperature increases. This is because the fraction of the molecules with energy that meets or exceeds some minimum requirement is greater at higher temperature.

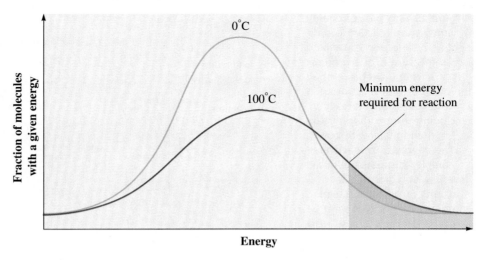

converted to products. The activated complex is generally designated by a chemical formula enclosed in brackets with a superscript double dagger ($\ddagger$), such as $O_3NO^{\ddagger}$ for the reaction of O_3 with NO in Figure 13.17. The activated complex is not a stable molecule that can be isolated. Rather, it is a very short-lived arrangement of molecules with the correct orientation and sufficient energy for reaction to proceed to products. Chemists have recently devised very sophisticated methods that use very fast ($\approx 10^{-15}$ s) laser spectroscopic techniques to investigate the strength of the interactions among the various atoms within the activated complex.

Compare the energy of the activated complex to the average energies of the reactants and products in Figure 13.18. The difference between the energy of the

Figure 13.17 The steps that must occur for a reaction to be completed, according to transition-state theory.

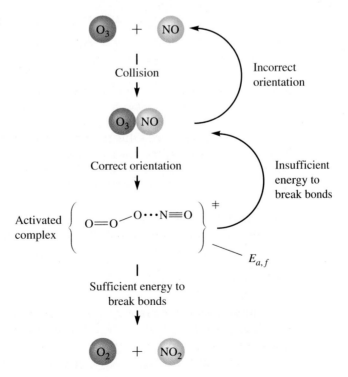

Figure 13.18 A plot of energy against the reaction coordinate. The reaction coordinate is a measure of the progress of a reaction. It represents all the changes that must occur in the course of the reaction, including the bending, breaking, and making of bonds. The energy of the activated complex is greater than the average energies of either reactants or products, creating an energy barrier that must be overcome for chemical reaction to occur.

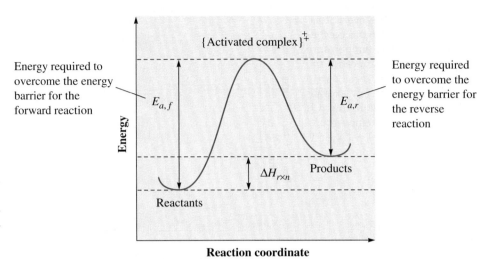

reactants and the much higher energy of the activated complex represents an *energy barrier* that must be overcome in order for reactants to be transformed to products. The energy that the reactants must have to overcome the barrier is called the **activation energy**, E_a. The activation energy, then, is the energy of the activated complex less the average energy of the reactants, as shown in Figure 13.18.

Why can we generally use the enthalpy change instead of the internal-energy change of reactions in discussions of the activation energy?

The activation energies associated with a reaction in the forward and the reverse directions are related by the enthalpy change (actually, the internal-energy change) of the reaction. To see how these qualities are related, let's look more closely at the activation energy for the reaction between ozone and nitric oxide, represented by the equation

$$O_3(g) + NO(g) \rightleftharpoons O_2(g) + NO_2(g)$$

The equation shows that this is an equilibrium reaction—it can proceed in either the forward or the reverse direction. We call the reaction between O_3 and NO the forward reaction and the reaction between O_2 and NO_2 the reverse reaction. There is an activation energy for both the forward reaction ($E_{a,f}$) and the reverse reaction ($E_{a,r}$), and these energies are not identical. The reaction of O_3 with NO has an activation energy of 10.7 kJ/mol in the forward direction and of 210.5 kJ/mol in the reverse direction. The difference in activation energies is due to the difference in average energies of the reactants and products. This difference equals the enthalpy of reaction, ΔH_{rxn}, as shown by the equation

$$\Delta H_{rxn} = E_{a,f} - E_{a,r}$$

For the ozone–nitric oxide reaction, the enthalpy of reaction can be calculated from the difference between the activation energies as follows:

$$\Delta H_{rxn} = 10.7 \text{ kJ/mol} - 210.5 \text{ kJ/mol} = -199.8 \text{ kJ/mol}$$

This relationship enables us to calculate any of these energy terms if we know the other two.

The ozone–nitric oxide reaction is an exothermic reaction. Thus it has a negative enthalpy of reaction, so the activation energy for the reverse reaction is greater than that for the forward reaction, as shown in Figure 13.19A. Conversely,

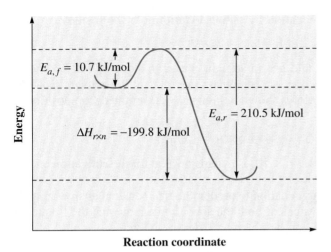

A. Exothermic reaction: $O_3 + NO \rightleftharpoons O_2 + NO_2$

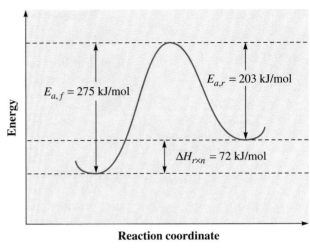

B. Endothermic reaction: $C_2H_5Cl \rightleftharpoons C_2H_4 + HCl$

Figure 13.19 For an exothermic reaction, the activation energy for the reverse process is greater than the activation energy for the forward process. For an endothermic reaction, the forward process has the greater activation energy.

in an endothermic reaction the enthalpy of reaction is positive, and the forward reaction has the greater activation energy. This is shown in Figure 13.19B for the decomposition of chloroethane,

$$C_2H_5Cl(g) \rightleftharpoons C_2H_4(g) + HCl(g)$$

Chloroethane, C_2H_5Cl, is a colorless, flammable gas at room temperature. Because its boiling point is only 12.3°C at 1 atm, it is easily condensed to a liquid. It has a characteristic etherlike odor. Chloroethane is used as a refrigerant and a solvent and in the manufacture of $Pb(C_2H_5)_4$, the "lead" in leaded gasoline.

The enthalpy of this reaction is 72 kJ/mol, and the activation energy in the forward direction is 275 kJ/mol. The activation energy for the reverse reaction is the difference between the two.

$$\Delta H_{rxn} = E_{a,f} - E_{a,r}$$
$$72 \text{ kJ/mol} = 275 \text{ kJ/mol} - E_{a,r}$$
$$E_{a,r} = 203 \text{ kJ/mol}$$

The Arrhenius Equation

How can we summarize the temperature dependence of reaction rates or rate constants without using extensive tables of data for each reaction?

Both collision theory and transition-state theory attribute the temperature dependence of the rate constant to an energy term. Both theories predict an exponential dependence on temperature, which is usually observed experimentally (Figure 13.14). This exponential dependence can be represented by a commonly used empirical equation, the **Arrhenius equation:**

$$k = Ae^{-E_a/RT}$$

In this equation, k is the rate constant for a reaction, A is the **frequency factor,** e is the base of the natural logarithms (approximately 2.718), E_a is the activation energy in units of joules per mole, R is the gas constant (8.314 J/mol K), and T is the absolute temperature. (A, the frequency factor, is related to the collision frequency and the molecular orientation. It depends slightly on temperature but can be considered constant over small temperature ranges.)

We can rearrange this equation by taking logarithms of both sides:

$$\ln k = \ln A - E_a/RT$$

A plot of $\ln k$ versus $1/T$ should give a straight line with slope of $-E_a/R$ and an intercept of $\ln A$. From this slope, we can obtain the value of the activation energy ($E_a = -\text{slope} \times R$). Such a plot is shown in Figure 13.20 for the decomposition of nitrogen dioxide,

The Arrhenius equation is named after Svante A. Arrhenius, who proposed it in 1889 on the basis of empirical observations. A Swedish chemist at Stockholm University, he was awarded the Nobel Prize in 1903 for his work on the theory of electrolyte dissociation.

$$2NO_2(g) \longrightarrow 2NO(g) + O_2(g)$$

The slope, with a value of -1.37×10^4 K, yields a value for the activation energy of 114 kJ/mol.

We can use the Arrhenius equation to predict the rate constant at any temperature, provided that we have values of both the activation energy and the frequency factor. The frequency factor can be obtained from the intercept of a plot of $\ln k$ against $1/T$. We can obtain the intercept by extrapolating the straight line to $1/T = 0$. The value obtained from the data in Figure 13.20 is 22.39. From this value, we can calculate a value of the frequency factor, because the intercept equals $\ln A$. The calculated frequency factor is 5.29×10^9 M^{-1} s^{-1}.

The Arrhenius equation can be used not only as a basis for a graphical determination of activation energies but also for an equation we can use to calculate activation energies from the values of the rate constant at two temperatures. We start with the logarithmic form of the Arrhenius equation for temperature T_1:

$$\ln k_1 = \ln A - E_a/RT_1$$

The same equation should be valid at a different temperature, T_2:

$$\ln k_2 = \ln A - E_a/RT_2$$

The frequency factor and the activation energy should have the same values at the two temperatures, provided that the temperature difference is not very large.

We now subtract the second equation from the first to eliminate the frequency factor.

$$\ln k_1 - \ln k_2 = \ln A - \ln A - E_a/RT_1 + E_a/RT_2$$

Figure 13.20 A plot of the Arrhenius equation for the reaction

$2NO_2(g) \longrightarrow 2NO(g) + O_2(g)$

The slope of the straight line is $-E_a/R$, and the intercept is $\ln A$. The measured value of the slope ($-13,700$ K^{-1}) and the value of the intercept (22.39) extrapolated from the straight line give values of 114 kJ/mol for the activation energy and 5.29×10^9 M^{-1} s^{-1} for the frequency factor.

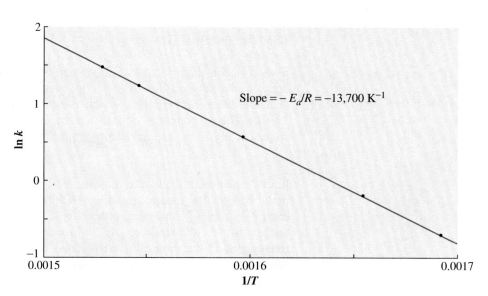

Slope $= -E_a/R = -13,700$ K^{-1}

Collecting terms gives

$$\ln(k_1/k_2) = (E_a/R)(1/T_2 - 1/T_1) \quad \text{or} \quad \ln(k_1/k_2) = (E_a/R)[(T_1 - T_2)/T_1T_2]$$

As we have said, if we know the rate constants at two temperatures, we can use this equation to calculate the activation energy. We can also use it to calculate the value of a rate constant at a certain temperature if we know the activation energy and the rate constant at another temperature.

EXAMPLE 13.5

The reaction of hydrogen peroxide with iodide ion,

$$H_2O_2(aq) + I^-(aq) \longrightarrow H_2O(l) + IO^-(aq)$$

follows the rate law

$$-\Delta[H_2O_2]/\Delta t = k[H_2O_2][I^-]$$

The rate constant is $0.0960\ M^{-1}\ s^{-1}$ at 318.2 K and $0.0230\ M^{-1}\ s^{-1}$ at 298.2 K. What is the activation energy for this reaction?

Solution   We use the equation derived from the Arrhenius equation:

$$\ln(k_1/k_2) = (E_a/R)[(T_1 - T_2)/T_1T_2]$$

Substituting values for the rate constants and temperatures yields

$$\ln(0.0960/0.0230) = (E_a/8.314)(318.2 - 298.2)/(298.2 \times 318.2)$$

Now we evaluate terms.

$$\ln(4.17) = (E_a/8.314) \times 20.0/(298.2 \times 318.2)$$
$$1.428 = (E_a/8.314) \times 0.000211$$

Finally, we solve for the activation energy.

$$E_a = 1.428 \times 8.314/0.000211 = 5.63 \times 10^4 \text{ J/mol} = 56.3 \text{ kJ/mol}$$

Practice Problem 13.5

The decomposition reaction of ethylene oxide,

$$C_2H_4O(g) \longrightarrow CH_4(g) + CO(g)$$

follows the rate law

$$-\Delta[C_2H_4O]/\Delta t = k[C_2H_4O]$$

The rate constant has been measured at several temperatures:

k (s^{-1}):	2.19×10^{-11}	69.2	1.02×10^6	1.23×10^8
T (K):	500.0	1000.0	1500.0	2000.0

What is the activation energy for this reaction?

Answer: 239 kJ/mol

The activation energy and the frequency factor can be used to summarize the effect of temperature on the rate constant for a reaction. Rather than listing rate constants at different temperatures, we need only list the activation energy and the frequency factor. With values of these two parameters, we can calculate the rate constant for a reaction at any temperature of interest by substituting their values into the Arrhenius equation.

EXAMPLE 13.6

The gas-phase decomposition of hydrogen peroxide,

$$2H_2O_2(g) \longrightarrow 2H_2O(g) + O_2(g)$$

follows the rate law

$$-\Delta[H_2O_2]/\Delta t = k[H_2O_2]^2$$

The frequency factor is $2.5 \times 10^{15}\ M^{-1}\ s^{-1}$, and the activation energy is 200.8 kJ/mol. What is the rate constant at 400.0°C?

Solution   We use the Arrhenius equation,

$$k = Ae^{-E_a/RT}$$

with the data

$$A = 2.5 \times 10^{15}\ M^{-1}\ s^{-1}$$
$$E_a = 200.8\ \text{kJ/mol} \times 1000\ \text{J/kJ} = 2.008 \times 10^5\ \text{J/mol}$$
$$R = 8.314\ \text{J/mol K}$$
$$T = 400.0°C + 273.2 = 673.2\ \text{K}$$

Inserting values gives

$$k = 2.5 \times 10^{15}\ M^{-1}\ s^{-1} \times e^{-(200800\ \text{J/mol})/(8.314\ \text{J/mol K} \times 673.2\ \text{K})}$$
$$= 2.5 \times 10^{15}\ M^{-1}\ s^{-1} \times e^{-35.88} = 2.5 \times 10^{15}\ M^{-1}\ s^{-1} \times 2.62 \times 10^{-16}$$
$$= 0.65\ M^{-1}\ s^{-1}$$

Practice Problem 13.6

The gas-phase decomposition of dinitrogen pentoxide,

$$2N_2O_5(g) \longrightarrow 4NO_2(g) + O_2(g)$$

follows the rate law

$$-\Delta[N_2O_5]/\Delta t = k[N_2O_5]$$

The frequency factor is $6.00 \times 10^{14}\ s^{-1}$ and the activation energy is 87.9 kJ/mol. What is the rate constant at 100.0°C?

Answer: 298 s⁻¹

How much can we expect reaction rates to increase with temperature? A rough rule is that the rate of a reaction increases by a factor of 2 for each 10°C rise in temperature. However, this rule is strictly applicable only for reactions that occur at 25°C and have an activation energy of 53 kJ/mol. In practice, many reactions increase in rate more rapidly or more slowly with temperature than this rule states.

Library and Museum Collections Endangered

It is the very composition of the paper itself that is causing the problem.

Reaction rates vary with concentrations, temperature, and the presence of catalysts. Generally, kinetic studies are used to reveal conditions under which a chemical reaction can be carried out efficiently and rapidly. At times, however, it is more important to prevent reactions than to carry them out rapidly. Examples are the reactions that cause paper in books to yellow and crumble, as shown in Figure 13.21. This type of deterioration is found in many millions of books and periodicals in libraries—in fact, in most of those printed since 1850.

What causes these reactions? It is not air pollution, as might be expected. It is the very composition of the paper itself that is causing the problem. Paper is made largely of natural fibers containing cellulose. Other substances have to be added to the cellulose to make paper suitable for printing. For example, some material is added as a sizing to keep the ink from spreading. Since 1850 the sizing has generally been aluminum sulfate. Aluminum sulfate gradually combines with moisture from air to form sulfuric acid.

$$Al_2(SO_4)_3(s) + 6H_2O(g) \longrightarrow 2Al(OH)_3(s) + 3H_2SO_4(l)$$

The sulfuric acid attacks the cellulose fibers, breaking them into smaller fragments, which ultimately leads to crumbling of the paper. Chemists at the Library of Congress have developed a process that neutralizes the acid formed in books. The acid-neutralizing substance is diethyl zinc gas, $Zn(C_2H_5)_2$. Because it is a gas, it can penetrate throughout even a closed book. The compound reacts with acid, releasing ethane gas and leaving behind a residue of zinc sulfate.

$$Zn(C_2H_5)_2(g) + H_2SO_4(l) \longrightarrow ZnSO_4(s) + 2C_2H_6(g)$$

Unfortunately, diethyl zinc is very reactive; it bursts into flames when brought into contact with oxygen in air and explodes when exposed to water. For this reason, it is necessary first to place the books into a vacuum chamber. The books are dried out by warming for three days, and all the moisture and air are pumped out of the chamber before the diethyl zinc is admitted. The ethane gas and unreacted diethyl zinc are then pumped away. Such deacidification chambers may well become a part of libraries in the future.

The problem of deterioration is not limited to libraries. It also arises in museums that have conserved plastic and synthetic fiber artifacts. For example, early 20th-century sculptures made of cellulose nitrate (or celluloid) are begin-

Figure 13.21 Most books printed since 1850 are slowly undergoing deterioration because components of the paper produce sulfuric acid. Books designed to last indefinitely are now being printed on acid-free paper.

ning to decompose. Chemicals, such as camphor, added to the celluloid to make it more pliable are slowly subliming from the plastic, causing cracks that expose more celluloid, which is then susceptible to deterioration. In the presence of light, cellulose nitrate decomposes to form nitrogen oxides that react with moisture and oxygen in the atmosphere to produce nitric acid. The nitric acid catalyzes the decomposition of cellulose nitrate. Similar problems are experienced with old movie film, which was also made of cellulose nitrate. To some extent, the deterioration of these artifacts can be countered by appropriate storage. For example, the Smithsonian Institution stores historic space suits at 5°C to slow their deterioration. Chemists at museums are gradually finding methods to slow down the disintegration of synthetic materials, but this disintegration cannot be completely stopped.

Questions for Discussion

1. List some examples of materials you use that are subject to deterioration. What do you do to prevent or inhibit this deterioration?
2. What value should be placed on old artifacts such as silent films? Is it worth spending much time and money to preserve these items?

(continued)

3. In the 1920s, celluloid was considered a ''miracle'' material. Today we have many other materials used for a variety of purposes. Is it likely that these modern materials will be found deficient in the future?

4. If neither paper nor plastics can be trusted to be permanent, how do we preserve our culture for the future?

13.4   Reaction Mechanisms

We represent chemical reactions with balanced chemical equations. These equations tell us the identity and number of the reactants and the products, but they do not provide any information about how the reaction takes place. This information is found in the **mechanism** of the reaction—the detailed pathway it follows, which often consists of a series of steps. A mechanism must be considered only as a postulate that is subject to change if new data are obtained. We propose a mechanism that is consistent with all known facts about a reaction, but we cannot prove that the mechanism is correct. It may be possible to propose more than one mechanism that can explain known facts. More experiments are then required to determine which mechanism is most likely.

How do we determine the mechanism of a reaction?

Rate laws can provide information that helps us predict the mechanism for a reaction. Any mechanism we propose must be consistent with the experimental rate law. However, rate laws do not prove that a particular mechanism is the one that occurs—only that it is possible. Often, more than one mechanism gives the same form of a rate law. Table 13.4 lists the stoichiometries and rate laws for several simple reactions. We will consider these reactions further in our examination of reaction mechanisms.

Elementary Reactions

The steps that make up a mechanism are called **elementary reactions.** These reactions cannot be further broken down into simpler steps. The mechanism of a reaction is often discussed in terms of **molecularity**—the number of reactant molecules or ions that participate in an elementary reaction. An elementary reaction with one reactant molecule is called **unimolecular,** a reaction with two is **bimolecular,** and a reaction with three is **termolecular.**

Acetonitrile, CH_3CN, is a colorless, volatile liquid that boils at 81.6°C at 1 atm. It has an etherlike odor and burns with a luminous flame. It is used as a starting material for a variety of organic syntheses and as a solvent, especially for polar or ionic substances. It is used to extract fatty materials from fish oil and vegetable oils.

Some simple reactions may occur in a single step, so the overall reaction may itself be an elementary reaction. Some of the reactions listed in Table 13.4, particularly those with a single reactant, fit this pattern. For example, consider the formation of CH_3CN from CH_3NC.

$$CH_3NC \longrightarrow CH_3CN$$

The overall reaction appears to be an elementary reaction, which occurs in a single step. In a single-step unimolecular reaction, the rate law depends only on the reactant concentration raised to the first power. *In the rate law for any elementary reaction, each reactant's concentration is raised to the power of its coefficient in the chemical equation.* Thus we can predict the rate law for elementary reactions, even though we cannot do so for other reactions. Here, because we assume that the

Table 13.4 Rate Laws for Some Simple Chemical Reactions

Reaction	Experimental Rate Law
$CH_3NC \longrightarrow CH_3CN$	Rate $= k[CH_3NC]$
$CH_3CHO \longrightarrow CH_4 + CO$	Rate $= k[CH_3CHO]^{3/2}$
$2N_2O_5 \longrightarrow 4NO_2 + O_2$	Rate $= k[N_2O_5]$
$H_2 + I_2 \longrightarrow 2HI$	Rate $= k[H_2][I_2]$
$H_2 + Br_2 \longrightarrow 2HBr$	Rate $= \dfrac{k[H_2][Br_2]^{3/2}}{[Br_2] + k'[HBr]}$
$2NO + O_2 \longrightarrow 2NO_2$	Rate $= k[NO]^2[O_2]$
$S_2O_8{}^{2-} + 2I^- \longrightarrow 2SO_4{}^{2-} + I_2$	Rate $= k[S_2O_8{}^{2-}][I^-]$
$Tl^{3+} + Hg_2{}^{2+} \longrightarrow Tl^+ + 2Hg^{2+}$	Rate $= k[Tl^{3+}][Hg_2{}^{2+}]/[Hg^{2+}]$
$BrO_3{}^- + 5Br^- + 6H^+ \longrightarrow 3Br_2 + 3H_2O$	Rate $= k[Br^-][BrO_3{}^-][H^+]^2$
$BrO_3{}^- + 6Fe^{2+} + 6H^+ \longrightarrow$ $Br^- + 6Fe^{3+} + 3H_2O$	Rate $= \dfrac{k[Fe^{2+}][BrO_3{}^-][H^+]}{k' + [H^+]}$

overall reaction is an elementary reaction, we raise the concentration of the reactant to the first power, which yields the following rate law:

$$\text{Rate} = k[CH_3NC]$$

We can test a proposed mechanism by comparing our predictions with observed kinetic behavior. The experimentally determined rate law for the disappearance of CH_3NC is

$$-\Delta[CH_3NC]/\Delta t = k[CH_3NC]$$

The predicted rate law and the experimental rate law match, so our mechanism is consistent with what we know about the reaction.

Bimolecular reactions—reactions that involve two reactant molecules—also have a tendency to follow a single-step mechanism. Let's look at the bimolecular reaction between ozone and nitric oxide.

$$O_3(g) + NO(g) \longrightarrow O_2(g) + NO_2(g)$$

If this bimolecular reaction is elementary, it should have a rate proportional to the concentration of each of the reactants:

$$\text{Rate} = k[O_3][NO]$$

The rate law indicates that the reaction is first-order in each reactant and second-order overall. It has been determined by experiment that the reaction does follow a second-order rate law:

$$-\Delta[O_3]/\Delta t = k[O_3][NO]$$

The proposal that this reaction is elementary is supported by the experimental evidence.

EXAMPLE 13.7

Write a rate law for the elementary reaction

$$H_2O_2(g) + OH(g) \longrightarrow H_2O(g) + HO_2(g)$$

Solution This reaction involves two molecules as reactants, so it is a bimolecular reaction. In an elementary reaction, the rate is proportional to the concentration of each reactant raised to the power of its coefficient, so this is a second-order reaction, first-order in each reactant.

$$Rate = k[H_2O_2][OH]$$

Practice Problem 13.7

Write a rate law for the elementary reaction

$$H_2O_2(g) \longrightarrow 2OH(g)$$

Answer: $Rate = k[H_2O_2]$

Not all reactions with a 1:1 stoichiometry are bimolecular elementary reactions. Examine the rate law for the hydrogen-bromine reaction in Table 13.4, for example. The rate law for the hydrogen-bromine reaction is much more complex than we would predict for an elementary reaction, indicating that the reaction occurs by a more complex mechanism.

Multistep Mechanisms

Elementary reactions may also be termolecular, involving the reaction of three molecules in a single step. However, a termolecular elementary reaction is unusual because it requires that three molecules collide simultaneously with proper orientation and energy, a highly improbable event. The mechanism for termolecular reactions is usually thought to consist of several elementary reactions occurring in sequence.

Let's consider the oxidation of iodide ion by peroxydisulfate ion.

$$S_2O_8^{2-} + 2I^- \longrightarrow 2SO_4^{2-} + I_2$$

Potassium persulfate, $K_2S_2O_8$, exists as white or colorless crystals. It decomposes gradually at room temperature and more rapidly at higher temperatures, losing oxygen in the process. This powerful oxidizing agent is used to bleach fabrics and soaps. In photography, it is employed to remove traces of thiosulfate from photographic plates and paper.

The experimental rate law is first-order in each reactant.

$$Rate = k[S_2O_8^{2-}][I^-]$$

Only one I^- ion appears in the rate law, but two appear as reactants in the balanced equation. Such a rate law indicates a possible elementary step in which one I^- ion reacts with one $S_2O_8^{2-}$ ion. Because the reaction is not completed in this elementary reaction, there must be at least one other elementary reaction in which the second I^- ion reacts. When some of the concentrations do not appear in the rate law, the mechanism must have more than one step.

In general, one step in a multistep mechanism is slower than the others; this is called the **rate-determining step.** The rate of this step governs the overall rate of the reaction, in the same way that the rate of doing the dishes is governed by the slower of the washing and drying steps. If the dish washer is slower than the dish drier, the rate at which the dishes are completed is determined by the rate of washing.

In a multistep mechanism, not only the rate of reaction but also the form of the rate law is determined by the identity of the rate-determining step. Reactant molecules that are consumed in the rate-determining step or in steps preceding it appear in the rate law. Reactants that are consumed after the rate-determining step do not appear in the rate law because they do not affect the rate.

In the reaction of I^- with $S_2O_8^{2-}$, one I^- ion must react in the rate-determining step, or in earlier steps, and the other in a later step. A possible mechanism is

$S_2O_8^{2-} + I^- \longrightarrow 2SO_4^{2-} + I^+$ (slower; rate-determining, with rate constant k)

$I^+ + I^- \longrightarrow I_2$ (faster; cannot measure its rate constant in this reaction)

In this mechanism, the first step is the slow step. It produces product SO_4^{2-} ion as well as a small amount of the species I^+, which is a reactant in the second, fast step. The species I^+ is called an **intermediate,** which is a species that is produced in one step and then reacts in a later step. It is not a final product of the reaction. (In our dish-washing analogy, the clean, wet dishes correspond to the intermediate.) Intermediates usually react readily in one of two ways. Either they react with another substance to go on to products, or they return to reactants by the reverse of the first step. In the proposed mechanism, the intermediate I^+ rapidly reacts with I^- to form the final product.

The identity of the intermediate, I^+, is a guess, because the rate law reflects only the elementary reactions that occur up to the rate-determining step. The form of the rate law is set by the rate-determining step—the first step, in this case. This step is bimolecular, so it should be second-order:

$$\text{Rate} = k[S_2O_8^{2-}][I^-]$$

Remember that reactants consumed in or before the slow, rate-determining step appear in the rate law, whereas reactants consumed after the rate-determining step do not. If the second step in a multistep mechanism is the rate-determining step, it is possible for all the reactant concentrations to appear in the rate law. Once again, this information can help us propose a reaction mechanism.

Let's consider the reaction of nitric oxide with oxygen.

$$2NO(g) + O_2(g) \longrightarrow 2NO_2(g)$$

The experimentally determined rate law for the reaction is

$$-\Delta[NO]/\Delta t = k[NO]^2[O_2]$$

This rate law is consistent with a single-step termolecular reaction, but because we know that termolecular elementary reactions are rare, we'll propose an alternative mechanism for this reaction. Any such alternative mechanism must involve at least two steps, each of which occurs at a characteristic rate.

In the nitric oxide–oxygen reaction, two NO molecules and one O_2 molecule must be consumed before or during the rate-determining step, because they appear in the rate law. Our mechanism, or set of elementary reactions, should consist of a sequence of unimolecular or bimolecular elementary reactions. Thus we must consume two of these molecules in one step and the third in another step. The first step generally consumes two molecules; if it did not, the second step would have to be termolecular.

Develop an analogy to a rate-determining step using toll-road traffic as an example.

This mechanism is not the only possibility, because all the rate law reveals is that one $S_2O_8^{2-}$ ion and one I^- ion are reactants in the first step. Propose another mechanism consistent with the observed rate law.

The second step in the sequence is the slower, rate-determining step. The first, rapid step is usually reversible. If it were not, we could actually detect two distinct reactions with observable products for each reaction. In the absence of such observations, we must conclude that there is an intermediate that does not accumulate enough to be identified as a product. This situation arises from rapid reversible reactions in which the forward rate equals the reverse rate—they are essentially equilibrium reactions.

We can propose a mechanism consistent with these requirements as follows:

$$NO + O_2 \rightleftharpoons NO_3 \qquad \text{(fast, with rate constants } k_1 \text{ and } k_{-1})$$
$$NO_3 + NO \longrightarrow 2NO_2 \qquad \text{(slow, rate-determining step, with rate constant } k_2)$$

As part of the mechanism, we must designate which step is rate-determining and which steps are fast. We also assign a rate constant to each step, including reverse steps, up to and including the rate-determining step. The rate constants are usually given a subscript to indicate the step with which they are associated: k_1 is the rate constant for the first step, k_{-1} is the rate constant for the reverse of the first step, and k_2 is the rate constant for the second step.

To determine whether our proposed mechanism is consistent with the experimental rate law, we must predict a rate law for it. To do so, we combine the rate laws for the elementary reactions to give an overall rate law for the proposed mechanism. We can predict the rate of each elementary reaction as we did for single-step reactions. We must then decide how to combine these rate laws. In this example, the second step is rate-determining, so its rate determines the overall rate of the reaction. Because this elementary reaction is biomolecular, it has a second-order rate law:

$$\text{Rate} = k_2[NO_3][NO]$$

The concentration of the intermediate, NO_3, does not appear in the experimental rate law, however, so we must somehow eliminate it from the predicted rate law. We can do so by considering the rapid first step. The forward reaction is bimolecular and has a rate constant k_1, so its rate law is

$$\text{Forward rate} = k_1[NO][O_2]$$

The reverse reaction is unimolecular and has a rate constant k_{-1}, so its rate law is

$$\text{Reverse rate} = k_{-1}[NO_3]$$

Assuming that these two rates are equal, we can combine them and derive the relationship

$$k_1[NO][O_2] = k_{-1}[NO_3]$$

This relationship can be used to obtain an expression for the concentration of the intermediate.

$$[NO_3] = (k_1/k_{-1})[NO][O_2]$$

We can eliminate the concentration of the intermediate NO_3 from the rate law by substituting the equivalent expression for $[NO_3]$.

$$\text{Rate} = k_2[NO_3][NO] = k_2(k_1/k_{-1})[NO][O_2][NO]$$
$$= (k_2k_1/k_{-1})[NO]^2[O_2]$$

This is just the same as the rate law that was determined experimentally, but here the experimental rate constant, k, is equal to the combination of rate constants for the elementary reactions, k_2k_1/k_{-1}.

EXAMPLE 13.8

Identify the intermediate in the following reaction mechanism for the reaction between thallium(III) and iron(II) ions.

$$Tl^{3+} + Fe^{2+} \rightleftharpoons Tl^{2+} + Fe^{3+} \quad \text{(fast)}$$
$$Tl^{2+} + Fe^{2+} \longrightarrow Tl^+ + Fe^{3+} \quad \text{(slow)}$$

Solution Both Tl^{2+} and Fe^{3+} are produced in the first step, but only Tl^{2+} reacts further in the second step. Fe^{3+} is one of the final products of the reaction. Thus only Tl^{2+} meets the criteria for an intermediate.

Practice Problem 13.8

Identify the intermediate in the following reaction mechanism for the reaction between dinitrogen pentoxide and nitric oxide to produce nitrogen dioxide.

$$N_2O_5 \rightleftharpoons NO_3 + NO_2 \quad \text{(fast)}$$
$$NO_3 + NO \longrightarrow 2NO_2 \quad \text{(slow)}$$

Answer: NO_3

We have looked at two examples of multistep mechanisms. In one the reaction orders matched the stoichiometry; in the other they did not match. In each case, the rate law was fairly simple. Simple rate laws arise when the first step in the mechanism is rate-determining or when each reactant order matches the stoichiometric coefficient of that substance. When these conditions are not met, the rate law can take quite complex forms. An example of a slightly more complex system is the reaction between thallium(III) and mercury(I) ions.

$$Tl^{3+} + Hg_2^{2+} \longrightarrow Tl^+ + 2Hg^{2+}$$

The experimentally determined rate law is

$$\text{Rate} = k[Tl^{3+}][Hg_2^{2+}]/[Hg^{2+}]$$

This rate law is complex in that it contains a denominator term. Species that appear in the numerator of a rate law behave as reactants in the rate-determining step or in earlier steps. Species that appear in the denominator of a rate law are involved as products of the reaction, but they must be formed prior to the rate-determining step, because the rate law contains only species involved in steps up to and including the rate-determining step.

Mercury(I) exists in the form of the colorless dimeric ion Hg_2^{2+}. This ion also exists in mercury(I) salts, such as the halides, nitrate, and sulfate. Except for the yellow iodide salt, these salts are all colorless or white. Mercury(I) chloride, also known as calomel, is almost insoluble in water. It is decomposed by light or alkali-metal iodide, bromide, or cyanide solutions, forming mercury(0) and mercury(II) chloride.

A possible mechanism for the reaction between thallium(III) and mercury(I) is

$$Hg_2^{2+} \rightleftharpoons Hg^0 + Hg^{2+} \quad \text{(fast, rate constants } k_1 \text{ and } k_{-1})$$
$$Tl^{3+} + Hg^0 \longrightarrow Tl^+ + Hg^{2+} \quad \text{(slow, rate constant } k_2)$$

Following the same procedure as we used for the rate law of the $NO-O_2$ reaction, we can predict the following rate law:

$$\text{Rate} = (k_1k_2/k_{-1})[Tl^{3+}][Hg_2^{2+}]/[Hg^{2+}]$$

This is the same form as the experimentally determined rate law, with $k = k_1k_2/k_{-1}$. A concentration in the denominator of a rate law always indicates that there are at least two steps in the mechanism and that the second step is rate-determining.

13.5 Catalysis

We have seen that the rate of reaction may change with concentration and temperature. There is yet another method for influencing the rate of reaction: the addition of a catalyst to the reaction system. A catalyst generally operates by changing the mechanism to a new pathway that has a lower activation energy, as shown in Figure 13.22. Following this alternative mechanism speeds up the reaction. For example, the dissociation of hydrogen iodide,

$$2HI(g) \longrightarrow H_2(g) + I_2(g)$$

normally has an activation energy of 192 kJ/mol. When the reaction is carried out in the presence of gold metal, however, the dissociation becomes faster, and the activation energy is only 105 kJ/mol; in the presence of platinum metal, the activation energy is lowered to 59 kJ/mol. Catalysts can modify the activation energy by lowering the energy required to break bonds. They can also modify the entropy of a system by helping reactants get close enough together to react, and possibly even by forming a catalyst–reactant bond.

Let's look more closely at how the addition of a catalyst can change a reaction mechanism. Ethylene is an organic compound containing a carbon–carbon double bond. It can be converted to ethane, C_2H_6, by reaction with hydrogen—a process called hydrogenation. This process occurs very slowly in aqueous solution, possibly by a one-step mechanism:

$$H_2 + H_2C{=}CH_2 \longrightarrow H_3C{-}CH_3$$
$$\text{Rate} = k[H_2][H_2C{=}CH_2]$$

When a copper salt is added to the solution, the rate becomes much higher, and the rate law takes a new form:

$$\text{Rate} = k[Cu^{2+}][H_2]$$

Note that the ethylene concentration has disappeared from the rate law, which indicates that a new mechanism has taken over that does not involve ethylene until

Hydrogen iodide is a colorless, nonflammable gas that fumes in moist air and decomposes into its elements when exposed to light. It is extremely soluble in water, forming hydriodic acid, HI(aq), which is colorless when first made but rapidly turns yellow on exposure to air and light because of the formation of iodine. It is used in the manufacture of iodide salts, pharmaceuticals, and disinfectants.

One liter of ethylene gas dissolves in about 9 L of water at 25°C. How could the concentration of ethylene in solution be increased?

Figure 13.22 A catalyst lowers the activation energy by making available a different reaction pathway with lower minimum energy requirements for the reactants. This results in an increase in reaction rate.

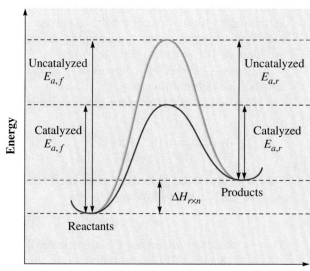

after the rate-determining step. The following two-step mechanism has been proposed for the catalyzed reaction:

$$Cu^{2+} + H_2 \longrightarrow CuH^+ + H^+ \qquad \text{(rate-determining)}$$
$$CuH^+ + H^+ + H_2C{=}CH_2 \longrightarrow Cu^{2+} + H_3C{-}CH_3 \quad \text{(fast)}$$

Note that the catalyst helps to break the H–H bond by forming a bond to one of the hydrogen atoms. Catalysis has added a step to the mechanism, but the total of the new steps is still faster than the one-step reaction in the uncatalyzed process. Note that Cu^{2+} appears as both a reactant and a product, so it is not consumed by the reaction; this is consistent with the definition of a catalyst.

The dissociation of hydrogen iodide and the hydrogenation of ethylene represent the major types of catalysis: heterogeneous and homogeneous. A **heterogeneous catalyst** (usually a solid, such as Au and Pt in the dissociation example) is in a different physical phase than that of the reaction system. In contrast, a **homogeneous catalyst,** such as Cu^{2+} in the hydrogenation of ethylene, exists in the same physical phase as the reaction system.

Homogeneous Catalysis

Many different homogeneous catalysts are used to prepare commercially important compounds. For example, many transition-metal compounds in low oxidation states behave as catalysts for reactions of carbon monoxide, or hydrogen, or ethylene. These transition-metal compounds can form chemical bonds to these molecules, thereby helping them both to weaken the chemical bonds that already exist in the molecules and to assemble the reactants.

The decomposition of hydrogen peroxide, for example, is catalyzed by a variety of metal ions, by the enzyme peroxidase, and by some nonmetal ions. The essential feature of a catalyst for this system is ease of oxidation and reduction. Hydrogen peroxide itself can undergo either oxidation to O_2 or reduction to H_2O. In fact, the decomposition reaction is a combination of these two processes.

$$2H_2O_2 \longrightarrow 2H_2O + O_2$$

Bromine or salts of bromide ion catalyze this reaction by cycling through successive oxidation and reduction processes.

$$H_2O_2 + 2Br^- + 2H^+ \longrightarrow 2H_2O + Br_2$$
$$H_2O_2 + Br_2 \longrightarrow 2H^+ + O_2 + 2Br^-$$

A typical example of homogeneous catalysis in the gas phase is the use of nitrogen oxides to catalyze the oxidation of sulfur dioxide by oxygen. The resulting sulfur trioxide is used in the production of sulfuric acid. Because these reactions are carried out in lead-lined chambers, the process is known as the lead chamber process.

$$2NO(g) + O_2(g) \longrightarrow 2NO_2(g)$$
$$\underline{2[SO_2(g) + NO_2(g) \longrightarrow SO_3(g) + NO(g)]}$$
$$2SO_2(g) + O_2(g) \longrightarrow 2SO_3(g)$$

The lead chamber process has fallen out of favor for the production of sulfuric acid because of high operating costs. It has largely been replaced by the contact process, which uses a heterogeneous catalyst. This process was described in more detail in Chapter 4.

Heterogeneous Catalysis

Although homogeneous catalysts are important for chemical syntheses in solution, many industrially important substances are prepared by reactions of gaseous molecules. Catalysts for such reactions are usually solids, and the reactions take place on the solid surface. In the contact process just mentioned, either V_2O_5 or metallic platinum coated onto silica gel is used to effect the oxidation of SO_2. The Haber process for production of ammonia uses iron or an iron–titanium alloy as a heterogeneous catalyst.

$$N_2(g) + 3H_2(g) \xrightarrow{\text{Fe}} 2NH_3(g)$$

The oxidation of ammonia by oxygen in air, a part of the Ostwald process for making nitric acid, uses a platinum gauze catalyst, as shown in Figure 13.23.

$$4NH_3(g) + 7O_2(g) \xrightarrow{\text{Pt}} 4NO_2(g) + 6H_2O(g)$$

The hydrogenation of multiple bonds, such as $C{=}C$, $C{\equiv}C$, $C{=}O$, and $C{\equiv}N$, is catalyzed by Raney nickel, finely divided nickel prepared from nickel–aluminum alloys.

$$H_2C{=}CH_2 + H_2 \xrightarrow{\text{Ni}} H_3C{-}CH_3$$

The catalytic activity of surfaces undoubtedly involves the phenomenon known as **chemisorption.** In this process, molecules are adsorbed, or attached, to solid surfaces. The molecules are chemically activated because of the weakening or breaking of bonds. Several steps must be involved in chemisorption: diffusion of reactant molecules to the surface; adsorption of reactant molecules onto the surface; reaction between adsorbed molecules; desorption, or release, of product molecules from the surface; and diffusion of product molecules from the surface. These steps are shown in Figure 13.24 for the reaction between carbon monoxide and oxygen gases.

Figure 13.23 When heated platinum metal is placed in the ammonia vapors above a concentrated solution of aqueous ammonia, it glows brightly because of the heat generated by the reaction of ammonia with oxygen that occurs on the surface of the platinum.

Figure 13.24 Heterogeneous catalysis in the reaction of carbon monoxide with oxygen proceeds in several steps: *(A)* Oxygen molecules are adsorbed on the metal surface. *(B)* Oxygen molecules break apart into oxygen atoms. *(C)* Carbon monoxide molecules are adsorbed on the metal surface. *(D)* A carbon monoxide molecule forms a bond to an oxygen atom. *(E)* The carbon dioxide product diffuses from the metal surface.

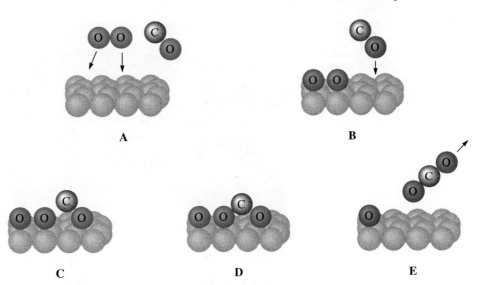

A   B

C   D   E

Catalysts in Automobiles

Catalysts have become important to the proper functioning of automobiles because gasoline does not burn smoothly in the absence of catalysts. Gasoline is a mixture of flammable hydrocarbons. However, the rate at which hydrocarbons burn varies. The cycle of fuel intake, combustion, and expulsion of gaseous products does not remain synchronized if the combustion process does not occur at the proper rate. When the cycle is unsynchronized, the engine runs roughly. Although using the appropriate hydrocarbons would solve this problem, it is less expensive to adjust the rate of combustion with various additives. For many years, tetraethyl lead, $Pb(C_2H_5)_4$, was used as an additive to cause gasoline to burn smoothly. More recently, automobiles have been required to run on "unleaded" gasoline, which contains other additives, such as MTBE, or methyl-tertiary-butyl ether, $(CH_3)_3C-O-CH_3$. This change eliminates toxic lead from the environment, but other factors also influenced the change to MTBE. These factors involve the catalytic converter.

Why do automobiles have catalytic converters?

The burning of hydrocarbons, such as those in gasoline, in air at high temperatures produces two types of pollutants—incompletely combusted carbon compounds (C_xH_y and CO) and nitrogen oxides (NO and NO_2). To be rendered harmless, the carbon materials must be further oxidized to CO_2. However, in order to eliminate their harmful effects on the atmosphere, the nitrogen oxides must be reduced back to the N_2 from which they were formed. Thus removing these pollutants requires both an oxidation and a reduction process. We could increase the amount of air in the fuel mixture to cause the combustion to be more nearly complete. But the resulting decrease in carbon monoxide and unburned hydrocarbons would be offset by an increase in the nitrogen oxides formed.

The current method of removing air pollutants from automobile exhaust uses the catalytic converter, which is a heterogeneous catalyst mounted in the exhaust system (Figure 13.25). The catalytic converter contains a mixture of two catalysts, one to facilitate further oxidation of hydrocarbons and the other to help reduce nitrogen oxides.

The oxidation of C_xH_y and CO is generally catalyzed by transition-metal oxides and noble metals such as Pt or Pd. A metal oxide such as Cr_2O_3 adsorbs O_2

Figure 13.25 Catalytic converters are used in automobile exhaust systems to catalyze the complete oxidation of hydrocarbons and carbon monoxide and the reduction of nitrogen oxides to nitrogen and oxygen gases.

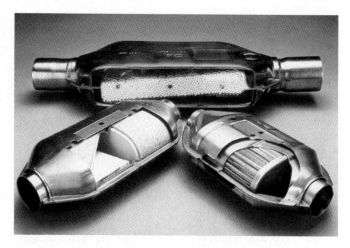

from air added to the exhaust gases and activates the oxygen molecule. Then carbon monoxide can react with oxygen atoms on the surface to form CO_2, as shown in Figure 13.24. Hydrocarbon oxidation proceeds in a similar fashion.

The second process, reduction of nitrogen oxides, is a thermodynamically favorable process, but it is slow.

$$2NO \longrightarrow N_2 + O_2$$

Transition-metal oxides and noble metals also catalyze this process. However, the transition-metal oxides that are the most effective catalysts for combustion are not the most effective catalysts for nitrogen oxide reduction. For this reason, catalytic converters contain a mixture of metal oxides, such as Cr_2O_3 and CuO, or Cr_2O_3 and ZnO, or a mixture of noble metals such as Pd, Pt, and Rh.

The catalysts currently in use are generally poisoned by $Pb(C_2H_5)_4$, tetraethyl lead. The lead is deposited on the catalyst surface, rendering the catalyst inoperative. Because of this effect, cars built in the United States since 1975 have been designed to discourage the use of ''leaded'' gasoline.

Enzymes

Enzymes are the most important catalysts to us, because we could not exist without them. Their action makes possible many complex chemical processes that could not otherwise occur at the relatively low temperature of the body. Enzymes are very specific catalysts, taking part in very limited reactions with only certain reactants. Thus an enormous number of enzymes—about 25,000—are required to catalyze all the processes necessary to the proper functioning of the body, such as digestion, respiration, cell synthesis, and energy transport.

Enzymes are large protein molecules with very high molar masses of 12,000–40,000 g/mol for the simplest enzymes and much higher molar masses for some of the more complex molecules. The intricate three-dimensional structures of enzymes (such as that shown in Figure 13.26 for lysozyme, which catalyzes the destruction of bacterial cell walls) contain depressions, or holes, that vary in shape and size according to the identity of the enzyme. It is these holes, which are called *active sites*, that make the enzymes such specific catalysts. In order for the catalytic process to occur, reactant molecules must fit into the structure of the enzyme at the active site. Interactions (such as covalent, ionic, or hydrogen bonding)

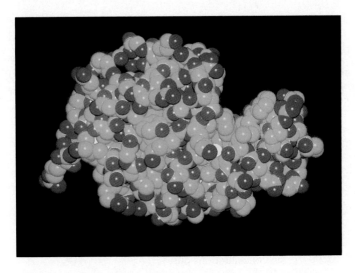

Figure 13.26 The three-dimensional structure of lysozyme is typical of the structures of enzymes. The active site in this enzyme is the vertical open groove in the upper right corner.

between the enzyme and the reactant help to modify the reactant in such a way that the conversion to product is enhanced.

Summary

The rates of reactions vary widely, depending on the identity of the reactants and other factors. The reaction rate is defined as the negative change in concentration of a reactant per unit of time or as the positive change in concentration of a product per unit of time. The concentrations of the species participating in the reaction affect the reaction rate. This effect is described by the rate law, which equates the rate to some function of the concentrations multiplied by a proportionality constant called the rate constant. Usually, the concentration of each substance that appears in the rate law is raised to some integral power called the order of that substance. The rate law can be determined from the variation of initial rates with concentration or from measurements of concentration at various times over the course of the reaction, evaluated graphically. The time required for the concentration of a reactant to be reduced to half its value is called the half-life of the reaction.

Reaction rates and rate constants generally increase with increasing temperature. Several theories, including collision theory and transition-state theory, seek to explain this effect. The variation of rate constants with temperature can be evaluated mathematically via the Arrhenius equation.

The stepwise manner in which reactions occur is called the reaction mechanism. The mechanism is made up of one or more elementary reactions, reactions that cannot be further broken down. Many mechanisms consist of a sequence of elementary reactions. One step in a multistep mechanism, called the rate-determining step, is slower than the others and determines the form of the rate law.

Catalysts also change the rate of reaction. They participate in a chemical reaction but can be recovered unchanged when the reaction is complete. Catalysts operate by decreasing the activation energy for a reaction and by changing the reaction mechanism.

Key Terms

activated complex (13.3)
activation energy
    (E_a) (13.3)
Arrhenius
    equation (13.3)
average rate (13.1)
bimolecular
    reaction (13.4)
chemisorption (13.5)
collision theory (13.3)

elementary
    reaction (13.4)
enzyme (13.5)
frequency factor (13.3)
half-life (13.2)
heterogeneous
    catalyst (13.5)
homogeneous
    catalyst (13.5)
initial rate (13.1)

instantaneous
    rate (13.1)
intermediate (13.4)
kinetics (p. 529)
mechanism (13.4)
molecularity (13.4)
rate constant (13.2)
rate-determining
    step (13.4)
rate equation (13.2)

rate law (13.2)
reaction order (13.2)
reaction rate (13.1)
termolecular
    reaction (13.4)
transition state (13.3)
transition-state
    theory (13.3)
unimolecular
    reaction (13.4)

Exercises

Reaction Rates

13.1   Why are reaction rates defined in such a way that they always have positive values?

13.2   Distinguish between initial rates and instantaneous rates.

13.3   Show how stoichiometric coefficients are used to relate the rates of different species involved in a reaction.

13.4   What is a rate law?

13.5   The reaction of ozone with ethylene follows the rate law

$$-\Delta[O_3]/\Delta t = k[O_3][C_2H_4]$$

with a rate constant of $k = 800\ M^{-1}\ s^{-1}$ at 25°C. What is the rate of reaction if the concentrations are $1.00 \times 10^{-8}\ M$ O_3 and $4.50 \times 10^{-6}\ M$ C_2H_4?

13.6   At an initial concentration of N_2O_5 of $1.510\ M$ at 45°C, the concentration decreased by $0.131\ M$ in 2.32 h. Calculate the initial rate of the reaction in units of $M\ s^{-1}$.

$$2N_2O_5 \longrightarrow 4NO_2 + O_2(g)$$

13.7   In the reaction of nitrogen with hydrogen gas (the Haber process),

$$N_2(g) + 3H_2(g) \longrightarrow 2NH_3$$

the rate of reaction was measured to be $-\Delta[H_2]/\Delta t = 2.50 \times 10^{-4}\ M\ s^{-1}$ in a particular experiment. What is the rate expressed in terms of the concentration (a) of nitrogen and (b) of ammonia?

13.8   Express the rate of reaction in terms of each of the species in the reaction

$$3I^- + IO_2^- + 4H^+ \longrightarrow 2I_2 + 2H_2O$$

13.9   At an initial concentration of $0.250\ M$, the concentration of O_3 decreased by $0.0313\ M$ in 100.0 s. Calculate the initial rate of the reaction

$$2O_3 \longrightarrow 3O_2$$

Reaction Order from Concentration–Time Data

13.10 (a) What is the order with respect to a reactant? (b) How is this related to the overall reaction order?

13.11 How does the rate of a reaction generally change with time?

13.12 What is the relationship between the order with respect to a reactant and the coefficient of that reactant in the balanced equation for the reaction?

13.13 Explain why reaction orders are almost always whole numbers.

13.14 The reaction of hydrocyanic acid with water is described by the equation

$$HCN(aq) + 2H_2O(l) \longrightarrow NH_4^+(aq) + HCO_2^-(aq)$$

The concentrations of HCN at various times follow. Make appropriate plots of these data to (a) determine the order of the reaction with respect to HCN and (b) determine the value of the rate constant.

Concentration of HCN (M)	Time (s)
0.100	0.00
0.0984	2.00×10^5
0.0961	5.00×10^5
0.0923	1.00×10^6
0.0851	2.00×10^6
0.0785	3.00×10^6
0.0724	4.00×10^6
0.0668	5.00×10^6
0.0617	6.00×10^6
0.0569	7.00×10^6
0.0525	8.00×10^6
0.0484	9.00×10^6
0.0447	1.00×10^7
0.0199	2.00×10^7
0.0089	3.00×10^7
0.0040	4.00×10^7

13.15 The decomposition of hypobromite ion follows the equation

$$3BrO^- \longrightarrow BrO_3^- + 2Br^-$$

The rate law is $-\Delta[BrO^-]/\Delta t = k[BrO^-]^n$. According to the following concentration–time data, what are (a) the reaction order and (b) the value of the rate constant?

Concentration of BrO– (M)	Time (s)
0.750	0
0.528	10
0.408	20
0.332	30
0.280	40
0.242	50
0.213	60
0.190	70
0.172	80
0.157	90
0.144	100

13.16 The hydrolysis of methyl chloride in water,

$$CH_3Cl + H_2O \longrightarrow CH_3OH + HCl$$

follows the rate law

$$-\Delta[CH_3Cl]/\Delta t = k[CH_3Cl]^n$$

According to the following concentration–time data, measured at 90°C, what are (a) the reaction order and

(b) the value of the rate constant?

Concentration of CH_3Cl (*M*)	Time (h)
0.100	0
0.0823	40
0.0678	80
0.0558	120
0.0460	160
0.0378	200
0.0311	240
0.0256	280
0.0211	320
0.0174	360

Rate Laws from Rate–Concentration Data

13.17 Why do rates depend on concentrations?

13.18 How can we make a reaction go faster?

13.19 Describe how rates can be measured and used to determine the rate law for a reaction.

13.20 (a) Determine the rate law for the reaction

$$2UO_2^+(aq) + 4H^+(aq) \longrightarrow$$
$$U^{4+}(aq) + UO_2^{2+}(aq) + 2H_2O(l)$$

using the following rate data.

Initial Concentration of UO$_2$+ (*M*)	H+ (*M*)	Rate (*M* s^{-1})
0.0010	0.20	2.65×10^{-5}
0.0010	0.80	1.04×10^{-4}
0.0030	0.20	2.34×10^{-4}

(b) Also calculate the value of the rate constant.

13.21 The thermal decomposition of acetaldehyde was studied at 800 K.
$$CH_3CHO \longrightarrow CH_4 + CO$$
The following rates were determined for different concentrations of acetaldehyde.

Rate (*M* s^{-1})	Concentration of CH_3CHO (*M*)
9.0×10^{-7}	0.100
36.0×10^{-7}	0.200
14.4×10^{-6}	0.400

Determine (a) the rate law and (b) the value of the rate constant.

13.22 The ozonization of pentene was studied in CCl_4 solvent at 25°C.

$$C_5H_{10} + O_3 \longrightarrow C_5H_{10}O_3$$

The following rates were determined for different concentrations of pentene and ozone.

Rate (*M* s^{-1})	Initial Concentration of C_5H_{10} (*M*)	O_3 (*M*)
2.2	0.010	0.0028
1.1	0.0050	0.0028
4.4	0.010	0.0056

Determine (a) the rate law and (b) a value of the rate constant.

13.23 The reaction of NO_2 with O_3 is first-order in each reactant. The rate of this reaction is 9.00×10^{-13} *M* s^{-1} when the concentration of NO_2 is 3.00×10^{-9} *M* and the concentration of O_3 is 1.00×10^{-8} *M*. What is the value of the second-order rate constant?

13.24 The decomposition of ozone dissolved in carbon tetrachloride follows the equation

$$2O_3 \longrightarrow 3O_2$$

The following rates were measured at 71°C for different initial concentrations of ozone.

Initial Concentration of O_3 (*M*)	Initial Rate (*M* s^{-1})
5.00×10^{-4}	6.15×10^{-8}
2.50×10^{-4}	3.06×10^{-8}
1.25×10^{-4}	1.55×10^{-8}

Determine the rate law, including the value of the rate constant, assuming that the rate is dependent only on the concentration of ozone.

13.25 (a) Determine the rate law for the reaction

$$2I^-(aq) + 2VO_2^+(aq) + 4H^+(aq) \longrightarrow$$
$$I_2(aq) + 2VO^{2+}(aq) + 2H_2O(l)$$

using the following rate data.

Initial Concentration of I– (*M*)	VO$_2$+ (*M*)	H+ (*M*)	Rate (*M* s^{-1})
0.00200	0.0100	0.100	2.60×10^{-8}
0.00400	0.0100	0.100	5.21×10^{-8}
0.00200	0.0200	0.100	5.19×10^{-8}
0.00200	0.0100	0.0500	6.50×10^{-9}

(b) Also calculate the value of the rate constant.

13.26 The reaction of hypochlorite ion with iodide ion was studied in aqueous solution containing OH^- at 25°C.

$$ClO^-(aq) + I^-(aq) \longrightarrow IO^-(aq) + Cl^-(aq)$$

The following rates were determined for different concentrations of hypochlorite, iodide, and hydroxide ions.

$-\Delta[I-]/\Delta t$ (*M* s^{-1})	Initial Concentration of ClO– (*M*)	I– (*M*)	OH– (*M*)
1750	0.0017	0.0017	1.00
3500	0.0034	0.0017	1.00
3450	0.0017	0.0034	1.00
3520	0.0017	0.0017	0.50

Determine (a) the rate law and (b) the value of the rate constant.

Half-Life

13.27 What is a half-life?

13.28 How can the half-life be used to calculate a value of the rate constant for a first-order reaction?

13.29 Derive the expression that relates the half-life to the rate constant and the initial concentration in a zero-order reaction.

13.30 Derive the expression that relates the half-life to the rate constant and the initial concentration in a second-order reaction.

13.31 The first-order decomposition of N_2O_5 has a half-life of 270 minutes at 45°C. What is the rate constant for this reaction?

13.32 The isotope Na-22 has a half-life of 2.60 years and decays by first-order kinetics. (a) How much of a 1.00-g sample would be left after 1.00 year? (b) What is the rate constant for the decay reaction?

13.33 The isotope I-131 has a half-life of 8.05 days and decays by first-order kinetics. If 10 mg of I_2 is injected into the thyroid gland to test for thyroid disease, how much will remain after 14 days?

13.34 The isotope S-35 has a half-life of 86.7 days and decays by first-order kinetics. How long will it take for 90.0% of a sample to disappear?

13.35 What is the half-life of a radioactive isotope if one-sixth of it disappears by first-order kinetics in 30.0 minutes?

13.36 The hydrolysis of HCN to give ammonium formate,

$$HCN(aq) + 2H_2O(l) \longrightarrow NH_4HCO_2(aq)$$

follows the rate law $-\Delta[HCN]/\Delta t = k[HCN]$, with $k = 8.06 \times 10^{-8} \text{ s}^{-1}$ at 65°C. The initial concentration of HCN is 0.0800 M. (a) What is the half-life of this reaction? (b) How long does it take to reduce the HCN concentration to 0.0600 M?

13.37 The gas-phase decomposition of ozone,

$$2O_3(g) \longrightarrow 3O_2(g)$$

follows the rate law

$$-\Delta[O_3]/\Delta t = k[O_3]^2$$

with $k = 1.40 \times 10^{-2} \text{ } M^{-1} \text{ s}^{-1}$ at 80°C. (a) If the initial ozone concentration is $6.00 \times 10^{-3} M$, what is the half-life of this reaction? (b) How long will it take for 90% of the ozone to decompose?

Rate Law

13.38 What is meant by molecularity of a reaction?

13.39 How is molecularity related to reaction order?

13.40 How do we predict the rate law of an elementary reaction?

13.41 Assume that each of the following reactions is an elementary reaction. Write a rate law for each reaction.
   a. $N_2O_5 \rightarrow NO_3 + NO_2$
   b. $NO_3 \rightarrow NO + O_2$
   c. $NO + NO_3 \rightarrow 2NO_2$
   d. $2NO_2 \rightarrow N_2O_4$

13.42 The reaction

$$2NO(g) + O_2(g) \longrightarrow 2NO_2(g)$$

is second-order in NO and first-order in O_2. (a) Write a rate law for the disappearance of O_2. (b) If concentrations are in units of moles per liter (mol/L), what are the units of the rate constant? (c) Write an equation for the rate of appearance of NO_2. (d) Does the rate constant in this equation have the same units as in part (b)? (e) The same value?

13.43 The reaction

$$5Br^-(aq) + BrO_3^-(aq) + 6H^+(aq) \longrightarrow$$
$$3Br_2(aq) + 3H_2O(l)$$

is first-order in Br^- and in BrO_3^- and second-order in H^+. (a) Write a rate law for this equation. (b) When $-\Delta[BrO_3^-]/\Delta t$ is $1.25 \times 10^{-3} \text{ } M \text{ s}^{-1}$, what are the values of $-\Delta[Br^-]/\Delta t$, $-\Delta[H^+]/\Delta t$, and $\Delta[Br_2]/\Delta t$?

13.44 The reaction

$$5Cl^-(aq) + BrO_3^-(aq) + 6H^+(aq) \longrightarrow$$
$$2Cl_2(aq) + BrCl(aq) + 3H_2O(l)$$

follows the rate law

$$-\Delta[BrO_3^-]/\Delta t = k[BrO_3^-][Cl^-][H^+]^2$$

Find (a) the order of the reaction with respect to each reactant and (b) the overall order of the reaction.

13.45 The addition of bromine to ethylene in carbon tetrachloride solution,

$$C_2H_4 + Br_2 \longrightarrow C_2H_4Br_2$$

is first-order in each reactant. The rate constant is $1.80 \times 10^{-2} \text{ } M^{-1} \text{ s}^{-1}$ at 0°C. If the reaction solution initially contains 0.00125 M C_2H_5 and 0.00142 M Br_2, what is the initial rate of this reaction?

Activation Energy and Temperature

13.46 How does collision theory explain the magnitude of reaction rates?

13.47 What is activation energy?

13.48 Contrast collision theory and transition-state theory.

13.49 Why do reactions usually go faster at higher temperature?

13.50 How is the enthalpy change of a reversible reaction related to the activation energies for the forward and reverse reactions?

13.51 At 300°C, the reaction between hydrogen molecules and bromine atoms,

$$H_2(g) + Br(g) \rightleftharpoons HBr(g) + H(g)$$

has an activation energy of 73.6 kJ/mol in the forward direction and an activation energy of 4.1 kJ/mol in the reverse direction. What is the enthalpy of reaction under these conditions?

13.52 The second-order reaction

$$H_2(g) + I_2(g) \longrightarrow 2HI(g)$$

has a value of the frequency factor (A) of 3.4×10^{10}

$M^{-1} s^{-1}$ and an activation energy (E_a) of 178 kJ/mol. What is the value of the rate constant at 25°C?

13.53 What activation energy corresponds to the rule of thumb that the rate of a reaction near 25°C doubles for each 10°C rise in temperature?

13.54 Hydrogen iodide decomposes according to the following equation:

$$2HI(g) \longrightarrow H_2(g) + I_2(g) \qquad \Delta H = 60 \text{ kJ}$$

The activation energy for the forward reaction is 183 kJ. (a) Plot the energy versus the reaction coordinate for this reaction. (b) Give the enthalpy change and activation energy for the reverse reaction,

$$H_2(g) + I_2(g) \longrightarrow 2HI(g)$$

(c) What is the minimum value of the activation energy for an endothermic reaction?

13.55 (a) If it takes an egg 3 minutes to hard-boil in 100°C water, how long would you expect it to take in a pressure cooker at 120°C? (b) On a mountain top at 80°C? (c) Explain your answers.

13.56 The hydrolysis of an ester, shown in the following equation, is catalyzed by hydrogen ions.

$$CH_3CO_2C_2H_5 + H_2O \longrightarrow CH_3CO_2H + C_2H_5OH$$

The rate law for this reaction is

$$\text{Rate} = k[CH_3CO_2C_2H_5][H^+]$$

Values of the rate constant are $4.51 \times 10^{-5} M^{-1} s^{-1}$ at 25°C, $1.09 \times 10^{-4} M^{-1} s^{-1}$ at 35°C, and $2.47 \times 10^{-4} M^{-1} s^{-1}$ at 45°C. (a) What is the activation energy (in kJ/mol) and (b) the frequency factor (in $M^{-1} s^{-1}$) for this reaction? (c) What is the rate constant at 65°C?

13.57 The reaction between ethylene and hydrogen,

$$C_2H_4 + H_2 \longrightarrow C_2H_6$$

is first-order in each reactant. The rate constant is $4.7 \times 10^{-3} M^{-1} s^{-1}$ at 460°C, and the activation energy is 180 kJ/mol. What is the rate constant at 200°C?

13.58 Benzene reacts with chlorine according to the equation

$$C_6H_6 + Cl_2 \longrightarrow C_6H_5Cl + HCl$$

The rate law is $-\Delta[C_6H_6]/\Delta t = k[C_6H_6][Cl_2]$. The rate constant is $1.48 \times 10^{-6} M^{-1} s^{-1}$ at 24.0°C and $3.90 \times 10^{-6} M^{-1} s^{-1}$ at 34.2°C. What is the value of the rate constant at 50.0°C?

Reaction Mechanism

13.59 What is meant by a reaction mechanism?

13.60 What is an elementary reaction?

13.61 Can we prove that a reaction mechanism is correct?

13.62 What is a rate-determining step?

13.63 Explain the relationship between the identity of the rate-determining step and the form of the rate law.

13.64 The decomposition of dinitrogen pentoxide is described by the equation

$$2N_2O_5(g) \longrightarrow 2N_2O_4(g) + O_2(g)$$

A possible mechanism is

$$N_2O_5 \longrightarrow NO_3 + NO_2$$
$$NO_3 \longrightarrow NO + O_2$$
$$NO + NO_3 \longrightarrow 2NO_2$$
$$2NO_2 \longrightarrow N_2O_4$$

Indicate the molecularity of each step in this mechanism.

13.65 The reaction between arsenic acid and iodide ion in aqueous solution,

$$2I^-(aq) + H_3AsO_4(aq) + 2H^+(aq) \longrightarrow$$
$$I_2(aq) + H_3AsO_3(aq) + H_2O(l)$$

follows the rate law

$$-\Delta[I^-]/\Delta t = k[I^-][H_3AsO_4][H^+]$$

Which step in the following mechanism is the rate-determining step?

$$H_3AsO_4 + H^+ \rightleftharpoons H_4AsO_4^+$$
$$H_4AsO_4^+ + I^- \longrightarrow H_3AsO_3 + HOI$$
$$HOI + I^- + H^+ \longrightarrow I_2 + H_2O$$

13.66 What is the rate law for the following mechanism?

$$Tl^+ + Co^{3+} \rightleftharpoons Tl^{2+} + Co^{2+} \qquad \text{(fast)}$$
$$Tl^{2+} + Co^{3+} \longrightarrow Tl^{3+} + Co^{2+} \qquad \text{(slow)}$$

13.67 What is the rate law for the following mechanism?

$$V^{3+} + Tl^{3+} + 2H_2O \longrightarrow$$
$$VO_2^+ + Tl^+ + 4H^+ \qquad \text{(slow)}$$
$$V^{3+} + VO_2^+ \longrightarrow 2VO^{2+} \qquad \text{(fast)}$$

13.68 What is the rate law for the following mechanism?

$$Cr^{2+} + UO_2^{2+} \rightleftharpoons CrUO_2^{4+} \qquad \text{(fast)}$$
$$CrUO_2^{4+} + Cr^{2+} \longrightarrow 2Cr^{3+} + UO_2 \qquad \text{(slow)}$$
$$UO_2 + 4H^+ \longrightarrow U^{4+} + 2H_2O \qquad \text{(fast)}$$

13.69 The reaction between iron(II) ions and iodine in aqueous solution,

$$2Fe^{2+}(aq) + I_2(aq) \longrightarrow$$
$$2Fe^{3+}(aq) + 2I^-(aq)$$

follows the rate law

$$-\Delta[Fe^{2+}]/\Delta t = k[Fe^{2+}][I_2]$$

A possible mechanism is

$$Fe^{2+} + I_2 \rightleftharpoons Fe^{3+} + I_2^-$$
$$Fe^{2+} + I_2^- \longrightarrow Fe^{3+} + 2I^-$$

Which step in this mechanism is rate-determining?

Catalysis

13.70 What is a catalyst?

13.71 How can a catalyst be involved in a reaction but not be used up during the reaction?

13.72 How does a catalyst increase the rate of a reaction?

13.73 The hydrogenation of ethylene,

$$C_2H_4 + H_2 \longrightarrow C_2H_6$$

is slow at room temperature but is accelerated considerably by finely divided nickel metal. Nickel strongly adsorbs hydrogen and weakens the H–H bond. (a) Outline a reaction sequence that explains the catalytic activity of nickel. (b) Why is the metal used in a finely divided state rather than as a foil or bar?

13.74 The decomposition of acetaldehyde has the following mechanism in the presence of iodine:

$$CH_3CHO + I_2 \longrightarrow CH_3I + CO + HI$$
$$CH_3I + HI \longrightarrow CH_4 + I_2$$

(a) What is the catalyst in this mechanism? (b) Identify a reaction intermediate.

13.75 Identify a catalyst in the following mechanism.

$$Ag^+ + S_2O_8^{2-} \longrightarrow Ag^{3+} + 2SO_4^{2-}$$
$$3Ag^{3+} + 2Cr^{3+} + 7H_2O \longrightarrow$$
$$3Ag^+ + Cr_2O_7^{2-} + 14H^+$$

Additional Exercises

13.76 Draw an energy-versus-reaction coordinate diagram for the reaction

$$NO(g) + NO_3(g) \rightleftharpoons 2NO_2(g)$$

The activation energy for the forward reaction is 7.1 kJ/mol and for the reverse reaction is 100.0 kJ/mol. Is the forward reaction endothermic or exothermic? What is the value of ΔH for the forward reaction?

13.77 The reaction of fluorine with chlorine dioxide was studied in the gas state at 250 K.

$$F_2(g) + 2ClO_2(g) \longrightarrow 2FClO_2(g)$$

The following rates were determined for different concentrations of fluorine and chlorine dioxide.

| | **Initial Concentration of** | |
Rate (M s^{-1})	F_2 (M)	ClO_2 (M)
1.2×10^{-3}	0.100	0.0100
4.8×10^{-3}	0.100	0.0400
2.4×10^{-3}	0.200	0.0100

Determine the rate law, including a value of the rate constant.

13.78 Unlike chemical reactions, the rate of radioactive decay is independent of temperature. What must be the activation energy for radioactive decay?

13.79 The reaction

$$5Br^-(aq) + BrO_3^-(aq) + 6H^+(aq) \longrightarrow$$
$$3Br_2(aq) + 3H_2O(l)$$

is first-order in Br^- and in BrO_3^- and second-order in H^+. (a) Write a rate law for the disappearance of Br^-. (b) If concentrations are in units of moles per liter, what are the units of the rate constant? (c) Write an equation for the rate of appearance of Br_2. (d) Does the rate constant in this equation have the same units as in part (b)? (e) The same value?

13.80 Hydrogen peroxide solutions are normally reasonably stable, but when metal ions such as Fe^{3+} are added, the hydrogen peroxide decomposes.

$$2H_2O_2(aq) \longrightarrow 2H_2O(l) + O_2(g)$$

You may have noted the evolution of gas bubbles when using H_2O_2 to disinfect a cut. The iron catalyst is in the form of the enzyme peroxidase in this case. In a solution containing 0.025 M $FeCl_3$, the concentration of H_2O_2 varies as follows as a function of time.

Concentration of H_2O_2 (M)	Time (s)
0.80	0
0.72	27
0.64	52
0.56	86
0.48	121
0.40	166
0.32	218
0.24	307
0.00	∞

Determine (a) the order of the reaction with respect to H_2O_2 and (b) the value of the rate constant, assuming the equation

$$-\Delta[H_2O_2]/\Delta t = k[H_2O_2]^n$$

13.81 The reaction of ozone with ethylene follows the rate law

$$-\Delta[O_3]/\Delta t = k[O_3][C_2H_4]$$

The rate constant is 800 M^{-1} s^{-1} at 25°C. If the rate of reaction is 1.60×10^{-16} M s^{-1}, and the concentration of ozone is 1.00×10^{-8} M, what is the ethylene concentration?

13.82 The isotope Au-198 has a half-life of 64.8 hours and decays by first-order kinetics. (a) How much of a 0.010-g sample is left after 1.00 day? (b) What is the rate constant for the decay reaction?

13.83 (a) Determine the rate law for the reaction

$$HCHO(g) \longrightarrow H_2(g) + CO(g)$$

using the following rate data.

Concentration of HCHO (M)	Rate (M s^{-1})
0.00500	9.02×10^{-11}
0.0100	3.61×10^{-10}
0.0200	1.43×10^{-9}

(b) Also calculate the value of the rate constant.

13.84 The reaction that takes place upon the addition of bromine to ethylene in carbon tetrachloride solution,

$$C_2H_4 + Br_2 \longrightarrow C_2H_4Br_2$$

is first-order in each reactant. The rate constant is 1.80×10^{-2} M^{-1} s^{-1} at 0°C and 1.70×10^{-4} M^{-1} s^{-1} at 25°C. What is the activation energy for this reaction?

13.85 The gas-phase decomposition of oxygen difluoride,

$$2OF_2 \longrightarrow 2F_2 + O_2$$

follows the rate law

$$-\Delta[OF_2]/\Delta t = k[OF_2]^2$$

with $k = 1.39 \times 10^{-2}$ M^{-1} s^{-1} at 250°C. (a) If the initial oxygen difluoride concentration is 4.50×10^{-2} M, what is the half-life of this reaction? (b) How long will it take for 85.0% of the oxygen difluoride to decompose?

* 13.86 The C-14 content of carbon from the heartwood of a giant sequoia tree is 70.0% of that of wood from the outer bark of the tree. If C-14 has a half-life of 5730 years, how old is the tree? Assume that the rate of incorporation of C-14 by growing wood was constant throughout the lifetime of the tree.

* 13.87 The reaction of chromium(II) with iron(III) was studied at 25°C in aqueous solution.

$$Cr^{2+}(aq) + Fe^{3+}(aq) \longrightarrow Cr^{3+}(aq) + Fe^{2+}(aq)$$

With Fe^{3+} in great excess, the reaction is first-order in Cr^{2+}: Rate = $k'[Cr^{2+}]$. Values of k' were determined for several iron(III) concentrations. The rate law is of the form

$$\text{Rate} = k[Cr^{2+}][Fe^{3+}]^n$$

Determine the value of n and the value of k.

Concentration of Fe^{3+} (M)	Pseudo-First-order Rate Constant, k' (s^{-1})
0.00100	2.00
0.00200	4.00
0.00375	7.50
0.00500	10.0

* 13.88 (a) Determine the rate law for the reaction

$$Sn^{2+}(aq) + 2Fe^{3+}(aq) \longrightarrow Sn^{4+}(aq) + 2Fe^{2+}(aq)$$

using the following rate data.

Initial Concentration of			
Sn^{2+} (M)	Fe^{3+} (M)	H$^+$ (M)	Rate (M s^{-1})
0.00200	0.0400	1.00	4.79×10^{-8}
0.00400	0.0400	1.00	9.60×10^{-8}
0.00200	0.0200	1.00	2.41×10^{-8}
0.00200	0.0200	0.500	9.63×10^{-8}

(b) Also calculate the value of the rate constant.

* 13.89 The reaction of ferricyanide ion with iodide ion was studied in aqueous solution at 25°C.

$$2Fe(CN)_6^{3-}(aq) + 2I^-(aq) \longrightarrow$$
$$2Fe(CN)_6^{4-}(aq) + I_2(aq)$$

The following rates were determined for different concentrations of ferricyanide, iodide, and ferrocyanide ions.

	Initial Concentration of		
$\Delta[I_2]/\Delta t$ (M s^{-1})	Fe(CN)$_6^{3-}$ (M)	I$^-$ (M)	Fe(CN)$_6^{4-}$ (M)
1.0×10^{-3}	0.0010	0.0010	0.0010
4.0×10^{-3}	0.0020	0.0010	0.0010
1.0×10^{-3}	0.0010	0.0020	0.0020
4.0×10^{-3}	0.0020	0.0020	0.0020
8.0×10^{-3}	0.0020	0.0040	0.0020

Determine the rate law, including a value of the rate constant.

* 13.90 The North American snowy tree cricket is popularly known as the "thermometer cricket," because it is possible to determine the approximate temperature in degrees Fahrenheit by counting the number of times the cricket chirps in 15 s and adding 40. According to this formula, the rate of emission of chirps is 2.47 chirps/s at 25°C and 4.27 chirps/s at 40°C. Determine the apparent activation energy for the rate of chirping, assuming that the Arrhenius equation for rate constants is followed.

* 13.91 The decomposition of ozone in carbon tetrachloride solution,

$$2O_3 \longrightarrow 3O_2$$

is first-order in the concentration of ozone:

$$-\Delta[O_3]/\Delta t = k[O_3]$$

Write a mechanism consistent with this result.

* 13.92 Nitric oxide reacts with chlorine in the gas phase according to the equation

$$2NO(g) + Cl_2(g) \longrightarrow 2NOCl(g)$$

The rate law is

$$-\Delta[Cl_2]/\Delta t = k[NO]^2[Cl_2]$$

Suggest a mechanism consistent with this result.

* 13.93 In the contact process for the manufacture of sulfuric acid, the oxidation of SO_2 to SO_3 by O_2 is catalyzed by Pt or V_2O_5. Outline a reaction sequence that explains the catalytic process.

* 13.94 A kinetics study was carried out on the hydrolysis reaction

$$Cr(NH_3)_5Cl^{2+}(aq) + H_2O(l) \longrightarrow$$
$$Cr(NH_3)_5(H_2O)^{3+}(aq) + Cl^-(aq)$$

Starting with a solution that contains 0.02 M $Cr(NH_3)_5Cl^{2+}$, investigators determined the order with respect to this ion. (a) Explain what data are needed to establish the reaction order, and describe the procedure used. (b) Explain why it is not possible to determine the order with respect to H_2O.

▶ ▶ ▶ Chemistry in Practice

13.95 When 1,3-butadiene gas, $H_2C\!=\!CH\!-\!CH\!=\!CH_2$, is heated, the pressure of the gas gradually decreases as a result of the formation of another substance. A sample of butadiene is injected into a 500.0-mL evacuated flask at 340.0°C. The initial pressure is 378 torr. The pressure decreases as indicated in the following table.

Time (min)	Pressure in Flask (torr)
0.00	378
20.0	318
40.0	286
60.0	267
80.0	255
100.0	245
140.0	233
180.0	225
250.0	216
300.0	212
400.0	207
500.0	203
1000.0	197
∞	189

The product of the reaction contains 88.82% C and 11.12% H by mass. When it is condensed to a liquid by cooling, it is found to weigh 0.2674 g.
    a. Identify the product of the reaction.
    b. Write a balanced equation for the reaction.
    c. Determine the rate law for the disappearance of butadiene.
    d. Calculate a value of the rate constant for the reaction.
    e. Propose a possible mechanism for the reaction.

Chemical Equilibrium

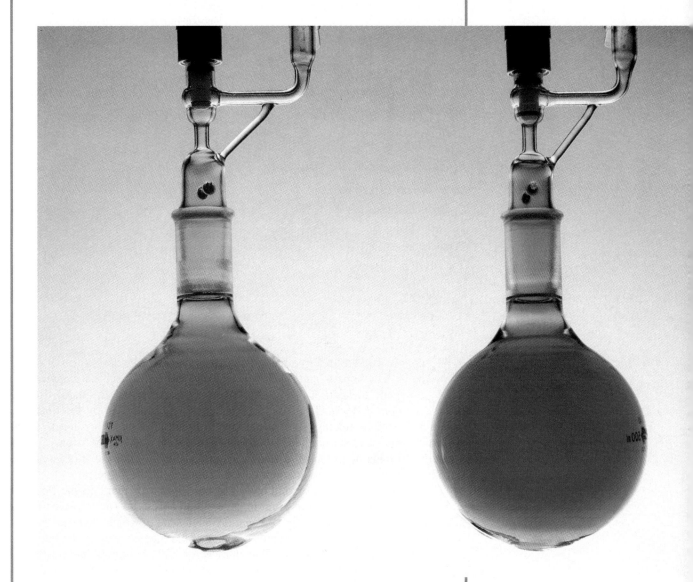

Nitrogen dioxide shown immediately after expanding into the vessel
(left) and after reaching a state of equilibrium (right).

CHAPTER OUTLINE

Encountering Chemistry

Suppose you are in a gymnasium, and you have a large basket full of tennis balls. You hope to start each ball bouncing by throwing it hard against the floor and to keep as many balls bouncing as possible, to qualify for the *Guinness Book of World Records*. You start throwing balls, but by the time you have half of them bouncing, some are stopping. Eventually, though you rush about as fast as you can, starting dead balls bouncing again, the proportion of balls bouncing to balls at rest on the floor remains constant. The two processes, the starting and the stopping of bouncing, are occurring at the same rate.

Now a friend brings a second basket of balls and rolls them on the floor for you. You no longer have to run as far to get to a ball at rest, so you are able to keep more balls bouncing. Eventually you again reach a state in which the proportion of balls bouncing to balls at rest remains constant. This may or may not be the same proportion as before, but the absolute number of balls bouncing has increased.

These opposing processes—the starting and stopping of bouncing—are similar to those involved in the vaporization of a liquid, described in Chapter 10. When a liquid evaporates in a closed container, molecules with sufficient kinetic energy overcome intermolecular forces and escape from the liquid to the gaseous state.

However, as the number of gaseous molecules increases, some of them start colliding with the surface of the liquid and are trapped in the liquid state again. Eventually, the two processes—evaporation and condensation—reach a state in which both continue at the same rate, so there is no overall change in the relative amounts of the substance existing in the liquid and gaseous states. This, you may recall, is a state of *dynamic equilibrium*.

Chemical reactions, too, can reach a state of dynamic equilibrium. A chemical equilibrium consists of reversible reactions occurring at equal rates, and this chemical equilibrium has the same characteristics as the physical equilibrium represented by the vaporization–condensation cycle. For example, the reddish gas nitrogen dioxide can combine with itself to give the colorless gas dinitrogen tetroxide.

$$2NO_2(g) \rightleftharpoons N_2O_4(g)$$

A change in the equilibrium concentrations results in changes in color intensity. When a gaseous mixture at high pressure is rapidly expanded into an evacuated flask (shown on the left in the chapter-opening photo), the color is less intense because the concentration of both molecules has decreased. On the right, we see that over a period of time, the color intensity increased as the position of equilibrium shifted toward the colored gas molecules.

This chapter examines both qualitative and quantitative aspects of chemical reactions in a state of dynamic equilibrium. It also discusses the factors that influence this state, such as the rates of the forward and reverse reactions, partial pressure and concentration, temperature, and free-energy change. On the basis of these factors, we can devise ways to shift a reaction to a different state of equilibrium. Methods are presented for determining the concentrations or pressures of reactants and products at equilibrium. Finally, the principles of equilibrium are applied to some gas-phase chemical systems used in industry.

14.1 The Equilibrium Constant

In Chapter 13, we saw how the concentrations of reactants and products change with time. Although when we were discussing kinetics, we generally considered reactions that go to completion, the same principles apply to reversible reactions that go to equilibrium. Using the rate constants for the forward and the reverse reactions, we can determine the concentrations of reactants and products at equilibrium.

Relationship Between Equilibrium and Kinetics

How is the position of equilibrium related to the rate constants for a reversible reaction?

The rate of a reversible reaction that comes to a state of equilibrium usually depends on the concentration of reactants. To understand how this dependence works and how we can use rate constants to determine the equilibrium composition of a mixture of reactants and products, we will first examine a very simple system. Consider an *isomerization* reaction, in which there is no change in the composition of the substance but only in the arrangement of atoms. The rearrangement of butane, C_4H_{10}, is such a reaction. As you may recall from Chapter 8, butane can

Figure 14.1 In a reversible reaction such as the isomerization of butane, concentrations of the reactant ([Butane]) and product ([i-Butane]) change with time. They ultimately remain constant at their equilibrium values when the rates of the forward and reverse reactions become equal.

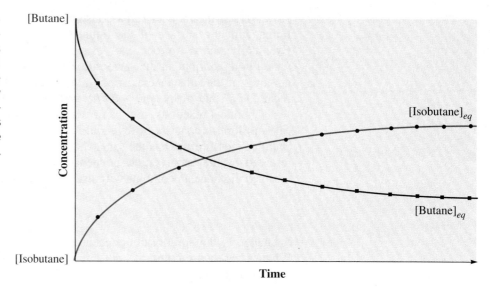

exist either as a straight chain of carbon atoms or as a branched chain.

$$CH_3CH_2CH_2CH_3 \rightleftharpoons CH_3CHCH_3$$
$$| \atop CH_3$$

normal butane   isobutane

To simplify the notation, we will symbolize these molecules as But and iBut. We will assume that initially the system contains all But and no iBut. As the reaction proceeds, the concentration of But decreases with time, and the concentration of iBut increases with time, as shown in Figure 14.1. Ultimately, the concentrations of the species reach constant values represented by the symbols $[But]_{eq}$ and $[iBut]_{eq}$, and a state of equilibrium is reached. The changes that occur in this system are shown in Figure 14.2.

In this state of equilibrium, chemical reaction is still occurring, but the rate of change of But into iBut is exactly equal to the rate of change of iBut into But. If we assume that both the forward and the reverse reactions are unimolecular elementary reactions, for which the rate is proportional to the concentration, then the

Figure 14.2 A reversible reaction reaches a state of equilibrium. (*A*) Initially, only butane molecules are present, so reaction proceeds only in the forward direction. (*B*) As formation of isobutane begins, the reverse reaction can also occur. It begins slowly, because not much isobutane is present yet. (*C*) Ultimately, the concentrations of butane and isobutane are such that the rate of formation of isobutane equals the rate of formation of butane. This is a state of equilibrium, in which there are no further changes in rate.

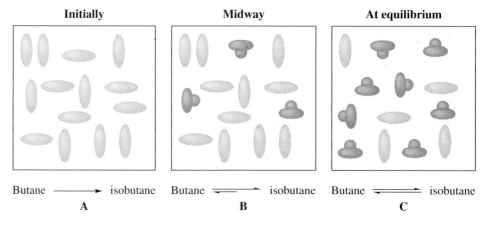

forward reaction should follow this rate law:

$$\text{Rate}_f = k_f[\text{But}]$$

and the reverse reaction should follow this rate law:

$$\text{Rate}_r = k_r[\text{iBut}]$$

In a state of equilibrium, the forward and reverse reactions must proceed at the same rate, because no net chemical change is occurring, and the concentrations have their equilibrium values.

$$\text{Rate}_f = \text{rate}_r$$
$$k_f[\text{But}]_{eq} = k_r[\text{iBut}]_{eq}$$

We know the rate constants have constant values at a given temperature, so we can rearrange this equation to obtain a ratio of rate constants, which must also be a constant.

$$\frac{k_f}{k_r} = \frac{[\text{iBut}]_{eq}}{[\text{But}]_{eq}}$$

The rate constant for the forward reaction, k_f, divided by the rate constant for the reverse reaction, k_r, is equal to a third constant, called the **equilibrium constant** (*K*).

$$K_c = \frac{k_f}{k_r} = \frac{[\text{iBut}]_{eq}}{[\text{But}]_{eq}}$$

In this case, we use the symbol K_c to indicate that the amounts of reactants and products are given in concentration units. This equation describes the concentrations of species in a state of equilibrium.

EXAMPLE 14.1

The following reversible reaction occurs in solution:

$$2NO_2 \underset{k_r}{\overset{k_f}{\rightleftharpoons}} N_2O_4$$

At 25°C, the rate laws are

$$\text{Rate}_f = k_f[NO_2]^2 \qquad k_f = 1.42 \times 10^7 \ M^{-1} \ s^{-1}$$
$$\text{Rate}_r = k_r[N_2O_4] \qquad k_r = 6.40 \times 10^4 \ s^{-1}$$

Calculate the value of the equilibrium constant.

Solution   At equilibrium, $\text{rate}_f = \text{rate}_r$. Substituting the rate laws, we get

$$k_f[NO_2]_{eq}^2 = k_r[N_2O_4]_{eq}$$

Now we rearrange to isolate a ratio of rate constants, which equals the equilibrium constant, K_c.

$$K_c = k_f/k_r = [N_2O_4]_{eq}/[NO_2]_{eq}^2$$

Inserting values for the rate constants gives us the value of the equilibrium constant.

$$K_c = (1.42 \times 10^7)/(6.40 \times 10^4) = 222$$

(continued)

The equilibrium constant determined by measurement of values of $[N_2O_4]_{eq}$ and $[NO_2]_{eq}$ is 218.

Practice Problem 14.1

Calculate the value of the equilibrium constant (K_c) at 400°C for the reaction

$$H_2(g) + I_2(g) \overset{k_f}{\underset{k_r}{\rightleftharpoons}} 2HI(g)$$

The rate constants measured at 400°C are $k_f = 5.00 \times 10^{-4}\ M^{-1}\ s^{-1}$ and $k_r = 1.00 \times 10^{-5}\ M^{-1}\ s^{-1}$.

Answer: 50.0

Our simple example assumed that the forward and reverse reactions were both elementary reactions. But the same principles apply to multistep forward and reverse reactions. Recall the reaction of thallium(III) and mercury(I) discussed in Chapter 13:

$$Tl^{3+} + Hg_2^{2+} \rightleftharpoons Tl^+ + 2Hg^{2+}$$

Thallium is a bluish-white, soft, easily melted metal. It oxidizes in air, forming a coating of black thallium(I) oxide. Alloyed with mercury, it is used in low-temperature switches. It is also used in semiconductors. Rodent poisons are made from its salts. In small doses, thallium salts cause weakness, pain in the extremities, and loss of hair. (These properties formed the basis for the Agatha Christie novel The Pale Horse.)

The mechanism proposed for this reaction was

$$Hg_2^{2+} \rightleftharpoons Hg^0 + Hg^{2+}$$
$$Tl^{3+} + Hg^0 \rightleftharpoons Tl^+ + Hg^{2+}$$

with the second step the rate-determining step. The rate law determined experimentally for the forward reaction is

$$\text{Rate}_f = k_f[Tl^{3+}][Hg_2^{2+}]/[Hg^{2+}]$$

The rate law for the reverse reaction, also found from experimentation, is

$$\text{Rate}_r = k_r[Tl^+][Hg^{2+}]$$

These two rate laws combine when equilibrium is reached.

$$\text{Rate}_r = \text{rate}_f$$
$$k_r[Tl^+][Hg^{2+}] = k_f[Tl^{3+}][Hg_2^{2+}]/[Hg^{2+}]$$

Rearranging gives an expression for K_c.

$$K_c = k_f/k_r = [Tl^+][Hg^{2+}]^2/[Tl^{3+}][Hg_2^{2+}]$$

How can we predict the expression for the equilibrium constant without knowing the rate laws for the forward and reverse reactions?

In a reversible reaction, no matter what the mechanism, the forward and reverse rate laws always combine in this manner to give an expression for the equilibrium constant that is predictable from the reaction stoichiometry.

Law of Mass Action

Examining combinations of rate laws for reversible reactions led the Norwegian chemists Cato Guldberg and Peter Waage to propose the **law of mass action** in 1867. This law states that at a given temperature a reaction will reach the same state of equilibrium, described by the same value of K_c, no matter what the starting conditions. In other words, no matter what concentrations of reactants and products we combine at, say, 25°C, when this mixture reaches a state of equilibrium, it has concentrations that give the value of K_c characteristic of that reaction at 25°C.

The dependence of the equilibrium constant on the equilibrium concentrations can be predicted from the stoichiometry of the reaction. Consider the general reversible reaction between substances A and B to give products C and D, represented as follows:

$$aA + bB \rightleftharpoons cC + dD$$

The equilibrium constant expression takes the form

$$K_c = [C]^c[D]^d/[A]^a[B]^b$$

The numerator of the expression is made up of the concentrations of the products, each raised to the power of its coefficient in the balanced equation, multiplied together. The denominator is made up of the concentrations of the reactants, each raised to the power of its coefficient, multiplied together. Although the actual value of each equilibrium concentration may vary from one experiment to another, the ratio of equilibrium concentrations always has the same value, provided the temperature is constant. It doesn't matter what the initial concentrations of reactants or products are, or whether the reaction is started with all reactants and no products or all products and no reactants. The reaction is reversible, and it will reach the same equilibrium position, which is governed by the value of the equilibrium constant, no matter where we start it.

EXAMPLE 14.2

Write the expression for K_c for the following reversible reaction:

$$CO(g) + 2H_2(g) \rightleftharpoons CH_3OH(g)$$

Solution   The concentration equilibrium constant is equal to the concentration of the products divided by the concentrations of the reactants, each concentration having first been raised to the power of its stoichiometric coefficient. For this reaction, the equation is thus

$$K_c = [CH_3OH]/[CO][H_2]^2$$

Note that the concentration of hydrogen is squared.

Practice Problem 14.2

Write the expression for K_c for the following reversible reaction:

$$2NO_2(g) \rightleftharpoons N_2O_4(g)$$

Answer: $K_c = [N_2O_4]/[NO_2]^2$

Determining K_c

Although the value of the equilibrium constant can be obtained from the rate constants for the forward and the reverse reactions, as we noted earlier, this is neither the simplest nor the most accurate method. Better results can be obtained by taking advantage of the law of mass action. The usual procedure involves preparing a reaction system under the desired conditions, allowing it to come to a state of equilibrium, and measuring the concentrations of the reactants and products. The measured values are then substituted into the equilibrium expression, and the equilibrium constant is calculated.

Suppose, for example, we allowed a mixture of NO, O_2, and N_2 gases to come to equilibrium at 1200°C.

$$N_2(g) + O_2(g) \rightleftharpoons 2NO(g)$$

The concentrations measured at equilibrium are 1.54×10^{-2} M N_2, 4.08×10^{-3} M O_2, and 2.64×10^{-5} M NO. The equilibrium constant expression is

$$K_c = [NO]^2/[N_2][O_2]$$

We need only substitute measured values for each equilibrium concentration into this equation to obtain the value of the equilibrium constant.

$$K_c = (2.64 \times 10^{-5})^2/(1.54 \times 10^{-2})(4.08 \times 10^{-3}) = 1.11 \times 10^{-5}$$

EXAMPLE 14.3

A sample of hydrogen iodide gas is placed in a vessel at 450°C and allowed to come to a state of equilibrium.

$$2HI(g) \rightleftharpoons H_2(g) + I_2(g)$$

The following concentrations of species are measured: [HI] = 0.195 mol/L, [H_2] = 0.0275 mol/L, and [I_2] = 0.0275 mol/L. What is the value of K_c?

Solution   We write the expression for K_c,

$$K_c = [H_2][I_2]/[HI]^2$$

and substitute the equilibrium concentrations into this expression.

$$K_c = (0.0275)(0.0275)/(0.195)^2 = 0.0199$$

Practice Problem 14.3

A mixture of hydrogen and carbon dioxide gases is allowed to come to equilibrium at 150°C.

$$H_2(g) + CO_2(g) \rightleftharpoons H_2O(g) + CO(g)$$

The following concentrations of species are measured: [H_2] = 0.171 mol/L, [CO_2] = 0.269 mol/L, [H_2O] = 0.232 mol/L, and [CO] = 0.325 mol/L. What is the value of K_c?

Answer: 1.64

Finding Equilibrium Constants by Using Partial Pressures

For gas-phase reactions, the equilibrium constant expression can be developed in two ways, which differ in how the amount of gas is expressed. We can express the amount of gas in terms of concentration in moles per liter (M), as we did earlier, in which case the equilibrium constant is symbolized K_c. For example, consider the reaction

$$H_2(g) + Cl_2(g) \rightleftharpoons 2HCl(g)$$

The expression for K_c is

$$K_c = [HCl]^2/[H_2][Cl_2]$$

Show how pressure and concentration are related for an ideal gas.

Because the pressure of a gas is proportional to its concentration, we can also express the amount of each gas in terms of partial pressures. The equilibrium constant is then symbolized K_p. The expression for K_p is like that for K_c except that the concentrations are replaced with the corresponding partial pressures:

$$K_p = P_{HCl}^2/P_{H_2}P_{Cl_2}$$

These expressions differ only in the use of partial pressures versus concentrations. The species and the powers to which the terms are raised remain the same.

Measuring K_p for a reaction is much like measuring K_c except that partial pressures must be used instead of concentrations. ▶ Suppose, for example, that we have allowed the mixture of NO_2 and N_2O_4 gases described in the chapter introduction to come to equilibrium at 25°C.

$$2NO_2(g) \rightleftharpoons N_2O_4(g)$$

We measure partial pressures of reactant and product and obtain the following values:

$$P_{NO_2} = 0.150 \text{ atm}$$
$$P_{N_2O_4} = 0.160 \text{ atm}$$

We write the expression for K_p:

$$K_p = \frac{P_{N_2O_4}}{P_{NO_2}^2}$$

Then we substitute the equilibrium partial pressures into this expression to obtain the value of the equilibrium constant.

$$K_p = 0.160/(0.150)^2 = 7.11$$

Units of the Equilibrium Constant

How do we write equilibrium expressions for heterogeneous equilibria?

You may have noticed that none of the equilibrium constants used so far have been assigned units. Although the ratios of product and reactant concentrations or pressures have units, the equilibrium constants do not. In a more sophisticated development of chemical equilibrium, the unitless nature of the equilibrium constant is generated by relating concentrations to the standard state, which is 1 atm for gases and 1 M for dissolved substances. When these standard-state quantities are included in the expression, they cancel out the units but do not change the numerical values of the pressures or concentrations. For this presentation, it is sufficient to realize that the absence of units on equilibrium constants is normal. In this book, we will always measure the quantities of materials in atmospheres when K_p is used and in molarity when K_c is used.

The Equilibrium Expression for Heterogeneous Equilibria

So far, we have been considering only *homogeneous equilibria,* in which all reactants and products are in the same physical state (either in the gaseous state or in solution). *Heterogeneous equilibria* involve more than one state of matter. In equilibrium expressions for heterogeneous equilibria, gases are represented in terms of partial pressures or molar concentrations, and dissolved substances are represented by their molar concentrations, as before. But liquids and solids are omitted from the equilibrium constant expression, because the concentrations of these substances do not change as long as some of the substances are present.

Consider the hydrolysis of bismuth(III) chloride, shown in Figure 14.3.

$$BiCl_3(aq) + H_2O(l) \rightleftharpoons BiOCl(s) + 2HCl(aq)$$

The expression for the equilibrium constant is

$$K_c = [HCl]^2/[BiCl_3]$$

The solid (BiOCl) and the liquid (H_2O) are not included because their concentrations are constant.

The same rule applies to heterogeneous equilibria involving gases. Solid ammonium carbamate, for example, dissociates as follows:

$$NH_4CO_2NH_2(s) \rightleftharpoons 2NH_3(g) + CO_2(g)$$

The equilibrium constant expressed in terms of partial pressures does not include an ammonium carbamate term.

$$K_p = P_{NH_3}^2 P_{CO_2}$$

When both gases and dissolved substances are involved in a reaction, it is common practice to use only K_c, as shown in Example 14.4.

Figure 14.3 Bismuth(III) chloride, $BiCl_3$, forms white to yellowish crystals. It hydrolyzes readily in moist air, releasing HCl gas. It is soluble in acidic solutions, but when diluted it forms white bismuthyl chloride, BiOCl, which is nearly insoluble in water and is used in face powders, pigments, and artificial pearls.

EXAMPLE 14.4

Write the expression for K_p or K_c (as appropriate) for the following reversible reactions.

a. $CaCO_3(s) \rightleftharpoons CaO(s) + CO_2(g)$
b. $3HNO_2(aq) \rightleftharpoons H^+(aq) + NO_3^-(aq) + 2NO(g) + H_2O(l)$

Solution

a. In this example only carbon dioxide is included; the other two substances are solids. Because carbon dioxide is a gas, we can use either K_p or K_c.

$$K_p = P_{CO_2} \qquad K_c = [CO_2]$$

b. The second example involves dissolved substances and a gas, which are included in the expression, and water, which is not. We use only K_c in this case, because not all the species in the expression are gases.

$$K_c = [NO]^2[H^+][NO_3^-]/[HNO_2]^3$$

Practice Problem 14.4

Write the expression for K_p and for K_c for the following reversible reaction.

$$C(s) + H_2O(g) \rightleftharpoons CO(g) + H_2(g)$$

Answer: $K_p = P_{H_2}P_{CO}/P_{H_2O} \qquad K_c = [H_2][CO]/[H_2O]$

Manipulation of Equilibrium Constants

Because reactions that come to a state of equilibrium are reversible, it is legitimate to write them with either set of species as reactants, depending on the starting conditions in the experiment. Consider the dissociation of phosphorus pentachloride.

Phosphorus pentachloride, PCl_5, is a white to pale yellow, fuming solid. It sublimes at about 100°C at atmospheric pressure but melts at 148°C and boils at 160°C when under pressure. It is hydrolyzed readily by water to form phosphoric and hydrochloric acids.

$$PCl_5(g) \rightleftharpoons PCl_3(g) + Cl_2(g)$$

The equilibrium constant for this reaction is given by the equation

$$K_c = [PCl_3][Cl_2]/[PCl_5]$$

The same equilibrium can be reached by reacting phosphorus trichloride and chlorine gases:

$$PCl_3(g) + Cl_2(g) \rightleftharpoons PCl_5(g)$$

Phosphorus trichloride, PCl_3, is a colorless, clear, fuming liquid that boils at 76°C. It is formed by the reaction of red phosphorus and chlorine gas. In excess chlorine, it forms phosphorus pentachloride.

In this case, we obtain a different equilibrium constant, which we designate with the notation K_c' to distinguish it from the original equilibrium constant.

$$K_c' = [PCl_5]/[PCl_3][Cl_2]$$

Note that when the reaction is reversed, the expression for the equilibrium constant is just inverted. Further, the value of the equilibrium constant for the reversed reaction is just the inverse of the original equilibrium constant, provided the temperature is the same for both.

$$K_c' = 1/K_c$$

The same manipulations can be carried out with K_p, because the same principles are involved.

EXAMPLE 14.5

The equilibrium constant K_c for decomposition of hydrogen iodide gas into its component elements is 0.0199 at 450°C. Determine the value of the equilibrium constant K_c' at 450°C for the reverse of this reaction, which is described by the equation

$$H_2(g) + I_2(g) \rightleftharpoons 2HI(g)$$

Solution   The expression for K_c' is given by the equation

$$K_c' = [HI]^2/[H_2][I_2]$$

This is just the inverse of the expression for K_c:

$$K_c = [H_2][I_2]/[HI]^2$$

The two equilibrium constants are related by the expression

$$K_c' = 1/K_c$$

Inserting the value of K_c, we obtain

$$K_c' = 1/0.0199 = 50.3$$

Practice Problem 14.5

A mixture of water and carbon monoxide gases is allowed to come to equilibrium at 150°C.

$$H_2O(g) + CO(g) \rightleftharpoons H_2(g) + CO_2(g)$$

Find the value of K_c for this reaction, given that the equilibrium constant for the reverse reaction has the value 1.64 at 150°C.

Answer: 0.610

Where do we get values of equilibrium constants?

Values of equilibrium constants can be calculated from measured values of equilibrium concentrations or partial pressures. Many of these equilibrium constant values can be found in handbooks. When the numerical value of an equilibrium constant is not already tabulated, it may sometimes be obtained from the equilibrium constants for reactions that can be added or subtracted to yield the desired reaction, provided that all equilibrium constants are at the same temperature.

Suppose we need to know the equilibrium constant (K_1) for the following reaction at 225°C, but the value is not readily available.

$$SO_2(g) + CO_2(g) \rightleftharpoons SO_3(g) + CO(g)$$

We do know values of the equilibrium constants for the following reactions at 225°C.

$$SO_2(g) + \tfrac{1}{2}O_2(g) \rightleftharpoons SO_3(g) \qquad K_2 = [SO_3]/[SO_2][O_2]^{\frac{1}{2}} = 2.74 \times 10^7$$
$$CO_2(g) \rightleftharpoons CO(g) + \tfrac{1}{2}O_2(g) \qquad K_3 = [CO][O_2]^{\frac{1}{2}}/[CO_2] = 1.05 \times 10^{-27}$$

(Although we do not commonly use fractional stoichiometric coefficients, in some cases they help us focus on the species of interest, such as the sulfur and carbon oxides in these reactions.) Note that when added together, these two equations give the equation of interest, because the oxygen cancels out.

$$SO_2(g) + \cancel{\tfrac{1}{2}O_2(g)} + CO_2(g) \rightleftharpoons SO_3(g) + CO(g) + \cancel{\tfrac{1}{2}O_2(g)}$$

How does this help us obtain the desired equilibrium constant value? Consider what happens when we multiply the two equilibrium constant expressions together and get the equation

$$K_2 \times K_3 = [SO_3]/[SO_2][O_2]^{\tfrac{1}{2}} \times [CO][O_2]^{\tfrac{1}{2}}/[CO_2]$$

Combining terms yields the desired equilibrium constant expression.

$$K_2 \times K_3 = [SO_3][CO]/[SO_2][CO_2] = K_1$$
$$K_1 = 2.74 \times 10^7 \times 1.05 \times 10^{-27} = 2.88 \times 10^{-20}$$

When we add two or more equations together, we get the equilibrium constant for the combined reaction by multiplying their equilibrium constants together. Similarly, when we subtract an equation from another equation, we divide the constant for the first reaction by the constant for the second to get the overall constant. Finally, when we multiply coefficients in a reaction equation by a constant, the new equation has an equilibrium constant equal to the old equilibrium constant raised to the power of the constant multiplier. This is equivalent to adding the same reaction an appropriate number of times. Consider the reaction

$$2SO_2(g) + O_2(g) \rightleftharpoons 2SO_3(g) \qquad K_1 = [SO_3]^2/[SO_2]^2[O_2]$$

This is just the reaction we saw earlier with the coefficients multiplied by 2.

$$SO_2(g) + \tfrac{1}{2}O_2(g) \rightleftharpoons SO_3(g) \qquad K_2 = [SO_3]/[SO_2][O_2]^{\tfrac{1}{2}} = 2.74 \times 10^7$$

The first equilibrium constant is just the second equilibrium constant raised to the second power.

$$K_1 = [SO_3]^2/[SO_2]^2[O_2] = K_2{}^2$$
$$= (2.74 \times 10^7)^2 = 7.51 \times 10^{14}$$

EXAMPLE 14.6

Consider the production of metallic iron from iron(II) oxide at 1000°C. Two reactions that are coupled in this process are

$$C(s) + CO_2(g) \rightleftharpoons 2CO(g) \qquad K_1 = 167.5$$
$$FeO(s) + CO(g) \rightleftharpoons Fe(s) + CO_2(g) \qquad K_2 = 0.403$$

Adding these two reactions together gives a description of the net chemical changes occurring as a result of the coupled reactions.

$$FeO(s) + C(s) \rightleftharpoons Fe(s) + CO(g) \qquad K_3 = ?$$

What is the value of the equilibrium constant for this overall reaction?

Solution   Two reaction equations were added, so their equilibrium constants may be multiplied to obtain the new equilibrium constant.

$$K_3 = K_1 \times K_2$$
$$= 167.5 \times 0.403 = 67.5$$

(continued)

Practice Problem 14.6

Consider the reaction of steam with coal to give carbon monoxide and hydrogen, a mixture called water gas. Equilibrium constants are known for the following reactions at 1000 K:

$$C(s) + CO_2(g) \rightleftharpoons 2CO(g) \qquad K_{p,1} = 1.75$$
$$H_2O(g) + CO(g) \rightleftharpoons H_2(g) + CO_2(g) \qquad K_{p,2} = 1.44$$

What is the equilibrium constant for the water-gas reaction at 1000 K?

$$C(s) + H_2O(g) \rightleftharpoons H_2(g) + CO(g) \qquad K_{p,3} = ?$$

Answer: 2.52

Relationship of K_p and K_c

When do K_p and K_c have the same value for a reaction?

For reactions involving gases, the values of K_p and K_c are not necessarily identical. Consider, for example, the decomposition reaction that occurs when phosgene is subjected to high temperatures.

$$COCl_2(g) \rightleftharpoons CO(g) + Cl_2(g)$$

Phosgene, $COCl_2$, is a colorless, highly toxic gas that has been used in gas warfare. When it is highly diluted with air, it has a suffocating odor similar to that of moldy hay. The liquid is colorless and boils at 8.2°C. It is used in the preparation of many organic chemicals.

At 727°C, the value of K_p is 26.3 and the value of K_c is 0.320.

It is a straightforward procedure to convert K_p to K_c, or vice versa, by using the ideal gas law:

$$PV = nRT$$

If we rearrange the equation to solve for pressure, and if we recognize that n/V is just the concentration in molarity (moles per liter), we obtain the following relationship between pressure and concentration for a gas:

$$P = MRT$$

The expression for K_p for the decomposition of phosgene is

$$K_p = P_{CO}P_{Cl_2}/P_{COCl_2}$$

Substituting the relationship for pressure obtained from the ideal gas law into this equation yields

$$K_p = ([CO]RT)([Cl_2]RT)/([COCl_2]RT)$$
$$= ([CO][Cl_2]/[COCl_2]) \times RT$$
$$= K_c \times RT$$

The gas constant, R, has a value of 0.08206 L atm/mol K, and the absolute temperature is 727°C + 273 = 1000 K. We can calculate K_p from K_c using this equation.

$$K_p = 0.320 \times 0.08206 \times 1000 = 26.3$$

This result is generalized for any gas-phase reaction in the equation

Show that this equation is true for the gas-phase reactions used in several examples earlier in this section.

$$K_p = K_c \times (RT)^{\Delta n}$$

In this equation, Δn is the number of molecules of *gaseous* products minus the number of molecules of *gaseous* reactants appearing in the balanced equation. The phosgene decomposition reaction has two gaseous product molecules and one gaseous reactant molecule, so $\Delta n = 1$ for that reaction.

Sometimes there is no change in the number of gas molecules during the reaction, as in the water-gas shift reaction.

$$CO(g) + H_2O(g) \rightleftharpoons CO_2(g) + H_2(g)$$

In this case, the values of K_p and K_c are the same because $\Delta n = 0$.

$$K_p = K_c \times (RT)^0 = K_c$$

EXAMPLE 14.7

Solid ammonium carbamate, used as an ammoniating agent, dissociates into ammonia and carbon dioxide upon evaporation.

$$NH_4CO_2NH_2(s) \rightleftharpoons 2NH_3(g) + CO_2(g)$$

The equilibrium constant for this reaction is $K_p = 2.31 \times 10^{-4}$ at 25°C. What is the value of K_c?

Solution   There are three gaseous product molecules—two ammonia molecules and one carbon dioxide molecule—and no gaseous reactant molecules, so the change in the number of gaseous molecules is

$$\Delta n = 2 + 1 - 0 = 3$$

Because Δn equals 3, the relationship between K_p and K_c is

$$K_p = K_c \times (RT)^3$$

The gas constant has a value of 0.08206, and the absolute temperature is $25 + 273 = 298$ K. Substituting values into this equation, we obtain

$$2.31 \times 10^{-4} = K_c \times (0.08206 \times 298)^3$$
$$K_c = 2.31 \times 10^{-4}/14{,}600 = 1.58 \times 10^{-8}$$

Practice Problem 14.7

The preparation of ammonia by the Haber process uses the equilibrium reaction

$$N_2(g) + 3H_2(g) \rightleftharpoons 2NH_3(g)$$

The equilibrium constant for this reaction is $K_c = 0.00237$ at 727°C. What is the value of K_p?

Answer: 3.52×10^{-7}

14.2   Free Energy and Equilibrium

We have seen that we can obtain values of equilibrium constants from rate constants for the forward and reverse reactions, by direct measurements of concentrations in equilibrium systems, or by combining equilibrium constants for other reactions. We can also calculate values by using thermodynamic functions.

Do we have to tabulate values
of equilibrium constants for all
possible reactions?

In Chapter 12, the free-energy change of a process was used to predict spontaneity. This is possible because the free-energy change depends on the exact conditions under which the process is being carried out, as described by the equation

$$\Delta G = \Delta G° + RT \ln Q$$

where $R = 8.314$ J/mol K and Q is the reaction quotient.

Recall that the expression for the reaction quotient is just the same as the expression for the equilibrium constant, except that the pressures or concentrations are not equilibrium values. The reaction quotient is a measure of the extent of reaction. It has a value of zero when no products are present and an infinitely large value when no reactants are present.

The value of the free-energy change, then, depends in part on the extent of reaction, as described by Q. When the value of ΔG is negative initially, the reaction proceeds spontaneously in the forward direction, with reactant pressures decreasing and product pressures increasing. As a result, the value of Q increases until ultimately it becomes sufficiently large to cause ΔG to equal zero. The system ceases to undergo any net change at this point and exists in a state of equilibrium:

$$\Delta G = 0 = \Delta G° + RT \ln Q_{eq}$$

Under this special condition, Q (or Q_{eq}) is the same as the equilibrium constant, K. Thus equilibrium constants can be calculated from standard free-energy changes. For gases, the relationship is

$$\Delta G° = -RT \ln K_p$$

The relationship for reactions occurring in solution is

$$\Delta G° = -RT \ln K_c$$

Because the standard free energy pertains to substances in their standard states, gases must be measured in terms of partial pressures, and the equation cannot be used to calculate K_c for gaseous reactions.

EXAMPLE 14.8

The standard free-energy change for the following reaction is $\Delta G° = 140.0$ kJ/mol O_2.

$$2SO_3(g) \rightleftharpoons 2SO_2(g) + O_2(g)$$

What is the equilibrium constant for this reaction at 25.0°C?

Solution   This is a gas-phase reaction, so we apply the equation for K_p:

$$\Delta G° = -RT \ln K_p$$

with $R = 8.314$ J/mol K and $T = 298.2$ K.

$$\Delta G° = -8.314 \text{ J/mol K} \times 298.2 \text{ K} \times \ln K_p$$

Using the value for the standard free-energy change in units of joules per mole, we get

$$1.400 \times 10^5 \text{ J/mol} = -8.314 \text{ J/mol K} \times 298.2 \text{ K} \times \ln K_p$$

Rearranged, this equation becomes

$$\ln K_p = 1.400 \times 10^5/(-8.314 \times 298.2) = -56.47$$

Taking the antilog gives

$$K_p = e^{-56.47} = 3.0 \times 10^{-25}$$

Practice Problem 14.8

The standard free-energy change for the following reaction is $\Delta G° = -55.60$ kJ/mol O_2.

$$2NO(g) + O_2(g) \rightleftharpoons 2NO_2(g)$$

What is the equilibrium constant for this reaction at 227.0°C?

Answer: 6.4×10^5

Equilibrium Constants and Spontaneity

How can we predict in which direction a reversible reaction will be spontaneous?

We have used free-energy changes to predict the spontaneity of a reaction. If we know the value of an equilibrium constant, however, we can determine spontaneity simply by examining the relative values of the equilibrium constant and the reaction quotient. When these values are the same, the system is at equilibrium and no net reaction, forward or backward, occurs. When these values are not equal, reaction goes in the direction that will cause the system to reach a state of equilibrium. If the reaction quotient is smaller than the equilibrium constant, the reaction proceeds spontaneously in the forward direction. If the reaction quotient is larger than the equilibrium constant, the reverse reaction proceeds spontaneously until a state of equilibrium is reached. These principles are summarized in Table 14.1.

EXAMPLE 14.9

Consider the dissociation of nitrosyl bromide.

$$2NOBr(g) \rightleftharpoons 2NO(g) + Br_2(g)$$

The equilibrium constant, K_p, for this reaction is 0.0225 at 350°C. If a mixture contains 0.50 atm NOBr, 0.40 atm NO, and 0.20 atm Br_2, will there be any net reaction? If so, in which direction?

Solution   First we calculate the value of Q from the equation.

$$Q = \frac{P_{NO}^2 P_{Br_2}}{P_{NOBr}^2}$$

$$= (0.40)^2(0.20)/(0.50)^2 = 0.13$$

Because $K = 0.0225$, it is smaller than the reaction quotient ($K_p < Q$), so the reaction should proceed spontaneously in the reverse direction, ultimately producing an equilibrium mixture that contains more NOBr and less NO and Br_2. In fact, the equilibrium pressures will be 0.64 atm NOBr, 0.26 atm NO, and 0.13 atm Br_2.

Practice Problem 14.9

Consider the dissociation of phosgene.

$$COCl_2(g) \rightleftharpoons CO(g) + Cl_2(g)$$

The equilibrium constant, K_p, is 0.329 at 1000 K. If a mixture contains 0.18 atm $COCl_2$, 0.25 atm CO, and 0.15 atm Cl_2, will there be any net reaction? If so, in which direction?

Answer: $Q = 0.21$, so the reaction proceeds spontaneously in the forward direction.

Table 14.1 Relative Values of K and Q Related to Reaction Spontaneity

Free-Energy Change	Relative Values of K and Q	Spontaneity
$\Delta G = 0$	$K = Q$	Equilibrium state
$\Delta G < 0$	$K > Q$	Spontaneous process
$\Delta G > 0$	$K < Q$	Nonspontaneous process (spontaneous in reverse direction)

14.3   Le Chatelier's Principle

▶ Remember the tennis-ball scenario described in the introduction to this chapter? When more tennis balls were added, an equilibrium was re-established in which more balls were resting on the gymnasium floor *and* more balls were bouncing. The addition of tennis balls represents the application of an external stress—that is, a change in the experimental conditions. The response of a system at equilibrium to an external stress is described by **Le Chatelier's principle:** If a system at equilibrium is subjected to an external stress, it shifts to a new equilibrium position to partially remove that stress. We say that the position of equilibrium—the relative amounts of reactants and products in the system—changes.

What kinds of stress can we place on an equilibrium reaction?

Let's look at a common example of a physical equilibrium, the solution of a gas in a liquid. A bottled soft drink contains carbon dioxide dissolved under pressure, with some carbon dioxide remaining in the gas phase above the liquid. The gas in the vapor phase and the gas in the liquid (dissolved) phase are in equilibrium with one another. When the bottle is opened, however, the pressure is reduced to atmospheric pressure, which is less than the original equilibrium pressure in the sealed bottle. Gas escapes from the container, and the liquid fizzes as carbon dioxide is removed from solution, as shown in Figure 14.4. Sufficient gas escapes from solution to create a new position of equilibrium.

Draw a series of pictures showing what happens at the molecular level when the equilibrium in the soft drink bottle is upset by reducing the pressure.

In this example, a change in pressure supplies the external stress that changes the position of equilibrium. Several other types of stress have similar effects.

Figure 14.4 Carbon dioxide dissolved under high pressure escapes from solution when the pressure is reduced.

These factors include the concentration of reactants and products and the temperature.

Concentration Changes

Adding a reactant or product to a chemical system at equilibrium disturbs the equilibrium state. In response, either a forward or a reverse chemical reaction occurs, shifting the position of equilibrium in such a way that some of the added substance is consumed. On the other hand, if the concentration of one of the reactants or products is decreased, reaction occurs to shift the position of equilibrium so that some of the removed substance is produced. None of these situations causes the value of the equilibrium constant to change. The value of K_c remains the same, and the concentrations adjust such that the ratio of product to reactant concentrations (each raised to the appropriate power) once again becomes equal to the equilibrium constant.

Predict what will happen to the position of equilibrium when more reactant is added to a system in a state of equilibrium.

Let's examine these effects in a real system. The formation of carbon dioxide from carbon monoxide and oxygen is described by the equation

$$2CO(g) + O_2(g) \rightleftharpoons 2CO_2(g)$$

First, we'll consider the effect of an increase in the concentration of CO. The system responds by reducing the amount of CO. It does this by reacting CO with O_2 to form more CO_2. When the system reaches a new state of equilibrium, the concentration of O_2 will have decreased, and the concentration of CO_2 will have increased. Thus the addition of CO results in a shift to a new equilibrium position that removes some of the CO we added.

Relate the relative values of Q and K to changes in the amounts of reactants and products.

We can also examine this result by considering the effect that increasing the concentration of CO has on the reaction quotient, Q. Before CO is added, Q is equal to K because the system is at equilibrium. After CO is added, the reaction quotient has a lower value than the equilibrium constant, because the concentration of CO, a reactant, is increased and this concentration appears in the denominator of the expression

$$[CO_2]^2/[CO]^2[O_2] = Q < K$$

Because Q has a lower value than K, the position of equilibrium shifts in the direction that increases the value of Q. (The value of K cannot change, of course, because the equilibrium constant must remain constant at a constant temperature.) The value of Q is increased by an increase in numerator terms and a decrease in denominator terms—that is, by a forward reaction. When Q again equals K, the system has reached a new state of equilibrium.

If the concentration of CO is decreased, the equilibrium position shifts back toward reactants to form more CO. Some of the CO_2 decomposes to form CO and O_2. A decrease in the concentration of CO causes Q to have a higher value than K, so the system must respond in such a way as to decrease the value of Q; this is accomplished by occurrence of the reverse reaction. The reverse reaction continues until Q equals K, and at that point a new equilibrium position is reached.

Because O_2 is also a reactant, addition or removal of O_2 has the same effect as addition or removal of CO. The addition or removal of CO_2 also causes the equilibrium to shift. Adding CO_2 causes the position of equilibrium to shift toward reactants, using up some of the added CO_2. In this case, the value of Q increases and is higher than the value of K. The reverse reaction decreases the numerator and increases the denominator terms of the expression for Q, resulting in a decrease in the value of Q. Conversely, the removal of some CO_2 causes the reaction to proceed in a forward direction, forming more CO_2.

EXAMPLE 14.10

The following system is in equilibrium.

$$O_2(g) + 2H_2(g) \rightleftharpoons 2H_2O(g)$$

Describe how each of the following concentration changes will affect the equilibrium.

a. The concentration of H_2 is decreased.
b. The concentration of O_2 is increased.
c. The concentration of H_2O is decreased.

Solution

a. If the concentration of H_2 is decreased, the reaction must shift in a direction that will increase the H_2 concentration. Thus the reaction will proceed in the reverse direction, forming more H_2 (and O_2) until a new equilibrium position is achieved.
b. An increase in the concentration of O_2 will cause the reaction to proceed in the forward direction, decreasing the concentration of O_2.
c. If the concentration of H_2O is decreased, the reaction will proceed in the forward direction, consuming H_2 and O_2 and thus increasing the concentration of H_2O, until a new state of equilibrium is reached.

Practice Problem 14.10

The following system is in equilibrium.

$$2NO(g) + O_2(g) \rightleftharpoons 2NO_2(g)$$

In what direction will equilibrium shift after each of the following concentration changes?

a. The concentration of NO_2 is increased.
b. The concentration of O_2 is decreased.
c. The concentration of NO is increased.

Answer: a. reverse, b. reverse, c. forward

Improving Reaction Yields   The existence of equilibrium states in chemical systems often causes problems for chemists who wish to produce a particular material. If the reaction being used reaches a state of equilibrium rather than proceeding to completion, the yield of product is reduced from 100%. The discussion of the effect of concentration changes on equilibrium suggests common methods for coping with this situation. For example, the amount of one of the reactants can be raised in order to cause the reaction to shift toward products.

Another approach is the continuous removal of one of the products from the reaction mixture. Because removal of a product causes the equilibrium to shift forward, more product will be formed. When product is removed continuously, equilibrium is never reached, and reversible reactions can proceed to completion.

Consider, for example, the reversible reaction

$$CaCO_3(s) \rightleftharpoons CaO(s) + CO_2(g)$$

The heating of limestone (calcium carbonate) is an important method for the preparation of lime (calcium oxide), which is used extensively in the production of

Figure 14.5 Chips of limestone *(Left)* heated in an open vessel are converted completely to powdered lime *(Right)*.

cement. When we carry out this reaction in a closed container, it reaches a state of equilibrium, and not all limestone can be converted to lime. However, when we remove the CO_2 from the system as it is formed (simply by allowing the container to remain open to the air), the $CaCO_3$ is completely converted to CaO, as shown in Figure 14.5.

Pressure Changes

Pressure is another factor that affects equilibrium when gases are involved. For an equilibrium system in which only liquids or solids are involved, pressure generally has a negligible effect, because the volumes of liquids and solids change only slightly with changes in pressure. However, the effect of changes in pressure on the equilibrium of a gaseous system can be substantial. When the pressure of such a system is changed, the concentrations change in such a way as to establish a new equilibrium position that offsets the change in pressure.

Pressure changes can be accomplished in three ways. A reactant or product can be added, with no change in volume; this is equivalent to changing a reactant or product concentration, as already described. Another method of changing pressure involves adding an inert gas—one not involved in the reaction—also with no change in volume. Such an increase in pressure has no effect on the position of equilibrium. We can see this by considering the reaction quotient expression. The partial pressure of the inert gas does not appear in this expression, so it cannot change the value of Q, which remains equal to K. The final method involves a change in volume. An increase in volume decreases the pressure, whereas a decrease in volume increases pressure.

Do pressure changes affect all equilibrium reactions in the same way?

Let's examine the effect of decreasing the volume on the following gaseous equilibrium.

$$2CO_2(g) \rightleftharpoons 2CO(g) + O_2(g)$$

This volume decrease causes an increase in the total pressure of the gases in the system. The system responds to the volume decrease (pressure increase) by shifting in a way that results in a decrease in pressure. Assuming that we maintain the new volume and keep the temperature constant, the pressure can be decreased only by a change in the number of moles of gas molecules, as shown in Figure 14.6. Note that 2 mol of gas appear as reactants on the left side of the equation and 3 mol of gas appear as products on the right side. A pressure decrease can be

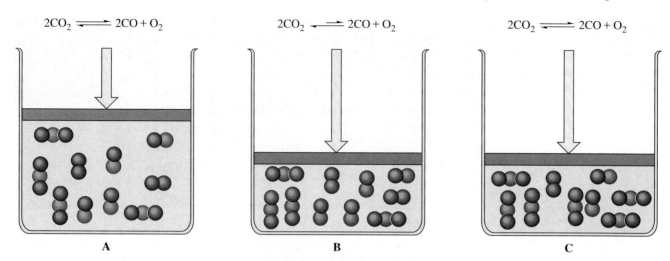

Figure 14.6 When the volume of an equilibrium mixture of CO_2, CO, and O_2 gases is decreased, the equilibrium shifts to form more CO_2. (*A*) The system is at equilibrium. Note the number of molecules of each gas. (*B*) The volume is decreased, causing the pressure to increase. The system is no longer at equilibrium, because the rates of the forward and reverse reactions are affected differently. (*C*) The system acts in such a way as to lower the pressure by decreasing the total number of molecules thus tending to restore equilibrium. In this case, the reaction shifts toward more carbon dioxide.

In the Mond process, solid nickel metal reacts with carbon monoxide gas to form a very toxic compound known as tetracarbonylnickel(0), $Ni(CO)_4$, which contains coordinate covalent bonds. This colorless liquid is quite volatile, boiling at 43°C. When heated at low pressure, the compound decomposes to give back nickel metal and carbon monoxide gas. Converting impure nickel to the gaseous compound and then decomposing the compound is a method for purifying nickel.

accomplished by shifting the reaction in the direction that leads to fewer gas molecules. In this case, the reaction will shift back toward reactants, producing more CO_2 and consuming CO and O_2. Conversely, a decrease in pressure, caused by an increase in volume, shifts the equilibrium in the forward direction, where there are more gas molecules.

The effect of changes in pressure and volume on gas-phase equilibria, then, depends on the change in number of gas molecules from reactants to products. If the number of gaseous product molecules is less than the number of gaseous reactant molecules, the effects we just observed are reversed. Consider, for example, the formation of tetracarbonylnickel(0).

$$Ni(s) + 4CO(g) \rightleftharpoons Ni(CO)_4(g)$$

A decrease in volume, with the accompanying increase in pressure, causes the reaction to shift in the forward direction, because the product contains fewer gaseous molecules than the reactants. The pressure increase can be offset by a decrease in the number of gas molecules.

If the total number of gas molecules is the same on both sides of a reaction equation, then the equilibrium position does not change when the pressure is altered by a change in volume. For example, in the reaction

$$H_2(g) + I_2(g) \rightleftharpoons 2HI(g)$$

the reactants and products both involve 2 mol of gas molecules. Thus an increase in pressure caused by a decrease in volume has no effect on the position of equilibrium.

The pressure effect can also be understood in terms of the effect of changes in volume and pressure on the reaction quotient. ▶ Suppose, for example, we start with the equilibrium system described in the chapter introduction:

$$N_2O_4(g) \rightleftharpoons 2NO_2(g) \qquad K_c = 0.0245 \text{ at } 55°C$$

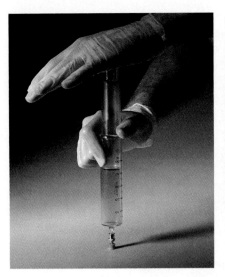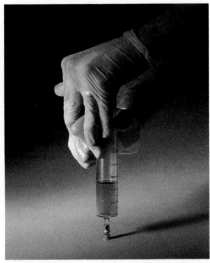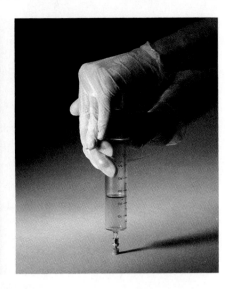

Figure 14.7 The effect of changes in volume and pressure on the equilibrium between red nitrogen dioxide and colorless dinitrogen tetroxide. *Left:* A syringe containing an equilibrium mixture of the gases. *Center:* Immediately after the volume is decreased to approximately one-half its original value, the concentration of nitrogen dioxide is doubled, as shown by the darker color. *Right:* By the time equilibrium is reached again, the red color has decreased in intensity because the equilibrium position now favors dinitrogen tetroxide.

The system contains 0.123 mol/L N_2O_4 and 0.0549 mol/L NO_2. We can double the pressure in the system by decreasing the volume to one-half its initial value. Decreasing the volume to one-half doubles the concentration of each species: $[N_2O_4] = 2 \times 0.123\ M = 0.246\ M$, and $[NO_2] = 2 \times 0.0549\ M = 0.110\ M$. The system is no longer at equilibrium, as we can see from the reaction quotient, which no longer equals K_c.

$$Q = [NO_2]^2/[N_2O_4] = [0.110]^2/[0.246] = 0.0492 \qquad K_c = 0.0245$$

The value of Q is larger than the value of K_c, so Q must be reduced to reach a new state of equilibrium.

A reduction in Q can be accomplished by a decrease in the concentration of NO_2 and an increase in the concentration of N_2O_4. These changes in concentration occur if the reaction proceeds in the reverse direction, which is the direction required to reduce the number of molecules and thus the pressure of the system. The new equilibrium concentrations are $[N_2O_4] = 0.261\ M$ and $[NO_2] = 0.0800\ M$. The concentration of N_2O_4 has indeed increased and that of NO_2 has decreased upon attainment of the new equilibrium. These changes can be observed as color changes, as shown in Figure 14.7.

EXAMPLE 14.11

Describe the effect of a decrease in pressure caused by an increase in volume on the equilibrium reaction

$$SO_2Cl_2(g) \rightleftharpoons SO_2(g) + Cl_2(g)$$

(continued)

Solution   There are 2 mol of gas molecules on the product side of the reaction and only 1 mol on the reactant side. A decrease in pressure causes the reaction to shift in the direction that increases pressure. That is, the reaction shifts in such a way as to form more molecules. Thus the new equilibrium position is reached by formation of more products.

Practice Problem 14.11

Describe the effect of an increase in pressure caused by a decrease in volume on the equilibrium reaction

$$2NOBr(g) \rightleftharpoons 2NO(g) + Br_2(g)$$

Answer: Equilibrium shifts toward the reactants.

In a heterogeneous reaction system, pressure changes affect primarily the gases, because pressure has little effect on liquids and solids. An example of a heterogeneous reaction is the reduction of copper oxide by hydrogen gas.

$$CuO(s) + H_2(g) \rightleftharpoons Cu(s) + H_2O(g)$$

Copper(II) oxide, CuO, is a black to brownish-black solid that occurs in nature as the minerals tenorite and paramelaconite. It is used as a pigment in glass, ceramics, enamel paints, porcelain glazes, and artificial gems.

A change in pressure has no effect on the position of equilibrium in this reaction; there is 1 mol of gas molecules on each side of the equation. On the other hand, a change in pressure does affect the reaction

$$H_2O(l) + CO_2(g) \rightleftharpoons H_2CO_3(aq)$$

When the pressure is increased, the system responds by shifting in the direction that decreases the number of gas molecules. The only gas in this system is CO_2, so an increase in pressure causes more CO_2 gas to dissolve and react with water to form aqueous carbonic acid. When the pressure is lowered, H_2CO_3 dissociates, and CO_2 comes out of solution to replace some of the CO_2 that has been removed. The fizzing of a carbonated beverage when the container is opened is a result of this effect.

Temperature Changes

A change in temperature is one more type of stress that can be placed on an equilibrium. Unlike changes in concentration and pressure, however, changes in temperature change the value of the equilibrium constant. In fact, the equilibrium constant depends only on the chemical reaction that is occurring and on the temperature.

Let's look at the dissociation of carbon dioxide, an endothermic reaction, as an example.

$$2CO_2(g) \rightleftharpoons 2CO(g) + O_2(g)$$

At 298 K, CO_2 has nearly no tendency to dissociate. At 1000 K and 1 atm pressure, only about $2 \times 10^{-5}\%$ of the CO_2 is dissociated. At 2000 K, the dissociation is about 2% complete, and at 3000 K, CO_2 is about 5% dissociated. As the temperature increases, then, the dissociation of CO_2 increases, and the concentrations of CO and O_2 increase. Because all these systems are in a state of equilibrium, the concentration changes indicate that as the temperature increases, the value of the equilibrium constant K_c increases.

$$K_c = [CO]^2[O_2]/[CO_2]^2$$

In fact, the values of K_c are 3.6×10^{-25} at 1000 K, 1.1×10^{-10} at 2000 K, and 6.0×10^{-5} at 3000 K. The equilibrium constant for endothermic reactions, such as this one, always increases as the temperature increases.

We can examine the effect of temperature on the equilibrium constant by using thermodynamics. Recall two equations that relate the standard free-energy change to other quantities:

Why does temperature affect the position of equilibrium and the value of the equilibrium constant?

$$\Delta G° = -RT \ln K \quad \text{and} \quad \Delta G° = \Delta H° - T\Delta S°$$

We can set these two equations equal to one another:

$$-RT \ln K = \Delta H° - T\Delta S°$$

and rearrange to give the equation

$$\ln K = -\Delta H°/RT + \Delta S°/R$$

From this equation, we see that an increase in temperature will give a larger (less negative) value of $\ln K$ when the reaction is endothermic ($\Delta H° > 0$) and a smaller value when the reaction is exothermic ($\Delta H° < 0$). If we examine this equation at two temperatures, we can combine equations to eliminate $\Delta S°$.

$$\ln\left(\frac{K_1}{K_2}\right) = -\Delta H°/R\left(\frac{1}{T_1} - \frac{1}{T_2}\right)$$

We can use this equation to calculate a value of the equilibrium constant at a new temperature if we know the value at one temperature and the value of the standard enthalpy change for the reaction.

Now let's examine the endothermic reaction from a more qualitative viewpoint. Heat is being absorbed—in other words, the enthalpy change is positive. The value of ΔH here is $+566$ kJ/mol O_2. We can represent the absorption of heat in equation form by treating heat as a reactant:

$$566 \text{ kJ} + 2CO_2(g) \rightleftharpoons 2CO(g) + O_2(g)$$

The form of this equation enables us to use Le Chatelier's principle to predict the effect of temperature changes on equilibrium. When heat is added by an increase in temperature, the equilibrium position shifts in such a way as to remove heat. Thus the position of equilibrium in the dissociation reaction of CO_2 shifts toward products when we raise the temperature, with a corresponding increase in the value of K_c. If the temperature is lowered, the position of equilibrium shifts toward the reactants, replacing some of the heat removed from the system by the temperature decrease.

Exothermic reactions, which release heat, show the opposite effects. Adding more heat by increasing the temperature causes the equilibrium to shift toward reactants, thus using up some of the added heat. This effect is illustrated by the change in position of equilibrium with change in temperature for the exothermic formation of dinitrogen pentoxide from nitrogen dioxide. ▶ This reaction,

$$2NO_2(g) \rightleftharpoons N_2O_4(g) + 57.2 \text{ kJ}$$

is shown in Figure 14.8 and was also described in the chapter introduction. For exothermic reactions, the value of the equilibrium constant decreases as the temperature increases.

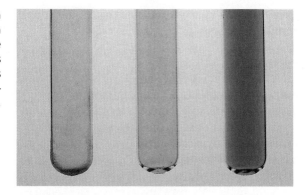

EXAMPLE 14.12

Predict the effect of a decrease in temperature on the position of equilibrium in the following reaction:

$$2SO_2(g) + O_2(g) \rightleftharpoons 2SO_3(g) \qquad \Delta H = -194 \text{ kJ/mol}$$

Solution   The reaction is exothermic; heat is given off. Thus, heat can be considered a product of the reaction:

$$2SO_2(g) + O_2(g) \rightleftharpoons 2SO_3(g) + 194 \text{ kJ/mol}$$

A decrease in temperature results in the removal of heat, so the reaction responds by shifting to a new equilibrium position that generates more heat. Thus, the reaction shifts forward at the lower temperature.

Practice Problem 14.12

Predict the effect of an increase in temperature on the position of equilibrium in the following reaction:

$$4HCl(g) + O_2(g) \rightleftharpoons 2Cl_2(g) + 2H_2O(g) \qquad \Delta H = -113 \text{ kJ/mol}$$

Answer: The equilibrium shifts toward reactants.

The Effect of Catalysts

Why is the position of equilibrium not affected by a catalyst?

A catalyst, you may recall, is a substance that increases the rate of a reaction but is not permanently changed by the reaction. For example, iron acts as a catalyst for the reaction between nitrogen and hydrogen.

$$N_2(g) + 3H_2(g) \xrightarrow{\text{Fe}} 2NH_3(g)$$

Because a catalyst may change the rate of a reaction, it is reasonable to ask whether it might shift the position of equilibrium. The answer is that it does not. Any substance that is a catalyst for the forward reaction in a reversible system must also be a catalyst for the reverse reaction, which in this case is

$$2NH_3(g) \xrightarrow{\text{Fe}} N_2(g) + 3H_2(g)$$

The catalyst does not change the energies of the reactants and products, so it

▶ ▶ ▶ WORLDS OF CHEMISTRY

Respiratory Diseases: Equilibria Gone Awry

An average adult breathes more than 12,000 L of air daily.

Diseases that interfere with respiration cause much suffering, ranging from fatigue to painful gasping for breath. These may be diseases such as tuberculosis caused by foreign organisms, diseases of the lungs such as emphysema, malfunctions of the blood such as the various anemias (iron deficiency anemia, pernicious anemia, acquired hemolytic anemia, and so on), or genetic disorders such as sickle cell anemia. An average adult breathes more than 12,000 L of air daily. This is the largest intake of any substance by the body and also one of the most important. We can go for long times without food or water, but life ends within minutes when we have no air. Each of these diseases interferes in some way with the equilibria that occur during respiration. But how does respiration work?

When food is oxidized in the human body, some of it is converted to CO_2, which must be eliminated. The excess CO_2 can be transported to the lungs through the bloodstream. However, the amount that can get into the blood from simple dissolution is only about 15% of the amount that must be eliminated. The rest is transported in the form of HCO_3^-.

Where does the HCO_3^- come from? When carbon dioxide dissolves, it forms carbonic acid.

$$CO_2(g) + H_2O(l) \rightleftharpoons H_2CO_3(aq)$$

The carbonic acid reacts with deoxygenated hemoglobin (Hb) in the red blood cells (shown in Figure 14.9), resulting in the transfer of a hydrogen ion.

$$H_2CO_3(aq) + Hb \rightleftharpoons HCO_3^-(aq) + HHb^+$$

When the blood reaches the lungs, the hemoglobin ion picks up an oxygen molecule.

$$HHb^+ + O_2(g) \rightleftharpoons HHbO_2^+$$

This weakens the bond to hydrogen ion, so the hemoglobin donates the hydrogen ion back to a bicarbonate ion.

$$HHbO_2^+ + HCO_3^-(aq) \rightleftharpoons HbO_2 + H_2CO_3(aq)$$

The resulting carbonic acid dissociates and releases CO_2 to be exhaled from the lungs.

$$H_2CO_3(aq) \rightleftharpoons H_2O(l) + CO_2(g)$$

The oxygenated hemoglobin is carried back to the cells, where the oxygen is released for use in food metabolism.

$$HbO_2 \rightleftharpoons Hb + O_2(aq)$$

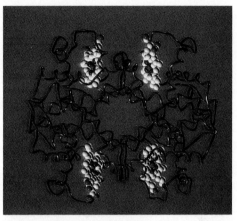

Figure 14.9 Hemoglobin is a large, complex, organic molecule that contains iron ions bonded to nitrogen atoms. Oxygen gas can bond to the iron in the hemoglobin, which is located in the red blood cells. Thus it is transported to cells in body tissue, where it is used for oxidation processes. The hemoglobin molecule contains four heme units, where oxygen bonds to iron.

The entire cycle of reversible equilibrium reactions can then be repeated—this is the chemical basis for our respiratory systems. Note that the reactions occurring in the cells are essentially just the reverse of the reactions occurring in the lungs. The direction in which these reactions proceed, according to Le Chatelier's principle, is dictated by whether there is a high concentration of CO_2 or of O_2 present. The cells have a high concentration of CO_2 resulting from food metabolism; the lungs have a high concentration of O_2 from breathing.

The respiratory diseases mentioned either interfere with getting oxygen into the lungs and available for respiration or reduce the amount or effectiveness of the hemoglobin that carries the oxygen to the cells.

Questions for Discussion

1. Hemoglobin carries oxygen from the lungs through the bloodstream. How does the oxygen get from the hemoglobin into the cells where it is needed for energy production?
2. Carbon monoxide binds strongly to hemoglobin. Why does exposure to carbon monoxide cause drowsiness and eventual impairment of mental and physical processes?

(continued)

3. What treatment would you suggest for a patient who has been exposed to excessive carbon monoxide?

4. Genetic engineering could result in cures for such genetic diseases as sickle cell anemia. But its misuse or carelessness in its use could cause harm. Should research in genetic engineering be carried out in universities? In other locations?

5. A condition called the blue-baby syndrome in infants, and more generally called methemoglobinemia, is an oxygen-deficiency disease. This condition results when iron in hemoglobin is oxidized from the +2 state to the +3 state. Why does this result in respiratory problems? What effect does it have on the equilibria involved in respiration?

cannot change the relative stabilities of these species and the position of equilibrium. A catalyst merely causes the state of equilibrium to be reached more quickly.

14.4 Calculations with Equilibrium Constants

Chapter 4 described various stoichiometric calculations that enable us to determine how much product results from a chemical reaction. In those calculations, it was assumed—though never explicitly stated—that the reactions went to completion, the limiting reagent being totally converted to products. Now we are considering reversible reactions, which do not go to completion but reach a state of equilibrium. In principle, all reactions are reversible—that is, at least a small amount of reactant always remains when a state of equilibrium is reached. In practice, we can distinguish between reactions that essentially go 100% to completion and reactions in which measurable amounts of reactants and products coexist when equilibrium is reached. The calculations required to determine reaction yield and related factors are different for these two types of reactions. This section illustrates some common types of calculations involving reactions that reach equilibrium.

What two laws must we apply to calculate the amounts of chemical substances present in a system at equilibrium?

One Unknown Equilibrium Pressure or Concentration

Among the simplest calculations are those in which we know the equilibrium constant and all but one of the equilibrium pressures or concentrations and we wish to know the pressure or concentration of the one unknown species. We can find this out simply by rearranging the mass law equation and solving for the unknown pressure or concentration.

It is always good problem-solving technique to determine whether the final answer is consistent with the information given. For equilibrium problems, an easy way to check for correctness is first to calculate the reaction quotient, Q, by substituting the answer back into the equilibrium-constant expression and then to compare the value of Q with the value of the equilibrium constant. If Q equals K, the answer is correct.

EXAMPLE 14.13

In the following reaction, $K_p = 7.13$ at 25°C.

$$2NO_2(g) \rightleftharpoons N_2O_4(g)$$

At equilibrium, the partial pressure of NO_2 is 0.150 atm. What is the partial pressure of N_2O_4?

Solution   First we write an expression for the equilibrium constant.

$$K_p = \frac{P_{N_2O_4}}{P_{NO_2}^2}$$

Next we substitute in the values for K_p (7.13) and P_{NO_2} (0.150).

$$7.13 = \frac{P_{N_2O_4}}{(0.150)^2}$$

Solving for the unknown pressure gives

$$P_{N_2O_4} = 7.13 \times (0.150)^2 = 7.13 \times 0.0225 = 0.160 \text{ atm}$$

To check for the correctness of this answer, we substitute it back into the reaction quotient expression and compare the result with the known value of the equilibrium constant.

$$Q = \frac{P_{N_2O_4}}{P_{NO_2}^2} = \frac{0.160}{(0.150)^2} = \frac{0.160}{0.0225} = 7.11$$

The agreement between $Q = 7.11$ and $K_p = 7.13$ is very good, indicating that no errors were made during the calculation.

Practice Problem 14.13

At 2000 K, ozone and oxygen are in equilibrium, with $K_p = 4.17 \times 10^{14}$.

$$2O_3(g) \rightleftharpoons 3O_2(g)$$

If the equilibrium partial pressure of O_2 is 0.205 atm, what is the equilibrium partial pressure of O_3?

Answer: 4.55×10^{-9} atm

It is often more convenient to work with concentrations and a value of K_c than to work with pressures and a value of K_p. The methods for solving the problem and checking the answer are just the same for concentrations as for partial pressures. The only difference is the use of molarities.

Known Initial Concentrations

Suppose we prepared a mixture of known proportions of reactants and then allowed it to proceed to a state of equilibrium. We could easily calculate the initial concentration of each species from the amount of that species added to the mixture. Then we could calculate the concentrations at equilibrium with the help of the equilibrium-constant expression.

When solving problems like this, it is usually helpful to make a table that includes the initial concentrations, the changes in concentration, and the equilibrium concentrations. As an example, consider the following reaction:

$$3H_2(g) + N_2(g) \rightleftharpoons 2NH_3(g)$$

Suppose the initial concentrations are 1.00 M H_2, 1.00 M N_2, and no NH_3. To determine the changes in concentration, we start by examining the coefficients in the balanced equation for the reaction. We can call the concentration of N_2 that reacts x. The concentration of H_2 that reacts must then be $3x$ because of the $3:1$ ratio of coefficients, and the amount of ammonia produced must be $2x$. The change in concentration between initial conditions and equilibrium will be given by $-x$ for N_2 (because it should disappear), $-3x$ for H_2, and $+2x$ for NH_3. The equilibrium concentrations will be the initial concentrations plus the changes in concentration. Now we arrange these facts in a short table.

Substance	N_2	H_2	NH_3
Initial concentration (M)	1.00	1.00	0
Change in concentration (M)	$-x$	$-3x$	$+2x$
Equilibrium concentration (M)	$1.00 - x$	$1.00 - 3x$	$0 + 2x$

We substitute the expressions for the equilibrium concentrations into the expression for the equilibrium constant.

$$K_c - [NH_3]^2/[N_2][H_2]^3$$
$$= \frac{(2x)^2}{(1.00 - x)(1.00 - 3x)^3}$$

With a known value of K_c, we can solve this equation for the value of x and then substitute that value back into the expressions in the table to determine the equilibrium concentration of each species. This approach to solving equilibrium problems is outlined in Table 14.2.

When setting up problems of this sort, we should always look for ways to simplify the mathematics. Ways of doing this include looking for equations that contain perfect squares and ignoring small unknown quantities. If we cannot simplify the mathematics, we may need to solve for x by using a quadratic equation.

Table 14.2 An Approach for Solving Equilibrium Problems

1. Write a balanced chemical equation.
2. Select one of the concentration changes and call it x.*
3. Use the stoichiometry to determine all the concentration changes in terms of x.
4. Make a table containing the substances, their initial concentrations, their changes in concentration, and their equilibrium concentrations (calculated from the initial concentrations and the concentration changes).
5. Write the equilibrium-constant expression.
6. Insert the equilibrium concentrations from the table into the equilibrium-constant expression.
7. Solve the equation for x.
8. Check any simplifications for validity.
9. Substitute the value of x into the expressions for the equilibrium concentrations and determine their values.
10. Use the equilibrium concentrations to calculate the reaction quotient, and compare it with the equilibrium constant to verify the accuracy of the answers.

*The same procedure works for partial pressures; simply substitute *pressure* for *concentration* in this list of steps.

Equations Containing Perfect Squares When an equation involves a perfect square, taking the square root of each side simplifies the mathematics. Always be on the alert for perfect squares when setting up problems.

EXAMPLE 14.14

At 440°C the equilibrium constant, K_c, for the following reaction has a value of 49.5.

$$H_2(g) + I_2(g) \rightleftharpoons 2HI(g)$$

Initially, we prepare a flask containing 0.0200 M I_2 and 0.0200 M H_2. What are the equilibrium concentrations of all three species?

Solution First we write an expression for the equilibrium constant.

$$K_c = [HI]^2/[H_2][I_2] = 49.5$$

We'll assume that when equilibrium has been established, the concentration of H_2 has decreased by x. Then the concentration of I_2 has also decreased by x, and the concentration of HI has increased by $2x$. We state these results in tabular form:

Substance	H_2	I_2	HI
Initial concentration (M)	0.0200	0.0200	0
Change in concentration (M)	$-x$	$-x$	$+2x$
Equilibrium concentration (M)	$0.0200 - x$	$0.0200 - x$	$0 + 2x$

The equilibrium concentrations of all the species can now be substituted into the equation for the equilibrium constant.

$$49.5 = \frac{(2x)^2}{(0.0200 - x)(0.0200 - x)}$$

Although this equation is a quadratic equation in x, it is also a perfect square, which is easier to solve than a quadratic equation. We start by taking the square root of each side.

$$(49.5)^{\frac{1}{2}} = \{[2x/(0.0200 - x)]^2\}^{\frac{1}{2}}$$

$$7.04 = \frac{2x}{0.0200 - x}$$

Rearranging this equation enables us to solve for x.

$$2x = 7.04(0.0200 - x) = 0.141 - 7.04x$$

$$9.04x = 0.141$$

$$x = 0.0156$$

The concentration of each species can now be calculated from the relationship of each concentration to x.

$$[H_2]_{eq} = 0.0200 - x = 0.0200 - 0.0156 = 0.0044 \ M$$
$$[I_2]_{eq} = 0.0200 - x = 0.0200 - 0.0156 = 0.0044 \ M$$
$$[HI]_{eq} = 2x = 2(0.0156) = 0.0312 \ M$$

As usual, it is a good idea to verify the results by substituting them into the reaction-quotient expression and then comparing the result with the known value of the equilibrium constant.

$$Q = [HI]^2/[H_2][I_2]$$

$$= \frac{(0.0312)^2}{(0.0044)(0.0044)} = 50$$

(continued)

The value of Q is nearly the same as that of K_c, 49.5, indicating that the calculations were carried out correctly. Differences of this magnitude commonly arise from rounding errors.

Practice Problem 14.14

The equilibrium constant for the water-gas shift reaction has a value of 0.227 at 2000 K.

$$CO(g) + H_2O(g) \rightleftharpoons CO_2(g) + H_2(g)$$

Suppose 0.0500 mol CO and 0.0500 mol H_2O are placed in a 2.00-L flask at 2000 K. What are the equilibrium concentrations of all species?

Answer: 0.0169 M CO, 0.0169 M H_2O, 0.00806 M CO_2, 0.00806 M H_2

Whatever mathematical method we use, we need not approach equilibrium from the reactant side. We can just as easily start with the products, in which case we think of the reaction as proceeding in the reverse direction to a state of equilibrium by the conversion of products to reactants.

Ignoring Small Unknown Quantities   A common method for simplifying calculations involves ignoring any negligibly small quantity in a sum or difference of quantities, such as the change of concentration, x, that is added to or subtracted from an initial concentration. When we encounter an equation that cannot be solved by use of perfect squares, we should always be on the lookout for physical conditions that enable us to simplify the equation.

Consider a case in which the equilibrium concentration is the initial concentration less the change in concentration, x: $C_{eq} = C - x$. If we assume that x is so small compared with the initial concentration that C remains essentially unchanged, we can replace the difference with the initial concentration. This eliminates one occurrence of x in the equilibrium-constant expression and simplifies the solution of the problem.

But how do we know whether it is safe to make this assumption? One approach is to examine the magnitude of the equilibrium constant. If it is quite small ($< 10^{-5}$, for instance), we can reasonably assume that there will be only a small amount of reaction in the forward direction. With a very large equilibrium constant, we expect a large change in the concentration of a reactant, so we might not want to try making this assumption. A second approach involves making the assumption blindly and then checking the result: Ignore x in one or more differences or sums, solve for x in the simplified equation, and then check whether the assumption was appropriate. Normally, if the value of x determined in this way is less than about 5% of the initial concentration from which it would have been subtracted, an acceptably small error results. If x is greater than 5% of the initial concentration, the equation must be solved without the simplifying assumption of small x.

EXAMPLE 14.15

Consider the decomposition of water at 500°C, where the equilibrium constant is $K_c = 6.00 \times 10^{-28}$.

$$2H_2O(g) \rightleftharpoons 2H_2(g) + O_2(g)$$

If 2.00 mol H_2O is placed in a 5.00-L container, what are the equilibrium concentrations of all species?

Solution   First we calculate the initial concentration of each species:

$$[H_2O] = 2.00 \text{ mol}/5.00 \text{ L} = 0.400 \ M$$

$$[H_2] = 0 \text{ mol}/5.00 \text{ L} = 0 \ M$$

$$[O_2] = 0 \text{ mol}/5.00 \text{ L} = 0 \ M$$

The changes in concentration are $+x$ for O_2, $+2x$ for H_2, and $-2x$ for H_2O. We summarize these results in tabular form.

Substance	H_2O	H_2	O_2
Initial concentration (M)	0.400	0	0
Change in concentration (M)	$-2x$	$+2x$	$+x$
Equilibrium concentration (M)	$0.400 - 2x$	$0 + 2x$	$0 + x$

Then we substitute these concentrations into the equilibrium-constant expression.

$$K_c = 6.00 \times 10^{-28} = [H_2]^2[O_2]/[H_2O]^2$$

$$6.00 \times 10^{-28} = \frac{(2x)^2(x)}{(0.400 - 2x)^2}$$

This is a cubic equation, which is rather difficult to solve, but some assumptions we can make will simplify it. If x is less than 5% of the number from which it is to be subtracted, it is safe to assume that it is negligible. We start by assuming that we can ignore x in the expression $0.400 - 2x$, replacing this expression with 0.400, and thereby simplifying the equation to

$$6.00 \times 10^{-28} = \frac{(2x)^2(x)}{(0.400)^2}$$

Now we can rearrange this equation and solve it readily.

$$4x^3 = (0.400)^2 \times 6.00 \times 10^{-28} = 9.60 \times 10^{-29}$$

$$x = (9.60 \times 10^{-29}/4)^{1/3} = 2.88 \times 10^{-10} \ M$$

After we have dropped unknown terms, as here, the first thing we must do after solving the equation is make sure that the assumption was valid. Because $2x$ is 5.76×10^{-10}, this assumption was indeed a very good one.

$$0.400 - 2x = 0.400 - 5.76 \times 10^{-10} = 0.400$$

If the assumption had not been valid, we would have had to solve the more complex equation.

Now we can go on to calculate the equilibrium concentrations on the basis of the value we have calculated for x.

$$[H_2O] = 0.400 - 2x = 0.400 \ M$$

$$[H_2] = 2x = 2(2.88 \times 10^{-10}) = 5.76 \times 10^{-10} \ M$$

$$[O_2] = x = 2.88 \times 10^{-10} \ M$$

Finally, we use these values to calculate the reaction quotient and verify the result.

$$Q = [H_2]^2[O_2]/[H_2O]^2$$

$$= \frac{(5.76 \times 10^{-10})^2(2.88 \times 10^{-10})}{(0.400)^2} = 5.97 \times 10^{-28}$$

This agrees very well with the value of K_c, 6.00×10^{-28}, so the calculations were performed correctly.

(continued)

Practice Problem 14.15

The equilibrium constant, K_c, at 100°C for the decomposition of phosgene is 2.18×10^{-10}.

$$COCl_2(g) \rightleftharpoons CO(g) + Cl_2(g)$$

Suppose 1.00 mol $COCl_2$ is placed in a 1.00-L container. What are the equilibrium concentrations of all species?

Answer: $[COCl_2] = 1.00\ M$, $[CO] = [Cl_2] = 1.48 \times 10^{-5}\ M$

Equilibrium Problems Involving Quadratic Equations   As noted, it is always wise to try to solve a problem by making simplifying assumptions and then examining the answer. But when no simplifying assumption will work, it may be necessary to use a quadratic equation.

Any algebraic equation that can be arranged into the form

$$ax^2 + bx + c = 0$$

is called a quadratic equation. In this general form, the parameters a, b, and c are positive or negative numbers generated by the problem being solved. The value of x in a quadratic equation is obtained from the following relation:

$$x = \frac{-b \pm \sqrt{(b^2 - 4ac)}}{2a}$$

The square root term must be either added to or subtracted from the term $-b$. The correct procedure—addition or subtraction—is usually apparent, because generally only one choice gives a positive answer, and a concentration must have a positive value. In other cases, when there are two positive answers, one is generally greater than the initial concentration, so it is not a valid answer.

EXAMPLE 14.16

At 500°C the equilibrium constant, K_c, for the following reaction is 0.0224.

$$PCl_5(g) \rightleftharpoons PCl_3(g) + Cl_2(g)$$

Suppose 0.100 mol PCl_5 is placed in a 2.00-L flask at 500°C. What are the equilibrium concentrations of all species?

Solution   First we calculate the initial concentration of PCl_5.

$$[PCl_5] = 0.100\ \text{mol}/2.00\ \text{L} = 0.0500\ M$$

When equilibrium is established, the concentration of PCl_5 will be decreased by x, and the concentrations of PCl_3 and Cl_2 will both be increased by x.

Substance	PCl_5	PCl_3	Cl_2
Initial concentration (M)	0.0500	0	0
Change in concentration (M)	$-x$	$+x$	$+x$
Equilibrium concentration (M)	$0.0500 - x$	$0 + x$	$0 + x$

The expression for the equilibrium constant is

$$K_c = [PCl_3][Cl_2]/[PCl_5] = 0.0224$$

Substituting the values for the equilibrium concentrations into this equation gives

$$K_c = \frac{(x)(x)}{(0.0500 - x)} = 0.0224$$

If the value of x is much smaller than 0.0500 M, then it is valid to neglect x in this equation. We test this possibility as follows:

$$\frac{(x)(x)}{(0.0500 - x)} = 0.0224$$

$$\frac{(x)(x)}{(0.0500)} = 0.0224$$

$$x^2 = 0.0500 \times 0.0224 = 0.00112$$

$$x = 0.0335 \; M$$

This value is nearly 70% of the initial concentration of 0.0500 M for PCl_5. Thus x is not negligible, and the equation must be solved as a quadratic equation.

The full equation to be solved is

$$\frac{(x)(x)}{(0.0500 - x)} = 0.0224$$

We multiply through by the denominator term.

$$x^2 = 0.0224(0.0500 - x) = 0.00112 - 0.0224x$$

Then we rearrange this equation into the quadratic form.

$$x^2 + 0.0224x - 0.00112 = 0$$

This form of the quadratic equation has $a = 1$, $b = 0.0224$, and $c = -0.00112$. We substitute these values into the formula for solution of the quadratic equation.

$$x = \frac{-b \pm \sqrt{(b^2 - 4ac)}}{2a}$$

$$= \frac{-0.0224 \pm \sqrt{(0.0224)^2 - 4(1)(-0.00112)}}{2}$$

$$= \frac{-0.0224 \pm 0.0706}{2}$$

Two possible answers result:

$$x = (-0.0224 + 0.0706)/2 = 0.0241$$
$$x = (-0.0224 - 0.0706)/2 = -0.0465$$

Because we defined x as the increase in concentration of PCl_3 and of Cl_2, it must have a positive value, so

$$x = 0.0241 \; M$$

The equilibrium concentrations are

$$[PCl_5] = 0.0500 - x = 0.0500 - 0.0241 = 0.0259 \; M$$
$$[PCl_3] = 0 + x = 0.0241 \; M$$
$$[Cl_2] = 0 + x = 0.0241 \; M$$

Again we check the answer by calculating the reaction quotient.

$$Q = [PCl_3][Cl_2]/[PCl_5] = (0.0241)(0.0241)/0.0259 = 0.0224$$

This value agrees exactly with the value of the equilibrium constant, $K_c = 0.0224$.

(continued)

Practice Problem 14.16

The equilibrium constant, K_c, at 55°C for the dissociation of dinitrogen tetroxide is 0.0245.

$$N_2O_4(g) \rightleftharpoons 2NO_2(g)$$

Suppose 1.50 mol N_2O_4 is placed in a 10.0-L container. What are the equilibrium concentrations of the species involved?

Answer: $[N_2O_4] = 0.123\ M$, $[NO_2] = 0.0548\ M$

Addition of a Substance to a System at Equilibrium

As we have seen, adding a reactant or product to a system at equilibrium shifts the position of equilibrium. We may wish to find out the new equilibrium concentrations—for example, if we are trying to determine the most desirable conditions for achieving a required yield of one of the species. The general procedure for solving such problems is the same as for the examples already considered, except that the initial concentrations must be those that were present in the initial *equilibrium* situation.

EXAMPLE 14.17

At 500°C the equilibrium constant, K_c, for the following reaction is 0.0224.

$$PCl_5(g) \rightleftharpoons PCl_3(g) + Cl_2(g)$$

Suppose 0.100 mol PCl_5 is placed in a 2.00-L flask at 500°C. The equilibrium concentrations of each species (calculated in Example 14.16) are $[PCl_5] = 0.0259\ M$, $[PCl_3] = 0.0241\ M$, and $[Cl_2] = 0.0241\ M$. An additional 0.0200 mol/L Cl_2 is added to this mixture. What are the new equilibrium concentrations?

Solution   The equilibrium has been upset by the addition of more Cl_2. The reaction shifts to remove some of the Cl_2, because its concentration is now too high for the equilibrium condition to be maintained. The concentrations of the species before any reaction occurs are:

$$[PCl_5] = 0.0259\ M$$
$$[PCl_3] = 0.0241\ M$$
$$[Cl_2] = 0.0241\ M + 0.0200\ M = 0.0441\ M$$

The concentration of Cl_2 will decrease by x, so the concentration of PCl_3 must also decrease by x and the concentration of PCl_5 must increase by x.

Substance	PCl_5	PCl_3	Cl_2
Initial concentration (*M*)	0.0259	0.0241	0.0441
Change in concentration (*M*)	$+x$	$-x$	$-x$
Equilibrium concentration (*M*)	$0.0259 + x$	$0.0241 - x$	$0.0441 - x$

The equilibrium expression is still valid:

$$K_c = [PCl_3][Cl_2]/[PCl_5] = 0.0224$$

Substituting the values for the equilibrium concentrations into this equation gives

$$K_c = \frac{(0.0241 - x)(0.0441 - x)}{(0.0259 + x)} = 0.0224$$

If x were small compared to 0.0241, we could ignore it, but then all the x terms would be eliminated from the equation and we would have nothing left to solve. For this reason, we cannot make simplifying assumptions. We must solve the full equation.

Expanding this equation gives

$$x^2 - 0.0441x - 0.0241x + 0.00106 = 0.0224x + 0.000580$$

Upon collecting terms, this equation becomes

$$x^2 - 0.0906x + 0.00048 = 0$$

This is a form of the quadratic equation with $a = 1$, $b = -0.0906$, and $c = 0.00048$. The solution to this equation is found via the quadratic formula.

$$x = \frac{-b \pm \sqrt{(b^2 - 4ac)}}{2a}$$

$$= \frac{0.0906 \pm \sqrt{(0.0906)^2 - 4(1)(0.00048)}}{2}$$

$$= 0.0057 \; M$$

Note that x is, in fact, not small compared with the initial concentrations.
   The new equilibrium concentrations are

$$[PCl_5] = 0.0259 + x = 0.0259 + 0.0057 = 0.0316 \; M$$

$$[PCl_3] = 0.0241 - x = 0.0241 - 0.0057 = 0.0184 \; M$$

$$[Cl_2] = 0.0441 - x = 0.0441 - 0.0057 = 0.0384 \; M$$

To check the answer, we compare the reaction quotient with the equilibrium constant.

$$Q = [PCl_3][Cl_2]/[PCl_5] = (0.0184)(0.0384)/(0.0316) = 0.0224$$

This value agrees exactly with the value of the equilibrium constant, $K_c = 0.0224$.

Practice Problem 14.17

The equilibrium constant, K_c, at 55°C for the dissociation of dinitrogen tetroxide is 0.0245.

$$N_2O_4(g) \rightleftharpoons 2NO_2(g)$$

Suppose 1.50 mol N_2O_4 is placed in a 10.0-L container. The equilibrium concentrations of the species are $[N_2O_4] = 0.123 \; M$ and $[NO_2] = 0.0548 \; M$. Now 0.0200 mol/L NO_2 is added to the mixture. What are the new equilibrium concentrations?

Answer: $[N_2O_4] = 0.132 \; M$, $[NO_2] = 0.0568 \; M$

14.5   Gas-Phase Equilibrium Reactions in Industry

A number of important industrial processes involve gas-phase reactions that reach a state of equilibrium. Clever application of the principles we have been discussing can circumvent the problems that arise from incomplete reaction in these systems.

Why does chemical equilibrium pose problems for industry?

Steam Reforming

Hydrogen is widely used as an industrial gas, a great deal of it in the manufacture of ammonia, an important component of fertilizers. Other uses include the synthesis of a variety of chemical compounds, such as methanol, cyclohexanol, and

Figure 14.10 Edible oils consist of unsaturated carboxylic acids, which contain carbon–carbon double bonds and the functional group —CO_2H. The carbon–carbon double bonds tend to make these fatty acids liquid. Hydrogenation of the double bonds converts them to single bonds, which tend to make the substances solids. Hydrogenation can take place at low temperatures and pressures in the presence of a nickel catalyst.

isooctane. It is also used in the hydrogenation of edible oils to make solid fats (Figure 14.10).

Hydrogen is obtained mostly from carbonaceous materials. The most common method of production is the steam-hydrocarbon reforming process, often called the *steam-reforming process* (Figure 14.11). A mixture of superheated steam and vaporized hydrocarbons is reacted over a nickel catalyst at high temperatures (about 800°C).

$$C_nH_m(g) + nH_2O(g) \rightleftharpoons nCO(g) + (\tfrac{1}{2}m + n)H_2(g)$$

This reforming reaction is highly endothermic, as shown here for methane (natural gas).

$$CH_4(g) + H_2O(g) \rightleftharpoons CO(g) + 3H_2(g) \qquad \Delta H° = 207 \text{ kJ}$$

Because the reaction uses heat, the formation of products is favored at high temperatures. Thus some methane (or other fuel) is burned to supply the heat needed to maintain high temperatures. And because there are more gaseous product molecules than gaseous reactant molecules, low pressures also favor the formation of products, so the reaction is run at relatively low pressures. Steam is generally used in excess of the stoichiometric amount; raising the partial pressure of a reactant helps to shift the equilibrium toward products.

Some of the excess steam reacts with carbon monoxide, a product of the steam reforming reaction.

$$CO(g) + H_2O(g) \rightleftharpoons CO_2(g) + H_2(g) \qquad \Delta H° = -42 \text{ kJ}$$

This is the *water-gas shift reaction*. It is desirable here, because it improves the yield of hydrogen gas and produces CO_2, which is easier to remove from the gaseous mixture than CO. However, it is not favored by the same conditions as the steam-reforming reaction. Being exothermic, it is favored by low temperature. The numbers of gaseous reactant and product molecules are identical, so pressure has no effect on it.

Addition of excess steam, a reactant, does help force the reaction to completion, however. To encourage the water-gas shift reaction, additional steam is added to the gases from the reforming reaction, the mixture is cooled to about 350°C, and an iron oxide/chromium oxide catalyst is used to speed up the reaction. Under these conditions, about 80–95% of the CO is converted to CO_2 at equilib-

Figure 14.11 A flowchart for the steam-reforming process.

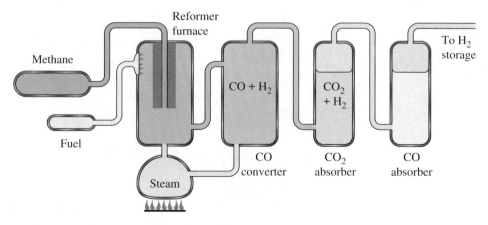

Ethanolamine, HO(CH₂)₂NH₂, is a viscous, hygroscopic liquid with an odor much like that of ammonia. It is used to remove carbon dioxide and hydrogen sulfide gases from natural gas and other gases. It is also used in polishes and hair waving solutions, as a softener for hides, and as a dispersing agent for agricultural chemicals.

rium. The hydrogen gas is isolated when the CO_2 reacts with a basic aqueous solution, such as an ethanolamine solution:

$$HO(CH_2)_2NH_2(aq) + CO_2(g) + H_2O(l) \rightleftharpoons HO(CH_2)_2NH_3^+HCO_3^-(aq)$$

The Haber Process

Ammonia is a colorless, poisonous gas with a characteristic choking odor. Ammonia is very soluble in water; 1 L H_2O dissolves over 700 L of NH_3 at room temperature and atmospheric pressure. About 80% of the ammonia produced by industry goes into the production of fertilizers, either anhydrous ammonia, which is pumped directly into the soil (Figure 14.12), or ammonium salts (Figure 14.13). Most of the remaining 20% of the ammonia produced is used in fibers, resins, plastics, nitric acid, explosives, and other products.

In the **Haber process** for synthesizing ammonia, air is added to the products of the steam-reforming reaction (CO and H_2) to introduce nitrogen gas. This mixture of gases is then subjected to the water-gas shift reaction and to purification as described for the steam reforming process. A mixture of 3 parts of hydrogen and 1 part of nitrogen results. This mixture comes to an equilibrium state that depends on the pressure and the temperature (Figure 14.14).

$$N_2(g) + 3H_2(g) \rightleftharpoons 2NH_3(g) \qquad \Delta H° = -92 \text{ kJ}$$

We expect the production of ammonia to be favored at high pressure, because four reactant molecules are converted to two product molecules, and increased pressure shifts the equilibrium to the side with fewer molecules. Measured yields of ammonia support this expectation, as shown in Figure 14.14A.

The yield of ammonia decreases with increasing temperature, as shown in Figure 14.14B. This is consistent with the fact that the reaction is exothermic. The equilibrium constant for exothermic reactions decreases as temperature increases. Some representative values are 6.85×10^5 at 25°C and 2.5×10^{-5} at 450°C. Note

Figure 14.12 Liquid ammonia is stored in a large tank and injected into soil as a fertilizer.

Figure 14.13 Ammonia and various ammonium salts derived from it are used in fertilizers.

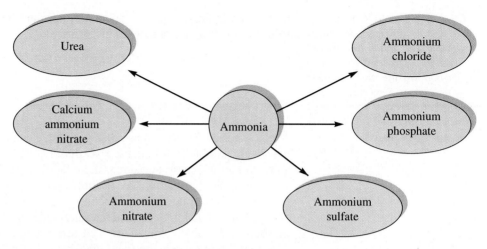

in Figure 14.14B that the maximum concentration of ammonia obtainable at a pressure of 200 atm is about 65% at 300°C, 35% at 400°C, and only 20% at 500°C. A higher yield would be favored by low temperatures, but the reaction is extremely slow at temperatures of 300°C or less, even in the presence of a catalyst. A compromise between acceptable rates and acceptable yields is achieved when the reaction is carried out at about 500°C and 200 atm. But here, less than 20% of the reactants are converted to ammonia.

Further conversion is possible when the product is removed so the equilibrium mixture can shift further toward product. Removal of ammonia involves

Figure 14.14 The position of equilibrium in the Haber process is affected by pressure and temperature. (A) At a given temperature, the yield of ammonia increases as the pressure increase. (B) At a given pressure, the yield of ammonia decreases as the temperature increases.

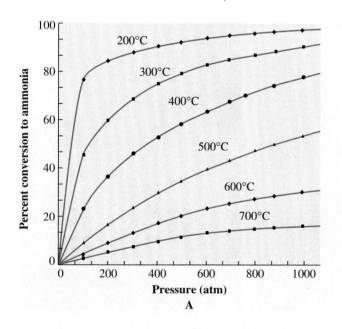

A

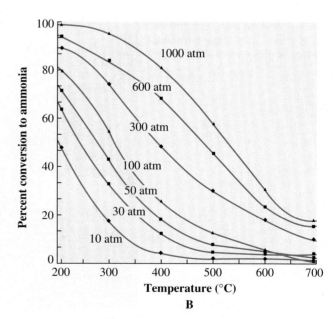

B

cooling the gaseous mixture to liquefy the ammonia. The uncondensed gases, primarily N_2 and H_2, are recycled to the reaction chamber for further conversion.

Synthesis of Methanol

Formaldehyde, H_2CO, is a flammable, colorless gas that is very soluble in water. It is usually sold as a solution commonly known as formalin. Formaldehyde is used as a disinfectant, germicide, fungicide, and insecticide. It is also used in the manufacture of a variety of materials, including plywood and particle board, artificial silk, dyes, glass mirrors, and explosives. Some evidence indicates that it may cause cancer in humans.

Methanol (or methyl alcohol, also known as wood alcohol) is a colorless liquid used as a solvent, as an additive in gasoline, and as a starting material for the synthesis of other materials, such as formaldehyde and acetic acid. The synthesis of methanol, diagrammed in Figure 14.15, involves passing a mixture of carbon monoxide and hydrogen gases, obtained from the steam reforming reaction, over a suitable catalyst at high pressures.

$$2H_2(g) + CO(g) \rightleftharpoons CH_3OH(g) \qquad \Delta H° = -103 \text{ kJ}$$

The reaction is exothermic, so the equilibrium constant decreases with increasing temperature. (Some typical values are 1.7×10^{-2} at 200°C, 1.3×10^{-4} at 300°C, and 1.1×10^{-5} at 400°C.) Thus it is necessary to carry out the reaction at as low a temperature as is consistent with a reasonable reaction rate. Generally, a temperature of about 300°C is used. The reaction rate is increased by use of a catalyst, usually copper or silver mixed with oxides of zinc, chromium, manganese, or aluminum.

Because the number of gaseous molecules decreases during the reaction, formation of methanol is favored by high pressure. At 300°C, a pressure of 50 atm gives an 8.0% conversion to methanol at equilibrium, whereas the conversion is 24.2% at 100 atm, 48.7% at 200 atm, and 62.3% at 300 atm. In practice, a pressure of about 300 atm is used, but the reaction is not generally allowed to come to equilibrium. Rather than wait for the reaction to reach equilibrium, it is more economical to move the gases rapidly through the reactor, cool the gases to liquefy and remove methanol, and then recycle the unreacted mixture of H_2 and CO. Generally, the yield is only about 12–15% for each pass over the catalyst in the reactor.

Figure 14.15 A flowchart for the synthesis of methanol from carbon monoxide and hydrogen.

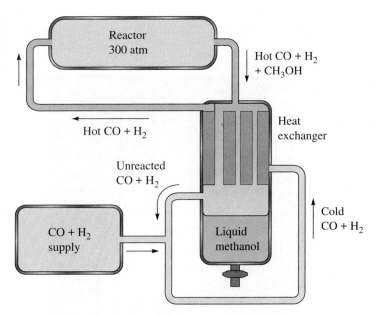

The Contact Process

Sulfuric acid is a colorless, transparent, oily liquid. Solutions of SO_3 in H_2SO_4 are called fuming sulfuric acid, or *oleum*. Oleum is extremely corrosive when brought into contact with moist air, where it forms a white mist of sulfuric acid, as shown in Figure 14.16.

$$H_2SO_4 \cdot SO_3(l) + H_2O(g) \rightleftharpoons 2H_2SO_4(l)$$

More sulfuric acid is produced than any other industrial chemical, and the commonest use for sulfuric acid is in the synthesis of superphosphate fertilizer from phosphate rock.

$$CaF_2 \cdot 3Ca_3(PO_4)_2(s) + 3H_2O(l) + 7H_2SO_4(l) \rightleftharpoons$$
$$3Ca(H_2PO_4)_2 \cdot H_2O(s) + 2HF(g) + 7CaSO_4(s)$$

Other uses include the synthesis of metal sulfates, mineral acids, organic acids, titanium dioxide (for paint pigments), and other chemicals; the cleaning of iron and steel before tinning or galvanizing; the refining of heavy metals; and electroplating. Indeed, few chemical products do not involve sulfuric acid somewhere in their manufacture.

Almost all the sulfuric acid produced for industry is prepared through the contact process, which was discussed in detail in Chapter 4. We will consider only one step here, the oxidation of sulfur dioxide by oxygen to form sulfur trioxide.

$$2SO_2(g) + O_2(g) \rightleftharpoons 2SO_3(g) \qquad \Delta H° = -196 \text{ kJ}$$

Because this reaction is exothermic, the formation of product is favored by low temperatures, as shown in Figure 14.17. The equilibrium constant is 1.6×10^5 at 400°C, 2.3×10^3 at 500°C, 91 at 600°C, 6.9 at 700°C, 0.84 at 800°C, and 0.034 at 1000°C. A reasonable reaction rate is achieved only at temperatures over 600°C, with about a 70% conversion of reactants to product.

To make the reaction more productive at lower temperatures, such as below 500°C, a V_2O_5 catalyst is used. The gases are introduced over the catalyst at 410–430°C. The temperature is allowed to increase as the reaction proceeds, and equilibrium is reached near 600°C. At this point, conversion to products is 60–70% complete. The gas mixture is then cooled to below 430°C and passed over

Figure 14.16 Oleum, which contains sulfur trioxide dissolved in sulfuric acid, fumes when exposed to moist air.

Figure 14.17 The position of equilibrium in the contact process for conversion of sulfur dioxide to sulfur trioxide shifts toward reactants as the temperature is increased.

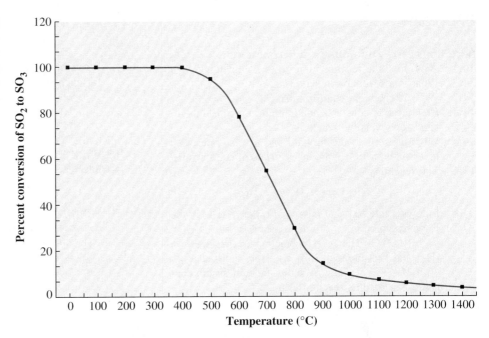

more catalyst. In some manufacturing plants, the equilibrium mixture of gases is then passed through an SO_3-absorbing tower to remove SO_3 from the mixture. The resulting mixture of gases, largely reactants, is then passed over the V_2O_5 catalyst again. When this method is used, overall conversions of 99.5% can be achieved.

Summary

A reversible reaction consists of a forward and a reverse reaction and so can proceed in either direction. When the forward and reverse reactions occur at the same rate, a state of equilibrium results. Combining the rate laws for the forward and reverse reactions gives an expression for the concentrations of species at equilibrium. The ratio of the rate constants is a new constant, the equilibrium constant, K. According to the law of mass action, this equilibrium-constant expression is valid no matter what the starting conditions at a given temperature. The standard free energy for a reaction can be used to calculate a value of the equilibrium constant at a given temperature. Like free-energy changes, equilibrium constants can be used to predict the spontaneity of reactions.

According to Le Chatelier's principle, when an external stress is placed on a system at equilibrium, the system shifts to a new position of equilibrium that removes some of the stress. The stress can take the form of a change in concentration, pressure, or temperature.

The calculations to determine reaction yields, though they vary depending on the type of reaction, can be carried out with a common procedure.

The principles of chemical equilibrium can be applied to improve the yields of industrial processes that involve reversible reactions. Some representative processes are steam reforming and the water-gas shift reaction for production of hydrogen, the Haber process for production of ammonia, the synthesis of methanol from carbon monoxide and hydrogen, and the contact process for production of sulfuric acid.

Key Terms

equilibrium constant
  (**K**) (14.1)
Haber process (14.5)

law of mass
  action (14.1)

Le Chatelier's
  principle (14.3)

Exercises

e. $2H_2O(g) \rightleftharpoons 2H_2(g) + O_2(g)$
f. $2NO_2(g) \rightleftharpoons N_2(g) + 2O_2(g)$
g. $2NO_2(g) \rightleftharpoons 2NO(g) + O_2(g)$

Rate Constants and Equilibrium Constants

14.1 Why do some reactions not go to completion?

14.2 What is a reversible reaction?

14.3 What is chemical equilibrium?

14.4 How are rate constants and equilibrium constants related for reversible reactions?

14.5 What happens to the forward and reverse rates as a reversible reaction approaches a state of equilibrium?

14.6 The reaction

$$2HI(g) \rightleftharpoons H_2(g) + I_2(g)$$

follows a rate law that is second-order in [HI] in the forward direction, with rate constant $k_f = 1.75\ M^{-1}$ s^{-1} at 600°C. The reverse reaction is first-order in $[H_2]$ and first-order in $[I_2]$, with rate constant $k_r = 66.8\ M^{-1}\ s^{-1}$ at 600°C. What is the equilibrium constant, K_c, for this reaction at 600°C?

14.7 The gas-phase reaction

$$2NO_2(g) \rightleftharpoons 2NO(g) + O_2(g)$$

has a rate constant for the forward reaction of $153\ M^{-1}\ s^{-1}$ and an equilibrium constant $K_c = 5.00 \times 10^{-7}$ at 520°C. What is the value of the rate constant for the reverse reaction?

14.8 Consider the reaction between vanadium(IV), VO^{2+}, and the iridium(IV) ion, $IrCl_6^{2-}$.

$$VO^{2+}(aq) + IrCl_6^{2-}(aq) + H_2O(l) \rightleftharpoons$$
$$VO_2^+(aq) + IrCl_6^{3-}(aq) + 2H^+(aq)$$

This reaction has a rate law in the forward direction of $k_f[VO^{2+}][IrCl_6^{2-}]/[H^+]$. Recognizing that the liquid, water, will appear in neither the rate law nor the expression for the equilibrium constant, predict a possible rate law for the reverse reaction.

Equilibrium-Constant Expressions

14.9 What is the relationship between the stoichiometry of a reaction and the form of the equilibrium-constant expression?

14.10 Why is the concentration of a pure liquid or solid in units of moles per liter always constant at constant temperature?

14.11 Why are the concentrations of pure solids or liquids not included in the equilibrium-constant expression?

14.12 What is the relationship between the equilibrium-constant expression and the reaction-quotient expression?

14.13 Write the expression for the equilibrium constant, K_p, for each of the following reactions.
   a. $3H_2(g) + N_2(g) \rightleftharpoons 2NH_3(g)$
   b. $2NH_3(g) \rightleftharpoons 3H_2(g) + N_2(g)$
   c. $\frac{3}{2}H_2(g) + \frac{1}{2}N_2(g) \rightleftharpoons NH_3(g)$
   d. $2HF(g) \rightleftharpoons H_2(g) + F_2(g)$

14.14 Write the expression for the equilibrium constant, K_c, for each of the following reactions.
   a. $2CO_2(g) \rightleftharpoons 2CO(g) + O_2(g)$
   b. $NH_4HS(s) \rightleftharpoons NH_3(g) + H_2S(g)$
   c. $N_2(g) + Cl_2(g) + 4H_2(g) \rightleftharpoons 2NH_4Cl(s)$
   d. $C(s) + CO_2(g) \rightleftharpoons 2CO(g)$

14.15 Write the expression for K_p and for K_c for the reversible reaction

$$2N_2O_5(g) \rightleftharpoons 4NO_2(g) + O_2(g)$$

14.16 Write the equilibrium-constant expression for K_p and K_c for each reaction.
   a. $Zn(s) + CO_2(g) \rightleftharpoons ZnO(s) + CO(g)$
   b. $NH_4Cl(s) \rightleftharpoons NH_3(g) + HCl(g)$
   c. $CaCO_3(s) \rightleftharpoons CaO(s) + CO_2(g)$

Equilibrium Constant from Equilibrium Concentrations and Pressures

14.17 Explain how the value of an equilibrium constant can be obtained from the concentrations of reactants and products at equilibrium.

14.18 Equilibrium reactions occurring in the gas phase are described by either K_p or K_c, but reactions in solution are described only by K_c. Explain this difference.

14.19 How can we use the value of the equilibrium constant and measured pressures or concentrations to decide whether any reaction will take place?

14.20 An equilibrium mixture at a certain temperature for the reaction

$$2H_2S(g) \rightleftharpoons 2H_2(g) + S_2(g)$$

has 2.00 atm pressure of H_2S, 0.400 atm H_2, and 1.60 atm S_2. What is K_p for this reaction?

14.21 When the reversible reaction

$$2A + B \rightleftharpoons 2C$$

reached equilibrium, the following concentrations were measured: $[A] = 0.40\ M$, $[B] = 0.30\ M$, and $[C] = 0.55\ M$. What is the value of K_c for this reaction?

14.22 An equilibrium mixture of PCl_5, PCl_3, and Cl_2 at a certain temperature in a 10.0-L container consists of 1.00 mol PCl_5, 0.30 mol PCl_3, and 0.80 mol Cl_2. Calculate K_c for the reaction

$$PCl_5(g) \rightleftharpoons PCl_3(g) + Cl_2(g)$$

14.23 The reaction

$$2NO(g) + O_2(g) \rightleftharpoons 2NO_2(g)$$

which occurs in auto exhausts, is allowed to come to a state of equilibrium at 160°C, and the partial pressures of the components are found to be as follows: $P_{NO} = 0.0052$ atm, $P_{NO_2} = 0.40$ atm, $P_{O_2} = 0.60$ atm. What is the value of K_p at 160°C?

14.24 Consider the reaction

$$2A + 3B \rightleftharpoons C + 2D$$

A mixture is prepared from 2.00 mol of A and 2.00 mol of B in a 1.00-L container. When the mixture reaches a state of equilibrium, 0.500 mol of D has been produced. What is the equilibrium constant for this reaction?

14.25 The reaction

$$H_2(g) + I_2(g) \rightleftharpoons 2HI(g)$$

has an equilibrium constant, K_p, of 55.3 at 699 K. (a) If a mixture consists of 0.70 atm HI, 0.020 atm H_2, and 0.020 atm I_2 at 699 K, will there be any net reaction? (b) If so, will HI be consumed or produced?

Combining Equilibrium Constants

14.26 Outline a procedure for calculating the value of the equilibrium constant for a reaction from the equilibrium constants known for two reactions that add together to give the reaction of interest.

14.27 How is Hess's law related to the combining of equilibrium constants for several reactions?

14.28 When combining equilibrium constants for two reactions to determine the equilibrium reaction for another reaction, why is it necessary for us to ensure that all the equilibrium constants are at the same temperature?

14.29 How can we calculate the equilibrium constant for a reaction written in reverse?

14.30 Given the following equilibrium constants:

$$H_2(g) + CO_2(g) \rightleftharpoons H_2O(g) + CO(g) \quad K_1 = 0.62$$
$$FeO(s) + H_2(g) \rightleftharpoons Fe(s) + H_2O(g) \quad K_2 = 0.42$$

find the equilibrium constant, K, at the same temperature for the reaction

$$FeO(s) + CO(g) \rightleftharpoons Fe(s) + CO_2(g)$$

14.31 Write expressions for the equilibrium constants for the following reactions:

$$N_2(g) + 3H_2(g) \rightleftharpoons 2NH_3(g)$$
$$\tfrac{1}{2}N_2(g) + \tfrac{3}{2}H_2(g) \rightleftharpoons NH_3(g)$$

  a. If $K_c = 0.49$ for the first reaction, what is the value of K_c for the second reaction at the same temperature?

  b. Determine the value of the equilibrium constant for the following reaction at the same temperature:

$$2NH_3(g) \rightleftharpoons N_2(g) + 3H_2(g)$$

14.32 The equilibrium constants are given for the following reactions:

$$2NO(g) \rightleftharpoons N_2(g) + O_2(g) \qquad K_c = 2.4 \times 10^{-18}$$
$$NO(g) + \tfrac{1}{2}Br_2(g) \rightleftharpoons NOBr(g) \quad K_c = 1.4$$

Using this information, determine the equilibrium constant for the following reaction at the same temperature:

$$\tfrac{1}{2}N_2(g) + \tfrac{1}{2}O_2 + \tfrac{1}{2}Br_2(g) \rightleftharpoons NOBr(g)$$

Relationship Between K_p and K_c

14.33 Describe a procedure for converting between K_p and K_c.

14.34 Under what conditions are the values of K_p and K_c the same for a gas-phase equilibrium reaction?

14.35 Calculate K_p for the dissociation of PCl_5 at 250°C.

$$PCl_5(g) \rightleftharpoons PCl_3(g) + Cl_2(g)$$

At this temperature, K_c is 0.00900.

14.36 At a certain temperature, the equilibrium constant K_p has a value of 2.4×10^{-18} for the reaction

$$2NO(g) \rightleftharpoons N_2(g) + O_2(g)$$

Calculate a value for K_c for this reaction.

14.37 K_c has a value of 12.9 for the following reaction at a temperature of 1550 K.

$$N_2(g) + 3H_2(g) \rightleftharpoons 2NH_3(g)$$

What is the value of K_p for this reaction?

14.38 For the reaction

$$CO_2(g) + H_2(g) \rightleftharpoons CO(g) + H_2O(g)$$

the equilibrium constant K_p is 0.10 at 690 K. What is the value of K_c at that temperature?

Free Energy and Equilibrium Constants

14.39 Write the equation that can be used to calculate a value of the equilibrium constant from the standard free energy of the reaction.

14.40 If the standard free-energy change for a reaction is needed to calculate the equilibrium constant for that reaction, but a value has not been measured directly, how can we calculate a value from other data?

14.41 Use the value of $\Delta G°$ to calculate the value of the equilibrium constant for each of the following reactions at 25.0°C.

  a. $CH_4(g) + I_2(g) \rightleftharpoons CH_3I(g) + HI(g)$
     $\Delta G° = 67.4$ kJ
  b. $2SbCl_3(g) + 3CO(g) \rightleftharpoons 2Sb(s) + 3COCl_2(g)$
     $\Delta G° = 401$ kJ
  c. $CH_4(g) + 2H_2S(g) \rightleftharpoons CS_2(g) + 4H_2(g)$
     $\Delta G° = 185$ kJ
  d. $4NH_3(g) + 3O_2(g) \rightleftharpoons 2N_2(g) + 6H_2O(g)$
     $\Delta G° = -1305$ kJ
  e. $Br_2(g) + 2HI(g) \rightleftharpoons I_2(g) + 2HBr(g)$
     $\Delta G° = -111$ kJ

14.42 Calculate the value of $\Delta G°$ and the value of the equilibrium constant for each of the following reactions at 25.0°C.

    a. $CO_2(g) + NO(g) \rightleftharpoons CO(g) + NO_2(g)$
    b. $2Ag(s) + H_2S(g) \rightleftharpoons H_2(g) + Ag_2S(s)$
    c. $2H_2S(g) + 3O_2(g) \rightleftharpoons 2H_2O(g) + 2SO_2(g)$
    d. $4NH_3(g) + 5O_2(g) \rightleftharpoons 4NO(g) + 6H_2O(g)$
    e. $CH_4(g) + 2H_2O(g) \rightleftharpoons CO_2(g) + 4H_2(g)$

Calculation of Equilibrium Concentrations

14.43 Outline a general procedure for determining the concentrations of reactants and products for a reaction at equilibrium.

14.44 What simplifying assumptions should we examine when calculating equilibrium concentrations?

14.45 The reaction

$$2NO(g) + O_2(g) \rightleftharpoons 2NO_2(g)$$

has an equilibrium constant $K_c = 100$ at a certain temperature.

    a. What is the concentration of O_2 in equilibrium with 0.00100 M NO and 0.0100 M NO_2?
    b. What is the concentration of NO in equilibrium with 0.100 M O_2 and 0.0100 M NO_2?

14.46 For the reaction

$$2NaHCO_3(s) \rightleftharpoons Na_2CO_3(s) + CO_2(g) + H_2O(g)$$

the equilibrium constant K_p is 3.90×10^{-4} at 50°C. A 5.0-g sample of $NaHCO_3$ is placed in a closed evacuated flask, and the temperature is raised to 50°C. What will be the total gas pressure at equilibrium?

14.47 For the reaction

$$CO_2(g) + H_2(g) \rightleftharpoons CO(g) + H_2O(g)$$

the equilibrium constant K_c is 0.10 at 690 K. What is the equilibrium pressure of each substance in the reaction that results from mixing 1.00 mol CO_2 and 1.00 mol H_2 in a 10.0-L flask at 690 K?

14.48 For the reaction

$$CO(g) + H_2O(g) \rightleftharpoons CO_2(g) + H_2(g)$$

the equilibrium constant is $K_p = 4.00$ at a certain temperature. Suppose 1.00 atm CO and 2.00 atm H_2O are introduced into a vessel. What is the pressure of CO_2 at equilibrium?

14.49 The equilibrium constant K_c for the following reaction at 520°C is 5.00×10^{-7}

$$2NO_2(g) \rightleftharpoons 2NO(g) + O_2(g)$$

Suppose 1.00 mol NO_2 is placed in a 1.00-L container at 520°C. What will be the final equilibrium concentration of each substance in the mixture?

14.50 For the reaction

$$Br_2(g) \rightleftharpoons 2Br(g)$$

the equilibrium constant, K_c, is 7.76 at 4000 K. Suppose 0.0100 mol Br_2 is confined to a 1.00-L flask and heated to 4000 K.

    a. How many moles of Br are present at equilibrium?
    b. What fraction of the initial Br_2 dissociates into atoms?

14.51 For the reaction

$$C_4H_{10}(g) \rightleftharpoons C_4H_6(g) + 2H_2(g)$$

the equilibrium constant K_c is 1.00×10^{-6} at 600°C. Suppose 1.00 mol C_4H_{10} and 1.00 mol H_2 are placed in a 1.00-liter flask at 600°C. How many moles of C_4H_6 will be formed?

14.52 The equilibrium constant for the formation of hydrazine is $K_c = 5.00 \times 10^{-3}$.

$$N_2(g) + 2H_2(g) \rightleftharpoons N_2H_4(g)$$

Suppose 2.00 mol H_2 and 1.00 mol N_2 are placed in a 1.00-L container. Calculate (a) the equilibrium concentration of each substance and (b) the equilibrium constant for the reverse reaction.

14.53 Calculate the equilibrium concentrations of all components when 2.00 mol HI is placed in a 1.00-L container at 25°C.

$$2HI(g) \rightleftharpoons H_2(g) + I_2(g) \qquad K_c = 1.23 \times 10^{-3}$$

Also calculate the total pressure in this system at 25°C.

14.54 At 986°C, the equilibrium constant for the reaction

$$H_2(g) + CO_2(g) \rightleftharpoons H_2O(g) + CO(g)$$

is $K_c = 1.60$. Calculate the equilibrium concentrations of all components of this system at 986°C when the initial concentrations of H_2 and CO_2 are both 1.00 mol/L, and H_2O and CO are initially absent.

Le Chatelier's Principle

14.55 What is Le Chatelier's principle?

14.56 Explain several ways in which we might drive a reaction in a desired direction by changing conditions.

14.57 Explain why the effect of a change in temperature on the position of equilibrium is different for exothermic and endothermic reactions.

14.58 Describe two conditions under which a change in pressure has no effect on the position of equilibrium in a reversible reaction.

14.59 Champagne is poured cold into chilled glasses because it goes "flat"—that is, loses dissolved CO_2—when poured into warm glasses. The temperature dependence of the position of equilibrium is the same for all processes of the following type:

$$\text{Gas} + \text{liquid} \rightleftharpoons \text{solution}$$

On the basis of this information and Le Chatelier's principle, determine whether the dissolution of a gas in a liquid is an exothermic or an endothermic process. Explain your answer.

14.60 Consider the following reaction at equilibrium.

$$NO(g) + SO_3(g) \rightleftharpoons NO_2(g) + SO_2(g)$$

Indicate the effect of the following stresses on the equilibrium.

a. a decrease in the pressure of NO
b. an increase in the pressure of SO_3
c. a decrease in the pressure of NO_2
d. an increase in the pressure of NO_2
e. an increase in the volume of the container by a factor of 2
f. a doubling of the pressure of NO and a doubling of the pressure of SO_2

14.61 The equilibrium constant K_c is 0.080 at 400°C and 0.41 at 600°C for the reaction

$$CO_2(g) + H_2(g) \rightleftharpoons CO(g) + H_2O(g)$$

Is the reaction endothermic or exothermic?

14.62 Consider the reaction

$$4NH_3(g) + 5O_2(g) \rightleftharpoons 4NO(g) + 6H_2O(g)$$

Based on Le Chatelier's principle, predict whether each of the following changes should favor the forward or the reverse direction.

a. removal of NO
b. addition of H_2O
c. addition of a catalyst
d. removal of O_2
e. a decrease in pressure (an increase in volume)

14.63 Consider the reaction

$$2H_2O_2(g) \rightleftharpoons 2H_2O(g) + O_2(g) + heat$$

Based on Le Chatelier's principle, predict whether each of the following changes should favor the forward or the reverse direction.

a. removal of O_2
b. addition of H_2O
c. addition of a catalyst
d. an increase in temperature
e. a decrease in pressure (an increase in volume)

14.64 In which direction will each of the following reactions shift after the specified stress is applied?

a. $N_2O_4(g) \rightleftharpoons 2NO_2(g)$
   a decrease in the total pressure (an increase in volume)
b. $2NOBr(g) \rightleftharpoons 2NO(g) + Br_2(g)$
   an increase in the total pressure (a decrease in volume)
c. $2Cl_2(g) + 2H_2O(g) \rightleftharpoons 4HCl(g) + O_2(g)$
   an increase in temperature; $\Delta H° = +113$ kJ
d. $H_2(g) + CO_2(g) \rightleftharpoons H_2O(g) + CO(g)$
   an increase in the concentration of CO_2
e. $2SO_2(g) + O_2(g) \rightleftharpoons 2SO_3(g)$
   addition of more catalyst
f. $CaCO_3(s) \rightleftharpoons CaO(s) + CO_2(g)$
   removal of some of the CO_2 formed
g. $H_2(g) + Cl_2(g) \rightleftharpoons 2HCl(g)$
   an increase in the concentration of HCl
h. $PCl_5(g) \rightleftharpoons PCl_3(g) + Cl_2(g)$
   addition of a small amount of gaseous helium
i. $2NO(g) \rightleftharpoons N_2(g) + O_2(g)$
   a decrease in the concentration of N_2
j. $NH_4HS(s) \rightleftharpoons NH_3(g) + H_2S(g)$
   an increase in the amount of NH_4HS

Industrial Chemistry

14.65 List some gas-phase industrial reactions that attain a state of equilibrium.

14.66 Why can equilibrium be a problem in a reaction that is being used industrially to produce a desired compound?

14.67 Describe some approaches that might be used to achieve a greater yield in a reaction that reaches a state of equilibrium.

14.68 Why is decreasing the temperature not always the appropriate method for increasing the yield in an exothermic reaction that reaches a state of equilibrium?

14.69 Write a balanced equation to describe the steam reforming of propane, C_3H_8. This reaction is endothermic. Discuss some general conditions that would help improve the yield of products.

14.70 a. What would be the percent NH_3 in a Haber synthesis carried out at 600 atm total pressure and 300°C?
   b. At 600 atm and 500°C?

Additional Exercises

14.71 For the reaction

$$N_2(g) + O_2(g) \rightleftharpoons 2NO(g)$$

the equilibrium constant is 1.0×10^{-6} at 1500 K and 6.2×10^{-4} at 2000 K. Is the reaction exothermic or endothermic?

14.72 Consider the reversible gas-phase reaction

$$SO_2Cl_2(g) \rightleftharpoons SO_2(g) + Cl_2(g)$$

The value of K_p is 22.5 at 380 K.

a. Write the expression for K_p for the equilibrium.
b. What is the relationship between the values of K_p and K_c for the reaction?

14.73 The gaseous compound NOBr decomposes according to the equation

$$NOBr(g) \rightleftharpoons NO(g) + \tfrac{1}{2}Br_2(g)$$

The equilibrium constant, K_p, has a value of 0.15 at 350 K. If 1.00 atm NOBr, 0.500 atm NO, and 0.300 atm Br_2 are mixed at this temperature, will there be a net reaction?

14.74 Consider the reaction

$$Heat + Cl_2(g) + 2O_2(g) \rightleftharpoons 2ClO_2(g)$$

Based on Le Chatelier's principle, predict whether each of the following changes should favor the forward or the reverse direction.
   a. removal of O_2
   b. addition of ClO_2
   c. an increase in the size of the container
   d. an increase in temperature
   e. an increase in pressure (a decrease in volume)

14.75 The reaction

$$2NO(g) + O_2(g) \rightleftharpoons 2NO_2(g)$$

has an equilibrium constant $K_c = 100.0$ at a certain temperature. What is the concentration of NO_2 in equilibrium with 0.00100 M NO and 0.0100 M O_2?

14.76 The decomposition of $CaSO_4$ takes place according to the equation

$$CaSO_4(s) \rightleftharpoons CaO(s) + SO_3(g) \qquad \Delta H = 400 \text{ kJ}$$

Suggest two ways to increase the yield of SO_3.

14.77 Calculate the value of the equilibrium constant for the reaction

$$2NO(g) + O_2(g) \rightleftharpoons N_2O_4(g) \qquad K_1 = ?$$

given the following equilibrium constants at the same temperature:

$$2NO(g) + O_2(g) \rightleftharpoons 2NO_2(g) \qquad K_2 = 2.5 \times 10^2$$
$$2NO_2(g) \rightleftharpoons N_2O_4(g) \qquad K_3 = 3.0 \times 10^3$$

14.78 How is the solubility of $O_2(g)$ in water affected by an increase in the pressure of O_2?

14.79 At 1483°C, bromine gas dissociates into atoms.

$$Br_2(g) \rightleftharpoons 2Br(g)$$

The equilibrium constant, K_c, is 4.0×10^{-4}. Suppose 1.00 mol Br_2 is injected into a 1.00-L container at 1483°C. What are the equilibrium concentrations?

14.80 Suggest four ways in which the concentration of SO_3 can be increased in a closed vessel for the reaction

$$SO_2(g) + \tfrac{1}{2}O_2(g) \rightleftharpoons SO_3(g) \quad \Delta H = -98 \text{ kJ/mol}$$

14.81 Suppose 5.00 g PCl_5 is placed in a 1.00-L flask and the PCl_5 is allowed to dissociate at 250°C according to the equation

$$PCl_5(g) \rightleftharpoons PCl_3(g) + Cl_2(g)$$

It is found that 1.25 g Cl_2 has formed. What is the value of K_c at 250°C?

14.82 Write the expression for K_p and for K_c for the reversible reaction

$$2N_2O_5(g) \rightleftharpoons 4NO_2(g) + O_2(g)$$

14.83 The equilibrium constant K_c is 0.030 at 760°C and 44.6 at 400°C for the reaction

$$PCl_3(g) + Cl_2(g) \rightleftharpoons PCl_5(g)$$

Is the reaction endothermic or exothermic?

14.84 During the Napoleonic wars, the tin buttons used on Russian army uniforms disintegrated in the winter when temperatures fell below 18°C. The culprit was "tin pest," a chemical process described by the equation

$$Sn_{(white\ metal)} \rightleftharpoons Sn_{(gray\ powder)}$$

Is this process exothermic or endothermic? Explain your answer.

14.85 Formamide decomposes into ammonia and carbon monoxide according to the equation

$$HCONH_2(g) \rightleftharpoons NH_3(g) + CO(g)$$

The equilibrium constant, K_c, has a value of 4.84 at 400 K. Suppose 10.0 g of formamide is allowed to dissociate in a 1.00-L flask at 400 K. What is the concentration of each species at equilibrium?

14.86 Calculate the value of $\Delta G°$ and the value of the equilibrium constant for each of the following reactions at 25.0°C.
   a. $NH_4Cl(s) \rightleftharpoons NH_3(g) + HCl(g)$
   b. $4NH_3(g) + 5O_2(g) \rightleftharpoons 4NO(g) + 6H_2O(g)$
   c. $2N_2(g) + O_2(g) \rightleftharpoons 2N_2O(g)$
   d. $N_2(g) + Cl_2(g) + 4H_2(g) \rightleftharpoons 2NH_4Cl(s)$
   e. $4HCl(g) + O_2(g) \rightleftharpoons 2H_2O(g) + 2Cl_2(g)$

14.87 The proper maintenance of a swimming pool requires control of both the chlorine level and the acidity. Chlorine is usually added in the form of a hypochlorite salt that forms chlorine by the following reversible reaction:

$$HOCl(aq) + Cl^-(aq) + H^+(aq) \rightleftharpoons$$
$$Cl_2(aq) + H_2O(l)$$

The chloride is generally present naturally in the water and is formed by photochemical decomposition of chlorine.
   a. Predict what effect the addition of more chloride to the water would have on the concentration of Cl_2.
   b. Predict what effect a decrease in the H^+ concentration would have on the concentration of Cl_2. Explain your answers.

14.88 Hydrogen and iodine react at 699 K according to the equation

$$H_2(g) + I_2(g) \rightleftharpoons 2HI(g)$$

Suppose 1.00 mol H_2 and 1.00 mol I_2 are placed in a 1.00-L container and allowed to react.

a. What molar concentration of HI is present at equilibrium? At 699 K, the value of K_c is 55.3.

b. What is the value of K if the amounts of the species are expressed in units of atmospheres?

14.89 a. Determine what relationship exists between K_1 and K_2, given the following equations:

$$2H_2(g) + O_2(g) \rightleftharpoons 2H_2O(g) \qquad K_1$$
$$H_2(g) + \tfrac{1}{2}O_2(g) \rightleftharpoons H_2O(g) \qquad K_2$$

b. What relationship exists between K_c and K_p for the reaction

$$N_2O_4(g) \rightleftharpoons 2NO(g) + O_2(g)$$

14.90 Calculate K_c for the reaction

$$2NH_3(g) \rightleftharpoons N_2(g) + 3H_2(g)$$

Assume that the equilibrium concentrations are 1.00 M N_2, 2.00 M H_2, and 2.00 M NH_3.

* 14.91 The equilibrium constant for the following reaction is $K_p = 2.40$ at a certain temperature.

$$SO_2Cl_2(g) \rightleftharpoons SO_2(g) + Cl_2(g)$$

Suppose 1.53 atm SO_2Cl_2 is placed in a container and allowed to dissociate until equilibrium is reached.

a. What are the equilibrium partial pressures of the gases?

b. Suppose 1.00 atm Cl_2 is then added to the container, and the reaction proceeds to a new equilibrium position. What is the partial pressure of each gas?

* 14.92 The poisonous gas phosgene dissociates according to the equation

$$COCl_2(g) \rightleftharpoons CO(g) + Cl_2(g)$$

The value of K_p is 0.0444 at a certain temperature. If the total pressure of an equilibrium mixture of these gases is 1.00 atm, what is the partial pressure of each gas at equilibrium, assuming only $COCl_2$ was present initially?

* 14.93 The reaction

$$2NO(g) + O_2(g) \rightleftharpoons 2NO_2(g)$$

has an equilibrium constant $K_c = 100.0$ at a certain temperature. What is the concentration of NO_2 produced when a mixture of 2.00 mol NO and 1.00 mol O_2 is placed in a 1.00-L flask and allowed to reach a state of equilibrium?

* 14.94 The equilibrium constant for the reaction

$$2A + 3B \rightleftharpoons C + 2D$$

is $K_c = 4.00 \times 10^{-3}$ at a certain temperature. Suppose 2.00 mol of A and 2.00 mol of B are placed in a 1.00-L container. How much C will be produced when the system reaches a state of equilibrium?

▶ ▶ ▶ Chemistry in Practice

14.95 Lanthanum(III) chloride undergoes hydrolysis in the presence of hot water vapor.

$$LaCl_3(s) + H_2O(g) \rightleftharpoons LaOCl(s) + 2HCl(g)$$

A sample of 0.5000 g $LaCl_3(s)$ was placed in a 250.0-mL flask, which was then evacuated to remove all air and was heated to 620.0°C. Water vapor at 620.0°C was admitted into the flask until the pressure was 210.90 torr. The pressure then continued to rise slowly because of the hydrolysis reaction, eventually reaching a constant value of 420.40 torr. The gas was pumped out of the flask and was absorbed by 50.00 mL of 0.1000 M

NaOH solution, which reacted with the HCl(g). This solution was then titrated with 0.1036 M HCl(aq), requiring 30.12 mL for complete reaction.

a. What is K_p for the hydrolysis reaction at 620.0°C?

b. How much of the original sample of 0.5000 g $LaCl_3(s)$ was converted to $LaOCl(s)$?

c. What initial pressure of $H_2O(g)$ should have been placed in the flask to convert 99.0% of the $LaCl_3(s)$ to $LaOCl(s)$?

15

▶ ▶ ▶ ▶

Acid–Base Equilibria

A variety of household products are either acids or bases.

CHAPTER OUTLINE

Encountering Chemistry

▶ ▶ ▶ ▶

Acids and bases are familiar substances. Many products used around the home contain one or the other. Most foods are acidic, for example, and many chemicals used to prepare foods are also acids. Vinegar, a solution of acetic acid, inhibits bacterial growth. Acids in baking powder release carbon dioxide from the sodium bicarbonate also present, causing bubbles in batter. Acids and bases also occur in cleaning products. Toilet bowl cleaners contain either hydrochloric acid or sodium hydroxide, a base. Drain cleaners often are a mixture of sodium hydroxide and aluminum chips that react in water to form hydrogen gas that helps dislodge greasy clogs. Most glass cleaners contain aqueous ammonia, another base. Many medicines are acids and bases. Aspirin and vitamin C, for example, are acids. Common antacid tablets contain calcium carbonate, and laxatives often contain magnesium hydroxide, both bases.

Acids and bases are important not only in the preparation of foods, but also in the growing of plants used as foods. The acidity of soils strongly affects the way plants grow, because it helps regulate the availability of plant nutrients and the activity of soil bacteria. In some cases, controlled acidity may help prevent the growth of weeds and may interfere in the life cycle of some destructive insects. Some plants grow well only in slightly acidic soils, whereas excess acidity prevents good growth in other plants. In alkaline soils, such as those in desert regions, the amounts of nitrogen, phosphorus, iron, and other nutrients are too low for good plant growth, and special procedures such as neutralization by acid treatment must be carried out to make the soil suitable for agriculture.

This chapter presents several models for the nature of acids and bases, emphasizing their behavior in aqueous solutions. Acids and bases can be strong or weak, depending on the degree to which they ionize. The chapter applies the concepts of chemical equilibrium to solutions of weak acids and bases, as well as to water, which itself undergoes ionization. The chapter explains how to calculate the acidity of solutions of acids and bases and, finally, discusses some industrially important acids and bases.

15.1   Properties of Acids and Bases

▶ Although we commonly speak of certain materials mentioned in the chapter introduction, like lemon juice and vinegar, as being acidic and of others, like household ammonia and milk of magnesia, as being basic, we often use these terms inexactly. To use these terms properly, we must adopt one of several acid–base theories. For example, according to the Arrhenius theory, any substance that generates an excess of hydrogen ions in aqueous solution is an **acid,** and a substance that generates an excess of hydroxide ions in aqueous solution is an **Arrhenius base.**

How do acids differ from bases?

Acids share a number of characteristic properties. They taste sour (as does citric acid in lemon juice and lactic acid in sour milk). They sting on the skin (as does formic acid, HCO_2H, which ants inject when they bite). Acids are corrosive—that is, they react with certain metals (such as zinc and iron) to produce solutions of the metal ions, in the process releasing hydrogen gas (Figure 15.1). They react with carbonate compounds to release carbon dioxide gas. They change the color of certain plant dyes called indicators (Figure 15.2). For example, litmus is red in the presence of acid; phenolphthalein is colorless. Acids react with bases to form salts.

Bases feel slippery, taste bitter, and change the color of some plant dyes (litmus turns blue in the presence of a base; phenolphthalein turns pink). Bases can neutralize acids. When an acid reacts with a base, the characteristic properties of

Figure 15.1 Active metals, such as iron, react with acids to release hydrogen gas and form solutions containing the ions formed by oxidation of the metal.

Figure 15.2 Certain plant extracts, known as indicators, change color in solutions of acids and bases. The left pair contains phenolphthalein in acid and in base, whereas the right pair contains bromcresol blue.

both the acid and the base disappear. For example, a common problem in automobile maintenance is the accumulation of battery acid on the top and terminals of the car's battery. This causes corrosion and shorting of the battery. A cure is neutralizing the acid by sprinkling solid sodium carbonate, a base, onto the battery. As shown by the following equation, the acid is consumed by the sodium carbonate, forming carbon dioxide gas and sodium sulfate solution, which can be washed away easily.

$$H_2SO_4(aq) + Na_2CO_3(s) \longrightarrow Na_2SO_4(aq) + H_2O(l) + CO_2(g)$$

This section explores three theories of acids and bases: the Arrhenius, Brønsted–Lowry, and Lewis theories.

Svante Arrhenius (1859–1927) was a Swedish chemist who received the 1903 Nobel Prize in chemistry for his electrolytic-dissociation theory, which explained the nature of the ionization of acids and bases in aqueous solution. He first proposed the theory in his doctoral dissertation. Because the theory was so controversial, the dissertation was almost rejected. He is also responsible for the equation, named after him, that describes the dependence of rate constants on temperature.

Arrhenius Theory of Acids and Bases

The first definition of the chemical nature of acids and bases was proposed by the Swedish chemist Svante Arrhenius. He defined an acid as a substance that produces protons (H^+, or hydrogen ions) when dissolved in aqueous solution, and he defined a base as a substance that donates hydroxide ions (OH^-) when dissolved in aqueous solution. The characteristic properties of acids and bases are due to the presence in aqueous solution of these ions, and the strength of an acid or base depends on the amount of ions present. The neutralization of an acid by a base involves reaction between hydrogen ions and hydroxide ions to form water.

$$H^+ + OH^- \longrightarrow H_2O$$

Metal ions are hydrated in aqueous solution, just as are the hydrogen and hydroxide ions. What symbols can we use for all these ions to indicate that they are hydrated, but that the waters of hydration are not the center of focus of their chemical behavior?

Problems with the Arrhenius Theory

Although it is useful in many common situations, the Arrhenius concept of acids and bases is a rather limited one with some fundamental difficulties. The hydrogen ion or proton, H^+, with a radius of about 10^{-13} cm, has an extremely concentrated

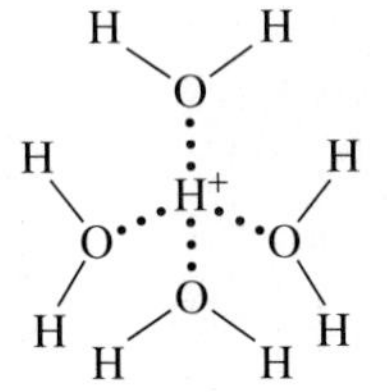

Figure 15.3 The average structure of the aqueous hydrogen ion has the proton hydrogen bonded to four water molecules.

ionic charge. As a result, simple unhydrated protons are unlikely to exist in aqueous solution. In fact, the hydration reaction

$$H^+ + H_2O \longrightarrow H_3O^+$$

is exothermic by about 1250 kJ/mol, which is such a large change in enthalpy that essentially no unhydrated protons exist in the presence of water. Evidence from X-ray diffraction patterns indicate that the proton is typically associated through hydrogen bonding with four water molecules in solution, as shown in Figure 15.3, giving an average formula for aqueous hydrogen ions of $H_9O_4^+$. We normally represent the aqueous hydrogen ion as H_3O^+ or $H^+(aq)$. The H_3O^+ ion, called the **hydronium ion,** is not part of the Arrhenius theory but is consistent with other theories of acids and bases.

Similarly, the hydroxide ion in aqueous solution participates in hydrogen bonding to water molecules. The typical structure of the aqueous hydroxide ion, shown in Figure 15.4, indicates that the average formula is $OH(H_2O)_3^-$ or $H_7O_4^-$, not the OH^- of Arrhenius theory. Other theories generally use the designation $OH^-(aq)$ for the aqueous hydroxide ion.

A further difficulty with the simple Arrhenius concept is that it is based solely on aqueous acid–base systems. However, substances other than hydroxide ion have basic properties, and acidic and basic properties are observed for substances existing in solvents other than water, such as in liquid ammonia.

Brønsted–Lowry Theory of Acids and Bases

The Brønsted–Lowry theory, which builds on the Arrhenius concept and deals with its shortcomings, was proposed independently in 1923 by J. N. Brønsted, a Danish chemist, and T. M. Lowry, a British chemist. The Brønsted–Lowry concept defines an acid as any substance that can donate a proton to another substance and a base as any substance that can accept a proton from another substance. This definition includes all Arrhenius acids and bases as well as substances other than $H^+(aq)$ and $OH^-(aq)$. ▶ The substances mentioned in the chapter introduction are all such **Brønsted–Lowry acids** or **Brønsted–Lowry bases.** For example, ammonium ion is an acid, because it can transfer a proton to water in solution.

$$NH_4^+(aq) + H_2O(l) \rightleftharpoons NH_3(aq) + H_3O^+(aq)$$

The carbonate ion is a base, because it can accept a proton from water.

$$H_2O(l) + CO_3^{2-}(aq) \rightleftharpoons OH^-(aq) + HCO_3^-(aq)$$

Note that the solvent, such as water, can either donate or accept the proton.

Conjugate Acids and Bases   The ionization of a Brønsted–Lowry acid or base can be considered an equilibrium reaction between two acid–base pairs. For example, hydrofluoric acid can donate a proton to water.

$$\underset{\text{acid}}{HF(aq)} + \underset{\text{base}}{H_2O(l)} \longrightarrow H_3O^+(aq) + F^-(aq)$$

Hydrogen fluoride is a colorless gas that fumes in air, forming hydrofluoric acid. This extremely corrosive acid forms burns on skin that are very slow to heal. It reacts with glass and silicon dioxide to form gaseous silicon tetrafluoride, so it is often used to etch glass.

The reverse reaction can also occur, wherein hydronium ion donates a proton to fluoride ion.

$$\underset{\text{acid}}{H_3O^+(aq)} + \underset{\text{base}}{F^-(aq)} \longrightarrow HF(aq) + H_2O(l)$$

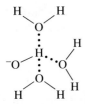

Figure 15.4 The average structure of the aqueous hydroxide ion is $H_7O_4^-$ or $OH(H_2O)_3^-$. The ion is hydrogen bonded to an average of three water molecules.

These reactions can be written as an equilibrium.

$$HF(aq) + H_2O(l) \rightleftharpoons H_3O^+(aq) + F^-(aq)$$
$$\text{acid 1} \qquad \text{base 2} \qquad\quad \text{acid 2} \qquad\quad \text{base 1}$$

Note that the base F^- can be formed from the acid HF by transfer of a proton to another substance, whereas the acid H_3O^+ can be formed from the base H_2O by transfer of a proton from another substance. Substances connected by the lines beneath the equilibrium equation are called **conjugate acid–base pairs.** Each member of a conjugate acid–base pair can be formed from the other by the gain or loss of a proton.

Conjugate acid–base pairs can be found for substances exhibiting acidic or basic properties in other solvents as well. Consider again the example of hydrofluoric acid dissolved in liquid ammonia.

$$HF + NH_3(l) \rightleftharpoons NH_4^+ + F^-$$
$$\text{acid 1} \quad \text{base 2} \qquad\quad \text{acid 2} \quad \text{base 1}$$

Hydrogen fluoride is an acid in water as well as in liquid ammonia. Its conjugate base in either case is formed by the loss of a proton.

According to the Brønsted–Lowry definition, the neutralization reaction of any acid–base pair is just a proton transfer reaction.

$$H_3O^+(aq) + OH^-(aq) \rightleftharpoons H_2O(l) + H_2O(l)$$
$$\text{acid 1} \qquad \text{base 2} \qquad\quad \text{acid 2} \qquad \text{base 1}$$

Amphoterism   Note that in the reverse reaction, water is acting as both an acid and a base. Substances that have both acidic and basic properties, no matter what definition of acids and bases is used, are called **amphoteric substances.** Thus by the Brønsted-Lowry definition, amphoteric substances can either gain or lose a proton. Perhaps the most important amphoteric substance is water. Pure water undergoes partial ionization (about 1 of every 555 million molecules ionizes), as shown by the reverse of the neutralization reaction.

$$H_2O(l) + H_2O(l) \rightleftharpoons H_3O^+(aq) + OH^-(aq)$$
$$\text{acid 1} \qquad \text{base 2} \qquad\quad \text{acid 2} \qquad \text{base 1}$$

Other substances can also behave in this way. For example, the bicarbonate ion, HCO_3^-, can react with a proton from another acid to form carbonic acid, in which case it acts as a base.

$$H_3O^+(aq) + HCO_3^-(aq) \rightleftharpoons H_2CO_3(aq) + H_2O(l)$$
$$\text{acid 1} \qquad\quad \text{base 2} \qquad\qquad \text{acid 2} \qquad \text{base 1}$$

Or it can donate a proton to water to behave as an acid.

$$HCO_3^-(aq) + H_2O(l) \rightleftharpoons H_3O^+(aq) + CO_3^{2-}(aq)$$
$$\text{acid 1} \qquad \text{base 2} \qquad\quad \text{acid 2} \qquad\quad \text{base 1}$$

Monoprotic and Polyprotic Acids   Brønsted–Lowry acids can also be categorized as **monoprotic** or **polyprotic,** depending on whether they can donate one proton or more than one. Monoprotic acids have only one proton to donate, as in hydrochloric acid.

$$HCl(aq) + H_2O(l) \rightleftharpoons H_3O^+(aq) + Cl^-(aq)$$

Are all hydrogen atoms in a substance acidic?

Polyprotic acids have more than one proton to donate and can be further categorized as **diprotic, triprotic,** and so on. A diprotic acid can donate two protons.

$$H_2SO_4(aq) + 2H_2O(l) \rightleftharpoons 2H_3O^+(aq) + SO_4^{2-}(aq)$$

Sulfur trioxide, SO_3, is a colorless, irritating gas that results from the combustion of sulfur or sulfur-containing materials. Although it is often encountered as an air pollutant, it can be condensed to volatile solids or liquids at room temperature. It combines violently with water to form sulfuric acid.

Acidic and Nonacidic Hydrogen Atoms The distinction between a monoprotic and a polyprotic acid is important in determining the proton-donating power of an acid, but it cannot always be made on the basis of the number of hydrogen atoms in the acid molecule. In such common acids as HCl, HBr, HNO_3, and $HClO_4$, which contain only one hydrogen atom, the hydrogen atom must be acidic. However, in *oxoacids* (acids that contain oxygen), only the hydrogens bonded to oxygen are acidic. Because of the relatively high electronegativity of oxygen, hydrogens bonded to an oxygen form a very polar bond, which can ionize in water to give acidic properties. For example, examine the Lewis electron-dot formulas for HNO_3, H_2SO_4, and CH_3CO_2H in Figure 15.5. Each hydrogen in nitric acid and sulfuric acid is bonded directly to an oxygen. Each of these hydrogen atoms can be transferred to a base when the acid is dissolved in water. In acetic acid, however, only one hydrogen atom is bonded to oxygen; the other three hydrogen atoms are bonded to carbon. In acetic acid, only the hydrogen atom bonded to oxygen is capable of exhibiting acidic behavior.

$$CH_3CO_2H(aq) + H_2O(l) \rightleftharpoons H_3O^+ + CH_3CO_2^-(aq)$$

Thus acetic acid is a monoprotic acid.

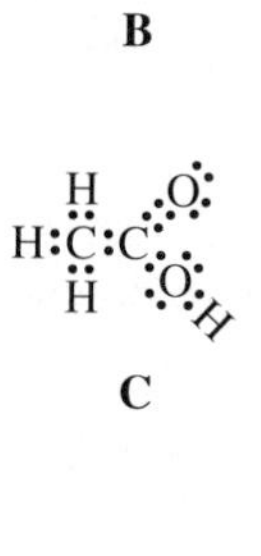

Strengths of Acids and Bases The strength of a Brønsted–Lowry acid is measured in terms of its tendency to donate protons. Correspondingly, the strength of a base reflects its tendency to accept protons from an acid. The strengths of acids and bases are important in determining what reactions will occur when they are mixed.

Normally we measure the strength, or proton-donating tendency, of an acid (or the proton-accepting tendency of a base) relative to the solvent. Substances considered to be strong acids—including HCl (hydrochloric acid), HBr (hydrobromic acid), HI (hydriodic acid), $HClO_4$ (perchloric acid), H_2SO_4 (sulfuric acid, first proton), and HNO_3 (nitric acid)—transfer their protons essentially completely when dissolved in water. The reaction for hydrochloric acid is

$$\underset{\text{acid}}{HCl(aq)} + \underset{\text{base}}{H_2O(l)} \longrightarrow \underset{\text{acid}}{H_3O^+(aq)} + \underset{\text{base}}{Cl^-(aq)}$$

The reverse reaction occurs only to a very small extent, so hydrochloric acid is a very strong acid—even stronger than the hydronium ion, itself a very strong acid. *The transfer of a proton tends to occur from a stronger acid to a stronger base, forming a weaker acid and a weaker base.* In this case, water is a stronger base than chloride ion. In fact, chloride is such a weak base in water that it exhibits essentially no basic properties (in the Brønsted–Lowry sense).

In contrast, weak acids, such as CH_3CO_2H (acetic acid), HSO_4^- (bisulfate ion), and HCN (hydrocyanic acid), only partially donate their protons to a base. That is, such reactions are dynamic equilibria, giving rise to an equilibrium mixture of conjugate acids and bases.

$$CH_3CO_2H(aq) + H_2O(l) \rightleftharpoons H_3O^+(aq) + CH_3CO_2^-(aq)$$

Figure 15.5 Lewis formulas of some oxoacids. *(A)* nitric acid *(B)* sulfuric acid *(C)* acetic acid.

Nitric acid

$$\text{H}:\ddot{\text{O}}:\text{N}::\ddot{\text{O}}:$$

A

Sulfuric acid

B

Acetic acid

C

Acetic acid is not as strong an acid as hydronium ion, so the equilibrium lies toward the reactants. Acetate ion is a stronger base than water; this is consistent with an equilibrium in which reactants predominate. Here again, the stronger acid reacts with the stronger base to form predominantly the weaker acid and weaker base—the reaction lies primarily to the left.

A ranking of some common Brønsted–Lowry acids (and their conjugate bases) in aqueous solution according to strength is given in Table 15.1. All the acids above H_3O^+ in the table are considered strong acids. All the bases above H_2O fall into the category of very weak bases, so weak that they show no basic properties in water. The strongest bases are those below OH^-; the weakest acids (which have essentially no acidic properties) are those below H_2O. The strongest acid that can exist in aqueous solution is H_3O^+, the conjugate acid of water. Any stronger acid simply reacts with the water to form H_3O^+. This phenomenon is known as the **leveling effect** of the solvent: The strongest acid that can exist in a solvent is the conjugate acid of that solvent. Similarly, the strongest base that can exist in a solvent is the conjugate base of that solvent (OH^- in the case of water).

In general, as you can see in the table, the stronger an acid, the weaker its conjugate base, and the stronger a base, the weaker its conjugate acid. This con-

How can we predict which acids and bases will react in a mixture?

Table 15.1 Relative Strengths of Some Brønsted–Lowry Acids and Their Conjugate Bases

	Acid	Base	
Strongest acids	$HClO_4$	ClO_4^-	Weakest bases
	H_2SO_4	HSO_4^-	
	HI	I^-	
	HBr	Br^-	
	HCl	Cl^-	
	HNO_3	NO_3^-	
	H_3O^+	H_2O	
	HSO_4^-	SO_4^{2-}	
	H_2SO_3	HSO_3^-	
	H_3PO_4	$H_2PO_4^-$	
	HNO_2	NO_2^-	
	HF	F^-	
	CH_3CO_2H	$CH_3CO_2^-$	
	H_2CO_3	HCO_3^-	
	H_2S	HS^-	
	NH_4^+	NH_3	
	HCN	CN^-	
	HCO_3^-	CO_3^{2-}	
	HS^-	S^{2-}	
	H_2O	OH^-	Strongest bases
Weakest acids	NH_3	NH_2^-	
	OH^-	O^{2-}	

Hydrogen sulfide is a colorless, flammable, poisonous gas with the smell of rotten eggs. It results from the decay of organic matter containing sulfur and from the reaction of metal sulfides with strong acids. In aqueous solution it forms hydrosulfuric acid, a weak diprotic acid. Hydrosulfuric acid solutions are often used to precipitate metal ions from solution.

clusion is consistent with the idea that the strength of an acid is a measure of its tendency to donate protons. If an acid has a strong tendency to donate protons, then its conjugate base, formed from the acid by the loss of protons, cannot have a strong tendency to accept protons.

As we have noted, all proton-transfer reactions occur in such a way as to form predominantly the weaker acid and the weaker base by reaction of the stronger acid with the stronger base. This rule enables us to decide what reaction is likely to occur in a mixture of acids and bases. For example, if hydrosulfuric acid, hydrogen sulfide ion, fluoride ion, and hydrofluoric acid are mixed, the following equilibrium is possible:

$$H_2S(aq) + F^-(aq) \rightleftharpoons HS^-(aq) + HF(aq)$$

To decide which set of species, the reactants or the products, predominates in such a mixture, we examine their relative acid and base strengths. According to Table 15.1, H_2S is a weaker acid than HF, and HS^- is a stronger base than F^-. The stronger acid and the stronger base should react to form the weaker acid and the weaker base. Thus the reverse reaction should predominate.

$$HS^-(aq) + HF(aq) \longrightarrow H_2S(aq) + F^-(aq)$$

EXAMPLE 15.1

For the reaction

$$HSO_4^-(aq) + HS^-(aq) \rightleftharpoons SO_4^{2-}(aq) + H_2S(aq)$$

which species, the reactants or the products, will predominate at equilibrium?

Solution   The two acids in this reaction are HSO_4^- and H_2S. Of these, HSO_4^- is the stronger. Of the two bases, HS^- is stronger than SO_4^{2-}. The stronger acid reacts predominantly with the stronger base to form the weaker acid and the weaker base. Thus the products predominate.

Practice Problem 15.1

For the reaction

$$HNO_2(aq) + NO_3^-(aq) \rightleftharpoons NO_2^-(aq) + HNO_3(aq)$$

which species, the reactants or the products, will predominate at equilibrium?

Answer: reactants

Oxides and Hydroxides   Substances containing hydrogen atoms are not the only compounds that show acidic or basic properties. Oxides of the elements have acid–base properties in the sense that they can neutralize Brønsted–Lowry acids or bases. For example, sulfur trioxide has acidic properties, because it reacts with a base in a neutralization reaction.

$$Ca(OH)_2(s) + SO_3(g) \longrightarrow CaSO_4(s) + H_2O(l)$$

Outline a procedure for finding the formula of an oxoacid or hydroxide base from the formula of an anhydride.

Calcium oxide, on the other hand, has basic properties because it reacts with an acid in a neutralization reaction.

$$CaO(s) + H_2SO_4(aq) \longrightarrow CaSO_4(s) + H_2O(l)$$

Phosphoric acid, H_3PO_4, exists as unstable colorless crystals that melt at 42°C or as a clear, syrupy liquid. It is commonly sold as a concentrated aqueous solution. Phosphoric acid is used in the manufacture of fertilizers, phosphate salts, and polyphosphates for detergents. It is used to provide acidity and flavor in carbonated soft drinks.

The oxides can be considered **anhydrides**—forms not containing water—of the oxoacids and hydroxide bases. Accordingly, they undergo reactions analogous to those characteristic of the corresponding acids or bases. How do we recognize the correspondence between the acids and bases and their anhydrides? We can remove the components of water (two hydrogens and one oxygen) from the acid or base until no hydrogen remains; the resulting formula is that of the acid or base anhydride. For example, if one H_2O is removed from sulfuric acid, sulfur trioxide results.

$$H_2SO_4 - H_2O = SO_3$$

Similarly, the removal of H_2O from calcium hydroxide gives calcium oxide.

$$Ca(OH)_2 - H_2O = CaO$$

For an acid or base that contains the components of more than one H_2O unit, we continue to remove H_2O from the formula until no hydrogen atoms remain. With phosphoric acid, upon removal of one H_2O from the formula, we get

$$H_3PO_4 - H_2O = HPO_3$$

A hydrogen remains, so we must remove more H_2O to get the formula of the oxide. Because this formula has only one hydrogen, we must remove H_2O from two formula units.

$$2HPO_3 - H_2O = P_2O_5$$

This is the simplest formula of the anhydride of phosphoric acid. To find the formula of the oxoacid of a nonmetal oxide, we perform this process in reverse.

EXAMPLE 15.2

Predict the formulas of the anhydrides of the following substances: a. H_3AsO_4; b. $Al(OH)_3$.

Solution In each case, we remove two hydrogens and one oxygen from the formula successively until no hydrogens remain. If there are an odd number of hydrogens in the formula, we must combine two molecules of the acid or base to obtain the formula of the anhydride.

a. $H_3AsO_4 - H_2O = HAsO_3$

Because one hydrogen remains, one more H_2O unit must be removed from two of these formulas.

$$2HAsO_3 - H_2O = As_2O_5$$

b. $Al(OH)_3 - H_2O = AlO(OH)$

Again there is an odd number of hydrogens, so one more molecule of water must be removed from two of the last species.

$$2AlO(OH) - H_2O = Al_2O_3$$

Practice Problem 15.2

Predict the formulas of the anhydrides of the following substances: a. $Th(OH)_4$; b. $Fe(OH)_3$; c. H_3PO_3

Answer: a. ThO_2, b. Fe_2O_3, c. P_2O_3

How do we recognize acids and bases among the oxides?

Metal oxides are generally bases, and nonmetal oxides are generally acids. These oxides exhibit the usual properties of acids and bases, especially their ability to neutralize one another by chemical reactions. Earlier, we saw that sulfur trioxide reacts with calcium hydroxide. This reaction is analogous to the acid–base neutralization reaction between sulfuric acid and calcium hydroxide.

$$Ca(OH)_2(aq) + H_2SO_4(aq) \longrightarrow CaSO_4(s) + 2H_2O(l)$$
$$Ca(OH)_2(aq) + SO_3(g) \longrightarrow CaSO_4(s) + H_2O(l)$$

We also saw that calcium oxide reacts with sulfuric acid. This reaction is also analogous to that of calcium hydroxide with sulfuric acid.

$$CaO(s) + H_2SO_4(aq) \longrightarrow CaSO_4(s) + H_2O(l)$$

Acid–base neutralization reactions also occur between metal oxides and nonmetal oxides, forming salts.

$$CaO(s) + SO_3(g) \longrightarrow CaSO_4(s)$$

All these reactions differ only in the amount of water produced. The reaction of oxides is an acid–base neutralization reaction, even though it does not involve a proton transfer and thus does not fit the Brønsted–Lowry definition (because the basic properties of calcium oxide and the acidic properties of sulfur trioxide have disappeared).

RULES OF REACTIVITY

A metal oxide reacts with a nonmetal oxide to form a metal oxoanion salt.

The oxides are the anhydrides of acids and bases, so it seems logical that we can generate normal Brønsted–Lowry acids and bases by dissolving the oxides in water. In fact, many oxides react with water in hydration reactions to form bases or acids. Metal oxides react with water to form bases.

$$MgO + H_2O \longrightarrow Mg(OH)_2$$

Nonmetal oxides react with water to form oxoacids.

$$N_2O_5 + H_2O \longrightarrow 2HNO_3$$

Why should these substances react in this way? To explain their behavior, consider a general reaction for the hydration of an oxide of an element, E, given by the following equation:

$$E-O + H_2O \longrightarrow E-O-H \quad \text{(not balanced)}$$

The O—H linkage is characteristic of both hydroxides and oxoacids. The acid–base behavior of these species arises from the breaking of one of the bonds, $E-O$ or $O-H$, upon dissolution in water.

$$E-O-H(aq) \rightleftharpoons E-O^-(aq) + H^+(aq) \qquad \text{acid}$$
$$E-O-H(aq) \rightleftharpoons E^+(aq) + OH^-(aq) \qquad \text{base}$$

Some species may undergo both types of ionization and so exhibit amphoteric behavior.

Figure 15.6 shows the acid–base properties of the oxides of the elements as a function of their position in the periodic table. If the element E has a low electronegativity, the $E-O$ bond will have a high degree of ionic character, and the substance will be primarily basic. Thus the metal oxides tend to be basic. If the element E is highly electronegative, then electron density is moved toward E in $E-O-H$ and away from H, so the O—H bond becomes very polar, and the substance is acidic. The nonmetal oxides tend to be acidic. The general tendency is

Legend:
A = acid
B = base
Am = amphoteric

VS = very strong
S = strong
M = moderate
W = weak
VW = very weak

1	2	3	4	5	6	7	8	9	10	11	12	13	14	15	16	17	18
H_2O																	
Li_2O VSB	BeO Am											B_2O_3 VWA	CO_2 WA; CO none	N_2O_5 SA; N_2O_4 WA; N_2O_3 WA			
Na_2O VSB	MgO WB											Al_2O_3 Am	SiO_2 VWA	P_4O_{10} MA; P_4O_6 MA	SO_3 SA; SO_2 WA	Cl_2O_7 VSA; ClO_3 SA; ClO_2 WA; Cl_2O VWA	
K_2O VSB	CaO MB	Sc_2O_3 WB	TiO_2 WA; Ti_2O_3 B; TiO B	V_2O_5 AmA; VO_2 Am; V_2O_3 B	CrO_3 SA; Cr_2O_3 AmB	Mn_2O_7 SA; MnO_2 AmA; Mn_2O_3 WB; MnO WB	Fe_2O_3 AmB; FeO AmB	CoO_2 WA; CoO AmB	NiO B	CuO AmB; Cu_2O Am	ZnO Am	Ga_2O_3 AmA	GeO_2 AmA; GeO VWA	As_4O_{10} MA; As_4O_6 AmA	SeO_3 SA; SeO_2 WA	Br_2O_7 SA; Br_2O WA	
Rb_2O VSB	SrO SB	Y_2O_3 MB	ZrO_2 AmB	Nb_2O_5 AmA	MoO_3 MA	Tc_2O_7 SA			PdO B	Ag_2O AmB	CdO AmB	In_2O_3 AmB	SnO_2 AmA; SnO Am	Sb_4O_{10} AmA; Sb_4O_6 Am-WA	TeO_3 WA; TeO_2 Am	I_2O_5 SA	
Cs_2O VSB	BaO SB	La_2O_3 MB–SB	HfO_2 AmB	Ta_2O_5 AmA; TaO_2 A	WO_3 A	Re_2O_7 SA; ReO_2 A		IrO_2 B	PtO_2 Am; Pt_2O_3 WA	Au_2O_3 AmA	HgO WB	Tl_2O_3 VWB; Tl_2O SB	PbO_2 AmA; PbO AmB	Bi_2O_3 WB			

Figure 15.6 Acid–Base Properties of Oxides

for the acidic character of the oxides to increase across the periodic table from left to right and from the bottom of the table to the top, paralleling the trend in electronegativity of the elements.

Lewis Theory of Acids and Bases

One other theory of acids and bases is worth considering at this time. It is that of G. N. Lewis, also known for the Lewis or electron-dot structures used to explain bonding in compounds. This theory brings the acid–base concept another step

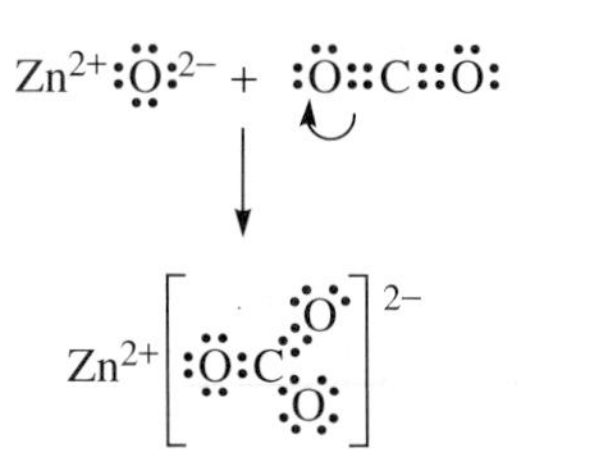

$Zn^{2+}\,:\!\ddot{O}\!:^{2-} + \;:\!\ddot{O}\!::C\!::\!\ddot{O}:$

$Zn^{2+}\left[\,:\!\ddot{O}\!:C\!\begin{smallmatrix}\ddot{O}\cdot\\ \ddot{O}:\end{smallmatrix}\right]^{2-}$

Figure 15.7 The reaction of a metal oxide with a nonmetal oxide involves the donation, to the nonmetal, of a share in an electron pair from the oxygen in the metal oxide.

Zinc oxide, ZnO, is also known as flowers of zinc, philosopher's wool, and zinc white. It is a white solid that occurs naturally as the mineral zincite *and can also be formed by the reaction of vaporized metal with hot air. It is used as a pigment in white paints, cosmetics, whitewall tires, white inks, and porcelains.*

closer to being universal—it includes even more compounds than the other theories. Lewis started with the Brønsted–Lowry concept but examined the sharing of electrons rather than the transfer of a proton. Consider the following reaction:

$$H_3N:(aq) + H_3O^+(aq) \rightleftharpoons H_3N:H^+(aq) + H_2O(aq)$$
$$\text{base} \qquad\quad \text{acid} \qquad\qquad\quad \text{acid} \qquad\quad \text{base}$$

Here a proton is transferred from hydronium ion to ammonia. At the same time, the proton in the hydronium ion is accepting a share in the electron pair on the nitrogen in ammonia, and that nitrogen atom is donating a share in its pair of electrons.

Embracing the viewpoint that acid–base reactions involve the donation and acceptance of a share in a pair of electrons, Lewis proposed that such reactions need not be limited to protons. A **Lewis acid,** then, is any species that will accept a share in a pair of electrons (an electron-pair acceptor). A **Lewis base** is any species that will share its electron pair with another substance (an electron-pair donor). These definitions, of course, include substances that transfer protons. However, they also include substances, such as oxides, that do not have protons. For example, zinc oxide reacts with carbon dioxide to form zinc carbonate according to the following equation:

$$ZnO(s) + CO_2(g) \rightleftharpoons ZnCO_3(s)$$

As can be seen in the Lewis formulas shown in Figure 15.7, this reaction involves the donation of a share in a pair of electrons on the oxide ion to the carbon in carbon dioxide.

The general form of a Lewis acid–base reaction is

$$\text{Acid} + \text{base} \rightleftharpoons \text{compound containing a coordinate covalent bond}$$

The formation of a *coordination compound* (sometimes called an **adduct**) is a Lewis acid–base neutralization reaction. A coordination compound, you may recall, is a compound that contains a coordinate covalent bond—a covalent bond in which both electrons were donated by one of the atoms. An example is the reaction between boron trifluoride and ammonia.

$$BF_3 + :NH_3 \rightleftharpoons F_3B:NH_3$$
$$\text{acid} \quad\;\; \text{base} \qquad\quad \text{adduct}$$

Proton-transfer reactions can be represented in the same way.

$$H^+ + :NH_3 \rightleftharpoons H:NH_3^+$$
$$\text{acid} \quad\;\; \text{base} \qquad\quad \text{adduct}$$

EXAMPLE 15.3

Identify the Lewis acid and base in the reaction

$$Cu^{2+}(aq) + 4NH_3(aq) \rightleftharpoons Cu(NH_3)_4^{2+}(aq)$$

Solution Each of the ammonia molecules has an unshared pair of electrons on the nitrogen, $:NH_3$. The copper ion can accept these electrons to form coordinate covalent bonds, $Cu:NH_3$. Thus the copper ion is an electron-pair acceptor, or acid, and the ammonia is an electron-pair donor, or base. The product, $Cu(NH_3)_4^{2+}$, is an adduct or coordination compound.

Practice Problem 15.3

Identify the Lewis acid and base in the reaction

$$Cr(s) + 6CO(g) \rightleftharpoons Cr(CO)_6(s)$$

Answer: Cr is an acid; CO is a base.

15.2 Strong Acids, Strong Bases, and the Ionization of Water

How can we determine the extent to which an acid donates its protons to solvent water?

The strengths of acids and bases, as we have said, are usually measured relative to the strength of the solvent. We can quantify these strengths by determining the percent ionization. This section examines ways to measure the extent of ionization of acids and bases relative to that of water.

Water, a covalent compound, ionizes to a small extent. This process is called **autoionization,** or self-ionization, because the water is the source of both the acidic proton and the basic hydroxide ion.

$$H_2O(l) + H_2O(l) \rightleftharpoons H_3O^+(aq) + OH^-(aq)$$

Autoionization is an equilibrium process, so it is possible to write an equilibrium-constant expression for it. As usual, the pure liquid is not included in the expression because its concentration is constant. This equilibrium constant, called the **ion-product constant of water,** is so widely used that it is given a special symbol, K_w.

$$K_w = [H_3O^+][OH^-]$$

At 25°C, K_w has a value of 1.008×10^{-14}. It is common to round this value to two or three significant digits.

$$K_w = [H_3O^+][OH^-] = 1.0 \times 10^{-14}$$

In pure water, with no other source of aqueous hydrogen ions (hydronium ions) or hydroxide ions, the concentrations of these ions must be equal.

$$[H_3O^+] = [OH^-]$$

Consider the equation for the ion-product constant:

$$[H_3O^+][OH^-] = 1.0 \times 10^{-14}$$

We can substitute the equality between the two concentrations into this equation:

$$[H_3O^+]^2 = 1.0 \times 10^{-14}$$

and solve for the hydronium-ion concentration, which equals the hydroxide-ion concentration.

$$[H_3O^+] = [OH^-] = 1.0 \times 10^{-7} \, M$$

Thus, in pure water, the concentration of hydronium ion and the concentration of hydroxide ion are both equal to $1.0 \times 10^{-7} \, M$ at 25°C. At other temperatures, the value of K_w is different, so the concentrations of $H_3O^+(aq)$ and of $OH^-(aq)$ have other values, but they are still equal to each other. For example, at 100°C, $K_w = 5.00 \times 10^{-13}$, and $[H_3O^+] = [OH^-] = 7.07 \times 10^{-7} \, M$.

Acidic, Basic, and Neutral Solutions

Pure water, as well as solutions containing only dissolved substances that have no acidic or basic properties, are neutral; hydronium ions and hydroxide ions come only from the autoionization of water, and $[H_3O^+] = [OH^-] = 1.0 \times 10^{-7}\ M$ at 25°C. When acidic or basic substances are dissolved in water, these concentrations are no longer identical, nor are they equal to $1.0 \times 10^{-7}\ M$. If the solution is acidic, the concentration of hydronium ion is greater than $10^{-7}\ M$, and the hydroxide-ion concentration is smaller than $10^{-7}\ M$. If the solution is basic, the reverse is true. But the concentrations of these two ions are still related by the ion-product expression for water, whenever the solution reaches a state of equilibrium.

We can calculate just what these concentrations are in much the same way we calculated the concentrations of $H_3O^+(aq)$ and $OH^-(aq)$ in pure water. If the acid or base is strong (if it ionizes completely) and the solution is not too dilute, the calculation is straightforward. In this case, two assumptions are made: (1) The acid or base is completely ionized. (2) The amount of $H_3O^+(aq)$ or $OH^-(aq)$ arising from the ionization of water can be neglected. The general approach assumes that the concentration of $H_3O^+(aq)$ equals the concentration of the strong acid. Then the concentration of $OH^-(aq)$ can be calculated from the ion-product expression.

$$K_w = [H_3O^+][OH^-] = 1.0 \times 10^{-14}$$

As we have said, this approach assumes that autoionization does not contribute significantly to the concentration of $H_3O^+(aq)$. We can check this assumption easily. Because the amount of $H_3O^+(aq)$ that would come from autoionization of water must equal the amount of $OH^-(aq)$ from that source, and because there is no other source of $OH^-(aq)$, the assumption must be valid if the concentration of $OH^-(aq)$ is much smaller than the total concentration of $H_3O^+(aq)$.

EXAMPLE 15.4

What are the concentrations of $H_3O^+(aq)$ and $OH^-(aq)$ in a solution prepared from 0.0050 mol HCl dissolved in 1.00 L H_2O?

Solution   HCl is a strong acid, so it is completely ionized in water, giving a hydronium-ion concentration arising from HCl of 0.0050 mol/1.00 L, or

$$[H_3O^+] = 5.0 \times 10^{-3}\ M$$

The autoionization of water produces H_3O^+ and OH^-, but we will assume that the amount of $H_3O^+(aq)$ coming from autoionization can be neglected. The hydroxide-ion concentration is calculated from the ion-product expression.

$$[H_3O^+][OH^-] = 1.0 \times 10^{-14}$$

Inserting the value of the hydronium-ion concentration, we get the equation

$$(5.0 \times 10^{-3})[OH^-] = 1.0 \times 10^{-14}$$

We can rearrange this equation and solve for the concentration of hydroxide ion.

$$[OH^-] = (1.0 \times 10^{-14})/(5.0 \times 10^{-3}) = 2.0 \times 10^{-12}\ M$$

This concentration is small compared to the hydronium-ion concentration, so our assumption was correct.

Practice Problem 15.4

What are the concentrations of $H_3O^+(aq)$ and $OH^-(aq)$ in a solution prepared from 0.0040 mol HCl dissolved in 2.00 L of water?

Answer: 2.0×10^{-3} M $H_3O^+(aq)$; 5.0×10^{-12} M $OH^-(aq)$

We can calculate the concentrations of $H_3O^+(aq)$ and $OH^-(aq)$ in solutions of strong bases in exactly the same way we just calculated their concentrations in solutions of strong acids. In this case, the $OH^-(aq)$ concentration is just the concentration of the strong base, and the $H_3O^+(aq)$ concentration is obtained from the ion-product constant of water.

In these calculations, we assumed that in a solution of a strong acid, the amount of $H_3O^+(aq)$ arising from the ionization of water could be neglected. Similarly, the $OH^-(aq)$ from the autoionization of water was neglected in a solution of a strong base. However, this assumption is not valid for very dilute solutions of acids or bases. When the concentration of acid or base becomes less than 10^{-6} M, the contribution from the autoionization of water becomes significant.

The pH, pOH, and Other pX Scales

We have been expressing the acidity of substances in terms of the $H_3O^+(aq)$ concentration, in moles per liter. Because the acidity of many substances is quite small, it is often more convenient to express the acidity of aqueous solutions on a *pH scale*. The **pH** of a solution is defined as the negative logarithm (base 10) of the hydronium-ion concentration.

$$pH = -\log [H_3O^+]$$

These quantities are also related by the inverse relationship.

$$[H_3O^+] = 10^{-pH}$$

For example, if $[H_3O^+] = 1.00 \times 10^{-2}$ M, then the pH is 2.000.

$$pH = -\log [H_3O^+]$$
$$= -\log(1.00) + -\log(10^{-2}) = -0.000 + 2 = 2.000$$

The larger the value of the pH, the less $H_3O^+(aq)$ exists in solution. Figure 15.8 gives the pH of various solutions formed from common substances.

We can also define a *pOH scale*. The pOH of a solution is the negative logarithm of the hydroxide-ion concentration.

$$pOH = -\log [OH^-]$$

The pH and pOH values for a given solution are related by the ion-product expression for water.

$$K_w = [H_3O^+][OH^-] = 1.0 \times 10^{-14}$$

Taking the negative logarithm of both sides of the equation gives

$$-\log([H_3O^+][OH^-]) = -\log(1.0 \times 10^{-14})$$
$$-\log [H_3O^+] - \log [OH^-] = -\log(1.0 \times 10^{-14})$$

The pH concept was originated in 1909 by the Danish biochemist S. P. L. Sørensen while he was working on the control of acidity during the brewing of beer. The p in pH stands for puissance *(in French),* potenz *(in German), or* power *in English; the H stands for hydrogen ion.*

Figure 15.8 The pH values of these common substances range from 0 (very acidic) to 14 (very basic).

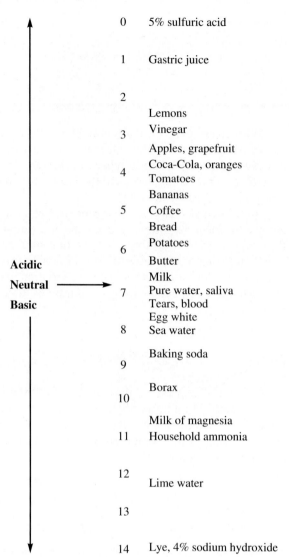

0	5% sulfuric acid
1	Gastric juice
2	
	Lemons
3	Vinegar
	Apples, grapefruit
	Coca-Cola, oranges
4	Tomatoes
	Bananas
5	Coffee
	Bread
	Potatoes
6	Butter
	Milk
7	Pure water, saliva
	Tears, blood
	Egg white
8	Sea water
	Baking soda
9	
	Borax
10	
	Milk of magnesia
11	Household ammonia
12	
	Lime water
13	
14	Lye, 4% sodium hydroxide

Acidic

Neutral

Basic

Because $pH = -\log [H_3O^+]$ and $pOH = -\log [OH^-]$, this relationship becomes

$$pH + pOH = 14.00$$

With a pH of 2.00, the pOH is

$$pOH = 14.00 - pH = 14.00 - 2.00 = 12.00$$

Both pH and pOH values correlate with the acidity of a solution. A neutral solution has a pH and a pOH value of 7.00; in such a solution, $[H_3O^+] = [OH^-] = 1.0 \times 10^{-7}$. If the pH is less than 7, or the pOH is greater than 7, the solution is acidic. Conversely, a basic solution has a pH greater than 7 and a pOH less than 7.

The use of significant figures can become confusing in work involving logarithms. A logarithm consists of two parts, the mantissa and the characteristic. The *mantissa* is the numerical part, and the *characteristic* is the whole-number power of 10. The characteristic is an exact number, because it simply indicates the position of the decimal point in the number. The digits in front of the decimal point in the logarithm are exact digits and are not considered when counting significant figures. Thus a pH of 2.00 has two significant figures, arising from the two significant figures in the concentration, 1.0×10^{-2}, from which it was calculated.

The "p" scales are used for other purposes besides describing the acidity of solutions. A general definition of such scales is that they involve the negative logarithm of whatever quantity is involved.

$$pX = -\log X$$

Thus, for example, we can express the values of equilibrium constants in terms of pK. If $K = 1.0 \times 10^6$, then $pK = -6.00$, and if $K = 5 \times 10^{-4}$, then $pK = 3.3$.

EXAMPLE 15.5

What are the pH and the pOH of a 0.050 M solution of HCl?

Solution   Recall that for solutions of strong acids that are more concentrated than 10^{-7} M, the $H_3O^+(aq)$ concentration is considered equal to the concentration of the acid—in this case, 0.050 M.

$$[H_3O^+] = 0.050 \ M$$

With solutions of this concentration, it is safe to assume that the amount of H_3O^+ from water is negligible. The pH is then calculated as follows:

$$pH = -\log [H_3O^+] = -\log(5.0 \times 10^{-2}) = 1.30$$

Because pH + pOH = 14.00, we simply subtract the value of pH from 14.00 to find pOH.

$$pOH = 14.00 - 1.30 = 12.70$$

Practice Problem 15.5

What are the pH and the pOH of a 0.0020 M solution of HCl?

Answer: pH = 2.70; pOH = 11.30

Calculation of the pH of a solution of a base is simpler when the pOH is calculated first, because the pOH can be directly related to the hydroxide-ion concentration. The pH is then calculated by subtracting the pOH from 14.00.

EXAMPLE 15.6

What is the pH of a 0.0020 M $Ba(OH)_2$ solution?

Solution $Ba(OH)_2$ is an ionic compound and a strong base, so it is completely dissociated into Ba^{2+} and OH^-. The concentration is so much greater than 10^{-7} that the hydroxide ion arising from ionization of water can be ignored. The $OH^-(aq)$ concentration is then equal to twice the $Ba(OH)_2$ concentration, because two $OH^-(aq)$ are produced per $Ba(OH)_2$ that dissociates.

$$[OH^-] = 2 \times 0.0020 = 0.0040 \ M, \ or \ 4.0 \times 10^{-3} \ M$$

(continued)

The pOH is calculated from this concentration.

$$pOH = -\log [OH^-] = -\log(4.0 \times 10^{-3}) = 2.40$$

The pH is obtained by subtraction of pOH from 14.00.

$$pH + pOH = 14.00$$

$$pH = 14.00 - 2.40 = 11.60$$

Practice Problem 15.6

What is the pH of a 0.015 M NaOH solution?

Answer: 12.18

Although it is convenient to express the acid or base concentration of a solution in terms of its pH or pOH value, the concentration values in units of molarity are often necessary in numerical applications. In such cases, we can calculate the concentrations from the pH or pOH values. This conversion involves taking the antilogarithm of the pH or pOH.

EXAMPLE 15.7

The pH of a solution is 3.60. What is the $H_3O^+(aq)$ concentration?

Solution   Because pH $= -\log [H_3O^+]$, $[H_3O^+] = 10^{-pH}$.

$$[H_3O^+] = 10^{-3.60} = 2.5 \times 10^{-4} \, M$$

Practice Problem 15.7

The pH of a solution is 5.65. What is the $H_3O^+(aq)$ concentration?

Answer: $2.2 \times 10^{-6} \, M$

Measurement of pH

We have considered methods for expressing the acidity or basicity of solutions. But how do we determine this property experimentally? There are several methods for determining the pH of a solution. One of the most familiar techniques for measuring pH is the litmus paper test. An acidic solution turns blue litmus red, and a basic solution turns red litmus blue, as shown in Figure 15.9. However, this very elementary and inaccurate measurement determines only whether the pH is greater or less than 7. Other pH-indicating papers enable us to estimate pH to about ±0.5 to 1 unit. Mixtures of indicators, which are often colored plant extracts, change color with pH, as shown in Figure 15.10, and can be used to determine pH colorimetrically. Several instrumental colorimetric methods and electrical methods are also available. In one electrical method, a voltage proportional to the pH develops when appropriate electrodes are dipped into a solution, and the pH is displayed on a pH meter (Figure 15.11). For solutions of fairly high concentration, the most

Figure 15.9 Red litmus paper turns blue when wet with a basic solution *(Left)*, and blue litmus paper turns red when wet with an acidic solution *(Right)*.

accurate technique for determining the acid or base concentration is titration, which was discussed in Chapter 4.

pH and Body Chemistry

▶ As we noted in the chapter introduction, acidity is important for proper plant growth. It is also important in animals, including humans. For example, our body chemistry normally controls the acidity of blood within a very narrow range. Serious illness and even death may result if the blood acidity varies by more than about 0.5 units from the normal pH of 7.3–7.5. Acidity greater than about 5×10^{-8} M $H^+(aq)$ (pH less than 7.3) results in a condition known as acidosis, whereas acidity less than about 3.5×10^{-8} M $H^+(aq)$ (pH greater than 7.45) leads to alkalosis. Acidosis may result from hypoventilation, diabetes mellitus, or ingestion of acidic substances. Alkalosis may be caused by hyperventilation, excitement, trauma, high temperatures, ingestion of basic substances, kidney disease, vomiting, or diarrhea.

Figure 15.10 Solutions of different pH create different colors in this mixture of plant extracts, which is known as a universal indicator.

Figure 15.11 A pH meter determines the pH of a solution
by measuring the voltage that
develops when electrodes are
dipped into the solution.

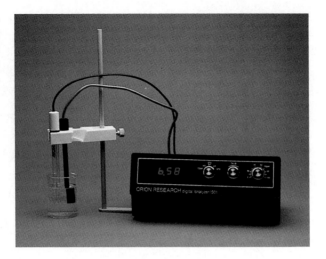

If the acidity of stomach acid increases much above the normal value, about 0.1 M $H^+(aq)$, the result is an acid stomach, with accompanying symptoms of heartburn and indigestion. Many over-the-counter antacids are used to treat these symptoms. However, most of these products are based on a relatively small number of antacid ingredients, all of which act by neutralizing stomach acid. Common antacid ingredients include sodium hydrogen carbonate (sodium bicarbonate), calcium carbonate, dihydroxyaluminum sodium carbonate, aluminum hydroxide, magnesium hydroxide, magnesium oxide, magnesium carbonate, magnesium trisilicate, sodium tartrate, and sodium citrate. The neutralizing chemistry of some of these compounds, illustrated in Figure 15.12, is summarized by the following equations:

$$CaCO_3(s) + 2HCl(aq) \longrightarrow CaCl_2(aq) + H_2O(l) + CO_2(g)$$
$$NaAl(OH)_2CO_3(s) + 4HCl(aq) \longrightarrow NaCl(aq) + AlCl_3(aq) + 3H_2O(l) + CO_2(g)$$
$$Mg(OH)_2(s) + 2HCl(aq) \longrightarrow MgCl_2(aq) + 2H_2O(l)$$
$$Mg_2Si_3O_8(s) + 4HCl(aq) \longrightarrow 2MgCl_2(aq) + 2H_2O(l) + 3SiO_2(s)$$

Figure 15.12 All antacids work
by neutralizing stomach acid,
roughly 0.1 M HCl. Here a suspension of magnesium hydroxide is neutralizing a
hydrochloric acid solution.

15.3   Weak Acids

Strong acids ionize completely into their component ions when dissolved in aqueous solution. Because of this, as we have seen, calculating the $H^+(aq)$ concentration in solutions of strong acids is relatively straightforward. However, there are not many strong acids. ▶ Most acids, including most of the household chemicals mentioned in the chapter introduction, are weak and do not ionize completely into their component ions when dissolved in solution. An example is nitrous acid, HNO_2. When HNO_2 dissolves in water, it exists in equilibrium with H_3O^+ and NO_2^- ions.

$$HNO_2(aq) + H_2O(l) \rightleftharpoons H_3O^+(aq) + NO_2^-(aq)$$

How can we succinctly express the strength of a weak acid?

This section considers ways of measuring and expressing the strengths of weak acids, trends in these strengths, and methods to determine the concentrations of all species in a solution of a weak acid.

Acid-Ionization Constants

By following the procedures developed for gas-phase equilibria and for the auto-ionization of water, we can write an equilibrium-constant expression for the ionization of a weak acid. The equilibrium constant, K_a, is called the **acid-ionization constant.** Consider the equation for ionization of nitrous acid:

$$HNO_2(aq) + H_2O(l) \rightleftharpoons H_3O^+(aq) + NO_2^-(aq)$$

The expression for the equilibrium constant is

Nitrous acid, formed by the reaction of strong acids with nitrite salts, is known only in solution, where it is pale blue. It is a weak acid that decomposes in aqueous solution, forming nitrogen monoxide and nitric acid. It forms stable, water-soluble nitrite salts with the alkali metals and the alkaline earth metals.

$$K_a = \frac{[H_3O^+][NO_2^-]}{[HNO_2]}$$

As usual, in developing the expression, we omit the concentration of pure water. For HNO_2 at 25°C, K_a equals 4.6×10^{-4}. The value of the acid-ionization constant is a measure of the strength of the acid—the larger the value, the stronger the acid. The acid-ionization constants for a variety of weak acids are given in Table 15.2. These constants fall in a rather wide range of values, which reflect the tendency for the various acids to produce H_3O^+ in aqueous solution, as was summarized in Table 15.1.

Acid Strength and the Structure of Acids

The strength of an acid is a measure of its tendency to ionize and is reflected in the magnitude of K_a. The larger the value of K_a, the more the acid ionizes in solution. The strength of an acid depends on the strength of the bond to hydrogen, which in turn depends on the size and electronegativity of the atom to which the proton is attached as well as on the oxidation numbers of the atoms in the molecule.

Oxoacids   Let's consider the oxoacids, wherein a central atom is surrounded by oxygen atoms and at least some of the oxygen atoms are bonded to hydrogen atoms. In a series of oxoacids that have the same central atom, the acid strength

Table 15.2 Ionization
Constants of Some Weak Acids
at 25°C

Name	Formula	K_{a1}	K_{a2}	K_{a3}
Acetic	CH_3COOH	1.75×10^{-5}		
Arsenic	H_3AsO_4	6.0×10^{-3}	1.05×10^{-7}	3.0×10^{-12}
Arsenious	H_3AsO_3	6.0×10^{-10}	3.0×10^{-14}	
Benzoic	C_6H_5COOH	6.14×10^{-5}		
Boric	H_3BO_3	5.83×10^{-10}		
Butanoic	C_3H_7COOH	1.51×10^{-5}		
Carbonic	H_2CO_3	4.45×10^{-7}	4.7×10^{-11}	
Chloric	$HClO_3$	5×10^2 (strong)		
Chloroacetic	$ClCH_2COOH$	1.36×10^{-3}		
Chlorous	$HClO_2$	1.1×10^{-2}		
Formic	$HCOOH$	1.77×10^{-4}		
Hydrazoic	HN_3	1.90×10^{-5}		
Hydrocyanic	HCN	4.0×10^{-10}		
Hydrofluoric	HF	7.0×10^{-4}		
Hydrogen peroxide	H_2O_2	2.7×10^{-12}		
Hydrosulfuric	H_2S	5.7×10^{-8}	1.2×10^{-13}	
Hypochlorous	$HOCl$	3.0×10^{-8}		
Iodic	HIO_3	1.67×10^{-1}		
Nitrous	HNO_2	4.6×10^{-4}		
Oxalic	$HOOCCOOH$	5.90×10^{-2}	6.40×10^{-5}	
Periodic	H_5IO_6	2.4×10^{-2}	5.0×10^{-9}	
Phenol	C_6H_5OH	1.00×10^{-10}		
Phosphoric	H_3PO_4	7.52×10^{-3}	6.23×10^{-8}	2.2×10^{-13}
Phosphorous	H_3PO_3	1.00×10^{-2}	2.6×10^{-7}	
Propanoic	C_2H_5COOH	1.34×10^{-5}		
Sulfamic	H_2NSO_3H	1.03×10^{-4}		
Sulfuric	H_2SO_4	Strong	1.20×10^{-2}	
Sulfurous	H_2SO_3	1.72×10^{-2}	6.43×10^{-8}	
Trichloroacetic	Cl_3CCOOH	1.29×10^{-4}		

Perchloric acid, $HClO_4$, is a strong acid. When anhydrous, it is a colorless, volatile, very hygroscopic liquid that explodes readily in the presence of reducing agents, such as organic material. Because of its instability in the pure state, it is usually sold as a 60–70% by mass solution in water.

increases with increasing oxidation number of the central atom. This is reflected in the chlorine oxoacids, for example, by the increasing values of the acid-ionization constants.

Oxoacid	$HOCl$	< $HClO_2$	< $HClO_3$	< $HClO_4$
Oxidation number of Cl	+1	+3	+5	+7
K_a	3.0×10^{-8}	1.1×10^{-2}	5×10^2	$\approx 10^8$

This effect shows periodic trends. The acid strength of oxoacids generally increases as the central atom changes from left to right and from bottom to top in the periodic table. For example, compare the strengths of the following acids, in which the central atom varies from left to right across a period.

$$
\begin{array}{cccc}
 & H_3BO_3 & < \ H_2CO_3 & < \ HNO_3 \\
K_a & 5.83 \times 10^{-10} & 4.45 \times 10^{-7} & \text{very large}
\end{array}
$$

We observe a similar effect when we go from lower to higher elements in a group.

$$
\begin{array}{cccc}
 & H_3AsO_4 & < \ H_3PO_4 & < \ HNO_3 \\
K_a & 6.0 \times 10^{-3} & 7.52 \times 10^{-3} & \text{very large}
\end{array}
$$

These variations in strength parallel the changes in electronegativity of the central atom. Recall that electronegativity increases from bottom to top and from left to right in the periodic table. The greater its electronegativity, the more the central atom (X) draws to itself the electron density associated with the covalent bonds, $X : O : H$. Consequently, the bond to the H atom becomes weaker and the proton is more readily ionized from the OH group.

The attraction of electrons to the central atom is also responsible for the effect of oxidation number on acid strength. The greater the oxidation number, the stronger the attraction for electrons, causing electrons to shift from the $O : H$ bond, making this bond more ionic.

Polyprotic Acids Another trend can be observed with polyprotic acids, which contain more than one acidic hydrogen per molecule. Consider phosphoric acid, which has three protons that can be donated to a base.

$$
\begin{array}{ll}
H_3PO_4(aq) + H_2O(l) \rightleftharpoons H_3O^+(aq) + H_2PO_4^-(aq) & K_{a1} = 7.52 \times 10^{-3} \\
H_2PO_4^-(aq) + H_2O(l) \rightleftharpoons H_3O^+(aq) + HPO_4^{2-}(aq) & K_{a2} = 6.23 \times 10^{-8} \\
HPO_4^{2-}(aq) + H_2O(l) \rightleftharpoons H_3O^+(aq) + PO_4^{3-}(aq) & K_{a3} = 2.2 \times 10^{-13}
\end{array}
$$

The ionization constants for the successive losses of protons decrease by a factor of very close to 10^5 in this example and for many other polyprotic acids. There are exceptions, but this rule is useful for making estimates when experimental values of acid-ionization constants are not available. The trend is consistent with the nature of the molecule or anion releasing the proton. A neutral molecule is more apt to release a proton than is a negatively charged ion. Thus H_3PO_4 will release a proton more readily than $H_2PO_4^-$, and $H_2PO_4^-$ will release a proton more readily than HPO_4^{2-}. This trend is typical of all polyprotic acids, including binary acids: Acidity decreases with increasing negative charge.

Binary Acids Within a given period, the trends observed for oxoacids can be extended to binary acids, which contain only hydrogen and another element. Values of acid-ionization constants for some binary acids are listed in Table 15.3. The strength of the acids increases from left to right across a period. For example, the following relative strengths are found for some Period 2 binary acids:

$$CH_4 < NH_3 < H_2O < HF$$
$$NH_4^+ < H_3O^+ < H_2F^+$$

Again, acid strength increases as electronegativity increases. The more electronegative the atom bonded to hydrogen, the more ionic the character of the bond, and the easier it is for the molecule to ionize in solution.

Although this approach works within periods, where the sizes of the central atom are nearly constant, it does not hold for comparisons within groups. As

Table 15.3 Acid-Ionization Constants of Some Binary Acids

CH_4	NH_3	H_2O	HF
10^{-58}	10^{-39}	10^{-14}	10^{-3}
SiH_4	PH_3	H_2S	HCl
10^{-35}	10^{-27}	10^{-7}	10^2
GeH_4	AsH_3	H_2Se	HBr
10^{-25}	10^{-19}	10^{-4}	10^3
SnH_4	SbH_3	H_2Te	HI
10^{-20}	10^{-15}	10^{-3}	10^5

indicated in the following comparison, the acid strength increases from the top to the bottom of a group.

$$HF < HCl < HBr < HI$$

Because the electronegativity decreases as we go down a group, this trend is opposite what is expected from changes in electronegativity. What causes this apparent anomaly? The increase in size down a group is an important factor. The larger size of the halogen atom leads to a lower charge density and a weaker bond to hydrogen, so the ionization is enhanced by size.

Calculations Involving Weak Acids

A more quantitative comparison of the acidity of weak acids is possible if we know the concentrations of all species in their solutions. A weak acid does not ionize completely, so the H_3O^+ concentration is less than the total concentration of acid placed in the solution.

Calculation of the H_3O^+ concentration follows the general approach used for solving equilibrium-system problems for gas-phase reactions (Chapter 14). In principle, the solutions involve third-order equations because there are two competing equilibria. The equilibria are the ionization of the weak acid, HA, and the autoionization of water.

$$HA(aq) + H_2O(l) \rightleftharpoons H_3O^+(aq) + A^-(aq)$$
$$2H_2O(l) \rightleftharpoons H_3O^+(aq) + OH^-(aq)$$

However, the equations can be reduced to a quadratic form if it is safe to assume that the weak acid contributes much more H_3O^+ than does water. Except for very dilute solutions and for very weak acids, this assumption is generally valid.

A second simplifying assumption is sometimes possible. To consider this assumption in more detail, we must first define the percent ionization of a weak acid, HA, as follows:

$$\text{Percent ionization} = \frac{[H_3O^+]}{[HA]} \times 100$$

If the percent ionization is small enough (generally 5% or less), the actual concentration of nonionized HA can be considered the same as the total concentration

(C_{HA}) that was placed in solution. From the stoichiometry of the reaction, we get the following relationship:

$$[HA] = C_{HA} - [H_3O^+]$$

If there is not much ionization, $[H_3O^+]$ is small enough to ignore, and the concentration of HA can be set equal to the known value of C_{HA}. Like all assumptions, this one must be checked after a solution is obtained. If it is not valid, the problem must be solved again without this assumption.

A general approach to solving problems of this sort involves creating a table of initial concentrations, changes in concentration needed to achieve equilibrium, and equilibrium concentrations. For example, suppose we have a 0.500 M solution of nitrous acid, HNO_2 ($K_a = 4.6 \times 10^{-4}$), which ionizes according to the equation

$$HNO_2(aq) + H_2O(l) \rightleftharpoons H_3O^+(aq) + NO_2^-(aq)$$

The initial concentrations are 0.500 M for HNO_2 and 0 M for the ions. If some amount of HNO_2—call it x—ionizes, then the balanced equation indicates that we will have x M of H_3O^+ and x M of NO_2^- at equilibrium. A table can be constructed from this information.

Substance	HNO_2	H_3O+	NO_2-
Initial concentration, M	0.500	0.0	0.0
Change in concentration, M	$-x$	$+x$	$+x$
Equilibrium concentration, M	$0.500 - x$	$0.0 + x$	$0.0 + x$

The expressions for the equilibrium concentrations are substituted into the expression for the equilibrium constant.

$$K_a = \frac{[H_3O^+][NO_2^-]}{[HNO_2]} = \frac{(x)(x)}{0.500 - x} = 4.6 \times 10^{-4}$$

We can expand this equation and solve it for x. If we can assume that the amount of ionization, x, is small compared with the total concentration, 0.500 M, we can simplify the equation as follows:

$$x^2/0.500 = 4.6 \times 10^{-4}$$

Solving for x gives the value 1.5×10^{-2} M. Because this value is 3% of 0.500, the assumption that x is small was valid. In general, differences as large as 5% give satisfactory results. Now that we know the value of x, we can calculate other concentrations from the relationships given in the table. The percent ionization is readily calculated from the value of x and the total acid concentration.

$$\text{Percent ionization} = \frac{1.5 \times 10^{-2}}{0.500} \times 100 = 3.0\%$$

The pH is calculated from the H_3O^+ concentration.

EXAMPLE 15.8

What are the concentrations of all species present in a 0.100 M hydrazoic acid (HN_3) solution at 25°C? What is the percent ionization? What is the pH? The value of the acid-ionization constant for hydrazoic acid is 1.90×10^{-5}.

(continued)

Solution   The equation for the ionization of hydrazoic acid is

$$HN_3(aq) + H_2O(l) \rightleftharpoons H_3O^+(aq) + N_3^-(aq)$$

Assuming that some amount, x, of HN_3 is ionized at equilibrium, the concentrations can be summarized as follows:

Substance	HN_3	H_3O^+	N_3^-
Initial concentration, M	0.100	0.0	0.0
Change in concentration, M	$-x$	$+x$	$+x$
Equilibrium concentration, M	$0.100 - x$	$0.0 + x$	$0.0 + x$

Substituting into the equilibrium-constant expression gives

$$K_a = \frac{[H_3O^+][N_3^-]}{[HN_3]} = \frac{(x)(x)}{0.100 - x} = 1.90 \times 10^{-5}$$

This equation can be expanded to assume a quadratic form, but we can simplify it by making the assumption that the quantity x is small compared with the initial concentration, 0.100. The equation then becomes

$$x^2/0.100 = 1.90 \times 10^{-5}$$

Multiplying through gives

$$x^2 = 1.90 \times 10^{-6}$$

Taking the square root gives a value for x.

$$x = 1.38 \times 10^{-3} \ M$$

At this point, it is essential to test the assumption that x is small compared with 0.100. We can do this by calculating the percent ionization based on these results.

$$\% \text{ ionization} = \frac{[H_3O^+]}{[HN_3]} \times 100 = \frac{x}{[HN_3]_{initial}} \times 100$$

$$= \frac{1.38 \times 10^{-3}}{0.100} \times 100 = 1.38\%$$

The value of x is less than 5% of the initial concentration, so only a small error is introduced by dropping x from the denominator term. Now the concentrations can be calculated by reference to the table.

$$[HN_3] = 0.100 - x = 0.100 - 0.00138 = 0.099 \ M$$
$$[H_3O^+] = x = 1.38 \times 10^{-3} \ M$$
$$[N_3^-] = x = 1.38 \times 10^{-3} \ M$$

The pH is based on the H_3O^+ concentration.

$$pH = -\log [H_3O^+] = -\log(1.38 \times 10^{-3}) = 2.860$$

Practice Problem 15.8

What are the concentrations of all species present in a 0.250 M acetic acid (CH_3CO_2H) solution at 25°C? What is the percent ionization? What is the pH? The value of the acid-ionization constant for acetic acid is 1.75×10^{-5}.

Answer: 0.248 M CH_3CO_2H; $2.09 \times 10^{-3} \ M$ $CH_3CO_2^-$; $2.09 \times 10^{-3} \ M$ H_3O^+; 0.836%, pH = 2.680

The weaker the acid, the more valid the assumption that x, the amount of the acid that will ionize, is negligible. For somewhat stronger acids, this assumption may not be valid, and the quadratic equation may have to be solved. The most practical approach is still to make the assumption, solve the simpler equation, and then test the assumption. Only if the assumption fails do we need to solve the quadratic equation.

EXAMPLE 15.9

What is the pH of a solution of hydrofluoric acid that has a concentration of 0.0100 M? The value of K_a is 7.0×10^{-4}.

Solution We follow the usual procedure. The reaction is

$$HF(aq) + H_2O(l) \rightleftharpoons H_3O^+(aq) + F^-(aq)$$

The concentrations can be summarized as follows:

Substance	HF	H_3O^+	F^-
Initial concentration, M	0.0100	0.0	0.0
Change in concentration, M	$-x$	$+x$	$+x$
Equilibrium concentration, M	$0.0100 - x$	$0.0 + x$	$0.0 + x$

Substituting into the equilibrium-constant expression yields

$$K_a = \frac{[H_3O^+][F^-]}{[HF]} = \frac{(x)(x)}{0.0100 - x} = 7.0 \times 10^{-4}$$

We make the simplifying assumption that x is small compared with 0.010 to get the equation

$$x^2/0.0100 = 7.0 \times 10^{-4}$$
$$x^2 = 0.0100 \times 7.0 \times 10^{-4} = 7.0 \times 10^{-6}$$

Taking the square root gives

$$x = 2.6 \times 10^{-3} \, M$$

In this case, the assumption that x is negligible is not valid. In fact, x is 26% of the initial concentration, so appreciable ionization has occurred. The equation must thus be solved in its quadratic form without the simplifying assumption.

$$\frac{x^2}{0.0100 - x} = 7.0 \times 10^{-4}$$

Expanding the equation into its quadratic form gives

$$x^2 + (7.0 \times 10^{-4})x - (7.0 \times 10^{-6}) = 0$$

Solving the quadratic equation gives

$$x = \frac{-(7.0 \times 10^{-4}) + \sqrt{(7.0 \times 10^{-4})^2 + 4(7.0 \times 10^{-6})}}{2}$$
$$= 2.3 \times 10^{-3} \, M$$

The concentrations can be calculated from the relationships in the table.

$$[HF] = 0.0100 - x = 0.0100 - 0.0023 = 7.7 \times 10^{-3} \, M$$
$$[H_3O^+] = x = 2.3 \times 10^{-3} \, M$$
$$[F^-] = x = 2.3 \times 10^{-3} \, M$$

The pH is

$$pH = -\log [H_3O^+] = -\log(2.3 \times 10^{-3}) = 2.64$$

(continued)

We can check our calculations by substituting back into the expression for the reaction quotient.

$$Q = \frac{[H_3O^+][F^-]}{[HF]} = \frac{(2.3 \times 10^{-3})(2.3 \times 10^{-3})}{7.7 \times 10^{-3}} = 6.9 \times 10^{-4}$$

This agrees well with the value of the acid-ionization constant, 7.0×10^{-4}, so the calculations were carried out correctly.

Practice Problem 15.9

What is the pH of a solution of formic acid (HCO_2H) that has a concentration of 0.0050 M? The value of K_a is 1.77×10^{-4}.

Answer: 3.07

Values of acid-ionization constants are available in tables such as Table 15.2. These values are determined by a variety of measurements. The simplest involves measuring the pH of a solution of a weak acid of known total concentration. Calculating the acid-ionization constant from the pH is very similar to working the examples just discussed. The percent ionization, which may be determined by measuring colligative properties (discussed in Chapter 11) if the ionization is sufficiently large, also provides a mean of calculating the acid-ionization constant. We assume that the percent of the acid that has ionized equals the amount of H_3O^+ formed.

EXAMPLE 15.10

A 0.25 M solution of HCN has a pH of 5.00. What is the value of K_a for HCN?

Solution   The equilibrium reaction is

$$HCN(aq) + H_2O(l) \rightleftharpoons H_3O^+(aq) + CN^-(aq)$$

The acid-ionization constant expression for this reaction is

$$K_a = \frac{[H_3O^+][CN^-]}{[HCN]}$$

The initial concentrations of species are 0.25 M HCN, 0 M H_3O^+, and 0 M CN^-. At equilibrium, some amount x of HCN has ionized. The initial and equilibrium concentrations of species can be summarized as follows:

Substance	HCN	H_3O^+	CN^-
Initial concentration, M	0.25	0.0	0.0
Change in concentration, M	$-x$	$+x$	$+x$
Equilibrium concentration, M	$0.25 - x$	$0.0 + x$	$0.0 + x$

Substituting into the equilibrium-constant expression gives

$$K_a = \frac{(x)(x)}{0.25 - x}$$

The H_3O^+ concentration, assumed to be equal to x, is obtained from the pH.

$$pH = 5.00 = -\log [H_3O^+]$$
$$[H_3O^+] = 1.0 \times 10^{-5} \, M$$

Substituting into the equation for x gives

$$K_a = \frac{(1.0 \times 10^{-5})(1.0 \times 10^{-5})}{0.25 - (1.0 \times 10^{-5})}$$

Because 1.0×10^{-5} is much less than 0.25, this equation can be simplified.

$$K_a = \frac{(1.0 \times 10^{-5})(1.0 \times 10^{-5})}{0.25} = 4.0 \times 10^{-10}$$

Practice Problem 15.10

A 0.10 M solution of HOI has a pH of 5.82. What is K_a for HOI?

Answer: 2.3×10^{-11}

Calculations Involving Polyprotic Weak Acids

Polyprotic acids further complicate calculating the composition of solutions. For example, in a solution of the diprotic acid sulfurous acid, two equilibrium reactions occur in addition to the ionization of water. Each releases H_3O^+ ions into solution. Thus the H_3O^+ concentration is affected by both equilibria:

$$H_2SO_3(aq) + H_2O(l) \rightleftharpoons H_3O^+(aq) + HSO_3^-(aq)$$
$$HSO_3^-(aq) + H_2O(l) \rightleftharpoons H_3O^+(aq) + SO_3^{2-}(aq)$$

The acid-ionization constants for these equilibria (numbered to indicate which proton is being released) are

$$K_{a1} = \frac{[H_3O^+][HSO_3^-]}{[H_2SO_3]} = 1.72 \times 10^{-2}$$

$$K_{a2} = \frac{[H_3O^+][SO_3^{2-}]}{[HSO_3^-]} = 6.43 \times 10^{-8}$$

Generally we can assume that the first ionization equilibrium is not affected by the second (or later) equilibrium reaction. This is a safe assumption so long as the ratio of equilibrium constants is as large as in this example, because the large ratio means the extent of the second equilibrium is slight compared with the first.

Using the simplifying assumption, we solve for the equilibrium concentrations by solving the first ionization equilibrium as though it were completely independent of the second ionization equilibrium. The solution to this first ionization equilibrium provides the starting concentrations for the second ionization equilibrium. We then solve each successive equilibrium independently, using the results of the previous steps.

EXAMPLE 15.11

What are the concentrations of H_3O^+, H_2SO_3, HSO_3^-, and SO_3^{2-} in a 0.100 M solution of H_2SO_3? The ionization constants are $K_{a1} = 1.72 \times 10^{-2}$ and $K_{a2} = 6.43 \times 10^{-8}$.

(continued)

Solution   It is necessary to consider both equilibrium reactions, but we can consider them one at a time because the equilibrium constants have such different values. We start with the first equilibrium and approach the problem in the usual way. The equilibrium reaction is

$$H_2SO_3(aq) + H_2O(l) \rightleftharpoons H_3O^+(aq) + HSO_3^-(aq)$$

The ionization-constant expression is

$$K_{a1} = \frac{[H_3O^+][HSO_3^-]}{[H_2SO_3]} = 1.72 \times 10^{-2}$$

The sulfurous acid will ionize to the extent of some amount, x, giving the following concentrations:

Substance	H_2SO_3	H_3O^+	HSO_3^-
Initial concentration, M	0.100	0.0	0.0
Change in concentration, M	$-x$	$+x$	$+x$
Equilibrium concentration, M	$0.100 - x$	$0.0 + x$	$0.0 + x$

Substituting into the equilibrium-constant expression gives

$$K_{a1} = \frac{(x)(x)}{0.100 - x} = 1.72 \times 10^{-2}$$

If x is much less than 0.100, the equation becomes

$$x^2 = 0.100 \times 1.72 \times 10^{-2}$$

We solve for x.

$$x = 0.0415 \ M$$

The assumption clearly was not valid. The result shows that x is 42% of the initial concentration, so the quadratic equation must be solved.

$$x^2 + 0.0172x - 0.00172 = 0$$

$$x = \frac{-0.0172 + \sqrt{(0.0172)^2 + 4(0.00172)}}{2} = 0.0338 \ M$$

The concentrations are thus

$$[H_2SO_3] = 0.100 - x = 0.100 - 0.0338 = 0.066 \ M$$
$$[H_3O^+] = x = 0.0338 \ M$$
$$[HSO_3^-] = x = 0.0338 \ M$$

These concentrations are the starting points for the second equilibrium reaction.

$$HSO_3^-(aq) + H_2O(l) \rightleftharpoons H_3O^+(aq) + SO_3^{2-}(aq)$$

$$K_{a2} = \frac{[H_3O^+][SO_3^{2-}]}{[HSO_3^-]} = 6.43 \times 10^{-8}$$

Substance	HSO_3^-	H_3O^+	SO_3^{2-}
Initial concentration, M	0.0338	0.0338	0.0
Change in concentration, M	$-x$	$+x$	$+x$
Equilibrium concentration, M	$0.0338 - x$	$0.0338 + x$	$0.0 + x$

Substituting into the equilibrium-constant expression gives

$$K_{a2} = \frac{(0.0338 + x)(x)}{0.0338 - x} = 6.43 \times 10^{-8}$$

(continued)

▶ ▶ ▶   WORLDS OF CHEMISTRY

Rain Isn't Right Anymore

The most acidic rainfall recorded in the United States was at Wheeling, West Virginia. The value was pH 1.4, which is more acidic than lemon juice.

In the 1950s, trout and salmon in Scandinavian lakes and rivers were killed by an invisible agent. In the 1960s, some similar agent killed fish in hundreds of lakes in the Adirondack Mountains of New York and left the lakes overgrown with moss. The surrounding trees also began to die (Figure 15.13) and crops were affected. In the late 1970s, similar effects began to show up in California and Colorado. What caused this destruction? Rain!

The pH of pure water is 7.0, exactly neutral. We have already seen that most water is not pure, however. Depending on the substances dissolved in the water and on the materials with which it has been in contact, the pH may have some other value. Water that is in contact with air, including rainwater, dissolves small quantities of the gases in the air. If the air is unpolluted, the only gas that dissolves in significant quantities in rainwater is carbon dioxide, which is the anhydride of carbonic acid.

The pH of water is lowered to about 5.6 by dissolved carbon dioxide, and this pH is the benchmark against which the pH of rainwater is measured. Rainwater more acidic than pH 5.6 is considered to be *acid rain*.

The average pH of the rainfall in the United States, especially in the eastern states, is much lower than the expected value of 5.6. Averages of 4.5–5.6 were common during the 1950s. In the 1970s, averages in the same areas dropped to 4.3–5.0. In the 1980s, values as low as 3.0–4.0 were recorded routinely. The most acidic rainfall recorded in the United States was at Wheeling, West Virginia; the value was pH 1.4, which is more acidic than lemon juice.

Most fish die at a pH of 4.5–5.0, which explains the effects of acid rain on lakes. However, the pH of waterways does not always match that of the rainfall. Once on the ground, water can be neutralized at least in part by some soils. Many soils, for example, contain limestone ($CaCO_3$), which can neutralize acids.

$$CaCO_3(s) + 2H_3O^+(aq) \longrightarrow Ca^{2+}(aq) + 3H_2O(l) + CO_2(g)$$

However, even the neutralizing capacity of soils is overcome if acid rain continues long enough. A negative effect of the neutralizing ability of calcium carbonate is the dam-

Figure 15.13 Fish and trees have died, leaving sterile lakes in central Ohio and elsewhere. The cause is thought to be acid rain.

age done by acid rain to limestone and marble buildings and statues (Figure 15.14).

The sources of acid rain are not known with certainty. Some are natural: lightning, decomposition of vegetable matter, volcanic eruptions, and sea salt sprays. However, it is believed that the combustion of fossil fuels is largely responsible for the increase in atmospheric pollution by acidic species.

The increased acidity is thought to come primarily from sulfur oxides and nitrogen oxides. Sulfur oxides are produced primarily by the burning of coal and petroleum that contain sulfur. As energy requirements have increased, so have sulfur oxide emissions. Nitrogen oxides, produced primarily by gasoline-powered vehicles and electrical power plants, have also increased.

The chemical behavior of the sulfur and nitrogen oxides in the atmosphere is not fully understood. Scientists are still trying to determine the mechanism by which these pollutants are converted to H_2SO_4 and HNO_3 and to find ways to prevent their generation in the atmosphere.

(continued)

Figure 15.14 Acid rain corrodes marble statues. The calcium carbonate reacts with acid in rainwater, and the statue gradually disappears. This gargoyle is on the Notre Dame Cathedral in Paris.

Questions for Discussion

1. Coal mined in the eastern United States contains more sulfur than coal mined in western states. Most of the worst incidents of acid rain have occurred in the eastern states. Do these observations support the hypothesis that acid rain arises from burning fossil fuels? Are there other hypotheses that could explain these same observations?

2. Should we continue to use coal that contains sulfur in electric power plants? Are there any other options?

3. Many metal ores, such as those of copper, are sulfides. When exposed to air by mining, they can weather and be oxidized to produce sulfur oxides. Smelting of sulfide ores also produces sulfur oxides. Discuss practical ways of coping with the potential air pollution that can arise from our need for metals.

4. Use any source of information you need to find out more about nitrogen oxides arising from automobile exhausts and their role in the formation of smog.

Assuming that x is much less than 0.0338 enables us to simplify the equation.

$$\frac{(0.0338)(x)}{0.0338} = 6.43 \times 10^{-8}$$

The two values of 0.0338 cancel, leaving

$$x = 6.43 \times 10^{-8} \ M$$

This time, the assumption that x is small compared with the initial concentration is indeed valid. The concentrations are thus

$$[H_2SO_3] = 0.066 \ M$$
$$[H_3O^+] = 0.0338 + x = 0.0338 \ M$$
$$[HSO_3^-] = 0.0338 - x = 0.0338 \ M$$
$$[SO_3^{2-}] = x = 6.43 \times 10^{-8} \ M$$

Practice Problem 15.11

What are the concentrations of H_3O^+, H_2S, HS^-, and S^{2-} in a 0.100 M solution of H_2S? The ionization constants are $K_{a1} = 5.7 \times 10^{-8}$ and $K_{a2} = 1.2 \times 10^{-13}$.

Answer: $7.5 \times 10^{-5} \ M \ H_3O^+$; $7.5 \times 10^{-5} \ M \ HS^-$; 0.100 $M \ H_2S$; $1.2 \times 10^{-13} \ M \ S^{2-}$

15.4   Weak Bases

So far, we have considered primarily the ionization of acids. However, weak bases form ions in a similar way, except that $OH^-(aq)$ is produced instead of $H_3O^+(aq)$,

Table 15.4 Ionization Constants of Some Weak Bases at 25°C

Name	Formula	K_b
Ammonia	NH_3	1.76×10^{-5}
Aniline	$C_6H_5NH_2$	3.94×10^{-10}
Butylamine	$C_4H_9NH_2$	4.0×10^{-4}
Dimethylamine	$(CH_3)_2NH$	5.9×10^{-4}
Ethanolamine	$HOC_2H_4NH_2$	3.18×10^{-5}
Ethylamine	$C_2H_5NH_2$	4.28×10^{-4}
Hydrazine	H_2NNH_2	1.3×10^{-6}
Hydroxylamine	$HONH_2$	1.07×10^{-8}
Methylamine	CH_3NH_2	4.8×10^{-4}
Pyridine	C_5H_5N	1.7×10^{-9}
Trimethylamine	$(CH_3)_3N$	6.25×10^{-5}

as shown in the following equation:

$$B(aq) + H_2O(l) \rightleftharpoons BH^+(aq) + OH^-(aq)$$

How is the behavior of bases in water similar to that of acids?

The equilibrium-constant expression for this reaction is

$$K_b = \frac{[BH^+][OH^-]}{[B]}$$

K_b is called the **base-ionization constant.** The ionization constants of some common weak bases are given in Table 15.4.

Calculations involving strong bases were discussed in Section 15.2 and involve the assumption that the OH^- concentration equals its stoichiometric concentration from the strong base. Calculations involving concentrations in solutions of weak bases, which do not ionize completely, are much like those for weak acids. Of course, we use K_b instead of K_a, and we find a value for the hydroxide-ion concentration rather the concentration of hydronium ion. If we wish to find the pH of a basic solution, we can do so most simply by deriving it from the value of pOH, which is calculated from the concentration of OH^-. Alternately, the concentration of H_3O^+ can be calculated from the concentration of OH^- and the value of K_w, and the pH can be calculated from the H_3O^+ concentration.

EXAMPLE 15.12

What is the pH of a 0.150 M solution of NH_3? The value of the base-ionization constant for ammonia is $K_b = 1.76 \times 10^{-5}$ at 25°C.

Solution The equilibrium reaction for the ionization of ammonia is

$$NH_3(aq) + H_2O(l) \rightleftharpoons NH_4^+(aq) + OH^-(aq)$$

The equilibrium-constant expression is

$$K_b = \frac{[NH_4^+][OH^-]}{[NH_3]} = 1.76 \times 10^{-5}$$

(continued)

To simplify the calculations, we assume that $OH^-(aq)$ from the ionization of water can be ignored. Now suppose that some amount, x, of $NH_3(aq)$ undergoes ionization. Then the initial and equilibrium concentrations can be summarized as follows:

Substance	NH_3	OH^-	NH_4^+
Initial concentration, M	0.150	0.0	0.0
Change in concentration, M	$-x$	$+x$	$+x$
Equilibrium concentration, M	$0.150 - x$	$0.0 + x$	$0.0 + x$

Substituting these equilibrium values into the equilibrium-constant expression gives

$$K_b = \frac{[NH_4^+][OH^-]}{[NH_3]} = \frac{(x)(x)}{0.150 - x} = 1.76 \times 10^{-5}$$

Assuming that x is much less than 0.150 enables us to simplify to

$$x^2/0.150 = 1.76 \times 10^{-5}$$
$$x^2 = 2.64 \times 10^{-6}$$

Taking the square root gives

$$x = 1.62 \times 10^{-3} \ M$$

Because x is only slightly more than 1% of the initial concentration of 0.150, it was valid to assume that x could be ignored in the denominator of the equation. The concentrations can be obtained from the relationships in the table.

$$[NH_3] = 0.150 - x = 0.150 - 0.00162 = 0.148 \ M$$
$$[OH^-] = x = 1.62 \times 10^{-3} \ M$$
$$[NH_4^+] = x = 1.62 \times 10^{-3} \ M$$

We can calculate the pH in either of two ways. We can use the ion-product constant of water to calculate the H_3O^+ concentration and derive the pH from that.

$$K_w = [H_3O^+][OH^-] = 1.00 \times 10^{-14}$$
$$[H_3O^+] = K_w/[OH^-] = (1.00 \times 10^{-14})/(1.62 \times 10^{-3}) = 6.17 \times 10^{-12} \ M$$
$$pH = -\log(6.17 \times 10^{-12}) = 11.210$$

Or we can calculate the pOH from the OH^- concentration.

$$pOH = -\log [OH^-] = -\log(1.62 \times 10^{-3}) = 2.790$$

and derive the pH from the following relationship.

$$pH + pOH = 14.000$$
$$pH = 14.000 - 2.790 = 11.210$$

Practice Problem 15.12

What is the pH of a 0.500 M solution of methylamine, CH_3NH_2? The value of the base-ionization constant for methylamine is $K_b = 4.8 \times 10^{-4}$ at 25°C.

Answer: 12.19

Determination of Base-Ionization Constants

As with weak acids, the value of the base-ionization constant for a weak base can be determined from the measured pH of a solution of that weak base. The technique is similar to that described for acid-ionization constants.

There is a second means of obtaining base-ionization constants, provided that the acid-ionization constant of the conjugate acid has been determined. Consider the acid HCN and its conjugate base CN^-. The equilibrium ionization reactions and the equilibrium-constant expressions are

$$HCN(aq) + H_2O(l) \rightleftharpoons H_3O^+(aq) + CN^-(aq)$$

$$K_a = \frac{[H_3O^+][CN^-]}{[HCN]}$$

$$CN^-(aq) + H_2O(l) \rightleftharpoons HCN(aq) + OH^-(aq)$$

$$K_b = \frac{[HCN][OH^-]}{[CN^-]}$$

Now see what happens when the two equilibrium expressions are multiplied together:

$$\frac{[H_3O^+][CN^-]}{[HCN]} \times \frac{[HCN][OH^-]}{[CN^-]} = K_a \times K_b$$

Canceling terms leaves

$$[H_3O^+][OH^-] = K_a K_b$$

The product $[H_3O^+][OH^-]$ is just the ion-product constant for water, so

$$K_a K_b = K_w$$

This is a general relationship for all conjugate acid–base pairs. If we know one of the ionization constants, we can calculate the other.

EXAMPLE 15.13

Determine the value of the base-ionization constant for the phenolate ion, which reacts as follows:

$$C_6H_5O^-(aq) + H_2O(l) \rightleftharpoons C_6H_5OH(aq) + OH^-(aq)$$

The acid-ionization constant for phenol is 1.0×10^{-10} at 25°C. The value of the ion-product constant of water is 1.0×10^{-14}.

Solution The value of K_b can be obtained from the value of K_a for the conjugate acid on the basis of the relationship

$$K_a K_b = K_w$$

Rearranging this equation, we get

$$K_b = K_w/K_a$$

Inserting values yields

$$K_b = (1.0 \times 10^{-14})/(1.0 \times 10^{-10}) = 1.0 \times 10^{-4}$$

The expression for K_b for this reaction is

$$K_b = \frac{[C_6H_5OH][OH^-]}{[C_6H_5O^-]} = 1.0 \times 10^{-4}$$

(continued)

Practice Problem 15.13

Determine the value of the base-ionization constant for the hypochlorite ion. The acid-ionization constant for hypochlorous acid is 3.0×10^{-8} at 25°C.

Answer: 3.3×10^{-7}

15.5 Titration Curves

So far, we have considered solutions of acids or bases prepared by dissolving or diluting a pure acid or a pure base. Also of interest is the situation in which solutions of an acid and a base are reacted together:

$$HCl(aq) + NH_3(aq) \rightleftharpoons NH_4^+(aq) + Cl^-(aq)$$

Titration, a common technique discussed in Chapter 4, uses reactions such as these for chemical analyses. For example, suppose we need to know how much acid there is in a chrome-plating bath (Figure 15.15). We simply react the bath solution with a solution of NaOH of known concentration until exactly equivalent amounts of acid and base are present in the mixture. From the volumes required to get equivalent amounts of acid and base, we can calculate the unknown concentration of the acid.

RULES OF REACTIVITY

In aqueous solution, acids react with bases in neutralization reactions to form salts and water.

How does an acid–base indicator work?

In order to add exactly equivalent amounts of reagents in a titration, we must be able to detect the **equivalence point,** the point at which the two reagents are present in chemically equivalent amounts. Generally, this point is sensed by the use of an *indicator*—a chemical substance whose color depends on the pH and changes noticeably over a short pH range. The point at which the color change occurs is called the *endpoint* of the titration. If we select the indicator carefully, the endpoint and the equivalence point will occur at the same volume of added reagent.

Figure 15.15 A chrome-plating bath contains CrO_3 dissolved in a sulfuric acid solution. The acidity must be controlled, or electrochemical reduction of chromium(VI) may result in deposition of $Cr(OH)_3$ instead of chromium metal.

Most acid–base indicators are complex molecules that themselves undergo acid–base reactions, as shown here for phenolphthalein. The acidic form is color-less. The basic form is pink.

Phenolphthalein, $C_{20}H_{14}O_4$, exists as small white or yellowish-white crystals. The sodium salt, $Na_2C_{20}H_{12}O_4$, is a reddish-brown or pale red powder. Phenolphthalein is almost insoluble in water but dissolves in ethanol, whereas the sodium salt is soluble in water. Besides its use as an acid–base indicator, phenolphthalein is used in laxatives.

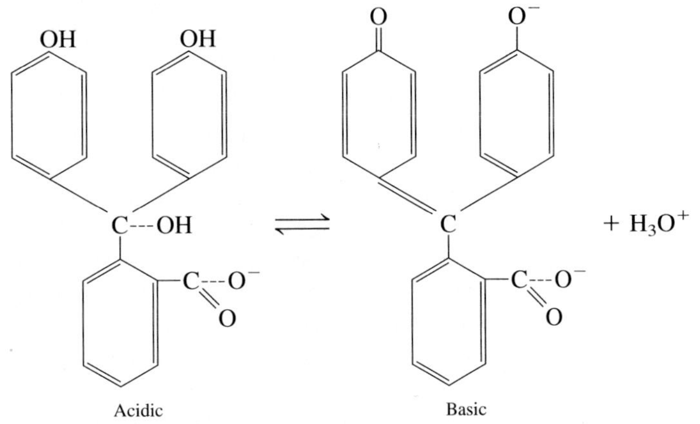

Acidic   Basic

The transition between the two forms occurs at pH values in the range 8.2–10. The pH at which an indicator changes color is close to its own pK value. Other indicators exhibit different colors and change color in other pH ranges, as shown in Table 15.5.

The best indicator to use for a given titration depends on the pH at the equivalence point, which depends on the strength of the acid and of the base used in the titration. To determine the pH at the equivalence point, we can follow the course of the titration with a pH meter, which gives us values of the pH of the solution as the titration proceeds. These data can then be examined in the form of a titration curve, a plot of pH against volume of added reagent that enables us to determine the pH at the equivalence point.

Table 15.5 Some Acid–Base Indicators

Indicator	Acid Color	pH Range of Color Change	Base Color
Thymol blue	Red	1.2–2.8	Yellow
Methyl orange	Red	3.1–4.5	Yellow
Bromcresol green	Yellow	3.8–5.5	Blue
Methyl red	Red	4.8–6.0	Yellow
Litmus	Red	5.0–8.0	Blue
Bromthymol blue	Yellow	6.0–7.6	Blue
Thymol blue	Yellow	8.0–8.6	Blue
Phenolphthalein	Colorless	8.2–10.0	Red
Alizarin yellow	Yellow	10.0–12.1	Lavender

Strong-Acid–Strong-Base Titrations

The titration curve determined for a strong-acid–strong-base titration is shown in Figure 15.16. The pH values in this figure were measured for a 0.100 M HCl solution as it is titrated with 0.100 M NaOH, according to the equation

$$HCl(aq) + NaOH(aq) \rightleftharpoons NaCl(aq) + H_2O(l)$$

At the equivalence point, exactly equal molar amounts of HCl and NaOH have been mixed and have reacted. The solution contains only dissolved NaCl at this point. Neither $Na^+(aq)$ nor $Cl^-(aq)$ has acid–base properties, so the pH is 7.00. Before any NaOH is added, the pH is 1.00, the value for 0.100 M HCl. When NaOH is first added, the pH increases slowly because acid is being neutralized by the added base. Near the equivalence point, most of the acid has been neutralized and the pH increases very rapidly, quickly passing pH 7 and reaching about pH 10. Further addition of NaOH increases the pH more slowly again as the hydroxide-ion concentration builds up.

Although we can measure the pH during a titration with a pH meter, it is not necessary actually to make measurements to see what the titration curve looks like. We can calculate the value of the pH for any combination of volumes of HCl and NaOH solutions. A calculated curve that agrees exactly with the measured titration curve can be constructed from the results of a number of such calculations. A point on the titration curve is determined by calculating the number of moles of each reactant from its concentration and volume, determining the amount of excess reactant, and converting this back to a concentration, using the total volume at the point of interest. Because the acid and base are strong, we need not take into account the dissociation of these species. They will be completely dissociated into ions.

EXAMPLE 15.14

Suppose we add 30.0 mL of 0.100 M NaOH to 50.0 mL of 0.100 M HCl. What is the pH of this mixture?

Figure 15.16 Titration curve for a strong-acid–strong-base titration. A 0.100 M solution of NaOH is being added to 50.0 mL of a 0.100 M solution of HCl. Near the equivalence point, the pH changes rapidly from about 3 to about 11 with the addition of a very small volume of NaOH solution. The color change for phenolphthalein occurs within the region where the pH is changing rapidly, so it is a good indicator for this reaction, giving an endpoint very close to the equivalence point.

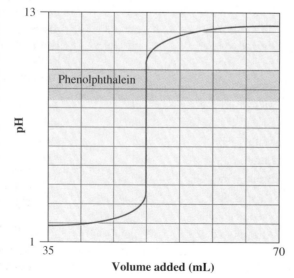

Volume added (mL)

Solution   First we calculate how many moles of each reagent have been added to the mixture.

$$\text{mol NaOH} = 30.0 \text{ mL} \times 1 \text{ L}/1000 \text{ mL} \times 0.100 \text{ mol/L} = 0.00300 \text{ mol}$$

$$\text{mol HCl} = 50.0 \text{ mL} \times 1 \text{ L}/1000 \text{ mL} \times 0.100 \text{ mol/L} = 0.00500 \text{ mol}$$

Now we consider the stoichiometry of the reaction to determine how much of each reagent reacts.

$$\text{HCl}(aq) + \text{NaOH}(aq) \longrightarrow \text{NaCl}(aq) + \text{H}_2\text{O}(l)$$

The acid and base react in a 1:1 ratio. Only 0.00300 mol of NaOH is present in the mixture, so only 0.00300 mol of each substance will have reacted. This leaves excess HCl.

$$\text{mol HCl in excess} = 0.00500 \text{ mol} - 0.00300 \text{ mol} = 0.00200 \text{ mol}$$

The volume at this point is the sum of the volumes of the added reagents:

$$\text{Total volume} = 30.0 \text{ mL} + 50.0 \text{ mL} = 80.0 \text{ mL or } 0.0800 \text{ L}$$

The concentration of excess acid is given by the moles divided by the volume.

$$[\text{HCl}] = 0.00200 \text{ mol}/0.0800 \text{ L} = 0.0250 \text{ } M$$

Because HCl is a strong acid, it is completely ionized, so the H_3O^+ concentration equals the hydrochloric acid concentration.

$$[\text{H}_3\text{O}^+] = 0.0250 \text{ } M$$

The pH is the negative log of this concentration.

$$\text{pH} = -\log [\text{H}_3\text{O}^+] = -\log(0.0250) = 1.602$$

Practice Problem 15.14

Suppose we add 60.0 mL of 0.100 M NaOH to 50.0 mL of 0.100 M HCl. What is the pH of this mixture?

Answer: 11.959

RULES OF REACTIVITY

Salts containing the conjugate acid of a strong base or the conjugate base of a strong acid do not undergo hydrolysis reactions in aqueous solution.

The indicator used for the experimental determination of the equivalence point should change color in a pH range close to the pH at the equivalence point, which is 7.00 for strong-acid–strong-base titrations. The pH changes rapidly with very small added amounts of sodium hydroxide at this point, so any indicator that changes color in the range of pH 6–10 is probably suitable. Phenolphthalein is often used for strong-acid–strong-base titrations; it changes color in the range of pH 8.2–10, as shown in Figure 15.16. Because this is in the range of rapid pH change, a very small excess of base changes phenolphthalein from colorless to pink.

Weak-Acid–Strong-Base Titrations

In a reaction involving a weak acid and strong base, the pH is not 7 at the equivalence point. An example is the titration of 0.100 M $\text{CH}_3\text{CO}_2\text{H}$ ($K_a = 1.75 \times 10^{-5}$) with 0.100 M NaOH, according to the equation

$$\text{CH}_3\text{CO}_2\text{H}(aq) + \text{OH}^-(aq) \longrightarrow \text{CH}_3\text{CO}_2^-(aq) + \text{H}_2\text{O}(l)$$

The change in pH during the course of the titration is shown in Figure 15.17. The pH at the equivalence point is 8.73, a value that can be calculated readily by using

Figure 15.17 Titration curve for a weak-acid–strong-base titration. A 0.100 M solution of NaOH is being added to 50.0 mL of a 0.100 M solution of CH_3CO_2H. Near the equivalence point, the pH changes rapidly from about 7 to about 11 with the addition of a very small volume of NaOH solution. Although this range is smaller than that for a strong-acid–strong-base titration, the color change for phenolphthalein nevertheless occurs within the region where the pH is changing rapidly. Thus phenolphthalein is a good indicator for this titration.

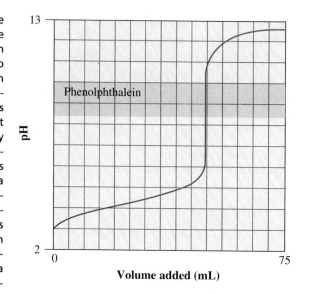

RULES OF REACTIVITY

Salts containing the conjugate base of a weak acid hydrolyze in aqueous solution to form a small amount of the weak acid and aqueous hydroxide ion.

the procedure for weak bases outlined in Section 15.4. At the equivalence point, where equal volumes of the two solutions have been mixed, the result is a 0.0500 M $NaCH_3CO_2$ solution (the concentration is halved because the volume has doubled). The pH of the reaction mixture can be calculated as the pH of the weak-base solution, 0.0500 M $CH_3CO_2^-(aq)$.

EXAMPLE 15.15

Calculate the pH of a mixture of 50.0 mL of 0.100 M CH_3CO_2H and 50.0 mL of 0.100 M NaOH.

Solution   First, we calculate the number of moles of each reactant.

mol CH_3CO_2H = 50.0 mL × 1 L/1000 mL × 0.100 mol/L = 0.00500 mol

mol NaOH = 50.0 mL × 1 L/1000 mL × 0.100 mol/L = 0.00500 mol

These substances react in a 1:1 ratio, as shown by the following equation:

$$CH_3CO_2H(aq) + NaOH(aq) \longrightarrow NaCH_3CO_2(aq) + H_2O(l)$$

As a result, no $CH_3CO_2H(aq)$ or NaOH(aq) remains in solution, and 0.00500 mol of $NaCH_3CO_2$ is formed. The concentration of $NaCH_3CO_2$ is calculated from the number of moles and the total volume (50.0 mL + 50.0 mL = 100.0 mL).

$$[NaCH_3CO_2] = \frac{0.00500 \text{ mol}}{100.0 \text{ mL}} \times \frac{1000 \text{ mL}}{1 \text{ L}} = 0.0500 \ M$$

Next we consider the reaction of acetate ion with water.

$$CH_3CO_2^-(aq) + H_2O(l) \rightleftharpoons CH_3CO_2H(aq) + OH^-(aq)$$

The equilibrium constant for this reaction can be calculated from the acid-ionization constant for acetic acid:

$$K_b = K_w/K_a = (1.00 \times 10^{-14})/(1.75 \times 10^{-5}) = 5.71 \times 10^{-10}$$

For simplicity, we assume that the amount of hydroxide ion coming directly from the ionization of water is negligible compared with the amount coming from the reaction of acetate ion with water. On the basis of that assumption, we can summarize the concentrations initially and at equilibrium as follows:

Substance	$CH_3CO_2^-$	OH^-	CH_3CO_2H
Initial concentration, M	0.0500	0.0	0.0
Change in concentration, M	$-x$	$+x$	$+x$
Equilibrium concentration, M	$0.0500 - x$	$0.0 + x$	$0.0 + x$

Substituting into the equilibrium-constant expression gives

$$K_b = \frac{[CH_3CO_2H][OH^-]}{[CH_3CO_2^-]} = \frac{(x)(x)}{0.0500 - x} = 5.71 \times 10^{-10}$$

To simplify the mathematics, we assume that x is much less than 0.0500.

$$x^2/0.0500 = 5.71 \times 10^{-10}$$
$$x^2 = 2.86 \times 10^{-11}$$
$$x = 5.35 \times 10^{-6}\ M$$

The assumption was indeed valid, so the concentrations can be calculated by replacing the value of x in the table.

$$[OH^-] = [CH_3CO_2H] = 5.35 \times 10^{-6}\ M$$
$$[CH_3CO_2^-] = 0.0500 - 5.35 \times 10^{-6} = 0.0500\ M$$
$$pOH = -\log [OH^-] = -\log(5.35 \times 10^{-6}) = 5.272$$
$$pH = 14.000 - pOH = 14.000 - 5.272 = 8.728$$

Practice Problem 15.15

Calculate the pH of a mixture of 25.0 mL of 0.200 M CH_3CO_2H and 25.0 mL of 0.200 M NaOH.

Answer: 8.879

In titrating a weak acid, such as acetic acid, with a strong base, we use an indicator that changes color at pH values slightly in the basic range. Phenolphthalein, which is used for strong-acid–strong-base titrations, is also suitable here, as shown in Figure 15.17. It changes from colorless to pink in the range of pH 8.2–10.0, which includes the pH value at the equivalence point in the acetic acid–sodium hydroxide titration.

A noteworthy feature of the titration curve for weak-acid–strong-base reactions is the relatively narrow transition range near the equivalence point. Compare this with the fairly wide range for strong-acid–strong-base titrations. It is possible to use any but the first and last indicators in Table 15.5 for almost any strong-acid–strong-base titration. Certainly phenolphthalein always works for such titrations, no matter which acid and base are used. For a weak-acid–strong-base titration, however, no single indicator works for every titration. The pH change is not as large, and the pH at the equivalence point varies from one weak acid to another. The indicator must be selected to change color near the pH of the equivalence point of a particular weak acid.

Strong-Acid–Weak-Base Titrations

It should not be surprising that in a strong-acid–weak-base titration, the pH at the equivalence point is less than 7; the solution will be slightly acidic. Consider the titration of 0.100 M HCl with a 0.100 M solution of the weak base $NH_3(aq)$, for which K_b is 1.76×10^{-5}. The reaction is

$$H_3O^+(aq) + NH_3(aq) \longrightarrow NH_4^+(aq) + H_2O(l)$$

Figure 15.18 shows the change in pH during the course of this titration. Note that the pH at the equivalence point is on the acidic side, pH 5.27. At the equivalence point, there are equal volumes of the two solutions in this example, producing a 0.0500 M solution of NH_4Cl. We can calculate the pH of this solution by considering the acid ionization of ammonium ion, as discussed in Section 15.3.

The pH at the equivalence point in the hydrochloric acid–aqueous ammonia titration does not fall within the range of the phenolphthalein color change, as shown in Figure 15.18. Thus we cannot use phenolphthalein here, because a large excess of $NH_3(aq)$ would have to be added before the pH would rise to a value in the color-change range, and the endpoint would occur much past the equivalence point. Instead, the indicator used must change color near pH 5. A common choice is methyl red, which changes from red to yellow in the range of pH 4.8–6.0. Again note the relatively narrow transition range near the equivalence point. The indicator must be selected with care.

Weak-Acid–Weak-Base Titrations

Titrations involving a weak acid and a weak base are usually not practical. The transition range is very narrow, making it very difficult to achieve reliable results. In addition, calculation of the pH involves mathematically complex calculations.

Figure 15.18 Titration curve for a strong-acid–weak-base titration. A 0.100 M solution of $NH_3(aq)$ is being added to 50.0 mL of a 0.100 M solution of HCl. Near the equivalence point, the pH changes rapidly from about 5 to about 7 with the addition of a very small volume of ammonia solution. A considerable excess of $NH_3(aq)$ must be added before phenolphthalein changes color, so it is not a good indicator for this titration. Methyl red does change color in the region of rapid pH change, so it can be used as an indicator for this titration.

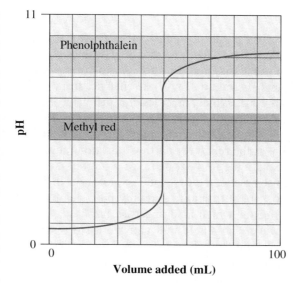

15.6 Industrial Acids and Bases

▶ The chapter introduction mentioned a number of important acids and bases used as household chemicals. Acids and bases are also high on the list of important industrial chemicals. Among the sixteen chemicals produced in the United States in excess of 10 billion pounds per year are seven acids and bases: sulfuric acid, lime (CaO, the basic anhydride of calcium hydroxide), ammonia, sodium hydroxide, phosphoric acid, nitric acid, and sodium carbonate. Hydrochloric acid is also produced in high volume, though not quite at this level. These chemicals are very useful themselves—for example, as cleaning products, such as toilet bowl cleaners (HCl), glass cleaners (NH_3), drain cleaners (NaOH), and rust removers (H_3PO_4)—and are also involved in the production of many other chemicals. This section examines the chemical properties of some important acids and bases.

Why are acids and bases produced in such large quantities?

Sulfuric Acid

Sulfuric acid is manufactured primarily by the contact process, which is described in Sections 4.6 and 14.5. Because it is inexpensive, sulfuric acid is often used when an acid is needed in a process. About two-thirds of the sulfuric acid produced is used to make phosphoric acid, which in turn is used in the production of fertilizers. Sulfuric acid is also important in the production of other fertilizer chemicals.

In concentrated solutions, especially when hot, sulfuric acid is a good oxidizing agent. As such, it can oxidize many nonmetals, such as bromide ion, shown in Figure 15.19.

$$2Br^- + H_2SO_4 \longrightarrow Br_2 + H_2O + SO_3^{2-}$$

Concentrated sulfuric acid has a very strong affinity for water, so it is used as a dehydrating agent. Sulfuric acid is so good at dehydrating that it can remove water from carbohydrates, which do not contain water, but do contain hydrogen and oxygen in a 2 : 1 ratio. As shown in Figure 15.20, sulfuric acid can convert sugar, $C_{12}H_{22}O_{11}$, to carbon. The dehydrating power of sulfuric acid is used to

Figure 15.19 Hot, concentrated sulfuric acid is a good oxidizing agent. It will oxidize a bromide salt *(Left)* to produce elemental bromine *(Right)*.

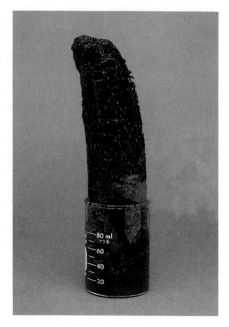

Figure 15.20 When concentrated sulfuric acid is added to sugar, it removes the components of water and forms carbon.

advantage in equilibrium reactions that produce water. Addition of sulfuric acid removes the water, thereby causing the reactions to go to completion in accordance with Le Chatelier's principle. For example, ethyl alcohol can be dehydrated in the presence of concentrated sulfuric acid to form diethyl ether, once widely used as an anaesthetic.

$$2C_2H_5OH \rightleftharpoons C_2H_5—O—C_2H_5 + H_2O$$

Concentrated sulfuric acid, which is not very volatile, releases volatile acids from their salts. For example, as shown in Figure 15.21, concentrated sulfuric acid converts sodium chloride to hydrogen chloride.

$$H_2SO_4(l) + NaCl(s) \longrightarrow NaHSO_4(s) + HCl(g)$$

The hydrogen chloride can then be dissolved in water to form hydrochloric acid. In fact, this was one of the earliest industrial methods of making hydrochloric acid.

Hydrochloric Acid

A great deal of hydrochloric acid is produced as a by-product of the chlorination of hydrocarbons.

$$C_6H_6 + Cl_2 \longrightarrow C_6H_5Cl + HCl$$

Hydrochloric acid is also a by-product of the preparation of sodium hydroxide, a process that involves the electrolysis of sodium chloride solutions.

$$2NaCl(aq) + 2H_2O(l) \xrightarrow{\text{electrolysis}} 2NaOH(aq) + H_2(g) + Cl_2(g)$$

The mixture of hydrogen and chlorine reacts if heated or exposed to light to form

Figure 15.21 When concentrated sulfuric acid is added to sodium chloride, hydrogen chloride gas is formed. The hydrogen chloride gas causes moist bromthymol blue paper to turn yellow.

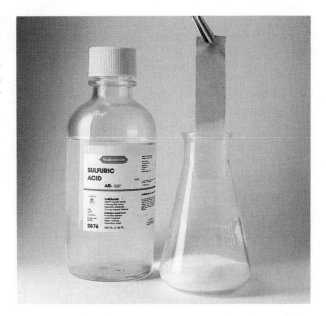

hydrogen chloride, which is dissolved in water to form hydrochloric acid.

$$H_2(g) + Cl_2(g) \longrightarrow 2HCl(g)$$

A saturated solution of hydrogen chloride at room temperature contains about 36% HCl by mass.

Hydrochloric acid is extremely corrosive to metals, so it is stored and handled in glass or plastic containers. Because it has little effect on clay-based brick but attacks concrete, it is often used to clean mortar from masonry. Hydrochloric acid is often injected into old oil wells to increase production by dissolving some of the subterranean limestone.

$$CaCO_3(s) + 2HCl(aq) \longrightarrow CaCl_2(aq) + CO_2(g) + H_2O(l)$$

RULES OF REACTIVITY

Metal carbonates react with acids that have a pK_a less than 6 to form carbon dioxide gas, water, and a metal salt.

Commercial hydrochloric acid, also called *muriatic acid,* is usually a brownish-yellow color because of iron impurities. This impure acid is widely used for pickling iron (removing rust) in preparation for plating or for recovery of scrap iron. It is also used to adjust the pH of swimming pools.

Phosphoric Acid

Phosphoric acid is made from naturally occurring phosphate rock, which is largely calcium phosphate, $Ca_3(PO_4)_2$, but also contains small amounts of calcium fluoride, CaF_2, and silicon dioxide. If the phosphoric acid need not be pure, it can be prepared directly from phosphate rock by reaction with sulfuric acid.

$$Ca_3(PO_4)_2(s) + 3H_2SO_4(l) + 6H_2O(l) \longrightarrow 3CaSO_4 \cdot 2H_2O(s) + 2H_3PO_4(l)$$

Phosphoric acid of greater purity is prepared from phosphorus. In the first step of the preparation of phosphoric acid, phosphorus (shown in Figure 15.22) is obtained from phosphate rock by heating with silicon dioxide and carbon.

$$2Ca_3(PO_4)_2(s) + 6SiO_2(l) \longrightarrow 6CaSiO_3(s) + P_4O_{10}(s)$$
$$P_4O_{10}(s) + 10C(s) \longrightarrow P_4(s) + 10CO(g)$$

Figure 15.22 Phosphorus exists in several allotropic forms, the most common being red phosphorus *(Left)* and white phosphorus *(Right)*, both of which are too reactive to occur naturally in a free state. White phosphorus, which occurs in the form of tetrahedral P_4 molecules, is more reactive than red phosphorus, which consists of chains of P_4 groups. White phosphorus must be stored under water, in which it is insoluble, to prevent its reaction with oxygen gas.

Next the oxide, P_4O_{10}, is formed by reaction with oxygen.

$$P_4(s) + 5O_2(g) \longrightarrow P_4O_{10}(s)$$

Finally, phosphoric acid is obtained from its anhydride, P_4O_{10}, which reacts quite vigorously with water.

$$P_4O_{10}(s) + 6H_2O(l) \longrightarrow 4H_3PO_4(aq)$$

Phosphoric acid is not very strong, so it does not find extensive use as an acid. Some phosphoric acid is used in naval jelly, a gelatinous mixture used to remove rust. The phosphoric acid dissolves the iron oxide and forms insoluble iron(III) phosphate, which makes a thin protective coating on the iron.

$$Fe_2O_3(s) + 2H_3PO_4(l) \longrightarrow 3H_2O(l) + 2FePO_4(s)$$

The primary use of phosphoric acid is in the preparation of inorganic fertilizers. The elements that fertilizers most commonly replace in the soil are nitrogen, phosphorus, and potassium. All these elements promote growth. In addition, nitrogen increases the size of leaves, phosphorus promotes root development, and potassium confers resistance to drought and disease. Phosphoric acid can be converted into salts that contain not only phosphorus but also the other two essential elements.

$$H_3PO_4(l) + NH_4OH(aq) \longrightarrow NH_4H_2PO_4(s) + H_2O(l)$$
$$H_3PO_4(l) + KOH(s) \longrightarrow KH_2PO_4(s) + H_2O(l)$$

These salts are mixed to prepare fertilizers.

Nitric Acid

Industrially, nitric acid is produced by the Ostwald process, in which ammonia is burned in the presence of a platinum–rhodium gauze catalyst.

$$4NH_3(g) + 5O_2(g) \longrightarrow 4NO(g) + 6H_2O(g)$$

The nitrogen monoxide reacts further with oxygen to give nitrogen dioxide.

$$2NO(g) + O_2(g) \longrightarrow 2NO_2(g)$$

The nitrogen dioxide is then reacted with water to produce nitrogen monoxide and a solution containing about 70% by mass of nitric acid.

$$3NO_2(g) + H_2O(l) \longrightarrow 2HNO_3(aq) + NO(g)$$

The nitrogen monoxide produced in this step is recycled.

Nitric acid is a strong monoprotic acid that reacts readily with metals and metal oxides to form soluble metal nitrate salts.

$$CuO(s) + 2HNO_3(aq) \longrightarrow Cu(NO_3)_2(aq) + H_2O(l)$$

Nitric acid is also a strong oxidizing agent, so it also reacts with less reactive metals. For example, in concentrated solutions it reacts with copper as follows:

$$Cu(s) + 4HNO_3(aq) \longrightarrow Cu(NO_3)_2(aq) + 2NO_2(g) + 2H_2O(l)$$

The products are different in dilute solution.

$$3Cu(s) + 8HNO_3(aq) \longrightarrow 3Cu(NO_3)_2(aq) + 2NO(g) + 4H_2O(l)$$

Very reactive metals reduce nitric acid all the way to ammonia or ammonium salts.

$$4Zn(s) + 10HNO_3(aq) \longrightarrow 4Zn(NO_3)_2(aq) + NH_4NO_3(aq) + 3H_2O(l)$$

Some metals, such as iron, chromium, and aluminum, do not react well with nitric acid because of the formation of a protective oxide coating; nitric acid is sometimes used to make these metals "passive" to corrosion.

Calcium Hydroxide

The cheapest and most readily available base is calcium hydroxide, also known as slaked lime, $Ca(OH)_2$. Slaked lime is prepared by addition of water to quicklime, CaO.

$$CaO(s) + H_2O(l) \longrightarrow Ca(OH)_2(s)$$

CaO results from the heating of calcium carbonate to 900–1200°C.

$$CaCO_3(s) \rightleftharpoons CaO(s) + CO_2(g)$$

RULES OF REACTIVITY

Metal carbonates, except for those of the alkali metals, decompose to metal oxides and carbon dioxide when heated.

The primary sources of calcium carbonate are limestone and its harder crystalline form, marble. Ultimately, both of these materials come from shellfish, which extract the ingredients from sea water. Deposits of shells subjected to high pressure become limestone, which is converted by heat and pressure to marble.

Limestone is used extensively in the iron and steel industry and in cement. Mortar used in cement contains slaked lime, which is gradually converted to calcium carbonate by reaction with atmospheric carbon dioxide.

$$Ca(OH)_2(s) + CO_2(g) \longrightarrow CaCO_3(s) + H_2O(l)$$

This is part of the process of hardening of cement.

As solids, calcium hydroxide and calcium oxide are widely used as agricultural lime to neutralize acid soils. Solid calcium oxide is also used to remove sulfur dioxide from smokestack gases.

$$CaO(s) + SO_2(g) \longrightarrow CaSO_3(s)$$

Calcium hydroxide is also useful in the form of suspensions and solutions. The addition of calcium oxide to an excess of water gives a white suspension of calcium hydroxide known as milk of lime. With sufficient water, a clear solution of calcium hydroxide results. The solution is known as lime water. Lime water is used for precipitating insoluble hydroxides of iron and magnesium from water. It is also used in the paper industry to prepare calcium hydrogen sulfite solutions.

$$Ca(OH)_2(aq) + 2SO_2(g) \longrightarrow Ca(HSO_3)_2(aq)$$

These solutions are used to dissolve lignin from wood fibers, leaving behind the cellulose needed for paper.

Sodium Hydroxide

Another common industrial base is sodium hydroxide. It is much more expensive than calcium hydroxide, so it is used industrially only where a sodium salt is required or where it is necessary to have the base in solution.

Most sodium hydroxide is prepared by electrolysis of brine (sodium chloride solutions).

$$2NaCl(aq) + 2H_2O(l) \xrightarrow{\text{electrolysis}} 2NaOH(aq) + H_2(g) + Cl_2(g)$$

Major industrial uses of sodium hydroxide include the manufacture of soaps and detergents, petroleum refining, and pulp and paper manufacture.

Aqueous Ammonia

Aqueous ammonia, also called ammonium hydroxide, is a weak base. It is prepared by the dissolution of ammonia gas in water. Ammonia is very soluble in water; under normal conditions, 1 volume of liquid water dissolves about 800 volumes of ammonia gas, yielding a 15 M solution. When diluted with enough water to give an ammonia content of about 10% by mass, aqueous ammonia solutions are used extensively as cleaning agents for windows, glassware, and porcelain ware (Figure 15.23). When mixed into a thick paste with chalk or other abrasive materials, aqueous ammonia makes an excellent polish for many metals such as brass and silver. Applied to an insect bite, aqueous ammonia can reduce pain and swelling by neutralizing the formic acid, HCOOH, that the insect has injected into the tissues.

Figure 15.23 Many glass cleaners contain dissolved ammonia, often mixed with a detergent.

Summary

S ubstances classified as acids have a number of properties in common, as do substances classified as bases. Acids and bases can neutralize one another, leading to a loss of the characteristic properties. Several theories define the chemical nature of acids and bases. These are the Arrhenius, Brønsted–Lowry, and Lewis theories.

The Brønsted–Lowry theory is the most widely used. According to this theory, an acid is a substance that can donate a proton to another substance, and a base is a proton acceptor. Acids and bases can be ranked according to their strength: their tendency to donate or accept protons. Acid–base neutralization reactions occur to form predominantly the weaker acid and weaker base by reaction of the stronger acid with the stronger base. The strengths of acids and bases can be quantified in terms of the extent to which they ionize when dissolved in water, the most common solvent. At 25°C, a neutral solution contains $1.00 \times 10^{-7}\ M\ H_3O^+(aq)$ and $1.00 \times 10^{-7}\ M\ OH^-(aq)$. An acidic solution contains more $H_3O^+(aq)$ than $OH^-(aq)$, and a basic solution contains more $OH^-(aq)$ than $H_3O^+(aq)$. The acidity of solutions is commonly expressed in terms of pH, the negative logarithm of $[H_3O^+]$.

Weak acids do not ionize completely when dissolved in solution but reach an equilibrium state. The equilibrium constant, called the acid-ionization constant, is a measure of the acid strength. Values of the acid-ionization constant and the concentration of acid dissolved in solution can be used to calculate the concentrations of all species in the solution. Base strength is measured in terms of the value of the base-ionization constant.

Acid–base reactions are often used in titrations to analyze the concentration of solutions. The change in pH as one reagent is added to the other follows a characteristic S-shaped curve called a titration curve. The pH at the equivalence point on the curve can be used to select an appropriate acid–base indicator.

Some common acids and bases—such as H_2SO_4, HCl, H_3PO_4, HNO_3, $Ca(OH)_2$, NaOH, and NH_3—are among the most widely used industrial chemicals. These substances are prepared by a variety of reactions, including acid–base reactions and oxidation–reduction reactions.

Key Terms

acid-ionization constant (K_a) (15.3)
adduct (15.1)
amphoteric substance (15.1)
anhydride (15.1)
Arrhenius acid (15.1)
Arrhenius base (15.1)

autoionization (15.2)
base-ionization constant (K_b) (15.4)
Brønsted–Lowry acid (15.1)
Brønsted–Lowry base (15.1)
conjugate acid–base

pair (15.1)
diprotic acid (15.1)
equivalence point (15.5)
hydronium ion (15.1)
ion-product constant of water (K_w) (15.2)
leveling effect (15.1)
Lewis acid (15.1)

Lewis base (15.1)
monoprotic acid (15.1)
pH (15.2)
polyprotic acid (15.1)
triprotic acid (15.1)

Exercises

Acids and Bases

15.1 What are a Brønsted–Lowry acid and a Brønsted–Lowry base?

15.2 Compare and contrast the Arrhenius, Brønsted–Lowry, and Lewis models for acids and bases.

15.3 Distinguish between monoprotic and polyprotic acids.

15.4 Give examples of monoprotic, diprotic, and triprotic acids.

15.5 What is a conjugate acid–base pair?

15.6 Indicate whether each of the following substances is an acid, a base, or neither.

a. Na_2SO_3	g. H_3AsO_4
b. HCl	h. $CaCO_3$
c. NaCl	i. NH_4Cl
d. $Ca(OH)_2$	j. NaOH
e. H_2CO_3	k. CH_3CO_2H
f. CH_4	l. $NaCH_3CO_2$

15.7 Indicate whether each of the following acids is monoprotic, diprotic, or triprotic.

a. HCl	e. CH_3CO_2H
b. $HClO_4$	f. HNO_3
c. H_2SO_3	g. H_3PO_2
d. H_2SO_4	h. H_3PO_4

15.8 Identify the conjugate base of each of the following acids.
   a. HCl
   b. NH_4^+
   c. NaHS
   d. HNO_2
   e. HCN
   f. H_3PO_4

15.9 Identify the conjugate acid of each of the following bases.
   a. NH_3
   b. PO_4^{3-}
   c. NaHS
   d. NaF
   e. NaOH
   f. OCN^-

15.10 Identify the Lewis acid and the Lewis base in each of the following reactions.
   a. $Ni^{2+}(aq) + 6NH_3(aq) \rightleftharpoons Ni(NH_3)_6^{2+}(aq)$
   b. $NF_3(g) + HBr(g) \rightleftharpoons NF_3HBr(s)$
   c. $NH_3(g) + BF_3(g) \rightleftharpoons H_3NBF_3(g)$

Strength of Acids and Bases

15.11 Distinguish between a strong and a weak acid or base.

15.12 How is the strength of an acid related to the strength of its conjugate base?

15.13 How can we predict whether the reactants or the products will predominate in a reaction between an acid and a base?

15.14 Indicate whether each of the following acids is strong or weak.
   a. HCl
   b. HBr
   c. HF
   d. HNO_3
   e. H_2SO_4
   f. H_2SO_3
   g. HSO_4^-
   h. H_3PO_4
   i. HNO_2
   j. HOCl
   k. $HClO_3$
   l. $HClO_4$

15.15 Indicate whether each of the following bases is strong or weak.
   a. NaOH
   b. $Ca(OH)_2$
   c. $Ba(OH)_2$
   d. NH_4OH
   e. $(CH_3)NH_3OH$

15.16 For the reaction

$$NH_4^+(aq) + F^-(aq) \rightleftharpoons NH_3(aq) + HF(aq)$$

indicate which species, the reactants or the products, will predominate at equilibrium.

15.17 What acid–base reaction occurs when each of the following species is dissolved in water?
   a. NH_4Br
   b. NaCN
   c. $NaHCO_3$
   d. NaBr
   e. NH_4NO_2

Oxides

15.18 What is a. an acid anhydride? b. a base anhydride?

15.19 What is an amphoteric substance?

15.20 Where in the periodic table would you find a. acidic oxides? b. basic oxides? c. amphoteric oxides?

15.21 Indicate whether each of the following oxides is acidic, basic, or amphoteric.
   a. Na_2O
   b. Br_2O_5
   c. CaO
   d. N_2O_5
   e. CO_2
   f. P_4O_{10}
   g. Al_2O_3
   h. Fe_2O_3
   i. B_2O_3
   j. Cr_2O_3
   k. CrO_3
   l. SO_2
   m. GeO_2

15.22 Write the formula of the oxide that is the anhydride of each of the following acids and bases.
   a. HNO_3
   b. $HClO_4$
   c. H_2SO_4
   d. NaOH
   e. $Ca(OH)_2$
   f. H_3BO_3
   g. H_2CrO_4
   h. $HBrO_3$
   i. $Zn(OH)_2$
   j. H_3PO_4
   k. $Fe(OH)_3$

pH and Acidity

15.23 What is pH?

15.24 How is pH measured?

15.25 Under what circumstances can the pH of a solution be used to determine how much acid was dissolved in that solution?

15.26 What is the pH of each of the following solutions?
   a. Milk, $[H_3O^+] = 3.2 \times 10^{-7}\ M$
   b. Orange juice, $[H_3O^+] = 3.2 \times 10^{-4}\ M$
   c. Seven-Up, $[H_3O^+] = 2.5 \times 10^{-4}\ M$
   d. Pickle juice, $[H_3O^+] = 2.0 \times 10^{-4}\ M$
   e. Beer, $[H_3O^+] = 3.2 \times 10^{-5}\ M$
   f. Blood, $[H_3O^+] = 4.0 \times 10^{-8}\ M$

15.27 What is the concentration of $H_3O^+(aq)$ in each of the following solutions?
   a. Lime juice (pH = 1.9)
   b. Tomato juice (pH = 4.2)
   c. Saliva (pH = 7.0)
   d. Kitchen cleaning solution (pH = 9.3)

15.28 What are the pH and the pOH of 0.156 M $HClO_4$?

15.29 What are the pH and the pOH of 0.00450 M NaOH?

15.30 What are the pH and the pOH of 0.0256 M HCl?

15.31 What is the pH a. of a 0.100 M solution of HNO_3? b. of a 0.100 M solution of HCl?

15.32 What are the pH and the pOH of 0.0652 M $Ba(OH)_2$?

15.33 What is the pH of each of the following HCl solutions?
   a. 1.00 M HCl
   b. 0.100 M HCl
   c. $1.00 \times 10^{-2}\ M$ HCl
   d. $1.00 \times 10^{-3}\ M$ HCl
   e. $1.00 \times 10^{-9}\ M$ HCl

15.34 What are the concentrations of $H_3O^+(aq)$ and $OH^-(aq)$ in a solution prepared from 0.100 mol of HNO_3 dissolved in 125 mL of water?

15.35 What are the concentrations of $H_3O^+(aq)$ and $OH^-(aq)$ in a solution prepared from 0.0250 mol of $Ba(OH)_2$ dissolved in 105 mL of water?

15.36 What are the concentrations of $H_3O^+(aq)$ and $OH^-(aq)$ in a solution prepared from 10.00 g of NaOH dissolved in 375 mL of water?

Acid and Base Ionization Constants

15.37 What is the product of the hydronium-ion concentration and the hydroxide-ion concentration in any aqueous solution at 25°C?

15.38 What is autoionization?

15.39 How is the acid-ionization constant defined?

15.40 How can an acid-ionization constant be measured?

15.41 How is the pH scale related to the ion-product constant of water?

15.42 A 0.500 M solution of pyridine contains 2.6×10^{-5} M OH^-. What is the base-ionization constant of pyridine?

15.43 The hydroxide-ion concentration of a 0.100 M solution of a weak acid HX is 5.0×10^{-10} M. What is the acid-ionization constant of HX?

15.44 A 1.00 M solution of acrylic acid has a hydrogen-ion concentration of 7.40×10^{-3} M. What is the acid-ionization constant of acrylic acid?

15.45 The pH of a 0.0400 M hypobromous acid solution is 4.00. What is the acid-ionization constant of hypobromous acid?

15.46 A weak acid HX is 0.150% ionized in a 0.0200 M solution. What is the acid-ionization constant of HX?

15.47 The pH of a 0.100 M solution of hydrazoic acid, HN_3, is 2.75. What is the acid-ionization constant of HN_3?

15.48 A 0.200 M solution of hydrazine, N_2H_4, has a hydroxide-ion concentration of 4.00×10^{-4} M. What is the base-ionization constant of hydrazine?

15.49 The pOH of a 0.100 M solution of aqueous ammonia, $NH_3(aq)$ or NH_4OH, is 2.87. What is the percent ionization of this solution of ammonia? What is the base-ionization constant?

15.50 The pH of a 0.100 M solution of a weak base is 9.63. What is the base-ionization constant of this base?

15.51 The pOH of a 0.120 M solution of formic acid, HCO_2H, is 11.67.
    a. What is the acid ionization constant of formic acid?
    b. What is the percent ionization of formic acid in this solution?

15.52 The pH of 0.0200 M aniline is 8.45. What is the base-ionization constant of aniline?

pH of Weak Acids and Bases

15.53 What assumptions can we make when calculating the concentrations of ions in a solution of a weak acid?

15.54 How are the pK_a and pK_b of a conjugate acid–base pair related?

15.55 The acid-ionization constant of acetic acid is 1.75×10^{-5}. What is the molar concentration of an acetic acid solution that has a pH of 3.45?

15.56 What is the pH of a 0.0100 M solution of hydrocyanic acid? The acid-ionization constant of HCN is 4.0×10^{-10}.

15.57 What is the pOH of a 0.100 M solution of aniline, $C_6H_5NH_2$? The base-ionization constant is 3.94×10^{-10}.

15.58 What is the pH of a solution prepared from 5.25 g of benzoic acid, $C_6H_5CO_2H$, dissolved in 500.0 mL of water, if the acid-ionization constant is 6.14×10^{-5}?

15.59 The ionization constant for HF is 7.0×10^{-4}. What is the pH of a 0.100 M solution of hydrofluoric acid?

15.60 The ionization constant for HOCl is 3.0×10^{-8}. What is the OCl^- concentration in a 0.0350 M solution of hypochlorous acid?

15.61 The ionization constant of HNO_2 is 4.6×10^{-4}. What molar concentration of HNO_2 is required to give a solution with a pH of 2.75?

15.62 The ionization constant for HCN is 4.0×10^{-10}. What is the CN^- concentration in a 0.125 M solution of hydrocyanic acid?

15.63 The ionization constant for HOBr is 2.2×10^{-9}. What is the pH of a 0.0250 M solution of hypobromous acid?

pH of Polyprotic Acid Solutions

15.64 Describe the procedure for calculating the pH of a polyprotic acid.

15.65 Why is the second ionization constant usually smaller than the first ionization constant for a polyprotic acid?

15.66 Calculate the concentration of HCO_3^- and of CO_3^{2-} in a 1.00 M solution of carbonic acid. The acid-ionization constants of H_2CO_3 are $K_{a1} = 4.45 \times 10^{-7}$ and $K_{a2} = 4.7 \times 10^{-11}$.

15.67 What is the oxalate ion ($C_2O_4^{2-}$) concentration in a 0.250 M solution of oxalic acid? The acid-ionization constants of $H_2C_2O_4$ are $K_{a1} = 5.90 \times 10^{-2}$ and $K_{a2} = 6.40 \times 10^{-5}$.

15.68 What are the concentrations of HS^- and S^{2-} in a 0.0100 M solution of H_2S? The acid-ionization constants of hydrosulfuric acid are $K_{a1} = 5.7 \times 10^{-8}$ and $K_{a2} = 1.2 \times 10^{-13}$.

pH in Titrations

15.69   What is a titration?

15.70   What is the general shape of a titration curve?

15.71   Is the endpoint of a weak-acid–strong-base titration at an acidic or a basic pH? Explain your answer.

15.72   What is the pH of the mixture resulting from the reaction of 50.0 mL of 0.150 M NaOH and 50.0 mL of 0.150 M HCl?

15.73   What is the pH of the mixture resulting from the reaction of 50.0 mL of 0.100 M NaOH and 50.0 mL of 0.200 M HCl?

15.74   What is the pH of the mixture resulting from the reaction of 25.0 mL of 0.0500 M NH_4OH and 25.0 mL of 0.0500 M HNO_3?

15.75   What is the pH of the mixture resulting from the reaction of 25.0 mL of 0.200 M KOH and 25.0 mL of 0.200 M CH_3CO_2H?

Additional Exercises

15.76   A weak acid is 0.0300% ionized in a 0.100 M solution. To what extent will it ionize in a 0.0100 M solution?

15.77   The pOH of a 0.100 M solution of hydrofluoric acid, HF, is 11.91.
    a. What is the acid-ionization constant of hydrofluoric acid?
    b. What is the percent ionization of hydrofluoric acid in this solution?

15.78   Identify the Lewis acid and the Lewis base in each of the following reactions.
    a. $Ag^+(aq) + 2NH_3(aq) \rightleftharpoons Ag(NH_3)_2{}^+(aq)$
    b. $NH_3(g) + HCl(g) \rightleftharpoons NH_4Cl(s)$
    c. $NCl_3(g) + BCl_3(g) \rightleftharpoons Cl_3NBCl_3(g)$

15.79   When lemon juice is added to tea, the tea changes color. Propose an explanation.

15.80   What is the pH of a solution prepared from 3.28 g of aniline, $C_6H_5NH_2$, dissolved in 300.0 mL of water? The base-ionization constant is 3.94×10^{-10}.

15.81   What is the pOH of a 0.200 M solution of pyridine, C_5H_5N? The base-ionization constant is 1.7×10^{-9}.

15.82   The ionization constant for NH_3 is 1.76×10^{-5}. What is the $NH_4{}^+$ concentration in a 0.0250 M solution of ammonia?

15.83   The ionization constant for HOCN is 1.20×10^{-4}. What is the OCN^- concentration in a 0.0275 M solution of cyanic acid?

15.84   The pH of a 0.100 M solution of trimethylamine, $(CH_3)_3N$, is 11.39. What is the base-ionization constant?

15.85   What is the pH of the mixture resulting from the reaction of 25.0 mL of 0.300 M NH_4OH with 50.0 mL of 0.150 M HCl?

15.86   What is the pH of a 0.100 M solution of nitrous acid, HNO_2? The acid ionization constant is 4.6×10^{-4}.

15.87   The pH of 0.300 M formic acid is 2.13. What is the acid-ionization constant of formic acid?

15.88   What is the pH of a 0.100 M solution of phosphoric acid? The acid-ionization constants are $K_{a1} = 7.52 \times 10^{-3}$, $K_{a2} = 6.22 \times 10^{-8}$, and $K_{a3} = 2.2 \times 10^{-13}$.

15.89   Dilute instant tea forms an orange color (midway between yellow and red) with methyl red. What is the approximate pH of the tea?

15.90   The pH of a 0.0100 M solution of pyridine, C_5H_5N, is 8.62. What is the acid-ionization constant of pyridinium ion, $C_5H_4NH^+$?

15.91   The ionization constant of HOCN is 1.20×10^{-4}. What molar concentration of HOCN will give a solution with a pH of 2.42?

15.92   The ionization constant of NH_3 is 1.76×10^{-5}. What molar concentration of NH_3 will give a solution with a pH of 11.50?

15.93   Write net ionic equations and equilibrium-constant expressions for each of the following reactions in aqueous solution.
    a. The acid ionization of the dimethylammonium ion, $(CH_3)_2NH_2{}^+$
    b. The base ionization of sulfate ion, $SO_4{}^{2-}$
    c. The acid ionization of hydrogen phosphate ion, $HPO_4{}^{2-}$
    d. The base ionization of hypochlorite ion, OCl^-
    e. The acid ionization of sulfurous acid, H_2SO_3

15.94   What are the pH and the pOH of 0.245 M KOH?

15.95   A sample of decolorized Gatorade® is tested for its pH as follows. When thymol blue is added, the color is yellow. When methyl orange is added, the color is red. Estimate a pH range in which the pH of Gatorade® is likely to be found.

15.96   For the reaction

$$HCO_3{}^-(aq) + NO_2{}^-(aq) \rightleftharpoons CO_3{}^{2-}(aq) + HNO_2(aq)$$

indicate which species, the reactants or the products, will predominate at equilibrium.

* 15.97   Calculate the concentrations of all ions in a 0.0100 M H_3PO_4 solution.

*15.98 Calculate the pH at the endpoint of the titration of 50.0 mL of 0.100 M hydroxybutyric acid ($HC_4H_7O_3$, $pK_a = 4.39$) with 0.100 M $NH_3(aq)$ ($pK_b = 4.84$).

*15.99 Starting with a 1.00 M solution of $HCl(aq)$, prepare a series of solutions by diluting the starting solution by a factor of 10 and then diluting the new solution by a factor of 10. Repeat the dilution process to give a total of eight solutions, each more dilute than the previous solution by a factor of 10. Calculate the pH of each solution.

*15.100 Calculate the concentrations of H_3O^+, $HCOOH$, $HCOO^-$, and OH^- in a solution prepared from 0.0150 mol of HCl, 0.0500 mol of HCOOH, and enough water to make 1.00 L of solution. The ionization constant, K_a, of formic acid (HCOOH) is 1.77×10^{-4}.

▶ ▶ ▶ Chemistry in Practice

15.101 A substance extracted from the resin of Styrax trees from Sumatra is found to have antifungal properties and is suitable for use as a food preservative. We wish to use the compound as a preservative for tomato juice, which has a pH of 4.20. The substance contains the elements carbon, hydrogen, and oxygen. A 0.3076-g sample of the substance, when burned in excess oxygen gas, produces 0.7758 g of CO_2 and 0.1361 g of H_2O. A sample was sublimed at 115.0°C into an evacuated 245-mL flask. The pressure in the flask was 228 torr, and the flask increased in mass by 0.2818 g. The substance is somewhat soluble in water. A 75.00-mL portion of a saturated solution was evaporated to dryness; this sample contained 0.2551 g of the substance. The substance is an acid; the pH of a saturated solution was measured to be 2.888. A 25.00-mL sample of the saturated solution was titrated with 0.03502 M NaOH and required 19.88 mL to reach the endpoint.

a. What is the empirical formula of the compound?
b. What is the molar mass?
c. What is the molecular formula?
d. How many acidic protons does the compound contain?
e. What is the value of pK_a for the compound?
f. To be effective as a preservative, the solution must contain 1.00×10^{-3} M of the nonionized form of the acid. What mass of the compound must be dissolved in each liter of tomato juice at pH 4.20 to act as a preservative?

16

▶ ▶ ▶ ▶

Ions and Ionic Equilibria

Iron(III) in different environments (clockwise from top): $Fe(OH)_3$ suspension; $Fe(NO_3)_3$ in solution; $Fe(NO_3)_3 \cdot 6H_2O$; and $FeCl_3 \cdot 6H_2O$.

Encountering Chemistry

▶ ▶ ▶ ▶

The form that ions take in aqueous solution is often different from their form in solid salts. The hydrated iron(III) ion, $Fe(H_2O)_6^{3+}$, is light pink, but this color is rarely seen. It can be observed in a few solids, such as hydrated iron(III) nitrate and iron(III) perchlorate. When iron(III) salts are dissolved in water, though, the resulting solutions have a yellowish-brown color much like that of hydrated iron(III) chloride. Why are these solutions not pale pink? Because they do not in fact contain the $Fe(H_2O)_6^{3+}$ ion, but rather other ions that arise from *hydrolysis* of the water bonded to the iron ion.

$$Fe(H_2O)_6^{3+}(aq) + H_2O(l) \rightleftharpoons Fe(H_2O)_5OH^{2+}(aq) + H_3O^+(aq)$$
$$\text{pink} \qquad\qquad\qquad\qquad \text{yellow}$$

As a result of this hydrolysis reaction, solutions of iron(III) salts are somewhat acidic. Solutions as concentrated as 1 *M* undergo further hydrolysis and eventually deposit a reddish-brown precipitate of $Fe(OH)_3$.

$$Fe(H_2O)_6^{3+}(aq) \rightleftharpoons Fe(OH)_3(s) + 3H_3O^+(aq)$$

Recall from Chapter 4 that hydrolysis is a reaction in which water is split into its component ions, OH^- and H_3O^+, which is also represented as $H^+(aq)$. Although hydrolysis may change the nature of substances that we dissolve in solution, we can control the form of substances in solution to some extent. For example, we can control the form of the iron(III) ion in an aqueous system by adjusting the pH of the system. Addition of acid shifts the equilibrium to one or more of the soluble ions, such as the hydrated iron(III) ion.

$$Fe(OH)_3(s) + 3H_3O^+(aq) \rightleftharpoons Fe(H_2O)_6^{3+}(aq)$$

The addition of many ionic salts, including iron(III) salts, to water or to aqueous solutions of other substances affects the pH of the solution. In fact, addition of the proper substances can be used to control the pH of solutions. We often control the pH of a solution with *buffers,* which are just mixtures of weak acids and their conjugate bases. You have probably heard of buffered medications, such as aspirin and the antacids used to control gastric acid.

Many metal ions, including iron(III) ions, precipitate from solution only within a certain range of pH values. The deposition of many minerals, especially those containing oxide, hydroxide, sulfide, and carbonate salts, depends on the pH of the solution that is in contact with these minerals. The deposition of such minerals may also depend on other ions present in the solution. For example, when a chloride salt is added to a solution containing iron(III) ions, the solution turns yellow, much like the color of hydrated iron(III) chloride, $FeCl_3 \cdot 6H_2O$. In both cases, chloride ions are bonded to the iron(III) to give species called *complex ions.* For example, the solid consists of $[Fe(H_2O)_4Cl_2]Cl \cdot 2H_2O$, which contains the complex ion $[Fe(H_2O)_4Cl_2]^+$. The pH at which iron(III) hydroxide precipitates from a chloride solution is different from the pH required in the absence of chloride ions.

This chapter considers various factors related to the iron(III) chemistry just discussed and to the chemical behavior of other ionic substances. Depending on the nature of ions in solution, they may undergo hydrolysis to change the pH. In appropriate buffer mixtures, this effect can be used to control pH. The formation of complex ions is also important in the chemical behavior of metal ions in solution. These factors, along with the concentrations of dissolved substances, help determine the solubility of ionic salts.

16.1 Hydrolysis

Why do solutions of many metal ions have a pH less than 7?

In Chapter 15, we saw that solutions of acids and bases have pH values different from that of pure water (7.00) and that the value depends on the concentration and the strength of the acid or base. Substances other than the common acids and bases can also affect the pH of solutions. Some salts, when dissolved in water, produce a solution that is acidic or basic. For example, when ammonium chloride is dissolved in water, the solution becomes slightly acidic, whereas the dissolution of sodium acetate gives a slightly basic solution, as shown in Figure 16.1. Other salts, such as NaCl, dissolve to yield a neutral solution.

The pH changes are due to *hydrolysis,* a Brønsted–Lowry acid–base reaction of a substance with water, in which either aqueous hydrogen ions or aqueous hydroxide ions are split out of the water molecule. The production of one of these ions causes the change in pH. Let's consider in more detail what happens when an

Figure 16.1 *Left:* When methyl red is added to a 1 *M* solution of ammonium chloride, the solution is red, indicating that the pH is less than 5 and the solution is acidic. *Right:* When bromthymol blue is added to a 1 *M* solution of sodium acetate, the solution is blue, indicating that the pH is 8 or greater and the solution is basic.

ionic salt dissolves in water. Generally, such salts first dissociate into their component ions, as shown by the following equation:

$$NH_4Cl(s) \xrightarrow{H_2O} NH_4^+(aq) + Cl^-(aq)$$

Once in solution, the ions may undergo a further reaction with water—hydrolysis. However, as indicated by the following equations, not all ions undergo hydrolysis, or *hydrolyze*.

$$NH_4^+(aq) + H_2O(l) \longrightarrow H_3O^+(aq) + NH_3(aq)$$
$$Cl^-(aq) + H_2O(l) \longrightarrow \text{no reaction}$$

What causes this difference in behavior? The acid or base strength of the ions is the deciding factor.

Some cations, such as NH_4^+, are the conjugate acids of weak bases, so they are weak acids. Some anions, such as $CH_3CO_2^-$, are the conjugate bases of weak acids, so they are weak bases. The hydrolysis reactions are just Brønsted–Lowry acid-ionization or base-ionization reactions, and these ions can undergo such ionization reactions in aqueous solution, just like any other weak acid or base.

What about the cations and anions that do not hydrolyze? Any anion that is produced by ionization of a strong acid does not hydrolyze in water, because the anion is an extremely weak base. Thus the anions Cl^-, Br^-, I^-, ClO_4^-, NO_3^-, and HSO_4^- do not hydrolyze in water.

Cations that do not hydrolyze in water are those that form ionic hydroxide salts that dissociate completely in water. This group includes all the metal ions in Groups IA (1) and IIA (2) of the periodic table, except for Be^{2+}. ▶ Many other 2+ cations and all cations of charge greater than 2+, such as the iron(III) ions mentioned in the chapter introduction, hydrolyze in water. These cations generally exist as hydrates; the $Cr(H_2O)_6^{3+}$ shown in Figure 16.2 is an example. These hydrated cations can hydrolyze by transferring a proton from one of the water molecules bonded to the cation to a solvent water molecule.

$$Cr(H_2O)_6^{3+}(aq) + H_2O(l) \rightleftharpoons Cr(H_2O)_5(OH)^{2+}(aq) + H_3O^+(aq)$$

Other common cations that hydrolyze in this manner include Fe^{3+}, Al^{3+}, Cu^{2+}, and Zn^{2+}.

Ionic salts can be placed in four categories on the basis of the strength of the acids and bases from which they were formed: (1) salts of strong bases and strong

Figure 16.2 The aqueous chromium(III) ion, $Cr(H_2O)_6^{3+}$, has a blue-gray color. This ion, like most 3+ ions, undergoes hydrolysis in solution, forming ions that have hydroxide bonded to the metal.

acids, such as NaCl; (2) salts of strong bases and weak acids, such as $NaCH_3CO_2$; (3) salts of weak bases and strong acids, such as NH_4Cl; and (4) salts of weak bases and weak acids, such as $NH_4CH_3CO_2$. Salts in each of these categories produce aqueous solutions that have different ranges of pH.

Hydrolysis of Salts of Strong Bases and Strong Acids

Sodium chloride and salts like it can be considered to be formed by the reaction of a strong base with a strong acid.

$$NaOH(aq) + HCl(aq) \longrightarrow Na^+(aq) + Cl^-(aq) + H_2O(l)$$

The net reaction is just the formation of water from hydroxide ion and hydronium ion.

$$Na^+(aq) + OH^-(aq) + H_3O^+(aq) + Cl^-(aq) \longrightarrow Na^+(aq) + Cl^-(aq) + 2H_2O(l)$$

Because the aqueous sodium ion arises from the dissociation of a strong base (NaOH), it must be a very weak acid. Thus it does not undergo significant hydrolysis.

$$Na^+(aq) + H_2O(l) \longrightarrow \text{no reaction}$$

Similarly, aqueous chloride ion comes from a strong acid (HCl), so it is a very weak base and does not hydrolyze.

$$Cl^-(aq) + H_2O(l) \longrightarrow \text{no reaction}$$

Cesium nitrate consists of white lustrous crystals that are very soluble in water. It is used for the synthesis of other cesium salts, which are used for prisms and X-ray fluorescent screens.

When any salt formed from a strong acid and a strong base is dissolved in water, the resulting solution is neutral, with a pH of 7.00. Other examples include $BaCl_2$, KBr, and $CsNO_3$.

Hydrolysis of Salts of Strong Bases and Weak Acids

Salts containing cations of strong bases and anions of weak acids dissolve in water to yield a slightly basic solution. Consider an aqueous solution of sodium fluoride, which dissociates into Na^+ and F^- ions. Cations such as sodium ion, which are extremely weak acids, do not undergo further reaction with water. However, the fluoride ion, F^-, is the conjugate base of a weak acid, HF. It is a moderately weak base, so it reacts with water.

$$F^-(aq) + H_2O(l) \rightleftharpoons HF(aq) + OH^-(aq)$$

Because a small amount of hydroxide ion is formed as a result of hydrolysis of fluoride ion, the solution is basic. All solutions resulting from the hydrolysis of salts that contain a cation of a strong base and an anion of a weak acid are basic, with a pH greater than 7.

How much hydrolysis of the anion will occur in such reactions? We can calculate this from the concentration of the salt and the equilibrium constant for the reaction, following the same calculation procedure used in Chapter 15 for the ionization of a weak base. The equilibrium constant for this hydrolysis reaction is identical in form to a base-ionization constant.

$$F^-(aq) + H_2O(l) \rightleftharpoons HF(aq) + OH^-(aq)$$

$$K_b = \frac{[HF][OH^-]}{[F^-]}$$

The equilibrium constant for a hydrolysis reaction is often referred to as the **hydrolysis constant,** symbolized K_h. Because they are identical to the K_b values, values of hydrolysis constants can be derived from the acid-ionization constant of the conjugate acid, HF in this case.

$$F^-(aq) + H_2O(l) \rightleftharpoons HF(aq) + OH^-(aq)$$

$$K_h = K_b = K_w/K_a = (1.0 \times 10^{-14})/(7.0 \times 10^{-4}) = 1.4 \times 10^{-11}$$

The procedure for determining the extent of hydrolysis in a reaction such as this is identical to that used to determine the extent of ionization of a weak base in Example 15.12 in Section 15.4 and Example 15.15 in Section 15.5. Such a calculation for a 0.10 M solution of NaF indicates that the pH will be 8.08 and that the percent hydrolysis will be 0.0012%.

Hydrolysis of Salts of Weak Bases and Strong Acids

What pH is expected for the solution of a salt that contains a cation that is a weak acid?

When a salt containing the cation derived from a weak base and the anion of a strong acid dissolves in water and dissociates into its ions, the cation undergoes hydrolysis, forming the weak base and H_3O^+. For example, NH_4Cl dissolves in water to give NH_4^+ and Cl^- ions.

$$NH_4Cl(s) \xrightarrow{H_2O} NH_4^+(aq) + Cl^-(aq)$$

Because the Cl^- ion is the anion of a strong acid, it does not react further with water. However, being the conjugate acid of a weak base, NH_4^+ does react with water.

$$NH_4^+(aq) + H_2O(l) \rightleftharpoons NH_3(aq) + H_3O^+(aq)$$

$$K_h = K_a = K_w/K_b = (1.0 \times 10^{-14})/(1.76 \times 10^{-5}) = 5.7 \times 10^{-10}$$

The hydrolysis of NH_4^+ produces $H_3O^+(aq)$, so the solution is acidic, with a pH less than 7.

Finding the extent of hydrolysis of a cation of a weak base is identical to the procedure described for the ionization of a weak acid in Examples 15.8 and 15.9

in Section 15.3. For example, we can calculate the pH of a 0.010 M NH_4Cl solution to be 5.62 and the percent ionization to be 0.024%.

Hydrolysis of Salts of Weak Bases and Weak Acids

When both the cation and the anion in a salt hydrolyze, how do we predict whether its solution will be acidic or basic?

When a salt contains both the cation of a weak base and the anion of a weak acid, both ions undergo hydrolysis when the salt is dissolved in water. The solution of such a salt may be acidic, basic, or neutral. If the cation hydrolyzes more than the anion hydrolyzes, more $H_3O^+(aq)$ than $OH^-(aq)$ is produced, and the solution is acidic. If the anion hydrolyzes to the greater extent, the solution is basic. If both hydrolyze to the same extent, the solution is neutral.

In a salt of this type, the cation and the anion are present in equal initial concentrations, so a comparison of the hydrolysis constants is all that is needed to determine the relative degrees of hydrolysis. If the hydrolysis constant of the cation is greater, the solution is acidic; if that of the anion is greater, the solution is basic.

Consider the example of ammonium cyanate, NH_4OCN. In aqueous solution, both ions in this salt hydrolyze.

$$NH_4^+(aq) + H_2O(l) \rightleftharpoons NH_3(aq) + H_3O^+(aq)$$
$$OCN^-(aq) + H_2O(l) \rightleftharpoons HOCN(aq) + OH^-(aq)$$

Ammonium cyanate, NH_4OCN, exists as very soluble white crystals that decompose when heated above 60°C. Cyanic acid is a colorless liquid with a strong, acrid odor that brings tears to the eyes. It decomposes slowly in water to form ammonia and carbon dioxide. Cyanate salts are used to inhibit the sickling of red blood cells (the condition known as sickle cell anemia).

The ammonium ion is the conjugate acid of aqueous ammonia, so its hydrolysis constant is

$$K_h = K_a \text{ of } NH_4^+ = K_w/K_b \text{ of } NH_3 = (1.0 \times 10^{-14})/(1.76 \times 10^{-5}) = 5.7 \times 10^{-10}$$

The cyanate ion is the conjugate base of cyanic acid, so its hydrolysis constant is

$$K_h = K_b \text{ of } OCN^- = K_w/K_a \text{ of } HOCN = (1.0 \times 10^{-14})/(3.46 \times 10^{-4}) = 2.9 \times 10^{-11}$$

Because NH_4^+ has a larger hydrolysis constant than OCN^-, it hydrolyzes to a greater extent. As shown in Figure 16.3, more $H_3O^+(aq)$ than $OH^-(aq)$ is produced, and the solution is acidic.

EXAMPLE 16.1

Predict whether a solution of ammonium nitrite, NH_4NO_2, will be acidic, basic, or neutral. The value of K_a for HNO_2 is 4.6×10^{-4}, and the value of K_b for NH_3 is 1.76×10^{-5}.

Solution When ammonium nitrite dissolves, it dissociates into NH_4^+ and NO_2^- ions. The ions are derived from a weak base and a weak acid, so both hydrolyze.

$$NH_4^+(aq) + H_2O(l) \rightleftharpoons NH_3(aq) + H_3O^+(aq)$$
$$NO_2^-(aq) + H_2O(l) \rightleftharpoons HNO_2(aq) + OH^-(aq)$$

To determine whether the solution will be acidic or basic, we must determine the values of the hydrolysis constants.

$$K_h = K_a \text{ of } NH_4^+ = K_w/K_b \text{ of } NH_3 = (1.0 \times 10^{-14})/(1.76 \times 10^{-5}) = 5.7 \times 10^{-10}$$
$$K_h = K_b \text{ of } NO_2^- = K_w/K_a \text{ of } HNO_2 = (1.0 \times 10^{-14})/(4.6 \times 10^{-4}) = 2.2 \times 10^{-11}$$

Because K_a is greater than K_b, the ammonium ion hydrolyzes to a greater extent than the nitrite ion. Thus more $H_3O^+(aq)$ than $OH^-(aq)$ is generated when the system reaches a state of equilibrium, and the solution is acidic.

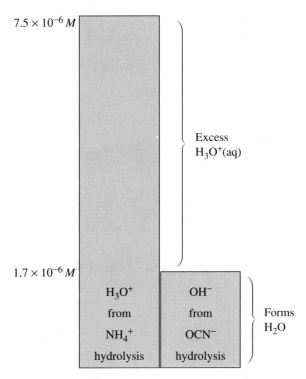

Figure 16.3 The hydrolysis of the ions in a 0.100 M NH$_4$OCN solution produces more H$_3$O$^+$(aq) than OH$^-$(aq), so the solution is acidic.

Practice Problem 16.1

Predict whether a solution of pyridinium hypochlorite, C_5H_5NHOCl, will be acidic, basic, or neutral. The value of K_a for HOCl is 3.0×10^{-8}, and the value of K_b for C_5H_5N is 1.7×10^{-9}.

Answer: slightly acidic

Summary of Hydrolysis of Salts

We have seen that the effect of the hydrolysis of a salt on the pH of a solution depends on the type of salt. The salts can be characterized according to the strength of the acids and bases from which they were derived. The classes of salts and the ions undergoing hydrolysis are summarized in Table 16.1. These classifications can be used to predict whether a solution will be acidic, basic, or neutral.

Hydrolysis of Metal Ions

▶ As noted in the chapter introduction, solutions of some metal ions are acidic. It is not always obvious that these metal ions are the conjugate acids of weak bases, because the hydroxide salts of most highly charged metal ions are insoluble. Metal ions that give acidic aqueous solutions usually are small and highly charged. Examples of such cations include Fe^{3+}, Cr^{3+}, Al^{3+}, Bi^{3+}, Cu^{2+}, Zn^{2+}, and Be^{2+}.

Table 16.1 Hydrolysis of Salts

Salt Derived from:	Ions Undergoing Hydrolysis	pH	Examples
Strong base, strong acid	Neither	Neutral, pH = 7	$NaCl$, KNO_3, $BaCl_2$, $CaBr_2$
Strong base, weak acid	Anion	Basic, pH > 7	$LiCN$, KNO_2, CaF_2 $NaCH_3CO_2$
Weak base, strong acid	Cation	Acidic, pH < 7	NH_4Cl, $Al(NO_3)_3$, $(CH_3)_3NHBr$
Weak base, weak acid	Both	Acidic if $K_b < K_a$; neutral if $K_b = K_a$; basic if $K_b > K_a$	NH_4NO_2 $NH_4CH_3CO_2$ NH_4CN

We usually represent these ions in aqueous solution as $M^{n+}(aq)$. For example, the iron(III) ion is written as $Fe^{3+}(aq)$. These ions are hydrated when in aqueous solution, the hydrated ion being bonded to the oxygen of water molecules by coordinate covalent bonds.

$$Fe^{3+} + 6H_2O \rightleftharpoons Fe(H_2O)_6{}^{3+}$$

This is a Lewis acid–base reaction in which oxygen donates a pair of electrons to the iron(III) ion: $Fe:OH_2$. The 3+ ions usually have six coordinated water molecules, whereas the 2+ ions usually have four. For some hydrated metal ions, the stability of the coordinate covalent bonds is sufficiently great that they persist when the ion is precipitated from solution. ▶ For example, iron(III) nitrate, discussed in the chapter introduction, has a formula that is usually represented as $Fe(NO_3)_3 \cdot 6H_2O$. In fact, though, this solid contains hydrated iron(III) ions and should be represented as $[Fe(H_2O)_6](NO_3)_3$.

When a water molecule is bonded to a metal ion by coordinate covalent bonds, changes occur in the electron distribution of the water molecule. The positively charged cation attracts electrons away from the oxygen atom. This in turn reduces the electron density around the hydrogen atoms in the water molecule, giving the O—H bond more ionic character. The net effect is to make the hydrogen ion easier to remove from the water molecule. The water molecule is more acidic when bonded to the metal ion than it was as an unbound molecule. This is the basis for the hydrolysis of metal ions. It is the bound water molecules that release hydrogen ions.

$$Fe(H_2O)_6{}^{3+}(aq) + H_2O(l) \rightleftharpoons Fe(H_2O)_5(OH)^{2+}(aq) + H_3O^+(aq)$$

The acid-ionization constants for several hydrated metal ions are given in Table 16.2. Note that the value of K_a generally increases as the charge on the ion increases and as the size of the ion decreases. This is consistent with the idea that a more concentrated charge on the metal ion pulls more electron density from the O—H bond and increases the acidity of the bound water molecules.

Table 16.2 Acid-Ionization Constants for Some Metal Ions

Metal Ion	Ionic Radius (pm)	K_a
Na^+	95	3.3×10^{-15}
Li^+	60	1.5×10^{-14}
Be^{2+}	31	3.2×10^{-7}
Mg^{2+}	65	3.8×10^{-12}
Ca^{2+}	99	2.0×10^{-13}
Sr^{2+}	113	6.6×10^{-14}
Ba^{2+}	135	1.5×10^{-14}
Fe^{2+}	76	7.9×10^{-11}
Ni^{2+}	72	4.0×10^{-10}
Cu^{2+}	70	3.0×10^{-8}
Zn^{2+}	74	2.5×10^{-10}
Al^{3+}	50	7.2×10^{-6}
Cr^{3+}	69	9.8×10^{-5}
Fe^{3+}	64	6.5×10^{-3}
Bi^{3+}	102	2.6×10^{-2}
Zr^{4+}	78	6.0×10^{-1}

16.2 The Common-Ion Effect

What happens to the pH when we prepare a solution from a mixture of salts?

The dissolution and hydrolysis of an ionic salt in water can shift the pH from the neutral value. Similarly, the addition of some salts to solutions of weak acids or bases can shift the pH. Salts that can do this contain the conjugate base or conjugate acid of the weak acid or base in solution.

We can understand the effect by recalling that a solution of a weak acid or weak base is an equilibrium system. When a reactant or product is added to a system at equilibrium, the reaction shifts in such a way as to reduce the stress caused by the added substance, and a new equilibrium condition is established, in accordance with Le Chatelier's principle. This principle was discussed in Chapter 14 as it applied to gas-phase equilibria, but it pertains to aqueous equilibria as well. The shift in equilibrium caused by the addition of a salt to an aqueous solution of a weak acid or weak base is called the **common-ion effect** when an ion produced by ionization of the weak acid or weak base is also contained in the salt.

Let's look at an example: the addition of some sodium acetate, $NaCH_3CO_2$, to a solution of acetic acid, CH_3CO_2H. The solution of acetic acid is initially in a state of equilibrium.

$$CH_3CO_2H(aq) + H_2O(l) \rightleftharpoons H_3O^+(aq) + CH_3CO_2^-(aq)$$

The solution contains a mixture of nonionized $CH_3CO_2H(aq)$ and its conjugate base $CH_3CO_2^-(aq)$, along with $H_3O^+(aq)$. Now we add $NaCH_3CO_2$, a strong electrolyte that dissociates completely to form Na^+ and $CH_3CO_2^-$ ions upon dissolution in water.

$$NaCH_3CO_2(aq) \xrightarrow{H_2O} Na^+(aq) + CH_3CO_2^-(aq)$$

Sodium acetate exists in an anhydrous form, $NaCH_3CO_2$, and in a hydrated form, $NaCH_3CO_2 \cdot 3H_2O$. Both are white solids used in the preparation of buffer solutions.

Addition of the salt to the acetic acid solution increases the concentration of product acetate ion to a value greater than the equilibrium value. The solution now contains $CH_3CO_2H(aq)$, $CH_3CO_2^-(aq)$, $H_3O^+(aq)$, and $Na^+(aq)$. Acetate ion is produced by both acetic acid and sodium acetate, so it is the **common ion.** Sodium ion is a spectator ion and does not participate in the equilibrium involving the other species. According to Le Chatelier's principle, addition of a reaction product causes the position of equilibrium to shift back toward reactants. Here acetate ion combines with aqueous hydronium ion until the reaction again reaches a state of equilibrium. The result of a shift in the equilibrium position is a decrease in the amount of aqueous hydronium ion. The solution becomes less acidic and the pH increases as a result of the addition of acetate ion, as shown in Figure 16.4. The net effect of the addition of sodium acetate to an acetic acid solution is suppression of the ionization of acetic acid.

We can calculate the quantitative effect of adding a common ion by using the general approach developed earlier for calculating the pH of a weak-acid or weak-base solution. The pH of a solution of a weak acid containing added conjugate base is calculated in the usual manner, except that the initial concentration of the conjugate base is not zero.

EXAMPLE 16.2

Calculate the pH of 0.100 M CH_3CO_2H before and after the addition of 0.0500 mol/L of $NaCH_3CO_2$. The value of K_a for acetic acid is 1.75×10^{-5}.

Solution   The equation for the ionization of acetic acid is

$$CH_3CO_2H(aq) + H_2O(l) \rightleftharpoons H_3O^+(aq) + CH_3CO_2^-(aq)$$

The equilibrium-constant expression is

$$K_a = \frac{[H_3O^+][CH_3CO_2^-]}{[CH_3CO_2H]} = 1.75 \times 10^{-5}$$

The concentrations for the system at equilibrium before addition of sodium acetate are summarized in the following table:

Substance	CH_3CO_2H	$CH_3CO_2^-$	H_3O^+
Initial concentration, M	0.100	0.0	0.0
Change in concentration, M	$-x$	$+x$	$+x$
Equilibrium concentration, M	$0.100 - x$	$0.0 + x$	$0.0 + x$

The expressions for the equilibrium concentrations can now be substituted into the equilibrium-constant expression.

$$K_a = \frac{(x)(x)}{0.100 - x} = 1.75 \times 10^{-5}$$

As usual, we make the assumption that x is much less than 0.100 to simplify the equation.

$$x^2/0.100 = 1.75 \times 10^{-5}$$

This equation can then be solved.

$$x^2 = 1.75 \times 10^{-6}$$
$$x = 1.32 \times 10^{-3} \, M$$

Figure 16.4 When sodium acetate is added to a solution of acetic acid, the solution becomes less acidic, as shown by the change in color of methyl orange from its orange acidic color (*Left*) to its yellow basic color (*Right*).

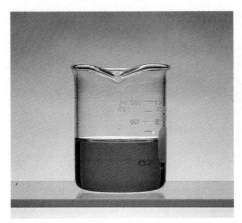 

The assumption is valid, because x is only about 1% of the initial concentration of acid. Substituting into the expression for the hydronium-ion concentration gives

$$[H^+] = x = 1.32 \times 10^{-3}\ M$$

The pH is obtained from this concentration.

$$pH = -\log [H_3O^+]$$
$$= -\log(1.32 \times 10^{-3}) = 2.879$$

The pH of the solution after the addition of sodium acetate is calculated in exactly the same way. When sodium acetate is dissolved in the solution, it dissociates completely.

$$NaCH_3CO_2 \xrightarrow{H_2O} Na^+(aq) + CH_3CO_2^-(aq)$$

The acetate ion, being a common ion, suppresses the ionization of acetic acid. The new concentrations can be summarized in a table, as before, only now the acetate ion concentration is not zero at the start.

Substance	CH_3CO_2H	$CH_3CO_2^-$	H_3O^+
Initial concentration, M	0.100	0.0500	0.0
Change in concentration, M	$-x$	$+x$	$+x$
Equilibrium concentration, M	$0.100 - x$	$0.0500 + x$	$0.0 + x$

The equilibrium concentrations are substituted into the equilibrium-constant expression.

$$K_a = \frac{[H_3O^+][CH_3CO_2^-]}{[CH_3CO_2H]} = \frac{(x)(0.0500 + x)}{0.100 - x} = 1.75 \times 10^{-5}$$

Again, we assume that x is small enough for the following relationships to be true.

$$0.100 - x = 0.100$$
$$0.0500 + x = 0.0500$$

Then

$$(x)(0.0500)/(0.100) = 1.75 \times 10^{-5}$$
$$x = 1.75 \times 10^{-5} \times 0.100/0.0500 = 3.50 \times 10^{-5}$$

The assumption that x is small compared with 0.100 and 0.0500 was valid. From the table, we derive the concentration of hydronium ion.

$$[H_3O^+] = x = 3.50 \times 10^{-5}$$

(continued)

The pH is given by

$$pH = -\log [H_3O^+]$$
$$= -\log(3.50 \times 10^{-5}) = 4.456$$

Thus the pH of the acetic acid solution is much higher after addition of acetate ion to the solution. The ionization of acetic acid was suppressed by addition of acetate ion.

Practice Problem 16.2

Calculate the pH of a solution of 0.200 M CH_3CO_2H before and after the addition of 0.150 mol/L of $NaCH_3CO_2$. The value of K_a for acetic acid is 1.75×10^{-5}.

Answers: 2.728 before and 4.633 after

In general, then, adding a salt containing an ion produced by the weak acid or weak base causes a shift in equilibrium toward the reactant. We have already seen that addition of the common ion to a solution of a weak acid increases the pH by decreasing the degree of ionization. Similarly, addition of the appropriate common ion to a solution of a weak base decreases the pH of the solution. Consider the ionization of the weak base ammonia, for example:

$$NH_3(aq) + H_2O(l) \rightleftharpoons NH_4^+(aq) + OH^-(aq)$$

The addition of NH_4^+ ions in the form of a salt such as NH_4Cl causes the equilibrium to shift back toward nonionized aqueous ammonia. The hydroxide-ion concentration is lowered as a result, so the hydronium-ion concentration is increased and the pH is decreased. The quantitative effect of addition of a common ion can be calculated in exactly the same way as when it is added to a weak acid.

16.3 Buffer Solutions

How can we keep the pH of a solution nearly constant, even if more acid or base is added?

We have just investigated the effect of the addition of a common ion on the pH of a solution of a weak acid or a weak base. One application of the common-ion effect in acid- or base-ionization reactions is the preparation of buffer solutions. A **buffer** solution is a solution prepared from nearly equal amounts of a weak acid and its conjugate base or of a weak base and its conjugate acid. Such a solution can consume a small amount of added strong acid or base with only a small accompanying change in the pH of the solution, so buffer solutions are often used to maintain nearly constant pH in reacting systems.

A typical example of a simple buffer system consists of a solution of acetic acid to which sodium acetate has been added. Acetic acid and acetate ion exist together in an equilibrium state.

$$CH_3CO_2H(aq) + H_2O(l) \rightleftharpoons H_3O^+(aq) + CH_3CO_2^-(aq)$$

Why does a buffer solution consume added acid?

Adding a small amount of a strong acid such as $HCl(aq)$ introduces more of the common ion $H_3O^+(aq)$ into the system. The addition of a product, remember, causes a shift in the equilibrium position, which in turn causes some of the added product to be consumed. The added $H_3O^+(aq)$, then, causes the reaction to shift back toward acetic acid. Some of the $H_3O^+(aq)$ is used up, along with some $CH_3CO_2^-(aq)$. Addition of the strong acid does lower the pH of the solution.

However, the same amount of hydrochloric acid would have had a much greater effect on an acetic acid solution with no added acetate ion, as shown in Figure 16.5.

Similarly, added bases react with the weak acid present in the buffer solution. If a small amount of a strong base such as NaOH is added to the acetic acid/acetate mixture, the OH$^-$ reacts with the acetic acid.

$$CH_3CO_2H(aq) + OH^-(aq) \longrightarrow CH_3CO_2^-(aq) + H_2O(l)$$

Almost all of the hydroxide ion is consumed, so the pH does not shift as much as if the same amount of sodium hydroxide had been added to pure water or to a solution of sodium acetate.

Buffering Action in Blood

Buffers are found in all body fluids and help to maintain the pH of these fluids at the proper level. In Chapter 15, we saw that blood normally has a pH in the range of 7.35–7.45 and that health problems arise if the pH shifts out of this range.

The pH of blood is controlled by three buffer systems, each a mixture of a weak acid and its salt.

Bicarbonate buffer: $H_2CO_3(aq) + H_2O(l) \rightleftharpoons H_3O^+(aq) + HCO_3^-(aq)$

Phosphate buffer: $H_2PO_4^-(aq) + H_2O(l) \rightleftharpoons H_3O^+(aq) + HPO_4^{2-}(aq)$

Various protein buffers: $HPr(aq) + H_2O(l) \rightleftharpoons H_3O^+(aq) + Pr^-(aq)$

Buffers act to maintain nearly constant pH by reacting with added acids or bases. If extra acid is added to blood, it is removed by reaction with HCO$_3^-$ in the bicarbonate buffer system (or the analogous anion in the other buffer systems).

$$H_3O^+(aq) + HCO_3^-(aq) \longrightarrow H_2CO_3(aq) + H_2O(l)$$

If acid is removed from blood, it can be replenished by H$_2$CO$_3$.

$$H_2CO_3(aq) + H_2O(l) \longrightarrow H_3O^+(aq) + HCO_3^-(aq)$$

As long as a mixture of H$_2$CO$_3$ and HCO$_3^-$ exists, the buffering capacity keeps the pH relatively constant.

Sodium bicarbonate, or sodium hydrogen carbonate, NaHCO$_3$, is a white crystalline solid. It begins to lose CO$_2$ if heated to 50°C, and it loses all its CO$_2$ when heated to 100°C. It is used to manufacture many other sodium salts by reaction with the appropriate acid. It is also used in baking powder and some antacid preparations, in fire extinguishers, and in cleaning powders.

Figure 16.5 The pH changes when 1.0 mL of 1.0 *M* HCl is added to 100 mL of 0.1 *M* CH$_3$CO$_2$H (*Left:* before, *Left center:* after) or 0.1 *M* CH$_3$CO$_2$H/0.1 *M* NaCH$_3$CO$_2$ buffer. (*Right center:* before, *Right:* after). The pH change is greater for the acetic acid solution than for the buffer solution.

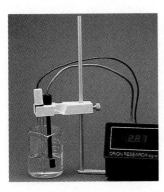
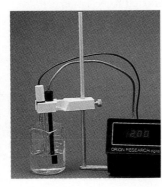
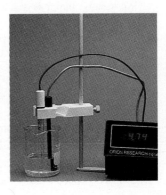
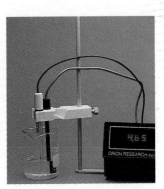

Calculating the pH of Buffer Systems

To calculate the pH of a buffer system before a strong acid or base is added, we follow the procedure outlined in Example 16.2. In this procedure, we assume that the amount of aqueous hydrogen ion formed is small compared with the initial concentrations of the acid and its conjugate base. This assumption is generally valid if the buffer components are present in sufficient concentration, say $0.1\ M$ or greater. The simplest method for calculating the pH after addition of a small quantity of a strong acid or base to the buffer system involves assuming complete reaction to remove all the added acid or base, followed by re-establishment of equilibrium with the new initial concentrations.

EXAMPLE 16.3

Consider a buffer system consisting of $1.00\ M$ CH_3CO_2H and $1.00\ M$ $NaCH_3CO_2$, with $K_a = 1.75 \times 10^{-5}$ for acetic acid. This buffer solution has a pH of 4.757. Calculate the pH of this buffer after the addition of 0.0100 mol HCl per liter of buffer solution and after the addition of 0.0500 mol NaOH per liter of buffer solution.

Solution   When $0.0100\ M$ HCl is added, the HCl ionizes completely into $H_3O^+(aq)$ and $Cl^-(aq)$ ions. To determine new initial conditions, we assume that the added $H_3O^+(aq)$ reacts completely with acetate ions.

$$H_3O^+(aq) + CH_3CO_2^-(aq) \longrightarrow CH_3CO_2H(aq) + H_2O(l)$$

Because 0.0100 mol/L of acetate is consumed in the reaction and 0.0100 mol/L of acetic acid is formed, the resulting solution has the initial concentrations shown in the following table. The change in concentration, x, is related to the equilibrium equation

$$CH_3CO_2H(aq) + H_2O(l) \rightleftharpoons H_3O^+(aq) + CH_3CO_2^-(aq)$$

The concentrations are as follows:

Substance	CH_3CO_2H	$CH_3CO_2^-$	H_3O^+
Initial concentration, M	1.01	0.99	0.0
Change in concentration, M	$-x$	$+x$	$+x$
Equilibrium concentration, M	$1.01 - x$	$0.99 + x$	$0.0 + x$

The expressions for the equilibrium concentrations are substituted into the equilibrium-constant expression.

$$K_a = \frac{[H_3O^+][CH_3CO_2^-]}{[CH_3CO_2H]} = \frac{(x)(0.99 + x)}{1.01 - x} = 1.75 \times 10^{-5}$$

We assume that x is small enough to be ignored.

$$\frac{(x)(0.99)}{1.01} = 1.75 \times 10^{-5}$$

Solving for x gives

$$x = 1.75 \times 10^{-5} \times \frac{1.01}{0.99} = 1.79 \times 10^{-5}$$

The assumption that x is small compared with 1.00 is valid, so the hydronium-ion concentration is

$$[H_3O^+] = x = 1.79 \times 10^{-5}$$

The pH is given by

$$pH = -\log [H_3O^+] = -\log(1.79 \times 10^{-5}) = 4.747$$

Note that the pH has decreased only slightly, by 0.01 pH unit. By contrast, the pH of a 0.0100 M HCl solution would have been 2.00 in the absence of the buffer, a change of 5 pH units from that of pure water.

The pH after the addition of a small amount of NaOH to the buffer solution is calculated similarly. When 0.0500 M NaOH is added, the NaOH dissociates completely into Na$^+$(aq) and OH$^-$(aq) ions. To determine the new initial conditions, we assume that the added OH$^-$(aq) reacts completely with acetic acid as follows:

$$OH^-(aq) + CH_3CO_2H(aq) \longrightarrow CH_3CO_2^-(aq) + H_2O(l)$$

Because 0.0500 mol/L of acetic acid is consumed in the reaction and 0.0500 mol/L of acetate ion is formed, the resulting solution has the following concentrations:

Substance	CH_3CO_2H	$CH_3CO_2^-$	H_3O^+
Initial concentration, M	0.95	1.05	0.0
Change in concentration, M	$-x$	$+x$	$+x$
Equilibrium concentration, M	$0.95 - x$	$1.05 + x$	$0.0 + x$

Only the initial concentrations of the buffer components have changed. The expressions for the equilibrium concentrations can be substituted into the equilibrium-constant expression.

$$K_a = \frac{[H_3O^+][CH_3CO_2^-]}{[CH_3CO_2H]} = \frac{(x)(1.05 + x)}{0.95 - x} = 1.75 \times 10^{-5}$$

We assume that x is small enough to be ignored.

$$\frac{(x)(1.05)}{0.95} = 1.75 \times 10^{-5}$$

We solve for x.

$$x = 1.75 \times 10^{-5} \times \frac{0.95}{1.05} = 1.58 \times 10^{-5}$$

The assumption that x is small compared with 1.05 and 0.95 was valid, so the hydronium-ion concentration is

$$[H_3O^+] = x = 1.58 \times 10^{-5}$$

The pH is given by

$$pH = -\log [H_3O^+] = -\log(1.58 \times 10^{-5}) = 4.801$$

The pH increased very little (less than 0.05 pH units) upon the addition of 0.0500 mol/L of NaOH (although the change was somewhat larger than that caused by the addition of the smaller quantity of HCl). The buffer system consumed most of the hydroxide ion added. Again, in the absence of a buffer, the change would have been much greater. For example, addition of 0.0500 M NaOH shifts the pH of pure water to 12.70.

Practice Problem 16.3

Consider a buffer system consisting of 0.500 M CH$_3$CO$_2$H and 1.00 M NaCH$_3$CO$_2$, with $K_a = 1.75 \times 10^{-5}$ for acetic acid. Calculate the pH of this buffer initially, after the addition of 0.0500 mol HCl per liter of buffer solution, and after the addition of 0.0200 mol NaOH per liter of buffer solution.

Answer: 5.058, 4.996, 5.084

Buffering Capacity and pH

Under what conditions does a mixture of a weak acid and its conjugate base act as a buffer?

Not all mixtures of a weak acid and its conjugate base make good buffers. Two conditions must be met. The first condition results from the fact that buffer solutions are most effective when the concentration of the weak acid and that of its conjugate base are approximately equal. Unequal concentrations work only if the ratio of acid concentration to base concentration falls in a certain range. Specifically, one component must be no more than ten times the other:

$$0.1 < [HA]/[A^-] < 10$$

Outside this range, a mixture of a weak acid and its conjugate base has very little buffering action.

The second condition that must be met arises from the limited capacity of a buffer to neutralize added acid or base. A buffer solution must be more concentrated than the amount of acid or base it is expected to neutralize. A buffer does not work if such large amounts of acid or base are added to the solution that one of the components is mostly converted to the other. The capacity of a buffer to consume a strong acid or base is limited to about 90% of the buffer concentration. For example, a buffer composed of 1.00 M HA and 1.00 M NaA consumes only 0.90 mol of a strong acid or base per liter of buffer solution. This means that the amount of the buffer components must be high compared with the amount of strong acid or base added.

The change in the hydronium-ion concentration caused by the addition of 0.00100 mol HCl per liter of various buffers is shown in Table 16.3. These results show that both the ratio of acid to conjugate base and the concentration of the buffer are important in determining the buffer's effectiveness. The buffering action is improved when the concentration of the buffer is increased and when the ratio of acid to conjugate base is close to a value of 1.

Table 16.3 Effect of Addition of 0.00100 mol of HCl to 1 L of $CH_3CO_2H/CH_3CO_2^-$ Buffer

Concentration of CH_3CO_2H, M	Concentration of $CH_3CO_2^-$, M	Ratio	pH Before Addition	pH After Addition	% Change in Concentration of $H_3O^+(aq)$
1.00	0.01	100	2.757	2.711	10
1.00	0.10	10	3.757	3.752	1.0
1.00	1.00	1	4.757	4.756	0.2
0.01	1.00	0.01	6.757	6.715	10
0.10	1.00	0.1	5.757	5.752	1.0
1.00	1.00	1	4.757	4.756	0.2
0.01	0.01	1	4.757	4.670	20
0.10	0.10	1	4.757	4.748	2.0
1.00	1.00	1	4.757	4.756	0.2

The Henderson–Hasselbalch Equation and Buffer Ranges

A convenient way to determine the pH of a buffer solution is provided by the **Henderson–Hasselbalch equation.** Consider the general case of a solution containing a weak acid (HA) and the anion that is the conjugate base (A^-):

$$HA(aq) + H_2O(l) \rightleftharpoons H_3O^+(aq) + A^-(aq)$$

The ionization constant is given by

$$K_a = [H_3O^+][A^-]/[HA]$$

This equation may be rearranged to give

$$[H_3O^+] = K_a[HA]/[A^-]$$

Taking the negative logarithm of both sides of the equation and then rearranging it yields a more useful form:

$$-\log [H_3O^+] = -\log K_a + \log [A^-]/[HA]$$
$$pH = pK_a + \log [A^-]/[HA]$$

What assumptions do we make implicitly when using the Henderson–Hasselbalch equation to calculate the pH of a buffer solution?

This final form of the equation is the Henderson–Hasselbalch equation. When a mixture qualifies as a buffer solution, we can use the Henderson–Hasselbalch equation to calculate its pH from the value of the acid-ionization constant and the ratio of the equilibrium concentration of the conjugate base and acid components of the buffer.

EXAMPLE 16.4

Use the Henderson–Hasselbalch equation to calculate the pH of a buffer containing 0.100 M CH_3CO_2H and 0.150 M $NaCH_3CO_2$. The value of K_a is 1.75×10^{-5} for acetic acid.

Solution The pH of the solution is given by

$$pH = pK_a + \log [CH_3CO_2^-]/[CH_3CO_2H]$$

The value of pK_a is

$$pK_a = -\log K_a = -\log(1.75 \times 10^{-5}) = 4.757$$

The values of all the parameters can now be substituted into the Henderson–Hasselbalch equation.

$$pH = 4.757 + \log(0.150/0.100)$$
$$= 4.757 + \log 1.50 = 4.757 + 0.176 = 4.933$$

Practice Problem 16.4

Use the Henderson–Hasselbalch equation to calculate the pH of a buffer containing 2.50 M CH_3CO_2H and 1.25 M $NaCH_3CO_2$. The value of K_a is 1.75×10^{-5} for acetic acid.

Answer: 4.456

We can also use the Henderson–Hasselbalch equation to calculate the pH of a buffer after the addition of a strong acid or base. We first calculate the changes in concentrations caused by the addition, assuming, as before, that there is a complete reaction to remove all the added acid or base. Then we use the new concentrations in the Henderson–Hasselbalch equation.

If we wish to prepare a buffer of a particular pH, we can use the Henderson–Hasselbalch equation to calculate the appropriate ratio of concentrations for the weak acid and conjugate base. Because the equation gives only the ratio of concentrations, we must set one concentration in advance. Then we need only substitute appropriate values into the equation.

EXAMPLE 16.5

A buffer of pH 4.230 is to be prepared by addition of $NaCH_3CO_2$ to a 1.00 M solution of CH_3CO_2H. What concentration of $NaCH_3CO_2$ is needed? The value of K_a is 1.75×10^{-5} for acetic acid, and the pK_a is 4.757.

Solution   The pH of the solution is given by

$$pH = pK_a + \log [CH_3CO_2^-]/[CH_3CO_2H]$$

All the parameters are known except the acetate concentration. That value can be obtained by substituting into the Henderson–Hasselbalch equation.

$$4.230 = 4.757 + \log([CH_3CO_2^-]/1.00)$$
$$\log [CH_3CO_2^-] = 4.230 - 4.757 = -0.527$$
$$[CH_3CO_2^-] = 10^{-0.527} = 0.297 \; M$$

Practice Problem 16.5

A buffer of pH 4.950 is to be prepared by addition of $NaCH_3CO_2$ to a 1.00 M solution of CH_3CO_2H. What concentration of $NaCH_3CO_2$ is needed? The value of K_a is 1.75×10^{-5} for acetic acid, and the pK_a is 4.757.

Answer: 1.56 M

We can use the Henderson–Hasselbalch equation in yet another way: to determine which weak acid and conjugate base to select for a buffer at a particular pH and to examine the range of concentrations that give good buffering action for the selected buffer system. Recall that the maximum buffering action occurs when the concentrations of the weak acid and its conjugate base are equal. Under these conditions, the $H_3O^+(aq)$ concentration equals the value of K_a, and the pH equals the pK_a, as can be seen in the following equations:

$$[H_3O^+] = K_a[HA]/[A^-] = K_a \times 1 = K_a$$
$$pH = pK_a + \log [A^-]/[HA] = pK_a + \log 1 = pK_a + 0 = pK_a$$

This result indicates that the best choice of a conjugate acid–base pair for a buffer system is one in which the weak acid has a pK_a equal to the desired pH.

It is not always possible to find an acid–base system that offers a perfect match between the pK_a and the desired pH. In such cases, we must find as close a match as possible and adjust the concentration ratio to give the desired pH. But how far apart can the pK_a and pH be before the buffer system no longer works

Figure 16.6 Relationship between pH and percent of each acetate species for an acetic acid/acetate ion buffer. Note that CH_3CO_2H predominates at low pH, whereas $CH_3CO_2^-$ predominates at high pH. There are equal amounts of each species at pH = pK_a. Buffering action is observed within the buffer range. This range lies between the dashed lines, which are one pH unit on either side of pK_a.

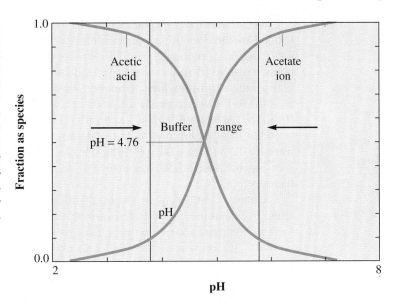

Show that this equation for the pH range for a buffer is consistent with a ratio of conjugate-base concentration to weak-acid concentration between 0.1 and 10.

properly? Recall that a buffer works properly even when the concentrations are not equal, provided that the concentration ratio falls in the range $0.1 < [A^-]/[HA] < 10$. In terms of the Henderson–Hasselbalch equation, this range converts to

$$\text{pH range} = pK_a \pm 1$$

This equation indicates that a particular buffer system is effective when it can achieve a pH that is within one unit of the pK_a. For example, mixtures of acetic acid and sodium acetate, with pK_a equal to 4.757, can be used to prepare buffers in the range of pH 3.757–5.757. If we want to obtain a pH outside this range, we must use a different buffer system. This range is often referred to as the **buffer range.** Figure 16.6 shows the distribution of species in acetic acid/acetate ion buffer systems. Note that outside the buffer range, one form—either the acid or the base—predominates, so little buffering action remains.

We have been examining buffer systems made up of a weak acid and its conjugate base, but as we noted earlier, a buffer can also be prepared from a mixture of a weak base and its conjugate acid. A common example is $NH_3(aq)$ and its salt, NH_4Cl, which is described by the following equilibrium.

$$NH_3(aq) + H_2O(l) \rightleftharpoons NH_4^+(aq) + OH^-(aq)$$

Calculations for this type of buffer system are similar to those we have been discussing. In fact, if the equilibrium between the weak base and its conjugate acid is written in terms of acid ionization (the reverse of this reaction), the calculations are identical.

16.4 Complex-Ion Equilibria

We have examined a number of equilibrium reactions in solution that involve some type of reaction between Brønsted–Lowry acids and bases, such as neutralization or hydrolysis. However, other types of equilibrium reactions can occur in solution. In this section, we will examine a Lewis acid–base reaction.

Recall that species containing one or more coordinate covalent bonds—bonds formed by the donation of an electron pair from one atom to another—are called coordination compounds. ▶ This group of compounds includes the iron compounds discussed in the chapter introduction. Within this general class is a subgroup called **complex ions.** Complex ions are species with a net ionic charge that are formed between a metal ion or atom and other molecules or ions called **ligands** (from the Latin word *ligare,* "to bind"). For example, when sufficient aqueous KI is added to a solution of mercury(II) ion, Hg^{2+}, a complex ion is formed from mercury(II) ion and four I^- ions: HgI_4^{2-}. As shown in Figure 16.7, this complex ion has the iodide ions arranged at the corners of a tetrahedron around the mercury ion.

Complex ions play an important role in many chemical and biological systems. For example, as indicated earlier, metal ions in aqueous solution exist in the form of hydrated ions—complex ions with water as a ligand. An example is the titanium(III) ion (Figure 16.8), which exists in aqueous solution as $Ti(H_2O)_6^{3+}$, with the oxygen atom of each water molecule donating its electron pair to the titanium ion. Complex ions are used chemically as catalysts, for the extraction of gold and silver from their ores, and in chemical analysis. Some complex ions of biological importance include complexes of iron, cobalt, and magnesium. The iron complex hemoglobin (Figure 16.9) is responsible for carrying oxygen in the bloodstream. Vitamin B_{12}, a complex of cobalt, is involved in the synthesis of red blood corpuscles. Magnesium forms complex ions in the form of chlorophyll, which is vital to the synthesis of carbohydrates from carbon dioxide and water in plants—photosynthesis.

Can we extend the principles of chemical equilibrium to Lewis acid–base reactions?

Why can hydrated metal ions be considered complex ions?

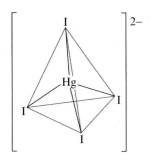

Figure 16.7 HgI_4^{2-} has mercury(II) surrounded by four iodide ions arranged at the corners of a tetrahedron.

Mercury(II) iodide is a heavy, scarlet-red powder that decomposes upon prolonged exposure to light. When heated, it turns yellow at 130°C; it becomes red again when cooled.

Formation Constants for Complex Ions

Complex ions with more than one ligand are often formed in a series of steps, with one ligand at a time added to the central metal. Each of these can be considered to reach a state of equilibrium. For example, the complex ion HgI_4^{2-} is formed in four steps.

$$Hg^{2+}(aq) + I^-(aq) \rightleftharpoons HgI^+(aq) \qquad K_{f1} = 7.9 \times 10^{12}$$
$$HgI^+(aq) + I^-(aq) \rightleftharpoons HgI_2(aq) \qquad K_{f2} = 1.0 \times 10^{11}$$
$$HgI_2(aq) + I^-(aq) \rightleftharpoons HgI_3^-(aq) \qquad K_{f3} = 5.0 \times 10^3$$
$$HgI_3^-(aq) + I^-(aq) \rightleftharpoons HgI_4^{2-}(aq) \qquad K_{f4} = 2.5 \times 10^2$$

Figure 16.8 Titanium(III) exists in aqueous solution as pink $Ti(H_2O)_6^{3+}$.

Figure 16.9 Hemoglobin is a large protein molecule that contains four heme units. The heme unit is a coordination compound containing iron(III) bonded to four nitrogen atoms by coordinate covalent bonds.

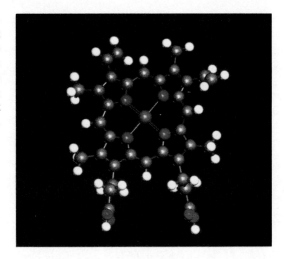

Some of these steps are shown in Figure 16.10. Each step reaches an equilibrium characterized by an equilibrium constant (K_{f1}, K_{f2}, K_{f3}, and K_{f4}) called a **formation constant.**

The overall formation process for HgI_4^{2-} is the sum of these four stepwise reactions and is represented as follows:

$$Hg^{2+}(aq) + 4I^-(aq) \rightleftharpoons HgI_4^{2-}(aq)$$

HgI_4^{2-} exists in a state of equilibrium with aqueous Hg^{2+} and I^- ions, as well as with the intermediate complex ions.

The formation constant, K_f, for the overall reaction is given by the following equation and has a value equal to the product of the formation constants for the individual steps.

$$K_f = \frac{[HgI_4^{2-}]}{[Hg^{2+}][I^-]^4} = 1.0 \times 10^{30}$$

$$K_f = K_{f1} \times K_{f2} \times K_{f3} \times K_{f4}$$

Table 16.4 lists the formation constants for some common complex ions.

Figure 16.10 When iodide ion is added to a solution of mercury(II) nitrate, it forms various complex ions. *Left:* Although some HgI_2 exists in solution, this substance is not very soluble, so it forms a red solid under the conditions of this experiment. *Right:* Addition of more iodide ion causes the solid HgI_2 to redissolve, because other complex ions are formed.

Table **16.4** Formation Constants for Some Complex Ions in Water at 25°C

Species	Reaction	K_f
$Ag(NH_3)_2^+$	$Ag^+ + 2NH_3 \rightleftharpoons Ag(NH_3)_2^+$	1.7×10^7
$Ag(CN)_2^-$	$Ag^+ + 2CN^- \rightleftharpoons Ag(CN)_2^-$	1.0×10^{21}
$Cu(NH_3)_4^{2+}$	$Cu^{2+} + 4NH_3 \rightleftharpoons Cu(NH_3)_4^{2+}$	1.2×10^{12}
$CdCl_4^{2+}$	$Cd^{2+} + 4Cl^- \rightleftharpoons CdCl_4^{2+}$	7.1×10^1
$Cd(NH_3)_4^{2+}$	$Cd^{2+} + 4NH_3 \rightleftharpoons Cd(NH_3)_4^{2+}$	4.0×10^6
$Cd(CN)_4^{2+}$	$Cd^{2+} + 4CN^- \rightleftharpoons Cd(CN)_4^{2+}$	7.1×10^{16}
$ZnCl_4^{2-}$	$Zn^{2+} + 4Cl^- \rightleftharpoons ZnCl_4^{2-}$	7.9×10^{-1}
$Zn(NH_3)_4^{2+}$	$Zn^{2+} + 4NH_3 \rightleftharpoons Zn(NH_3)_4^{2+}$	6.3×10^8
$Co(CN)_6^{4-}$	$Co^{2+} + 6CN^- \rightleftharpoons Co(CN)_6^{4-}$	1.2×10^{19}
$Co(CN)_6^{3-}$	$Co^{3+} + 6CN^- \rightleftharpoons Co(CN)_6^{3-}$	1.0×10^{64}
$Co(NH_3)_6^{2+}$	$Co^{2+} + 6NH_3 \rightleftharpoons Co(NH_3)_6^{2+}$	8.3×10^4
$Co(NH_3)_6^{3+}$	$Co^{3+} + 6NH_3 \rightleftharpoons Co(NH_3)_6^{3+}$	4.5×10^{33}
$Ni(NH_3)_6^{2+}$	$Ni^{2+} + 6NH_3 \rightleftharpoons Ni(NH_3)_6^{2+}$	1.8×10^8
$Fe(CN)_6^{3-}$	$Fe^{3+} + 6CN^- \rightleftharpoons Fe(CN)_6^{3-}$	1.0×10^{42}
$Fe(CN)_6^{4-}$	$Fe^{2+} + 6CN^- \rightleftharpoons Fe(CN)_6^{4-}$	1.0×10^{35}
$HgCl_4^{2-}$	$Hg^{2+} + 4Cl^- \rightleftharpoons HgCl_4^{2-}$	1.6×10^{15}
HgI_4^{2-}	$Hg^{2+} + 4I^- \rightleftharpoons HgI_4^{2-}$	1.0×10^{30}
AlF_6^{3-}	$Al^{3+} + 6F^- \rightleftharpoons AlF_6^{3-}$	5.0×10^{19}
$SnCl_4^{2-}$	$Sn^{2+} + 4Cl^- \rightleftharpoons SnCl_4^{2-}$	2.5×10^1

Equilibrium Concentrations

Generally, we calculate equilibrium concentrations in solutions of complex ions just as we do in solutions of weak acids or bases. The stepwise dissociation of complex ions is analogous to the stepwise ionization of polyprotic acids. However, the relative values of the stepwise constants are different for the two types of reactions. Because successive acid-ionization constants usually have quite different values, we were able to treat each step as an independent equilibrium when solving for the equilibrium concentrations. This is not the case for stepwise complex-ion formation constants, which may have quite similar values. For HgI_4^{2-}, the first two constants are only about a factor of 10 apart, as are the last two. This results in the coexistence of three mercury(II) species over certain concentration ranges, as shown in Figure 16.11. Under these conditions, it is not possible to treat the stepwise formation equilibria independently. However, if we have a sufficiently large concentration of I^-, most of the mercury(II) will be in the form of HgI_4^{2-}, and we can calculate its concentration. Under conditions where only one complex ion is present in significant quantities, the calculations are analogous to those for weak acids.

Figure 16.11 The distribution of the various ions and compounds formed between mercury(II) and iodide ions indicates that in some ranges of iodide-ion concentration, three mercury species coexist in a state of equilibrium.

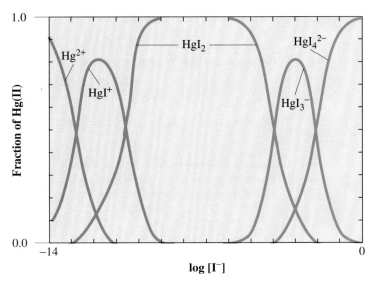

EXAMPLE 16.6

Ammonia is added to a 0.00100 M AgNO$_3$ solution to give a total NH$_3$(aq) concentration of 0.100 M. The formation constant has a value of $K_f = 1.7 \times 10^7$ for Ag(NH$_3$)$_2{}^+$. What are the concentrations of all the species in this solution? Assume that the intermediate complex ion AgNH$_3{}^+$ is not present in significant concentrations.

Solution The reaction involves the equilibrium

$$\text{Ag}^+(aq) + 2\text{NH}_3(aq) \rightleftharpoons \text{Ag(NH}_3)_2{}^+(aq)$$

The concentrations of these three species can be summarized in the form of a table.

Substance	Ag$^+$	NH$_3$	Ag(NH$_3$)$_2{}^+$
Initial concentration, M	0.00100	0.100	0.00
Change in concentration, M	$-x$	$-2x$	$+x$
Equilibrium concentration, M	$0.00100 - x$	$0.100 - 2x$	$0.00 + x$

The equilibrium-constant expression is given by

$$K_f = \frac{[\text{Ag(NH}_3)_2{}^+]}{[\text{Ag}^+][\text{NH}_3]^2} = 1.7 \times 10^7$$

Expressions for the equilibrium concentrations can be substituted into this equation.

$$K_f = \frac{x}{(0.00100 - x)(0.100 - 2x)^2} = 1.7 \times 10^7$$

If all the silver ion reacted, x would have a value of about 0.001 M, so it is reasonable to assume that $2x$ is much less than 0.100. This assumption allows us to simplify the equation.

$$\frac{x}{(0.00100 - x)(0.100)^2} = 1.7 \times 10^7$$

(continued)

The equation can be rearranged.

$$x = (1.7 \times 10^7)(0.100)^2(0.00100 - x)$$

Collecting terms and solving for x yields

$$x(1 + 1.7 \times 10^5) = 1.7 \times 10^2$$
$$x = (1.7 \times 10^2)/(1 + 1.7 \times 10^5) = 1.0 \times 10^{-3}$$

The concentration of the complex ion is thus

$$[Ag(NH_3)_2{}^+] = x = 1.0 \times 10^{-3}\ M$$

Because the concentration of the complex ion equals the concentration of silver(I) initially in the solution, the reaction proceeds essentially to completion.

The ammonia concentration is given by

$$[NH_3] = 0.100 - 2x = 0.100 - 2(0.0010) = 0.098\ M$$

The silver ion concentration cannot be obtained from the relations set out in the table. These will give a value of 0, because the number of significant figures is too small to yield an answer by subtraction. Instead, the concentrations of $Ag(NH_3)_2{}^+$ and NH_3 can be substituted back into the formation-constant expression, which can then be solved for the silver ion concentration.

$$\frac{[Ag(NH_3)_2{}^+]}{[Ag^+][NH_3]^2} = 1.7 \times 10^7$$

$$\frac{(1.0 \times 10^{-3})}{[Ag^+](0.098)^2} = 1.7 \times 10^7$$

Upon rearrangement, this becomes

$$[Ag^+] = (1.0 \times 10^{-3})/(0.098)^2(1.7 \times 10^7) = 6.1 \times 10^{-9}\ M$$

Only a very small concentration of silver ion is left in this solution.

Practice Problem 16.6

Sodium cyanide (NaCN) is added to a $0.00100\ M$ $AgNO_3$ solution to give a total $CN^-(aq)$ concentration of $0.200\ M$. The following reaction occurs:

$$Ag^+(aq) + 2CN^-(aq) \rightleftharpoons Ag(CN)_2{}^-(aq)$$

The formation constant for $Ag(CN)_2{}^-$ has a value of $K_f = 1.0 \times 10^{21}$. What are the concentrations of all the species in this solution? Assume that the intermediate species $AgCN(aq)$ is not present in significant concentrations.

Answer: $0.00100\ M$ $Ag(CN)_2{}^-$; $0.198\ M$ CN^-; $2.6 \times 10^{-23}\ M$ Ag^+

16.5   Solubility and Solubility Product

So far, we have considered hydrolysis, acid–base neutralization, and complex-ion-formation reactions in ionic solutions. Another reaction that can occur when two ionic substances are dissolved in solution is precipitation. If any two of the ions in solution form an insoluble salt, a precipitate may separate from the solution. In this section, we will investigate equilibrium reactions between ionic solids and ions in solution. Understanding such equilibria enables us to determine the conditions under which solids dissolve or ions precipitate.

Can we apply the principles of chemical equilibrium to precipitation reactions?

Solubility

Silver chloride is a white powder that turns black when exposed to light because it decomposes to metallic silver. It is used in silver plating and in antiseptic silver preparations.

Recall that solubility is the equilibrium amount of a substance that dissolves in a given quantity of solvent at a specific temperature. Ionic salts vary considerably in their solubility in water. For example, 100 g of water dissolves 34 g KCl, 36 g NaCl, or 79 g LiCl but only 1.9×10^{-4} g AgCl. Furthermore, the solubility of ionic salts does not follow simple, predictable trends. A number of factors are involved, including the lattice energies of the salts and the heats of hydration of the ions. However, some generalizations can be made, as in the solubility rules for ionic salts given in Appendix B. It should be noted that the definitions of *soluble* and *insoluble* are not very precise. However, for practical purposes, solutes with a solubility less than about 0.1 g/100 g solvent are considered insoluble or slightly soluble.

Insoluble Salts and the Solubility Product Constant

Mercury(II) sulfide exists in a black and a red form, the latter occurring naturally as the mineral cinnabar. Both forms are used as pigments.

Even a salt classified as insoluble, such as AgCl, does have a limited solubility in water. Indeed, there is no completely insoluble salt. There are always some ions in solution when a salt and water are in contact and when they reach a state of equilibrium.

The solubility of salts classified as insoluble is best considered in terms of the equilibrium between undissolved and dissolved salt. For example, the most nearly insoluble salt known is mercury(II) sulfide, HgS. In contact with water, this salt dissolves to give 1.3×10^{-27} M Hg^{2+} and 1.3×10^{-27} M S^{2-} in solution. (This means that it requires almost 1300 L of water to contain a single pair of ions in solution at equilibrium.) HgS(s) in equilibrium with Hg^{2+} and S^{2-} ions is represented as follows:

$$HgS(s) \rightleftharpoons Hg^{2+}(aq) + S^{2-}(aq)$$

A solution such as this one, in which a salt has reached a state of equilibrium with its ions, is called a *saturated solution*.

The equilibrium constant for such solutions is called the **solubility product constant** (K_{sp}). For the mercury(II) sulfide solution, the solubility product constant is given by the following equation:

$$K_{sp} = [Hg^{2+}][S^{2-}]$$

The value of the solubility product constant depends only on the temperature and not on the amounts of solvent and salt used to prepare the solution. The solubility product expression is the product of the concentration of each ion raised to the power of its subscript in the ionic formula. These subscripts are equal to the coefficients of the ions in the equation for dissolution of the solid. For example, the dissolution of calcium phosphate, $Ca_3(PO_4)_2$, follows the equation

$$Ca_3(PO_4)_2(s) \rightleftharpoons 3Ca^{2+}(aq) + 2PO_4^{3-}(aq)$$

The expression for the solubility product constant is

$$K_{sp} = [Ca^{2+}]^3[PO_4^{3-}]^2$$

Like other equilibrium constants, K_{sp} is not given units. Table 16.5 lists the K_{sp} values for some common ionic salts. As we have said, the value of the solubility product constant should depend only on temperature, as is the case for other

Table 16.5 Values of Some Solubility Product Constants for the Dissolution of Ionic Salts in Pure Water

Compound	K_{sp}	Compound	K_{sp}
$Al(OH)_3$	1.9×10^{-33}	PbS	8.4×10^{-28}
$BaCO_3$	8.1×10^{-9}	$Mg(OH)_2$	1.2×10^{-11}
$BaCrO_4$	2.4×10^{-10}	MgC_2O_4	8.6×10^{-5}
BaF_2	1.73×10^{-6}	$Mn(OH)_2$	4.5×10^{-14}
$BaSO_4$	1.5×10^{-9}	MnS	5.6×10^{-16}
CdS	3.6×10^{-29}	Hg_2Cl_2	1.1×10^{-18}
$CaCO_3$	9.0×10^{-9}	HgS	1.6×10^{-54}
CaF_2	1.70×10^{-10}	NiS	2.0×10^{-21}
$CaSO_4$	2.00×10^{-4}	$AgCH_3CO_2$	2.3×10^{-3}
CoS	5.9×10^{-21}	Ag_2CO_3	8.2×10^{-12}
CuS	8.7×10^{-36}	$AgCl$	1.70×10^{-10}
Cu_2S	1.6×10^{-48}	$AgBr$	5.0×10^{-13}
$Fe(OH)_2$	2.0×10^{-15}	AgI	8.5×10^{-17}
$Fe(OH)_3$	1.1×10^{-36}	Ag_2CrO_4	9.0×10^{-12}
FeC_2O_4	2.1×10^{-7}	$AgCN$	1.6×10^{-14}
FeS	3.7×10^{-19}	Ag_2S	1.0×10^{-51}
$PbCl_2$	1.6×10^{-5}	$Sn(OH)_2$	5.0×10^{-26}
$PbCrO_4$	1.8×10^{-14}	SnS	8.0×10^{-29}
PbC_2O_4	2.7×10^{-11}	$Zn(OH)_2$	4.5×10^{-17}
$PbSO_4$	1.80×10^{-8}	ZnS	1.2×10^{-23}

Calcium phosphate, $Ca_3(PO_4)_2$, is a white solid that is practically insoluble in water but dissolves in acids because of the formation of protonated forms of the phosphate ion. Calcium phosphate is used in the manufacture of phosphorus, phosphoric acid, fertilizers, and various phosphorous compounds, as a polishing agent in toothpastes and in enamels for porcelain and pottery.

equilibrium constants. However, there can also be so-called medium effects, in which the behavior of ions in solution depends somewhat on the presence of other dissolved materials. As a result, values of solubility product constants often vary slightly according to the ionic medium in which they were measured. Thus values of K_{sp} obtained from different sources often differ somewhat.

In comparing the values of K_{sp} for two substances, we must remember that only salts that have the same number of ions can be compared directly. Thus the relative solubilities of AgCl, with two ions, and $PbCl_2$, with three ions, cannot be determined by a simple examination of the values of K_{sp} for the two salts. For salts that do have the same number of ions, however, the relative values of K_{sp} provide a basis for comparing solubilities. For example, the values of K_{sp} for AgCl and AgCN are 1.7×10^{-10} and 1.6×10^{-14}, respectively, and AgCl is more soluble than AgCN. The larger the value of K_{sp}, the more soluble the salt, so long as the salts are made up of the same number of ions.

Determining K_{sp} from Solubilities

To relate the solubility product constant to solubility, we must derive the solubility of ions in solution from the molar solubility of the dissolving substance, using the

RULES OF REACTIVITY

A precipitation reaction occurs when a chloride salt solution is mixed with a solution of a metal ion, if that metal ion forms an insoluble chloride. All chlorides are soluble except $AgCl$, Hg_2Cl_2, and $PbCl_2$. ($PbCl_2$, though insoluble in cold water, is soluble in hot water.)

How are solubilities of salts related to their solubility product constants?

Lead(II) chloride, $PbCl_2$, is a white crystalline solid used in the manufacture of various lead pigments.

coefficients from the balanced dissolution equation as conversion factors. For a 1:1 salt such as $AgCl(s)$, the molar concentration of $Ag^+(aq)$ and of $Cl^-(aq)$ in solution equals the molar solubility of AgCl—the number of moles of AgCl that dissolve in 1 L of solution—because the dissolution of one formula unit of the salt produces one of each ion:

$$AgCl(s) \rightleftharpoons Ag^+(aq) + Cl^-(aq)$$

We will give the molar solubility of an ionic salt the symbol S. The molar solubility of silver chloride is related to the concentrations of ions in solution as follows:

$$S = [Ag^+] \qquad S = [Cl^-]$$

Other salts have other relationships, determined by the stoichiometry of the dissolution reaction. For example, the dissolution of lead chloride is represented by the equation

$$PbCl_2(s) \rightleftharpoons Pb^{2+}(aq) + 2Cl^-(aq)$$

The molar solubility of lead chloride equals the lead-ion concentration but is only one-half the chloride-ion concentration because of the 2:1 ratio of ions:

$$S = [Pb^{2+}] \qquad 2S = [Cl^-]$$

The relationships between ion concentrations and solubility must be established carefully if the value of the solubility product constant is to be correctly calculated.

For salts in which the cation and the anion occur in equal amounts, then, the value of the solubility product constant is given by the square of the molar solubility, as shown for silver chloride in the equation

$$K_{sp} = [Ag^+][Cl^-] = (S)(S) = S^2$$

For other salts, however, this is not the case. For these other salts, we must use the appropriate relationships between the concentrations of the ions and the solubility of the salt. For $PbCl_2$, we have

$$K_{sp} = [Pb^{2+}][Cl^-]^2 = (S)(2S)^2 = 4S^3$$

EXAMPLE 16.7

Calculate the K_{sp} of Ag_2CrO_4, assuming that the solubility is 1.30×10^{-4} mol/L.

Solution The equilibrium established when solid Ag_2CrO_4 dissolves in water is described by the equation

$$Ag_2CrO_4(s) \rightleftharpoons 2Ag^+(aq) + CrO_4^{2-}(aq)$$

The solubility product expression is

$$K_{sp} = [Ag^+]^2[CrO_4^{2-}]$$

The concentrations of the Ag^+ and CrO_4^{2-} ions must be calculated from the molar solubility. Because one silver chromate formula unit generates one chromate ion and two silver ions, the relationships are

$$[CrO_4^{2-}] = S = 1.30 \times 10^{-4} \, M$$
$$[Ag^+] = 2S = 2 \times 1.30 \times 10^{-4} = 2.60 \times 10^{-4} \, M$$

(continued)

We can calculate the value of K_{sp} by inserting the concentrations into the solubility product expression.

$$K_{sp} = [Ag^+]^2[CrO_4^{2-}] = (2S)^2(S)$$
$$= (2.60 \times 10^{-4})^2(1.30 \times 10^{-4}) = 8.79 \times 10^{-12}$$

This calculation can also be set up with the table we have used for other equilibrium calculations. Note that the coefficient of the solubility comes directly from the equation for the dissolution reaction.

$$Ag_2CrO_4(s) \rightleftharpoons 2Ag^+(aq) + CrO_4^{2-}(aq)$$

Substance	Ag^+	CrO_4^{2-}
Initial concentration, M	0.0	0.0
Change in concentration, M	2S	S
Equilibrium concentration, M	0.0 + 2S	0.0 + S

Substituting the equilibrium concentrations into the expression for the solubility product constant yields the same result we obtained earlier.

Practice Problem 16.7

Calculate the K_{sp} of $PbCl_2$, assuming that the solubility is 1.59×10^{-2} mol/L.

Answer: 1.61×10^{-5}

Determining Solubilities from K_{sp}

Just as we can calculate K_{sp} from solubility, we can calculate solubility from K_{sp}. The procedure is just the reverse of that used to calculate K_{sp} from solubilities. Again, the appropriate relationships between concentrations of ions and molar solubility must be used.

EXAMPLE 16.8

Calculate the solubility of calcium fluoride in water, assuming that $K_{sp} = 1.70 \times 10^{-10}$ for CaF_2.

Solution   The dissolution of calcium fluoride is described by the equation

$$CaF_2(s) \rightleftharpoons Ca^{2+}(aq) + 2F^-(aq)$$

The expression for the solubility product constant is given by

$$K_{sp} = [Ca^{2+}][F^-]^2$$

We can relate the concentrations of the ions to the solubility by using the reaction stoichiometry.

$$CaF_2(s) \rightleftharpoons Ca^{2+}(aq) + 2F^-(aq)$$

Substance	Ca^{2+}	F^-
Initial concentration, M	0.0	0.0
Change in concentration, M	S	2S
Equilibrium concentration, M	0.0 + S	0.0 + 2S

(continued)

▶ ▶ ▶ WORLDS OF CHEMISTRY

Chemical Equilibrium in the Underworld

The magnificent formations found in caves are due to rather simple chemical changes caused by shifts in the concentration of CO_2.

Many caves are composed primarily of limestone and often contain wondrous natural formations—stalactites, stalagmites, columns, curtains, flowstones, and other shapes, as shown in Figure 16.12. What causes these fantastic structures? Water. But water alone could not reshape limestone in this way. The other sculptor is carbon dioxide.

Limestone is calcium carbonate, which is only slightly soluble in water, with a solubility product constant of 9.0×10^{-9}.

$$CaCO_3(s) \rightleftharpoons Ca^{2+}(aq) + CO_3^{2-}(aq)$$

Though calcium carbonate is only slightly soluble in pure water, it does dissolve in acidic solutions.

$$CaCO_3(s) + 2H^+(aq) \longrightarrow Ca^{2+}(aq) + H_2O(l) + CO_2(g)$$

Calcium carbonate dissolves readily in solutions of strong acids and to a lesser extent in solutions of weak acids. The acid involved in the chemistry of caves is carbonic acid, H_2CO_3. Carbonic acid is formed when carbon dioxide in air dissolves in rainwater. As the rainwater soaks into the ground, it absorbs more CO_2, which is present in soil as a product of the decay of plant and animal matter. The carbon dioxide forms carbonic acid when it dissolves.

$$CO_2(g) + H_2O(l) \rightleftharpoons H_2CO_3(aq)$$

The acidic rainwater percolates down through cracks in layers of limestone, dissolving some limestone.

$$CaCO_3(s) + H_2CO_3(aq) \rightleftharpoons Ca^{2+}(aq) + 2HCO_3^-(aq)$$

If the limestone contains a hollow space—a cave—drops of calcium bicarbonate solution collect on the roof. Some water may evaporate from these drops, and they also lose CO_2 to the air in the cave. A decrease in the dissolved CO_2 concentration results in calcium carbonate being precipitated from the water drops.

$$Ca(HCO_3)_2(aq) \rightleftharpoons CaCO_3(s) + CO_2(g) + H_2O(l)$$

This calcium carbonate is deposited as a small solid ring around the drop. As more calcium bicarbonate solution percolates into the cave, the ring becomes thicker, forming a stalactite. As the initially hollow stalactite grows longer, the ring eventually closes into a solid cone. Drops that fall

Figure 16.12 Stalactites hang from the ceiling and stalagmites rise from the floor in Luray Caverns, Virginia.

to the floor release CO_2 as they splash and also deposit $CaCO_3$, forming stalagmites. If the two formations meet, a solid column is created. These formations take a long time to grow. It commonly takes about 1000 years for a stalactite to lengthen by 6 cm.

If the calcium carbonate were pure, the limestone formations would be white. Instead, they often occur in subtle shades of yellow, orange, or red as a result of iron impurities that the water has extracted from the soil.

The magnificent formations found in caves are due to rather simple chemical changes caused by shifts in the concentration of CO_2. A reversible reaction lies toward soluble calcium bicarbonate in the limestone and shifts toward insoluble calcium carbonate in the air.

Questions for Discussion

1. It takes many thousands of years for stalactites and stalagmites to grow. Is it reasonable for people to remove these formations from caves to make pen holders, jewelry, and other ornaments?

2. How did the calcium carbonate that makes the limestone deposits out of which caves are fashioned get in the ground?

3. Acid rain may kill fish in one lake that exists over a granite bedrock, whereas it may leave fish unharmed in a nearby lake that is located over a mass of limestone. Explain the chemical basis for the difference.

4. Carbon dioxide emissions may be contributing to the "greenhouse effect" that is suspected to be causing a gradual warming of the Earth. The vast quantities of

(continued)

cement that are used in buildings and roads is composed in part of calcium oxide, which is formed by decomposing calcium carbonate at high temperatures. If the green-

house effect is real and is a threat to our long-term well-being, should we stop using cement and concrete? If so, what should we use instead?

These relationships can be substituted into the solubility product expression.

$$K_{sp} = [Ca^{2+}][F^-]^2 = (S)(2S)^2 = 4S^3 = 1.70 \times 10^{-10}$$

And this equation can be solved for the molar solubility.

$$S = (1.70 \times 10^{-10}/4)^{\frac{1}{3}} = 3.49 \times 10^{-4} \ M$$

Practice Problem 16.8

Calculate the solubility of silver sulfide in water, assuming that $K_{sp} = 1.00 \times 10^{-51}$ for Ag_2S.

Answer: $6.30 \times 10^{-18} \ M$

Questions 71 through 82 in the Test Bank correspond to this section.

16.6   Modification of Solubility with Ionic Salts

The solubility of salts can be modified in a number of ways. ▶ For example, the chapter introduction mentioned that the insoluble salt iron(III) hydroxide can be dissolved by addition of acid. It is also possible to modify the solubility of some salts by adding other salts to the solution. For example, hard water contains calcium, magnesium, iron, carbonate, and sulfate ions. Adding sodium carbonate to a sample of hard water results in the precipitation of calcium carbonate from the solution, reducing the hardness of the water. A hydrated form of sodium carbonate, known as washing soda, has been used in detergents for just this purpose.

Solutions found in nature rarely contain only one substance. How does the presence of a second substance in solution affect the solubility of other substances?

The Effect of Common Ions on Solubility

The calcium carbonate example illustrates a general phenomenon. The solubility of an ionic salt is reduced in a solution that contains one of the salt's component ions, if the ions have come from a source other than this ionic salt. This is the common-ion effect, which has already been described for the ionization of weak acids and bases.

Consider another example. As a result of the common-ion effect, silver chloride is less soluble in a solution containing sodium chloride than in pure water. The dissolution of silver chloride in water reaches a state of equilibrium, as described by the equation

$$AgCl(s) \rightleftharpoons Ag^+(aq) + Cl^-(aq)$$

Addition of sodium chloride, which dissociates completely in solution into aqueous sodium ions and chloride ions, changes the concentration of a product in this

Figure 16.13 *Left:* A saturated solution of AgCl contains some dissolved $Ag^+(aq)$ and $Cl^-(aq)$. *Right:* When NaCl is added, some of the dissolved AgCl precipitates from solution.

equilibrium reaction. In accordance with Le Chatelier's principle, the equilibrium shifts back toward the reactant, partially undoing the added concentration of chloride ion. In this way, addition of a soluble salt containing either Ag^+ or Cl^- to a saturated solution of AgCl results in a shift of the equilibrium to the left, causing the precipitation of some silver chloride, as shown in Figure 16.13.

We can also consider this effect in terms of the solubility product expression. Addition of one of the ions from another source, such as NaCl or HCl, increases the product of the ionic concentrations above the value of K_{sp}. The ion product must be reduced in order for a new state of equilibrium to be reached. Precipitation of silver chloride, which reduces the concentrations of the ions, accomplishes this reduction.

We can further examine the common-ion effect by calculating the solubility of an ionic salt in pure water and in a solution containing a common ion. The calculation is exactly like those already outlined, except that we must take into account the fact that the concentration of the common ion comes from more than one source.

EXAMPLE 16.9

Calculate the solubility of silver chloride in water and in a solution containing 0.0100 *M* NaCl. The K_{sp} of AgCl is 1.70×10^{-10}.

Solution   The equilibrium established by dissolution of silver chloride is described by the equation

$$AgCl(s) \rightleftharpoons Ag^+(aq) + Cl^-(aq)$$

The solubility product expression is

$$K_{sp} = [Ag^+][Cl^-]$$

(continued)

We can collect the various concentration terms in a table, using the symbol S to represent the solubility of silver chloride. First we consider the solubility of AgCl in water.

Substance	Ag^+	Cl^-
Initial concentration, M	0.0	0.0
Change in concentration, M	S	S
Equilibrium concentration, M	$0.0 + S$	$0.0 + S$

The equilibrium concentrations are then substituted into the solubility product expression, and the resulting equation is solved for the molar solubility.

$$K_{sp} = [Ag^+][Cl^-] = (S)(S) = 1.70 \times 10^{-10}$$
$$S = (1.70 \times 10^{-10})^{\frac{1}{2}} = 1.30 \times 10^{-5}\ M$$

In the solution that already contains some chloride ion, the initial concentration of chloride ion equals the sodium chloride concentration.

Substance	Ag^+	Cl^-
Initial concentration, M	0.0	0.010
Change in concentration, M	S	S
Equilibrium concentration, M	$0.0 + S$	$0.0100 + S$

Substituting these equilibrium concentrations into the solubility product expression gives a different equation from the one we obtained for the solubility in pure water.

$$K_{sp} = [Ag^+][Cl^-] = (S)(0.0100 + S) = 1.70 \times 10^{-10}$$

The solubility of AgCl in water was $1.30 \times 10^{-5}\ M$. Because we expect the solubility in water to be greater than the solubility in sodium chloride solution (S), we can safely assume that S is small compared with 0.0100.

$$S \ll 0.0100 \quad\text{or}\quad (0.0100 + S) = 0.0100$$

If this assumption is valid, the equation relating solubility and solubility product becomes

$$(S)(0.0100) = 1.70 \times 10^{-10}$$

This equation is solved to give

$$S = (1.70 \times 10^{-10})/(0.0100) = 1.70 \times 10^{-8}\ M$$

The solubility is indeed very small compared with 0.0100 M, so the assumption was valid. Note that the solubility is considerably reduced from the value in pure water.

Practice Problem 16.9

Calculate the solubility of silver iodide in water and in a solution containing 0.100 M $AgNO_3$. The K_{sp} of AgI is 8.5×10^{-17}.

Answer: $9.2 \times 10^{-9}\ M;\ 8.5 \times 10^{-16}\ M$

Precipitation of Salts

We have examined the solubility of insoluble salts and the effect of common ions on this solubility. In the examples considered so far, the saturated solutions were formed by dissolution of the insoluble salt. However, the equilibrium between an insoluble salt and its solution can also be established by mixing solutions of two soluble salts that contain the component ions of the insoluble salt. If the concentrations are appropriate, a precipitation reaction occurs and a saturated solution of the salt is formed.

How can we precipitate a salt from a solution that contains its component ions?

The conditions for the precipitation of a salt from solution can be determined from the value of the solubility product constant and the ion concentrations initially placed in solution. The ion concentrations can be examined simply in terms of the reaction quotient, the product of the ion concentrations raised to appropriate powers. Three relationships of Q_{sp} to K_{sp} are possible, each associated with a specific condition of the solution.

1. $Q_{sp} > K_{sp}$: Precipitation occurs until $Q_{sp} = K_{sp}$.
2. $Q_{sp} = K_{sp}$: Equilibrium exists (the solution is saturated).
3. $Q_{sp} < K_{sp}$: No precipitation occurs; if solid is added, it dissolves until $Q_{sp} = K_{sp}$.

A test of these relationships can be used to determine the conditions under which a precipitation reaction will occur.

EXAMPLE 16.10

Will a precipitate of $BaCO_3$ form when 0.20 L of 1.0×10^{-6} M $BaCl_2$ is added to 1.00 L of 1.2×10^{-6} M Na_2CO_3? The value of K_{sp} for $BaCO_3$ is 8.1×10^{-9}.

Solution   After the two solutions are combined, the total volume is 0.20 L + 1.00 L = 1.20 L. First we calculate the initial concentrations of ions—the concentrations before any precipitation occurs. The number of moles of Ba^{2+} in 0.20 L of 1.0×10^{-6} M $BaCl_2$ solution is given by

$$\text{mol } Ba^{2+} = (1.0 \times 10^{-6} \text{ mol/L})(0.20 \text{ L}) = 2.0 \times 10^{-7} \text{ mol}$$

The concentration of Ba^{2+} in 1.20 L of the mixture is

$$[Ba^{2+}] = (2.0 \times 10^{-7} \text{ mol})/(1.20 \text{ L}) = 1.7 \times 10^{-7} \text{ } M$$

We calculate the moles and initial concentration of CO_3^{2-} in the same way.

$$\text{mol } CO_3^{2-} = (1.2 \times 10^{-6} \text{ mol/L})(1.00 \text{ L}) = 1.2 \times 10^{-6} \text{ mol}$$
$$[CO_3^{2-}] = (1.2 \times 10^{-6} \text{ mol})/(1.2 \text{ L}) = 1.0 \times 10^{-6} \text{ } M$$

The reaction quotient is given by

$$Q_{sp} = [Ba^{2+}][CO_3^{2-}] = (1.7 \times 10^{-7})(1.0 \times 10^{-6}) = 1.7 \times 10^{-13}$$

A comparison with $K_{sp} = 8.1 \times 10^{-9}$ indicates that $Q_{sp} < K_{sp}$, so no $BaCO_3$ will precipitate from this mixture.

Practice Problem 16.10

Will a precipitate of $CaCO_3$ form when 0.50 L of 2.0×10^{-3} M $CaCl_2$ is added to 1.00 L of 1.5×10^{-2} M Na_2CO_3? The value of K_{sp} for $CaCO_3$ is 9.0×10^{-9}.

Answer: Yes ($Q_{sp} = 6.7 \times 10^{-6}$)

Frequently, testing a specific solution, as in the preceding example, is less useful than determining the conditions under which precipitation will occur. The same factors come into play, and the calculations are very similar. A typical calculation involves determining the minimum amount of a solution that would have to be added to initiate precipitation of an ionic salt. The critical condition is achievement of the set of concentrations for which Q_{sp} just becomes equal to K_{sp}.

EXAMPLE 16.11

What concentration of chloride ion must be present in a 0.100 M solution of lead(II) ion in order for $PbCl_2$ to just begin precipitating? (We will assume that the solution does not become supersaturated, which does happen with $PbCl_2$.) To simplify the calculations, assume that adding chloride ion produces no volume change. The value of K_{sp} for $PbCl_2$ is 1.6×10^{-5}.

Solution   Precipitation will begin when the reaction quotient just equals the solubility product constant. The lead-ion concentration is 0.100 M, and the chloride-ion concentration can be calculated from the relationship

$$K_{sp} = [Pb^{2+}][Cl^-]^2 = 1.6 \times 10^{-5}$$
$$(0.100)[Cl^-]^2 = 1.6 \times 10^{-5}$$
$$[Cl^-]^2 = 1.6 \times 10^{-5}/0.100 = 1.6 \times 10^{-4}$$
$$[Cl^-] = (1.6 \times 10^{-4})^{\frac{1}{2}} = 1.3 \times 10^{-2}\ M$$

Any concentration of Cl^- above $1.3 \times 10^{-2}\ M$ will cause the precipitation of $PbCl_2$.

Practice Problem 16.11

What concentration of chloride ion must be present in a $2.00 \times 10^{-5}\ M$ solution of silver ion in order for AgCl to precipitate? Assume that adding chloride ion produces no volume change. The value of K_{sp} for AgCl is 1.70×10^{-10}.

Answer: $8.50 \times 10^{-6}\ M$

Precipitation reactions can be used to separate two ions. By carefully controlled addition of another ion, it is possible to precipitate one ion selectively from a mixture, leaving the other in solution. Thus, by adding fluoride ion, we could remove all the calcium ion from solution while leaving barium ion still dissolved. The amount of fluoride ion must be carefully controlled in relation to the concentrations of calcium and barium ions so that in one case Q_{sp} is greater than K_{sp}, whereas in the other case Q_{sp} is less than K_{sp}.

EXAMPLE 16.12

A solution contains both Ca^{2+} and Ba^{2+}, each present at a concentration of 0.0100 M. We wish to remove the calcium ion from solution, while leaving all the barium ion still dissolved, by adding solid NaF. What concentration of F^- is necessary for CaF_2 but not BaF_2 to precipitate? What concentration of calcium ion will be left in solution when the barium ion just begins to precipitate if more fluoride ion is added? The K_{sp} values are 1.70×10^{-10} for CaF_2 and 1.73×10^{-6} for BaF_2.

Solution   First we determine the concentrations of F^- that are necessary to precipitate CaF_2 and BaF_2. The solubility product equations are

$$K_{sp} = [Ca^{2+}][F^-]^2 = 1.70 \times 10^{-10}$$
$$K_{sp} = [Ba^{2+}][F^-]^2 = 1.73 \times 10^{-6}$$

The concentrations of $[Ca^{2+}]$ and $[Ba^{2+}]$ are both 0.0100 M, and the necessary fluoride-ion concentrations are as follows.

For CaF_2 precipitation:

$$[F^-] = \sqrt{(1.70 \times 10^{-10})/(0.0100)} = 1.30 \times 10^{-4} \ M$$

For BaF_2 precipitation:

$$[F^-] = \sqrt{(1.73 \times 10^{-6})/(0.0100)} = 1.32 \times 10^{-2} \ M$$

Thus, in order for CaF_2 but not BaF_2 to precipitate, the concentration of F^- must be greater than $1.30 \times 10^{-4} \ M$ but not more than $1.32 \times 10^{-2} \ M$.

The concentration of the calcium ion left in solution when barium ion begins precipitating is given by the solubility product equation, where $[F^-] = 1.32 \times 10^{-2} \ M$.

$$K_{sp} = [Ca^{2+}][F^-]^2 = 1.70 \times 10^{-10}$$
$$[Ca^{2+}](1.32 \times 10^{-2})^2 = 1.70 \times 10^{-10}$$
$$[Ca^{2+}] = (1.70 \times 10^{-10})/(1.32 \times 10^{-2})^2 = 9.76 \times 10^{-7} \ M$$

Thus the calcium ion is reduced from 0.0100 M to $9.76 \times 10^{-7} \ M$ at this point.

Practice Problem 16.12

A solution contains both Pb^{2+} and Ca^{2+}, each present at a concentration of 0.0100 M. We wish to remove the lead ion from solution, while leaving all the calcium ion still dissolved, by adding solid Na_2SO_4. What concentration of SO_4^{2-} is necessary to precipitate $PbSO_4$ but not $CaSO_4$? What concentration of lead ion will be left in solution when the calcium ion just begins to precipitate if more sulfate ion is added? The K_{sp} values are 2.00×10^{-4} for $CaSO_4$ and 1.80×10^{-8} for $PbSO_4$.

Answer: $1.80 \times 10^{-6} \ M < [SO_4^{2-}] < 2.00 \times 10^{-2} \ M$; remaining $[Pb^{2+}] = 9.00 \times 10^{-7} \ M$

Solubility and pH

When one of the ions in an insoluble salt is acidic or basic, the solubility of the salt depends on the pH of the solution. ▶ The most straightforward examples of this are the metal hydroxides, such as the iron(III) hydroxide described in the chapter introduction. Most of these compounds, except the alkali metal hydroxides, are insoluble or only slightly soluble in water. Consider magnesium hydroxide, which has a limited solubility. Suspensions of this salt, called milk of magnesia, are used as antacid preparations. The dissolution equation is

$$Mg(OH)_2(s) \rightleftharpoons Mg^{2+}(aq) + 2OH^-(aq)$$

with $K_{sp} = 1.2 \times 10^{-11}$. From this value, the solubility of $Mg(OH)_2$ in water can be calculated in the usual way to be $1.4 \times 10^{-4} \ M$. Because two OH^- ions are produced for each $Mg(OH)_2$ formula unit that dissolves, the OH^- concentration is twice the $Mg(OH)_2$ concentration, or $2.8 \times 10^{-4} \ M$.

When OH^- is one of the ions in a salt, as here, the solubility is pH-dependent. The pH in a saturated solution of $Mg(OH)_2$ is

$$pOH = -\log [OH^-] = -\log(2.8 \times 10^{-4}) = 3.55$$
$$pH = 14 - pOH = 14.00 - 3.55 = 10.45$$

What types of salts have solubilities that are pH-dependent?

RULES OF REACTIVITY

A precipitation reaction occurs when hydrolysis of a metal ion produces an insoluble hydroxide. All hydroxides are insoluble except NaOH, KOH, NH_4OH, and $Ba(OH)_2$. $Ca(OH)_2$ is slightly soluble.

This solution is decidedly basic. Changing the pH of the solution, which we could do by adding an acid or a base, is equivalent to introducing a common ion. If the pH is raised, the concentration of OH^- is increased and the solubility is decreased, in line with the predictions of Le Chatelier's principle. On the other hand, if the pH is lowered, the solubility increases. In fact, if the solution is made acidic, the solubility becomes characteristic of the magnesium salt of the acid's anion. Calculation of the solubility at various pH values follows the usual procedure, except that the OH^- concentration must be calculated from the pH.

EXAMPLE 16.13

The molar solubility of magnesium hydroxide in water is 1.44×10^{-4} mol/L, and $K_{sp} = 1.2 \times 10^{-11}$. The pH of this solution is 10.45. What is the solubility of $Mg(OH)_2$ in a solution buffered to pH 9.00?

Solution   The solubility product constant is

$$K_{sp} = [Mg^{2+}][OH^-]^2 = 1.2 \times 10^{-11}$$

In the absence of a common-ion effect, the relationships between ionic concentrations and solubility for magnesium hydroxide dissolved in water are

$$[Mg^{2+}] = S \qquad [OH^-] = 2S$$

The pH of this solution is 10.45, as stated.

If the pH of the solution is changed, so are the OH^- concentration and the relationship between this concentration and the solubility. We can calculate the OH^- concentration directly from the following equation and the pH of 9.00.

$$pOH = -\log [OH^-] = 14.00 - pH = 14.00 - 9.00 = 5.00$$

We obtain the hydroxide-ion concentration from the equation

$$[OH^-] = 10^{-pOH} = 10^{-5.00} = 1.0 \times 10^{-5} \, M$$

The magnesium-ion concentration, and thus the solubility of magnesium hydroxide, can be calculated from the expression for the solubility product constant.

$$K_{sp} = [Mg^{2+}][OH^-]^2 = (S)(1.0 \times 10^{-5})^2 = 1.2 \times 10^{-11}$$
$$S = 1.2 \times 10^{-11}/(1.0 \times 10^{-5})^2 = 0.12 \, M$$

Note that because this hydroxide-ion concentration comes from an external source and not from the magnesium hydroxide, it is *not* doubled. From this calculation, we see that at pH 9.00, which is more acidic than the solution resulting from dissolution of $Mg(OH)_2$ in water, the solubility is increased.

Practice Problem 16.13

Manganese hydroxide, $Mn(OH)_2$, has a K_{sp} of 4.5×10^{-14}. What is the solubility of this salt in water? What is the pH of this solution? What is the solubility of $Mn(OH)_2$ in a solution buffered to pH 8.00?

Answer: $2.2 \times 10^{-5} \, M$; pH 9.64; 0.045 M

So far, we have examined the pH-dependent solubility of only hydroxide salts. However, the same factors come into play for any ion that is an acid or a base. Thus basic anions—such as F^-, $CH_3CO_2^-$, CN^-, and S^{2-}—form salts whose solubility is pH-dependent. Consider calcium fluoride, which exists in nature as the mineral fluorspar, for example. The dissolution equilibrium is

$$CaF_2(s) \rightleftharpoons Ca^{2+}(aq) + 2F^-(aq)$$

If a saturated solution of calcium fluoride is made acidic, the concentration of F^- is reduced because HF, a weak acid, is formed.

$$H_3O^+(aq) + F^-(aq) \rightleftharpoons HF(aq) + H_2O(l)$$

The removal of F^- to form HF forces the dissolution equilibrium to shift to the right, and the solubility of CaF_2 increases.

Complex-Ion Formation and Solubility

Solubility can be affected by the presence of some Lewis acids and bases, which, as you recall, may form complex ions. The formation of a complex ion in a saturated solution of an insoluble salt generally increases the solubility of that salt. For example, lead chloride precipitates out of a solution containing Pb^{2+} ion upon addition of hydrochloric acid. However, if enough hydrochloric acid is added, the precipitate redissolves, as shown in Figure 16.14. This phenomenon is caused by the formation of a soluble complex ion at high chloride-ion concentrations.

$$PbCl_2(s) + 2Cl^-(aq) \rightleftharpoons PbCl_4^{2-}(aq)$$

The dissolution of precipitates by complex-ion formation does not necessarily involve a common ion, as in this example. Addition of other substances can have the same effect if they result in the formation of complex ions. An example involves silver chloride, an insoluble salt that forms silver ion and chloride ion in solution.

$$AgCl(s) \rightleftharpoons Ag^+(aq) + Cl^-(aq)$$

Figure 16.14 *Left:* A saturated solution of lead(II) chloride. *Center:* When chloride ion is added, solid dissolves, and a colorless complex ion is formed (*Right*).

Figure 16.15 (*A*) A saturated solution of AgCl is in contact with undissolved solid. (*B*) Addition of concentrated aqueous ammonia results in the formation of a complex ion and (*C*) in dissolution of solid AgCl.

When silver ion is removed from this system, the equilibrium should shift to the right, and more silver chloride should dissolve. This effect can be produced by addition of aqueous ammonia, which forms a silver ammine complex.

$$Ag^+(aq) + 2NH_3(aq) \rightleftharpoons Ag(NH_3)_2^+$$

When the ammonia concentration reaches about 5 *M*, silver chloride becomes highly soluble, as shown in Figure 16.15. Removal of silver ion is so nearly complete that the solubility equilibrium shifts far to the right, and the silver chloride dissolves.

Summary

Hydrolysis, often of salts derived from acids or bases, can affect the pH of a solution. Some metal ions, particularly those with charges greater than 2+, also undergo hydrolysis by loss of a hydrogen ion from a water molecule covalently bonded to the metal ion. The pH of a solution of a weak acid or a weak base can be changed by addition of a salt that contains the conjugate base or conjugate acid. This phenomenon is the common-ion effect. Solutions prepared from mixtures of a weak acid and its conjugate base, or mixtures of a weak base and its conjugate acid, resist changes in pH when strong acids or bases are added to the mixture. Such solutions, called buffer solutions, are very important in regulating the pH of body fluids and other systems. The pH of a buffer solution can be calculated from the Henderson–Hasselbalch equation: $pH = pK_a + \log [A^-]/[HA]$.

Complex ions formed between a metal ion and a ligand are important in many chemical and biological processes. In many cases, more than one complex ion can be formed by a series of successive equilibrium reactions.

The solubility of insoluble salts is best described in terms of an equilibrium between the solid salt and dissolved ions. The equilibrium constant for such a reaction is called the solubility product constant. The solubility of an insoluble salt can be modified by addition of other salts, adjustment of the pH in some cases, or formation of a complex ion.

Key Terms

buffer (16.3)
buffer range (16.3)
common ion (16.2)
common-ion

effect (16.2)
complex ion (16.4)
formation constant (K_f) (16.4)

Henderson–Hasselbalch equation (16.3)
hydrolysis constant (K_h) (16.1)

ligand (16.4)
solubility product constant (K_{sp}) (16.5)

Exercises

Hydrolysis

16.1 What is hydrolysis?

16.2 How can we decide whether a solution of a salt will be acidic, neutral, or basic?

16.3 Why are solutions of some metal ions acidic?

16.4 How can we obtain values of the hydrolysis constant of anions?

16.5 Would you expect the percentage hydrolysis of sodium acetate to increase or decrease as the solution is diluted? Explain.

16.6 Calculate the pH of a 0.250 M solution of NH_4NO_3. The K_b for NH_3 is 1.76×10^{-5}.

16.7 A solution made up to be 0.200 M NaOCl exhibits a pH of 10.41 at equilibrium. Calculate K_a for HOCl.

16.8 Calculate the pH of a 0.150 M $LiNO_2$ solution. The K_a of HNO_2 is 4.6×10^{-4}.

16.9 Calculate the pH of a 0.200 M solution of ammonium chloride, NH_4Cl. The base-ionization constant of ammonia is 1.76×10^{-5}.

16.10 Calculate the hydronium-ion concentration in a 0.225 M solution of sodium chlorite, $NaClO_2$. The acid-ionization constant of chlorous acid is 1.1×10^{-2}.

16.11 Calculate the hydrolysis constant for dimethylammonium chloride, $(CH_3)_2NH^+Cl^-$. The base-ionization constant for dimethylamine is 5.9×10^{-4}.

16.12 Calculate the pH of 250.0 mL of a solution containing 23.2 g of sodium acetate. The acid-ionization constant of acetic acid is 1.75×10^{-5}.

16.13 Calculate the pH and percent hydrolysis for each of the following solutions.

   a. 0.0500 M NH_4Br; K_b of $NH_3 = 1.76 \times 10^{-5}$
   b. 2.0×10^{-3} M NH_4Cl; K_b of $NH_3 = 1.76 \times 10^{-5}$
   c. 0.0100 M KCN; K_a of HCN = 4.0×10^{-10}
   d. 0.250 M NaCN; K_a of HCN = 4.0×10^{-10}

Buffer Solutions

16.14 What is a buffer?

16.15 What are buffers used for?

16.16 What is the significance of the buffer range?

16.17 How can we pick the appropriate compounds to use for a buffer?

16.18 Explain why buffers resist changes in pH.

16.19 A buffer solution can be made from acetic acid and sodium acetate dissolved in water.

   a. Write a balanced equation to describe the equilibrium reaction that prevents the pH from changing when small quantities of acid are added to the solution.
   b. Write a balanced equation to describe the equilibrium reaction that prevents the pH from changing when small quantities of base are added to the solution.

16.20 What is the molar ratio of ammonium ion to ammonia in a buffer solution with a pH of 9.00? The K_b of ammonia is 1.76×10^{-5}.

16.21 Calculate the pH of a solution of 0.0500 M nitrous acid, HNO_2, that also contains 0.0200 M sodium nitrite, $NaNO_2$. The K_a of HNO_2 is 4.6×10^{-4}.

16.22 A solution contains 0.0750 M $NH_3(aq)$ and 0.100 M NH_4Cl. Calculate the pH of the solution, assuming that K_b of $NH_3 = 1.76 \times 10^{-5}$.

16.23 Calculate the pH of a buffer solution prepared from 0.0500 mol CH_3CO_2H and 0.0250 mol $NaCH_3CO_2$ dissolved in enough water to yield 500.0 mL of solution. The K_a of CH_3CO_2H is 1.75×10^{-5}.

16.24 A solution is prepared from 0.200 mol $NaHCO_2$ and 0.250 mol HCO_2H dissolved in approximately 200 mL of water. Calculate the concentrations of $H_3O^+(aq)$ and $OH^-(aq)$ in this solution. The acid-ionization constant of formic acid is 1.77×10^{-4}.

16.25 Calculate the pH of a solution prepared by addition of 30.0 mL of 0.100 M HCl to 245 mL of a solution containing 0.0500 M CH_3CO_2H and 0.125 M $NaCH_3CO_2$. The value of K_a for acetic acid is 1.75×10^{-5}.

16.26 Calculate the pH of a solution that is 0.125 M in acetic acid and 0.100 M in sodium acetate. The value of K_a for acetic acid is 1.75×10^{-5}.

16.27 What mass of sodium formate, $NaHCO_2$, must be dissolved in 250.0 mL of 0.1000 M formic acid (HCO_2H) to yield a buffer solution of pH 3.88? The value of K_a for formic acid is 1.77×10^{-4}.

16.28 You are instructed to prepare a buffer solution with a pH of 3.75, using formic acid and sodium formate. If the value of K_a for HCO_2H is 1.77×10^{-4}, what ratio of formic acid to formate ion is necessary to prepare this buffer?

16.29 What concentration of NH_4Cl is required to adjust the pH of a 0.100 M $NH_3(aq)$ solution to 8.00? The value of K_b for $NH_3(aq)$ is 1.76×10^{-5}.

16.30 A solution contains 0.100 M acetic acid and 0.100 M sodium acetate.

   a. What is the pH of the solution if the value of K_a for acetic acid is 1.75×10^{-5}?
   b. What is the change in pH if 0.0200 mol HCl is added to 1.00 L of this solution?
   c. What is the change in pH if 0.0200 mol NaOH is added to 1.00 L of this solution? Assume there is no volume change.

16.31 A solution contains 0.100 M $NH_3(aq)$ and 0.100 M NH_4Cl.

   a. What is the pH of the solution if the value of K_b for $NH_3(aq)$ is 1.76×10^{-5}?

    b. What change in pH is produced when 0.0100 mol NaOH is added to 500.0 mL of this solution?

16.32 Suppose 0.00100 mol HCl is added to 500.0 mL of an acetic acid/sodium acetate buffer solution with a pH of 4.74. What is the new pH of the solution? The concentrations of acetic acid and sodium acetate were both initially 0.100 M.

16.33 How much 10.0 M NaOH would have to be added to 500.0 mL of an acetic acid/sodium acetate buffer solution to raise the pH by 0.20 pH units from the initial pH of 4.74? The concentrations of acetic acid and sodium acetate were both initially 0.500 M.

Complex-Ion Equilibria

16.34 What is a complex ion?

16.35 Write an equation for a typical complex-ion formation reaction.

16.36 Write the formulas of some complex ions that are important in nature.

16.37 When ammonia is added to a solution containing blue $Cu^{2+}(aq)$, the solution turns a deeper blue because of formation of a complex ion. What is the concentration of Cu^{2+} in a solution containing 0.0100 M $Cu(NH_3)_4^{2+}$ and 0.400 M excess NH_3? The formation constant of $Cu(NH_3)_4^{2+}$ is 1.2×10^{12}.

16.38 If the formation constant of $Ag(NH_3)_2^+$ is 1.7×10^7, what is the concentration of Ag^+ in a 0.100 M $Ag(NH_3)_2^+$ solution that also contains 0.250 M excess NH_3?

16.39 Enough KCN is added to a 0.010 M Ni^{2+} solution to form the complex ion $Ni(CN)_4^{2-}$ and give a concentration of free CN^- of 0.020 M. What is the concentration of Ni^{2+} in this solution if K_f of $Ni(CN)_4^{2-}$ is 5.6×10^{13}?

Solubility Product Constant

16.40 Write the expression for the solubility product.

16.41 Why is the concentration of the insoluble salt not included in the expression for the solubility product?

16.42 Write equilibrium chemical equations and solubility product constant expressions for the dissolution of the following slightly soluble salts.
    a. Ag_2CrO_4 d. AgI
    b. $PbCrO_4$ e. $Cr(OH)_3$
    c. PbI_2 f. $Ca_3(PO_4)_2$

16.43 Write the expression for the solubility product constant, K_{sp}, for each of the following.
    a. $PbCl_2$ d. $Cr_2(CO_3)_3$
    b. $Cr(OH)_3$ e. Ag_2S
    c. MnS

16.44 Calculate the solubility product constant for each of the following substances.
    a. $Mg(OH)_2$, with a solubility of 0.0084 g/L
    b. $PbSO_4$, with a solubility of 0.0300 g/L
    c. Ag_2CO_3, with a solubility of 0.00317 g/100 mL
    d. $BaSO_4$, with a solubility of 0.00220 g/L
    e. $Al(OH)_3$, with a solubility of 0.0308 g/L

16.45 Calculate the K_{sp} of CuBr if it dissolves to the extent of 0.0104 g/L in water at 25°C.

16.46 Aluminum fluoride is less toxic than many other fluoride salts because of its low solubility. If the solubility of AlF_3 is 0.559 g per 100.0 mL of water at 25°C, what is the value of K_{sp}?

16.47 Cobalt(II) cyanide is a deep-blue solid that turns red when hydrated. It is used as a catalyst. If the solubility of $Co(CN)_2$ is 0.00415 g per 100.0 mL of water at 25°C, what is the value of K_{sp}?

16.48 Iron(II) hydroxide is a pale green crystalline solid that is oxidized to iron(III) hydroxide upon exposure to moist air. When finely divided, it spontaneously ignites in air. If the concentration of hydroxide ion in a saturated solution of $Fe(OH)_2$ is 3.20×10^{-5} M, what is the K_{sp} of iron(II) hydroxide?

Solubility

16.49 How is solubility related to the solubility product constant?

16.50 What is necessary for a slightly soluble salt to be in a state of equilibrium when placed in water?

16.51 Calculate the concentrations of the ions present in saturated solutions of each of the following ionic salts.
    a. ZnS ($K_{sp} = 1.2 \times 10^{-23}$)
    b. $Fe(OH)_2$ ($K_{sp} = 2.0 \times 10^{-15}$)
    c. $SrSO_4$ ($K_{sp} = 2.8 \times 10^{-7}$)
    d. $CaCO_3$ ($K_{sp} = 9.0 \times 10^{-9}$)
    e. $Fe(OH)_3$ ($K_{sp} = 1.1 \times 10^{-36}$)

16.52 Calculate the solubility in grams per liter for each of the following substances.
    a. $AgBr$ ($K_{sp} = 5.0 \times 10^{-13}$)
    b. FeS ($K_{sp} = 3.7 \times 10^{-19}$)
    c. BaF_2 ($K_{sp} = 1.73 \times 10^{-6}$)
    d. $SrSO_4$ ($K_{sp} = 2.8 \times 10^{-7}$)
    e. $PbCl_2$ ($K_{sp} = 1.6 \times 10^{-5}$)

16.53 Radium sulfate was once used to make luminous paints for watch dials, but such use was discontinued because of its radioactivity. How many grams of Na_2SO_4 can be dissolved in 3.00 L of 0.0250 M $RaCl_2$ solution without the formation of a precipitate? K_{sp} of $RaSO_4$ is 3.85×10^{-15}.

16.54 Strontium chromate is a yellow powder used in pigments. What mass of strontium chromate will dissolve in 125 mL of water if the K_{sp} of $SrCrO_4$ is 3.6×10^{-5}?

16.55 Magnesium fluoride is a colorless crystalline solid used in the manufacture of certain types of glass and ceramics. Calculate the molar solubility of MgF_2 if the K_{sp} is 6.4×10^{-9}.

16.56 Silver chromate is a dark reddish-brown crystalline solid. A solution is 0.100 M in $AgNO_3$. What is the molar solubility of Ag_2CrO_4 in this solution? K_{sp} of $Ag_2CrO_4 = 9.0 \times 10^{-12}$.

16.57 Calculate the concentration of Cu^+ in a saturated solution of $CuBr$ if the solution also contains 0.100 M NaBr. The K_{sp} of $CuBr$ is 5.26×10^{-9}.

Precipitation of Salts

16.58 What is a precipitation reaction?

16.59 How can a metal ion be precipitated from solution when it is in equilibrium with its insoluble salt?

16.60 Explain how two metal ions could be separated from one another by precipitation.

16.61 What concentration of S^{2-} ions is necessary to start the precipitation of HgS from a solution that is 0.0200 M in $Hg(NO_3)_2$? The value of K_{sp} for HgS is 1.6×10^{-54}.

16.62 Predict whether a precipitate would form in each of the following mixtures.
   a. 300.0 mL of 0.100 M $CuCl_2$ and 200.0 mL of 0.200 M KIO_3; K_{sp} of $Cu(IO_3)_2 = 1.4 \times 10^{-7}$
   b. 20.0 mL of 0.100 M KCl and 50.0 mL of 0.100 M $Pb(NO_3)_2$; K_{sp} of $PbCl_2 = 1.6 \times 10^{-5}$
   c. 50.0 mL of 0.000100 M $BaCl_2$ and 20.0 mL of 0.00100 M Na_2SO_4; K_{sp} of $BaSO_4 = 1.5 \times 10^{-9}$

16.63 Lead chromate is a yellow to yellow-orange solid used as a pigment in oil and water colors, fabric printing, and ceramics. What concentration of Pb^{2+} is necessary to start the precipitation of $PbCrO_4$ from a solution that is 0.0100 M in Na_2CrO_4? The K_{sp} of $PbCrO_4$ is 1.8×10^{-14}.

16.64 Lead iodate, $Pb(IO_3)_2$, is a slightly soluble salt with a K_{sp} of 2.6×10^{-13}. To 35.0 mL of a 0.150 M $Pb(NO_3)_2$ solution, 15.0 mL of a 0.800 M KIO_3 solution is added, and a precipitate of $Pb(IO_3)_2$ is formed. What concentration of Pb^{2+} and IO_3^- are left in solution when the system reaches equilibrium?

16.65 Solid NaF is added slowly to a solution that is 0.0100 M in Ca^{2+} and 0.0100 M in Pb^{2+}.

   a. Which solid, CaF_2 or PbF_2, will precipitate first?
   b. Calculate the percentage of Ca^{2+} or Pb^{2+} that precipitates just before the second compound begins to precipitate. The value of K_{sp} is 1.70×10^{-10} for CaF_2 and 3.70×10^{-8} for PbF_2.

Solubility and pH

16.66 The solubility of some salts is dependent on the pH. What types of salts are these?

16.67 How could buffers be used to control the solubility of some salts?

16.68 How would you separate two metal ions by adjusting the pH?

16.69 Calculate K_{sp} for $Ca(OH)_2$. The pH of a saturated solution is 12.40.

16.70 Cadmium hydroxide is a colorless solid used in the electrodes of NiCad batteries. What concentration of Cd^{2+} can exist in a solution of pH 8.00? The K_{sp} of $Cd(OH)_2$ is 1.2×10^{-14}.

16.71 Aluminum hydroxide is used in municipal water treatment to help remove suspended solid particles. At what pH will $Al(OH)_3$ start to precipitate from a solution that contains 0.0100 M Al^{3+}? The K_{sp} of $Al(OH)_3$ is 1.9×10^{-33}.

Solubility and Complex-Ion Formation

16.72 Describe the bonding in complex ions.

16.73 What is the general form of the expression for a complex-ion formation constant?

16.74 How can complex-ion formation be used to control the solubility of slightly soluble salts?

16.75 Silver iodide is a light yellow solid used in rainmaking to "seed" clouds. The slightly soluble compound AgI reacts with CN^- to form the stable complex ion $Ag(CN)_2^-$.
   a. Write an equation for this reaction.
   b. What is the equilibrium constant for this reaction in terms of K_{sp} and K_f?

16.76 What mass of AgBr will dissolve in 500.0 mL of 4.00 M NH_3 solution? The formation constant of $Ag(NH_3)_2^+$ is 1.7×10^7, and the K_{sp} of AgBr is 5.0×10^{-13}.

16.77 What mass of AgCl will dissolve in 1.00 L of 1.00 M NH_3 solution? The formation constant of $Ag(NH_3)_2^+$ is 1.7×10^7, and the value of K_{sp} for AgCl is 1.7×10^{-10}.

16.78 Calculate the solubility of AgI in 2.00 M aqueous NH_3. The value of K_{sp} for AgI is 8.5×10^{-17}, and the value of K_f for $Ag(NH_3)_2^+$ is 1.7×10^7.

Additional Exercises

16.79 a. Calculate the molar solubility of $PbCl_2$ in water at 25°C, assuming that $K_{sp} = 1.6 \times 10^{-5}$.

   b. Also calculate the solubility in a solution containing 0.100 mol/L NaCl.

16.80 Into 1.00 L of a solution of 0.250 M HCl is added 0.600 mol $NaCH_3CO_2$. Assume that no volume change occurs, and calculate the concentrations of $CH_3CO_2H(aq)$, $CH_3CO_2^-(aq)$, $H_3O^+(aq)$, and $OH^-(aq)$.

16.81 Calculate the molar solubility of Ag_3PO_4. Its K_{sp} is 1.77×10^{-18}.

16.82 Arrange the following metal sulfides in order of increasing solubility in water: CoS, CuS, MnS, NiS.

16.83 A sample of 0.200 M HCN is neutralized with an equal volume of 0.200 M NaOH. Calculate the pH at the equivalence point. The value of K_a for HCN is 4.0×10^{-10}.

16.84 Barium sulfate is so insoluble that it is used to take stomach X rays even though barium is toxic. Calculate the value of K_{sp} for barium sulfate. The molar solubility is 3.87×10^{-5} mol/L.

16.85 Lead sulfate is formed on the electrodes of an automobile battery. The solubility product constant of lead sulfate, $PbSO_4$, is 1.80×10^{-8}. Calculate the solubility of $PbSO_4$ in each of the following.

   a. pure water

   b. 0.100 M $Pb(NO_3)_2$ solution

   c. 0.00100 M Na_2SO_4 solution

16.86 Predict the effect on the solubility of $Cr(OH)_3$ of adding aqueous solutions of the following salts. Explain your predictions.

   a. $CrCl_3$ b. NaOH   c. HCl   d. NaCl

16.87 What is the pH of a buffer prepared from 0.250 mol of hydrazine, N_2H_4, and 0.350 mol of hydrazinium chloride, N_2H_5Cl, dissolved in 250.0 mL of solution? The value of K_b for hydrazine is 1.3×10^{-6}.

16.88 Will a precipitate form in the following solutions?

   a. A solution containing 0.025 M $CaCl_2$ and 0.050 M Na_2CO_3 ($K_{sp} = 9 \times 10^{-9}$ for $CaCO_3$)

   b. A solution containing 0.0050 M $Pb(NO_3)_2$ and 0.020 M $CaCl_2$ ($K_{sp} = 1.6 \times 10^{-5}$ for $PbCl_2$)

16.89 Silver sulfide is a grayish-black solid used in ceramics. Calculate the molar solubility of silver sulfide in 0.100 M $AgNO_3$ solution. The K_{sp} for Ag_2S is 1.0×10^{-51}.

16.90 Consider the chemical equilibrium

$$AgCl(s) + 2NH_3(aq) \rightleftharpoons Ag(NH_3)_2^+(aq) + Cl^-(aq)$$

Will the solubility of $AgCl(s)$ increase, decrease, or stay the same as a result of each of the following changes?

   a. A decrease in the concentration of Cl^-

   b. An increase in the concentration of NH_3

   c. An increase in the amount of AgCl

16.91 Solid silver nitrate is slowly added to a solution containing 0.100 M NaBr, 0.0500 M Na_2CrO_4, and 0.00100 M Na_3PO_4. In what order will AgBr, Ag_2CrO_4, and Ag_3PO_4 precipitate? The values of K_{sp} are 5.0×10^{-13} for AgBr, 9.0×10^{-12} for Ag_2CrO_4, and 1.77×10^{-18} for Ag_3PO_4.

16.92 Write balanced net ionic equations to describe any precipitation reactions that occur when the following pairs of aqueous solutions are mixed.

   a. $Ni(NO_3)_2 + Na_2S$

   b. $Na_2CO_3 + CaCl_2$

   c. $CrCl_3 + Na_3PO_4$

   d. $Al(NO_3)_3 + NaOH$

   e. $MgS + Hg(NO_3)_2$

   f. $HCl + Pb(NO_3)_2$

   g. $MgSO_4 + Ba(OH)_2$

   h. $Fe_2(SO_4)_3 + BaCl_2$

   i. $Zn(NO_3)_2 + (NH_4)_2SO_4$

   j. $Sr(NO_3)_2 + H_2SO_4$

   k. $KCl + CuSO_4$

16.93 Bismuth hydroxide is a yellowish-white amorphous powder used in the separation of plutonium from uranium. If the solubility of $Bi(OH)_3$ is 0.00140 g per 1.00 L water at 25°C, what is the value of K_{sp} for $Bi(OH)_3$?

16.94 The pH of a solution that is 0.200 M CH_3CO_2H and 0.200 M $NaCH_3CO_2$ is 4.76. What is the pH of a solution that is 0.0500 M CH_3CO_2H and 0.0500 M $NaCH_3CO_2$?

16.95 A buffer solution can be made from ammonia and ammonium chloride dissolved in water. Write a balanced equation to describe the equilibrium reaction that prevents the pH from changing when small quantities of acid are added to the solution. Write a balanced equation to describe the equilibrium reaction that prevents the pH from changing when small quantities of base are added to the solution.

* 16.96 To a solution containing 0.100 M Ca^{2+} and 0.100 M Ba^{2+}, sodium sulfate is added slowly. The solubility product constants of $CaSO_4$ and $BaSO_4$ are 2.00×10^{-5} and 1.5×10^{-9}, respectively.

   a. What is the SO_4^{2-} concentration at the instant the first solid precipitates?

   b. What is that solid? Neglect dilution and calculate the Ba^{2+} concentration when the first precipitation of $CaSO_4$ occurs.

   c. Is it possible to separate Ca^{2+} and Ba^{2+} by selective precipitation of the sulfates?

*16.97 A solution that contains 0.0100 M each of Ag^+, Fe^{3+}, and Zn^{2+} is adjusted to a pH of 4.35. The values of K_{sp} are 2.00×10^{-8} for AgOH, 1.10×10^{-36} for $Fe(OH)_3$, and 4.5×10^{-17} for $Zn(OH)_2$.
  a. What precipitates, if any, are formed?
  b. At what pH does each of the metal ions start to precipitate?

c. Is it possible to separate these metal ions by selective precipitation brought about by changing the pH? Justify your answer by calculating the concentration of the precipitating ion that is still in solution when the next ion begins to precipitate.

▶ ▶ ▶ Chemistry in Practice

16.98 In the United States, daily direct personal use (not including industrial use) of purified water is about 270 L. For a municipality of 50,000 people, this amounts to 13.5 million liters daily. It is common to add fluoride ion (F^-) in amounts of 2.0 ppm (or mg/L) to municipal water to help prevent tooth decay. Calcium fluoride, CaF_2, with a K_{sp} of 1.70×10^{-10}, is a possible source of F^-. Of concern, however, is the natural hardness of the water, because a high concentration of Ca^{2+} could suppress the solubility of CaF_2 and lower the F^- concentration. The following experiments were done to test the hardness of the water: A standard solution of $CaCl_2$ was prepared by weighing 4.015 g $CaCO_3$ into a volumetric flask, reacting it with a slight excess of HCl(aq), and diluting to 500.0 mL. A 50.00-mL portion of this $CaCl_2$ solution was titrated with an EDTA solution; 39.72 mL of EDTA was required. The EDTA solution was then used to titrate 50.00 mL of the municipal water; 14.92 mL of EDTA was required.
  a. What is the concentration of Ca^{2+} in the municipal water?

b. Is it possible to maintain a concentration of 0.8 ppm of F^- in this water by dissolving CaF_2?
c. If CaF_2 is not sufficiently soluble under these conditions, how much Na_2CO_3 should be added daily to the water supply to lower the concentration of Ca^{2+} enough for it to be possible to maintain this F^- concentration? The solubility product constant of $CaCO_3$ is 9.0×10^{-9}.
d. If CaF_2 is sufficiently soluble in the municipal water, what is the maximum concentration of Ca^{2+} that can exist in this water supply without removing any fluoride ion?
e. How much CaF_2 would this municipality need each year?
f. The municipality wants to manufacture its own CaF_2 by the reaction between $CaCO_3$ and HF. How much of each reactant compound would be needed annually if the reaction gave a 78% yield?

Oxidation–Reduction Reactions

Zinc reduces tin(II) chloride in solution to form a tin tree.

CHAPTER OUTLINE

Encountering Chemistry

A white solid, tin(II) chloride, is dissolved in water and forms a clear, colorless solution. A strip of zinc is dropped into this solution. In a few minutes, metallic crystals begin to grow on the surface of the zinc. Eventually the crystals fill the metal surface, protruding like the branches of a tree. This is a displacement reaction (discussed in Chapter 4), in which a metal ion is displaced from solution by another, more active metal, which in turn goes into solution in the form of ions. The reaction is represented by the following equation:

$$Zn(s) + Sn^{2+}(aq) \longrightarrow Zn^{2+}(aq) + Sn(s)$$

Displacement reactions such as this are but one type of *oxidation–reduction reaction,* in which the oxidation number of one element increases and that of another element decreases.

In this chapter we examine oxidation–reduction reactions, the assignment of oxidation numbers, the balancing of oxidation–reduction reaction equations, and methods for predicting the outcomes of oxidation–reduction reactions. These principles are applied to a variety of reactions, including those involved in recovering metals from their ores.

17.1 Oxidation and Reduction

In Chapter 4 we considered oxidation–reduction reactions as one example of the class of displacement reactions. ▶ Oxidation–reduction reactions, or *redox reactions,* may involve a metal and ions of another element, as shown for zinc and

Figure 17.1 When solutions containing colorless $Fe^{2+}(aq)$ and yellow-orange $Cr_2O_7^{2-}(aq)$ are mixed, they undergo an oxidation–reduction reaction to form $Fe^{3+}(aq)$ and $Cr^{3+}(aq)$. This mixture of ions is green.

The term oxidation originally arose from investigations of reactions that involve O_2, oxidation being the addition of oxygen to a species. How can this definition be extended to include a much wider variety of reactions?

tin(II) ions in the introduction to this chapter. They may also involve reactions between two substances, neither of them in the elemental state. For example, iron(II) in solution reacts with chromium(VI) in solution, in the presence of aqueous hydrogen ions, to form iron(III) and chromium(III) in solution, as shown in Figure 17.1 and summarized by the equation

$$6Fe^{2+}(aq) + Cr_2O_7^{2-} + 14H^+(aq) \longrightarrow 6Fe^{3+}(aq) + 2Cr^{3+}(aq) + 7H_2O(l)$$

To simplify the balancing of equations, we will use the designation $H^+(aq)$ to represent the aqueous hydrogen ion in this chapter—rather than $H_3O^+(aq)$, which we used in Chapters 15 and 16.

    Oxidation–reduction reactions involve a transfer of electrons from one substance to another, with the results outlined in Figure 17.2. As a result of the electron transfer, there is a change in the number of electrons assigned to one of the component elements in each substance. These elements undergo a change in oxidation number. Among the simplest electron-transfer reactions is the loss of

Figure 17.2 The oxidation or reduction of an element corresponds to a change in oxidation number and to a change in the number of electrons assigned to the element.

Oxidation

Decrease in number of electrons
(loss of electrons)
Increase in oxidation number

$\longrightarrow$

Oxidation number: −3   −2   −1   0   +1   +2   +3

$\longleftarrow$

Increase in number of electrons
(gain of electrons)
Decrease in oxidation number

Reduction

electrons by a metal atom to form a metal ion, as shown here for manganese.

$$Mn \longrightarrow Mn^{2+} + 2e^- \quad \text{(where } e^- \text{ symbolizes the electron)}$$

This representation is known as a **half-reaction,** in which only half the electron transfer (oxidation or reduction) is shown. The metal atom is oxidized in this half-reaction; it loses electrons and increases its oxidation number.

Electrons cannot just be released; they must be transferred to some other atom or ion. For example, aqueous hydrogen ion can react with the electrons to form gaseous molecular hydrogen, as shown by the following reduction half-reaction.

$$2H^+(aq) + 2e^- \longrightarrow H_2(g)$$

This reaction involves a gain of electrons and a decrease in oxidation number; the ion is reduced. Oxidation–reduction reactions, then, involve the oxidation of one substance and the reduction of another.

$$Mn(s) + 2H^+(aq) \longrightarrow Mn^{2+}(aq) + H_2(g)$$

Manganese is a steel-gray, hard, brittle metal that burns with a bright light when heated in air. It reacts slowly with cold water, and more rapidly with hot water, to release hydrogen gas. It reacts readily with dilute acids to release hydrogen gas and form manganese(II) solutions. Manganese is used in several alloys, the most common being steel.

In this reaction, two electrons are transferred from the manganese, which is oxidized, to two aqueous hydrogen ions, which are reduced. The substance that is oxidized during the reaction is called the **reducing agent,** because it brings about the reduction of another substance. The substance that is reduced is called the **oxidizing agent.**

We can think of oxidation–reduction reactions as involving a competition of two substances for electrons. One substance has a greater attraction for additional electrons and becomes the oxidizing agent; the other substance loses electrons and becomes the reducing agent. Oxidizing and reducing agents vary in strength. The stronger oxidizing agents are those substances with a greater affinity for extra electrons, and the stronger reducing agents are those substances with the least affinity for the electrons they already have.

Activity Series

The tendencies of various metals to give up their electrons are ranked in the **chemical activity series** shown in Figure 17.3. An activity series was first introduced in Section 4.2. The metals are arranged in order of decreasing reducing strength. The metals at the top of the series lose their electrons and form cations most readily, and those at the bottom lose their electrons with the greatest difficulty. Note that the most reactive metals are the strongest reducing agents.

Cadmium is a silver-white, lustrous metal with a blue tinge. It is so soft it can be easily cut with a knife. It does not react with water but does react slowly with dilute nitric acid and hot hydrochloric acid. Cadmium is used in low-melting alloys, such as those employed in solders and in dental amalgam.

This activity series can also provide information on the strengths of some oxidizing agents: the cations formed when the metals are oxidized. The cations of the metals at the bottom of the series gain electrons to form the metal more readily than the cations of the metals higher in the series. We can see this more easily if we expand the entries in the activity series as a series of reduction half-reactions, as shown in Table 17.1. The metal ions at the top of the series are the weakest oxidizing agents; those at the bottom are the strongest. From this table we can tell, for example, that manganese is a stronger reducing agent than cadmium but the Cd^{2+} ion is a stronger oxidizing agent than the Mn^{2+} ion. The numbers listed in the rightmost column of Table 17.1 provide a quantitative measure of the relative oxidizing and reducing strength of the elements and ions in the table. They will be discussed in detail in Section 17.5.

Figure 17.3 A chemical activity series.

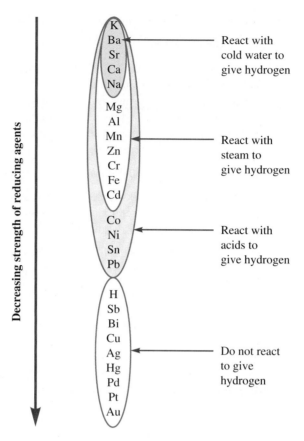

Decreasing strength of reducing agents

React with cold water to give hydrogen

React with steam to give hydrogen

React with acids to give hydrogen

Do not react to give hydrogen

How can chemical reactions be used to establish an activity series?

Nickel is a lustrous, white, hard metal. It is ferromagnetic—that is, it is attracted to a magnet. It is stable in air at ordinary temperatures but forms nickel oxide when heated in oxygen. It does not react with water but decomposes steam at high temperatures (red heat). It is slowly attacked by dilute hydrochloric and sulfuric acids and more rapidly by dilute nitric acid. Nickel is used for plating other metals to provide a high shine and to protect from corrosion, as well as in various alloys such as German silver, Monel metal, and stainless steels.

How do we know where substances fit in an activity series? We can determine the relative strengths of oxidizing and reducing agents—their reactivity—in a number of ways. Here we examine two ways: the displacement of H_2 from water or acids and the displacement of metal ions from solution. An additional method will be discussed in Section 17.5.

Liberation of Hydrogen Gas   One way of determining the reducing strength of metals involves examining the displacement of H_2 from a solution. As shown in Figure 17.4, very active metals, those with the greatest ability to donate electrons, can displace H_2 from water.

$$2Na(s) + 2H_2O(l) \longrightarrow H_2(g) + 2NaOH(aq)$$

Some metals, which are weaker electron donors, are not quite so active; they displace H_2 from water only when the water is heated to form steam.

$$Mn(s) + 2H_2O(g) \xrightarrow{\Delta} H_2(g) + Mn(OH)_2(s)$$

Metals that are even less active can release H_2 from aqueous solution but also require the presence of a better source of hydrogen ions, an acid.

$$Ni(s) + 2HCl(aq) \longrightarrow H_2(g) + NiCl_2(aq)$$

Still less active metals—those below hydrogen in the activity series—cannot displace hydrogen under any reaction conditions.

Table 17.1 Relative Reducing and Oxidizing Strengths of Common Metals and Their Cations

Reduction Half-Reaction		Standard Reduction Potential (V)
Weakest oxidizing agent	Strongest reducing agent	
$K^+(aq) + e^- \longrightarrow K(s)$		−2.93
$Ba^{2+}(aq) + 2e^- \longrightarrow Ba(s)$		−2.90
$Sr^{2+}(aq) + 2e^- \longrightarrow Sr(s)$		−2.89
$Ca^{2+}(aq) + 2e^- \longrightarrow Ca(s)$		−2.87
$Na^+(aq) + e^- \longrightarrow Na(s)$		−2.71
$Mg^{2+}(aq) + 2e^- \longrightarrow Mg(s)$		−2.37
$Al^{3+}(aq) + 3e^- \longrightarrow Al(s)$		−1.66
$Mn^{2+}(aq) + 2e^- \longrightarrow Mn(s)$		−1.18
$Zn^{2+}(aq) + 2e^- \longrightarrow Zn(s)$		−0.76
$Cr^{3+}(aq) + 3e^- \longrightarrow Cr(s)$		−0.74
$Fe^{2+}(aq) + 2e^- \longrightarrow Fe(s)$		−0.44
$Cd^{2+}(aq) + 2e^- \longrightarrow Cd(s)$		−0.40
$Co^{2+}(aq) + 2e^- \longrightarrow Co(s)$		−0.28
$Ni^{2+}(aq) + 2e^- \longrightarrow Ni(s)$		−0.25
$Sn^{2+}(aq) + 2e^- \longrightarrow Sn(s)$		−0.14
$Pb^{2+}(aq) + 2e^- \longrightarrow Pb(s)$		−0.13
$2H^+(aq) + 2e^- \longrightarrow H_2(g)$		0.00
$SbO^+(aq) + 2H^+(aq) + 3e^- \longrightarrow Sb(s) + H_2O(l)$		+0.21
$AsO^+(aq) + 2H^+(aq) + 3e^- \longrightarrow As(s) + H_2O(l)$		+0.25
$BiO^+(aq) + 2H^+(aq) + 3e^- \longrightarrow Bi(s) + H_2O(l)$		+0.32
$Cu^{2+}(aq) + 2e^- \longrightarrow Cu(s)$		+0.34
$Ag^+(aq) + e^- \longrightarrow Ag(s)$		+0.80
$Hg^{2+}(aq) + 2e^- \longrightarrow Hg(l)$		+0.85
$Pd^{2+}(aq) + 2e^- \longrightarrow Pd(s)$		+0.99
$Pt^{2+}(aq) + 2e^- \longrightarrow Pt(s)$		+1.20
$Au^{3+}(aq) + 3e^- \longrightarrow Au(s)$		+1.50
Strongest oxidizing agent	Weakest reducing agent	

How can we use the activity series to predict redox reactions?

The location of a metal in the activity series, then, can be used to predict the circumstances under which it will produce hydrogen gas. Metals from potassium through sodium will react even with cold water to release hydrogen. And, of course, they will also release hydrogen in reaction with steam or aqueous acid. Metals from magnesium through cadmium will react with steam or aqueous acids

Figure 17.4 Sodium reacts with water to release hydrogen gas. This reaction also produces sodium hydroxide, a base, as shown by the change in color of the acid–base indicator, thymol blue.

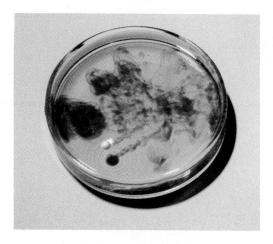

to release hydrogen. Metals from cobalt through lead will react only with aqueous acids to release hydrogen. Finally, metals below hydrogen in the series (antimony through gold) do not release hydrogen from aqueous systems under any circumstances.

EXAMPLE 17.1

Predict the reactivity of aluminum and of tin with water, steam, and aqueous acid.

Solution   Aluminum, located in the second group of metals in the activity series, will react with steam or aqueous acid to release hydrogen, but not with cold water.

Tin is located in the third group of metals in the activity series. It will react only with aqueous acid to release hydrogen.

Practice Problem 17.1

Predict the reactivity of calcium and of copper with water, steam, and aqueous acid.

Answer: Ca will react with cold water, steam, or acid to release hydrogen gas; copper will not release hydrogen gas under any circumstances.

Displacement of Metal Ions from Solution   Displacement reactions involving a metal and a solution that contains the cation of another metal can also be used to determine relative reactivity. ▶ A more reactive metal can displace a less reactive metal from a solution of its salt, as shown for zinc and tin in the chapter introduction. Consider another example: Zn can displace Pb from a solution containing Pb(II) ions, as shown in Figure 17.5 and summarized in the equation

$$Zn(s) + Pb^{2+}(aq) \longrightarrow Zn^{2+}(aq) + Pb(s)$$

From this observation, we can conclude that Zn is more reactive than Pb. The information on relative reducing and oxidizing strengths given in Figure 17.3 and Table 17.1 can thus be used to predict whether certain reactions will occur.

Note that the reaction will always occur between the stronger reducing agent and the stronger oxidizing agent to produce the weaker reducing agent and the weaker oxidizing agent. You may recall that a similar rule pertained to acid–base

Figure 17.5 Zinc is a more active metal than lead, so zinc displaces lead from a solution of lead(II) chloride. *Left:* A strip of zinc metal is immersed in a lead(II) chloride solution. *Right:* Over a period of time, lead crystals are deposited on the zinc strip.

reactions, wherein the stronger acid and base react to produce predominantly the weaker acid and base.

EXAMPLE 17.2

Predict whether any reaction will occur between Fe metal and a solution containing Al^{3+}.

Solution Iron is lower in the activity series than aluminum, so iron will not reduce aluminum(III) ions in aqueous solution. The reverse reaction will occur, however—aluminum metal will reduce Fe^{2+} to iron metal.

Practice Problem 17.2

Predict whether any reaction will occur between Mg metal and a solution containing Al^{3+}.

Answer: Mg will react with Al^{3+} to form Al metal and Mg^{2+}.

17.2 Oxidation Numbers

Besides the direct transfer of electrons, in what other ways could redox reactions occur?

We have seen the oxidation–reduction reactions involve, at least formally, the transfer of one or more electrons between two species. In some reactions, it seems fairly clear that electrons have been transferred from one species to another. For example, consider the reaction between aqueous iron(III) and chromium(II) ions shown in Figure 17.6:

$$Fe^{3+}(aq) + Cr^{2+}(aq) \rightleftharpoons Cr^{3+}(aq) + Fe^{2+}(aq)$$

The iron(III) ion loses charge, so it must gain an electron, and the increase in charge exhibited by the chromium(II) ion must be due to the loss of an electron. In other examples, such as the reduction of NO_3^- by SO_2, it is not so clear that the net changes are brought about by the transfer of electrons.

$$NO_3^-(aq) + SO_2(aq) \rightleftharpoons SO_3(aq) + NO_2^-(aq)$$

Because we cannot always easily tell that electron transfer is taking place in an oxidation–reduction reaction, we need some method to keep track of the number

How are oxidation numbers assigned?

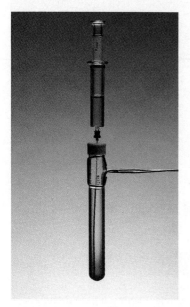

Figure 17.6 When pale yellow $Fe^{3+}(aq)$ (in the test tube) and sky-blue $Cr^{2+}(aq)$ (in the syringe) are mixed, they rapidly form colorless $Fe^{2+}(aq)$ and blue-green $Cr^{3+}(aq)$ by the transfer of an electron.

of electrons around the atoms in each substance so that we can see where the changes are occurring. This method is supplied by oxidation numbers, which we first encountered in Chapter 3.

Ideally, the oxidation number of an atom in a molecule or ion should represent the actual charge on that atom. However, precise electron distributions are not known for many molecules, and even when they are, the idea of an atom's actual charge is often ambiguous. Oxidation numbers, then, are simply charges assigned according to a sometimes arbitrary method and used primarily to keep track of electrons during reactions.

The oxidation number of an atom in its elemental state is 0. Its oxidation number in any other state is determined by comparison with the elemental state. If it has fewer electrons than in the elemental state, its oxidation number is positive; if it has more electrons, its oxidation number is negative. Simply counting electrons is readily applied to monatomic ions, but a more complex procedure is needed for polyatomic compounds or ions that contain covalent bonds. The oxidation number of an atom in a molecule or polyatomic ion can be obtained from the following equation:

$$\text{Oxidation number} = \begin{array}{c}\text{number of valence } e^- \\ \text{in the element}\end{array} - \begin{array}{c}\text{number of valence } e^- \text{ assigned} \\ \text{to the atom in the molecule or ion}\end{array}$$

To assign electrons to an atom in a covalent compound or polyatomic ion, we first draw a Lewis electron-dot formula for the substance and then assign all electrons in each covalent bond to the more electronegative atom sharing the electrons. When like atoms (atoms of the same electronegativity) are joined, the electrons are divided equally between them.

A set of secondary rules for assigning oxidation numbers was given in Section 3.4 as Table 3.1. These rules work as a short cut for many common ions and molecules, but not for all substances. The secondary rules usually work for compounds that contain hydrogen, oxygen, halogens, alkali metals, or alkaline-earth metals. If necessary, review the examples presented in Section 3.4. When the secondary rules do not apply, we must write out the Lewis electron-dot formula and use the primary procedure to determine the oxidation numbers.

EXAMPLE 17.3

What are the oxidation numbers of the elements in SCN^-?

Solution   It is not possible to use any of the secondary rules, so we must examine the electron-dot structure of the thiocyanate ion.

Electronegativities can be found in Figure 7.20. From this figure, we see that sulfur and carbon have the same electronegativity and that nitrogen has a greater electronegativity. The electron pair that sulfur and carbon share is divided equally between them. Sulfur thus has seven valence electrons, compared with the six valence electrons it has as an atom, so sulfur has an oxidation number of -1. Nitrogen is more electronegative than carbon, so the six electrons they share are assigned to nitrogen. Nitrogen thus has eight valence electrons, compared with its normal five, so it has an oxidation number of -3. According to these assignments, carbon is left with one electron, compared with its normal four, so carbon has an oxidation number of $+3$. Because the sum of the oxidation numbers (-1, $+3$, and -3) equals the ionic charge ($1-$), the assignment is internally consistent.   *(continued)*

▶ ▶ ▶ **WORLDS OF CHEMISTRY**

Images with a Grain of Salt

A number of chemical reactions must be used to convert exposed photographic film to a visible image.

Black-and-white photography makes use of a series of processes involving oxidation–reduction reactions, as well as solubility equilibria and complex-ion formation.

Photographic film and paper contain a thin layer of silver halide crystals mixed with gelatin to form an emulsion. The silver halides—silver chloride, silver bromide, and silver iodide—are light-sensitive. Silver chloride, the least sensitive silver halide, is normally used on printing paper, where longer exposure times are possible. The bromide is used for film and the iodide for very fast film.

When the silver halide (AgX) emulsion is exposed to light, a few halide ions lose their electrons.

$$X^- \longrightarrow X + e^-$$

The electrons eventually react with silver ions within the crystal to form a small number of silver atoms within the crystal.

$$Ag^+ + e^- \longrightarrow Ag$$

The number of silver atoms formed in an exposed film is quite small—too small to leave an image—but these silver atoms sensitize the silver halide crystal for further reduction to metallic silver. The sensitivity of any part of the emulsion to further reduction is proportional to the number of silver atoms in that part and thus to the intensity of light to which it was exposed.

A number of chemical reactions must be used to convert exposed photographic film to a visible image. For example, further reduction of silver halide to silver is carried out with a developing agent, which is a weak reducing agent (Figure 17.7). A typical developer contains hydroquinone, HO—C_6H_4—OH (two hydroxyl groups bonded to a benzene ring). Under appropriate conditions, including the right length of time, the hydroquinone reduces the silver

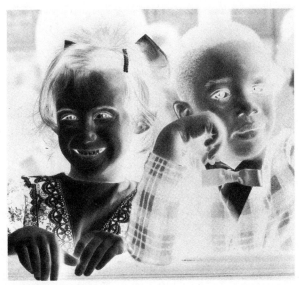

Figure 17.7 The chemistry of photography. *(A)* The objects being photographed reflect light, and the camera lens focuses this light onto light-sensitive film, where it converts silver halide units into silver and halogen atoms. Development of the film produces silver metal by reaction of silver halide with a weak reducing agent such as hydroquinone. A fixing bath, which contains moderately weak acid, is used to stop

the reduction so that the film does not become overdeveloped. The fixing solution contains sodium thiosulfate, which forms complex ions with silver ions, thereby dissolving unreacted silver halide. *(B)* The negative film is washed to remove any residual chemicals, is dried, and then is used to prepare positive prints, a process that involves similar chemical changes.

(continued)

halide crystals that have been sensitized by exposure to light.

Next, the reduction of silver halide must be stopped and the developed image must be fixed. The fixing bath usually contains a moderately weak acid such as $KHSO_4$ to stop the reduction and sodium thiosulfate (fixer) to remove excess silver halide. Thiosulfate ions form stable complex ions with silver ions, so the insoluble undeveloped silver halide crystals dissolve and are no longer susceptible to deposition on the film by reduction.

$$AgX(s) + 2S_2O_3^{2-}(aq) \longrightarrow$$
$$Ag(S_2O_3)_2^{3-}(aq) + X^-(aq)$$

These soluble ions can then be washed away with water.

This process produces a negative. A positive image is obtained by exposing photographic paper to light through the negative film and repeating the chemical processes of developing and fixing.

Questions for Discussion

1. In photochromic sunglasses, the lenses automatically darken when exposed to bright light. These sunglasses are made from glass in which silver chloride has been dissolved. Explain how these sunglasses work.

2. Daguerreotypes, used in nineteenth century photography, were made by plating a thin layer of silver on a sheet of copper and then exposing the silver to iodine vapor in the dark. Upon exposure to light in a camera, an image developed on the surface. Daguerreotypes, however, gradually darkened after exposure, so they did not provide permanent images. Explain why they darkened, whereas modern photographic film does not.

3. What happens to all the silver that is washed from photographic film? Is this an environmental hazard? Is it recycled?

4. How does color photography work?

Practice Problem 17.3

What are the oxidation numbers of the elements in HCN?

Answer: H, +1; C, +2; N, −3

17.3   Common Oxidizing and Reducing Agents

Why do metals normally have positive oxidation numbers, whereas nonmetals have both positive and negative oxidation numbers?

The ability to recognize common oxidizing and reducing agents is helpful. All the elements, with the exception of the light noble gases, exist in stable compounds with one or more oxidation numbers in addition to the oxidation number 0, which is found in the uncombined element. The oxidation numbers commonly found for the elements are listed in the periodic table in Figure 17.8. Note that the metallic elements normally exhibit positive oxidation numbers, whereas most of the nonmetallic elements have both positive and negative oxidation numbers. The maximum positive and negative values correspond to the numbers of electrons that would have to be lost or gained for the element to achieve a noble-gas configuration. It is unusual for an element to lose or gain more electrons than this.

Depending on what oxidation numbers it can have, an element may be found in substances that can behave as oxidizing agents, reducing agents, or both. For example, consider the oxidation numbers available to nitrogen, shown in Table 17.2. Nitrogen can have an oxidation number as high as +5 (as in HNO_3, NO_3^-, and N_2O_5) and as low as −3 (as in NH_3 and NH_4^+). A compound containing nitrogen with oxidation number +5, at the top of the range, can act only as an oxidizing agent. At the other extreme, a compound containing nitrogen with oxidation number −3 can act only as a reducing agent. Substances containing nitrogen with intermediate oxidation numbers can act as either oxidizing or reducing

1 IA	2 IIA	3 IIIB	4 IVB	5 VB	6 VIB	7 VIIB	8 ⎯VIIIB⎯	9	10	11 IXB	12 XB	13 IIIA	14 IVA	15 VA	16 VIA	17 VIIA	18 VIIIA
H +1 −1																	He
Li +1	Be +2											B +3	C +4 +2 −4	N +5 +4 +3 +2 +1 −3	O −1 −2	F −1	Ne
Na +1	Mg +2											Al +3	Si +4 −4	P +5 +3 −3	S +6 +4 +2 −2	Cl +7 +5 +3 +1 −1	Ar
K +1	Ca +2	Sc +3	Ti +4 +3 +2	V +5 +4 +3 +2	Cr +6 +3 +2	Mn +7 +6 +4 +3 +2	Fe +3 +2	Co +3 +2	Ni +2	Cu +2 +1	Zn +2	Ga +3	Ge +4 +2 −4	As +5 +3 −3	Se +6 +4 −2	Br +7 +5 +1 −1	Kr +4 +2
Rb +1	Sr +2	Y +3	Zr +4	Nb +5 +4	Mo +6 +4 +3	Tc +7 +6 +4	Ru +8 +6 +4 +3	Rh +4 +3 +2	Pd +4 +2	Ag +1	Cd +2	In +3	Sn +4 +2	Sb +5 +3 −3	Te +6 +4 −2	I +7 +5 +1 −1	Xe +6 +4 +2
Cs +1	Ba +2	La +3	Hf +4	Ta +5	W +6 +4	Re +7 +6 +4	Os +8 +4	Ir +4 +3 +2	Pt +4 +2	Au +3 +1	Hg +2 +1	Tl +3 +1	Pb +4 +2	Bi +5 +3	Po +2	At −1	Rn

Figure 17.8 Common oxidation numbers of the elements.

agents, depending on what other substances are present and on reaction conditions, such as pH.

Oxidizing Agents

As we have noted, a substance that contains an element with its highest oxidation number can function only as an oxidizing agent, because it can only gain electrons. Compounds that contain elements with intermediate oxidation numbers can also function as oxidizing agents. Some common oxidizing agents, along with the substances produced when they are reduced, are listed in Table 17.3. The strongest oxidizing agent is F_2, which always forms the F^- ion in its reactions and reacts with most substances. The fluorine molecule is such a strong oxidizing agent that no stable aqueous solutions of it can be prepared. Indeed, fluorine oxidizes water so vigorously that the reaction mixture bursts into flame.

Fluorine exists as a pale yellow gas containing F_2 molecules. It is the most reactive of the nonmetals and combines, either directly or indirectly, with all the elements except the lighter noble gases.

$$2F_2(aq) + 2H_2O(l) \longrightarrow 4HF(aq) + O_2(g)$$

Table 17.2 Possible Oxidation Numbers for Nitrogen Species

Oxidation Number[*]	Common Species
+5	HNO_3, NO_3^-, N_2O_5
+4	NO_2, N_2O_4
+3	NO_2^-, HNO_2, N_2O_3
+2	NO
+1	N_2O
0	N_2
−1	NH_2OH
−2	N_2H_4
−3	NH_3, NH_4OH, NH_4^+

[*]Oxidation increases the oxidation number; reduction decreases it.

Molecular oxygen is another strong oxidizing agent. It reacts with most of the elements to form oxides and with some to form peroxides or superoxides. Molecular oxygen also reacts with many compounds. In these reactions, it gives the same oxides as would be formed from the component elements. For example, hydrocarbons are oxidized to carbon dioxide and water.

$$C_3H_8(g) + 5O_2(g) \rightarrow 3CO_2(g) + 4H_2O(g)$$

Propane, C_3H_8, is a colorless gas found in natural gas and crude petroleum. It is odorless when pure and burns with a luminous, smoky flame. It is widely used as a fuel gas.

When oxygen is combined with other elements, as we have seen, it usually forms oxides and has an oxidation number of −2. These oxides can also behave as oxidizing agents, but it is usually the other element that gives the compound its oxidizing power. In redox reactions of the oxides, oxygen generally keeps its

Table 17.3 Some Common Oxidizing Agents and Their Products

Oxidizing Agent	Product of Its Reduction
O_2	O^{2-}
H_2O_2	H_2O
F_2, Cl_2, Br_2, I_2	F^-, Cl^-, Br^-, I^-
ClO_3^-	Cl^-
BrO_3^-	Br^-
H_2SO_4 (concentrated)	SO_2, H_2S, or S
SO_3^{2-}	$S_2O_3^{2-}$, S, or H_2S
HNO_3 (concentrated)	NO_2, NO, N_2, or NH_3
NO_3^-	NO_2, NO, N_2, or NH_3
MnO_4^- (base)	MnO_2
MnO_4^- (acid)	Mn^{2+}
CrO_4^{2-} (base)	$Cr(OH)_3$
$Cr_2O_7^{2-}$ (acid)	Cr^{3+}

oxidation number of -2, and the other element undergoes a change in oxidation number. For example, when sulfur trioxide is reduced to sulfur dioxide, the oxygen retains its -2 oxidation number, while the sulfur undergoes a change in oxidation number from $+6$ to $+4$.

$$SO_3(g) + 2Br^-(aq) + 2H^+(aq) \longrightarrow SO_2(g) + Br_2(aq) + H_2O(l)$$

Oxoacids—acids that contain oxygen—are usually good oxidizing agents, because the nonmetal central atom has a high oxidation number, as in H_2SO_4, HNO_3, and $HClO_4$. Oxoanions derived from these acids, as well as oxoanions of other nonmetals that have high oxidation numbers, tend to be good oxidizing agents. Examples include BrO_3^-, ClO_3^-, and NO_3^-. Oxides of transition metals that have their maximum oxidation numbers also are excellent oxidizing agents. Species such as CrO_4^{2-} or $Cr_2O_7^{2-}$, FeO_4^{2-}, and MnO_4^-, for instance, are commonly used as oxidizing agents in the laboratory.

A number of species in which a component element has an oxidation number lower than the maximum but higher than the minimum are also useful as oxidizing agents. For example, the elemental halogens are commonly used as oxidizing agents, as in the following reaction with methane, where chlorine changes its oxidation number from 0 to -1.

$$Cl_2(g) + CH_4(g) \longrightarrow CH_3Cl(g) + HCl(g)$$

Another common oxidizing agent with an intermediate oxidation number is hydrogen peroxide, H_2O_2, in which oxygen has an oxidation number of -1. Hydrogen peroxide oxidizes iron(II), for example, the oxidation number of oxygen decreasing to -2.

$$2Fe^{2+}(aq) + H_2O_2(aq) + 2H^+(aq) \longrightarrow 2Fe^{3+}(aq) + 2H_2O(l)$$

Hydrogen peroxide can also act as a reducing agent; in such reactions, the oxidation number of oxygen increases from -1 to 0.

Reducing Agents

Why are metals commonly used as reducing agents?

A substance that contains an element with its lowest oxidation number can function only as a reducing agent, because it can only lose electrons. The metallic elements provide many examples of common reducing agents. The strongest reducing agents are the elemental metals with the lowest electronegativities—alkali metal and alkaline-earth metal elements—because they have the greatest tendency to give up electrons. All of the metals above hydrogen in the activity series shown in Figure 17.3 behave as reducing agents when placed in water, where they release hydrogen gas. For example, calcium metal reacts with water to form aqueous calcium hydroxide and hydrogen gas.

$$Ca(s) + 2H_2O(l) \longrightarrow Ca^{2+}(aq) + 2OH^-(aq) + H_2(g)$$

These metals, along with those below hydrogen in the activity series, can also act as reducing agents when reacted with other substances. Of these metals, Fe, Zn, Al, Sn, Na, and Mg are most often used as reducing agents. ▶ For example, we saw in the chapter introduction that zinc metal reduces Sn^{2+} in solution.

$$Zn(s) + Sn^{2+}(aq) \longrightarrow Zn^{2+}(aq) + Sn(s)$$

Nonmetals with low oxidation numbers can also function as reducing agents. The strongest reducing agent of this type is the hydride ion, H^-. Hydrogen gas can

Potassium permanganate, $KMnO_4$, is a dark purple solid that is stable in air. It decomposes slowly in water to deposit MnO_2, a dark brown solid. In the presence of some MnO_2, the decomposition is much faster. It is used for bleaching various fibers and skins, for dyeing wood brown and printing on fabrics, for purifying water, and as an antiseptic agent.

Figure 17.9 When hydrogen gas is passed over heated white SnO_2, tin metal (which appears black when finely divided) and invisible water vapor are produced. The H_2 is the reducing agent, and the SnO_2 is the oxidizing agent.

be used as a reducing agent as well, although it is not so strong as hydride ion. Hydrogen gas is often used to reduce metal oxides to the metal, as shown in Figure 17.9 for tin(IV) oxide.

$$SnO_2(s) + 2H_2(g) \xrightarrow{\Delta} Sn(s) + 2H_2O(g)$$

Iodide ion, I^-, is the only halide commonly used as a reducing agent. Among the Group VIA (16) elements, S^{2-} is often used as a reducing agent; SO_2 and SO_3^{2-} are used less commonly. Some common reducing agents, along with their oxidation products, are listed in Table 17.4.

Table 17.4 Some Common Reducing Agents and Their Products

Reducing Agent	Product of Its Oxidation
H_2	H^+
H_2O_2	O_2
SO_3^{2-}	SO_4^{2-}
NO_2^-	NO_3^-
I^-	I_2
S^{2-}	S
H_2S	S
NH_3	N_2
N_2H_4	N_2
Cr^{2+}	Cr^{3+}
Sn^{2+}	Sn^{4+}
Fe^{2+}	Fe^{3+} (acid)
Alkali metals	+1 ions (such as Na^+)
Alkaline-earth metals	+2 ions (such as Ca^{2+})
Al	Al^{3+}
Sn	Sn^{2+}
Zn	Zn^{2+}
C (as coke)	CO or CO_2

Identification of Oxidizing and Reducing Agents

We can reliably identify the oxidizing agent and reducing reagent in a redox reaction by assigning oxidation numbers to all the component elements. The substance that contains an element whose oxidation number decreases is the oxidizing agent, and the substance that contains an element whose oxidation number increases is the reducing agent.

EXAMPLE 17.4

Identify the oxidizing and reducing agents in the following reaction between oxalate and iodate ions.

$$5C_2O_4^{2-}(aq) + 2IO_3^-(aq) + 12H^+(aq) \longrightarrow 10CO_2(g) + I_2(aq) + 6H_2O(l)$$

Solution   We assign oxidation numbers to the elements according to the approach developed in Section 17.2. The oxidation numbers are

Element	Reactants	Products
C	+3	+4
O	−2	−2
I	+5	0
H	+1	+1

Note that the oxidation numbers of oxygen and hydrogen do not change during the reaction. Carbon undergoes an increase in oxidation number, so oxalate ion is the reducing agent. Iodine undergoes a decrease in oxidation number, so iodate ion is the oxidizing agent.

Practice Problem 17.4

Identify the oxidizing and reducing agents in the reaction

$$10VO^{2+}(aq) + 2BrO_3^-(aq) + 4H_2O(l) \longrightarrow 10VO_2^+(aq) + Br_2(aq) + 8H^+(aq)$$

Answer: VO^{2+} is the reducing agent. BrO_3^- is the oxidizing agent.

Some familiarity with the common oxidizing and reducing agents enables us in some cases to identify them by inspection. For example, if we see that the chromate ion, CrO_4^{2-}, is involved in a reaction, we can readily identify it as the oxidizing agent, because chromium has its highest oxidation number, +6, in this ion.

17.4 Balancing Equations

We turn now to balancing equations for oxidation–reduction reactions. Some are so simple that they can be balanced by inspection, as we balanced equations in Chapter 4. For example, in the oxidation of magnesium with oxygen to form magnesium oxide, the unbalanced equation is

Earlier we balanced chemical equations by inspection. Why do most redox reactions require a more complex approach?

$$Mg(s) + O_2(g) \longrightarrow MgO(s)$$

We can balance this equation simply by balancing atoms.

$$2Mg(s) + O_2(g) \longrightarrow 2MgO(s)$$

Most oxidation–reduction reactions are more complex, however, and require a more systematic approach. Two commonly used methods are the half-reaction method and the oxidation-number-change method. For both, it is necessary to know not only the identity of the reactants and products but also whether the reaction is carried out in acidic or basic solution.

The Half-Reaction Method

The **half-reaction method** (also called the **ion–electron method**) develops the oxidation process and the reduction process as separate half-reactions. The half-reactions are balanced independently and then combined in a way that equalizes the loss and gain of electrons and gives the overall balanced equation for the oxidation–reduction reaction.

A procedure for applying the half-reaction method is summarized in Table 17.5. Let's apply this procedure to the oxidation of hydrogen sulfide by nitrate ion in acidic solution to form sulfur and nitric oxide. The unbalanced equation is

$$H_2S(aq) + NO_3^-(aq) \longrightarrow S(s) + NO(g)$$

The steps will be labeled to match those in Table 17.5.

Table 17.5 Balancing Oxidation–Reduction Reactions by the Half-Reaction Method

Step 1: Write an unbalanced half-reaction for either the oxidation or the reduction process.

Step 2: Balance the half-reaction according to steps a through e.
  (a) Balance all atoms except H and O.
  (b) Balance O by adding H_2O to the appropriate side of the equation.
  (c) Balance H by adding H^+ to the appropriate side of the equation.
  (d) Balance ionic charges by adding the proper number of electrons to the appropriate side of the equation.
  (e) If the reaction is occurring in basic solution, add as many OH^- ions to each side of the equation as there are H^+ ions on one side. On the side of the equation that now contains both H^+ and OH^-, convert these ions to an equal number of H_2O molecules. Cancel any H_2O molecules that appear on both sides of the equation.

Step 3: Write an unbalanced half-reaction for the other process.

Step 4: Balance the half-reaction according to steps a through e.

Step 5: Equalize the number of electrons lost and the number gained by multiplying each coefficient in the balanced oxidation half-reaction and in the balanced reduction half-reaction by appropriate constants.

Step 6: Add the two half-reactions, and cancel equal amounts of any substance that occurs on both sides of the overall equation.

Step 7: Make a final check of atom and charge balances.

Step 1. In this reaction, sulfur in hydrogen sulfide is undergoing an increase in oxidation number. The unbalanced half-reaction for the oxidation is thus

$$H_2S(aq) \longrightarrow S(s)$$

Step 2a. The sulfur is already balanced.

Step 2b. There are no oxygen atoms to balance in this half-reaction.

Step 2c. The hydrogen atoms are balanced by addition of two H^+ ions to the right-hand side of the equation.

$$H_2S(aq) \longrightarrow S(s) + 2H^+(aq)$$

Step 2d. The right-hand side of the equation contains a net charge of $2+$. There are no charged species on the left-hand side of the equation. We add two electrons to the right-hand side to obtain a net charge of 0, thereby balancing the ionic charges.

$$H_2S(aq) \longrightarrow S(s) + 2H^+(aq) + 2e^-$$

The half-reaction for oxidation of hydrogen sulfide is now balanced; identical numbers of both charges and atoms appear on each side of the equation.

Step 2e. The reaction is occurring in acidic solution, so this step is not relevant.

Step 3. Nitrate ion is reduced to nitric oxide in this half-reaction. The unbalanced half-reaction is

$$NO_3^-(aq) \longrightarrow NO(g)$$

Step 4a. The numbers of nitrogen atoms are the same on both sides of the equation.

Step 4b. The left-hand side of the equation has three oxygen atoms, whereas the right-hand side has only one, so two water molecules must be added to the right-hand side to balance oxygen atoms.

$$NO_3^-(aq) \longrightarrow NO(g) + 2H_2O(l)$$

Step 4c. The right-hand side of the equation contains four hydrogen atoms, and the left-hand side has none. Four hydrogen ions must be added to the left-hand side of the equation to balance hydrogen atoms.

$$NO_3^-(aq) + 4H^+(aq) \longrightarrow NO(g) + 2H_2O(l)$$

Step 4d. The left-hand side of the equation has a net charge of $3+$, whereas there is no charged species on the right. Addition of three electrons to the left-hand side balances the charges:

$$NO_3^-(aq) + 4H^+(aq) + 3e^- \longrightarrow NO(g) + 2H_2O(l)$$

The reduction half-reaction is now balanced.

Step 4e. The reaction is occurring in acidic solution, so this step is not relevant.

Step 5. The balanced half-reactions are

$$H_2S(aq) \longrightarrow S(s) + 2H^+(aq) + 2e^-$$
$$NO_3^-(aq) + 4H^+(aq) + 3e^- \longrightarrow NO(g) + 2H_2O(l)$$

To equalize the numbers of electrons lost and gained, we can multiply the oxidation half-reaction by 3 and the reduction half-reaction by 2.

$$3H_2S(aq) \longrightarrow 3S(s) + 6H^+(aq) + 6e^-$$
$$2NO_3^-(aq) + 8H^+(aq) + 6e^- \longrightarrow 2NO(g) + 4H_2O(l)$$

Step 6. First we add the two half-reactions.

$$3H_2S(aq) \longrightarrow 3S(s) + 6H^+(aq) + 6e^-$$
$$2NO_3^-(aq) + 8H^+(aq) + 6e^- \longrightarrow 2NO(g) + 4H_2O(l)$$

$$3H_2S(aq) + 2NO_3^-(aq) + 8H^+(aq) + 6e^- \longrightarrow$$
$$3S(s) + 6H^+(aq) + 6e^- + 2NO(g) + 4H_2O(l)$$

The six electrons on each side cancel out, as do six H^+ ions on each side.

$$3H_2S(aq) + 2NO_3^-(aq) + 2H^+(aq) \longrightarrow 3S(s) + 2NO(g) + 4H_2O(l)$$

Step 7. To make a final check of atom and charge balances, we set up a table.

Element	Left-hand Side of Equation	Right-hand Side of Equation
H	8	8
S	3	3
N	2	2
O	6	6
Charge	0	0

Atoms and charges balance, so the equation is properly balanced.

In this example, only hydrogen and oxygen had to be balanced. Now we consider a different situation.

The procedure for balancing redox reactions makes use of a number of chemical principles we discussed in earlier chapters. Cite some of these principles.

EXAMPLE 17.5

Balance the equation for the following reaction in acidic solution by using the half-reaction method.

$$Cr_2O_7^{2-}(aq) + Br^-(aq) \longrightarrow Cr^{3+}(aq) + Br_2(aq)$$

Solution   Oxidation half-reaction:

Step 1: $Br^-(aq) \longrightarrow Br_2(aq)$
Step 2a: The bromine atoms are not balanced.

$$2Br^-(aq) \longrightarrow Br_2(aq)$$

Step 2b: There are no oxygen atoms.
Step 2c: There are no hydrogen atoms.
Step 2d: $2Br^-(aq) \longrightarrow Br_2(aq) + 2e^-$
Step 2e: The reaction is occurring in acidic solution.

Reduction half-reaction:

Step 3: $Cr_2O_7^{2-}(aq) \longrightarrow Cr^{3+}(aq)$
Step 4a: The chromium atoms are not balanced.

$$Cr_2O_7^{2-}(aq) \longrightarrow 2Cr^{3+}(aq)$$

Step 4b: $Cr_2O_7^{2-}(aq) \longrightarrow 2Cr^{3+}(aq) + 7H_2O(l)$
Step 4c: $Cr_2O_7^{2-}(aq) + 14H^+(aq) \longrightarrow 2Cr^{3+}(aq) + 7H_2O(l)$
Step 4d: $Cr_2O_7^{2-}(aq) + 14H^+(aq) + 6e^- \longrightarrow 2Cr^{3+}(aq) + 7H_2O(l)$
Step 4e: The reaction is occurring in acidic solution.

Overall reaction:

Step 5: $6Br^-(aq) \longrightarrow 3Br_2(aq) + 6e^-$
$\qquad Cr_2O_7{}^{2-}(aq) + 14H^+(aq) + 6e^- \longrightarrow 2Cr^{3+}(aq) + 7H_2O(l)$
Step 6: $Cr_2O_7{}^{2-}(aq) + 14H^+(aq) + 6e^- + 6Br^-(aq) \longrightarrow$
$\qquad\qquad\qquad\qquad\qquad 3Br_2(aq) + 6e^- + 2Cr^{3+}(aq) + 7H_2O(l)$
$\qquad Cr_2O_7{}^{2-}(aq) + 14H^+(aq) + 6Br^-(aq) \longrightarrow 3Br_2(aq) + 2Cr^{3+}(aq) + 7H_2O(l)$

Step 7: There are two Cr atoms, seven O atoms, fourteen H atoms, six Br atoms, and a net charge of 6+ on each side of the equation, so the equation is balanced.

Practice Problem 17.5

Balance the equation for the following reaction in acidic solution by using the half-reaction method.

$$Cl^-(aq) + MnO_4{}^-(aq) \longrightarrow Cl_2(aq) + Mn^{2+}(aq)$$

Answer: $10Cl^-(aq) + 2MnO_4{}^-(aq) + 16H^+(aq) \longrightarrow 5Cl_2(aq) + 2Mn^{2+}(aq) + 8H_2O(l)$

The reactions we have examined so far occurred in acidic solution, so it was not necessary to apply Steps 2e and 4e. These steps provide a method for balancing an equation in basic solution as though the reaction occurred in acidic solution and the acid were neutralized with hydroxide ion. For example, suppose the balanced half-reaction in acidic solution is

$$H_2O_2(aq) \longrightarrow O_2(g) + 2H^+(aq) + 2e^-$$

Explain how the procedure used for basic solutions is analogous to an acid–base titration.

To balance this equation in basic solution, we can add two hydroxide ions to each side of the equation. Because the OH^- ions are added to both sides of the equation, neither the atomic balance nor the charge balance is upset. The equation at this stage is

$$H_2O_2(aq) + 2OH^-(aq) \longrightarrow O_2(g) + 2H^+(aq) + 2OH^-(aq) + 2e^-$$

Combining $H^+(aq)$ with $OH^-(aq)$ to give $H_2O(l)$ results in the balanced equation.

$$H_2O_2(aq) + 2OH^-(aq) \longrightarrow O_2(g) + 2H_2O(l) + 2e^-$$

EXAMPLE 17.6

Balance the equation for the following reaction occurring in basic solution.

$$Bi_2O_3(s) + ClO^-(aq) \longrightarrow BiO_3{}^-(aq) + Cl^-(aq)$$

Solution Oxidation half-reaction:

Step 1: $Bi_2O_3(s) \longrightarrow BiO_3{}^-(aq)$
Step 2a: The bismuth atoms are not balanced.

$$Bi_2O_3(s) \longrightarrow 2BiO_3{}^-(aq)$$

Step 2b: $Bi_2O_3(s) + 3H_2O(l) \longrightarrow 2BiO_3{}^-(aq)$
Step 2c: $Bi_2O_3(s) + 3H_2O(l) \longrightarrow 2BiO_3{}^-(aq) + 6H^+(aq)$
Step 2d: $Bi_2O_3(s) + 3H_2O(l) \longrightarrow 2BiO_3{}^-(aq) + 6H^+(aq) + 4e^-$
Step 2e: We add six OH^- ions to each side to neutralize the H^+ ions.

$$Bi_2O_3(s) + 3H_2O(l) + 6OH^-(aq) \longrightarrow 2BiO_3{}^-(aq) + 6H^+(aq) + 6OH^-(aq) + 4e^-$$

(continued)

Then we combine H^+ and OH^- ions to form water.

$$Bi_2O_3(s) + 3H_2O(l) + 6OH^-(aq) \longrightarrow 2BiO_3^-(aq) + 6H_2O(l) + 4e^-$$

We cancel the water molecules that appear on both sides of the equation.

$$Bi_2O_3(s) + 6OH^-(aq) \longrightarrow 2BiO_3^-(aq) + 3H_2O(l) + 4e^-$$

Reduction half-reaction:

Step 3: $ClO^-(aq) \longrightarrow Cl^-(aq)$
Step 4a: Chlorine atoms are already balanced.
Step 4b: $ClO^-(aq) \longrightarrow Cl^-(aq) + H_2O(l)$
Step 4c: $ClO^-(aq) + 2H^+(aq) \longrightarrow Cl^-(aq) + H_2O(l)$
Step 4d: $ClO^-(aq) + 2H^+(aq) + 2e^- \longrightarrow Cl^-(aq) + H_2O(l)$
Step 4e: We add two OH^- ions to each side to neutralize the H^+ ions.

$$ClO^-(aq) + 2H^+(aq) + 2OH^-(aq) + 2e^- \longrightarrow Cl^-(aq) + H_2O(l) + 2OH^-(aq)$$

We combine H^+ and OH^- ions to form water.

$$ClO^-(aq) + 2H_2O(l) + 2e^- \longrightarrow Cl^-(aq) + H_2O(l) + 2OH^-(aq)$$

We cancel the water molecules that appear on both sides of the equation.

$$ClO^-(aq) + H_2O(l) + 2e^- \longrightarrow Cl^-(aq) + 2OH^-(aq)$$

Overall reaction:

Step 5: The reduction equation must be multiplied by 2 to equalize electron gain and loss.

$$Bi_2O_3(s) + 6OH^-(aq) \longrightarrow 2BiO_3^-(aq) + 3H_2O(l) + 4e^-$$
$$2ClO^-(aq) + 2H_2O(l) + 4e^- \longrightarrow 2Cl^-(aq) + 4OH^-(aq)$$

Step 6: $Bi_2O_3(s) + 6OH^-(aq) + 2ClO^-(aq) + 2H_2O(l) + 4e^- \longrightarrow$
$$2Cl^-(aq) + 4OH^-(aq) + 2BiO_3^-(aq) + 3H_2O(l) + 4e^-$$

We cancel equal numbers of any species that appear on both sides of the equation:

$$Bi_2O_3(s) + 2OH^-(aq) + 2ClO^-(aq) \longrightarrow 2Cl^-(aq) + 2BiO_3^-(aq) + H_2O(l)$$

Step 7: There are two Bi atoms, seven O atoms, two H atoms, two Cl atoms, and a net charge of $4-$ on each side of the equation, so the equation is balanced.

Practice Problem 17.6

Balance the equation for the following reaction occurring in basic solution.

$$MnO_4^-(aq) + H_2O_2(aq) \longrightarrow MnO_2(s) + O_2(g)$$

Answer: $2MnO_4^-(aq) + 3H_2O_2(aq) \longrightarrow 2MnO_2(s) + 3O_2(g) + 2OH^-(aq) + 2H_2O(l)$

The Oxidation-Number-Change Method

The **oxidation-number-change method** uses the changes in oxidation number of the elements undergoing oxidation and reduction to balance the equation. The total increase in oxidation number is balanced against the total decrease in oxidation number. The procedure is summarized in Table 17.6. This method tends to take fewer steps than the half-reaction method. However, it does not work for half-reactions, which will be used extensively in Chapter 18. The following example illustrates the application of the oxidation-number-change method to a reaction in basic solution.

Table 17.6 Balancing Oxidation–Reduction Reactions by the Oxidation-Number-Change Method

Step 1: Assign oxidation numbers to the elements.

Step 2: Balance the atoms undergoing a change in oxidation number such that the same number of each appears on both sides of the equation.

Step 3: Determine the *total* change in oxidation number for all atoms being oxidized and all atoms being reduced.

Step 4: Equalize the oxidation number gain and loss:
(a) Find the lowest common multiple for the total increase in oxidation number and the total decrease in oxidation number.
(b) Multiply the coefficients of the species involved in the oxidation process by the following ratio:

   lowest common multiple/total increase in oxidation number

(c) Multiply the coefficients of the species involved in the reduction process by the following ratio:

   lowest common multiple/total decrease in oxidation number

Step 5: Balance other atoms as necessary, except O and H.

Step 6: Balance the ionic charges by adding the appropriate number of H^+ (in acidic solution) or OH^- (in basic solution) to the appropriate side of the equation.

Step 7: Complete the balancing by adding sufficient H_2O to the appropriate side of the equation to balance H and O.

Step 8: Make a final check of charge balance and mass balance.

EXAMPLE 17.7

Use the oxidation-number-change method to balance the equation for the following reaction occurring in basic solution.

$$Al(s) + OH^-(aq) \longrightarrow Al(OH)_4^-(aq) + H_2(g)$$

Solution   We follow the steps outlined in Table 17.6.

Step 1:   0 -2 $+1$ $+3$ -2 $+1$ 0
$$Al(s)   +   OH^-(aq) \longrightarrow   Al(OH)_4^-(aq) + H_2(g)$$

Step 2: The aluminum atoms are balanced, but there is one more hydrogen atom on the right-hand side of the equation (in H_2) than on the left-hand side (in OH^-), so we change the coefficient of OH^- to 2.

$$Al(s) + 2OH^-(aq) \longrightarrow Al(OH)_4^-(aq) + H_2(g)$$

(We momentarily disregard the H in $Al(OH)_4^-$ because these H atoms are not changing the oxidation number.)

Step 3: Aluminum increases in oxidation number by 3. Hydrogen decreases by 1 per atom, a total of 2 for two atoms.

increase of 3

$$Al(s) + 2OH^-(aq) \longrightarrow Al(OH)_4^-(aq) + H_2(g)$$

decrease of 2

(continued)

Step 4: The lowest common multiple for the total increase of 3 and the total decrease of 2 in oxidation number is 6. The aluminum species must be multiplied by 6/3, or 2, and the hydrogen species must be multiplied by 6/2, or 3.

$$2Al(s) + 6OH^-(aq) \longrightarrow 2Al(OH)_4^-(aq) + 3H_2(g)$$

Step 5: The aluminum is already balanced, so no changes are required.

Step 6: There is a total charge of $6-$ on the left-hand side of the equation and $2-$ on the right-hand side of the equation. Because the reaction occurs in basic solution, OH^- must be added to the right-hand side to balance the charges.

$$2Al(s) + 6OH^-(aq) \longrightarrow 2Al(OH)_4^-(aq) + 3H_2(g) + 4OH^-(aq)$$

We cancel the OH^- that occurs on both sides of the equation.

$$2Al(s) + 2OH^-(aq) \longrightarrow 2Al(OH)_4^-(aq) + 3H_2(g)$$

Step 7: The equation contains two O atoms and two H atoms on the left and eight O atoms and fourteen H atoms on the right. The difference is six O atoms and twelve H atoms, corresponding to six H_2O molecules, which must be added to the left-hand side of the equation:

$$2Al(s) + 2OH^-(aq) + 6H_2O(l) \longrightarrow 2Al(OH)_4^-(aq) + 3H_2(g)$$

Step 8: We check the balance by counting atoms and charges:

Element	Left-hand Side	Right-hand Side
Al	2	2
O	8	8
H	14	14
Charge	$2-$	$2-$

Atoms and charges balance, so the equation is properly balanced.

Practice Problem 17.7

Use the oxidation-number-change method to balance the equation for the following reaction occurring in basic solution.

$$Zn(s) + NO_3^-(aq) \longrightarrow Zn(OH)_4^{2-}(aq) + NH_3(aq)$$

Answer: $4Zn(s) + NO_3^-(aq) + 7OH^-(aq) + 6H_2O(l) \longrightarrow 4Zn(OH)_4^{2-}(aq) + NH_3(aq)$

17.5   Spontaneous Oxidation–Reduction Reactions

Suppose we place a piece of lead in a solution that contains Ag^+ ions. ▶ In the resulting oxidation–reduction reaction, which is similar to the zinc/tin(II) reaction described in the chapter introduction, silver metal is deposited and lead metal is oxidized to Pb^{2+}, as shown in Figure 17.10. The changes in each element can be represented by the following half-reactions:

$$Pb(s) \longrightarrow Pb^{2+}(aq) + 2e^-$$
$$Ag^+(aq) + e^- \longrightarrow Ag(s)$$

The overall reaction is

$$Pb(s) + 2Ag^+(aq) \longrightarrow Pb^{2+}(aq) + 2Ag(s)$$

Figure 17.10 The displacement of Ag⁺ from aqueous solution by lead metal. The lead wires are partially immersed in a silver nitrate solution. Silver crystals are formed on the wire in contact with the solution.

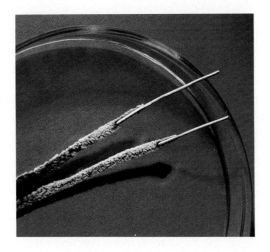

How can we predict whether a possible redox reaction will be spontaneous?

This oxidation–reduction reaction occurs spontaneously, releasing energy. The reverse of this reaction, however, does not. Thus, if a piece of silver is added to a solution containing Pb^{2+} ions, no metallic lead is deposited unless energy is added to the system. This section examines methods of predicting whether an oxidation–reduction reaction will occur spontaneously.

Activity Series and Spontaneity

In any oxidation–reduction reaction for which we can write an equation, either the forward net reaction or its reverse is spontaneous. We can tell which by comparing the reactivities of the chemical species involved. To make such comparisons, we can use an activity series such as the one in Figure 17.3. This activity series reveals, for example, that lead is more active, and thus a stronger reducing agent, than silver. That means that the oxidized form of lead, Pb^{2+}, is a weaker oxidizing agent than the oxidized form of silver, Ag^+.

$$Pb(s) \; + \; 2Ag^+(aq) \; \rightleftharpoons \; Pb^{2+}(aq) \; + \; 2Ag(s)$$

| stronger reducing agent | stronger oxidizing agent | weaker oxidizing agent | weaker reducing agent |

Silver is a white, malleable, ductile metal. It is an excellent conductor of heat and electricity. It does not react with water or oxygen in the air but does form a black coating when exposed to ozone, hydrogen sulfide, or sulfur. It does not react with most acids except dilute nitric acid and hot concentrated sulfuric acid. It is used for coins, mirrors, silverware, and jewelry and as an ingredient in dental amalgam.

Section 17.1 pointed out that an oxidation–reduction reaction always occurs between the stronger reducing agent and the stronger oxidizing agent to produce predominantly the weaker oxidizing agent and the weaker reducing agent. This, then, is the direction in which the reaction proceeds spontaneously.

Although the activity series can tell us about the relative oxidizing or reducing strengths of two substances, it cannot tell us the magnitude of the difference. To make more quantitative comparisons of oxidizing and reducing strength, we need to consider a property of oxidizing and reducing agents that can be measured and assigned a numerical value.

Electromotive Force

We have seen that thinking of oxidation–reduction reactions as oxidation half-reactions and reduction half-reactions provides a convenient method for balancing equations. Although neither can occur alone, these half-reactions can be carried out with the reactants in separate containers linked only by electrical connections, as shown in Figure 17.11. If the oxidation–reduction process is spontaneous, electrons flow from the reducing agent to the oxidizing agent through the connecting wire. As a result, an electrical current is generated in the circuit. This electrical current is characterized by its voltage, a measure of the **electromotive force (emf)** or **electrical potential.** The electromotive force, usually measured in units of volts, is a driving force that pushes electrons through the electrical circuit. In the case of the lead/silver-ion reaction, the electromotive force generated in the circuit is +0.93 V. Every spontaneous oxidation–reduction reaction generates a characteristic electromotive force. Under standard thermodynamic conditions, the electromotive force is called the *standard emf, $E°$.*

Lead is a silvery-gray metal. Although it has a high luster when cut, it quickly tarnishes in air, forming an oxide coating. It is very soft and has a low melting point. Lead is used for tank linings, pipes, and other equipment designed for the handling of corrosive gases and liquids. It is also used to protect against atomic radiation and X rays.

$$Pb(s) + 2Ag^+(aq) \longrightarrow Pb^{2+}(aq) + 2Ag(s) \qquad E° = +0.93 \text{ V}$$

The driving force for a redox reaction can be considered to be the sum of two driving forces: the tendency of the reducing agent to donate electrons and the tendency of the oxidizing agent to accept electrons. In the same way, the electromotive force generated by a redox reaction can be considered to be the sum of two electromotive forces, one associated with the oxidation half-reaction and one associated with the reduction half-reaction. We call these electromotive forces the **oxidation potential, $E°_{ox}$,** and the **reduction potential, $E°_{red}$.** The overall *reaction potential* (or emf), $E°_{rxn}$, is the sum of these half-reaction potentials:

$$E°_{rxn} = E°_{ox} + E°_{red}$$

Values of the half-reaction potentials provide a method for quantifying the oxidizing and reducing strength of substances. However, it is not possible to carry out a single half-reaction—they must be carried out in pairs. And we can measure only the voltage generated by an overall oxidation–reduction reaction, not that associated with each half-reaction.

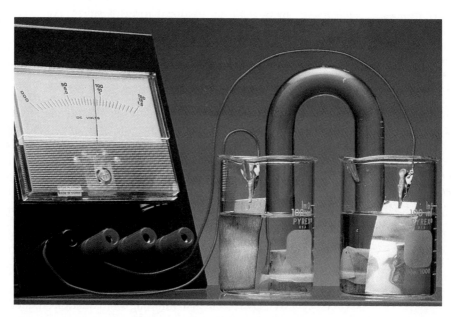

Figure 17.11 Oxidation and reduction half-reactions can be carried out in separate containers linked by a tube to allow the passage of ions and by a wire to allow the passage of electrons. Here, lead metal is oxidized to Pb^{2+}, and Ag^+ is reduced to silver metal. The voltmeter registers a voltage of 0.93 V.

To solve this problem, all half-reaction potentials are assigned values relative to a standard, which is provided by the following half-reaction:

$$2H^+(aq) + 2e^- \longrightarrow H_2(g)$$

Under standard conditions (1 atm H_2, 1 M H^+) at 25°C, this reaction is assigned a half-reaction potential, E°_{red}, of 0 V. Values for some other half-reactions can be determined from the voltage generated by the oxidation of a substance by aqueous hydrogen ions.

Consider, for example, the oxidation of lead metal by $H^+(aq)$ under standard conditions.

$$Pb(s) + 2H^+(aq) \longrightarrow Pb^{2+}(aq) + H_2$$

The electromotive force generated by this reaction, 0.13 V, is the sum of the oxidation potential and the reduction potential.

$$E^\circ_{rxn} = E^\circ_{ox} + E^\circ_{red}$$

Recall that we assigned a value of 0.00 V to the reduction potential for H^+/H_2.

$$0.13 \text{ V} = E^\circ_{ox} + 0.00 \text{ V}$$

Thus, the oxidation potential of lead is given by:

$$E^\circ_{ox} = 0.13 \text{ V} - 0.00 \text{ V} = +0.13 \text{ V}$$

These results can in turn be used to determine other potentials. We have seen that the electromotive force generated by the reaction between lead metal and silver ions is 0.93 V.

$$Pb(s) + 2Ag^+(aq) \longrightarrow Pb^{2+}(aq) + 2Ag(s) \qquad E^\circ = +0.93 \text{ V}$$

We can use this information, along with the oxidation potential of Pb, to calculate the reduction potential of Ag^+.

$$E^\circ_{rxn} = E^\circ_{ox} + E^\circ_{red}$$
$$0.93 \text{ V} = 0.13 \text{ V} + E^\circ_{red}$$
$$E^\circ_{red} = 0.93 \text{ V} - 0.13 \text{ V} = +0.80 \text{ V}$$

Note that even though we may have to multiply a half-reaction by an integer to balance the oxidation–reduction reaction, we do not multiply its potential by that integer. The rationale for this procedure will be explored in Chapter 18.

EXAMPLE 17.8

E°_{rxn} for the following reaction is +1.08 V.

$$Co(s) + 2Ag^+(aq) \longrightarrow Co^{2+}(aq) + 2Ag(s)$$

E°_{red} for the following half-reaction is +0.80 V.

$$Ag^+(aq) + e^- \longrightarrow Ag(s)$$

Calculate E°_{ox} for the other half-reaction.

$$Co(s) \longrightarrow Co^{2+}(aq) + 2e^-$$

(continued)

Solution   The overall reaction can be divided into two half-reactions:

$$Ag^+(aq) + e^- \longrightarrow Ag(s) \qquad E_{red}^\circ = +0.80 \text{ V}$$
$$Co(s) \longrightarrow Co^{2+}(aq) + 2e^- \qquad E_{ox}^\circ = ?$$

The sum of the two half-reaction potentials equals the overall reaction potential.

$$E_{rxn}^\circ = E_{ox}^\circ + E_{red}^\circ$$

We substitute known values to get

$$1.08 \text{ V} = E_{ox}^\circ + 0.80 \text{ V}$$

We rearrange this equation to obtain the desired oxidation potential.

$$E_{ox}^\circ = 1.08 \text{ V} - 0.80 \text{ V} = +0.28 \text{ V}$$

Practice Problem 17.8

E_{rxn}° for the following reaction is +4.37 V.

$$3Ca(s) + 2Au^{3+}(aq) \longrightarrow 3Ca^{2+}(aq) + 2Au(s)$$

E_{red}° for the following half-reaction is +1.50 V.

$$Au^{3+}(aq) + 3e^- \longrightarrow Au(s)$$

Calculate E_{ox}° for the other half-reaction.

$$Ca(s) \longrightarrow Ca^{2+}(aq) + 2e^-$$

Answer: +2.87 V

Reduction Potentials and Spontaneity

When a half-reaction is reversed, what happens to the value of its electromotive force?

Zinc is a bluish-white, lustrous metal. It is stable in air because it forms a white coating of basic zinc carbonate, $Zn(OH)_2 \cdot ZnCO_3$, which prevents further corrosion. When heated in air, it burns with a bluish-green flame. Zinc reacts with dilute acids to release hydrogen gas.
Zinc is used in galvanized sheet iron, where a coating of zinc on the surface of the iron prevents corrosion of the iron. It is also an ingredient in alloys such as bronze, brass, and German silver.

Because we can assign values for oxidation and reduction potentials, we can arrange the elements in the order of their ability to reduce or oxidize other elements. Such an arrangement, called the **electromotive force series,** is given in Table 17.1. This series is the same as the activity series discussed in Section 17.1 but has more focus on the emf values. By convention, half-reaction potentials are normally reported as reduction potentials. In Table 17.1 the metal ion with the greatest oxidizing power, Au^{3+}, appears at the bottom of the table. The corresponding element, Au, is the weakest reducing agent in the table, with a standard reduction potential of +1.50 V. The strongest reducing agents are the elements at the top of the table. More reduction potentials can be found in Appendix G.

The standard reduction potentials given in Table 17.1 measure the tendency of a half-reaction to occur as a reduction. The more positive the value of the reduction potential, the greater the tendency for reduction to occur. The more negative the reduction potential, the greater the tendency for the reverse oxidation reaction to occur. In fact, a negative reduction potential corresponds to a positive oxidation potential for the oxidation half-reaction obtained by reversing a reduction half-reaction.

Consider the reduction half-reaction for zinc.

$$Zn^{2+}(aq) + 2e^- \longrightarrow Zn(s) \qquad E_{red}^\circ = -0.76 \text{ V}$$

Reversing the half-reaction is like changing the direction of electron flow in an electrical circuit, or reversing the electrical polarity. Thus the potential for the oxidation half-reaction is just opposite in sign to the potential for the corresponding reduction half-reaction.

$$Zn(s) \longrightarrow Zn^{2+}(aq) + 2e^- \qquad E^\circ_{ox} = +0.76 \text{ V}$$

We have already seen that the standard oxidation and reduction potentials for two half-reactions can be summed to give the standard emf for a reaction consisting of those two half-reactions. This provides us with a means of determining whether a reaction will be spontaneous, because spontaneous reactions generate a positive emf.

▶ Consider, for example, the possible reactions between zinc, tin, and their ions, described in the chapter introduction.

$$Zn(s) + Sn^{2+}(aq) \longrightarrow Zn^{2+}(aq) + Sn(s)$$
$$Zn^{2+}(aq) + Sn(s) \longrightarrow Zn(s) + Sn^{2+}(aq)$$

Only one of these reactions can be spontaneous. The reduction half-reactions for the two possible reactions are

$$Zn^{2+}(aq) + 2e^- \longrightarrow Zn(s) \qquad E^\circ_{red} = -0.76 \text{ V}$$
$$Sn^{2+}(aq) + 2e^- \longrightarrow Sn(s) \qquad E^\circ_{red} = -0.14 \text{ V}$$

We must combine these half-reactions into an oxidation–reduction reaction. To do so, we must reverse one of them to make it an oxidation half-reaction. When we reverse the half-reaction, we must also reverse the sign of the reduction potential. Because the overall reaction should have a positive potential, we want to reverse the half-reaction with the lower reduction potential, so that combining its potential with that of the other half-reaction gives a positive total. In this case, we reverse the zinc half-reaction.

$$Zn(s) \longrightarrow Zn^{2+}(aq) + 2e^- \qquad E^\circ_{ox} = +0.76 \text{ V}$$

Combining it with the tin reduction half-reaction, we obtain the following redox reaction:

$$Zn(s) + Sn^{2+}(aq) \longrightarrow Zn^{2+}(aq) + Sn(s)$$

The potential generated by this oxidation–reduction reaction can be calculated from the two half-reaction potentials.

$$Sn^{2+}(aq) + 2e^- \longrightarrow Sn(s) \qquad E^\circ_{red} = -0.14 \text{ V}$$
$$Zn(s) \longrightarrow Zn^{2+}(aq) + 2e^- \qquad E^\circ_{ox} = +0.76 \text{ V}$$

The reaction potential is the sum of the two half-reaction potentials.

$$E^\circ_{rxn} = E^\circ_{ox} + E^\circ_{red}$$
$$= +0.76 \text{ V} + -0.14 \text{ V} = +0.62 \text{ V}$$

The positive potential generated by this reaction confirms that it will proceed spontaneously as written. ▶ This result is supported by the experiment described in the chapter introduction.

EXAMPLE 17.9

Will copper metal spontaneously reduce Mg^{2+} in aqueous solution?

$$Cu(s) + Mg^{2+}(aq) \longrightarrow Cu^{2+}(aq) + Mg(s) \qquad \textit{(continued)}$$

Solution   The half-reactions and their potentials from Table 17.1 are

$$Mg^{2+}(aq) + 2e^- \longrightarrow Mg(s) \qquad E^\circ_{red} = -2.37 \text{ V}$$
$$Cu^{2+}(aq) + 2e^- \longrightarrow Cu(s) \qquad E^\circ_{red} = +0.34 \text{ V}$$

Copper metal is oxidized in the reaction, so the copper half-reaction must be reversed. The sign on the potential must also be changed.

$$Cu(s) \longrightarrow Cu^{2+}(aq) + 2e^- \qquad E^\circ_{ox} = -0.34 \text{ V}$$

The reaction potential is the sum of the two half-reaction potentials.

$$E^\circ_{rxn} = E^\circ_{ox} + E^\circ_{red}$$
$$= -0.34 \text{ V} + (-2.37 \text{ V}) = -2.71 \text{ V}$$

Because E°_{rxn} is negative, the oxidation–reduction reaction will not occur as written. However, the reverse reaction will occur spontaneously.

$$Cu^{2+}(aq) + Mg(s) \longrightarrow Cu(s) + Mg^{2+}(aq)$$

Practice Problem 17.9

Will aluminum metal spontaneously reduce Zn^{2+} in aqueous solution?

$$2Al(s) + 3Zn^{2+}(aq) \longrightarrow 2Al^{3+}(aq) + 3Zn(s)$$

Answer: yes

Reduction potentials can be used not only to predict whether a redox reaction will be spontaneous but also to predict the product of reaction between two substances, in cases where an element has several stable oxidation numbers. Consider, for example, the reaction between VO_2^+ and excess vanadium metal. What will be the product of this reaction? Appendix G contains the following possible reaction half-reactions and their standard potentials for vanadium having various oxidation numbers, which are shown in Figure 17.12.

$$VO_2^+ + 2H^+ + e^- \longrightarrow VO^{2+} + H_2O \qquad E^\circ_{red} = +1.00 \text{ V}$$
$$VO^{2+} + 2H^+ + e^- \longrightarrow V^{3+} + H_2O \qquad E^\circ_{red} = +0.36 \text{ V}$$
$$V^{3+} + e^- \longrightarrow V^{2+} \qquad E^\circ_{red} = -0.26 \text{ V}$$
$$V^{2+} + 2e^- \longrightarrow V \qquad E^\circ_{red} = -1.20 \text{ V}$$

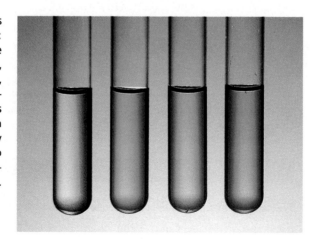

Figure 17.12 Vanadium forms four ions in acidic solution: yellow $VO_2^+(aq)$, bright blue $VO^{2+}(aq)$, blue-gray $V^{3+}(aq)$, and purple $V^{2+}(aq)$. In each, the oxidation number of vanadium is different. $V^{2+}(aq)$ is rapidly oxidized by oxygen in air, whereas $V^{3+}(aq)$ slowly reduces water. The other two ions are stable for long periods in solution exposed to air.

To determine the reaction product, we need merely consider the possible oxidation or reduction steps that each reactant can undergo and examine the standard emf when they are combined. (Of course, the conclusions reached will be valid only under standard conditions.) To simplify this process, we initially consider only the oxidation of vanadium metal to V^{2+}, according to the half-reaction

$$V \longrightarrow V^{2+} + 2e^- \qquad E^\circ_{ox} = +1.20 \text{ V}$$

Because V is present in excess, this assumption will be valid if vanadium with a higher oxidation number is reduced to V^{2+}. If it is not, we must consider the further oxidation of V^{2+}. Having excess vanadium metal also means that there is sufficient V to reduce VO_2^+ to any of the vanadium ions of lower oxidation number, assuming that these reductions are spontaneous.

We examine the combination of this oxidation half-reaction with each of the reduction half-reactions, starting with the reduction of VO_2^+. If the resulting reaction potential is positive, the reaction will occur. The potential for the first step is positive, so that reaction is spontaneous.

$$2VO_2^+(aq) + V(s) + 4H^+(aq) \longrightarrow 2VO^{2+}(aq) + V^{2+}(aq) + 2H_2O(l)$$
$$E^\circ_{rxn} = 1.00 \text{ V} + 1.20 \text{ V} = +2.20 \text{ V}$$

We examine the possible reduction of VO^{2+} in the same way and find a positive reaction potential, so this reaction is also spontaneous.

$$2VO^{2+}(aq) + V(s) + 4H^+(aq) \longrightarrow 2V^{3+}(aq) + V^{2+}(aq) + 2H_2O(l)$$
$$E^\circ_{rxn} = 0.36 \text{ V} + 1.20 \text{ V} = +1.56 \text{ V}$$

Repeating the procedure for the product, V^{3+}, we find another spontaneous reaction.

$$2V^{3+}(aq) + V(s) \longrightarrow 3V^{2+}(aq)$$
$$E^\circ_{rxn} = -0.26 \text{ V} + 1.20 \text{ V} = +0.94 \text{ V}$$

Thus vanadium metal will reduce VO_2^+ completely to V^{2+} and will itself be oxidized to V^{2+}.

$$2VO_2^+(aq) + 3V(s) + 8H^+(aq) \longrightarrow 5V^{2+}(aq) + 4H_2O(aq)$$

This prediction agrees with experimental results, as shown in Figure 17.13.

Figure 17.13 When yellow $VO_2^+(aq)$ (*Left*) is reduced with excess vanadium metal, purple $V^{2+}(aq)$ (*Right*) results.

A similar procedure can be used to determine the reaction products of redox reactions between two different elements.

EXAMPLE 17.10

What is the product of the reaction between excess I_2 and vanadium metal? The half-reactions and their reduction potentials of vanadium were given earlier.

$$I_2(aq) + 2e^- \longrightarrow 2I^-(aq) \qquad E^\circ_{red} = +0.54 \text{ V}$$

Solution Iodine is present in excess, so we must determine the oxidation number to which vanadium metal can be oxidized. We combine the iodine/iodide reduction half-reaction with each vanadium oxidation half-reaction to determine whether the reaction potential is positive.

$$V(s) + I_2(aq) \longrightarrow V^{2+}(aq) + 2I^-(aq) \qquad E^\circ_{rxn} = 1.20 \text{ V} + 0.54 \text{ V} = +1.74 \text{ V}$$

$$2V^{2+}(aq) + I_2(aq) \longrightarrow 2V^{3+}(aq) + 2I^-(aq) \qquad E^\circ_{rxn} = 0.26 \text{ V} + 0.54 \text{ V} = +0.80 \text{ V}$$

$$2V^{3+}(aq) + I_2(aq) + 2H_2O(l) \longrightarrow 2VO^{2+}(aq) + 2I^-(aq) + 4H^+(aq)$$
$$E^\circ_{rxn} = -0.36 \text{ V} + 0.54 \text{ V} = +0.18 \text{ V}$$

$$2VO^{2+}(aq) + I_2(aq) + 2H_2O(l) \longrightarrow 2VO_2^+(aq) + 2I^-(aq) + 4H^+(aq)$$
$$E^\circ_{rxn} = -1.00 \text{ V} + 0.54 \text{ V} = -0.46 \text{ V}$$

Iodine can oxidize vanadium metal stepwise up to VO^{2+}, because positive reaction potentials result for each of these steps. However, iodine is not a strong enough oxidizing agent to overcome the 1.0 V required to oxidize VO^{2+} to VO_2^+. The overall reaction with excess iodine is obtained by adding the spontaneous reactions in such a way that the vanadium ions having intermediate oxidation numbers cancel.

$$V(s) + 2I_2(aq) + H_2O(l) \longrightarrow VO^{2+}(aq) + 4I^-(aq) + 2H^+(aq)$$

Practice Problem 17.10

Which iodine species can be oxidized by excess aqueous chlorine? What is the product of oxidation in each case? The reduction half-reactions and their potentials are

$$H_5IO_6(aq) + H^+(aq) + 2e^- \longrightarrow IO_3^-(aq) + 3H_2O(l) \qquad E^\circ_{red} = +1.70 \text{ V}$$

$$2IO_3^-(aq) + 12H^+(aq) + 10e^- \longrightarrow I_2(aq) + 6H_2O(l) \qquad E^\circ_{red} = +1.20 \text{ V}$$

$$I_2(aq) + 2e^- \longrightarrow 2I^-(aq) \qquad E^\circ_{red} = +0.54 \text{ V}$$

$$Cl_2(aq) + 2e^- \longrightarrow 2Cl^-(aq) \qquad E^\circ_{red} = +1.36 \text{ V}$$

Answer: I^- and I_2 will be oxidized to IO_3^- by excess Cl_2.

Instability of Species in Solution

Another application of reduction potentials is in predicting the stability of species in solution. We can make such predictions simply by examining reduction half-reactions and their potentials. Here, we look at three types of instability: disproportionation, oxidation or reduction of water (or H^+), and oxidation by O_2 from air.

Disproportionation Reactions   Some species are not stable in solution but spontaneously form two new substances, one of them having a higher oxidation number than the original substance and the other a lower oxidation number. This reaction is known as **disproportionation.** In general, for example, an ion will disproportionate if the reduction potential for the half-reaction that forms it is smaller than the reduction potential for the half-reaction that consumes it. This criterion is identical to that used earlier to predict spontaneity: The sum of the oxidation potential and the reduction potential for the two half-reactions is positive for a spontaneous reaction. The only difference in a disproportionation reaction is that the same element participates in both half-reactions.

Consider two half-reactions that involve the ruthenium(II) ion:

$$RuO_4^{2-}(aq) + 8H^+(aq) + 4e^- \longrightarrow Ru^{2+}(aq) + 4H_2O(l) \qquad E^\circ_{red} = +0.30 \text{ V}$$
$$Ru^{2+}(aq) + 2e^- \longrightarrow Ru(s) \qquad E^\circ_{red} = +0.45 \text{ V}$$

We can combine these two half-reactions so that the only ruthenium species in the reactants is ruthenium(II) by reversing the first half-reaction and adding it to the second.

$$Ru^{2+}(aq) + 4H_2O(l) \longrightarrow RuO_4^{2-}(aq) + 8H^+(aq) + 4e^- \qquad E^\circ_{ox} = -0.30 \text{ V}$$
$$Ru^{2+}(aq) + 2e^- \longrightarrow Ru(s) \qquad E^\circ_{red} = +0.45 \text{ V}$$

The products are RuO_4^{2-} and $Ru(s)$.

$$3Ru^{2+}(aq) + 4H_2O(l) \longrightarrow RuO_4^{2-}(aq) + 2Ru(s) + 8H^+(aq)$$
$$E^\circ_{rxn} = E^\circ_{ox} + E^\circ_{red} = -0.30 \text{ V} + 0.45 \text{ V} = +0.15 \text{ V}$$

Because the reaction potential is positive, this disproportionation reaction is spontaneous, and ruthenium(II) is not stable in aqueous solution. We would have to examine all possible half-reactions involving Ru^{2+} to see whether this is the only possible disproportionation reaction.

Reaction with Water   A second type of instability may characterize a species in aqueous solution. A sufficiently strong reducing agent may spontaneously undergo oxidation by aqueous hydrogen ion (or water) to release hydrogen gas, as was shown for metals in the chemical activity series in Figure 17.3. Very active metals produce H_2 in pure water. Less active metals produce $H_2(g)$ from 1 M $H^+(aq)$ solution. Any metal above hydrogen in the activity series displaces hydrogen gas from acid solution. Metals with lower activity do not produce hydrogen at all.

Reduction potentials provide a quantitative method for deciding whether a substance can displace hydrogen gas from water in 1 M $H^+(aq)$ solution. Recall that the standard emf for the hydrogen half-reaction in 1 M $H^+(aq)$ solution is 0 V.

$$2H^+(aq) + 2e^- \longrightarrow H_2(g) \qquad E^\circ_{red} = 0.000 \text{ V}$$

To obtain the positive reaction potential necessary for a spontaneous reaction, we need only have an oxidation half-reaction with a positive oxidation potential. This criterion is met by any substance listed in Table 17.1 (or Appendix G) that has a negative reduction potential. All such substances are capable of releasing H_2 spontaneously from 1 M $H^+(aq)$ solution.

Consider, for example, the half-reactions for vanadium.

$$VO_2^+(aq) + 2H^+(aq) + e^- \longrightarrow VO^{2+}(aq) + H_2O(l) \qquad E^\circ_{red} = +1.00 \text{ V}$$
$$VO^{2+}(aq) + 2H^+(aq) + e^- \longrightarrow V^{3+}(aq) + H_2O(l) \qquad E^\circ_{red} = +0.36 \text{ V}$$
$$V^{3+}(aq) + e^- \longrightarrow V^{2+}(aq) \qquad E^\circ_{red} = -0.26 \text{ V}$$
$$V^{2+}(aq) + 2e^- \longrightarrow V(s) \qquad E^\circ_{red} = -1.20 \text{ V}$$

The latter two reduction half-reactions have negative potentials, so V and V^{2+} should be able to reduce $H^+(aq)$ to $H_2(g)$, as shown by the positive reaction potentials.

$$V(s) + 2H^+(aq) \longrightarrow V^{2+}(aq) + H_2(g) \qquad E^\circ_{rxn} = 1.20\ V + 0.00\ V = +1.20\ V$$
$$2V^{2+}(aq) + 2H^+(aq) \longrightarrow 2V^{3+}(aq) + H_2(g) \qquad E^\circ_{rxn} = 0.26\ V + 0.00\ V = +0.26\ V$$

Although these results are consistent with thermodynamic criteria for spontaneity, not all possible reactions occur rapidly in practice. The liberation of H_2 from acid solution occurs at an appreciable rate only if E°_{rxn} is greater than 0.4–0.5 V. Thus vanadium metal dissolves in 1 M $H^+(aq)$ solution, but $V^{2+}(aq)$ is oxidized to $V^{3+}(aq)$ only very slowly.

EXAMPLE 17.11

Consider the following reduction half-reactions and their potentials.

$$Cr_2O_7^{2-} + 14H^+(aq) + 6e^- \longrightarrow 2Cr^{3+}(aq) + 7H_2O(l) \qquad E^\circ_{red} = +1.33\ V$$
$$Cr^{3+}(aq) + e^- \longrightarrow Cr^{2+}(aq) \qquad E^\circ_{red} = -0.41\ V$$
$$Cr^{2+}(aq) + 2e^- \longrightarrow Cr(s) \qquad E^\circ_{red} = -0.91\ V$$

Which chromium species, if any, can reduce 1 M $H^+(aq)$ to release H_2 gas?

Solution   Only Cr and Cr^{2+} are produced from higher oxidation states with negative reduction potentials, so these two species should release H_2 from 1 M $H^+(aq)$ solution.

$$Cr(s) + 2H^+(aq) \longrightarrow Cr^{2+}(aq) + H_2(g) \qquad E^\circ_{rxn} = 0.91\ V + 0.00\ V = +0.91\ V$$
$$2Cr^{2+}(aq) + 2H^+(aq) \longrightarrow 2Cr^{3+}(aq) + H_2(g) \qquad E^\circ_{rxn} = 0.41\ V + 0.00\ V = +0.41\ V$$

In fact, chromium metal dissolves readily in 1 M acid solution, but chromium(II) salts, though they are very good reducing agents, do not quickly reduce acid.

Practice Problem 17.11

Consider the following reduction half-reactions and their potentials.

$$TiO^{2+}(aq) + 2H^+(aq) + e^- \longrightarrow Ti^{3+}(aq) + H_2O(l) \qquad E^\circ_{red} = +0.10\ V$$
$$Ti^{3+}(aq) + e^- \longrightarrow Ti^{2+}(aq) \qquad E^\circ_{red} = -0.37\ V$$
$$Ti^{2+}(aq) + 2e^- \longrightarrow Ti(s) \qquad E^\circ_{red} = -1.63\ V$$

Which titanium species, if any, can reduce 1 M $H^+(aq)$ and release H_2 gas?

Answer: Ti ($E^\circ_{rxn} = 1.63$ V) and Ti^{2+} ($E^\circ_{rxn} = 0.37$ V) react spontaneously.

A sufficiently strong oxidizing agent may be able to oxidize water to release oxygen gas. The reduction potentials for the oxidizing agent must be strong enough to overcome the reduction potential for oxygen gas in 1 M $H^+(aq)$ solution. The reduction half-reaction and its potential are

$$O_2(g) + 4H^+(aq) + 4e^- \longrightarrow 2H_2O(l) \qquad E^\circ_{red} = +1.23\ V$$

The oxidation of water occurs by the reverse of this half-reaction:

$$2H_2O(l) \longrightarrow O_2(g) + 4H^+(aq) + 4e^- \qquad E^\circ_{ox} = -1.23\ V$$

A possible oxidizing agent must have a reduction potential greater than +1.23 V in 1 M $H^+(aq)$ to release $O_2(g)$ from solution. It should be noted that, like the release of H_2 from water, the production of O_2 usually occurs at an appreciable rate only if the net reaction potential is greater than 0.4–0.5 V.

Consider the reduction half-reactions and potentials for silver:

$$AgO^+(aq) + 2H^+(aq) + e^- \longrightarrow Ag^{2+}(aq) + H_2O(l) \qquad E^{\circ}_{red} = +2.10 \text{ V}$$
$$Ag^{2+}(aq) + e^- \longrightarrow Ag^+(aq) \qquad E^{\circ}_{red} = +1.98 \text{ V}$$
$$Ag^+(aq) + e^- \longrightarrow Ag(s) \qquad E^{\circ}_{red} = +0.80 \text{ V}$$

Silver normally exists in solution as the aqueous silver(I) ion, Ag^+(aq), a colorless substance. However, under very strongly oxidizing conditions, silver can be oxidized to Ag^{2+}(aq) or AgO^+(aq), which are short-lived because they oxidize water.

We can predict that AgO^+ and Ag^{2+} should oxidize water in 1 M $H^+(aq)$ solution, whereas Ag^+ should not. We need only calculate the reaction potentials to see whether they are positive.

$$4AgO^+(aq) + 4H^+(aq) \longrightarrow 4Ag^{2+}(aq) + 2H_2O(l) + O_2(g)$$
$$E^{\circ}_{rxn} = -1.23 \text{ V} + 2.10 \text{ V} = +0.87 \text{ V}$$

$$4Ag^{2+}(aq) + 2H_2O(l) \longrightarrow 4Ag^+(aq) + O_2(g) + 4H^+(aq)$$
$$E^{\circ}_{rxn} = -1.23 \text{ V} + 1.98 \text{ V} = +0.75 \text{ V}$$

Both these values are well above the 0.4–0.5-V limit needed for oxygen to evolve at an appreciable rate.

Oxidation by O_2 in Air   Finally, we examine instability involving the oxidation of substances by oxygen in air. The half-reaction of interest here, occurring in 1 M $H^+(aq)$ solution, is

$$O_2(g) + 4H^+(aq) + 4e^- \longrightarrow 2H_2O(l) \qquad E^{\circ}_{red} = +1.23 \text{ V}$$

Any substance with an oxidation potential greater than -1.23 V should be oxidized by oxygen gas. Note that this is the reverse of the oxidation of water.

EXAMPLE 17.12

Consider the following reduction half-reactions for chromium.

$$Cr_2O_7^{2-} + 14H^+(aq) + 6e^- \longrightarrow 2Cr^{3+}(aq) + 7H_2O(l) \qquad E^{\circ}_{red} = +1.33 \text{ V}$$
$$Cr^{3+}(aq) + e^- \longrightarrow Cr^{2+}(aq) \qquad E^{\circ}_{red} = -0.41 \text{ V}$$
$$Cr^{2+}(aq) + 2e^- \longrightarrow Cr(s) \qquad E^{\circ}_{red} = -0.91 \text{ V}$$

Which chromium species can be oxidized by O_2 from air?

Solution   The reduction potentials suggest that both Cr metal and Cr^{2+} should be oxidized by oxygen gas.

$$2Cr(s) + O_2(g) + 4H^+(aq) \longrightarrow 2Cr^{2+} + 2H_2O(l)$$
$$E^{\circ}_{rxn} = 0.91 \text{ V} + 1.23 \text{ V} = +2.14 \text{ V}$$

$$4Cr^{2+}(aq) + O_2(g) + 4H^+(aq) \longrightarrow 4Cr^{3+}(aq) + 2H_2O(l)$$
$$E^{\circ}_{rxn} = 0.41 \text{ V} + 1.23 \text{ V} = +1.64 \text{ V}$$

However, when chromium metal is oxidized by O_2, it forms a tight protective coating of oxide, so it is not particularly susceptible to air oxidation. (Because of this oxide, we can use chromium to protect steel on objects such as faucets.) Chromium(II) solutions are extremely susceptible to air oxidation. Chromium(III) is not oxidized by air, however, which is consistent with the negative reaction potential (-1.33 V + 1.23 V = -0.10 V) for its possible oxidation.

(continued)

Practice Problem 17.12

Consider the following reduction half-reactions and their potentials.

$$VO_2^+(aq) + 2H^+(aq) + e^- \longrightarrow VO^{2+}(aq) + H_2O(l) \qquad E^\circ_{red} = +1.00 \text{ V}$$
$$VO^{2+}(aq) + 2H^+(aq) + e^- \longrightarrow V^{3+}(aq) + H_2O(l) \qquad E^\circ_{red} = +0.36 \text{ V}$$
$$V^{3+}(aq) + e^- \longrightarrow V^{2+}(aq) \qquad E^\circ_{red} = -0.25 \text{ V}$$
$$V^{2+}(aq) + 2e^- \longrightarrow V(s) \qquad E^\circ_{red} = -1.20 \text{ V}$$

Which vanadium species can be oxidized by O_2 from air?

Answer: V and V^{2+} readily; V^{3+} slowly; and VO^{2+} possibly, but so slowly that it is generally not observed.

17.6 Hydrometallurgy

Since ancient times, metals have been produced from their ores by *pyrometallurgy:* high-temperature processes for melting and reducing metal salts. Toward the end of the nineteenth century, new processes began to be introduced. These processes, known as **hydrometallurgy,** involve aqueous solutions at moderate temperatures. Hydrometallurgical processes usually require less energy and produce less air pollution than pyrometallurgical processes, and they are especially well suited to low-grade ores and ores that contain several metals with similar properties. Hydrometallurgy is now used to produce several metals, including gold and silver, as well as salts of other metals. And it has begun to compete economically with pyrometallurgy in the production of copper, zinc, and nickel.

In hydrometallurgical processes, ores are leached into solutions, sometimes with the aid of oxygen from air and of complexing agents. The extracted metals are deposited from solution by application of oxidation–reduction reactions and various principles of chemical equilibrium.

Leaching of Ores

Which chemical principles are invoked in attempts to dissolve metal ores?

In the first step of any hydrometallurgical process, the ore must be dissolved in solution, or *leached,* so it can be processed further to recover the metal. A variety of ores can be leached by treatment with various reagents, such as water, acids, bases, and aqueous salt solutions. Water can be used to dissolve soluble salts, such as chlorides and sulfates, but acids are the most commonly used leaching agents. A variety of ores, particularly oxides, can be leached with acids.

$$CuO(s) + 2H^+(aq) \longrightarrow Cu^{2+}(aq) + H_2O(l)$$
$$2UO_2(s) + 4H^+(aq) + O_2(g) \longrightarrow 2UO_2^{2+}(aq) + 2H_2O(l)$$

Sulfuric acid is often used, because it is cheapest. However, because some sulfates are not soluble, hydrochloric acid and nitric acid are frequently used instead. For example, uranium ores and phosphate rock, which may contain radium, leave behind deposits of radioactive, insoluble radium sulfate when treated with sulfuric acid. If radium is to be recovered by a hydrometallurgical process, it must be in solution, so sulfuric acid cannot be used for this ore.

Bases can also be used as leaching agents, particularly for amphoteric metals. For example, an early hydrometallurgical process known as the Bayer process is

Uranium dioxide, UO_2, is a black solid that occurs in nature as the minerals uraninite and pitchblende. It is insoluble in water and dilute acids but dissolves in concentrated acids to give a solution containing green U^{4+}(aq). This ion is oxidized by oxygen in air to give yellow uranium(VI), UO_2^{2+}(aq).

used to recover alumina (Al_2O_3) from bauxite, an ore containing hydrated Al_2O_3. The bauxite is leached with sodium hydroxide solution to produce a soluble hydroxy aluminum(III) complex ion.

$$Al_2O_3(s) + 2OH^-(aq) + 3H_2O(l) \rightleftharpoons 2Al(OH)_4{}^-(aq)$$

Shifting the pH to a more acidic region causes aluminum hydroxide to precipitate.

$$Al(OH)_4{}^-(aq) + H^+(aq) \rightleftharpoons Al(OH)_3(s) + H_2O(l)$$

The aluminum hydroxide can be dehydrated thermally to produce alumina.

$$2Al(OH)_3(s) \xrightarrow{\Delta} Al_2O_3(s) + 3H_2O(g)$$

This alumina can then be used to obtain aluminum metal.

Oxidizing Agents and Sulfide Ores

In order to dissolve some ores, such as sulfides, it may be necessary to have an oxidizing agent in the system as well as a leaching reagent. The reaction of oxygen gas with insoluble sulfides spontaneously forms soluble sulfates, for example, as shown by the following reaction potentials.

$$4H^+ + 2S^{2-} + O_2 \longrightarrow 2S + 2H_2O \qquad E°_{rxn} = +1.09 \text{ V}$$
$$\underline{2S + 3O_2 + 2H_2O \longrightarrow 4H^+ + 2SO_4{}^{2-}} \qquad E°_{rxn} = +0.87 \text{ V}$$
$$S^{2-} + 2O_2 \longrightarrow SO_4{}^{2-} \qquad E°_{rxn} = +0.93 \text{ V}$$

A typical example is insoluble zinc sulfide, which can be converted to soluble zinc sulfate by reaction with oxygen from air.

$$ZnS(s) + 2O_2(aq) \longrightarrow ZnSO_4(aq)$$

Complexing Agents

It is also often helpful to have a *complexing agent* present in the system to dissolve the ore. Formation of a complex ion converts the metal ion in solution to a different form, so the solubility is increased in accordance with Le Chatelier's principle. This is the case in one of the first hydrometallurgical processes introduced, the cyanide process for extracting gold from ores via oxidation of the metal by oxygen gas in the presence of cyanide ion, which complexes the gold(I) ions produced.

$$4Au(s) + 8CN^-(aq) + O_2(g) + 2H_2O(l) \longrightarrow 4Au(CN)_2{}^-(aq) + 4OH^-(aq)$$

This is a spontaneous reaction ($E°_{rxn} = +1.00$ V). Gold(I) ions are not stable in aqueous solution unless they are stabilized by formation of a complex ion such as $Au(CN)_2{}^-$.

Reduction of Ores

Once a metal salt has been made soluble by leaching, it can be purified if necessary and then must be recovered from solution. Several methods are available, but a reduction process must be used if a pure metal is to be recovered. For example, recovery of metallic gold from the $Au(CN)_2{}^-$ solution just mentioned involves

Aluminum hydroxide is a bulky white solid. It dissolves in acidic and basic solutions but is almost insoluble in water. On contact with water, it forms gels. It is used in water-treatment plants to remove suspended particles from the water and is an ingredient in antiperspirants, dentifrices, and antacids. It is also used in the manufacture of glass, paper, pottery, and inks.

Figure 17.14 Copper is leached from large beds of low-grade copper ore by a process in which dilute sulfuric acid is sprayed over the ore beds and the solution is allowed to filter through the ore.

reduction with a metal or other substance that is a stronger reducing agent than gold. Zinc powder is commonly used.

$$2Au(CN)_2^-(aq) + Zn(s) \longrightarrow 2Au(s) + Zn(CN)_4^{2-}(aq) \qquad E_{rxn}^\circ = +0.66 \text{ V}$$

This process is still used worldwide to extract gold from its ores.

Hydrometallurgical processes are also used to recover copper from low-grade ores. The ores are leached with sulfuric acid, which produces a dilute copper(II) sulfate solution (Figure 17.14). This solution is then poured over scrap iron, and copper metal is deposited.

$$Cu^{2+}(aq) + Fe(s) \longrightarrow Cu(s) + Fe^{2+}(aq)$$

Other reducing agents that have been used include molecular hydrogen, iron(II) salts, and sulfur dioxide. Finally, electrolysis—the passage of an electrical current through the solution—can effect the reduction of a metal ion to a metal.

$$Cu^{2+}(aq) + 2e^- \longrightarrow Cu(s)$$
$$Ni^{2+}(aq) + 2e^- \longrightarrow Ni(s)$$

Electrolysis will be explored in detail in Chapter 18.

Summary

Oxidation–reduction reactions, which involve the transfer of electrons from the reducing agent to the oxidizing agent, can be separated into oxidation and reduction half-reactions. Reducing agents can be arranged in order of their reducing power in an activity series, which can be used to predict the oxidation–reduction reactions of these substances and their ions.

Many elements form various compounds in which they have different oxidation numbers. We can recognize substances as oxidizing or reducing agents by assigning oxidation numbers to the elements in the reactants and products and determining which are increased or decreased during an oxidation–reduction reaction. Substances that contain an element with its highest oxidation number can act only as oxidizing agents, whereas those that contain an element with its lowest oxidation number can act only as reducing agents. Although equations for some oxidation–reduction reactions can be balanced by inspection, most are complex enough to require a more systematic approach, such as the half-reaction method and the oxidation-number-change method.

The spontaneity of an oxidation–reduction reaction can be predicted from an activity series. The stronger reducing agent reacts spontaneously with the stronger oxidizing agent to form a weaker oxidizing agent and a weaker reducing agent. A more quantitative approach uses values of the electromotive force associated with half-reactions. Addition of the half-reaction potentials results in a positive reaction potential if the oxidation–reduction reaction is spontaneous.

Key Terms

chemical activity series (17.1)	**(emf)** (17.5)	**hydrometallurgy** (17.6)	**potential** (17.5)
disproportionation (17.5)	**electromotive force series** (17.5)	**ion–electron method** (17.4)	**oxidizing agent** (17.1)
electrical potential (17.5)	**half-reaction** (17.1)	**oxidation-number-change method** (17.4)	**reduction potential** (17.5)
electromotive force	**half-reaction method** (17.4)	**oxidation**	**reducing agent** (17.1)

Exercises

Oxidizing and Reducing Agents

17.1 What is meant by oxidizing agent and by reducing agent?

17.2 Define oxidation and reduction.

17.3 What is an activity series?

17.4 Why is oxidation always coupled with reduction?

17.5 Pick the strongest reducing agent and the weakest reducing agent from the following list: Zn, Pb, Ag, H_2, Li

17.6 Pick the strongest oxidizing agent and the weakest oxidizing agent from the following list: Ag^+, Na^+, H^+, Pb^{2+}, Fe^{3+}

17.7 Indicate whether each of the following reactions is an oxidation–reduction reaction.

a. $BaCl_2(aq) + H_2SO_4(aq) \longrightarrow$
$BaSO_4(s) + 2HCl(aq)$

b. $3H_2(g) + N_2(g) \longrightarrow 2NH_3(g)$

c. $H_2CO_3(aq) \longrightarrow H_2O(l) + CO_2(g)$

d. $AgNO_3(aq) + NaCl(aq) \longrightarrow$
$AgCl(s) + NaNO_3(aq)$

e. $2Na(s) + 2H_2O(l) \longrightarrow 2NaOH(aq) + H_2(g)$

f. $H_2(g) + F_2(g) \longrightarrow 2HF(g)$

g. $C(s) + H_2O(s) \longrightarrow CO(g) + H_2(g)$

17.8 Consider the reaction

$6Fe^{2+}(aq) + Cr_2O_7^{2-}(aq) + 14H^+(aq) \longrightarrow$
$6Fe^{3+}(aq) + 2Cr^{3+}(aq) + 7H_2O(l)$

a. Which substance is oxidized?

b. Which substance is reduced?

c. Which substance is the oxidizing agent?

d. Which substance is the reducing agent?

17.9 Identify the oxidizing agent and reducing agent in the following equation for the preparation of sodium chlorite, an industrial bleaching agent.

$4NaOH(aq) + Ca(OH)_2(aq) + C(s) + 4ClO_2(g) \longrightarrow$
$4NaClO_2(aq) + CaCO_3(s) + 3H_2O(l)$

17.10 In each of the following reactions, identify the oxidizing agent, the reducing agent, the element oxidized, and the element reduced.

a. $4Na(s) + O_2(g) \longrightarrow 2Na_2O(s)$

b. $Ni(s) + Br_2(l) \longrightarrow NiBr_2(s)$

c. $C(s) + O_2(g) \longrightarrow CO_2(g)$

d. $2Na(s) + H_2(g) \longrightarrow 2NaH(s)$

17.11 In each of the following oxidation–reduction reactions, identify the oxidizing agent, the reducing agent, and the number of electrons transferred.

a. $I_2(aq) + 2OH^-(aq) \longrightarrow$
$I^-(aq) + IO^-(aq) + H_2O(l)$

b. $Cr(s) + 2H^+(aq) \longrightarrow Cr^{2+}(aq) + H_2(g)$

c. $2Cr_2O_7^{2-}(aq) + 16H^+(aq) \longrightarrow$
$4Cr^{3+}(aq) + 3O_2(g) + 8H_2O(l)$

d. $3Fe^{3+}(aq) + Al(s) \longrightarrow$
$3Fe^{2+}(aq) + Al^{3+}(aq)$

e. $6I^-(aq) + BrO_3^-(aq) + 6H^+(aq) \longrightarrow$
$3I_2(aq) + Br^-(aq) + 3H_2O(l)$

f. $5Br^-(aq) + BrO_3^-(aq) + 6H^+(aq) \longrightarrow$
$3Br_2(aq) + 3H_2O(l)$

g. $XeF_4(s) + 2H_2O(l) \longrightarrow$
$Xe(g) + 4HF(aq) + O_2(g)$

Oxidation Numbers

17.12 What is an oxidation number?

17.13 Describe two methods for assigning oxidation numbers.

17.14 How do oxidation numbers change during oxidation and during reduction?

17.15 What is the oxidation number of phosphorus in each of the following oxides?

a. P_4O_{10} d. P_4O_8

b. P_4O_7 e. P_4O_9

c. P_4O_6

17.16 Determine the oxidation number of bromine in each of these bromine oxides.

a. Br_2O d. BrO_2

b. BrO_3 e. BrO

c. Br_2O_3 f. Br_2O_5

17.17 Indicate the oxidation number of phosphorus in each of the following compounds.

a. $AlPO_4$ d. H_3PO_2

b. PF_5 e. PH_3

c. H_3PO_4 f. H_3PO_3

17.18 Determine the oxidation number of chlorine in each of the following compounds.
- a. NaCl
- b. Cl_2
- c. ClO_2
- d. Cl_2O_7
- e. $KClO_4$
- f. $Ca(ClO_2)_2$
- g. $HClO_2$
- h. $AlCl_3$
- i. ClF

17.19 Indicate the oxidation number of sulfur in each of the following compounds.
- a. $SOCl_2$
- b. H_2S
- c. SO_3
- d. SO_2Cl_2
- e. S_8
- f. $Fe_2(SO_4)_3$

17.20 Indicate the oxidation number of iron in each of the following compounds.
- a. $FeCl_2$
- b. $FeSO_4$
- c. $Fe_2(C_2O_4)_3$
- d. $Fe(H_2O)_6Cl_3$
- e. $Fe(NO_3)_3$
- f. $Fe_2(SO_4)_3$

17.21 Determine the oxidation number of each atom in the following compounds.
- a. ClO_2
- b. CaF_2
- c. H_2TeO_3
- d. Na_2O_2
- e. $Fe(NO_3)_3$
- f. Sc_2O_3

17.22 Assign an oxidation number to iodine in each of the following species.
- a. IF_7 b. IO_3^- c. I^- d. ICl_3 e. $IOCl_3$

17.23 Determine the oxidation numbers of the underlined elements in each of the following compounds.
- a. $K_2\underline{Cr}O_4$
- b. $Na_2\underline{Se}O_4$
- c. $NH_4\underline{N}O_3$
- d. $Ca\underline{H}PO_4$
- e. $Ca\underline{H}_2$

17.24 Determine the oxidation number of the central atom in each of the following ions.
- a. NO_2^-
- b. $P_2O_7^{4-}$
- c. $AgCl_2^-$
- d. SO_3^{2-}
- e. CO_3^{2-}
- f. ClO_2^-
- g. $H_2PO_4^-$
- h. $FeCl_6^{3-}$
- i. $Ni(CN)_4^{2-}$
- j. $Co(NH_3)_6^{3+}$
- k. SiF_6^{2-}
- l. AsO_4^{3-}

Balancing Oxidation–Reduction Equations

17.25 What is a half-reaction?

17.26 What are three methods for balancing oxidation–reduction reactions?

17.27 Why are oxidation–reduction reactions not usually balanced by inspection?

17.28 How are oxidation numbers used to help balance oxidation–reduction equations?

17.29 Balance the following half-reactions by adding $H^+(aq)$, $H_2O(l)$, and e^- as appropriate.
- a. $Ba(s) \longrightarrow Ba^{2+}(aq)$
- b. $I_2(aq) \longrightarrow I^-(aq)$
- c. $HNO_2(aq) \longrightarrow NO(g)$
- d. $H_2O(l) \longrightarrow O_2(g)$

- e. $H_2O_2(aq) \longrightarrow H_2O(l)$
- f. $MnO_4^-(aq) \longrightarrow Mn^{2+}(aq)$
- g. $Cr^{3+}(aq) \longrightarrow Cr_2O_7^{2-}(aq)$

17.30 Balance the following equations using the oxidation-number-change method.
- a. $Cu(s) + H^+(aq) + NO_3^-(aq) \longrightarrow$
  $Cu^{2+}(aq) + NO(l) + H_2O(l)$
- b. $Zn(s) + H^+(aq) + NO_3^-(aq) \longrightarrow$
  $Zn^{2+}(aq) + NH_4^+(aq) + H_2O(l)$
- c. $H^+(aq) + I^-(aq) + SO_4^{2-}(aq) \longrightarrow$
  $H_2S(aq) + I_2(aq) + H_2O(l)$
- d. $Cr^{3+}(aq) + ClO_3^-(aq) + OH^-(aq) \longrightarrow$
  $CrO_4^{2-}(aq) + Cl^-(aq) + H_2O(l)$
- e. $Cl_2(aq) + OH^-(aq) \longrightarrow$
  $Cl^-(aq) + ClO_3^-(aq) + H_2O(l)$

17.31 Balance the following equations using the half-reaction method.
- a. $H^+(aq) + Br^-(aq) + SO_4^{2-}(aq) \longrightarrow$
  $Br_2(aq) + SO_2(g) + H_2O(l)$
- b. $Al(s) + H_2O(l) + OH^-(aq) \longrightarrow$
  $Al(OH)_4^-(aq) + H_2(g)$
- c. $H_2O_2(aq) + I^-(aq) + H^+(aq) \longrightarrow$
  $I_2(aq) + H_2O(l)$
- d. $H_2O_2(aq) + MnO_4^-(aq) + H^+(aq) \longrightarrow$
  $O_2(g) + Mn^{2+}(aq) + H_2O(l)$
- e. $Pb(s) + PbO_2(s) + H^+(aq) + SO_4^{2-}(aq) \longrightarrow$
  $PbSO_4(s) + H_2O(l)$

17.32 Balance the following equations using either the oxidation-number-change method or the half-reaction method.
- a. $H_2O_2(aq) + PbS(s) \longrightarrow PbSO_4(s) + H_2O(l)$
- b. $KMnO_4(aq) + H_2SO_4(aq) + NaCl(aq) \longrightarrow$
  $Cl_2(aq) + NaHSO_4(aq) + K_2SO_4(aq)$
  $+ MnSO_4(aq) + H_2O(l)$
- c. $Na_2SO_3(aq) + I_2(aq) + H_2O(l) \longrightarrow$
  $Na_2SO_4(aq) + HI(aq)$
- d. $HNO_2(aq) + HI(aq) \longrightarrow$
  $NO(g) + I_2(aq) + H_2O(l)$
- e. $Hg_2Cl_2(aq) + NaOH(aq) \longrightarrow$
  $Hg(l) + HgO(s) + H_2O(l) + NaCl(aq)$

17.33 Balance the following equations, assuming that the reactions occur in basic solution.
- a. $H_2O_2(aq) + ClO_2(aq) \longrightarrow ClO_2^-(aq) + O_2(g)$
- b. $MnO_4^-(aq) + I^-(aq) \longrightarrow MnO_2(s) + I_2(aq)$
- c. $Cl_2(aq) \longrightarrow ClO^-(aq) + Cl^-(aq)$

17.34 Balance the following equations.
- a. iron(III) sulfate + potassium iodide $\longrightarrow$
  iron(II) sulfate + potassium sulfate + iodine
- b. arsenic(III) oxide + iodine + water $\longrightarrow$
  hydrogen iodide + arsenic(V) oxide
- c. nitric acid + sulfur dioxide $\longrightarrow$
  sulfuric acid + nitrogen dioxide
- d. potassium chlorate + hydrogen iodide +
  sulfuric acid $\longrightarrow$ hydrogen chloride +
  potassium hydrogen sulfate + iodine + water

e. ammonium dichromate $\longrightarrow$
chromium(III) oxide + nitrogen + water

17.35 Complete and balance the following equations, assuming that the oxidation–reduction reactions occur in acidic solution.

a. $H_2S(aq) + Cr_2O_7{}^{2-}(aq) \longrightarrow$
$$S(s) + Cr^{3+}(aq)$$

b. $V^{2+}(aq) + MnO_4{}^-(aq) \longrightarrow$
$$VO_2{}^+(aq) + Mn^{2+}(aq)$$

c. $Fe^{2+}(aq) + ClO_3{}^-(aq) \longrightarrow$
$$Fe^{3+}(aq) + Cl^-(aq)$$

d. $Cu(s) + NO_3{}^-(aq) \longrightarrow Cu^{2+}(aq) + NO(g)$

e. $C_2O_4{}^{2-}(aq) + MnO_4{}^-(aq) \longrightarrow$
$$CO_2(g) + Mn^{2+}(aq)$$

f. $ClO_3{}^-(aq) + Cl^-(aq) \longrightarrow Cl_2(aq)$

g. $S_2O_3{}^{2-}(aq) + I_2(aq) \longrightarrow$
$$SO_4{}^{2-}(aq) + I^-(aq)$$

17.36 Complete and balance the following equations, assuming that the oxidation–reduction reactions occur in basic solution.

a. $S^{2-}(aq) + CrO_4{}^{2-}(aq) \longrightarrow$
$$S(s) + Cr(OH)_3(s)$$

b. $MnO_4{}^-(aq) + CN^-(aq) \longrightarrow$
$$MnO_4{}^{2-}(aq) + CNO^-(aq)$$

c. $Al(s) + OH^-(aq) \longrightarrow$
$$Al(OH)_4{}^-(aq) + H_2(g)$$

d. $Br_2(aq) \longrightarrow Br^-(aq) + BrO^-(aq)$

e. $NH_3(aq) + ClO^-(aq) \longrightarrow$
$$N_2H_4(aq) + Cl^-(aq)$$

f. $Cr(OH)_4{}^-(aq) + HO_2{}^-(aq) \longrightarrow$
$$CrO_4{}^{2-}(aq) + H_2O(l)$$

g. $P_4(s) \longrightarrow PH_3(g) + H_2PO_2{}^-(aq)$

Reaction Potentials

17.37 What is a reaction potential or electromotive force?

17.38 What are standard conditions?

17.39 How do we measure the value of the reduction potential for a half-reaction?

17.40 How are the potentials for a reaction and its reverse reaction related?

17.41 What does the sign of a reaction potential mean?

17.42 Calculate the standard reaction potential for each of the following reactions.

a. $2Al(s) + 6H^+(aq) \longrightarrow 2Al^{3+}(aq) + 3H_2(g)$

b. $3Cl_2(aq) + 2Cr^{3+}(aq) + 7H_2O(l) \longrightarrow$
$$6Cl^-(aq) + Cr_2O_7{}^{2-}(aq) + 14H^+(aq)$$

c. $Mg(s) + 2H^+(aq) \longrightarrow Mg^{2+}(aq) + H_2(g)$

d. $5Fe^{2+}(aq) + MnO_4{}^-(aq) + 8H^+(aq) \longrightarrow$
$$5Fe^{3+}(aq) + Mn^{2+}(aq) + 4H_2O(l)$$

e. $Pb(s) + PbO_2(s) + 2H_2SO_4(aq) \longrightarrow$
$$2PbSO_4(s) + 2H_2O(l)$$

f. $2Au^{3+}(aq) + 3Zn(aq) \longrightarrow$
$$2Au(s) + 3Zn^{2+}(aq)$$

17.43 On the basis of standard reaction potentials, decide whether each of the following reactions is spontaneous.

a. $Cr(s) + 2H^+(aq) \longrightarrow Cr^{2+}(aq) + H_2(g)$

b. $3Fe^{3+}(aq) + Al(s) \longrightarrow 3Fe^{2+}(aq) + Al^{3+}(aq)$

c. $6I^-(aq) + BrO_3{}^-(aq) + 6H^+(aq) \longrightarrow$
$$3I_2(aq) + Br^-(aq) + 3H_2O(l)$$

d. $5Br^-(aq) + BrO_3{}^-(aq) + 6H^+(aq) \longrightarrow$
$$3Br_2(aq) + 3H_2O(l)$$

e. $2ClO_2{}^-(aq) \longrightarrow ClO_3{}^-(aq) + ClO^-(aq)$

17.44 The reduction potential, E_{red}°, is given for each of the following half-reactions.

$$Al^{3+}(aq) + 3e^- \longrightarrow Al(s) \qquad E_{red}^\circ = -1.66 \text{ V}$$
$$Zn^{2+}(aq) + 2e^- \longrightarrow Zn(s) \qquad E_{red}^\circ = -0.76 \text{ V}$$
$$Cu^{2+}(aq) + 2e^- \longrightarrow Cu(s) \qquad E_{red}^\circ = +0.34 \text{ V}$$

Will Cu react with either Al^{3+} or Zn^{2+}? If so, write a balanced equation for the reaction.

17.45 Use standard reduction potentials to verify the following statements.

a. Cobalt(III) ion, $Co^{3+}(aq)$, is capable of oxidizing water to $O_2(g)$.

b. Chromous ion, $Cr^{2+}(aq)$, must be protected from atmospheric oxygen to prevent its oxidation to $Cr^{3+}(aq)$.

c. Fluorine (F_2) will oxidize cobalt(II) ion to cobalt(III) ion, but chlorous acid ($HClO_2$) will not.

d. Titanium metal will dissolve in 1 M HCl.

e. Iron(II) salts dissolved in water will not reduce the water in acid solution.

f. Gold can be oxidized to $Au^{3+}(aq)$ by HOCl.

17.46 Given the reduction potentials that follow, predict the product of

a. the reaction between $Fe^{2+}(aq)$ and $MnO_4{}^-(aq)$ in acidic solution

b. the reaction between Zn and $MnO_4{}^-(aq)$ in acidic solution

c. the reaction between Zn and $Fe^{3+}(aq)$ in acidic solution

$$MnO_4{}^-(aq) + e^- \longrightarrow MnO_4{}^{2-}(aq)$$
$$E_{red}^\circ = +0.564 \text{ V}$$

$$MnO_4{}^{2-}(aq) + 4H^+(aq) + 2e^- \longrightarrow$$
$$MnO_2(s) + 2H_2O(l) \qquad E_{red}^\circ = +2.26 \text{ V}$$

$$MnO_2(s) + 4H^+(aq) + e^- \longrightarrow$$
$$Mn^{3+}(aq) + 2H_2O(l) \qquad E_{red}^\circ = +0.95 \text{ V}$$

$$Mn^{3+}(aq) + e^- \longrightarrow Mn^{2+}(aq) \quad E_{red}^\circ = +1.51 \text{ V}$$
$$Mn^{2+}(aq) + 2e^- \longrightarrow Mn(s) \quad E_{red}^\circ = -1.18 \text{ V}$$
$$Zn^{2+}(aq) + 2e^- \longrightarrow Zn(s) \quad E_{red}^\circ = -0.763 \text{ V}$$
$$Fe^{3+}(aq) + e^- \longrightarrow Fe^{2+}(aq) \quad E_{red}^\circ = +0.771 \text{ V}$$
$$Fe^{2+}(aq) + 2e^- \longrightarrow Fe(s) \quad E_{red}^\circ = -0.44 \text{ V}$$

17.47 On the basis of the following reduction potentials in acidic solution, comment on the stability of WO_2 and of $W^{3+}(aq)$ with respect to disproportionation.

$$W_2O_5 + 2H^+(aq) + 2e^- \longrightarrow 2WO_2 + H_2O$$
$$E^\circ_{red} = -0.04 \text{ V}$$

$$WO_2 + 4H^+(aq) + e^- \longrightarrow W^{3+}(aq)$$
$$E^\circ_{red} = -0.15 \text{ V}$$

$$W^{3+}(aq) + 3e^- \longrightarrow W + 2H_2O(l)$$
$$E^\circ_{red} = -0.11 \text{ V}$$

17.48 Considering the following reduction half-reactions and their potentials in acidic and basic solution, predict the stability of the peroxide species with respect to disproportionation. In acidic solution:

$$O_2(g) + 2H^+(aq) + 2e^- \longrightarrow H_2O_2(aq)$$
$$E^\circ_{red} = +0.67 \text{ V}$$

$$H_2O_2(aq) + 2H^+(aq) + 2e^- \longrightarrow 2H_2O(l)$$
$$E^\circ_{red} = +1.77 \text{ V}$$

In basic solution:

$$O_2(g) + H_2O(l) + 2e^- \longrightarrow HO_2^-(aq) + OH^-(aq)$$
$$E^\circ_{red} = -0.08 \text{ V}$$

$$HO_2^-(aq) + H_2O(l) + 2e^- \longrightarrow 3OH^-(aq)$$
$$E^\circ_{red} = +0.88 \text{ V}$$

17.49 Considering the half-reactions and reduction potentials for chromium in acidic and basic solution, predict the stability of the various chromium species with respect to disproportionation. In acidic solution:

$$Cr_2O_7^{2-}(aq) + 14H^+(aq) + 6e^- \longrightarrow$$
$$2Cr^{3+}(aq) + 7H_2O(l) \quad E^\circ_{red} = +1.33 \text{ V}$$

$$Cr^{3+}(aq) + e^- \longrightarrow Cr^{2+}(aq) \quad E^\circ_{red} = -0.41 \text{ V}$$

$$Cr^{2+}(aq) + 2e^- \longrightarrow Cr(s) \quad E^\circ_{red} = -0.91 \text{ V}$$

In basic solution:

$$CrO_4^{2-}(aq) + 4H_2O(l) + 3e^- \longrightarrow$$
$$Cr(OH)_3(s) + 5OH^-(aq) \quad E^\circ_{red} = -0.13 \text{ V}$$

$$Cr(OH)_3(s) + e^- \longrightarrow Cr(OH)_2(s) + OH^-$$
$$E^\circ_{red} = -1.1 \text{ V}$$

$$Cr(OH)_2(s) + 2e^- \longrightarrow Cr(s) + 2OH^-$$
$$E^\circ_{red} = -1.4 \text{ V}$$

17.50 Considering the half-reactions and reduction potentials for sulfur in acidic solution, predict the stability of the various sulfur species with respect to disproportionation.

$$2SO_4^{2-}(aq) + 4H^+(aq) + 2e^- \longrightarrow$$
$$S_2O_6^{2-}(aq) + 2H_2O(l) \quad E^\circ_{red} = -0.22 \text{ V}$$

$$S_2O_6^{2-}(aq) + 4H^+(aq) + 2e^- \longrightarrow$$
$$2H_2SO_3(aq) \quad E^\circ_{red} = +0.57 \text{ V}$$

17.51 Which manganese species, if any, will reduce $H^+(aq)$ to H_2 in 1 M $H^+(aq)$ solution?

$$MnO_4^-(aq) + e^- \longrightarrow MnO_4^{2-}(aq)$$
$$E^\circ_{red} = +0.564 \text{ V}$$

$$MnO_4^{2-}(aq) + 4H^+(aq) + 2e^- \longrightarrow$$
$$MnO_2(s) + 2H_2O(l) \quad E^\circ_{red} = +2.26 \text{ V}$$

$$MnO_2(s) + 4H^+(aq) + e^- \longrightarrow$$
$$Mn^{3+}(aq) + 2H_2O(l) \quad E^\circ_{red} = +0.95 \text{ V}$$

$$Mn^{3+}(aq) + e^- \longrightarrow Mn^{2+}(aq) \quad E^\circ_{red} = +1.51 \text{ V}$$

$$Mn^{2+}(aq) + 2e^- \longrightarrow Mn(s) \quad E^\circ_{red} = -1.18 \text{ V}$$

17.52 Which bismuth species, if any, will reduce $H^+(aq)$ to H_2 in 1 M $H^+(aq)$ solution?

$$BiO_3^-(aq) + 4H^+(aq) + 2e^- \longrightarrow$$
$$BiO^+(aq) + 2H_2O(l) \quad E^\circ_{red} = +1.62 \text{ V}$$

$$BiO^+(aq) + 2H^+(aq) + 3e^- \longrightarrow Bi(s) + H_2O(l)$$
$$E^\circ_{red} = +0.32 \text{ V}$$

17.53 Discuss the stability of the various chromium and manganese species in acidic solution with respect to disproportionation, to oxidation by O_2, and to reduction of $H^+(aq)$ to H_2.

$$Cr_2O_7^{2-}(aq) + 14H^+(aq) + 6e^- \longrightarrow$$
$$2Cr^{3+}(aq) + 7H_2O(l) \quad E^\circ_{red} = +1.33 \text{ V}$$

$$Cr^{3+}(aq) + e^- \longrightarrow Cr^{2+}(aq) \quad E^\circ_{red} = -0.41 \text{ V}$$

$$Cr^{2+}(aq) + 2e^- \longrightarrow Cr(s) \quad E^\circ_{red} = -0.91 \text{ V}$$

$$MnO_4^-(aq) + e^- \longrightarrow MnO_4^{2-}(aq)$$
$$E^\circ_{red} = +0.564 \text{ V}$$

$$MnO_4^{2-}(aq) + 4H^+(aq) + 2e^- \longrightarrow$$
$$MnO_2(s) + 2H_2O(l) \quad E^\circ_{red} = +2.26 \text{ V}$$

$$MnO_2(s) + 4H^+(aq) + e^- \longrightarrow$$
$$Mn^{3+}(aq) + 2H_2O(l) \quad E^\circ_{red} = +0.95 \text{ V}$$

$$Mn^{3+}(aq) + e^- \longrightarrow Mn^{2+}(aq) \quad E^\circ_{red} = +1.51 \text{ V}$$

$$Mn^{2+}(aq) + 2e^- \longrightarrow Mn(s) \quad E^\circ_{red} = -1.18 \text{ V}$$

17.54 What is the product of the reaction between $Cr_2O_7^{2-}(aq)$ and excess chromium metal? The reduction potentials are

$$Cr_2O_7^{2-}(aq) + 14H^+(aq) + 6e^- \longrightarrow$$
$$2Cr^{3+}(aq) + 7H_2O(l) \quad E^\circ_{red} = +1.33 \text{ V}$$

$$Cr^{3+}(aq) + e^- \longrightarrow Cr^{2+}(aq) \quad E^\circ_{red} = -0.41 \text{ V}$$

$$Cr^{2+}(aq) + 2e^- \longrightarrow Cr(s) \quad E^\circ_{red} = -0.91 \text{ V}$$

17.55 Which iodine species, if any, will disproportionate and what will be the products?

$$H_5IO_6(aq) + H^+(aq) + 2e^- \longrightarrow$$
$$IO_3^-(aq) + 3H_2O(l) \quad E^\circ_{red} = +1.70 \text{ V}$$

$$IO_3^-(aq) + 5H^+(aq) + 4e^- \longrightarrow$$
$$HOI(aq) + 2H_2O(l) \quad E^\circ_{red} = +1.14 \text{ V}$$

$2HOI(aq) + 2H^+(aq) + 2e^- \longrightarrow$
$\qquad I_2(aq) + 2H_2O(l) \qquad E^\circ_{red} = +1.45 \text{ V}$

$I_2(aq) + 2e^- \longrightarrow 2I^-(aq) \qquad E^\circ_{red} = +0.54 \text{ V}$

17.56 Consider the following reduction potentials.

$CuO^+(aq) + 2H^+(aq) + e^- \longrightarrow$
$\qquad Cu^{2+}(aq) + H_2O(l) \qquad E^\circ_{red} = +1.80 \text{ V}$

$Cu^{2+}(aq) + e^- \longrightarrow Cu^+(aq) \qquad E^\circ_{red} = +0.15 \text{ V}$

$Cu^+(aq) + e^- \longrightarrow Cu(s) \qquad E^\circ_{red} = +0.52 \text{ V}$

Which copper species, if any, can oxidize water in 1 M $H^+(aq)$ solution to release O_2 gas?

Additional Exercises

17.57 Pick the strongest reducing agent and the weakest reducing agent from the following list: Na^+, Cu, Mg, H_2, F^-

17.58 Balance the following equations, adding $H^+(aq)$, $OH^-(aq)$, and water where necessary:
   a. $Au(s) + CN^-(aq) + O_2(g) \longrightarrow$
   $\qquad Au(CN)_2^-(aq) + OH^-(aq)$
   b. $Fe(CN)_6^{3-}(aq) + Cr_2O_3(s) \longrightarrow$
   $\qquad Fe(CN)_6^{4-}(aq) + CrO_4^{2-}(aq)$
   c. $N_2H_4(l) + Cu(OH)_2(s) \longrightarrow N_2(g) + Cu(s)$

17.59 Identify the oxidizing and reducing agents and the oxidation and reduction half-reactions in each of the following reactions.
   a. $Tl^+(aq) + 2Ce^{4+}(aq) \longrightarrow$
   $\qquad 2Ce^{3+}(aq) + Tl^{3+}(aq)$
   b. $2NO_3^-(aq) + 2Br^-(aq) + 4H^+(aq) \longrightarrow$
   $\qquad Br_2(l) + 2NO_2(g) + 2H_2O(l)$

17.60 a. Balance the following equation, assuming that the reaction occurs in basic solution.

$Cr^{2+}(aq) + H_2O_2(aq) \longrightarrow Cr(OH)_4^-(aq)$

Also identify b. the oxidizing agent, c. the reducing agent, d. the element oxidized, and e. the element reduced.

17.61 Indicate the oxidation number of nitrogen in each of the following compounds.
   a. NH_3 d. NH_2OH
   b. N_2H_4 e. $Fe(NO_3)_3$
   c. NF_3 f. HNO_2

17.62 Balance the following equations, assuming that the reactions occur in basic solution.
   a. $CoCl_2(s) + Na_2O_2(aq) \longrightarrow$
   $\qquad Co(OH)_3(s) + Cl^-(aq) + Na^+(aq)$
   b. $Bi_2O_3(s) + ClO^-(aq) \longrightarrow$
   $\qquad BiO_3^-(aq) + Cl^-(aq)$
   Also identify the oxidizing agent, the reducing agent, the element oxidized, and the element reduced in each reaction.

17.63 Calculate the reaction potential, E°_{rxn}, for the reaction

$3Fe^{2+}(aq) + 2Cr(s) \longrightarrow 3Fe(s) + 2Cr^{3+}(aq)$

17.64 Denitrification occurs when nitrogen in the soil is lost to the atmosphere. One way in which denitrification can occur in acidic soils rich in plant matter is given in the following equation.

$C_6H_{12}O_6(aq) + NO_3^-(aq) \longrightarrow CO_2(g) + N_2(g)$

Balance this equation, adding $H^+(aq)$ and H_2O as necessary.

17.65 Which bismuth species, if any, can be oxidized by oxygen from air in 1 M $H^+(aq)$ solution?

$BiO_3^-(aq) + 4H^+(aq) + 2e^- \longrightarrow$
$\qquad BiO^+(aq) + 2H_2O(l) \qquad E^\circ_{red} = +1.62 \text{ V}$
$BiO^+(aq) + 2H^+(aq) + 3e^- \longrightarrow$
$\qquad Bi(s) + H_2O(l) \qquad E^\circ_{red} = +0.32 \text{ V}$

17.66 Aqua regia is a mixture of concentrated HNO_3 and concentrated HCl. It is often used for cleaning laboratory glassware, and it also dissolves platinum. Balance the equation for the dissolution of platinum metal.

$Pt(s) + NO_3^-(aq) + Cl^-(aq) \longrightarrow$
$\qquad PtCl_4^{2-}(aq) + NO_2(g)$

17.67 When iodine is oxidized in concentrated HNO_3, it produces the white solid iodic acid, HIO_3.

$I_2(s) + NO_3^-(aq) \longrightarrow HIO_3(s) + NO_2(g)$

Balance this equation, adding $H^+(aq)$ and H_2O as necessary.

17.68 Explain why bromide ion can function only as a reducing agent and why permanganate ion can function only as an oxidizing agent.

17.69 a. Balance the following equation, assuming that the reaction occurs in acidic solution.

$NH_4^+(aq) + NO_3^-(aq) \longrightarrow N_2O(g) + H_2O(l)$

Also identify b. the oxidizing agent, c. the reducing agent, d. the element oxidized, and e. the element reduced.

17.70 What is the black image on exposed and developed photographic film?

17.71 Pick the strongest oxidizing agent from the following list: Mg^{2+}, Na, H^+, $Cr_2O_7^{2-}$, I^-

17.72 What is the product of the reaction between $Cr_2O_7^{2-}(aq)$ and $H_2SO_3(aq)$ in acidic solution?

$$Cr_2O_7^{2-}(aq) + 14H^+(aq) + 6e^- \longrightarrow$$
$$2Cr^{3+}(aq) + 7H_2O(l) \qquad E^\circ_{red} = +1.33 \text{ V}$$

$$Cr^{3+}(aq) + e^- \longrightarrow Cr^{2+}(aq) \qquad E^\circ_{red} = -0.41 \text{ V}$$

$$Cr^{2+}(aq) + 2e^- \longrightarrow Cr(s) \qquad E^\circ_{red} = -0.91 \text{ V}$$

$$2SO_4^{2-}(aq) + 4H^+(aq) + 2e^- \longrightarrow$$
$$S_2O_6^{2-}(aq) + 2H_2O(l) \qquad E^\circ_{red} = -0.22 \text{ V}$$

$$S_2O_6^{2-} + 4H^+(aq) + 2e^- \longrightarrow 2H_2SO_3$$
$$E^\circ_{red} = +0.57 \text{ V}$$

* 17.73 Assuming that elements 116 through 120 were synthesized, what would you predict to be the most common oxidation number for each element?

* 17.74 Which of the following substances can act as either an oxidizing agent or a reducing agent, depending on the identity of the other reactant?

    a. $S_2^{2-}(aq)$ f. $H_2SO_4(aq)$
    b. $Al^{3+}(aq)$ g. $Na(s)$
    c. $Br_2(aq)$ h. $H_3AsO_3(aq)$
    d. $C(s)$ i. $H_2(g)$
    e. $Cr^{3+}(aq)$

* 17.75 Balance the following equation, assuming that the reaction occurs in basic solution.

$$CrI_3(s) + Cl_2(g) \longrightarrow$$
$$CrO_4^{2-}(aq) + IO_4^-(aq) + Cl^-(aq)$$

* 17.76 Balance the following equation, assuming that the reaction occurs in acidic solution.

$$FeAsS(s) + HNO_3(aq) \longrightarrow$$
$$Fe^{3+}(aq) + H_3AsO_4(aq) + HSO_4^-(aq) + NO_2(g)$$

* 17.77 People sometimes kill algae in swimming pools by adding $CuSO_4$ as a fungicide. Copper sulfate can be prepared by addition of hot H_2SO_4 to Cu metal.

$$Cu(s) + H_2SO_4(aq) \longrightarrow Cu^{2+}(aq) + SO_2(g)$$

    a. Balance this equation, adding $H^+(aq)$ and H_2O as necessary.
    b. Is this reaction spontaneous at 25°C under standard conditions?
    c. What mass of 98% H_2SO_4 is needed to react with 35.0 g Cu?
    d. What volume of $SO_2(g)$ at STP can be formed from 35.0 g Cu?

▶ ▶ ▶ Chemistry in Practice

17.78 An iron ore contains Fe_2O_3, Fe_3O_4, and inert silicate minerals. To determine whether this ore is an economically feasible source of iron, it is necessary to determine its composition. A 2.1250-g sample of the ore is treated with acid to dissolve the iron oxides. An excess of sodium iodide is added to reduce all the iron ions to Fe^{2+}; the product I_2 is quite soluble in the presence of excess I^-. The solution is diluted to 100.0 mL with water. A 25.00-mL portion of the solution is titrated with a 0.05012 M solution of sodium thiosulfate, $Na_2S_2O_3$, requiring 43.42 mL. This reaction converts the iodine to iodide ion and the thiosulfate ion to tetrathionate ion, $S_4O_6^{2-}$. Another 25.00-mL sample of the solution is mixed with $CCl_4(l)$, which extracts the iodine. The aqueous layer is titrated with $KMnO_4$ solution, requiring 35.95 mL. In its reaction with iron(II), manganese(VII) is converted to manganese(II). Because potassium permanganate solution is not stable, it is standardized by titrating 0.1820 g $Na_2C_2O_4$,

which requires 36.41 mL. This reaction produces $CO_2(g)$ and Mn^{2+}.

    a. Write balanced equations for all the reactions used in these experiments.
    b. What is the concentration of the $KMnO_4$ solution?
    c. What percent by mass of the ore is iron?
    d. What percent by mass of the ore does each of the iron oxides represent?
    e. Fe_3O_4 has a composition equivalent to an equimolar mixture of FeO and Fe_2O_3. Is the formation of Fe_3O_4 from the other two oxides an exothermic or an endothermic process? Is it likely to be spontaneous?
    f. If the ore is to be reduced to metallic iron by reaction with carbon at high temperatures, which is preferable, a composition with more Fe_2O_3 or with more Fe_3O_4?
    g. If the silicates dissolved in acid to form SiO_3^{2-}, would this ion have been reduced by the excess sodium iodide?

Electrochemistry

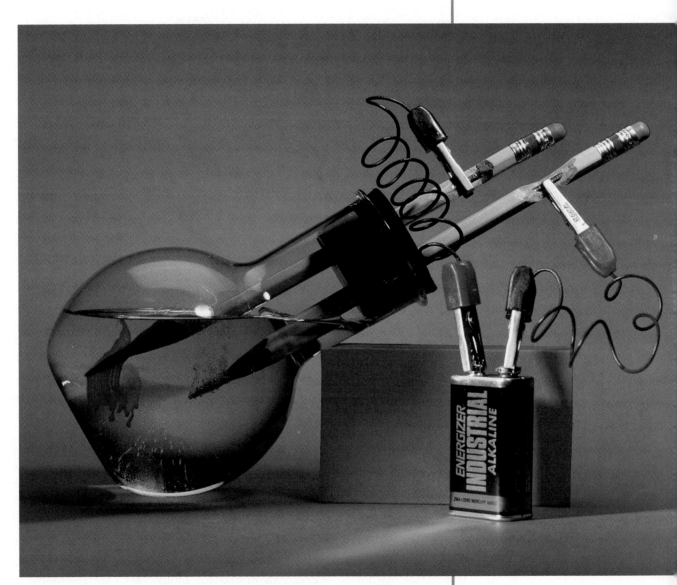

Electrolysis of potassium iodide solution produces elemental iodine (top electrode) and elemental hydrogen (bottom electrode).

CHAPTER OUTLINE

Encountering Chemistry

▶ ▶ ▶ ▶

A white salt, NaI, is dissolved in water, forming a colorless solution. Two pieces of pencil lead, made of graphite and clay, are wired to a battery and placed in this solution. Bubbles of H$_2$(g) are formed at one piece. The solution near the other piece turns brown from the formation of dissolved molecular iodine, I$_2$(aq), and the I$_3^-$(aq) ion formed from iodine and iodide ion. Electricity has been used to produce two elements, I$_2$ and H$_2$.

This process involves electrolysis, the passage of electrical current through a substance, which supplies the energy needed to cause nonspontaneous oxidation–reduction reactions to occur. Sodium iodide in solution does not decompose spontaneously, but it can be forced to decompose through electrolysis, as shown in the following equation:

$$2I^-(aq) + 2H_2O(l) \xrightarrow{\text{electrolysis}} I_2(aq) + H_2(g) + 2OH^-(aq)$$

Electrolysis of salts led to the discovery of several new elements, as well as to new methods of isolating a variety of previously known elements and compounds of these elements. Many elements and compounds are still isolated by this method. Electrolysis is a primary method of producing such important metals as aluminum, magnesium, calcium, sodium, potassium, pure copper, and pure zinc, for example.

This chapter describes the production of electricity by oxidation–reduction reactions in *voltaic cells*. Half-reaction potentials are used to calculate the voltage delivered by a voltaic cell and to predict the spontaneity of oxidation–reduction reactions under various conditions. Then the chapter examines the use of electricity to carry out nonspontaneous oxidation–reduction reactions in *electrolysis cells*. These topics make up the field known as *electrochemistry*.

18.1 Electrochemical Cells

How can chemical energy be converted into electrical energy?

Electrochemistry involves the study of spontaneous oxidation–reduction reactions that generate an electrical current and the study of nonspontaneous oxidation–reduction reactions that are forced to occur by the passage of an electrical current. In both cases, the conversion between chemical and electrical energy is carried out by electrochemical cells. The energy released in a spontaneous oxidation–reduction reaction is transformed into electrical energy in a **voltaic cell,** or **galvanic cell,** a device in which electrons are passed through an external pathway rather than directly between reactants. ▶ Conversely, in an **electrolytic cell,** or **electrolysis cell,** such as that described in the chapter introduction, electrical current causes a nonspontaneous oxidation–reduction reaction to occur. The discussion of electrochemical cells in this section involves primarily voltaic cells, but the same principles apply to electrolytic cells.

To understand the conversion between chemical and electrical energy, consider this simple oxidation–reduction reaction:

$$Zn(s) + Cu^{2+}(aq) \longrightarrow Zn^{2+}(aq) + Cu(s)$$

RULES OF REACTIVITY

In a spontaneous oxidation–reduction reaction, the stronger oxidizing agent reacts with the stronger reducing agent to produce a weaker reducing agent and a weaker oxidizing agent.

The more active metal and stronger reducing agent, zinc, reduces copper ions to deposit the less active metal and weaker reducing agent, copper. In the process, the zinc is oxidized to become zinc ions. The oxidation–reduction reaction has an E°_{rxn} of +1.10 V and therefore occurs spontaneously. As shown in Figure 18.1, it is possible to carry out this process in two ways. In the first case, a bar of zinc metal is simply placed in a solution containing copper ions. During the simple chemical oxidation–reduction reaction that results, the zinc bar begins to dissolve, while reddish copper metal is deposited on the zinc bar. As the bar dissolves, the copper falls to the bottom of the container. The solution initially contains blue copper sulfate, but as the copper ions are reduced to copper metal, the solution loses its blue color. Ultimately the solution contains only colorless zinc and sulfate ions. It is not possible to harness electrical energy from a reaction carried out in this way. Electrons flow directly between the zinc bar and the copper ions in solution, and there is no way to use them to do useful work.

If the flow of electrons between zinc metal and copper ions is to be harnessed, it must be directed through an external circuit, as shown in Figure 18.1. The zinc bar is immersed in a solution of zinc sulfate, and a copper bar in a solution of copper sulfate. The two elements and their compounds are physically isolated from

Figure 18.1 *Left:* Chemical oxidation–reduction processes between zinc metal and copper sulfate solution result in the deposition of copper metal and the dissolution of zinc metal. Electrons flow directly from the zinc metal to the copper ions. *Right:* The electrochemical oxidation–reduction processes give the same result as the chemical oxidation–reduction processes. In this case, however, electron flow is directed through an external circuit, which allows the electricity that is generated to do other useful work.

one another so that no direct oxidation–reduction processes can occur. The oxidation–reduction reaction occurs only if the two metals, zinc and copper, are connected by an external wire and the two solutions are connected by a tube to allow ionic movement. The electron flow can then be harnessed to do electrical work.

The arrangement just described is a voltaic cell. Several characteristics of this cell are common to all electrochemical cells.

• The two half-reactions must be isolated from one another.
• The container in which each half-reaction is carried out must contain a conductive solid that can pass electrons into or out of the system, and the two halves of the system must be electrically connected through an external circuit so that electrons can flow from one half to the other.
• There must be some means of passing ions between the two halves of the system.

We will look more closely at each of these characteristics and then examine the electrical potential of cells.

Isolation of Half-Reactions

The oxidation and reduction half-reactions must occur in separate locations so that the oxidizing agent (Cu^{2+}) and the reducing agent (Zn) never make physical contact. The physical arrangement of each half-reaction is sometimes called a *half-cell.*

Passage of Electrons

A solid conducting material must be present to provide the site for each half-reaction. These materials are called the **electrodes.** The electrode at which oxidation occurs is called the **anode.** The other electrode, the **cathode,** is the site of the reduction half-reaction. In the system described earlier, the electrodes are zinc (the anode) and copper (the cathode). ▶ The electrodes are often reactants or products of the chemical reaction, but they can also be nonreactive parts, as in the electrochemical cell described in the chapter introduction.

In order for electrons to flow from one electrode to another, they must be connected by an external circuit. This circuit may simply be a wire, or it may be a motor or other device that uses the electrical energy produced by the flow of electrons. Because oxidation occurs at the anode, electrons must flow away from this electrode into the external circuit. As a result, the anode has a negative charge from the viewpoint of the external circuit, but removal of electrons gives it a positive charge from the viewpoint of the solution in the cell. Conversely, reduction occurs at the cathode, so electrons flow through the external circuit toward this electrode. It is positive externally and negative internally.

Passage of Ions

Because electrons flow away from the anode, it has a positive internal charge, and anions in the surrounding solution are attracted to it. Conversely, cations are attracted to the cathode. If there were no means to transport ions from one half-cell to the other, ions of a given charge would build up in each half-cell. However, because ionic systems must be electrically neutral, an electrochemical cell must contain some means for ions to pass from one side of the cell to the other. A common method for accomplishing this involves a **salt bridge.** The salt bridge may simply be a porous glass barrier, or it may be a more complex system such as a gelatinous material, like agar, that contains an ionic salt. In either case, the salt bridge allows the passage of ions so the system can remain electrically neutral.

Cell Potentials

The electrical energy developed by a voltaic cell is characterized by the voltage of the cell. A voltage-measuring device, such as a voltmeter, placed in the external electrical circuit measures a voltage difference between the two electrodes. This voltage difference is the electromotive force (emf) discussed in Chapter 17; it is also called the **cell potential.** It is the driving force that pushes electrons through the external circuit. In the Zn/Cu^{2+} example, an emf of $+1.10$ volts is generated at 25°C when the concentrations of Zn^{2+} and Cu^{2+} are 1.00 M. The voltage varies if either of the concentrations or the temperature is changed.

Half-Reaction Potentials   We can treat cell potentials just as we treated overall reaction potentials in Chapter 17. The emf of a cell can be thought of as the sum of two **half-reaction potentials,** that for oxidation and that for reduction. Very often, cell potentials and half-reaction potentials are reported under standard-state conditions. We can express the relationships between standard-state potentials just as we did in Chapter 17:

How can we measure standard reduction potentials?

$$E°_{cell} = E°_{ox} + E°_{red}$$

As mentioned in Chapter 17, values for a half-reaction potential are determined relative to a point of reference—the **standard hydrogen electrode,** shown in Figure 18.2. The reduction half-reaction that occurs at this electrode is

$$2H^+(aq) + 2e^- \longrightarrow H_2(g) \qquad E°_{red} = 0.00 \text{ V}$$

The standard half-reaction potential for this reference reaction is *defined* as zero. To measure any other standard half-reaction potential, we need only measure the standard cell potential involving the standard hydrogen electrode and the other

Figure 18.2 The standard hydrogen electrode consists of a glass envelope to contain hydrogen gas at 1 atm pressure, which is bubbled over a platinum electrode immersed in an acid solution containing 1.00 M H$^+$(aq). Platinum black, PtO$_2$, is coated on the platinum electrode as a catalyst to make the reaction between gaseous hydrogen and aqueous hydrogen ions proceed faster.

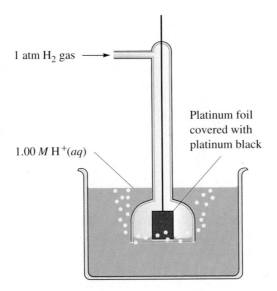

half-reaction. For example, suppose we combine the half-reaction potential for the Zn/Zn^{2+} half-reaction with the standard hydrogen electrode, as shown in Figure 18.3. When these two half-cells are connected, a spontaneous oxidation–reduction reaction occurs, with electrons flowing from the Zn/Zn^{2+} half-cell to the standard hydrogen half-cell, and the measured cell potential is +0.76 V.

$$Zn(s) + 2H^+(aq) \longrightarrow Zn^{2+}(aq) + H_2(g) \qquad E^{\circ}_{cell} = 0.76 \text{ V}$$

As related in Chapter 17, this potential can be used to calculate the standard Zn^{2+}/Zn reduction potential.

$$Zn^{2+}(aq) + 2e^- \longrightarrow Zn(s) \qquad E^{\circ}_{red} = -0.76 \text{ V}$$

The hydrogen electrode is difficult to work with experimentally, so it is not commonly used to measure other half-reaction potentials. When we know the

Figure 18.3 A cell consisting of a Zn/Zn^{2+} half-cell connected to a standard hydrogen half-cell can be used to determine the half-cell potential of the Zn/Zn^{2+} half-reaction. The standard hydrogen electrode is assigned a half-cell potential of exactly zero at 25°C.

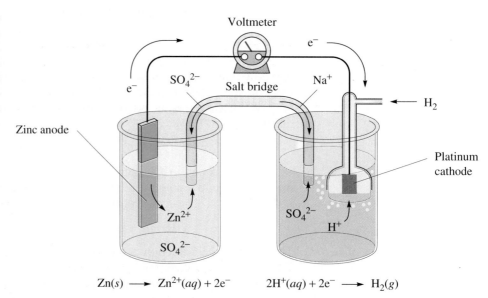

$$Zn(s) \longrightarrow Zn^{2+}(aq) + 2e^- \qquad 2H^+(aq) + 2e^- \longrightarrow H_2(g)$$

Mercury(I) chloride, also known as calomel, has the formula Hg_2Cl_2. It is a white solid that is practically insoluble in water. It decomposes slowly in sunlight into mercury(II) chloride and mercury metal. Mercury(I) chloride is used in standard electrodes, is applied as a fungicide, and is mixed with gold for painting on porcelain.

potential of a half-reaction, we can use it to determine the potentials of other half-reactions. Some stable electrodes that are easily handled experimentally are the silver/silver chloride electrode and the calomel electrode, which use the following reduction half-reactions:

$$AgCl(s) + e^- \longrightarrow Ag(s) + Cl^-(aq) \qquad E°_{red} = +0.220 \text{ V}$$
$$Hg_2Cl_2(s) + 2e^- \longrightarrow 2Hg(l) + 2Cl^-(aq) \qquad E°_{red} = +0.2802 \text{ V}$$

Some typical values of standard reduction potentials measured in electrochemical cells are listed in Table 18.1, and others are given in Table 17.1 and in Appendix G. Following the procedures used in Chapter 17, we can use standard reduction potentials such as the ones in these tables to calculate the standard cell potential for any oxidation–reduction reaction.

18.2   Voltaic Cells

We have seen that a spontaneous oxidation–reduction reaction carried out in a voltaic cell can be used to generate an electrical current at a specific voltage. Voltaic cells are highly useful because of their portability. The ''batteries'' we buy to use in flashlights, radios, and the like are voltaic cells. (Strictly speaking, they are not batteries unless several are used together; a **battery** is a series of voltaic cells.) Cells have the disadvantage of being relatively expensive. Electricity from the cells commonly used in flashlights, for example, may cost 500 to 1000 times more per kilowatt hour than electricity from a power plant.

Voltaic cells used in consumer products generally have some features in common. The anode is usually a metal, and the cathode usually contains a material with a high oxidation number, which can be reduced. The voltaic cell contains an electrolyte—generally an aqueous solution or a moist paste—to allow ions to migrate between the cathode and the anode. Theoretically, any spontaneous oxidation–reduction reaction could be used in a voltaic cell to generate electricity, but adaptation to a rugged, compact unit requires considerable ingenuity. For a cell to be portable, it should have a high energy output per unit of mass.

How can voltaic cells be designed to provide practical sources of electrical energy?

Voltaic cells can be classified as *primary cells* or as *secondary cells;* the latter are commonly known as *rechargeable* cells. Primary cells are especially inefficient energy sources, because the chemicals can only be used once. A rechargeable cell, which can be reused, requires a chemical reaction that is rapid and readily reversible. Many types of primary and secondary cells are available, and others are in various stages of development.

As shown in Table 18.2, voltaic cells can also be classified according to the type of electrolyte system they employ: acidic or alkaline aqueous solution, organic liquid, molten salt, or solid-state electrolyte. The materials separated by a slash in the table are the primary anodic and cathodic materials in the cells. We will examine some of these cells in more detail.

Acidic Dry Cells

A dry cell is a voltaic cell whose contents cannot be spilled—they are in the form of a solid or a paste that does not flow readily. Dry cells can be either acidic or alkaline, depending on the nature of the electrolyte.

Table 18.1 Standard Reduction Potentials at 25°C

Reduction Half-Reaction	E°_{red} (V)
$Li^+(aq) + e^- \longrightarrow Li(s)$	-3.05
$K^+(aq) + e^- \longrightarrow K(s)$	-2.93
$Na^+(aq) + e^- \longrightarrow Na(s)$	-2.71
$Mg^{2+}(aq) + 2e^- \longrightarrow Mg(s)$	-2.37
$Al^{3+}(aq) + 3e^- \longrightarrow Al(s)$	-1.66
$Mn^{2+}(aq) + 2e^- \longrightarrow Mn(s)$	-1.18
$2H_2O(l) + 2e^- \longrightarrow H_2(g) + 2OH^-(aq)$	-0.83
$Zn^{2+}(aq) + 2e^- \longrightarrow Zn(s)$	-0.76
$Cr^{3+}(aq) + 3e^- \longrightarrow Cr(s)$	-0.74
$Fe^{2+}(aq) + 2e^- \longrightarrow Fe(s)$	-0.44
$Cd^{2+}(aq) + 2e^- \longrightarrow Cd(s)$	-0.40
$PbSO_4(s) + 2e^- \longrightarrow Pb(s) + SO_4^{2-}(aq)$	-0.35
$Co^{2+}(aq) + 2e^- \longrightarrow Co(s)$	-0.28
$V^{3+}(aq) + e^- \longrightarrow V^{2+}(aq)$	-0.26
$Ni^{2+}(aq) + 2e^- \longrightarrow Ni(s)$	-0.25
$Pb^{2+}(aq) + 2e^- \longrightarrow Pb(s)$	-0.13
$2H^+(aq) + 2e^- \longrightarrow H_2(g)$	0.00
$Sn^{4+}(aq) + 2e^- \longrightarrow Sn^{2+}(aq)$	$+0.15$
$SO_4^{2-}(aq) + 4H^+(aq) + 2e^- \longrightarrow H_2SO_3(aq) + H_2O(l)$	$+0.17$
$AgCl(s) + e^- \longrightarrow Ag(s) + Cl^-(aq)$	$+0.22$
$Cu^{2+}(aq) + 2e^- \longrightarrow Cu(s)$	$+0.34$
$VO^{2+}(aq) + 2H^+(aq) + e^- \longrightarrow V^{3+}(aq) + H_2O(l)$	$+0.34$
$Cu^+(aq) + e^- \longrightarrow Cu(s)$	$+0.52$
$I_2(s) + 2e^- \longrightarrow 2I^-(aq)$	$+0.54$
$O_2(g) + 2H^+(aq) + 2e^- \longrightarrow H_2O_2(aq)$	$+0.68$
$Fe^{3+}(aq) + e^- \longrightarrow Fe^{2+}(aq)$	$+0.77$
$Ag^+(aq) + e^- \longrightarrow Ag(s)$	$+0.80$
$2Hg^{2+}(aq) + 2e^- \longrightarrow Hg_2^{2+}(aq)$	$+0.91$
$VO_2^+(aq) + 2H^+(aq) + e^- \longrightarrow VO^{2+}(aq) + H_2O(l)$	$+1.00$
$Br_2(l) + 2e^- \longrightarrow 2Br^-(aq)$	$+1.07$
$IO_3^-(aq) + 6H^+(aq) + 6e^- \longrightarrow I^-(aq) + 3H_2O(l)$	$+1.09$
$MnO_2(s) + 4H^+(aq) + 2e^- \longrightarrow Mn^{2+}(aq) + 2H_2O(l)$	$+1.23$
$O_2(g) + 4H^+(aq) + 4e^- \longrightarrow 2H_2O(l)$	$+1.23$
$Cr_2O_7^{2-}(aq) + 14H^+(aq) + 6e^- \longrightarrow 2Cr^{3+}(aq) + 7H_2O(l)$	$+1.33$
$Cl_2(g) + 2e^- \longrightarrow 2Cl^-(aq)$	$+1.36$
$BrO_3^-(aq) + 6H^+(aq) + 6e^- \longrightarrow Br^-(aq) + 3H_2O(l)$	$+1.44$
$Ce^{4+}(aq) + e^- \longrightarrow Ce^{3+}(aq)$	$+1.61$
$PbO_2(s) + SO_4^{2-}(aq) + 4H^+(aq) + 2e^- \longrightarrow PbSO_4(s) + 2H_2O(l)$	$+1.69$
$F_2(g) + 2e^- \longrightarrow 2F^-(aq)$	$+2.87$

Table 18.2 Some Voltaic Cells*

Aqueous Solution		Organic Liquid	Molten Salt	Solid State
Acidic	**Alkaline**			
Mg/AgCl	Al/air	Doped $(CH)_x$	$Li+Al/FeS_2$	Ca/NiF_2
Mg/MnO_2	Cd/Ag_2O	Li/Br_2	Li/Cl_2	Cu/TiS_2
Pb/PbO_2	Cd/air	$Li/NbSe_2$	Na/S	Li/TiS_2
Zn/Br_2	Fe/air	Li/S	$Na/SbCl_3$	
Zn/Cl_2	Li/air	Li/TiO_2		
Zn/MnO_2	Ni/Cd	Li/TiS_2		
	Ni/Fe			
	Ni/H_2			
	Ni/Zn			
	Zn/Ag_2O			
	Zn/air			

*The materials separated by a slash in this table are the primary anodic and cathodic materials in the cells.

The voltaic cell frequently used in flashlights, radios, and toys is an acidic dry cell (also known as the Leclanché cell, after the man who invented it in 1866). This dry cell, shown in Figure 18.4, uses a zinc container as the anode, where the zinc is oxidized.

$$Zn(s) \longrightarrow Zn^{2+}(aq) + 2e^- \qquad E° = 0.763 \text{ V}$$

A carbon rod, surrounded by a mixture of manganese dioxide and graphite, serves as the cathode. The manganese dioxide is an oxidizing agent, and the graphite is used to increase the conductivity. The electrolyte consists of an aqueous NH_4Cl–$ZnCl_2$ solution made into a paste with flour or starch. The ammonium chloride acts as a weak acid in this system. The cathode reaction is not known exactly but

Figure 18.4 The Leclanché dry cell.

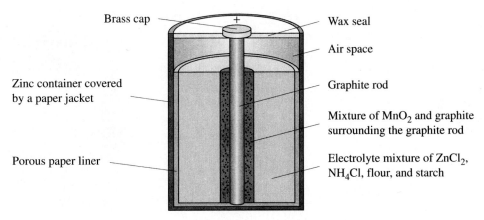

Brass cap

+

Wax seal

Air space

Zinc container covered by a paper jacket

Graphite rod

Mixture of MnO_2 and graphite surrounding the graphite rod

Porous paper liner

Electrolyte mixture of $ZnCl_2$, NH_4Cl, flour, and starch

appears to involve reduction of Mn(IV) to Mn(III), as in the following half-reaction:

$$2MnO_2(s) + 2NH_4^+(aq) + 2H_2O(l) + 2e^- \longrightarrow$$
$$2NH_4OH(aq) + 2MnO(OH)(s) \qquad E° = 0.5 \text{ V}$$

Manganese(IV) oxide, commonly known as manganese dioxide, is a black solid that occurs in nature as the mineral pyrolusite. It is insoluble in water but reacts slowly with hydrochloric acid, forming molecular chlorine and aqueous $MnCl_2$. It acts as a catalyst for the decomposition of hydrogen peroxide into molecular oxygen and water. Manganese dioxide is used in the manufacture of manganese steel, as an oxidizing agent, in dry cells, and for making amethyst glass.

The combination of these two half-reactions gives the following overall reaction:

$$Zn(s) + 2MnO_2(s) + 2NH_4^+(aq) + 2H_2O(l) \longrightarrow$$
$$Zn^{2+}(aq) + 2NH_4OH(aq) + 2MnO(OH)(s) \qquad E° = 1.26 \text{ V}$$

The voltage calculated from the half-cell potentials is that predicted for the standard state. The actual voltage is 1.5 V. These two values do not agree because the cell contains substances with concentrations higher than standard-state concentrations.

The Leclanché dry cell is a primary cell characterized by a short life, high cost, and low production of energy per unit of mass (0.066 kwh/kg). The zinc container is easily corroded. It is not unusual for an appliance, such as a flashlight, to be ruined when the zinc corrodes and the contents of the cell leak out.

Variations on the Leclanché dry cell include the magnesium dry cell, which is constructed much like the Leclanché cell, except that magnesium replaces the zinc and the electrolyte is $MgBr_2$ or $Mg(ClO_4)_2$. Because magnesium is lighter than zinc, magnesium dry cells produce more energy per unit mass (0.088 kwh/kg). Magnesium dry cells also can be stored longer than zinc dry cells.

Alkaline Dry Cells

Why is the zinc container not corroded so readily by OH^- as by NH_4^+?

In alkaline dry cells (Figure 18.5), potassium hydroxide is used as the electrolyte. The zinc container is not corroded so readily by OH^- as by NH_4^+. The anode is surrounded by a paste of zinc mixed with potassium hydroxide. The chemical reactions are similar to those in the Leclanché dry cell. The standard cell potential for an alkaline dry cell is 1.54 V. The alkaline dry cell can be stored longer than the Leclanché dry cell and can be recharged, although not indefinitely.

Figure 18.5 The alkaline dry cell.

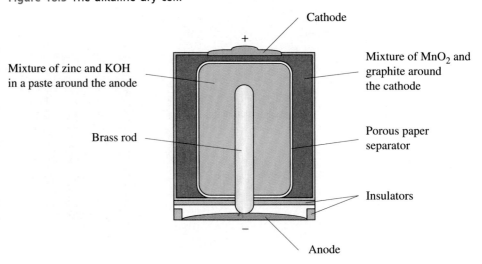

Lead Storage Battery: Pb/PbO₂

The lead storage battery—extensively used in automobiles for ignition, lights, and so on—is the most common of the secondary, or rechargeable, voltaic cell systems. A lead storage battery is shown in Figure 18.6. Because of the high density of lead, this battery delivers relatively low energy per unit mass (0.022 kwh/kg). The normal 12-V storage battery consists of six cells, each containing anodes made of lead and cathodes made of lead coated with lead dioxide. Partially diluted sulfuric acid is used as the electrolyte. As the cells discharge, the anodes and cathodes become coated with insoluble lead sulfate. The reaction at the cathode is

$$PbO_2(s) + 4H^+(aq) + SO_4^{2-}(aq) + 2e^- \longrightarrow PbSO_4(s) + 2H_2O(l) \qquad E° = 1.69 \text{ V}$$

At the anode, the reaction is

$$Pb(s) + SO_4^{2-}(aq) \longrightarrow PbSO_4(s) + 2e^- \qquad E° = 0.35 \text{ V}$$

The overall reaction is

$$Pb(s) + PbO_2(s) + 2SO_4^{2-} + 4H^+(aq) \longrightarrow 2PbSO_4(s) + 2H_2O(l) \qquad E° = 2.04 \text{ V}$$

At the sulfuric acid concentration normally used (6 M), this cell delivers 2.0 V. As the cell reactions occur, the sulfuric acid concentration decreases. Auto mechanics measure the density of the battery fluid (the sulfuric acid solution) with a hydrometer to determine how much sulfuric acid is present and thus how much energy can still be delivered.

Lead storage batteries can be recharged by an electrical current, which is passed through the battery to drive the net chemical reaction backwards.

$$2PbSO_4(s) + 2H_2O(l) \longrightarrow Pb(s) + PbO_2(s) + 2SO_4^{2-}(aq) + 4H^+(aq)$$

In an automobile, the electrical current is supplied by a generator or an alternator driven by the engine. The battery can be recharged only if the lead sulfate and the lead dioxide adhere to the lead electrodes; if they do not, they cannot be involved in the oxidation–reduction reaction. If the battery is recharged too rapidly, water in the cell can be dissociated into hydrogen and oxygen gases, which may dislodge these lead compounds from the electrodes, thereby shortening the life of the battery.

Lead dioxide, PbO₂, is a dark brown powder that decomposes when heated, forming PbO and O₂. It is insoluble in water but reacts with hydrochloric acid to form Cl₂. It is used as an oxidizing agent, in pyrotechnics, in the manufacture of pigments, and in lead storage batteries. Lead sulfate, PbSO₄, is an insoluble white solid. It is used as a white pigment and in lead storage batteries.

Figure 18.6 The lead storage battery.

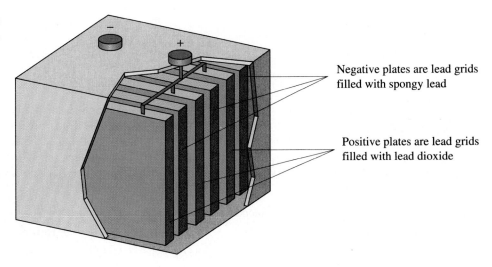

Negative plates are lead grids filled with spongy lead

Positive plates are lead grids filled with lead dioxide

NiCad Cell: Cd/NiO$_2$

Cadmium hydroxide, Cd(OH)$_2$, is a colorless solid that dehydrates when heated to form brown cadmium oxide. It is insoluble in water but dissolves in dilute acids. It is used in voltaic cells. Nickel(IV) oxide is a black, insoluble solid that is a strong oxidizing agent. It decomposes when heated to form green NiO.

The nickel–cadmium secondary cell, known as a NiCad cell, is widely used in portable rechargeable power tools and appliances such as drills, hedge clippers, shavers, and toothbrushes. The anode is cadmium, the cathode is nickel coated with nickel oxides, and the electrolyte is concentrated potassium hydroxide solution. The cell is sealed to prevent leakage of potassium hydroxide. The NiCad cell has a long life, produces a reasonable amount of energy per unit mass (0.033 kwh/kg), and can be recharged readily because the reaction products adhere to the electrodes. The electrode reactions are

Anode: $Cd(s) + 2OH^- \longrightarrow Cd(OH)_2 + 2e^-$ $E° = 0.81$ V

Cathode: $NiO_2(s) + 2H_2O + 2e^- \longrightarrow Ni(OH)_2(s) + 2OH^-$ $E° = 0.49$ V

The overall reaction is

$Cd(s) + NiO_2(s) + 2H_2O \longrightarrow Cd(OH)_2 + 2Ni(OH)_2(s)$ $E° = 1.30$ V

Zinc/Silver Oxide Cell

Silver(II) oxide, AgO, is a charcoal-gray solid that is a very strong oxidizing agent. It is insoluble in water but reacts with dilute acids to evolve O$_2$ and with dilute NH$_4$OH to evolve N$_2$. It dissolves in concentrated acid to form intensely colored solutions— brown in HNO$_3$ and olive green in H$_2$SO$_4$. Silver(I) oxide, Ag$_2$O, is a brownish-black solid that decomposes into its constituent elements when heated or exposed to sunlight. It is insoluble in water but dissolves in acids. It is used as a catalyst, in glass to obtain a yellow color, and in the purification of water.

The Zn/AgO cell is very expensive. However, it delivers a high amount of energy per unit of mass (0.18 kwh/kg), so it is particularly useful in applications where weight is a problem, such as in satellite power systems. The cell is also used in cameras, hearing aids, and watches because of its light weight. The cell can be sealed reliably to prevent loss of the 20–40% KOH electrolyte, so it is safe to use in these devices, which would be ruined if the cell leaked. The reactions are

Anode: $Zn(s) + 2OH^- \longrightarrow Zn(OH)_2 + 2e^-$ $E° = 1.25$ V

Cathode: $2AgO(s) + H_2O + 2e^- \longrightarrow Ag_2O(s) + 2OH^-$ $E° = 0.61$ V

Overall: $Zn(s) + 2AgO(s) + H_2O \longrightarrow Zn(OH)_2 + Ag_2O(s)$ $E° = 1.86$ V

After the silver(II) oxide has been completely consumed, the cell can continue to function by reduction of Ag$_2$O to Ag, but at a lower cell potential.

$Zn(s) + Ag_2O(s) + H_2O \longrightarrow Zn(OH)_2 + 2Ag(s)$ $E° = 1.58$ V

These cells are rechargeable as long as they have not been discharged by more than about 50% of capacity.

Mercury Cell: Zn/HgO

Mercury(II) oxide, HgO, exists in a red and a yellow solid form. The yellow form converts to the red form upon heating. Both forms are insoluble in water but dissolve in acids. When heated sufficiently, HgO decomposes into its constituent elements. Mercury(II) oxide is used in marine paints, as a catalyst in organic reactions, and for diluting pigments used to paint on porcelain.

The Zn/HgO cell, shown in Figure 18.7, is widely used where high reliability and long life are required; typical applications are digital watches and pacemakers. This is another of the alkaline cells that is sealed to prevent loss of the KOH electrolyte. The anode (amalgamated zinc mixed with KOH) and the cathode [mercury(II) oxide mixed with graphite] are both compressed powders. The reactions are

Anode: $Zn(s) + 2OH^- \longrightarrow Zn(OH)_2 + 2e^-$ $E° = 1.25$ V

Cathode: $HgO(s) + H_2O + 2e^- \longrightarrow Hg(l) + 2OH^-$ $E° = 0.098$ V

Overall: $Zn(s) + HgO(s) + H_2O \longrightarrow Zn(OH)_2 + Hg(l)$ $E° = 1.35$ V

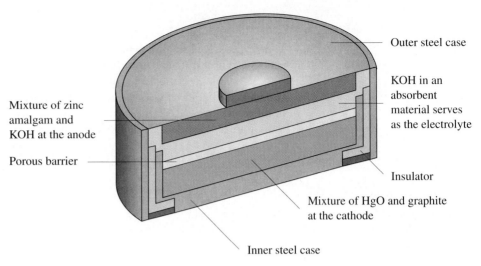

Outer steel case

KOH in an absorbent material serves as the electrolyte

Mixture of zinc amalgam and KOH at the anode

Porous barrier

Insulator

Mixture of HgO and graphite at the cathode

Inner steel case

Figure 18.7 The mercury cell.

High-Temperature Batteries

Much research and development work has been devoted to creating electric automobiles that can compete with gasoline-powered vehicles in expense and driving range. Central to this development is the availability of appropriate batteries to power the automobiles. The battery must be easily and quickly rechargeable, must be light, and must deliver enough energy to drive the car about 300 miles without recharging. Lead storage batteries are not appropriate; it would take 500 lb of lead storage batteries to provide as much energy as 1 gallon of gasoline!

If the electricity for charging batteries continues to be generated by the burning of fossil fuels, will this advancement really solve our energy problems? Will it reduce air pollution?

A great deal of research has centered on the use of very light metals—particularly alkali metals—in batteries. It is necessary to use nonaqueous electrolytes with these metals because they react with water, so solid or molten-salt electrolytes at high temperatures are generally used. One such battery under development is the sodium/sulfur battery, which uses beta-alumina ($NaAl_{11}O_{17}$) as the electrolyte and operates at 300°C. The reactions are

$$\text{Anode:} \quad Na \longrightarrow Na^+ + e^-$$
$$\text{Cathode:} \quad 3Na_2S_5 + 4Na^+ + 4e^- \longrightarrow 5Na_2S_3$$
$$\text{Overall:} \quad 3Na_2S_5 + 4Na \longrightarrow 5Na_2S_3 \qquad E° = 2.0 \text{ V}$$

The similar lithium/sulfur battery operates at 330°C and delivers 2.2 V. These systems are troubled with corrosion problems. Furthermore, it is difficult to maintain the required high temperatures.

Lithium is a soft, silvery-white metal, the lightest of the alkali metals. It becomes yellowish upon exposure to moist air. It reacts with water to form the hydroxide and release H_2 gas. It is used in alloys, greases, and missile fuels and as a catalyst.

Similar problems plague the lithium/chlorine system, which operates at 650°C. Molten lithium chloride is used as the electrolyte in this battery. The reactions are

$$\text{Anode:} \quad Li \longrightarrow Li^+ + e^-$$
$$\text{Cathode:} \quad Cl_2 + 2e^- \longrightarrow 2Cl^-$$
$$\text{Overall:} \quad 2Li + Cl_2 \longrightarrow 2Li^+ + 2Cl^- \qquad E° = 3.5 \text{ V}$$

Air Batteries

Another approach to the production of lightweight batteries for electric automobiles is the use of air as the cathode material. Zinc has generally formed the anode in these batteries, but aluminum, lithium, iron, and cadmium have also been used. The cathode is made of a porous ceramic material containing a Pt catalyst. Air is admitted to the system by convection or is blown in by fans. The electrolyte is potassium hydroxide. The chemical reactions are

$$
\begin{array}{lll}
\text{Anode:} & \text{Zn}(s) + 2\text{OH}^- \longrightarrow \text{Zn(OH)}_2 + 2e^- & E° = 1.25 \text{ V} \\
\text{Cathode:} & \text{O}_2 + 2\text{H}_2\text{O} + 4e^- \longrightarrow 4\text{OH}^- & E° = 0.40 \text{ V} \\
\text{Overall:} & 2\text{Zn} + \text{O}_2 + 2\text{H}_2\text{O} \longrightarrow 2\text{Zn(OH)}_2 & E° = 1.65 \text{ V}
\end{array}
$$

Zinc hydroxide, $Zn(OH)_2$, is an amphoteric, colorless solid. It dissolves in acidic or basic aqueous solutions. Upon heating, it is converted to white zinc oxide, which is used as a white pigment in paints and cosmetics.

The voltage generated under actual operating conditions is about 1.4 V. Zinc/air batteries are available for use in hearing aids, replacing mercury batteries or silver batteries.

Fuel Cells

In fuel cells, both reactants are stored outside the cell and are introduced as needed at the electrodes. Fuel cells are light, so they are widely used in space travel where weight is a primary concern. As shown in Figure 18.8, the electrodes are porous graphite (or metal) plates containing appropriate metallic catalysts separated by a concentrated potassium hydroxide electrolyte. In the most common fuel cell, hydrogen is the fuel and the other reactant is oxygen. The reactions are

$$
\begin{array}{lll}
\text{Anode:} & \text{H}_2 + 2\text{OH}^- \longrightarrow 2\text{H}_2\text{O} + 2e^- & E° = 0.83 \text{ V} \\
\text{Cathode:} & \text{O}_2 + 2\text{H}_2\text{O} + 4e^- \longrightarrow 4\text{OH}^- & E° = 0.40 \text{ V} \\
\text{Overall:} & 2\text{H}_2 + \text{O}_2 \longrightarrow 2\text{H}_2\text{O} & E° = 1.23 \text{ V}
\end{array}
$$

The cell is generally only about 60% efficient and generates about 0.9–1.0 V. Other fuels, such as CH_4 or ammonia, can also be used.

Figure 18.8 A hydrogen/oxygen fuel cell.

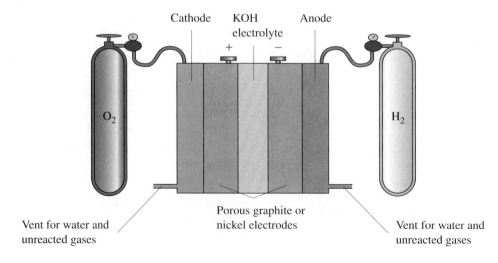

▶ ▶ ▶ **WORLDS OF CHEMISTRY**

Rusting: Buildings at Risk

Replacing materials that have corroded costs billions of dollars annually, so the protection of metals from corrosion is extremely important.

Corrosion is the deterioration of metals from their interaction with components of the environment. It involves electrochemical processes in which an electrochemical cell is set up on the surface of the metal. Familiar signs of corrosion include the red stains of rust, $Fe_2O_3 \cdot xH_2O$, on iron and steel; the green coating, $Cu_2(OH)_2CO_3$, on copper and brass; and the black coating, Ag_2S, on silver. Replacing materials that have corroded costs billions of dollars annually, so the protection of metals from corrosion is extremely important.

Rusting of iron is an especially harmful corrosion process that involves both water and atmospheric oxygen. The rate of rusting can also be increased by stresses in the metal, salts in the water, and low pH.

When iron rusts, a part of the metal acts as an anode, where an oxidation process occurs.

$$Fe \longrightarrow Fe^{2+} + 2e^-$$

A reduction process occurs simultaneously in another location, which acts as a cathode, with molecular oxygen the oxidizing agent.

$$O_2 + 2H_2O + 4e^- \longrightarrow 4OH^-$$

Electrons released by the oxidation process migrate through the metal to the site where the oxygen atoms in molecular oxygen gain electrons. In the presence of water, the oxidized metal ions migrate to the cathodic site and combine with hydroxide ions to form insoluble hydroxide salts, which react further with oxygen to form iron(III). Rust flakes off the iron and exposes fresh metal, as shown in Figure 18.9, so iron continues to corrode, ultimately suffering serious damage.

Many methods that are used to protect iron from corrosion involve setting up a less destructive electrochemical cell. These methods include cathodic protection and plating.

When two metals are placed in contact in a corrosive environment, the more active of the two metals corrodes and the other does not. This principle forms the basis for the cathodic protection of iron. Iron pipes or tanks buried underground are often connected to a more active metal, such as magnesium or zinc (Figure 18.10). The trans-Alaska pipeline, for example, was connected to buried strips of

Figure 18.9 Iron corrodes badly because the iron oxide coating flakes off, exposing fresh metal that can be oxidized.

zinc cable for corrosion protection. The zinc forms the anode, where oxidation occurs.

$$Zn \longrightarrow Zn^{2+} + 2e^-$$

This converts the iron to the cathode, where reduction of oxygen occurs.

$$O_2 + 2H_2O + 4e^- \longrightarrow 4OH^-$$

The iron acts only as a conductor of electrons and does not take part in either the oxidation half-reaction or the reduction half-reaction. Thus the iron stays completely uncorroded as long as the zinc remains. It is simpler to replace the zinc cable with a fresh piece than to dig up and replace the pipe. Magnesium is often used to protect buried gasoline tanks and ships' hulls.

A variation on cathodic protection involves plating the iron with a coating of a more active metal. Galvanized iron has a coating of zinc, for example. Again, it is the zinc that corrodes, because zinc becomes the anode in an electrochemical cell, and iron becomes the cathode, where reduction of molecular oxygen occurs. Oxidation of zinc occurs slowly, because a protective coating of basic zinc carbonate, $Zn_2(OH)_2CO_3$, forms on the zinc. Even if some iron is exposed by a scratch or gouge, it does not corrode because the zinc reacts instead.

Another type of plating involves covering a more active metal with a less active metal that corrodes only very slowly. Iron is often covered with tin, as in a tin can, for

(continued)

Figure 18.10 Cathodic protection may involve attaching a more active metal to an iron pipe, thus preventing the iron from participating in oxidation processes. In the trans-Alaska pipeline, a buried zinc cable provides the cathodic protection.

instance. The low activity of the tin prevents corrosion. However, if the tin coating is scratched to expose some iron, the iron corrodes very rapidly at the exposed position, and the tin soon flakes off, resulting in even more extensive corrosion.

Questions for Discussion

1. If rusting is such a major problem, why is iron so widely used as a structural metal? What other metals might be used instead?
2. Steel is sometimes rustproofed by dipping it into a solution containing phosphoric acid and small amounts of zinc or manganese ions. Propose a hypothesis to explain why this treatment prevents rusting.
3. Some metal parts on automobiles are being replaced by plastic parts. What advantages does this offer? What are the disadvantages?
4. Iron dissolves in dilute nitric acid, but it becomes "passive" when placed in concentrated nitric acid. Propose an explanation.

Plastic Batteries

One of the most recent developments in the search for a lightweight rechargeable battery is the use of thin, stretchable films of plastics such as polyacetylene. Acetylene, $HC\equiv CH$, can react with itself to form long-chain hydrocarbons.

$$n \ HC\equiv CH \longrightarrow \ (HC=CH)_n$$

These films can be partially oxidized or reduced to give them some charge. The films then have a shiny surface not unlike that of a metal in appearance and an electrical conductivity in the range typical of metals. When two films of opposite charge are immersed in a solution of lithium perchlorate dissolved in an organic solvent and are connected together in a cell, an electrical current is produced. Voltages of about 0.5 V have been produced. Such a plastic battery can be recharged and reused almost indefinitely. Because polyacetylene has a very low density, these batteries are very light.

18.3 Thermodynamics of Oxidation–Reduction Reactions

We have been using the value of the cell potential, $E°$, as a measure of the spontaneity of an oxidation–reduction reaction. The more positive the value of $E°$, the greater the tendency for the reaction to occur spontaneously. A negative value of $E°$ means the reaction will not occur spontaneously. In earlier chapters, we devel-

How are the various criteria for spontaneity of redox reactions related?

oped criteria for spontaneity based on thermodynamics—the free-energy change and the equilibrium constant. This section examines the connection between these criteria for spontaneity and shows how free-energy changes lead to a method for predicting spontaneity under nonstandard conditions.

Recall from Chapter 12 that the free-energy change, ΔG, for a spontaneous process is negative. The free-energy change for an oxidation–reduction reaction (or for a half-reaction) can be related to the cell potential, E, by the following equation, because electrical energy is equivalent to electrical charge multiplied by voltage.

$$\Delta G = -n\mathscr{F}E$$

Under standard-state conditions, the equation is

$$\Delta G° = -n\mathscr{F}E°$$

In these equations, n is the number of moles of electrons transferred in the reaction, and $\mathscr{F}$ is the **Faraday constant,** which is equal to the charge on 1 mol of electrons.

$$1\ \mathscr{F} = 9.650 \times 10^4\ \text{coul/mol e}^- = 9.650 \times 10^4\ \text{J/V mol e}^-$$

Note that a positive cell potential corresponds to a negative-free energy change; both are criteria for a spontaneous reaction.

EXAMPLE 18.1

Calculate the standard free-energy change associated with the half-reaction

$$Cu^{2+}(aq) + 2e^- \longrightarrow Cu(s) \quad E°_{red} = 0.34\ \text{V}$$

Solution The standard free-energy change is related to the standard half-cell potential by the equation

$$\Delta G° = -n\mathscr{F}E°$$

In this half-reaction, two electrons are transferred for each copper ion reduced, so $n = 2$ mol. The value of $\mathscr{F}$ is 9.650×10^4 J/V mol. Inserting values gives

$$\Delta G° = -2\ \text{mol} \times 9.650 \times 10^4\ \text{J/V mol} \times 0.34\ \text{V} = -6.6 \times 10^4\ \text{J, or } -66\ \text{kJ}$$

Practice Problem 18.1

Calculate the standard free-energy change associated with the half-reaction

$$Li^+(aq) + e^- \longrightarrow Li(s) \quad E°_{red} = -3.05\ \text{V}$$

Answer: 294 kJ

Free Energy and the Additivity of Half-Reaction Potentials

In Chapter 12 we saw that when two reactions are added, the free-energy change for the new reaction equals the sum of the free-energy changes for the component reactions. This relationship holds for half-reactions as well, and it is the reason we

Chromium(III) ion is blue-gray and stable in aqueous solution. Chromium(II) ion is sky-blue and reacts readily with O_2 in air to form a green chromium(III) ion, $Cr_2(OH)_2^{4+}$.

can add oxidation and reduction half-reaction potentials to obtain the total reaction potential. We can see the relationships by calculating the free-energy changes for the following half-reactions:

$$
\begin{array}{lll}
2Br^-(aq) \longrightarrow Br_2(l) + 2e^- & & E^\circ_{ox} = -1.07 \text{ V} \\
Cl_2(g) + 2e^- \longrightarrow 2Cl^-(aq) & & E^\circ_{red} = 1.36 \text{ V} \\
\hline
2Br^-(aq) + Cl_2(g) \longrightarrow Br_2(l) + 2Cl^-(aq) & & E^\circ_{cell} = 0.29 \text{ V}
\end{array}
$$

The equation $\Delta G^\circ = -n\mathscr{F}E^\circ$ can be used in all the calculations.

$$\Delta G^\circ_{ox} = -2(9.650 \times 10^4)(-1.07 \text{ J}) = 2.06 \times 10^5 \text{ J}$$
$$\Delta G^\circ_{red} = -2(9.650 \times 10^4)(1.36 \text{ J}) = -2.62 \times 10^5 \text{ J}$$
$$\Delta G^\circ_{cell} = -2(9.650 \times 10^4)(0.29 \text{ J}) = -5.6 \times 10^4 \text{ J}$$

Because free-energy change is additive, we can also calculate ΔG° for the reaction as follows:

$$\Delta G^\circ_{cell} = \Delta G^\circ_{ox} + \Delta G^\circ_{red} = 2.06 \times 10^5 \text{ J} - 2.62 \times 10^5 \text{ J} = -5.6 \times 10^4 \text{ J}$$

According to both criteria, this reaction should be spontaneous. We can verify this result experimentally, as shown in Figure 18.11.

EXAMPLE 18.2

Calculate the standard free-energy change for the reaction

$$Fe(s) + F_2(g) \longrightarrow Fe^{2+}(aq) + 2F^-(aq)$$

Solution   First we identify the component half-reactions.

$$
\begin{array}{l}
F_2(g) + 2e^- \longrightarrow 2F^-(aq) \\
Fe(s) \longrightarrow Fe^{2+}(aq) + 2e^- \\
\hline
Fe(s) + F_2(g) \longrightarrow Fe^{2+}(aq) + 2F^-(aq)
\end{array}
$$

Next we find the appropriate reduction potentials in Table 18.1.

$$
\begin{array}{ll}
F_2(g) + 2e^- \longrightarrow 2F^-(aq) & E^\circ_{red} = 2.87 \text{ V} \\
Fe^{2+}(aq) + 2e^- \longrightarrow Fe(s) & E^\circ_{red} = -0.44 \text{ V}
\end{array}
$$

The second reduction half-reaction must be reversed. Then the two half-reaction potentials can be combined to give the overall cell potential.

$$
\begin{array}{lll}
F_2(g) + 2e^- \longrightarrow 2F^-(aq) & & E^\circ_{red} = 2.87 \text{ V} \\
Fe(s) \longrightarrow Fe^{2+}(aq) + 2e^- & & E^\circ_{ox} = 0.44 \text{ V} \\
\hline
Fe(s) + F_2(g) \longrightarrow Fe^{2+}(aq) + 2F^-(aq) & & E^\circ_{cell} = 3.31 \text{ V}
\end{array}
$$

The free-energy change is calculated from the cell potential.

$$\Delta G^\circ = -n\mathscr{F}E^\circ$$
$$\Delta G^\circ = -2 \text{ mol} \times 9.650 \times 10^4 \text{ J/V mol} \times 3.31 \text{ V} = -6.39 \times 10^5 \text{ J, or } -639 \text{ kJ}$$

Practice Problem 18.2

Calculate the standard free-energy change for the reaction

$$Mg(s) + I_2(g) \longrightarrow Mg^{2+}(aq) + 2I^-(aq)$$

Answer: -562 kJ

Figure 18.11 Colorless bromide ion mixed with molecular chlorine is oxidized to red molecular bromine, and Cl_2 is reduced to chloride ion. Chloroform is added as a separate layer in these photos because it extracts the halogens and makes their colors easier to see.

We have been adding two half-reactions to get the overall oxidation–reduction reaction. However, it is also possible to combine two half-reactions to get a new half-reaction. Consider the following half-reactions for chromium, which are found in Appendix G.

$$Cr^{3+}(aq) + e^- \longrightarrow Cr^{2+}(aq) \qquad E^\circ_{red} = -0.41$$
$$Cr^{2+}(aq) + 2e^- \longrightarrow Cr(s) \qquad E^\circ_{red} = -0.91$$
$$Cr^{3+}(aq) + 3e^- \longrightarrow Cr(s) \qquad E^\circ_{red} = -0.74$$

The first two half-reactions can be added to yield a new half-reaction.

$$Cr^{3+}(aq) + e^- + Cr^{2+}(aq) + 2e^- \longrightarrow Cr^{2+}(aq) + Cr(s)$$

Canceling and collecting terms, we get the third half-reaction.

$$Cr^{3+}(aq) + 3e^- \longrightarrow Cr(s)$$

Why are half-reaction potentials additive some times but not other times?

You can see that the standard reduction potentials are not additive for these half-reactions. To see why half-reaction potentials are additive when we combine half-reactions to give an overall reaction, but not when we combine them to give another half-reaction, consider the relationships among the free-energy changes (ΔG°_1 and ΔG°_2) and the potentials (E°_1 and E°_2) for the two half-reactions and the results of their addition (ΔG°_3 and E°_3). Free-energy changes are additive in either case, so we have the following equation:

$$\Delta G^\circ_3 = \Delta G^\circ_1 + \Delta G^\circ_2$$

Substituting the relation $\Delta G^\circ = -n\mathscr{F}E^\circ$ into this equation gives

$$(-n_3\mathscr{F}E^\circ_3) = (-n_1\mathscr{F}E^\circ_1) + (-n_2\mathscr{F}E^\circ_2)$$

Collecting terms yields the following equation for the combined potential.

$$E^\circ_3 = (n_1 E^\circ_1 + n_2 E^\circ_2)/n_3$$

In an overall balanced oxidation–reduction reaction, the number of electrons transferred (n_3) always equals the number of electrons transferred in each of the component half-reactions (n_1 and n_2), so these terms cancel out and the equation simplifies to

$$E_3^\circ = E_1^\circ + E_2^\circ$$

This is exactly the relationship we have been using to combine oxidation and reduction half-reaction potentials to give cell potentials.

When two half-reactions are combined to give another half-reaction, however, the n_1, n_2, and n_3 terms do not cancel out, because the numbers of electrons involved in the reactions are not the same. In this case, we must use the unsimplified equation to find the potential. Consider again the half-reactions for chromium given earlier. We can use the known potentials for the first two half-reactions to calculate the new potential.

$$E_3^\circ = (n_1E_1^\circ + n_2E_2^\circ)/n_3$$
$$= (1 \times -0.41 \text{ V} + 2 \times -0.91 \text{ V})/3 = -0.74 \text{ V}$$

This result agrees with the value in Appendix G.

EXAMPLE 18.3

Calculate the potential for the reduction of thallium(III) to thallium(I). The potentials for reduction of both oxidation states to the metal are known:

$$\text{Tl}^+(aq) + \text{e}^- \longrightarrow \text{Tl}(s) \qquad E^\circ = -0.336 \text{ V}$$
$$\text{Tl}^{3+}(aq) + 3\text{e}^- \longrightarrow \text{Tl}(s) \qquad E^\circ = 0.721 \text{ V}$$

Solution   The reaction of interest is

$$\text{Tl}^{3+}(aq) + 2\text{e}^- \longrightarrow \text{Tl}^+(aq)$$

To obtain this half-reaction, we reverse the first reduction half-reaction.

$$\text{Tl}(s) \longrightarrow \text{Tl}^+(aq) + \text{e}^- \qquad E^\circ = 0.336 \text{ V}$$

We then add the resulting oxidation half-reaction to the second reduction half-reaction.

$$\text{Tl}^{3+}(aq) + 3\text{e}^- \longrightarrow \text{Tl}(s) \qquad E^\circ = 0.721 \text{ V}$$

The result is the half-reaction of interest.

$$\text{Tl}^{3+}(aq) + 2\text{e}^- \longrightarrow \text{Tl}^+(aq)$$

The potential for this half-reaction is given by

$$E_3^\circ = (n_1E_1^\circ + n_2E_2^\circ)/n_3$$
$$= (1 \times 0.336 \text{ V} + 3 \times 0.721 \text{ V})/2 = 1.25 \text{ V}$$

Practice Problem 18.3

Calculate the potential for the reduction of copper(II) to copper(I). The potentials are known for the following reduction half-reactions.

$$\text{Cu}^+(aq) + \text{e}^- \longrightarrow \text{Cu}(s) \qquad E^\circ = 0.52 \text{ V}$$
$$\text{Cu}^{2+}(aq) + 2\text{e}^- \longrightarrow \text{Cu}(s) \qquad E^\circ = 0.34 \text{ V}$$

Answer: 0.16 V

Cell Potential and Equilibrium Constant

Another criterion used to determine whether a reaction should occur spontaneously is the value of the equilibrium constant for the reaction. Large values of the equilibrium constant, K, correspond to a negative value of the standard free-energy change and to a spontaneous reaction in the forward direction. These quantities are related by the equation

$$\Delta G° = -RT \ln K = -2.303RT \log K$$

Because the free-energy change is also related to the cell potential, a relationship between the cell potential and the equilibrium constant can be developed.

$$\Delta G° = -n\mathscr{F}E°$$
$$-n\mathscr{F}E° = -2.303RT \log K$$
$$E° = (2.303RT/n\mathscr{F}) \log K$$

The more positive the value of the cell potential, the larger the value of the equilibrium constant.

It is convenient to evaluate the collection of constants in this equation for the common condition of 25°C.

$$T = 273.2 + 25.0°C = 298.2 \text{ K}$$
$$R = 8.314 \text{ J/mol K}$$
$$\mathscr{F} = 9.650 \times 10^4 \text{ J/V mol e}^-$$
$$2.303 \, RT/\mathscr{F} = 2.303 \times 8.314 \times 298.2/9.650 \times 10^4 = 0.05916$$

The following equation, then, is valid for 25.0°C:

$$E° = (0.05916/n) \log K$$

This equation enables us to use the standard cell potential for an oxidation–reduction reaction to derive the equilibrium constant.

EXAMPLE 18.4

Calculate the equilibrum constant at 25°C for the reaction

$$3Tl^+(aq) \longrightarrow Tl^{3+}(aq) + 2Tl(s) \qquad E° = -1.59 \text{ V}$$

Solution   In this equation two electrons are transferred, because Tl^+ loses two electrons to form Tl^{3+} and each of the two Tl^+ ions gains an electron for a total of two electrons to form $2Tl$. It is necessary only to insert values into the equation.

$$E° = \frac{0.05916}{n} \log K$$

$$-1.59 = \frac{0.05916}{2} \log K$$

This equation can be rearranged:

$$\log K = -1.59 \times \frac{2}{0.05916} = -53.8$$

and then solved for the equilibrium constant:

$$K = 10^{-53.8} = 2 \times 10^{-54}$$

(continued)

Practice Problem 18.4

Calculate the equilibrium constant at 25°C for the reaction

$$Cu^{2+}(aq) + Cu(s) \longrightarrow 2Cu^{+}(aq) \qquad E° = -0.36 \text{ V}$$

Answer: 8×10^{-7}

Nonstandard Conditions and the Nernst Equation

We have been assuming in our calculations that standard-state conditions applied, but cells are generally not operated under such conditions. To correct for nonstandard concentrations of dissolved materials or nonstandard pressures of gases, we can use for the free-energy change a relationship developed from thermodynamics (see Section 12.7).

What effect does concentration or pressure have on the cell potential?

$$\Delta G = \Delta G° + RT \ln Q = \Delta G° + 2.303RT \log Q$$

Recall that Q is the reaction quotient.

The equation relating ΔG to $\Delta G°$ can be combined with the relation between free energy and reaction potential.

$$\Delta G = -n\mathscr{F}E$$
$$\Delta G° = -n\mathscr{F}E°$$
$$\Delta G = \Delta G° + 2.303RT \log Q$$
$$-n\mathscr{F}E = -n\mathscr{F}E° + 2.303RT \log Q$$
$$E = E° - (2.303RT/n\mathscr{F}) \log Q$$

This equation, which is called the **Nernst equation,** takes the following form at 25°C:

$$E = E° - (0.05916/n) \log Q$$

We can use this equation to calculate cell potentials or half-reaction potentials for nonstandard concentrations of reactants and products.

EXAMPLE 18.5

Consider the following reaction at 25°C.

$$Zn(s) + Cu^{2+}(aq) \longrightarrow Zn^{2+}(aq) + Cu(s) \qquad E°_{cell} = 1.10 \text{ V}$$

What is the cell potential for this system when the concentrations are 5.00 M Cu^{2+} and 0.050 M Zn^{2+}?

Solution   The reaction quotient is

$$Q = [Zn^{2+}]/[Cu^{2+}]$$

By examining the changes in charge, we can see that $n = 2$. We apply the Nernst equation as follows:

$$E = E° - (0.05916/n) \log Q$$
$$= E° - (0.05916/n) \log\{[Zn^{2+}]/[Cu^{2+}]\}$$
$$= 1.10 \text{ V} - (0.05916/2) \log(0.050 \ M/5.00 \ M)$$

$$= 1.10 \text{ V} - (0.05916/2) \log 0.010 = 1.10 - (0.05916/2) \times -2.00$$
$$= 1.10 + 0.0592 = 1.16 \text{ V}$$

Practice Problem 18.5

Consider the following reaction at 25°C.

$$Zn(s) + Cu^{2+}(aq) \longrightarrow Zn^{2+}(aq) + Cu(s) \qquad E^{\circ}_{cell} = 1.10 \text{ V}$$

What is the cell potential for this system when the concentrations are 0.0150 M Cu^{2+} and 1.25 M Zn^{2+}?

Answer: 1.04 V

We can make a number of observations about the relationships in the Nernst equation. If reactant and product concentrations are not standard (1.00 M), and they do not cancel to give $Q = 1$, the cell potential (E) will not equal the standard cell potential (E°). When the system has a higher concentration of reactants than of products ($Q < 1$), E is more positive than E°. When the cell "runs down" to give more products than reactants, the value of E decreases accordingly. Thus, as the reactants are depleted in a cell, the voltage developed by that cell decreases. At the point where reactants and products are in a state of equilibrium ($Q = K$), then $E = 0$.

$$E = E^{\circ} - (0.05916/n) \log K$$
$$E = E^{\circ} - E^{\circ} = 0$$

These relationships can be summarized as follows:

$$Q < 1 \qquad E > E^{\circ}$$
$$Q > 1 \qquad E < E^{\circ}$$
$$Q = 1 \qquad E = E^{\circ}$$
$$Q = K \qquad E = 0$$

The Nernst equation forms the basis for some analytical techniques used to monitor the concentrations of ions. For example, the pH meter discussed in Chapter 15 measures the voltage between two electrodes, one of which develops a potential that is dependent on the H_3O^+ concentration. Other systems are based on ion-selective electrodes, which develop a potential that depends on the concentration of a specific ion, such as fluoride ion.

18.4 Electrolysis

What happens to atoms or ions during electrolysis?

So far we have been considering primarily voltaic cells, in which spontaneous chemical reactions are used to generate electricity. In this section, we turn our attention to electrolysis cells, in which, as you recall, electrical energy is used to cause nonspontaneous oxidation–reduction reactions to occur. Electrolysis is useful for isolating active elements, purifying metals, and electroplating metals. ▶ For example, in the experiment described in the chapter introduction, electrolysis is used to isolate elemental iodine from an iodide salt. Here, we examine how to predict the products of electrolysis reactions, as well as how to predict the amount of product formed.

Electrolysis of Pure Compounds

Sodium occurs in a wide variety of minerals: NaCl, $Na_2B_4O_7$ (borax), Na_3AlF_6 (cryolite), $NaNO_3$ (Chile saltpeter), and Na_2CO_3 (soda ash). Sodium is a silvery-white metal with a low melting point (97.8°C) and density (0.97 g/mL). It is used as a coolant in nuclear reactors and as a reducing agent in the preparation of titanium and sodium peroxide. Because of its high reactivity, sodium is normally stored under kerosene or dry nitrogen gas.

Sir Humphrey Davy first isolated sodium in 1807 by electrolysis of sodium hydroxide. The primary modern method of electrolysis of molten sodium chloride dates back to Faraday's work in 1833. However, this method was not commercially successful until the Downs cell was introduced by the du Pont Chemical Company in 1921.

The simplest application of electrolysis involves a chemical compound in its pure form—for example, an ionic salt such as sodium chloride. In this process, the salt is melted, two inert electrodes are inserted into the molten salt, and electricity is passed through it, as shown in Figure 18.12. The current causes one of the electrodes to become negatively charged and causes the other to become positively charged. The electrodes attract ions, which are free to move around in the molten salt. The negatively charged electrode (the cathode) attracts cations, such as Na^+ in molten NaCl. The sodium ions can be reduced if the voltage applied to the cell is high enough to overcome the half-reaction potential for that reduction, as shown in Figure 18.13A. At the same time, the positively charged electrode (the anode) attracts chloride ions, which are oxidized, as shown in Figure 18.13B. The proportion of cations reduced must be the same as the proportion of anions oxidized in order for charge balance to be maintained. In a system as simple as molten NaCl, there is only one possible reduction and one possible oxidation.

$$Na^+(l) + e^- \longrightarrow Na(l) \qquad E° = -2.71 \text{ V}$$
$$2Cl^-(l) \longrightarrow Cl_2(g) + 2e^- \qquad E° = -1.36 \text{ V}$$

Electrolysis of molten salts is the primary commercial method of production of a number of metals, including sodium. Much sodium metal is produced in Downs cells, such as that shown in Figure 18.14, which provide for the separation and removal of sodium and chlorine gas as they are formed. The component reactions are simple.

$$\begin{aligned} \text{Cathode:} \quad & Na^+ + e^- \longrightarrow Na \\ \text{Anode:} \quad & 2Cl^- \longrightarrow Cl_2 + 2e^- \\ \hline & 2NaCl(l) \longrightarrow 2Na(l) + Cl_2(g) \end{aligned}$$

The anode is made of graphite and the cathode of iron, each in a separate compartment to facilitate removal of the products. Sodium chloride melts at 804°C. Because temperatures this high can lead to ignition of sodium vapor, the process is

Figure 18.12 A molten salt contains negative and positive ions that are free to move around relatively independently of one another. The cations are attracted to the cathode, where they are reduced. The anions are attracted to the anode, where they are oxidized.

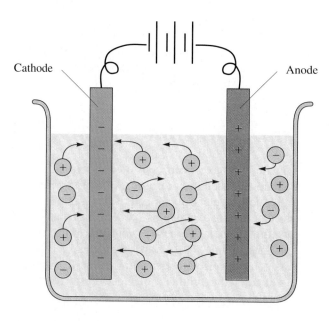

Cathode   Anode

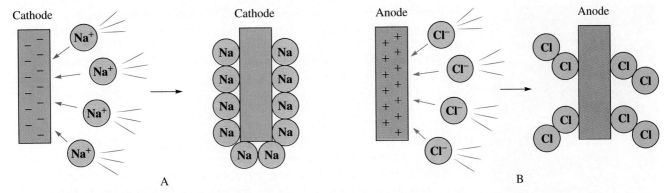

A   B

Figure 18.13 Electrostatic attraction between opposite charges causes migration of the sodium ions to the cathode and migration of chloride ions to the anode. (*A*) The sodium ions take on electrons to become neutral sodium atoms, which are deposited onto the cathode. (*B*) Chloride ions give up electrons to become neutral chlorine atoms. These atoms combine to form diatomic molecules.

Why does addition of calcium chloride lower the melting point of sodium chloride?

normally carried out at 575–600°C, and calcium chloride (about 60% by mass) is added to lower the melting point of the sodium chloride.

Like molten salts, water can conduct electricity, at least to some extent, because it contains small quantities of ions from autoionization, and it can be decomposed by electrolysis. However, electrolysis of water is a very slow process unless some ionic substance is dissolved in the water to help conduct electrical current. To effect the electrolysis of water, shown in Figure 18.15, dilute sulfuric acid is usually added to the water to enhance electrical conductivity. Water undergoes dissociation into H_2 and O_2 gases in the same 2:1 proportion as exists in the water molecules.

Hydrogen forms, by reduction, at the cathode.

$$2H_2O(l) + 2e^- \longrightarrow H_2(g) + 2OH^-(aq) \qquad E° = -0.83 \text{ V}$$

Note that -0.83 V is a standard potential, corresponding to the half-reaction that occurs in a solution containing 1.00 M OH^- (pH 14). Using the Nernst equation, we can calculate a potential of -0.41 V for this half-reaction at pH 7, corresponding to pure water. In 0.5 M H_2SO_4, which contains 1.00 M $H^+(aq)$, the potential is 0.00 V.

Oxygen is formed at the anode.

$$2H_2O(l) \longrightarrow O_2(g) + 4H^+(aq) + 4e^- \qquad E° = -1.23 \text{ V}$$

Figure 18.14 In the Downs cell for the production of sodium and chlorine, molten sodium chloride is electrolyzed. Chlorine is produced at the graphite anode, where it rises and is vented and collected. The sodium metal is produced at the iron cathode. It is captured in an iron container, from which it is siphoned out of the system.

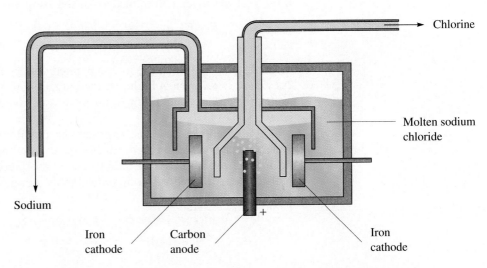

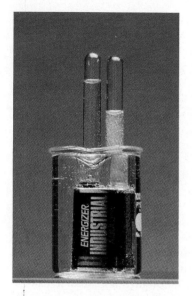

Figure 18.15 The electrolysis of water produces hydrogen and oxygen in a 2:1 ratio. The terminals on the dry cell perform the function of electrodes in the electrolysis cell.

This standard potential corresponds to the half-reaction that occurs in a solution containing 1.00 M $H^+(aq)$ (pH 0). We can use the Nernst equation to calculate a potential of -0.82 V at pH 7.

When we add the half-reaction potentials to obtain the cell potential, we obtain a value of -1.23 V at either pH.

$$\text{At pH 7: } E_{cell} = -0.41 \text{ V} + -0.82 \text{ V} = -1.23 \text{ V}$$
$$\text{At pH 0: } E_{cell} = -0.00 \text{ V} + -1.23 \text{ V} = -1.23 \text{ V}$$

This means it is necessary to apply a voltage greater than $+1.23$ V to this cell to cause the decomposition reaction to occur.

To maintain electrical balance, the reduction reaction must occur twice as often as the oxidation reaction, giving the expected 2:1 ratio of H_2 to O_2.

Electrolysis of Aqueous Solutions

Predicting the result of electrolyzing a pure compound is a straightforward task, because only one oxidation and one reduction are possible. But water can be electrolyzed, so predicting the products of the electrolysis of salts in aqueous solution is more complicated. Normally only one of the substances (salt or water) undergoes oxidation, and only one undergoes reduction. The oxidation and reduction potentials for the possible half-reactions determine which substance is oxidized and which is reduced. These potentials are affected by the concentration of the dissolved salt, as described by the Nernst equation.

Let's consider an example of electrolysis in an aqueous solution. Suppose the ionic salt sodium sulfate is added to water. Experimentally, we find that the electrolysis of this solution yields H_2 and O_2 in a 2:1 ratio. However, there are two possible reactions at each electrode. At the cathode, either water molecules or sodium ions could be reduced, as described by the following half-reactions.

$$2H_2O(l) + 2e^- \longrightarrow H_2(g) + 2OH^-(aq) \qquad E° = -0.83 \text{ V}$$
$$Na^+(l) + e^- \longrightarrow Na(l) \qquad E° = -2.71 \text{ V}$$

The reduction of sodium ion requires an applied potential of 2.71 V at all pH levels. The reduction of water requires 0.83 V at pH 14, 0.41 V at pH 7, and 0.00 V at pH 0, as determined from the Nernst equation (assuming 1 atm H_2).

$$E = E° - (0.05916/2) \log P_{H_2} [OH^-]^2$$
$$= -0.83 + 0.05916 \text{ pOH} = -0.83 + 0.05916 (14 - \text{pH})$$

The substance requiring the lowest applied voltage is the one most easily reduced. As we see here, even at high pH, the reduction of water requires significantly less voltage than the reduction of sodium, so water is reduced and Na^+ remains unreacted.

At the anode, we must consider the possible oxidation of sulfate ions and water. As can be seen in Appendix G, the only sulfur-containing oxoanion that can be formed by oxidation of sulfate ion is the peroxydisulfate ion, $S_2O_8^{2-}$. The half-reactions for oxidation of sulfate and water are given by the following equations:

$$2H_2O(l) \longrightarrow O_2(g) + 4H^+(aq) + 4e^- \qquad E° = -1.23 \text{ V}$$
$$2SO_4^{2-}(aq) \longrightarrow S_2O_8^{2-}(aq) + 2e^- \qquad E° = -2.01 \text{ V}$$

This standard potential corresponds to the half-reaction that occurs in a solution containing 1.00 M $H^+(aq)$, or pH 0. We can use the Nernst equation to calculate a potential at pH 7 of -0.82 V for the water half-reaction and of -2.01 V for the sulfate half-reaction. No matter what the pH, it takes less voltage to oxidize water than to oxidize sulfate ion, so water is oxidized at the anode to produce oxygen gas. The potentials for the possible half-reactions at each electrode allowed us to predict correctly the products of electrolysis of this aqueous solution.

▶ Different results are obtained when sodium iodide is dissolved in water, as shown in the introduction to this chapter. Here, electrolysis produces hydrogen gas and dissolved iodine. The potentials for the reduction of water and sodium ion at the cathode are the same as for the sodium sulfate solution, so again we predict that water is reduced to give hydrogen gas at the cathode. At the anode, we have the following possible half-reactions.

Sodium iodide is a colorless solid. It is highly soluble in water, where it is readily oxidized to form aqueous I_2. This reaction occurs slowly in the presence of O_2 from air.

$$2I^-(aq) \longrightarrow I_2(aq) + 2e^- \qquad E° = -0.54 \text{ V at all pH}$$
$$2H_2O(l) \longrightarrow O_2(g) + 4H^+(aq) + 4e^- \qquad E° = -1.23 \text{ V at pH 0}$$
$$2H_2O(l) \longrightarrow O_2(g) + 4H^+(aq) + 4e^- \qquad E = -0.82 \text{ V at pH 7}$$

Assuming we start with a 1.00 M solution of NaI at pH 7, so that we can use the standard half-reaction potential, we see that it requires 0.54 V to oxidize iodide ion and 0.82 V to oxidize water. Because it requires less voltage to oxidize iodide ion than to oxidize water, the iodide ion should be oxidized at the anode, in agreement with experimental results.

EXAMPLE 18.6

What reactions occur at each electrode in the electrolysis of aqueous copper fluoride?

Solution The possible cathode and anode reactions are

Cathode: $\quad Cu^{2+}(aq) + 2e^- \longrightarrow Cu(s) \qquad E° = 0.34 \text{ V}$

$\qquad\qquad 2H_2O(l) + 2e^- \longrightarrow H_2(g) + 2OH^-(aq) \qquad E° = -0.83 \text{ V at pH 14}$

$\qquad\qquad 2H_2O(l) + 2e^- \longrightarrow H_2(g) + 2OH^-(aq) \qquad E = -0.41 \text{ V at pH 7}$

Anode: $\quad 2F^-(aq) \longrightarrow F_2(g) + 2e^- \qquad E° = -2.87 \text{ V}$

$\qquad\qquad 2H_2O(l) \longrightarrow O_2(g) + 4H^+(aq) + 4e^- \qquad E° = -1.23 \text{ V at pH 0}$

$\qquad\qquad 2H_2O(l) \longrightarrow O_2(g) + 4H^+(aq) + 4e^- \qquad E = -0.82 \text{ V at pH 7}$

At the cathode, reduction of copper actually generates some voltage, while reduction of water uses electrical energy. Because of this difference in potential, copper metal should be deposited at the cathode. At the anode, oxygen gas is evolved, because less voltage is required to oxidize water than to oxidize fluoride ion.

Practice Problem 18.6

What reactions occur at each electrode in the electrolysis of aqueous vanadium(IV) iodide?

Answer: Reduction of VO^{2+} to V^{3+} at the cathode; oxidation of I^- to I_2 at the anode.

Overvoltage Some electrolysis reactions in aqueous solution require more voltage than the standard half-reaction potentials predict. This effect is due to several factors, which are generally described collectively as **overvoltage.** Overvoltage, then, is the difference between the potential actually required to carry out an electrolysis and the standard potential.

Factors that can cause overvoltage include formation of a protective coating on an electrode, which slows down the electrolysis reaction. Extra voltage must be applied to cause the reaction to occur as rapidly as it would on a clean electrode. The overvoltage for the deposition of metals on an electrode is usually very small.

The largest overvoltages—often as large as several tenths of a volt—are observed for electrochemical reactions in which a gas is produced. The overvoltage for production of gaseous oxygen is usually about 0.5 volts, for example. We will generally ignore the overvoltage effect, because it is hard to predict reliably.

Faraday's Law and the Amount of Electrolysis Product

So far, we have used reaction potentials to decide what reaction will occur at each electrode in an electrolysis cell. We would also like to be able to calculate what amount of electrolysis product will be produced by passage of a given amount of electrical current. The relationship needed to carry out such calculations is provided by **Faraday's law:** The quantity of a substance produced by a given quantity of direct current is proportional to the equivalent weight of the substance. The **equivalent weight** of a substance being oxidized or reduced is equal to the molar mass divided by the number of electrons being transferred in the oxidation–reduction reaction. Thus, the equivalent weight is the mass of a substance that can be oxidized or reduced by 1 mol of electrons.

Suggest an experiment that would make use of Faraday's law to determine a value of Avogadro's number.

The relationship between the amount of electrical current passed through a cell and the mass of the product formed at an electrode is illustrated by the series of electrolysis cells shown in Figure 18.16. When 1 mol of electrons is passed through this system, different quantities of material are generated at each of the cathodes. The quantities of reduced materials are equal to the equivalent weights of the cations being reduced.

Cation	Ag^+	Cu^{2+}	Au^{3+}	H^+
Molar mass (g/mol)	107.8682	63.546	196.9665	1.0079
Equivalent weight (g/equiv)	107.8682	31.773	65.6555	1.0079
Mass deposited (g)	107.868	31.773	65.6555	1.0079
Amount deposited (mol)	1 mol Ag	$\frac{1}{2}$ mol Cu	$\frac{1}{3}$ mol Au	$\frac{1}{2}$ mol H_2

Figure 18.16 Amounts of materials deposited by the passage of 1 mol of electrons (1 Faraday) through a series of connected electrolysis cells. Deposited at the cathodes:

107.868 g Ag	31.773 g Cu	65.6555 g Au	1.0079 g H_2
(1 mol)	($\frac{1}{2}$ mol)	($\frac{1}{3}$ mol)	($\frac{1}{2}$ mol)

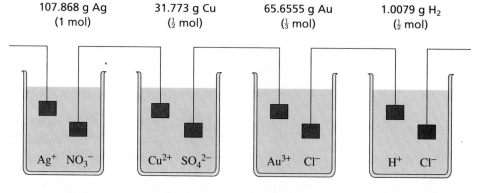

We can explain these results by examining the reduction half-reactions, which show the number of electrons required for complete reduction of the substance in solution.

$$Ag^+(aq) + e^- \longrightarrow Ag(s)$$
$$Cu^{2+}(aq) + 2e^- \longrightarrow Cu(s)$$
$$Au^{3+}(aq) + 3e^- \longrightarrow Au(s)$$
$$2H^+(aq) + 2e^- \longrightarrow H_2(g)$$

The amount of reduced material produced at the various cathodes corresponds exactly to the stoichiometric ratio of oxidizing agent to electrons. Silver ion requires one electron for reduction, so 1 mol of electrons produces 1 mol of metallic silver. Copper ion requires two electrons for complete reduction, so 1 mol of electrons can produce only $\frac{1}{2}$ mol of copper metal. The other results are similarly consistent with the number of electrons required for complete reduction.

We can, then, calculate the amount of material produced by the passage of 1 mol of electrons through a cell. But how much charge corresponds to 1 mol of electrons? This relationship is given by the Faraday constant:

$$\text{Charge on 1 mol electrons} = 1\ \mathscr{F} = 9.650 \times 10^4\ \text{coul}$$

A coulomb (coul) is the amount of charge passing a point when a current of 1 ampere (A) flows for 1 second (s):

$$1\ \text{coul} = 1\ \text{A s}$$

Thus if we know the amperage and the time, we can calculate the amount of charge in coulombs.

$$\text{Charge in coul} = \text{current in amps} \times \text{time in seconds} \times 1\ \text{coul/A s}$$

From this quantity, we can determine the moles of electrons passed through the cell.

$$\text{Moles of electrons} = \text{charge in coul} \times 1\ \text{mol}/(9.650 \times 10^4\ \text{coul})$$

Using the stoichiometry of the balanced half-reaction, we can relate moles of electrons to moles of substance reduced. Finally, we can convert moles of substance to grams by using the molar mass of the substance.

EXAMPLE 18.7

Suppose the electrolysis of molten magnesium chloride is carried out with a current of 8.00×10^4 A. What masses of magnesium and chlorine are produced in exactly 1 h?

Solution The overall reaction is

$$MgCl_2(l) \longrightarrow Mg(l) + Cl_2(g)$$

We break this reaction down into the two half-reactions.

$$Mg^{2+}(l) + 2e^- \longrightarrow Mg(l)$$
$$2Cl^-(l) \longrightarrow Cl_2(g) + 2e^-$$

Now we calculate the amount of charge passed through the electrolysis cell.

$$\text{Charge in coul} = \text{current in amperes} \times \text{time in seconds} \times 1\ \text{coul/A s}$$
$$\text{Charge} = 8.00 \times 10^4\ \text{A} \times 1\ \text{h} \times 3600\ \text{s/h} \times 1\ \text{coul/A s} = 2.88 \times 10^8\ \text{coul}$$

(continued)

Then we calculate the number of moles of electrons required to carry this amount of charge.

$$\text{Moles of electrons} = \text{charge in coul} \times 1 \text{ mol}/(9.650 \times 10^4 \text{ coul})$$

$$\text{Moles of electrons} = 2.88 \times 10^8 \text{ coul} \times 1 \text{ mol}/(9.650 \times 10^4 \text{ coul})$$

$$= 2.98 \times 10^3 \text{ mol}$$

We convert moles of electrons to moles of product using the balanced half-reactions and moles of product to grams of product using the molar masses:

$$\text{mol Mg} = 2.98 \times 10^3 \text{ mol e}^- \times 1 \text{ mol Mg}/2 \text{ mol e}^- = 1.49 \times 10^3 \text{ mol Mg}$$

$$\text{g Mg} = 1.49 \times 10^3 \text{ mol Mg} \times 24.305 \text{ g/mol Mg} = 3.62 \times 10^4 \text{ g Mg}$$

$$\text{mol Cl}_2 = 2.98 \times 10^3 \text{ mol e}^- \times 1 \text{ mol Cl}_2/2 \text{ mol e}^- = 1.49 \times 10^3 \text{ mol Cl}_2$$

$$\text{g Cl}_2 = 1.49 \times 10^3 \text{ mol Cl}_2 \times 70.906 \text{ g/mol Cl}_2 = 1.06 \times 10^5 \text{ g Cl}_2$$

Practice Problem 18.7

Suppose the electrolysis of molten magnesium chloride is carried out with a current of 2.50×10^3 A. What masses of magnesium and chlorine are produced in 2.50 h?

Answer: 2.82×10^3 g Mg; 8.23×10^3 g Cl$_2$

By adding a few more relationships, we can calculate the amount of energy and power required to carry out an electrolysis. Energy is normally measured in units of joules.

$$1 \text{ J} = 1 \text{ coul V}$$

Power, measured in watts, is the rate of energy use and is expressed as 1 joule per second.

$$1 \text{ W} = 1 \text{ J/s}$$

Electrical energy is usually measured in units of kilowatt hours (kwh)—that is, the rate of energy usage (power) in kilowatts multiplied by the time in hours. Thus we need a conversion between kilowatt hours and joules (or coul V), which we can obtain from the definitions of units just given.

$$1 \text{ kwh} = 1000 \text{ W} \times 1 \text{ h} \times \frac{1 \text{ J/s}}{1 \text{ W}} \times \frac{3600 \text{ s}}{1 \text{ h}} \times \frac{1 \text{ coul V}}{1 \text{ J}}$$

From this series of conversions, we obtain the desired conversion.

$$1 \text{ kwh} = 3.6 \times 10^6 \text{ coul V}$$

If the amount of charge is calculated in units of coulombs and the applied voltage is known, then it is possible to calculate the amount of electrical energy required from this conversion equation.

$$\underset{\text{in kwh}}{\text{Electrical energy}} = \underset{\text{in coul}}{\text{charge}} \times \underset{\text{in V}}{\text{voltage}} \times 1 \text{ kwh}/(3.6 \times 10^6 \text{ coul V})$$

EXAMPLE 18.8

Suppose the electrolysis of molten magnesium chloride is carried out with an applied voltage of 6.00 V and a current of 8.00×10^4 A. How much electrical energy is consumed in exactly 1 h?

Solution   As in the previous example, we must calculate the amount of charge passed through the electrolysis cell.

$$\text{Charge} = 8.00 \times 10^4 \text{ A} \times 1 \text{ h} \times 3600 \text{ s/h} \times 1 \text{ coul/A s} = 2.88 \times 10^8 \text{ coul}$$

The electrical energy can then be calculated.

$$\text{Electrical energy in kwh} = 2.88 \times 10^8 \text{ coul} \times 6.00 \text{ V} \times 1 \text{ kwh}/(3.6 \times 10^6 \text{ coul V})$$
$$= 4.80 \times 10^2 \text{ kwh}$$

Thus, theoretically, it takes 480 kwh of electricity to produce 36.2 kg of magnesium metal.

Practice Problem 18.8

Suppose the electrolysis of molten magnesium chloride is carried out with a current of 2.50×10^3 A and a potential of 8.00 V. How much electrical energy is consumed in exactly 2 h?

Answer: 40.0 kwh

18.5   Electrolytic Processes with Metals

A variety of materials can be manufactured by electrochemical processes. Whether this is the most economical means of production depends on whether inexpensive electricity is readily available, as it is in regions such as Niagara Falls, where hydroelectric plants are located. A listing of some of the materials produced electrochemically is given in Table 18.3. Here we will look at a few of these processes in detail.

What are the advantages and the disadvantages of using an electrolytic process, rather than a chemical reduction process, to prepare metals?

Aluminum

Aluminum is the third most abundant element in the earth's crust (8.13% by mass) and occurs widely dispersed in the form of complex silicates such as $KAlSi_3O_8$ in

Table 18.3 Some Metals Made Electrochemically

Metal	Process
Aluminum	Electrolytic reduction of alumina
Cadmium	Electrolytic precipitation from dissolved Zn–Pb residues
Calcium	Electrolytic reduction of molten salt
Copper	Electrolytic refining
Gold	Electrolytic refining
Lead	Electrolytic refining
Magnesium	Electrolysis of molten $MgCl_2$
Sodium	Electrolysis of molten NaCl
Zinc	Electrolytic precipitation

Aluminum has a low density (2.70 g/mL) and a relatively low melting point (660°C), and it is a good conductor of heat and electricity. Aluminum has been used for electrical wiring, but it may melt if too much current is passed, giving rise to a fire hazard, so it is no longer used for this purpose. Though it is a relatively weak metal, aluminum can be strengthened by alloying with other elements, chiefly magnesium, copper, zinc, and lead.

clays and as bauxite, $Al_2O_3 \cdot nH_2O$. The oxide is the more important ore for recovery of the metal. Clay is more plentiful than bauxite, but it is more difficult to recover Al_2O_3 from clay at a reasonable cost.

The formation of a stable oxide of aluminum makes it difficult to recover aluminum from its ores, because the oxide is difficult to reduce. The existence of this oxide enables us to use aluminum in ways that would not otherwise be feasible, however. For an element to be used as a structural metal, it must not only be strong but must also be unreactive with water and air. Although aluminum is one of the most chemically active metals, with an extremely strong affinity for oxygen, it is surpassed only by steel in its use as a structural metal. A freshly produced aluminum surface does react rapidly with atmospheric oxygen, but once a layer of Al_2O_3 is formed, the metal becomes impervious to further attack by O_2 as well as by many other corrosive agents, although it is dissolved by most strong acids and bases.

The oxide coating is only 5–10 nm thick, but it can be thickened to about 25 μm by an electrolysis process called *anodizing,* in which the metal is made the anode in a cell so that oxidation of the surface can take place. The anodized coating can be produced in a variety of permanent colors when the electrolysis is carried out in the presence of certain dyes, so this technique is used to produce colored aluminum objects, such as those shown in Figure 18.17.

The first pure aluminum was prepared by the French chemist Henri Sainte-Claire Deville in 1854, who used a chemical method in which sodium metal is reacted with aluminum(III) chloride. At this time, aluminum was more expensive than gold; Napoleon's most favored guests were served with aluminum utensils, whereas his lesser guests had to make do with gold and silver.

A usable electrolytic process for producing aluminum, which led to dramatic decreases in cost, was discovered nearly simultaneously in 1886 by Charles Martin Hall and Paul Héroult. The Hall–Héroult process is basically the electrolysis of a 5% by mass solution of Al_2O_3 in molten cryolite, Na_3AlF_6, at about 1000°C. Because Al_2O_3 melts at 2045°C, which is too high a temperature to be practical, the electrolysis is carried out in a solution at a lower temperature. Cryolite was selected as the solvent because of its lower melting point. The electrolytic cell is shown in Figure 18.18. The cell consists of a steel box lined with carbon, which, along with molten aluminum, serves as the cathode. Large carbon anodes are immersed in the molten salt bath. These anodes not only serve as electrical con-

Although cryolite, Na_3AlF_6, occurs naturally in Greenland, most is now made synthetically by treating aluminum oxide and sodium carbonate with hydrofluoric acid.

$3Na_2CO_3(aq) + Al_2O_3(s) +$
$12HF(aq) \longrightarrow$
$2Na_3AlF_6(s) + 6H_2O(l) +$
$3CO_2(g)$

Figure 18.17 Anodized aluminum consists of aluminum metal coated with electrolytically deposited aluminum oxide, which can be dyed many colors, such as those of the plaque and the keys.

Figure 18.18 In the electrolytic cell used for aluminum production, the aluminum is reduced in a carbon-lined box containing a molten salt of cryolite in which aluminum oxide has been dissolved. The aluminum settles to the bottom, where it can be drained off.

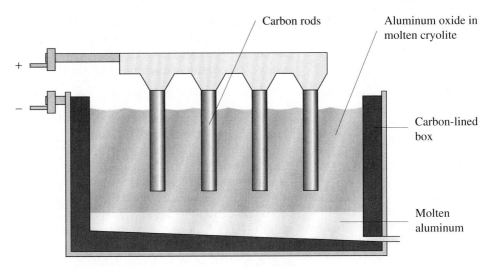

Carbon rods

Aluminum oxide in molten cryolite

Carbon-lined box

Molten aluminum

The electrolytic method for producing aluminum was discovered almost simultaneously by two 23-year-old men. One, Charles Martin Hall, a student at Oberlin College in Ohio, worked in a lab in a backyard woodshed. The other, Paul Héroult, worked in a borrowed lab in Paris.

ductors but also participate chemically in the process, as shown in the following equation for the overall cell reaction.

$$7Al_2O_3(soln) + 12C(s) \longrightarrow 14Al(l) + 3CO(g) + 9CO_2(g)$$

Because the anodes are consumed by this reaction, they must be replaced regularly. More Al_2O_3 is added as needed to keep its concentration near 5% by mass. The molten aluminum is periodically drained from the cell. The Hall–Héroult method uses tremendous amounts of electrical energy—by one estimate, 5% of the total energy output in the United States—not only for the electrolysis but also for heating the cryolite bath. In spite of its high energy requirements, though, this process is the primary method of obtaining aluminum from its ores. Recycling of aluminum requires only about 5% as much electrical energy because aluminum melts at a lower temperature (660°C), so much energy could be saved if all disposable aluminum objects were recycled.

Magnesium

Magnesium is a silvery-white metal with a low density (1.74 g/mL). It is the lightest of the structural metals and is widely used to produce strong, light alloys with other metals. Magnesium is a very active metal, but it resists chemical attack to some extent by forming a protective oxide coating. Because of its high chemical activity, magnesium is used as an anode for protecting other metals against corrosion and as a reducing agent in the production of other metals, such as titanium, zirconium, uranium, and beryllium.

Magnesium is the eighth most abundant element in the earth's crust (2.1% by mass). It occurs widely in nature as the ores dolomite ($CaCO_3 \cdot MgCO_3$), magnesite ($MgCO_3$), and carnallite ($KMgCl_3 \cdot 6H_2O$). It also occurs in the dissolved state as $MgCl_2$ in seawater and underground brines.

Magnesium was first prepared as an amalgam in 1808 by Sir Humphrey Davy via the electrolytic reduction of MgO at a mercury cathode. In 1833 Michael Faraday produced magnesium by the electrolytic reduction of magnesium chloride, a process now widely used.

$$MgCl_2(l) \xrightarrow{\text{electrolysis}} Mg(l) + Cl_2(g)$$

The reactions occurring at the electrodes are

$$\text{Cathode:} \quad Mg^{2+} + 2e^- \longrightarrow Mg$$
$$\text{Anode:} \quad 2Cl^- \longrightarrow Cl_2 + 2e^-$$

In 1941, the Dow Chemical Company developed a process using seawater, an essentially inexhaustible source of magnesium in the form of $Mg^{2+}(aq)$. Today the

Dow process is responsible for two-thirds of the magnesium produced in the United States each year. Dolomite ($CaCO_3 \cdot MgCO_3$) or oyster shells (largely $CaCO_3$) are heated to form an alkaline oxide ($CaO \cdot MgO$ or CaO). When added to seawater, the oxide causes magnesium hydroxide to precipitate.

Discuss the precipitation of magnesium hydroxide from the standpoint of the equilibria involved.

$$Mg^{2+}(aq) + CaO(s) + H_2O(l) \longrightarrow Mg(OH)_2(s) + Ca^{2+}(aq)$$

The $Mg(OH)_2$ is removed and treated with hydrochloric acid to form a concentrated solution of magnesium chloride.

$$Mg(OH)_2(s) + 2HCl(aq) \longrightarrow MgCl_2(aq) + 2H_2O(l)$$

Michael Faraday (1791–1867) was a British chemist and physicist who was a professor of chemistry at the Royal Institution in London, where he started as the laboratory assistant of Sir Humphrey Davy.

The aqueous magnesium chloride is evaporated to yield a solid containing 73% $MgCl_2$ and 27% water. This solid is fed into a steel electrolysis cell that uses consumable graphite anodes. This salt is heated to 710°C and electrolyzed at 6 V and 80–100 kA. The magnesium, which melts at 651°C, floats to the top of the cell and is withdrawn periodically. The chlorine that is produced can be converted to hydrochloric acid

$$2Cl_2(g) + 2H_2O(g) \xrightarrow{\Delta} 4HCl(g) + O_2(g)$$

and recycled through the process.

Copper

Copper is widely used as an electrical conductor, as well as in the alloys brass and bronze. It can be found in nature as the free metal (Figure 18.19) and so was one of the first metals used by ancient peoples. It is also widely distributed in over 165 copper minerals, but only six of these figure in 95% of commercial copper-mining operations. Sulfide ores such as chalcopyrite, $CuFeS_2$, and bornite, Cu_3FeS_3, and carbonate ores such as malachite, $CuCO_3 \cdot Cu(OH)_2$, and azurite, $2CuCO_3 \cdot Cu(OH)_2$, are of primary importance.

Extraction of copper from its ores is largely concerned with concentrating the ore, which typically contains less than 1–2% copper compounds. After the ore is concentrated, it is processed by pyrometallurgical or hydrometallurgical methods (discussed in Section 17.6), which yield copper that is about 98% pure and contains a variety of metals, including gold, silver, platinum, selenium, tellurium, and nickel.

Figure 18.19 Copper can be found in nature in the form of interesting dendritic structures.

Figure 18.20 Copper cathodes, which start as thin sheets of pure copper, are combined with impure copper anodes in an electrolysis cell. Copper is transferred from the anode to the cathode, and impurities fall to the bottom of the cell.

Because of its appearance, this impure copper is known as blister copper. Molten blister copper is poured into molds to be cast into anodes for electrolytic refining. The anodes are suspended in a solution containing $CuSO_4$ and H_2SO_4. The cathodes are very thin sheets of copper that is more than 99.9% pure (Figure 18.20). Upon electrolysis, the copper is dissolved from the impure anode and deposited on the pure cathode, as shown in the following equations.

$$\text{Anode:} \quad Cu(s) \longrightarrow Cu^{2+}(aq) + 2e^-$$
$$\text{Cathode:} \quad Cu^{2+}(aq) + 2e^- \longrightarrow Cu(s)$$

The impurities in the blister copper either dissolve or fall to the bottom of the tank as a "slime" that can later be treated to recover these metals. Copper refining in the United States yields 25% of the molybdenum, 93% of the selenium, 76% of the tellurium, 32% of the gold, and 28% of the silver produced.

Electroplating

So far, we have seen that electrolysis can be used to recover metals from their salts and to purify metals. Electrolysis can also be used to deposit one metal on another, a process called **electroplating.** Often, thin coatings of one metal are plated on another metal to protect it from corrosion. At one time, for example, bumpers on automobiles were steel electroplated with chromium. Many metal parts are coated with chromium or nickel in this way to protect them from corrosion. The metal part is immersed in an electroplating bath that contains an electrolyte and a salt of chromium (or nickel). Passage of electricity, with the metal part as the cathode, results in reduction of the chromium ion and deposition of chromium onto the metal part.

Why are metal parts plated with other metals such as chromium instead of just being made of chromium in the first place?

18.6 Industrial Chemicals from Seawater

Seawater contains a wealth of minerals, and electrolysis can be used to recover at least some of them, including several compounds, pure nonmetals, and metals. We have already seen how magnesium is recovered from seawater in a process that uses electrolysis. ▶ In the chapter introduction, we saw how iodine could be recovered from an aqueous iodide solution. Here we examine the recovery of two

What problems arise when
electrolysis is carried out with
aqueous solutions instead of
with molten salts?

industrial chemicals, sodium hydroxide and chlorine. Other chemicals, such as
sodium chlorate, can also be recovered from seawater.

Sodium hydroxide is used in the manufacturing of soap, paper, dyes, and
textiles. Chlorine is used for water purification, the bleaching of paper pulp, and
the preparation of a variety of materials such as PVC plastic (polyvinyl chloride).
These two substances, together called chlor-alkali, are prepared in a single process, which involves the electrolysis of brine (NaCl solution). Essentially all
chemical industries are involved, either as producers or as consumers in chlor-
alkali processes, which have a market of a few billion dollars a year. Hydrogen gas
is an important by-product of this process.

The half-reactions that occur at the electrodes in the chlor-alkali process are
given by the following equations.

$$\text{Cathode:} \quad 2H_2O + 2e^- \longrightarrow H_2 + 2OH^-$$
$$\text{Anode:} \quad 2Cl^- \longrightarrow Cl_2 + 2e^-$$

These half-reactions add up to the net ionic reaction

$$2Cl^- + 2H_2O \longrightarrow 2OH^- + Cl_2 + H_2$$

Written in terms of starting materials and products, this equation becomes

$$2NaCl(aq) + 2H_2O(l) \longrightarrow 2NaOH(aq) + Cl_2(g) + H_2(g)$$

The process involves one major difficulty: keeping the products separated
from one another and from the brine. Chlorine must be kept separated from the
sodium hydroxide solution because it will disproportionate in basic solution, as
shown in the equation

$$Cl_2 + 2OH^- \longrightarrow Cl^- + ClO^- + H_2O$$

In addition, sodium hydroxide is very soluble, so it is difficult to separate from
brine in a pure condition.

One type of cell for chlor-alkali production uses mercury as the cathode. In
this cell, shown in Figure 18.21, sodium ions are reduced to sodium metal, which

Figure 18.21 The mercury cell
for chlor-alkali production.

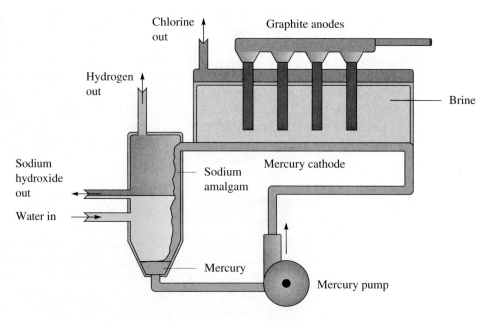

dissolves in mercury to form sodium amalgam, Na(Hg).

$$Na^+ + e^- + Hg \longrightarrow Na(Hg)$$

The mercury is passed through a tank of water, and the sodium reacts to give sodium hydroxide and hydrogen gas.

$$2Na(Hg) + 2H_2O(l) \longrightarrow 2NaOH(aq) + H_2(g) + 2Hg(l)$$

Because the sodium hydroxide is produced at a site separated from the brine, there is no problem with contamination. Chlorine gas is produced at the carbon anodes, where it is removed. This method is coming into disfavor, however, because of environmental pollution by mercury.

Other methods for electrolysis use a diaphragm cell or a polymeric membrane cell (Figure 18.22) to keep the oxidation and reduction products separated. In the diaphragm cell, sodium ions and chloride ions can pass through the wet asbestos diaphragm, but hydrogen and chlorine gases cannot. A positive pressure of the solution in the anode compartment prevents the passage of hydroxide ions to this side of the diaphragm. The diaphragm method consumes the least power but produces a sodium hydroxide solution of relatively low concentration (11%, compared with 50% in a mercury cell) and with a high residual brine content (15%).

Figure 18.22 Cells for chlor-alkali production.

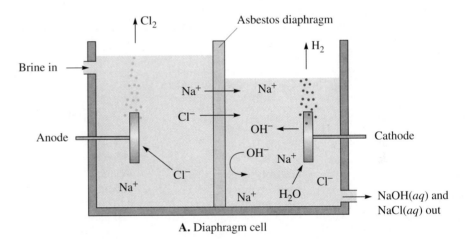

A. Diaphragm cell

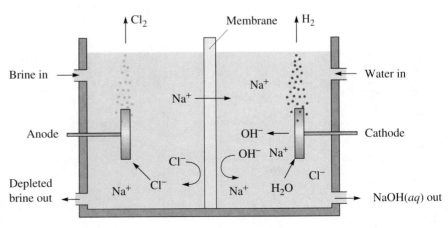

B. Membrane cell

The membrane cell, which consumes more power, produces a more concentrated sodium hydroxide solution (28%) with only a small brine residual (50 ppm). In this cell, only sodium ions can pass through the ion-exchange membrane, so the sodium hydroxide product produced at the cathode is not contaminated by sodium chloride.

Summary

Oxidation and reduction half-reactions can be carried out separately in the compartments of an electrochemical cell. Voltaic cells generate electrical energy, and electrolysis cells use electricity to cause a nonspontaneous oxidation–reduction to occur. Numerous voltaic cells, exploiting a variety of oxidation–reduction reactions, have been designed for use as portable power supplies.

The free-energy change associated with an oxidation–reduction reaction or a half-reaction is related to the potential by the equation $\Delta G = -n\mathscr{F}E$. When half-reactions are added to give a new half-reaction, the free-energy changes can be added but the potentials cannot. The potentials must be multiplied by the number of electrons involved in the half-reactions, as described by the equation $E_3 = (n_1E_1 + n_2E_2)/n_3$. The standard cell potential can be related to the equilibrium constant for the oxidation–reduction reaction by the equation $E° = (2.303RT/n\mathscr{F}) \log K$. Under nonstandard conditions, the cell potential is different from the standard cell potential. They are related by the Nernst equation: $E = E° - (2.303RT/n\mathscr{F}) \log Q$.

The products of electrolysis reactions of pure substances can be predicted from the composition of the substances. For electrolysis reactions in aqueous solutions, it is necessary to consider the magnitude of the half-reaction potentials for all possible half-reactions. The component of the solution that will be oxidized at the anode or reduced at the cathode is the one that requires the least voltage. The amount of electrolysis product can be predicted from the amount of electrical current passed through the cell and the equivalent weight of the reactant, and vice versa. Various metals, including aluminum, magnesium, and sodium, can be produced by electrolysis. Electrolysis can be used to purify metals, such as copper. And it can be used to convert seawater to various important chemicals.

Key Terms

anode (18.1)
battery (18.2)
cathode (18.1)
cell potential (18.1)
electrode (18.1)

electrolysis cell (18.1)
electrolytic cell (18.1)
electroplating (18.5)
equivalent weight (18.4)
Faraday constant (18.3)

Faraday's law (18.4)
galvanic cell (18.1)
half-reaction
    potential (18.1)
Nernst equation (18.3)

overvoltage (18.4)
salt bridge (18.1)
standard hydrogen
    electrode (18.1)
voltaic cell (18.1)

Exercises

Cell Potentials and Reduction Potentials

18.1   What is a voltaic cell?

18.2   How is a standard cell potential related to standard reduction potentials?

18.3   Define the terms *cathode* and *anode*.

18.4   Describe the components of an electrochemical cell.

18.5   Distinguish between a voltaic cell and an electrolytic cell.

18.6   An iron bar placed in a solution containing nickel(II) ions will be plated with nickel metal. Calculate the standard cell potential for the reaction

$$Fe(s) + Ni^{2+}(aq) \rightleftharpoons Fe^{2+}(aq) + Ni(s)$$

18.7   An electrical current is generated in a voltaic cell by the reaction of Zn with Cu^{2+} to produce Zn^{2+} and Cu. Write equations for the half-reactions and for the overall cell reaction. Under standard conditions, what voltage does this cell generate?

18.8   An electrical current is generated in a voltaic cell by the reaction of Cd with Ag^+ to produce Cd^{2+} and Ag. Write equations for the half-reactions and for the overall cell reaction. Under standard conditions, what voltage does this cell generate?

18.9   Calculate the standard cell potential for the reaction

$$Zn(s) + Cl_2(aq) \rightleftharpoons Zn^{2+}(aq) + 2Cl^-(aq)$$

18.10 Solutions of potassium iodate are commonly used to analyze solutions that contain a reducing agent. Calculate the standard cell potential for the reaction

$$10Br^-(aq) + 2IO_3^-(aq) + 12H^+(aq) \rightleftharpoons$$
$$5Br_2(aq) + I_2(aq) + 6H_2O(l)$$

18.11 The dichromate ion is a strong oxidizing agent used in chemical analysis. Calculate the standard half-reaction potential for the following half-reaction, using data available from a table of half-reaction potentials.

$$Cr_2O_7^{2-}(aq) + 14H^+(aq) + 12e^- \longrightarrow$$
$$2Cr(s) + 7H_2O(l)$$

18.12 The reduction of aqueous tin(IV) solutions usually results in tin(II) ions, but further reduction can produce tin metal. Calculate the standard half-reaction potential for the following half-reaction, using data available from a table of half-reaction potentials.

$$Sn^{4+}(aq) + 4e^- \longrightarrow Sn(s)$$

18.13 Silver can be recovered from solutions of its salts by reduction with a more active metal. Calculate the standard cell potential for the following reaction, which might be used to recover silver.

$$Mg(s) + 2Ag^+(aq) \rightleftharpoons Mg^{2+}(aq) + 2Ag(s)$$

18.14 Although the reduction of copper(II) ions in solution normally results in the deposition of copper metal, it is possible to reduce copper(II) to copper(I). Calculate the standard half-reaction potential for the following half-reaction, using data available from a table of half-reaction potentials.

$$Cu^{2+}(aq) + e^- \longrightarrow Cu^+(aq)$$

Free Energy, Cell Potential, and Equilibrium Constant

18.15 Write equations that relate the following:
   a. $\Delta G°$ and $E°_{cell}$ b. $\Delta G°$ and K
   c. $E°_{cell}$ and K

18.16 Describe various criteria that can be used to decide whether a reaction will be spontaneous.

18.17 What is the equation that relates the potentials for two half-reactions to the potential of a third half-reaction that is the sum of the other two?

18.18 If we multiply the coefficients in a half-reaction by 2, what happens to the value of the half-reaction potential?

18.19 Iron metal may be corroded in the presence of a variety of oxidizing agents, including aqueous tin(IV) ions. Calculate a. the standard free energy and b. the equilibrium constant at 25.0°C for the reaction

$$Fe(s) + Sn^{4+}(aq) \rightleftharpoons Fe^{2+}(aq) + Sn^{2+}(aq)$$

18.20 Calculate a. the cell potential, b. the standard free energy, and c. the equilibrium constant at 25.0°C for the reaction

$$Zn(s) + Pb^{2+}(aq) \rightleftharpoons Zn^{2+}(aq) + Pb(s)$$

   d. Would this reaction provide a means of producing lead metal from its salts?

18.21 Is the discharge reaction for the lead storage battery thermodynamically spontaneous? Show that the standard free-energy change is consistent with your answer.

$$Pb(s) + 4H^+(aq) + 2SO_4^{2-}(aq) + PbO_2(s) \rightleftharpoons$$
$$2PbSO_4(s) + 2H_2O(l)$$

18.22 Calculate the standard free-energy change for the reaction

$$Fe(s) + 2Ag^+(aq) \rightleftharpoons Fe^{2+}(aq) + 2Ag(s)$$

18.23 Calculate a. the standard free-energy change and b. the standard cell potential for the reaction

$$Mg(s) + 2Ag^+(aq) \rightleftharpoons Mg^{2+}(aq) + 2Ag(s)$$

18.24 Calculate a. the standard cell potential, b. the standard free energy, and c. the equilibrium constant at 25.0°C for the reaction

$$2Fe^{2+}(aq) + 2H^+(aq) \rightleftharpoons 2Fe^{3+}(aq) + H_2(g)$$

   d. Will iron(II) ions be oxidized to iron(III) ions in acidic aqueous solution under standard conditions?

18.25 Calculate a. the standard free energy and b. the equilibrium constant at 25.0°C for the reaction

$$Zn(s) + Co^{2+}(aq) \rightleftharpoons Zn^{2+}(aq) + Co(s)$$

Nernst Equation

18.26 How can we predict whether an oxidation–reduction reaction occurring in a cell is spontaneous under standard conditions?

18.27 How can we predict whether an oxidation–reduction reaction occurring in a cell is spontaneous under nonstandard conditions?

18.28 What is the Nernst equation?

18.29 Under what conditions will the following relationships between the cell potential and the standard cell potential be observed?
   a. $E = E°$ b. $E > E°$
   c. $E < E°$ d. $E = 0$

18.30 Calculate the half-reaction potential at 25.0°C for the reaction

$$Zn^{2+}(aq) + 2e^- \longrightarrow Zn(s)$$

The concentration of Zn^{2+} is 0.0120 M.

18.31 What is the concentration of $Mg^{2+}(aq)$ in the following reaction if the half-reaction potential at 25.0°C is −2.42 V?

$$Mg^{2+}(aq) + 2e^- \longrightarrow Mg(s)$$

18.32 Calculate the half-reaction potential at 25.0°C for the reaction

$$Cl_2(aq) + 2e^- \longrightarrow 2Cl^-(aq)$$

The concentration of $Cl_2(aq)$ is 5.00×10^{-4} M, the concentration of $Cl^-(aq)$ is 0.0100 M, and $E° = 1.36$ V.

18.33 Calculate the cell potential at 25.0°C for the reaction

$$Co(s) + Sn^{4+}(aq) \rightleftharpoons Co^{2+}(aq) + Sn^{2+}(aq)$$

The concentration of $Cl_2(aq)$ is 5.00×10^{-4} M, the concentration of $Cl^-(aq)$ is 0.0100 M, and $E° = 1.36$ V.

18.34 Magnesium metal reacts vigorously with aqueous acids. What is the concentration of $Mg^{2+}(aq)$ in the following reaction if the concentration of $H^+(aq)$ is 0.200 M, the pressure of H_2 is 1.00 atm, and the cell potential at 25.0°C is 2.46 V?

$$Mg(s) + 2H^+(aq) \rightleftharpoons Mg^{2+}(aq) + H_2(g)$$

18.35 Calculate the cell potential at 25.0°C for the reaction

$$Zn(s) + Cl_2(aq) \rightleftharpoons Zn^{2+}(aq) + 2Cl^-(aq)$$

The concentrations of $Cl_2(aq)$ and $Cl^-(aq)$ are 0.0100 M, and the concentration of $Zn^{2+}(aq)$ is 5.00×10^{-4} M.

18.36 The standard cell potential for the reaction

$$Zn(s) + 2H^+(aq) \rightleftharpoons Zn^{2+}(aq) + H_2(g)$$

is 0.76 V. What is the standard cell potential at 25.0°C for the following reaction?

$$Zn(s) + 2H_2O(l) \rightleftharpoons Zn^{2+}(aq) + 2OH^-(aq) + H_2(g)$$

Voltaic Cells and Batteries

18.37 What is a battery?

18.38 Why don't the voltages generated by various voltaic cells usually match the cell potential calculated from the standard half-reaction potentials?

18.39 List some voltaic cells and their commercial uses.

18.40 Distinguish between primary and secondary voltaic cells.

18.41 Describe the difference between a Leclanché dry cell and an alkaline dry cell.

18.42 a. Write the reactions that occur at the anode and the cathode in a lead storage battery.
   b. Write the overall reaction that occurs when this battery is being charged.

18.43 a. Draw a diagram of a Leclanché dry cell and label the cathode and anode.
   b. Write balanced equations to describe the processes that occur at each electrode.

18.44 Write the reactions that occur at the anode and the cathode in a mercury cell.

18.45 Write the reactions that occur at the anode and the cathode in a zinc/silver oxide cell.

18.46 Draw a diagram of a voltaic cell based on the following oxidation–reduction reaction:

$$Ni(s) + 2CuBr(s) \rightleftharpoons Ni^{2+}(aq) + 2Br^-(aq) + 2Cu(s)$$

   a. Label the cathode and anode, and draw arrows to show the direction of movement of cations, anions, and electrons.
   b. Indicate whether each electrode gains or loses mass during the electrochemical process.
   c. Write balanced equations to describe the processes that occur at each electrode.

18.47 Draw a diagram of a voltaic cell based on the following oxidation–reduction reaction:

$$4Al(s) + 3O_2(g) + 12H^+ \rightleftharpoons 4Al^{3+} + 6H_2O$$

   a. Label the cathode and anode, and draw arrows to show the direction of movement of cations, anions, and electrons.
   b. Write balanced equations to describe the processes that occur at each electrode.
   c. What voltage would this cell produce under standard conditions?

Electrolysis

18.48 What is electrolysis?

18.49 a. Describe how electrolysis is used to refine impure copper.
   b. How could this process be modified to refine nickel?

18.50 Why is electrolysis of a molten salt generally simpler than electrolysis of an aqueous solution of that salt?

18.51 Describe what happens at each electrode during the electrolysis of molten sodium chloride.

18.52 Describe the preparation of aluminum metal by electrolysis.

18.53 Compare the electrolysis of molten and of aqueous sodium chloride.

18.54 Indicate the major products of the electrolysis of the following aqueous solutions.
   a. sodium iodide, NaI
   b. sodium sulfate, Na_2SO_4
   c. copper sulfate, $CuSO_4$
   d. sodium chloride, NaCl
   e. hydrochloric acid, HCl
   f. aluminum chloride, $AlCl_3$
   g. zinc sulfate, $ZnSO_4$
   h. lead nitrate, $Pb(NO_3)_2$

18.55 What are most likely to be the products of electrolysis of a mixture of $CuSO_4(aq)$ and $HI(aq)$?

18.56 What are most likely to be the products of electrolysis of a mixture of $FeCl_2(aq)$ and $ZnCl_2(aq)$?

18.57 What are most likely to be the products of electrolysis of a mixture of $CuSO_4(aq)$ and $ZnSO_4(aq)$?

18.58 What chemical compounds can be prepared by electrolysis of seawater? Write balanced equations to describe the electrolysis reactions.

Electrolysis and Electrical Energy

18.59 What is a Faraday?

18.60 What is Faraday's law?

18.61 How can Faraday's law be used to determine how much electricity is needed to deposit a specific amount of a metal in an electrolysis process?

18.62 Three electrolytic cells are connected in series so that the same amount of current passes through each. The three cells contain aqueous solutions of $AgNO_3$, $CuCl_2$, and $CrCl_3$. If enough current is passed through the cells to deposit 1.00 mol of copper, a. how many moles of silver and b. how many moles of chromium will be deposited?

18.63 How many grams of silver are deposited by the passage of 0.75 Faraday of electricity through a solution of $AgNO_3$?

18.64 When a current of 10.0 A is passed through a solution of $AuCl_3$ for 1.00 min, what mass of gold, in grams, will be deposited at the cathode?

18.65 During the electrolysis of molten $CaCl_2$, 125 mL of chlorine gas is produced at STP. What mass of calcium is also produced?

18.66 How many grams of sodium are deposited by the passage of 2.85 Faraday of electricity through molten NaCl?

18.67 When a current of 0.250 A is passed through a solution of $CuSO_4$ for 10.0 min, how much copper, in grams, will be deposited at the cathode?

18.68 What current, in amps, is required to deposit 125 g Cu from a solution of $CuSO_4$ in 1.00 h?

18.69 How long must a current of 1.00 A be passed through an aqueous solution of sodium chloride to release 1.25 g H_2 gas?

18.70 What is the equivalent weight of a metal if a current of 0.50 A deposits 0.80 g of the metal in 30.0 min?

18.71 What volume of H_2 at STP is released in the electrolysis of aqueous NaCl carried out with a current of 0.250 A for 2.50 h?

18.72 How many moles of Mg are deposited at the cathode during the electrolysis of molten $MgCl_2$ with a current of 0.65 A for 95 min?

18.73 When 250 mL of 0.20 M $CuSO_4$ solution is electrolyzed for 25 min with a current of 0.15 A, what will be the concentration of copper ions left in solution?

Additional Exercises

18.74 An electrical current is generated in a voltaic cell by the reaction of Cd with Ag^+ to produce Cd^{2+} and Ag.
   a. Write equations for the half-reactions and for the overall cell reaction.
   b. Under standard conditions, what voltage does this cell generate?

18.75 Calculate a. the cell potential, b. the standard free energy, and c. the equilibrium constant at 25.0°C for the following reaction in acidic solution.

$$3Br_2(aq) + 3H_2O(l) \rightleftharpoons$$
$$BrO_3^-(aq) + 5Br^-(aq) + 6H^+(aq)$$

18.76 Write the reactions that occur at the anode and the cathode in a NiCad cell.

18.77 Calculate the half-reaction potential at 25.0°C for the reaction

$$Ni^{2+}(aq) + 2e^- \longrightarrow Ni(s)$$

The concentration of $Ni^{2+}(aq)$ is 0.050 M.

18.78 Calculate the standard cell potential for the reaction

$$Fe(s) + Sn^{4+}(aq) \rightleftharpoons Fe^{2+}(aq) + Sn^{2+}(aq)$$

18.79 Describe the preparation of sodium metal by electrolysis.

18.80 Two electrolysis cells are connected in series, one containing $AgNO_3(aq)$ and the other $CuSO_4(aq)$. If 5.38 g Ag is deposited in the cell containing $AgNO_3$, how much Cu will be deposited in the cell containing $CuSO_4$?

18.81 Calculate the standard cell potential for the reaction

$$Fe(s) + 2Ag^+(aq) \rightleftharpoons Fe^{2+}(aq) + 2Ag(s)$$

18.82 When a current of 1.25 A is passed through a solution of $NiCl_2$ for 35.0 min, how many grams of nickel will be deposited at the cathode?

18.83 Calculate the standard half-reaction potential for the following half-reaction, using data available from a table of half-reaction potentials.

$$Fe^{3+}(aq) + 3e^- \longrightarrow Fe(s)$$

18.84 Calculate the standard free-energy change for the reaction

$$10Br^-(aq) + 2IO_3^-(aq) + 12H^+(aq) \rightleftharpoons$$
$$5Br_2(aq) + I_2(aq) + 6H_2O(l)$$

18.85 How many grams of copper are deposited by the passage of 1.25 Faraday of electricity through a solution of $CuSO_4$?

18.86 Calculate the half-reaction potential at 25.0°C for the reaction

$$Fe(s) \longrightarrow Fe^{2+}(aq) + 2e^-$$

The concentration of $Fe^{2+}(aq)$ is 0.25 M.

18.87   a. Draw a diagram of a voltaic cell based on the following oxidation–reduction reaction.

$$Sn^{2+}(aq) + 2Fe^{3+}(aq) \rightleftharpoons Sn^{4+}(aq) + 2Fe^{2+}(aq)$$

Label the cathode and anode, and draw arrows to show the direction of movement of cations, anions, and electrons.

b. Write balanced equations to describe the processes that occur at each electrode.

18.88 What is the equivalent weight of a metal if 4.18 g of the metal are deposited from solution by passage of a current of 3.00 A through the solution for 0.75 h?

18.89 Calculate the cell potential at 25.0°C for the reaction

$$Fe(s) + Ni^{2+}(aq) \rightleftharpoons Fe^{2+}(aq) + Ni(s)$$

The concentration of $Ni^{2+}(aq)$ is 0.0500 M, and the concentration of $Fe^{2+}(aq)$ is 0.250 M.

18.90 Write the reactions that occur at the anode and the cathode in a hydrogen/oxygen fuel cell.

18.91 What are most likely to be the products of electrolysis of a mixture of $KCl(aq)$, $NaCl(aq)$, $CuSO_4(aq)$, and $H_2SO_4(aq)$?

18.92 Calculate the standard free energy and the equilibrium constant at 25.0°C for the reaction

$$Fe(s) + Ni^{2+}(aq) \rightleftharpoons Fe^{2+}(aq) + Ni(s)$$

18.93 Iron is a more active metal than copper. A common error of amateur plumbers is to connect iron and copper pipes directly; the pipe eventually corrodes at the junction if it is used with impure water or in moist soil. Which metal corrodes? Explain.

18.94 If aluminum siding is fastened to a house with common iron nails, the siding may fall down as a result of corrosion.

a. Considering the relative activity of these metals, determine which metal corrodes.

b. Would corrosion have occurred if aluminum nails had been used? Explain your answers.

* 18.95 An electrochemical cell has one compartment that contains 1.00 M NaCl and an electrode composed of silver coated with silver chloride. The other compartment has an electrode composed of manganese coated with manganese dioxide. Show how this cell could be used to measure the pH of any solution placed in the second compartment.

* 18.96 A type of voltaic cell called a concentration cell has the same components in both cell compartments; the cell fluids differ only in concentration. Consider a cell in which a mixture of $Ce^{4+}(aq)$ and $Ce^{3+}(aq)$ is present in both compartments. The electrodes are platinum wires, which don't participate in the half-reactions but serve only to conduct electrical current. What voltage would be developed by this cell at 25.0°C if one compartment contained 1.00 M $Ce^{4+}(aq)$ and 1.00 M $Ce^{3+}(aq)$ and the other compartment contained 1.00 M $Ce^{4+}(aq)$ and 0.0100 M $Ce^{3+}(aq)$?

* 18.97 Calculate the solubility product constant for silver chromate at 25.0°C, given the following standard reduction potentials:

$$Ag^+(aq) + e^- \longrightarrow Ag(s) \qquad E° = 0.799 \text{ V}$$
$$Ag_2CrO_4(s) + 2e^- \longrightarrow 2Ag(s) + CrO_4^{2-}(aq)$$
$$E° = 0.446 \text{ V}$$

* 18.98 Two electrolysis cells are connected in series; one contains $AgNO_3(aq)$ and the other consists of a solution of a salt that contains molybdenum. The oxidation number of molybdenum is not known. If 1.475 g Ag and 0.328 g Mo are deposited, what is the oxidation number of molybdenum in the solution?

▶ ▶ ▶ Chemistry in Practice

18.99 Complex ions, such as $Cr(H_2O)_5Cl^{2+}$ (abbreviated $CrCl^{2+}$), undergo substitution reactions in aqueous solution. One such reaction is *aquation*:

$$Cr(H_2O)_5Cl^{2+}(aq) + H_2O(l) \longrightarrow$$
$$Cr(H_2O)_6^{3+}(aq) + Cl^-(aq)$$

Understanding the mechanism of such reactions requires careful study of the reaction kinetics. A concentrated solution of $CrCl^{2+}$ is prepared by dissolving $CrCl_3 \cdot 6H_2O$ in 1.0 M $HClO_4$ and puri-

fying it via ion exchange. To analyze the solution, we make 1.00 mL of it basic and then treat it with excess hydrogen peroxide, which oxidizes the chromium(III).

$$2Cr(OH)_3(s) + 3HO_2^-(aq) + OH^-(aq) \longrightarrow$$
$$2CrO_4^{2-}(aq) + 5H_2O(l)$$

The solution is then heated to destroy any excess hydrogen peroxide, after which it is reacidified. This solution is titrated with a solution of iron(II) chloride prepared by reacting 2.9822 g of iron powder with excess dilute hydrochloric acid and diluting to 500.0 mL. The titration of the oxidized chromium solution requires 30.49 mL of the iron(II) solution.

$$CrO_4^{2-}(aq) + 3Fe^{2+}(aq) + 8H^+(aq) \longrightarrow$$
$$Cr^{3+}(aq) + 3Fe^{3+}(aq) + 4H_2O(l)$$

For the kinetics study, an electrochemical cell is set up with two compartments connected by a salt bridge. Each compartment has a Ag/AgCl electrode. The reference compartment is filled with 1.000 M HCl. The other compartment contains 100.0 mL of a solution made by adding 9.24 mL of the $CrCl^{2+}$ solution to 90.76 mL of 1.000 M

$HClO_4$. The voltage developed by this cell is measured periodically. The following table lists results at various times.

Time (h)	E (V)	Time (h)	E (V)
1	0.232	600	0.0760
10	0.173	800	0.0713
50	0.132	1000	0.0681
100	0.115	1200	0.0659
200	0.0984	1400	0.0643
400	0.0836	1600	0.0631

Use these data to determine the reaction order and the rate constant for the aquation reaction.

19

▶ ▶ ▶ ▶

Coordination Chemistry

Compounds of copper(II): Cu(NH$_3$)$_4$$^{2+}$(aq) and Cu(H$_2$O)$_4$$^{2+}$(aq) (top from left); CuSO$_4$·5H$_2$O, Cu(NH$_3$)$_4$SO$_4$·H$_2$O, and CuSO$_4$ (bottom from left).

Encountering Chemistry

Copper sulfate pentahydrate is a blue hydrated solid, $CuSO_4 \cdot 5H_2O$, that dissolves readily in water to form a light blue solution. When solid copper sulfate pentahydrate is heated, it loses water to form a very pale blue (almost white) solid containing no water—anhydrous copper sulfate, $CuSO_4$. Adding colorless aqueous ammonia to anhydrous copper sulfate causes the solid to dissolve. The resulting solution has an intense royal blue color. If acetone is added to make the solvent less polar, a royal blue solid precipitates. This solid has the composition $Cu(NH_3)_4SO_4 \cdot H_2O$.

What causes these changes in color? We have seen that such changes can accompany oxidation–reduction reactions, but that is not the case here, because each of the copper compounds contains copper(II). Sulfate ions are always colorless, so the color must come from the copper(II) ions. The environment in which the copper(II) ions find themselves must have considerable bearing on the color. The environment is provided by the substances that are covalently bonded to the copper: water, ammonia, or no bonded substances (in the anhydrous salt). The formation of copper(II) compounds in which molecules such as water or ammonia are bonded to copper involves Lewis acid–base reactions in which electron pairs are donated from a Lewis base (electron-pair donor) to a Lewis acid (electron-pair acceptor), with the formation of a coordinate covalent bond. The formation of adducts containing coordinate covalent bonds was discussed in Sections 7.6 and 15.1.

Let's consider a specific case. The royal-blue color in both solid $Cu(NH_3)_4SO_4 \cdot H_2O$ and its solution arises from the ion $Cu(NH_3)_4{}^{2+}$. Both copper(II) ion and ammonia are capable of independent existence. Their association can be explained only in terms of electron-pair sharing, ammonia being the electron-pair donor in a Lewis acid–base reaction.

$$CuSO_4(s) + 4NH_3(aq) \longrightarrow Cu(NH_3)_4{}^{2+}(aq) + SO_4{}^{2-}(aq)$$

Copper(II) ion in the anhydrous copper sulfate is a Lewis acid in this reaction; aqueous ammonia is a Lewis base.

The color is due to the Lewis adduct, such as $Cu(H_2O)_4^{2+}$ and $Cu(NH_3)_4^{2+}$. The difference in color is due to the difference in the nature of the bond to the Lewis base. This chapter investigates compounds that contain such ions, the structures of such substances, their reactions, and the manner in which Lewis bases are bonded to Lewis acids. Once the nature of the coordinate covalent bond is understood, we will be in a position to understand the colors of Lewis adducts, as well as other characteristics such as magnetic properties.

19.1 Coordination Compounds

What types of molecules or ions are typically found in coordination compounds?

Many compounds contain coordinate covalent bonds. Such compounds are called *coordination compounds*. Most coordination compounds are formed from a metal atom or ion acting as the Lewis acid and another species (atom, molecule, or ion) acting as the Lewis base. The Lewis acid is often called the *central metal ion* (or atom), because it is surrounded by the Lewis bases. Although the central metal ion or atom may be a main-group metal, most are transition metals, because transition metals have unfilled low-energy *d* orbitals that can accept electron pairs from the Lewis base. The Lewis bases are called the *ligands*. A ligand is simply a species covalently bonded to the central metal atom. The name is derived from the Latin *ligare*, ''to bind.'' ▶ In both $Cu(NH_3)_4SO_4 \cdot H_2O$ and $Cu(NH_3)_4^{2+}$, which were mentioned in the chapter introduction, copper(II) ion is the central metal ion and ammonia is the ligand. In the compound, sulfate ion is ionically bonded to the $Cu(NH_3)_4^{2+}$ ion.

We have noted that coordinate covalent bonds can be found both in compounds and in ions. Ions that contain a metal ion or atom bonded to ligands by coordinate covalent bonds are called *complex ions*. A coordination compound, then, consists of a neutral coordinate-covalent compound or of a complex ion and ions of opposite charge. For example, the coordination compound $Cu(NH_3)_4SO_4 \cdot H_2O$ contains the complex cation $Cu(NH_3)_4^{2+}$ and the anion SO_4^{2-}. It should be noted that the water molecule in $Cu(NH_3)_4SO_4 \cdot H_2O$ is not a ligand; it is hydrogen-bonded to the sulfate ion in this compound. The complex ion can also be an anion. The coordination compound K_2CuCl_4, for example, contains the complex ion $CuCl_4^{2-}$. Coordination compounds that contain complex ions are sometimes called *metal complexes*.

Chlorophyll, the green pigment of plants, consists of several coordination compounds in which magnesium(II) is surrounded by four nitrogen atoms that are part of a large organic molecule. Solutions of chlorophyll are shades of green. Chlorophyll is used to color soaps, oils, waxes, cosmetics, perfumes, and liquors, as well as for dying leather and as an ingredient in deodorizers.

The formula for a complex ion is often enclosed in square brackets to indicate which species are ligands and which are not. Thus we could write the formulas $[Cu(NH_3)_4]SO_4 \cdot H_2O$, $[Cu(NH_3)_4]^{2+}$, and $K_2[CuCl_4]$.

Coordination compounds are very common and important species. Many have biochemical significance. Heme, for example, is a complex ion of iron that is responsible for carrying oxygen in the bloodstream. Vitamin B_{12}, a complex of cobalt, is involved in the synthesis of red blood cells. Chlorophyll, which is involved in the photosynthesis of carbohydrates from carbon dioxide and water in plants, is a complex of magnesium. Coordination compounds are also widely employed as catalysts for various industrial processes used to synthesize organic compounds. Complex ions are often used in electroplating baths to increase efficiency and to yield a smooth and adherent deposit on the metal surface. Further, complex ions are involved in the extraction of metals from their ores. The use of

Potassium dicyanoargentate(I), K[Ag(CN)₂], exists as white crystals that are sensitive to light. When dissolved in acidic solutions, silver cyanide precipitates. The coordination compound is used as a bactericide and in silver-plating baths.

As far as we know, the first synthetic coordination compound was Prussian blue, KFeFe(CN)₆, which contains the ferricyanide ion, Fe(CN)₆³⁻. This compound was prepared by an artist's color maker named Diesbach in about 1700 and is still used as a pigment. Diesbach produced the substance accidentally by heating animal wastes and sodium carbonate in an iron container. The true beginning of coordination chemistry, however, occurred in 1798, when the French chemist B. M. Taessert isolated orange crystals of Co(NH₃)₆Cl₃ by allowing a mixture of CoCl₂ and aqueous ammonia to stand in air. This compound was so unusual that it inspired numerous studies of similar species.

cyanide complexes in the recovery of silver as $[Ag(CN)_2]^-$ and of gold as $[Au(CN)_4]^-$ was described in Section 17.6.

As we have noted, coordination compounds can contain complex ions in which ligands are bonded to the central metal ion by coordinate covalent bonds and the complex ion is bonded to ions of opposite charge by ionic bonds. But how do we know the type of bonds present? For example, how do we know that $K[Ag(CN)_2]$ consists of K^+ and $[Ag(CN)_2]^-$ ions and not a mixture of K^+, Ag^+, and CN^- ions? Some simple experiments can provide evidence for the presence or absence of ions in coordination compounds.

Many of the compounds prepared in the early history of coordination chemistry involved the metallic elements cobalt and chromium. These elements form a wide variety of species that are usually quite stable and colorful and can be studied both in solution and in the solid state. Consider, for example, the compounds formed among cobalt, chloride, and ammonia. These compounds can be found with four, five, or six ammonia groups: $CoCl_3 \cdot 4NH_3$, $CoCl_3 \cdot 5NH_3$, and $CoCl_3 \cdot 6NH_3$. And they all have different colors, as shown in Figure 19.1. It is reasonable to suppose that the ammonia molecules are ligands, because they would not otherwise normally be found in these compounds. But what about the chloride ions? Are they ligands, or are they simply ions of charge opposite to that of the complex ion? One way to determine how the chlorides are bonded in these compounds involves the reaction of solutions of the compounds with silver nitrate solution. If the chloride is present as free chloride ion, it precipitates out of solution as white silver chloride, AgCl, almost immediately. If the chloride is covalently bonded to the cobalt, however, this precipitation reaction happens only very slowly because the chloride is not present in ionic form. The amount of silver chloride formed immediately after addition of excess silver nitrate solution, then, indicates how many of the chlorides are ionic. As shown in Figure 19.2, the number of ionic chlorides varies for the compounds $CoCl_3 \cdot 4NH_3$, $CoCl_3 \cdot 5NH_3$, and $CoCl_3 \cdot 6NH_3$. The numbers of free chloride ions deduced in this way for these compounds

Figure 19.1 Coordination compounds of cobalt, ammonia, and chloride have different colors, depending on the number of ammonia molecules present. *Top and right:* CoCl₃·4NH₃ (in two different forms). *Bottom:* CoCl₃·5NH₃. *Left:* CoCl₃·6NH₃.

Figure 19.2 Addition of excess silver nitrate solution to solutions of *Left:* $CoCl_3 \cdot 4NH_3$, *Center:* $CoCl_3 \cdot 5NH_3$, and *Right:* $CoCl_3 \cdot 6NH_3$ produces different amounts of solid silver chloride.

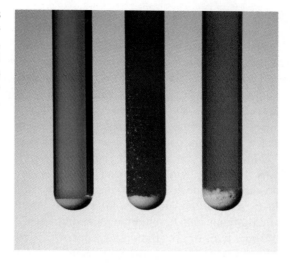

are listed in Table 19.1. Note that the number of ligands—ammonia molecules and nonionic chlorides—is always six in these compounds. Six is the most common number of ligands found in coordination compounds. We will see later that the coordination compounds of every metal have a characteristic number of ligands, called the *coordination number*.

Another way to determine the number of ligands and the number of free ions in a coordination compound involves the measurement of ionic charge on the complex ion: The more anions that serve as ligands, the lower the charge on the complex ion. We can measure the charge by using ion-exchange resins similar to those described in Section 11.5 for softening hard water. Ions can be absorbed on an ion-exchange resin contained in a tube or column. The lower the charge on the ion, the farther down the column the ion moves. Different regions in the column then correspond to ions of different charge.

In what other ways could we determine the charge of a complex ion?

Consider the chromium compound $CrCl_3 \cdot 6H_2O$. This compound exists in several forms that have different colors—blue-gray, dark green, and bright green. The differences in color arise from the identity of the ligands in each compound. Figure 19.3 shows what happens when the different forms of these compounds are absorbed in a column of ion-exchange resin. A blue-gray band stays near the top of the column. A dark green band moves down the column below the blue-gray band, and a bright green band moves even further down the column. The positions of the complex ions on the column indicate that they are blue-gray $[Cr(H_2O)_6]^{3+}$, dark green $[Cr(H_2O)_5Cl]^{2+}$, and bright green $[Cr(H_2O)_4Cl_2]^+$, respectively.

Table 19.1 Compositions and Formulas of Compounds of Cobalt, Chloride, and Ammonia

Composition	Color	Number of Cl^- Precipitated	Formula
$CoCl_3 \cdot 6NH_3$	Yellow	3	$[Co(NH_3)_6]Cl_3$
$CoCl_3 \cdot 5NH_3$	Purple	2	$[Co(NH_3)_5Cl]Cl_2$
$CoCl_3 \cdot 4NH_3$	Green	1	$[Co(NH_3)_4Cl_2]Cl$
$CoCl_3 \cdot 4NH_3$	Violet	1	$[Co(NH_3)_4Cl_2]Cl$

Figure 19.3 The extent of movement of a complex ion down a column of yellow ion-exchange resin is an indication of the charge on the complex ion. Here the top band is blue-gray $[Cr(H_2O)_6]^{3+}$, the middle band is dark green $[Cr(H_2O)_5Cl]^{2+}$, and the bottom band is bright green $[Cr(H_2O)_4Cl_2]^+$.

19.2 Werner's Coordination Theory

Although many coordination compounds were synthesized and studied in the 18th and 19th centuries, it was not until 1893 that a plausible explanation for the behavior and composition of coordination compounds was proposed. Alfred Werner was awarded the Nobel Prize for this work in 1913. The essence of Werner's theory remains valid today and forms the basis for our present understanding of coordination compounds of metals.

What determines the composition of a coordination compound?

Oxidation Number, Coordination Number, and Coordination Sphere

Werner proposed that the nature of a coordination compound is determined by two factors, which we now call the oxidation number and the **coordination number,** the number of ligands that surround a central metal atom. According to Werner's theory, the coordination number is a constant for each metal. That is, whenever the metal has a particular oxidation number and forms a coordination compound, it is surrounded by the same number of ligands. The ligands may be anions, neutral molecules, or both. How many ligands are necessary to satisfy the needs of a metal cation? The oxidation number imposes one requirement on the composition of the coordination compound. Enough anions must be present to neutralize the charge of the metal cation. These anions may be ligands, but if there are not enough anions present as ligands to neutralize the charge, others must be present as free ions. The charge on the complex ion is determined by the number of anions that are ligands, but the coordination compound itself must bear no charge.

The metal atom or ion and its covalently bonded ligands are often collectively referred to as the first **coordination sphere** of the metal. The first coordination sphere, then, contains the complex ion. The first coordination spheres of an anionic, a cationic, and an uncharged complex are shown in Figure 19.4. In the formulas of coordination compounds, the portion enclosed in square brackets de-

Figure 19.4 The first coordination sphere of a coordination compound includes the metal ion and its ligands.

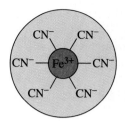

A. $[Fe(CN)_6]^{3-}$

coordination
number = 6

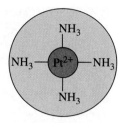

B. $[Pt(NH_3)_4]^{2+}$

coordination
number = 4

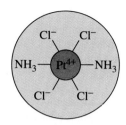

C. $[Pt(NH_3)_2Cl_4]$

coordination
number = 6

scribes the first coordination sphere of the coordination compound: the complex ion. In $K_3[Fe(CN)_6]$, for example, the complex ion is $[Fe(CN)_6]^{3-}$. The first coordination sphere includes iron(III) and six cyanide-ion ligands, as shown in Figure 19.4A. Three of the ligands must be present to neutralize the 3+ charge of iron. The additional ligands are necessary to satisfy the typical coordination number of 6 for iron. Because this results in a complex anion, the potassium ions are present to neutralize the charge in the coordination compound. The first coordination sphere of $[Pt(NH_3)_4]Cl_2$ includes the platinum(II) ion with its four ammonia ligands (Figure 19.4B). The chlorides present in the coordination compound neutralize the charge of the complex cation. In neutral coordination compounds such as $[Pt(NH_3)_2Cl_4]$, the entire compound makes up the first coordination sphere (Figure 19.4C).

Potassium hexacyanoferrate(III), $K_3[Fe(CN)_6]$, forms ruby-red crystals and a yellowish-red solution. It is used to prepare a blue coloring material for blueprints, wood stains, wool dyes, and fabric printing.

Some characteristic coordination numbers are 6 for Co(III), Cr(III), Fe(II), Fe(III), and Pt(IV); 4 for Cu(II), Zn(II), Pt(II), Pd(II), and Au(III); 2 for Cu(I), Ag(I), and Au(I); and 8 for Zr(IV) and Th(IV).

EXAMPLE 19.1

Determine the oxidation number and the coordination number of the metal in $[Cr(NH_3)_4Cl_2]Cl$.

Solution   Chloride ion has a charge of $1-$, whereas ammonia is uncharged. The three chloride ions give a total anionic charge of $3-$, so the chromium must have an oxidation number of $+3$ to neutralize the anionic charge. There are four ammonia molecules and two chloride ions covalently bonded to the chromium(III), so the coordination number is 6, the total of the numbers of molecules and ions bonded to the chromium.

Practice Problem 19.1

Determine the oxidation number and the coordination number of the metal in $[Zn(NH_3)_2Cl_2]$.

Answer: oxidation number, $+2$; coordination number, 4

Coordination Number and Geometry

Another of Werner's postulates was that the geometry of the first coordination sphere is correlated with the coordination number. That is, the ligands around the central metal ion point in particular and predictable directions. For example, six ligands around a central metal ion point to the corners of an octahedron. On this basis, Werner was able to predict, for example, that there should be two different forms of the coordination compound $[Co(NH_3)_4Cl_2]Cl$, with the two chlorides in the first coordination sphere either next to one another or opposite one another in the octahedron. The two forms are the green and violet forms listed in Table 19.1 and shown in Figure 19.1.

The most common coordination numbers are 4 and 6. However, all coordination numbers from 2 through 10 have been found in various coordination compounds. The coordination numbers and their geometries are outlined in Table 19.2. Complex ions such as $[AgCl_2]^-$, which have a coordination number of 2, are linear. Complex ions such as $[Ag(P(CH_3)_3)_3]^+$, which have a coordination number of 3, have a triangular planar structure. This coordination number is relatively rare. For a coordination number of 4, two geometries are possible: square planar and tetrahedral. Complexes of Pt(II), Pd(II), and Au(III), such as $[PtCl_4]^{2-}$, have a coordination number of 4 and are square planar. Complex ions of Zn(II) and Co(II), as well as a number of other metal ions with a coordination number of 4, are tetrahedral. Examples include $[Zn(NH_3)_4]^{2+}$ and $[CoCl_4]^{2-}$. Two possible shapes are also associated with a coordination number of 5: trigonal bipyramidal and square pyramidal. Trigonal bipyramidal complexes are more common than square pyramidal ones. A variety of shapes fall between these two ideal shapes, but they do not vary greatly in energy or stability. For a coordination number of 6,

How can we predict the geometry of complex ions?

Draw pictures of all the shapes described here.

Table 19.2 Common Coordination Numbers and Their Geometries

Coordination Number	Geometry	Example	Structure
2	Linear	$[Ag(NH_3)]_2^+$	$H_3N—Ag—NH_3$
4	Square planar	$[PtCl_4]^{2-}$	
	Tetrahedral	$[CuCl_4]^{2-}$	
6	Octahedral	$[Co(NH_3)_6]^{3+}$	

complex ions have octahedral geometry. An octahedron has eight faces but only six corners. The six ligands point to the corners of the octahedron. The very common Co(III) and Cr(III) complex ions all have this shape.

EXAMPLE 19.2

What are the possible geometries of $[Zn(NH_3)_2Cl_2]$?

Solution   The coordination number of zinc in this compound is 4, because there are two ammonia molecules and two chloride ions in the coordination sphere. The possible geometries for a coordination number of 4 are tetrahedral and square planar. (The actual structure is tetrahedral.)

Practice Problem 19.2

What are the possible geometries of $[Cr(NH_3)_4Cl_2]Cl$?

Answer: octahedral

Ligands

So far, we have seen in various examples that H_2O, NH_3, Cl^-, CN^-, and $P(CH_3)_3$ can behave as ligands toward metal ions. There are many other ligands as well. Some of the more common ones are listed in Table 19.3, along with the names used for these molecules or ions when they are ligands. In Section 19.3 we will see how these names are formed.

Monodentate and Polydentate Ligands   Most ligands occupy only one site in the coordination sphere of a metal. These ligands donate a single pair of electrons to the metal. A ligand that occupies only one coordination site is called **monodentate,** from Latin roots meaning "one-toothed."

Why does a molecule have to be long to serve as a polydentate ligand?

Some ligands have enough electron pairs available and are large enough that they can occupy more than one coordination site. These ligands are called **polydentate.** An example is ethylenediamine, abbreviated *en*, which has the formula $H_2\ddot{N}{-}CH_2{-}CH_2{-}\ddot{N}H_2$. This molecule has two unshared electron pairs, one on each nitrogen, that can be donated to a metal. It is long enough to stretch across two adjacent coordination sites on most metal ions. Thus ethylenediamine can occupy two coordination sites; it is **bidentate.** It is possible, for example, to form cobalt(III) complexes that contain one or more ethylenediamine molecules. In these complexes, the coordination number of 6 can be satisfied by replacing two monodentate ligands with one ethylenediamine. Cobalt(III) complexes are possible that contain either ammonia or ethylenediamine or both:

Ethylenediamine, $H_2NCH_2CH_2NH_2$, is a colorless, viscous liquid with the odor of ammonia. It often has a yellow color due to the presence of impurities. Besides its use as a chelating ligand, it is used to stabilize rubber latex and to serve as a solvent for shellac, sulfur, and albumin and as a rust inhibitor in antifreezes.

$$[Co(NH_3)_6]^{3+} \qquad [Co(NH_3)_4(en)]^{3+} \qquad [Co(NH_3)_2(en)_2]^{3+} \qquad [Co(en)_3]^{3+}$$

The structures of some complex ions containing ethylenediamine are shown in Figure 19.5. The ethylenediamine wraps around the metal ion from one octahedral site to another, making a ring containing the metal ion.

A widely used polydentate ligand is ethylenediaminetetraacetic acid, $(HOOC{-}CH_2)_2{-}\ddot{N}{-}CH_2{-}CH_2{-}\ddot{N}{-}(CH_2{-}COOH)_2$, abbreviated EDTA.

Table 19.3 Formulas and Names of Some Common Ligands

Formula	Name
H_2O	aqua
NH_3	ammine
CO	carbonyl
NO	nitrosyl
$H_2NC_2H_4NH_2$	ethylenediamine
OH^-	hydroxo
O^{2-}	oxo
F^-	fluoro
Cl^-	chloro
Br^-	bromo
I^-	iodo
CN^-	cyano
$—NCS^-$	isothiocyanato*
$—SCN^-$	thiocyanato*
SO_4^{2-}	sulfato
SO_3^{2-}	sulfito
NO_3^-	nitrato
$—NO_2^-$	nitro*
$—ONO^-$	nitrito*
CO_3^{2-}	carbonato

*In these ligands two forms are known; they differ in the atom that donates the electron pair to the metal ion.

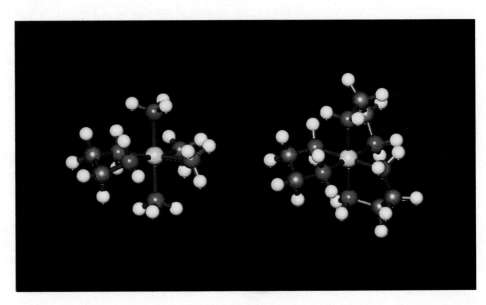

Figure 19.5 Ethylenediamine is a bidentate ligand, as can be seen from the structures of $[Co(NH_3)_4(en)]^{3+}$ and $[Co(en)_3]^{3+}$.

This molecule (and the anions that can be formed by loss of the acidic hydrogens) has six atoms with unshared electron pairs that can be donated to a metal: the two nitrogen atoms and one oxygen atom in each of the four —COOH (carboxylic acid) groups. Depending on how many of these electron pairs are donated to the metal, complex ions formed with EDTA may be tetradentate, pentadentate, or hexadentate.

When a ligand is polydentate, it wraps around the metal like a claw. The resulting complex is called a **chelate**, from the Greek word for "claw." EDTA is a commonly used *chelating agent*—a ligand that forms a chelate with a metal ion. EDTA forms a chelate with a wide variety of metal ions. For example, it can be used to form a chelate with calcium ions in water. The calcium ions are tied up, so they cannot be precipitated from solution as a carbonate salt. This results in soft water. EDTA is also used in bathroom cleansers to help remove deposits of calcium carbonate. EDTA can be used to form a chelate with heavy-metal ions to help remove these poisonous metals (most commonly lead) from the body. It is used in liquid plant fertilizers to form a chelate with iron. This "solubilized" iron does not hydrolyze in soil, so it can be absorbed by plants. If the iron were present as a salt such as $FeCl_3$ instead of as a chelate, the iron(III) ions would hydrolyze and precipitate as $Fe(OH)_3$ in the soil. In this form, the iron would not be available for uptake by plants. EDTA is also used to remove an iron taste from mayonnaise; iron is present because mayonnaise is slightly acidic and is prepared in steel containers.

Why does $FeCl_3$ hydrolyze but the EDTA complex of iron(III) does not?

19.3   Nomenclature of Coordination Compounds

Why can't we use the same rules for naming coordination compounds as for naming other compounds?

The nomenclature rules developed in Chapter 3 for inorganic compounds do not work with coordination compounds, which have more complex structures and compositions. Thus we must use a different set of rules for coordination compounds. We will deal here only with rules pertaining to some relatively simple complexes. These rules, devised by Werner and later modified somewhat, are outlined in Table 19.4. If a coordination compound, such as $[Co(NH_3)_6][Fe(CN)_6]$, contains both a cation and an anion that are complex ions, each is named by these rules. If one of the ions is not a complex ion, it is named by the normal rules of nomenclature. In addition to naming complex ions in coordination compounds, we can use these rules to name complex ions as they exist in solution.

Let's consider first a cation with ligands of two different types, an anion and a neutral molecule.

EXAMPLE 19.3

Name the complex cation $[Co(NH_3)_5Br]^{2+}$.

Solution   We consider the ligands first and the central atom last, as stated in Rule 3a. The ligands are NH_3, named *ammine* (Rule 3c), and Br^-, named *bromo* (Rule 3d). They are placed in alphabetical order (Rule 3b). There are five ammine groups, so ammine should be given the prefix *penta-* (Rule 3e). The central metal takes its usual name, *cobalt*, because the complex ion is a cation (Rule 4). The net ionic charge is 2+ and there is one anionic

Table 19.4 Nomenclature of Coordination Compounds

1. The cation is named first in ionic compounds, then the anion.
2. Nonionic compounds are given a one-word name.
3. The following rules pertain to the names of ligands.
   a. The ligands are named first and the central atom last.
   b. Ligands are named in alphabetical order by their root name.
   c. Neutral ligands are named the same as the molecule, except for a few such as H_2O (*aqua*) and NH_3 (*ammine*), which have special names.
   d. Anionic ligands are named by adding *-o* to the stem of the usual name, such as *chloro* for Cl^- and *sulfato* for SO_4^{2-}. (A number of these names are listed in Table 19.3.)
   e. The name of each ligand is preceded by a Latin prefix (*di-*, *tri-*, *tetra-*, *penta-*, *hexa-*, etc.) if more than one of that ligand is bonded to the central atom. For example, the ligands in $PtCl_4^{2-}$ are named *tetrachloro*, and the ligands in $Co(NH_3)_4Cl_2^+$ are named *tetraamminedichloro*.

   If the ligand is polydentate, as in ethylenediamine, the number of ligands bonded to the central atom is indicated by the corresponding Greek prefixes (*bis-*, *tris-*, *tetrakis-*, *pentakis-*, *hexakis-*, etc.). For example, the ligands in $Co(en)_3^{3+}$ are named *trisethylenediamine*. A Greek prefix is also used when a Latin prefix forms a part of the name of the ligand, as in triethylamine, $N(CH_3)_3$. In this case, the ligand name is enclosed in parentheses. For example, the ligands in $[Co(N(CH_3)_3)_4]^{2+}$ are named *tetrakis(triethylamine)*.

4. For a cationic complex ion or a nonionic compound, the central atom is given its ordinary name followed by its oxidation number in Roman numerals, enclosed in parentheses. For example, $[Cr(H_2O)_5Cl]^{2+}$ is named pentaaquachlorochromium(III) ion, and $[Cr(NH_3)_3Cl_3]$ is named triamminetrichlorochromium(III).
5. For anionic complex ions, the suffix *-ate* is added to the name of the central atom, followed by the oxidation number in Roman numerals, enclosed in parentheses. For example, $[Cr(CN)_6]^{3-}$ is named hexacyanochromate(III) ion.

ligand, Br^-, giving a total ligand charge of $1-$, so the oxidation number is found as follows:

$$\text{Net ionic charge} = \text{oxidation number of metal} + \text{ionic charges of ligands}$$
$$2+ = \text{oxidation number} + (1-)$$
$$\text{Oxidation number} = +3$$

The oxidation number is written in Roman numerals enclosed in parentheses (Rule 4). The name of the complex ion is thus pentaamminebromocobalt(III) ion.

Practice Problem 19.3

Name the complex cation $[Cr(NH_3)_4F_2]^+$.

Answer: tetraamminedifluorochromium(III) ion

Neutral coordination compounds are named in exactly the same way that a complex cation is named. For example, the coordination compound $[Pt(NH_3)_2Cl_2]$ is named diamminedichloroplatinum(II). If the complex ion is an anion, the name of the central atom must end with *-ate*. Otherwise, the previous procedure applies.

EXAMPLE 19.4

Name the complex anion $[Pt(NH_3)Cl_3]^-$.

Solution   The ligands are *ammine* and *chloro*, in alphabetical order. There are three chlorides, so the prefix *tri-* should precede *chloro*. Platinum is called *platinate* here, because the complex ion is an anion. There are three Cl^- ions and a net ionic charge of $1-$. Thus the oxidation number of platinum is $+2$. The name is amminetrichloroplatinate(II) ion.

Practice Problem 19.4

Name the complex anion $[CuCl_4]^{2-}$.

Answer: tetrachlorocuprate(II) ion

If the complex ion contains more than two ligands, the procedure is the same: Each ligand must be named, and the names must be arranged in alphabetical order. For example, the complex ion $[Cr(NH_3)_2(H_2O)_2BrCl]^+$ is named diamminediaquabromochlorochromium(III) ion.

When the ligands are polydentate or when their names already contain number prefixes, the Latin prefixes *di-*, *tri-*, and so on are replaced by Greek prefixes (*bis-*, *tris-*, etc.). This is true, for example, of the ligand ethylenediamine (*en*), which is polydentate, and of trimethylamine ($N(CH_3)_3$) and triphenylphosphine ($P(C_6H_5)_3$), which begin with *tri*. The names of these ligands are enclosed in parentheses when they are preceded by a Greek number prefix. Thus the coordination compound $[Pt(P(C_6H_5)_3)_4]$ is named tetrakis(triphenylphosphine)platinum(0), and the complex ion $[Co(en)_2(H_2O)_2]^{3+}$ is named diaquabis(ethylenediamine)cobalt(III) ion.

When naming a coordination compound that contains both a cation and an anion, we name each ion separately. The name of the cation is given first, then the name of the anion. If one of these ions is not a complex ion, it is named according to normal rules of nomenclature. The coordination compound $[Cr(NH_3)_4F_2]ClO_4$ is called tetraamminedifluorochromium(III) perchlorate.

Suppose we know the name of a compound but not the formula. We can find the formula simply by reversing the procedure used to name coordination compounds.

EXAMPLE 19.5

Write the formula for the compound tetraaquadicyanochromium(III) perchlorate.

Solution   Consider the cation first. It contains four water molecules (H_2O) and two cyanide ions (CN^-) as ligands. Chromium has an oxidation number of $+3$, so the complex ion must have a net charge of $1+$. The formula of the complex ion, then, is $[Cr(H_2O)_4(CN)_2]^+$. The anion is the perchlorate ion, which has the formula ClO_4^-. Because the cation has one unit of positive charge and the anion has one unit of negative charge, the final formula for the compound must contain equal numbers of these species: $[Cr(H_2O)_4(CN)_2]ClO_4$.

Practice Problem 19.5

Write the formula for triammineaquadichlorocobalt(III) sulfate.

Answer: $[Co(NH_3)_3(H_2O)Cl_2]_2SO_4$

19.4 Geometry and Isomerism

How can compounds that differ in such obvious properties as color have the same chemical formula?

As can be seen in Figure 19.1, $[Co(NH_3)_6]Cl_3$ has only one form (yellow) and $[Co(NH_3)_5Cl]Cl_2$ has only one form (red), whereas $[Co(NH_3)_4Cl_2]Cl$ has two forms, one green and the other violet. Many coordination compounds exist in two or more forms that have the same chemical composition. Such compounds are called **isomers.** Isomers can be placed into two general classes: **stereoisomers,** in which the bonding in the compounds is identical, and **structural isomers,** in which the bonding is different but the overall composition is the same.

Stereoisomers

Stereoisomers, in which the bonding is the same, contain exactly the same ligands in their first coordination spheres. They differ, however, in the arrangement of those ligands. There are two types of stereoisomers, **geometrical isomers** and **optical isomers.**

Geometrical Isomers   Geometrical isomers differ in the arrangement of two or more different ligands with respect to one another. The two forms of $[Co(NH_3)_4Cl_2]Cl$ are examples. In this compound, the central metal atom is surrounded by six ligands, so the coordination number is 6. Recall from Table 19.2 that this coordination number is associated with an octahedral arrangement of the ligands. In a first coordination sphere of this geometry, where there are at least two different ligands and at least two of each kind, there can be more than one arrangement of the ligands. Two arrangements of the ligands are possible in $[Co(NH_3)_4Cl_2]^+$, as shown in Figure 19.6. In one, called the *cis* isomer, the two chloride ligands are next to one another. In the other, the *trans* isomer, the two chloride ligands are opposite one another. For any complex ion with a formula of MA_4B_2, these are the only two forms possible. Other arrangements may look

Figure 19.6 Geometrical isomers of $[Co(NH_3)_4Cl_2]^+$. *(A)* Orientation of the ligands in the *cis* isomer. *(B)* Orientation of the ligands in the *trans* isomer. *(C)* and *(D)* are computer-generated models.

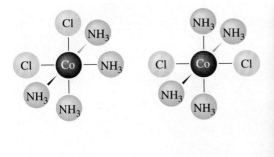

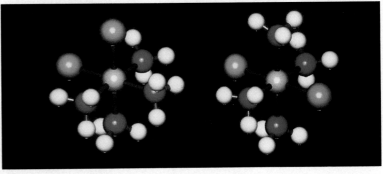

A   B   C   D

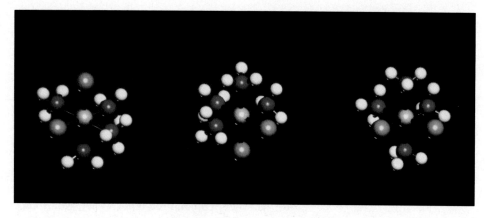

different, but they can be made identical to one of these two arrangements simply by rotation around one of the principal axes. For example, several different orientations of the *cis* isomer of [Co(NH$_3$)$_4$Cl$_2$]Cl are shown in Figure 19.7. Note that all of these arrangements have the two chloride ligands next to one another and differ only in the directions in which the chlorides are pointing.

The *cis* and *trans* designations can be included in the names and formulas used to identify the isomers. The green form of [Co(NH$_3$)$_4$Cl$_2$]Cl contains *trans*-[Co(NH$_3$)$_4$Cl$_2$]$^+$, which is called *trans*-tetraamminedichlorocobalt(III) ion. The violet form contains *cis*-[Co(NH$_3$)$_4$Cl$_2$]$^+$, called *cis*-tetraamminedichlorocobalt(III) ion.

Other geometries also give rise to *cis* and *trans* isomers. Here we will discuss only one more—square planar geometry. Metal complexes with a coordination number of 4 have two possible geometries, tetrahedral and square planar. In the tetrahedral geometry, every ligand is adjacent to all the other ligands, as shown in Table 19.2, so it is not possible to have different geometrical arrangements. With the square planar geometry, however, two arrangements are possible if there are two different ligands, each present in equal number, and these arrangements have like ligands either across from one another (*trans*) or adjacent to one another (*cis*). An example is [Pt(NH$_3$)$_2$Cl$_2$], shown in Figure 19.8. Geometrical isomerism can occur in any square planar complex of the general formula MA_2B_2.

Figure 19.8 Geometrical isomers of [Pt(NH$_3$)$_2$Cl$_2$]. *(A)* Orientation of the ligands in the *cis* isomer. *(B)* Orientation of the ligands in the *trans* isomer. (C) and (D) are computer-generated models.

Optical Isomerism   A second type of stereoisomerism is optical isomerism, which exists when the mirror image of a structure cannot be made identical to the original image by rotation of the molecule or ion. Optical isomers are related to one another

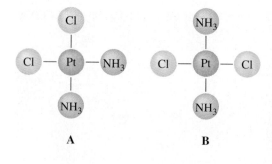

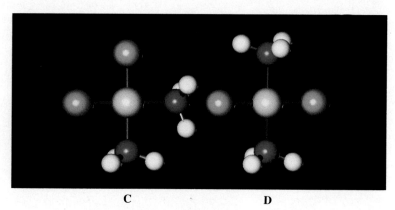

A   B

C   D

in the same way a right hand is related to a left hand. They are not superimposable, even though they have essentially the same structural features. When two optical isomers exist for a molecule or ion, they are called **enantiomers,** or opposite forms.

How can we distinguish between optical isomers?

Enantiomers are often chemically similar and have identical physical properties, with one exception—their optical properties. The isomers interact differently with plane-polarized light. Light normally consists of waves oriented in all directions. When a polarizing filter such as that used in Polaroid sunglasses is placed in a light beam, the light waves that pass through the filter are oriented in a single plane, producing plane-polarized light. Only light waves oriented in the plane parallel to that of the filter pass through the filter, as shown in Figure 19.9. Plane-polarized light that is passed through a solution containing a single enantiomer is rotated by some specific angle, depending on the concentration of the solution and on the identity of the coordination compound. This angle, α, is determined by using a second polarizing filter, as shown in Figure 19.9. No light passes through this filter unless it is aligned with the plane of the rotated light. If the solution contains the other enantiomer instead, the plane-polarized light is rotated the same amount, but in the opposite direction. If the solution contains an equal mixture of the two optical isomers (called a *racemic mixture*), their effects cancel each other, and the plane of the light is not rotated.

How do we know which coordination compounds will have optical isomers? We have seen that the isomers cannot be superimposable by rotation. However, it is not necessary to examine all compounds to see if they have superimposable isomers. We can eliminate some compounds.

Let's consider coordination compounds that have a coordination number of 6. Optical isomers do not exist for octahedral species that contain only two different ligands unless one of the ligands is bidentate. Thus, for example, neither *cis*- nor *trans*-$[Co(NH_3)_4Cl_2]^+$ has optical isomers; but *cis*-$[Co(en)_2Cl_2]^+$, which contains bidentate ligands, does have optical isomers. The isomers of *cis*-$[Co(en)_2Cl_2]^+$ are pictured in Figure 19.10, which also shows that *trans*-$[Co(en)_2Cl_2]^+$ does not have optical isomers (the mirror image of its structure is identical to the original image). Neither are optical isomers found in an octahedral complex ion that contains only one bidentate ligand, such as $[Co(en)(CN)_4]^-$, because the mirror images of the structures are always superimposable. However, complex ions with three bidentate ligands, such as $[Co(en)_3]^{3+}$, have nonsuperimposable mirror images and thus exist as enantiomers.

Optical isomers are usually not found in coordination compounds with a coordination number of 4. Because a square planar complex can always be flipped over, it is never possible to find optical isomers with this geometrical arrangement. Tetrahedral complexes give optical isomers only if all four ligands are different.

Figure 19.9 Interaction of plane-polarized light with optical isomers.

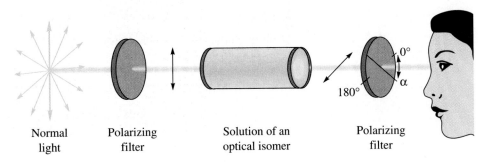

| Normal light | Polarizing filter | Solution of an optical isomer | Polarizing filter |

Figure 19.10 Optical isomers exist for *cis*-[Co(en)$_2$Cl$_2$]$^+$ but not for *trans*-[Co(en)$_2$Cl$_2$]$^+$. In the structures, N N represents ethylenediamine.

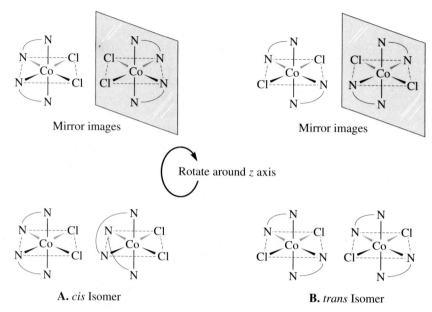

A. *cis* Isomer   **B.** *trans* Isomer

Structural Isomers

Structural isomers differ in either the identity of the ligands or the way in which the ligands are bonded. Thus these isomers involve more than just a difference in the spatial orientation of the ligands. Types of structural isomers include ionization isomers, hydration isomers, and linkage isomers.

Ionization isomers differ in the identity of the ion in the coordination sphere. Consider the isomers [Co(NH$_3$)$_4$(NO$_2$)Cl]Cl and [Co(NH$_3$)$_4$Cl$_2$]NO$_2$. Nitrite is a ligand in the first isomer, but in the second isomer, nitrite is the anion that neutralizes the charge of the complex cation. Nitrite and chloride have exchanged places in the first coordination sphere of the cobalt complex ion.

EXAMPLE 19.6

Write the formula of an ionization isomer of [Co(NH$_3$)$_4$Br$_2$]Cl.

Solution   Ionization isomers involve the exchange of anionic ligands and free anions, in this case bromide and chloride. Another isomer would have bromide as the anion:

$$[Co(NH_3)_4BrCl]Br$$

Practice Problem 19.6

Write the formula of an ionization isomer of [Pt(NH$_3$)$_3$Cl]F.

Answer: [Pt(NH$_3$)$_3$F]Cl

Hydration isomers are similar to ionization isomers, except that it is water that may either be coordinated to the central metal atom or be outside the coordination sphere. The compounds [Co(NH$_3$)$_4$(H$_2$O)Cl]Cl$_2$ and [Co(NH$_3$)$_4$Cl$_2$]Cl·H$_2$O,

Geometrical Isomers and Cancer

...cancer is a disease in which cells undergo rapid, uncontrolled division. If cell division could be prevented by addition of a platinum compound, it might be possible to control cancerous tumors.

In 1964 a group of scientists at Michigan State University, headed by Barnett Rosenberg, discovered that bacteria did not undergo cell division in the presence of platinum electrodes. Instead, they just grew into larger, elongated cells. By individually varying each component of the electrochemical cell, Rosenberg and his colleagues eventually discovered that the important components were platinum metal, oxygen gas, ammonia, and chloride ion. They then added various platinum(IV) and platinum(II) compounds containing ammonia or ammonium ion and chloride ion, such as $(NH_4)_2[PtCl_6]$ and $(NH_4)_2[PtCl_4]$. Most of the compounds had no effect. The only compound that prevented cell division was $[Pt(NH_3)_2Cl_2]$; and only the *cis* isomer of that compound worked. This study was important because cancer is a disease in which cells undergo rapid, uncontrolled division. If cell division could be prevented by addition of a platinum compound, it might be possible to control cancerous tumors. When *cis*-$[Pt(NH_3)_2Cl_2]$, now called *cisplatin*, was administered to mice with tumors, the tumors were destroyed. Ultimately, the compound was developed as an anticancer drug for humans. It works very well in the treatment of testicular, ovarian, and bladder cancers. Unfortunately, like compounds of most heavy metals, platinum compounds are toxic to the kidneys. To combat this, patients must drink lots of water so that the platinum compound is diluted as it passes through the kidneys.

It is thought that the *cis*, but not the *trans*, isomer works because it can lose two adjacent ligands and form coordinate covalent bonds to close positions in a DNA molecule, as shown in Figure 19.11. Researchers continue to

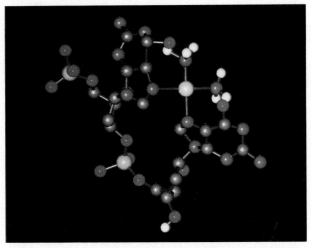

Figure 19.11 Cisplatin works as an anticancer agent by binding to two nitrogen atoms in DNA.

search for other *cis* platinum and palladium complexes that might also be effective against cancer but have lower kidney toxicity.

Questions for Discussion

1. How can a metal both be toxic and be effective as a cure?
2. It has been said that "The dose is the poison." What is meant by this statement?
3. Why do so many drugs (medicines) have undesirable side effects?
4. It takes many years—often decades—of testing for a new pharmaceutical drug to meet the stringent requirements of the United States Food and Drug Administration (FDA). Only then may it be made available to patients. Should the FDA loosen its restrictions so that drugs can become available sooner?

shown in Figure 19.12, are hydration isomers. Here, water is a ligand in the first isomer and is water of hydration in the second isomer.

Some ligands, which are called *ambidentate*, can bond to a metal at more than one atomic site. For example, the cyanide ion, CN^-, can bond to a metal either by

Figure 19.12 The hydration isomers Co[(NH$_3$)$_4$(H$_2$O)Cl]Cl$_2$ *(Left)* and [Co(NH$_3$)$_4$Cl$_2$]Cl · H$_2$O *(Right)* have different colors.

donating the electron pair on the carbon or (infrequently) by donating the electron pair on the nitrogen. The cyanide ion can bond in either way with chromium, forming the isomers [Cr(H$_2$O)$_5$CN]$^{2+}$ and [Cr(H$_2$O)$_5$NC]$^{2+}$, which are called *linkage isomers*. The first isomer is pentaaquacyanochromium(III) ion, and the second is pentaaquaisocyanochromium(III) ion. Other ligands that can give linkage isomers include thiocyanate (—SCN$^-$ and —NCS$^-$) and nitrite (—NO$_2$$^-$ and —ONO$^-$). Thus we have pentaammineisothiocyanatocobalt(III) ion, [Co(NH$_3$)$_5$NCS]$^{2+}$, and pentaamminethiocyanatocobalt(III) ion, [Co(NH$_3$)$_5$SCN]$^{2+}$, as well as triamminenitroplatinum(II) ion, [Pt(NH$_3$)$_3$NO$_2$]$^+$, and triamminenitritoplatinum(II) ion, [Pt(NH$_3$)$_3$ONO]$^+$.

19.5   Reactions of Coordination Compounds

In which ways are the reactions of complex ions different from those of other ions?

Coordination compounds can undergo essentially three types of reactions: oxidation–reduction reactions, which involve changes in the oxidation number of the central metal atom; substitution reactions, which involve changes in the coordination sphere; and various reactions of the coordinated ligands.

Oxidation–Reduction Reactions

As a result of oxidation–reduction reactions, the central metal atom may change its oxidation number by transferring one or more electrons from one coordination compound to another. When coordination compounds undergo oxidation or reduction, the coordination sphere may remain intact or may change drastically as a result of the oxidation–reduction. On the basis of this difference, oxidation–reduction reactions are classified into two types: outer-sphere and inner-sphere reactions.

In *outer-sphere reactions,* the coordination spheres of the reactants do not change during the reaction. Usually, the coordination spheres of the products are the same as those of the reactants. The only change is a difference in the oxidation number of reactants and products. An example is given by the reaction of hexacyanoferrate(II) ion with hexachloroiridate(IV) ion.

$$\text{Fe(CN)}_6{}^{4-}(aq) + \text{IrCl}_6{}^{2-}(aq) \longrightarrow \text{Fe(CN)}_6{}^{3-}(aq) + \text{IrCl}_6{}^{3-}(aq)$$

The reactants contain Fe(II) and Ir(IV); the products contain Fe(III) and Ir(III). The iron is oxidized and the iridium reduced. The ligands bonded to the reactants remain bonded to the same metals in the products. The transfer of an electron in this oxidation–reduction reaction can be seen readily by examining the component half-reactions.

$$Fe(CN)_6^{4-}(aq) \longrightarrow Fe(CN)_6^{3-}(aq) + e^-$$
$$IrCl_6^{2-}(aq) + e^- \longrightarrow IrCl_6^{3-}(aq)$$

Reactions such as this can, in fact, be carried out in separate compartments of an electrochemical cell.

In *inner-sphere reactions*, a change in the coordination sphere of at least one reactant accompanies the transfer of an electron. One ligand simultaneously bonds to the metals in both reactants, acting as a bridge to help the electron transfer from the reducing agent to the oxidizing agent. Frequently, inner-sphere reactions are accompanied not only by a change in oxidation numbers but also by the transfer of a ligand from one metal to the other. One of the first examples discovered was the reaction of chromium(II) ion with pentaamminechlorocobalt(III) ion, shown in Figure 19.13.

$$Cr(H_2O)_6^{2+}(aq) + Co(NH_3)_5Cl^{2+}(aq) + 5H_2O(l) \longrightarrow$$

$$Cr(H_2O)_5Cl^{2+}(aq) + Co(H_2O)_6^{2+}(aq) + 5NH_3(aq)$$

In this reaction, the oxidation number of chromium changes from +2 to +3, so it is oxidized, whereas the oxidation number of cobalt is reduced from +3 to +2. Before electron transfer occurs, chromium(II) loses one of its water molecules, and that position in its coordination sphere is taken by the chloride ion already bonded to cobalt(III). After the electron transfer takes place, the chloride ion remains bonded to the newly formed chromium(III), and water molecules replace all ligands bonded to cobalt(II).

Could an inner-sphere reaction be carried out in separate compartments of an electrochemical cell? If a reaction occurred under these conditions, would it be the same reaction?

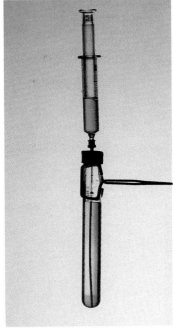

Figure 19.13 Blue $Cr^{2+}(aq)$ reacts with red $Co(NH_3)_5Cl^{2+}(aq)$ to form green $Cr(H_2O)_5Cl^{2+}(aq)$ and light red $Co^{2+}(aq)$. The chloride ion is transferred from the cobalt(III) to the chromium(III).

Substitution Reactions

Reactions in which one ligand is removed from the coordination sphere and replaced with another are called *substitution reactions*. We have just seen an example of a substitution reaction accompanying an inner-sphere oxidation–reduction reaction. Substitution reactions are used to prepare a wide variety of coordination compounds. For example, salts of pentaammineaquacobalt(III) ion can be prepared by an oxidation–reduction reaction in which oxygen in air is used to oxidize cobalt(II) in a solution of aqueous ammonia, as shown in Figure 19.14 for the chloride salt. This reaction forms solid $[Co(NH_3)_5H_2O]Cl_3$, which can be used as a starting material to prepare another coordination compound in a substitution reaction. As shown in Figure 19.15, heating this compound drives water out of the coordination sphere, giving a new coordination compound.

$$[Co(NH_3)_5H_2O]Cl_3(s) \xrightarrow{\Delta} [Co(NH_3)_5Cl]Cl_2(s) + H_2O(g)$$

This method can be used to prepare compounds containing complex ions of the form $[Co(NH_3)_5X]^{2+}$ with almost any anion that does not decompose when heated. When water is driven off by heating, a vacancy is left in the coordination sphere of cobalt(III). Because cobalt(III) strongly prefers a coordination number of 6, the anion substitutes into the coordination sphere to fill the vacancy.

Substitution reactions also take place in solution. A common example is *sol-vation*, a substitution reaction in which a ligand is replaced by a solvent molecule. When the solvent is water, as is most often the case, this reaction is also known as *aquation*. In the following reaction, for example, water substitutes for the chloride ligand in pentaamminechlorocobalt(III) ion.

$$[Co(NH_3)_5Cl]^{2+}(aq) + H_2O(l) \longrightarrow [Co(NH_3)_5H_2O]^{3+}(aq) + Cl^-(aq)$$

Depending on the conditions of concentration and temperature, it is possible to carry out the reverse of this reaction, called *anation*, in which an anion in solution substitutes for ligand water.

$$[Co(NH_3)_5H_2O]^{3+}(aq) + NCS^-(aq) \longrightarrow [Co(NH_3)_5NCS]^{2+}(aq) + H_2O(l)$$

Another type of substitution reaction in solution is *direct substitution*. Here, neither the incoming nor the outgoing ligand is the solvent. Such reactions are most commonly found in the chemistry of platinum(II) complexes—in the synthesis of *cis*-[Pt(NH_3)_2Cl_2], for example. This compound is prepared by the successive replacement of a chloride ligand in $[PtCl_4]^{2-}$ by aqueous ammonia.

$$[PtCl_4]^{2-}(aq) + NH_3(aq) \longrightarrow [Pt(NH_3)Cl_3]^-(aq) + Cl^-(aq)$$
$$[Pt(NH_3)Cl_3]^-(aq) + NH_3(aq) \longrightarrow cis\text{-}[Pt(NH_3)_2Cl_2](aq) + Cl^-(aq)$$

Reactions of Ligands

The great utility of complex ions as catalysts in a variety of reactions, especially synthetic organic processes, is a result of the reactivity of ligands. For example, Zeise's salt, $K[PtCl_3(H_2C{=}CH_2)]$, has ethylene bonded to platinum, under which conditions it is more reactive than free ethylene. Zeise's salt is used in hydrogenation and polymerization reactions of ethylene because of the enhanced reactivity of the ligand. Ligands in coordination compounds can undergo a variety of reactions, some of which will be reviewed here.

A molecule or ion and a metal joined in a coordination compound react differently than they would as separate entities. Consider, for example, the reactivity of water. In the pure liquid, water ionizes into hydronium and hydroxide ions.

$$2H_2O(l) \rightleftharpoons H_3O^+(aq) + OH^-(aq)$$

This reaction has an equilibrium constant (K_w) of 1.008×10^{-14} at 25°C. The position of this equilibrium reaction is changed dramatically when water is bonded to a metal ion. For example, many aquametal ions in solution are acidic because of the enhanced ionization of ligand water, as shown here for hexaaquairon(III) ion.

$$[Fe(H_2O)_6]^{3+}(aq) + H_2O(l) \rightleftharpoons [Fe(H_2O)_5OH]^{2+}(aq) + H_3O^+(aq)$$

This reaction has an equilibrium constant of 10^{-3}, which is considerably greater than the equilibrium constant for ionization of pure water. The yellow color associated with solutions of iron(III) salts is due to the presence of $[Fe(H_2O)_5OH]^{2+}(aq)$, as well as other iron(III) complexes; the unhydrolyzed ion, $[Fe(H_2O)_6]^{3+}(aq)$, is light pink in color (Figure 19.16). Like the equilibrium constant for hydrolysis of $[Fe(H_2O)_6]^{3+}(aq)$, the ionization constants for many metal ions (listed in Table 16.2) are larger than that for pure water, because the positive metal ion bonded to the oxygen draws electrons away from the O—H bond, rendering it more ionic.

Coordinated ligands can also undergo reactions other than hydrolysis, such as oxidation–reduction. Coordinated thiocyanate can be oxidized with hydrogen peroxide to give ligand ammonia, for example.

$$[Co(NH_3)_5NCS]^{2+}(aq) + 4H_2O_2(aq) + 2H_2O(l) \longrightarrow$$
$$[Co(NH_3)_6]^{3+}(aq) + 3H_3O^+(aq) + CO_3^{2-}(aq) + SO_4^{2-}(aq)$$

Figure 19.16 Iron(III) perchlorate hydrolyzes in solution, producing a yellow color. The solid has the pink color of the $[Fe(H_2O)_6]^{3+}$ ion because this ion is not hydrolyzed.

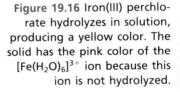

The oxidation–reduction reaction just mentioned involved the making and breaking of covalent bonds within the ligand. Ligands can also undergo other processes that involve bond breaking and bond making within the ligand but that don't involve oxidation–reduction. For example, ligand water can form bonds to other molecules, such as carbon dioxide.

$$[Co(NH_3)_5OH_2]^{3+}(aq) + CO_2(aq) + 2H_2O(l) \longrightarrow$$
$$[Co(NH_3)_5OCO_2]^+(aq) + 2H_3O^+(aq)$$

Here, the O—H bonds are broken and an O—C bond is formed. Ligand water is transformed to ligand carbonate.

19.6   The Coordinate Covalent Bond

When Werner started investigating coordination compounds, many of their features were quite perplexing. Indeed, their very existence made little sense. The bonding in transition-metal coordination compounds is difficult to explain in terms of the concepts used to explain other compounds. Ionic main-group compounds form in such a way as to satisfy the electroneutrality of the compound, and covalent main-group compounds usually form so as to achieve a noble-gas configuration. In contrast, the transition elements form compounds in which more atoms or ions are bonded to the metal than are needed to maintain electroneutrality, and they frequently do not achieve a noble-gas configuration. The differences in bonding are accompanied by differences in properties. For example, with only a few exceptions, compounds of the main-group elements are not attracted into a magnetic field; they are *diamagnetic*. Many transition-metal compounds are *paramagnetic;* that is, they are attracted into a magnetic field. This property is unusual, because it indicates the presence of unpaired electrons. Another property characteristic of transition-metal coordination compounds is color. ▶ As indicated in the chapter introduction, these compounds show a wide range of colors, whereas most main-group compounds are colorless.

A bonding theory for transition-metal coordination compounds should explain not only the existence of the compounds but also some of their properties, including the characteristic magnetism and color. We will consider two bonding theories: *valence bond theory* and *crystal field theory*. Each has some merits and some limitations.

Why do metals form more bonds in coordination compounds than are necessary to maintain electroneutrality?

Why are coordination compounds usually paramagnetic and colored?

Valence Bond Theory

The valence bond approach assumes that the geometry of a complex ion or an uncharged coordination compound is related to the orbitals on the metal atom used to form coordinate covalent bonds between the metal and the ligands. Hybridized orbitals are invoked to explain the structure and magnetic characteristics of the complex ion. (Valence bond theory was applied to covalent compounds of main-group elements in Section 8.3.)

According to valence bond theory, when complex ions are formed from metal ions and ligands, electron pairs from the donor atom of the ligand are donated into low-energy, empty valence orbitals on the metal ion. When we represent the bonding, we select empty orbitals that can be hybridized to give the observed geometry and magnetic properties of the complex ion. If there are not enough orbitals available on the metal to hold the ligand electron pairs, any unpaired metal electrons are paired up, thereby freeing some orbitals to accept electron pairs from the ligands. The empty orbitals are then hybridized to yield the appropriate geometry. The common geometries and their hybridization schemes are

octahedral d^2sp^3
tetrahedral sp^3
square planar dsp^2

As in Chapter 8, we will use boxes to represent the valence orbitals and will use arrows to represent electrons and their spins. Each orbital (box) can hold two electrons (arrows) of opposite spin (direction of arrow). Let's consider a few common cases to illustrate the use of the valence bond approach to explain the structure and magnetism of complex ions.

First we'll consider the iron atom, which has the electronic configuration $1s^22s^22p^63s^23p^64s^23d^6$, with four unpaired electrons. The valence electrons are represented as follows:

	3d	4s	4p
Fe atom:	↑↓ ↑ ↑ ↑ ↑	↑↓	☐ ☐ ☐

This configuration agrees with the magnetic data, which indicate the presence of four unpaired electrons in the iron atom. Now let's consider the structure, bonding, and magnetism of iron(III) complex ions. Before we can see which orbitals are available to accept the ligand electron pairs, we must examine the electronic configuration of the Fe^{3+} ion. This ion is known to have five unpaired electrons. When a transition-metal atom such as iron is ionized, electrons are removed first from the outer s orbital, then from completely filled d orbitals. The electronic configuration of the Fe^{3+} ion is thus

	3d	4s	4p
Fe^{3+} ion:	↑ ↑ ↑ ↑ ↑	☐	☐ ☐ ☐

To form a complex ion of iron(III), say $[Fe(CN)_6]^{3-}$, electron pairs from six cyanide ions must be donated into six empty orbitals. As we have seen, complex ions with a coordination number of 6 have octahedral geometry. This geometry arises from d^2sp^3 hybridization, as shown in Figure 19.17. Thus, for the ion to exhibit octahedral geometry, the unshared electron pairs in $:CN^-$ must be donated into two 3d, one 4s, and three 4p orbitals on Fe^{3+}. Magnetic data for $[Fe(CN)_6]^{3-}$ indicate that it has one unpaired electron—it is paramagnetic. Thus it is necessary to pair up the unpaired 3d electrons. This costs energy, giving an Fe(III) ion in an excited (higher-energy) state, but this energy will be regained from the formation of coordinate covalent bonds. The electronic configuration of the excited Fe^{3+} ion is

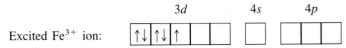

	3d	4s	4p
Excited Fe^{3+} ion:	↑↓ ↑↓ ↑	☐ ☐	☐	☐ ☐ ☐

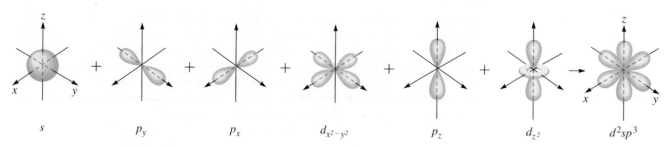

Figure 19.17 Combination of two *d* orbitals, one *s* orbital, and three *p* orbitals results in six equivalent d^2sp^3 hybrid orbitals. These orbitals point to the corners of an octahedron.

This gives the two empty 3*d* orbitals needed for the hybridization scheme. Now the appropriate number of carbon electron pairs can be donated into vacant orbitals.

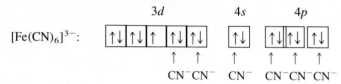

Valence bond theory can explain the octahedral structure corresponding to d^2sp^3 hybridization and the paramagnetism of this complex ion.

Now let's consider a different complex ion of iron(III) that requires a different valence bond approach to explain its structure and magnetism. This ion is octahedral $[FeF_6]^{3-}$, which has five unpaired electrons. Here, the 3*d* electrons cannot be paired up to give two empty *d* orbitals; all five must remain unpaired to match the observed paramagnetism. The electron pairs from the fluoride ions are donated into the 4*s* and the three 4*p* orbitals. In addition, two 4*d* orbitals must be used.

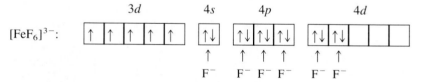

In spite of the unavailability of the 3*d* orbitals, it is still possible to achieve a hybridization scheme that gives an octahedral geometry and is consistent with the observed magnetic properties of the complex ion.

These two iron(III) ions, $[Fe(CN)_6]^{3-}$ and $[FeF_6]^{3-}$, represent two types of complex ions. They are called **low-spin complexes** or **high-spin complexes,** depending on whether they have the minimum or the maximum number of unpaired electrons. Depending on whether inner or outer *d* orbitals are used to bond to the ligands, they are also called *inner-orbital* or *outer-orbital* complexes.

An interesting example of high-spin and low-spin complexes is provided by hemoglobin, a complex of iron(II) that is involved in respiration. Hemoglobin consists of a large protein molecule containing four heme groups, which have iron(II) surrounded by a large molecule called a porphyrin. As was shown in Figure 16.9, the iron is covalently bonded to four nitrogen atoms in this complex. A fifth coordination site is used to bind to the protein, and a sixth site is normally occupied by an oxygen molecule, a HCO_3^- ion, or a water molecule. When this site contains an oxygen molecule, the complex is called oxyhemoglobin and is bright red. The iron ions have a low-spin electronic configuration. Oxyhemoglobin is used to transport oxygen to the cells and is found in red arterial blood. When the oxygen has been given up to the cells, high-spin deoxyhemoglobin is formed. This complex is bluish-red and is the cause of the bluish color in the veins. A pronounced version of this bluish color is seen in people who are choking and in people who have been poisoned by carbon monoxide, which binds to the heme groups and prevents their bonding to oxygen.

As the examples of iron complexes illustrate, valence bond theory can explain the observed structure and magnetic properties of complex ions. Similar explanations can be developed for tetrahedral complexes such as $[Zn(NH_3)_4]^{2+}$ and for square planar complexes such as $[PtCl_4]^{2-}$. However, valance bond theory cannot predict the magnetic properties of Fe(III) in different coordination compounds. We had to know that $[Fe(CN)_6]^{3-}$ has one unpaired electron and that $[FeF_6]^{3-}$ has five unpaired electrons in order to construct a valence bond representation of the bonding and electronic structure of these complex ions. If we had not known the number of unpaired electrons, we could not have decided whether to use inner or outer orbitals to bond to the ligands. The inability of valence bond theory to make such predictions is a serious shortcoming.

Valence bond theory has some other shortcomings:

1. It provides only qualitative explanations; it cannot explain why one complex ion is more stable than another.
2. It does not account for color and electronic spectra.
3. It cannot explain the relative stabilities of different structural isomers.
4. It resorts to artificial explanations, such as the distinction between inner- and outer-orbital complexes, to make things fit. However, it remains a useful theory for some applications because of its inherent simplicity.

Potassium tetrachloroplatinate(II), $K_2[PtCl_4]$, is a ruby-red solid used in acid toning baths in photography.

Crystal Field Theory

Another bonding theory for transition-metal complex ions, **crystal field theory,** assumes that the interaction between metal ion and ligands is purely electrostatic. The bonding is assumed to arise from attractions between the positive charge on the metal and the negative charge on the electrons provided by the ligands.

Crystal field theory is based on the premise that ligands can be regarded as negative electronic charges distributed at points in space in a geometrical arrangement around a metal ion. The name is based on early applications of the theory to metal ions in crystals. The bonding is considered a purely electrostatic attraction between the positively charged metal ion and the charges of the anions concentrated at the geometrical points. The relative energies of electrons around the metal ion can be calculated from the effect of repulsions between the ligand electron pairs and the outermost metal electrons, the $(n - 1)$ *d*-orbital electrons. From this information, it is possible to predict the structure, magnetic properties, and color of complex ions.

Geometry and *d*-Orbital Splitting   To see how ligand electronic charges interact with metal electrons, let's consider the arrangement of the five *d* orbitals around a metal, shown in Figure 19.18. Two of these orbitals ($d_{x^2-y^2}$ and d_{z^2}) point along (are located along) the *xyz* axes, whereas three of these orbitals point between the axes (d_{xy}, d_{xz}, and d_{yz}). In a free ion or in an ion in an electrical field that is evenly

Figure 19.18 Orientation of *d*
orbitals around a metal.

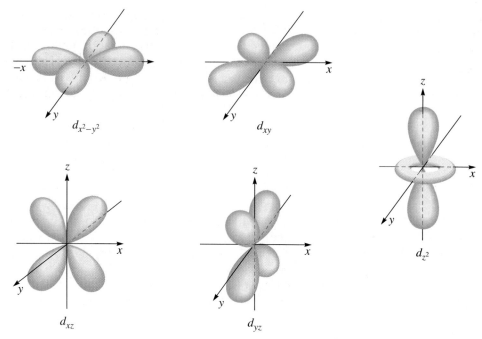

$d_{x^2-y^2}$ d_{xy} d_{z^2}

d_{xz} d_{yz}

distributed around a sphere surrounding the ion, all five *d* orbitals have the same
energy. However, in the presence of ligands, the electrical field is no longer
spherical. The electrical field now has specific areas of concentration, and the
energies of the orbitals are no longer equal. Some of the orbitals are closer to the
electronic charges of the ligands, whereas others are farther away. Greater repulsions between charges in the closer orbitals cause them to have higher energy than
the other orbitals.

*How does the splitting of the
energy of the d orbitals explain
the color and magnetic
properties of transition-metal
complex ions?*

To see this effect in more detail, first consider an octahedral complex ion,
with six ligands located along the *x*, *y*, and *z* axes, as shown in Figure 19.19. This
arrangement provides an electrical field around the metal called an *octahedral*

Figure 19.19 Location of the
six ligands in an octahedral
crystal field.

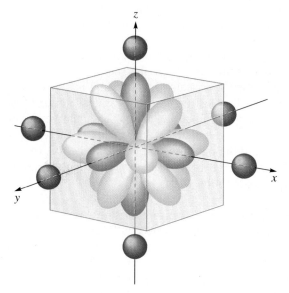

Figure 19.20 Crystal field splitting energy in an octahedral complex.

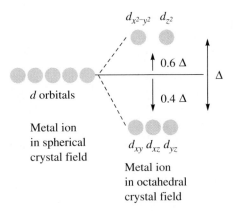

crystal field. The two orbitals that point along the axes also point directly at the ligands. The other three orbitals point between the ligands. The repulsions between the ligands and the *d* orbitals pointing along the axes are greater than the repulsions between the ligands and the *d* orbitals pointing between the axes. This leads to a difference in energy; the d_{xy}, d_{xz}, and d_{yz} orbitals are lower in energy than the $d_{x^2-y^2}$ and d_{z^2} orbitals. This difference, shown in Figure 19.20, is called the **crystal field splitting energy** and is symbolized Δ.

Now consider a tetrahedral complex ion. The ligands are located at alternate corners of the cube enclosing the axes, as shown in Figure 19.21. The ligands are closer to the *d* orbitals pointing at the centers of the edges of the cube than to the *d* orbitals pointing at the centers of the faces. Thus a *tetrahedral crystal field* also causes splitting in the energy of the *d* orbitals, but here three orbitals are higher in energy and two orbitals are lower in energy, as shown in Figure 19.22.

What *d* orbitals are occupied by electrons in complex ions that have these geometries? This depends on the magnitude of the crystal field splitting energy, which varies with the geometry, the identity of the metal, and the identity of the ligands. The crystal field splitting energy for a tetrahedral complex is generally about 45% of the crystal field splitting energy for an octahedral complex. The crystal field splitting energy also depends on the principal energy level and on the charge of the metal ion. The splitting for 4*d* orbitals is about 1.45 times as large as

Figure 19.21 Location of the four ligands in a tetrahedral crystal field.

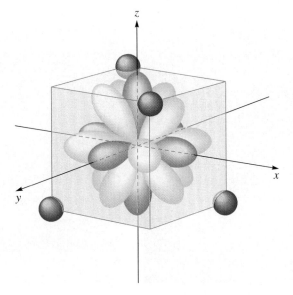

Figure 19.22 Crystal field splitting energy in a tetrahedral complex.

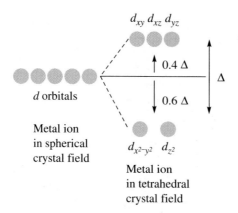

$d_{xy}\ d_{xz}\ d_{yz}$

$\uparrow$ 0.4 Δ

Δ

$\downarrow$ 0.6 Δ

d orbitals

Metal ion in spherical crystal field

$d_{x^2-y^2}\ d_{z^2}$

Metal ion in tetrahedral crystal field

Aquacobalt(III) ion, Co^{3+}(aq), is an extremely strong oxidizing agent. It does not survive long in aqueous solution because it oxidizes water. It can be prepared in low concentrations by electrochemical oxidation of cobalt(II) solutions.

Potassium hexacyanocobaltate(III), $K_3[Co(CN)_6]$, is a faintly yellow solid that dissolves readily in water to form a pale yellow solution. It decomposes slowly in solution—more rapidly when exposed to light—releasing HCN.

the splitting for 3*d* orbitals, and the splitting for 5*d* orbitals is about 1.75 times the splitting for 3*d* orbitals. Thus the crystal field splitting energy of Ir(III) is greater than that of Rh(III), which in turn is greater than that of Co(III). The splitting for 3+ metal ions is 40–80% larger than for 2+ metal ions. For example, the value of Δ for $[Co(H_2O)_6]^{3+}$ is 220 kJ/mol, whereas that for $[Co(H_2O)_6]^{2+}$ is 110 kJ/mol. There is also variation with the particular metal forming the complex ion. For example, compare the values of Δ for $[Fe(H_2O)_6]^{3+}$ (167 kJ/mol) and $[Fe(H_2O)_6]^{2+}$ (130 kJ/mol) with the corresponding values for ions of cobalt. Finally, the magnitude of Δ depends on the ligands surrounding the metal ion. For example, CN^- generates a larger crystal field splitting energy than NH_3. The value of Δ for $[Co(CN)_6]^{3-}$ is 390 kJ/mol, whereas that for $[Co(NH_3)_6]^{3+}$ is 260 kJ/mol. We will explore these variations in greater detail shortly.

Color and Electronic Transitions   Crystal field theory was proposed in an attempt to explain the colors imparted to gems by transition-metal-ion impurities. For example, small amounts of chromium(III) cause emerald to be green and ruby to be red (Figure 19.23). Chromium(III) is also the color-causing impurity in a variety of gems and minerals, including jade (green), tourmaline (green), spinel (yellow or greenish brown), topaz (red), and alexandrite (red or green). The differ-

Figure 19.23 The green color of emerald arises from a chromium(III) impurity in $Be_3Al_2Si_6O_{18}$, whereas the red color of ruby arises from a chromium(III) impurity in Al_2O_3.

Hexaamminecobalt(III) chloride, [Co(NH₃)₆]Cl₃, is a wine-red to brownish-orange solid. It decomposes slowly in aqueous solution, releasing NH₃. It is used to analyze for phosphate, with which it forms an insoluble precipitate.

ences in color are attributed to the ligands surrounding the Cr^{3+} ion. For example, ruby is primarily Al_2O_3, and emerald is primarily $Be_3Al_2Si_6O_{18}$.

Differences in color are reflected in the electronic spectrum. The spectra of ruby and emerald are shown in Figure 19.24, along with a scale relating energy, wavelength, and color. Note that ruby transmits a small amount of blue light and red light, whereas emerald transmits light in the blue-green region. Some other chromium(III) complex ions and their colors are listed in Table 19.5. The electronic spectra of these complex ions are similar in appearance to those of Cr^{3+} in ruby and emerald, but the colors of maximum absorbance and transmission occur at different wavelengths.

Crystal field theory provides an explanation for these differences in color—the color depends on the crystal field splitting energy. When light is absorbed by a complex ion, an electron is promoted from the lower-energy *d* orbitals to the higher-energy *d* orbitals, as shown in Figure 19.25. Conversely, when an electron moves from the higher-energy to the lower-energy *d* orbitals, the complex ion emits light. Because the wavelength of the light absorbed must be in the same range as the value of the crystal field splitting energy to shift an electron from one orbital to another, light is not absorbed at all wavelengths. The variation of light absorption with wavelength gives rise to spectra such as those shown in Figure 19.24. If the absorbed light is in the visible region of wavelengths, the complex ion is colored. Not all complex ions are colored, however. Where the *d* orbitals are either empty or completely filled, there can be no electronic transitions, and the complex ion is colorless. (Such a complex ion may have a faint color that is due to other mechanisms for generating color).

Figure 19.24 Electronic spectra of ruby and emerald. *(A)* The spectra. *(B)* The relationship among color, wavelength, and energy. From *Lapidary Journal,* August 1975, p. 924. Copyright © 1975. Used by permission.

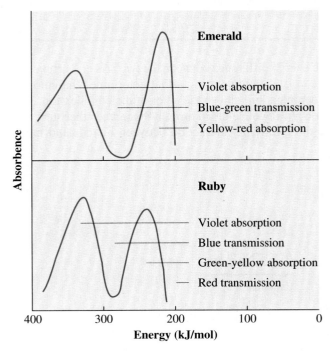

A

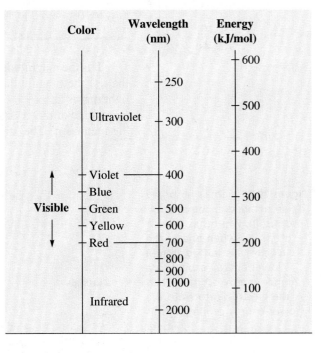

B

Table 19.5 Colors of Some
Chromium(III) Complex Ions

Complex Ion	Color
$Cr(H_2O)_5(H_2O)^{3+}$	Blue-gray
$Cr(H_2O)_5(NH_3)^{3+}$	Purple
$Cr(H_2O)_5CN^{2+}$	Pink
$Cr(H_2O)_5F^{2+}$	Blue-green
$Cr(H_2O)_5Cl^{2+}$	Grass green
$Cr(H_2O)_5Br^{2+}$	Kelly green
$Cr(H_2O)_5I^{2+}$	Light green

EXAMPLE 19.7

Predict whether any complex ions of Cr^{3+} and of Sc^{3+} should be colored.

Solution   Chromium(III) has an electronic configuration of $4s^0 3d^3$. Because it can transfer electrons between the lower-energy and higher-energy d orbitals, as shown in Figure 19.25, it should be colored. Scandium(III) has an electronic configuration of $4s^0 3d^0$. It has no electrons in d orbitals, so no electrons can be transferred between d orbitals, and it should be colorless.

Practice Problem 19.7

Predict whether any complex ions of Ti^{4+} and of Mn^{4+} should be colored.

Answer: Ti^{4+} should be colorless, but Mn^{4+} should be colored.

It is the magnitude of the energy difference between the two sets of d orbitals—that is, the magnitude of the crystal field splitting—that determines the color of light that is absorbed by the complex ion, and so the color of the complex ion. The magnitude of the crystal field splitting depends in turn on the ligands that surround the metal ion. The stronger the electronic interaction between ligands and metal

Figure 19.25 Light is absorbed by an octahedral complex ion of Cr(III), which has three d electrons, when one of the electrons is shifted from a lower-energy orbital to a higher-energy orbital. When the electron returns to the lower-energy orbital, light is emitted.

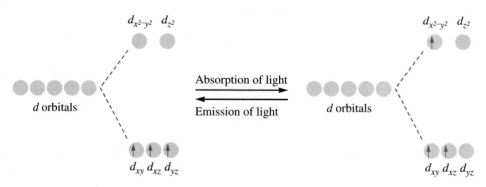

ion, the greater the splitting between the two sets of orbitals. This effect can be summarized by the **spectrochemical series,** which lists ligands in order of decreasing crystal field splitting energy, as in Table 19.6. Note that ligands tend to group together in the spectrochemical series according to the atom that donates the electron pair to the metal ion:

$$C > N > O > F > Cl > S > Br > I$$

Thus the identity of this atom appears to be primarily responsible for the position of the ligand in the series. ▶ The change in ligands is responsible for the change in color observed for the copper compounds described in the chapter introduction. The correlation of these ligands with the wavelength of maximum absorption—the position of the peak in the spectrum, which is related to color—is reasonably good. For example, $[Co(NH_3)_5CN]^{2+}$, with maximum absorption at 440 nm, is yellow, whereas $[Co(NH_3)_5Cl]^{2+}$, with maximum absorption at 533 nm, is red. Coordination compounds that contain these complex ions are shown in Figure 19.26. Crystal field theory does, then, explain the colors of complex ions, and it has some predictive capability as well.

Table 19.6 Spectrochemical Series of Ligands

Ligand	Wavelength in nm of Peak in Spectrum of $Co(NH_3)_5X^*$ (nm)	Crystal Field Splitting Energy for $Co(NH_3)_5X$ (kJ/mol)	
CO			
CN^-	440	272	
$—NO_2^-$	458	261	
en			
$—NC^-$			Large Δ
NH_3	475	252	
$—NCS^-$	496	241	
OH_2	490	244	
$C_2O_4^{2-}$			
			Pairing energy
$—ONO^-$	491	244	
OH^-	503	238	
$—ONO_2^-$	500	239	
F^-	510	235	
Cl^-	533	224	Small Δ
$—SCN^-$			
S^{2-}			
Br^-	550	217	
I^-	580	206	

*This wavelength is that at which there is a maximum absorption of light in the spectrum of the cobalt(III) complex ion.

Figure 19.26 The color of [Co(NH$_3$)$_5$X] complex ions varies with the identity of the ligand. *Left:* [Co(NH$_3$)$_5$CN](ClO$_4$)$_2$. *Right:* [Co(NH$_3$)$_5$Cl]Cl$_2$.

EXAMPLE 19.8

Which complex ion, [Cr(NH$_3$)$_5$NO$_2$]$^{2+}$ or [Cr(NH$_3$)$_5$ONO]$^{2+}$, should have a peak in its spectrum—its point of maximum absorption—at a higher wavelength?

Solution   To determine the relative positions of peaks in the spectra of two complex ions that differ in only one ligand, we consult the spectrochemical series in Table 19.6. The higher the ligand is in this series, the lower the wavelength of the absorbance peak. Nitrogen-bonded nitrite ion is higher in the spectrochemical series than oxygen-bonded nitrite ion. Therefore, [Cr(H$_2$O)$_5$ONO]$^{2+}$ should have a peak in the spectrum at a higher wavelength.

Practice Problem 19.8

Which complex ion, [Cr(H$_2$O)$_5$NC]$^{2+}$ or [Cr(H$_2$O)$_5$CN]$^{2+}$, should have a peak in its spectrum at a higher wavelength?

Answer: [Cr(H$_2$O)$_5$NC]$^{2+}$

Why are electrons paired up in one complex ion of a metal but unpaired in another?

Magnetism and Crystal Field Splitting   The splitting of the *d* orbitals into sets with different energy provides an explanation for the magnetic properties of complex ions. Furthermore, unlike valence bond theory, crystal field theory provides a basis for predictions. Specifically, crystal field theory enables us to predict how many unpaired electrons are present in a complex ion, yielding an explanation of such properties as color and magnetism.

Let's consider again complex ions of Fe(III) having an octahedral structure. Recall that iron(III) has five valence electrons in *d* orbitals. In an octahedral crystal field, three orbitals have lower energy than the other two orbitals. The first three iron(III) valence electrons go into separate *d* orbitals of the lower-energy set, because electrons repel one another and it costs energy to pair them up in a single orbital (recall Hund's rule from Chapter 6). The fourth and fifth electrons can pair up with electrons in the lower-energy *d* orbitals, paying the energy cost demanded by pairing, or they can go into separate higher-energy *d* orbitals, with an energy

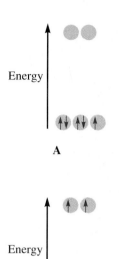

Figure 19.27 The electronic configuration of iron(III) complex ions depends on the relative values of the pairing energy and the crystal field splitting energy. *(A)* When the pairing energy is less than Δ, electrons occupy only lower-energy orbitals. *(B)* When the pairing energy is greater than Δ, electrons occupy all orbitals.

cost equal to the crystal field splitting energy. These two cases are shown in Figure 19.27. Which orbitals are occupied by electrons depends on which is greater, the pairing energy or the crystal field splitting energy. If the splitting energy is lower, the electrons occupy both lower-energy and higher-energy d orbitals. Because this electronic configuration gives the maximum number of unpaired electrons, such a complex ion is called a high-spin complex. If the splitting energy is greater, the electrons pair up in the lower-energy d orbitals. Such complex ions are called low-spin complexes, because they have the minimum number of unpaired electrons. For iron(III) octahedral complex ions with five d electrons, the magnetic properties can correspond to either one or five unpaired electrons, depending on the magnitude of the crystal field splitting energy.

The size of the crystal field splitting energy is determined, for a given metal ion, by the nature of its ligands. In Table 19.6, ligands were divided into two groups, those with large Δ and those with small Δ. Those ligands associated with high splitting are called *strong-field ligands,* and those that result in low splitting energies are called *weak-field ligands.* A strong-field ligand produces a low-spin complex, and a weak-field ligand produces a high-spin complex.

Earlier we considered two iron(III) complex ions, $[Fe(CN)_6]^{3-}$ and $[FeF_6]^{3-}$. On the basis of the spectrochemical series, we classify cyanide ion as a strong-field ligand and fluoride ion as a weak-field ligand. For a strong-field ligand in a complex ion with five d electrons, we predict a low-spin complex ion with one unpaired electron. This result matches that observed for $[Fe(CN)_6]^{3-}$. For a weak-field ligand on iron(III), we predict a high-spin complex with five unpaired electrons, just as is observed for $[FeF_6]^{3-}$.

EXAMPLE 19.9

What is the number of unpaired electrons expected for low-spin and for high-spin octahedral complex ions of cobalt(II) that contain seven d electrons?

Solution In either case, the first three electrons go into the lower-energy orbitals. For the low-spin complex, the next three electrons also enter these orbitals and the seventh electron enters a higher-energy d orbital, giving only one unpaired electron.

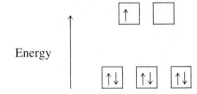

In the high-spin complex, the fourth and fifth electrons enter the higher-energy d orbitals, and then the sixth and seventh pair up with electrons already in the lower-energy d orbitals. This leaves one unpaired electron in the lower-energy orbitals and two unpaired electrons in the higher-energy orbitals, for a total of three unpaired electrons.

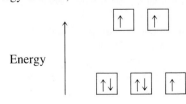

(continued)

Practice Problem 19.9

What is the number of unpaired electrons expected for low-spin and for high-spin octahedral complex ions of manganese(II) that contain five *d* electrons?

Answer: For the low-spin complex, one unpaired electron; for the high-spin complex, five unpaired electrons.

The distinction between high-spin and low-spin complexes holds only for octahedral complex ions containing four, five, six, or seven *d* electrons. The other electronic configurations have identical orbital occupancy.

The crystal field splitting energy for a tetrahedral complex ion is less than half that for an octahedral complex ion. For this reason, ligands must produce a stronger electrical field to generate sufficient crystal field splitting energy to produce a low-spin complex ion with tetrahedral geometry than with octahedral geometry. In fact, there are no ligands that provide a sufficiently strong crystal field to give low-spin tetrahedral complexes. All tetrahedral complex ions are high-spin.

We have seen that crystal field theory makes it possible to explain and predict color and magnetic properties of complex ions. Using calculations beyond the scope of this discussion, we can also predict and explain a number of other properties of complex ions. These properties include the stability of various oxidation numbers, the preference of a given ion for one ligand over another, and trends in thermodynamic properties such as the heat of hydration of metal ions.

Summary

Many coordination compounds of transition metals contain a complex ion that consists of a central metal ion surrounded by ligands. Enough anions and neutral ligands must be present to bring about electroneutrality and to satisfy the coordination number of the complex ion, which is most commonly 6 or 4. Coordination compounds are named with a special set of rules.

Many coordination compounds exist as isomers—more than one form that have the same formula. Isomers are classified as stereoisomers (geometrical or optical) or structural isomers. Geometrical isomers have ligands arranged in different positions relative to one another. In what are called *cis* geometrical isomers, two like ligands are adjacent; in *trans* isomers, two like ligands are opposite one another. Optical isomers exist when rearranging ligands yields a mirror image of a complex ion that cannot be superimposed on the original. Structural isomers differ either in the identity of the ligands bonded to the central metal ion or in the way these ligands are bonded.

Coordination compounds can undergo oxidation–reduction reactions, substitution reactions, or reactions of ligands. Ligands can undergo a variety of reactions, including dissociation, hydrolysis, and oxidation–reduction.

Several theories have been advanced to explain the bonding and properties of complex ions. Valence bond theory uses hybridized orbitals that match the structure of the complex ion to hold the metal *d* electrons and the electron pairs donated by the ligands. This theory can explain the structure and magnetic properties of complex ions, but it cannot predict or explain the colors of complex ions. Crystal field theory treats ligands as points of negative electrical charge surrounding the metal ion in a geometrical arrangement. The presence of the ligands causes the metal *d* orbital energies to split to an extent called the crystal field splitting energy. Shifts of electrons from the lower-energy orbitals to the higher-energy orbitals cause absorption of light, often in the visible region, and result in color in complex ions. Magnetic properties can also be predicted and explained by feeding electrons into the split *d* orbitals in patterns determined by the strength of the ligands.

Key Terms

bidentate ligand (19.2)
chelate (19.2)
coordination
  number (19.2)
coordination
  sphere (19.2)
crystal field splitting

energy (Δ) (19.6)
crystal field
  theory (19.6)
enantiomers (19.4)
geometrical
  isomers (19.4)
high-spin

complex (19.6)
isomers (19.4)
low-spin complex (19.6)
monodentate
  ligand (19.2)
optical isomers (19.4)
polydentate

ligand (19.2)
spectrochemical
  series (19.6)
stereoisomers (19.4)
structural
  isomers (19.4)

Exercises

Properties of Complex Ions

19.1 Distinguish between a complex ion and a coordination compound.

19.2 What is the difference between a covalent bond and a coordinate covalent bond?

19.3 Discuss why both coordination number and oxidation number must be used to account for the composition of coordination compounds.

19.4 What is the first coordination sphere of a complex ion?

19.5 Which elements are typically found as the central atom in complex ions?

19.6 What is a ligand?

19.7 Distinguish between a monodentate and a polydentate ligand.

19.8 What geometries are found for the common coordination numbers of complex ions?

19.9 Give the oxidation number of the central metal ion in each of the following complexes.
  a. $[Co(CN)_6]^{3-}$ c. $[CoF_6]^{3-}$
  b. $[Co(NH_3)_6]^{3+}$ d. $[Cr(CN)_6]^{4-}$

19.10 Give the oxidation number and the coordination number of the central metal ion in each of the following coordination compounds.
  a. $K_3[Fe(CN)_5CO]$
  b. $[Cr(NH_3)_3(H_2O)_3]Cl_3$
  c. $[Cr(NH_3)_2(H_2O)Cl_3]$
  d. $[Co(NH_3)_5NO_2](NO_3)_2$
  e. $[Cr(H_2O)_5CN](ClO_4)_2$
  f. $[Pt(NH_3)_2(NCS)_2]$
  g. $K_2[NiCl_4]$

19.11 What are the possible geometries of the complexes $[Pt(NH_3)_2Br_2]$ and $[CuCl_4]^{2-}$?

19.12 A compound has the empirical formula $CrCl_3 \cdot 6H_2O$. Reaction of 1 mol of the compound with excess silver nitrate gives an immediate precipitate of 1 mol of silver chloride. Write a formula for the coordination compound.

19.13 Suggest a simple test for distinguishing between the isomers $[Cr(NH_3)_4(H_2O)(SO_4)]Cl$ and $[Cr(NH_3)_4(SO_4)Cl] \cdot H_2O$.

Nomenclature and Formulas

19.14 Describe the conventions for writing formulas of coordination compounds.

19.15 Outline the main features of the rules for naming coordination compounds.

19.16 What is each of the following ions called when it is a ligand?
  a. IO_3^- d. CrO_4^{2-}
  b. S^{2-} e. F^-
  c. SO_4^{2-} f. OH^-

19.17 Name the following coordination compounds.
  a. $[Cr(NH_3)_3(H_2O)Cl_2]Cl$
  b. $[Cr(NH_3)_3(H_2O)_2Cl]Cl_2$
  c. $[Cr(NH_3)(H_2O)_3Cl_2]Cl$
  d. $[Cr(NH_3)_3(H_2O)_3]Cl_3$
  e. $[Cr(NH_3)_2(H_2O)Cl_3]$

19.18 Name the following complexes.
  a. $[Co(NH_3)_3(H_2O)_2NO_2]Cl_2$
  b. $K_4[Fe(CN)_6]$
  c. $[Co(H_2O)_6]^{2+}$
  d. $[Cr(H_2O)_5Cl][Ni(CN)_4]$

19.19 Name the following coordination compounds.
  a. $[Co(en)_2Cl_2]NO_3$
  b. $[Pt(en)_2]Cl_2$
  c. $[Co(en)(NH_3)_2(H_2O)Cl](ClO_4)_2$
  d. $[Cu(en)_2][CuCl_4]$

19.20 Write formulas for the following complex ions.
  a. hexacyanoruthenate(III)
  b. tetrachlorocuprate(II)
  c. triamminechloroplatinum(II)
  d. triamminetrichloroiridium(IV)
  e. pentaaquachloroiron(III)

19.21 Write formulas for the following coordination compounds.
   a. sodium hexanitrocobaltate(II)
   b. diamminesilver(I) nitrate
   c. zinc hexachloroplatinate(IV)
   d. sodium aquapentacyanoferrate(II)
   e. tetraamminechloronitritocobalt(III) chloride

19.22 Write formulas for the following coordination compounds.
   a. tris(ethylenediamine)chromium(III) tetrabromoplatinate(II)
   b. triammineaquacarbonatocobalt(III) carbonate
   c. potassium tetracyanoethylenediamine-cobaltate(III)
   d. dichlorobis(ethylenediamine)chromium(III) tetrachloroaurate(III)

Isomerism

19.23 What is an isomer?

19.24 Distinguish between stereoisomers and structural isomers.

19.25 What are the common geometrical isomers?

19.26 For which common geometries is it possible to have geometrical isomers?

19.27 How can we tell whether a compound can exist as optical isomers?

19.28 List some types of structural isomers.

19.29 How do ionization isomers differ from hydration isomers?

19.30 What is a linkage isomer?

19.31 Write the formula of a linkage isomer of each of the following coordination compounds.
   a. $[Co(NH_3)_5NO_2](NO_3)_2$
   b. $[Cr(H_2O)_5CN](ClO_4)_2$
   c. $[Pt(NH_3)_2(NCS)_2]$

19.32 Write the formula of an ionization isomer of each of the following coordination compounds.
   a. $[Co(NH_3)_4(NO_2)Cl]Cl$
   b. $[Cr(NH_3)_5Br]Cl_2$
   c. $[Ir(NH_3)_4Cl_2]NCS$

19.33 Draw all the stereoisomers of the following complexes. Indicate whether each is a geometrical or an optical isomer.
   a. $[Co(en)_3]^{3+}$
   b. $[Cr(NH_3)_4Cl_2]^+$
   c. $[Cr(H_2O)_2(CN)_4]^-$
   d. $[Co(NH_3)_4(en)]^{3+}$
   e. $[Pt(NH_3)_2Cl_2]$ (square planar)
   f. $[Ni(NH_3)_2Cl_2]^+$ (tetrahedral)

19.34 Write the formula of a hydration isomer of each of the following coordination compounds.
   a. $[Co(NH_3)_5(H_2O)]Cl_3$
   b. $[Cr(H_2O)_5Cl]Cl_2$
   c. $[Pt(NH_3)_2Cl_2] \cdot H_2O$

19.35 Indicate the type of isomerism shown by each of the following pairs of complexes.
   a. cis-$[Cr(NH_3)_4Cl_2]^+$ and $trans$-$[Cr(NH_3)_4Cl_2]^+$
   b. $[Cr(NH_3)_4(H_2O)(SO_4)]Cl$ and $[Cr(NH_3)_4(SO_4)Cl] \cdot H_2O$
   c. $[Co(NH_3)_5NO_2]SO_4$ and $[Co(NH_3)_5ONO]SO_4$
   d. $[Co(NH_3)_5NO_2]SO_4$ and $[Co(NH_3)_5SO_4]NO_2$

Valence Bond Approach

19.36 Describe the valence bond approach to bonding in complex ions.

19.37 What are some limitations of valence bond theory?

19.38 Write the valence bond description of the bonding in tetrahedral $[Ni(NH_3)_4]^{2+}$.

19.39 For each of the following complexes, describe the bonding in terms of valence bond theory. Indicate the orbitals used for the coordinate covalent bonds, the magnetic behavior (including the number of unpaired electrons), and the geometry.
   a. $[Ag(NH_3)_2]^+$ c. $[Cu(H_2O)_4]^{2+}$
   b. $[CoCl_4]^{2-}$ d. $[ZnCl_4]^{2-}$

19.40 For each of the following complexes, describe the bonding in terms of valence bond theory. Indicate the orbitals used for the coordinate covalent bonds, the magnetic behavior (including the number of unpaired electrons), and the geometry.
   a. $[Cr(NH_3)_6]^{3+}$ c. $[Mn(CN)_6]^{4-}$
   b. $[Ti(H_2O)_6]^{3+}$ d. $[Pt(NH_3)_4]^{2+}$

Crystal Field Theory

19.41 Describe the crystal field approach to bonding in complex ions.

19.42 What advantages does crystal field theory have over valence bond theory for describing the bonding in complex ions?

19.43 What is crystal field splitting energy?

19.44 Why are transition-metal complex ions colored?

19.45 What is a spectrochemical series?

19.46 Distinguish between low-spin and high-spin complexes.

19.47 How does the crystal field splitting energy of tetrahedral complexes compare to that of octahedral complexes for the same metal?

19.48 Discuss the effect of each of the following factors on the crystal field splitting energy.
   a. Charge on the metal ion
   b. Geometrical structure of the complex ion
   c. Nature of the ligands
   d. Pairing energy
   e. Principal energy level of the outermost d orbitals

19.49 Using crystal field theory, draw a *d*-orbital splitting diagram for each of the following complex ions. Indicate the number of unpaired electrons.
  a. $[Cr(H_2O)_6]^{2+}$ d. $[V(H_2O)_6]^{3+}$
  b. $[Mn(CN)_6]^{3-}$ e. $[Ti(H_2O)_6]^{3+}$
  c. $[V(CN)_6]^{3-}$

19.50 Determine the number of unpaired electrons in a low-spin and a high-spin octahedral complex that contains five *d* electrons.

19.51 Given the following complexes and their wavelength of maximum absorption, arrange the ligands in order of decreasing crystal field splitting energy: $[MoX_6]^{3-}$, 525 nm; $[MoY_6]^{3-}$, 705 nm; $[MoZ_6]^{3-}$, 482 nm.

19.52 Decide, on the basis of crystal field splitting patterns, whether each of the following compounds should be diamagnetic or paramagnetic.
  a. $[Co(CN)_6]^{3-}$ c. $[CoF_6]^{3-}$
  b. $[Co(NH_3)_6]^{3+}$ d. $[Cr(CN)_6]^{3-}$

19.53 Use crystal field theory to decide whether each of the following complexes should be paramagnetic.
  a. $[Pt(NH_3)_4Cl_2]^{2+}$ c. $[Co(H_2O)_6]^{2+}$
  b. $[Co(NH_3)_5Cl]^{2+}$ d. $[Pt(CN)_6]^{2-}$

19.54 Predict whether each of the following metal ions should have colored complex ions.
  a. Sc^{3+} d. Mn^{3+}
  b. Cr^{3+} e. Fe^{2+}
  c. Zn^{2+} f. Ni^{2+}

19.55 Determine the number of unpaired electrons in a tetrahedral complex that contains eight *d* electrons. (These are always high-spin complexes.)

19.56 On the basis of crystal field theory, decide whether each of the following complexes should be low-spin or high-spin complexes.
  a. $[Co(NH_3)_6]^{3+}$ c. $[Fe(CN)_6]^{4-}$
  b. $[FeBr_6]^{3-}$ d. $[CoCl_4]^{2-}$ (tetrahedral)

19.57 Using the spectrochemical series, predict which of each pair of complex ions should absorb visible radiation at the longer wavelength.
  a. $[Cr(NH_3)_5CN]^{2+}$ or $[Cr(NH_3)_5Cl]^{2+}$
  b. $[Cr(H_2O)_4Cl_2]^{+}$ or $[Cr(H_2O)_4(NH_3)_2]^{3+}$
  c. $[PtCl_4]^{2-}$ or $[Pt(NH_3)_4]^{2+}$
  d. $[Co(NH_3)_4(H_2O)NO_2]^{2+}$ or $[Co(NH_3)_4(H_2O)Br]^{2+}$

Additional Exercises

19.58 Name the compound $K_3[Fe(CN)_5CO]$.

19.59 Draw a diagram showing the crystal field splitting energy for the Co(III) ion in a strong octahedral field.

19.60 Write formulas for the following coordination compounds.
  a. bromocyanobis(ethylenediamine)chromium(III) chloride
  b. triamminebromochloronitrorhodium(III)

19.61 Give examples of the following types of isomers.
  a. optical isomers   d. ionization isomers
  b. stereoisomers   e. hydration isomers
  c. linkage isomers

19.62 On the basis of crystal field theory, decide whether the following complexes should be low-spin or high-spin complexes.
  a. $[Co(H_2O)_6]^{3+}$ b. $[Mn(CN)_6]^{3-}$

19.63 Name the following complexes.
  a. $Na_3[MnF_6]$ b. $[Ru(NH_3)_6][Mn(CN)_6]$

19.64 Determine the number of unpaired electrons in a low-spin and a high-spin octahedral complex that contains eight *d* electrons.

19.65 Draw a diagram showing the crystal field splitting energy for a high-spin tetrahedral complex that contains four *d* electrons.

19.66 Write the valence bond description of the bonding in square planar $[Cu(NH_3)_4]^{2+}$.

19.67 Decide, on the basis of crystal field splitting patterns, whether each of the following compounds should be diamagnetic or paramagnetic.
  a. $[Cr(NH_3)_6]^{3+}$ c. $[Fe(CN)_6]^{4-}$
  b. $[Fe(CN)_6]^{3-}$ d. $[Fe(H_2O)_6]^{3+}$

19.68 Predict whether each of the following metal ions should have colored complex ions.
  a. Ti^{4+} d. Co^{3+}
  b. V^{3+} e. Fe^{6+}
  c. Cu^{+} f. Cu^{2+}

19.69 Determine the number of unpaired electrons in a tetrahedral complex that contains five *d* electrons. (These are always high-spin complexes.)

19.70 Using the spectrochemical series, predict which of each pair of complex ions should absorb visible radiation at the shorter wavelength.
  a. $[Co(NH_3)_4BrCl]^{+}$ or $[Co(NH_3)_4Br(H_2O)]^{2+}$
  b. $[Cr(CN)_6]^{3-}$ or $[Cr(CN)_5(H_2O)]^{2-}$
  c. $[Fe(H_2O)_6]^{3+}$ or $[Fe(CN)_6]^{3-}$
  d. $[Mn(H_2O)_6]^{3+}$ or $[MnF_6]^{3-}$

19.71 Write formulas for the following complex ions.
  a. tetrahydroxozincate(II)
  b. tris(ethylenediamine)nickel(II)

19.72 Give the oxidation number of the central metal ion in each of the following complexes.

    a. $[Cr(NH_3)_6]^{3+}$ c. $[Fe(CN)_6]^{4-}$

    b. $[Fe(CN)_6]^{3-}$ d. $[Fe(H_2O)_6]^{2+}$

19.73 Use crystal field theory to decide whether each of the following complexes should be paramagnetic.

    a. $[Fe(H_2O)_6]^{2+}$

    b. $[Ru(CN)_6]^{4-}$

    c. $[Rh(NH_3)_6]^{3+}$

19.74 Using crystal field theory, draw a *d*-orbital splitting diagram for the following complex ions. Indicate the number of unpaired electrons.

    a. $[Mn(H_2O)_6]^{3+}$ b. $[Mn(H_2O)_6]^{2+}$

19.75 What is each of the following ions called when it is a ligand?

    a. NO_3^- d. SCN^-

    b. O^{2-} e. I^-

    c. PO_4^{3-} f. ClO_2^-

19.76 Indicate the oxidation number and coordination number of the central metal ion in each of the following coordination compounds.

    a. $[Cr(NH_3)_5Br]Cl_2$

    b. $[Ir(NH_3)_4Cl_2]NCS$

    c. $[Co(NH_3)_5(H_2O)]Cl_3$

    d. $[Cr(H_2O)_5Cl]Cl_2$

    e. $[Pt(NH_3)_2Cl_2] \cdot H_2O$

    f. $K_4[NiCl_4]$

19.77 What are the possible geometries of the complexes $[Zn(NH_3)_4]^{2+}$ and $[PtCl_6]^{2-}$?

19.78 Excess silver nitrate is added to 100.00 mL of 0.100 M $PtCl_4 \cdot 5NH_3$, producing 4.300 g of AgCl. What is the formula of the complex ion in this coordination compound?

* 19.79 The formula $Co(NH_3)_4ClCO_3$ could represent three isomers. Write the structures of these isomers, and suggest experiments that could be used to distinguish between them.

* 19.80 The pentacyanoiodocobalt(III) ion can undergo a direct-substitution reaction with cyanide ion, as shown in the equation

$$[Co(CN)_5I]^{3-}(aq) + CN^-(aq) \longrightarrow [Co(CN)_6]^{3-}(aq) + I^-(aq)$$

Addition of cobalt(II) chloride to this solution greatly accelerates the reaction. Suggest an explanation for this observation. (*Note:* When cobalt(II) chloride is dissolved in a solution containing cyanide ions, the complex ion $[Co(CN)_5]^{3-}$ is formed.)

* 19.81 The dark green solid $CrCl_3 \cdot 6H_2O$ contains a complex of chromium(III).

    a. Write formulas for four possible complexes that are consistent with this empirical formula.

    b. Explain how an ion-exchange column could be used to determine which of these complexes is actually present.

* 19.82 Tripropylphosphine, $P(C_3H_7)_3$, is often abbreviated PPr$_3$. It can be used as a ligand, for example, with platinum(II). Two compounds with the formula $[Pt(PPr_3)_2Cl_2]$ have been formed. One is pale yellow and has a melting point of 85°C, and the second is colorless and melts at 151°C. Both are soluble in benzene, where the freezing-point depression indicates that the molar mass of each is 586 g/mol. When dissolved in nitrobenzene, both complexes are nonelectrolytes. The yellow form has a dipole moment of 0 D, and the colorless form has a dipole moment of 11.5 D. What are the structures of these two complexes?

▶ ▶ ▶ Chemistry in Practice

19.83 After 5.48 g of $CoCl_2 \cdot 6H_2O$ has been dissolved in water, the solution is made basic with NH_4OH, and air is bubbled through. After cooling, 3.96 g of red crystals precipitate from the solution. After the crystals are removed by filtration, NaOH is added to the remaining solution, and it is boiled. All color is removed from the solution, and 0.482 g of Co_3O_4 precipitates. The following experiments were carried out on the red crystals. When 0.1466 g of crystals is dissolved in 10.00 mL of water, the freezing point is −0.292°C. An excess of $AgNO_3$ is added to this solution, and 0.1800 g of AgCl precipitates immediately. After standing for a long time, 0.0900 g more of AgCl precipitates. A 0.1466 g sample of the red crystals is boiled with excess NaOH solution, releasing ammonia gas. The ammonia gas is absorbed by 25.00 mL of 0.2015 M HCl. This solution is then titrated with 0.1074 M NaOH solution, and 23.51 mL of the solution is required. Suggest, for the red compound, a structure that is consistent with all these results.

Nuclear Chemistry

Nuclear fuel elements glow with a blue color under water.

CHAPTER OUTLINE

Encountering Chemistry

▶ ▶ ▶ ▶

S olid cylinders are lowered into a vat of water. Water in the vicinity of the cylinders glows with an eerie blue color and becomes warm. Where do the light and the heat come from?

The cylinders contain uranium used in nuclear power plants. When the cylinders are no longer effective as a nuclear fuel, they are stored under water. Here, they continue to lose some of their remaining radioactivity until they can be stored safely elsewhere. The energy arising from this loss of radioactivity results in the observed light and heat.

This chapter examines nuclear chemistry, the changes that occur in the nuclei of atoms that exhibit radioactivity. The chapter also examines radioactivity and its uses, energy changes in nuclear reactions, the use of such reactions as energy sources, and possible mechanisms for the formation of the elements.

20.1 Radioactivity

The chemical elements exist in widely varying amounts, which differ for the universe as a whole and for Earth. Figure 20.1 shows the relative abundances of many elements in the universe, and Table 20.1 lists the more abundant elements in the universe, Earth, and the sun. Hydrogen is the most abundant element in the uni-

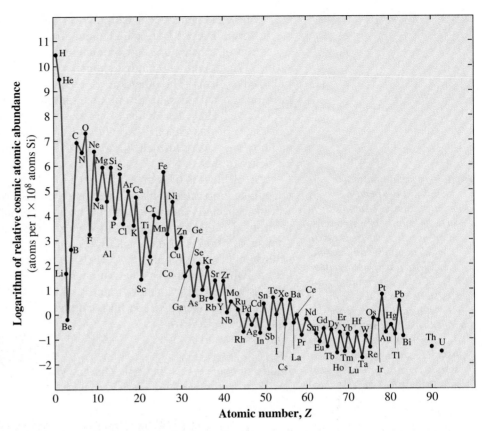

Figure 20.1 Relative abundances of elements in the universe. A logarithmic scale is used because there is tremendous variation in the abundances of the elements. Absolute numbers are not known, so the abundances are given relative to that of silicon, which occurs throughout the universe.

Do atoms of all the elements last indefinitely?

verse; nearly 91% of all atoms are hydrogen. In Earth's crust, the most abundant element is oxygen, which accounts for over 62% of the atoms. At the other extreme, some of the synthetic elements have been prepared in amounts of only a few atoms, which exist for only fractions of a second before they destroy themselves by radioactive decay—the emission of radiation.

Table 20.1 Distribution of Elements in Parts of the Universe

Universe		Sun		Earth (total)		Earth's Crust		
Element	Atom %	Element	Volume %	Element	% by mass	Element	% by mass	Atom %
H	90.7%	H	92.5%	Fe	35.4%	O	46.6%	62.6%
He	9.1	He	7.3	O	27.8	Si	27.2	21.2
O	0.057	O	0.068	Mg	17.0	Al	8.13	6.47
N	0.042	C	0.037	Si	12.6	Fe	5.00	1.92
C	0.021	N	0.0093	S	2.7	Ca	3.63	1.94
Ne	0.003	Ne	0.0074	Ni	2.7	Na	2.83	2.64
Si	0.003	Mg	0.0044	Ca	0.6	K	2.59	1.42
Mg	0.002	Si	0.0029	Al	0.44	Mg	2.09	1.84
Fe	0.002	Fe	0.00185			Ti	0.63	0.28

Ninety elements occur naturally on Earth, and of these, 81 have at least one stable form. These 81 are the elements from hydrogen through bismuth, with the exception of technetium and promethium, which have only unstable radioactive forms. The remaining 9 natural elements, polonium through uranium, exist only in radioactive forms. The other known elements are all synthetic, although they may have once occurred naturally in small amounts. These synthetic elements too are radioactive. Thus the elements are not all equally stable or equally capable of existence. In this chapter, we will consider the relative stability of the elements and how they were formed. First, however, we must examine the structure of the elements further.

An element is of course composed of atoms. An atom can be subdivided, but the unique properties of that element are then lost. In discussing nuclear chemistry, we are especially interested in the nuclear particles, or **nucleons:** protons and neutrons. Recall from Chapter 2 that the atomic number, symbolized Z, equals the number of protons in the nucleus. The number of neutrons in the nucleus is the neutron number, symbolized N. The sum of these two quantities is the mass number, A, which equals the total number of nucleons.

As we noted in Chapter 2, atoms with the same atomic number, Z, but different neutron numbers, N, and mass numbers, A, are called *isotopes*. An isotope is represented as follows:

$$_Z^A E^{n+}$$

In this symbol, $n+$ is the ionic charge. When we are particularly interested in the nuclei of isotopes of the elements, we often call them *nuclides*.

Numerous additional subatomic particles have been found by atomic physicists. Among them are the pions, or pi mesons, which are thought to act as a "nuclear glue" holding the nuclear particles together. Because proton–proton repulsions would destabilize a nucleus, strong cohesive nonelectrostatic forces must be operating at very short distances to maintain nuclear stability. These strong forces are attributed to pions, which act between all nuclear particles with a magnitude 30 to 40 times that of proton–proton repulsions.

Stability of Nuclides

There are somewhat more than 1700 known nuclides. Of these, only 264 are stable. ▶ The rest decompose over some period of time, like the uranium described in the chapter opening, emitting radiation in the process. That is, they exhibit **radioactivity.** The stable nuclides are found in a rather restricted range of combinations of Z and N values, as shown in Figure 20.2. The stable nuclides in this narrow "island of stability" have equal numbers of neutrons and protons ($N/Z = 1$) in the lighter elements ($Z = 1$–20) and a slight excess of neutrons ($N/Z \leq 1.6$) in the heavier elements. No nuclides heavier than $_{83}^{209}\text{Bi}$ are stable. All nuclides found outside the stability region are radioactive. They undergo nuclear transformations that convert them into stable isotopes of the same element or into different elements, though not always directly or rapidly.

Note in Figure 20.2 that the isotopes found in the region of stability are arranged in a zig-zag pattern to some extent. Even numbers of protons and neutrons have greater stability than odd numbers, as shown in Table 20.2. Of the stable nuclides, about 60% have an even number of protons and an even number of neutrons. Elements with an even atomic number have more stable isotopes than elements with an odd atomic number. Looking back at Figure 20.1, you can see that elements with even atomic numbers are more abundant in the universe than elements with odd atomic numbers. An important exception to this rule is the most abundant element, hydrogen, where the isotope $_1^1\text{H}$ predominates.

Which nuclides are stable and which are radioactive?

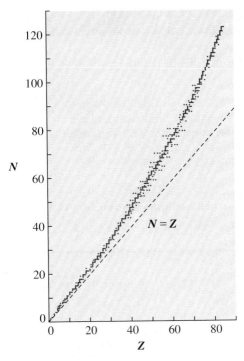

Figure 20.2 Relative numbers of neutrons and protons in stable nuclides.

Why are some nuclear compositions more stable than others?

According to one model of the nucleus, the *shell model*, neutrons and protons—like electrons—have energy levels. Special stability is associated with certain numbers of electrons (2, 10, 18, 36, 54, and 86) that correspond to filled electronic shells; this results in the unusual chemical inertness of the noble gases. A similar inertness is noted for nuclei that have certain numbers of protons or neutrons called *magic numbers*: 2, 8, 20, 28, 50, 82, and 126. The most stable isotopes have magic numbers of both protons and neutrons.

The most common example of this is the helium nucleus, $^4_2$He, which has two protons and two neutrons. The most stable isotope of the relatively abundant element oxygen is $^{16}_8$O, which has a magic number of protons (8) and of neutrons (8). Similarly, the most stable isotope of calcium, the fifth most abundant element in Earth's crust, is $^{40}_{20}$Ca, with magic numbers of protons (20) and neutrons (20).

Table 20.2 Types of Stable Nuclides

Number of Protons	Number of Neutrons	Number of Stable Nuclides
Even	Even	157
Even	Odd	52
Odd	Even	50
Odd	Odd	5

EXAMPLE 20.1

Decide which of the following two nuclides is stable and which is radioactive: $^{124}_{50}Sn$ and $^{125}_{50}Sn$.

Solution   The isotope $^{124}_{50}Sn$ has 50 protons and 74 neutrons, whereas $^{125}_{50}Sn$ has 50 protons and 75 neutrons. Both nuclides have 50 protons, a magic number, but neither has a magic number of neutrons. An even number of nucleons is more stable than an odd number of nucleons, so $^{124}_{50}Sn$ is stable and $^{125}_{50}Sn$ is radioactive.

Practice Problem 20.1

Decide which of the following two nuclides is stable and which is radioactive: $^{76}_{31}Ga$ and $^{45}_{21}Sc$.

Answer: $^{45}_{21}Sc$ is stable and $^{76}_{31}Ga$ is radioactive.

Radiation

Nuclear reactions involve changes in the composition of the nucleus, generally accompanied by the emission of radiation. Natural radiation associated with radioactivity can be placed in three classes: alpha particles, beta particles, and gamma rays. Other particles, such as protons and neutrons, may also be involved in nuclear reactions.

At one time many people built backyard bomb shelters in order to survive possible attack by atomic bombs. Why were these shelters built of concrete?

Alpha particles, usually designated as α or $^4_2\alpha$, contain two protons and two neutrons; they are the nuclei of helium atoms. We often omit the charge of 2+ on the helium nucleus from the symbol for the alpha particle, because it is not necessary to account for electrons when examining nuclear reactions. Alpha particles are the least penetrating of the three types of radiation and so are the least harmful to animal tissue upon external exposure. Alpha particles can be stopped by a thickness of aluminum metal greater than 10^{-3} cm, by a piece of paper, or by skin cells. Thus wrapping an alpha-emitting source with aluminum foil can successfully contain alpha particles, and they are not particularly harmful to human tissue because they do not penetrate the skin.

Beta particles are small charged particles that can be emitted from atoms at speeds approaching the speed of light. Beta particles are more penetrating than alpha particles; it takes about 0.05–0.1 cm of aluminum metal to stop them. They can travel as far as 10 feet through air and can easily be stopped by several sheets of paper or by a sheet of aluminum foil. Beta particles are commonly emitted by television sets. There are two types of beta particles, electrons and positrons. **Positrons** have the same mass as electrons but the opposite charge. Electrons are designated as $^{\,0}_{-1}\beta^-$ or $^{\,0}_{-1}e^-$ and positrons as $^0_1\beta^+$.

Gamma rays are the most penetrating type of radiation; a thickness of 5–11 cm of aluminum metal is required to stop them. They are therefore the most damaging type of natural radiation. Gamma rays can also be stopped by other types of heavy shielding, such as lead or thick layers of concrete. Gamma rays are so penetrating because they have no mass to slow them down. They are high-energy electromagnetic radiation—energy with no charge or mass.

Other particles are also commonly emitted during nuclear reactions. Two of these have already been discussed: protons (p^+ or ^1_1p or $^1_1H^+$) and neutrons (n or

$_0^1$n). Other particles are also involved, but because they have essentially no mass or charge, we do not need to consider them in our introductory treatment of nuclear reactions. Examples include the neutrino ($_0^0\nu$) and the antineutrino ($_0^0\bar{\nu}$), which always accompany the emission of beta particles.

20.2   Nuclear Reactions

Radioactivity arises from a change in the composition of a nucleus—the transformation of one nuclear particle into another or the emission of a particle from the nucleus. We can write equations for these transformations just as we wrote equations for chemical reactions. However, the procedure for writing these equations is somewhat different.

Equations for Nuclear Reactions

How do we write balanced equations to represent nuclear reactions?

Two conditions must be met for the equation for a nuclear reaction to be balanced: (1) conservation of mass number and (2) conservation of nuclear charge (atomic number). We can identify the particles emitted in a particular nuclear transformation by checking these two conservation conditions.

Alpha Particle Emission   When an alpha particle is emitted by a nucleus, both its atomic number and its neutron number decrease by 2 and its mass number decreases by 4. For example, thorium-232 undergoes radioactive decay by alpha particle emission.

$$_{90}^{232}\text{Th} \longrightarrow {}_{88}^{228}\text{Ra} + {}_2^4\alpha$$

Thorium is a grayish-white, lustrous metal. It burns in air to form thorium(IV) oxide, reacts with concentrated acids, and reacts with the halogens at red heat. It is used in nuclear reactors, along with ^{235}U, because the natural isotope ^{232}Th forms fissionable ^{233}U when struck by neutrons.

The atomic number decreases by 2 when the nuclide changes from thorium to radium; meanwhile the mass number decreases by 4. The emission of an alpha particle conserves the mass number ($232 = 228 + 4$) and the atomic number ($90 = 88 + 2$).

Beta Particle (Electron) Emission   The emission of a beta particle (electron) results in no change in the mass number, but the atomic number increases by 1; this is equivalent to the conversion of a neutron into a proton. The number of protons is increased, resulting in an increase in the atomic number. Because the number of neutrons and protons remains the same, the mass number does not change. Thorium-231 decays by beta emission.

$$_{90}^{231}\text{Th} \longrightarrow {}_{91}^{231}\text{Pa} + {}_{-1}^{0}\beta^-$$

The mass number stays the same, so it is conserved. The atomic number increases by 1, so the atomic number is also conserved ($90 = 91 + -1$).

EXAMPLE 20.2

Determine the identity of the unknown particle X in the following nuclear reaction, thereby balancing the equation.

$$_6^{14}\text{C} \longrightarrow X + {}_{-1}^{0}\beta^-$$

(continued)

Solution The emitted electron has a mass number of 0, so the particle X must have a mass number of 14, equal to that of the reactant. With a charge of $1-$ on the emitted electron, the product X must have a nuclear charge greater by 1 than the carbon reactant, or a value of 7. Atomic number 7 indicates nitrogen, so X must be ^{14}N.

$$^{14}_{6}C \longrightarrow {}^{14}_{7}N + {}^{0}_{-1}\beta^{-}$$

Practice Problem 20.2

Determine the identity of the unknown particle X in the following nuclear reaction, thereby balancing the equation.

$$^{234}_{90}Th \longrightarrow X + {}^{0}_{-1}\beta^{-}$$

Answer: $^{234}_{91}Pa$

Beta Particle (Positron) Emission A positron has a mass number of 0 and a charge of $1+$. Thus the emission of a positron by a radioactive nucleus must result in no change in the mass number and a decrease of 1 in the atomic number, equivalent to the conversion of a proton into a neutron. This result is just the reverse of the result of emission of an electron. An example of positron emission is the decay of magnesium-23.

$$^{23}_{12}Mg \longrightarrow {}^{23}_{11}Na + {}^{0}_{1}\beta^{+}$$

Because mass number $(23 = 23 + 0)$ and atomic number $(12 = 11 + 1)$ are both conserved, this nuclear equation is balanced.

Electron Capture Emission of a positron by a nucleus transforms that nucleus into the nucleus of another element with the same mass number. The new element has an atomic number less by 1 than that of the original element. This transformation can take place in another way as well—through **electron capture.** The transformation of a nucleus by electron capture involves the capturing of an inner-shell electron by the nucleus. When an electron is added to the nucleus, a proton combines with the electron to form a neutron. This process, then, accomplishes the same nuclear change as positron emission. For example, beryllium-7 undergoes radioactive decay by electron capture.

$$^{7}_{4}Be + {}^{0}_{-1}e^{-} \longrightarrow {}^{7}_{3}Li$$

The mass number stays the same $(7 + 0 = 7)$, but the atomic number is decreased by 1 $(4 - 1 = 3)$.

What happens to the energy released during a spontaneous nuclear reaction?

Gamma Ray Emission During all the radioactive decay processes already discussed, the nucleus changes from a state of higher energy to a state of lower energy. Excess energy can be carried off by the emission of an alpha or beta particle with a high kinetic energy. Energy can also be emitted in the form of high-energy gamma rays.

In many cases, the product of a nuclear reaction is produced in an excited state; it has an excess of energy. The excess energy is released by the product nucleus in the form of gamma rays—pure electromagnetic radiation. These gamma rays have characteristic energies for particular nuclides. A given nuclide may emit anywhere from none to more than 100 different gamma ray frequencies. The frequencies of the gamma rays emitted help us identify the products of nuclear reactions. An example of a reaction that produces gamma rays is the transformation of excited radium-224 to ground-state radium-224, wherein the radium is produced by the spontaneous alpha decay of thorium-228:

$$^{228}_{90}\text{Th} \longrightarrow {}^{224}_{88}\text{Ra}^* + {}^4_2\alpha$$
$$^{224}_{88}\text{Ra}^* \longrightarrow {}^{224}_{88}\text{Ra} + \gamma$$

The gamma decay process involves no change in the mass number or in the atomic number.

Bombardment Reactions   So far we have examined only the nuclear reactions that occur when one nucleus spontaneously changes into another. Nuclear reactions may also occur when a nucleus is bombarded with another particle. When vanadium-51 is bombarded with deuterium atoms, a neutron is transferred from the deuterium nucleus to the vanadium nucleus, and a proton is emitted.

Are all nuclear reactions spontaneous?

$$^{51}_{23}\text{V} + {}^2_1\text{H} \longrightarrow {}^{52}_{23}\text{V} + {}^1_1\text{H}$$

The mass number $(51 + 2 = 52 + 1)$ and the atomic number $(23 + 1 = 23 + 1)$ are conserved. It is possible to identify an unknown particle in a nuclear bombardment reaction by mass and charge balance if all other particles are known.

Spontaneous Nuclear Decay Reactions

We have examined a number of different types of nuclear decay reactions in which one nuclide is transformed into another. But why does one nuclide decay by emission of an alpha particle whereas another decays by emission of a beta particle? The nuclear reactions that occur in naturally radioactive nuclides can be accounted for in terms of a tendency for the N/Z ratio to move toward the "island of stability" shown in Figure 20.3. Changes in the composition of the nucleus can be accomplished by alpha decay, beta (electron or positron) decay, and electron capture. Excited nuclei can change to a lower-energy state by emitting gamma rays. The characteristics of these types of nuclear changes are listed in Table 20.3. The consequences of the processes with respect to nuclear stability are shown in Figure 20.3.

Radioactive nuclides convert spontaneously, over time, to stable nuclides. This conversion may occur in a single step, but more commonly several successive nuclear reactions are required before a stable state within the island of stability is reached.

Decay of Heavier Elements   As we have said, all isotopes of the elements beyond atomic number 83 (bismuth) are radioactive and decay to become nuclides with lower mass and lower atomic number. To do so, these elements generally decay by a combination of alpha and beta decay processes. ▶ The alpha decay gets rid of excess mass via the loss of two neutrons and two protons, as in the decay of uranium-238 described in the chapter opening.

$$^{238}_{92}\text{U} \longrightarrow {}^{234}_{90}\text{Th} + {}^4_2\alpha$$

An unstable neutron/proton ratio is not adjusted significantly in heavy elements through this process, however. The thorium-234 produced in the decay of uranium-238 is still unstable. Beta decay processes are also necessary to convert a heavy, unstable nuclide to a more stable one. Beta decay converts a neutron to a

Figure 20.3 Nuclear changes in natural radioactive decay. Nuclides are transformed to other nuclides along lines parallel or perpendicular to the line that represents $N/Z = 1$.
  o These unstable nuclides decay through both alpha and beta (electron) emission. The alpha decay relieves the nucleus of excess mass, whereas the beta decay keeps the N/Z ratio from rising too high.
  + These unstable nuclides need to convert a neutron to a proton to enter the island of nuclear stability. This is accomplished by beta decay.
  * These unstable nuclides need to convert a proton to a neutron. They accomplish this by ejecting a positron or by electron capture.

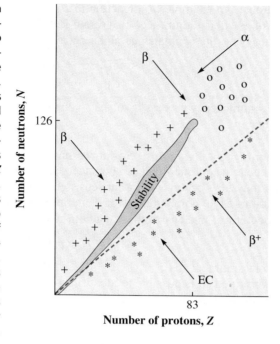

Uranium is a silver-white, lustrous, radioactive metal. It is used as a nuclear fuel and in atomic and hydrogen bombs. It burns in air to form U_3O_8 and reacts slowly with water when finely divided.

proton, so it decreases the neutron/proton ratio and increases the atomic number, as in the decay of thorium-234.

$$^{234}_{90}\text{Th} \longrightarrow ^{234}_{91}\text{Pa} + ^{0}_{-1}\beta^{-}$$

Thus, as shown in Figure 20.3, alpha decay adjusts the nucleus along a line parallel to $N/Z = 1$, whereas beta decay adjusts the nucleus along a line perpendicular to this. Positron decay has the opposite effect: Protons are converted to neutrons,

Table 20.3 Natural Radioactive Decay Processes

Reason for Nuclear Instability	Radioactive Process	Emitted Radiation	Nuclear Change	Change in N/Z Ratio
Excess mass	α decay	$^{4}_{2}\alpha$	Loss of 2 protons and 2 neutrons occurs.	Slight increase
N/Z too high	β^{-} decay	$^{0}_{-1}\beta$	A neutron is converted into a proton and an electron.	Decrease
N/Z too low	β^{+} decay	$^{0}_{+1}\beta$	A proton is converted into a neutron and a positron.	Increase
N/Z too low	Electron capture	Neutrino	A proton combines with an inner-shell electron to become a neutron.	Increase
Energetically excited	γ emission	Gamma ray	Loss of excess nuclear energy occurs.	None

so the N/Z ratio is increased and the atomic number is decreased, as in the decay of fluorine-18.

$$^{18}_{9}F \longrightarrow {}^{18}_{8}O + {}^{0}_{+1}\beta^{+}$$

Decay of Lighter Elements For the lighter elements, where stable nuclides have $N/Z = 1$, it is fairly simple to predict the mode of decay. Radioactivity in lighter nuclides results from an unacceptable N/Z ratio. The ratios may be adjusted through beta decay, positron emission, or electron capture.

If the N/Z ratio is too high, it is necessary to convert a neutron to a proton by beta decay (emission of electrons). For example, carbon-14 has too high a ratio: $N/Z = (14 - 6)/6 = 8/6 = 1.33$. A stable state is achieved by beta decay.

$$^{14}_{6}C \longrightarrow {}^{14}_{7}N + {}^{0}_{-1}e^{-}$$

The product, nitrogen-14, has a ratio of $N/Z = (14 - 7)/7 = 7/7 = 1$.

On the other hand, nitrogen-13 has a low ratio: $N/Z = (13 - 7)/7 = 0.86$. Positron decay results in a higher ratio.

$$^{13}_{7}N \longrightarrow {}^{13}_{6}C + {}^{0}_{+1}\beta$$

The ratio in the product of this transformation, carbon-13, is not perfect either: $N/Z = (13 - 6)/6 = 7/6 = 1.17$. Nevertheless, carbon-13 is the only stable isotope of carbon other than carbon-12, which has the perfect $N/Z = 1$. It is not possible to get closer to a ratio of 1 by beta emission or electron capture.

It is also possible to raise the N/Z ratio by electron capture, as in the decay of beryllium-4.

$$^{7}_{4}Be + {}^{0}_{-1}e^{-} \longrightarrow {}^{7}_{3}Li$$

Beryllium is a gray metal with chemical properties similar to those of aluminum. The metal resists attack by acids because of the formation of an adherent oxide coating. When finely divided, it does react with aqueous acids and bases to release hydrogen gas. It is used as a neutron deflector and a neutron moderator in nuclear reactors. It is also used in aerospace structures and inertial guidance systems.

The occurrence of electron capture cannot be predicted readily. However, among the first 20 elements, this process apparently occurs in only four cases besides that just shown: the decay of $^{22}_{11}Na$, $^{37}_{18}Ar$, $^{40}_{19}K$, and $^{41}_{20}Ca$.

EXAMPLE 20.3

Predict the method of radioactive decay of the unstable nuclide $^{24}_{10}Ne$.

Solution The N/Z ratio in this nuclide is $(24 - 10)/10 = 14/10 = 1.4$. This ratio is higher than the ideal ratio of 1, so radioactive decay is needed to reduce it. This means that a neutron should be converted to a proton, which can be accomplished by emission of an electron (beta decay).

$$^{24}_{10}Ne \longrightarrow {}^{24}_{11}Na + {}^{0}_{-1}e^{-}$$

This sodium nuclide is also radioactive, because its N/Z ratio is also greater than 1.

Practice Problem 20.3

Predict the method of radioactive decay of the unstable nuclide $^{18}_{10}Ne$.

Answer: β^{+} decay

Radioactive Decay Series As noted earlier, some radioactive nuclides decay to stable nuclides in a single step. However, frequently the product of radioactive decay is itself radioactive, especially in the heavier elements. In such cases, sequences of alpha and beta decay steps ultimately lead to stable nuclides. Three such radioactive decay series, which involve only alpha and beta decay, are found among the natural elements and account for most of the radioactivity among elements 83 through 92. These are the uranium series (shown in Figure 20.4), the actinium series, and the thorium series. An additional series, the neptunium series, involves the synthetic element neptunium.

Neptunium is a silvery metal prepared by nuclear reactions. It exists as a number of radioactive isotopes, the most stable of which has a half-life of about 2 million years. It exhibits five oxidation numbers, which can be obtained in aqueous solution by oxidation–reduction processes. The most stable is +5, which is found in the ion NpO_2^+.

Each of these series consists of members whose mass numbers differ by units of 4, because mass is lost only through loss of alpha particles with mass number 4. Atomic number is decreased by alpha decay and increased by beta decay. In the decay series, the number of alpha particles lost is determined by the change in mass number. Any discrepancy between the change in atomic number that would be consistent with this number of alpha particles and the change that actually is observed is accounted for by beta decay.

Nuclear Bombardment Reactions

There are three types of nuclear reactions besides spontaneous decay processes: fission, fusion, and nuclear bombardment reactions. **Fission** is the splitting of a

Figure 20.4 The uranium series.

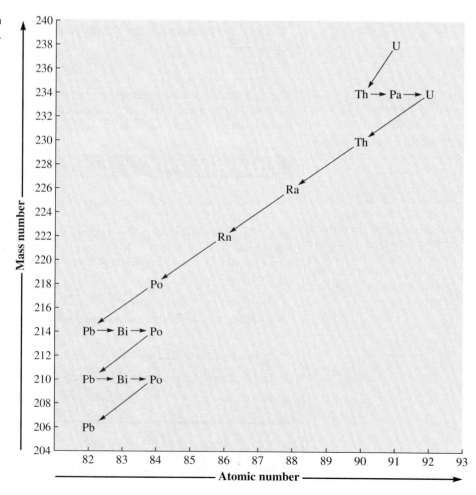

heavy nucleus into two or more lighter nuclei and some number of neutrons. **Fusion** is the combination of light nuclei to form heavier nuclei. These two processes are discussed in separate sections. Like fission and fusion, the third type of nuclear reaction, nuclear bombardment, is an induced nuclear reaction.

In **nuclear bombardment reactions,** nuclei are bombarded with other nuclei or with nuclear particles. For example, the bombardment of $^{238}_{92}\text{U}$ with neutrons gives $^{239}_{92}\text{U}$, which spontaneously decays into $^{239}_{93}\text{Np}$:

$$^{238}_{92}\text{U} + {}^{1}_{0}\text{n} \longrightarrow {}^{239}_{93}\text{Np} + {}^{0}_{-1}\text{e}^{-}$$

All elements with atomic numbers above 92, the atomic number of uranium, are artificial elements that do not exist naturally on Earth. There are also lighter elements that have never been found in a natural state. For a long time, there were "holes" in the periodic table at elements 43 (Tc), 61 (Pm), and 85 (At). Promethium has been identified among the fission products of uranium, but technetium and astatine were synthesized via nuclear bombardment reactions.

Technetium was first formed in 1936 by the bombardment of molybdenum with deuterium atoms.

$$^{97}_{42}\text{Mo} + {}^{2}_{1}\text{H} \longrightarrow {}^{97}_{43}\text{Tc} + 2{}^{1}_{0}\text{n}$$

It has also been formed by bombardment of molybdenum with neutrons. All isotopes of technetium are unstable; most of them undergo radioactive decay in a matter of a few minutes to a few days, but three of the isotopes survive for a few million years.

Astatine was formed in 1940 by the bombardment of bismuth with alpha particles.

$$^{209}_{83}\text{Bi} + {}^{4}_{2}\alpha \longrightarrow {}^{211}_{85}\text{At} + 2{}^{1}_{0}\text{n}$$

All isotopes of astatine are highly radioactive, the most stable, $^{210}_{85}\text{At}$, having a half-life of only 8.3 hours. No more than about 0.05 μg of astatine has ever been produced in a single preparation.

All of the transuranium elements have been synthesized by nuclear bombardment reactions. We saw earlier that the bombardment of uranium with neutrons yields neptunium. Other nuclear particles or other nuclei can be used to bombard uranium or other heavy nuclei, as in the following examples:

$$^{238}_{92}\text{U} + {}^{4}_{2}\alpha \longrightarrow {}^{239}_{94}\text{Pu} + 3{}^{1}_{0}\text{n}$$
$$^{238}_{92}\text{U} + {}^{12}_{6}\text{C} \longrightarrow {}^{246}_{98}\text{Cf} + 4{}^{1}_{0}\text{n}$$
$$^{238}_{92}\text{U} + {}^{14}_{7}\text{N} \longrightarrow {}^{247}_{99}\text{Es} + 5{}^{1}_{0}\text{n}$$

Nuclear scientists continue to use bombardment reactions to prepare heavier elements. By bombarding bismuth-209 with iron-58 nuclei for one week, for example, scientists synthesized one atom of element 109, which has mass number 266.

$$^{209}_{83}\text{Bi} + {}^{58}_{26}\text{Fe} \longrightarrow {}^{266}_{109}X + {}^{1}_{0}\text{n}$$

The atom decayed by alpha particle emission to element 107, which had previously been prepared in a greater amount (6 atoms). Element 107 decayed further by alpha emission to $^{258}_{105}\text{Unp}$, which underwent electron capture to form $^{258}_{104}\text{Unq}$, which decayed by fission to smaller nuclei.

The lifetimes of the transuranium elements generally decrease as the atomic numbers increase. Thus the heavier undiscovered elements, if they are synthesized, are expected to have extremely short lifetimes. However, it is possible that islands of stability will be found among the "superheavy" elements. As shown in

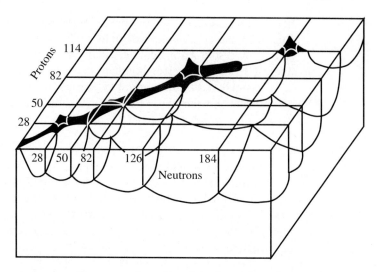

Figure 20.5 Known and predicted regions of nuclear stability. Areas that rise above the surface of the grid indicate combinations of protons and neutrons that are expected to be stable.

Figure 20.5, it is expected that elements around atomic number 114 will have greater stability than adjacent elements. Attempts to synthesize nuclides in this region will focus on bombardment of transuranium elements with medium-heavy nuclei. For example, it is possible that element 116 may result from the bombardment of curium-248 with calcium-48.

$$^{248}_{96}Cm + {}^{48}_{20}Ca \longrightarrow {}^{296}_{116}X$$

Another region suitable for exploration involves isotopes with around 126 protons and 184 neutrons. This nuclide has magic numbers of protons and neutrons, so it is expected to be more stable than nearby nuclides.

20.3   Radioactive Decay

Nuclear reactions are characterized by the emission of radiation. The type of radiation and its properties can be used to identify a nuclear reaction. The emitted radiation has a characteristic energy, and it is produced at a characteristic rate.

Detection of Radiation

How can we tell that a nuclide is radioactive? Many different techniques can be used to study the characteristics of the radiation produced by radioactive nuclides. The simplest method of detecting radiation involves the exposure of photographic film. Indeed, individuals who work around radiation often must wear film badges, which are developed on a regular schedule to determine the extent of exposure to radiation. However, the use of photographic film is neither very accurate nor very rapid.

► ► ►   WORLDS OF CHEMISTRY

Radon: Is Your House Radioactive?

Several surveys indicate that the U.S. national average for radon in houses is about 1.4 pCi/L, but some houses were found to register as high as 2700 pCi/L.

Although concerns about air pollution are directed mostly to outside air, indoor air pollution is also a health risk. But what could be in the air in our houses that could cause problems? Of course, second-hand cigarette smoke and carbon monoxide from a faulty burner are possibilities. But surprisingly, a rather rare noble gas, radon, has been implicated as a possible cause of lung cancer. This element is rare because all of its isotopes are radioactive and decay to other elements in relatively short times. The isotope with the longest lifetime is ^{222}Rn, which has a half-life of 3.8 days. Radon is one of the intermediate products of the radioactive decay of uranium and thorium. Uranium is widely present in soil in components of granite and shale, and thorium is present in granite and gneiss. Radon-222 decays further to polonium-218.

$$^{222}\text{Rn} \longrightarrow {}^{218}\text{Po} + \alpha$$

Breathing radon is no more harmful than breathing any other noble gas, unless a radon atom undergoes radioactive decay while in the lungs. The polonium and all decay products after it are solids, which can become attached to tissue in the lungs, where the ensuing radiation can cause cell damage.

The amount of damage radon causes depends on just how much radon is in a particular house. Measurements are made in terms of the radioactivity in picocuries per liter, where a curie (Ci) corresponds to 3.7×10^{10} radioactive disintegrations per second. Several surveys indicate that the U.S. national average for radon in houses is about 1.4 pCi/L, but some houses were found to register as high as 2700 pCi/L. The national average corresponds to about 56% of the average radiation dosage experienced annually by residents of the United States. Estimates suggest that people living in homes with radon levels above 8 pCi/L have a 2% increase in lung cancer risk and that about 10,000 lung cancer deaths a year can be attributed to radon.

The radon comes from the soil, but how does it get into our houses? No one is absolutely certain, but radon has become a much greater problem since we started to build houses that are airtight in attempts to save on energy used for heating and/or air conditioning. Although the air in our

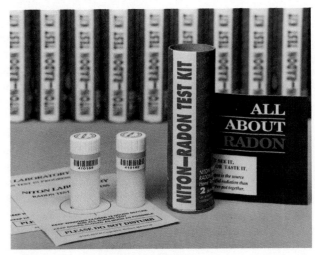

Figure 20.6 Test kits are available to determine the radon level in buildings.

houses is exchanged with outside air once or twice an hour, this is less frequent than it used to be, and radon concentrations are higher as a result.

What can be done about radon in our houses? First, it is necessary to determine whether the radon level in a house is unusually high. Kits, such as those shown in Figure 20.6, can be used to measure radon levels. These kits consist of either a special piece of plastic or a canister of activated charcoal. The plastic is damaged by passage of α particles, so microscopic examination reveals the number of penetrations that occurred during a set time period. The charcoal adsorbs radon atoms. In both cases, the kit must be returned to a laboratory for evaluation. In some cases, unusually high radon levels have been cured by relatively simple approaches such as placing blowers under a house to remove accumulations of radon in the vicinity of the house.

Questions for Discussion

1. Should you have the air in your house tested for radon?
2. The risk to the general population from radon is about 100–1000 times higher than the risk from carbon tetrachloride or benzene, which have been tightly controlled by the U.S. Environmental Protection Agency. Should the EPA generate regulations for radon in houses?

(continued)

3. Why do toxic chemicals of human origin generally get more attention in the news media than toxic chemicals of natural origin?

4. Radon is estimated to cause 10,000 lung cancer deaths

per year, compared to about 110,000 attributed to the smoking of tobacco. Why is tobacco not more heavily regulated?

One of the earliest instruments used to measure radiation was the *Geiger–Müller counter*, shown in Figure 20.7. Radiation enters a tube through a thin glass or plastic window. The tube contains argon gas, which is ionized when it interacts with the entering radiation. A potential of 1000–2000 V is applied between the walls of the tube and a wire that passes into the center. When the gas is ionized, electrons are released and become accelerated by the electrical potential, ionizing other argon atoms. The result is a pulse of electrical current, which is passed through a digital counter or through a speaker to make a clicking sound. A Geiger–Müller counter can detect alpha, beta, and gamma radiation.

Another common detection device is a *scintillation counter*, which contains a tube coated with a material such as zinc sulfide that fluoresces or gives a flash of light when struck by radiation. Flashes of light that result from the impact of radiation cause pulses of electricity to pass through a photoelectric tube, where the electrical signals are amplified to activate a digital counter so that the intensity of the radiation can be measured in units of counts per minute. Because a scintillation counter can be tuned to detect radiation at different energies, it is able to determine the energy associated with radiation from decay of a particular nuclide.

The amount of radiation that can be produced by a sample of radioactive material is measured by its **activity,** which is the number of nuclei that disintegrate per unit time. Activity is usually measured in units of **curies,** abbreviated Ci. A substance with an activity of 1 Ci is a very powerful and dangerous source of radiation—this is the activity of 1 g of ^{226}Ra. One curie corresponds to an activity of 3.7×10^{10} disintegrations per second.

Figure 20.7 Geiger–Müller counter.

Kinetics of Radioactive Decay

So far, we have been concerned primarily with the types of particles involved in nuclear reactions. Another important characteristic of these reactions is the rate at which they proceed. All nuclear decay reactions follow first-order kinetics, described in Section 13.2, so the half-life ($t_{\frac{1}{2}}$) for radioactive decay is constant throughout the process. It takes the same time for a fresh sample to decay to one-half the original number of atoms of that nuclide as it does for one-half to decay to one-fourth, and so on. The shorter the half-life, the more radioactive the nuclide and the more intense the radiation, because it is being produced at a greater rate.

The intensity of radiation can be measured as the number of nuclear disintegrations per minute per gram of the nuclide. For example, carbon-14 undergoes beta decay with a radioactivity of 15.3 disintegrations per minute per gram of total carbon. The intensity of radiation is related to the number of radioactive nuclides present in the sample. Normally, we do not measure the disintegration rate directly but rather use the response of some detector, such as a scintillation counter. The detector response is measured over a convenient time period, and the radiation intensity is reported in terms of counts per minute (cpm). Figure 20.8 shows the intensity of radiation as a function of time for the beta decay of gold-198, which has $t_{\frac{1}{2}} = 2.7$ days.

The number of atoms of a given nuclide is exponentially related to time according to the first-order kinetic equation

$$N = N°e^{-kt}$$

Here N is the number of atoms present at time t, $N°$ is the number of atoms at time $t = 0$, and k is the first-order rate constant for the reaction.

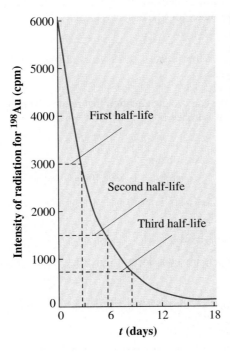

Figure 20.8 As shown by a radioactive decay curve for ^{198}Au, the intensity of radiation decreases by a factor of 2 during each successive half-life. The number of ^{198}Au atoms also decreases by a factor of 2 during each half-life.

Why are radioactive decay rates listed in terms of half-lives instead of rate constants, as was done for chemical reactions?

The rate of a radioactive decay process is usually given in terms of the half-life for the process. The relationship between rate constant and half-life was discussed in Chapter 13. Briefly, when the time equals one half-life, the number of atoms is decreased to one-half its original value.

$$N = \tfrac{1}{2}N° \qquad \text{when } t = t_{\frac{1}{2}}$$

Inserting these relationships into the first-order equation gives

$$\ln N = \ln N° - 0.693t/t_{\frac{1}{2}}$$

Thus the half-life for the nuclear process can be obtained from the slope of a plot of $\ln N$ against time. The half-lives measured in this way for the nuclides involved in the uranium decay series are given in Table 20.4. Half-lives for a variety of other nuclides are listed in Appendix H.

The first-order rate equation for radioactive decay can be used to carry out a variety of calculations. A common calculation determines the amount of radioactive nuclide remaining after a specific time. This calculation is useful in a number of situations. For example, radioactivity is employed to treat some serious cancers. In some cases, the intensity of the radiation is so high that the patient is a radiation risk to other individuals and has to be kept in isolation. In such cases it is desirable to know what the radioactivity level will be after various periods of time. As long as we know the half-life and the specific time interval of interest, we can calculate the fraction of nuclide remaining (in terms of $N/N°$) by solving the equation for N.

EXAMPLE 20.4

Strontium-90 is a common product found in the fallout of atomic bomb explosions. It is particularly harmful because it may end up associated with the calcium in cows' milk. Strontium-90 undergoes beta decay to give yttrium-90.

$$^{90}_{38}\text{Sr} \longrightarrow \, ^{90}_{39}\text{Y} + \, ^{0}_{-1}\beta^{-}$$

The half-life of the process is 28 years. What fraction of the strontium-90 is left after exactly 10 years?

Solution   We can use the equation that relates the number of nuclide atoms remaining (N) to the original number $N°$:

$$N/N° = e^{-0.693t/t_{\frac{1}{2}}}$$

Inserting values, we obtain

$$N/N° = e^{-0.693 \times 10 \text{ years}/28 \text{ years}} = e^{-0.248} = 0.78$$

Thus, 78% of the strontium-90 is left after 10 years.

Practice Problem 20.4

Gold-128 undergoes beta decay to give mercury-128:

$$^{128}_{79}\text{Au} \longrightarrow \, ^{128}_{80}\text{Hg} + \, ^{0}_{-1}\beta^{-}$$

The half-life is 2.7 days. What fraction of the gold-128 is left after 5.0 days?

Answer: 28%

Table 20.4 Half-Lives for the Nuclides in the Uranium Decay Series

Nuclide	Method of Decay	Half-life
$^{238}_{92}U$	Alpha	4.46×10^9 years
$^{234}_{90}Th$	Beta	24.1 days
$^{234}_{91}Pa$	Beta	6.75 hours
$^{234}_{92}U$	Alpha	2.45×10^5 years
$^{230}_{90}Th$	Alpha	7.54×10^4 years
$^{226}_{88}Ra$	Alpha	1.62×10^3 years
$^{222}_{86}Rn$	Alpha	3.82 days
$^{218}_{84}Po$	Alpha	3.11 minutes
$^{214}_{82}Pb$	Beta	26.8 minutes
$^{214}_{83}Bi$	Beta	19.7 minutes
$^{214}_{84}Po$	Alpha	1.6×10^{-4} seconds
$^{210}_{82}Pb$	Beta	22.3 years
$^{210}_{83}Bi$	Beta	5.01 days
$^{210}_{84}Po$	Alpha	138.4 days
$^{206}_{82}Pb$	—	—

We can also calculate the time it would take for the radioactivity level to be reduced by some specific amount. In this case, the $N/N°$ ratio is known. If the half-life is also known, the first-order equation can be solved for the elapsed time.

EXAMPLE 20.5

Potassium-40 undergoes two simultaneous decay processes. Electron capture converts 11.0% of the K-40 to Ar-40.

$$^{40}_{19}K + {}^{0}_{-1}\beta^- \longrightarrow {}^{40}_{18}Ar$$

Beta decay converts 89.0% of the K-40 to Ca-40:

$$^{40}_{19}K \longrightarrow {}^{40}_{20}Ca + {}^{0}_{-1}\beta^-$$

The half-life for these processes is 1.27×10^9 years. How long does it take for $\frac{2}{3}$ of a sample of K-40 to decay?

Solution We can apply the equation relating the number of atoms remaining to the original number.

$$N/N° = e^{-0.693t/t_{\frac{1}{2}}}$$

The number of K-40 atoms remaining is $1/3 \, N°$. Inserting values, we obtain

$$N/N° = 0.333 = e^{-0.693 \times t/1.27 \times 10^9 \text{ years}}$$

Taking the natural logarithm yields

$$\ln(0.333) = -1.10 = -0.693 \times t/1.27 \times 10^9 \text{ years}$$

Rearranging gives

$$t = 1.27 \times 10^9 \text{ years} \times 1.10/0.693 = 2.02 \times 10^9 \text{ years}$$

(continued)

Practice Problem 20.5

Strontium-90 undergoes beta decay to give yttrium-90.

$$^{90}_{38}\text{Sr} \longrightarrow \, ^{90}_{39}\text{Y} + \, ^{0}_{-1}\beta^-$$

The half-life is 28 years. How long does it take for $\frac{1}{8}$ of a sample of Sr-90 to decay?

Answer: 5.4 years

Lifetime and Radioactive Storage

Knowing the half-lives of various nuclides is necessary in several important applications. ▶ Radioactive materials produced in reactors must be stored in safe locations until their radioactivity is reduced sufficiently, as described in the chapter opening. The half-life of the nuclides in the material determines the required storage time. For example, chromium-51 decays by electron capture and emits weak gamma radiation. The half-life for decay of Cr-51 is 27.8 days. This nuclide is often used to trace the fate of chromium in solution reactions of coordination compounds of chromium. All solutions containing Cr-51 must be stored in shielded locations after use. Generally, the substance's radioactivity reaches an acceptable level after 10 half-lives (at which point the Cr-51 is 99.9% decayed). Thus it is necessary to shield these solutions for 10×27.8, or 278, days.

How long is it necessary to store radioactive wastes?

Archeological Dating

The decay of radioactive nuclides is also used to measure time on an archeological or geological scale. For example, artifacts are dated according to the amount of carbon-14 they contain—a technique called **radiocarbon dating.** C-14 is continuously being produced in the upper atmosphere by the reaction of nitrogen-14 with neutrons from outer space (cosmic rays).

$$^{14}_{7}\text{N} + \, ^{1}_{0}\text{n} \longrightarrow \, ^{14}_{6}\text{C} + \, ^{1}_{1}\text{H}$$

For purposes of C-14 dating, we assume that the rate of production has been fairly constant over very long times, although there may have been some fluctuations associated with phenomena such as solar flares. The C-14 produced in this way is eventually incorporated as carbon dioxide into plant tissues during photosynthesis and into animal tissue when plants are consumed.

As long as the plant or animal is alive, its C-14 content should match that in the atmosphere. After it dies, the C-14 content decreases because of beta decay.

$$^{14}_{6}\text{C} \longrightarrow \, ^{14}_{7}\text{N} + \, ^{0}_{-1}\beta^-$$

This process has a half-life of 5730 years. Measurement of the amount of C-14 in the carbon of a dead substance, then, can be used to determine the time elapsed since the tissues died. The method has been calibrated by measurement of the C-14 content of wood in individual tree rings of very old bristle-cone pine. Thus corrections can be made for fluctuations in the intensity of cosmic rays.

Radiocarbon dating has been used on many different substances, such as wood, grains, cloth, bone, shells, peat, charcoal, organic mud, and carbonates. Living tissue contains enough carbon-14 to give about 15.3 disintegrations per minute per gram of carbon (disintegrations $min^{-1} g^{-1}$). Measurement of C-14 radioactivity has a limit of detection of about 0.03 disintegrations $min^{-1} g^{-1}$. This limit is reached in about 9 half-lives, so we can use radiocarbon dating to determine the age of objects younger than about 50,000 years.

EXAMPLE 20.6

A wooden bowl found in an archeological dig had a C-14 content corresponding to 6.29 disintegrations $min^{-1} g^{-1}$. If living tissue gives 15.3 disintegrations $min^{-1} g^{-1}$, and the half-life of C-14 is 5730 years, how old is the bowl?

Solution   The intensity of radiation is proportional to the number of C-14 atoms, so we can use the equation

$$I/I° = N/N° = e^{-0.693t/t_{\frac{1}{2}}}$$

The measurements give $I/I° = 6.29/15.3 = 0.411$. Substituting this value into the equation yields

$$0.411 = e^{-0.693t/5730} \text{ years}$$
$$\ln(0.411) = -0.899 = -t(0.693/5730)$$
$$t = 0.889 \times 5730/0.693 = 7350 \text{ years}$$

Practice Problem 20.6

A leather thong found in an archeological dig had a C-14 content corresponding to 12.8 disintegrations $min^{-1} g^{-1}$. If living tissue gives 15.3 disintegrations $min^{-1} g^{-1}$, and the half-life of C-14 is 5730 years, how old is the leather?

Answer: 1470 years

Geological Dating

Geologists date their samples in much the same way as archeologists date samples containing C-14, but they use other nuclides. One such method involves the ratio of K-40 to Ar-40 in minerals containing potassium. Potassium-40 undergoes two decay processes: 11.0% undergoes electron capture to produce Ar-40, and 89.0% undergoes beta decay to produce calcium-40.

$$^{40}_{19}K + ^{\ 0}_{-1}\beta^- \longrightarrow ^{40}_{18}Ar$$
$$^{40}_{19}K \longrightarrow ^{40}_{20}Ca + ^{\ 0}_{-1}\beta^-$$

The half-life for these processes is 1.27×10^9 years. Measurements based on the K/Ar ratio have been made on many different potassium-containing minerals. Terrestrial rocks have been dated to ages as great as 2 billion (2×10^9) years. The K/Ar method has also been applied to meteorites, resulting in measurements of ages as great as 4.5 billion (4.5×10^9) years. Geological samples can also be dated with other long-lived radioactive elements. Such investigation have given $(2.5–3.0) \times 10^9$ years as the ages of the oldest terrestrial rocks and 4.5×10^9 years as the ages of a number of meteorites.

20.4   Energy Changes in Nuclear Reactions

In earlier chapters, we discussed the energy changes associated with chemical reactions. Values were usually around 100–1000 kJ/mol. Here, we examine the energy changes associated with nuclear reactions, which are much larger when compared on the basis of mass. It is these very large energy changes that make nuclear reactions interesting to chemists. We begin by examining the change in mass that accompanies a nuclear reaction.

When specifying the identity of a given nuclide, we used the atomic number (Z) as a measure of the number of protons and the mass number (A) as a measure of the number of protons plus neutrons, or nucleons. The mass of a proton is 1.00728 amu, of a neutron 1.00867 amu, and of an electron 0.0005486 amu. Because the masses of neutrons and protons are slightly greater than 1 amu, we would expect the mass of a nuclide to be slightly greater than its mass number. However, the nuclide is always lighter than expected. This discrepancy in mass is called the **mass defect.** Consider $^{20}_{10}\text{Ne}$, which contains 10 protons, 10 electrons, and 10 neutrons. When determining the mass defect of $^{20}_{10}\text{Ne}$, we are dealing with the following hypothetical reaction:

Why is the mass of the nucleus less than the mass of its component particles?

$$10\text{p}^+ + 10\text{n}^\circ + 10\text{e}^- \longrightarrow {}^{20}_{10}\text{Ne}$$

The sum of the masses of the component particles is determined as follows:

$$\begin{aligned} \text{Mass}_{\text{particles}} &= 10 \times 1.00728 \text{ amu} + 10 \times 1.00867 \text{ amu} + 10 \times 0.0005486 \text{ amu} \\ &= 10.0728 + 10.0867 + 0.005486 = 20.1650 \text{ amu} \end{aligned}$$

However, the mass of the $^{20}_{10}\text{Ne}$ atom is 19.9924 amu. The mass defect is calculated as the difference in masses.

$$\text{Mass defect} = 20.1650 \text{ amu} - 19.9924 \text{ amu} = 0.1726 \text{ amu}$$

When an atom is formed from its component particles, then, mass is lost. This loss—the mass defect—corresponds to a lowering in the energy of the system. The atom is more stable than the separate particles by an amount of energy corresponding to the mass defect. This energy, called the **nuclear binding energy,** is the amount of energy that would be released if the appropriate numbers of subnuclear particles were combined to give an atom. To calculate the magnitude of this energy, we use Einstein's equation for the mass–energy equivalence:

$$E = mc^2$$

where m is the mass and c is the speed of light. If we express mass in kilograms (1 amu $= 1.66056 \times 10^{-27}$ kg) and the speed of light in meters per second (2.9979×10^8 m/s), then the energy has units of kg m^2 s^{-2}, or joules. To get more manageable numbers, it is common to express nuclear binding energies in units of million electron volts, or MeV, where 1 MeV $= 1.6022 \times 10^{-13}$ J.

Now we can calculate the nuclear binding energy for $^{20}_{10}\text{Ne}$.

$$\begin{aligned} E &= 0.1726 \text{ amu} \times 1.66056 \times 10^{-27} \text{ kg/amu} \times (2.9979 \times 10^8 \text{ m/s})^2 \\ &= 2.576 \times 10^{-11} \text{ J} \end{aligned}$$

This value can be converted to units of MeV.

$$E = 2.576 \times 10^{-11} \text{ J} \times 1 \text{ MeV}/1.6022 \times 10^{-13} \text{ J} = 160.8 \text{ MeV}$$

This is a small amount of energy compared with the chemical reaction energies we have been dealing with, but it is the amount of energy associated with one atom. A mole of $^{20}_{10}$Ne atoms has the following nuclear binding energy:

$$E = 2.576 \times 10^{-11} \text{ J/atom} \times 6.022 \times 10^{23} \text{ atoms/mol} = 1.551 \times 10^{13} \text{ J/mol}$$

The square of the speed of light and the conversions from amu to kg and from J to MeV occur in every mass–energy problem. When working problems, it is simpler to collect these terms into one equivalence expression:

$$1 \text{ amu} = 931.48 \text{ MeV}$$

Then we can use a simple, one-step procedure to calculate nuclear binding energies from mass defects.

EXAMPLE 20.7

Calculate the nuclear binding energy for $^{40}_{20}$Ca, given that the mass of this atom is 39.9626 amu.

Solution   We are dealing with the following hypothetical reaction:

$$20p^+ + 20n^\circ + 20e^- \longrightarrow {}^{40}_{20}\text{Ca}$$

The mass of the subnuclear particles is

$$\begin{aligned} \text{Mass} &= 20 \times 1.00728 \text{ amu} + 20 \times 1.00867 \text{ amu} + 20 \times 0.000549 \text{ amu} \\ &= 40.33000 \text{ amu} \end{aligned}$$

The mass defect is then

$$\text{Mass defect} = 40.3300 \text{ amu} - 39.9626 \text{ amu} = 0.3674 \text{ amu}$$

The nuclear binding energy is given by the equation

$$E \text{ in MeV} = \text{mass defect in amu} \times 931.48 \text{ MeV/1 amu}$$

Substituting the value of the mass defect gives the required energy.

$$E \text{ in MeV} = 0.3674 \text{ amu} \times 931.48 \text{ MeV/1 amu} = 342.2 \text{ MeV}$$

Thus 342.2 MeV of energy is given off in the formation of $^{40}_{20}$Ca from its component particles.

Practice Problem 20.7

Calculate the nuclear binding energy for $^{46}_{22}$Ti, given that the mass of this atom is 45.9526 amu.

Answer: 398.4 MeV

We can use nuclear binding energies to compare the stabilities of various nuclides. The greater the nuclear binding energy, the lower the relative energy of the nuclide and the more stable it is. To do this, we divide the calculated nuclear binding energy by the number of nucleons in the nuclide. This procedure enables us to compare the amount of nuclear stability per nucleon, thereby correcting for the contribution of each nucleon to the stability. For $^{40}_{20}$Ca, for example, we obtain 342.2 MeV/40 nucleons = 8.56 MeV/nucleon. For $^{20}_{10}$Ne we obtain 160.8 MeV/20 nucleons = 8.04 MeV/nucleon. A comparison of the nuclear binding energies per nucleon for various nuclides is shown in Figure 20.9, which displays a binding-energy curve over a wide range of mass numbers.

Figure 20.9 Binding energy per nucleon plotted against mass number.

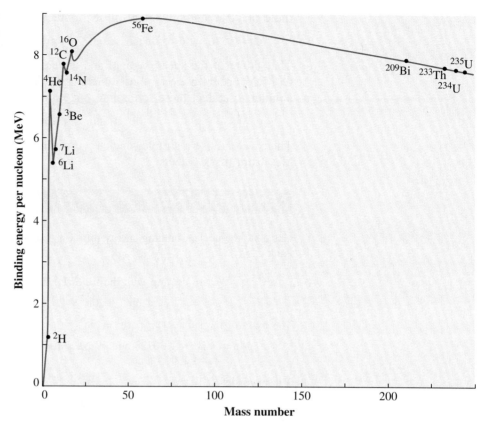

Note that certain nuclides have unusually high binding energies. In particular, observe the relative values of ^4_2He, $^{12}_6\text{C}$, and $^{16}_8\text{O}$. These nuclides all have even numbers of neutrons and of protons, and all have N equal to Z. As we noted earlier, these are all conditions for stability. In contrast, ^6_3Li, and $^{14}_7\text{N}$, with odd numbers of protons and neutrons, have binding energies per nucleon lower than their neighbors on the binding-energy curve. Elements with atomic numbers in the range of 20–30 have the greatest binding energy; the maximum occurs around $^{56}_{26}\text{Fe}$. The relative stability measured in terms of binding energy is reflected in the distribution of the elements in Earth's crust. For example, no element with a mass number greater than that of iron is found in the crust in amounts greater than 1%.

The binding-energy curve can also be used to predict the types of nuclear reactions the elements might undergo. Lighter nuclides are susceptible to transmutation by bombardment reactions—they can undergo fusion processes to yield heavier nuclides with higher binding energies. The combination of hydrogen nuclei to form helium nuclei is the source of the energy in the sun, for example. In contrast, the heavier nuclides, with smaller binding energies, are prone to undergo fission, creating lighter nuclides with higher binding energies. For example, when elements such as uranium undergo fission, they generally produce elements with atomic numbers roughly in the range of 40–50. These elements have higher binding energies and are thus more stable.

Fission, fusion, and bombardment reactions that result in products that have greater average binding energies than the reactants result in the loss of mass and the production of energy. To calculate the energy accompanying these nuclear reactions, we need only calculate the mass change for the process and convert this to its energy equivalent (1 amu = 931.5 MeV).

EXAMPLE 20.8

Calculate the energy change for the following fission reaction:

$$^{235}_{92}U + ^{1}_{0}n \longrightarrow ^{94}_{40}Zr + ^{140}_{58}Ce + 2^{1}_{0}n + 6^{0}_{-1}e^{-}$$

Solution   First we need to know the masses of these particles. They are

$^{235}_{92}U$, 235.04392 amu	$^{140}_{58}Ce$, 139.90543 amu
$^{94}_{40}Zr$, 93.9063 amu	$^{1}_{0}n$, 1.00867 amu
$^{0}_{-1}e^{-}$, 0.000549 amu	

Next we calculate the mass loss accompanying the process.

Mass loss = 235.04392 amu + 1.00867 amu − 93.9063 amu − 139.90543 amu
$$- 2 \times 1.00867 \text{ amu} - 6 \times 0.000549 \text{ amu} = 0.2202 \text{ amu}$$

Then we convert this mass loss into its energy equivalent.

$$E = 0.2202 \text{ amu} \times 931.48 \text{ MeV/amu} = 205.1 \text{ MeV}$$

Practice Problem 20.8

Calculate the energy change for the following reaction:

$$^{10}_{5}B + ^{1}_{0}n \longrightarrow ^{7}_{3}Li + ^{4}_{2}\alpha$$

The masses of these particles are

$^{10}_{5}B$, 10.01294 amu	$^{7}_{3}Li$, 7.01600 amu
$^{1}_{0}n$, 1.00867 amu	$^{4}_{2}\alpha$, 4.00260 amu

Answer: 2.80 MeV

20.5   Nuclear Fission

Some heavy radioactive nuclides are so unstable, because of relatively low binding energies, that they undergo spontaneous fission. In other cases the heavy nuclide must be irradiated or bombarded with some appropriate particle to cause fission. ▶ A fissionable nuclide of particular importance is uranium-235, which is extensively used in nuclear power plants, as described in the chapter opening.

Uranium-235 Fission

The fission of uranium-235 is depicted in Figure 20.10. When U-235 is struck by a slow neutron—one having a low kinetic energy—it absorbs the neutron and becomes the even less stable U-236 nuclide. The U-236 nuclide then splits into two lighter nuclei and also emits neutrons and gamma radiation. Many different fission reactions may occur when U-235 is bombarded with neutrons. Indeed, over

Figure 20.10 U-235 fission can be induced by neutron bombardment. The intermediate product is unstable U-236, which splits into lighter nuclides and emits neutrons that can cause other U-235 atoms to undergo fission.

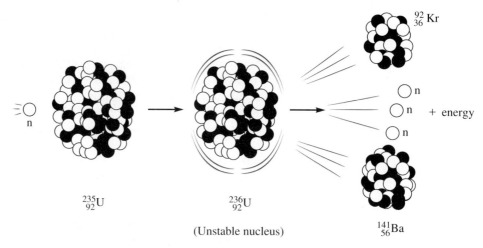

200 different fission products of this reaction have been identified. As the following examples show, these fission reactions differ only in the manner in which the very unstable U-236 splits up.

$$^{235}_{92}\text{U} + ^{1}_{0}\text{n} \longrightarrow ^{92}_{36}\text{Kr} + ^{141}_{56}\text{Ba} + 3^{1}_{0}\text{n}$$

$$^{235}_{92}\text{U} + ^{1}_{0}\text{n} \longrightarrow ^{90}_{38}\text{Sr} + ^{143}_{54}\text{Xe} + 3^{1}_{0}\text{n}$$

$$^{235}_{92}\text{U} + ^{1}_{0}\text{n} \longrightarrow ^{94}_{40}\text{Zr} + ^{140}_{58}\text{Ce} + 2^{1}_{0}\text{n} + 6^{0}_{-1}\text{e}$$

Chain Reactions

Why does U-235 fission continue to occur after neutron bombardment is stopped?

The fission of U-235 can be started by a single slow neutron, but all the resulting reactions produce more neutrons than are consumed. Overall, on average, 2.4 neutrons are produced for each neutron used. If all the product neutrons initiate more fission steps, the process continuously gets faster. Such a process is called a **chain reaction**—a reaction in which a product of one step is a reactant in another step.

Several conditions must be met for a chain reaction to sustain itself. The neutrons produced by the fission reaction must be of appropriate energy to cause other nuclei to undergo fission. Generally, the energy of the neutrons produced by U-235 fission is higher than is optimal for inducing further fission steps. The neutrons can lose energy in collisions, but if they escape from the uranium sample before they have slowed down sufficiently, they cannot induce fission. In order for the chain reaction to sustain itself, then, the amount and shape of the sample of fissionable material must be such that the neutrons will not simply escape. The smallest amount of fissionable material necessary to support a continuing chain reaction is called the **critical mass.**

Another important factor is how the fission progresses. Ideally, to sustain itself, the process should maintain a constant rate, becoming neither faster nor slower. If too few of the emitted neutrons are absorbed by other nuclei, the chain reaction gets slower and eventually stops. If too many of the emitted neutrons are absorbed by other nuclei, the process becomes faster and faster and culminates in an explosion. This is the basis of the atomic bomb, which consists of several pieces of uranium, each having a mass less than the critical mass. A conventional explosion is detonated to force the pieces together to form a critical mass, thereby activating the bomb.

Fission Reactors

How can fission be controlled?

Fission can be used to produce not only atomic bombs but also electrical energy. Obviously, in this case, it is necessary to maintain very careful control of the rate of the chain reaction. Neutron absorbers such as cadmium and boron make such control possible. Normally, movable cadmium or boron rods called *control rods* are inserted into a core of uranium fuel in fission reactors (Figure 20.11). The material in the rods reacts with and removes neutrons.

$$^{10}_{5}B + ^{1}_{0}n \longrightarrow ^{7}_{3}Li + ^{4}_{2}\alpha$$

The control rods are adjusted so that the chain reaction proceeds at just the constant rate necessary to maintain the desired energy output.

It is not necessary to have pure U-235 to support fission. Normally, the uranium fuel is enriched in U-235 from its natural abundance of 0.7% to about 2 or 3%. The fuel does not have to be uranium metal; it usually consists of U_3O_8.

The uranium oxide U_3O_8 is the chief component of the uranium ore pitchblende. Other uranium oxides are black UO_2, pale yellow $UO_4 \cdot 2H_2O$, and red UO_3.

As we have said, the neutrons produced by U-235 fission are usually too fast (too energetic) to produce further fission steps efficiently, so they must be slowed down. A *moderator* such as normal water (H_2O), heavy water (D_2O), or graphite is included in the reactor design to slow down the neutrons (Figure 20.11). When the neutrons collide with the moderator, they lose energy.

The fission of U-235 produces about 200 MeV of energy per atom, or about 2×10^{10} kJ/mol, in the form of kinetic energy in the reaction products. To be usable, the energy must be converted to another form. This conversion normally involves circulating a coolant around the reactor core (Figure 20.12). The hot coolant (as hot as 310°C) is pressurized to prevent boiling and then passed through a steam generator.

Steam generators are massive structures containing thousands of thin-walled tubes for heat transfer. The hot primary coolant is pumped through the tubes, and the secondary coolant is pumped around the tubes and converted to steam. The steam is used to drive a turbine, which operates a generator, producing electricity. The steam must be condensed and cooled so that it can be pumped back through

Figure 20.11 Reactor core in a fission reactor.

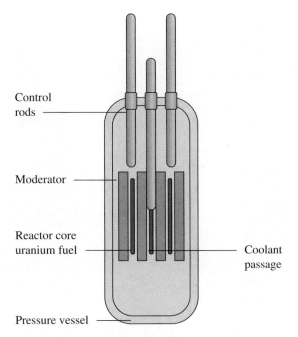

Control rods

Moderator

Reactor core uranium fuel

Coolant passage

Pressure vessel

Figure 20.12 Heat-exchange system in a pressurized fission reactor.

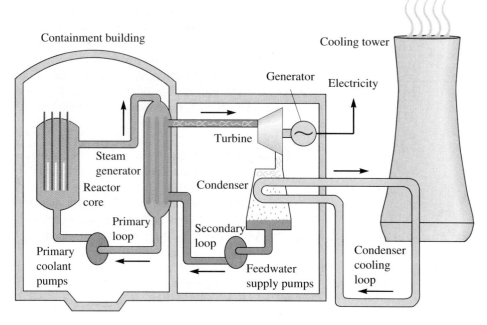

the steam generator. This is often accomplished by circulating a third coolant through a condenser and a cooling tower. Heat is removed from the secondary coolant in the condenser and then released to the environment in the massive cooling towers (Figure 20.13) that have become the hallmark of nuclear power plants.

For protection in case of leaks, the reactor core is located in a massive containment building, often built in a spherical shape (Figure 20.13). In the event of a reactor leak, the radioactive material would, in principle, be kept within the containment building, and other areas would not become contaminated. The nuclear disaster that occurred in 1986 at the nuclear power plant in Chernobyl, Rus-

Figure 20.13 Cooling tower at the Trojan nuclear power plant in Oregon.

sia, released a considerable amount of radioactive nuclides into the surrounding environment, some of which spread throughout the world. A major difference exists between Russian and U.S. nuclear reactors. The Russian reactors do not include the containment building, which is designed to prevent the release of radioactive material.

In most cases, the immediate products of U-235 fission are themselves radioactive. For example, $^{92}_{36}Kr$ undergoes four successive beta decay steps, with half-lives of 3 s, 5 s, 3 h, and 3.5 h, to yield $^{92}_{40}Zr$, and $^{141}_{56}Ba$ decays to yield $^{141}_{59}Pr$ via three successive beta emissions that have half-lives of 18 min, 4 h, and 33 days. The radioactivity of the products of fission poses a major challenge. Once the fission fuel (U-235) has been used up, a complex mixture of radioactive substances remains. This radioactive waste, as well as other contaminated materials, must be stored so that it can become inactive without causing damage to living things. The best method for disposing of radioactive waste has not yet been determined, but many methods are under investigation. These include reprocessing the fuel and storing it in liquid or solid form in steel tanks or underground.

20.6   Nuclear Fusion

Another method for nuclear energy production is fusion, the combination of light nuclei to form heavier nuclei. Fusion occurs continuously in the sun:

$$4^1_1H \longrightarrow {}^4_2He + 2{}^{\ 0}_{+1}\beta^+$$

About 25 MeV of energy per fusion event evolves through this process. Fusion requires considerable energy to overcome the strong internuclear repulsions that occur as the nuclei get enough together for nuclear forces to cause the nuclei to fuse together. In the sun, the energy is provided by extremely high temperatures in excess of 10^7 K. In a fusion bomb, the high temperatures and pressures are provided by the explosion of a fission bomb. The temperature required to initiate a fusion reaction is called the ignition temperature.

Fission reactors suffer from a limited supply of fissionable material and from the necessity of storing highly radioactive products. These shortcomings can be avoided in fusion reactors. At present, no fusion reactor has been designed to actually produce power, but considerable research is going on in this area.

A variety of fusion reactions could be used in a commercial fusion reactor. The following reactions are some of the possibilities:

$$^2_1H + {}^3_1H \longrightarrow {}^4_2He + {}^1_0n$$
Energy yield = 17.6 MeV
Ignition temperature = 1×10^8 K (Ignition energy = 10 keV)
Energy gain (energy out/energy in) = 1760

$$^2_1H + {}^2_1H \longrightarrow {}^3_1H + {}^1_1p^+$$
Energy yield = 4.0 MeV
Ignition temperature = 5×10^8 K (Ignition energy = 50 keV)
Energy gain = 80

$$^2_1H + {}^3_2He \longrightarrow {}^4_2He + {}^1_1p^+$$
Energy yield = 18.3 MeV
Ignition temperature = 1×10^9 K (Ignition energy = 100 keV)
Energy gain = 183

Fusion Reactors

What advantages do fusion reactors offer, compared to fission reactors?

The first of the reactions just given, between deuterium and tritium, is the most likely candidate for the first commercial fusion reactor, because the energy gain from this reaction is much greater than for the other reactions listed. Furthermore, deuterium is present in water in sufficient quantities that it will be available for millions of years. The technology for inexpensively separating deuterium from water has already been developed. Eight gallons of sea water contains about 1 gram of deuterium, which during fusion can release an amount of energy equivalent to the burning of about 2500 gallons of gasoline. Tritium is a radioactive isotope of hydrogen, which emits beta rays with a half-life of 12.3 years. Although tritium occurs naturally to some extent, it is not present in nature in sufficient quantity to be used as a fuel. However, tritium can be prepared synthetically by neutron bombardment of lithium, which is available in the earth and in the seas in sufficient quantities to provide at least a thousand-year supply.

In a fusion reactor (Figure 20.14), the reaction chamber would be surrounded by a lithium "blanket," which would transfer the heat generated to a reactor coolant. Heat would then be transferred to a steam generator, as in fission reactors. The lithium would also absorb excess neutrons formed in the fusion process, thereby producing more tritium.

$$\begin{aligned} {}_{3}^{7}\text{Li} + {}_{0}^{1}\text{n (fast)} &\longrightarrow {}_{1}^{3}\text{H} + {}_{2}^{4}\text{He} + {}_{0}^{1}\text{n (slow)} \\ {}_{3}^{6}\text{Li} + {}_{0}^{1}\text{n (slow)} &\longrightarrow {}_{1}^{3}\text{H} + {}_{2}^{4}\text{He} + 4.8 \text{ MeV} \end{aligned}$$

In fact, it is possible to create more tritium than is used, and the excess tritium could be used to fuel other fusion reactors. A reactor that produces fuel that can be used in other reactors is called a *breeder reactor*.

The major problems to be solved before fusion can be used as an energy source include the creation and control of a *plasma*, an ionized gas, at temperatures of about 10^8 K. In order to achieve fusion, it is necessary to condense the gaseous reactants to a small volume at high temperatures. This is the purpose of the plasma, which must be kept at a very high temperature for longer than 1 second at a density of about 10^{13} atoms/cm^3. At this high temperature, the plasma will

Figure 20.14 Probable layout of a fusion reactor.

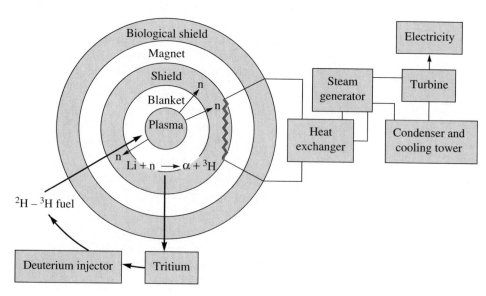

melt any normal container material. Two approaches are under investigation—magnetic and inertial confinement. Most work on magnetic confinement has concentrated on doughnut-shaped machines called *tokamaks*, in which magnetic fields are used to contain the plasma. The magnetic fields also induce an electrical current in the plasma to heat it. Temperatures over 7×10^7 K have been achieved by this method. The second method, inertial confinement, uses hydrogen fuel contained in tiny pellets, about 0.1 mm in diameter, made of glass or metal. The fuel pellet is bombarded with laser beams or electron beams. The beams compress and heat the fuel to form a very dense plasma in which it will undergo fusion. Whatever method is ultimately determined to be better, nuclear fusion is unlikely to provide commercial energy for many years.

Fusion Reactions in the Stars

Although fusion reactions have not yet been harnessed for producing energy in nuclear power plants, we do use energy created in fusion processes. The sun and other active stars are centers of fusion activity. The fusion reactions in the stars are the basis for an explanation of why elements heavier than hydrogen exist.

From where did the elements come? The **big bang theory** holds that the universe came into being at some definite time in the past when densely packed nuclear particles reached a temperature and pressure that caused them to explode outward and disperse matter throughout space. This matter eventually grouped itself into various stellar systems. Using various measurements, astronomers estimate the age of the universe to be about 1.2×10^{10} years, the age of our sun to be about 5×10^9 years, and the age of solidified material, such as planets and meteorites, in our solar system to be about 4.6×10^9 years.

On the basis of our current understanding of the fusion processes that are occurring in the stars, it is difficult to explain the amount of helium in the universe, which exceeds the amount that could be produced in the stars over a period of 10^{10} years. It is necessary to invoke some special event, such as a "big bang," to explain the amount of helium. During the big bang, initial temperatures of about 10^{12} K would have dropped in about 5 minutes to about 10^9 K. At this temperature, a number of sequential nuclear fusion reactions would have occurred within 30 to 60 minutes.

$$^1_1\text{H} + ^1_0\text{n} \longrightarrow ^2_1\text{H}$$

$$^2_1\text{H} + ^1_1\text{H} \longrightarrow ^3_2\text{He}$$

$$^3_2\text{He} + ^1_0\text{n} \longrightarrow ^4_2\text{He}$$

$$^4_2\text{He} + ^1_0\text{n} \longrightarrow ^5_2\text{He}$$

The sequence of reactions stops at this point, because the ^5_2He has a half-life of only 2×10^{-21} s and decays back to ^4_2He. The temperature would soon have dropped, probably down to about 4×10^7 K at the end of one day, and all these reactions would have stopped, leaving large amounts of helium. Some fairly simple models of the big bang come close to predicting the existing amount of helium in the universe.

Calculations involving the rates of numerous nuclear reactions suggest that the big bang could not have produced significant amounts of elements heavier than helium (except perhaps ^7_3Li). The heavier elements are generally thought to have originated in the intensely hot interiors of stars.

Stars proceed through an evolutionary series of stages: hydrogen burning, helium burning, carbon burning, oxygen burning, and silicon burning. The density, temperature, elemental composition, and elemental transformations in the stars differ for each stage. The various stellar processes that give rise to the elements are summarized in Figure 20.15.

The initial process involved in the synthesis of elements in stars is called *hydrogen burning*. Gravitational forces acting on the hydrogen that has condensed to form the star raise the density to about 100 g/cm^3 and the temperature to at least 5×10^6 K and more commonly to about $1–3 \times 10^7$ K. A variety of processes are thought to be involved in hydrogen burning, but the net result is the formation of helium. One possible set of processes is

$$^1_1H + ^1_1H \longrightarrow ^2_1H + ^{\ 0}_{+1}\beta^+$$

$$^2_1H + ^1_1H \longrightarrow ^3_2He$$

$$^3_2He + ^3_2He \longrightarrow ^4_2He + 2^1_1H$$

Figure 20.15 Summary of nucleosynthesis in the stars.

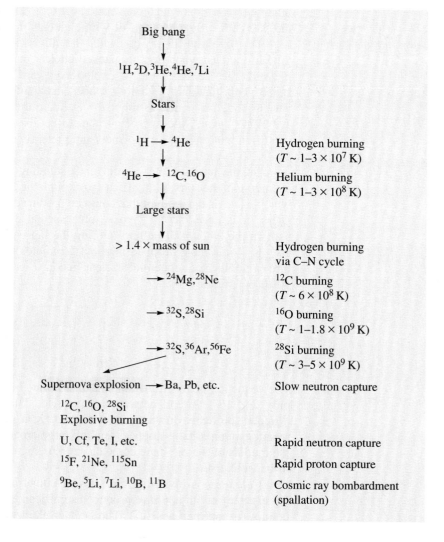

Big bang	
^{1}H,^{2}D,^{3}He,^{4}He,^{7}Li	
Stars	
^{1}H $\longrightarrow$ ^{4}He	Hydrogen burning ($T \sim 1–3 \times 10^7$ K)
^{4}He $\longrightarrow$ ^{12}C,^{16}O	Helium burning ($T \sim 1–3 \times 10^8$ K)
Large stars	
> 1.4 × mass of sun	Hydrogen burning via C–N cycle
$\longrightarrow$ ^{24}Mg,^{28}Ne	^{12}C burning ($T \sim 6 \times 10^8$ K)
$\longrightarrow$ ^{32}S,^{28}Si	^{16}O burning ($T \sim 1–1.8 \times 10^9$ K)
$\longrightarrow$ ^{32}S,^{36}Ar,^{56}Fe	^{28}Si burning ($T \sim 3–5 \times 10^9$ K)
Supernova explosion $\longrightarrow$ Ba, Pb, etc.	Slow neutron capture
^{12}C, ^{16}O, ^{28}Si Explosive burning	
U, Cf, Te, I, etc.	Rapid neutron capture
^{15}F, ^{21}Ne, ^{115}Sn	Rapid proton capture
^{9}Be, ^{5}Li, ^{7}Li, ^{10}B, ^{11}B	Cosmic ray bombardment (spallation)

Other processes produce $_4^7$Be, $_3^7$Li, $_4^8$Be, and $_5^8$B. In larger, somewhat hotter stars, cycles involving reactions of protons with carbon, nitrogen, and oxygen are also involved. (These elements have apparently condensed from material that has come from other stars in other stages of development.) These reactions produce $_7^{13}$N, $_6^{13}$C, $_7^{14}$N, $_8^{15}$O, $_7^{15}$N, and $_6^{12}$C.

Upon exhaustion of the hydrogen fuel, the density increases to about 10^5 g/cm^3 and the temperature to about 3×10^8 K. Under these conditions, *helium burning* commences. Helium burning involves a sequence of reactions of $_2^4$He to form $_4^8$Be, $_6^{12}$C, and $_8^{16}$O.

When the helium is exhausted, gravitational contraction again occurs, raising the temperature to about $8-11 \times 10^8$ K. At this point *carbon burning* occurs, involving the conversion of $_6^{12}$C to $_{10}^{20}$Ne and $_{11}^{23}$Na. Small amounts of $_{12}^{23}$Mg and $_{12}^{24}$Mg may also be formed at this stage.

The next major nuclear burning stage is reached when the supply of carbon-12 is exhausted and the temperature rises to about 2×10^9 K. Now *oxygen burning* starts, involving the conversion of $_8^{16}$O to $_{14}^{28}$Si, $_{15}^{31}$P, $_{16}^{31}$S, and $_{16}^{32}$S. Some argon and chlorine may also result at this stage by reaction of the product elements with alpha particles, protons, and neutrons.

The next major nuclear process, which occurs at $3-5 \times 10^9$ K and a density of about 10^8 g/cm^3, is *silicon burning*. Here heavier nuclei, such as $_{12}^{24}$Mg, $_{16}^{32}$S, and $_{18}^{36}$Ar, are formed from $_{14}^{28}$Si. Such processes account for the elements through iron, the element at the maximum in the binding-energy curve (Figure 20.9). The fusion process continues to release energy up to the formation of $_{26}^{56}$Fe. Further increases in density and temperature at this point lead not to further synthesis of heavier elements but rather to disintegration of the elements, to stellar core collapse, and to supernova explosions.

Neutron Capture Processes

How are elements heavier than $_{26}^{56}$Fe formed? Two types of neutron capture processes are possible. In the first type, "s" or *slow neutron capture*, the elements are built up from iron by successive captures of single neutrons. These processes tend to produce the lighter, proton-rich isotopes of the elements.

$$_{26}^{58}\text{Fe} + _0^1\text{n} \longrightarrow _{26}^{59}\text{Fe} \longrightarrow _{27}^{59}\text{Co} + _{-1}^0\text{e}^-$$

The heaviest isotope produced in this manner is $_{83}^{209}$Bi. The other neutron capture process, "r" or *rapid neutron capture*, occurs in environments of high neutron density, where several neutrons may be captured in a single step. This type of process leads to neutron-rich isotopes—for example,

$$_{26}^{56}\text{Fe} + 13\ _0^1\text{n} \longrightarrow _{26}^{69}\text{Fe} \longrightarrow _{27}^{69}\text{Co} + _{-1}^0\text{e}^-$$

Other processes involved in the synthesis of elements include rapid proton capture and spallation reactions. Rapid proton capture produces $_7^{15}$N, $_8^{18}$O, $_9^{19}$F, $_{10}^{21}$Ne, and many nuclides heavier than $_{26}^{56}$Fe that are not produced by neutron capture processes. Spallation reactions involve the bombardment of nuclides with heavy cosmic ray particles, which causes the nuclides to break up into lighter nuclides. None of the processes discussed earlier yield the nuclides $_1^2$H, $_3^6$Li, $_4^9$Be, $_5^{10}$B, and $_5^{11}$B. These nuclides are probably produced by spallation reactions.

20.7   Particle Accelerators

How are bombardment
reactions carried out?

In earlier sections we spoke rather glibly about bombarding nuclides with subatomic particles or with other nuclides, but generating and manipulating these
particles require sophisticated and expensive equipment. It is particularly important to increase the velocity of the particles and thereby make them energetic
enough to induce a nuclear change. Particle accelerators have been essential in the
study of the structure of the nucleus, as well as in the application of bombardment
reactions to the synthesis of new elements. Recent efforts in this area have created
new elements with atomic numbers greater than 103, although these elements are
usually produced in amounts of only a few atoms.

Several different types of particle accelerators have been developed. The
most common are linear accelerators and circular accelerators.

In a linear accelerator, an ion beam or electron beam is directed into a tube, as
shown in Figure 20.16. Increments of energy are fed to the charged particles as
they pass through a series of cavities. The voltage applied to the cavities is alter-
nated at an appropriate rate so that the cavities repel the charged particles from
behind and attract them from the front. Because the particles go faster as they
proceed through the accelerator, the farther the tubes are from the source, the
longer they must be.

Most of the accelerators in use are based on a circular design. These *cyclo-
trons* consist of two D-shaped cavities, as shown in Figure 20.17. Ions are generated in the center of the cyclotron. A magnetic field perpendicular to the cavities
constrains the ions to follow a circular path through the "dees." Every time the
particles pass through the gap between the dees, an alternating voltage across the
dees is changed to accelerate the particles. As the particles move faster, they also
move outward in the accelerator. After sufficient acceleration, the particles leave
the outer edge of the cyclotron and are directed onto an external target.

Perhaps the most successful accelerator—and certainly the largest—is the
synchrotron, which also uses a circular path for the accelerating particles
(Figure 20.18). The radius of the larger synchrotrons exceeds 1 km. The synchrotron uses electromagnets to produce a variable magnetic field that maintains the
path of the particles at a given radius as they are accelerated by voltage increments.
The electromagnets are enclosed within the hollow ring through which the charged
particles pass, so it is not necessary to have an electromagnet with radius of 1 km.
In addition to being able to accelerate particles to enormous energies, which ex-
ceed 500 GeV for protons, synchrotrons radiate considerable electromagnetic energy (in the range of 1 to 10^5 eV).

Accelerators are typically used to prepare new nuclides. In some cases, the
nuclides are new elements; in others, the nuclides are isotopes of existing elements. For example, the element technetium is not found naturally. It was first

Figure 20.16 Linear accelerator. An alternating current
changes the charges on the
tubes as the particle passes
through, causing a continual
acceleration.

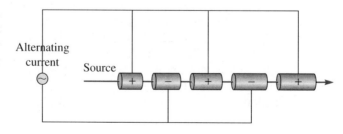

Figure 20.17 Cyclotron.

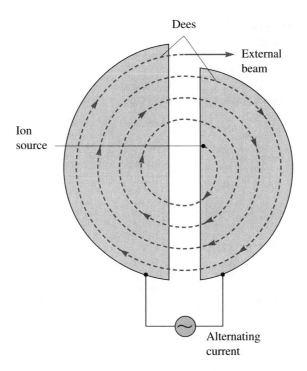

Dees

External beam

Ion source

Alternating current

prepared by accelerating deuterium nuclei in a cyclotron and bombarding a piece of molybdenum. This bombardment reaction yields technetium-97 and a neutron.

$$^{96}_{42}\text{Mo} + ^{2}_{1}\text{H} \longrightarrow ^{97}_{43}\text{Tc} + ^{1}_{0}\text{n}$$

Figure 20.18 Synchrotron. (*A*) The synchrotron at Fermilab in Batavia, Illinois, is built within a circular ring 2.0 km in diameter. (*B*) Inside the ring, electromagnets are used to keep the accelerating ions moving in a circular path.

Technetium, the first artificially produced element, is prepared by the bombardment of a molybdenum plate by a beam of deuterons for a few months. It is used for medical scanning of the brain, thyroid, and other organs.

What are some ways in which radioisotopes affect your life?

20.8   Applications of Isotopes

Nuclear chemistry has been extensively applied to the solution of problems in various fields, from archeology to art to medicine. We have already discussed several applications: Radioactive isotopes (radioisotopes) can be used to date objects of archeological or geological significance and to generate electricity. Unfortunately, the fission and fusion processes used for power generation can deliberately be allowed to run out of control, creating bombs with devastating power to destroy. This section considers other applications of radioisotopes.

Power Generators

Akin to the huge fission reactors now in use and to the fusion reactors expected in the next century are lightweight portable power packs that use radioactive isotopes as fuel. In these power packs, radioactive nuclides generate thermoelectric power to operate instruments that are not readily accessible for maintenance. These power supplies operate for many years at nearly constant output. For example, ^{242}Cm, ^{244}Cm, and ^{210}Po have been used to power instruments in space vehicles and in remote terrestrial sites, such as the polar regions. Pacemakers powered by ^{238}Pu power devices can be implanted surgically and need be replaced only infrequently.

Radioactive Tracers

Many radioactive nuclides are used as tracers. These nuclides can be detected and measured by their characteristic radiation. The radioactive labeling of portions of molecules can help to determine the mechanism of chemical reactions. It can be determined, for example, that in the following reaction, the hydroxide bonded to the product chromium(III) originated from that bound to cobalt(III) rather than from water.

$$Co(NH_3)_5{}^*OH^{2+}(aq) + Cr^{2+}(aq) + 5H_2O(l) \longrightarrow$$
$$Cr(H_2O)_5{}^*OH^{2+}(aq) + Co^{2+}(aq) + 5NH_3(aq)$$

Here, hydroxide bonded to the cobalt is labeled with some radioactive oxygen (^{18}O). The radioactivity ends up in the hydroxide that is bonded to the chromium, showing that the hydroxide could not come from the water, which had no radioactivity.

Medical Diagnoses

Tracers find extensive use in medical studies. The isomer of technetium-99 known as ^{99m}Tc decays to ^{99}Tc by emission of gamma radiation with a half-life of 6.0 hours. When ingested in an appropriate chemical form, this nuclide can help doctors locate tumors in the spleen, liver, brain, and thyroid. The general approach for medical tracers is to employ a radioactive nuclide that is used in some way by the tissues to be investigated. Cancerous cells are characterized by unusually high metabolism and cell replication, so the tracer becomes concentrated in those cells.

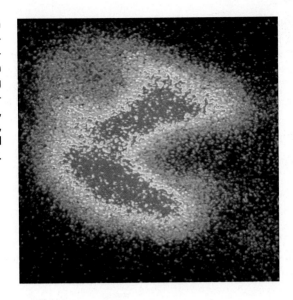

Figure 20.19 A scintillation counter is used to scan a patient's thyroid after the patient has ingested a sodium iodide sample containing iodine-123. The scan is displayed on a screen or film, where normal thyroid tissue, now radioactive, is displayed as the brighter areas.

For example, thyroid problems can also be diagnosed via the patient's ingestion of radioactive ^{131}I or radioactive ^{123}I (which decays to ^{123}Te by electron capture with a half-life of 13.3 h). Scanning the distribution of radiation in the thyroid gland reveals the existence of malignant growths, as shown in Figure 20.19. The scan for radioactivity creates a picture of the thyroid gland. A tumor shows up either as an area of higher-than-average radioactivity or as an area of low radioactivity, depending on the nature of the tumor. The thyroid shown in the figure was diagnosed as having nodular goiter. Many other radioactive nuclides are used in a similar fashion to detect other disorders, especially cancerous tissue.

Cancer Therapy

Radioactive nuclides can be used to treat cancer as well as to detect it. Isotopes that emit high-energy gamma rays can destroy young cancer cells if the cells absorb the nuclides. Because cancer cells have abnormally high growth rates, they preferentially absorb cell nutrients. If these nutrients contain a gamma-emitting component, the radioactivity becomes concentrated in the cancerous cells, which are destroyed in greater numbers than are normal cells. For example, ^{131}I destroys thyroid tumors. Gold, as ^{198}Au, has been used to treat lung cancer, and phosphorus as ^{32}P has been used for eye tumors.

Positron Emission Tomography

One of the newer medical diagnostic tools is called PETT (or PET), for positron emission transaxial tomography. This tool detects abnormalities in living tissues without disrupting the tissue. Some radioactive nuclides are incorporated into chemicals that are normally used by the tissues that are being investigated. The isotopes used are short-lived positron emitters such as ^{11}C ($t_{\frac{1}{2}} = 20.4$ min), ^{18}F (110 min), ^{15}O (2.05 min), and ^{13}N (9.98 min). The radioactive chemical, such as ^{11}C-labeled carbon dioxide or glucose, is administered by inhalation or injection.

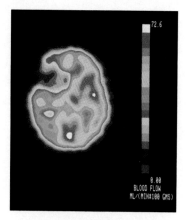

Figure 20.20 Image of a brain constructed after a PET scan on a patient who suffered a stroke. The image was taken after the patient received an injection of blood labeled with ^{15}O–water. The absence of color in the upper-left region indicates the area in which the stroke occurred.

The patient is placed into a cylindrical gamma ray detector. Positrons emitted within the patient collide with electrons in nearby tissue. Positrons react with electrons to emit gamma rays, which are detected by the gamma ray detector. A computer reconstructs the location of many positron-electron reactions, thereby creating an image of the radionuclide's distribution in the tissues being scanned. Such a computer-constructed image is shown in Figure 20.20.

Chemical Analysis

The detection and measurement of characteristic radiation provide a means of tracing particular elements, so radiochemistry finds numerous applications in chemical analysis. A highly useful analytical technique is neutron activation analysis. This method is essentially nondestructive, so it can be used for purposes such as the authentication of old paintings and other valuable works of art. The object is bombarded with a fine beam of neutrons, causing transformations such as

$$^{208}_{82}Pb + {}^{1}_{0}n \longrightarrow {}^{209}_{82}Pb$$

The new isotope of the element is usually radioactive and gives off either a beta particle or gamma radiation. Evaluation of this radiation is used to characterize the elements present and to determine their relative amounts. The distribution of elements in the paint can be compared to that in those paints that the supposed artist is known to have used.

Comment on the statement that a painting found to contain chloride could not have been painted in A.D. 1648, as claimed, because the element chlorine was not discovered until 1774.

How does radiation damage cells?

Biological Effects of Radiation

We have seen that radioisotopes can be used to diagnose and to cure cancer. Unfortunately, they can also cause cancer. Radiation can cause tissue damage of varying degrees, depending on the intensity and length of exposure to the radiation.

Tissue damage to living animals and humans is a matter of serious concern. We are exposed to radiation continually from a variety of sources, not just from medical exposure and nuclear power plants. Granite contains radioactive elements, as do soil, water, food, and air. Brick and concrete used in construction contain radioactive nuclides. Some radioactive nuclides, such as potassium-40, are contained in our bodies. Some irradiation occurs by exposure to cosmic rays, which are particularly intense during high-altitude airplane flights. Inhalation of radon that accumulates in poorly ventilated houses has received much attention as a potential source of cancer.

Irradiation can have one of four effects on the proper functioning of a cell.

1. The radiation can pass through the cell without doing any damage.
2. The radiation can be absorbed by the cell and damage it, but the cell subsequently repairs the damage and resumes normal functioning.
3. The cell can be damaged so severely that it cannot repair itself. New cells formed from this cell are abnormal. This mutant cell can ultimately cause cancer if it continues to proliferate. Such damage can take years to show up as a tumor.
4. The cell can be so severely damaged that it dies.

Table 20.5 Sources of Average Exposure of the U.S. Population to Radiation

Source	Dose (mrem/yr)
Cosmic rays	50
From the earth	47
From building materials	3
In human tissue (as ^{40}K)	21
From air	5
Medical (X rays, radiodiagnosis, radiotherapy)	61
Nuclear power	0.3
Radioactive fallout	4
Consumer products such as TV tubes and watch dials	0.04

The extent of cell damage in a human depends on the level of exposure to radiation. Exposure rates, or doses, are measured in units called *rems* (roentgen equivalent man). A rem is a measure of the amount of energy absorbed by a living system. It is equal to a *rad* (radiation absorbed dose) multiplied by a factor called the *RBE*, or relative biological effectiveness.

$$1 \text{ rem} = 1 \text{ rad} \times 1 \text{ RBE}$$

A rad is an amount of radiation that results in the absorption of 1×10^{-5} J of energy per gram of irradiated material. The RBE indicates the relative damage caused by different types of radiation. The RBEs for some common types of radiation are 1 for X rays, gamma rays, and beta rays; 2.5 for slow neutrons; 10 for alpha particles, protons, and fast neutrons; and 20 for heavy ions. Low levels of radiation are often measured in units of 10^{-3} rem, or millirems (mrem).

Normal exposure to radiation in an average area in the United States is about 200 mrem per year (Table 20.5). This amount of radiation has no observable effects on the average person. More important is the effect of dose levels for a single exposure to radiation (lasting for a few minutes to a few hours). A single dose of radiation of 0–25 rem causes no observable physical effects. However, radiation levels of 25–100 rem cause some temporary changes in blood cells. Higher levels of 100–300 rem cause symptoms of radiation sickness, which include nausea, diarrhea, fever, anemia, and achiness. Such exposure results in an increased statistical risk of cancer. Doses of 400–600 rem cause radiation sickness and result in a 50% chance of death within 30 days. Exposure to 1000 rem and higher levels of radiation results in a nearly 100% chance of death within 30 days. The longer the exposure to radiation, the greater the chance that cancer will result. It is because of these risks that such great care is taken in the design of nuclear power plants and in other uses of radionuclides.

The destructive effects of radiation on plant and animal cells are being investigated (and, to a limited extent, used) as means to prevent the spoilage of foods,

Figure 20.21 The potatoes on the right are moldy and sprouting roots after two weeks of cold storage, whereas those on the left, which were irradiated with gamma rays, are still fresh.

as shown in Figure 20.21. Exposure to appropriate levels of radiation kills bacteria, molds, worms, and other parasites without causing significant changes in the food itself. If this irradiation is carried out on food stored in sealed containers, spoilage can be prevented without recourse to refrigeration or freezing.

Summary

Most nuclides of the elements are radioactive—they decompose to other elements and in so doing emit radiation. The stable nuclides are all found within a narrow island of stability in which the nuclear composition includes nearly equal numbers of protons and neutrons. The heavier radioactive elements decay by combinations of alpha and beta decay processes to form nuclides with lower mass and lower atomic number, eventually reaching the island of stability. The lighter radioactive nuclides decay by beta (electron) decay, beta (positron) decay, or electron capture. Nuclear bombardment reactions are induced nuclear reactions in which nuclides are bombarded with other particles to create new nuclides.

Nuclear decay reactions all follow first-order kinetics, and all radioactive nuclides have characteristic half-lives. Half-lives are used to determine proper storage times for nuclear waste and to date objects that are of archeological or geological interest.

A nuclide is always lighter than its component particles. This mass defect can be used to calculate the nuclear binding energy, which can be used to explain the relative stability and abundance of nuclides, to predict nuclear reactions, and to calculate the energy accompanying a nuclear reaction.

Fission is a nuclear reaction in which a nuclide splits to form two other nuclides and, usually, small particles such as neutrons. Fusion, the combination of light nuclides to form heavier nuclides, is a potential source of electrical energy that has not yet been realized. The existence and origin of the elements can be explained by a variety of fusion reactions that occur in different types of stars.

Nuclear reactions are used for portable power generators, radioactive tracers, medical diagnosis, cancer therapy, positron emission tomography, and chemical analyses. However, too much radiation exposure can cause tissue damage to living things.

Key Terms

activity (20.3)
big bang theory (20.6)
chain reaction (20.5)
critical mass (20.5)
curie (20.3)

electron capture (20.2)
fission (20.2)
fusion (20.2)
mass defect (20.4)
nuclear

bombardment (20.2)
nuclear binding
  energy (20.4)
nucleon (20.1)
positron (20.1)

radioactivity (20.1)
radiocarbon
  dating (20.3)

Exercises

Nuclear Stability

20.1 What is radioactivity?

20.2 What is the island of stability?

20.3 How is the island of stability used to predict the type of radioactive decay a given nuclide might undergo?

20.4 What are the different types of particles or rays that may result from radioactive decay?

20.5 Predict the relative stability of the three isotopes $^{16}_{8}O$, $^{17}_{8}O$, and $^{18}_{8}O$.

20.6 Predict the relative stability of the three isotopes $^{28}_{14}Si$, $^{29}_{14}Si$, and $^{30}_{14}Si$.

20.7 Predict what type of radioactive decay each of the following unstable nuclides undergoes.

    a. $^{14}_{6}C$ d. $^{15}_{8}O$

    b. $^{234}_{90}Th$ e. $^{16}_{7}N$

    c. $^{234}_{92}U$

Nuclear Reaction Equations

20.8 What conservation rules are used to balance the equations that represent nuclear reactions?

20.9 Distinguish between fission and fusion.

20.10 What is a nuclear bombardment reaction?

20.11 What product results from electron emission by $^{99}_{42}Mo$?

20.12 What is the product of beta (electron) decay of $^{234}_{90}Th$?

20.13 Bombardment of $^{250}_{98}Cf$ by $^{11}_{5}B$ yields 4 neutrons. What is the other product?

20.14 What product results from electron capture by $^{44}_{22}Ti$?

20.15 What product results from alpha emission by $^{226}_{88}Ra$?

20.16 Write an equation for the alpha decay of $^{228}_{90}Th$.

20.17 Complete and balance the following equations.

    a. $^{238}_{92}U \longrightarrow \ ^{234}_{90}Th + ?$

    b. $^{234}_{90}Th + ? \longrightarrow \ ^{234}_{91}Pa$

    c. $^{234}_{92}U \longrightarrow \ ^{4}_{2}He + ?$

    d. $^{214}_{82}Pb + ? \longrightarrow \ ^{214}_{83}Bi$

    e. $^{210}_{83}Bi + \ ^{0}_{-1}e^{-} \longrightarrow \ ?$

20.18 Write balanced equations for the following radioactive decay reactions.

    a. alpha decay by $^{221}_{87}Fr$

    b. positron emission by $^{18}_{9}F$

    c. electron capture by $^{133}_{56}Ba$

    d. beta (electron) emission by $^{114}_{47}Ag$

    e. bombardment of $^{96}_{42}Mo$ with protons, resulting in neutron emission

Half-life

20.19 How is radiation detected?

20.20 What is the half-life of a radioactive nuclide?

20.21 What uses are made of the half-life of a radioactive nuclide?

20.22 How are the half-lives of radioactive nuclides measured?

20.23 The half-life of ^{24}Na is 15.0 h. What fraction of the isotope will remain after 60.0 h?

20.24 The half-life of the radioactive isotope of hydrogen known as tritium, $^{3}_{1}H$, is 12.3 years. How much of a 100.0-g sample of tritium will be left after a period of 36 years?

20.25 One method used to date moon rocks returned by the Apollo 11 flight involved measuring the ratio of ^{40}K to ^{40}Ar in the rocks. Potassium-40 decays to argon-40 by positron emission with a half-life of 1.3 billion years. Assuming that no ^{40}Ar was present initially and that none escaped, how old is a rock that had a ratio of $^{40}K/^{40}Ar$ of 0.17?

20.26 The half-life of $^{35}_{16}S$ is 86.7 days. How long does it take for 35% of its activity to disappear?

20.27 Technetium-99, with a half-life of 6.0 h, is used for a number of diagnostic purposes in medicine. For example, ^{99}Tc introduced into the bloodstream is not normally absorbed by brain cells. But it is absorbed by a tumor, and once it has been absorbed, a brain scan can locate the tumor. When a sample of ^{99}Tc is introduced into a body in this way, what percentage of its radioactivity remains after 24 h?

20.28 A sample of a radioactive nuclide has an activity of 3500 counts/min. If 45.0 min later the activity has been reduced to 1500 counts/min, what is the half-life of the nuclide?

20.29 Barley and wheat grains recovered from an ancient Egyptian burial chamber were analyzed for their carbon-14 content. The amount of ^{14}C was found to be 42% of that present in living plants. The half-life of ^{14}C is 5730 years. How old are these grains?

20.30 The nuclide $^{22}_{11}Na$ decays by positron emission. If 76.6% of the original activity still remains after 1.00 year, what is the half-life for this reaction?

20.31 The half-life of the nuclide $^{211}_{85}At$ is 7.5 h. If 0.100 mg of this nuclide is administered for thyroid treatment, how much will remain exactly 1 day later?

20.32 The half-life of $^{35}_{16}S$ is 86.7 days. If 80.0 mg is absorbed by an orange, how much will remain in the orange after a period of exactly 100 days?

20.33 What fraction of the $^{237}_{93}Np$ formed at the time that Earth originated still exists today? The age of Earth is about 3.5×10^{9} years, and the half-life of $^{237}_{93}Np$ is 2.1×10^{6} years.

20.34 To investigate metabolic pathways, a laboratory rat is injected with a sample containing phosphorus-32, which has a half-life of 14.3 days. Assuming that none of the ^{32}P is excreted, how much of the ^{32}P activity would be expected to remain in the rat after 2 months?

Mass–Energy Equivalence

20.35 What is mass defect?

20.36 What equation is used to calculate the energy change for a nuclear reaction?

20.37 How does the energy change in a nuclear reaction compare to that in a chemical reaction?

20.38 How much energy is produced in the following nuclear reaction?

$$^{6}_{3}Li + ^{1}_{0}n \longrightarrow ^{4}_{2}He + ^{3}_{1}H$$

The masses of the nuclides, reading from left to right, are 6.01512 amu, 1.00867 amu, 4.00260 amu, and 3.01605 amu.

20.39 The nuclide $^{213}_{85}At$ (mass 212.9931 amu) decays to $^{209}_{83}Bi$ (mass 208.9804 amu) by emission of alpha particles (mass 4.00260 amu). What is the energy released by this process?

20.40 What is the energy released by positron emission from $^{18}_{10}Ne$ (mass 18.00572 amu) to give $^{18}_{9}F$ (mass 18.00095 amu)? The mass of a positron is 0.0005486 amu.

20.41 The decay of $^{27}_{14}Si$ to $^{27}_{13}Al$ (mass 26.98154 amu) by positron (mass 0.0005486 amu) emission produces 3.80 MeV of energy per decay. What is the mass of $^{27}_{14}Si$?

20.42 Calculate the binding energy for $^{235}_{92}U$ if the mass is 235.0439 amu. Other masses are as follows: neutron, 1.008665 amu; proton, 1.007277 amu; electron, 0.0005486 amu.

20.43 Given the masses that follow, calculate the energy change for the fission reaction

$$^{235}_{92}U + ^{1}_{0}n \longrightarrow ^{94}_{38}Sr + ^{139}_{54}Xe + 3^{1}_{0}n$$

$^{235}_{92}U$, 235.0439 amu	$^{94}_{38}Sr$, 93.9154 amu
$^{139}_{54}Xe$, 138.9187 amu	$^{1}_{0}n$, 1.008665 amu

20.44 Calculate the binding energy per nucleon for $^{15}_{7}N$ (mass 15.000108 amu) and for $^{14}_{7}N$ (mass 14.003074 amu). Other masses are as follows: neutron, 1.008665 amu; proton, 1.007277 amu; electron, 0.0005486 amu.

20.45 How much energy is produced in the following nuclear reaction?

$$^{2}_{1}H + ^{1}_{1}H \longrightarrow ^{3}_{2}He$$

The masses of the nuclides, reading from left to right, are 2.014102 amu, 1.007825 amu, and 3.016030 amu.

20.46 Calculate the binding energy per nucleon for $^{56}_{26}Fe$ (mass 55.9349 amu). Other masses are as follows: neutron, 1.008665 amu; proton, 1.007277 amu; electron, 0.0005486 amu.

20.47 How much energy is produced in the following nuclear reaction?

$$^{14}_{7}N + ^{2}_{1}H \longrightarrow ^{4}_{2}He + ^{12}_{6}C$$

The masses of the nuclides, reading from left to right, are 14.003074 amu, 2.0140 amu, 4.002604 amu, and 12.000000 amu.

Fission

20.48 What are the main components of a fission reactor?

20.49 What are the composition and function of (a) a moderator in a nuclear fission reactor? (b) a control rod?

20.50 Why is the heat from fission reactions not used directly to create steam to operate a turbine?

Fusion

20.51 What fuels are likely to be used as a power source in fusion reactors?

20.52 Discuss the problems that must be overcome before nuclear fusion can become a useful energy source for peaceful purposes.

20.53 Why is carbon burning not an important source of energy in cooler stars?

20.54 Why is helium burning not considered likely to be a source of elements heavier than iron?

20.55 Outline the nuclear reactions that are the primary source of energy in our sun.

Additional Exercises

20.56 How much energy is produced in the following nuclear reaction?

$$^{21}_{10}Ne + ^{4}_{2}He \longrightarrow ^{1}_{0}n + ^{24}_{12}Mg$$

The masses of the nuclides, reading from left to right, are 20.993849 amu, 4.002604 amu, 1.008665 amu, and 23.985045 amu.

20.57 Complete and balance the following equations.

a. $^{7}_{3}Li + ? \longrightarrow 2^{4}_{2}He$
b. $^{10}_{5}B + ^{2}_{1}H \longrightarrow ? + ^{1}_{0}n$
c. $^{27}_{13}Al + ^{1}_{0}n \longrightarrow ? + ^{4}_{2}He$
d. $^{63}_{29}Cu + ? \longrightarrow ^{64}_{29}Cu + ^{1}_{1}H$

e. $^{12}_{6}C + ? \longrightarrow ^{13}_{7}N$
f. $? + ^{2}_{1}H \longrightarrow ^{64}_{30}Zn + ^{1}_{0}n$

20.58 Describe the process of radioactive dating.

20.59 What product will result from neutron capture by $^{238}_{92}U$?

20.60 An old piece of wood has a carbon-14 activity of 0.011 times that of a freshly cut piece of wood. What is the age of the old wood if the half-life of ^{14}C is 5730 years?

20.61 Calculate the binding energy per nucleon for $^{64}_{31}Ga$ if the mass is 63.93674 amu. Other masses are as follows: neutron, 1.008665 amu; proton, 1.007277 amu; electron, 0.0005486 amu.

20.62 Predict what type of radioactive decay each of the following unstable nuclides undergoes.

    a. $^{17}_{9}F$ d. $^{19}_{10}Ne$

    b. $^{21}_{9}F$ e. $^{15}_{6}C$

    c. $^{216}_{84}Po$

20.63 Carbon-12 can be produced by the bombardment of beryllium-9 with alpha particles. Predict any other products of this reaction, and write a balanced equation to describe the reaction.

20.64 What product results from positron emission by $^{15}_{8}O$?

*20.65 Polonium-210 is an alpha-particle emitter with a half-life of 138 days. Static-eliminator brushes for cleaning dust from camera lenses contain a small amount of ^{210}Po at the base of the bristles. Suppose you are in charge of maintaining inventory at a camera store and you are offered a very good buy on a three-year supply of brushes. Should you maintain a stock this size? To help you decide, consider how much ^{210}Po is left at the end of each year during this period.

*20.66 Although potassium-40 decays to produce calcium-40 with a half-life of 1400 million years, rocks cannot be dated by measuring the ratio of ^{40}K to ^{40}Ca. Explain this statement.

*20.67 Suppose that the entire mass of Earth (6.0×10^{27} g) was originally composed of neptunium's longest-lived isotope, ^{237}Np, which has a half-life of 2.14×10^6 years. If this were true, how many atoms of ^{237}Np would be surviving today? Earth is 3.35×10^9 years old.

*20.68 The predominant isotopes of uranium, their abundances, and their half-lives are given in the following table.

Isotope	Mass	Current Abundance	Half-Life
^{234}U	234.0409 amu	0.005%	2.45×10^5 years
^{235}U	235.0439 amu	0.720%	7.04×10^8 years
^{238}U	238.0508 amu	99.275%	4.46×10^9 years

    a. What is the atomic weight of uranium?

    b. Assuming that ^{234}U has remained constant at its present abundance (0.005 g per 100 g of uranium) and that the Earth is 3.35×10^9 years old, what was the abundance of uranium at the beginning of Earth compared to today?

    c. What was the average atomic weight of uranium at the beginning of Earth?

    d. Why must we assume that ^{234}U has remained constant during this time period?

▶ ▶ ▶ Chemistry in Practice

20.69 Chloride-ion concentrations in solutions are usually determined by titration with a silver nitrate solution of known concentration. Some sodium chromate is usually added to make the endpoint visible. After the silver chloride is precipitated, red silver chromate starts to precipitate, turning the white precipitate slightly red. The following alternative method, which does not require visual detection of the endpoint, was developed. A solution of silver nitrate was prepared by dissolving 7.4752 g of anhydrous $AgNO_3$ in water in a volumetric flask, adding a small amount of dilute radioactive $^{110}AgNO_3$ solution, and diluting to 1.000 L. The radioactive silver solution was so dilute that it did not change the concentration of the silver nitrate solution, but it added 0.3 mCi of radioactivity to the solution. A 10.00-mL pipet was used to deliver an unknown potassium chloride solution into a 50.00-mL volumetric flask. About 10 mL of water was added, and the flask was warmed to 60°C so that any silver chloride would form a coagulated precipitate rather than a colloidal suspension. The radioactive $AgNO_3$ so-lution of known concentration was then dropped into the solution, with stirring, until a total of 10.00 mL was added. The solution was then diluted to 50.00 mL. The solution was filtered, and 20.00 mL of the solution was used to determine the radioactivity. First, background radiation was assessed with the scintillation counter, giving 101 counts per minute (cpm). This measurement determines what radioactivity (if any) is generally present in the room. Next, the radioactivity of the filtrate was determined to be 925 cpm. Finally, 5.00 mL of the standard $AgNO_3$ solution was diluted to 20.00 mL with water and used to calibrate the radioactivity, giving 7556 cpm.

    a. What is the concentration of the unknown chloride solution?

    b. If 25.00 mL of a 0.1000 M solution of KCl, containing 0.00100 M K_2CrO_4, is titrated with the standard $AgNO_3$ solution, what fraction of the Cl^- is still in solution when the Ag_2CrO_4 starts to precipitate? Solubility product constants are 1.7×10^{-10} for AgCl and 9.0×10^{-12} for Ag_2CrO_4.

21

Chemistry of the Main-Group Elements

▶ ▶ ▶ ▶

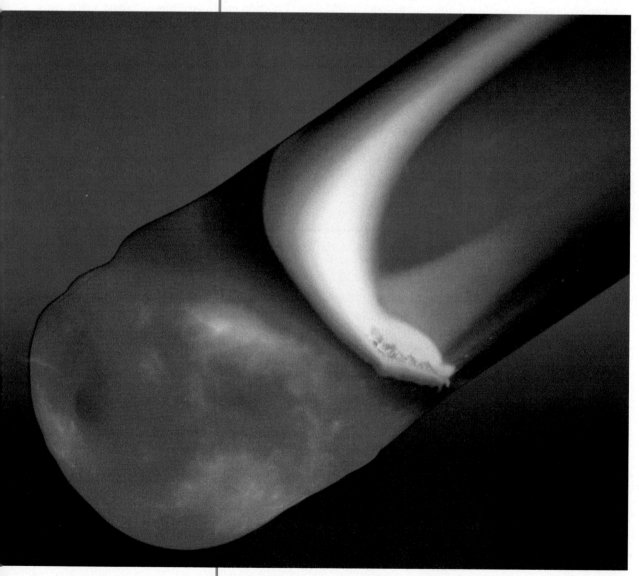

Magnesium reacts exothermically with silicon dioxide to form magnesium oxide and silicon.

CHAPTER OUTLINE

Encountering Chemistry

What do these elements have in common? How are they different?

A silvery-gray powder mixed with sand is placed in a test tube and heated with a Bunsen burner. The mixture begins to glow and continues to glow when the burner is extinguished. As the reaction proceeds, it gives off much heat, and two solids deposit on the test tube. One, a shiny silver film, is elemental silicon; the other, a white solid, is magnesium oxide. The reactants were magnesium metal and silicon dioxide. This reaction is summarized by the equation

$$SiO_2(s) + 2Mg(s) \longrightarrow Si(s) + 2MgO(s)$$

All the elements involved in this reaction are main-group elements: those elements, shown in Figure 21.1, in which the valence electrons are filling the s and p orbitals. These three elements are examples of the three types of main-group elements—oxygen is a nonmetal, magnesium is a metal, and silicon is a metalloid. As we have already seen, the elements on the left-hand side of the periodic table are metals, those on the right-hand side are nonmetals, and metalloids fall on a diagonal separating the other two classes. Many of the properties of the elements vary regularly from left to right across the periodic table, consistent with the change from metals to nonmetals.

Most nonmetals are volatile substances with low melting and boiling points. They have low density, are usually brittle or soft in the solid state, and have low thermal and electrical conductance. The nonmetal solids usually are dull in appearance, not lustrous or shiny as the metals are. Many nonmetals, such as O_2 and P_4, are made up of discrete molecules containing relatively few atoms. The atoms in these small molecules are joined by strong covalent bonds, but the intermolecular forces—van der Waals forces—between adjacent molecules are quite weak. The nature of these bonding and intermolecular forces is responsible for the characteristic physical properties of the nonmetals. The chemical properties of the nonmetals are characterized by their tendency to gain electrons—to be reduced—in reactions with metals, whereas the oxides of the nonmetals are usually acidic.

The metals generally are nonvolatile and have high melting and boiling points. Their density and their thermal and electrical conductance are high. The metals have a characteristic shine or luster in the solid state, where they exist in crystalline form. Metallic solids are malleable and ductile; they can be beaten into sheets and drawn into wires. The metals consist of large arrays of atoms that have long-range bonding forces (the metallic bond) in which metal cations are attracted to large numbers of relatively freely moving electrons. Chemically, metals tend to lose electrons—to be oxidized—in reactions, and their oxides are alkaline in water.

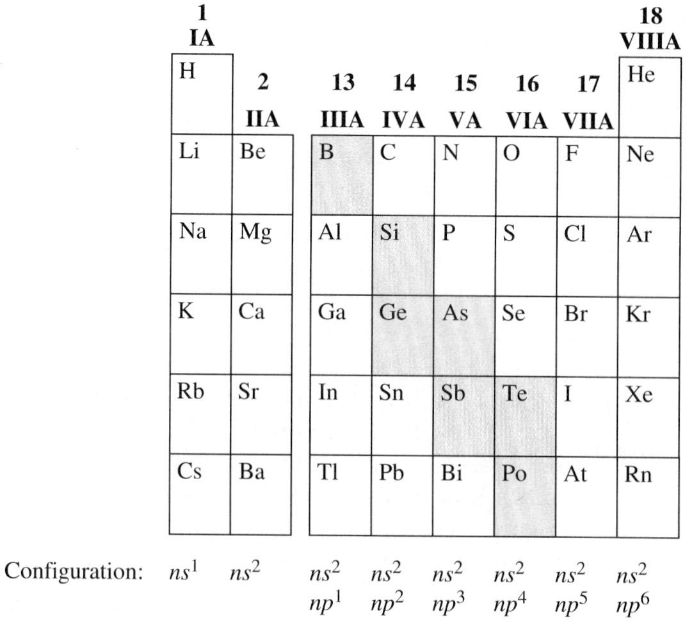

Graphite, a form of carbon, is an exception to the usual observation of low electrical conductance in nonmetals. Because graphite is a good electrical conductor, even though only in two dimensions, it is often used as an electrode for the electrolytic production of metals.

Why do the nonmetals have many properties in common?

Why are the properties of the metals different from those of the nonmetals?

Figure 21.1 The main-group elements are those in which the valence electrons are occupying the outermost s and p orbitals.

1 IA	2 IIA	13 IIIA	14 IVA	15 VA	16 VIA	17 VIIA	18 VIIIA
H							He
Li	Be	B	C	N	O	F	Ne
Na	Mg	Al	Si	P	S	Cl	Ar
K	Ca	Ga	Ge	As	Se	Br	Kr
Rb	Sr	In	Sn	Sb	Te	I	Xe
Cs	Ba	Tl	Pb	Bi	Po	At	Rn

Configuration: ns^1 ns^2 $ns^2 np^1$ $ns^2 np^2$ $ns^2 np^3$ $ns^2 np^4$ $ns^2 np^5$ $ns^2 np^6$

In their physical properties, metalloids are more like metals than nonmetals, but their chemical reactivity is more like that of nonmetals. Many of their atomic properties are intermediate between those of metals and of nonmetals.

This chapter examines the physical and chemical properties of some of the main-group elements, as well as trends in these properties. The properties of the elements can be accounted for largely in terms of the principles covered in earlier chapters. The chemistry of the noble gases was discussed in Section 6.7 and will not be repeated here.

The properties of the first element in each group are usually more distinctive, whereas the rest of the elements in a group tend to have similar properties. For this reason, the first element is usually considered separately from the other elements in a group. The unusual properties of the first element in a group can be explained on the basis of its unusually small size, which arises because the valence electrons are not shielded from the nucleus and the electrons are held relatively tightly in the atom.

21.1 Hydrogen

Where would you place hydrogen in the periodic table?

Because of its very small atomic radius, hydrogen tends to exhibit behavior that is uniquely its own. Its electronic configuration, $1s^1$, is consistent with that of the Group IA (1) elements. Because only one electron is needed to complete the valence shell of hydrogen, its electronic configuration is also similar to that of Group VIIA (17) elements, which are one electron short of a completed valence shell. Hydrogen has some properties consistent with our placing it in Group IA (1) and other properties that suggest placing it in Group VIIA (17). In many of its reactions, hydrogen is a reducing agent, attaining an oxidation number of $+1$, but in reactions with very active metals, hydrogen can be an oxidizing agent, forming H^- upon reduction. The physical properties of hydrogen were reviewed in Section 3.3

Preparation of Elemental Hydrogen

RULES OF REACTIVITY

Some metals react with water to form metal hydroxides and hydrogen gas. A vigorous reaction is observed with the Group IA (1) metals and the Group IIA (2) metals Ca, Sr, and Ba. A mild reaction is observed with Be, Mg, the Group IIIA (13) metals, and the lanthanide metals.

Elemental hydrogen is prepared by decomposing hydrogen-containing compounds such as water, acids, bases, and hydrocarbons. Very active metals such as sodium release H_2 from water; less active metals such as zinc and chromium (Figure 21.2) react with nonoxidizing acids to form hydrogen.

$$Zn(s) + 2H^+(aq) \longrightarrow Zn^{2+}(aq) + H_2(g)$$

Some metals react with strong bases to form hydrogen.

$$2Al(s) + 2OH^-(aq) + 6H_2O \longrightarrow 3H_2(g) + 2Al(OH)_4^-(aq)$$

An expensive method for producing hydrogen is the electrolysis of water.

$$2H_2O(l) \xrightarrow{\text{electrolysis}} 2H_2(g) + O_2(g)$$

The primary commercial sources of hydrogen gas are the reaction of hydrocarbons or coal with water. For example, hot coal reacts with steam to yield a mixture of carbon monoxide and hydrogen.

$$C(s) + H_2O(g) \longrightarrow CO(g) + H_2(g)$$

More hydrogen can be produced by reaction with more steam in the water–gas shift reaction, which was discussed in detail in Section 14.5.

$$CO(g) + H_2O(g) \longrightarrow CO_2(g) + H_2(g)$$

Uses of Hydrogen

Hydrogen is widely used as an industrial gas. Over one-third of the hydrogen produced is used in the manufacture of ammonia for the fertilizer industry. Most of the remainder is used in various other chemical processes, including the hydrogenation of edible oils and the synthesis of chemical compounds such as methanol, cyclohexanol, and octane. Heating an unsaturated oil or fat, which contains carbon–carbon double bonds, with hydrogen at high pressure in the presence of a catalyst such as nickel gives a saturated fat, which contains carbon–carbon single bonds.

$$-CH_2-CH{=}CH-CH_2- + H_2 \xrightarrow{\text{Ni}} -CH_2-CH_2-CH_2-CH_2-$$

Figure 21.2 Many metals, including chromium, react with hydrochloric acid to produce gaseous hydrogen. The equation for this reaction is $Cr(s) + 2H^+(aq) \longrightarrow Cr^{2+}(aq) + H_2(g)$

Vegetable oils are converted to margarine, and animal fats to shortening, in this way.

The fairly high bond energy of the H—H bond (436 kJ/mol) makes hydrogen relatively inert at room temperature, but it becomes reactive at high temperatures or in the presence of a catalyst. When enough energy is supplied in the form of heat, light, or a spark, hydrogen molecules dissociate into atoms, which are quite reactive, so hydrogen can be used as a fuel. For example, hydrogen is used as a gaseous fuel in the hydrogen–oxygen fuel cell, which was discussed in Section 18.2.

Reactions with Other Elements

Hydrogen reacts with many of the elements to form hydrides. These hydrides have different properties, depending on the metallic nature of the other element.

The more active metals react directly with hydrogen gas to form metal hydrides containing H^-. The Group IIA (2) metals, especially beryllium and magnesium, react so vigorously that they burst into flame.

$$Mg(s) + H_2(g) \xrightarrow{\Delta} MgH_2(s)$$

Many of these metal hydrides are white crystalline salts that react readily with water.

$$MgH_2(s) + H_2O(l) \longrightarrow Mg(OH)_2(aq) + H_2(g)$$

Magnesium hydride, MgH₂, is a white, nonvolatile, crystalline solid. It decomposes at 280°C under a vacuum. It is a strong reducing agent, reacting with oxygen in air to form MgO and water. It reacts violently with water to form hydrogen gas.

Metal hydrides, such as lithium aluminum hydride, $LiAlH_4$, and sodium borohydride, $NaBH_4$, are excellent reducing agents for organic compounds. The anions in these reducing agents contain hydrogen atoms covalently bonded to boron or aluminum.

Hydrides of the nonmetals and metalloids are largely covalent and generally form small gaseous molecules or volatile liquids, such as ammonia, which is

Lithium aluminum hydride, LiAlH₄, is a white powder that is stable in dry air at room temperature. It decomposes in moist air or when heated above 125°C. It reacts rapidly with water or alcohols, forming hydrogen gas, and is widely used as a reducing agent for a variety of organic compounds.

synthesized from nitrogen and hydrogen gases by the Haber process (Section 14.5). The nonmetals and metalloids from Group IIIA (13) through Group VIIA (17) all form hydrides. Examples include B_2H_6, CH_4, NH_3, H_2O, and HF. Some of the heavier metals in these groups, such as Sn, Pb, Sb, and Bi, form covalent hydrides typical of their groups. Thus SnH_4 and PbH_4 are covalent molecules with properties similar to those of the other Group IVA (14) hydrides, such as CH_4 and SiH_4. The hydrides of Group VIIA (17), especially HCl, HBr, and HI, form strong acids in polar solvents such as water.

21.2   Group IA (1): The Alkali Metals

Group IA (1) elements (Li, Na, K, Rb, Cs, Fr) are called the alkali metals. All occur widely distributed in Earth's crust as deposits primarily of chloride salts and in the oceans as a wide variety of salts. Francium occurs only in trace quantities.

Physical Properties

Cesium is a silver-white metal with a slight golden color. This golden sheen distinguishes it from the other alkali metals.

Would the alkali metals be used as reducing agents or oxidizing agents?

The alkali metals are all soft, silvery-white, reactive metals (Figure 21.3). All have low melting points (Table 21.1). The melting point of cesium, for example, is only 28.5°C, and it tends to supercool readily; thus it usually occurs as a liquid at room temperature. All the metals crystallize in a body-centered cubic lattice, and all exhibit a ns^1 valence-electron configuration. The electronegativities of the alkali metals are quite low, as are the values of the first ionization energies. Accordingly, the metals all readily form ions with a +1 oxidation number and the same number of electrons as the previous noble gas. The ions are all considerably smaller than the parent atoms, because ionization removes all the valence-shell electrons.

Figure 21.3 Sodium, like all the alkali metals, is quite soft and can be cut with a knife. The cut surface soon dulls, because the metal is reactive enough to form an oxide coating fairly quickly when exposed to air. The metal must be stored under kerosene or some other nonreactive liquid to protect it from such oxidation.

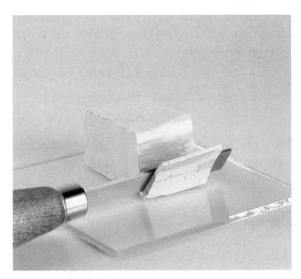

Property	Li	Na	K	Rb	Cs
Ionization energy (kJ/mol)	520	496	419	403	376
Electronegativity	1.0	0.9	0.8	0.8	0.7
Atomic radius (pm)	123	158	203	216	235
Ionic radius of 1+ ion (pm)	60	95	133	148	169
Density (g/cm³)	0.534	0.971	0.862	1.53	1.87
Melting point (°C)	187	97.5	63.65	38.89	28.5
Boiling point (°C)	1326	889	774	688	690
Reduction potential (V)	−3.05	−2.71	−2.92	−2.93	−2.92

Table 21.1 Some Properties of the Alkali Metals

Preparation of the Metals

Stannane, SnH₄, is a covalent gas that decomposes slowly at room temperature and very rapidly if heated to about 150°C. Plumbane, PbH₄, is also known but has not been very well characterized.

The free or uncombined alkali metals are prepared by the electrolysis of molten salts. For example,

$$2LiCl(l) \xrightarrow{\text{electrolysis}} 2Li(l) + Cl_2(g)$$

Details of the electrolysis process for preparing sodium were presented in Section 18.4.

Chemical Properties

The alkali metals are strong reducing agents with large negative reduction potentials (given in Table 21.1 for aqueous solution). No information is available on the chemistry of francium. Its most stable isotope has a half-life of only 22 minutes, so it has not been possible to accumulate enough of it to study extensively.

A low-melting metal would be useful in fuses. Why isn't sodium used for this purpose?

Reactions of the alkali metals with molecular hydrogen and oxygen were discussed in Section 4.3. The alkali metals react vigorously with water, either as a liquid or as water vapor in air, to form metal hydroxides and hydrogen, as shown here for potassium.

$$2K(s) + 2H_2O(l) \longrightarrow 2KOH(aq) + H_2(g)$$

They react in much the same way with acids.

$$2Li(s) + H_2SO_4(aq) \longrightarrow Li_2SO_4(aq) + H_2(g)$$

They can also absorb carbon dioxide from air in the presence of water vapor, as exemplified by the reaction of lithium.

$$2Li(s) + CO_2(g) + H_2O(g) \longrightarrow Li_2CO_3(s) + H_2(g)$$

Because of the reactivity of the alkali metals toward the components of air, they must be stored under a protective liquid such as kerosene or mineral oil.

The alkali metals react violently with the halogens, especially fluorine and chlorine, as shown in Figure 21.4 for the reaction between potassium and chlorine.

$$2K(s) + Cl_2(g) \longrightarrow 2KCl(s)$$

Figure 21.4 Potassium reacts vigorously with chlorine gas to form the white salt potassium chloride.

Uses

The alkali metals are used as reducing agents, especially for organic syntheses, and for the production of less reactive metals. For example, sodium can be used in the production of titanium.

$$TiCl_4 + 4Na \longrightarrow Ti + 4NaCl$$

An extremely wide variety of alkali-metal salts is produced for laboratory use, primarily because they are generally highly soluble and can be used to make concentrated solutions. Sodium salts are the most common because they are readily available and inexpensive. Sodium hydroxide and other alkali-metal hydroxides are widely used as bases for the neutralization of acids. Lithium carbonate is used in the treatment of manic-depressive illness. Sodium carbonate is widely used in making various glasses and as a water softener in detergents. Potassium compounds are used in fertilizers because potassium is essential for plant growth.

21.3 Group IIA (2): The Alkaline-Earth Metals

The name *alkaline-earth metals* arises from some of the properties of the elements in this group, Group IIA (2). The term *earth* was used in the Middle Ages to denote substances (primarily oxides) that did not melt or change into another substance upon application of fire. The alkaline-earth-metal oxides have basic properties and form hydroxides upon reaction with water, as shown in the following equation for calcium oxide.

$$CaO(s) + H_2O(l) \longrightarrow Ca(OH)_2(aq)$$

Group Properties

Although the alkali metals have fairly uniform properties, the same is not true of the alkaline-earth metals. This group of elements, which is characterized by a valence-electron configuration of ns^2, consists of beryllium, magnesium, calcium,

Beryllium chloride, BeCl₂, is a white to faintly yellow solid. It melts at 405°C and sublimes in a vacuum at 300°C, consistent with its covalent nature. It dissolves in water with the evolution of heat, forming an acidic solution as a result of the hydrolysis of beryllium(II) to give beryllium hydroxide species. It is used in the manufacture of beryllium metal and as a catalyst.

Why does beryllium have a greater tendency than the other metals to form covalent bonds?

Which oxides from other groups of the periodic table would you expect to be amphoteric?

strontium, barium, and radium. Beryllium has a noticeable tendency to form covalent compounds, such as $BeCl_2$, whereas the heavier members of the group form primarily ionic compounds. Some properties of these elements are summarized in Table 21.2.

The elements are typically metallic in appearance. However, calcium, strontium, and barium become covered with a white film of metal oxide upon standing in air. Beryllium and magnesium also develop an oxide film, but it adheres tightly and protects these metals from further rapid corrosion.

The alkaline-earth metals are all harder, higher-melting, and stronger than the alkali metals. These properties are consistent with the greater number of valence electrons that are available for bonding. Because of the greater effective nuclear charge, the atoms and ions of the alkaline earth metals are smaller than their alkali-metal counterparts, as discussed in Section 6.8. The metallic character of the elements increases down Group IIA (2), as can be seen from the electronegativity and the ionization energy, both of which decrease down the group. The covalent nature of beryllium is consistent with its small size and with its electronegativity of 1.5, which is rather high for a metal. Lower down the group, however, the elements have properties comparable to those of sodium and potassium. For example, barium hydroxide is a very strong base.

The alkaline-earth metals lose their valence electrons quite readily, forming 2+ ions with noble-gas electronic configurations. The metals are all strong reducing agents as a result of this tendency to lose electrons, as can be seen from the reduction potentials in acidic aqueous solution. The alkaline-earth oxides are definitely basic, except for that of beryllium, which is amphoteric.

Chemical Reactivity

The alkaline-earth metals are quite reactive, as indicated by the large negative reduction potentials of the 2+ ions. The chemical reactivity of the metals increases down the period, which is consistent with the trends in electronegativity, ionization energy, and reduction potential.

Table 21.2 Some Properties of the Alkaline-Earth Metals

Element	Be	Mg	Ca	Sr	Ba	Ra
Ionization energy (kJ/mol)	899	738	590	549	503	509
Electronegativity	1.5	1.2	1.0	1.0	0.9	0.9
Atomic radius (pm)	89	136	174	191	198	—
Ionic radius of 2+ ion (pm)	31	65	99	113	135	140
Density (g/cm³)	1.85	1.74	1.54	2.60	3.51	5
Melting point (°C)	1278	649	839	769	725	700
Boiling point (°C)	2970	1107	1484	1384	1640	1140
Acidic reduction potential (V)	−1.85	−2.37	−2.87	−2.89	−2.90	−2.92
Basic reduction potential (V)	−2.62	−2.69	−3.03	−2.99	−2.97	—

Beryllium and magnesium do not react with water, although magnesium reacts slowly with steam to release hydrogen gas.

$$Mg(s) + 2H_2O(g) \xrightarrow{\Delta} Mg(OH)_2(aq) + H_2(g)$$

Calcium reacts slowly with cold water and rather rapidly with hot water.

$$Ca(s) + 2H_2O(l) \longrightarrow Ca^{2+}(aq) + 2OH^-(aq) + H_2(g)$$

Calcium tends to form a protective oxide coating, so it can be stored for laboratory use in contact with moist air. It does react to form the oxide over extended periods of time, however. Strontium and barium react violently with water, somewhat as sodium and potassium do.

All the metals of Group IIA (2) react with oxygen upon heating, although beryllium must be finely divided to react rapidly. The oxides of the alkaline-earth metals all have unusually high melting points: BeO, 2750°C; MgO, 2800°C; CaO, 1728°C; SrO, 1635°C; BaO, 1475°C. Magnesium and beryllium oxides are commonly used as insulating materials because of their high melting points. ▶ Magnesium oxide, described in the chapter introduction, is used to shield the heating coils on electric ranges, because it is a good conductor of heat but a poor conductor of electricity.

Occurrence and Recovery

The primary minerals in which these elements are found are oxide, carbonate, silicate, and sulfate salts. Among the beryllium minerals, only beryl, $3BeO \cdot Al_2O_3 \cdot 6SiO_2$ (shown in Figure 21.5), is industrially significant. In pure crystalline form, beryl occurs in emerald and aquamarine. Beryllium is recovered from beryl by conversion to BeF_2, which is reduced to beryllium metal by reaction with magnesium metal.

The recovery of magnesium from its salts and from seawater was described in Section 18.5. Metallic calcium is produced by the electrolysis of molten calcium chloride.

Figure 21.5 Beryllium is extracted from crystalline beryl.

Strontium is often found with calcium minerals and may displace calcium in bones; this is the reason for concern when radioactive strontium-90 is released into the atmosphere during atomic tests and nuclear accidents such as the one that occurred at Chernobyl, Russia, in 1986. Strontium metal is produced by electrolysis of a molten $SrCl_2/KCl$ mixture.

Barite, $BaSO_4$, is the chief source of barium in the United States. Barite is first reduced to barium sulfide by carbon in a furnace.

$$BaSO_4 + 2C \longrightarrow BaS + 2CO_2$$

Barium sulfate can be precipitated as a fine powder by mixtures of any soluble barium and sulfate salts.

$$Ba^{2+}(aq) + SO_4^{2-}(aq) \longrightarrow BaSO_4(s)$$

Barium sulfide can be converted to a variety of other barium compounds. Barium oxide is prepared by heating barium carbonate with carbon.

$$BaCO_3 + C \xrightarrow{\Delta} BaO + 2CO(g)$$

This white salt is very insoluble, so in spite of its toxicity, it is used to produce an opaque background for X-ray photography of the gastrointestinal system. Barium sulfate is also used as a white pigment in paint, cosmetics, paper, leather, and rubber.

Barium metal is produced by reducing barium oxide with aluminum in a vacuum at high temperatures (1100–1200°C):

$$4BaO + 2Al \longrightarrow BaO \cdot Al_2O_3 + 3Ba(g)$$

Radium is present in small quantities in uranium ores, where it is formed by radioactive decay. Uranium ores are treated first with sulfuric acid and then with excess carbonate salt to recover radium and barium carbonates from the ores. After reaction with hydrochloric acid,

$$RaCO_3(s) + 2HCl(aq) \longrightarrow RaCl_2(aq) + CO_2(g) + H_2O(l)$$

the mixture of radium and barium ions is separated by repeated crystallization (Figure 21.6). Radium metal is prepared from its salts by electrolysis.

Uses

Which of the alkaline-earth metals would you expect to be used as structural metals?

The uses of some of the compounds of the alkaline-earth metals are listed in Table 21.3. The following section considers the uses of the elemental metals.

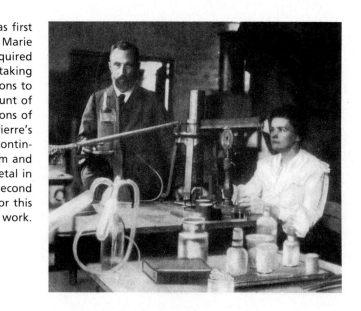

Figure 21.6 Radium was first identified in 1898 by Marie and Pierre Curie. It required many months of painstaking successive recrystallizations to separate a small amount of radium from several tons of pitchblende. After Pierre's death in 1906, Marie continued their work on radium and isolated the pure metal in 1910. She received her second Nobel Prize in 1911 for this work.

Table 21.3 Important Alkaline-Earth-Metal Compounds and Their Uses

BeO	High-temperature electrical insulators
MgO	Ceramics, rubber production, catalyst, uranium ore processing, decolorization of dry cleaning fluids, dentifrices, cosmetic powders, fertilizers, SO_2 recovery in paper making, petroleum additives
$Mg(OH)_2$	Antacid, laxative (milk of magnesia)
$MgSO_4 \cdot 7H_2O$	Laxative
$MgCO_3$	Filler in paper, paints, plastics, and rubber; antacid; added to table salt to prevent caking
$MgCl_2$	Electrolytic production of magnesium metal, fireproofing agent for wood, refrigerant brine
$CaCO_3$	Slag formation in the steel industry; manufacture of cement, mortar, plaster, glass, and ceramics
CaC_2	Preparation of acetylene
$Ca(HSO_3)_2$	Wood pulp processes
$CaCN_2$	Fertilizers, production of cyanides, production of ammonia, production of melamine $N_3C_3(NH_2)_3$, which reacts with formaldehyde to form a hard plastic
CaF_2	Flux, source of hydrogen fluoride for the preparation of freons and Teflon
$Ca_3(PO_4)_2$	Fertilizers, phosphate chemicals
$CaCl_2$	Drying agent, freezing-point lowering of solutions, prevention of ice formation on roads
$CaSO_4 \cdot 2H_2O$	Gypsum, used in wallboard, tile, and plasters
$CaSO_4 \cdot \frac{1}{2}H_2O$	Plaster of Paris
SrS	Depilatory (hair-removing) preparations, luminous paints
$BaCO_3$	Ceramics, optical glass
BaO	Lubricating oil detergents
$BaSO_4$	Primary starting material for other barium chemicals; X-ray photography; white pigment in paint, cosmetics, paper, leather, and rubber

Beryllium is often used with copper to form alloys that are as hard as many steels, are good electrical conductors, and resist corrosion. These alloys are used in electrical switches and airplane brakes. Because beryllium absorbs neutrons, it is used to keep neutrons from escaping from nuclear reactors.

Beryllium and its compounds are quite toxic to humans. Because of its small size and large charge as an ion, beryllium readily bonds to nitrogen atoms in proteins, thereby disrupting the proper functioning of the protein. Being a component of coal, beryllium dust causes the deadly disease of beryllosis in the lungs of miners.

Magnesium is used for structural purposes because of its strength and low density and in chemical processes because of its high reactivity. Strong, lightweight magnesium alloys are widely used in aircraft, spacecraft, automobile and truck bodies and engines, camera bodies, and sporting goods.

Magnesium is used as a reducing agent in the production of many other metals and in the production of various organic compounds. ▶ The chapter introduction described the use of magnesium to produce elemental silicon. Because of its reactivity, magnesium is used in several types of batteries, which are small and lightweight with high current output. As we noted in Chapter 18, magnesium is also used for corrosion protection of water heaters, underground pipelines and storage tanks, and ship hulls.

Calcium metal is used as an alloying agent for aluminum, lead, copper, and magnesium and as a reducing agent in the preparation of numerous metals. Strontium and barium metals and their salts are not widely used.

21.4　Boron

Boron, the first member of Group IIIA (13), is a black crystalline element that is extremely hard and brittle, has a low density (2.34 g/cm^3), and has a high melting point (2300°C) and boiling point (2550°C). Because boron has a low electrical conductivity, it is classified as a semiconductor and has become an important element for the semiconductor electronics industry. Boron is sometimes added to steel to increase its strength and to copper to increase its electrical conductivity.

Occurrence and Preparation

In spite of the rarity of boron in Earth's crust, it occurs in deposits of sufficient concentration to be mined easily. The best-known boron mineral is borax, $Na_2B_4O_7 \cdot 10H_2O$, which is found in dry lake beds, particularly in Death Valley, California.

Elemental boron is recovered from its ores by a process in which boron(III) oxide is formed by acidification.

$$B_4O_7^{2-}(aq) + 2H^+(aq) + (2n-1)H_2O(l) \longrightarrow 2B_2O_3 \cdot nH_2O(s)$$

Water is removed from the hydrated oxide by heating to about 600°C.

$$2B_2O_3 \cdot nH_2O(s) \xrightarrow{\Delta} 2B_2O_3(s) + nH_2O(g)$$

The oxide can then be reduced by heating with an active metal.

$$B_2O_3(s) + 3Mg(s) \xrightarrow{\Delta} 3MgO(s) + 2B(s)$$

Most methods of producing boron result in a product that contains small amounts of impurities. Boron pure enough for semiconductor applications is obtained by **zone refining** (Figure 21.7), which is now commonly used in the semiconductor industry.

Chemical Reactions of Boron

Why is the chemical behavior of boron not like that of aluminum?

Like all metalloids, boron resembles metals in its physical properties but is more like nonmetals chemically. However, the chemical behavior of boron is complex and highly unusual. The ionization energy of boron is 799 kJ—over 200 kJ higher

Figure 21.7 In zone refining, a glass or metal tube is filled with the impure boron (or other material to be purified). This tube is passed slowly through a narrow furnace so that only a small zone of material is molten at a time. The zone slowly passes down the cylinder of solid material. The impurities are more soluble in the molten boron than in the solid, so they become concentrated in the molten zone. After many such passes through the furnace, the impurities are collected at one end of the crystal, and most of the crystal is very pure.

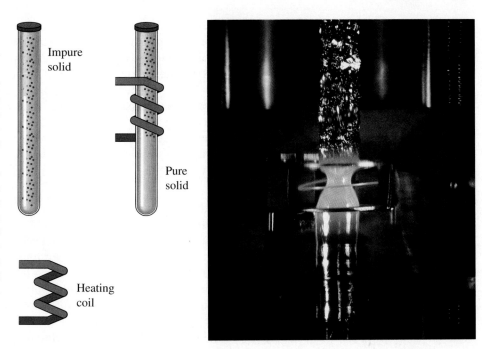

Impure solid

Pure solid

Heating coil

than that of the other Group IIIA (13) elements—which indicates that the formation of a cation is much more difficult for boron. The electronegativity is 2.0, which is higher than that of other elements in this group and higher than that of metals in general. The value is comparable to that of several nonmetals, such as hydrogen and carbon.

All the compounds formed between boron and other elements are covalent. With a valence-electron configuration of $2s^2 2p^1$, boron can form only three normal covalent bonds. However, this leaves an electron deficiency that makes boron a good Lewis acid, reacting with electron donors to form adducts. Oxidation number $+3$ is common, but other oxidation numbers are apparent in the boranes, a special class of compounds to be described shortly.

Boron reacts directly with fluorine and chlorine to give trihalides of boron(III) in direct synthesis reactions, as shown here for fluorine.

$$2B(s) + 3F_2(g) \longrightarrow 2BF_3(g)$$

Boron trifluoride combines with hydrogen fluoride in a Lewis acid–base reaction to form fluoboric acid.

$$BF_3(aq) + HF(aq) \longrightarrow H^+(aq) + BF_4^-(aq)$$

Boron has a great affinity for molecular oxygen. This property is used to purify metals such as copper, aluminum, or brass by addition of boron to the molten metal. The boron removes the oxygen from the metal oxide:

$$3MO + 2B \longrightarrow B_2O_3 + 3M$$

and the boron(III) oxide rises to the surface of the molten metal. Boron(III) oxide is acidic and dissolves slowly in water to give boric acid or rapidly in basic solution to give borate salts.

$$B_2O_3(s) + 3H_2O(l) \longrightarrow 2H_3BO_3(aq)$$
$$2B_2O_3(s) + 2NaOH(aq) \longrightarrow Na_2B_4O_7(aq) + H_2O(l)$$

Dilute boric acid solutions have long been used as eyewash solutions; they have germicidal properties and do not harm eye tissue.

Borax, or sodium tetraborate, is the most important of commercial boron(III) compounds. Besides acting as a raw material in the production of elemental boron, borax is widely used for water softening in washing products. Addition of borax to wash water removes hardness by precipitating insoluble calcium and magnesium borates.

Boron reacts directly with nitrogen at temperatures of about 1000°C to give the very stable nitride.

$$2B(s) + N_2(g) \xrightarrow{\Delta} 2BN(s)$$

The stability is due to its graphite-like structure, shown in Figure 21.8A. Because boron has only three valence electrons, it tends to bond with three nitrogens, using sp^2 hybrid orbitals. The nitrogen has five valence electrons, so to achieve an octet, it also bonds to three boron atoms, using sp^2 hybrid orbitals. This bonding pattern leads to sheets of boron–nitrogen rings, much like that in graphite. Lone pairs on the nitrogen atoms are attracted to the electron-deficient boron atoms in an adjacent layer, giving rise to fairly strong intermolecular forces between the layers. However, these forces are weak enough to allow the layers to slide over one another, giving boron nitride a slippery feeling like graphite. Because of this property, boron nitride is used as a high-temperature lubricant.

Graphitic boron nitride can be converted into a diamond-like structure, shown in Figure 21.8B, by application of high temperature and pressure. The unshared pair of electrons on a nitrogen atom in one layer is donated to the electron-deficient boron in an adjacent layer, forming four bonds around both elements. The resulting tetrahedral structure contains strong covalent bonds, so this form of boron nitride is extremely hard (nearly as hard as diamond) and is an important abrasive.

Boron forms a large number of hydrides, though not by direct reaction with hydrogen. The expected hydride with the formula BH_3 is not a stable species. The only monomeric boron–hydrogen species is the borohydride ion, which is formed by the reaction of excess sodium hydride and boron trifluoride.

$$BF_3(g) + 4NaH(s) \longrightarrow NaBH_4(s) + 3NaF(s)$$

Figure 21.8 Structure of two forms of boron nitride. (*A*) Graphite-like BN occurs in layers of hexagonal sheets. (Dashed lines were added to show the alignment of atoms.) (*B*) Diamond-like BN.

○ Nitrogen
● Boron

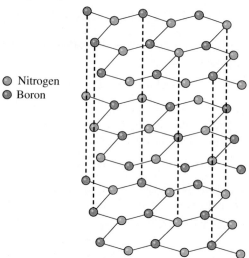

A. Graphite-like BN

○ Nitrogen
● Boron

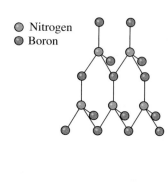

B. Diamond-like BN

Figure 21.9 Structure and bonding in some boranes.

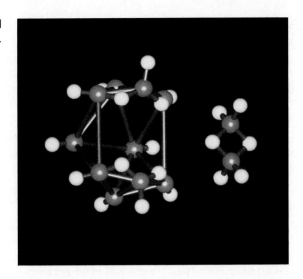

The simplest of the boron hydrides is the gas diborane, which is prepared by the reaction between boron trifluoride and lithium aluminum hydride in ether solution.

$$4BF_3 + 3LiAlH_4 \longrightarrow 2B_2H_6(g) + 3LiF(s) + 3AlF_3(s)$$

Diborane decomposes to other boranes and hydrogen gas when heated above 100°C.

$$2B_2H_6(g) \xrightarrow{\Delta} B_4H_{10}(g) + H_2(g)$$

The structure and bonding in the boranes are highly unusual, not at all like those found with other second-period elements. Consider diborane, with two boron atoms, six hydrogen atoms, and 12 valence electrons. A normal covalent bond consists of two shared electrons, so there are not enough electrons to bond these atoms in the usual fashion. Assuming that each boron–hydrogen bond used two electrons, no electrons would be left to bond the two BH_3 groups together. However, diborane does not contain a direct boron–boron bond. Instead, diborane has a highly unusual structure in which two of the hydrogens act as bridges between the boron atoms, as shown in Figure 21.9A. In each of these bridges, the B—H—B bonding arrangement has the three atoms sharing an electron pair and is called three-center bonding. As shown in Figure 21.9B, larger boranes such as $B_{10}H_{14}$ have even more complex structures. Because of the electron deficiencies that are characteristic of the boranes, these compounds are all highly reactive.

21.5   Aluminum and Group IIIA (13) Metals

Group IIIA (13) consists of boron, aluminum, gallium, indium, and thallium. Boron has already been discussed. The properties of the remaining elements, which are all soft metals, are compared in Table 21.4. The activity of the metals decreases down the group. Aluminum is a very active metal but is protected from reaction by its oxide coating. Indium also forms a protective oxide coating. The lower elements in the group are much less active than aluminum. The ionization energies and electronegativities of these elements are all very similar and are larger than those of Group IA (1) and Group IIA (2) elements.

Element	B	Al	Ga	In	Tl
Ionization energy (kJ/mol)	800	577	579	558	589
Electronegativity	2.0	1.5	1.6	1.7	1.8
Atomic radius (pm)	89	125	125	142	144
Ionic radius of 3+ ion (pm)	20	50	62	81	95
Density (g/cm^3)	2.34	2.70	5.91	7.30	11.85
Melting point (°C)	2300	660	29.8	157	304
Boiling point (°C)	2550	2467	2403	2080	1457

Table 21.4 Some Properties of the Group IIIA (13) Elements

The chemical behavior of the Group IIIA (13) metals is dominated by oxidation number +3. These metals all react with acid to release hydrogen gas. Aluminum and gallium also react readily with strong basic solutions to release hydrogen and form the 3+ ions. Although gallium has an intermediate oxidation number of +2 and indium has oxidation numbers of +2 and +1, these intermediate oxidation numbers are rather unstable with respect to disproportionation.

$$3Ga_2^{4+} \longrightarrow 2Ga + 4Ga^{3+}$$

Thallium, however, has a stable +1 oxidation number.

Aluminum

The stability of aluminum is often viewed negatively because of the accumulation of roadside litter. For what reasons might this stability be a positive characteristic of aluminum?

Aluminum is the most active metal in Group IIIA (13)—so reactive that it is found naturally only in combined states, primarily as oxides and silicates. In spite of this, aluminum structures have survived exposure to nature for over 80 years because of a strongly adherent layer of Al_2O_3 formed by reaction with atmospheric O_2. The properties and preparation of aluminum metal were discussed in detail in Section 18.5.

Recycling of Aluminum   The recycling of aluminum is becoming fairly common for three reasons. Aluminum is impervious to corrosion, so aluminum beverage cans (which have an expected lifetime of about 100 years) stay with us as litter. Even more serious is the depletion of usable ores of aluminum. Although aluminum is the third most abundant element in Earth's crust, the supply of aluminum ores of reasonably usable concentration is limited. The manufacture of aluminum from its ores also requires large amounts of electrical energy, which amounts to about 25% of the total cost of producing aluminum. The Aluminum Company of America, Alcoa, claims that remelting recycled aluminum saves 95% of the energy required to process aluminum ores.

Reactions of Aluminum   The reactions of aluminum are dominated by the oxidation of aluminum metal and by acid–base reactions of aluminum(III) ion. Aluminum metal reacts slowly with hydrochloric acid.

$$2Al(s) + 6HCl(aq) \longrightarrow 2AlCl_3(aq) + 3H_2(g)$$

Aluminum metal reacts violently with liquid bromine, producing flames and flashes of light.

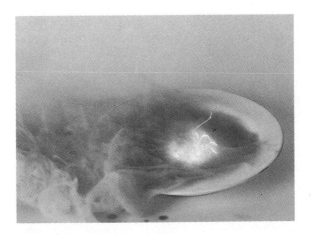

Similar reactions occur with other acids, including dilute oxoacids. However, aluminum does not react with concentrated oxoacids such as H_2SO_4 and HNO_3. These acids form on the aluminum surface a thin coat of aluminum(III) oxide, which prevents further reaction. Aluminum reacts with strong alkalies, which dissolve its protective oxide coating.

$$2Al(s) + 2OH^-(aq) + 6H_2O(l) \longrightarrow 2Al(OH)_4^- + 3H_2(g)$$

Aluminum(III), formed as a product of these oxidation–reduction reactions, is an amphoteric substance; $Al(OH)_3$ acts as either a base or an acid.

The reactivity of aluminum becomes evident at very high temperatures. At 1000°C and above, aluminum reduces many oxides. This is the basis for the *Goldschmidt reaction*, which is also known as the thermite reaction when the metal oxide is iron oxide.

$$2Al(l) + 3MO(s) \xrightarrow{\Delta} Al_2O_3(s) + 3M(l)$$

The Goldschmidt reaction is used to produce metals, such as chromium, manganese, and vanadium, that are otherwise difficult to recover from their oxides. ▶ This reaction is similar to the reduction of silicon dioxide by magnesium metal, described in the chapter introduction.

Aluminum forms compounds with a number of nonmetals. Aluminum reacts with all the halogens fairly violently (Figure 21.10), although iodine does not react unless moistened slightly. Aluminum fluoride is ionic, but the other aluminum halides are covalent compounds. These substances have low melting and boiling points and do not conduct electricity when molten. As gases, they exist as dimers, Al_2X_6, with two of the halogen atoms bridging between the two aluminum atoms.

Why do these properties of aluminum halides suggest that they are covalent?

21.6 Carbon

Carbon, the first member of Group IVA (14), forms a vast number of compounds—more than the total formed by all other elements except hydrogen. In this section we discuss the inorganic chemistry of carbon. Carbon occurs in three common forms: diamond, graphite, and an amorphous form generally taken to include any

Why do the properties of carbon vary with the form in which it is found?

form that is not diamond or graphite. In 1985, a new allotrope of carbon called buckminsterfullerene was produced by laser vaporization of graphite.

Diamond

Diamond is a clear, crystalline form of carbon and one of the hardest substances known. The structure and bonding in diamond were discussed in Sections 8.4 and 10.5. It occurs naturally in significant amounts only in a few places, most notably South Africa. The larger crystals of diamond are, of course, used as gems. A more practical use of diamonds is in grinding and sawing tools, where size and purity are not critical. About 40% of industrial diamonds, in sizes up to 2 mm, are synthetic. A mixture of carbon and a metal catalyst is heated sufficiently to melt the mixture, which is raised to a high pressure, where diamond is more stable than graphite. Small diamonds can be crystallized from the melt under these conditions.

Graphite

Pencil "lead" is formed primarily from the very soft natural graphite found in Sri Lanka. It is mixed and baked with clay in appropriate proportions to give it the desired hardness. Graphite was originally named plumbago *because of its resemblance to lead* (plumbum). *In the late 1700s it became known as graphite, from the Greek word* graphein, *which means "to write."*

Graphite is a slippery, gray-black solid. Its structure and bonding were discussed in Sections 8.4 and 10.5. Strong covalent bonds hold atoms together in each layer, but the layers are bonded only by weak van der Waals forces, so the layers slide across one another readily. This property accounts for the widespread use of graphite as a lubricant. Graphite is found widely distributed in Earth's crust. It is also synthesized from amorphous forms of carbon; indeed, this is the principal source of graphite in the United States. Graphite is used in crucibles, lubricants, pencils, nuclear reactors (to slow down fast neutrons), and electrodes for industrial electrolysis and for electric-arc furnaces.

Amorphous Carbon

Amorphous carbon exists in the forms of carbon blacks, charcoal, activated carbon, soot, and coke. These are all essentially microcrystalline forms of graphite—that is, they contain graphite-like fragments with no layering. The primary sources of these forms of carbon are the thermal decomposition or partial decomposition of various carbon-containing materials such as coal, petroleum, natural gas, and wood. When such materials are burned with an insufficient supply of oxygen, finely divided carbon results. For example, in a low-oxygen atmosphere, natural gas burns with a very sooty flame (Figure 21.11), producing carbon black.

$$CH_4(g) + O_2(g) \longrightarrow 2H_2O(g) + C(s)$$

Soot formed by burning oil is called lampblack. Heating coal in the absence of air produces coke, whereas charcoal results from heating wood in the absence of air.

Carbon black is commonly used as a filler in the manufacture of rubber tires, where it increases toughness and prevents brittleness. Lampblack is used in inks, paints, and the coating on carbon paper. Charcoal, in a form called activated charcoal, is used in filters to adsorb odors, in gas masks to adsorb poisonous gases, in the decolorizing of sugar, in water treatment, and in the reclamation of dry-cleaning solvents. Coke is widely used in the extraction of metals from their oxide ores.

Figure 21.11 Natural gas burns to form solid carbon when there is a low concentration of oxygen.

Buckminsterfullerene

Buckminsterfullerene, whose structure is affectionately known as a buckyball, is a form of carbon consisting of molecules of C_{60}. It is formed by the laser vaporization or the high-temperature carbon arc vaporization of graphite. This substance is one member of a class of relatively new forms of carbon called fullerenes, which consist of clusters containing even numbers of carbon atoms—from 44 to 84, and perhaps even much larger numbers. The C_{60} molecule exists as a truncated icosahedron, which contains 12 pentagonal faces and 20 hexagonal faces (Figure 21.12). This molecule seems to have a remarkable physical stability, but it is chemically reactive. Various molecules can be chemically combined on the surface of the buckminsterfullerene cage. It is also possible to encapsulate metal ions within the cage. These compounds are likely to have unique properties that may be of practical importance. One of the remarkable early findings about C_{60} is that it is a superconductor at relatively high temperatures.

Reactions of Carbon

Elemental carbon is relatively unreactive in all its forms at room temperature. It is insoluble in water, dilute acids and bases, and organic solvents. At high temperatures, however, carbon becomes highly reactive and combines directly with many elements.

Diamond does not begin to oxidize in oxygen or air until the temperature reaches about 800°C, graphite oxidizes above 450°C, and some amorphous forms of carbon oxidize at temperatures less than 450°C. Carbon forms two stable gaseous oxides, CO and CO_2, which were discussed in Section 9.9.

Carbon also reacts with oxygen in the form of oxides or oxoanions. Section 14.5 discussed the water–gas shift reaction, in which carbon reacts with the oxygen in water vapor.

$$C(s) + H_2O(g) \longrightarrow CO(g) + H_2(g)$$

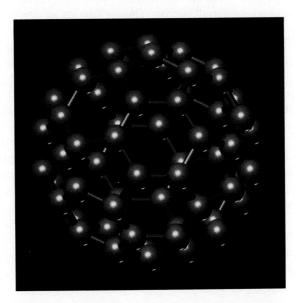

Figure 21.12 Buckminsterfullerene, a form of carbon containing C_{60} molecules.

Carbon also reduces metal oxides, as when coke is used to prepare iron from its oxide ores (Section 4.6).

$$Fe_2O_3(s) + 3C(s) \xrightarrow{\Delta} 2Fe(l) + 3CO(g)$$

▶ This reaction is similar to the reduction of silicon dioxide by magnesium metal described in the chapter introduction. As an example of the reduction of an oxoanion, carbon is used to reduce phosphate rock in the presence of silica for the commercial production of elemental phosphorus.

$$Ca_3(PO_4)_2 + 5C + 3SiO_2 \longrightarrow 3CaSiO_3 + 2P + 5CO$$

Around 1000°C, carbon reacts with hydrogen at high pressure to form small amounts of methane.

$$C(s) + 2H_2(g) \xrightarrow{\Delta} CH_4(g)$$

However, methane is usually obtained as the primary component of natural gas. Methane is quite stable in air but is combustible when heated.

$$CH_4(g) + 2O_2(g) \longrightarrow CO_2(g) + 2H_2O(g) \qquad \Delta H = -882 \text{ kJ}$$

This highly exothermic reaction becomes self-sustaining once it has started. The commercial production of methane and its use as a fuel were discussed in Section 9.9.

Carbon combines with elements that have an electronegativity lower than or similar to that of carbon to form binary compounds called carbides. These substances can be formed by direct reaction of the elements at temperatures above 2200°C, by heating metal oxides with carbon, or by heating the metal in a vapor of a hydrocarbon. Two typical reactions are

$$2CaO(s) + 5C(s) \xrightarrow{\Delta} 2CaC_2(s) + CO_2(g)$$

$$SiO_2(s) + 2C(s) \xrightarrow{\Delta} SiC(s) + CO_2(g)$$

Carbides can be divided into three major classes: salt-like, interstitial, and covalent. The salt-like carbides are formed with the most electropositive metals and are essentially ionic. These carbides are hydrolyzed by water or dilute acid at room temperature to give hydrocarbons. Perhaps the most familiar of these salt-like carbides is calcium carbide, a member of a group called acetylides (they contain C_2^{2-}), which produces acetylene upon reaction with water.

$$CaC_2(s) + 2H_2O(l) \longrightarrow Ca(OH)_2(s) + C_2H_2(g)$$

This reaction was used in miners' lamps for many years, because it provided a cheap, portable, and safe means of carrying acetylene to be burned for light.

The interstitial carbides are formed primarily with the transition metals. These substances are very hard and have very high melting points, high metallic conductivity, and metallic luster. These carbides consist essentially of a metal with carbon atoms located in some of the interstitial sites (or holes) in the metal structure. Some examples are W_2C, Fe_3C, and TaC.

The third class, the covalent carbides, includes the carbides of silicon and boron, which are close in size and electronegativity to carbon. These substances are completely covalent and form infinite network structures. They are exceptionally hard materials widely used as abrasives. The structure of silicon carbide is similar to that of diamond, but with every other carbon atom substituted by a silicon atom. Thus silicon carbide is exceptionally hard.

21.7 Silicon and Group IVA (14) Elements

The Group IVA (14) elements are carbon, silicon, germanium, tin, and lead. Some properties of these elements are summarized in Table 21.5. Carbon, which we discussed in the preceding section, is nonmetallic. Silicon is essentially nonmetallic, but its chemistry is only partially like that of carbon. Germanium is a metalloid, whereas tin and lead are metallic. The ionization energies and the melting points decrease going down the group, reflecting the change from nonmetallic to metallic nature. The behavior exhibited by carbon is dominated by the tendency of carbon to bond to itself, but the ability of the element to bond to itself diminishes on going down Group IVA (14). (For this reason, contrary to many early science fiction stories, there are no known life forms based on silicon.) This trend reflects the bond strength between two atoms of the element, which, because of the increasing size of the elements, decreases considerably on going down the group. The unique position of carbon in this group is also shown by the existence of carbon–carbon double and triple bonds, as in C_2H_4 and C_2H_2. No analogous bonds have been found for the other elements in the group.

Oxidation number +4 dominates the chemistry of the elements near the top of Group IVA (14), such as silicon, but oxidation number +2 becomes more stable down the group. Consider the dichlorides of these elements, which have oxidation number +2. The species CCl_2 and $SiCl_2$ have only fleeting existence because they are so reactive. However, $GeCl_2$, $SnCl_2$, and $PbCl_2$ are stable species, even though they can be oxidized. For example, they react with molecular chlorine to give MCl_4 species, as shown here for tin.

$$SnCl_2 + Cl_2 \longrightarrow SnCl_4$$

Why is carbon almost unique in its tendency to bond to itself? What effect does this property of carbon have on our lives?

Table 21.5 Some Properties of the Group IV Elements

Element	C (diamond, graphite)	Si	Ge	Sn	Pb
Ionization energy (kJ/mol)	1086	786	766	708	715
Electronegativity	2.5	1.8	1.8	1.8	1.8
Atomic radius (pm)	77	117	122	140	150
Ionic radius of 2+ ion (pm)	—	—	73	93	121
Density (g/cm³)	3.51, 2.25	2.33	5.35	7.28	11.35
Melting point (°C)	3570	1410	937	232	328
Boiling point (°C)	sublimes	2355	2830	2270	1750
Bond energies (kJ/mol)	356	210–240	190–210	105–145	—

Silicon

Silicon is second only to oxygen in abundance in Earth's crust, where it occurs in the form of silicon dioxide and over 800 silicate minerals. Elemental silicon is obtained by the reduction of silicon dioxide with carbon or calcium carbide at 1500°C in an electric furnace.

$$SiO_2(s) + C(s) \xrightarrow{\Delta} Si(l) + CO_2(g)$$

▶ As described in the chapter introduction, magnesium can also be used as the reducing agent. The primary use of silicon is in semiconductor devices, which require very high purity. To obtain high purity, the silicon is converted to trichlorosilane ($HSiCl_3$) by reaction with hydrogen chloride at about 225°C.

$$Si(s) + 3HCl(g) \xrightarrow{\Delta} HSiCl_3(l) + H_2(g)$$

The liquid product can be purified by fractional distillation and then reduced back to silicon by reaction with molecular hydrogen at about 1025°C.

$$HSiCl_3(g) + H_2(g) \xrightarrow{\Delta} Si(s) + 3HCl(g)$$

Properties and Uses Silicon is a brittle, gray-black, metallic-looking solid. It is quite hard, with a melting point of 1410°C. Silicon crystallizes in a diamond-like tetrahedral network structure. Silicon is inert at room temperature but becomes reactive at high temperatures. For example, it reacts with chlorine at 450°C.

$$Si(s) + 2Cl_2(g) \xrightarrow{\Delta} SiCl_4(g)$$

Silicon tetrachloride is used to prepare smoke screens, because it hydrolyzes readily in moist air, giving a smoke containing solid silicon dioxide.

$$SiCl_4(g) + 2H_2O(g) \longrightarrow SiO_2(s) + 4HCl(g)$$

Silicon reacts with hydrofluoric acid to form fluosilicic acid.

$$Si(s) + 6HF(aq) \longrightarrow H_2SiF_6(aq) + 2H_2(g)$$

Fluosilicic acid, a colorless fuming liquid that is quite soluble in water, is sold commercially as a 30% solution (stored in wax-covered bottles, because it attacks glass) used for glass etching.

At temperatures greater than 1710°C, silicon reacts completely with oxygen to form silicon dioxide. Silicon has a tendency to react with oxygen-containing substances to form Si—O—Si linkages with tetrahedral structures.

Compounds with Hydrogen Silicon hydrides, called silanes, are prepared by reaction of magnesium silicide with acids, generally in an atmosphere of hydrogen to prevent oxidation.

$$Mg_2Si(s) + 4H^+(aq) \longrightarrow 2Mg^{2+}(aq) + SiH_4(g)$$

Actually, this reaction produces a mixture of silanes: SiH_4, Si_2H_6, Si_3H_8, Si_4H_{10}, Si_5H_{12}, and Si_6H_{14}. These are analogous to the alkanes (discussed in Section 8.4), which are made up of carbon and hydrogen. The silanes are all quite reactive.

Only the first two are stable at room temperature; the others decompose to give SiH_4, Si_2H_6, and H_2. The silanes are spontaneously flammable in air, as illustrated here for Si_3H_8.

$$Si_3H_8(g) + 5O_2(g) \longrightarrow 3SiO_2(s) + 4H_2O(g)$$

Compounds with Carbon   We have seen that silicon carbide (SiC) results from the reaction of sand and coke at high temperatures. Another class of compounds containing silicon and carbon—in this case, along with oxygen—is the silicones. Their general formula is $(R_2SiO)_n$, where R is any hydrocarbon group, such as CH_3 or C_6H_5. These complex substances can be made as plastics, rubbers, oils, greases, and resins. All have excellent thermal stability, resist chemical attack, and repel water. Silicon rubbers are used in artificial heart valves, in electrical insulation, and for leakproof seals in machinery. The resins set to solids when heated and are used as glues and in nonstick surfaces. Silicone greases are used to ensure a chemically resistant seal between pieces of glass. Silicone oils are widely used in pumps, engines, transformers, and hydraulic systems (including brakes), as well as in hand lotions and in water-repellant preparations for shoes.

Silicon Dioxide   Although adjacent members of a group in the periodic table often have similar properties, the oxides of carbon and silicon are not at all alike. Carbon dioxide exists as a simple molecular substance (CO_2), whereas silicon dioxide exists in polymeric forms, $(SiO_2)_n$, all of which have silicon covalently bonded to four bridging oxygen atoms. For example, the form of silicon dioxide known as cristobalite has a diamond-like structure in which the silicon atoms are arranged in the tetrahedral diamond arrangement, with an oxygen atom halfway between each pair of silicon atoms (Figure 21.13). This extended covalent-bonding network causes silicon dioxide to be a hard, high-melting solid. In contrast, solid carbon dioxide (dry ice) is a soft, weak solid with a low melting point and high volatility, because it contains discrete carbon dioxide molecules held together only by weak intermolecular forces.

How would you expect the properties of CO_2 and SiO_2 to differ?

Glass   Because of its strength, transparency, and low chemical reactivity, glass is extremely useful in many applications. It is formed by heating together silicon

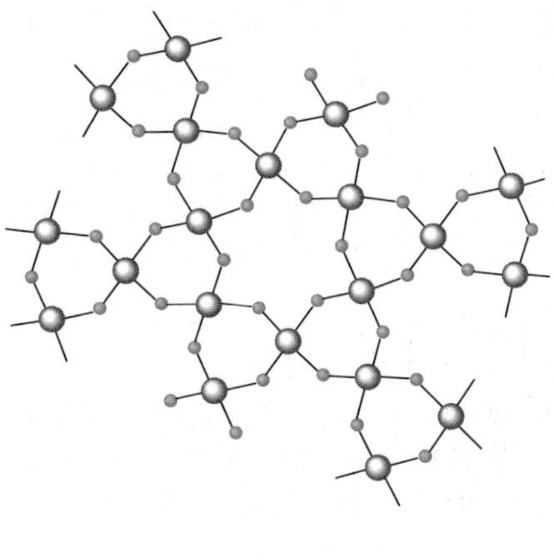

Figure 21.13 The structure of silicon dioxide in the cristobalite form.

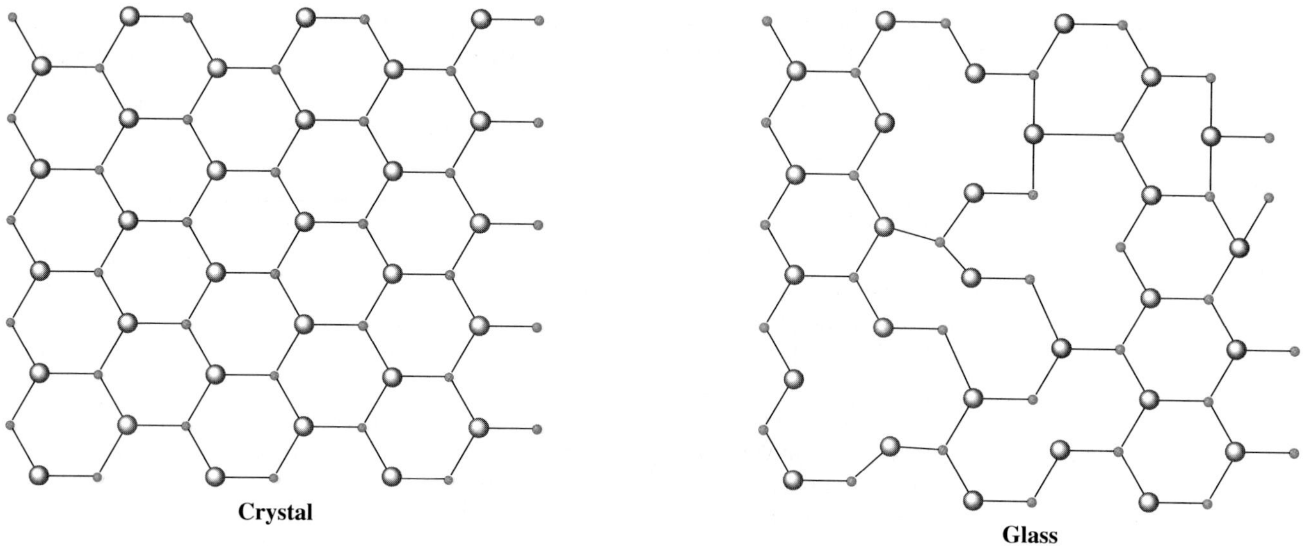

Crystal

Glass

Figure 21.14 Two-dimensional structures of a crystalline silicate and a silicate glass.

dioxide, alkali-metal and alkaline-earth-metal oxides, and sometimes other oxides. The particular properties of a glass depend on the materials of which it is composed, but the general properties are related to the silicate structure present in the glass. This structure is not regular, as in crystalline silicates, but is somewhat random (see Figure 21.14). In this structure, some of the bonds are under more strain than others, which causes the glass to melt over a range of temperatures rather than sharply at a characteristic temperature. This means that glass can be softened without melting, so it can be worked into various shapes.

Common window glass is prepared by melting together silicon dioxide and sodium carbonate (a major use for this chemical).

$$SiO_2(s) + Na_2CO_3(s) \xrightarrow{\Delta} Na_2SiO_3(l) + CO_2(g)$$

The water-soluble sodium silicate produced by this reaction, known as *water glass,* is sometimes used in washing powders and in fireproofing preparations. Addition of calcium oxide to water glass produces a fairly insoluble substance; the calcium oxide can be added in the form of calcium carbonate (from limestone).

$$CaCO_3(s) + Na_2CO_3(s) + 2SiO_2(s) \xrightarrow{\Delta} Na_2SiO_3(l) + CaSiO_3(l) + 2CO_2(g)$$

Upon cooling, the mixture of sodium and calcium silicates forms window glass, otherwise known as *soft glass* or *soda-lime glass.*

Silicates   We have seen that some silicates are synthesized in the preparation of glasses from silicon dioxide and metal oxides. Many others occur naturally. The basic structural unit in all these silicates is the SiO_4 tetrahedron, but these structural units can occur in several varieties: singly, in small groups sharing oxygen atoms, in small cyclic groups, in infinite chains, in double-stranded chains (or bands), and in infinite sheets. From these very few structures, many hundreds of minerals are formed by combination of the silicate anions with various metal cations. The cations lie between the silicate structures. The various types of silicates, their structures, and some common examples are listed in Figure 21.15.

Figure 21.15 Structures of silicate minerals.

Orthosilicates SiO_4^{4-}	Zircon, $ZrSiO_4$ Garnet, $Fe_3Al_2(SiO_4)_3$	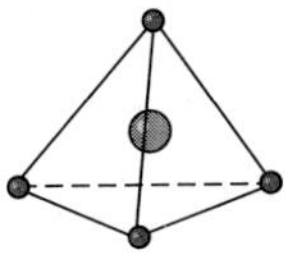
Pyrosilicates $Si_2O_7^{6-}$	Thortvetite, $Sc_2Si_2O_7$	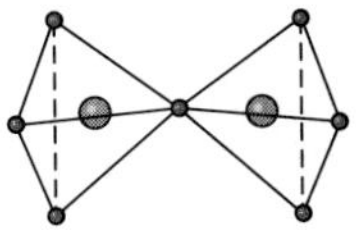
Cyclicsilicates $(SiO_3)_3^{6-}$	Benitoite, $BaTiSi_3O_9$	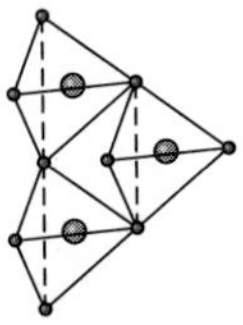
$(SiO_3)_6^{12-}$	Beryl, $Be_3Al_2Si_6O_{18}$ Emerald (beryl with traces of chromium)	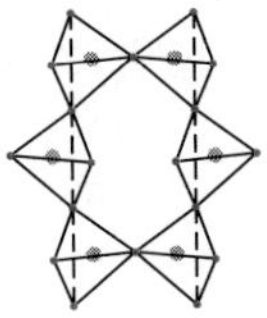
Chain silicates: single-stranded (pyroxenes) $(SiO_3)_n^{2n-}$	Jadeite, $NaAl(SiO_3)_2$	
Double-stranded (amphiboles) $(Si_4O_{11})_n^{6n-}$	Asbestos	
Infinite sheets $(Si_2O_5)_n^{2n-}$	Micas, $KAl_2(OH)_2(AlSi_3O_{10})$ Clay, $Al_4(OH)_8(Si_2O_5)_2$ Talc, $Mg_3(OH)_2(SiO_5)_2$	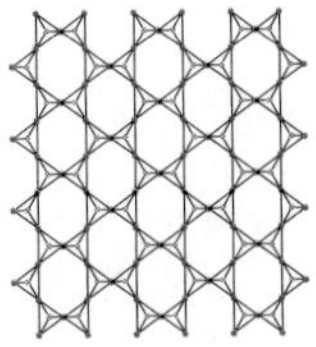
Three-dimensional framework	Feldspars, $KAlSi_3O_8$ Zeolites, $Na_2(Al_2Si_3O_{10}) \cdot 2H_2O$	Structure is like SiO_2 with some Si substituted by other elements.

The silicate structure in some minerals is manifested in the physical structure of the minerals. An example is the asbestos minerals, which contain double-stranded silicate structures. As shown in Figure 21.16, asbestos contains fibers.

Figure 21.16 Asbestos.

These are very strong parallel strands, which are connected by weaker links involving attractions to the intervening cations. The health hazards posed by some forms of asbestos are caused by small fibers breaking off and penetrating lung tissue, where they can cause cancer.

Another example of the internal structure being manifested in the physical structure is provided by mica (Figure 21.17), which contains layers of silicate in the form of infinite sheets, in which many single-stranded chains are connected together by covalent bonds to bridging oxygen atoms. The layers are connected by weaker interactions with the metal ions between the layers, so it is relatively easy to split mica into thin sheets.

Figure 21.17 A piece of mica can be separated into layers easily.

21.8 Nitrogen

Nitrogen is the first member of Group VA (15). Although the electronic configurations of the Group VA (15) elements are similar (ns^2np^3), and some of the simpler compounds have similar stoichiometries (such as NH_3, PH_3, and AsH_3), there is little resemblance between the chemistry of nitrogen and that of other elements in this group. Nitrogen is an important element because of its involvement in life processes. In its elemental form as the colorless, odorless, diatomic molecule N_2, it constitutes 78.1% by volume of dry air. Furthermore, nitrogen is an essential component of all living matter. Protein contains about 17% nitrogen in the form of amino acids. Growing plants need a suitable source of nitrogen to synthesize protein, so nitrogen compounds are important components of chemical fertilizers—75% of the NH_3 and urea manufactured ends up in fertilizers.

How does nitrogen get from the air into proteins?

Elemental nitrogen is used commercially to provide an inert atmosphere for operations in electronics and metals industries and as a freezing agent (liquid nitrogen). It is obtained primarily by the fractional distillation of liquid air.

Most uses of nitrogen involve not the element but its compounds, which are known with all possible oxidation numbers from -3 to $+5$.

Ammonia

The common compounds of nitrogen are either hydrides or oxides. We consider the hydrides first. The hydrides all contain nitrogen with a negative oxidation number. The most common of the hydrides of nitrogen is ammonia, NH_3, in which nitrogen has an oxidation number of -3. Ammonia is a colorless gas with a characteristic sharp, choking odor. A small amount of ammonia exists in the atmosphere, about 20% of it from industrial processes and the rest from natural sources, especially the hydrolysis of urea from animal urine. The chemistry of ammonia was discussed in Section 9.9, and details of the synthesis of ammonia were presented in Section 14.5.

Nitrogen Oxides

The positive oxidation numbers of nitrogen occur in the oxides, which were discussed briefly in Section 9.9. The six principal oxides are N_2O, NO, N_2O_3, NO_2, N_2O_4, and N_2O_5. In aqueous solution, N_2O_3 is converted to nitrous acid, HNO_2, and N_2O_5 is converted to nitric acid, HNO_3.

Dinitrogen oxide, or nitrous oxide, N_2O, is a naturally occurring, relatively unreactive component of the atmosphere that is formed, along with N_2, in the natural degradation of proteins by soil microorganisms. It can also be obtained by thermal decomposition of ammonium nitrate.

$$NH_4NO_3 \xrightarrow{\Delta} N_2O + 2H_2O$$

Nitrous oxide is used as an anesthetic (in laughing gas), but it must be mixed with oxygen because it does not support life.

Nitric oxide can be formed by reaction of copper metal with dilute aqueous nitric acid.

$$3Cu(s) + 8H^+(aq) + 2NO_3^-(aq) \longrightarrow 2NO(g) + 4H_2O(l) + 3Cu^{2+}(aq)$$

It is also formed in high-temperature combustion processes ($> 1200°C$) that involve air, such as occurs in automobile engines:

$$N_2(g) + O_2(g) \xrightarrow{\Delta} 2NO(g)$$

Nitrogen oxides formed in this way have been implicated in the formation of smog. The commercial preparation of NO involves oxidation of ammonia gas, as we noted in Section 15.6.

As shown in Figure 21.18, NO reacts rapidly with O_2 to form reddish-brown NO_2.

$$2NO(g) + O_2(g) \longrightarrow 2NO_2(g)$$

Dinitrogen trioxide results from the reaction between NO and NO_2. When condensed to a liquid, N_2O_3 is blue. In the gaseous state, N_2O_3 reacts with water to form gaseous nitrous acid, HNO_2. In the liquid state, this reaction also produces NO, N_2O_4, and NO_3^-. Nitrous acid is a weak acid that is blue in color and susceptible to decomposition to NO and NO_3^-.

Nitrogen dioxide, NO_2, is a poisonous, reddish-brown gas with an irritating odor. Nitrogen dioxide exists in equilibrium with the colorless dimer N_2O_4, which is formed when the unpaired electrons are shared between two nitrogen atoms.

$$O_2N \cdot + \cdot NO_2 \rightleftharpoons O_2N-NO_2$$

Dinitrogen pentoxide, N_2O_5, is a volatile white solid that melts at $41°C$. When dissolved in water, N_2O_5 forms the strong acid HNO_3, nitric acid, which is widely used industrially for the preparation of numerous products, including fertilizers and explosives, and for cleaning and etching metals. It is also employed to form nitrate salts of numerous metals, which are extensively used because they are soluble in water. Pure nitric acid can be formed by heating a mixture of solid $NaNO_3$ and pure H_2SO_4, but normally HNO_3 is encountered not as a pure substance but as a 68% solution in water. $HNO_3(aq)$ is colorless when properly stored, but it turns yellow upon standing in sunlight because some decomposition to NO_2 takes place.

$$4HNO_3(aq) \longrightarrow 4NO_2(aq) + O_2(g) + 2H_2O(l)$$

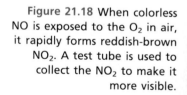

Figure 21.18 When colorless NO is exposed to the O_2 in air, it rapidly forms reddish-brown NO_2. A test tube is used to collect the NO_2 to make it more visible.

Nitric acid is a strong acid and a strong oxidizing agent. The product of reduction of nitric acid depends on the concentration. In dilute solutions, HNO_3 is reduced to NO, whereas at higher concentrations, NO_2 is also produced. With some reducing agents, such as zinc metal, HNO_3 can be reduced all the way to NH_3.

$$4Zn(s) + NO_3^-(aq) + 10H^+(aq) \longrightarrow 4Zn^{2+}(aq) + NH_4^+(aq) + 3H_2O(l)$$

21.9 Phosphorus and Group VA (15) Elements

We have said that there is little resemblance between the chemistry of nitrogen and that of the other elements in Group VA (15). The occurrence of the elements illustrates these differences. Whereas nitrogen is found in nature primarily as the unreactive N_2 molecule in the atmosphere, the other elements are found only in the combined state. For example, phosphorus is found primarily as phosphate salts. The uncombined elements, such as P_4, are too reactive to exist as such in nature.

What features of the structure and bonding of P_4 make it much more reactive than N_2?

Some properties of the Group VA (15) elements are summarized in Table 21.6. The metallic nature of the elements increases down the group. Phosphorus is essentially nonmetallic and typically forms covalent bonds. Its oxides are acidic. Arsenic exhibits properties between those of a nonmetal and those of a metalloid. Its oxides are amphoteric, though more acidic than basic. Antimony is mostly metallic but has some properties of a metalloid. Its oxides are amphoteric, but more basic than acidic. The last element, bismuth, is definitely metallic, with basic oxides. These trends are reflected in the dramatic decreases in ionization energy observed on going down the group.

The changes in metallic character are also reflected in trends in the oxidation numbers found for the members of Group VA (15). Nitrogen and phosphorus form compounds with oxidation number ranging from -3 to $+5$. Arsenic forms AsH_3, with oxidation number -3, but it is found primarily with oxidation numbers $+3$ and $+5$. Antimony is found mostly as $+3$ but occasionally as $+5$. Bismuth is almost exclusively found with oxidation number $+3$. Note that the lower, more metallic elements in this group do not have nearly so many stable oxidation numbers as nitrogen.

Table 21.6 Some Properties of the Group VA (15) Elements

Element	N	P	As	Sb	Bi
Ionization energy (kJ/mol)	1402	1012	947	834	703
Electronegativity	3.0	2.1	2.0	1.9	1.9
Atomic radius (pm)	70	110	121	141	157
Ionic radius of 3− ion (pm)	171	212	222	245	—
Density (g/cm³)	0.00125	1.83	5.73	6.68	9.80
Melting point (°C)	−210	44	817	631	271
Boiling point (°C)	−196	280	sub-limes	1750	1560

Phosphorus

Phosphorus is widely distributed in nature, primarily as phosphates. There are nearly 190 different phosphorus minerals, but only the apatite series of minerals, often in the form of phosphate rock, are an important commercial source of the element. These minerals are $Ca_5(PO_4)_3OH$ (apatite), $Ca_5(PO_4)_3F$ (fluorapatite), and $Ca_5(PO_4)_3Cl$ (chloroapatite).

Many phosphate deposits have a biological origin. Phosphates are an important constituent of all bone tissue, and phosphate in the form of adenosine triphosphate (ATP) is an important component of all tissue. Phosphate rock was deposited in ancient oceans from urine and animal skeletons.

Elemental phosphorus is recovered from phosphate rock by heating the rock with coke, as described in Section 21.6. The phosphorus is collected and stored under water because it ignites in air.

Properties and Uses   Phosphorus occurs in the form of 19 allotropes. The main allotropes are called white, red, and black phosphorus. The vapor consists of tetrahedral P_4 molecules in which each phosphorus atom is bonded to the other three. The liquid and solid phases of condensed phosphorus vapors are called white phosphorus and also consist of P_4 molecules. White phosphorus, a molecular solid, exhibits the softness typical of that class. The molecules are held together by weak intermolecular forces. In this form, phosphorus is quite poisonous, and contact with the solid causes painful skin burns. When exposed to air in the dark, white phosphorus glows with a greenish color and gives off white fumes. This glow, called phosphorescence, is the origin of the name of the element.

White phosphorus is quite reactive and has a tendency to ignite spontaneously in air. It can be stored and handled safely under water. Because it exists as nonpolar molecules, white phosphorus is soluble in many nonpolar organic solvents, such as hexane, but not in water.

As white phosphorus ages, it tends to develop a yellow or pinkish tinge that is due to the formation of red phosphorus. This amorphous substance can be formed more rapidly (though still slowly) from molten white phosphorus at about 200°C. Red phosphorus is not nearly so reactive as white phosphorus and can be stored in contact with air. It is not soluble in organic solvents. At temperatures above 460°C, amorphous red phosphorus can be converted to several crystalline forms. Red phosphorus is thought to involve P_4 tetrahedral molecules bonded to one another in long chains, but the structure has never been fully investigated.

The structure of black phosphorus has not been determined experimentally. Why would we think it might be graphite-like?

An even less reactive form is black phosphorus. It may be amorphous, or it may have a graphite-like structure that roughly resembles corrugated sheets of atoms stacked parallel to one another. Pressure of about 13,000 atm (or a catalyst) and temperatures of 220°C are required to form black phosphorus. It is metallic in appearance and is an electrical conductor.

About 95% of the pure phosphorus produced from phosphate rock is converted to phosphoric acid by burning to the oxide and subsequent dissolution of the oxide in water.

$$P_4(s) + 5O_2(g) \xrightarrow{\Delta} P_4O_{10}(s)$$
$$P_4O_{10}(s) + 6H_2O(l) \longrightarrow 4H_3PO_4(aq)$$

Phosphoric acid is used in soft drinks, metal cleaners, and liquid fertilizers, as well as in the preparation of other phosphorus chemicals.

The principal use of phosphorus, as phosphate salts in phosphate fertilizers, often does not involve the pure substance at all. Rather, fertilizers may be made directly from phosphate rock. The rock is treated with an acid to release the phosphate from the apatite structure. For example, sulfuric acid may be used.

$$2Ca_5(PO_4)_3F(s) + 7H_2SO_4(aq) + 3H_2O(l) \longrightarrow$$
$$7CaSO_4(s) + 3Ca(H_2PO_4)_2 \cdot H_2O(s) + 2HF(g)$$

The resulting product is called superphosphate. Such processes account for about two-thirds of the phosphate fertilizers sold. Most of the rest is ammonium phosphate, $(NH_4)_2HPO_4$, prepared by the acid–base neutralization reaction between ammonia and phosphoric acid.

Diammonium hydrogen phosphate, $(NH_4)_2HPO_4$, occurs as colorless crystals or powder. It is used in fireproofing paper, cloth, and wood, in dentifrices, in fertilizers, and in corrosion inhibitors.

The next largest application of phosphorus as phosphate salts is as a ''builder'' in synthetic detergents, the most common being sodium tripolyphosphate, $Na_5P_3O_{10}$. A builder is a substance added to raise the pH and so to assist in converting oil to colloidal particles, to buffer against pH changes, and to form complex ions with Ca^{2+} and Mg^{2+}, which might otherwise precipitate onto clothing as carbonate salts.

Sodium tripolyphosphate, $Na_5P_3O_{10}$, is a slightly hygroscopic solid. It is used in water softening to tie up calcium and magnesium ions without precipitating them. It is also an ingredient of drilling fluids, where it controls the viscosity of drilling mud in oil fields.

Compounds of Phosphorus Phosphorus usually has oxidation numbers of $+5$, $+3$, and -3. The $+5$ state is generally preferred. Phosphorus seldom forms monatomic ions, except for P^{3-}. Rather, it tends to bond covalently to other atoms to form molecules or polyatomic ions, such as those formed with hydrogen, oxygen, and the halogens.

Phosphorus forms a poisonous hydride, phosphine (PH_3), that has a structure and properties similar to those of ammonia, although it is a much weaker base than ammonia. There are two oxides of phosphorus, P_4O_{10} and P_4O_6, which can be formed by burning phosphorus in oxygen. Phosphorus pentoxide, named for its empirical formula P_2O_5, is a colorless solid that sublimes at temperatures above 360°C. Because it readily absorbs moisture from the atmosphere, as shown in Figure 21.19, it is often used as a drying agent for gases. Phosphorus trioxide, P_4O_6, is also a colorless solid and has a melting point of 24°C. Both oxides react with water to form acids—phosphoric acid (H_3PO_4) from P_4O_{10}, and phosphorous acid (H_3PO_3) from P_4O_6.

Figure 21.19 P_4O_{10} *(Left)* has such a high affinity for water that it removes water vapor from the atmosphere, *(Right)* forming phosphoric acid.

21.10   Oxygen

Oxygen, the first member of Group VIA (16), is the most abundant element in Earth's crust; it accounts for 49.5% of the crustal weight. As O_2 it makes up about 20.9% of the volume of the atmosphere, second only to nitrogen. Besides its occurrence in the atmosphere, where it sustains life and supports combustion, oxygen is a significant constituent of water, of most rocks and minerals, of fats, carbohydrates, and proteins, and of the common oxoacids.

Molecular Oxygen

Elemental oxygen exists as a colorless, odorless diatomic molecule with a bond strength of 494 kJ/mol, which is consistent with a double bond. The molecule is paramagnetic, however, as we noted in Section 8.5. In Section 8.6, we saw that the paramagnetism could be predicted by molecular orbital theory but not by valence bond theory.

Oxygen is obtained commercially by the fractional distillation of air. Some O_2 is obtained in a very pure state by electrolysis of aqueous solutions.

$$2H_2O(l) \xrightarrow{\text{electrolysis}} 2H_2(g) + O_2(g)$$

Oxygen is also prepared in small amounts in the laboratory by the thermal decomposition of oxoanions or peroxides.

$$2KNO_3 \xrightarrow{\Delta} 2KNO_2 + O_2$$

$$2BaO_2 \xrightarrow{\Delta} 2BaO + O_2$$

Oxygen is widely used as an industrial oxidizing agent in applications such as the conversion of coke to CO for the reduction of iron oxides, the oxyacetylene torch for high-temperature welding, and hydrogen–oxygen fuel cells. Liquid oxygen is used as an oxidizer for fuels such as kerosene and liquid hydrogen in rockets.

Ozone

Ozone (O_3) is an allotrope of oxygen (O_2). The ozone molecule is bent because the central oxygen atom has an unshared pair of electrons, as shown by the following Lewis structure:

$$:\!\ddot{O}::\!\ddot{O}:\!\ddot{O}: \quad \longleftrightarrow \quad :\!\ddot{O}:\!\ddot{O}::\!\ddot{O}:$$

The bonds are identical, with a bond angle of 117° and a bond length intermediate between those expected for a single bond and for a double bond.

Normally a gas, ozone exists at 1 atm pressure as a deep blue, explosive liquid when the temperature is below −111.9°C and as a solid below −192.5°C. Its characteristic pungent odor is associated with thunderstorms. Lightning, electric-arc discharges, and ultraviolet light turn O_2 into O_3. Ozone is a very strong and fast-working oxidizing agent, second in strength only to fluorine (F_2) among simple species. When ozone acts as an oxidizing agent, it is usually converted to O_2,

The Ozone Hole

Scientists believe that life on Earth has adapted to an environment strongly influenced by the absorption of UV light in the ozone layer and that changes in the ozone layer pose potential threats to the well-being of plants and animals.

Far above the Earth—8 to 30 miles from the surface—the atmosphere contains a layer of ozone. This layer doesn't contain much ozone: If contained at standard temperature and pressure, the layer would be only 3 mm thick! But this layer of ozone has a profound effect on our lives. It absorbs ultraviolet radiation from the sun, a phenomenon believed to be the main mechanism for heating the stratosphere. The temperature of the atmosphere decreases with increasing altitude up to a certain point. In the ozone layer, ozone converts the ultraviolet (UV) radiation to heat, so the temperature does not continue to drop as the altitude increases. The temperature of the stratosphere is apparently linked to the temperature close to Earth's surface. In addition, this temperature change slows down the loss of water vapor from the atmosphere. Because of the temperature inversion, water vapor cannot rise from Earth's surface indefinitely but rather condenses and falls back to the surface as rain or snow. Scientists believe that life on Earth has adapted to an environment strongly influenced by the absorption of UV light in the ozone layer and that changes in the ozone layer pose potential threats to the well-being of plants and animals. For example, UV light is known to cause skin cancer, so a reduction of ozone is expected to result in a rise in the incidence of skin cancer.

In 1974, chemists F. S. Rowland and M. J. Molina warned that the ozone layer was being threatened by the presence of fluorochlorocarbons—the freons that are used in refrigerators, air conditioners, and aerosol cans. Indeed, measurements made since 1979 have shown that the concentration of ozone in the stratosphere above the South Pole has been diminishing steadily. This low concentration of ozone, known as an ozone "hole," is especially pronounced during the Antarctic spring. The hole is attributed to high concentrations of chlorine and bromine species: ClO, $(ClO)_2$, and BrO. These substances, all of which are catalysts for the conversion of O_3 to O_2, arise from interaction of UV light with freons.

$$CCl_2F_2 \xrightarrow{\text{UV light}} CClF_2 + Cl$$

Oxygen in the atmosphere also dissociates when exposed to UV light.

$$O_2 \xrightarrow{\text{UV light}} 2O$$

The chlorine atoms produced by the decomposition of freons react with ozone molecules to form ClO molecules that behave as a catalyst.

$$Cl + O_3 \longrightarrow ClO + O_2$$
$$O + ClO \longrightarrow Cl + O_2$$

The interconversion of Cl and ClO can proceed in continuous cycles known as a chain mechanism. The net reaction is the conversion of an oxygen atom and an ozone molecule to two oxygen molecules.

$$O + O_3 \longrightarrow 2O_2$$

A single chlorine atom can destroy about 100,000 ozone molecules in this way. The amount of ozone depletion can be exactly accounted for by the concentration of chlorine and bromine species found in the ozone layer above Antarctica.

These findings have caused much consternation in government and industry. Governments of many countries have agreed to reduce the use of fluorochlorocarbons, and industries are frantically trying to synthesize new refrigerants to substitute for the freons. Some likely candidates are halogenated hydrocarbons in which the hydrogens have not been completely replaced by halogens, such as CH_2FCF_3 (HFC-134a) and $CHClF_2$ (HCFC-22).

Questions for Discussion

1. How would your life be affected if all existing freons were collected and destroyed tomorrow?
2. It has been estimated that the ozone destroyed by each supersonic transport (SST), such as the Concorde, results in 50 additional cases of skin cancer per year. Should we destroy all SSTs?
3. Some environmental groups blame chemical companies for the atmospheric damage caused by the freons that they discovered and produced. Is this accurate? Or should we blame the users of freon-containing products?
4. Could we cure the ozone problem by producing it on Earth and injecting it into the stratospheric ozone layer?

as in the following reaction:

$$2NO_2 + O_3 \longrightarrow N_2O_5 + O_2$$

*Reconcile the facts that ozone
is harmful in the air we breathe
but is a necessary component
of the atmosphere if we are to
remain healthy.*

Ozone is used as a germicide and as a bleaching agent for waxes, varnishes, and fats. It is sometimes used to deodorize sewage gases. Ozone is especially important as a component of the upper atmosphere, where it forms a protective layer that screens out harmful ultraviolet radiation. Excess O_3 at ground level is harmful to breathe and causes rubber, such as that in tires, to decompose.

Oxygen Compounds

The most common of the oxygen compounds are the oxides, in which oxygen has an oxidation number of -2. Oxygen forms stable oxides with most of the elements, as we noted in Section 4.3. The properties of the oxides were discussed in Section 3.3. Oxygen is also found in combined form as peroxides, with oxidation number -1, and as superoxides, with oxidation number $-\frac{1}{2}$, which were discussed in Section 4.3.

*Solutions of hydrogen peroxide,
such as those that can be
purchased in a drug store for
use as an antiseptic, normally
contain a stabilizer such as
acetanilide, $C_6H_5NHCOCH_3$,
to slow down the
decomposition by
disproportionation. The
stabilizer reacts with OH
groups that would cause
further decomposition.*

Some of the more reactive metals will form peroxides (containing O_2^{2-}) or superoxides (containing O_2^-) instead of oxides. The formation of these compounds was discussed in Section 4.3. An especially important peroxide is the most common example, hydrogen peroxide, which decomposes readily by disproportionation.

$$2H_2O_2(aq) \xrightarrow{\Delta} 2H_2O(l) + O_2(g)$$

Because it has an intermediate oxidation number, -1, hydrogen peroxide can act as either an oxidizing agent or a reducing agent.

21.11   Group VIA (16): The Chalcogens

Group VIA (16) consists of sulfur, selenium, tellurium, and polonium, in addition to oxygen. These elements are called **chalcogens,** from the Greek for "copper giver," because they are often found in copper ores. Some properties of these elements are summarized in Table 21.7. A predominant trend that affects other properties is the increase in metallic character down the group, as indicated by the decreases in ionization energy and electronegativity. However, nonmetallic character dominates in this group. Oxygen exists as diatomic molecules and is a nonmetal. Sulfur exists in a variety of covalently bonded polyatomic forms and is also a nonmetal. Selenium and tellurium are more metallic than sulfur and have metalloid character, but they nevertheless bear some resemblance to sulfur. Polonium is even more metallic, but its behavior is known only from studies in which polonium is a small impurity in a sample of another element, such as tellurium. Polonium is a rare, radioactive element, so it is not possible to obtain large samples.

*Why doesn't oxygen normally
exhibit positive oxidation
numbers?*

Although the chemistry of oxygen is dominated by oxidation number -2, the other Group VIA (16) elements exhibit oxidation numbers from -2 through $+6$. The higher oxidation numbers are quite common, especially in combination with oxygen.

Table 21.7 Some Properties of the Group VIA (16) Elements

Element	O	S	Se	Te	Po
Ionization energy (kJ/mol)	1314	999	941	869	813
Electronegativity	3.5	2.5	2.4	2.1	2.0
Atomic radius (pm)	66	104	117	137	—
Ionic radius of 2− ion (pm)	140	184	198	221	—
Density (g/cm³)	0.00143	2.07	4.81	6.25	9.4
Melting point (°C)	−218	113	217	450	254
Boiling point (°C)	−183	445	684	990	962

Sulfur

Figure 21.20 Free sulfur is found in large quantities in nature. Here rhombic sulfur crystals are embedded in calcite.

Sulfur is found in Earth's crust principally as sulfide and sulfate minerals and as the free element (often embedded in calcite, as shown in Figure 21.20). Sulfur is also a small but critical constituent of plant and animal tissue and is found as sulfur dioxide and sulfur trioxide in the atmosphere. Sulfur is cycled through a complex series of transformations in nature. These transformations are dominated by the biological synthesis and destruction of sulfur-containing amino acids, such as cysteine, $HSCH_2CH(NH_2)CO_2H$. In addition, bacteria cause transformations among sulfur, sulfide, and sulfate. In the atmosphere, transformations occur among sulfur dioxide, sulfur trioxide, and sulfuric acid, which has been implicated in acid rain (Chapter 15).

Elemental sulfur is obtained primarily by melting underground sulfur deposits and bringing them to the surface by the Frasch process, discussed in Section 4.6. Increasingly significant amounts are also obtained in the refinement of crude oils and coal. And sulfur is recovered as sulfuric acid from smelting operations involving sulfide ores.

Properties and Uses Elemental sulfur is a tasteless, odorless, combustible yellow solid that exists in a wide variety of allotropes under different conditions. The properties of the rhombic, monoclinic, and amorphous forms were discussed in Section 10.7. The various forms of sulfur have different molecular structures. Both the rhombic and monoclinic forms consist of eight-membered rings of sulfur atoms (S_8), but they occur in different crystal modifications. Liquid sulfur also consists primarily of eight-membered rings.

The major use of sulfur is in the preparation of sulfuric acid, discussed in Section 4.6. Sulfuric acid is in turn used to make phosphate fertilizers and impure phosphoric acid from phosphate rock. It is also used in the preparation of titanium dioxide as a paint pigment; in the preparation of detergents, soaps, cellulose, and various fibers; in the pickling of steel; and as battery acid. The synthesis of sulfuric acid was discussed in Sections 4.6 and 14.5. Reactions of sulfuric acid were discussed in Section 15.6.

Compounds of Sulfur Elemental sulfur is quite reactive, even at room temperature, though not so reactive as oxygen. Sulfur forms binary compounds with all the elements except iodine and the noble gases. With most of these elements, direct reaction with elemental sulfur is possible. For example, sulfur combines directly with hydrogen, carbon, and iron (as shown in Figure 21.21) to form sulfides.

$$S(s) + H_2(g) \rightleftharpoons H_2S(g)$$

$$2S(l) + C(s) \xrightarrow{\Delta} CS_2(l)$$

$$S(l) + Fe(s) \xrightarrow{\Delta} FeS(s)$$

Hydrogen sulfide (H$_2$S) is usually prepared by the reaction of a metal sulfide with an acid, such as hydrochloric acid.

$$FeS(s) + 2HCl(aq) \longrightarrow H_2S(g) + FeCl_2(aq)$$

Hydrogen sulfide is a gas under normal conditions. The gas is well known for its "rotten egg" odor and is extremely poisonous—even more toxic than hydrogen cyanide. Hydrogen sulfide is the largest source of sulfur in the atmosphere, arising primarily from the decay of organic matter, the biological reduction of sulfate, and volcanoes.

Hydrogen sulfide gas dissolves in water; a saturated solution is 0.1 M hydrosulfuric acid, H$_2$S(aq). Hydrogen sulfide and sulfide solutions form insoluble precipitates when added to many metal ion solutions, such as a solution of mercury(II) shown in Figure 21.22.

$$Hg^{2+}(aq) + H_2S(aq) \longrightarrow HgS(s) + 2H^+(aq)$$

Because hydrogen sulfide is a weak acid, the concentration of sulfide ion in solution depends on the pH. Thus some metal sulfides are insoluble in basic solution only (CoS, FeS, MnS, NiS, and ZnS), whereas less soluble metal sulfides are also insoluble in acidic solution (PbS, Bi$_2$S$_3$), CuS, CdS, HgS, As$_2$S$_3$, Sb$_2$S$_3$, and SnS$_2$). These differences in solubility form the basis for a common analytical method for separating metal ions. The values of K_{sp} for these sulfide salts can be found in Appendix E.

Sulfur reacts with oxygen to form two different oxides, sulfur dioxide and sulfur trioxide. These oxides form oxoanions (SO$_3^{2-}$ and SO$_4^{2-}$) and oxoacids (H$_2$SO$_3$ and H$_2$SO$_4$) by reaction with metal oxides or water. The chemistry of these oxides was discussed in Section 9.9.

Figure 21.21 When iron metal is heated with sulfur, a hard mass of iron(II) sulfide results.

Figure 21.22 When H₂S is bubbled through a solution containing mercury(II) ions *(Left)*, black HgS precipitates *(Right)*.

Another oxoanion of sulfur is thiosulfate ion, $S_2O_3^{2-}$, which is produced when sulfur dissolves in hot sulfite solutions.

$$SO_3^{2-}(aq) + S(s) \longrightarrow S_2O_3^{2-}(aq)$$

Formally, the oxidation number of sulfur in this ion is +2, but the two sulfur atoms are not equivalent, because one is attached to oxygen atoms and the other is not: $S-SO_3^{2-}$. Thiosulfate is used in photography as a complexing agent (photographic fixer) to dissolve silver bromide from photographic film.

$$AgBr(s) + 2S_2O_3^{2-}(aq) \longrightarrow Ag(S_2O_3)_2^{3-}(aq) + Br^-(aq)$$

In this context, sodium thiosulfate is known as *hypo*, from the old term *hyposulfite*. Thiosulfate is also used in analytical chemistry as a reducing agent to analyze iodine solutions. This reaction is shown in Figure 21.23 and is represented by the equation

$$I_2(aq) + 2S_2O_3^{2-}(aq) \longrightarrow S_4O_6^{2-}(aq) + 2I^-(aq)$$

Figure 21.23 Brown aqueous iodine reacts rapidly with colorless thiosulfate solution to form colorless iodide ions.

21.12 Group VIIA (17): The Halogens

How can we determine the properties of an element such as astatine that we cannot isolate in visible quantities?

The most common of the Group VIIA (17) elements, the halogens, are chlorine, bromine, and iodine. The other members of the group are fluorine, whose properties are atypical of the group, and astatine, which is radioactive and exists naturally in only very small amounts. As uncombined elements, the halogens exist as diatomic molecules. At room temperature, fluorine is a yellow gas, chlorine a pale green gas, bromine a red liquid, and iodine a purple solid. Some properties of the halogens are summarized in Table 21.8. Note the very high values of the ionization energies, which are typical of nonmetals.

Fluorine

Fluorine occurs in nature principally as CaF_2 (fluorspar), Na_3AlF_6 (cryolite), and $Ca_5(PO_4)_3F$ (fluorapatite). Molecular fluorine is prepared by the electrolysis of concentrated solutions of KF, anhydrous liquid HF being the solvent.

Fluorine is an extremely reactive gas. It reacts with all of the elements except oxygen and the lighter noble gases to form stable fluorides, as in the following examples:

$$S(s) + 3F_2(g) \longrightarrow SF_6(g)$$
$$Zn(s) + F_2(g) \longrightarrow ZnF_2(s)$$

Many of these reactions occur explosively; a mixture of hydrogen and fluorine ignites spontaneously, for example. Fluorine reacts explosively to remove hydrogen from all hydrocarbons (including wood and paper) and form HF and CF_4, as illustrated in the following equation for propane.

$$C_3H_8(g) + 10F_2(g) \longrightarrow 3CF_4(g) + 8HF(g)$$

If the hydrogen in organic material is replaced by fluorine atoms, as in the polymer teflon, $-(CF_2-CF_2)_n-$, the organic material becomes inert to fluorine.

Fluorine is normally contained in vessels made of certain metals, such as copper, that form a protective fluoride coating in much the same way that aluminum forms a protective oxide coating in air.

Fluorine is such a strong oxidizing agent that it can convert oxides, including water, to molecular oxygen. The product of reaction with water depends on the acidity of the solution.

$$2F_2(g) + 2H_2O(l) \longrightarrow O_2(g) + 4HF(aq)$$
$$2OH^-(aq) + 2F_2(aq) \longrightarrow OF_2(g) + H_2O(l) + 2F^-(aq)$$

Table 21.8 Some Properties of the Group VIIA (17) Elements

Element	F	Cl	Br	I	At
Ionization energy (kJ/mol)	1681	1256	1143	1009	—
Electronegativity	4.0	3.0	2.8	2.5	2.2
Atomic radius (pm)	64	99	114	133	—
Ionic radius of 1− ion (pm)	136	181	195	216	—
Density (g/cm³)	0.00181	0.00321	3.12	4.93	—
Melting point (°C)	−220	−101	−7.2	114	—
Boiling point (°C)	−188	−34.6	58.8	184	—

Chlorine

Chlorine exists principally as chlorides in seawater, salt lakes, and brine deposits. Chlorine gas is prepared industrially by the electrolysis of sodium chloride solutions, which also yields sodium hydroxide, as we noted in Section 18.6. Chlorine is also a byproduct of the preparation of metals by electrolysis of molten salts such as NaCl, $MgCl_2$, and $CaCl_2$. Most of the chlorine produced is used as a raw material in the production of over a thousand other chemicals. Chlorine is extensively used in the synthesis of herbicides and insecticides, in the bleaching of textiles and paper, in purifying drinking water, and in the production of plastics such as polyvinyl chloride (PVC).

Bromine

Bromine exists in very small quantities in Earth's crust primarily in the form of bromides coexisting with chlorides. Bromine is prepared by reacting a solution that contains bromide ion, such as seawater, with chlorine, as shown in Figure 21.24.

$$2Br^-(aq) + Cl_2(g) \longrightarrow Br_2(aq) + 2Cl^-(aq)$$

Bromine is used as a bleach and in the manufacture of bromide compounds, such as ethylene bromide, $C_2H_4Br_2$, which is used as an antiknock agent in gasoline.

Iodine

Iodine exists as iodide in brines and as iodates in deposits of sodium nitrate ($NaNO_3$, or Chile saltpeter). It may be recovered by oxidation of I^- with Cl_2 or by reduction of IO_3^- with HSO_3^-.

$$2I^-(aq) + Cl_2(g) \longrightarrow I_2(aq) + 2Cl^-(aq)$$
$$2IO_3^-(aq) + 5HSO_3^-(aq) \longrightarrow I_2(s) + 5SO_4^{2-}(aq) + H_2O(l) + 3H^+(aq)$$

Iodine is used as an antiseptic and disinfectant and as a reagent for chemical analysis.

Figure 21.24 When chlorine gas is bubbled through a solution of aqueous bromide ion (*Left*), they react to form aqueous bromine and chloride ion (*Right*).

Halogen Compounds

The most important compound of fluorine is hydrogen fluoride, HF, which freezes at −81.3°C and boils at 19.9°C. Because of very strong hydrogen bonding, which we discussed in Section 10.1, these temperatures are much higher than expected from the values for the other hydrogen halides. The HF molecule is also very polar because of the very high electronegativity of fluorine.

Also because of its strong hydrogen bonding and high polarity, liquid hydrogen fluoride is an exceptionally good solvent. Indeed, hydrogen fluoride comes as close as any substance to being a universal solvent. As a gas or a liquid or in aqueous solution, for example, it reacts with glass and is commonly used for etching glass, as shown in Figure 21.25.

Is HF really acting as a solvent when it etches glass?

$$SiO_2(s) + 4HF \longrightarrow SiF_4(g) + 2H_2O(l)$$

Hydrogen fluoride is formed by heating fluorspar with concentrated sulfuric acid.

$$CaF_2(s) + H_2SO_4(l) \longrightarrow CaSO_4(s) + 2HF(g)$$

Calcium fluoride, CaF₂, is a white powder that becomes luminous when heated. It dissolves in concentrated mineral acids to form HF. Known as fluorspar, it is the primary source of fluorine and its compounds. Because fluorspar transmits ultraviolet radiation, it is often used in the optical industry. It is also used to fluoridate drinking water.

In addition to the common −1 and 0 oxidation numbers, the halogens (except fluorine) exist with each positive oxidation number through +7, although some of the aqueous-solution species have only transitory existence. For example, many aqueous halogen species are susceptible to disproportionation. These include ClO_2, $HClO_2$, $HOCl$, Cl_2 (in base), $HOBr$, Br_2 (in base), HOI, IO^-, and I_2 (in base).

The halogens form a variety of species with oxygen. The oxides were discussed briefly in Section 3.3. Fluorine forms two very reactive substances: OF_2, a colorless gas, and O_2F_2, a yellow-orange solid that melts at −160°C to give a red liquid that decomposes upon heating above −100°C. Over a dozen oxides of the other halogens have been characterized. Halogen oxides are known with oxidation numbers as high as +7. Most of the halogen oxides are very strong oxidizing agents and tend to be rather unstable.

Figure 21.25 Hydrogen fluoride is used to etch glass, giving a frosted effect.

Why is fluorine not found with positive oxidation numbers?

All the halogens except fluorine form a wide variety of oxoacids and corresponding oxoanion salts. Some substances are known with oxidation numbers of $+1$, $+3$, $+5$, and $+7$.

Summary

Hydrogen has some properties like those of other Group IA (1) elements and other properties that are more like those of the Group VIIA (17) elements. The alkali metals are all soft, low-melting, silvery-white, reactive metals that form basic oxides.

Properties of the alkaline-earth metals vary down the group. Compounds of the lighter elements are somewhat covalent, whereas those of the heavier elements are primarily ionic. Beryllium and magnesium form protective oxide coatings, so they are stable in air, but the heavier elements oxidize and form a thick, non-protective coating of the oxide. The alkaline-earth metals are all strong reducing agents. They form basic oxides and hydroxides, except for those of beryllium, which are amphoteric.

Boron, a semiconductor, forms covalent compounds with various nonmetals. Many of these compounds are electron-deficient, which leads to unusual structures and to high chemical reactivity. Other members of Group IIIA (13) are soft metals. The activity of these metals decreases down the group, although aluminum and indium are stable in air because of a protective oxide coating.

Carbon forms a large number of compounds, primarily the hydrocarbons and their derivatives. The Group IVA (14) elements change from nonmetals at the top of the group through metalloids to metals at the bottom. Carbon forms a very strong bond to itself in compounds and so is characterized by the existence of many compounds containing C—C bonds. Bond strengths decrease down the group, so similar compounds are less prevalent for silicon and almost nonexistent for the lower elements.

Nitrogen is found in the atmosphere as N_2 molecules and is a constituent of all living matter. The Group VA (15) elements are generally found in combined form rather than as the free elements. The metallic nature of the elements increases down the group, from nonmetallic phosphorus to metallic bismuth.

Oxygen occurs in the atmosphere as a diatomic molecule, O_2. It is widely used as an oxidizing agent. An allotrope is ozone, O_3, which is present at low concentrations in the atmosphere. Oxygen also occurs in the form of oxides, peroxides, and superoxides. The Group VIA (16) elements increase in metallic character down the group.

The Group VIIA (17) elements, the halogens, occur as nonmetallic diatomic molecules. Their chemical reactivity decreases down the group. Fluorine is so reactive that it is rarely found except as fluorides. Chlorine, bromine, and iodine also exist in many compounds as halides, oxides, and oxoanions.

Key Terms

chalcogens (21.11)
zone refining (21.4)

Exercises

Main-Group Elements

21.1 Write the electronic configuration expected for members of each of the main-group elements from Group IA (1) to Group VIIA (17).

21.2 Identify the nonmetals in Groups IA (1) to VIIA (17).

21.3 Identify the metalloids in Groups IA (1) to VIIA (17).

21.4 Identify the metals in Groups IIIA (15) to VIIIA (18).

21.5 What types of bonding are expected for compounds formed from elements in each of the groups from Group IA (1) to Group VIIA (17)?

21.6 What are the common oxidation numbers for members of each of the groups from Group IA (1) to Group VIIA (17)?

21.7 Lithium and its compounds often resemble magnesium and its compounds. There is a similar resemblance between beryllium and aluminum. What property of these elements might give rise to such similarities?

21.8 Select four physical properties and compare them for metals, nonmetals, and metalloids.

21.9 Compare the general chemical properties of the metals, nonmetals, and metalloids.

21.10 Discuss the tendency toward covalent bond formation across the periodic table.

Hydrogen and Group IA (1) Elements

21.11 Discuss the placement of hydrogen as a member of Group IA (1) or Group VIIA (17) in the periodic table.

21.12 Describe several methods for the preparation of hydrogen gas.

21.13 What is the formula of the hydride that would be formed (a) from Se? (b) from Ge? (c) from Be?

21.14 The hydrides of calcium and gallium have melting points of 816°C and −21°C, respectively. Explain this difference.

21.15 Would you expect the hydride ion, H^-, to be a reducing agent or an oxidizing agent in aqueous solution?

21.16 Arrange the following hydrides in order of increasing ionic character: BeH_2, CH_4, MgH_2, NaH, SiH_4.

21.17 Describe the characteristic physical properties of the alkali metals.

21.18 Why are the Group IA (1) elements not found in the elemental state in nature?

21.19 Write a balanced equation to describe the electrolysis of molten sodium chloride.

21.20 The alkali metals are more likely than any other group of metals to produce a color when placed in the flame of a Bunsen burner. Explain this observation.

21.21 Write a balanced equation to describe the reaction that occurs when potassium metal is immersed in liquid water.

21.22 Even though rubidium metal is a stronger reducing agent than magnesium metal, the following reaction can be forced to completion at high temperatures:

$$Rb_2CO_3 + Mg \longrightarrow 2Rb + MgCO_3$$

Explain this result.

Group IIA (2) Elements

21.23 Describe the characteristic physical properties of the alkaline-earth metals.

21.24 Compare the following properties of beryllium with those of the other Group IIA (2) elements, and explain the differences.

a. Most binary compounds of Be are covalent.

b. Beryllium is a relatively weak reducing agent.

c. The oxide and hydroxide of beryllium are amphoteric.

d. The hydrated Be(II) cation hydrolyzes to give an acidic aqueous solution.

e. Molten $BeCl_2$ is a very poor conductor of electricity.

21.25 Which of the following compounds are likely to have high electrical conductivity when molten?

a. MgO   d. CaO

b. $MgCl_2$ e. $BaCl_2$

c. BeO   f. $BeCl_2$

21.26 Suggest a method for converting (a) $CaCO_3$ to $CaCl_2$, (b) $CaSO_4$ to Ca.

21.27 Calcium metal stored in air develops a white powder coating. Explain this result, and write an equation for the formation of the white powder.

21.28 Magnesium burns vigorously in air to produce a white ash. (a) What is this white substance? (b) What reaction will occur if this white substance is placed in water?

21.29 Write equations to describe the following conversion processes: (a) $BaCO_3$ to BaO (b) BaO to Ba

Boron and Group IIIA (13) Elements

21.30 Write the formula and describe or draw the structure of the simplest of the boron hydrides (boranes).

21.31 What is unique about the series of hydrides formed by boron?

21.32 Discuss the properties and structures of the two forms of boron nitride.

21.33 In the presence of excess fluoride ion, both BF_3 and AlF_3 dissolve to form aqueous solutions of a complex ion. Explain these results, and write balanced equations for the reactions that occur.

21.34 Aluminum does not corrode in moist air, even though it is very active. Explain this observation.

21.35 Addition of aqueous sodium hydroxide to $Al^{3+}(aq)$ gives a gelatinous precipitate. Continued addition of sodium hydroxide causes the precipitate to dissolve. Write balanced equations to describe these reactions. What property of aluminum(III) ion gives rise to this effect?

21.36 Evaporation of an aqueous solution of $AlCl_3$ gives a solid product that does not contain chlorine. Explain this result and identify the solid product.

Carbon and Group IVA (14) Elements

21.37 Identify the semiconducting elements in Group IVA (14).

21.38 Diamond is the hardest substance known, but graphite is soft and slippery and is used for lubrication. Both substances are pure carbon. Explain the difference between them.

21.39 Describe a laboratory preparation of carbon dioxide.

21.40 What products would you expect from the reaction between strontium carbide (SrC_2) and water?

21.41 Compare the heats of combustion of CH_4 and SiH_4 under standard conditions at 25.0°C.

21.42 Write an equation for the preparation of silicon from sand and coke.

21.43 Carbon dioxide is a gas, whereas silicon dioxide is a solid. Describe the differences in structure and bonding that give rise to this difference in physical state.

21.44 The fundamental unit in all silicate minerals is SiO_4. Explain how modifications of this unit give rise to the various types of silicate minerals.

21.45 Use balanced equations to outline methods for the following conversions.
   a. SiO_2 to Na_2SiF_6
   b. SiO_2 to Na_2SiO_3
   c. $SiCl_4$ to SiO_2

Nitrogen and Group VA (15) Elements

21.46 Arrange the following nitrides in order of increasing ionic character: BN, TiN, K_3N, NH_3.

21.47 Suggest a method for converting (a) Na_3N to NH_3, (b) P_4 to H_3PO_3.

21.48 Identify the products and write equations to describe the reaction of copper metal with dilute nitric acid and with concentrated nitric acid.

21.49 Describe the appearance and structures of the various allotropes of elemental phosphorus.

21.50 Write a reaction to describe how elemental phosphorus is recovered from phosphate rock.

21.51 Describe, using balanced equations, the production of phosphoric acid from elemental phosphorus. How would you modify this procedure to produce phosphorous acid?

Oxygen and Group VIA (16) Elements

21.52 Write a Lewis structure for molecular oxygen and discuss this structure in light of the fact that O_2 is paramagnetic.

21.53 Write Lewis structures for ozone and hydrogen peroxide and describe their shapes.

21.54 No ionic oxide simply dissolves in water. Explain this statement, using balanced chemical equations to illustrate your answer.

21.55 Describe the molecular structure of monoclinic sulfur.

21.56 Arrange the H_2X compounds for Group VIA (16) in order of increasing tendency to form hydrogen bonds: H_2O, H_2S, H_2Se, H_2Te.

The Halogens

21.57 Fluorine is the only halogen that forms compounds with the noble gases. What is special about fluorine that might cause this unique reactivity?

21.58 A common method of preparing the halogens is the oxidation of a halide salt with a chemical oxidizing agent. However, fluorine cannot be prepared by the chemical oxidation of fluoride. Discuss this discrepancy.

21.59 The melting and boiling points of the halogens increase down the group. Discuss the factors responsible for this behavior.

21.60 Compare F_2 and Cl_2 in terms of their reactivity with water, writing balanced equations for any reactions that occur and explaining any differences.

21.61 The halogens occur in nature primarily as halide salts. However, iodine can also be found as $NaIO_3$. Explain this difference.

21.62 In spite of its tremendous chemical reactivity, fluorine can be stored in a nickel or copper container. Explain why these two metals, but no others, can be used for fluorine containers.

21.63 Write balanced chemical equations to describe the reaction of water with each of the following.
   a. Cl_2 b. Cl_2O_7 c. Cl_2O d. F_2 e. I_2O_5

21.64 Write a Lewis structure for ClO_2 and decide whether this compound is diamagnetic or paramagnetic.

21.65 Bromine occurs in seawater in the form of dissolved bromides. Elemental bromine can be prepared by reacting seawater with chlorine gas. The only halogen that will work in this reaction is chlorine. Explain this observation.

21.66 The disproportionation of bromine proceeds further toward completion in basic solution than in acidic solution. Write balanced equations to describe these two processes, and explain the results.

21.67 Predict the products of the reaction of fluorine with sulfur.

21.68 Cyanogen, $(CN)_2$, is called a pseudohalogen because many of its chemical properties are similar to those of the halogens. On the basis of this observation, predict the products of the hydrolysis of cyanogen.

Additional Exercises

21.69 Calculate the percent by mass of boron in borax, $Na_2B_4O_7 \cdot 10H_2O$.

21.70 The hydrides of beryllium and barium have melting points of 125°C and 675°C, respectively. Explain this difference.

21.71 Write balanced equations to indicate what happens when a stream of the indicated gas is passed over the indicated hot metal.
   a. $H_2(g) + Ba(s)$ d. $Cl_2(g) + Mg(s)$
   b. $O_2(g) + Ba(s)$ e. $O_2(g) + Be(s)$
   c. $O_2(g) + Li(s)$

21.72 Potassium perchlorate, $KClO_4$, has a solubility product constant (K_{sp}) of 1.07×10^{-2}. What is the molar solubility of potassium perchlorate?

21.73 Aluminum forms a tightly bonded protective oxide coating upon reaction with oxygen gas or with an oxidizing acid. This coating protects the aluminum from corrosion. However, aluminum reacts with strong bases to release hydrogen gas. Why does the oxide coating not protect the aluminum from corrosion by strong bases?

21.74 K_{sp} for $BaCO_3$ is 8.1×10^{-9}. Will barium carbonate precipitate from a solution containing 2.0×10^{-4} M Ba^{2+} and 1.0×10^{-5} M $CO_3{}^{2-}$?

21.75 Magnesium metal reacts with nitrogen at high temperatures to form a compound that dissolves in water to give a basic solution. Write equations to explain these results.

21.76 Chlorine disproportionates slightly in water. Explain why the addition of Ag^+ to this solution drives the disproportionation reaction to completion.

21.77 Write balanced chemical equations to illustrate the acidic nature of the oxides of nitrogen, phosphorus, and sulfur.

*21.78 The equilibrium constant, K_c, for the dissociation of N_2O_4 into NO_2 is 0.0245 at 55°C. Suppose 0.100 mol N_2O_4 is placed in a 2.00-L vessel. What will be the equilibrium concentrations of the two species?

*21.79 The disproportionation of aqueous bromine gives bromide and hypobromous acid. The equilibrium constant at 25°C for this reaction is 7.2×10^{-9}. Suppose a 0.100 M solution of Br_2 is prepared in pure water. What is the concentration of HOBr at equilibrium?

*21.80 The etching of glass can be used to synthesize silicon tetrafluoride.

$$SiO_2(s) + 4HF(g) \longrightarrow SiF_4(g) + 2H_2O(l)$$

Calculate the standard enthalpy, entropy, and free-energy changes for this reaction at 25°C. Also calculate the equilibrium constant for the reaction.

*21.81 The combustion of white phosphorus is highly exothermic.

$$P_4(s) + 5O_2(g) \longrightarrow P_4O_{10}(s) \quad \Delta H = -2984 \text{ kJ}$$

How much heat is generated by the complete reaction of a mixture of 150.0 g P_4 with 250.0 g O_2?

*21.82 Write a balanced equation to describe the Goldschmidt reaction between PbO_2 and Al. Using a table of thermodynamic quantities, decide whether this reaction is endothermic or exothermic at 25.0°C. What is the standard free-energy change for this reaction at 25.0°C? Is the reaction spontaneous under these conditions?

*21.83 The reaction of boron with oxygen is exothermic at 600°C.

$$4B(s) + 3O_2(g) \longrightarrow 2B_2O_3(l) \quad \Delta H° = -2920 \text{ kJ}$$

How much heat is released during the reaction of 100.0 g of boron with excess oxygen?

▶ ▶ ▶ Chemistry in Practice

21.84 Magnesium boride, MgB_2, is treated with hydrochloric acid, producing a mixture of gaseous boron–hydrogen compounds. One compound is removed from the mixture by condensing it to a liquid at 18°C and solidifying it at −120°C. A 0.2490-g sample of this compound is then placed in a 25.00-mL sealed vessel and allowed to evaporate at 42.0°C. A pressure of 3672 torr is measured. This sample of the gas is transferred to an evacuated 2.00-L container and set aside at 42.0°C for a day. A dark solid presumed to be boron and weighing 0.2020 g is deposited; the pressure in the container increases to 229 torr. The gas in the container is identified as molecular hydrogen. In another experiment, 1.000 g of the compound is dissolved in 104.0 mL of water in a 2346-mL sealed container at 25.0°C. The pressure initially was 727 torr. Over a 24-hour period, the pressure rises and then levels off at 1435 torr. The solution contains boric acid. The pH of this solution is measured to be 4.681.

a. What is the molar mass of the gaseous compound?

b. What is the percent composition by mass of the compound?

c. What is the empirical formula of the compound?

d. What is the molecular formula of the compound?

e. Write a balanced equation for the reaction of this compound with water.

f. What is the value of the acid-dissociation constant of boric acid?

Chemistry of the Transition Metals

Vanadium as the metal and in solution as $V^{2+}(aq)$, $V^{3+}(aq)$, $VO^{2+}(aq)$, and $VO_2^+(aq)$ (left to right).

CHAPTER OUTLINE

Encountering Chemistry

Why are all these ions colored?

A yellow solution is treated with silvery-white zinc metal. The solution turns sky-blue, then blue-gray, then violet. A silvery-gray metal is placed in a hydrochloric acid solution. It dissolves slowly, forming a violet solution. On exposure to air, this solution slowly turns blue-gray. The metal is vanadium, and the colored solutions contain vanadium ions that have different oxidation numbers. The yellow ion, $VO_2^+(aq)$, contains vanadium(V), whereas the sky-blue ion, $VO^{2+}(aq)$, contains vanadium(IV). The blue-gray ion is $V^{3+}(aq)$ and the violet ion is $V^{2+}(aq)$. These ions are complex ions, with water as ligands. The ions exhibit behavior typical of the transition elements, including formation of complex ions, existence as stable substances with different oxidation numbers that are formed from one another by oxidation–reduction reactions, and changes in color with changes in oxidation number.

The transition elements are those elements in which the valence electrons are filling the *d* orbitals. These elements are located in the middle of the periodic table, in what is sometimes called the *d* block, between the two sets of main-group elements (the *s* block and the *p* block), as shown in Figure 22.1. Because all the transition elements are metals, they are often called the transition metals. Located in the middle of the transition metals, after La and Ac, are two other sets of elements, the lanthanides and actinides, which are called the inner-transition elements. Usually they are separated in the periodic table, in the *f* block, as shown in Figure 22.1, to keep the width of the table reasonable for printed media.

This chapter focuses on some of the transition metals in the fourth period, in which valence electrons are entering the 3*d* orbitals, and to some extent on the inner-transition metals. For the most part, we can understand the properties of the fifth- and sixth-period transition metals by referring to the fourth-period transition elements and to some general trends in properties. We will focus in particular on those fourth-period transition metals that have an extensive and varied aqueous-solution chemistry.

3	4	5	6	7	8	9	10	11	12
IIIB	**IVB**	**VB**	**VIB**	**VIIB**		— VIIIB —		**IB**	**IIB**
21 Sc	22 Ti	23 V	24 Cr	25 Mn	26 Fe	27 Co	28 Ni	29 Cu	30 Zn
39 Y	40 Zr	41 Nb	42 Mo	43 Tc	44 Ru	45 Rh	46 Pd	47 Ag	48 Cd
57 La	72 Hf	73 Ta	74 W	75 Re	76 Os	77 Ir	78 Pt	79 Au	80 Hg
89 Ac	104 Unq	105 Unp	106 Unh	107 Uns					

58 Ce	59 Pr	60 Nd	61 Pm	62 Sm	63 Eu	64 Gd	65 Tb	66 Dy	67 Ho	68 Er	69 Tm	70 Yb	71 Lu
90 Th	91 Pa	92 U	93 Np	94 Pu	95 Am	96 Cm	97 Bk	98 Cf	99 Es	100 Fm	101 Md	102 No	103 Lr

Figure 22.1 The transition and inner-transition elements.

This chapter examines the physical and chemical properties of selected transition metals, as well as trends in their properties. The properties of the elements can be explained in terms of principles covered in earlier chapters.

22.1   Properties of the Transition Metals

Which metal oxides would you expect to decompose upon heating?

The transition elements are widely distributed in Earth's crust. Most of them are not found in great quantities, however. Exceptions are iron (5.6% by mass of the crust), titanium (0.63%), and manganese (0.095%); the others are found in amounts of less than 0.01% of the crust. Most of the transition elements occur in nature as oxides, sulfides, carbonates, and silicates. Table 22.1 lists the minerals that are of greatest importance as commercial sources of these elements.

Four general methods for recovery of the transition metals from their minerals are summarized in Figure 22.2.

1. A few of the elements, including copper and gold, are found in the elemental state in nature, at least to a limited extent. Other noble metals occur in compounds that decompose readily when heated.

2. Some minerals can be converted to low-melting salts, which are melted and electrolyzed, much as in the methods used for recovery of alkali and alkaline-earth metals described in Section 18.5.

3. Other elements can be recovered by chemical reduction with a reducing agent such as carbon, magnesium, or aluminum.

4. Many of the transition metals occur as sulfide minerals, which can be heated with oxygen from air (a process called roasting) and then reduced with a chemical reducing agent, usually carbon.

The similarity in types of minerals of the transition metals and the few methods used to recover them indicate a characteristic of these elements—they have

Table 22.1 Commercially Important Minerals of the First-Series Transition Metals

Element	Mineral	Formula
Scandium	Thortveitite	$(Sc,Y)_2Si_2O_7$
Titanium	Rutile	TiO_2
	Ilmenite	$FeTiO_3$
Vanadium	Patronite	V_2S_5 occurring with sulfur
	Bravoite	$(Fe,Ni)S_2$ containing vanadium sulfide
Chromium	Chromite	$FeCr_2O_4$
	Croicite	$PbCrO_4$
Manganese	Pyrolusite	MnO_2
	Rhodochrosite	$MnCO_3$
Iron	Hematite	Fe_2O_3
	Magnetite	Fe_3O_4
	Iron pyrite	FeS_2
Cobalt	Cobaltite	$CoAsS$
	Smaltite	$CoAs_2$
Nickel	Pentlandite	$(Ni,Fe)_9S_8$
	Garnierite	$(Ni,Mg)_6Si_4O_{10}(OH)_2$
Copper	Chalcocite	Cu_2S
	Chalcopyrite	$CuFeS_2$
Zinc	Zinc blende	ZnS

very similar properties. These metals are relatively hard, have high melting points, and are more dense than most of the main-group metals. They have a typical metallic luster and are good electrical and thermal conductors. Most of the transition metals are strong, malleable, and ductile. Selected properties of the fourth-period transition metals are summarized in Table 22.2.

Many trends in the properties of the transition metals can be understood in terms of the electronic configurations of these metals, which were discussed in Sections 6.5 and 6.6 and listed in Table 6.4 and Figure 6.27. The electrons are successively added to the $3d$ orbitals across the first transition series. In accord-

Figure 22.2 Methods of recovering the transition elements from their minerals.

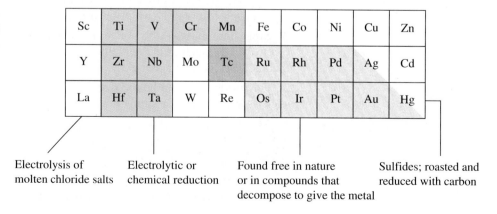

Sc	Ti	V	Cr	Mn	Fe	Co	Ni	Cu	Zn
Y	Zr	Nb	Mo	Tc	Ru	Rh	Pd	Ag	Cd
La	Hf	Ta	W	Re	Os	Ir	Pt	Au	Hg

Electrolysis of molten chloride salts

Electrolytic or chemical reduction

Found free in nature or in compounds that decompose to give the metal

Sulfides; roasted and reduced with carbon

Element	Sc	Ti	V	Cr	Mn	Fe	Co	Ni	Cu	Zn
Ionization energy (kJ/mol)	631	658	651	653	717	759	760	736	745	906
Electronegativity	1.3	1.5	1.6	1.6	1.5	1.8	1.8	1.8	1.9	1.6
Atomic radius (pm)	144	132	122	119	118	117	116	115	118	121
2+ ionic radius (pm)	—	94	88	89	80	74	72	69	72	74
3+ ionic radius (pm)	73	67	64	62	65	65	61	60	—	—
Density (g/cm³)	3.00	4.51	6.11	7.19	7.4	7.87	8.92	9.91	8.94	7.135
Melting point (°C)	1541	1660	1890	1857	1244	1525	1495	1453	1083	420
Boiling point (°C)	2831	3287	3380	2672	1962	2750	2870	2732	2567	907

Table 22.2 Some Properties of Fourth-Period Transition Metals

ance with Hund's rule, the electrons remain unpaired until each of the d orbitals is half filled. Only then does a second electron go into one of the d orbitals. Two of the elements, Cr and Cu, deviate from the smooth addition of electrons to the d orbitals. Because of the special stability of half-filled and filled subshells, they have a configuration in which one of the $4s$ electrons is shifted to a $3d$ orbital.

The magnetic properties of transition-metal complex ions were discussed in Chapter 19. The metals themselves also have magnetic properties. Many of the transition metals have relatively large numbers of unpaired electrons, and most transition metals are paramagnetic—that is, they exhibit a weak magnetism, which is characteristic of the presence of unpaired electrons. Three of the elements, iron, cobalt, and nickel, have a much stronger magnetism, called **ferromagnetism.** Here large numbers of the unpaired electrons spin in the same direction, resulting in an additivity of their magnetic fields and an enhanced magnetism. Ferromagnetic substances can be permanently magnetized by being exposed to another magnetic field. In addition to these three metals, some oxides of iron are also ferromagnetic.

Trends in the properties of the transition elements can be explained in terms of the interaction between the nucleus and the electrons. The d orbitals do not shield the outer s orbitals very well from the nuclear charge, so each successive transition metal across a period holds its outer electrons more tightly. However, the effect of poor shielding may be moderated somewhat by the crystal field splitting energy, introduced in Section 19.6, which modifies the energy of the electrons in complex ions. We will examine a number of properties of transition metals that vary across the periodic table and will try to understand these variations in terms of electronic stability.

Size

The atomic and ionic radii of the first transition series are listed in Table 22.3. The atomic radii decrease across the period, then become constant, and finally increase slightly. The ionic radii vary in a similar way but do not decrease so much as the atomic radii. These variations, shown in Figure 22.3, can be explained in terms of the effective nuclear charge, which determines how tightly the electrons are held to

Table 22.3 Aqueous Species of the First Transition-Metal Series*

	Sc	Ti	V	Cr	Mn	Fe	Co	Ni	Cu	Zn
d^0	Sc^{3+}	**TiO^{2+}**	VO_2^+	CrO_4^{2-}	MnO_4^-					
d^1		Ti^{3+}	**VO^{2+}**	H_3CrO_4	MnO_4^{2-}					
d^2		Ti^{2+}	V^{3+}	(Cr^{4+})	MnO_3^-	FeO_4^{2-}				
d^3			V^{2+}	**Cr^{3+}**	MnO_2					
d^4				Cr^{2+}	Mn^{3+}	(Fe^{IV})		NiO_4^{2-}		
d^5					**Mn^{2+}**	**Fe^{3+}**	CoO_2			
d^6						Fe^{2+}	Co^{3+}, CoL_6^{n+}	NiO_2		
d^7							**Co^{2+}**, CoL_6^{m+}			
d^8							(Co^+)	**Ni^{2+}**	CuO^+	
d^9									**Cu^{2+}**	
d^{10}									Cu^+	**Zn^{2+}**

*Formulas are empirical and do not always accurately represent the first coordination sphere of the ion. The formula printed in boldface represents the most stable oxidation number for that metal.

the nucleus and, thus, the size. The filling of the inner $3d$ orbitals begins with scandium, which shows a sharp drop in atomic radius compared with calcium. The electron in the $3d$ orbital is found primarily closer to the nucleus than the electrons in the $4s$ orbital. Because the $3d$ electrons do not shield the nucleus from the $4s$ electrons very well, the increase in nuclear charge from calcium to scandium is not totally canceled by the added $3d$ electron. The effective nuclear charge—the charge experienced by the valence electrons—is thus larger for Sc than for Ca, so the $4s$ electrons are pulled closer to the nucleus in Sc and the atomic size decreases. The atomic radius continues to decrease as additional electrons are added to the $3d$ orbitals, through chromium with its $4s^1 3d^5$ electronic configuration. After that, the atomic radius remains approximately constant. Although atomic size tends to decrease as atomic number increases across a period, you may notice a two-humped curve in Figure 22.3, one curve from scandium to manganese and the other from manganese to zinc. This is caused by the stability of certain electronic configurations resulting from the splitting of the d orbitals in a crystal field. There are similar effects in ionic radii, although the double-humped curve is not well defined in this case.

The atomic and ionic sizes vary not only across a period but also down a group. As we go down a group in the periodic table, we observe an average increase of 15–20 pm in these radii, smaller than the increase we observe on going down a group of main-group metals. Surprisingly, the second- and third-series transition metals are usually nearly the same size. For example, titanium has an atomic radius of 146 pm, and both zirconium and hafnium have an atomic radius of 157 pm. The near identity in size of the second- and third-series elements is due to an effect called the **lanthanide contraction.** The filling of the $4f$ orbitals in the lanthanides, which occurs within the third series of transition elements, causes

Figure 22.3 Atomic and ionic radii of the first transition series.

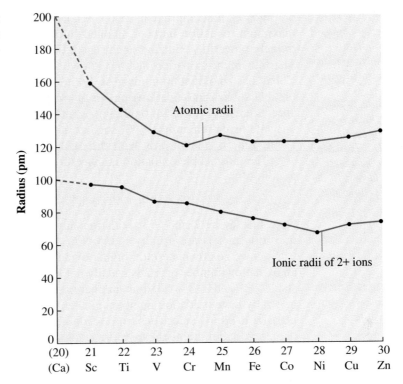

these transition metals to be smaller than expected. This effect occurs because the 4*f* orbitals are very poor nuclear shielders. The effective nuclear charge increases sufficiently that the third-series elements have nearly the same effective nuclear charge as the second-series elements—and thus nearly the same size. The unexpectedly small size for the second- and third-series transition metals causes them to hold electrons more tightly and thus to exhibit more covalent character in their bonding, to be less reactive than the first-series elements, and to have a high density. Because they have a similar size and charge, they tend to have very similar chemical properties.

Oxidation Number

Across a period, the number of stable species of an element that have different oxidation numbers increases. The maximum oxidation number increases to a value of $+7$ with manganese and then decreases, as shown in Table 22.3. The decreased oxidation numbers ($+2$, $+3$) of the post-manganese elements lends them critical roles in biological systems, such as iron(II)'s role in hemoglobin and that of cobalt(II) in vitamin B_{12}.

The stability of the maximum possible oxidation number decreases across the period. Consider the stability of oxides or oxoanions of the maximum oxidation number of some elements. Scandium(III) oxide, Sc_2O_3, is a stable oxide, as is titanium(IV) oxide, TiO_2, which occurs naturally as the mineral rutile. ▶ Vanadium(V) oxide and its oxoion, VO_2^+, are mild oxidizing agents, as shown in the chapter introduction. Chromium(VI) as CrO_4^{2-} or $Cr_2O_7^{2-}$ and manganese(VII) as MnO_4^- are powerful oxidizing agents, so they are not stable in the

Scandium(III) oxide, Sc_2O_3, is a fine white powder obtained by igniting the metal or its compounds. It dissolves readily in concentrated acids.

Vanadium pentoxide, V_2O_5, is a yellow to rust-brown solid prepared by heating ammonium vanadate, NH_4VO_3. It is used as a catalyst in the synthesis of sulfuric acid and in the manufacture of yellow glass.

presence of reducing agents. Iron is not known with oxidation number $+8$; the maximum oxidation number attained by iron is $+6$ in FeO_4^{2-}, which is an extremely powerful oxidizing agent and exists only in concentrated basic solution. The maximum oxidation number known for the other elements is less than $+6$. Cobalt has an oxidation number of $+4$ in CoO_2, but this compound is quite unstable. Nickel has an oxidation number of $+6$ in NiO_4^{2-} and $+4$ in NiO_2. Both these species are very powerful oxidizing agents. Copper is found occasionally with the unstable oxidation number $+3$, such as in CuO^+, which is a very strong oxidizing agent. Zinc is known only with the oxidation number $+2$.

The most stable oxidation numbers for each first-series transition element are as follows:

$$Sc^{III} \quad Ti^{IV} \quad V^{IV} \quad Cr^{III} \quad Mn^{II} \quad Fe^{III} \quad Co^{II} \quad Ni^{II} \quad Cu^{II,I} \quad Zn^{II}$$

Note that the $+2$ and $+3$ oxidation numbers predominate. Stable salts of the transition metals are usually found with oxidation numbers $+2$ and $+3$.

Why are higher oxidation numbers more stable going down a transition group?

As we go down a group, higher oxidation numbers tend to be more stable and an oxidation number of $+2$ is usually not stable. For example, although cobalt has stable $+2$ and $+3$ oxidation numbers, stable species for rhodium and iridium generally have oxidation numbers greater than $+3$. Because of their larger size, it is easier to remove electrons from second- and third-series transition elements, and higher oxidation numbers are more stable than for first-series transition elements.

Coordination Number

Many compounds of the transition elements are coordination compounds, in which the composition of the compound is determined not only by the oxidation number of the metal but also by its coordination number. As we saw in Chapter 19, there is a connection between these two numbers. Metal ions with a higher charge tend to have a higher coordination number. Cobalt(II) is usually found with coordination numbers of 4 or 6, for example, whereas cobalt(III) is always found with the coordination number 6.

As we go down a group, the coordination number tends to increase because of slightly increasing sizes of the ions. Coordination numbers from 6 through 10 are commonly found for the second- and third-series transitior elements. Coordination numbers as low as 4 are not usually found for these elements, with the exception of ions of these elements that have an electronic configuration of $ns^0(n-1)d^8$. Such ions commonly have coordination number 4 with square planar geometry. These species include Pd(II), Pt(II), Ir(I), Rh(I), and Au(III).

Chemical Reactivity

The chemical reactivity of the transition metals varies widely. For example, iron is quite reactive, as evidenced by its tendency to rust, whereas gold and platinum are exceptionally unreactive. Compounds of all the transition metals are known and can be formed fairly readily. The transition metals typically form complex ions, with a wide variety of coordination numbers and geometries. ▶ As discussed in Chapter 19 and illustrated in the chapter opening photograph, compounds of transition metals, especially the coordination compounds, are usually colored and their reactions are accompanied by color changes. The colors of many gemstones, for example, come from transition-metal ions.

22.2 Titanium

Titanium, the second member of the first transition period, is a strong, light metal that is silver-colored and very hard. With a density of 4.51 g/cm^3, it is 45% less dense than steel but just as strong. Titanium has a higher melting point (1660°C) than steel, but its strength drops rapidly above a temperature of 800°C, and it begins to absorb oxygen and nitrogen from air above 1300°C, which causes it to become brittle. To avoid brittleness, titanium must be welded under an inert atmosphere, such as argon. Titanium is quite resistant to corrosion. However, it burns in air and is the only transition element that burns in gaseous nitrogen.

Because of its strength and low density, titanium has been used in alloys for the construction of aircraft, missiles, spacecraft, and surgical instruments. Commercial aircraft contain approximately 7% titanium; military aircraft are about 27% titanium. The difference arises because titanium is a relatively expensive metal. Because titanium metal is not appreciably corroded by seawater, it is used for propeller shafts and other parts of ships that are exposed to seawater. This quality, along with its lightness, makes it well suited for masts and spars on racing yachts. Titanium is commonly used for artificial joints, such as ball-and-socket joints for hips, knees, wrists, and ankles, and in tooth implants. Besides being light and corrosion-resistant, it is not toxic and can be made porous so that the adjoining bone can attach itself readily to the replacement joint.

Titanium dioxide is the most common compound of titanium. It is a white amphoteric solid that reacts with both acids and bases. Titanium dioxide is formed by the hydrolysis of titanium tetrachloride, shown in Figure 22.4.

$$TiCl_4(l) + 2H_2O(l) \longrightarrow TiO_2(s) + 4HCl(aq)$$

Titanium tetrachloride, TiCl$_4$, is a colorless liquid that boils at 136°C. It fumes in moist air, producing a cloud of titanium dioxide and hydrochloric acid, so it has been used for smoke screens and skywriting.

Titanium dioxide has an extremely bright white color and is quite stable, so it is used extensively as a pigment in paints, enamels, paper, floor coverings, textiles, inks, plastics, ceramics, rubber, and cosmetics.

The oxidation–reduction properties of titanium in acidic solution are summarized by the following half-reactions.

$$TiO^{2+}(aq) + 2H^+(aq) + e^- \longrightarrow Ti^{3+}(aq) + H_2O(l) \qquad E^\circ_{red} = 0.1 \text{ V}$$

$$Ti^{3+}(aq) + e^- \longrightarrow Ti^{2+}(aq) \qquad E^\circ_{red} = -0.37 \text{ V}$$

$$Ti^{2+}(aq) + 2e^- \longrightarrow Ti \qquad E^\circ_{red} = -1.63 \text{ V}$$

Figure 22.4 When liquid titanium tetrachloride is added to water, solid titanium dioxide forms.

Figure 22.5 When titanium metal dissolves in hydrochloric acid *(Left)*, violet $Ti(H_2O)_6^{3+}(aq)$ forms *(Right)*.

Three aqueous titanium species exist. Ti(IV) exists as the colorless titanyl ion, TiO^{2+}. Solutions of titanium(IV) tend to deposit TiO_2 unless they are quite acidic. The +4 oxidation number is the most stable for titanium.

Ti(III) solutions are violet and contain the $Ti(H_2O)_6^{3+}$ or $Ti^{3+}(aq)$ ion, which reacts slowly with water or oxygen to produce Ti(IV). Solutions can be prepared by the reaction between titanium metal and hydrochloric acid, as shown in Figure 22.5.

Ti(II) solutions are extremely unstable. Because Ti^{2+} is a very strong reducing agent, it reduces water and does not exist very long in aqueous solution.

22.3   Vanadium

Vanadium is a light gray metal that is fairly soft and ductile, with a density of 6.11 g/cm³ and a melting point of 1890°C. The metal has good structural strength and resists corrosion by hydrochloric acid, sulfuric acid, alkalies, and salt water.

Most of the vanadium that is produced is alloyed with steel for applications such as rust-resistant materials, springs, and high-speed tools. Vanadium is also alloyed with aluminum and titanium for use in aircraft construction.

▶ Vanadium is found in compounds with oxidation numbers of +2, +3, +4, and +5, as mentioned in the chapter introduction. The most common compounds of vanadium that have these oxidation numbers are oxides and oxoion salts. Several oxides can be formed by heating vanadium metal in air. These are black VO, brownish black V_2O_3, blue-black VO_2, and red-orange V_2O_5. Vanadium pentoxide, as is typical for the oxide of a metal having its highest oxidation number, is amphoteric. With lower oxidation numbers, metal oxides are usually basic. We can explain this effect by considering that the higher the charge on the metal, the more it pulls electron density from the oxygen atoms, giving the metal–oxygen bond more covalent character. A hydroxide formed from such a metal oxide has more covalent character in the metal–oxygen bond than in the oxygen–hydrogen bond, resulting in more acidic character than is usually expected for a metal oxide. Because of its amphoteric character, vanadium(V) can be produced in different ionic forms. When V_2O_5 is heated with molten sodium hydroxide, the

Why would we expect V_2O_5 to be amphoteric?

vanadium(V) oxide acts as an acid to form sodium metavanadate, which also has the oxidation number +5.

$$V_2O_5(s) + 2NaOH(l) \xrightarrow{\Delta} 2NaVO_3(s) + H_2O(g)$$

If more sodium hydroxide is used, sodium pyrovanadate, another form of vanadium(V), forms.

$$V_2O_5(s) + 4NaOH(l) \xrightarrow{\Delta} Na_4V_2O_7(s) + 2H_2O(g)$$

Finally, sodium orthovanadate, which also contains vanadium(V), can be formed by addition of more sodium hydroxide.

$$V_2O_5(s) + 6NaOH(l) \xrightarrow{\Delta} 2Na_3VO_4(s) + 3H_2O(g)$$

When vanadium pentoxide is added to acids, it acts as a base and dissolves to give yellow VO_2^+.

$$V_2O_5(s) + 2H^+(aq) \longrightarrow 2VO_2^+(aq) + H_2O(l)$$

Ammonium vanadate, NH₄VO₃, is a white or slightly yellow powder. It is used as a dye for staining wool and wood black and for making indelible ink.

The VO_2^+ ion is probably actually $V(OH)_4^+$ or $VO_2(H_2O)_2^+$; the exact formula in solution has never been definitely determined.

Two vanadium compounds, vanadium pentoxide and ammonium vanadate (NH_4VO_3), are widely used as catalysts. An important catalytic application of V_2O_5 is in the contact process for sulfuric acid (Section 4.6), where it has largely replaced the use of the more expensive platinum oxide. Vanadium pentoxide is also used in ceramics, such as in yellow glass.

Vanadium occurs with a variety of oxidation numbers in acidic solution, as shown by the following reduction half-reactions:

$$VO_2^+ + 2H^+ + e^- \longrightarrow VO^{2+} + H_2O \qquad E^\circ_{red} = +1.00 \text{ V}$$
$$VO^{2+} + 2H^+ + e^- \longrightarrow V^{3+} + H_2O \qquad E^\circ_{red} = +0.36 \text{ V}$$
$$V^{3+} + e^- \longrightarrow V^{2+} \qquad E^\circ_{red} = -0.26 \text{ V}$$
$$V^{2+} + 2e^- \longrightarrow V \qquad E^\circ_{red} = -1.20 \text{ V}$$

Vanadium(V) is a strong oxidizing agent and can be selectively reduced to species that have +4, +3, or +2 oxidation numbers via reducing agents of appropriate strength. ▶ The ions in these half-reactions were discussed in the introduction to this chapter.

The most stable oxidation number for vanadium is +4, which is stable toward oxidation or reduction by water and to oxidation by oxygen in acidic solution. It can be prepared by the reduction of solutions of vanadium(V) with sulfur dioxide (sulfurous acid), as shown in Figure 22.6.

$$VO_2^+(aq) + H_2SO_3(aq) \longrightarrow VO^{2+}(aq) + H_2SO_4(aq)$$

Vanadium(III) exists as the hexaaqua ion $V(H_2O)_6^{3+}$. It can be prepared by the reduction of VO_2^+ (or VO^{2+}) with zinc metal.

$$VO_2^+(aq) + Zn(s) + 4H^+(aq) \longrightarrow V^{3+}(aq) + Zn^{2+}(aq) + 2H_2O(l)$$

Vanadium(III) is a good reducing agent that reacts slowly with oxygen from air to form VO^{2+}.

If vanadium(V), vanadium(IV), or vanadium(III) solution is reacted with zinc metal in the absence of air, violet vanadium(II) forms.

$$2V^{3+}(aq) + Zn(s) \longrightarrow 2V^{2+}(aq) + Zn^{2+}(aq)$$

Figure 22.6 Gaseous sulfur dioxide reacts with yellow VO_2^+ solutions *(Left)* to produce blue VO^{2+} solutions *(Right)*. *Center:* A green stage in the reduction process arises from mixtures of yellow and blue.

Vanadium(II), a strong reducing agent, slowly reduces water, releasing hydrogen gas, and it rapidly reduces oxygen gas, forming vanadium(III). For this reason, solutions of vanadium(II) must be prepared shortly before use and stored under an inert gas, such as nitrogen or argon.

22.4   Chromium

Why is the susceptibility of iron to corrosion blamed on its oxide, whereas the oxide of chromium is given credit for its resistance to corrosion?

Most alloys of chromium and nickel contain more nickel than chromium. For example, the heating element in products such as toasters is made of nichrome *wire, an alloy of about 80% Ni and 20% Cr.*

Barium chromate, BaCrO$_4$, is a poisonous yellow solid used as a pigment in artists' colors, glass, and porcelain. It is also used as an anticorrosion paste for covering joints between two metals.

Chromium is a hard, silvery-white, lustrous metal that takes a high polish and strongly resists corrosion by forming a very thin oxide layer. The protective oxide layer is formed by exposure to oxygen or to oxoacids such as nitric acid, chromic acid, and phosphoric acid.

Because pure chromium metal is not very ductile and is rather brittle, it is usually alloyed with a little nickel to improve its workability and its corrosion-resisting properties at high temperatures. Chromium is used most to increase the hardness and improve the corrosion resistance of alloys, such as stainless steel, which contains 10–25% chromium in addition to less than 1% carbon. Lesser quantities of chromium are added to other steels to improve their hardness. Steel containing chromium is used to make objects such as ball bearings and machine tools. Chromium is often electroplated over other metals, such as steel, to give a hard, shiny, corrosion-resistant surface. The chromium plate is so thin that the chromium on all the automobiles in the United States adds up to only about 100 tons.

The principal compounds of chromium are Na_2CrO_4, $Na_2Cr_2O_7$, CrO_3, $Cr_2(SO_4)_3$, and Cr_2O_3, which are shown in Figure 22.7, as well as many coordination compounds of chromium(III). Compounds containing water as a ligand are very soluble and difficult to recover from solution, but some of them, such as $[Cr(H_2O)_6]Cl_3$ and $[Cr(H_2O)_4Cl_2]Cl \cdot 2H_2O$, can be obtained as solids. Compounds that contain ammonia and similar ligands can be precipitated from

Figure 22.7 Compounds of chromium are highly colored. *Clockwise from bottom:* The compounds are Na_2CrO_4, $Cr_2(SO_4)_3$, $Na_2Cr_2O_7$, Cr_2O_3, and CrO_3.

solution readily. Some examples are $[Cr(NH_3)_5H_2O]Cl_3$, $[Cr(NH_3)_6]Cl_3$, $[Cr(NH_3)_5Cl]Cl_2$, and $K_3[Cr(CN)_6]$.

Compounds of chromium(VI), such as sodium chromate and sodium dichromate, are the most important chromium compounds. Sodium chromate is prepared by the reaction of chromite ore or of chromium(III) oxide with sodium carbonate in the presence of air.

$$4FeCr_2O_4 + 8Na_2CO_3 + 7O_2(g) \xrightarrow{\Delta} 8Na_2CrO_4 + 2Fe_2O_3 + 8CO_2(g)$$

Reaction of sodium chromate with sulfuric acid produces sodium dichromate.

$$2Na_2CrO_4(aq) + H_2SO_4(aq) \xrightarrow{\Delta} Na_2Cr_2O_7(aq) + Na_2SO_4(s) + H_2O(l)$$

In aqueous solution, the chromate and dichromate ions are in a state of equilibrium. As shown in Figure 22.8, the position of equilibrium depends on the pH. Other chromate salts are lead chromate, $PbCrO_4$, and barium chromate, $BaCrO_4$, both yellow solids used as pigments. These salts are highly insoluble and can be precipitated from solution, as shown in Figure 22.9.

Figure 22.8 Addition of acid to yellow chromate ion *(Left)* forms orange dichromate ion *(Right)*. The reaction is reversible, and chromate ion can be reformed by addition of base.

Figure 22.9 Addition of sodium chromate solution to lead nitrate solution results in the precipitation of lead chromate.

Chromium(III) oxide is used as a green pigment and as a starting material for chromium production. Another oxide is CrO_2, which exists as black ferromagnetic crystals. It is commonly used in magnetic recording tapes.

Compounds of chromium are known with all the oxidation numbers from -2 to $+6$, but only species with oxidation numbers of $+2$, $+3$, and $+6$ exist for any time in aqueous solution. The primary substances found in acidic aqueous solution are indicated in the following half-reactions.

Chromium(III) oxide, Cr_2O_3, is a dark green solid used in abrasives and as a pigment for printing fabrics and banknotes and for coloring green glass.

$$Cr_2O_7{}^{2-} + 14H^+(aq) + 6e^- \longrightarrow 2Cr^{3+}(aq) + 7H_2O(l) \qquad E^\circ_{red} = +1.33 \text{ V}$$
$$Cr^{3+}(aq) + e^- \longrightarrow Cr^{2+}(aq) \qquad E^\circ_{red} = -0.41 \text{ V}$$
$$Cr^{2+}(aq) + 2e^- \longrightarrow Cr(s) \qquad E^\circ_{red} = -0.91 \text{ V}$$

Chromium metal dissolves in acidic solution to give blue Cr^{2+} if no O_2 is present, as shown in Figure 22.10. Cr(II) ion can also be produced by reduction of solutions of Cr(III) salts with zinc.

$$2Cr^{3+}(aq) + Zn(s) \longrightarrow 2Cr^{2+}(aq) + Zn^{2+}(aq)$$

Figure 22.10 Chromium metal reacts with acid to produce chromium(II) ion in solution and hydrogen gas.

▶ ▶ ▶   **WORLDS OF CHEMISTRY**

Transition Metals and Cancer

Several factors determine whether an element is beneficial or detrimental to health.

Almost daily, we hear or read reports of chemicals that can cause cancer. Unfortunately, reports of chemicals that can cure cancer are not so frequent. Can't we just avoid the chemicals that might cause cancer? Not in every case. How do the transition metals fit into this scenario?

Although some transition metals are necessary to life, a number are also toxic or carcinogenic. Several factors determine whether an element is beneficial or detrimental to health. Among these are the method of exposure, the dose, and the compound of the element involved.

The fourth-period transition metals can be divided into several groups on the basis of carcinogenicity.

- Proven carcinogen: Cr, Co, Ni, Zn
- Suspected carcinogen: Sc, Ti, Mn
- Some complexes are carcinogens: Fe, Cu
- May be a carcinogen (no direct evidence): V

Chromium is a proven carcinogen and is involved primarily in lung cancer, which has been observed particularly among workers in chromium metallurgy and refining plants and in dichromate manufacturing plants, usually with a time lag of about 17 years after first exposure. Although the actual carcinogenic agent is not known, both Cr(VI) and chromite ore dust are considered agents. Chromate dusts, fumes, and mists have also been judged dangerous. No lung cancer was found among workers in a plant producing Cr(III) oxide and sulfate, and no cancer has been found to result from oral intake of chromium compounds. Thus it appears that the compound, the oxidation number, and the method of exposure are important.

Zinc is required by over 50 enzymes. No life can exist if zinc is absent. If the amount present is sufficient only to sustain life, children are severely retarded and small. If more than this (but still less than the optimal amount) is present, taste and smell are deficient. If more than the optimal amount of zinc is present, nausea results and the immune system is endangered. Even more zinc produces fume fever and granulomas in the lungs. Much more zinc produces death. Thus the dose of zinc is extremely important in its effect on health.

We often hear it said that nickel is carcinogenic. However, its carcinogenicity depends on the specific compound of nickel(II), as shown by the following list, where the numbers are the percent of rats that experienced sarcomas when they ingested the compound: NiS (100%), NiO (65%), NiSe (50%), Ni (CH$_3$CO$_2$)$_2$ (22%), NiCrO$_4$ (6%), and NiCl$_2$ (0%).

The studies cited indicate that we cannot make blanket statements about the carcinogenicity of an element or a compound. The extent of carcinogenicity depends on factors such as oxidation number, chemical composition, dose, and type of exposure.

Questions for Discussion

1. Judging on the basis of the information provided here, should your instructor ask you to work with a 0.100 M solution of potassium chromate in lab?
2. If nickel is carcinogenic, is it reasonable to ban it from instructional chemistry labs but to use it in coins?
3. Much of the knowledge we have of carcinogenicity comes from studies with rats. Why are rats used for these studies? Do they serve as good models for the action of chemicals on humans? Or are they just unattractive animals that people don't mind harming?
4. Find a recent newspaper article on cancer, especially one that involves chemicals in some way, and discuss it with your classmates.

Chromium(II), a strong reducing agent, reacts with many aqueous ions, such as iron(III) ion:

$$\text{Cr}^{2+}(aq) + \text{Fe}^{3+}(aq) \longrightarrow \text{Cr}^{3+}(aq) + \text{Fe}^{2+}(aq)$$

and cobalt(III) complex ions:

$$\text{Co(NH}_3)_5\text{Cl}^{2+}(aq) + \text{Cr}^{2+}(aq) \longrightarrow \text{CrCl}^{2+}(aq) + \text{Co}^{2+}(aq) + 5\text{NH}_3(aq)$$

Figure 22.11 Chromium(III) in solution *(Left)* is converted to solid $Cr(OH)_3$ *(Center)* when sodium hydroxide is added. Excess sodium hydroxide causes the solid to dissolve and form $Cr(OH)_4{}^-$ *(aq)* *(Right)*.

Chromium(III) exists in solution primarily as $[Cr(H_2O)_6]^{3+}$, which is usually abbreviated $Cr^{3+}(aq)$. In slightly basic solutions, this ion forms amphoteric chromium(III) hydroxide.

$$Cr^{3+}(aq) + 3OH^-(aq) \longrightarrow Cr(OH)_3(s)$$

As shown in Figure 22.11, this gray-green precipitate redissolves upon addition of excess hydroxide to give the green chromite ion.

$$Cr(OH)_3(s) + OH^-(aq) \longrightarrow Cr(OH)_4{}^-(aq)$$

In acidic solution, Cr(VI) is a very strong oxidizing agent and is extensively used as an analytical reagent to titrate oxidizable metal ions. For example, it can be used to analyze iron(II) solutions or salts.

$$6Fe^{2+}(aq) + Cr_2O_7{}^{2-}(aq) + 14H^+(aq) \longrightarrow 6Fe^{3+}(aq) + 2Cr^{3+}(aq) + 7H_2O(l)$$

22.5   Manganese

Manganese, a gray-white metal with a reddish tinge, is hard enough to scratch glass and so brittle that it cannot be machined readily. Manganese is used extensively in alloys, especially with iron. It improves the workability, strength, hardness, toughness, and wearability of steels. Manganese is added in small amounts ($< 1\%$) to aluminum and magnesium to increase their corrosion resistance and hardness.

The most stable compounds of manganese are $KMnO_4$, MnO_2, $MnCl_2$, $MnSO_4$, and $Mn(NO_3)_2$. These compounds contain manganese with an oxidation number of +7, +4, or +2.

What property of manganese dioxide makes it useful in dry cells?

Manganese dioxide is used in dry cells (discussed in Section 18.2), as a decolorizer for glass that is green because of iron(II) impurities, and as a black pigment in glass and in enamel paints. Manganese dioxide occurs naturally in the form of pyrolusite. It can also be produced by electrolysis; this source is generally used for dry cells. In the laboratory, it can be prepared by reaction of Mn(II) with Mn(VII), as shown in Figure 22.12.

$$2MnO_4{}^-(aq) + 3Mn^{2+}(aq) + 2H_2O(l) \longrightarrow 5MnO_2(s) + 4H^+(aq)$$

Figure 22.12 When a solution containing potassium permanganate is added to a manganese(II) solution, solid manganese dioxide precipitates.

Manganese dioxide is a good oxidizing agent and is used as such in a variety of chemical processes. For example, MnO_2 reacts with hot acidic chloride solutions to produce chlorine gas, a common method for the laboratory preparation of Cl_2.

$$MnO_2(s) + 4H^+(aq) + 2Cl^-(aq) \xrightarrow{\Delta} Mn^{2+}(aq) + 2H_2O(l) + Cl_2(g)$$

Potassium permanganate is a strong oxidizing agent widely used in analytical chemistry, water and air purification, bleaching, and disinfecting and as an oxidizer in organic syntheses such as that of saccharine and vitamin C. Potassium permanganate is a dark purple solid formed by air oxidation of manganese dioxide in molten potassium hydroxide. This reaction first produces potassium manganate, a dark green solid.

$$2MnO_2(s) + 4KOH(l) + O_2(g) \xrightarrow{\Delta} 2K_2MnO_4(s) + 2H_2O(g)$$

Potassium manganate reacts with water in the presence of carbon dioxide or other acids to form potassium permanganate.

$$3K_2MnO_4(aq) + 4H^+(l) \longrightarrow 2KMnO_4(aq) + 4K^+(aq) + MnO_2(s) + 2H_2O(aq)$$

Most of the aqueous chemistry of manganese is associated with oxidation numbers +2, +4, and +7. The aqueous-solution species of manganese and their interconversion by oxidation–reduction reactions are summarized by the following half-reactions.

$$MnO_4^-(aq) + 8H^+(aq) + 5e^- \longrightarrow Mn^{2+}(aq) + 4H_2O(l) \qquad E^\circ_{red} = 1.51 \text{ V}$$
$$MnO_4^-(aq) + e^- \longrightarrow MnO_4^{2-}(aq) \qquad E^\circ_{red} = 0.564 \text{ V}$$
$$MnO_4^-(aq) + 4H^+(aq) + 3e^- \longrightarrow MnO_2(s) + 2H_2O(l) \qquad E^\circ_{red} = 1.695 \text{ V}$$
$$MnO_4^{2-}(aq) + 4H^+(aq) + 2e^- \longrightarrow MnO_2(s) + 2H_2O(l) \qquad E^\circ_{red} = 2.26 \text{ V}$$
$$MnO_2(s) + 4H^+(aq) + e^- \longrightarrow Mn^{3+}(aq) + 2H_2O(l) \qquad E^\circ_{red} = 0.95 \text{ V}$$
$$MnO_2(s) + 4H^+(aq) + 2e^- \longrightarrow Mn^{2+}(aq) + 2H_2O(l) \qquad E^\circ_{red} = 1.23 \text{ V}$$
$$Mn^{3+}(aq) + e^- \longrightarrow Mn^{2+}(aq) \qquad E^\circ_{red} = 1.51 \text{ V}$$
$$Mn^{2+}(aq) + 2e^- \longrightarrow Mn(s) \qquad E^\circ_{red} = -1.18 \text{ V}$$

Mn(II) exists as faint pink Mn^{2+} in acidic solution, where it is the most stable manganese species, and it precipitates as an insoluble white hydroxide, $Mn(OH)_2$,

Figure 22.13 A paper towel contains components that are weak reducing agents. In slighly basic solution, violet permanganate ion is reduced by the paper towel to green manganate ion.

from basic solution. It forms a variety of six-coordinate complex ions such as $[Mn(CN)_6]^{4-}$.

Mn(III) can be prepared by electrolytic or chemical oxidation of Mn^{2+} in acidic solution, but it is unstable and tends to disproportionate.

$$2Mn^{3+}(aq) + 2H_2O(l) \longrightarrow MnO_2(s) + Mn^{2+}(aq) + 4H^+(aq)$$

Brown insoluble $Mn(OH)_3$ is formed by the oxidation of $Mn(OH)_2$ with air. In general, manganese(III) exists in a stable state only as insoluble compounds or as complex ions such as $[Mn(CN)_6]^{3-}$.

Mn(IV) exists primarily as insoluble brown MnO_2, which has a limited solubility in its hydrated forms and acts as a strong oxidizing agent in acidic solutions.

Mn(VI) exists as the green manganate ion, MnO_4^{2-}. This ion is stable only in alkaline solution. As shown in the following equation, it disproportionates in neutral and acidic solutions.

$$3MnO_4^{2-}(aq) + 4H^+(aq) \longrightarrow MnO_2(s) + 2MnO_4^-(aq) + 2H_2O(l)$$

The manganate ion slowly releases oxygen gas from aqueous solution at any pH. This species is often the product of the reduction of permanganate ion with weak reducing agents in basic solution. An example is shown in Figure 22.13.

Mn(VII) exists as the violet permanganate ion, MnO_4^-. The product of reduction of MnO_4^- depends on the strength of the reducing agent as well as on the pH of the solution. Possible products in acidic solution are shown in the half-reactions listed earlier. For most reducing agents, reduction in acidic solutions produces Mn^{2+}; in neutral solutions, MnO_2; and in basic solutions, MnO_4^{2-}.

22.6 Iron

Iron is a tough, malleable gray metal that is used in alloys in greater amounts than any other metal. This is because of its widespread occurrence, the ease with which it can be produced from its ores, and its desirable structural properties. It is used in magnets because of its strong ferromagnetism.

Pure iron metal is not used extensively. Most of the iron extracted from its ores is converted to steel, which contains small amounts of carbon and other elements. The basic process for steel production was discussed in Section 4.6. The properties of the steel depend on the elements present in the alloy. One series of steels, known as carbon steels, contains only iron and carbon, for example. Low-carbon steel, such as wrought iron, contains only a few tenths of a percent of carbon; such steels are soft, tough, and malleable. Steels with about 0.5% carbon are used for structural materials such as beams and rails. High-carbon steels, with 0.8–1.5% carbon, are very hard and are used for knife blades, razor blades, drill bits, and other cutting tools. Alloying iron with other elements produces a variety of steels called alloy steels. Carbon is usually also present in these steels. Stainless steel, for example, is an alloy steel that contains small amounts of carbon along with large amounts of chromium and nickel.

Iron is present in all mammalian cells, where it is essential for life. Best known in living systems is the occurrence of iron in the form of hemoglobin in blood, which is responsible for the transport of oxygen through the body. Iron is also found in a variety of enzymes that act as oxidation–reduction catalysts.

One of the unfortunate characteristics of iron is that it rusts, or oxidizes, in moist air—a characteristic that requires that iron be protected or replaced frequently. This characteristic arises because iron is a reactive element and a strong reducing agent. It slowly displaces hydrogen from water at room temperature and displaces it rapidly above 500°C. Iron reduces a number of aqueous ions (including Hg^{2+}, Sn^{2+}, Ni^{2+}, and Cu^{2+}) to the respective metals. Iron can be passivated, or rendered inactive, by treating it with concentrated nitric acid to form a thin protective coating of FeO on the surface. However, this film is easily destroyed by hammering or scratching or by reducing agents. The difference in reactivity of iron with dilute and concentrated nitric acid is shown in Figure 22.14.

A number of compounds of iron can be formed by reactions of iron metal. For example, iron reacts with non-oxidizing acids to release hydrogen gas.

$$Fe(s) + 2H^+(aq) \longrightarrow Fe^{2+}(aq) + H_2(g)$$

Iron combines with dry oxygen only at elevated temperatures, where it burns to form Fe_3O_4, which contains iron with oxidation numbers of +2 and +3.

$$3Fe(s) + 2O_2(g) \xrightarrow{\Delta} Fe_3O_4(s)$$

In the presence of moist air at ordinary temperatures, iron rusts to form hydrated iron(III) oxide, $Fe_2O_3 \cdot nH_2O$, which does not adhere well to the iron and so flakes

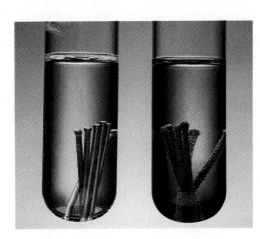

Figure 22.14 Iron does not react with concentrated nitric acid because a protective coating of FeO is formed *(Left)*. Dilute nitric acid does not form this coating and reacts with iron to release hydrogen gas *(Right)*. Nitric acid may oxidize the iron to $Fe^{3+}(aq)$.

off and continually provides a fresh surface to rust. The prevention of such corrosion is described in Chapter 18.

Iron(III) oxide exists as well in an anhydrous form, Fe_2O_3, which is found in nature as hematite. Iron(III) hydroxide may be prepared by addition of alkali-metal hydroxides or aqueous ammonia to a solution of an iron(III) salt.

$$Fe^{3+}(aq) + 3OH^-(aq) \longrightarrow Fe(OH)_3(s)$$

Iron(III) oxide, Fe_2O_3, is used to produce a wide range of pigments, from yellow ochre to rouge and Venetian red.

Unlike the hydroxides of some metal ions with a 3+ charge, such as Al^{3+} and Cr^{3+}, this red solid does not dissolve in excess hydroxide. Rather, it slowly converts to red-brown hydrated Fe_2O_3 in basic solution.

Iron forms a variety of salts with common anions, such as pale green $FeCl_2 \cdot 4H_2O$, blue-green $FeSO_4 \cdot 7H_2O$, and yellow $FeCl_3$. Iron(II) and iron (III) form a variety of six-coordinate octahedral complex ions. The most common of these have water or cyanide as ligands: $[Fe(H_2O)_6]^{2+}$, $[Fe(H_2O)_6]^{3+}$, $[Fe(CN)_6]^{4-}$, and $[Fe(CN)_6]^{3-}$.

The aqueous-solution species of iron and their oxidation–reduction properties are summarized in the following half-reactions.

$$FeO_4^{2-}(aq) + 8H^+(aq) + 3e^- \longrightarrow Fe^{3+}(aq) + 4H_2O(l) \qquad E^{\circ}_{red} = 2.20 \text{ V}$$
$$Fe^{3+}(aq) + e^- \longrightarrow Fe^{2+}(aq) \qquad E^{\circ}_{red} = 0.771 \text{ V}$$
$$Fe^{2+}(aq) + 2e^- \longrightarrow Fe(s) \qquad E^{\circ}_{red} = -0.440 \text{ V}$$

The aqueous chemistry of iron is dominated by oxidation numbers +2 and +3. Iron(III) is the most stable state for iron in aqueous solution, but it can be reduced readily to iron(II) by strong reducing agents.

Iron(III) exists in solution as the pale pink $[Fe(H_2O)_6]^{3+}$ ion. Solutions of Fe(III) salts are generally yellow, however, because of partial hydrolysis of the iron(III) ion, as shown in the equation

$$[Fe(H_2O)_6]^{3+}(aq) \rightleftharpoons [Fe(H_2O)_5OH]^{2+}(aq) + H^+(aq)$$

The yellow color may also arise from complex ions formed with the anion in the iron(III) salt, such as chloride ion.

$$[Fe(H_2O)_6]^{3+}(aq) + Cl^-(aq) \rightleftharpoons [Fe(H_2O)_5Cl]^{2+}(aq) + H_2O(l)$$

Iron(III) forms a characteristic blood-red complex ion (Figure 22.15) when reacted with a thiocyanate salt.

$$[Fe(H_2O)_6]^{3+}(aq) + NCS^-(aq) \rightleftharpoons [Fe(H_2O)_5NCS]^{2+}(aq) + H_2O(l)$$

Figure 22.15 Aqueous potassium thiocyanate reacts with aqueous iron(III) chloride to form a red isothiocyanato iron(III) complex ion.

Iron(III) chloride, FeCl₃, is a red or green solid, depending on whether it is viewed by transmitted or reflected light. As a hexahydrate, it is a brownish-yellow to orange solid. It is used in pigments, ink, and catalysts and for deodorizing sewage.

Iron(II) is a colorless or very pale green ion that exists in solution as $[Fe(H_2O)_6]^{2+}$. It is a mild reducing agent that is oxidized slowly when exposed to air. Because of its lower charge, $Fe^{2+}(aq)$ is not so acidic as $Fe^{3+}(aq)$ and does not undergo hydrolysis reactions such as those observed for $Fe^{3+}(aq)$.

22.7 Cobalt

Cobalt is a hard, silvery-white metal that resembles iron and nickel but has a slight pink tinge. Like those metals, it is ferromagnetic. Alloys of cobalt with iron and nickel along with other metals have exceptional magnetism. One such alloy is *Alnico,* which contains *al*uminum (12%), *ni*ckel (21%), *co*balt (5%), and iron (62%) and is often used in magnets.

About 80% of the cobalt produced is alloyed with other metals, either to make magnetic alloys or to make steels that exhibit improved resistance to wear and to oxidation and corrosion at high temperatures. These properties make cobalt steels useful, for example, in surgical instruments and high-speed cutting tools.

Most cobalt compounds have oxidation numbers of +2 or +3. Co(II) forms simple compounds with a variety of anions, such as oxide, hydroxide, sulfate, sulfide, fluoride, chloride, bromide, iodide, acetate, carbonate, phosphate, and nitrate ions. Many of these compounds exist as hydrates.

Co(II) forms the reddish-pink Co^{2+} ion in solution. Unless a ligand such as NH_3, CN^-, or NO_2^- is present, Co(II) is the most stable form of cobalt in aqueous solution. In the presence of strong-field ligands such as these, Co(II) becomes a strong reducing agent.

An important aspect of the chemistry of cobalt occurs in basic solutions. Addition of an alkali-metal hydroxide to an aqueous solution of a cobalt(II) salt produces blue Co(II) hydroxide.

$$Co^{2+}(aq) + 2OH^-(aq) \longrightarrow Co(OH)_2(s)$$

On standing, the blue solid slowly turns pink, probably as a result of hydration. A blue color for Co(II) species is characteristic of a coordination number of 4, and pink is associated with a coordination number of 6.

Ammine complexes are formed in a solution of a Co(II) salt made basic with ammonium hydroxide.

$$Co^{2+}(aq) + 6NH_3(aq) \rightleftharpoons [Co(NH_3)_6]^{2+}(aq)$$

This complex ion is very susceptible to air oxidation to $[Co(NH_3)_6]^{3+}$ or related Co(III) complex ions.

Heating cobalt(II) hydroxide in the absence of air produces black Co(II) oxide.

$$Co(OH)_2(s) \xrightarrow{\Delta} CoO(s) + H_2O(g)$$

This material is added to glass to give the deep blue color characteristic of cobalt glass.

What can you deduce on the basis of color about the structural environment of cobalt in cobalt glass?

Co(III) does not form simple hydrated salts because it is not stable in the presence of water. Co(III) is found most often as coordination compounds such as $[Co(NH_3)_6]Cl_3$, $[Co(NH_3)_5Cl]SO_4$, and $Na_3[Co(NO_2)_6]$.

Sodium cobaltinitrite, or sodium hexanitrito-cobaltate(III), $Na_3[Co(NO_2)_6]$, is a yellow to brownish-yellow solid. It is used to detect potassium ion, with which it forms an insoluble compound.

The aqueous chemistry of cobalt is dominated by oxidation numbers +2 and +3, as shown by the following half-reactions.

$$Co^{3+}(aq) + e^- \longrightarrow Co^{2+}(aq) \qquad E^\circ_{red} = 1.82 \text{ V}$$
$$Co^{2+}(aq) + 2e^- \longrightarrow Co(s) \qquad E^\circ_{red} = -0.277 \text{ V}$$

Co(III) can be prepared in aqueous solution as the Co^{3+} ion, which is actually $[Co(H_2O)_6]^{3+}$. This very strong oxidizing agent cannot be kept for long in water, because it oxidizes water to release O_2 gas.

$$4Co^{3+}(aq) + 2H_2O(l) \longrightarrow 4Co^{2+}(aq) + O_2(g) + 4H^+(aq)$$

22.8   Nickel

Nickel is a hard, silvery-white metal that is quite ductile and malleable, moderately strong and tough, and resistant to corrosion. It is also ferromagnetic, as shown in Figure 22.16, and is used in many magnets.

Especially pure nickel metal is produced via the Mond process. Nickel obtained by electrolysis is powdered and reacted with carbon monoxide under high pressures, such as 100–400 atm, at about 40–50°C.

$$Ni(s) + 4CO(g) \xrightarrow{\Delta} Ni(CO)_4(g)$$

Is the Mond process for purifying nickel consistent with Le Chatelier's principle?

The tetracarbonylnickel(0), a tetrahedral coordination compound that exists as a colorless volatile liquid at room temperature, is decomposed at 200°C.

$$Ni(CO)_4(g) \xrightarrow{\Delta} Ni(s) + 4CO(g)$$

Figure 22.16 Many Canadian coins are made of pure nickel metal, which is ferromagnetic and so is attracted to a magnet.

Small pellets of pure nickel are used as seeds in this process to grow larger pellets, such as those shown in Figure 22.17.

Nickel is often used to plate other metals because it is hard, is resistant to corrosion and wear, and takes a high polish. Chrome-plated objects usually have a nickel plating beneath their very thin chromium plating. Nickel is a component of most stainless steels. Other alloys include brass, nickel silver (Cu with 18% Ni and 18% Zn), nichrome (Ni with 20% Cr), German silver (Cu, Ni, and Zn), and coin silver. Most ''silver'' coins, such as dimes and quarters, are actually copper–nickel alloys containing little if any silver, and most nickel coins are also a copper–nickel alloy.

Nickel is similar to iron, cobalt, and copper in its reaction chemistry. It is less reactive than iron and about as reactive as cobalt. Like iron, it reacts with dilute acids but is inactivated by nitric acid. Unless powdered or heated to high temperatures, it does not react with oxygen in air or with steam. The halogens form Ni(II) halides. Fluorine, however, reacts with nickel to form an adherent protective coating, so it does not react further; nickel containers are commonly used to hold fluorine.

Most nickel compounds contain Ni(II), but other oxidation numbers are also known. Ni(IV) is found in the complex ion $[NiF_6]^{2-}$, which decomposes in water to give Ni(II) and O_2, as does the Ni(III) complex ion $[NiF_6]^{3-}$. Ni(I) is found in complexes such as $K_2[Ni_2(CN)_4]$. Several nickel(0) complexes are known, including $[Ni(CO)_4]$ and $[Ni(CN)_4]^{4-}$. Nickel(II) forms salts, usually hydrated, with all the common anions, such as green $NiBr_2 \cdot 6H_2O$, bright green $NiCl_2 \cdot 6H_2O$, emerald green $Ni(NO_3)_2 \cdot 6H_2O$, and blue-green $NiSO_4 \cdot 6H_2O$. All contain the pale green complex ion $[Ni(H_2O)_6]^{2+}$. Nickel also forms complex ions with many other ligands, adopting coordination numbers of 4 or 6. Examples are dark blue $[Ni(NH_3)_6]^{2+}$ and bright yellow $[Ni(CN)_4]^{2-}$.

Reactions of Ni(II) generally involve precipitation or complex formation reactions, such as those shown in Figure 22.18. For example, when aqueous ammonia or an alkali-metal hydroxide is added to a solution of $Ni^{2+}(aq)$, Ni(II) hydroxide precipitates as a gelatinous green solid.

$$Ni^{2+}(aq) + 2OH^-(aq) \longrightarrow Ni(OH)_2(s)$$

Excess aqueous ammonia dissolves the precipitate to form a blue complex ion.

$$Ni(OH)_2(s) + 6NH_3(aq) \longrightarrow [Ni(NH_3)_6]^{2+}(aq) + 2OH^-(aq)$$

Figure 22.17 Pellets of nickel produced by the Mond process.

Figure 22.18 When aqueous ammonia is added to a solution of a nickel(II) salt *(Left)*, nickel(II) hydroxide is precipitated *(Center)*. The solid redissolves and $[Ni(NH_3)_6]^{2+}$ is formed when more aqueous ammonia is added *(Right)*.

22.9 The Inner-Transition Metals

Why are the inner-transition elements displayed separately from the other elements on the periodic table?

The inner-transition elements consist of two groups, the lanthanides and the actinides. As we have seen, although these groups are normally displayed in the periodic table separated from the rest of the elements, the lanthanides really belong between lanthanum (La) and hafnium (Hf), and the actinides follow actinium (Ac). They are separated out for two reasons. One is that the periodic table would otherwise be too wide to be displayed easily on a printed page. The other is that many of the properties of the elements in each group are very similar, as though they occupied but a single position in the periodic table. For example, when examining periodicity of properties of the lanthanides, we find that lanthanum serves to represent all these elements. Both the lanthanides and the actinides are filling $(n-2)f$ orbitals and have two electrons in the ns orbital and zero, one, or two electrons in the $(n-1)d$ orbitals. The electronic configurations of these elements are summarized in Table 22.4.

Lanthanides

Why do the lanthanide elements occur together in minerals?

The lanthanide elements are known collectively as the **rare-earth elements.** The rare earths also include yttrium (Y) and lanthanum (La), whose properties are similar to those of the lanthanides. All of these elements except promethium (Pr) occur naturally.

The rare-earth elements all occur together in nature as oxide or oxoanion minerals in which one element can readily substitute for another in the mineral

Table 22.4 Valence Electronic Configurations of the Inner-Transition Metals

Lanthanides			Actinides		
Lanthanum	La	$6s^25d^14f^0$	Actinium	Ac	$7s^26d^15f^0$
Cerium	Ce	$6s^25d^14f^1$	Thorium	Th	$7s^26d^25f^0$
Praseodymium	Pr	$6s^25d^04f^3$	Protactinium	Pa	$7s^26d^15f^2$
Neodymium	Nd	$6s^25d^04f^4$	Uranium	U	$7s^26d^15f^3$
Promethium	Pm	$6s^25d^04f^5$	Neptunium	Np	$7s^26d^15f^4$
Samarium	Sm	$6s^25d^04f^6$	Plutonium	Pu	$7s^26d^05f^6$
Europium	Eu	$6s^25d^04f^7$	Americium	Am	$7s^26d^05f^7$
Gadolinium	Gd	$6s^25d^14f^7$	Curium	Cm	$7s^26d^15f^7$
Terbium	Tb	$6s^25d^04f^9$	Berkelium	Bk	$7s^26d^05f^9$
Dysprosium	Dy	$6s^25d^04f^{10}$	Californium	Cf	$7s^26d^05f^{10}$
Holmium	Ho	$6s^25d^04f^{11}$	Einsteinium	Es	$7s^26d^05f^{11}$
Erbium	Er	$6s^25d^04f^{12}$	Fermium	Fm	$7s^26d^05f^{12}$
Thulium	Tm	$6s^25d^04f^{13}$	Mendelevium	Md	$7s^26d^05f^{13}$
Ytterbium	Yb	$6s^25d^04f^{14}$	Nobelium	No	$7s^26d^05f^{14}$
Lutetium	Lu	$6s^25d^14f^{14}$	Lawrencium	Lw	$7s^26d^15f^{14}$

structure. Thus no specific mineral is associated with any specific rare-earth element. Although the term *rare earth* suggests that all the lanthanide elements are scarce, in fact they are reasonably abundant. The term came to be used because these elements were originally recovered as oxides (or *earths*) from relatively rare minerals. The most abundant of the lanthanides is cerium, which is 28th in abundance in Earth's crust; the least abundant is thulium, which ranks 63rd, ahead of gold, silver, iodine, mercury, and platinum.

The lanthanides are put to a variety of uses in industry and science. In the laser and optical industries, they are used for polishing lenses and as an additive to glass and ceramics. Addition of lanthanides (often as a mixture called mischmetal) to alloys of other metals, including steels, makes the alloys stronger. Mischmetal is also used to make flints for lighters, because it sparks when struck. Lanthanide oxides are used in the coating applied to color television screens, because they give off light when struck by an electron beam. Cerium salts are commonly used to analyze other minerals by means of redox titrations.

Examination of the electronic structures of the lanthanides (and actinides) in Table 22.4 reveals that more than one element in each group has an electronic configuration corresponding to f^0, f^7, and f^{14}, filled or half-filled subshells. As we have seen for the transition elements, such configurations have unusual stability. From our experience with the transition elements, we would expect the lanthanides to lose the $6s$ electrons to form 2+ ions. And we might expect that the elements with an electronic configuration of $6s^2 5d^1 4f^n$ would lose the $6s$ and $5d$ electrons to form 3+ ions. In fact, though, all the lanthanides form 3+ ions in solution and have a characteristic oxidation number of +3 in compounds. Two elements, Ce and Tb, can also be oxidized to the +4 state; Ce^{4+} has a configuration of $6s^0 5d^0 4f^0$, and Tb^{4+} has a configuration of $6s^0 5d^0 4f^7$—corresponding to empty and half-filled subshells, respectively. These ions are very good oxidizing agents. The 3+ ions of Ce and Tb are more stable than the 4+ ions. Praseodymium adopts oxidation number +4 in some compounds but has not been isolated in solution because it is such a strong oxidizing agent that it oxidizes water. Four of the lanthanides can be reduced to oxidation number +2 to form powerful reducing agents. Two of these are europium and ytterbium. Eu^{2+} and Yb^{2+} can be kept for some time in aqueous solution, because they have particularly stable electronic configurations, $6s^0 5d^0 4f^7$ for Eu^{2+} and $6s^0 5d^0 4f^{14}$ for Yb^{2+}. The other two elements that can adopt oxidation number +2 are samarium and thulium. Sm^{2+} and Tm^{2+} are much stronger reducing agents and rapidly reduce water in aqueous solution. They have the less stable electronic configurations $6s^0 5d^0 4f^6$ (Sm^{2+}) and $6s^0 5d^0 4f^{13}$ (Tm^{2+}).

Many of the lanthanide ions are highly colored because of a shift of electrons from one f orbital to another. However, those with no unpaired electrons and those with a stable half-filled shell are colorless. Solutions of these ions are shown in Figure 22.19.

Table 22.5 lists the metallic and ionic radii of the lanthanides. Note that the radii decrease slightly with increasing atomic number. This effect, called the lanthanide contraction, was discussed earlier. The lanthanides are all nearly the same size—somewhat larger than the transition metals and significantly smaller than the alkali metals and alkaline-earth metals. The similarity in size of all the lanthanides explains why they occur together in minerals; they are easily substituted for one another in crystal structures.

The lanthanides are all soft, gray metals. They are reactive, as we would expect of active metals that have low electronegativity. All the lanthanide metals

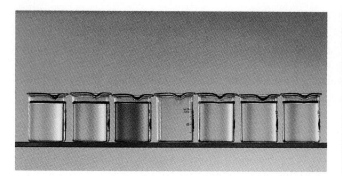 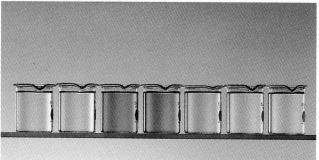

Figure 22.19 Most of the lanthanide(III) ions are colored. The solutions shown are arranged as in the periodic table, from Ce^{3+} *(Left)* to Lu^{3+} *(Right)*. The fourth beaker is empty because all isotopes of Pm are radioactive and therefore are not safe to handle.

Cerium(IV) oxide, CeO_2, is a white or pale yellow powder. It is used for polishing and decolorizing glass, in the preparation of Ce(IV) solutions used as an analytical reagent, and in catalysts.

(symbolized *Ln*) react slowly with oxygen gas at room temperature and more rapidly at high temperatures.

$$4Ln(s) + 3O_2(g) \longrightarrow 2Ln_2O_3(s)$$

Halogens react slowly with the lanthanides at room temperature, but the metals in contact with halogens burn at temperatures in excess of 200°C, as shown here for chlorine.

$$2Ln(s) + 3Cl_2(g) \longrightarrow 2LnCl_3(s)$$

Table 22.5 Metallic and Ionic Radii of the Lanthanides

Element	Metallic Radius (pm)	2+ Ionic Radius (pm)	3+ Ionic Radius (pm)	4+ Ionic Radius (pm)
La	188		106	
Ce	183		103	92
Pr	182		101	90
Nd	181		100	
Pm	181		98	
Sm	180	111	96	
Eu	204	109	95	
Gd	180		94	
Tb	178		92	84
Dy	177		91	
Ho	177		89	
Er	176		88	
Tm	175	94	87	
Yb	194	93	86	
Lu	173		85	

Lanthanides react with aqueous acids at room temperature to give 3+ lanthanide ions.

$$2Ln(s) + 6H^+(aq) \longrightarrow 2Ln^{3+}(aq) + 3H_2(g)$$

The lanthanides also form numerous complex ions, especially with polydentate ligands.

Actinides

The actinides can be divided into two groups in terms of their properties. The lighter actinides behave more like transition metals than like lanthanides. For example, thorium is more like hafnium than like cerium. Similarly, protactinium resembles tantalum, and uranium resembles tungsten. The second half of the actinide series has properties parallel to those of the lanthanides. The actinides are all radioactive.

The actinides exhibit a variety of oxidation numbers, in contrast to the overwhelming preference of the lanthanides for the oxidation number +3. Actinium is found as Ac^{3+}, thorium as Th^{4+}, and protactinium as protactinium(V). Uranium exhibits several oxidation numbers; uranium(IV) and uranium(VI) are most important, but uranium(III) and uranium(V) are also known. The most important oxidation numbers for neptunium and plutonium are +4, +5, and +6. Starting with americium and proceeding through the rest of the actinides (where enough element has been isolated to investigate its chemistry), the oxidation number is primarily +3, reflecting the similarity of these elements to the lanthanides. Like the lanthanides, the actinides undergo a decrease in radius with increasing atomic number.

Only a few of the actinides are found as minerals. Uranium and thorium are isolated from natural minerals. Actinium and protactinium are formed by the radioactive decay of uranium, so they are found with uranium in ores such as pitchblende, U_3O_8. The elements following uranium are all synthetic. The various nuclear transformations that give rise to these elements were described in Chapter 20.

Summary

Most transition metals occur in nature as oxide, sulfide, carbonate, or silicate minerals. The metals are recovered from their minerals by four general methods: thermal decomposition, molten salt electrolysis, chemical reduction, and roasting followed by reduction. The transition metals all have similar properties; they are hard, dense, and have high melting points. Most transition metals are paramagnetic, but three—Fe, Co, and Ni—are ferromagnetic. Atomic and ionic radii decrease slightly across the periodic table, become constant in the middle of the period, and then increase slightly. The size increases slightly down a group, but the fifth- and sixth-period elements usually have very similar sizes. The number of observed oxidation numbers increases from the left-hand side of the transition-metal groups to the center and then decreases toward the right-hand side. The magnitude of the maximum value of the oxidation number for the transition metals similarly increases from Sc to Mn and then decreases from Fe to Zn. The transition metals form many coordination compounds, which usually have a coordination number of 4 or 6.

The inner-transition metals consist of the lanthanides and the actinides, which are found together in a variety of ores. Isolation of the elements is difficult because they are similar in size and other properties. With a few exceptions, the lanthanides form only +3 ions. The radii of these elements decrease slightly with increasing atomic number. This effect is called the lanthanide contraction. The lighter actinides behave like transition metals, whereas the heavier actinides behave like lanthanides.

Key Terms

ferromagnetism (22.1)   lanthanide contraction (22.1)   rare-earth metals (22.9)

Exercises

Properties of Transition Metals

22.1   Which elements form the first transition series?

22.2   What general methods are used to obtain the transition metals from natural sources?

22.3   What characteristics do the transition metals have in common?

22.4   Discuss the magnetic properties of the transition metals.

22.5   What is the lanthanide contraction?

22.6   Arrange the following elements in order of increasing atomic radius: Co, Cr, Cu, Mn, Ti, V.

22.7   State the maximum known oxidation number for each of the following elements.

   a. Mn   f. Sc
   b. Fe   g. Cu
   c. Cr   h. Ni
   d. Co   i. Au
   e. V   j. Zn

22.8   State the most common oxidation number(s) for each of the following elements.

   a. Mn   b. Fe   c. Cr   d. Co   e. V   f. Sc

22.9   Explain why the properties of the second and third transition-metal series are more similar to one another than to the those of the first transition-metal series.

22.10 Explain the following trend in values of atomic radii: Ti, 132 pm; Zr, 145 pm; Hf, 144 pm.

22.11 Explain the following trend in values of atomic radii: Sc, 144 pm; Ti, 132 pm; V, 122 pm; Cr, 119 pm.

22.12 Explain the differences in the densities of the following elements: Cr, 7.19 g/cm^3; Mo, 10.2 g/cm^3; W, 19.3 g/cm^3.

22.13 Both zinc and calcium ionize to form 2+ ions by losing $4s$ electrons, but calcium is much more reactive than zinc. Explain this difference in reactivity.

Titanium

22.14 List some properties of titanium metal.

22.15 What properties of titanium metal make it useful for building airplanes?

22.16 Write formulas for titanium species that can be found in aqueous solution.

22.17 Write a balanced equation for the reaction that occurs when titanium tetrachloride is used for skywriting.

22.18 Why is $TiCl_4$ a colorless liquid rather than a solid at room temperature?

22.19 Why is it necessary to process or weld titanium metal under an atmosphere of argon?

22.20 Write equations to show that titanium dioxide is amphoteric.

22.21 Describe, with equations, a method for preparing Ti_2O_3 from TiO_2.

22.22 Why is $TiO^{2+}(aq)$ colorless and $Ti^{3+}(aq)$ purple?

Vanadium

22.23 Discuss the uses of vanadium metal and V_2O_5.

22.24 What are the common oxidation numbers found in vanadium compounds and ions?

22.25 Write equations to show that V_2O_5 is amphoteric.

22.26 Write a balanced chemical equation to describe the use of the thermite reaction (with hot aluminum) to convert vanadium(V) oxide to vanadium metal.

22.27 The zinc amalgam reduction of yellow $VO_2^+(aq)$ gives successively the colors green, sky-blue, blue-gray, and violet. Identify the species that cause these colors.

22.28 Indicate whether each of the following aqueous vanadium species is oxidized upon exposure to oxygen gas.

   a. V^{2+} d. V^{3+}
   b. VO^{2+} e. VO_4^{3-}
   c. VO_2^+

22.29 Sulfurous acid, or aqueous sulfur dioxide, reduces vanadium(V) in acidic aqueous solution. What is the vanadium product of this reaction?

22.30 Write balanced equations to indicate two methods for preparing aqueous solutions that contain V^{3+}.

Chromium

22.31 Discuss the major uses of elemental chromium.

22.32 Write formulas for all the oxides of chromium and arrange them in order of increasing acidity.

22.33 Write a formula for the most stable form of Cr(VI) in (a) acidic solution and (b) basic solution.

22.34 A small piece of chromium metal was reacted with 6 M HNO_3. The products were a sky-blue solution and a colorless, flammable gas. When the sky-blue solution reacts with Fe^{3+}, it turns blue-gray. Identify all substances described, and write balanced equations for the reactions.

22.35 Complete and balance the following equations.

   a. $Cr_2O_3(s) + OH^-(aq) \longrightarrow$
   b. $Cr^{2+}(aq) + CrO_4^{2-}(aq) + H^+(aq) \longrightarrow$
   c. $Cr(s) + HCl(aq) \longrightarrow$
   d. $K_2CrO_4(aq) + KI(aq) \longrightarrow$

22.36 Why does chromium react with concentrated hydrochloric acid but not with concentrated nitric acid?

Manganese

22.37 List the common ions of manganese.

22.38 Discuss the oxidation number of manganese in hausmannite, Mn_3O_4.

22.39 Manganese is commonly alloyed with other metals, such as iron. However, it is not much used as a pure metal. Explain.

22.40 Manganese metal reacts with hot water. Write a balanced chemical equation to describe this reaction.

22.41 Oxidation of MnO_2 by a mixture of $KClO_3$ and KOH under proper conditions produces a green solution of the manganate ion, MnO_4^{2-}. When the green solution is acidified, it turns pink and deposits a brown solid. Identify the products of this reaction and write a balanced equation.

22.42 Complete and balance the following equations.
a. $MnO_4^-(aq) + Fe^{2+}(aq) + H^+(aq) \longrightarrow$
b. $MnO_4^-(aq) + [Fe(CN)_6]^{4-}(aq) +$
$$OH^-(aq) \longrightarrow$$
c. $MnO_4^-(aq) + Mn^{2+}(aq) + H^+(aq) \longrightarrow$
d. $MnO_4^{2-}(aq) + H^+(aq) \longrightarrow$

22.43 Write a balanced chemical equation to describe the disproportionation of Mn^{3+} in aqueous solution.

22.44 Identify the usual products of reduction of MnO_4^- in a. acidic solution, b. neutral solution, and c. basic solution.

Iron

22.45 Discuss the reasons why iron is so widely used as a structural metal. What are the disadvantages of using iron as a structural metal?

22.46 What is galvanized iron? Why does it not corrode like ordinary iron?

22.47 What are the main components of steel? Explain the differences between carbon steels and alloy steels.

22.48 Complete and balance the following equations.
a. $Fe(NO_3)_2(aq) + O_2(g) + H^+(aq) \longrightarrow$

b. $Fe(s) + H_2SO_4(aq) \xrightarrow{\Delta}$
c. $Fe(s) + Fe_2(SO_4)_3(aq) \longrightarrow$
d. $FeCl_3(aq) + NaOH(aq) \longrightarrow$
e. $Fe(OH)_3(s) + OH^-(aq) \longrightarrow$

Cobalt

22.49 Discuss the uses of cobalt metal.

22.50 Discuss the coordination number of cobalt(II) in cobalt glass, which is deep blue because of the addition of CoO.

22.51 Which is more stable in aqueous solution, Co^{2+} or Co^{3+}? Explain how the less stable ion decomposes.

22.52 When a reddish-pink solution of $[Co(H_2O)_6]Cl_2$ is heated, it becomes blue. Identify the species that causes the blue color.

22.53 Discuss the relative stabilities of cobalt having oxidation number +2 or +3, when present in water and when present in aqueous ammonia.

22.54 What are the product(s) of the reaction between $AgNO_3(aq)$ and $[Co(NH_3)_5Br]Cl_2$?

Nickel

22.55 Discuss the primary uses of nickel metal.

22.56 What are some common compounds of nickel?

22.57 What is the product of the reaction of aqueous nickel(II) solutions with hydroxide ion?

22.58 What types of reactions are typical of nickel(II)?

22.59 Write balanced chemical equations to indicate the role of $Ni(CO)_4$ in the purification of nickel metal.

22.60 When aqueous ammonia is added to a solution of $NiCl_2$, a pale green precipitate forms. This precipitate redissolves when more aqueous ammonia is added. Write balanced chemical equations to explain these observations.

22.61 Write a balanced chemical equation to describe the reaction between NiS and hot nitric acid, which produces a solution of $NiSO_4$.

Inner-Transition Metals

22.62 Identify the rare-earth elements.

22.63 Explain why the rare-earth elements occur together in nature.

22.64 Compare the common oxidation numbers of the lanthanides and the actinides.

22.65 Explain the meaning of the term *lanthanide contraction*. Give examples of its consequences.

22.66 Discuss the oxidation numbers commonly found for the lanthanides in terms of their electronic configurations.

22.67 Why are cerium and europium relatively easy to separate from the rest of the lanthanide elements?

22.68 Discuss the colors of lanthanide 3+ ions in terms of their electronic configurations.

Additional Exercises

22.69 a. Predict the product of the reaction of chromium metal with fluorine gas.
b. Predict the product of the reaction of vanadium metal with fluorine gas.

22.70 When ammonium dichromate is ignited, a voluminous green powder is produced, along with a colorless, nonflammable, odorless gas and water vapor. Identify the products of the reaction and write a balanced equation to describe the decomposition reaction.

22.71 Adding iron(III) chloride to boiling water produces a reddish-brown precipitate. Identify this precipitate.

22.72 Explain why Mn(II) is a much poorer reducing agent than either Cr(II) or Fe(II).

22.73 Indicate which of the following transition-metal ions you would expect to be colorless and which you would expect to be colored.

 a. Cr^{2+} e. V^{3+}
 b. Fe^{2+} f. Ti^{3+}
 c. Cr^{3+} g. Ti^{4+}
 d. Sc^{3+}

*22.74 A solution of $Fe(H_2O)_6^{3+}$ is pale pink, nearly colorless. When the NCS^- ion is added to this solution, it turns blood red. Explain the difference in the colors of these solutions.

*22.75 Outline, with balanced equations, a procedure for converting chromite, $FeCr_2O_4$, to $K_2Cr_2O_7$.

*22.76 A sample of chromite ore, $FeCr_2O_4$, was heated with concentrated nitric acid. Part of the solution was made basic with concentrated aqueous ammonia, giving a green precipitate. This precipitate was heated in air with potassium carbonate, water was added, and the solution was adjusted to pH 5 and then mixed with aqueous lead(II) nitrate. A yellow precipitate resulted. Identify both the green and the yellow precipitates. Write equations for all the reactions that occurred.

▶ ▶ ▶ Chemistry in Practice

22.77 A newly discovered ore contains streaks of a black material that is found to contain copper and sulfur. The ore is analyzed as follows. A sample of 1.3832 g of crushed ore is treated with $HClO_4$ to dissolve the copper sulfide. The H_2S formed is completely driven off by warming the solution and is bubbled through a solution of $AgNO_3(aq)$. This solution precipitates Ag_2S, which is filtered, dried, and found to weigh 0.2503 g. Inert material from the ore is filtered from the $HClO_4$ solution, which is then diluted to 100.0 mL. A 25.00-mL portion of this solution is purged of air by bubbling N_2 through it. To this solution is added 25.00 mL of 0.09843 M $Cr(ClO_4)_2(aq)$, stored under a nitrogen atmosphere. Copper metal precipitates from the solution, and $Cr^{2+}(aq)$ is converted to $Cr^{3+}(aq)$. The excess $Cr^{2+}(aq)$ is titrated with 0.01217 M $KMnO_4$ solution, requiring 32.14 mL. The copper metal is then filtered from this solution, dried, and found to weigh 0.0321 g.

To process the ore, it will be roasted with air to form copper(II) oxide and sulfur dioxide. The sulfur dioxide will be collected, reacted with air in the presence of a vanadium(V) oxide catalyst to form sulfur trioxide, and reacted with water to form sulfuric acid. The copper(II) oxide will be reacted with hydrogen gas, H_2, to produce copper metal and water. The hydrogen gas will be formed by the reaction between zinc metal and sulfuric acid.

 a. What is the percent composition by mass of the copper sulfide found in the ore?

 b. What is the empirical formula of the copper sulfide in the ore?

 c. What is the percent by mass of copper sulfide in the ore?

 d. What is the percent by mass of copper in the ore?

 e. What mass of sulfuric acid will be produced as a by-product per metric ton (1000 kg) of ore processed?

 f. What volume of hydrogen at STP is needed per metric ton of ore?

 g. What mass of zinc is needed to produce the hydrogen necessary to process 1 metric ton of ore?

 h. Does this process produce sufficient sulfuric acid to be self-maintaining, or will additional sulfuric acid be required?

 i. Write equations to describe all the reactions used in these processes.

Organic Chemistry

Mixtures of acidic potassium dichromate with a primary, secondary, and tertiary alcohol (left to right).

CHAPTER OUTLINE

Encountering Chemistry

▶ ▶ ▶ ▶

An orange solution is added to three beakers containing colorless liquids. The liquid in the first beaker slowly turns blue-gray, the liquid in the second turns green, and the liquid in the third remains orange. The colorless liquids are all alcohols, and the orange solution contains potassium dichromate and sulfuric acid. Why don't these mixtures undergo the same color changes? The answer lies in the structures of the alcohols, which all have the formula C_4H_9OH but differ in the location of the OH group in the molecule.

The alcohol that turned blue-gray is 1-butanol, or butyl alcohol, which is oxidized to butanoic acid, C_3H_7COOH.

$$3CH_3CH_2CH_2CH_2OH + 2Cr_2O_7^{2-}(aq) + 16H^+(aq) \longrightarrow$$
$$3CH_3CH_2CH_2COOH + 4Cr^{3+}(aq) + 11H_2O(l)$$

The blue-gray color arises from the chromium(III) ion, $Cr^{3+}(aq)$.

The alcohol that turned green is 2-butanol, or *sec*-butyl alcohol, which is oxidized to 2-butanone, also called methyl ethyl ketone, $CH_3COCH_2CH_3$.

$$3CH_3\underset{\underset{OH}{|}}{CH}CH_2CH_3 + Cr_2O_7^{2-}(aq) + 8H^+(aq) \longrightarrow 3CH_3\underset{\overset{||}{O}}{C}CH_2CH_3 + 2Cr^{3+}(aq) + 7H_2O(l)$$

The green color arises from a mixture of orange $Cr_2O_7^{2-}(aq)$ and blue $Cr^{3+}(aq)$.

The alcohol that remained orange is 2-methyl-2-propanol, or *tert*-butyl alcohol, which did not react at all. The orange color is due to unreacted $Cr_2O_7^{2-}(aq)$. The structure of *tert*-butyl alcohol is

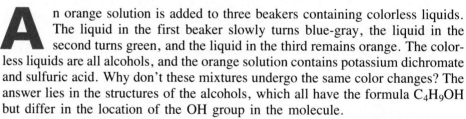

We have just seen three types of compounds (alcohols, carboxylic acids, and ketones) that are among those carbon-containing compounds classified as organic. This chapter considers the chemistry of organic compounds—their names, their classes, their properties, and their reactions.

23.1 Carbon Chemistry

Simple inorganic compounds of carbon are discussed in Section 21.6.

The element carbon is unique. In combination with about half a dozen other elements (primarily H, O, N, and S), it forms several million known compounds. The number of carbon compounds far exceeds that of any other element except hydrogen.

The chemistry of carbon can be considered in terms of three classes of materials. One is elemental carbon, which exists as diamond, graphite, and buckminsterfullerene. A second consists of the simple inorganic compounds of carbon which are much like those of other nonmetallic elements. These include carbon monoxide, carbon dioxide, and the carbides, carbonates, cyanides, and cyanates. The final class of carbon species is made up of the *organic compounds,* which contain carbon and hydrogen and often have other elements replacing one or more hydrogen atoms. ▶ Examples include the alcohols described in the chapter introduction. The study of organic compounds is called **organic chemistry.**

The term *organic* reflects the importance of carbon compounds in living systems. Before 1828, scientists believed that living systems possessed a vital force necessary to synthesize organic compounds. Thus a distinction was made between organic compounds, which were thought to be produced only by living systems, and inorganic compounds. However, Friedrich Wöhler (1800–1882), a German chemist at the University of Göttingen, proved that this distinction was artificial by synthesizing an organic compound from an inorganic compound. He was attempting to prepare ammonium cyanate, NH_4OCN, by heating together ammonium salts and various metal cyanates. Instead, he synthesized the organic compound urea, NH_2CONH_2, which he later found could be formed by heating ammonium cyanate.

$$NH_4^+OCN^- \xrightarrow{\Delta} NH_2CONH_2$$

This compound was considered organic, because it had previously been isolated from urine and therefore had been produced by a living system.

Urea, H_2NCONH_2, is a colorless solid that decomposes slowly, developing the odor of ammonia. It is used in fertilizers, animal feeds, and dentifrices and in the manufacture of resins and plastics.

Organic Chemistry and Life

Organic compounds need not come from living systems, but they are vital to all life processes. The profusion of carbon compounds and the involvement of carbon in life processes arise from the nature of carbon–carbon bonds. Consider the bond energy for some of the nonmetals bonded to themselves, and the reactivity of these bonds, listed in Table 23.1. This comparison reflects the relative thermodynamic and kinetic stability of compounds that contain these bonds. Of the compounds listed, C_2H_6 has the strongest bond and is the least reactive under conditions prevalent on Earth, such as temperatures of 0–30°C, atmospheric pressure, and the

Table 23.1 Some Bond Energies and Reactivities of Dimeric Hydrides

Bond	Average Bond Energy kJ/mol	Hydride	Reactivity of Hydride
C—C	347	H_3C—CH_3	Nonreactive in air
S—S	214	HS—SH	Extremely reactive
P—P	213	H_2P—PH_2	Spontaneously flammable in air
Si—Si	176	H_3Si—SiH_3	Spontaneously flammable in air
N—N	159	H_2N—NH_2	Extremely reactive
O—O	138	HO—OH	Reactive

presence of oxygen gas and water. Under these conditions, the C—C bond embodies a compromise between stability and reactivity that enables it to form and persist and thus to provide the basis for life. The high energy of the C—C bond gives carbon another quality: the ability to bond to itself to form long chains of carbon atoms. This ability accounts for the large number of organic compounds, which are all based structurally on the hydrocarbons considered in the next section.

23.2   The Hydrocarbons

What different kinds of hydrocarbons are known?

Hydrocarbons—compounds that contain only carbon and hydrogen—make up one of the principal classes of organic compounds and may be considered the basic structural components of most other classes of organic compounds.

Classes of Hydrocarbons

Hydrocarbons differ in the type of C—C bonds they contain and in whether the carbon atoms are arranged as open chains or as closed rings. The division of hydrocarbons into two major classes, the *aliphatics* and the *aromatics,* was described in Section 8.4 and outlined in Figure 8.25.

Aliphatic hydrocarbons are those in which the bonding is localized. These hydrocarbons may contain only carbon–carbon single bonds, as in the *alkanes* (such as ethane, H_3C—CH_3), in which case they are called **saturated hydrocarbons.** Other aliphatic hydrocarbons are **unsaturated hydrocarbons,** which contain one or more double or triple bonds between carbon atoms. Examples include the *alkenes,* such as H_2C=CH_2 (ethene or ethylene), and the *alkynes,* such as HC≡CH (ethyne or acetylene). These molecules are called unsaturated because the pi electrons of the multiple bond can be used to form more sigma bonds by reaction with other substances.

Aromatic hydrocarbons have delocalized bonding that can be explained by resonance, introduced in Section 7.5. This situation arises only in cyclic structures, such as benzene, which formally contain alternating single and double bonds.

Such alternating single and double bonds are said to be *conjugated*. The pi electrons in benzene are delocalized over the entire ring structure, giving the molecule special properties: The C—C bonds are all the same length and, unlike normal double bonds, are very unreactive. The structure is generally more accurately represented with the double-bond character denoted by dashed lines or a circle to indicate that it is delocalized.

There are two types of aliphatic hydrocarbons. The **acyclic aliphatic hydrocarbons** occur in the form of chains of carbon atoms, as in propane, H_3C—CH_2—CH_3, and possibly with side chains as in 4-ethylheptane:

$$\begin{array}{c} CH_2{-}CH_3 \\ | \\ CH_3{-}CH_2{-}CH_2{-}CH{-}CH_2{-}CH_2{-}CH_3 \end{array}$$

Cyclobutane, C_4H_8, is a colorless gas that burns with a luminous flame. It is prepared by the hydrogenation of cyclobutene in the presence of a nickel catalyst.

Cyclic aliphatic hydrocarbons can occur in the form of rings of carbon atoms, as in cyclobutane:

$$\begin{array}{c} H_2C{-}CH_2 \\ | \qquad | \\ H_2C{-}CH_2 \end{array}$$

23.3 Saturated Hydrocarbons

The simplest organic compounds are the saturated hydrocarbons, which consist of the alkanes and cycloalkanes. In this section, we review these compounds, their structures, and their nomenclature.

Alkanes

Alkanes are saturated hydrocarbons with the general formula C_nH_{2n+2}. The alkanes are obtained directly from petroleum or by *cracking* of petroleum, which breaks larger alkanes into smaller ones (and into alkenes). Where the value of n is 1 through 4, the alkanes are gases under normal conditions. Where n equals 5 through 18, they are liquids, and where n is greater than 18, they are solids. The melting points and boiling points increase as the number of carbon atoms increases. This is because the larger size is accompanied by a greater surface area and an increase in intermolecular forces. The shape of the molecule also has an effect on intermolecular forces.

Why does the physical state of the alkanes change as the number of carbon atoms in the molecule increases?

The alkanes are named by use of a prefix denoting the number of carbon atoms in the longest chain, followed by the ending -*ane*. The four lower alkanes—

methane (CH_4), ethane (C_2H_6), propane (C_3H_8), and butane (C_4H_{10})—have special names, whereas the longer alkanes have prefixes derived from the Greek or Latin numbers, such as octane for C_8H_{18}. The names of the first ten alkanes are given in Table 23.2.

Each successive alkane can be considered to be derived by insertion of a —CH_2— group between the carbon and one of the hydrogen atoms in a smaller alkane. For example, insertion of a —CH_2— group into methane gives ethane, CH_3—CH_2—H or CH_3—CH_3. The third alkane is propane, C_3H_8, which has a second —CH_2— group inserted to give CH_3—CH_2—CH_2—H or CH_3—CH_2—CH_3. The next molecule in this series is butane, C_4H_{10}.

$$CH_3—CH_2—CH_2—CH_2—H \quad \text{or} \quad CH_3—CH_2—CH_2—CH_3$$

However, unlike the first three alkanes, this molecule can be viewed as being assembled in two different ways, depending on the C—H bond into which the CH_2 group is inserted. The alternative arrangement has the —CH_2— group attached to the center carbon.

$$
\begin{array}{cc}
\text{H} & \\
| & \\
CH_2 & CH_3 \\
| & | \\
CH_3CH—CH_2—H \quad \text{or} \quad & CH_3—CH—CH_3
\end{array}
$$

Straight-chain alkanes are not really straight, because each carbon is at the center of a tetrahedron. A straight-chain alkane thus has a zig-zag structure. For this reason, some chemists call these alkanes continuous-chain alkanes or unbranched-chain alkanes.

Alkanes in which the carbon atoms form a continuous chain, with no branches, are called *normal alkanes* or *straight-chain alkanes*. The alternative arrangements are called *branched-chain alkanes* and are structural isomers of these compounds. Thus butane can have two structural isomers, which are often called normal butane (or just butane)—an unbranched chain of four carbon atoms—and isobutane—a chain of three carbon atoms with one carbon atom attached as a branch to the middle carbon atom.

The number of isomers increases rapidly as the number of carbon atoms increases. Pentane (C_5H_{12}) has 3 isomers, $C_{10}H_{22}$ has 75 isomers, and $C_{20}H_{42}$ has 366,319 isomers. Because of the large number of possible isomers, alkanes are

Table 23.2 Names of Simple Alkanes

Alkane Formula	Name
CH_4	Methane
C_2H_6	Ethane
C_3H_8	Propane
C_4H_{10}	Butane
C_5H_{12}	Pentane
C_6H_{14}	Hexane
C_7H_{16}	Heptane
C_8H_{18}	Octane
C_9H_{20}	Nonane
$C_{10}H_{22}$	Decane

named in accordance with a systematic nomenclature, the IUPAC (International Union of Pure and Applied Chemistry) nomenclature.

To use it, we start with the names of the groups (**alkyl groups**) derived from the alkanes by removal of one hydrogen atom, with a general formula C_nH_{2n+1}. The names of the alkyl groups are similar to those of the alkanes (which are given in Table 23.2), except that the *-ane* ending is replaced by *-yl*. Thus —CH$_3$ is methyl and —C$_2$H$_5$ is ethyl.

A few rules enable us to achieve consistent nomenclature for the alkanes.

1. Determine the *longest* continuous carbon chain in the hydrocarbon. The name of this alkane constitutes the parent name.
2. Number the carbon atoms in the parent hydrocarbon from the end nearer a branch, so that the branch will have the lowest possible position number.
3. Branches attached to the parent chain are named as alkyl groups.
4. The name consists of the position number, a dash, the branch name, and the straight-chain name. If there is more than one branch, they are named in alphabetical order. If the branches are the same alkyl group, the name is preceded by a prefix (*di-*, *tri-*, etc.), and the position numbers are separated by a comma.

Of course, the rules can be reversed to yield the structural formula from the IUPAC name.

EXAMPLE 23.1

Name isopentane according to the IUPAC nomenclature rules.

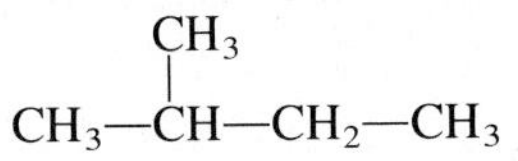

$$CH_3-\overset{\overset{\displaystyle CH_3}{|}}{CH}-CH_2-CH_3$$

Solution

Step 1: The longest continuous carbon chain has four carbons, so the name is based on butane.

Step 2: The carbons in this chain are numbered from the end closest to a branch. The only branch occurs at position 2.

Step 3: The branch occurring at position 2 is CH$_3$, or methyl.

Step 4: The name is 2-methylbutane.

Practice Problem 23.1

Give an IUPAC name for isobutane.

Answer: 2-methylpropane

If the structure contains more than one branch, each is named separately. Consider the molecule

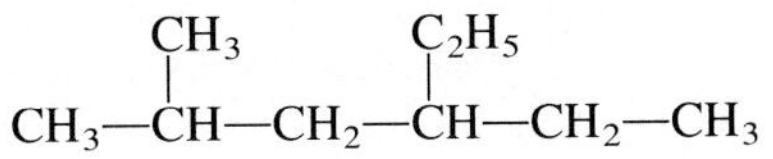

$$CH_3-\overset{\overset{\displaystyle CH_3}{|}}{CH}-CH_2-\overset{\overset{\displaystyle C_2H_5}{|}}{CH}-CH_2-CH_3$$

The longest chain is hexane. A methyl branch occurs at position 2, and an ethyl branch occurs at position 4. The molecule is given the name 4-ethyl-2-methylhexane.

EXAMPLE 23.2

Name the isomer of $C_{11}H_{24}$ that has the structure

$$CH_3—CH_2—\overset{\displaystyle C_2H_5}{\underset{\displaystyle CH_3}{C}}—CH_2—\overset{\displaystyle C_2H_5}{CH}—CH_3$$

Solution   The longest continuous chain has seven carbons, as shown by the following rearrangement of the structure, so the compound is named as a derivative of heptane.

$$CH_3—CH_2—\overset{\displaystyle C_2H_5}{\underset{\displaystyle CH_3}{C}}—CH_2—\overset{\displaystyle CH_3}{CH}—CH_2—CH_3$$

There are two methyl groups, one at carbon 3 and one at carbon 5, and there is one ethyl group at carbon 3. The ethyl group should be named first, because it occurs first alphabetically. The methyl groups can be named together, using the carbon numbers to indicate the position of each. The full name is thus 3-ethyl-3,5-dimethylheptane.

Practice Problem 23.2

Name the isomer of $C_{10}H_{22}$ that has the structure

$$CH_3—\overset{\displaystyle CH_3}{\underset{\displaystyle CH_3}{C}}—CH_2—\overset{\displaystyle C_2H_5}{CH}—CH_2—CH_3$$

Answer: 4-ethyl-2,2-dimethylhexane

Cycloalkanes

Cyclopropane, C_3H_6, is a colorless, flammable gas with a characteristic odor. It is explosive in air. It was once used as an anesthetic.

A cyclic hydrocarbon corresponds to the removal of a hydrogen from each end of a hydrocarbon chain and the bonding together of the two end carbons. Such hydrocarbons are called **cycloalkanes.** The general formula for these molecules is C_nH_{2n}. The simplest cycloalkane is cyclopropane:

$$\begin{array}{c} H_2C — CH_2 \\ \diagdown\diagup \\ CH_2 \end{array}$$

The cycloalkanes are named in the same way as the alkanes but have the prefix *cyclo-*. Many of these molecules (especially those in which the rings are small) are strained, because they cannot adopt the normal tetrahedral bond angles around carbon. Thus they tend to be somewhat more reactive than the alkanes. The most common of the cycloalkanes are cyclopentane and cyclohexane, which are low-boiling liquids often used as solvents.

WORLDS OF CHEMISTRY

Hydrocarbons: An Essential Product for Automobiles

Our automobiles would be completely immobile if hydrocarbons did not exist as liquids.

Hydrocarbons differ in their physical characteristics. For example, as the number of carbon atoms increases, the physical state changes from gas to liquid to solid. The liquid range is observed at common temperatures with hydrocarbons containing from five to eighteen carbon atoms. Our automobiles would be completely immobile if hydrocarbons did not exist as liquids. Gasoline, for example, is made up of the lighter hydrocarbons. Kerosene contains slightly heavier hydrocarbons, and lubricating oil is even heavier. Each of these products is a mixture of hydrocarbons. Their properties depend on the hydrocarbons they contain. An important characteristic of lubricating oil is its viscosity, or resistance to flow. The higher the viscosity, the less readily the oil flows, as shown in Figure 23.1. Oils of widely different viscosities are "tailored" for the different conditions under which they must work. For example, low-viscosity oil is used to lubricate machinery at low temperatures, whereas high-viscosity oil is used at high temperatures. The viscosity of an oil can be expressed in several ways. A common method is that used by the Society of Automotive Engineers (SAE). A typical motor oil for low temperatures is SAE 15W (where W stands for winter); SAE 40 is a summer motor oil. Often oils are made of mixtures of these viscosity ranges so that they include some components that will lubricate when the engine is cold in the winter and others that will lubricate after the engine has heated up. A typical mixture has a viscosity rating of SAE 10W-30.

The viscosity of an oil is determined by the number of carbon atoms in the hydrocarbon and by the relative proportions of straight-chain and branched-chain isomers. The larger the hydrocarbon, the greater the intermolecular forces between adjacent molecules, and the greater the viscosity. It is the intermolecular forces that cause liquids to flow with difficulty. Because intermolecular forces can be overcome to some extent by heating a liquid, the viscosity of a liquid generally decreases at higher temperatures.

A variety of lubricants is needed for automobiles: oil for the engine, oil for the transmission, and grease for the wheels. Lubrication of each part poses different requirements of viscosity and stability when heated. Different hy-

Figure 23.1 Marbles sink more rapidly in motor oil, the less viscous the oil. The viscosities range from SAE 40 (Left) to 5W (Right).

drocarbons can satisfy these various needs. But what happens if (or when) we run out of hydrocarbons with these properties? Chemists have been working for a long time to synthesize motor oils, and various products are already available, though usually at greater cost. Some of these synthetic oils are based on silicones, which contain silicon-oxygen chains and hydrocarbon branches. The ability to prepare desired hydrocarbons or their substitutes will ensure that our automobiles continue to operate, at least for some years to come.

Questions for Discussion

1. Why do motor oils have to be changed periodically?
2. A 5W oil gives about 1% better fuel economy than a 10W oil. What is the connection between oil viscosity and fuel economy?
3. In many areas, used motor oils must be collected for recycling. What has to be done to used oil in order to recycle it?
4. Some synthetic motor oils are based on silicone oils rather than mixtures of hydrocarbons. What advantages do these synthetic oils offer? Do they have any disadvantages?

Cyclohexane, C$_6$H$_{12}$, is a colorless, flammable liquid used as a solvent for lacquers and resins and in paint removers.

A shorthand notation is often used for cyclic hydrocarbons. Rather than writing all the carbon and hydrogen atoms, we draw the structure with lines to denote the bonds between carbons and leave out the atoms themselves. For example, the structure of cyclohexane is abbreviated as

23.4   Unsaturated Hydrocarbons

Some hydrocarbons contain carbon–carbon multiple bonds. In this section, we consider two such classes of hydrocarbons, alkenes and alkynes.

Where do we get unsaturated hydrocarbons?

Alkenes

Alkenes contain at least one C=C double bond and have the general formula C$_n$H$_{2n}$. They are generally obtained from petroleum by the cracking of alkanes, using heat and a catalyst. The simplest such reaction is

$$CH_3\text{—}CH_3(g) \xrightarrow{\Delta} H_2C\text{=}CH_2(g) + H_2(g)$$

The product of this reaction is named ethylene, or ethene.

The names of the alkenes include the alkane root, but the ending *-ene* replaces the alkane ending *-ane*. If there is more than one possible position for a double bond, a number is used to indicate the position. The carbon atoms are numbered from the end nearer the double bond. For the following compound, this simply involves numbering the carbon chain from left to right, which places the C=C bond between carbon atoms 1 and 2.

$$\overset{1}{H_2C}\text{=}\overset{2}{CH}\text{—}\overset{3}{CH_2}\text{—}\overset{4}{CH_3}$$

This molecule has four carbon atoms, so it is derived from butane. It has a double bond, so the *-ane* ending is replaced by *-ene*, giving it the name butene. The position of the double bond is designated by the lowest-numbered carbon involved in that bond, so this molecule is 1-butene. Similarly, the molecule CH$_3$—CH=CH—CH$_3$ is named 2-butene.

The names for side chains on an alkene follow the rules for naming alkanes: The name has the *-yl* ending, and a number designates the position of attachment. Again, numbering starts at the end nearest the double bond.

EXAMPLE 23.3

Name the following compound:

$$CH_3\text{—}\underset{\underset{CH_3}{|}}{CH}\text{—}CH\text{=}CH\text{—}CH_3$$

Solution   The longest chain has five carbons, so the compound is a derivative of pentane, making the name pentene (because of the double bond). The position of the double bond is after the second carbon from the right. Counting from that end, the methyl side chain is attached to the fourth carbon atom. The name is 4-methyl-2-pentene.

Practice Problem 23.3

Name the following compound:

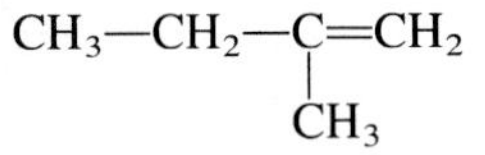

$$CH_3-CH_2-\underset{\underset{\displaystyle CH_3}{|}}{C}=CH_2$$

Answer: 2-methyl-1-butene

It is possible to have geometric isomers in the alkenes, because the double bond restricts rotation around the $C=C$ bond. As a result, the relative positions of side chains attached to the carbons involved in a double bond are fixed. The double bond is formed by the overlap of sp^2 hybrid orbitals on two carbon atoms and by the sideways overlap of the remaining p_z orbitals, as shown in Figure 23.2. Sufficient rotation of one of the carbon atoms with respect to the other would break the pi bond formed by overlap of the p_z orbitals. The bond energy of this pi bond thus acts as an energy barrier to rotation around the bond.

Geometric isomers must have atoms other than hydrogen attached to both of the carbon atoms involved in the double bond. The attached atoms may be on the same side of the molecule (a *cis* isomer) or on opposite sides (a *trans* isomer). This nomenclature is similar to that used for transition-metal coordination compounds, given in Chapter 19. A simple example is 2-butene.

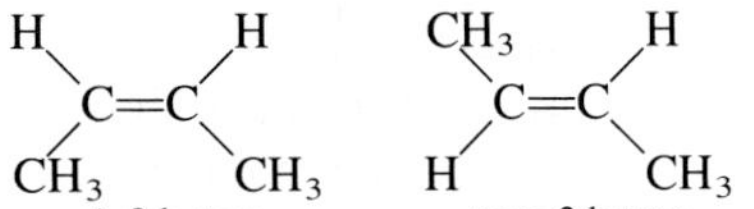

cis-2-butene   *trans*-2-butene

1,3-Butadiene, $CH_2=CHCH=CH_2$, is a colorless gas used in the manufacture of synthetic rubber.

A molecule may, of course, have more than one double bond. In such a case, an appropriate prefix (such as *di-*) is placed just before the *-ene* ending to indicate the number of double bonds. For example, $H_2C=CH-CH=CH_2$ is named 1,3-butadiene.

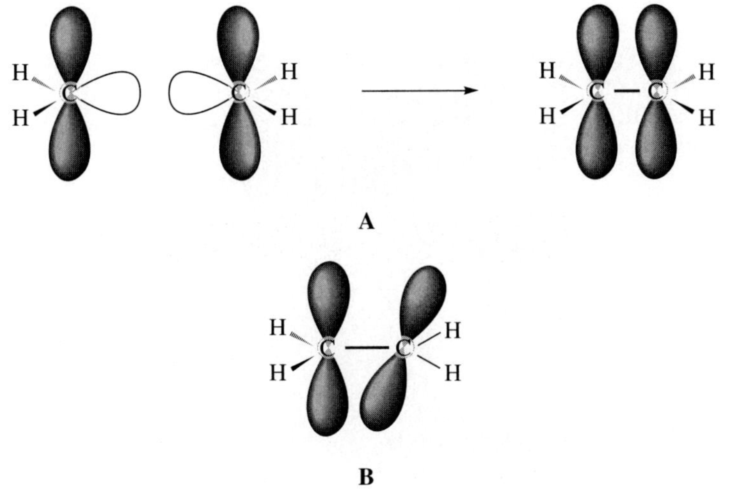

Figure 23.2 *(A)* The overlap of sp^2 orbitals forms a sigma bond between two carbon atoms. The sideways overlap of p_z orbitals forms a pi bond. *(B)* If one carbon atom is rotated with respect to the other, the overlap of the p_z orbitals is decreased. Rotation by as much as 90° breaks the pi bond. Thus molecules that contain a pi bond cannot undergo free internal rotation.

A

B

Figure 23.3 The overlap of *sp* orbitals forms a sigma bond between two carbon atoms. The sideways overlap of two *p* orbitals forms two pi bonds.

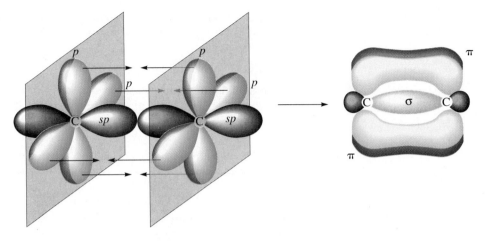

Cyclohexene, C_6H_{10}, is a colorless liquid used as a solvent, in the manufacture of a variety of organic compounds, and as a stabilizer in high-octane gasolines.

Cyclic alkenes are also known. A common example is cyclohexene.

$$
\begin{array}{c}
CH_2 \\
H_2C \qquad CH \\
\qquad \parallel \\
H_2C \qquad CH \\
CH_2
\end{array}
$$

Why do alkynes not have geometric isomers such as were found for alkenes?

Alkynes

Alkynes contain a C≡C triple bond and have the general formula C_nH_{2n-2}. Alkynes are named the same as alkenes but have the ending *-yne*. As shown in Figure 23.3, the triple bond arises from the overlap of an *sp* hybrid orbital to form a sigma bond and from the overlap of two *p* orbitals to form two pi bonds. Alkynes are also called acetylenes, from the common name of the simplest alkyne, HC≡CH. Acetylene (or ethyne) is generally prepared from calcium carbide, as shown in Figure 23.4.

$$CaC_2(s) + 2H_2O(l) \longrightarrow HC \equiv CH(g) + Ca(OH)_2(s)$$

Figure 23.4 Calcium carbide reacts with water to form acetylene, which burns with a sooty flame.

23.5   Aromatic Hydrocarbons

As we noted earlier, aromatic hydrocarbons are cyclic hydrocarbons with conjugated double bonds. The actual structure of aromatic hydrocarbons such as C_6H_6 has the electrons in the pi bonds delocalized over the entire ring system, as shown in Figure 23.5. Although isolated double bonds have a characteristic reactivity, pi bonds with electrons delocalized over the entire ring lack this reactivity. The resulting inertness relative to normal double bonds is referred to as aromatic character. The delocalized bonding in benzene is usually represented by a circle inside the structure.

The aromatic hydrocarbons are usually named as derivatives of the parent ring structure—benzene, in most cases. Many of these molecules have special names as well. The following molecule, for example, is called both methylbenzene and, more commonly, toluene. It has one hydrogen atom on the benzene ring substituted by a methyl group:

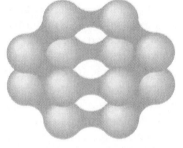

Figure 23.5 The pi electrons in benzene are located in delocalized orbitals that lie above and below the plane of the carbon atoms.

Toluene, $C_6H_5CH_3$, is a flammable, colorless liquid with an odor much like that of benzene. It is used as a solvent and in the manufacture of benzoic acid, benzaldehyde, dyes, explosives, and numerous organic compounds.

An aromatic group formed by removal of a hydrogen from a benzene ring is called an **aryl group,** which is the aromatic analogue of an alkyl group. The aryl group formed from benzene, C_6H_5, is called phenyl. For example, C_6H_5—C≡CH is commonly known as phenylacetylene.

If two positions on a benzene ring have substituents, the positions are indicated by numbers, counted from the location of the first substituent. If two methyl groups are substituents on benzene, the molecule is dimethylbenzene, or xylene. There are three isomers: 1,2-dimethylbenzene, 1,3-dimethylbenzene, and 1,4-dimethylbenzene. These and several other examples are summarized in Figure 23.6.

Figure 23.6 Some substituted benzene molecules.

Why do conjugated double bonds in an aromatic ring lack the reactivity of isolated double bonds?

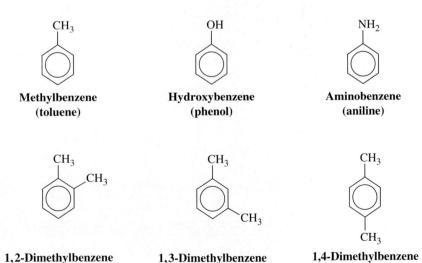

Methylbenzene (toluene)

Hydroxybenzene (phenol)

Aminobenzene (aniline)

1,2-Dimethylbenzene

1,3-Dimethylbenzene

1,4-Dimethylbenzene

23.6 Petroleum

Along with coal, petroleum is the primary source of organic chemicals. The major use of these materials is, of course, energy production, but around 10% of the coal and petroleum recovered is used to make the numerous industrial products that contain carbon.

1-butanethiol, which has a skunk-like odor, is added to natural gas as an odorant so that gas leaks can be detected in the home.

Petroleum is a primary source of carbon-containing compounds that show up in a variety of consumer products such as plastics. Should we continue to use most of our petroleum as a fuel?

Petroleum is generally found in two forms, natural gas and crude oil. Natural gas consists mainly of methane (CH_4), with smaller amounts of ethane (C_2H_6), propane (C_3H_8), and butane (C_4H_{10}). All of these are colorless, odorless, flammable gases. Crude oil is a dark, thick, odoriferous liquid composed of about 95% hydrocarbons, including alkanes, cycloalkanes, and aromatics.

The processing of petroleum takes place in a refinery and involves just a few basic processes. The first step is **fractional distillation,** or the separation of the petroleum into fractions. These fractions are then used directly, purified further, or converted into more desirable products. Fractional distillation is carried out in a *pipe still,* where the crude petroleum is heated in a furnace and fed under pressure through pipes to a *fractionating tower* (Figure 23.7). The fractionating tower is a vertical steel cylinder, usually 15 to 20 stories tall. The heated gaseous petroleum rises up the tower (Figure 23.8). The temperature decreases up the tower, so the lowest-boiling molecules rise higher in the tower than the higher-boiling molecules. Inside the tower, trays are located like floors in a building; these trays contain holes covered with bubble caps. The bubble caps cause the rising vapors to bubble through any condensed hydrocarbons already in the tray, thereby vaporizing any lower-boiling molecules present in the liquid. Overflow pipes return liquid to the next lowest tray, so the higher-boiling fractions work their way down the tower while the lower-boiling fractions work their way up. Heavy oils are drawn off the trays near the bottom of the tower. Lighter products, such as gas oil, kerosene, and naphtha, are removed at the middle and are usually further purified by heating in another tower with steam—a process called stripping. The lighter materials that are stripped off with steam are returned to the fractionating tower. Lighter fractions near the top of the tower are removed as gasoline. Some lighter fractions do not condense in the tower but are removed as vapors and piped to a

Figure 23.7 Fractionating tower.

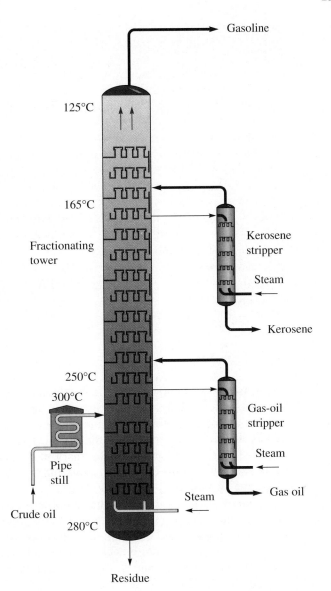

Figure 23.8 Diagram of a pipe still and fractionating tower. From CHEM 13 NEWS, March 1976, page 5, published by the Department of Chemistry, University of Waterloo, Waterloo, Ontario, Canada. Used by permission.

vapor recovery unit, where they are condensed and separated into gasoline, liquefied petroleum (LP), and chemical raw materials for further processing. The various fractions and their boiling ranges are listed in Table 23.3.

Most petroleum is used to make gasoline. Unfortunately, the fraction of petroleum that can be directly separated as gasoline is not sufficient to meet demand, because over half of the alkanes in petroleum have more than 13 carbon atoms, the longest chain found in the gasoline fraction. Larger amounts of gasoline can be obtained from the heavier fractions of petroleum by the process of **catalytic cracking.** In this process, larger alkanes from a higher-boiling fraction are broken or cracked in the presence of a catalyst to form smaller alkanes (and alkenes). In this way, for example, gas oil can be converted into gasoline and other products.

Isooctane, or 2,2,4-trimethylpentane, is a flammable, colorless liquid with the odor of gasoline. It is used as a standard in determining the octane rating of fuels.

$$\underset{\text{alkane}}{C_{14}H_{30}} \xrightarrow{\Delta} \underset{\text{alkene}}{C_6H_{12}} + \underset{\text{alkane}}{C_8H_{18}}$$

Table 23.3 Petroleum Fractions

Fraction	Boiling Range (°C)	Composition	Uses
Natural gas	< 20	CH_4—C_4H_{10}	Fuel, petrochemicals
Petroleum ether	20–60	C_5H_{12}, C_6H_{14}	Solvent
Ligroin, or naphtha	60–100	C_6H_{14}, C_7H_{16}	Solvent, raw material
Gasoline	40–220	C_4H_{10}—$C_{13}H_{28}$, mostly C_6H_{14}—C_8H_{18}	Motor fuel
Kerosene	175–325	C_8H_{18}—$C_{14}H_{30}$	Heating fuel
Gas oil	> 275	$C_{12}H_{26}$—$C_{18}H_{38}$	Diesel and heating fuel
Lubricating oils and greases	Viscous liquids	> $C_{18}H_{38}$	Lubrication
Paraffin	M.p. 50–60	$C_{23}H_{48}$—$C_{29}H_{60}$	Wax products
Asphalt, or petroleum coke	Solids	Residue	Roofing, paving, fuel, reducing agent

Gasoline as it comes from the fractionating tower or from the catalytic crack-ing unit is not suitable for use in motor vehicles because it has a low **octane rating.** The octane rating is a relative scale that measures the tendency of a fuel to knock. Knocking occurs when the gasoline–air mixture in an engine cylinder explodes before it is ignited by the spark plug. The octane rating scale is set at 0 for heptane (which knocks a great deal) and at 100 for isooctane, or 2,2,4-trimeth-ylpentane (which does not knock much).

The presence of chain branches, double bonds, and rings increases the octane number, thereby giving a gasoline that burns more smoothly. A process of **cata-lytic reforming,** carried out in the presence of a catalyst, is used to change straight-chain alkanes into branched-chain and cyclic alkanes and to convert some ring structures into aromatics. Some typical reforming reactions are

$$CH_3(CH_2)_8CH_3 \longrightarrow H_3C(CH_2)_6CH(CH_3)_2$$

Tetraethyllead is a colorless liquid once used in large amounts to prevent knocking when gasoline burns in combustion engines.

The octane number can also be increased by the addition of antiknock agents such as tetraethyllead, $Pb(C_2H_5)_4$. This compound usually increases the octane number by 15 to 20 units. However, because of the lead pollution of our air that has been caused by the use of this compound in gasoline, tetraethyllead is being phased out of gasoline. Other additives—such as MTBE, or methyl *tert*-butyl ether, CH_3—O—$C(CH_3)_3$—are used to increase the octane number in unleaded gasolines.

23.7   Functional Groups

How can we use the concept of functional groups to simplify the study of organic chemistry?

The study of organic compounds is considerably simplified by two aspects of organic chemistry: (1) Certain atomic linkages called functional groups give characteristic properties to an organic molecule. (2) When the functional group undergoes reactions, the remainder of the molecule generally remains intact. Here, we briefly describe the common functional groups and give a few examples of each.

A *functional group* is a reactive part of a molecule that undergoes characteristic reactions. The functional groups we have already considered include the $C=C$ double bond and the $C\equiv C$ triple bond. Others, such as —OH, —COOH, and —NH$_2$, contain elements (most commonly oxygen) in addition to carbon and hydrogen. The common functional groups are listed, and their properties summarized, in Table 23.4.

In dealing with the functional groups, it is customary to abbreviate the hydrocarbon part of the molecule as R— (R'— if there are two hydrocarbon parts), which can stand for any alkyl or aryl group. The official nomenclature for these classes of molecules replaces the alkane ending of the hydrocarbon chain(s) with the ending given in Table 23.4. Numbers are used, as necessary, to indicate the position of the functional group, just as they were used to indicate the positions of double bonds.

Alcohols

Ethylene glycol, or 1,2-ethanediol, HO—CH$_2$—CH$_2$—OH, is an alcohol that contains two —OH groups. This colorless liquid is commonly used as an antifreeze in automobile radiators. In a molecule containing more than one —OH group, a numerical prefix (di, tri, etc.) is used before the -ol to indicate the number of groups.

The *alcohols* are organic molecules that contain an —OH group, as in R—OH. Polarity is introduced by the —OH group, and so even the lower-molar-mass alcohols are liquids. The short-chain alcohols are water-soluble, but solubility decreases as the chain length increases. For example, C_4H_9OH is only slightly soluble in water, and alcohols containing five or more carbon atoms are essentially insoluble in water.

The common names of the alcohols are made up of the alkyl group name followed by the word *alcohol*, whereas the accepted IUPAC names use the alkane root and replace the *-ane* ending with *-anol*. Thus CH_3OH is methyl alcohol or methanol, C_2H_5OH is ethyl alcohol or ethanol, and C_3H_7OH is propyl alcohol or propanol.

An —OH group may be attached directly to an aromatic hydrocarbon group, as in C_6H_5—OH. This molecule is known as phenol (from phenyl alcohol). The chemistry of the —OH group is changed greatly when it is directly attached to an aromatic ring, and such compounds are classified as phenols and not as alcohols.

Alcohols can be classified as primary, secondary, or tertiary, depending on the number of carbon atoms attached to the same carbon as the —OH group. For example, CH_3CH_2OH (ethanol) is a primary alcohol, $(CH_3)_2CHOH$ (2-propanol) is a secondary alcohol, and $(CH_3)_3COH$ (*tert*-butanol or 2-methyl-2-propanol) is a tertiary alcohol. ▶ Differences in the reactivity of these types of alcohol were described in the introduction to this chapter.

Alcohols are useful substances, and some of them are among the top fifty chemicals produced in the United States. Alcohols produced in large amounts include methanol, ethylene glycol, phenol, isopropanol, and ethanol.

Methanol is a colorless toxic liquid; even its vapors, if breathed extensively, may cause blindness or death. Methanol is used as a solvent for shellac and various resins and in automobile windshield-washer solvents. It is often added to nonbev-

Name	Formula	Nomenclature	Example	
Alkane	$-\overset{\displaystyle \vert}{\underset{\displaystyle \vert}{C}}-\overset{\displaystyle \vert}{\underset{\displaystyle \vert}{C}}-$	—	CH_3CH_3	Ethane
Alkene	$\overset{\diagup}{\diagdown}C{=}C\overset{\diagdown}{\diagup}$	Replace *-ane* with *-ene*	$CH_2{=}CH_2$	Ethene (ethylene)
Alkyne	$-C{\equiv}C-$	Replace *-ane* with *-yne*	$CH{\equiv}CH$	Ethyne (acetylene)
Alcohol	$-\overset{\displaystyle \vert}{\underset{\displaystyle \vert}{C}}-OH$	Replace *-ane* with *-anol*	CH_3CH_2OH	Ethanol
Aldehyde	$\underset{C\quad H}{\overset{O}{\overset{\|}{C}}}$	Replace *-ane* with *-anal*	CH_3CHO	Ethanal (acetaldehyde)
Ketone	$\underset{C\quad C}{\overset{O}{\overset{\|}{C}}}$	Replace *-ane* with *-anone*	$CH_3\overset{O}{\overset{\|}{C}}CH_3$	Propanone (acetone)
Acid	$-\overset{O}{\overset{\|}{C}}-OH$	Replace *-ane* with *-anoic acid*	$CH_3\overset{O}{\overset{\|}{C}}OH$	Ethanoic acid (acetic acid)
Ether	$-\overset{\displaystyle \vert}{\underset{\displaystyle \vert}{C}}-O-\overset{\displaystyle \vert}{\underset{\displaystyle \vert}{C}}-$	Substituted ether	$CH_3CH_2-O-CH_3$	Ethyl methyl ether
Amine	$-NH_2$	Substituted amine	$CH_3CH_2NH_2$	Ethyl amine
Halide	$-\overset{\displaystyle \vert}{\underset{\displaystyle \vert}{C}}-X$	Halo derivative	CH_3CH_2Cl	Chloroethane

Table 23.4 The Common Functional Groups

erage ethanol to make it unfit for drinking, so that the ethanol can be sold for scientific purposes without the high federal taxes levied on ethanol used in beverages. In this role, methanol is called a *denaturant,* and the modified ethanol is called *denatured alcohol.*

Phenol, C_6H_5-OH, is a colorless solid so corrosive to skin that it causes blisters on contact, but dilute solutions, called carbolic acid, are used as a disinfectant. The major use of phenol is in the preparation of resins, which end up as adhesives, plastics, and fibers. In turn, the largest end use of these resins is in binders for plywood and particle board.

Ethers

Diethyl ether was once commonly used as an anesthetic. It is a colorless, flammable, volatile liquid that boils at 35°C.

The **ethers** are organic molecules that contain an —O— group bonded to two carbon atoms. They are related to the alcohols but have two hydrocarbon groups attached to an oxygen: R—O—R'. The names of the ethers include the names of the alkyl substituents, listed alphabetically, followed by the word *ether*, as in diethyl ether, C_2H_5—O—C_2H_5. The ethers tend to be more volatile than the corresponding alcohols. Replacement of the hydrogen on the oxygen in alcohols with a hydrocarbon in the ethers causes a loss of hydrogen bonding between the molecules, resulting in lower intermolecular forces in the ethers. Ethers are used primarily as solvents.

Aldehydes

The **aldehydes** are oxidation products of the alcohols. They have the general formula R—CHO, in which the hydrogen and the oxygen are both bonded to the carbon, and the oxygen–carbon bond is a double bond.

$$R-\overset{\overset{\textstyle O}{\|}}{C}-H$$

The most common aldehydes are formaldehyde (HCHO) and acetaldehyde (CH_3CHO). Formaldehyde, or methanal, is a colorless gas with an irritating odor. It is quite soluble in water and is commonly sold as a 37% solution known as *formalin*. Formalin causes coagulation of proteins, which makes it useful as a preservative of tissues. Formaldehyde is used with urea and with phenol to form resins, which are employed in adhesives and plastics. Urea–formaldehyde foam, shown in Figure 23.9, was widely used for house insulation until it was banned because of the toxic properties of formaldehyde. Phenol–formaldehyde resins are used as the adhesive in plywood and particle board.

Acetaldehyde, or ethanal, CH_3CHO, is a volatile, colorless liquid that is soluble in water and has a characteristic pungent odor much like that of cut green

Figure 23.9 Polyurethane foams used for insulation are formed by mixing two amber liquids, which react to form a polymer and a gas that causes the foam to expand. In a short time, the foam hardens to form a rigid solid that contains a lot of trapped gas.

apples. It is also toxic and flammable. Acetaldehyde is used in perfumes, flavors, aniline dyes, plastics, and synthetic rubber and as a reducing agent for silvering mirrors (Figure 23.10).

Ketones

Ketones resemble aldehydes in that both contain the carbonyl group, a carbon double-bonded to an oxygen, or $\diagup$C$=$O. In the aldehydes, this group is located on an end carbon. In ketones, however, alkyl groups lie on either side of the C$=$O group:

$$\underset{\displaystyle R-\overset{\textstyle O}{\overset{\|}{C}}-R'}{}$$

The simplest and most important of the ketones is acetone, $(H_3C)_2C$$=$$O$, whose official name is propanone. Acetone is a colorless, highly flammable liquid with a characteristic pungent odor. It is readily soluble in water. Acetone is widely used as a solvent for fats, oils, waxes, resins, rubber, plastics, lacquers, varnishes, and rubber cements, and it is the principal ingredient in fingernail polish remover.

Figure 23.10 In the Tollens test for aldehydes, diamminesilver(I) ion is reduced to metallic silver. Here, the reaction with formaldehyde creates a silver mirror on the flask. Ketones do not react in this way, so the Tollens test can be used to distinguish aldehydes from ketones.

Carboxylic Acids

▶ The oxidation of primary or secondary alcohols with weak oxidizing agents produces aldehydes or ketones, but under stronger oxidizing conditions, aldehydes can be further oxidized to **carboxylic acids,** as described in the chapter introduction. The characteristic functional group is a carbon double-bonded to an oxygen and also bonded to an —OH group, —COOH.

$$\underset{\displaystyle R-\overset{\textstyle O}{\overset{\|}{C}}-OH}{}$$

The simplest carboxylic acid is formic acid (methanoic acid), which was first obtained by the destructive distillation of red ants. It is the chemical agent that causes ant bites and bee stings to be so painful. Formic acid is a colorless, water-soluble liquid with a pungent odor. It is used in dyeing wool, dehairing and tanning hides, electroplating, coagulating rubber latex, and regenerating old rubber.

Like all the carboxylic acids, formic acid forms salts by reaction with metal hydroxides.

$$\text{HCOOH} + \text{NaOH} \longrightarrow \text{HCO}_2\text{Na} + \text{H}_2\text{O}$$

The most common of the carboxylic acids is acetic acid, CH_3COOH, which is produced as vinegar when cider ferments. Fermentation causes the sugar to form ethanol in the presence of catalytic yeast enzymes, and in the presence of air, the ethanol oxidizes further to acetic acid.

$$\text{CH}_3\text{CH}_2\text{OH} + \text{O}_2 \longrightarrow \text{CH}_3\text{CO}_2\text{H} + \text{H}_2\text{O}$$

Pure acetic acid is a liquid that has a pungent odor and burns the skin. It is completely soluble in water and is also an excellent solvent for many organic compounds. Acetic acid is used in the synthesis of rubber and various acetate plastics. It is used as a solvent for gums, resins, and oils and is used in printing calico and dyeing silk.

Esters

Esters are compounds with the general formula R—COO—R'. In these structures, one of the alkyl groups is directly attached to the carbon, the other is bonded to one of the oxygens, and the other oxygen is double-bonded to the carbon.

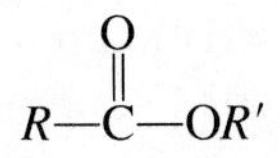

$$
\begin{array}{c}
\quad\ \ \overset{\displaystyle O}{\overset{\displaystyle \|}{}} \\
R-C-OR'
\end{array}
$$

The esters are formed by the reaction of a carboxylic acid with an alcohol. Water is removed in the process, as shown here for the reaction of acetic acid with ethanol.

$$CH_3COOH + HOC_2H_5 \rightleftharpoons CH_3COOC_2H_5 + H_2O$$

These reactions are reversible and usually involve catalysis by an acid. Addition of a base causes the reaction to shift in the reverse direction by neutralizing and thus removing the carboxylic acid. Most esters are volatile liquids with pleasant odors and flavors. In fact, the natural odors of flowers and plants often come from esters. Esters are often the flavoring agents in synthetic and processed foods. An ester that is used medicinally is aspirin—or, more properly, acetylsalicylic acid— $C_6H_4(COOH)(OOCCH_3)$.

Explain how these observations are consistent with Le Chatelier's principle.

The common animal and vegetable oils and fats are esters, mostly based on an alcohol variously called glycerol, glycerine, and 1,2,3-propanetriol. Glycerol has the following structure:

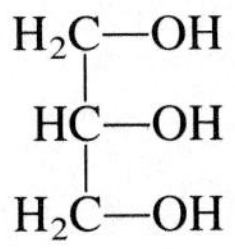

$$
\begin{array}{l}
H_2C-OH \\
HC-OH \\
H_2C-OH
\end{array}
$$

Esters that are formed by reaction of this alcohol with high-molar-mass fatty acids form liquid oils or solid fats, depending on the degree of unsaturation of the acid. Glyceryl tristearate is one such solid ester formed from glycerol and stearic acid, a saturated fatty acid, $C_{17}H_{35}COOH$.

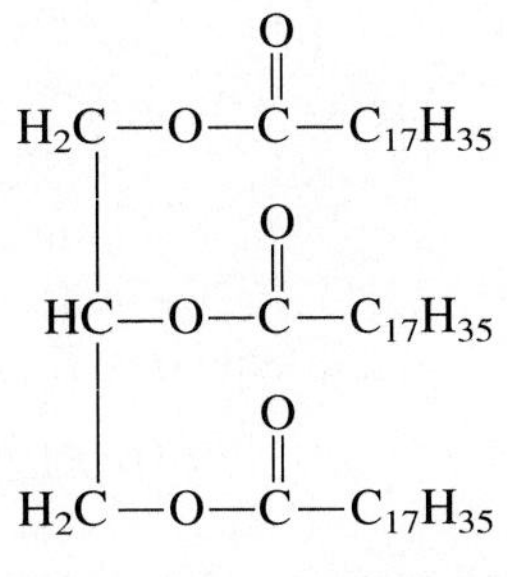

$$
\begin{array}{l}
\qquad\qquad\qquad\ \ O \\
\qquad\qquad\qquad\ \ \| \\
H_2C-O-C-C_{17}H_{35} \\
\ | \qquad\qquad\quad O \\
\ | \qquad\qquad\quad \| \\
HC-O-C-C_{17}H_{35} \\
\ | \qquad\qquad\quad O \\
\ | \qquad\qquad\quad \| \\
H_2C-O-C-C_{17}H_{35}
\end{array}
$$

The sodium salts of these carboxylic acids are used in soaps. The salts have polar ends that attract water and nonpolar ends that attract grease and oil, dispersing these materials into the water. Soap is prepared from animal fats by *saponification*—a reaction with sodium hydroxide (lye) that cleaves the ester linkages and converts the ester to sodium carboxylates and glycerol.

$$C_3H_5(OOCC_{17}H_{35})_3 + 3NaOH \xrightarrow{\Delta} 3NaOOCC_{17}H_{35} + C_3H_5(OH)_3$$

Alkyl Halides

Any nonaromatic saturated hydrocarbon containing a halogen in place of a hydrogen atom is called an **alkyl halide.** There are also aromatic halides, in which halogen atoms substitute for a hydrogen on a benzene ring.

Amines

Amines are the principal examples of organic bases. The amines are themselves based on ammonia, with alkyl groups substituting for one or more of the hydrogen atoms on ammonia. Much like alcohols, amines are classified as primary, secondary, or tertiary, depending on how many carbon atoms are bonded to the nitrogen. In addition, it is possible to make quaternary ammonium compounds such as tetramethyl ammonium chloride, $(CH_3)_4N^+Cl^-$. The covalent substances have a disagreeable fishy smell. There are also aromatic amines, such as aniline, $C_6H_5—NH_2$.

23.8   Reactions of Organic Compounds

Functional groups, as we have seen, are reactive portions of organic molecules that undergo predictable reactions. The reactions of the various functional groups can be broadly classified into a few types: displacement, addition–elimination, and oxidation–reduction. It is useful to review these broad classifications before looking at the reactions of each functional group.

What types of reactions are found for organic compounds?

In a *displacement reaction,* or *substitution reaction,* one functional group is displaced, or replaced, by another functional group.

$$CH_3CH_2OH + HBr \longrightarrow CH_3CH_2Br + H_2O$$

In an *addition reaction,* atoms are added to a functional group, and in an *elimination reaction,* atoms are removed from a functional group. These reactions, then, are the reverse of one another.

$$H_2C{=}CH_2 + Br_2 \underset{\text{elimination}}{\overset{\text{addition}}{\rightleftharpoons}} BrH_2C—CH_2Br$$

In *oxidation,* a functional group or an atom within a functional group attains a higher oxidation number, usually by addition of oxygen or removal of hydrogen. In *reduction,* the functional group achieves a lower oxidation number. ▶ An example of oxidation, discussed in the chapter introduction and shown in Figure 23.11, is the reaction of methanol (or of other alcohols) with dichromate ion in acidic aqueous solution.

$$3CH_3OH + 2Cr_2O_7{}^{2-} + 16H^+ \longrightarrow 3HCOOH + 4Cr^{3+} + 11H_2O$$

EXAMPLE 23.4

Identify the type of reaction that would lead to the conversion of C_2H_5Br to C_2H_5OH.

Solution   An alkyl halide is converted to an alcohol. The only change is in the functional group, where there is a direct replacement of the bromide group by the —OH group. This is a displacement reaction.

Figure 23.11 Methanol reacts with orange dichromate *(Left)* to form blue-gray chromium(III) and formic acid *(Right).*

Practice Problem 23.4

Identify the type of reaction that would lead to the conversion of $(CH_3)_2C{=}O$ to $(CH_3)_2CHOH$.

Answer: Reduction reaction

Reactions of Functional Groups

Although most organic reactions can be classified into the few types just described, the nature of the reactions varies somewhat for different functional groups. Here, we briefly review the reactions that are characteristic of some common functional groups.

Alkanes As shown in Figure 23.12, alkanes are characterized by chemical inertness. Only under forcing conditions of heat or light do the alkanes react. For

Figure 23.12 Typical reactions of the alkanes.

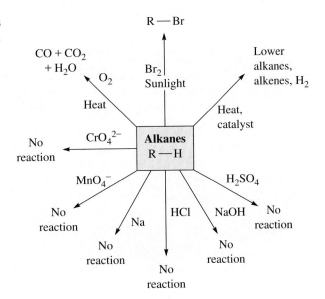

example, alkanes undergo **pyrolysis,** or cracking, when heated in the presence of a catalyst. Such reactions produce shorter-chain alkanes, cycloalkanes, alkenes, and molecular hydrogen.

Alkanes undergo oxidation in exothermic *combustion* reactions, in which they act as fuels and burn in the presence of oxygen gas to form carbon monoxide or carbon dioxide and water.

$$C_3H_8 + 5O_2 \xrightarrow{\Delta} 3CO_2 + 4H_2O$$

In the presence of sunlight or ultraviolet radiation, alkanes undergo **halogenation reactions,** which are essentially displacement reactions. As shown in Figure 23.13, light is needed as a source of energy to break the halogen–halogen bonds and initiate the reaction.

$$CH_4 + Cl_2 \xrightarrow{\text{light}} CH_3Cl + HCl$$

EXAMPLE 23.5

Predict the products of the reaction of propane with bromine in the dark at room temperature.

Solution   Alkanes do not react with bromine unless subjected to heat, sunlight, or ultraviolet radiation. Thus there would be no reaction.

Practice Problem 23.5

Predict the products of the reaction of an excess of butane with bromine in sunlight.

Answer: One or more of the isomers of C_4H_9Br

Alkenes and Alkynes   Alkenes typically undergo addition reactions across the double bond, as shown in Figure 23.14. To occur at an appreciable rate, most of these reactions require a catalyst. Alkynes undergo reactions similar to those of double bonds, as shown in Figure 23.15.

Figure 23.13 Hexane does not react with bromine unless exposed to bright light. *Left:* Mixture in the absence of light. *Right:* Mixture subjected to bright light.

Figure 23.14 Typical reactions of the alkenes.

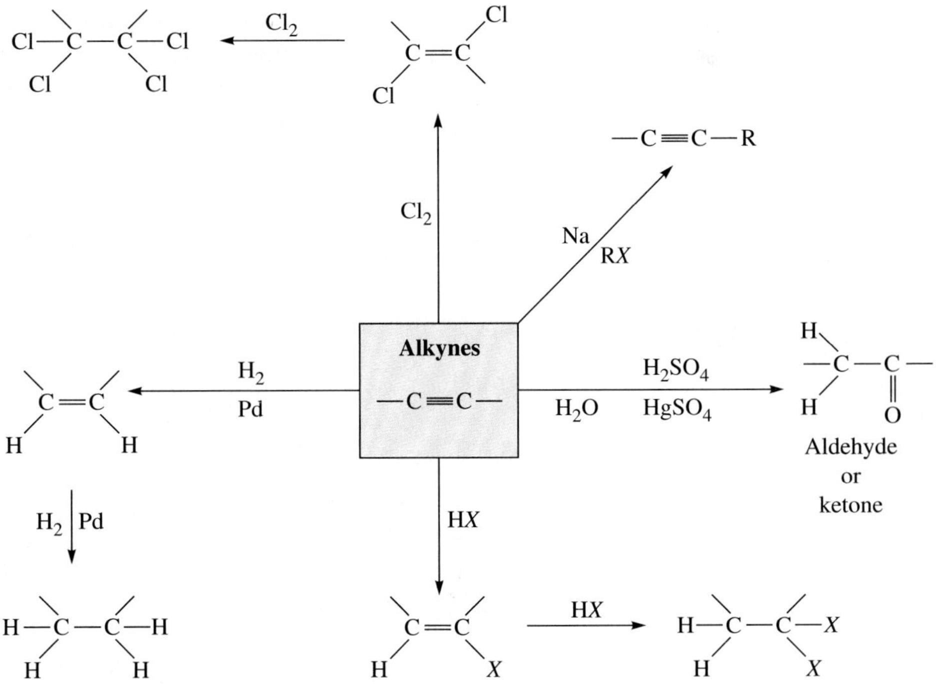

Figure 23.15 Typical reactions of the alkynes.

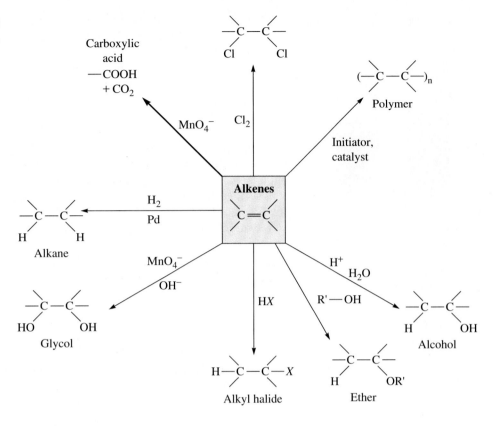

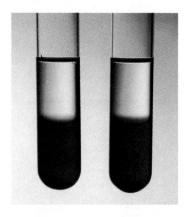

Alkenes can also undergo oxidation reactions with permanganate ion, as shown in Figure 23.16. If the solution is heated, the alkene is oxidized to a carboxylic acid and carbon dioxide. If the solution is cold and is made basic, it forms a glycol in which two —OH groups are added across the double bond.

$$-\underset{\underset{HO}{|}}{C}-\underset{\underset{OH}{|}}{C}-$$

Figure 23.16 Permanganate ion oxidizes cyclohexene *(Left)* but not cyclohexane *(Right).* Permanganate is converted to brown manganese dioxide by the reaction.

Aromatics   Benzene and other aromatic hydrocarbons lack the reactivity of the double bond in alkenes. Normally, benzene does not react at all, but with the right catalysts it can be made to undergo substitution reactions in which some functional group replaces one or more hydrogens on the benzene ring. For example, benzene reacts with chlorine in the presence of $FeCl_3$ to form chlorobenzene.

$$C_6H_6 + Cl_2 \xrightarrow{\text{FeCl}_3} C_6H_5Cl + HCl$$

An important displacement reaction, *alkylation*, occurs between benzene and alkyl halides in the presence of an aluminum halide catalyst. In a typical reaction, benzene reacts with chloromethane to form toluene.

$$C_6H_6 + CH_3Cl \xrightarrow{\text{AlCl}_3} C_6H_5CH_3 + HCl$$

Benzoic acid, C_6H_5COOH, is a colorless solid that sublimes in steam. It occurs naturally in most berries and is used for preserving foods and fruit juices.

Unlike alkenes and alkynes, benzene cannot be oxidized by permanganate or chromate ions. However, an alkyl substituent on a benzene ring—for example, the methyl group in toluene—can be oxidized to a carboxylic acid. For instance, toluene is oxidized to benzoic acid:

$$C_6H_5CH_3 \xrightarrow{\text{MnO}_4^-} C_6H_5COOH$$

Alcohols   Alcohols undergo a wide variety of reactions, some of which are shown in Figure 23.17. Many of these reactions involve the loss of water from the alcohol. Among these reactions is dehydration with sulfuric acid to form an ether.

$$2CH_3-OH \xrightarrow{\text{H}_2\text{SO}_4} CH_3-O-CH_3 + H_2O$$

Figure 23.17 Typical reactions of alcohols.

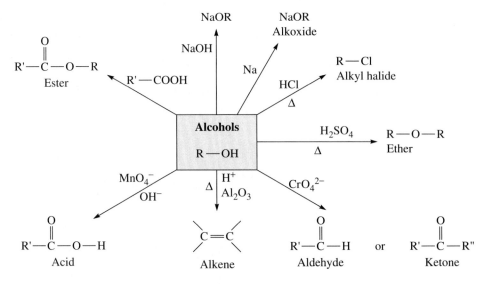

In the presence of alumina, sulfuric acid dehydrates an alcohol to an alkene.

$$CH_3—CH_2—CH_2—OH \xrightarrow{H_2SO_4, \ Al_2O_3} CH_3—CH=CH_2 + H_2O$$

Alcohols react with carboxylic acids to form esters.

$$CH_3OH + C_2H_5COOH \longrightarrow C_2H_5COOCH_3 + H_2O$$

▶ Some alcohols can be oxidized to aldehydes or ketones under mild conditions by permanganate or chromium(VI), as noted in the introduction to this chapter. The oxidation product depends on the type of alcohol. A primary alcohol is oxidized to an aldehyde (or further to a carboxylic acid), a secondary alcohol is oxidized to a ketone, and tertiary alcohols are not oxidized unless extreme conditions, such as high temperature, are used.

EXAMPLE 23.6

Predict the products of the oxidation of $CH_3CH_2CH_2OH$.

Solution   Primary alcohols are oxidized to aldehydes or, under more strongly oxidizing conditions, to carboxylic acids. The product of the oxidation of $CH_3CH_2CH_2OH$, then, is either CH_3CH_2CHO or CH_3CH_2COOH.

Practice Problem 23.6

Predict the products of the oxidation of $(CH_3)_2CHOH$.

Answer: acetone, a ketone: $(CH_3)_2C=O$

Aldehydes and Ketones   Aldehydes and ketones undergo primarily oxidation–reduction reactions at the $C=O$ functional group. Aldehydes can be oxidized to carboxylic acids with very strong oxidizing agents; the reaction resembles the oxidation of alcohols. Aldehydes and ketones can be reduced back to alcohols. The reducing agent is either hydrogen gas, with a palladium or platinum catalyst, or lithium aluminum hydride ($LiAlH_4$) or sodium borohydride ($NaBH_4$), all of which are very powerful reducing and hydrogenation agents.

$$(CH_3)_2C=O + H_2 \xrightarrow{Pd} (CH_3)_2CHOH$$

Another method for converting aldehydes and ketones to alcohols involves a *Grignard reagent*, a magnesium salt containing an alkyl group and a halide, $RMgX$. This salt adds across the carbonyl double bond.

$$CH_3MgCl + \overset{|}{\underset{|}{C}}=O \longrightarrow CH_3—\overset{|}{\underset{|}{C}}—OMgCl$$

Addition of water, usually in acidic solution, decomposes the product magnesium salt to give an alcohol.

$$CH_3—\overset{|}{\underset{|}{C}}—OMgCl + H_2O \longrightarrow CH_3—\overset{|}{\underset{|}{C}}—OH + Mg(OH)Cl$$

Carboxylic Acids and Esters   Carboxylic acids are weak acids, and they undergo partial ionization in aqueous solution to give a carboxylate ion and aqueous hydrogen ions. For example, acetic acid ionizes as follows:

$$CH_3-COOH(aq) \rightleftharpoons CH_3-COO^-(aq) + H^+(aq)$$

Other reactions of carboxylic acids are summarized in Figure 23.18. Analogous to the formation of esters with alcohols is the formation of a peptide bond with organic amines, which is important in the formation of proteins from amino acids.

Esters undergo only a few reactions (Figure 23.18). Among these are hydrolysis to re-form the alcohol and the carboxylic acid from which they were formed, which proceeds to completion in the presence of a hydroxide base.

Alkyl Halides   Most reactions of alkyl halides are displacement reactions in which the halide group is replaced by another group. Some displacement reactions involve reaction with a base, such as sodium ethoxide.

$$CH_3Cl + NaOC_2H_5 \longrightarrow CH_3-O-C_2H_5 + NaCl$$

Reaction of alkyl halides with hydroxide ion can result in two different products. Hydrolysis in basic solution produces alcohols.

$$C_2H_5Cl + OH^- \longrightarrow C_2H_5OH + Cl^-$$

When an alkyl halide is heated with concentrated potassium hydroxide, however, an elimination reaction occurs, producing an alkene.

$$CH_3-CH_2-CH_2-Cl + KOH \overset{\Delta}{\longrightarrow} CH_3-CH=CH_2 + KCl + H_2O$$

The elimination reaction is favored in strongly basic solutions, at high temperatures, and with more alkyl groups bonded to the carbon.

Figure 23.18 Typical reactions of carboxylic acids and esters.

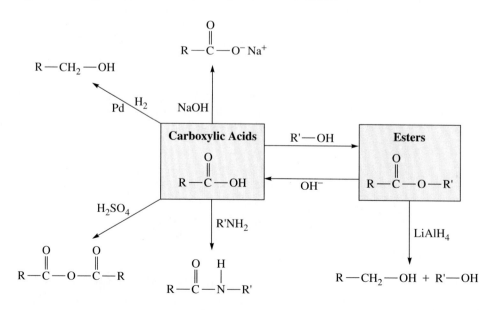

Summary

O rganic compounds contain carbon, hydrogen, and sometimes other elements. Organic compounds are found in great numbers and are involved in all life processes. The extensive occurrence of organic compounds is due to the great strength of carbon–carbon bonds. Hydrocarbons—compounds of carbon and hydrogen—form the basic structural components of organic compounds. Types of hydrocarbons include the alkanes, cycloalkanes, alkenes, cyclic alkenes, alkynes, and aromatic hydrocarbons. Petroleum, a primary source of organic compounds, is found as natural gas and crude oil. Petroleum is refined into useful products by fractional distillation in a pipe still and fractionating tower.

A functional group is a reactive part of a molecule that undergoes characteristic reactions. Various classes of organic compounds that contain different functional groups include alcohols, ethers, aldehydes, ketones, carboxylic acids, esters, alkyl halides, and amines. Reactions of an organic compound are characteristic of the functional groups found in that compound. Most reactions can be classified as displacement (or substitution), addition–elimination, or oxidation–reduction reactions. The nature of these types of reactions varies somewhat for different functional groups.

Key Terms

acyclic
 hydrocarbon (23.2)
aldehyde (23.7)
aliphatic
 hydrocarbon (23.2)
alkyl group (23.3)
alkyl halide (23.7)
amine (23.7)
aromatic

hydrocarbon (23.2)
aryl group (23.5)
carboxylic acid (23.7)
catalytic cracking (23.6)
catalytic
 reforming (23.6)
cycloalkane (23.3)
cyclic
 hydrocarbon (23.2)

ester (23.7)
ether (23.7)
fractional
 distillation (23.6)
halogenation
 reaction (23.8)
ketone (23.7)
octane rating (23.6)
organic

chemistry (23.1)
pyrolysis (23.8)
saturated
 hydrocarbon (23.2)
unsaturated
 hydrocarbon (23.2)

Exercises

Alkanes
23.1 Outline the different types of hydrocarbons.
23.2 What is an alkane?
23.3 Name the following compounds.
   a. $CH_3CH(CH_3)CH_2CH(CH_3)_2$

   b.
$$
\begin{array}{c}
\text{H}_2 \\
\text{C} \\
\diagup \quad \diagdown \\
\text{H}_3\text{C}-\text{CH} \quad \text{HC}-\text{CH}_3 \\
| \qquad\qquad | \\
\text{H}_2\text{C}----\text{CH}_2
\end{array}
$$

   c. $(CH_3)_3C-C(CH_3)_3$
23.4 Write structural formulas for the following compounds.
   a. isobutane
   b. 3,3-diethyl-2-methylpentane
   c. 1,2-dimethylcyclopropane
   d. 2,4-dimethyloctane
23.5 Complete the following reactions.

   a. $CH_3CH_2CH_3 + Cl_2 \xrightarrow{\Delta}$

   b. $(CH_3)_3CCH_2CH_2CH_3 \xrightarrow{\Delta}$

   c. $CH_3CH_3 + Br_2 \xrightarrow{light}$

   d. $CH_4 + O_2 \xrightarrow{\Delta}$

   e. $CH_3CH_2CH_3 + MnO_4^- \xrightarrow{OH^-}$
23.6 Identify the three isomers of C_5H_{12}.

Alkenes and Alkynes
23.7 Distinguish between alkenes and alkynes.
23.8 Draw the structures and name the isomers that might be formed when bromine reacts with acetylene.
23.9 Name the following compounds.
   a. $CH_3CH_2-C\equiv C-CH(CH_3)CH_2CH_3$

   b.
$$
\begin{array}{c}
\text{CH}_3 \qquad\quad \text{CH}_3 \\
\diagdown \qquad \diagup \\
\text{C}=\text{C} \\
\diagup \qquad \diagdown \\
\text{CH}_3 \qquad\quad \text{C}_3\text{H}_7
\end{array}
$$

   c. $(CH_3)_2C=CHCH_3$
   d. $CH_3C\equiv CC(CH_3)_3$
23.10 Write structural formulas for the following compounds.
   a. 3-methyl-1-butene
   b. propyne
   c. 1,3-hexadiene
   d. 2,3,3-trimethyl-1-hexene

23.11 Complete the following reactions.

a. $CH_3CH{=}CHCH_3 + HBr \longrightarrow$

b. $CH_3CH_2CH{=}CHCH_3 + Br_2 \longrightarrow$

c. $CH_3C{\equiv}CCH_3 + H_2 \xrightarrow{Pd}$

d. $(CH_3)_2CHCH{=}CHCH_3 + KMnO_4 \xrightarrow{\Delta}$

e. $CH_3CH{=}CHCH_3 + H_2O \xrightarrow{H^+}$

f. $CH_3CH{=}CH_2 + MnO_4^- \xrightarrow{OH^-}$

g. $HC{\equiv}CH + CH_3Br + Na \longrightarrow$

23.12 Suggest a simple test to distinguish $CH_3CH_2CH_2CH_3$ from $CH_2{=}CHCH_2CH_3$.

Aromatic Hydrocarbons

23.13 What feature is common to all aromatic hydrocarbons?

23.14 Name the following compounds.

23.15 Write structural formulas for the following compounds.

a. 1,2,4-trimethylbenzene

b. chlorobenzene

c. 4-chlorotoluene

d. 2,5-dichloro-1-ethylbenzene

e. pentachlorobenzoic acid

23.16 Complete the following reactions.

a. $C_6H_6 + Cl_2 \xrightarrow{FeCl_3}$

b. $C_6H_6 + CH_3Cl \xrightarrow{AlCl_3}$

c. $C_6H_6 + HCl \longrightarrow$

d. $C_6H_5CH_3 + MnO_4^- \xrightarrow{\Delta}$

e. $C_6H_5CH_3 + Cl_2 \xrightarrow{light}$

23.17 A compound of formula C_8H_{10} is oxidized with hot permanganate. The product is $1,2\text{-}C_6H_4(COOH)_2$. What is the structure of the compound?

23.18 Suggest a method for preparing each of the following compounds, starting with benzene.

a. C_6H_5COOH b. C_6H_5Cl c. $C_6H_5CH_2CH_3$

Alcohols

23.19 What is an alcohol?

23.20 Name the following compounds.

a. $(CH_3)_2C{=}CHCH_2CH_2OH$

b. $CH_3CH(OH)CH_2CH_3$

c. $HOCH_2CH_2CHOHCH_2CH_3$

d. $(CH_3)_2CHCH_2CH_2CH_2OH$

23.21 Write structural formulas for the following compounds.

a. 3-hexanol

b. 2,2-dimethyl-3-pentanol

c. phenol

23.22 What is the product of oxidation of 2-pentanol with CrO_3?

23.23 Complete the following reactions.

a. $CH_3CH_2CH(OH)CH_2CH_3 + H_2SO_4 \xrightarrow{\Delta}$

b. $CH_3CH_2OH + Na \longrightarrow$

c. $CH_3OH + (CH_3)_2CHCH_2COOH \xrightarrow{H^+}$

d. $CH_3CH_2CH_2OH + MnO_4^- \xrightarrow{OH^-}$

e. $CH_3CH(OH)CH_3 + H_2SO_4 \xrightarrow{Al_2O_3}$

f. $CH_3CH(OH)CH_2CH_3 + CrO_3 \longrightarrow$

23.24 Propose a method to synthesize $CH_3CHClCH_2Cl$ from $CH_3CH_2CH_2OH$.

23.25 Suggest a simple test to distinguish $CH_3CH_2CH{=}CHCH_3$ from $CH_3CH_2CHOHCH_2CH_3$.

23.26 Propose a method to convert cyclohexanol to 1,2-dibromocyclohexane.

23.27 Suggest methods to distinguish cyclohexane from cyclohexanol.

23.28 Classify the following as primary, secondary, or tertiary alcohols.

a. methanol

b. ethanol

c. 1-propanol

d. 2-propanol

e. 2,4-dimethyl-2-hexanol

f. cyclohexanol

Ethers

23.29 How do ethers differ from alcohols?

23.30 Name the following compounds.

a. $CH_3CH_2{-}O{-}C(CH_3)_3$

b. $CH_3{-}O{-}CH_2CH_3$

c. $CH_3CH_2{-}O{-}CH_2CH_3$

23.31 Write structural formulas for the following compounds.

a. ethyl methyl ether

b. ethyl phenyl ether

c. isopropyl methyl ether

23.32 Suggest a method for synthesizing $CH_3{-}O{-}CH_3$.

23.33 Name the ethers that could be formed if a mixture of ethanol and 1-propanol were heated with sulfuric acid.

23.34 Suggest a simple test to distinguish $CH_3CH_2CH_2{-}O{-}CH_3$ from $CH_3CH_2CH_2CH_2OH$.

Aldehydes and Ketones

23.35 What is the difference between an aldehyde and a ketone?

23.36 Name the following compounds.
    a. $CH_3CH_2CH_2CHO$
    b. $CH_3CH_2(CH=O)CH_2CH_3$
    c. $(CH_3)_2C=O$
    d. $H_2C=O$
    e. CH_3CHO

23.37 Write structural formulas for the following compounds.
    a. acetone
    b. propanone
    c. propanal
    d. 2-butenal
    e. ethanal
    f. acetaldehyde

23.38 Complete the following reactions.
    a. $(CH_3)_2C=O + H_2 \xrightarrow{Pd}$
    b. $(CH_3)_2C=O + LiAlH_4 \longrightarrow$
    c. $CH_3CHO + KMnO_4 \xrightarrow{\Delta}$

Carboxylic Acids and Esters

23.39 How does an ester differ from a carboxylic acid?

23.40 Name the following compounds.
    a. CH_3CH_2COOH
    b. $CH_3CH_2COOCH_3$
    c. $HCOOCH_3$
    d. $C_6H_5COOCH_2CH_3$
    e. Cl_3CCOOH
    f. $(CH_3)_2CHCOOH$

23.41 Write structural formulas for the following compounds.
    a. 1-propyl butyrate
    b. pentanoic acid
    c. methyl acetate
    d. formic acid
    e. propanoic acid

23.42 Show the structure and name the ester that is formed in the reaction of 2-propanol with acetic acid.

23.43 Complete the following reactions.
    a. $CH_3COOH + H_2 \xrightarrow{Pd}$
    b. $CH_3COOH + NaOH \longrightarrow$
    c. $CH_3COOH + CH_3CH_2OH \longrightarrow$
    d. $CH_3COOH + H_2SO_4 \longrightarrow$
    e. $CH_3COOCH_2CH_3 + OH^- \longrightarrow$
    f. $CH_3COOCH_2CH_3 + LiAlH_4 \longrightarrow$
    g. $CH_3COOH + CH_3NH_2 \longrightarrow$

23.44 Suggest methods for synthesizing acetic acid, starting from ethylene or from acetylene.

Alkyl Halides

23.45 What are some common alkyl halides?

23.46 Name the following compounds.
    a. $CH_2=CHCH_2Cl$
    b. $Cl_2CHCHBrCH_2CH_2CH_2Br$
    c. $(CH_3)_3CF$

23.47 Write structural formulas for the following compounds.
    a. 1,4-dichlorocyclohexane
    b. 2-chloro-3,3-dimethylpentane
    c. iodobenzene
    d. 1,1,1,2-tetrachloroethane

23.48 Complete these reactions.
    a. $CH_3CH_2CHBrCH_2CH_3 + KOH \xrightarrow{\Delta}$
    b. $CH_3CH_2Br + OH^- \longrightarrow$
    c. $CH_3CH_2Br + NaOCH_3 \longrightarrow$
    d. $CH_3Br + KCN \longrightarrow$

23.49 Give a method for the preparation of the following compounds from appropriate starting materials.
    a. $CH_3CH_2CH_2CHBrCH_2CH_3$
    b. 4-bromo-cyclohexene

Amines

23.50 What is an alkyl amine?

23.51 Name the following compounds.
    a. $(CH_3)_2NH$ c. $(CH_3CH_2)_4N^+Cl^-$
    b. $(CH_3)_2N(C_2H_5)$ d. $C_6H_5NH_2$

23.52 Write structural formulas for the following compounds.
    a. dimethylpropylamine
    b. aniline
    c. tetramethylammonium bromide
    d. dipropylamine
    e. methylethylamine

Additional Exercises

23.53 Give an example of a compound that has each of the following general formulas.
    a. C_nH_{2n-2} b. C_nH_{2n} c. C_nH_{2n+2} d. C_nH_n

23.54 Give the structural formulas of the following compounds.
    a. 2,2,4-trimethylpentane
    b. 3-ethyl-2-pentene
    c. 1,3-butadiene
    d. cyclobutane
    e. chlorobenzene
    f. toluene

23.55 Compare the structures of primary, secondary, and tertiary amines and alcohols.

23.56 Give an example of each of the following classes of
       compounds.
       a. ketone
       b. carboxylic acid
       c. ether
       d. aldehyde
       e. alcohol
       f. alkene
       g. ester
       h. cycloalkane
23.57 Identify the class of each of the following com-
       pounds.
       a. $CH_3CH_2CH_2CH_3$
       b. $(CH_3)_2CH—O—CH_3$
       c. CH_3CH_2COOH
       d. $(CH_3)_2CHCHO$

       e. $(CH_3)_3CCOOCH_2CH_3$
       f. $CH_3CH_2NHCH_3$
       g. $CH_3CHClCH_2CH_2CH_3$
       h. $(CH_3CH_2)_2CO$
23.58 Write structural formulas for four isomers of C_4H_9Cl.
23.59 Suggest a method for synthesizing CH_3CH_3 from
       CH_3CH_2OH.
23.60 Give examples of chlorination reactions that involve
       substitution and addition reactions.
*23.61 Which of the following compounds can exist as *cis*
       and *trans* isomers? Write structural formulas for
       those that can.
       a. propene   d. 2-pentene
       b. 1-butene   e. 2-pentyne
       c. 2-butene

▶ ▶ ▶ Chemistry in Practice

23.62 A colorless liquid that melts at 12.4°C and boils at
       211.9°C is formed from benzaldehyde. Upon ex-
       posure to air for some time, this colorless liquid
       begins to deposit a white solid. Over time, more
       and more of this solid settles to the bottom of the
       bottle. The solid is not soluble in water, but it does
       dissolve in alcohol, ether, acetone, and benzene.
       It also dissolves in dilute sodium hydroxide solu-
       tion, neutralizing the sodium hydroxide. The solid
       melts at 142°C. When suspended in water in a
       graduate cylinder, 4.85 g of the solid increases the
       volume of the water from 18.32 mL to 21.46 mL.
       A solution of 25.00 mL of 0.1005 M NaOH is
       neutralized by 0.3934 g of the solid. Combustion
       of 0.1570 g of the solid produces 0.3089 g CO_2
       and 0.0452 g H_2O. A 0.1227-g sample of the
       solid is oxidized with concentrated HNO_3, and the
       products are treated with excess $AgNO_3$ solution,
       producing 0.1123 g of AgCl precipitate. A 1.55-g
       sample of the solid is dissolved in 20.0 g of ben-
       zene. The solution freezes at 2.97°C. (The normal

       freezing point of benzene is 5.50°C, and its
       freezing-point constant is 5.12°C/kg/mol.)
       a. What is the density of the solid?
       b. What is the percent composition of the
          solid?
       c. What is the empirical formula of the solid?
       d. What is the molar mass of the solid?
       e. How many ionizable hydrogens are con-
          tained in the solid?
       f. Suggest a structural formula for the com-
          pound that is consistent with the available
          data.
       g. What uncertainties exist in your structural
          formula? How could these uncertainties be
          resolved by access to a handbook of molecu-
          lar properties?
       h. What is the colorless liquid originally
          formed from benzaldehyde?
       i. What reaction is forming the white solid
          from the colorless liquid?

Synthetic and Biological Polymers

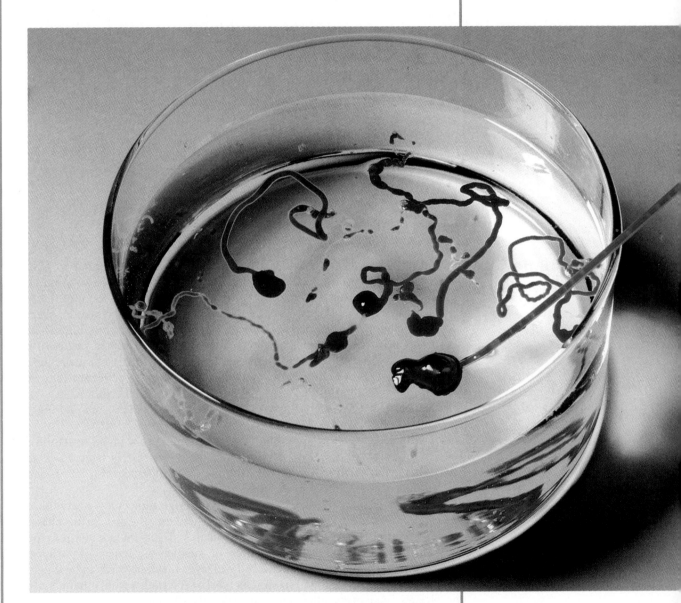

Rayon being produced by the cuprammonium process.

CHAPTER OUTLINE

Encountering Chemistry

Pieces of paper are placed in a blue solution, where they dissolve. This blue solution is squirted through a capillary tube into a colorless solution, and fibrous strands are produced. These strands can be used to make thread and can be woven into clothing—they contain rayon, a popular synthetic fiber. The blue solution contains the complex ion $[Cu(NH_3)_4]^{2+}$ and hydroxide ion, which help break up strong intermolecular forces between the fibers of cellulose in the paper. The colorless solution is dilute sulfuric acid, which neutralizes the basic components of the solution, allowing the cellulose fibers to be reconstituted in the form of rayon. Cellulose is a very large molecule with the formula $(C_6H_{10}O_5)_n$. It is the chief constituent of plant fibers, such as cotton, which is about 90% cellulose.

Large molecules, such as cellulose and proteins, occur in the structural materials of living systems. They have very different properties from those of the atoms, ions, and small molecules we examined in earlier chapters. In this chapter, we consider some of these properties.

Previous chapters have focused on certain large molecules—network solids such as diamond, graphite, and silicon dioxide, and the silicates. In this chapter we examine several types of large organic molecules. These include the **polymers**—giant molecules formed by combination of smaller molecules—that appear in many consumer products. Synthetic and extracted materials such as the huge variety of fibers that can be found in various fabrics, the rubbers and other elastomers, the plastics, and many other modern materials all consist of very large organic molecules.

We also consider the naturally occurring substances of life—proteins, carbohydrates, and nucleic acids, which are also polymers.

24.1 Natural and Synthetic Rubber

In Great Britain, pencil erasers are called rubbers. Rubber derives its name from this use. Joseph Priestley, famous for his work with oxygen, proposed this name when he noted that rubber was a much better material for rubbing out pencil marks than the bread crumbs that were then in use.

Among the consumer materials that are composed of large molecules is rubber. Rubber is used in automobile tires, which consume about 50% of all rubber, both natural (5.3×10^9 kg/y) and synthetic (1.0×10^{10} kg/y). The numerous other articles made of rubber include washers, nautical gear, balloons, artificial organs, and pencil erasers.

Rubber is a curious substance. Natural rubber is obtained from the sap of trees grown in Malaysia, Indonesia, and other Asian countries, as well as in some places in West Africa, South America, and Central America. When dried, this sap coagulates into a tough gum. A distinguishing property of rubber is its resilience. It bounces, and it can be stretched to five or ten times its original length but will return to the original length when released. Similarly, it can be squeezed or twisted and will return to its original shape. No other material behaves this way. Rubber is also flexible, waterproof, and airtight. It is highly resistant to abrasion—more so than any metal, including steel. It is not corroded by most chemicals. It can be bonded to metals, glass, plastics, and fabrics. These properties explain the use of rubber in products ranging from automobile tires to sporting goods.

In addition to its useful properties, rubber has two characteristics that make using it difficult. Rubber is sticky at room temperature and gets soft and stickier at higher temperatures. And at low temperatures, rubber becomes hard and stiff, losing its flexibility.

An early attempt to cope with the stickiness was made in 1823 by Charles Macintosh, a Scot, who sandwiched the sticky rubber between two layers of cotton fabric, producing the material used for the rainwear known as a mackintosh.

In 1839 the problem of stickiness was solved by an American, Charles Goodyear, who discovered the process of *vulcanization,* which consists of heating rubber to 140–150°C with 5–8% sulfur. After vulcanization, rubber loses its stickiness. The vulcanized rubber produced by Goodyear over a hundred years ago is very similar to modern rubber. Vulcanization spawned such an increase in the use of rubber that by the end of the nineteenth century, a worldwide rubber shortage had resulted. Even the establishment of new rubber plantations in Asia could not meet the demand. This led many researchers to focus their efforts on understanding the nature of rubber, with the goal of producing synthetic rubber.

As early as 1836, Michael Faraday, well known for his work in electrolysis, determined that natural rubber has the empirical formula C_5H_8. In 1860 the English chemist C. G. Williams recovered a volatile substance with the formula C_5H_8 by heating rubber to decompose it. He called this substance isoprene, but its IUPAC name is 2-methyl-1,3-butadiene.

$$CH_2 = \underset{\underset{CH_3}{|}}{C} - CH = CH_2$$

How does knowing the composition of rubber help us make synthetic rubber?

Because isoprene can also be prepared from more readily available substances such as turpentine, scientists attempted to recombine isoprene to form synthetic rubber. By 1887 several scientists in France, England, and Germany had converted isoprene back to a rubber-like substance. However, the process took several months and did not achieve complete conversion. Another 30 years passed before the process for synthesizing rubber became sufficiently practical to produce as much as 1800 tons in a year. By the end of World War II, the United States was producing over 700,000 tons of synthetic rubber per year. However, it was not possible to synthesize a rubber exactly like natural rubber until the discovery of an appropriate catalyst by Karl Ziegler of Germany and Guilio Natta of Italy around the middle of the present century.

24.2 Structure and Properties of Polymers

The giant molecules found in substances such as rubber are called *polymers*—from the Greek *poly* ("many") and *mer* ("parts")—because they consist of many connected smaller units. The individual units that combine to make a polymer are called **monomers** ("single parts"). Polymers are often called **plastics** because they can be molded or shaped. ▶ The polymers found in wide variety in consumer products may be natural materials, such as the cellulose used to make rayon as described in the chapter introduction, or they may be materials synthesized by industry in imitation of natural products.

How are polymers synthesized?

The synthesis of rubber and other **elastomers** (elastic polymers) involves building up the polymer by reacting a large number of monomers. Natural rubber consists of long molecules made up of isoprene molecules bonded together.

$$CH_2{=}\overset{\overset{\displaystyle CH_3}{|}}{C}{-}CH{=}CH_2 + CH_2{=}\overset{\overset{\displaystyle CH_3}{|}}{C}{-}CH{=}CH_2 + CH_2{=}\overset{\overset{\displaystyle CH_3}{|}}{C}{-}CH{=}CH_2 + \cdots \longrightarrow$$

$$-CH_2{-}\overset{\overset{\displaystyle CH_3}{|}}{C}{=}CH{-}CH_2{-}CH_2{-}\overset{\overset{\displaystyle CH_3}{|}}{C}{=}CH{-}CH_2{-}CH_2{-}\overset{\overset{\displaystyle CH_3}{|}}{C}{=}CH{-}CH_2{-}\cdots$$

It is impractical to represent a molecule containing 30,000–40,000 monomer units in this manner, however. Instead, we enclose the repeating unit in square brackets and place a subscript n outside the brackets to indicate that the exact number of repeating units is not known. Lines drawn to each end of this shorthand formula indicate that the polymer chain extends in either direction. Although this representation suggests linear chains, polymers actually have three-dimensional structures, which are discussed in more detail later in this section. The polymerization reaction can be represented in a simpler form by using the shorthand formula

$$nCH_2{=}\overset{\overset{\displaystyle CH_3}{|}}{C}{-}CH{=}CH_2 \longrightarrow \left[CH_2{-}\overset{\overset{\displaystyle CH_3}{|}}{C}{=}CH{-}CH_2 \right]_n$$

The difficulty in synthesizing an isoprene rubber with exactly the same properties as natural rubber arises because of the structure of the long rubber molecules. When a monomer, such as isoprene, that contains two double bonds (a diene) combines to form a polymer, one double bond remains in the polymer, which means there can be structural isomers. The long chains of polyisoprene stick together because intermolecular forces exist between them and because they become entangled with one another. The amount of stickiness depends on how the methyl groups are arranged around the double bonds. In natural rubber, the methyl groups are all on the same side of the molecule. Reproducing this arrangement was a major difficulty in synthesizing a natural polyisoprene with the same properties as rubber.

How can we change the stickiness of the synthetic rubber?

The properties of polymers can be adjusted to any desired degree of hardness, strength, rigidity, and plasticity. The identity of the monomer and of the conditions for polymerization dictate the properties of the polymer that results. For example, the stickiness of polyisoprene can be modified by preparing a rubber from monomers with slightly different structures. To see how this happens, let's consider the polymerization of 1,3-butadiene, which has no side chains but is otherwise similar to isoprene. There are two possible configurations about the double bond, as shown in Figure 24.1. These configurations involve either a *cis* or a *trans* arrangement of the hydrogen atoms (or the —CH₂— groups in the chain). The *cis* arrangement, called polybutadiene rubber, gives rubber-like properties

Figure 24.1 Possible configurations of polymerized butadiene.

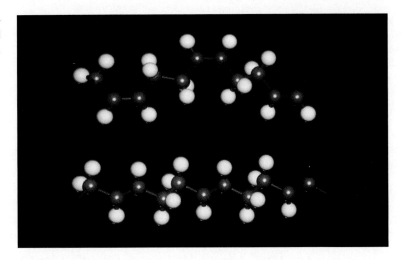

because it has flexibility; if this molecule is pulled, it can stretch. This rubber is used extensively in tires. The *trans* arrangement, called gutta-percha, is already in a somewhat stretched configuration and is more plastic and fibrous. It is used for deep-sea cable insulation, golf ball covers, and packing for root canals. Like rubber, gutta-percha is obtained from an exudate of various trees found in the Malayan Archipelago; it can also be synthesized. Comparison of rubber and gutta-percha indicates that the arrangement of groups around a double bond has a significant effect on the properties of the polymer.

Cross-Linking

Variation in structural isomers in a rubber provides some control over the properties of the rubber. As we have seen, the properties of rubber can also be changed by vulcanization. How does vulcanization change the properties of rubber? Before vulcanization, the large rubber molecules are not connected but interact with one another only by weak van der Waals forces or by physical entanglement. If force is applied, the molecules tend to straighten out, giving the effect of elasticity. If sufficient force is applied to a rubber object, however, the molecules may slide over one another, and the shape of the object can be altered permanently. In the vulcanization process, chains of sulfur atoms bond between polymer molecules, causing the polymer molecules to be permanently fixed in position relative to one another (Figure 24.2). The chains of sulfur atoms that connect the polymer molecules are called *cross-links,* and these cross-links cause major changes in the properties of the polymer. A tangle of polymer molecules becomes a three-dimensional network, with increased rigidity and elasticity, because the molecules no longer can move relative to one another and it is more difficult to straighten out a polymer chain. Thus vulcanized rubber is stronger and more elastic than unvulcanized rubber.

How do chain length and shape affect the properties of polymers?

The properties of polymers can be affected by other means in addition to vulcanization. Linear polymer chains are harder and more brittle if the chains are longer; this is due to increased intermolecular forces and to the physical entanglement of long chains. More force is required to move longer chains relative to one another. Thus polymers that contain shorter chains tend to be softer and more pliable.

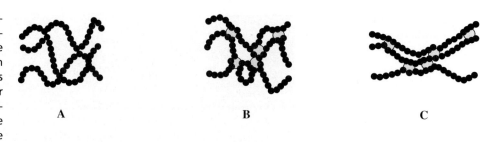

Figure 24.2 (*A*) In unvulcanized rubber, polymer molecules can slip past one another. (*B*) Vulcanization causes chains of sulfur atoms to bond between polymer molecules. (*C*) Vulcanized molecules cannot slide across one another when they are stretched, which makes them stronger and more elastic.

A　　　　　　　　　　B　　　　　　　　　　C

Branched Polymers

The properties of a polymer also vary with the number of side chains. A polymer that has side chains, such as methyl or longer hydrocarbon groups, has a greater tendency to become entangled, because the polymer molecule is not so smooth:

$$\left[CH_2-\underset{\underset{CH_3}{|}}{C}=CH-CH_2\right]_n$$

A polymer with side chains is harder and more rigid than one without side chains. The properties of a polymer can thus be adjusted by forming it from monomers that have side chains of various types. Several polymers with side chains are described in the next section. It is also possible to synthesize a polymer that has branches bonded as a cross-link between polymer chains. Such a network arrangement is harder and more rigid than a polymer that is not cross-linked. The effect is similar to that caused by vulcanization.

Thermoplastic and Thermosetting Polymers

Another property affected by the degree of branching and cross-linking is the reaction to heat. Some polymers, called **thermoplastics,** become soft (or plastic) upon heating and can be reformed into different shapes (Figure 24.3A). These polymers do not have extensive cross-linking or branching.

Upon heating, the intermolecular forces can be overcome easily and the molecules can be moved relative to one another. In contrast, polymers with extensive cross-linking, called **thermosetting polymers,** remain hard when heated (Figure 24.3B). It is not possible to reshape these polymers by heating. To allow movement, it would be necessary to heat them enough to sever the cross-linked chains, and they generally burn before they melt.

Figure 24.3 Thermoplastic and thermosetting polymers. (*A*) Thermoplastic polymers, which have few cross-links, can be molded because they become soft when heated. (*B*) Thermosetting polymers, which have many cross-links, cannot be remelted or reshaped with heat.

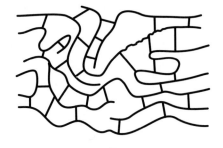

A　　　　　　　　　　　　　　　　B

Copolymers

Another method of controlling the properties of a polymer involves the polymerization of a mixture of two different monomers. Such a polymer is called a **copolymer.** The properties of the copolymer are somewhere between those of polymers generated from the pure monomers. This phenomenon provides another mechanism for controlling the properties of the synthesized polymer. An example is styrene butadiene rubber (or SRB), widely produced for automobile tires, which uses about 25% styrene (phenyl ethene) and about 75% 1,3-butadiene in the polymerization mixture. In the resulting polymer, the 25% styrene units are distributed randomly among the 75% butadiene units that make up the bulk of the molecular chains. This polymer might be represented as

$$\left[(CH_2-CH=CH-CH_2)_x(CH_2-\underset{\underset{C_6H_5}{|}}{CH})_y\right]_n$$

Styrene, $C_6H_5CH=CH_2$, is a colorless to yellowish, oily liquid. On heating to 200°C, it undergoes polymerization to polystyrene, a clear plastic with excellent insulating properties. It is used in the manufacture of plastics, synthetic rubber, resins, and insulation.

where $x = 3$ and $y = 1$. The values of x and y vary with the relative amounts of the two monomers. The more styrene in the mixture, the harder and less plastic the resulting polymer.

24.3 Formation of Polymers

We have seen that it is possible to synthesize a variety of rubbers by selecting the appropriate monomer. Most synthetic rubbers are not identical to natural rubber but are modified as appropriate to give improved properties. An important polymer with rubber-like properties is neoprene, which is polymerized chloroprene.

$$n CH_2=\underset{\underset{chloroprene}{}}{\underset{\underset{Cl}{|}}{C}}-CH=CH_2 \longrightarrow \left[CH_2-\underset{\underset{neoprene}{}}{\underset{\underset{Cl}{|}}{C}}=CH-CH_2\right]_n$$

What types of molecules can be used to form polymers?

Neoprene, a commercial name for the polymer polychloroprene, is more elastic, resists abrasion better, and is less affected by oil and gasoline than natural rubber. It is used in gasoline and oil hoses, auto and refrigerator parts, and electrical insulation. Various other rubbers or elastomers are summarized in Table 24.1.

Not all polymers have the properties of rubbers. Often, other properties are required. To obtain a polymer with the desired properties, different types of organic molecules can be polymerized. The simplest monomer is ethylene, which polymerizes to give polyethylene.

$$n CH_2=CH_2 \longrightarrow [CH_2-CH_2]_n$$

Depending on the reaction conditions and on the catalysts used, this material can be prepared with different numbers of side chains and in different lengths, as shown in the following representation of a branched polymer.

$$x CH_2=CH_2 \longrightarrow \left[CH_2-\underset{\underset{\underset{H}{|}}{\underset{(CH_2)_z}{|}}}{CH}\right]_y$$

Polyethylene is a thermoplastic, so it can be heated to soften and shape it and then cooled to give a plastic substance with the original properties. For example,

Table 24.1 The Major Elastomers

Styrene–butadiene rubber

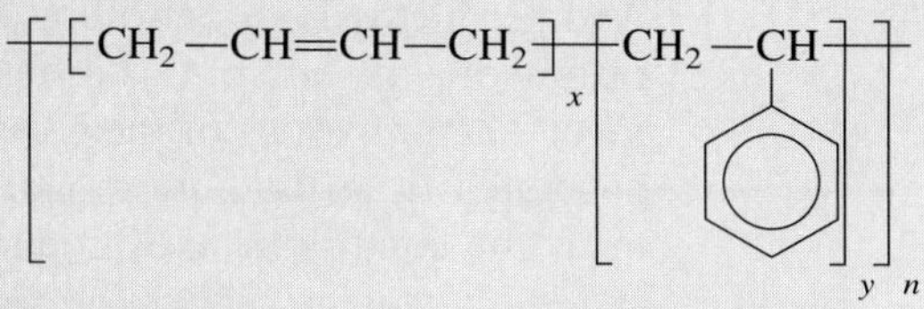

$$\left[\left[CH_2-CH=CH-CH_2\right]_x \left[CH_2-CH\right]_y\right]_n$$

(with phenyl group on the CH)

Polybutadiene rubber

$$\left[CH_2 \begin{array}{c} HC=CH \\ \end{array} CH_2-CH_2 \begin{array}{c} \\ HC=CH \end{array} CH_2-CH_2 \begin{array}{c} HC=CH \\ \end{array} CH_2\right]_n$$

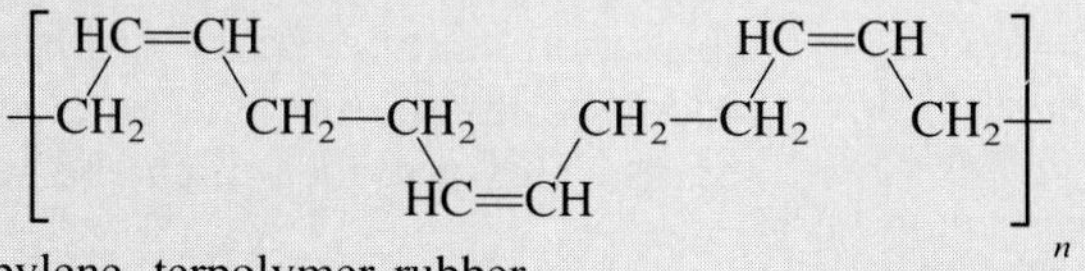

Ethylene–propylene–terpolymer rubber

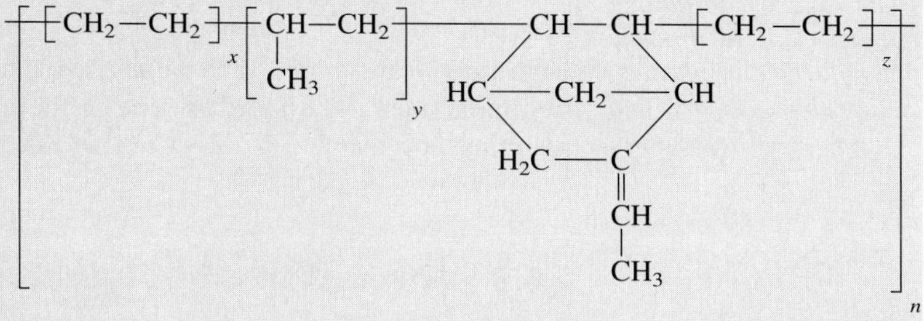

$$\left[\left[CH_2-CH_2\right]_x \left[\begin{array}{c} CH-CH_2 \\ | \\ CH_3 \end{array}\right]_y \begin{array}{c} CH-CH \\ HC-CH_2-CH \\ H_2C-C \\ \| \\ CH \\ | \\ CH_3 \end{array} \left[CH_2-CH_2\right]_z\right]_n$$

Butyl rubber (copolymer of isobutylene and isoprene)

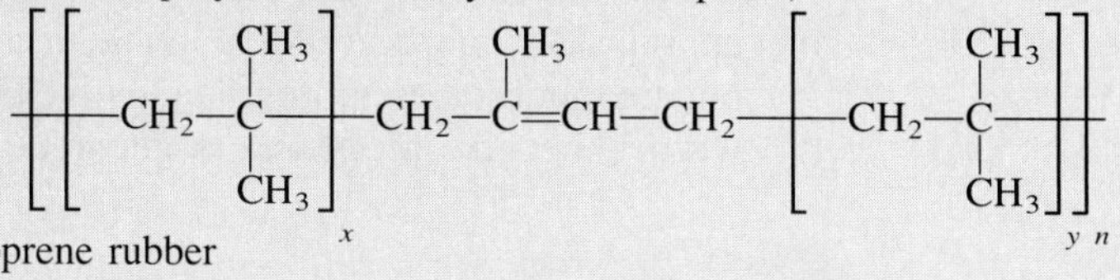

$$\left[\left[\begin{array}{c} CH_3 \\ | \\ -CH_2-C- \\ | \\ CH_3 \end{array}\right]_x \begin{array}{c} CH_3 \\ | \\ -CH_2-C=CH-CH_2- \end{array} \left[\begin{array}{c} CH_3 \\ | \\ -CH_2-C- \\ | \\ CH_3 \end{array}\right]_y\right]_n$$

Polyisoprene rubber

$$\left[\begin{array}{c} CH_3 \\ | \\ -CH_2-C=CH-CH_2- \end{array}\right]_n$$

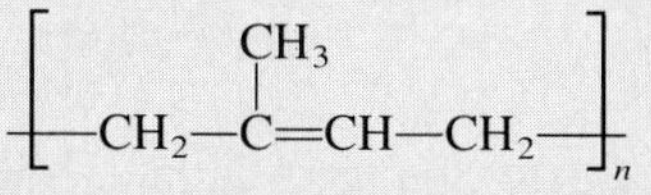

Polychloroprene rubber

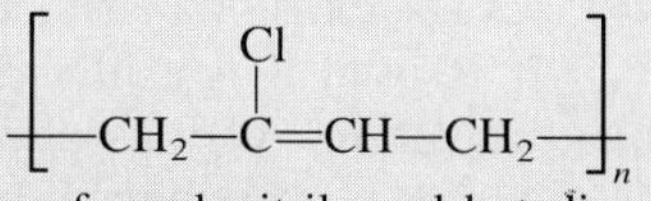

$$\left[\begin{array}{c} Cl \\ | \\ -CH_2-C=CH-CH_2- \end{array}\right]_n$$

Nitrile rubber (copolymer of acrylonitrile and butadiene)

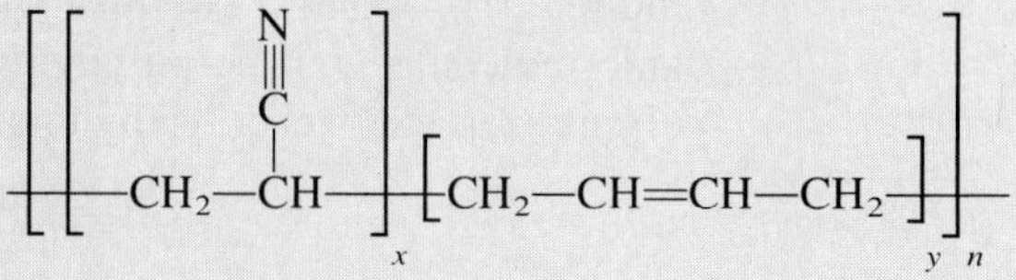

$$\left[\left[\begin{array}{c} N \\ \| \\ C \\ | \\ -CH_2-CH- \end{array}\right]_x \left[-CH_2-CH=CH-CH_2-\right]_y\right]_n$$

polyethylene bags can be heat-sealed. Polyethylene films hold in water vapor but allow the passage of oxygen. Polyethylene is a good electrical insulator. It is commonly used in plastic bottles, bags for fruits and vegetables, plastic wrap, wire coatings, and coatings for milk cartons, among many other uses (Figure 24.4).

There are two types of polyethylene: high-density polyethylene, used in rigid food containers such as milk bottles, and low-density polyethylene, used in plastic produce bags. High-density polyethylene has little branching, making it the more compact and more linear form. It is harder and more rigid because the polymer chains fit closer together, with little unoccupied space to allow flexing of the

Figure 24.4 Many consumer products are made of polyethylene.

molecules. Low-density polyethylene has more extensive branching, resulting in a less compact and less dense structure. The strands are kept apart by the side chains, so this polyethylene has more empty space and is more flexible than the high-density form.

Addition Polymerization

The formation of polyethylene and that of the other polymers discussed so far are examples of **addition polymerization** reactions, in which monomers add to one another without losing any atoms. The polymer is formed by the addition of monomers to the monomer double bond, as shown here for ethylene.

$$CH_2{=}CH_2 + CH_2{=}CH_2 + CH_2{=}CH_2 + \cdots \longrightarrow$$
$$-CH_2-CH_2-CH_2-CH_2-CH_2-CH_2- \cdots$$

Addition polymers require a monomer with at least one double bond.

What is the mechanism of polymerization reactions?

In order for an addition reaction to occur, an *initiator* is required. The initiator is usually a substance that will decompose to form *free radicals*—substances that have an unpaired electron—when exposed to heat or light or perhaps to a catalyst. Common initiators are organic or inorganic species that have a peroxide linkage, such as $R{-}O{-}O{-}R$, where R is an alkyl group. The *initiation step* is the decomposition of this species into free radicals, with an unpaired electron on the oxygen atom.

A common initiator is benzoyl peroxide, $C_6H_5(CO){-}O{-}O{-}(CO)C_6H_5$. This substance is a crystalline solid that may explode when heated. It is used as an oxidizer to bleach oils and flour, as well as in the plastics industry as a polymerization initiator.

$$R{-}O{-}O{-}R \longrightarrow 2R{-}O\cdot$$

The initiator free radical reacts with a molecule of monomer to form another free radical.

$$R{-}O\cdot + CH_2{=}CH_2 \longrightarrow R{-}O{-}CH_2{-}CH_2\cdot$$

The reaction then proceeds to the *propagation step*, which continues the chain growth by reaction between the current form of the free radical with another monomer molecule.

$$R{-}O{-}CH_2{-}CH_2\cdot + CH_2{=}CH_2 \longrightarrow R{-}O{-}CH_2{-}CH_2{-}CH_2{-}CH_2\cdot$$

This step continues the chain growth.

$$R{-}O{-}(CH_2{-}CH_2)_n\cdot + CH_2{=}CH_2 \longrightarrow R{-}O{-}(CH_2{-}CH_2)_{n+1}\cdot$$

Figure 24.5 Polystyrene is used in products ranging from hard plastic objects to styrofoam objects.

Unless the chain reaction is stopped in some manner, this chain building continues until all the monomer is gone. The chain building is stopped by a *termination step,* a reaction in which a free radical is destroyed. For example, one polymer chain can react with another polymer chain, producing a molecule with no unpaired electrons.

$$R—O—(CH_2—CH_2)_n \cdot + \cdot (CH_2—CH_2)_m—O—R \longrightarrow$$
$$R—O—(CH_2—CH_2)_n—(CH_2—CH_2)_m—O—R$$

A variety of other addition polymers is described in Table 24.2. Each of these addition polymers is formed from a substituted ethylene molecule. The substituent ends up as a side chain in the polymer, and its identity has a bearing on the properties and uses of the polymer. Let's consider a few common examples.

Propylene is polymerized similarly to ethylene, but the polymer (*polypropylene*) is harder and less flexible than polyethylene because of its methyl side chains.

$$n CH_2{=}\overset{\overset{\displaystyle CH_3}{|}}{CH} \longrightarrow \left[CH_2{-}\overset{\overset{\displaystyle CH_3}{|}}{CH} \right]_n$$

Polypropylene is translucent, thermoplastic, and chemically inert.

Another related material is the polymer of styrene (phenylethene, $C_6H_5—CH{=}CH_2$).

$$n CH_2{=}\overset{\overset{\displaystyle C_6H_5}{|}}{CH} \longrightarrow \left[CH_2{-}\overset{\overset{\displaystyle C_6H_5}{|}}{CH} \right]_n$$

Polystyrene is used in packaging, food and household containers, combs, coatings for television lead-in wire, and many styrofoam objects, some of which are shown in Figure 24.5. It is rigid, clear, and nontoxic, with good resistance to aqueous solutions. However, it has only limited resistance to organic solvents, as shown in Figure 24.6. It can be prepared in the form of expandable beads, which are used to make styrofoam. The beads are expanded in a mold, under which conditions they adhere to one another and adopt the shape of the mold.

Vinyl chloride polymerizes to give *poly(vinyl chloride),* or *PVC.*

$$n CH_2{=}\overset{\overset{\displaystyle Cl}{|}}{CH} \longrightarrow \left[CH_2{-}\overset{\overset{\displaystyle Cl}{|}}{CH} \right]_n$$

Figure 24.6 The intermolecular forces that hold polystyrene molecules together in styrofoam can be disrupted by solvents such as acetone. Evaporation of the acetone leaves a hard, clear plastic.

Monomer	Polymer	Polymer Name	Some Uses
$CH_2{=}CH_2$	$\left[\begin{array}{c} \text{H} \;\; \text{H} \\ -\text{C}-\text{C}- \\ \text{H} \;\; \text{H} \end{array}\right]_n$	Polyethylene	Plastic bags, bottles, electrical insulation, toys
$CH_2{=}CH-CH_3$	$\left[\begin{array}{c} \text{H} \;\; \text{H} \\ -\text{C}-\text{C}- \\ \text{H} \;\; \text{CH}_3 \end{array}\right]_n$	Polypropylene	Bottles, indoor–outdoor carpeting
$CH_2{=}CH-C_6H_5$	$\left[\begin{array}{c} \text{H} \;\; \text{H} \\ -\text{C}-\text{C}- \\ \text{H} \;\; \text{C}_6\text{H}_5 \end{array}\right]_n$	Polystyrene	Packaging, styrofoam packing materials, cups, insulation
$CH_2{=}CH-Cl$	$\left[\begin{array}{c} \text{H} \;\; \text{H} \\ -\text{C}-\text{C}- \\ \text{H} \;\; \text{Cl} \end{array}\right]_n$	Poly(vinyl chloride), PVC	Plastic wrap, Naugahyde, garden hoses, phonograph records
$CF_2{=}CF_2$	$\left[\begin{array}{c} \text{F} \;\; \text{F} \\ -\text{C}-\text{C}- \\ \text{F} \;\; \text{F} \end{array}\right]_n$	Poly(tetra-fluoroethylene) (*Teflon*)	Nonstick coatings, electrical insulation
$CH_2{=}CH-CN$	$\left[\begin{array}{c} \text{H} \;\; \text{H} \\ -\text{C}-\text{C}- \\ \text{H} \;\; \text{CN} \end{array}\right]_n$	Polyacrylonitrile (*Orlon, Acrilan, Creslan*)	Yarns, wigs, carpets
$CH_2{=}CH-O{-}\overset{\displaystyle O}{\overset{\|}{C}}{-}CH_3$	$\left[\begin{array}{c} \text{H} \;\; \text{H} \\ -\text{C}-\text{C}- \\ \text{H} \;\; \text{O}{-}\text{C}{-}\text{CH}_3 \\ \| \\ \text{O} \end{array}\right]_n$	Poly(vinyl acetate), PVA	Adhesives, paints, chewing-gum resin, textile coatings
$CH_2{=}\overset{\displaystyle CH_3}{\underset{}{C}}{-}\overset{\displaystyle O}{\overset{\|}{C}}{-}O{-}CH_3$	$\left[\begin{array}{c} \text{H} \;\; \text{CH}_3 \\ -\text{C}-\text{C}- \\ \text{H} \;\; \text{C}{-}\text{O}{-}\text{CH}_3 \\ \| \\ \text{O} \end{array}\right]_n$	Poly(methyl methacrylate) (*Lucite, Plexiglass*)	Bowling balls, glass substitutes

Table 24.2 Some Addition Polymers

PVC is used in floor tiles, heat-sealed food wraps, plumbing pipes, phonograph records, wire insulation, wading pools, and rainwear (Figure 24.7). When pressed onto a fabric backing, poly(vinyl chloride) is sold as Naugahyde, a leather substitute. PVC is long-wearing and a good electrical insulator, can be heat-sealed, and resists penetration by water. It is transparent with a slight bluish tint.

Figure 24.7 Numerous consumer products are made of poly(vinyl chloride).

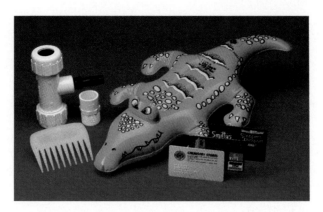

Polyacrylonitrile, a polymer of $CH_2{=}CH{-}CN$, is used to prepare fibers with properties similar to those of wool.

$$n CH_2{=}\underset{\underset{CN}{|}}{CH} \longrightarrow \left[CH_2{-}\underset{\underset{CN}{|}}{CH} \right]_n$$

Acrylonitrile, $CH_2{=}CHCN$, is an explosive, flammable, toxic, colorless liquid. It undergoes polymerization spontaneously when exposed to light or in the presence of concentrated alkali. It is used in acrylic fibers, plastics, surface coatings, and adhesives and as a fumigant for grain.

This polymer is used in knitwear (such as sweaters) and carpeting and is commonly available under the trade names *Orlon* and *Acrilan.*

Addition Copolymers

Addition polymerization can also involve the copolymerization of two monomers. Some examples of addition copolymers are listed in Table 24.3. A common example is *Saran* wrap, which uses chloroethene (vinyl chloride) and 1,1-dichloroethene as monomers.

$$n CH_2{=}CHCl + n CH_2{=}CCl_2 \longrightarrow [CH_2{-}CHCl{-}CH_2{-}CCl_2]_n$$

Table 24.3 Some Addition Copolymers

Monomers	Polymer Name	Some Uses
Vinyl chloride, vinyl acetate	Vinyl, vinylite	Shower curtains, rainwear, phonograph records
Vinyl chloride, vinylidene chloride	Saran	Saran wrap, auto seat cover fibers, pipes
Vinyl chloride, acrylonitrile	Vinyon, Dynel	Clothing fibers
Ethylene, propylene	Ethylene–propylene rubber (EPR)	Tires
Styrene, butadiene	Styrene–butadiene rubber (SBR)	Tires
Isobutylene, isoprene	Butyl rubber	Inner tubes
Acrylonitrile, butadiene, styrene	ABS	Luggage, crash helmets, pipes, battery cases, knobs, handles, telephones

Styrene and butadiene are also commonly copolymerized.

$$n\mathrm{CH_2}{=}\mathrm{CH(C_6H_5)} + n\mathrm{CH_2}{=}\mathrm{CH{-}CH}{=}\dot{\mathrm{C}}\mathrm{H_2} \longrightarrow$$
$$\mathrm{+CH_2{-}CH(C_6H_5){-}CH_2{-}CH}{=}\mathrm{CH{-}CH_2}\mathrm{\}_n}$$

These polymers can contain different proportions of the two monomers, giving styrene–butadiene rubbers (SBR) with different properties. They can also be polymerized in the form of small spheres suspended in water to form a colloidal emulsion. In this form, this polymer is used in latex paints.

Condensation Polymers

In contrast to addition polymers, **condensation polymers** are formed in polymerization reactions during which small molecules are split out of the monomers. These reactions involve two different monomers, and both must have two reactive functional groups. A variety of condensation polymers are listed in Table 24.4.

A familiar condensation polymer is *nylon,* a copolymer of adipic acid and hexamethylenediamine (or 1,6-diaminohexane). It is shown in Figure 24.8.

$$(n+1)\mathrm{HO\overset{O}{\overset{\|}{C}}{-}(CH_2)_4{-}\overset{O}{\overset{\|}{C}}OH} + (n+1)\mathrm{H_2N{-}(CH_2)_6{-}NH_2} \longrightarrow$$

$$\mathrm{HO{-}\overset{O}{\overset{\|}{C}}{-}(CH_2)_4{-}\overset{O}{\overset{\|}{C}}\Big[\overset{H}{\overset{|}{N}}{-}(CH_2)_6{-}\overset{H}{\overset{|}{N}}{-}\overset{O}{\overset{\|}{C}}{-}(CH_2)_4{-}\overset{O}{\overset{\|}{C}}\Big]_n\overset{H}{\overset{|}{N}}{-}(CH_2)_6{-}\overset{H}{\overset{|}{N}}{-}H} + (2n+1)\mathrm{H_2O}$$

Adipic acid, $HOOC(CH_2)_4COOH$, is a colorless solid found in beet juice. It can also be prepared by oxidizing cyclohexanol with nitric acid. It is used in the manufacture of nylon, resins, and urethane foams. It is also an ingredient in some baking powders and lubricating-oil additives.

As noted in Chapter 23, the reaction between an amine and a carboxylic acid to form an amide linkage is among the reactions typical of these functional groups. In this reaction, the —OH group from a carboxylic acid (—COOH) and a hydrogen from an amine group (—NH$_2$) form water. The original functional groups form a carbon–nitrogen bond.

$$\mathrm{{-}\overset{O}{\overset{\|}{C}}{-}OH + H{-}\overset{H}{\overset{|}{N}}{-}} \longrightarrow \mathrm{{-}\overset{O}{\overset{\|}{C}}{-}\overset{H}{\overset{|}{N}}{-}} + \mathrm{H_2O}$$

Polymers formed by the loss of water from a carboxylic acid and an amine are called *polyamides,* a class that includes nylon and proteins. The polyamide formed

Figure 24.8 Nylon results when a dicarboxylic acid reacts with a diamine. The reaction occurs at the interface between the two insoluble liquids. A nylon strand can be pulled from this interface until one of the liquids is completely consumed. A dye has been added to the aqueous layer to help distinguish the layers.

Table 24.4 Some Condensation Polymers

Polyester (from glycol and terephthalic acid)

$$\left[\begin{array}{c} \overset{O}{\overset{\|}{C}}-C_6H_4-\overset{O}{\overset{\|}{C}}-O-CH_2CH_2-O \end{array}\right]_n$$

Acetate (from cellulose and acetic anhydride)

Rayon (reconstituted cellulose)

Spandex (from diisocyanates, polyesters, and glycol)

$$\begin{array}{c} \overset{O}{\overset{\|}{}}\ \overset{H}{\overset{|}{}} \\ [\text{polyester}]\!-\!O\!-\!C\!-\!N\!-\!R \\ N\!-\!H \\ [\text{polyester}]\!-\!O\!-\!C\!=\!O \end{array}$$

Nylon (from diamines and di-carboxylic acids)

$$\left[\begin{array}{c} \overset{O}{\overset{\|}{C}}-(CH_2)_4-\overset{O}{\overset{\|}{C}}-\overset{H}{\overset{|}{N}}-(CH_2)_6-\overset{H}{\overset{|}{N}} \end{array}\right]_n$$

from adipic acid and hexamethylenediamine is called nylon 66, indicating that there are six carbon atoms in each monomer. Nylon is strong, tough, and resistant to abrasion. It is used primarily to make fibers, which end up in hosiery and other clothing, toothbrush bristles, surgical sutures, tennis racket string, fishing line, and automobile tire cord.

Dacron is made from ethylene glycol and terephthalic acid. Recall that an alcohol will react with a carboxylic acid to form an ester. Ethylene glycol has two alcohol functional groups, whereas terepthalic acid has two carboxylic acid groups.

$$(n + 1)HOCH_2CH_2OH + (n + 1)HO\overset{O}{\overset{\|}{C}}-C_6H_4-\overset{O}{\overset{\|}{C}}OH \longrightarrow$$

$$HOCH_2CH_2O\left[\overset{O}{\overset{\|}{C}}-C_6H_4-\overset{O}{\overset{\|}{C}}-O-CH_2CH_2-O\right]_n\overset{O}{\overset{\|}{C}}-C_6H_4\overset{O}{\overset{\|}{C}}OH + (2n + 1)H_2O$$

Terephthalic acid, HOOC—C₆H₄—COOH, is a colorless, crystalline solid that forms polyesters with glycols. These polyesters are used to prepare plastic films and sheets.

This is one of the general class of polymers known as *polyesters. Alkyd* paints, which include most enamel and other high-gloss paints, are based on this type of polymer. Polyesters are used to reinforce fiberglass in boat construction, in fishing rods, and in various household articles.

A similar reaction is used in the formation of *Bakelite*, the hard, brittle plastic once used for telephone bodies and still used for the handles of cooking pots and pans and for electrical insulators. This polymer is formed from phenol, C_6H_5OH, and formaldehyde, $H_2C=O$. However, in this case the reaction occurs in two steps. The first is the substitution of formaldehyde onto the benzene ring.

Phenol, C₆H₅OH, is a colorless, crystalline solid also known as carbolic acid. It is used as a disinfectant, especially for toilets, stables, cesspools, floors, and drains. It is also used to manufacture colorless or light-colored resins.

$$HO-C_6H_5 + H_2C=O \longrightarrow HO-C_6H_4-CH_2-OH$$

This substitution occurs primarily either adjacent to or opposite to the OH group. The —OH group attached to the —CH₂ group can then form water with a hydrogen from the benzene ring on another substituted phenol.

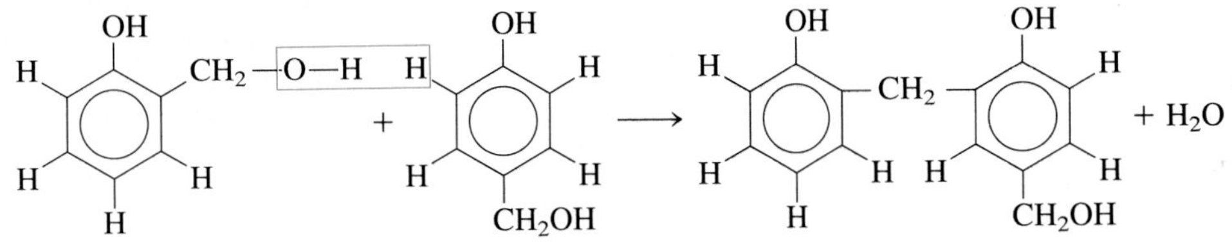

This type of substitution and formation of water continues, forming a highly cross-linked condensation polymer, which sets into a rigid structure (shown in Figure 24.9) upon heating. Resins made from phenol and formaldehyde are used for the adhesive in plywood and particle board.

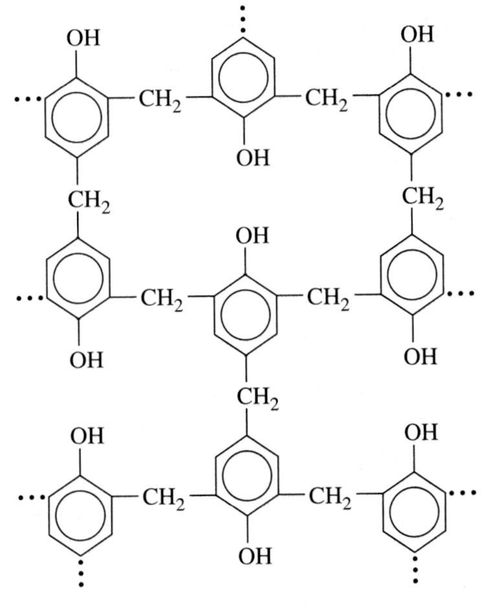

Figure 24.9 The highly cross-linked structure of Bakelite.

24.4   Proteins

The realm of biochemistry includes a few relatively simple substances and many very complex ones. Most of the species that are of biological importance are organic compounds containing several functional groups, but some inorganic substances are also significant in lesser amounts. Three important classes of biomolecules are considered essential food constituents. ▶ These are **proteins,** *carbohydrates* (such as cellulose, described in the chapter introduction), and *fats* (or *lipids*). In addition, *nucleic acids* carry genetic information and regulate the formation of proteins. All these molecules are very large and can be considered biological polymers. We look more closely at proteins here. Later in this chapter, we will discuss the other classes of biomolecules.

What types of molecules are involved in the chemistry of life?

Composition of Proteins

Proteins, in some form, make up most of body tissue. Enzymes, viruses, hormones, antibodies, genes, muscle, tendons, hair, and fingernails are all composed at least in part of proteins. Proteins are composed largely of carbon ($\approx 52\%$), hydrogen ($\approx 7\%$), oxygen ($\approx 23\%$), nitrogen ($\approx 16\%$), and lesser amounts of other elements such as sulfur, iodine, phosphorus, iron, and copper. They are very complex substances of high molar mass (about 15,000 to several million g/mol) composed of various combinations of 20 different amino acids.

The amino acids are small molecules that contain both the amine and the carboxylic acid functional groups. The amino acids of biological significance are called **alpha-amino acids** (α-amino acids). In these substances, —NH_2 and —COOH groups are both attached to the same carbon. The structures of the α-amino acids are shown in Figure 24.10. Note that each of the structures contains the group

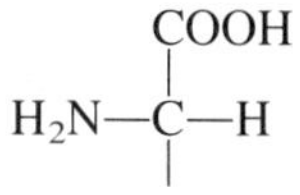

$$H_2N-\overset{\displaystyle COOH}{\underset{\displaystyle |}{\overset{\displaystyle |}{C}}}-H$$

Figure 24.10 Structures of the α-amino acids.

Glycine
(Gly)

L-Alanine
(Ala)

L-Valine
(Val)

L-Leucine
(Leu)

L-Isoleucine
(Ile)

L-Serine
(Ser)

L-Threonine
(Thr)

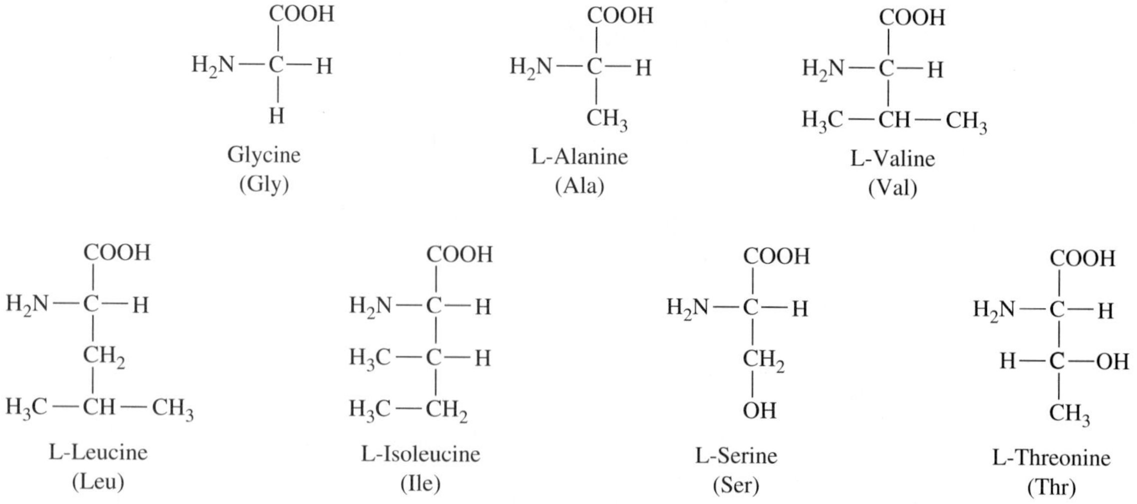

Figure 24.10 (cont'd)

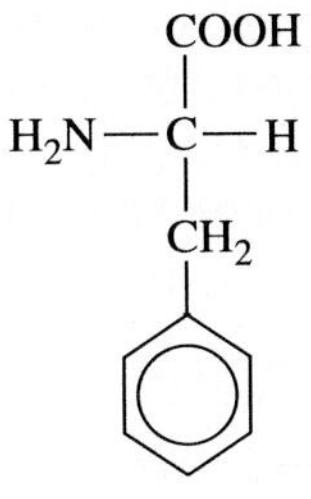

L-Phenylalanine
(Phe)

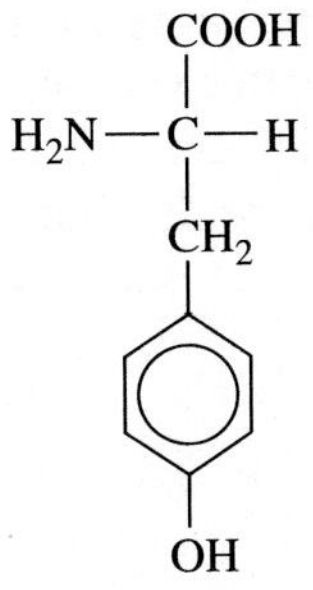

L-Tyrosine
(Tyr)

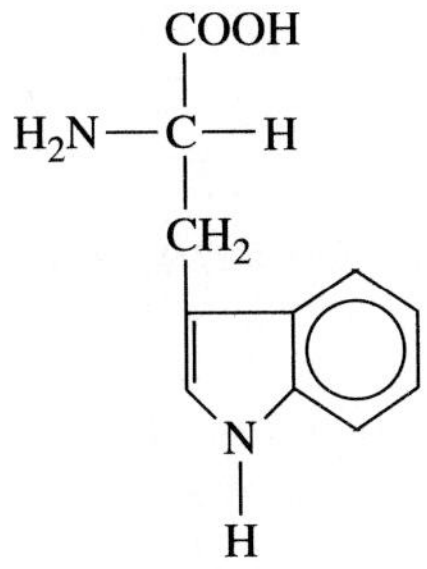

L-Tryptophan
(Trp)

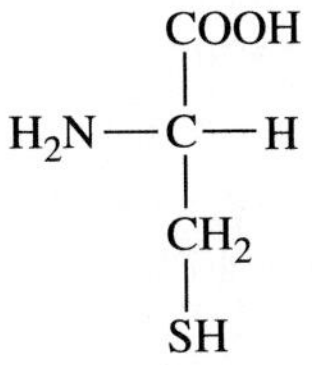

L-Cysteine
(Cys)

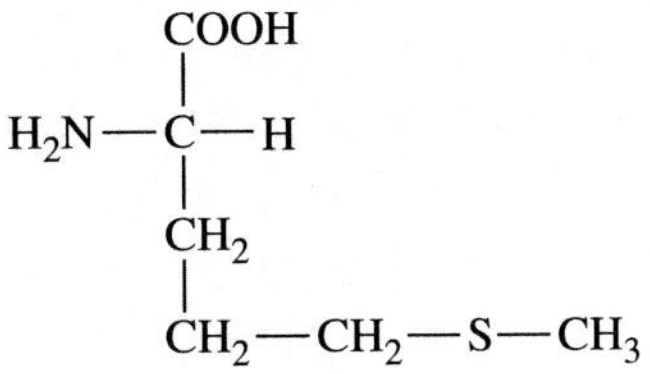

L-Methionine
(Met)

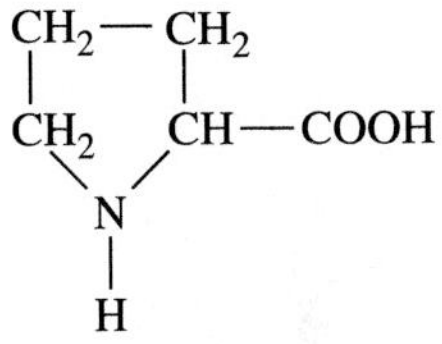

L-Proline
(Pro)

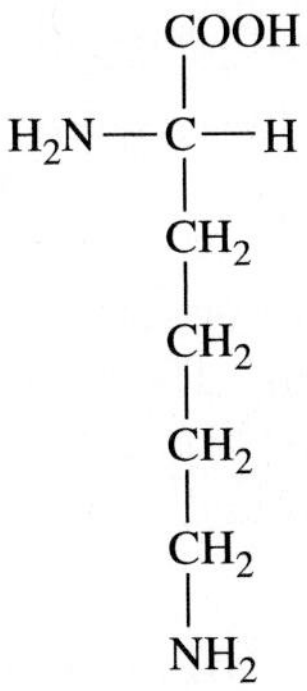

L-Lysine
(Lys)

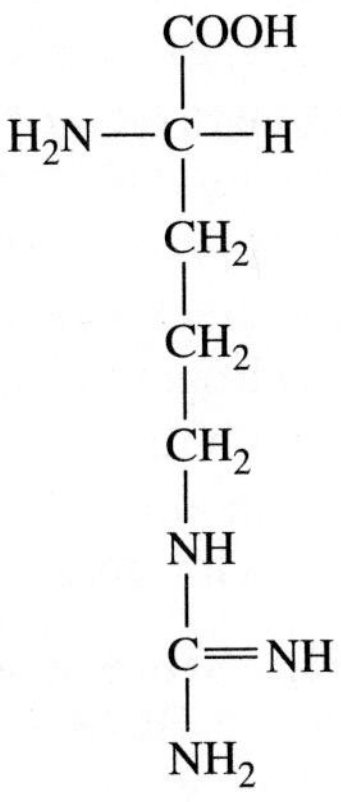

L-Arginine
(Arg)

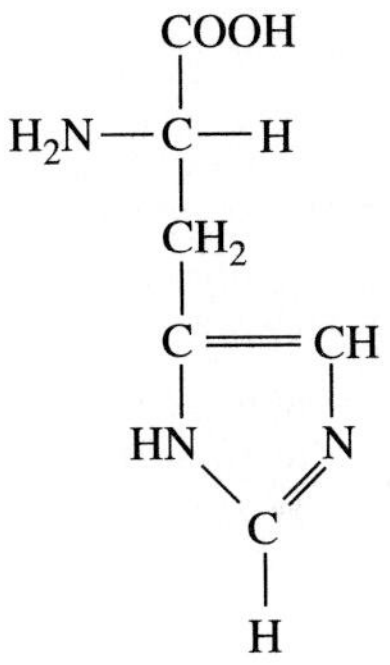

L-Histidine
(His)

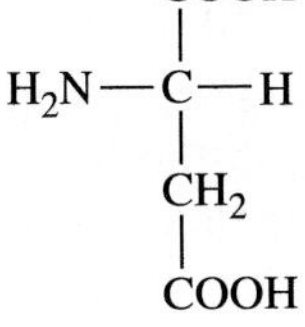

L-Aspartic acid
(Asp)

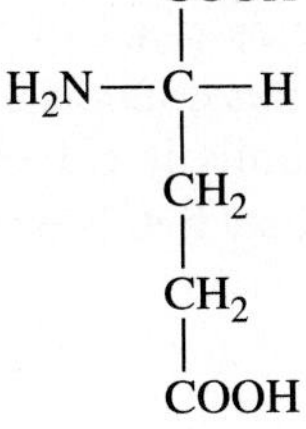

L-Glutamic acid
(Glu)

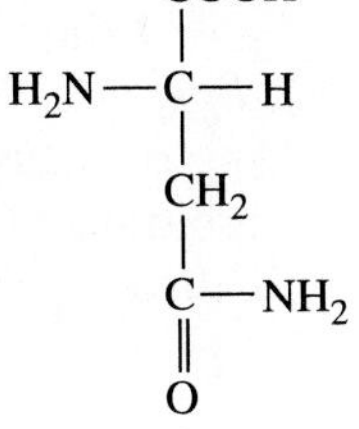

L-Asparagine
(Asn)

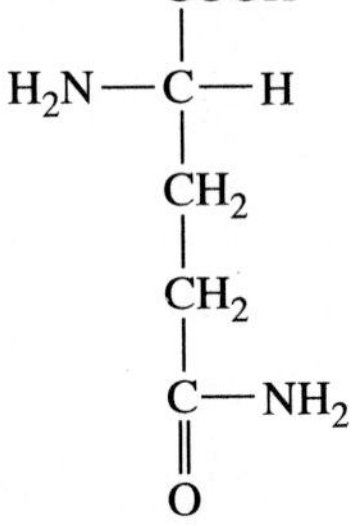

L-Glutamine
(Gln)

The other parts of the structures—side chains on this group—are all different. It is the identity of these side chains that determines the properties of the amino acids, such as their solubility and acidity. The names of the amino acids are abbreviated with three-letter symbols constructed of the first three unique letters in the names. For example, the abbreviation for glycine is Gly.

Glycine, NH_2CH_2COOH, is a nonessential amino acid found in gelatin. It is a colorless solid with a sweet taste that is used as a nutrient.

Amino acids—and thus proteins—contain both the acidic carboxyl group (—COOH) and the basic amine group (—NH_2).

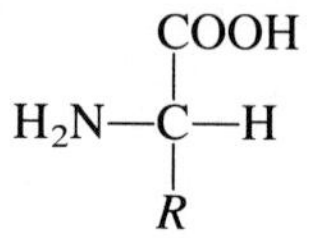

$$\begin{array}{c} COOH \\ | \\ H_2N-C-H \\ | \\ R \end{array}$$

Thus proteins are amphoteric; they can consume both OH^- (by converting —COOH to —COO^-) and H^+ (by converting —NH_2 to —NH_3^+). Proteins are built up from amino acids by the elimination of water during the condensation reaction of the —NH_2 group from one amino acid with the —COOH group from another to form a peptide linkage.

$$\begin{array}{cc} H & O \\ | & \| \\ -N-C- \end{array}$$

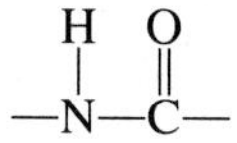

We previously encountered the peptide linkage in the polymer nylon. Consider the following reaction between two amino acids to form a peptide linkage.

$$HOOC—CH(CH_3)—NH_2 + HOOC—CH_2—NH_2 \longrightarrow$$
$$HOOC—CH(CH_3)—NH—CO—CH_2—NH_2 + H_2O$$

Using abbreviations, we can represent this reaction as follows:

$$Ala + Gly \longrightarrow Ala–Gly + H_2O$$

Alanine, $CH_3CH(NH_2)COOH$, is a nonessential amino acid that can be obtained by the hydrolysis of proteins, such as that in white silk. It is a colorless solid.

Note that a molecule of water is formed from an —OH group from the carboxylic acid and a hydrogen from the amine group. The condensation product has a feature in common with the reactant amino acids, because it has a carboxylic acid group at one end and an amine group at the other end. Thus this condensation product can react further with another amino acid.

$$HOOC—CH_2—NH_2 + HOOC—CH(CH_3)—NH—CO—CH_2—NH_2 \longrightarrow$$
$$HOOC—CH_2—NH—CO—CH(CH_3)—NH—CO—CH_2—NH_2 + H_2O$$
$$Gly + Ala–Gly \longrightarrow Gly–Ala–Gly + H_2O$$

This process can continue to form longer and longer condensation polymers. Such polymerization reactions are similar to what occurs in the formation of nylon (described in Section 24.3).

When two amino acids react to form a molecule, that molecule is called a *dipeptide*. Three amino acids condense to form a *tripeptide*, and a polymer consisting of many amino acid units is called a *polypeptide*. Although the borderline is arbitrary, a polypeptide that has a molar mass greater than about 5000 g/mol is called a protein.

Structure of Proteins

Many different proteins are known. Their properties and functions depend on their configurations, which can be described in terms of four types of structure: primary, secondary, tertiary, and quaternary.

Primary Structure   The sequence of amino acids in each different protein is characteristic of that protein and is responsible for some of the protein's properties. Figure 24.11, for example, shows the sequence of amino acids in the enzyme lysozyme, a protein of molar mass 14,600 g/mol. Lysozyme is found in tears, nasal mucus, milk, saliva, and blood serum, where it functions to hydrolyze the cell walls of bacteria. This enzyme consists of 20 of the amino acids in a chain of 129 subunits arranged in a specific and characteristic order. The sequence of amino acids in a protein is known as that protein's **primary structure.**

Figure 24.11 Amino-acid sequence of lysozyme.

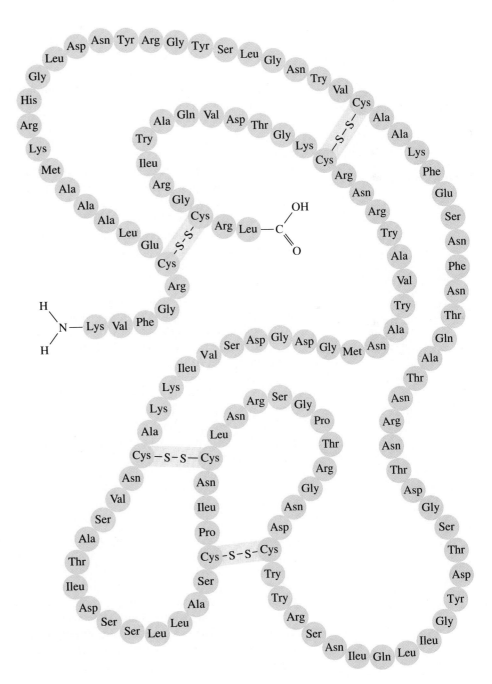

Figure 24.12 A folded-sheet-like secondary structure is found in some proteins, such as in silk and stretched hair. Protein chains are connected by the hydrogen bonds represented by dashed lines.

Figure 24.13 The alpha-helical secondary structure is found in many proteins, such as the keratin in hair, horn, and hoofs. Hydrogen bonds, represented by dashed lines, connect parts of the helix, giving some rigidity to the structure.

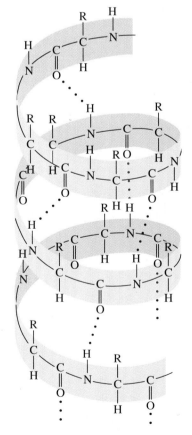

Amino acids can be recovered from a protein by heating it with aqueous acids or bases. The protein undergoes hydrolysis to form its component amino acids. This process can also occur in the presence of hydrolytic enzymes, which catalyze the hydrolysis process. By means of such enzymatic hydrolysis processes, the amino acids an organism needs to generate its specific proteins can be obtained from other protein sources. The digestion of proteins produces amino acids, which can be delivered to other parts of the body to build those proteins the body needs. The human body must obtain nine of the amino acids (His, Ile, Leu, Lys, Met, Phe, Thr, Trp, and Val) from the diet. These are called the *essential amino acids.* Another amino acid is essential only in the diets of children, who need it for proper growth, but it is present in sufficient quantities in adults. The remaining amino acids, the nonessential amino acids, are synthesized within the body.

Secondary Structure   The sequence of amino acids in a protein determines some of its properties and functioning. However, the shape and arrangement of the protein chains are also important. The spatial arrangement of amino acids near one another in the protein chain is called the protein's **secondary structure.** This structure is due to the presence of hydrogen bonds between peptide linkages. In many proteins, this local structure shows up as a regular, periodic structure. Some proteins have a structure something like a folded sheet, in which protein chains are lined up beside one another, as shown in Figure 24.12. The chains are held together by hydrogen bonds. Other proteins have an **alpha-helical structure,** in which the protein chain is twisted into a helix, as shown in Figure 24.13. This helical structure is maintained by hydrogen bonds between an amine group in one part of the chain and a carbonyl group in another position. The helical structures are found in fibrous proteins, such as hair and muscles, and are often bundled together, like a cable (Figure 24.14).

Tertiary Structure   The cylinder formed by the helical arrangement of a protein chain has further structure; it can be linear, or it can be folded back on itself in various ways. The nature of the folding of the helical chains or of nonhelical secondary structures is very specific for particular proteins and is called their **tertiary structure.** Look, for example, at the lysozyme chain shown in Figure 24.11. In four places, the chain is connected to itself by bridging disulfide groups. These bridges help to maintain the three-dimensional tertiary structure of the enzyme, which is necessary for its proper functioning. Another sort of tertiary structure is that of myoglobin, a protein with a molar mass of about 18,000 g/mol that is involved in respiratory processes as an oxygen carrier. The tertiary structure of myoglobin is shown in Figure 24.15. Note the extent to which this protein is folded back on itself. Proteins like myoglobin, which are folded into a rough spherical shape, are known as *globular proteins.*

The folding that constitutes the tertiary structure determines how proteins interact with other molecular species. In the case of enzymes, for example, the **active site** of the enzyme, the place where the catalytic activity takes place, is often on the inside of the molecule. In order for a molecular change to be catalyzed by an enzyme, the molecule to be changed must fit in the cavity around the active site, and the shape of the cavity is dictated by the tertiary structure of the enzyme. It is for this reason that enzymes are such specific catalysts.

Quaternary Structure Many proteins are composed of more than one polypeptide chain. For example, hemoglobin, the primary oxygen carrier in the blood of higher animals, contains pairs of two different polypeptide chains. The structure that results from the association in a protein of two or more polypeptide chains is called the **quaternary structure.** The functioning of a protein depends on this fourth aspect of its structure as well as on the other three.

Denaturation of Proteins When the structure of a protein is disrupted, the protein is said to be denatured, and it loses many of its characteristic properties. The term *denaturation* describes a number of structural changes in proteins, from the loss of tertiary structure to the uncoiling of chains. Generally, the protein is converted to a state of greater disorder. Denaturation can be accomplished by heating, adding strong acids, placing the protein in a nonaqueous solvent, or adding heavy-metal ions. A common example of denaturation is the boiling of an egg.

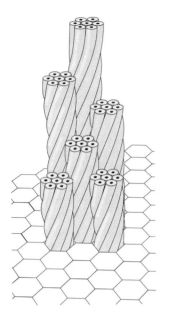

Figure 24.14 Fibrous proteins consist of bundles of helices. An example is collagen, found in growing bones, tendons, and muscle.

What happens to the protein in an egg when it is cooked in boiling water?

Figure 24.15 Tertiary structure of myoglobin. From Ebbing, *General Chemistry*, 4/e. Copyright © 1993 by Houghton Mifflin Company. Used with permission.

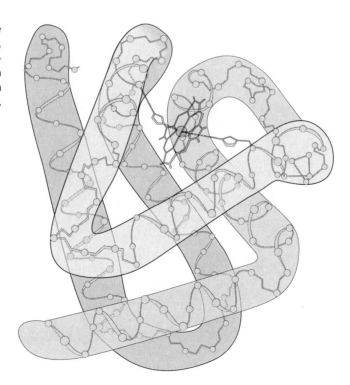

24.5   Nucleic Acids

The basic biological unit of all living matter is the cell. Many different types of cells fulfill many different functions in living systems. However, for all their variety, cells have many features in common. (A typical cell is shown in Figure 24.16.) All cells are enclosed by a *cell membrane*. Within the membrane is a gelatinous fluid called *cytoplasm,* which contains a number of smaller bodies. The *nucleus* is at the center of the cell. *Lysosomes* contain digestive enzymes that break down large molecules into smaller molecules. These smaller molecules are oxidized by enzymes in the *mitochondria,* the site of reactions that provide energy to the cell. The dots that line the *endoplasmic reticulum* are *ribosomes,* which are the sites of protein synthesis. The *centrosomes* are involved in the replication of chromosomes.

   Nucleic acids are found in both the nucleus and cytoplasm—in other words, everywhere except in the cell membrane. There are two classes of nucleic acids. **Deoxyribonucleic acid (DNA)** occurs in the cell nucleus and has a molar mass of 10^6 to 10^9 g/mol. **Ribonucleic acid (RNA)** is found in the cytoplasm and has a molar mass of 25,000 to 10^6 g/mol.

   Nucleic acids consist of three parts: an acid group (phosphates), a basic group (four possible organic bases), and a sugar unit. The sugars found in nucleic acids are shown in Figure 24.17. They are D-ribose, found in RNA, and D-2-deoxyribose, found in DNA. The bases (and their common abbreviations) are adenine (A), guanine (G), cytosine (C), and uracil (U) in RNA and adenine (A),

Figure 24.16 Diagram of a typical biological cell.

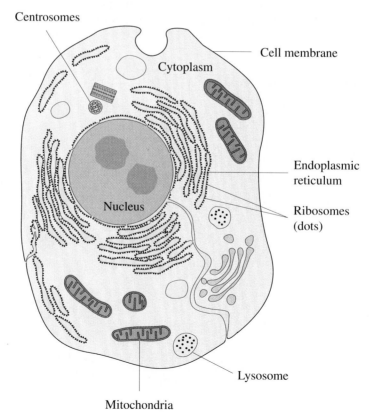

Centrosomes

Cell membrane

Cytoplasm

Endoplasmic
reticulum

Ribosomes
(dots)

Nucleus

Lysosome

Mitochondria

Figure 24.17 Structures of the sugars found in nucleic acids.

A. β–D–ribose

B. β–D–deoxyribose

guanine (G), cytosine (C), and thymine (T) in DNA. Note that uracil and thymine are found in one but not the other nucleic acid. Occasionally, other bases also occur. The structures of the primary bases are shown in Figure 24.18.

The fundamental unit of a nucleic-acid structure is called a **nucleotide.** Each nucleotide is made up of a sugar molecule to which is bonded both a phosphate group and a base. The nucleic acids are polymers, made up of chains of nucleotides linked together. The arrangements of nucleotides in RNA and DNA are shown in Figure 24.19. The sugar and phosphate units form the backbones of these polymeric chains, and the bases are bonded as side chains to the sugars.

Deoxyribonucleic Acid

How are the genetic codes stored in cells?

Deoxyribonucleic acid carries the genetic information in the cell. It consists of two interwoven helical strands in which the bases point toward the center of the helix, as shown in Figure 24.20, and the phosphate groups point to the outside. The sugar and phosphate units form the backbone of the helix. The bases pair up inside the double helix in very specific ways—adenine to thymine (A–T) and cytosine to guanine (C–G)—and are held together by hydrogen bonding, as shown in Figure 24.21. Thus the arrangement of the nucleotide units in the polymeric strand must enable one strand to match another strand to produce the appropriate base pairing.

The double-strand arrangement enables DNA to replicate itself. Replication occurs when the double helix starts to become unraveled. A new strand can grow on each separated strand, which serves as a template for the forming of DNA

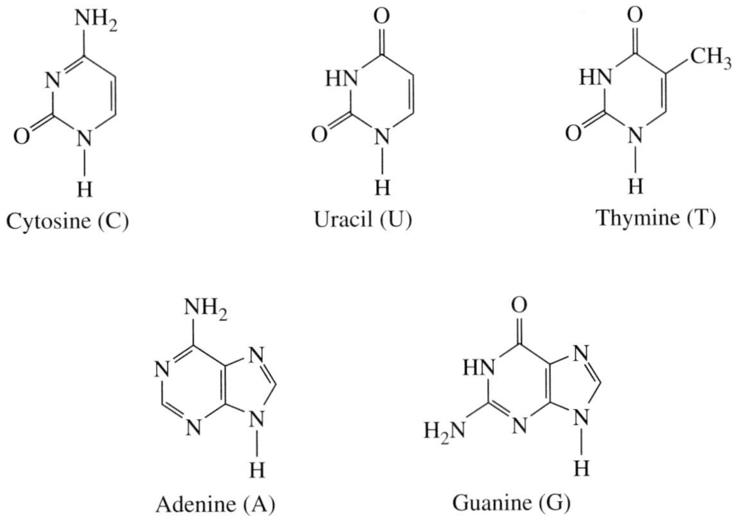

Figure 24.18 Structures of the bases found in nucleic acids and their one-letter abbreviations.

Cytosine (C)

Uracil (U)

Thymine (T)

Adenine (A)

Guanine (G)

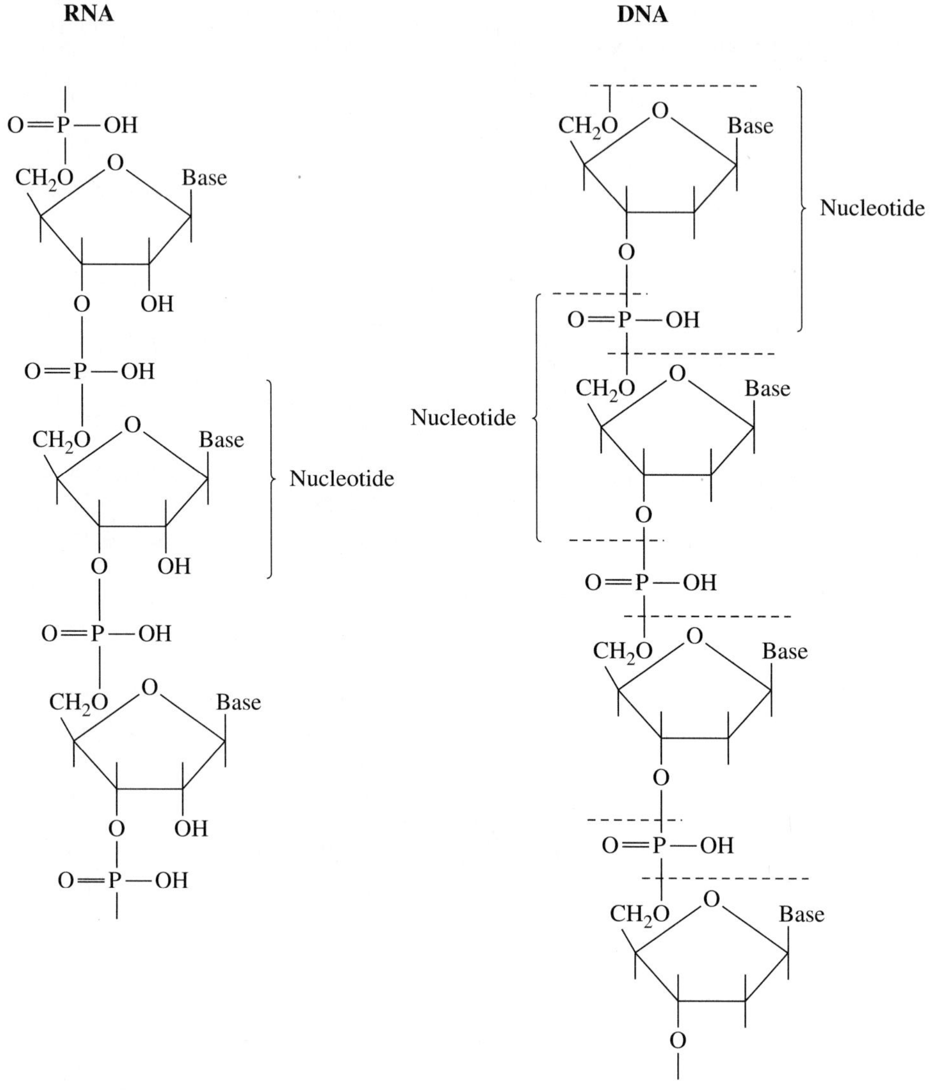

Figure 24.19 Polymeric arrangement of nucleotides in RNA and DNA.

Figure 24.19 Polymeric arrangement of nucleotides in RNA and DNA.

(Figure 24.22). Through this replication process, DNA is generated for transmission to new cells as they develop.

Genes, made up of DNA molecules, specify the kinds of proteins that are made in cells for different types of tissues. Recall that proteins are formed from up to 20 different amino acids and that the order of the amino acids in the polypeptide chain determines the functioning of the protein. The genetic information that determines the order of the amino acids in a protein is stored as the sequence of bases on the DNA strand. There are only four bases but 20 amino acids, so it requires more than one base to designate a particular amino acid. A sequence of two bases would provide 16 (4^2) possibilities, whereas a series of three bases provides 64 (4^3) possibilities. Because the DNA needs to be able to specify about 20 amino acids, it must use a series of at least three bases. Each meaningful sequence of three bases, then, provides a genetic code for a particular amino acid, and the code can be used to determine which amino acid must be added next when a protein chain is grow-

C:::G
A:::T
G:::C
C:::G
T:::A
G:::C
C:::G
A:::T
G:::C
Bases
Sugar–phosphate
backbone

Figure 24.20 Double-helical structure of DNA.

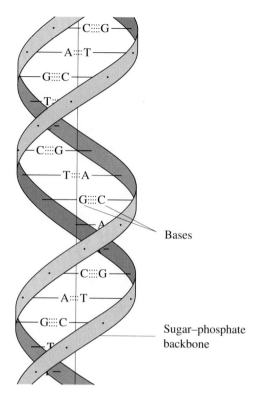

ing. This series of three bases is called a **codon.** An example is shown in Figure 24.23.

Ribonucleic Acid

The information stored by the codons, to be used in protein synthesis, has to be transferred from the nuclear DNA to the growing protein in the cytoplasm. This is where RNA comes into the picture. **Messenger RNA** (m-RNA) takes the genetic code from DNA and carries it to the *ribosomes,* the sites in the cell at which protein synthesis takes place. The bases in m-RNA are complementary to those in DNA—for example, where DNA has a cytosine, m-RNA has a guanine. As we

Thymine
Adenine
A–T pair
Cytosine
Guanine
G–C pair

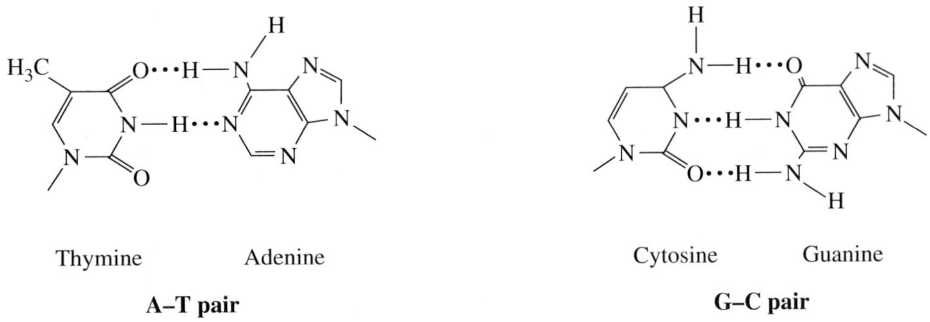

Figure 24.21 Hydrogen bonding in base pairs.

Figure 24.22 DNA is replicated by growing new strands on a partially unwound double helix. The order of bases on the unwound strands determines the order in which nucleotides are added to the growing strands, because the only allowed pairings are C–G and A–T.

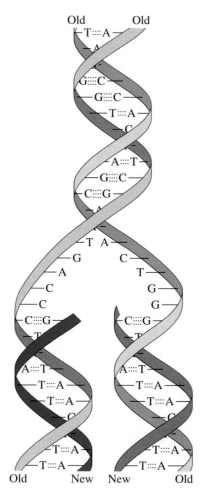

noted earlier, RNA has uracil instead of thymine. Thus a base sequence of AAT-CAGTT in DNA translates to UUAGUCAA in RNA, as Figure 24.24 shows. Once the m-RNA becomes attached to the ribosomes, the genetic information is available for synthesis of proteins.

Another form of RNA called **transfer RNA** (t-RNA) plays a role in bringing the appropriate amino acid in an activated form to the site of protein synthesis, the m-RNA in the ribosome. There is at least one kind of t-RNA for each amino acid. Transfer RNA consists of about 75 nucleotides in a strand that bends back on itself. The molecule has two regions of double-helical structure and some loops, giving an L-shape, as shown in Figure 24.25. The closed loop at the end of the L contains the coding of three bases associated with a particular amino acid, which is

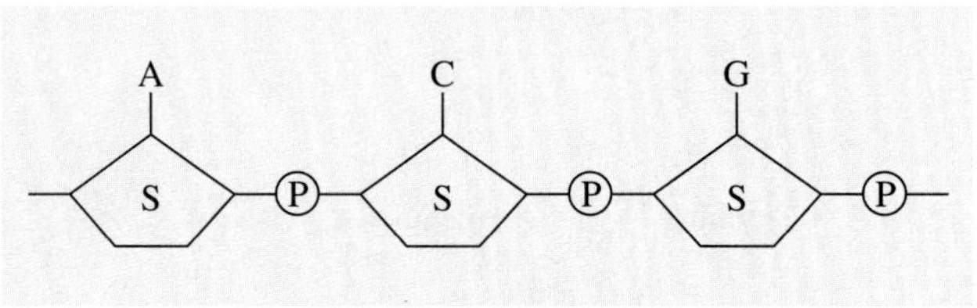

Figure 24.23 Diagram of a codon, which consists of a sugar–phosphate backbone with three bases attached. The identity of the three bases provides one item of information in the genetic code used to synthesize proteins.

Figure 24.24 Messenger RNA is complementary to DNA. From *The Genetic Code: II* by Marshall W. Nirenberg, *Scientific American,* March 1963. Copyright © 1963 by Scientific American, Inc. All rights reserved.

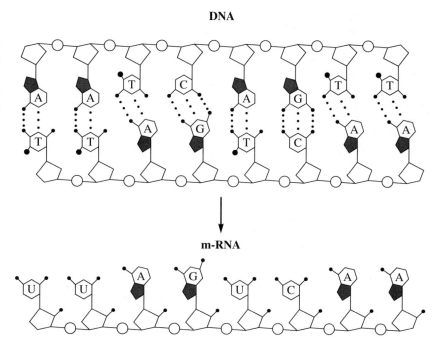

DNA

m-RNA

Figure 24.25 Structure of t-RNA bonded to an amino acid. From BIOCHEMISTRY 3/e by Lubert Stryer. Copyright © 1988 by Lubert Stryer. Reprinted with permission of W.H. Freeman and Company.

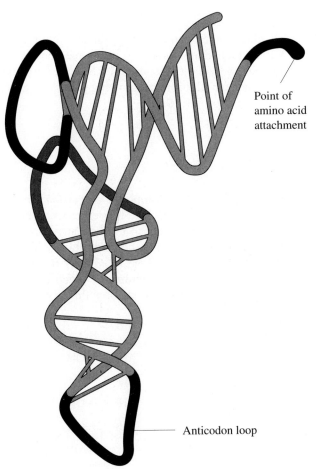

Point of amino acid attachment

Anticodon loop

Figure 24.26 Synthesis of protein molecules. Messenger RNA provides the codes needed to determine the sequence of amino acids to be added to the protein. Transfer RNA brings the necessary amino acid to the m-RNA in the ribosome. The t-RNA bonds to the m-RNA, and the amino acid is transferred to the growing protein chain. After transfer of the amino acid, the t-RNA is released from the m-RNA. From *The Genetic Code: III* by F.H. C. Crick, *The Scientific American,* October 1966, Vol. 215, No. 4, pp. 55–62. Copyright © 1966 by Scientific American, Inc. All rights reserved.

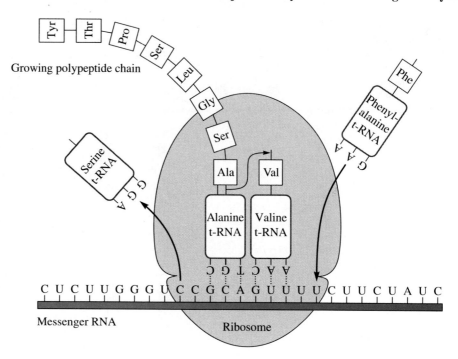

attached at the open end of the chain at the other end of the L. The base coding in t-RNA (called an *anticodon*) is complementary to that in m-RNA, so it can form a temporary hydrogen bond to m-RNA, thereby holding the amino acid in the proper sequence for addition to a growing protein chain. This process is shown schematically in Figure 24.26.

24.6   Carbohydrates

Carbohydrates are composed of carbon, hydrogen, and oxygen and have the general formula $C_n(H_2O)_m$. This empirical formula suggests the origin of the name *carbohydrate,* but it doesn't tell us anything about the structure of carbohydrate molecules. Carbohydrates contain multiple alcohol groups (—OH) and either aldehyde ($RCH{=}O$) or ketone ($R_2C{=}O$) functional groups or groups readily converted to these. The structures of glucose, a typical aldehyde carbohydrate, and fructose, a typical ketone carbohydrate, can be represented as linear, open-chain arrangements.

$$
\begin{array}{cc}
\text{CHO} & \text{CH}_2\text{OH} \\
\text{H—C—OH} & \text{C=O} \\
\text{HO—C—H} & \text{HO—C—H} \\
\text{H—C—OH} & \text{H—C—OH} \\
\text{H—C—OH} & \text{H—C—OH} \\
\text{CH}_2\text{OH} & \text{CH}_2\text{OH} \\
\text{D-glucose} & \text{D-fructose}
\end{array}
$$

Fighting Crime with DNA

The Human Genome Project is an international effort to locate and identify every gene in human DNA.

With a bit of skin scraped from a murderer by the fingernails of a murder victim, a biochemist can unambiguously identify the murderer—if a skin sample from the murderer is available for comparison. Given a drop of blood from a man and a child, a biochemist can establish with certainty whether the man is the father of the child. This technique also extends to the vindication of those falsely accused. A swatch of hair caught on a window screen at the scene of a burglary can be used to determine that an accused thief is *not* the individual who crawled through the window.

How do biochemists do this? The method is called DNA fingerprinting, which is beginning to be accepted by the courts as legitimate proof of identity. Every individual's DNA is unique, so identification of the exact makeup of the DNA can be used to identify the origin of a tissue sample by comparison with a known tissue sample from that person. The composition of the DNA is determined by breaking it into fragments with the assistance of enzymes or other chemical agents and then subjecting the mixture to chromatography in an electrical field—a process known as electrophoresis. An example of the resulting DNA fingerprint is shown in Figure 24.27. Comparison of such fingerprints from two DNA samples determines whether the two samples come from the same person.

Some states are now taking blood samples from convicted rapists and using them to establish a data bank of DNA fingerprints. This data bank could be used by police across the United States to trace a rapist who moves to another state and rapes again.

Even more ambitious efforts are under way. The Human Genome Project is an international effort to locate and identify every gene in human DNA. Such a genetic map could be used to detect and perhaps cure genetic diseases and could also be used for criminal identification. For example, a DNA analysis of a clue from a crime could be used to identify the criminal's eye color if scientists knew the location of the gene for eye color. If successful, the Human Genome Project could provide a powerful tool for law enforcement. But civil-liberties groups also fear that innocent people might be convicted with this new technology, since

Figure 24.27 Matching bars in a DNA fingerprint indicate that two DNA samples are from the same individual.

it is based on probabilities of fingerprint matches, rather than on absolute matching.

Questions for Discussion

1. How do we know that the DNA of each individual is unique?
2. What other uses could be made of DNA fingerprinting?
3. Biochemists are applying genetic-engineering techniques to isolate pieces of DNA and insert them into bacteria, where synthesis of biologically active proteins can occur. What use can be made of these techniques?
4. Gene therapy is being investigated as a medical tool. In this therapy, modified genetic material is introduced into human cells in attempts to cure genetic diseases. What are some advantages and some dangers associated with this approach?

Glucose, $C_6H_{12}O_6$, is a colorless solid also known as dextrose, blood sugar, grape sugar, and corn sugar. It is a main source of energy for living organisms and is found in fruits and other plant parts. It is prepared in commercial amounts by the hydrolysis of starch. It is about three-quarters as sweet as sucrose.

These molecules actually exist primarily in cyclic structures in which two carbon atoms are linked by an oxygen atom.

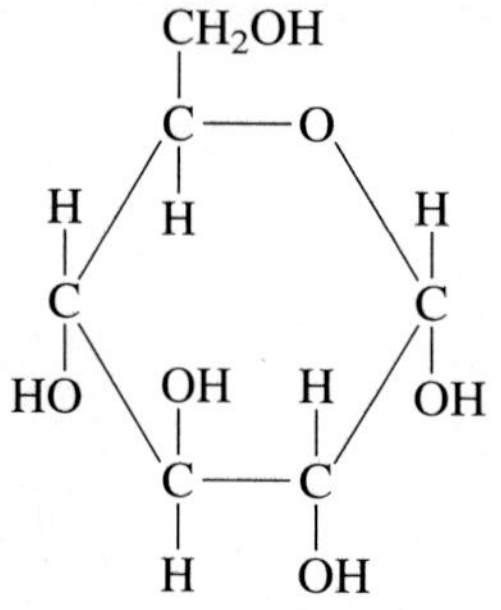

glucose 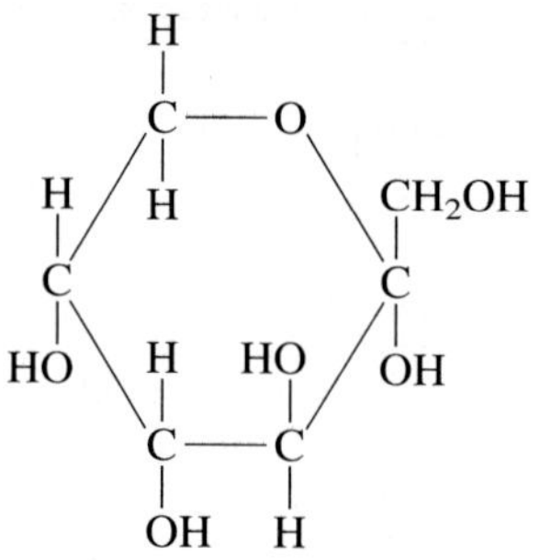   fructose

Although the predominant form of fructose is a six-membered ring, it also forms a five-membered ring, which is the form used in derivatives of fructose, such as sucrose (discussed later).

What types of molecules are included in the carbohydrates?

The simpler carbohydrates include a group that has a sweet taste, the *sugars,* which include glucose and fructose. These carbohydrates are often classified as *monosaccharides, disaccharides, oligosaccharides,* or *polysaccharides.*

The monosaccharides have just one five- or six-membered sugar ring, as in glucose, or blood sugar, $C_6H_{12}O_6$. This monosaccharide exists primarily in a cyclic form, which is produced by the addition of an alcohol group across the $C{=}O$ of the aldehyde group in the open-chain form.

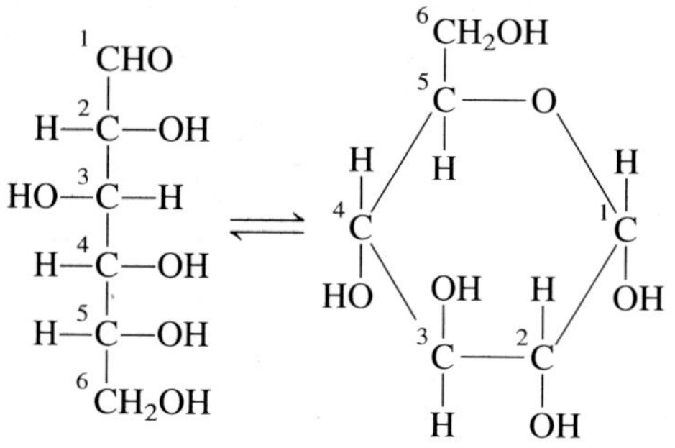

α-D-glucose

There are two ways in which the cyclic form could be formed from the open-chain form. To help visualize this, we have numbered the carbon atoms in these structures. The carbon numbered 1 is in the —CHOH group attached to the ether oxygen. The carbon numbered 6 is in the hydroxymethyl group, —CH$_2$OH, which is not part of the ring. The glucose ring is given a label of alpha when the —OH group on carbon 1 is on the opposite side of the ring from the hydroxymethyl

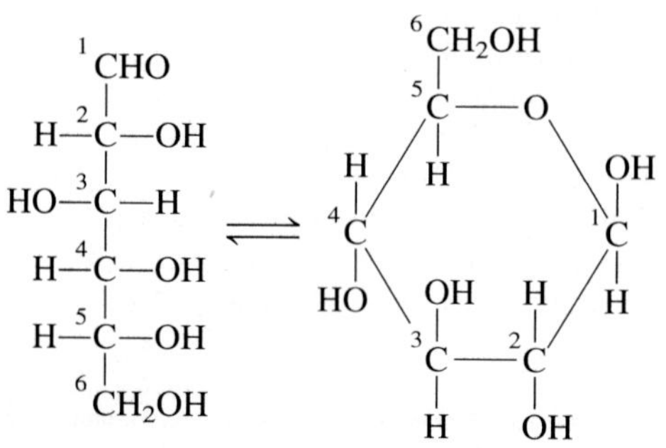

β-D-glucose

Fructose, $C_6H_{12}O_6$, also known as fruit sugar, is the sweetest of the sugars and is found in fruits and honey. The white solid is used to prevent sandiness in ice cream.

group. If the —OH group on carbon 1 is on the same side of the ring as the hydroxymethyl carbon, the ring is labeled beta.

Other monosaccharides include fructose (shown earlier), mannose, and ribose. These monosaccharides differ from glucose and fructose only in the orientation of the —OH groups on the five- or six-membered rings.

Disaccharides, which contain two sugar groups, include common table sugar, sucrose, $C_{12}H_{22}O_{11}$. It is formed (at least formally) by the bonding of a glucose unit with a fructose unit, accompanied by the loss of one water molecule.

$$C_6H_{12}O_6 + C_6H_{12}O_6 \longrightarrow C_{12}H_{22}O_{11} + H_2O$$

glucose   fructose   sucrose

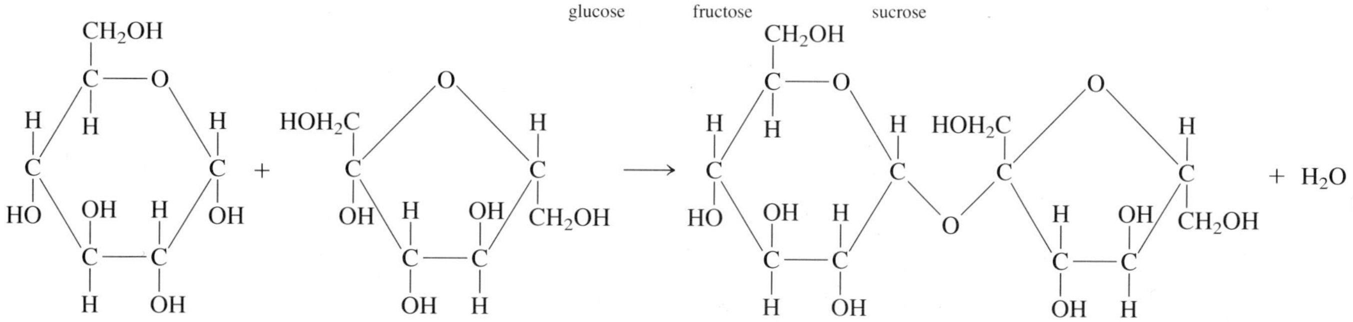

Oligosaccharides are carbohydrates that contain a few monosaccharide units in a chain, and polysaccharides are long-chain polymers of monosaccharide units. Included among the polysaccharides are starch, glycogen, and cellulose. These substances are composed of glucose monomer units but differ in how the sugar rings bond together. The ways in which rings can be connected are illustrated by the structures of some disaccharides in Figure 24.28. These apparently small variations in the bonding between two glucose molecules profoundly affect the properties of the polysaccharides that result from polymerization.

Note the labels given to these linkages in the figure. These labels have two parts. The numbers, such as 1,4 for the glucose rings linked in cellulose, indicate the numbers assigned to the carbons on the two rings involved in the formation of the ether linkage between the rings. Thus carbon 1 on the first ring loses its —OH group, while carbon 4 on the second ring loses a hydrogen from its —OH group to eliminate a water molecule and form the ether linkage. The second part of the

Figure 24.28 Structures of some disaccharides.

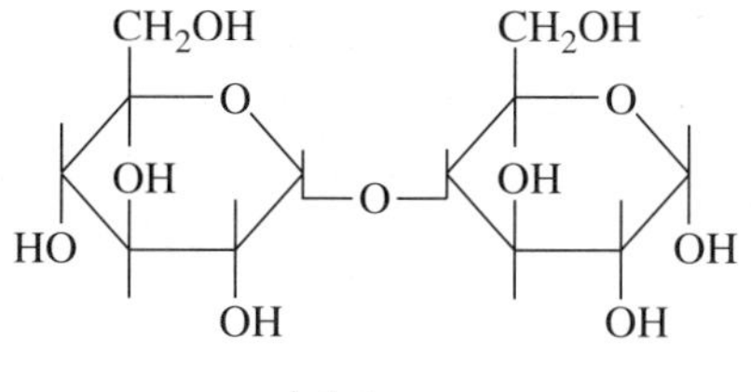

A 1,4-α

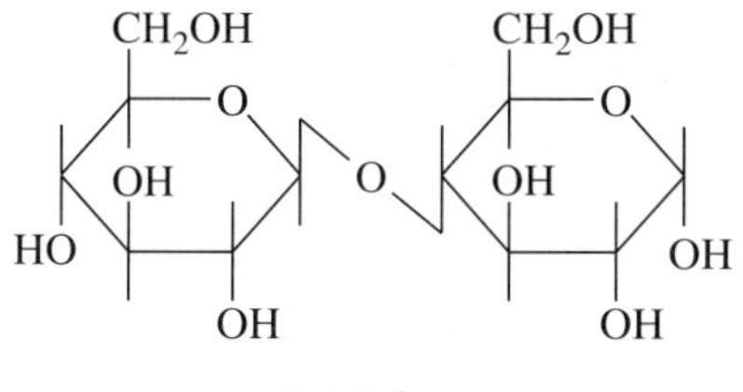

B 1,4-β

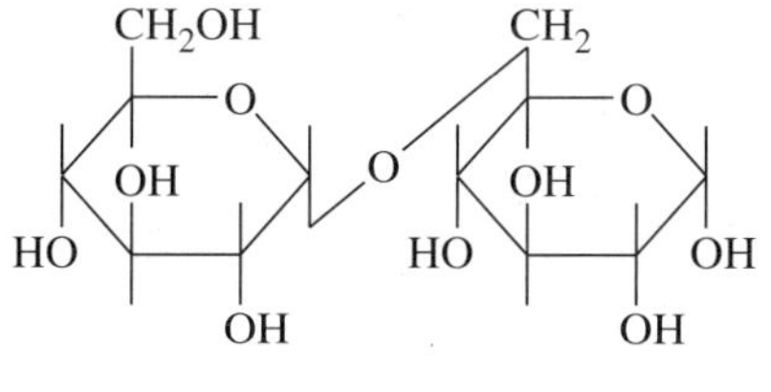

C 1,6-α

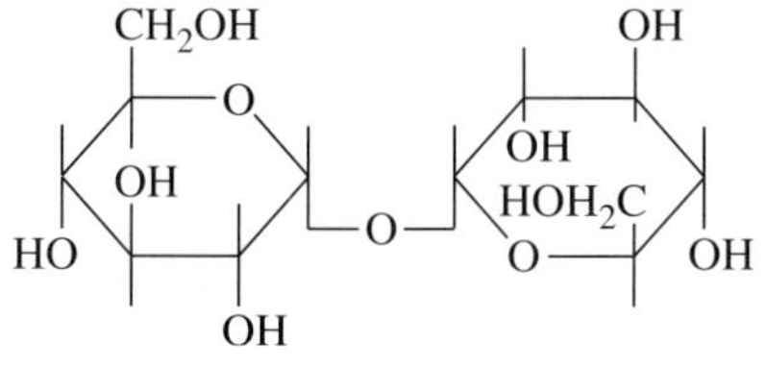

D 1,1-α

label, alpha or beta, indicates whether the —O— group on carbon 1 is on the opposite side (alpha) or the same side (beta) as the hydroxymethyl group.

Starch

Starch is a long chain polymer with the general formula $(C_6H_{10}O_5)_n$. Seeds, tubers, and fruits of plants are sources of natural starches, because starch is the principal store of carbohydrates in plants. It is used by young plants as a food source until they develop a leaf system and are able to start synthesizing their own foods by photosynthesis. It is also used as food by established plants during periods when they cannot synthesize carbohydrates. Animals use starch as an energy source by ingesting plant material.

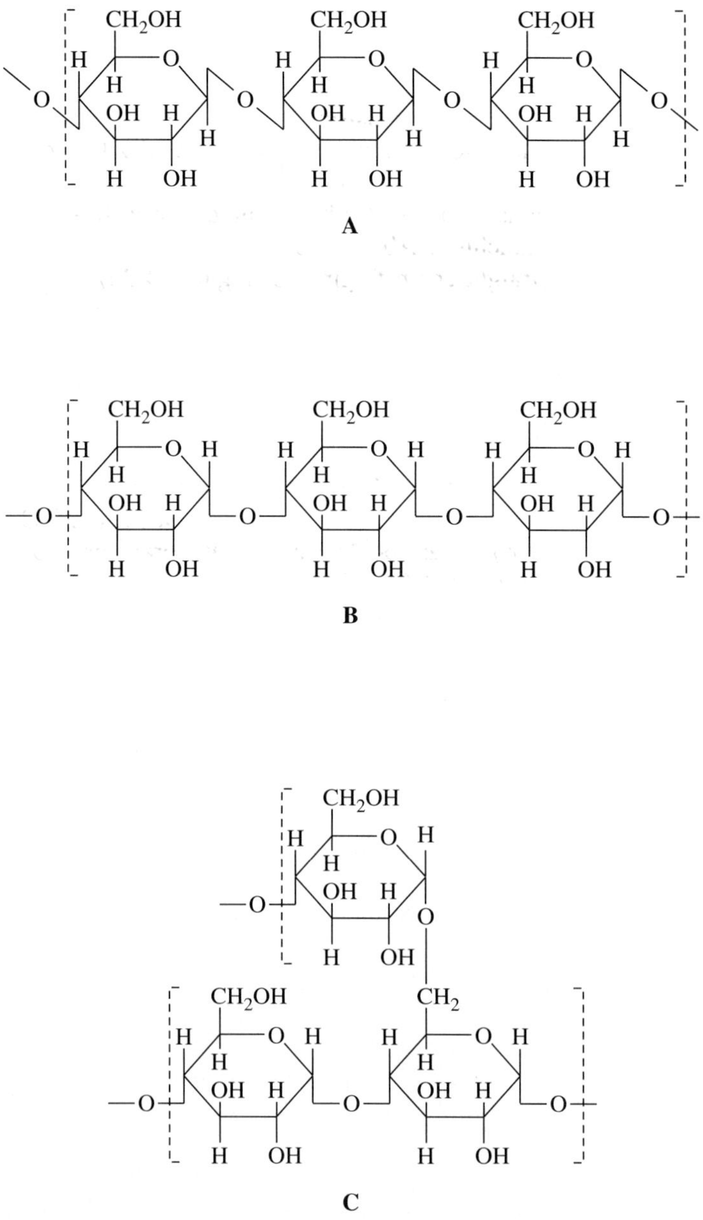

Figure 24.29 Structures of amylose, amylopectin, glycogen, and cellulose. *(A)* Cellulose contains 1,4-β linked glucose units. *(B)* Amylose contains 1,4-α linked glucose units. *(C)* Amylopectin and glycogen contain 1,6-α branches.

Figure 24.30 Branched structure of amylopectin and
glycogen.

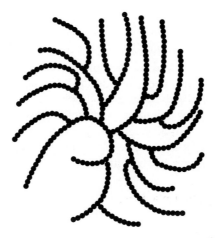

Starch consists of two types of carbohydrate, *amylose* and *amylopectin,* which are diagrammed in Figure 24.29. Amylose, a straight-chain polymer with 100 to 1000 glucose units, occurs in an open helical structure. Amylopectin contains about 5000 glucose units and differs from amylose in that it has some side branching. About 400 of its glucose units occur as branching points to connect straight-chain fragments (Figure 24.30).

Glycogen

Glycogen, $(C_6H_{10}O_5)_n$, has a structure much like that of amylopectin (Figures 24.29 and 24.30), with glucose units connected in a branched structure. Glycogen, though, has about twice as many branching points as amylopectin. Glycogen is the principal storage food of higher animals and of some microbes. It is found primarily in the liver and muscles, where it breaks down to provide energy for the muscles and other tissues.

Cellulose

▶ *Cellulose,* discussed in the introduction to this chapter, is the most important of the polysaccharides that play a structural role in plants. Its empirical formula is $(C_6H_{10}O_5)_n$. Cellulose is made up of glucose units, with molar masses in the range of 6×10^5 to 2×10^6 g/mol. Like amylose, it has a linear array of glucose units. It differs from amylose in the way the glucose units are attached to one another (Figure 24.29). Chains of cellulose are arranged in parallel bundles to form fibers, which are extremely insoluble. These fibers contain about 50% of the carbon found in plants and are the main structural material for plant cells. Cellulose cannot be digested by most animals but is important as one of the major sources of dietary fiber in humans. Most animals are unable to hydrolyze the ether linkages in cellulose. Exceptions are the ruminants, such as cows, sheep, and goats. These animals' stomachs contain bacteria that have the necessary enzyme to break cellulose down to smaller saccharide units that can be used as a source of energy.

24.7 Lipids

Lipids are a class of biochemical substances that can be extracted from cells by nonpolar solvents. Lipids consist of molecules much smaller than those of carbohydrates or proteins. They include fats, oils, waxes, long-chain carboxylic acids, steroids, phospholipids, and glycolipids.

The substances found in greatest quantity in living tissue are fats and oils. These substances are all *triglycerides,* esters of glycerol with long-chain carboxylic acids, or **fatty acids.** They have the following structure, where *R* represents the hydrocarbon part of the fatty acid.

What is the difference between fats and oils?

$$
\begin{array}{c}
\quad\quad\quad\quad\quad \text{O} \\
\quad\quad\quad\quad\quad \| \\
\text{H}_2\text{C}-\text{O}-\text{C}-R \\
\quad | \\
\quad\quad\quad\quad\quad \text{O} \\
\quad\quad\quad\quad\quad \| \\
\text{HC}-\text{O}-\text{C}-R \\
\quad | \\
\quad\quad\quad\quad\quad \text{O} \\
\quad\quad\quad\quad\quad \| \\
\text{H}_2\text{C}-\text{O}-\text{C}-R
\end{array}
$$

The fatty acids in a triglyceride need not all be the same, and they may be saturated or unsaturated. If the fatty acid is saturated, the triglyceride is a solid, or *fat;* and if the fatty acid is highly unsaturated, with many carbon–carbon double bonds, the triglyceride is a liquid, or *oil.* The greater the degree of unsaturation, the more liquid-like the triglyceride. Common fats include lard, tallow, and milk fat, and some oils are linseed oil, cottonseed oil, coconut oil, soybean oil, peanut oil, corn oil, and olive oil.

Fats and oils are used in the body to synthesize essential tissue constituents. The body also uses them as a source of energy. Because they are insoluble, they can be deposited in cells and can remain there inert until they are needed as a source of energy.

Summary

Rubber and other polymers are synthesized by combining large numbers of monomers. The properties of polymers depend on the presence and arrangement of side chains, on cross-linking between chains, and on the length of the chains. Polymers are formed by several types of reactions, the most common being addition and condensation reactions. Addition reactions are used to form polymers such as polyethylene, polypropylene, polystyrene, poly(vinyl chloride), and polyacrylonitrile. Condensation polymers are formed by splitting out a small molecule from the monomers, which then combine to form a larger molecule. A common example is nylon.

Most body tissue is made up of proteins, which are condensation polymers of the amino acids. The biologically significant amino acids are called alpha amino acids and have the —NH$_2$ and —COOH groups both attached to the same carbon. Protein structures are described in terms of four types: primary, secondary, tertiary, and quaternary.

Nucleic acids, which are made up of an acidic phosphate group, an organic base, and a sugar unit, are found everywhere in a biological cell except the cell membrane. There are two classes of nucleic acids, deoxyribonucleic acid (DNA) and ribonucleic acid (RNA). DNA carries the genetic information in the cell and exists in a double-helical structure. DNA, by means of the sequence of bases on its strands, carries codes used to direct the synthesis of proteins. Messenger RNA reads the genetic code on the DNA and carries it to ribosomes—sites in the cell where protein synthesis takes place. Transfer RNA brings the appropriate amino acid to the ribosome and matches up with the messenger RNA to transfer the amino acid to the growing protein strand.

Carbohydrates (molecules containing carbon, hydrogen, and oxygen) contain multiple alcohol functional groups as well as either an aldehyde group or a ketone group. The smallest carbohydrates are the sugars such as glucose and fructose. The sugar units polymerize to form condensation polymers by eliminating a water molecule by forming an ether linkage between two cyclic structures. Depending on how the sugar units are linked together, various biological polymers are formed. These include starch, glycogen, and cellulose.

Lipids comprise a variety of substances—fats, oils, waxes, steroids, phospholipids, and glycolipids—that can be extracted from cells with nonpolar solvents.

Key Terms

active site (24.4)

addition
  polymerization (24.3)
alpha-amino acid (24.4)
alpha helix (24.4)
carbohydrate (24.6)
codon (24.5)
condensation
  polymerization (24.3)

copolymer (24.1)
deoxyribonucleic acid
  (DNA) (24.5)
elastomer (24.1)
fatty acid (24.7)
lipid (24.7)
messenger RNA (24.5)
monomer (24.1)
nucleic acid (24.5)

nucleotide (24.5)
plastics (24.1)
polymer (p. 1000)
primary
  structure (24.4)
protein (24.4)
quaternary
  structure (24.4)
ribonucleic acid

(RNA) (24.5)
secondary
  structure (24.4)
tertiary structure (24.4)
thermoplastic (24.2)
thermosetting
  polymer (24.2)
transfer RNA (24.5)

Exercises

Rubber

24.1 What is rubber?

24.2 Explain why vulcanization increases the elasticity of rubber.

24.3 Isobutylene (2-methylpropene) polymerizes to form butyl rubber. Write a formula for butyl rubber, showing several repeating units.

24.4 Describe the differences in structure and properties between rubber and gutta-percha.

Polymers

24.5 Explain the difference between a monomer and a polymer.

24.6 Kodel is a polyester formed from terephthalic acid ($HOOC-C_6H_4-COOH$) and 1,4-di(hydroxymethyl)cyclohexane ($HO-CH_2-C_6H_{10}-CH_2-OH$). Write an equation to describe the polymerization reaction used to prepare Kodel.

24.7 Draw the structure of the repeating unit in one of the nylons.

24.8 Compare the properties of polyethylene, polypropylene, and polybutylene.

24.9 Vinyl acetate is a monomer with the structure $CH_2=CH-OOC-CH_3$. Draw the structure of poly(vinylacetate).

24.10 Explain why Dacron is called a polyester.

24.11 Discuss the effect of cross-linking on the properties of polymers.

24.12 Write an equation for the formation of a polyester.

24.13 Draw the structures of the monomers used to make each of the following polymers.
  a. polyethylene
  b. polybutadiene
  c. polypropylene
  d. polyacrylonitrile
  e. polystyrene
  f. poly(vinyl chloride)

24.14 Malonic acid is a dicarboxylic acid, $HOOC-CH_2-COOH$, and ethylene glycol has the structure $HO-CH_2-CH_2-OH$.
  a. Write an equation to describe the condensation polymerization reaction between these species.
  b. What kind of condensation polymer is formed?

24.15 Draw the structures of the polymers that result from the addition polymerization reactions of each of the following monomers.
  a. $CH_2=CHCl$
  b. $CH_2=C(CH_3)-CH=CH_2$
  c. $CH_2=CHNH_2$
  d. $CH_3-CH=CH-CH=CH-CH_3$
  e. $CH_2=CH-CH_2-CH_3$

24.16 Draw a structure for a copolymer that might result from the polymerization of styrene and acrylonitrile.

24.17 What features are necessary in a molecule if it is to be polymerized?

24.18 Write the structure of nylon 6,10.

24.19 Draw the structures of the polymers that would result from the condensation polymerization reactions of

each of the following monomers.
   a. H_2N-CH_2-COOH
   b. $HOOC-CH_2-CH=CH-CH_2-NH_2$
   c. $H_2N-CH(CH_3)-CH_2-COOH$
   d. $H_2N-CH_2-CH_2-NH_2$ and
      $HOOC-CH_2-COOH$
   e. $HO-CH_2-CH=CH-CH_2-OH$ and
      $HOOC-C_6H_4-COOH$

24.20 Identify the repeating unit in each of the following polymers.
   a. polyacrylonitrile
   b. polystyrene
   c. polypropylene
   d. polybutadiene
   e. styrene–butadiene polymer
   f. polytetrafluoroethylene

24.21 Draw six-carbon (or longer) sections of the polymer chains that would be formed from each of the following monomers. Give the chemical and/or common name for the polymer and indicate at least one use for this polymer.
   a. $CH_2=CHCl$
   b. $CH_2=CH-CN$
   c. $CH_2=CH-CH=CH_2$
   d. $C_6H_5-CH=CH_2$
   e. $CF_2=CF_2$
   f. $CH_3-CH=CH_2$

Proteins

   Draw the structures of five amino acids.

24.23 When a protein is hydrolyzed, what chemical species results?

24.24 Draw the structure of a polypeptide that contains only alanine.

24.25 Distinguish among the primary, secondary, tertiary, and quaternary structures of proteins.

24.26 What is the general structure of an alpha-amino acid?

24.27 How does the structure of myoglobin differ from that of wool (a fibrous protein)?

24.28 Discuss the relationship between the proper functioning of enzymes and the tertiary structure of proteins.

24.29 Describe several functions of proteins in living systems.

24.30 Write an equation to describe how amino acids combine to form proteins.

24.31 Draw the dipeptides that could be formed from alanine and leucine.

24.32 Indicate which amino acids are obtained by hydrolyzing the tripeptide

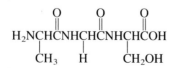

24.33 Draw the structure of the tetrapeptide Ala–Ser–Phe–Gly.

24.34 Describe the role of hydrogen bonds in developing the alpha-helical structure of proteins.

24.35 Both fibrous and globular proteins are formed from long polypeptide chains. Explain the differences between the tertiary structures of these proteins.

24.36 Describe the effect of denaturation on a protein and list some ways to denature proteins.

Nucleic Acids

24.37 What is the relationship among nucleic acids, amino acids, and proteins?

24.38 Describe the differences between RNA and DNA in terms of their constituent parts, their structures, and their functions.

24.39 Consider the structures of adenine, cytosine, thymine, and guanine. On the basis of hydrogen-bonding interactions between all possible combinations of these bases, explain why the only pairings found in DNA are between adenine and thymine and between cytosine and guanine.

24.40 How does DNA replicate itself?

24.41 Describe the parts of a nucleotide.

24.42 If the base sequence in a portion of a strand of DNA is A–C–T–A–A–G–A–T, what is the base sequence in the complementary strand?

24.43 If the base sequence in a portion of a strand of DNA is A–C–T–A–A–G–A–T, what is the base sequence in a strand of RNA formed from this DNA?

24.44 How is genetic information encoded by nucleic-acid molecules?

24.45 Why do m-RNA and t-RNA have complementary base coding?

Carbohydrates

24.46 Write a formula and give a name for a. a monosaccharide, b. a disaccharide, and c. a polysaccharide.

24.47 Why is cellulose called a carbohydrate?

24.48 What monosaccharides could be obtained from the hydrolysis of each of the following?
   a. sucrose   c. cellulose
   b. starch   d. glycogen

24.49 Draw the structure of glucose.

24.50 Describe the difference between alpha-glucose and beta-glucose.

24.51 Describe and explain any major differences between the properties of cellulose and those of starch.

Lipids

24.52 Discuss the structural differences between fats and oils.

24.53 Write a general formula to represent the chemical composition of fats and that of oils. How do they differ from one another?

Additional Exercises

24.54 How do thermosetting and thermoplastic polymers differ in structure and in properties?

24.55 Draw the structure of a polypeptide that contains only serine.

24.56 Name ten polymers that you encounter regularly.

24.57 Compare the primary, secondary, and tertiary structures of a protein with the structure of a telephone cord.

24.58 Draw the structure of the tripeptide Val–Cys–Tyr.

24.59 Write structures for the polymers that could be formed from each of the following monomers.

   a. $CF_2{=}CCl_2$

   b. $H_2N{-}(CH_2)_2{-}NH_2$ and $HOOC{-}(CH_2)_4{-}COOH$

   c. $H_2N{-}C_6H_4{-}COOH$

   d. $CH_2{=}CH{-}C_6H_4{-}CH_3$

24.60 Compare the functions of proteins and carbohydrates in living systems.

24.61 Write an equation to describe how proteins hydrolyze to form amino acids.

▶ ▶ ▶ Chemistry in Practice

24.62 A solid sample of hemoglobin is isolated from blood. A solution prepared by dissolving 0.2325 g of the hemoglobin in 25.00 mL of water developed an osmotic pressure of 2.660 torr at 25.0°C. Digestion of 22.548 g of the hemoglobin followed by alkali treatment precipitated 0.1109 g of Fe_2O_3.

   a. What is the molar mass of hemoglobin?

   b. How many iron atoms are present in each hemoglobin molecule?

Appendix A

▶ ▶ ▶ ▶

Mathematical Operations

Scientific Notation

Many numbers used in chemistry are either very large, such as 602,200,000,000,000,000,000,000 atoms in 12 g of carbon, or very small, such as 0.0000000001 cm per picometer. It is very easy to make mistakes with such numbers, so we express them in a shorthand notation called exponential or scientific notation. In scientific notation, the numbers used in those examples would be written as 6.022×10^{23} and 1×10^{-10}, respectively. The numbers are easier to write in this form, and it is easier to keep track of the position of the decimal point when carrying out calculations.

A number in scientific notation is expressed as $C \times 10^n$, where C is called the **coefficient** and n the **exponent.** The coefficient, C, is a number greater than or equal to 1 and less than 10 that is obtained by moving the decimal point the appropriate number of places. The exponent, n, is a positive or negative integer equal to the number of places the decimal point must be moved to give C. For numbers greater than 1, the decimal point is moved to the left and the exponent is positive. For numbers smaller than 1, the exponent is negative because the decimal point must be moved to the right.

In order to carry out mathematical operations with numbers written in scientific notation, we need to know how to operate on the exponential part of the number.

To multiply numbers that have exponents, we add the exponents, as in the following examples:

$$10^4 \times 10^7 = 10^{(4+7)} = 10^{11}$$
$$10^3 \times 10^{-5} = 10^{[3+(-5)]} = 10^{-2}$$
$$10^{-4} \times 10^4 = 10^{(-4+4)} = 10^0 = 1$$

To divide numbers written in exponential notation, we divide the coefficients and subtract the exponents.

$$10^6/10^4 = 10^{(6-4)} = 10^2$$
$$10^{-3}/10^{-8} = 10^{[-3-(-8)]} = 10^5$$

To raise an exponential number to a power, we multiply the exponent by the power.

$$(10^4)^3 = 10^{(4 \times 3)} = 10^{12}$$
$$(10^{-2})^{-3} = 10^{(-2 \times -3)} = 10^6$$

A-1

To extract the root of an exponential number, we divide the exponent by the root.

$$(10^8)^{1/2} = 10^{(8/2)} = 10^4$$
$$(10^{27})^{1/9} = 10^{(27/9)} = 10^3$$

If the exponent is not evenly divisible by the root, we rewrite the number in such a way that the exponent is evenly divisible.

$$(10^7)^{1/2} = (10 \times 10^6)^{1/2} = 10^{1/2} \times (10^6)^{1/2} = 3.16 \times 10^3$$

To add or subtract numbers written in exponential notation, we must first express both numbers with identical exponents. To do so, we shift the decimal point in one of the numbers so that its exponential part is the same as that of the other number. Then the coefficients are simply added or subtracted, and the exponential part is carried through unchanged. For example,

$$5.2 \times 10^4 + 7.0 \times 10^3 = 52. \times 10^3 + 7.0 \times 10^3 =$$
$$(52. + 7.0) \times 10^3 = 59. \times 10^3 = 5.9 \times 10^4$$

Error, Accuracy, and Precision

Every measurement has some degree of uncertainty associated with it. People make mistakes, and measuring devices are not perfect. The concepts of accuracy and precision are used to describe this uncertainty. *Accuracy* defines how close a measured value is to the true value. *Precision* describes the repeatability of a measurement.

Accuracy

Several terms are used to describe accuracy, and all of them are based on the idea of the *error* in a measurement. Error is defined as the difference between the measured value and the true value. The smaller the error, the more accurate the measurement. This difference can be expressed in several ways:

$$\text{Absolute error} = \text{measured value} - \text{true value}$$
$$\text{Relative error} = \frac{\text{absolute error}}{\text{true value}}$$
$$\% \text{ error} = \frac{\text{absolute error}}{\text{true value}} \times 100$$

Precision

The true value of the quantity we want to measure is not always known. In such a case, the error terms just described are not of much use. Instead, the same measurement is made repeatedly using the same procedure. Then the uncertainty can be expressed as the **precision,** or reproducibility, of the measurement. Suppose that a set of n repetitive measurements has been made of some value: x_1, x_2, x_3, . . . , x_n. The **average value** for this set of measurements is calculated by dividing the *sum* of all the values by the *number* of values.

$$\bar{x} = (x_1 + x_2 + x_3 + \cdots + x_n)/n$$

The average value, rather than the true value, is then used as the reference point. To determine the precision, it is necessary to calculate the **deviation** (d_i), which is the difference between an individual measured value and the average value.

$$d_i = x_i - \bar{x}$$

The precision of the set of measurements is then calculated as the **average deviation,** which is the sum of the absolute values of the individual deviations divided by the number of measurements.

$$\bar{d} = (|d_1| + |d_2| + |d_3| + \cdots + |d_n|)/n$$

The smaller the average deviation, the more precise the set of measurements. The terms **relative deviation** and **% deviation** are also often used in this context.

$$\text{Relative deviation} = \bar{d}/\bar{x}$$
$$\text{% deviation} = (\bar{d}/\bar{x}) \times 100$$

These definitions can be illustrated with an example. Suppose an object is weighed four times and the following results are obtained:

$$16.22 \text{ g} \qquad 16.15 \text{ g} \qquad 16.17 \text{ g} \qquad 16.26 \text{ g}$$

Also suppose that the true value, obtained by very careful measurements with a more accurate balance, is 16.26 g.

We wish to determine the error, deviation, % error, and % deviation for each measurement and the average deviation and % deviation for the set of measurements. First, we obtain the average value.

$$\text{Average mass} = (16.22 \text{ g} + 16.15 \text{ g} + 16.17 \text{ g} + 16.26 \text{ g})/4 = 16.20 \text{ g}$$

Next, we calculate the other quantities for the first value.

$$\text{Error} = 16.22 \text{ g} - 16.26 \text{ g} = -0.04 \text{ g}$$
$$\text{% error} = (-0.04 \text{ g}/16.26 \text{ g}) \times 100 = -0.25\%$$
$$\text{Deviation} = 16.22 - 16.20 = 0.02 \text{ g}$$
$$\text{% deviation} = (0.02 \text{ g}/16.20 \text{ g}) \times 100 = 0.12\%$$

Proceeding in the same manner for the other values yields the following results:

Value	Error	% Error	Deviation	% Deviation
16.22	−0.04	−0.25	0.02	0.12
16.15	−0.11	−0.68	−0.05	−0.31
16.17	−0.09	−0.55	−0.03	−0.19
16.26	0.00	0.00	0.06	0.37

The average deviation is then given by

$$\text{Average deviation} = \tfrac{1}{4}(|0.02| + |-0.05| + |-0.03| + |0.06|)$$
$$= \tfrac{1}{4}(0.02 + 0.05 + 0.03 + 0.06)$$
$$= \tfrac{1}{4}(0.16) = 0.04 \text{ g}$$

The average value is conveniently written as 16.20 ± 0.04 g, expressing the idea that if we measured this value again, we would expect it to fall within the range of 16.16 to 16.24 g. This range is actually somewhat smaller than the actual range of values observed (16.15 to 16.26 g).

Error Analysis: An Advanced Treatment

When carrying out mathematical operations with numbers that were determined experimentally, we need to be concerned with the **propagation of errors** in the calculated result. The simplest—though not a completely sufficient—approach to this problem is the consideration of significant figures in the result of a mathematical operation. As an example, consider the determination of the density of a liquid by the following procedure: (1) Weigh a beaker. (2) Deliver a given volume of liquid into the beaker with a pipet. (3) Weigh the beaker again. Each of the experimental measurements involved in this operation has some uncertainty associated with it. We want to know how much uncertainty will be generated in the calculated value of the density by virtue of uncertainty in these measurements. The calculation is carried out as follows:

$$\text{Density} = \frac{\text{mass of beaker and liquid} - \text{mass of empty beaker}}{\text{volume of liquid}}$$

By considering the proper number of significant figures in such a calculation, we can get an estimate of the uncertainty in the calculated result. However, a more exact treatment, which gives the maximum possible uncertainty in the calculated result, requires application of the following rules:

1. In *addition* and *subtraction,* the **absolute precision** of the answer is given by the *sum of the absolute precisions* of the component quantities.

   Example: $15 \pm 5 + 28 \pm 2 = 43 \pm 7$

2. In *multiplication* and *division,* the **relative precision** of the answer is given by the *sum of the relative precisions* of the component quantities.

   Example: $(15 \pm 5) \times (28 \pm 2) = 420 \pm 168$

This result arises from the sum of the relative precisions,

$$(5/15) + (2/28) = 0.33 + 0.07 = 0.40$$

which is converted to a deviation.

$$\text{Deviation}/420 = 0.40$$
$$\text{Deviation} = 420 \times 0.40 = 168$$

Let's consider the example of a density measurement in more detail, using the following data:

Mass of beaker and liquid	53.1638 g $\pm$ 0.0002 g
Mass of empty beaker	35.8266 g $\pm$ 0.0002 g
Mass of liquid	? g $\pm$? g
Volume of liquid	10.05 mL $\pm$ 0.01 mL
Density of liquid	? g/mL $\pm$? g/mL

We calculate the mass of the liquid as follows:

$$\text{Mass of liquid} = 53.1638 \text{ g} - 35.8266 \text{ g} = 17.3372 \text{ g}$$

Then the precision of the mass is given by the sum of the absolute precisions.

$$\text{Precision of mass} = 0.0002 \text{ g} + 0.0002 \text{ g} = 0.0004 \text{ g}$$

Therefore, the mass of the liquid is 17.3372 ± 0.0004 g. The validity of obtaining the precision of the mass of liquid by adding the absolute precisions of the component masses can be demonstrated by considering that 53.1638 ± 0.0002 g means that this mass lies between 53.1636 g and 53.1640 g, and that 35.8266 ± 0.0002 g means that the mass lies between 35.8264 g and 35.8268 g. By subtracting the extreme values, we see that the difference in mass must lie between $53.1640 - 35.8264 = 17.3376$ g and $53.1636 - 35.8268 = 17.3368$ g. Thus the average value of these extremes is 17.3372 g with a precision of ± 0.0004 g.

The density is then given by

$$d = (17.3372 \pm 0.0004 \text{ g})/(10.05 \pm 0.01 \text{ mL}) = 1.725 \pm 0.002 \text{ g/mL}$$

In this case, we calculate the precision by using the rule that the relative deviation of the answer is given by the sum of the relative deviations of the component numbers.

$$\text{rel. dev. of density} = \text{rel. dev. of mass} + \text{rel. dev. of volume}$$

$$\frac{\text{abs.dev. of density}}{\text{density}} = \frac{\text{abs. dev. of mass}}{\text{mass}} + \frac{\text{abs. dev. of volume}}{\text{volume}}$$

$$\frac{\text{abs. dev. of } d}{d} = \frac{0.0004}{17.3372} + \frac{0.01}{10.05} = 2 \times 10^{-5} + 1 \times 10^{-3} = 1 \times 10^{-3}$$

Thus the absolute deviation of the density is given by

$$\text{Absolute deviation} = 1.725 \times 1 \times 10^{-3} = 0.002 \text{ g/mL}$$

Again, the validity of this procedure can be demonstrated by considering that the mass lies between 17.3368 and 17.3376 g and that the volume lies between 10.04 and 10.06 mL. Calculating the density with the extreme values gives 17.3368 g/10.06 mL = 1.723 g/mL   and   17.3376 g/10.04 mL = 1.727 g/mL. The average of these two values is 1.725 g/mL, and the average deviation is then ± 0.002 g/mL.

Note that in subtractions, it is the absolute precision that is important, so even though each measurement may be made with good relative precision, the difference between two close values is very imprecise. For example, if the two masses had been 21.1637 ± 0.0002 g and 21.1629 ± 0.0002 g, the values would have had % precisions of only 0.001%, but the difference between the masses would have been 0.0008 ± 0.0004 g, with a precision of 50%. Small differences between large numbers are very unreliable and should be avoided.

Logarithms

We generally use numbers that can be expressed as a power, or exponent, of 10. **Common logarithms** are simply the exponent to which 10 must be raised to equal the number in question. Thus the logarithm of 1 is 0 because $1 = 10^0$, or log 1 = 0.
Similarly,

$$\log 10 = 1 \qquad \log 100 = 2 \qquad \log 0.1 = -1 \qquad \log 0.01 = -2$$

The logarithm of any number that is an integral power of 10 is a positive or a negative whole number. The logarithms of such numbers can be determined simply by inspection. One of the important aspects of logarithms is that because they are just exponents of 10, they can be manipulated in the same manner as pure exponential numbers. Recall that when multiplying two exponential numbers, we just add the exponents.

$$10^3 \times 10^4 = 10^{3+4} = 10^7$$

This relationship can be extended by considering that logarithms are exponents of 10, so the logarithm of a product is simply the sum of the logarithms of the numbers being multiplied.

$$\log(10^3 \times 10^4) = \log(10^3) + \log(10^4) = 3 + 4 = 7 = \log(10^7)$$

By using logarithms, it is possible to carry out multiplications and divisions via the less tedious operations of addition and subtraction. Of course, with integral multiples of 10, the processes of multiplication and division are so simple that logarithms are not of much use. The utility of logarithms is more evident when we consider that any number can be expressed in a purely exponential form. Some common nonintegral powers of 10 include the square root.

$$\sqrt{10} = 10^{1/2} = 10^{0.50} = 3.16$$

Thus the logarithm of 3.16 is 0.500 ($\log 3.16 = 0.500$). A number that falls between 1 and 10 has a logarithm which falls between $\log 1 = 0$ and $\log 10 = 1$, as in this example. Similarly, the logarithm of a number between 10 and 100 lies between 1 and 2; the logarithm of a number between 0.01 and 0.001 falls between -2 and -3; and so on. Tables of logarithms or the logarithm function on a calculator can be used to convert any number to its purely exponential form.

The logarithm is the exponent to which 10 must be raised to give the number of interest.

$$2 = 10^{0.3010}$$
$$200 = 10^{2.3010}$$
$$0.002 = 10^{-2.6990} = 10^{(0.3010-3)}$$

The logarithm of a number contains two parts, the fractional exponent (called the **mantissa**), which is the number found in log tables, and the whole-number exponent (called the **characteristic**). The mantissa gives the numerical value, and the characteristic establishes the position of the decimal point. For numbers greater than 1, the characteristic has a value one less than the number of digits to the left of the decimal point. With numbers less than 1, the logarithm is negative, and the characteristic has a numerical value of one more than the number of zeroes between the decimal point and the first nonzero digit to the right of the decimal point.

Having numbers in their pure exponential forms makes it convenient to carry out mathematical operations, because only addition and subtraction are required. However, when the answer is obtained, it is usually more convenient to have it in scientific notation. With an answer in the form of a logarithm, we want instead the number that corresponds to this logarithm—that is, the **antilogarithm.**

To find the antilogarithm, reverse the procedure for finding logarithms. For example, if the logarithm is 5.7340, then the antilogarithm is

$$\text{antilog } 5.7340 = 10^{5.7340} = 10^{0.7340} \times 10^5$$

To convert $10^{0.7340}$ to normal decimal notation, locate 0.7340 among the mantissa values listed in the log table, and read the number corresponding to it: 5.42. The number is thus

$$\text{antilog } 5.7340 = 10^{0.7340} \times 10^5 = 5.42 \times 10^5$$

Alternatively, calculators with the logarithm function either have an inverse button that activates the antilogarithm function or have a 10^x button. In either case, the antilog value is obtained directly by use of this button.

Logarithms are just exponents, so they follow the same mathematical rules as numbers written in exponential notation.

1. To multiply two numbers, add their logarithms and then take the antilog.
2. To divide two numbers, subtract their logarithms (log dividend minus log divisor) and then take the antilog.
3. To raise a number to a power, multiply the logarithm of the number by the power and then take the antilog.
4. To take a root of a number, divide the logarithm of the number by the value of the root and then take the antilog.

Quadratic Equations

Because many of the principles of chemistry can be represented with algebraic equations, finding the solutions to problems that involve these principles requires the ability to solve and manipulate such equations. Many problems can be solved by simple rearrangement of the equations. However, some equations are more complex.

For example, equations that contain terms with both x^2 and x can be put into the general form of a quadratic equation:

$$ax^2 + bx + c = 0$$

For such an equation, the value of the variable x is given by the relationship

$$x = [-b \pm (b^2 - 4ac)^{1/2}]/2a$$

There are two possible values for x that satisfy the equation, but in most chemical applications of such equations, one value will be much more reasonable than the other.

Graphs

A functional relationship between two variables x and y can be represented by means of an algebraic equation. It is often desirable to discover such relationships between variables in experimental data, in order that a large amount of data may be presented in the form of a single equation. The discovery of such relationships, or the demonstration that such relationships do indeed exist within a given set of data, is often facilitated by the proper construction of an appropriate graph.

Following these guidelines generally leads to a graph that presents data in a clear and easily interpreted manner.

1. Carefully examine the data to be plotted. Organizing the data in a table facilitates their transfer to the graph.

2. Clearly label each axis to show the variable that is to be plotted along it. Include the units of the variable, either in parentheses following the name of the

variable, or separated from the variable name by a comma: mass (g) or mass, g. The vertical axis, which is called the **ordinate,** is used for the y variable. This is the dependent variable—the variable measured experimentally. The horizontal axis, called the **abscissa,** is used for the x variable. This is the independent variable—the variable that is controlled and ''varied,'' or manipulated, by the experimenter.

3. Decide on the scales and on the limits of values to be plotted along each axis. The graph should fill approximately the entire available area of graph paper. The scale should be adjusted so that the data can be entered easily on the graph and subsequently can be read off the graph with about the same precision as that with which they were measured. The value of each major division on an axis is usually selected so that minor divisions represent workable numbers of units (use 1 unit, 5 units, 0.5 units, or 0.1 unit rather than 1.6 units, 0.28 unit, or 0.07 unit). To avoid using only a small area of the graph, it may sometimes not be desirable for the scale of each axis to begin at zero, or for each division to represent the same number of units along both axes.

4. After selecting the scales to be plotted along each axis, clearly mark the value of each major division along each axis. If the numbers to be plotted are very small (such as 0.0001) or very large (such as 10,000), there may not be enough space to mark the value of each major division. Expressing the numbers in exponential notation may not save much space. One solution is to label the major divisions with the coefficient of the number expressed in exponential notation (to use, for instance, the 5 from 5×10^{-4}) and then to label the axis as the quantity being plotted times the inverse of the exponential part of the number (such as $y \times 10^4$). In this example, the implication is that $y \times 10^4 = 5$, or that $y = 5/10^4 = 5 \times 10^{-4}$.

5. Place dots on the graph at appropriate intersecting y and x values to represent each data point. Each point plotted on the graph should have a circle (or square) about 1–2 mm in diameter drawn around it in order to emphasize its location. If several sets of data are to be plotted on the same graph, use different geometric figures, such as squares, triangles, and diamonds) to outline the other sets of data points.

6. A smooth line should be drawn through the data. To avoid obscuring the exact location of the experimental points, the line should not pass through the inside of the outlining circles but rather should be temporarily discontinued when it reaches a circle. Do not connect the points by *separate* straight-line segments. It is generally assumed that a measured variable varies *smoothly* with the independent variable and that when points do not fall exactly on a smooth line, this is due to experimental error. If the data appear to follow a straight line, use a straight edge to draw the line. If the data follow a curve, use a graphing aid such as a French curve or a flexible spline.

7. The completed graph should be identified with an appropriate title in a clear space near the top. Your name and the date should be included if necessary for identification.

These steps are illustrated by the following example, which involves plotting the volume of a gas as a function of the absolute temperature.

Volume (L)	Temperature (K)
2.20	300
2.37	325
2.59	350
2.95	400
3.30	450
3.71	500
4.37	600
5.17	700

The y axis should be used for volume and the x axis for temperature. Suppose we have a piece of graph paper that is four major divisions high and six major divisions wide. The y axis then has four major divisions, and we need to cover 5.17 L − 2.20 L = 2.97 L, so to keep even units for each minor division, we assign 1.0 unit per major division (rather than 2.97 L/4 = 0.74 L/major division). On the x axis, we have six major divisions to cover 700 K − 300 K = 400 K. Although we could assign 400 K/6 = 66.7 K per major division, the graphing procedure is simplified if we assign 100 K per major division. Proceeding according to the steps outlined, we obtain the accompanying graph.

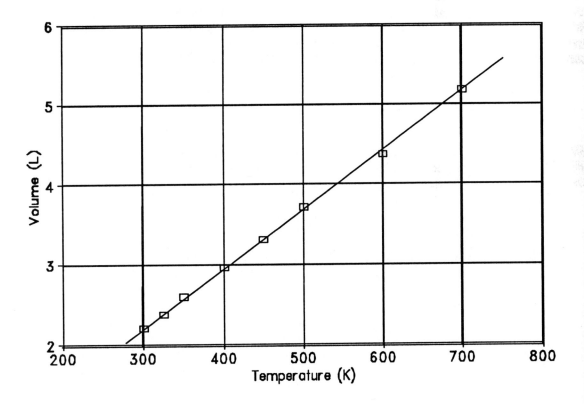

If the data are such that a straight line can be drawn through the points on the graph, then the data obey the general function

$$y = mx + b$$

where y is the dependent variable and x is the independent variable. The parameter

m is called the **slope.** The parameter b is the **intercept** on the y axis (the value of y when $x = 0$); it is also known as the y intercept. The slope is calculated from the coordinates of any two points (y_1, x_1) and (y_2, x_2) on the line. The slope is given by

$$m = (y_2 - y_1)/(x_2 - x_1)$$

Do not use two experimental points, because the line, if drawn properly, represents the average behavior of all the data, and some of the experimental points may not fall directly on the line. A properly drawn straight line will fall as close as possible to as many of the points as possible. Of those points that lie off the line, approximately as many should fall above the line as fall below the line. The intercept can be read directly from the graph if the scale of the x axis goes down to zero. Otherwise, having determined the value of m, we obtain the intercept by using any point (y_1, x_1) on the line.

$$b = y_1 - mx_1$$

If the data do not fit a straight line, it might be possible to obtain a linear graph by plotting some other function instead of y versus x. Some possibilities are

$$y \text{ versus } 1/x \qquad 1/y \text{ versus } x$$
$$\log y \text{ versus } x \qquad y \text{ versus } \log x$$
$$y \text{ versus } x^2 \qquad y^2 \text{ versus } x$$
$$\log y \text{ versus } 1/x$$

Appendix B

▶ ▶ ▶ ▶

Solubility Rules for Inorganic Salts in Water

NO_3^-	All nitrates are soluble.
$CH_3CO_2^-$	All acetates are soluble. ($AgCH_3CO_2$ is only moderately soluble.)
NO_2^-, ClO_4^-, ClO_3^-, MnO_4^-, $Cr_2O_7^{2-}$	Most are soluble.
Cl^-, Br^-, I^-, SCN^-	All chlorides, bromides, iodides, and thiocyanates are soluble, except AgX, Hg_2X_2, PbX_2 (but $PbCl_2$ is slightly soluble in cold water and moderately soluble in hot water), and HgI_2.
SO_4^{2-}	All sulfates are soluble, except $BaSO_4$, $SrSO_4$, and $PbSO_4$. [$CaSO_4$, Hg_2SO_4, and Ag_2SO_4 are only slightly soluble; the corresponding bisulfate (HSO_4^-) salts are more soluble.]
CO_3^{2-}, PO_4^{3-}, SO_3^{2-}, AsO_4^{3-} AsO_3^{3-}, BO_3^{3-}, SiO_3^{2-}	All carbonates, phosphates, sulfites, arsenates, arsenites, borates, and silicates are insoluble, except those of Na^+, K^+, and NH_4^+. [Many acid phosphates are soluble, such as $Mg(H_2PO_4)_2$ and $Ca(H_2PO_4)_2$.]
CrO_4^{2-}	Most chromates are insoluble, except those of K^+, Na^+, NH_4^+, Mg^{2+}, Ca^{2+}, Al^{3+}, and Ni^{2+}. Chromates are soluble in acid solution.
OH^-	All hydroxides are insoluble, except $NaOH$, KOH, NH_4OH, and $Ba(OH)_2$. [$Ca(OH)_2$ is slightly soluble.]
O^{2-}	Oxides are usually insoluble, except those of metal ions that form soluble hydroxides.
S^{2-}	All sulfides are insoluble, except those of Na^+, K^+, and NH_4^+ and those of the alkaline earths: Mg^{2+}, Ca^{2+}, Sr^{2+}, and Ba^{2+}. (Sulfides of Al^{3+} and Cr^{3+} react with water to precipitate the corresponding hydroxides.)
Na^+, K^+, NH_4^+	All salts of sodium, potassium, and ammonium ions are soluble, except some uncommon ones. ($KClO_4$ is only slightly soluble.)
Ag^+	All silver salts are insoluble, except $AgNO_3$ and $AgClO_4$. $AgCH_3CO_2$ and Ag_2SO_4 are only moderately soluble.

Appendix C

Thermodynamic Quantities

Substance	Standard Molar Enthalpies of Formation, $\Delta H^\circ_{f,298.15}$ (kJ/mol)	Standard Molar Free Energies of Formation, $\Delta G^\circ_{f,298.15}$ (kJ/mol)	Absolute Standard Entropies at 25°C and 1 atm, $S^\circ_{298.15}$ (J/K mol)
Ag(s)	0	0	42.55
Ag(g)	284.6	245.7	172.89
AgCl(s)	−127.1	−109.8	96.2
Ag$_2$O(s)	−31.0	−11.2	121
Ag$_2$S(s)	−32.6	−40.7	144.0
Al(s)	0	0	28.3
Al(g)	326	286	164.4
AlCl$_3$(s)	−704.2	−628.9	110.7
AlF$_3$(s)	−1504	−1425	66.44
Al$_2$O$_3$(s)	−1676	−1582	50.92
Al$_2$S$_3$(s)	−724	−492.4	—
Al$_2$(SO$_4$)$_3$(s)	−3440.8	−3100.1	239
As(s)	0	0	35
As(g)	303	261	174.1
As$_4$O$_6$(s)	−1313.9	−1152.5	214
AsCl$_3$(g)	−258.6	−245.9	327.1
AsH$_3$(g)	66.44	68.91	222.7
As$_2$S$_3$(s)	−169	−169	164
B(s)	0	0	5.86
B(g)	562.7	518.8	153.3
BCl$_3$(g)	−403.8	−388.7	290.0
B$_2$H$_6$(g)	36	86.6	232.0
BF$_3$(g)	−1137.3	−1120.3	254.0
B$_3$N$_3$H$_6$(l)	−541.0	−392.8	200
B$_2$O$_3$(s)	−1263.6	−1184.1	78.7
B(OH)$_3$(s)	−1094.3	−969.01	88.83
Ba(s)	0	0	66.9

Substance	Standard Molar Enthalpies of Formation, $\Delta H^\circ_{f,298.15}$ (kJ/mol)	Standard Molar Free Energies of Formation, $\Delta G^\circ_{f,298.15}$ (kJ/mol)	Absolute Standard Entropies at 25°C and 1 atm, $S^\circ_{298.15}$ (J/K mol)
Ba(g)	175.6	144.8	170.3
BaCl$_2$(s)	−860.06	−810.9	126
BaO(s)	−558.1	−528.4	70.3
BaSO$_4$(s)	−1465	−1353	132
Be(s)	0	0	9.54
Be(g)	320.6	282.8	136.17
BeO(s)	−610.9	−581.6	14.1
Bi(s)	0	0	56.74
Bi(g)	207	168	186.90
BiCl$_3$(s)	−379	−315	177
Bi$_2$O$_3$(s)	−573.88	−493.7	151
Bi$_2$S$_3$(s)	−143	−141	200
Br$_2$(l)	0	0	152.23
Br$_2$(g)	30.91	3.142	245.35
Br(g)	111.88	82.429	174.91
BrF$_3$(g)	−255.6	−229.5	292.4
C(s) (graphite)	0	0	5.740
C(s) (diamond)	1.887	2.900	2.38
C(g)	716.681	671.289	157.987
CCl$_4$(l)	−135.4	−65.27	216.4
CCl$_4$(g)	−102.9	−60.63	309.7
CH$_4$(g)	−74.81	−50.75	186.15
C$_2$H$_2$(g)	226.7	209.2	200.8
C$_2$H$_4$(g)	52.26	68.12	219.5
C$_2$H$_6$(g)	−84.68	−32.9	229.5
C$_3$H$_8$(g)	−103.85	−23.49	269.9
C$_6$H$_6$(g)	82.927	129.66	269.2
C$_6$H$_6$(l)	49.028	124.50	172.8
CHCl$_3$(l)	−134.5	−73.72	202
CHCl$_3$(g)	−103.1	−70.37	295.6
CH$_2$Cl$_2$(l)	−121.5	−67.32	178
CH$_2$Cl$_2$(g)	−92.47	−65.90	270.1
CH$_3$Cl(g)	−80.83	−57.40	234.5
C$_2$H$_5$Cl(l)	−136.5	−59.41	190.8
C$_2$H$_5$Cl(g)	−112.2	−60.46	275.9
CH$_3$OH(l)	−238.7	−166.4	126.8
CH$_3$OH(g)	−200.7	−162.0	239.7

Substance	Standard Molar Enthalpies of Formation, $\Delta H^\circ_{f,298.15}$ (kJ/mol)	Standard Molar Free Energies of Formation, $\Delta G^\circ_{f,298.15}$ (kJ/mol)	Absolute Standard Entropies at 25°C and 1 atm, $S^\circ_{298.15}$ (J/K mol)
$CH_3CHO(g)$	−166.4	−133.7	266
$C_2H_5OH(l)$	−277.7	−174.9	160.7
$C_2H_5OH(g)$	−235.1	−168.6	282.6
$CH_3CO_2H(l)$	−484.5	−390.0	159.8
$CH_3CO_2H(g)$	−432.25	−374.0	282.4
$CO(g)$	−110.52	−137.15	197.56
$CO_2(g)$	−393.51	−394.36	213.6
$CS_2(l)$	89.70	65.27	151.3
$CS_2(g)$	117.4	67.15	237.7
$Ca(s)$	0	0	41.6
$Ca(g)$	192.6	158.9	154.78
$CaCO_3(s)$ (calcite)	−1206.9	−1128.8	92.9
$CaO(s)$	−635.5	−604.2	39.7
$Ca(OH)_2(s)$	−986.59	−896.76	76.1
$CaSO_3 \cdot 2H_2O(s)$	−1762	−1565	184
$CaSO_4 \cdot 2H_2O(s)$	−2021.1	−1795.7	194.0
$CaSO_4(s)$	−1432.7	−1320.3	106.7
$Cd(s)$	0	0	51.76
$Cd(g)$	112.0	77.45	167.64
$CdCl_2(s)$	−391.5	−344.0	115.3
$CdO(s)$	−258	−228	54.8
$CdS(s)$	−162	−156	64.9
$CdSO_4(s)$	−933.28	−822.78	123.04
$Cl_2(g)$	0	0	222.96
$Cl(g)$	121.68	105.70	165.09
$ClF(g)$	−54.48	−55.94	217.8
$ClF_3(g)$	−163	−123	281.5
$Cl_2O(g)$	80.3	97.9	266.1
$Co(s)$	0	0	30.0
$Co(NO_3)_2(s)$	−420.5	—	—
$CoO(s)$	−237.9	−214.2	52.97
$Co_3O_4(s)$	−891.2	−774.0	103
$Cr(s)$	0	0	23.8
$Cr(g)$	397	352	174.4
$Cr_2O_3(s)$	−1140	−1058	81.2
$CrO_3(s)$	−589.5	—	—
$(NH_4)_2Cr_2O_7(s)$	−1807	—	—

Substance	Standard Molar Enthalpies of Formation, $\Delta H^\circ_{f,298.15}$ (kJ/mol)	Standard Molar Free Energies of Formation, $\Delta G^\circ_{f,298.15}$ (kJ/mol)	Absolute Standard Entropies at 25°C and 1 atm, $S^\circ_{298.15}$ (J/K mol)
$Cu(s)$	0	0	33.15
$Cu(g)$	338.3	298.5	166.3
$Cu(NO_3)_2(s)$	−303	—	—
$Cu_2O(s)$	−169	−146	93.14
$CuO(s)$	−157	−130	42.63
$Cu_2S(s)$	−79.5	−86.2	121
$CuS(s)$	−53.1	−53.6	66.5
$CuSO_4(s)$	−771.36	−661.9	109
$F_2(g)$	0	0	202.7
$F(g)$	78.99	61.92	158.64
$F_2O(g)$	−22	−4.6	247.3
$Fe(s)$	0	0	27.3
$Fe(g)$	416	371	180.38
$FeO(s)$	−272	—	—
$Fe_2O_3(s)$	−824.2	−742.2	87.40
$Fe_3O_4(s)$	−1118	−1015	146
$Fe(OH)_2(s)$	−569.0	−486.6	88
$Fe(OH)_3(s)$	−823.0	−696.6	107
$FeS(s)$	−100	−100	60.29
$H_2(g)$	0	0	130.57
$H(g)$	217.97	203.26	114.60
$HBr(g)$	−36.4	−53.43	198.59
$HCl(g)$	−92.307	−95.299	186.80
$HCN(l)$	108.9	124.9	112.8
$HCN(g)$	135	124.7	201.7
$HCO_2H(g)$	−362.6	−335.7	251
$HF(g)$	−271	−273	173.67
$HI(g)$	26.5	1.7	206.48
$HNO_3(l)$	−174.1	−80.79	155.6
$HNO_3(g)$	−135.1	−74.77	266.2
$H_2O(l)$	−285.83	−237.18	69.91
$H_2O(g)$	−241.82	−228.59	188.71
$H_2O_2(l)$	−187.8	−120.4	109.6
$H_2O_2(g)$	−136.3	−105.6	233
$H_3PO_2(s)$	−604.6	—	—
$H_3PO_3(s)$	−964.4	—	—
$H_3PO_4(s)$	−1279	−1119	110.5

Substance	Standard Molar Enthalpies of Formation, $\Delta H^\circ_{f,298.15}$ (kJ/mol)	Standard Molar Free Energies of Formation, $\Delta G^\circ_{f,298.15}$ (kJ/mol)	Absolute Standard Entropies at 25°C and 1 atm, $S^\circ_{298.15}$ (J/K mol)
$H_3PO_4(l)$	−1267	—	—
$H_2S(g)$	−20.6	−33.6	205.7
$H_2SO_4(l)$	−813.989	−690.101	156.90
$H_2Se(g)$	30	16	218.9
$H_2SiO_3(s)$	−1189	−1092	130
$H_4SiO_4(s)$	−1481	−1333	190
$Hg(l)$	0	0	76.02
$Hg(g)$	61.317	31.85	174.8
$Hg_2Cl_2(s)$	−265.2	−210.78	192
$HgCl_2(s)$	−224	−179	146
$HgO(s)$ (red)	−90.83	−58.555	70.29
$HgO(s)$ (yellow)	−90.46	−57.296	71.1
$HgS(s)$ (red)	−58.16	−50.6	82.4
$HgS(s)$ (black)	−53.6	−47.7	88.3
$I_2(s)$	0	0	116.14
$I_2(g)$	62.438	19.36	260.6
$I(g)$	106.84	70.283	180.68
$IBr(g)$	40.8	3.7	258.66
$ICl(g)$	17.8	−5.44	247.44
$IF(g)$	−95.65	−118.5	236.1
$IF_7(g)$	−943.9	−818.4	346
$K(s)$	0	0	63.6
$KCl(s)$	−435.87	−408.32	82.68
$Li(s)$	0	0	28.0
$Li(g)$	155.1	122.1	138.67
$LiCl(s)$	−435.9	−408.3	390.5
$Li_2CO_3(s)$	−1215.6	−1132.4	90.4
$LiF(s)$	−612.1	−584.1	35.9
$LiH(s)$	−90.42	−69.96	25
$LiOH(s)$	−487.23	−443.9	50.2
$Mg(s)$	0	0	32.5
$Mg(g)$	149	115	148.5
$MgO(s)$	−601.83	−569.57	27
$Mg(OH)_2(s)$	−924.66	−833.75	63.14
$MgCl_2(s)$	−641.83	−592.33	89.5
$MgSO_4(s)$	−1278	−1174	91.6
$Mn(s)$	0	0	32.0

Substance	Standard Molar Enthalpies of Formation, $\Delta H^\circ_{f,298.15}$ (kJ/mol)	Standard Molar Free Energies of Formation, $\Delta G^\circ_{f,298.15}$ (kJ/mol)	Absolute Standard Entropies at 25°C and 1 atm, $S^\circ_{298.15}$ (J/K mol)
Mn(g)	281	238	173.6
MnO(s)	−385.2	−362.9	59.71
Mn$_2$O$_3$(s)	−959.0	−881.2	110
MnO$_2$(s)	−520.03	−465.18	53.05
Mn$_3$O$_4$(s)	−1388	−1283	156
N$_2$(g)	0	0	191.5
N(g)	472.704	455.579	153.19
NH$_3$(g)	−46.11	−16.5	192.3
N$_2$H$_4$(l)	50.63	149.2	121.2
N$_2$H$_4$(g)	95.4	159.3	238.4
NH$_4$Br(s)	−270.8	−175	113
NH$_4$Cl(s)	−314.4	−201.5	94.6
NH$_4$I(s)	−201.4	−113	117
NH$_4$NO$_3$(s)	−365.6	−184.0	151.1
N$_2$O(g)	82.05	104.2	219.7
NO(g)	90.25	86.57	210.65
NO$_2$(g)	33.2	51.30	239.9
N$_2$O$_3$(g)	83.72	139.4	312.2
N$_2$O$_4$(g)	9.16	97.82	304.2
N$_2$O$_5$(g)	11	115	356
Na(s)	0	0	51.0
Na(g)	108.7	78.11	153.62
NaBr(s)	−359.9	−347	83.7
NaCl(s)	−411.00	−384.03	72.38
NaF(s)	−569.0	−541.0	58.6
NaI(s)	−288.0	−282	91.2
NaOH(s)	−426.7	−381	64
Na$_2$O(s)	−415.9	−377	72.8
Na$_2$O$_2$(s)	−504.1	−451	95
O$_2$(g)	0	0	205.03
O(g)	249.17	231.75	160.95
O$_3$(g)	143	163	238.8
P(s)	0	0	41.1
P(g)	58.91	24.5	280.0
P$_4$(g)	314.6	278.3	163.08
PCl$_3$(g)	−287	−268	311.7
PCl$_5$(g)	−375	−305	364.5

Substance	Standard Molar Enthalpies of Formation, $\Delta H^\circ_{f,298.15}$ (kJ/mol)	Standard Molar Free Energies of Formation, $\Delta G^\circ_{f,298.15}$ (kJ/mol)	Absolute Standard Entropies at 25°C and 1 atm, $S^\circ_{298.15}$ (J/K mol)
$PH_3(g)$	5.4	13	210.1
P_4O_6	−1640	—	—
P_4O_{10}	−2984	−2698	228.9
$Pb(s)$	0	0	64.81
$Pb(g)$	195	162	175.26
$PbCl_2(s)$	−359.4	−314.1	136
$Pb(NO_3)_2(s)$	−451.9	—	—
$PbO(s)$ (yellow)	−217.3	−187.9	68.70
$PbO(s)$ (red)	−219.0	−188.9	66.5
$PbO_2(s)$	−277	−217.4	68.7
$Pb(OH)_2(s)$	−515.9	—	—
$PbS(s)$	−100	−98.7	91.2
$PbSO_4(s)$	−918.39	−811.2	147
$S(s)$ (rhombic)	0	0	31.8
$S(g)$	278.80	238.27	167.75
$SF_4(g)$	−774.9	−731.4	291.9
$SF_6(g)$	−1210	−1105	291.7
$SO_2(g)$	−296.83	−300.19	248.1
$SO_3(g)$	−395.7	−371.1	256.6
$Sb(s)$	0	0	45.69
$Sb(g)$	262	222	180.2
$SbCl_3(s)$	−382.2	−323.7	184
$SbCl_3(g)$	−314	−301	337.7
$SbCl_5(g)$	−394.3	−334.3	401.8
$Sb_4O_6(s)$	−1441	−1268	221
$Sb_2S_3(s)$	−175	−174	182
$Si(s)$	0	0	18.8
$Si(g)$	455.6	411	167.9
$SiC(s)$	−65.3	−62.8	16.6
$SiCl_4(l)$	−687.0	−619.90	239.7
$SiCl_4(g)$	−657.01	−617.01	330.6
$SiF_4(g)$	−1614.9	−1572.7	282.4
$SiH_4(g)$	34	56.9	204.5
$SiO_2(s)$	−910.94	−856.67	41.84
$Sn(s)$	0	0	51.55
$Sn(g)$	302	267	168.38
$SnCl_4(l)$	−511.2	−440.2	259

Substance	Standard Molar Enthalpies of Formation, $\Delta H^{\circ}_{f,298.15}$ (kJ/mol)	Standard Molar Free Energies of Formation, $\Delta G^{\circ}_{f,298.15}$ (kJ/mol)	Absolute Standard Entropies at 25°C and 1 atm, $S^{\circ}_{298.15}$ (J/K mol)
$SnCl_4(g)$	−471.5	−432.2	366
$SnO(s)$	−286	−257	56.5
$SnO_2(s)$	−580.7	−519.7	52.3
$Ti(s)$	0	0	30.6
$Ti(g)$	469.9	425.1	180.19
$TiCl_4(l)$	−804.2	−737.2	252.3
$TiCl_4(g)$	−763.2	−726.8	354.8
$TiO_2(s)$	−944.7	−889.5	50.33
$W(s)$	0	0	32.6
$W(g)$	849.4	807.1	173.84
$WO_3(s)$	−842.87	−764.08	75.90
$Zn(s)$	0	0	41.6
$Zn(g)$	130.73	95.178	160.87
$ZnCl_2(s)$	−415.1	−369.43	111.5
$ZnCO_3(s)$	−812.78	−731.57	82.4
$ZnO(s)$	−348.3	−318.3	43.64
$Zn(OH)_2(s)$	−642.2	—	—
$ZnS(s)$	−206.0	−201.3	57.7
$ZnSO_4(s)$	−982.8	−874.5	120

Acid-Ionization Constants and Base-Ionization Constants

Acid	Formula	K_a at 25°C
acetic	CH_3COOH	1.75×10^{-5}
arsenic	H_3AsO_4	6.0×10^{-3}
	$H_2AsO_4^-$	1.05×10^{-7}
	$HAsO_4^{2-}$	3.0×10^{-13}
arsenious	H_3AsO_3	6.0×10^{-10}
boric	H_3BO_3	5.83×10^{-10}
carbonic	H_2CO_3	4.45×10^{-7}
	HCO_3^-	4.7×10^{-11}
cyanic	$HOCN$	3.46×10^{-4}
formic	$HCOOH$	1.77×10^{-4}
hydrazoic	HN_3	1.90×10^{-5}
hydrocyanic	HCN	4.0×10^{-10}
hydrofluoric	HF	7.0×10^{-4}
hydrogen peroxide	H_2O_2	2.7×10^{-12}
hydroselenic	H_2Se	1.7×10^{-4}
	HSe^-	1×10^{-10}
hydrosulfuric	H_2S	5.7×10^{-8}
	HS^-	1.2×10^{-13}
hydrotelluric	H_2Te	2.3×10^{-3}
	HTe^-	1×10^{-5}
hypobromous	$HOBr$	2.2×10^{-9}
hypochlorous	$HOCl$	3.0×10^{-8}
hypoiodous	HOI	2.3×10^{-11}
nitrous	HNO_2	4.6×10^{-4}
oxalic	$H_2C_2O_4$	5.90×10^{-2}
	$HC_2O_4^-$	6.40×10^{-5}
phosphoric	H_3PO_4	7.52×10^{-3}
	$H_2PO_4^-$	6.23×10^{-8}
	HPO_4^{2-}	2.2×10^{-13}

phosphorous	H_3PO_3	1.00×10^{-2}
	$H_2PO_3^-$	2.6×10^{-7}
sulfuric	H_2SO_4	very strong
	HSO_4^-	1.20×10^{-2}
sulfurous	H_2SO_3	1.72×10^{-2}
	HSO_3^-	6.43×10^{-8}

Base	**Formula**	K_b **at 25°C**
ammonia	NH_3	1.76×10^{-5}
aniline	$C_6H_5NH_2$	3.94×10^{-10}
butylamine	$C_4H_9NH_2$	4.0×10^{-4}
dimethylamine	$(CH_3)_2NH$	5.9×10^{-4}
ethanolamine	$HOC_2H_4NH_2$	3.18×10^{-5}
ethylamine	$C_2H_5NH_2$	4.28×10^{-4}
hydrazine	H_2NNH_2	1.3×10^{-6}
hydroxylamine	$HONH_2$	1.07×10^{-8}
methylamine	CH_3NH_2	4.8×10^{-4}
pyridine	C_5H_5N	1.7×10^{-9}
trimethylamine	$(CH_3)_3N$	6.25×10^{-5}

Appendix E
Solubility Product Constants

Substance	K_{sp} at 25°C
AgBr	5.0×10^{-13}
AgCl	1.70×10^{-10}
Ag_2CO_3	8.2×10^{-12}
AgCN	1.6×10^{-16}
Ag_2CrO_4	9.0×10^{-12}
AgI	8.5×10^{-17}
AgSCN	1.03×10^{-12}
AgOH	2.0×10^{-8}
Ag_3PO_4	8.88×10^{-17}
Ag_2S	1.0×10^{-51}
Ag_2SO_4	1.18×10^{-5}
$Al(OH)_3$	1.9×10^{-33}
$BaCO_3$	8.1×10^{-9}
$BaSO_4$	1.50×10^{-9}
$BaCrO_4$	2.4×10^{-10}
BaF_2	1.84×10^{-7}
$Ba(OH)_2 \cdot 8H_2O$	2.55×10^{-4}
$Ba_3(PO_4)_2$	1.3×10^{-29}
BiO(OH)	1.0×10^{-12}
BiOCl	7.0×10^{-9}
Bi_2S_3	1.82×10^{-99}
$CaCO_3$	9.0×10^{-9}
$CaC_2O_4 \cdot H_2O$	2.34×10^{-9}
CaF_2	1.70×10^{-10}
$Ca(OH)_2$	4.68×10^{-6}
$Ca_3(PO_4)_2$	2.07×10^{-33}
$CaSO_4 \cdot 2H_2O$	2.4×10^{-5}
$CdCO_3$	6.18×10^{-12}
$Cd(OH)_2$	5.27×10^{-15}
CdS	1.40×10^{-29}
$CoCO_3$	1.0×10^{-12}
$Co(OH)_2$	1.1×10^{-16}

Substance	K_{sp} at 25°C
$Co(OH)_3$	2.5×10^{-43}
CoS	5.9×10^{-21}
$Cr(OH)_3$	6.7×10^{-31}
$CuBr$	6.27×10^{-9}
$CuCl$	1.72×10^{-7}
$CuCO_3$	1.37×10^{-10}
CuI	1.27×10^{-12}
$CuNCS$	1.77×10^{-13}
$Cu(OH)_2$	5.6×10^{-20}
Cu_2S	1.6×10^{-48}
CuS	8.7×10^{-36}
$FeCO_3$	3.07×10^{-11}
$Fe(OH)_2$	2.0×10^{-15}
$Fe(OH)_3$	1.1×10^{-36}
FeS	3.7×10^{-19}
Hg_2Br_2	6.41×10^{-23}
Hg_2Cl_2	1.1×10^{-18}
Hg_2CO_3	3.67×10^{-17}
Hg_2CrO_4	2.0×10^{-9}
Hg_2I_2	5.33×10^{-29}
$Hg_2O \cdot H_2O$	1.6×10^{-23}
HgS	1.6×10^{-54}
Hg_2S	1.0×10^{-45}
Hg_2SO_4	7.99×10^{-7}
$KClO_4$	1.05×10^{-2}
$KHC_4H_4O_6$	3.0×10^{-4}
K_2PtCl_6	7.48×10^{-6}
$MgCO_3 \cdot 3H_2O$	2.38×10^{-6}
MgC_2O_4	8.6×10^{-5}
MgF_2	7.42×10^{-11}
$MgNH_4PO_4$	2.5×10^{-13}
$Mg(OH)_2$	1.2×10^{-11}
$MnCO_3$	2.24×10^{-11}
$Mn(OH)_2$	4.5×10^{-14}
MnS	5.6×10^{-16}
$NiCO_3$	1.42×10^{-7}
$Ni(OH)_2$	5.47×10^{-16}
NiS	2.0×10^{-21}
$PbBr_2$	6.6×10^{-6}
$PbCl_2$	1.6×10^{-5}
$PbCO_3$	1.46×10^{-13}
$PbCrO_4$	1.8×10^{-14}

Substance	K_{sp} at 25°C
PbF_2	7.2×10^{-7}
PbI_2	8.49×10^{-9}
$Pb(OH)_2$	1.42×10^{-20}
$Pb_3(PO_4)_2$	3.0×10^{-44}
PbS	8.4×10^{-28}
$PbSO_4$	1.82×10^{-8}
$Sn(OH)_2$	5.0×10^{-26}
$Sn(OH)_4$	1.0×10^{-56}
SnS	8.0×10^{-29}
$SrCO_3$	5.60×10^{-10}
$SrC_2O_4 \cdot H_2O$	5.61×10^{-8}
$SrCrO_4$	3.6×10^{-5}
$Sr(OH)_2 \cdot 8H_2O$	3.2×10^{-4}
$SrSO_4$	3.44×10^{-7}
$TlCl$	1.9×10^{-4}
$TlNCS$	5.8×10^{-4}
$Tl(OH)_3$	1.5×10^{-44}
Tl_2S	1.2×10^{-24}
$ZnCO_3$	1.99×10^{-10}
$Zn(OH)_2$	4.5×10^{-17}
ZnS	1.2×10^{-23}

Appendix F

Complex-Ion Formation Constants

Equilibrium	K_f
$Al^{3+} + 6F^- \rightleftharpoons [AlF_6]^{3-}$	5.0×10^{19}
$Cd^{2+} + 4NH_3 \rightleftharpoons [Cd(NH_3)_4]^{2+}$	4.0×10^6
$Cd^{2+} + 4CN^- \rightleftharpoons [Cd(CN)_4]^{2-}$	7.1×10^{16}
$Co^{2+} + 6NH_3 \rightleftharpoons [Co(NH_3)_6]^{2+}$	8.3×10^4
$Co^{3+} + 6NH_3 \rightleftharpoons [Co(NH_3)_6]^{3+}$	4.5×10^{33}
$Cu^+ + 2CN^- \rightleftharpoons [Cu(CN)_2]^-$	1.0×10^{16}
$Cu^{2+} + 4NH_3 \rightleftharpoons [Cu(NH_3)_4]^{2+}$	1.2×10^{12}
$Fe^{2+} + 6CN^- \rightleftharpoons [Fe(CN)_6]^{4-}$	1.0×10^{35}
$Fe^{3+} + 6CN^- \rightleftharpoons [Fe(CN)_6]^{3-}$	1.0×10^{42}
$Fe^{3+} + 6NCS^- \rightleftharpoons [Fe(NCS)_6]^{3-}$	3.2×10^3
$Hg^{2+} + 4Cl^- \rightleftharpoons [HgCl_4]^{2-}$	1.6×10^{15}
$Ni^{2+} + 6NH_3 \rightleftharpoons [Ni(NH_3)_6]^{2+}$	1.8×10^8
$Ag^+ + 2Cl^- \rightleftharpoons [AgCl_2]^-$	2.5×10^5
$Ag^+ + 2CN^- \rightleftharpoons [Ag(CN)_2]^-$	1.0×10^{21}
$Ag^+ + 2NH_3 \rightleftharpoons [Ag(NH_3)_2]^+$	1.7×10^7
$Zn^{2+} + 4CN^- \rightleftharpoons [Zn(CN)_4]^{2-}$	1.0×10^{19}
$Zn^{2+} + 4OH^- \rightleftharpoons [Zn(OH)_4]^{2-}$	2.9×10^{15}

Appendix G
Standard Reduction Potentials at 25°C

Reduction Half-Reaction	E°_{red}, volts
$Li^+(aq) + e^- \longrightarrow Li(s)$	-3.05
$K^+(aq) + e^- \longrightarrow K(s)$	-2.93
$Rb^+(aq) + e^- \longrightarrow Rb(s)$	-2.93
$Ra^{2+}(aq) + 2e^- \longrightarrow Ra(s)$	-2.92
$Ba^{2+}(aq) + 2e^- \longrightarrow Ba(s)$	-2.90
$Sr^{2+}(aq) + 2e^- \longrightarrow Sr(s)$	-2.89
$Ca^{2+}(aq) + 2e^- \longrightarrow Ca(s)$	-2.87
$Na^+(aq) + e^- \longrightarrow Na(s)$	-2.71
$La^{3+}(aq) + 3e^- \longrightarrow La(s)$	-2.52
$SiO_3^{2-}(aq) + 3H_2O(l) + 4e^- \longrightarrow Si(s) + 6OH^-(aq)$	-2.52
$Ce^{3+}(aq) + 3e^- \longrightarrow Ce(s)$	-2.48
$Nd^{3+}(aq) + 3e^- \longrightarrow Nd(s)$	-2.44
$Sm^{3+}(aq) + 3e^- \longrightarrow Sm(s)$	-2.41
$Gd^{3+}(aq) + 3e^- \longrightarrow Gd(s)$	-2.40
$Mg^{2+}(aq) + 2e^- \longrightarrow Mg(s)$	-2.37
$Y^{3+}(aq) + 3e^- \longrightarrow Y(s)$	-2.37
$Am^{3+}(aq) + 3e^- \longrightarrow Am(s)$	-2.32
$Lu^{3+}(aq) + 3e^- \longrightarrow Lu(s)$	-2.25
$H_2(g) + 2e^- \longrightarrow 2H^-(aq)$	-2.25
$Sc^{3+}(aq) + 3e^- \longrightarrow Sc(s)$	-2.08
$Pu^{3+}(aq) + 3e^- \longrightarrow Pu(s)$	-2.07
$Th^{4+}(aq) + 4e^- \longrightarrow Th(s)$	-1.90
$Np^{3+}(aq) + 3e^- \longrightarrow Np(s)$	-1.86
$Be^{2+}(aq) + 2e^- \longrightarrow Be(s)$	-1.85
$U^{3+}(aq) + 3e^- \longrightarrow U(s)$	-1.80
$Hf^{4+}(aq) + 4e^- \longrightarrow Hf(s)$	-1.70
$Al^{3+}(aq) + 3e^- \longrightarrow Al(s)$	-1.66
$Ti^{2+}(aq) + 2e^- \longrightarrow Ti(s)$	-1.63
$Zr^{4+}(aq) + 4e^- \longrightarrow Zr(s)$	-1.53
$Cr(OH)_3(s) + 3e^- \longrightarrow Cr(s) + 3OH^-(aq)$	-1.30
$Zn(OH)_2(s) + 2e^- \longrightarrow Zn(s) + 2OH^-(aq)$	-1.25
$CdS(s) + 2e^- \longrightarrow Cd(s) + S^{2-}(aq)$	-1.21

Reduction Half-Reaction	E°_{red}, volts
$V^{2+}(aq) + 2e^- \longrightarrow V(s)$	-1.20
$ZnS(s) + 2e^- \longrightarrow Zn(s) + S^{2-}(aq)$	-1.18
$Mn^{2+}(aq) + 2e^- \longrightarrow Mn(s)$	-1.18
$FeS(s) + 2e^- \longrightarrow Fe(s) + S^{2-}(aq)$	-1.01
$PbS(s) + 2e^- \longrightarrow Pb(s) + S^{2-}(aq)$	-0.95
$SnS(s) + 2e^- \longrightarrow Sn(s) + S^{2-}(aq)$	-0.94
$Cr^{2+}(aq) + 2e^- \longrightarrow Cr(s)$	-0.91
$Fe(OH)_2(s) + 2e^- \longrightarrow Fe(s) + 2OH^-(aq)$	-0.88
$SiO_2(s) + 4H^+(aq) + 4e^- \longrightarrow Si(s) + 2H_2O(l)$	-0.86
$NiS(s) + 2e^- \longrightarrow Ni(s) + S^{2-}(aq)$	-0.83
$2H_2O(l) + 2e^- \longrightarrow H_2(g) + 2OH^-(aq)$	-0.83
$Zn^{2+}(aq) + 2e^- \longrightarrow Zn(s)$	-0.76
$Cr^{3+}(aq) + 3e^- \longrightarrow Cr(s)$	-0.74
$HgS(s) + 2e^- \longrightarrow Hg(l) + S^{2-}(aq)$	-0.72
$Ga^{3+}(aq) + 3e^- \longrightarrow Ga(s)$	-0.53
$H_3PO_2(aq) + H^+(aq) + e^- \longrightarrow P(s) + 2H_2O(l)$	-0.51
$S(s) + 2e^- \longrightarrow S^{2-}(aq)$	-0.48
$Fe^{2+}(aq) + 2e^- \longrightarrow Fe(s)$	-0.44
$Cr^{3+}(aq) + e^- \longrightarrow Cr^{2+}(aq)$	-0.41
$Cd^{2+}(aq) + 2e^- \longrightarrow Cd(s)$	-0.40
$Se(s) + 2H^+(aq) + 2e^- \longrightarrow H_2Se(aq)$	-0.40
$ClO_4^-(aq) + H_2O(l) + 2e^- \longrightarrow ClO_3^-(aq) + 2OH^-(aq)$	-0.36
$PbSO_4(s) + 2e^- \longrightarrow Pb(s) + SO_4^{2-}(aq)$	-0.35
$In^{3+}(aq) + 3e^- \longrightarrow In(s)$	-0.34
$Tl^+(aq) + e^- \longrightarrow Tl(s)$	-0.34
$Co^{2+}(aq) + 2e^- \longrightarrow Co(s)$	-0.28
$V^{3+}(aq) + e^- \longrightarrow V^{2+}(aq)$	-0.26
$Ni^{2+}(aq) + 2e^- \longrightarrow Ni(s)$	-0.25
$Sn^{2+}(aq) + 2e^- \longrightarrow Sn(s)$	-0.14
$CrO_4^{2-}(aq) + 4H_2O(l) + 3e^- \longrightarrow Cr(OH)_3(s) + 5OH^-(aq)$	-0.13
$Pb^{2+}(aq) + 2e^- \longrightarrow Pb(s)$	-0.13
$P(s) + 3H^+(aq) + 3e^- \longrightarrow PH_3(g)$	-0.065
$MnO_2(s) + 2H_2O(l) + 2e^- \longrightarrow Mn(OH)_2(s) + 2OH^-(aq)$	-0.05
$2H^+(aq) + 2e^- \longrightarrow H_2(g)$	0.00
$NO_3^-(aq) + H_2O(l) + 2e^- \longrightarrow NO_2^-(aq) + 2OH^-(aq)$	$+0.01$
$TiO^{2+}(aq) + 2H^+(aq) + e^- \longrightarrow Ti^{3+}(aq) + H_2O(l)$	$+0.10$
$S(s) + 2H^+(aq) + 2e^- \longrightarrow H_2S(aq)$	$+0.14$
$Cu^{2+}(aq) + e^- \longrightarrow Cu^+(aq)$	$+0.15$
$Sn^{4+}(aq) + 2e^- \longrightarrow Sn^{2+}(aq)$	$+0.15$
$Co(OH)_3(s) + e^- \longrightarrow Co(OH)_2(s) + OH^-(aq)$	$+0.17$
$SO_4^{2-}(aq) + 4H^+(aq) + 2e^- \longrightarrow H_2SO_3(aq) + H_2O(l)$	$+0.17$
$AgCl(s) + e^- \longrightarrow Ag(s) + Cl^-(aq)$	$+0.22$

Reduction Half-Reaction	E°_{red}, volts
$Hg_2Cl_2(s) + 2e^- \longrightarrow 2Hg(l) + 2Cl^-(aq)$	+0.27
$ClO_3^-(aq) + H_2O(l) + 2e^- \longrightarrow ClO_2^-(aq) + 2OH^-(aq)$	+0.33
$Cu^{2+}(aq) + 2e^- \longrightarrow Cu(s)$	+0.34
$VO^{2+}(aq) + 2H^+(aq) + e^- \longrightarrow V^{3+}(aq) + H_2O(l)$	+0.34
$O_2(g) + 2H_2O(l) + 4e^- \longrightarrow 4OH^-(aq)$	+0.40
$NiO_2(s) + 2H_2O(l) + 2e^- \longrightarrow Ni(OH)_2(s) + 2OH^-(aq)$	+0.49
$Cu^+(aq) + e^- \longrightarrow Cu(s)$	+0.52
$TeO_2(s) + 4H^+(aq) + 4e^- \longrightarrow Te(s) + 2H_2O(l)$	+0.53
$I_2(s) + 2e^- \longrightarrow 2I^-(aq)$	+0.54
$MnO_4^-(aq) + 2H_2O(l) + 3e^- \longrightarrow MnO_2(s) + 4OH^-(aq)$	+0.59
$2HgCl_2(aq) + 2e^- \longrightarrow Hg_2Cl_2(s) + 2Cl^-(aq)$	+0.63
$ClO_2^-(aq) + H_2O(l) + 2e^- \longrightarrow ClO^-(aq) + 2OH^-(aq)$	+0.66
$O_2(g) + 2H^+(aq) + 2e^- \longrightarrow H_2O_2(aq)$	+0.68
$Fe^{3+}(aq) + e^- \longrightarrow Fe^{2+}(aq)$	+0.77
$Hg_2^{2+}(aq) + 2e^- \longrightarrow 2Hg(l)$	+0.79
$Ag^+(aq) + e^- \longrightarrow Ag(s)$	+0.80
$Hg^{2+}(aq) + 2e^- \longrightarrow Hg(l)$	+0.85
$HO_2^-(aq) + H_2O(l) + 2e^- \longrightarrow 3OH^-(aq)$	+0.88
$ClO^-(aq) + H_2O(l) + 2e^- \longrightarrow Cl^-(aq) + 2OH^-(aq)$	+0.89
$2Hg^{2+}(aq) + 2e^- \longrightarrow Hg_2^{2+}(aq)$	+0.91
$NO_3^-(aq) + 3H^+(aq) + 2e^- \longrightarrow HNO_2(aq) + H_2O(l)$	+0.94
$NO_3^-(aq) + 4H^+(aq) + 3e^- \longrightarrow NO(g) + 2H_2O(l)$	+0.96
$Pd^{2+}(aq) + 2e^- \longrightarrow Pd(s)$	+0.99
$HNO_2(aq) + H^+(aq) + e^- \longrightarrow NO(g) + H_2O(l)$	+1.00
$VO_2^+(aq) + 2H^+(aq) + e^- \longrightarrow VO^{2+}(aq) + H_2O(l)$	+1.00
$Br_2(l) + 2e^- \longrightarrow 2Br^-(aq)$	+1.07
$IO_3^-(aq) + 6H^+(aq) + 6e^- \longrightarrow I^-(aq) + 3H_2O(l)$	+1.09
$ClO_4^-(aq) + 2H^+(aq) + 2e^- \longrightarrow ClO_3^-(aq) + H_2O(l)$	+1.19
$Pt^{2+}(aq) + 2e^- \longrightarrow Pt(s)$	+1.20
$2IO_3^-(aq) + 12H^+(aq) + 10e^- \longrightarrow I_2(aq) + 6H_2O(l)$	+1.20
$ClO_3^-(aq) + 3H^+(aq) + 2e^- \longrightarrow HClO_2(aq) + H_2O(l)$	+1.21
$MnO_2(s) + 4H^+(aq) + 2e^- \longrightarrow Mn^{2+}(aq) + 2H_2O(l)$	+1.23
$O_2(g) + 4H^+(aq) + 4e^- \longrightarrow 2H_2O(l)$	+1.23
$Cr_2O_7^{2-}(aq) + 14H^+(aq) + 6e^- \longrightarrow 2Cr^{3+}(aq) + 7H_2O(l)$	+1.33
$Cl_2(g) + 2e^- \longrightarrow 2Cl^-(aq)$	+1.36
$BrO_3^-(aq) + 6H^+(aq) + 6e^- \longrightarrow Br^-(aq) + 3H_2O(l)$	+1.44
$2HOI(aq) + 2H^+(aq) + 4e^- \longrightarrow I_2(aq) + 2H_2O(l)$	+1.45
$HOCl(aq) + H^+(aq) + 2e^- \longrightarrow Cl^-(aq) + H_2O(l)$	+1.49
$Au^{3+}(aq) + 3e^- \longrightarrow Au(s)$	+1.50
$MnO_4^-(aq) + 8H^+(aq) + 5e^- \longrightarrow Mn^{2+}(aq) + 4H_2O(l)$	+1.51
$2BrO_3^-(aq) + 12H^+ + 10e^- \longrightarrow Br_2(aq) + 6H_2O(l)$	+1.52
$Ce^{4+}(aq) + e^- \longrightarrow Ce^{3+}(aq)$	+1.61

Reduction Half-Reaction	E°_{red}, volts
$2HOCl(aq) + 2H^+ + 2e^- \longrightarrow Cl_2(aq) + 2H_2O(l)$	+1.63
$HClO_2(aq) + 2H^+(aq) + 2e^- \longrightarrow HOCl(aq) + H_2O(l)$	+1.64
$Au^+(aq) + e^- \longrightarrow Au(s)$	+1.68
$NiO_2(s) + 4H^+(aq) + 2e^- \longrightarrow Ni^{2+}(aq) + 2H_2O(l)$	+1.68
$PbO_2(s) + SO_4^{2-}(aq) + 4H^+(aq) + 2e^- \longrightarrow PbSO_4(s) + 2H_2O(l)$	+1.69
$H_2O_2(aq) + 2H^+(aq) + 2e^- \longrightarrow 2H_2O(l)$	+1.77
$Co^{3+}(aq) + e^- \longrightarrow Co^{2+}(aq)$	+1.82
$S_2O_8^{2-}(aq) + 2e^- \longrightarrow 2SO_4^{2-}(aq)$	+2.01
$F_2(g) + 2e^- \longrightarrow 2F^-(aq)$	+2.87

Appendix H

Half-Lives of Some Radioactive Isotopes

Isotope	Half-Life	Type of Emission*
^{14}C	5730 y	β^-
^{13}N	10.0 min	β^+
^{24}Na	15.0 h	β^-
^{32}P	14.3 d	β^-
^{40}K	1.3×10^9 y	β^- or EC
^{60}Co	5.3 y	β^-
^{87}Rb	4.7×10^{10} y	β^-
^{90}Sr	28 y	β^-
^{115}In	4.4×10^{14} y	β^-
^{131}I	8.04 d	β^-
^{142}Ce	5×10^{15} y	α
^{198}Au	64.6 h	β^-
^{208}Tl	3.1 min	β^-
^{210}Pb	22.3 y	β^-
^{212}Pb	10.6 h	β^-
^{214}Pb	26.8 min	β^-
^{206}Bi	6.2 d	β^+ or EC
^{210}Bi	5.01 d	β^-
^{212}Bi	60.5 min	α or β^-
^{214}Bi	19.7 min	β^-
^{207}Po	5.8 h	α, β^+, or EC
^{210}Po	138.4 d	α
^{212}Po	3×10^{-7} s	α
^{214}Po	1.6×10^{-4} s	α
^{216}Po	0.15 s	α
^{218}Po	3.11 min	α or β^-
^{215}At	10^{-14} s	α
^{218}At	1.3 s	α
^{220}Rn	54.5 s	α
^{222}Rn	3.82 d	α
^{224}Rn	3.64 d	α
^{226}Ra	1620 y	α

Isotope	Half-Life	Type of Emission*
^{228}Ra	6.7 y	β^-
^{228}Ac	6.13 h	β^-
^{228}Th	1.91 y	α
^{230}Th	7.54×10^4 y	α
^{232}Th	1.39×10^{10} y	α, β^-, or SF
^{233}Th	22.3 min	β^-
^{234}Th	24.1 d	β^-
^{223}Pa	27 d	β^-
^{233}U	1.59×10^5 y	α
^{234}U	2.45×10^5 y	α or SF
^{235}U	7.04×10^8 y	α or SF
^{238}U	4.46×10^9 y	α or SF
^{239}U	23.5 min	β^-
^{239}Np	2.35 d	β^-
^{239}Pu	2.41×10^4 y	α or SF
^{240}Pu	6.54×10^3 y	α or SF
^{241}Pu	14.4 y	α or β^-
^{241}Am	432 y	α
^{242}Cm	163 d	α or SF
^{243}Bk	4.5 h	α or EC
^{245}Cf	43.6 min	α or EC
^{253}Es	20.5 d	α or SF
^{254}Fm	3.24 h	SF
^{255}Fm	20.1 h	α
^{256}Md	76 min	EC
^{254}No	55 s	α
^{257}Lr	0.65 s	α
^{263}(106)	0.8 s	α

*EC = electron capture; SF = spontaneous fission

Answers to Selected Even Exercises

Note: Your answers may differ from those given here in the last significant figure.

Chapter 1

1.6 **a.** chemical **b.** physical **c.** physical **d.** physical **e.** chemical **f.** physical **g.** physical (possibly chemical) **1.8** **a.** pure **b.** solution **c.** heterogeneous **d.** pure **e.** heterogeneous **f.** heterogeneous **g.** heterogeneous **1.10** Limestone is a compound. We can't classify lime. **1.18** **a.** 6.2 cm **b.** 2.2 h **c.** 3.2 cm^2 **d.** 3.2 g/mL **e.** 2.04 kg **1.20** **a.** 91 mm **b.** 0.30 min **c.** 1400 mL or 1.4×10^3 mL **d.** 0.018 kg **e.** 2.34×10^6 cm^3 **f.** 0.55 m **1.22** **a.** 5.93 g/mL **b.** 2.70 g/mL **c.** 1.43 g/L **d.** 1.2 g/mL **e.** 19.3 g/mL **1.24** 9.19 mL **1.26** 0.789 g/mL **1.28** 394 K **1.32** **a.** 2 **b.** 2 or 3 **c.** 3 **d.** 4 **e.** 1 **f.** 3 **g.** 4 **1.34** **a.** 6.0 **b.** 1.03 **c.** 4.0×10^{10} **d.** 13 **1.38** 3.8×10^7 min **1.40** 55.8 kg **1.42** 28.4 g **1.44** 88 km h^{-1} **1.46** 5.0×10^2 s **1.50** 9.79×10^2 kg, 2.16×10^3 lb **1.54** **a.** chemical **b.** chemical **c.** physical **d.** physical **e.** chemical **f.** chemical **g.** physical **1.60** **a.** 8.3 cm **b.** 2.4 min **c.** 1.0×10^1 cm^3 **d.** 0.90 g mL^{-1} **e.** 1.99 g

Chapter 2

2.10 Calculated from mass data: 67.0% zinc, 33% S, matching the values for zinc blende. Thus the composition is independent of the source of the substance. **2.14** **a.** 4, 4, 5 **b.** 11, 11, 12 **c.** 12, 12, 14 **d.** 2, 3, 4 **e.** 10, 9, 10 **2.16** $^{35}_{17}Cl$ 17 protons, 17 electrons, 18 neutrons; $^{37}_{17}Cl$ 17 protons, 17 electrons, 20 neutrons **2.18** **a.** 14, 6, 8 **b.** 32, 15, 17 **c.** 192, 77, 115 **d.** 27, 13, 14 **e.** 209, 84, 125 **2.22** **a.** Fe **b.** Pb **c.** Ag **d.** Au **e.** Sb **f.** Cu **g.** Hg **h.** Sn **i.** Na **j.** W **2.28** hydrogen, boron, carbon, nitrogen, oxygen, fluorine, phosphorus, sulfur, potassium, vanadium, yttrium, iodine, tungsten, uranium **2.34** **a.** $BaCl_2$ **b.** $FeBr_3$ **c.** $Ca_3(PO_4)_2$ **d.** $Cr_2(SO_4)_3$ **e.** Na_3N **f.** $CsClO_3$ **g.** $Ti_2(CO_3)_3$ **2.36** **a.** FeO, Fe_2O_3 **b.** $FeCl_2$, $FeCl_3$ **2.38** ^{35}Cl has 17 protons, 17 electrons, and 18 neutrons; ^{37}Cl has 17 protons, 17 electrons, and 20 neutrons. **2.40** 1 ($^{19}_9F$) **2.42** 24.31 amu **2.46** **a.** 472.09 amu **b.** 174.26 amu **c.** 146.054 amu **d.** 208.85 amu **e.** 120.060 amu **f.** 98.960 amu **g.** 342.301 amu **h.** 150.903 **2.48** 34.4 g/mol **2.52** 10^{14} atoms **2.54** 1.7×10^{21} molecules **2.56** 2.6 g **2.58** Na **2.60** 9.76 g **2.62** 179 g **2.64** **a.** 0.0999 mol **b.** 0.293 mol **c.** 0.127 mol **d.** 0.0831 mol **2.66** 1.254×10^{23} molecules **2.68** 357.3 g **2.72** **a.** 16.48% **b.** 26.19% **c.** 87.41% **d.** 63.65% **e.** 36.85% **2.74** **a.** 79.86% **b.** 57.48% **c.** 79.89% **d.** 55.31% **2.76** 63.92% **2.82** Ca_3SiO_5 **2.84** $C_8H_{17}N$ **2.86** $C_7H_5O_6N_3$ **2.88** **a.** N_2O_3 **b.** NO_2 **2.92** **a.** C_3H_2Cl **b.** C_6H_5Cl **c.** N_2O_5 **d.** NO_2 **e.** HCO_2 **f.** CH_2O **2.94** N_2O_4 **2.96** $C_6H_{14}N_4$ **2.100** 156 g **2.102** 20.17 amu **2.106** 22 mol **2.110** 13.59% Na, 35.51% C, 4.77% H, 8.28% N, 37.84% O **2.112** law of conservation of mass **2.114** 4.08 g

Chapter 3

3.2 **a.** Ar, argon **b.** Na, sodium **c.** Sm, samarium **d.** Cr, chromium **e.** Cl, chlorine **f.** Ba, barium **g.** Na, sodium **h.** Gd, gadolinium **3.4** **a.** halogens **b.** alkaline-earth metals **c.** noble gases **d.** alkali metals **e.** noble gases **f.** halogens **3.6** **a.** main group **b.** transition **c.** transition **d.** inner transition **e.** transition **f.** main group **g.** main group **h.** inner transition **i.** main group **3.8** **a.** lithium **b.** aluminum **c.** barium **d.** germanium **e.** graphite **3.14** mp 6°C, bp 75°C **3.18** **a.** 2+ **b.** 3+ **c.** 2+ **d.** 2+, 3+ **e.** 1+ **f.** 2+, 4+ **3.20** **a.** Li_2O **b.** Ca_3N_2 **c.** PbO or PbO_2 **d.** RbCl **e.** Al_4C_3 **f.** BaS **3.22** **a.** $MgCl_2$ **b.** MgS **c.** AlN **d.** NaH **e.** Na_2S **f.** Mg_3N_2 **g.** SnH_4 **h.** Al_2O_3 **i.** Be_2C **3.24** **a.** ClO_3^- or NO_3^- **b.** SO_3^{2-} or CO_3^{2-} **c.** ClO_4^- **d.** SO_4^{2-} **e.** PO_4^{3-} **f.** SiO_4^{4-} **3.26** **a.** AsO_3^{3-} **b.** SiO_4^{4-} **c.** PO_4^{3-} **d.** BO_3^{3-} **e.** NO_2^- **f.** ClO_2^- **3.28** **a.** MgO **b.** Tl_2O_3 **c.** SO_2 **d.** I_2O_5 **e.** Cr_2O_3 **f.** H_2O **3.30** CO_2, CO, N_2O_4, N_2O_3, NO_2, NO, N_2O, SO_2, F_2O, Cl_2O_5, ClO_2, and Cl_2O **3.32** H_2S **3.36** **a.** -2 **b.** $+1$ **c.** $+2$ **d.** $+3$ **e.** -3 **f.** -1 **3.38** **a.** -1 **b.** $+1$ **c.** $+5$ **d.** -1 **e.** $+1$ **f.** $+6$ **g.** $+4$ **h.** $+2$ **i.** $+6$ **3.40** **a.** phosphorus pentafluoride **b.** phosphorus trifluoride **c.** carbon monoxide **d.** carbon dioxide **e.** sulfur dioxide **f.** sulfur trioxide **g.** dinitrogen tetroxide **h.** nitrogen dioxide **i.** carbon disulfide **3.42** **a.** hypochlorite ion **b.** hydrogen sulfate ion **c.** borate ion **d.** hydrogen sulfide ion **e.** peroxide ion **f.** nitrate ion **g.** perbromate ion **h.** carbonate ion **i.** phosphate ion **3.44** **a.** ammonium sulfide **b.** aluminum oxide **c.** silver(I) oxide **d.** copper(II) bromide **e.** barium cyanide **f.** zinc chloride **g.** potassium carbonate **h.** potassium hydroxide **i.** magnesium iodide **3.50** **a.** copper(II) sulfate pentahydrate **b.** magnesium sulfate heptahydrate **c.** cobalt(II) chloride hexahydrate **d.** copper(II) sulfate hydrate **3.52** **a.** sodium nitrate **b.** barium perchlorate **c.** copper(II) sulfate **d.** chromium(III) phosphate **e.** sodium hydrogen sulfate **f.** thorium(IV) chlorate **g.** calcium carbonate **h.** sodium hypobromite **i.** sodium arsenate **3.54** **a.** hydrobromic acid **b.** bromic acid **c.** hypobromous acid **d.** perbromic acid **e.** bromous acid **f.** phosphoric acid **g.** boric acid **h.** carbonic acid **i.** nitrous acid **3.60** **a.** ionic **b.** covalent **c.** covalent **d.** ionic **e.** ionic **3.62** **a.** binary acid **b.** not an acid **c.** oxoacid **d.** not an acid **e.** oxoacid **f.** not an acid **g.** not an acid **h.** oxoacid **i.** oxoacid **3.64** **a.** oxide **b.** not an oxide **c.** not an oxide **d.** oxide **e.** not an oxide **f.** not an oxide **3.68** N_2, O_2, H_2, Ar, He **3.72** ammonia **3.74** carbonates, halides, oxides, phosphates, silicates, sulfates, sulfides **3.76** **a.** ionic **b.** covalent **c.** covalent **d.** ionic **e.** covalent **f.** covalent **3.80** phosphorus or nitrogen **3.82** **a.** $AgClO_4$ **b.** ZnI_2 **c.** SiF_4 **d.** CS_2 **e.** $Zn(NO_3)_2$ **f.** Na_2O **g.** $Ba(OH)_2$ **h.** Na_3PO_4 **i.** CaC_2 **3.92** ammonium chloride

Chapter 4

4.12 **a.** $2NaHCO_3(s) \xrightarrow{\Delta} Na_2CO_3(s) + CO_2(g) + H_2O(g)$

b. $Na_2B_4O_7 \cdot 10H_2O(s) \xrightarrow{\Delta} Na_2B_4O_7(s) + 10H_2O(g)$

c. $2Cu(NO_3)_2(s) \xrightarrow{\Delta} 2CuO(s) + 4NO_2(g) + O_2(g)$

d. $2HgO(s) \xrightarrow{\Delta} 2Hg(l) + O_2(g)$

e. $(NH_4)_2Cr_2O_7(s) \xrightarrow{\Delta} Cr_2O_3(s) + N_2(g) + 4H_2O(g)$
f. $3HBrO(aq) \longrightarrow HBrO_3(aq) + 2HBr(aq)$
4.14 **a.** $CaCO_3(s) + H_2SO_4(aq) \longrightarrow$
$$CaSO_4(s) + CO_2(g) + H_2O(l)$$
b. $2SbCl_3(aq) + 3H_2S(aq) \longrightarrow Sb_2S_3(s) + 6HCl(aq)$
c. $3NaOH(aq) + H_3PO_4(aq) \longrightarrow Na_3PO_4(aq) + 3H_2O(l)$
d. $H_2SO_4(aq) + 2KOH(aq) \longrightarrow K_2SO_4(aq) + 2H_2O(l)$
4.16 DS = direct synthesis, D = decomposition, SD = single-displacement, DD = double-displacement.
a. $CaCl_2(aq) + Na_2SO_4(aq) \longrightarrow$
$$CaSO_4(s) + 2NaCl(aq) \quad DD$$
b. $Ba(s) + 2HCl(aq) \longrightarrow BaCl_2(aq) + H_2(g)$ SD

c. $4Al(s) + 3O_2(g) \xrightarrow{\Delta} 2Al_2O_3(s)$ DS

d. $FeO(s) + CO(g) \xrightarrow{\Delta} Fe(s) + CO_2(g)$ DD
e. $CaO(s) + H_2O(l) \longrightarrow Ca(OH)_2(aq)$ DS
f. $Na_2CrO_4(aq) + Pb(NO_3)_2(aq) \longrightarrow$
$$PbCrO_4(s) + 2NaNO_3(aq) \quad DD$$
g. $2KI(aq) + Cl_2(g) \longrightarrow 2KCl(aq) + I_2(aq)$ SD

h. $2NaHCO_3(s) \xrightarrow{\Delta} Na_2CO_3(s) + CO_2(g) + H_2O(g)$ D

i. $CaO(s) + SO_3(g) \xrightarrow{\Delta} CaSO_4(s)$ DS

j. $PCl_5(g) \xrightarrow{\Delta} PCl_3(g) + Cl_2(g)$ D

k. $Ca_3N_2(s) + 6H_2O(l) \xrightarrow{\Delta} 3Ca(OH)_2(aq) + 2NH_3(g)$ DD
4.24 $Cu(s) + 2H_2SO_4(aq) \longrightarrow CuSO_4(aq) + SO_2(g) + 2H_2O(l)$ **a.** 0.50 mol **b.** 80 g **c.** 0.50 mol **d.** 3.0×10^{23} molecules **4.26** 21.0 g **4.28** **a.** 1.53×10^8 g **b.** 6.67×10^7 g **4.30** 8.72 g **4.32** 17.8 g **4.34** **a.** 92.67 g **b.** 162.1 g **c.** 106.5 g **d.** 163.3 g **4.36** **a.** $C_6H_{12}O_6 + 6KNO_3 \longrightarrow 6CO_2 + 3H_2O + 6KOH + 3N_2O$ **b.** 6.11 g **4.38** **a.** 2.70 g **b.** 176 mmol **4.40** 107 kg **4.42** 9.39 g **4.44** 63.4 g **4.46** 2.14×10^3 g **4.48** 0.278 g **4.54** **a.** 2.03 M **b.** 0.540 M **c.** 3.20 M **4.56** **a.** 1.00 L **b.** 0.0833 L or 83.3 mL **c.** 0.167 L or 167 mL **d.** 0.333 L or 333 mL **4.58** 23.3 mL **4.60** 0.02607 M **4.62** 0.0210 M

4.64 15.6 mL **4.66** 0.1815 M NaOH **4.68** 0.0918 M **4.74** **a.** H_2O **b.** CO or CO_2 **c.** SO_2 or SO_3 **d.** P_4O_6 or P_4O_{10} **e.** SeO_2 **f.** B_2O_3
4.76 **a.** $Na_2O(s) + H_2O(l) \longrightarrow 2NaOH(aq)$
b. $CaO(s) + H_2SO_4(aq) \longrightarrow CaSO_4(s) + H_2O(l)$
c. $2NaOH(aq) + Cl_2O(g) \longrightarrow 2NaOCl(aq) + H_2O(l)$
d. $Ba(OH)_2(aq) + 2HNO_3(aq) \longrightarrow Ba(NO_3)_2(aq) + 2H_2O(l)$
e. $Al_2O_3(s) + 6HCl(aq) \longrightarrow 2AlCl_3(aq) + 3H_2O(l)$
f. $Al_2O_3(s) + 2NaOH(aq) + 3H_2O(l) \longrightarrow 2NaAl(OH)_4(aq)$
g. $BaO(s) + CO_2(g) \longrightarrow BaCO_3(s)$
h. $2NaOH(aq) + SO_2(g) \longrightarrow Na_2SO_3(aq) + H_2O(l)$
i. $2NaOH(aq) + H_2SO_3(aq) \longrightarrow Na_2SO_3(aq) + 2H_2O(l)$
j. $Na_2O(s) + SO_2(g) \longrightarrow Na_2SO_3(s)$
4.86 14.1 g **4.88** 41.4 g **4.90** 0.1141 M **4.94** 82.4% **4.98** 0.0543 M **4.100** 14.0 g

Chapter 5

5.6 3.47×10^4 cal **5.8** 114 Cal **5.12** 5.47×10^3 J **5.14** **a.** 22.6 J/mol°C **b.** 20.1 J/°C **5.16** 0.129 J/g°C **5.18** 392 g **5.20** 7.80×10^5 J **5.22** -210 J **5.28** 39.0°C **5.30** 54.8°C **5.34** -5.17×10^4 J **5.36** 1.11×10^4 J **5.38** 4.11×10^4 J **5.40** -6.20×10^4 J **5.42** 4.69×10^6 J **5.48** -1.5×10^5 kJ **5.50** -2.27×10^3 kJ **5.52** 31.4 kJ **5.54** 25.3 g **5.56** -4.87×10^3 kJ **5.60** -202.5 kJ **5.62** $+41$ kJ **5.64** -790.6 kJ **5.66** -3351.0 kJ **5.70** -957.0 kJ **5.72** -157.0 kJ **5.74** -2855.60 kJ **5.76** 4.6 kJ/mol **5.78** -511 kJ **5.80** -408 kJ **5.82** -293.0 kJ **5.86** -136.2 kJ **5.90** 2.13×10^4 J/mol **5.96** -890.3 kJ

Chapter 6

6.6 3.31×10^{-19} J **6.8** **a.** 1.26×10^{-6} m or 1260 nm **b.** 2.38×10^{14} s^{-1} **6.10** **a.** 1.50×10^{22} s^{-1} **b.** 9.94×10^{-12} J **6.12** 6.35×10^{14} s^{-1} **6.18** 7.90×10^4 m s^{-1} **6.28** 5 **6.30** 3 **6.32** **a.** 18 **b.** 18 **c.** 8 **d.** 10 **e.** 8 **f.** 18 **6.34** $+\frac{1}{2}$ or $-\frac{1}{2}$ **6.42** **a.** $[Kr]5s^24d^{10}5p^3$ **b.** $[Ar]4s^23d^1$ **c.** $[Ar]4s^23d^{10}4p^5$ **d.** $[He]2s^22p^3$ **e.** $[Ar]4s^2$ **f.** $[Ar]4s^23d^5$ **6.44** **a.** $[Ne]3s^23p^6$ **b.** $[Ar]4s^23d^{10}4p^6$ **c.** $[Ar]4s^23d^{10}4p^6$ **d.** $[He]2s^22p^6$ **e.** $[He]2s^22p^6$ **f.** $1s^2$ **6.46** **a.** Sc **b.** Se **c.** Cr **d.** I **e.** Sn **6.54** Ar < Sc < Ca < K **6.56** $Sc^{3+} < Y^{3+} < La^{3+}$ **6.60** $N^{3-} < Cl^- < S^{2-} < Br^- < P^{3-}$ **6.62** **a.** K **b.** Sr **c.** C **d.** O **6.66** aluminum < silicon < sulfur < phosphorus **6.72** **a.** O **b.** S **c.** Li **d.** Cs **e.** Si **6.74** N < C < Si < O < S **6.84** Na^+, Mg^{2+}, Al^{3+}, Si^{4+} or Si^{4-}, P^{3-}, S^{2-}, Cl^-, Ar^0 (no ion)

Chapter 7

7.4 **a.** no **b.** yes **c.** no **d.** yes **e.** no **7.10** **a.** covalent **b.** ionic **c.** covalent **d.** covalent **e.** ionic between Ca^{2+} and C_2^{2-}; covalent in C_2^{2-} **f.** covalent **7.14** **a.** $:\!\ddot{K}r\!:$ **b.** $:\!\dot{S}b\,\cdot$ **c.** $:\!\ddot{F}\cdot$ **d.** $\cdot\dot{I}n\cdot$ **e.** $Ba\cdot$ **f.** $\cdot\dot{S}i\cdot$ **7.20** **a.** Li^+ $:\!\ddot{C}l\!:^-$ **b.** Ba^{2+} $(:\!\ddot{C}l\!:^-)_2$ **c.** $(Na^+)_2$ $:\!\ddot{O}\!:^{2-}$ **d.** Li^+ $:\!\ddot{O}\!:H^-$ **e.** Cs^+ $:\!\ddot{F}\!:^-$ **f.** Ba^{2+} $:\!\ddot{S}\!:^{2-}$ **7.24** -1101 kJ/mol

7.30 a. H:Ö:N::Ö **b.** H:N̈:H⁺ **c.** H:C::Ö:

with H above and below in b.

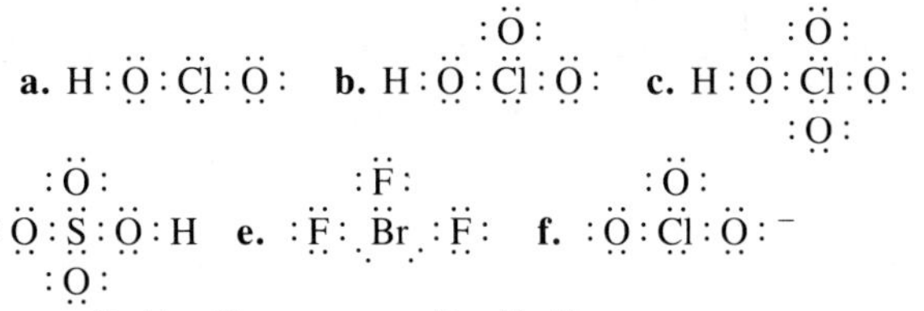

d. H:C̈:C̈:C̈:H **e.** :Ö:H⁻ **f.** H:C̈:N::Ö:

with H O H arrangement in d, and H below in f.

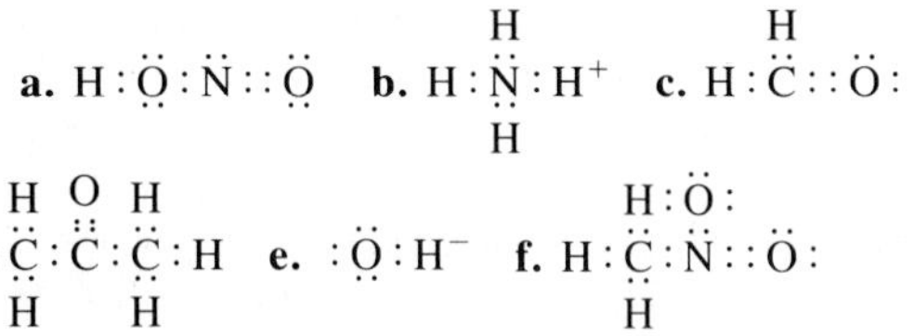

7.32 normal covalent

7.34 a. H:C:::N: **b.** H:C̈:C:::N: **c.** H:C̈:Br̈:
d. H:C:::C:H **e.** H:C̈::C̈:H **f.** H:C̈:C̈:H

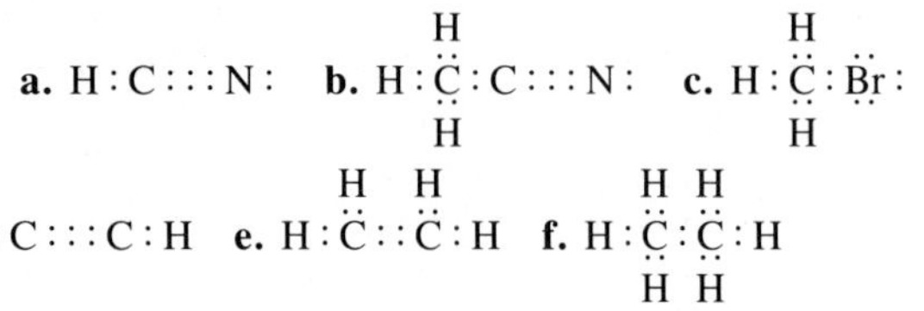

7.36 a. H:Ö:C̈l:Ö: **b.** H:Ö:C̈l:Ö: **c.** H:Ö:C̈l:Ö:
with :Ö: below in c.

d. H:Ö:S̈:Ö:H **e.** :F̈:Br̈:F̈: **f.** :Ö:C̈l:Ö:⁻
with :Ö: below in d, and :F: above in e.

7.40 a. :Ö:S::Ö: ⟷ :Ö::S:Ö:

b. :Ö:S::Ö: ⟷ :Ö::S:Ö: ⟷ :Ö:S:Ö:

c. :Ö::C::Ö: ⟷ :Ö:C:::O: ⟷ :O:::C:Ö:

d. :Ö:C̈::Ö:²⁻ ⟷ :Ö::C̈:Ö:²⁻ ⟷ :Ö:C̈:Ö:²⁻

e. :Ö:N::Ö:⁻ ⟷ :Ö::N:Ö:⁻ ⟷ :Ö:N:Ö:⁻

f. :Ö:N::Ö: ⟷ :Ö::N:Ö:

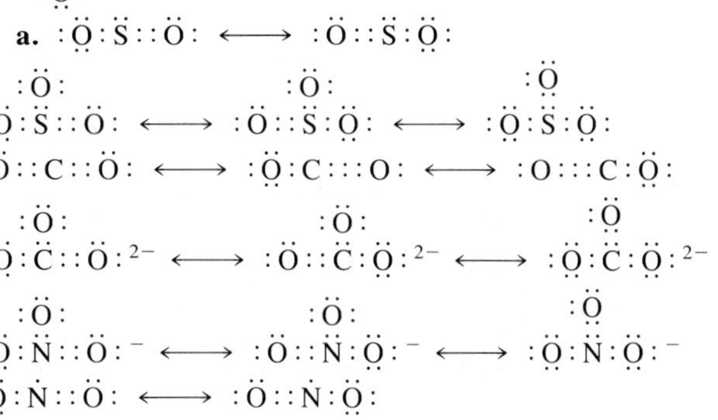

7.42 a. +1 **b.** −1 **c.** +1 **d.** −2 **e.** 0 **f.** +1 **7.46 a.** yes **b.** yes **c.** no, F has 7 electrons **d.** no, S has 10 electrons **e.** yes **f.** no, H has 2 electrons **7.54 a.** HF **b.** ICl **c.** HI **7.60 a.** −24 kJ **b.** 184 kJ **c.** −2818 kJ **d.** −216 kJ **e.** −558 kJ **7.62** 268 kJ **7.64** HBr **7.66 a.** −826 kJ **b.** −834 kJ **c.** 53 kJ **d.** −7674 kJ **e.** −1278 kJ **7.72 a.** HF **b.** FCl **c.** BO **7.80 a.** covalent **b.** ionic **c.** covalent **d.** covalent **e.** ionic **f.** ionic **7.86 a.** F 1004, Cl 802, Br 734, I 652 **b.** Mullikan $\chi = 250 \times$ Pauling χ

Chapter 8

8.4 a. S **b.** Be **c.** C **d.** S **e.** Xe **f.** S **g.** P **h.** S **8.6 a.** trigonal pyramidal **b.** trigonal bipyramidal **c.** bent **d.** distorted tetrahedral **e.** octahedral **f.** linear **g.** square planar **h.** trigonal pyramidal **8.10 a.** ≈ 107° **b.** ≈ 105° **c.** 180° **d.** 180° **e.** 120° **f.** 120° **g.** ≈ 107° **h.** 90°, 180°, and 120° **8.16 a.** CH_2Cl_2 **b.** CH_3F **c.** HCl **d.** NF_3 **e.** H_2O **8.30 a.** d^3sp^3 **b.** sp^2 **c.** sp^3 **d.** sp^3 **e.** dsp^3 **f.** sp **8.34** 1 sigma bond and 2 pi bonds in C≡C; 2 sigma bonds in C—H **8.58** CN^+ (2) < CN (2½) < CN^- (3) **8.62** BN^+ (1½) < BN (2) < BN^- (2½) **8.64 a.** paramagnetic **b.** paramagnetic **c.** diamagnetic **d.** diamagnetic **e.** paramagnetic

8.72 a. BN $(\sigma_{1s})^2(\sigma_{1s}^*)^2(\sigma_{2s})^2(\sigma_{2s}^*)^2[(\pi_{2p_y})^2(\pi_{2p_z})^2](\sigma_{2p_x})^0$ $[(\pi_{2p_y}^*)^0(\pi_{2p_z}^*)^0](\sigma_{2p_x}^*)^0$
b. BO $(\sigma_{1s})^2(\sigma_{1s}^*)^2(\sigma_{2s})^2(\sigma_{2s}^*)^2[(\pi_{2p_y})^2(\pi_{2p_z})^2](\sigma_{2p_x})^1$ $[(\pi_{2p_y}^*)^0(\pi_{2p_z}^*)^0](\sigma_{2p_x}^*)^0$
c. CN $(\sigma_{1s})^2(\sigma_{1s}^*)^2(\sigma_{2s})^2(\sigma_{2s}^*)^2[(\pi_{2p_y})^2(\pi_{2p_z})^2](\sigma_{2p_x})^1$ $[(\pi_{2p_y}^*)^0(\pi_{2p_z}^*)^0](\sigma_{2p_x}^*)^0$
d. CN^- $(\sigma_{1s})^2(\sigma_{1s}^*)^2(\sigma_{2s})^2(\sigma_{2s}^*)^2[(\pi_{2p_y})^2(\pi_{2p_z})^2](\sigma_{2p_x})^2$ $[(\pi_{2p_y}^*)^0(\pi_{2p_z}^*)^0](\sigma_{2p_x}^*)^0$
8.76 a. trigonal pyramidal **b.** bent **c.** T-shape **d.** square pyramidal **e.** bent **f.** bent **g.** bent **h.** octahedral **8.88** $F_2 < O_2 < N_2 < C_2$

Chapter 9

9.6 a. 81.6 torr **b.** 0.100 torr **c.** 658 torr **d.** 74.5 torr **e.** 2.18 atm **9.8** 182 atm **9.10** 0.472 L **9.14 a.** 418 torr **b.** 673 torr **c.** 212 torr **d.** 2.25 atm **e.** 1.49 atm **9.16 a.** 201.8 K, −71.4°C **b.** 1430 K, 1160°C **c.** 5.10×10^2 K **d.** 105 K **e.** 902 K, 629°C **9.22** 0.407 mL **9.24 a.** 546 K, 273°C **b.** 69 K, −204°C **c.** 224 K **d.** 0.1493 K, −273.00°C **e.** 356 K, 83°C **9.26 a.** 1.16 L **b.** 1.33 L **c.** 9.10×10^3 torr **d.** 468 torr **e.** 588 K **f.** 77.8 K, −195.3°C **9.28** 4840 torr **9.32 a.** 0.40 mol, 9.0 L **b.** 1.5 mol, 34 L **c.** 0.050 mol, 1.1 L **d.** 0.20 mol, 4.5 L **e.** 2.2 mol, 49 L **9.34 a.** 0.40 mol, 0.82 L **b.** 1.5 mol 3.1 L **c.** 0.050 mol, 0.10 L **d.** 0.20 mol, 0.41 L **e.** 2.2 mol, 4.5 L **9.40 a.** 0.672 g/L **b.** 1.11 g/L **c.** 1.74 g/L **b.** 1.19 g/L **e.** 1.82 g/L **9.42 a.** 61.22 g/mol **b.** 21.2 g/mol **c.** 33.30 g/mol **d.** 86.70 g/mol **e.** 42.87 g/mol **9.44** 131 g/mol **9.48** 32.3 g/mol **9.52** 20.8 g/mol **9.62** 726 torr **9.66** 8.7×10^9 L **9.68** 54.0 L CO_2, 85.5 L O_2 **9.70** 1680 g **9.72** 24 L H_2 **9.74** 547 L **9.80** 1.74 L **9.82** 30 L SO_3 **9.84** 0.355 of the original volume **9.86** 5.07 atm **9.88** 55.1 atm gas, 33.6 atm N_2, 21.5 atm Ar **9.90** 30.9 L **9.92** 2.13×10^3 cm/s **9.94** 2.53×10^7 L **9.96** 7.06 L **9.100 a.** 5.165 g/L **b.** 137.0 g/mol **c.** $CFCl_3$

Chapter 10

10.6 a. London dispersion forces **b.** ionic bond **c.** London dispersion forces **d.** dipole–dipole **e.** hydrogen bond **f.** dipole–dipole forces **g.** London dispersion forces **h.** covalent bonding **10.8 a.** dipole–induced-dipole force **b.** London dispersion force **c.** London dispersion force **d.** hydrogen bonding **e.** dipole–induced-dipole force **f.** hydrogen bonding **10.10 a.** yes **b.** yes **c.** no **d.** no **e.** yes **10.12** He < N_2 < I_2 < H_2O < $CaBr_2$ **10.18** CH_3OH < C_2H_5OH < C_3H_7OH **10.20** CH_4 < SiH_4 < GeH_4 < SnH_4 **10.38 a.** CH_3OH **b.** CH_3OH **c.** NH_3 **d.** Cl_2 **e.** PH_3 **10.52** They are identical structures. **10.54** 4 **10.56** 359 pm **10.58** 191 pm **10.60** 10.5 g/cm³ **10.64** $4S^{2-}$, $4Zn^{2+}$ **10.66** C_3A **10.74** metallic solid **10.76 a.** ionic **b.** molecular **c.** metallic **d.** network covalent **e.** molecular **10.78 a.** yes **b.** no **c.** no **d.** no **e.** no **10.84** diamond **10.88** Cl_2 < Br_2 < I_2 **10.96** 16.3 g/cm³ **10.98** −679 kJ/mol

Chapter 11

11.6 a. 33.22% **b.** 0.1629 **c.** 10.81 *m* **d.** 7.800 *M* **11.8** 0.854 *m* **11.10** 2.87 *M* **11.12** 8.9 *M* **11.14 a.** 0.185 *M* **b.** 0.250 *M* **c.** 1.31 *M* **11.16 a.** 21 g **b.** 44 g **c.** 19 g **11.24 a.** $CH_3CH_2CH_3$ **b.** $CH_3CH_2CH_2CH_2OH$ **c.** CH_4 **d.** $CH_3CH_2CH_3$ **e.** $CH_3CH_2CH_2CO_2H$ **f.** H_2S **11.26** 0.500 *M* **11.28** 17.1 g **11.30** 45 g **11.34 a.** electrolyte **b.** electrolyte **c.** electrolyte **d.** nonelectrolyte **e.** nonelectrolyte **f.** nonelectrolyte (or very weak) **11.40**
a. $Sr(NO_3)_2(aq) + H_2SO_4(aq) \longrightarrow SrSO_4(s) + 2HNO_3(aq)$
$Sr^{2+}(aq) + SO_4^{2-}(aq) \longrightarrow SrSO_4(s)$
b. $Zn(NO_3)_2 + Na_2SO_4 \longrightarrow$ no reaction
c. $CuSO_4(aq) + BaS(aq) \longrightarrow CuS(s) + BaSO_4(s)$
$Cu^{2+}(aq) + SO_4^{2-}(aq) + Ba^{2+}(aq) + S^{2-}(aq) \longrightarrow$
$$CuS(s) + BaSO_4(s)$$
d. $3(NH_4)_2CO_3(aq) + 2CrCl_3(aq) \longrightarrow$
$$6NH_4Cl(aq) + Cr_2(CO_3)_3(s)$$
$3CO_3^{2-}(aq) + 2Cr^{3+}(aq) \longrightarrow Cr_2(CO_3)_3(s)$
e. $Al(OH)_3(s) + 3HCl(aq) \longrightarrow AlCl_3(aq) + 3H_2O(l)$
$Al(OH)_3(s) + 3H^+(aq) \longrightarrow Al^{3+}(aq) + 3H_2O(l)$
11.46 a. $Ca^{2+}(aq) + 2Na^+(resin) \longrightarrow$
$$Ca^{2+}(resin) + 2Na^+(aq)$$
b. $Ca(HCO_3)_2(aq) \xrightarrow{\Delta} CaCO_3(s) + H_2O(l) + CO_2(g)$
c. $3Ca^{2+} + 2PO_4^{2-}(aq) \longrightarrow Ca_3PO_4(s)$
d. $Ca(HCO_3)_2(aq) + CaO(s) \longrightarrow 2CaCO_3(s) + H_2O(l)$
11.54 100.78°C **11.56** 1.8×10^2 g/mol **11.58** 35.0 g/mol **11.60** 0.675 atm **11.62** 98 g/mol **11.64** 6.00 torr C_2H_5OH, 8.32 torr CCl_4 **11.66** 6.67×10^6 g/mol **11.70 a.** solution **b.** solution **c.** colloid **d.** solution **e.** colloid **f.** colloid **g.** suspension **h.** colloid **i.** solution **11.72** -25°C, -13°F **11.76** 667 g **11.78** 71 g **11.80** 9.98 *M* **11.82** 3×10^3 g **11.84 a.** 54.91% **b.** 0.183 **c.** 12.4 *m* **11.86** 9.62×10^5 g/mol **11.90** 60.8 g/mol **11.92 a.** 0.2074 **b.** 4.117 *m* **c.** 28.2 *M*

Chapter 12

12.6 1.86×10^3 J **12.8** -2046.44 kJ **12.14** $Hg(s) < Hg(l) < Hg(g)$ **12.16 a.** increase **b.** decrease **c.** increase **d.** decrease **e.** decrease **f.** increase **12.18** 143 J/mol K **12.20 a.** 88.75 J/K **b.** -146.53 J/K **c.** 170 J/K **d.** -188.0 J/K **e.** 152.3 J/K **12.28 a.** -750 kJ **b.** exothermic **c.** spontaneous **12.30 a.** 313 kJ **b.** endothermic **c.** not spontaneous **12.32 a.** -29.4 kJ, -24.8 kJ, 15.3 J/K **b.** -141.8 kJ, -197.7 kJ, -188.0 J/K **c.** -81.82 kJ, -132.5 kJ, -169.6 J/K **d.** 84.4 kJ, 87.2 kJ, 8.5 J/K **e.** -990.4 kJ, -1036.1 kJ, -152.9 J/K **12.36** enthalpy **12.38 a.** spontaneous **b.** not spontaneous **c.** not spontaneous **d.** spontaneous **e.** spontaneous **12.40** < 1550 K **12.42 a.** $Q = P_{CO_2}^2/P_{CO}^2 P_{O_2}$ **b.** $Q = P_{C_2H_6}/P_{C_2H_2}P_{H_2}^2$ **c.** $Q = P_{C_2H_6}/P_{C_2H_4}P_{H_2}$ **d.** $Q = P_{SiF_4}P_{H_2O}^2/P_{HF}^4$ **e.** $Q = P_{O_2}$ **12.46** -28.62 kJ **12.52** 0.031 **12.54** >1 **12.56 a.** 94.1 kJ, 3.3×10^{-17} **b.** -75.98 kJ, 2.0×10^{13} **c.** -990.4 kJ, 3×10^{173} **d.** 146.0 kJ, 2.7×10^{-26} **e.** -221.94 kJ, 7.3×10^{38} **12.58 a.** decrease **b.** decrease

c. decrease **d.** decrease **e.** decrease **12.60** 228.05 kJ, 694.3°C **12.62** **a.** -514.42 kJ **b.** -242.1 kJ **c.** -141.1 kJ **d.** -27.00 kJ **e.** -81 kJ **f.** -233.6 kJ **g.** -452.3 kJ **12.64 a.** not spontaneous **b.** spontaneous **c.** spontaneous **d.** not spontaneous e. spontaneous **12.66** > 1750 K **12.68** 2×10^{140} **12.76 a.** -800.79 kJ, -802.34 kJ, -5.2 J/K **b.** -430.2 kJ, -413.6 kJ, 54.3 J/K **c.** -207.8 kJ, -295.4 kJ, -294.3 J/K **d.** 194.0 kJ, 237.6 kJ, 146.18 J/K **e.** -233.6 kJ, -196.1 kJ, 125.6 J/K **12.78** spontaneous **12.82** -206.4 kJ, spontaneous

Chapter 13

13.6 7.84×10^{-6} *M* s^{-1} **13.8** Rate $= -1/3\ \Delta[I^-]/\Delta t = -\Delta[IO_2^-]/\Delta t = -1/4\ \Delta[H^+]/\Delta t = 2\Delta[I_2]/\Delta t = 2\Delta[H_2O]/\Delta t$ **13.14 a.** first order **b.** 8.07×10^{-8} s^{-1} **13.16 a.** first order **b.** 0.00486 h^{-1} **13.20 a.** $-\frac{1}{2}\Delta[UO_2^+]/\Delta t = k[UO_2^+]^2[H^+]$ **b.** 130 M^{-2} s^{-1} **13.22 a.** Rate $= k[C_5H_{10}][O_3]$ **b.** 7.9×10^4 M^{-1}s^{-1} **13.24** Rate $= k[O_3]$, 1.24×10^{-4} s^{-1} **13.26 a.** Rate $= k[ClO^-][I^-][OH^-]^{-1}$ **b.** 6.1×10^8 s^{-1} **13.32 a.** 0.766 g **b.** 0.267 y^{-1} **13.34** 288 d **13.36 a.** 8.60×10^6 s **b.** 3.56×10^6 s **13.42 a.** $-\Delta[O_2]/\Delta t = k[NO]^2[O_2]$ **b.** M^{-2}s^{-1} **c.** $\frac{1}{2}\Delta[NO_2]/\Delta t = k[NO]^2[O_2]$ **d.** yes **e.** yes **13.44 a.** Order with respect to $Cl^{-1} = 1$, $BrO_3^- = 1$, $H^+ = 2$ **b.** 4 **13.52** 2.3×10^{-2} M^{-1}s^{-1} **13.56 a.** 64 kJ/mol **b.** 7.3×10^6 M^{-1}s^{-1} **c.** 9.1×10^{-4} M^{-1}s^{-1} **13.58** 1.56×10^{-5} M^{-1}s^{-1} **13.64** 1, 1, 2, 2 **13.66** Rate $= (k_2k_1/k_{-1})[Tl^+][Co^{3+}]^2/[Co^{2+}]$ **13.68** Rate $= (k_2k_1/k_{-1})[Cr^{2+}]^2[UO_2^+]$ **13.74 a.** I_2 **b.** HI **13.78** 0 **13.80 a.** first order **b.** 0.00412 s^{-1} **13.84** 130 kJ/mol **13.86** 2950 y **13.88 a.** Rate $= k[Sn^{2+}][Fe^{3+}][H^+]^{-2}$ **b.** 6.01×10^{-4} *M* s^{-1} **13.90** 28.3 kJ/mol

Chapter 14

14.6 0.0262 **14.8** $k_r[VO_2^+][IrCl_6^{3-}][H^+]$ **14.14 a.** $K_c = [CO]^2[O_2]/[CO_2]^2$ **b.** $K_c = [NH_3][H_2S]$ **c.** $K_c = 1/[N_2][Cl_2][H_2]^4$ **d.** $K_c = [CO]^2/[CO_2]$ **14.16 a.** $K_p = P_{CO}/P_{CO_2}$, $K_c = [CO]/[CO_2]$ **b.** $K_p = P_{HCl}P_{NH_3}$, $K_c = [HCl][NH_3]$ **c.** $K_p = P_{CO_2}$, $K_c = [CO_2]$ **14.20** 0.0640 **14.22** 0.024 **14.24** 0.0142 **14.30** 0.67 **14.32** 9.1×10^8 **14.36** 2.4×10^{-18} **14.38** 0.10 **14.42 a.** 221.95 kJ, 1.32×10^{-39} **b.** -7.1 kJ, 20 **c.** -990.4 kJ, 3×10^{173} **d.** -959.3 kJ, 1×10^{168} **e.** 113.57 kJ, 1.27×10^{-20} **14.46** 3.94×10^{-2} atm **14.48** 0.845 atm **14.50 a.** 0.0199 mol **b.** 0.995 (or 99.5%) **14.52 a.** 0.98 *M* N_2, 1.96 *M* H_2, 0.0200 *M* N_2H_4 **b.** 2.00×10^2 **14.54** 0.44 *M* CO_2, 0.44 *M* H_2, 0.558 *M* CO, 0.558 *M* H_2O **14.60 a.** shift toward reactants **b.** shift toward products **c.** shift toward products **d.** shift toward reactants **e.** no effect **f.** no effect **14.62 a.** forward **b.** reverse **c.** no effect **d.** reverse **e.** forward **14.64 a.** forward **b.** reverse **c.** forward **d.** forward **e.** no effect **f.** forward **g.** reverse **h.** no effect **i.** forward **j.** no effect **14.70 a.** $\approx 80\%$ **b.** $\approx 40\%$ **14.78** increased **14.84** exothermic **14.86 a.** 89.7 kJ, 2×10^{-16} **b.** -959.3 kJ, 1×10^{164} **c.** 208.4 kJ, 3.1×10^{-37}

d. -403.0 kJ, 4.13×10^{70} **e.** -75.98 kJ, 2.05×10^{13}
14.88 1.58 M **b.** 55.3 **14.90** 2.00

Chapter 15

15.4 $HCl(aq)$, $H_2SO_4(aq)$, $H_3PO_4(aq)$ **15.6** **a.** base
b. acid **c.** neither **d.** base **e.** acid **f.** neither **g.** acid
h. base **i.** acid **j.** base **k.** acid **l.** base **15.8** **a.** Cl^-
b. NH_3 **c.** S^{2-} **d.** NO_2^- **e.** CN^- **f.** $H_2PO_4^-$ **15.10**
a. $Ni^{2+}(aq)$, $NH_3(aq)$ **b.** $HBr(g)$, $NF_3(g)$ **c.** $BF_3(g)$, $NH_3(g)$
15.14 **a.** strong **b.** strong **c.** weak **d.** strong **e.** strong
f. weak **g.** weak **h.** weak **i.** weak **k.** strong **l.** strong
15.16 reactants **15.22** **a.** N_2O_5 **b.** Cl_2O_7 **c.** SO_3
d. Na_2O **e.** CaO **f.** B_2O_3 **g.** CrO_3 **h.** Br_2O_5 **i.** ZnO
j. P_2O_5 **k.** Fe_2O_3 **15.26** **a.** 6.49 **b.** 3.49 **c.** 3.60 **d.** 3.70
e. 4.49 **f.** 7.40 **15.28** 0.807, 13.193 **15.30** 1.592, 12.408
15.32 13.1153, 0.8847 **15.34** 0.800 M, 1.25×10^{-14} M
15.36 1.50×10^{-14} M, 0.667 M **15.42** 1.4×10^{-9} **15.44**
15.44 5.52×10^{-5} **15.46** 4.50×10^{-8} **15.48** $8.00 \times$
10^{-7} **15.50** 1.8×10^{-8} **15.52** 3.9×10^{-10} **15.56** 5.70
15.58 2.638 **15.60** 3.2×10^{-5} M **15.62** 7.1×10^{-6} M
15.66 6.67×10^{-4} M, 4.7×10^{-11} M **15.68** $2.4 \times$
10^{-5} M, 1.2×10^{-13} M **15.72** 7.00 **15.74** 5.424 **15.76**
0.0949% **15.80** 8.832 **15.82** 6.63×10^{-4} M **15.84**
6.4×10^{-5} **15.86** 2.18 **15.88** 1.622 **15.94** 13.389,
0.611 **15.98** 6.78 **15.100** 0.0156 M H_3O^+, 0.0494 M
$HCOOH$, 5.68×10^{-4} M $HCOO^-$, 6.41×10^{-13} M OH^-

Chapter 16

16.6 4.924 **16.8** 8.26 **16.10** 2.2×10^{-8} **16.12** 9.405
16.20 1.8 **16.22** 9.121 **16.24** 2.21×10^{-4} M, $4.52 \times$
10^{-11} M **16.26** 4.660 **16.28** 1.0 **16.30** **a.** 4.757
b. -0.17 **c.** 0.16 **16.32** 4.72 **16.38** 9.4×10^{-8} M
16.42 **a.** $Ag_2CrO_4(s) \rightleftharpoons 2Ag^+(aq) + CrO_4^{2-}(aq)$
$$K_{sp} = [Ag^+]^2[CrO_4^{2-}]$$
b. $PbCrO_4(s) \rightleftharpoons Pb^{2+}(aq) + CrO_4^{2-}(aq)$
$$K_{sp} = [Pb^{2+}][CrO_4^{2-}]$$
c. $PbI_2(s) \rightleftharpoons Pb^{2+}(aq) + 2I^-(aq)$ $K_{sp} = [Pb^{2+}][I^-]^2$
d. $AgI(s) \rightleftharpoons Ag^+(aq) + I^-(aq)$ $K_{sp} = [Ag^+][I^-]$
e. $Cr(OH)_3(s) \rightleftharpoons Cr^{3+}(aq) + 3OH^-(aq)$
$$K_{sp} = [Cr^{3+}][OH^-]^3$$
f. $Ca_3(PO_4)_2(s) \rightleftharpoons 3Ca^{2+}(aq) + 2PO_4^{3-}(aq)$
$$K_{sp} = [Ca^{2+}]^3[PO_4^{3-}]^2$$
16.44 **a.** 1.1×10^{-11} **b.** 9.8×10^{-9} **c.** 6.08×10^{-12}
d. 9.70×10^{-11} **e.** 6.57×10^{-13} **16.46** 5.31×10^{-4}
16.48 1.64×10^{-14} **16.52** **a.** 1.33×10^{-4} g/L **b.** $5.4 \times$
10^{-8} g/L **c.** 1.3 g/L **d.** 0.097 g/L **e.** 4.4 g/L **16.54**
0.15 g **16.56** 9.0×10^{-10} M **16.62** **a.** yes **b.** yes **c.** yes
16.64 2.9×10^{-10} M Pb^{2+}, 0.030 M IO_3^- **16.70** $1.2 \times$
10^{-2} M **16.76** 1.1 g **16.78** 7.6×10^{-5} M **16.80**
0.250 M CH_3CO_2H, 0.350 M $CH_3CO_2^-$, 1.25×10^{-5} M
H_3O^+, 8.00×10^{-10} M OH^- **16.82** $CuS < NiS < CoS <$
MnS **16.84** 1.50×10^{-9} **16.86** **a.** decrease **b.** decrease
c. increase **d.** no effect **16.88** **a.** yes **b.** no **16.90**
a. increase **b.** increase **c.** no change **16.94** 4.76 **16.96**
a. 1.5×10^{-8} M **b.** $BaSO_4$, 1.5×10^{-9} **c.** yes

Chapter 17

17.6 Ag^+, Na^+ **17.8** **a.** Fe^{2+} **b.** $Cr_2O_7^{2-}$ **c.** $Cr_2O_7^{2-}$
d. Fe^{2+} **17.10** **a.** O_2, Na, Na, O_2 **b.** Br_2, Ni, Ni, Br_2
c. O_2, C, C, O_2 **d.** H_2, Na, Na, H_2 **17.16** **a.** $+1$ **b.** $+6$
c. $+3$ **d.** $+4$ **e.** $+2$ **f.** $+5$ **17.18** **a.** -1 **b.** 0 **c.** $+4$
d. $+7$ **e.** $+7$ **f.** $+3$ **g.** $+3$ **h.** -1 **i.** $+1$ **17.20** **a.** $+2$
b. $+2$ **c.** $+3$ **d.** $+3$ **e.** $+3$ **f.** $+3$ **17.22** **a.** $+7$ **b.** $+5$
c. -1 **d.** $+3$ **e.** $+5$ **17.24** **a.** $+3$ **b.** $+5$ **c.** $+1$ **d.** $+4$
e. $+4$ **f.** $+3$ **g.** $+5$ **h.** $+3$ **i.** $+2$ **j.** $+3$ **k.** $+4$ **l.** $+5$
17.30 **a.** $3Cu(s) + 8H^+(aq) + 2NO_3^-(aq) \longrightarrow$
$$3Cu^{2+}(aq) + 2NO(l) + 4H_2O(l)$$
b. $4Zn(s) + 10H^+(aq) + NO_3^-(aq) \longrightarrow$
$$4Zn^{2+}(aq) + NH_4^+(aq) + 3H_2O(l)$$
c. $10H^+(aq) + 8I^-(aq) + SO_4^{2-}(aq) \longrightarrow$
$$H_2S(aq) + 4I_2(aq) + 4H_2O(l)$$
d. $2Cr^{3+}(aq) + ClO_3^-(aq) + 10OH^-(aq) \longrightarrow$
$$2CrO_4^{2-}(aq) + Cl^-(aq) + 5H_2O(l)$$
e. $3Cl_2(aq) + 6OH^-(aq) \longrightarrow$
$$5Cl^-(aq) + ClO_3^-(aq) + 3H_2O(l)$$
17.32 **a.** $4H_2O_2(aq) + PbS(s) \longrightarrow PbSO_4(s) + 4H_2O(l)$
b. $2KMnO_4(aq) + 13H_2SO_4(aq) + 10NaCl(aq) \longrightarrow$
$$5Cl_2(aq) + 10NaHSO_4(aq) + K_2SO_4(aq)$$
$$+ 2MnSO_4(aq) + 8H_2O(l)$$
c. $Na_2SO_3(aq) + H_2O(l) + I_2(aq) \longrightarrow$
$$Na_2SO_4(aq) + 2HI(aq)$$
d. $2HNO_2(aq) + 2HI(aq) \longrightarrow 2NO(g) + I_2(aq) + 2H_2O(l)$
e. $Hg_2Cl_2(aq) + 2NaOH(aq) \longrightarrow$
$$HgO(s) + Hg(l) + 2NaCl(aq) + H_2O(l)$$
17.34 **a.** $Fe_2(SO_4)_3 + 2KI \longrightarrow 2FeSO_4 + K_2SO_4 + I_2$
b. $As_2O_3 + 2H_2O + 2I_2 \longrightarrow As_2O_5 + 4HI$
c. $SO_2 + 2HNO_3 \longrightarrow H_2SO_4 + 2NO_2$
d. $KClO_3 + H_2SO_4 + 6HI \longrightarrow$
$$HCl + KHSO_4 + 3H_2O + 3I_2$$
e. $(NH_4)_2Cr_2O_7 \longrightarrow Cr_2O_3 + N_2 + 4H_2O$
17.36 **a.** $2CrO_4^{2-}(aq) + 3S^{2-}(aq) + 8H_2O(l) \longrightarrow$
$$2Cr(OH)_3(s) + 3S(s) + 10OH^-(aq)$$
b. $2MnO_4^-(aq) + CN^-(aq) + 2OH^-(aq) \longrightarrow$
$$2MnO_4^{2-}(aq) + CNO^-(aq) + H_2O(l)$$
c. $2Al(s) + 6H_2O(l) + 2OH^-(aq) \longrightarrow$
$$2Al(OH)_4^-(aq) + 3H_2(g)$$
d. $Br_2(aq) + 2OH^-(aq) \longrightarrow BrO^-(aq) + Br^-(aq) + H_2O(l)$
e. $ClO^-(aq) + 2NH_3(aq) \longrightarrow Cl^-(aq) + N_2H_4(aq) + H_2O(l)$
f. $2Cr(OH)_4^-(aq) + 3HO_2^-(aq) \longrightarrow$
$$2CrO_4^{2-}(aq) + 5H_2O(aq) + OH^-(aq)$$
g. $P_4(s) + 3H_2O(l) + 3OH^-(aq) \longrightarrow$
$$PH_3(g) + 3H_2PO_2^-(aq)$$
17.42 **a.** $+1.66$ V **b.** $+0.03$ V **c.** $+2.37$ V **d.** $+0.74$ V
e. $+2.04$ V **f.** $+2.26$ V **17.44** no **17.46** **a.** Fe^{3+} and
Mn^{2+} **b.** Zn^{2+} and Mn^{2+} **c.** Zn^{2+} and Fe **17.48** H_2O_2
should disproportionate in acidic solution, and HO_2^- should disproportionate in basic solution. **17.50** SO_4^{2-} cannot disproportionate; $S_2O_6^{2-}$ should disproportionate. **17.52** none
17.54 Cr^{2+} **17.56** CuO^+
17.58 **a.** $4Au(s) + 8CN^-(aq) + O_2(g) + 2H_2O(l) \longrightarrow$
$$4Au(CN)_2^-(aq) + 4OH^-(aq)$$

b. $Cr_2O_3(s) + 6Fe(CN)_6{}^{3-}(aq) + 10OH^-(aq) \longrightarrow$
$$2CrO_4{}^{2-}(aq) + 6Fe(CN)_6{}^{4-}(aq) + 5H_2O(l)$$
c. $N_2H_4(l) + 2Cu(OH)_2(s) \longrightarrow N_2(g) + 2Cu(s) + 4H_2O(l)$
17.60 **a.** $2Cr^{2+}(aq) + H_2O_2(aq) + 6OH^-(aq) \longrightarrow$
$$2Cr(OH)_4{}^-(aq)$$
b. H_2O_2 **c.** Cr^{2+} **d.** Cr **e.** O
17.62 **a.** $2CoCl_2(s) + 2OH^-(aq) + Na_2O_2(aq) +$
$2H_2O(l) \longrightarrow 2Co(OH)_3(s) + 2Na^+(aq) + 4Cl^-(aq)$
$$Na_2O_2, CoCl_2, Co, O$$
b. $Bi_2O_3(s) + 2OH^-(aq) + 2ClO^-(aq) \longrightarrow$
$$2BiO_3{}^-(aq) + H_2O(l) + 2Cl^-(aq)$$
$$ClO^-, Bi_2O_3, Bi, Cl$$
17.64 $5C_6H_{12}O_6(aq) + 24NO_3{}^-(aq) + 24H^+(aq) \longrightarrow$
$$30CO_2(g) + 12N_2(g) + 42H_2O(l)$$
17.66 $Pt(s) + 4Cl^-(aq) + 2NO_3{}^-(aq) + 4H^+(aq) \longrightarrow$
$$PtCl_4{}^{2-}(aq) + 2NO_2(g) + 2H_2O(l)$$
17.72 Cr^{3+} and $SO_4{}^{2-}$
17.74 **a.** yes **b.** no **c.** yes **d.** yes **e.** yes **f.** no **g.** no
h. yes **i.** yes
17.76 $FeAsS(s) + 14HNO_3(aq) + 2H^+(aq) \longrightarrow$
$$Fe^{3+}(aq) + H_3AsO_4(aq) + HSO_4{}^-(aq)$$
$$+ 14NO_2(g) + 6H_2O(l)$$

Chapter 18

18.6 $+0.19$ V **18.8** $+1.20$ V **18.10** 0.13 V **18.12**
$+0.005$ V **18.14** $+0.16$ V **18.20** **a.** $+0.63$ V
b. -1.2×10^5 J/mol **c.** 7×10^{20} **d.** yes **18.22** $-2.39 \times$
10^5 J/mol **18.24** **a.** -0.77 V **b.** 1.5×10^5 J/mol **c.** $3 \times$
10^{-27} **d.** no **18.30** -0.82 V **18.32** $+1.38$ V **18.34**
3.6×10^{-5} M **18.36** -0.07 V
18.42 **a.** cathode: $PbO_2(s) + 4H^+(aq) + SO_4{}^{2-}(aq)$
$$+ 2e^- \longrightarrow PbSO_4(s) + 2H_2O(l)$$
anode: $Pb(s) + SO_4{}^{2-}(aq) \longrightarrow PbSO_4(s) + 2e^-$
b. $2PbSO_4(s) + 2H_2O(l) \longrightarrow$
$$Pb(s) + PbO_2(s) + 2SO_4{}^{2-}(aq) + 4H^+(aq)$$
18.44 anode: $Zn(s) + 2OH^- \longrightarrow Zn(OH)_2 + 2e^-$
cathode: $HgO(s) + H_2O + 2e^- \longrightarrow Hg(l) + 2OH^-$
18.54 **a.** I_2 and H_2 **b.** H_2 and O_2 **c.** Cu and O_2 **d.** H_2 and
O_2 **e.** H_2 and O_2 **f.** H_2 and O_2 **g.** H_2 and O_2 **h.** Pb and
HNO_2 **18.56** H_2 (or maybe Fe) and O_2 **18.62** **a.** 2 mol
b. 0.67 mol **18.64** 0.408 g **18.66** 65.5 g **18.68** 106 A
18.70 86 g/equiv **18.72** 0.019 mol
18.74 **a.** $Cd(s) \longrightarrow Cd^{2+}(aq) + 2e^-$
$Ag^+(aq) + e^- \longrightarrow Ag(s)$
$Cd(s) + 2Ag^+(aq) \longrightarrow Cd^{2+}(aq) + 2Ag(s)$
b. $+1.20$ V
18.76 anode: $Cd(s) + 2OH^- \longrightarrow Cd(OH)_2 + 2e^-$
cathode: $NiO_2(s) + 2H_2O + 2e^- \longrightarrow$
$$Ni(OH)_2(s) + 2OH^-$$
18.78 $+0.59$ V **18.80** 1.58 g **18.82** 0.798 g **18.84**
-1.3×10^5 J/mol **18.86** $+0.46$ V **18.88** 50 g/equiv
18.96 0.12 V **18.98** $+4$

Chapter 19

19.8 coordination number 2: linear; coordination number 4:
tetrahedral or square planar; coordination number 6: octahedral

19.10 **a.** $+2, 6$ **b.** $+3, 6$ **c.** $+3, 6$ **d.** $+3, 6$ **e.** $+3, 6$
f. $+2, 4$ **g.** $+2, 4$ **19.12** $[Cr(H_2O)_4Cl_2]Cl \cdot 2H_2O$ **19.16**
a. iodato **b.** sulfido **c.** sulfato **d.** chromato **e.** fluoro
f. hydroxo **19.18** **a.** triamminediaquanitrocobalt(III) chloride
b. potassium hexacyanoferrate(II) **c.** hexaaquacobalt(II) ion
d. pentaaquachlorochromium(III) tetracyanonickelate(II)
19.20 **a.** $Ru(CN)_6{}^{3-}$ **b.** $CuCl_4{}^{2-}$ **c.** $Pt(NH_3)_3Cl^+$
d. $Ir(NH_3)_3Cl_3{}^+$ **e.** $Fe(H_2O)_5Cl^{2+}$ **19.22** **a.** $[Cr(en)_3]_2[PtBr_4]_3$
b. $[Co(NH_3)_3(H_2O)CO_3]_2CO_3$ **c.** $K[Co(en)(CN)_4]$
d. $[Cr(en)_2Cl_2][AuCl_4]$ **19.32** **a.** $[Co(NH_3)_4Cl_2]NO_2$
b. $[Cr(NH_3)_5Cl]BrCl$ **c.** $[Ir(NH_3)_4(NCS)Cl]Cl$
19.34 **a.** $[Co(NH_3)_5Cl]Cl_2 \cdot H_2O$ **b.** $[Cr(H_2O)_4Cl_2]Cl \cdot H_2O$
c. $[Pt(NH_3)_2(H_2O)Cl]Cl$ **19.38** $[Ni(NH_3)_4]^{2+}$:

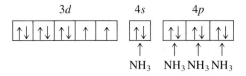

19.50 low-spin, 1 unpaired electron; high-spin, 5 unpaired
electrons **19.52** **a.** diamagnetic **b.** paramagnetic **c.** paramagnetic **d.** paramagnetic **19.54** **a.** colorless **b.** colored
c. colorless **d.** colored **e.** colored **f.** colored **19.56**
a. high-spin **b.** high-spin **c.** low-spin **d.** High-spin and low-
spin have the same configuration. **19.58** potassium carbonyl-
pentacyanoferrate(II) **19.60** **a.** $[Cr(en)_2(CN)Br]Cl$
b. $[Rh(NH_3)_3ClBrNO_2]$ **19.62** **a.** high-spin **b.** low-spin
19.64 Low-spin and high-spin have 2 unpaired electrons.
19.68 **a.** colorless **b.** colored **c.** colorless **d.** colored
e. colored **f.** colored **19.70** **a.** $[Co(NH_3)_4Br(H_2O)]^{2+}$
b. $[Cr(CN)_6]^{3-}$ **c.** $[Fe(CN)_6]^{3-}$ **d.** $[Mn(H_2O)_6]^{3+}$ **19.72**
a. $+3$ **b.** $+3$ **c.** $+2$ **d.** $+2$ **19.78** $[Pt(NH_3)_5Cl]^{3+}$

Chapter 20

20.6 $^{28}_{14}Si$ should be the most stable, $^{30}_{14}Si$ the next most
stable, and $^{29}_{14}Si$ the least stable. **20.12** $^{234}_{90}Th \longrightarrow$
$^{234}_{91}Pa + \beta^-$ **20.14** $^{44}_{21}Sc$ **20.16** $^{228}_{90}Th \longrightarrow ^{224}_{88}Ra + \alpha$
20.18 **a.** $^{221}_{87}Fr \longrightarrow \alpha + ^{217}_{85}At$ **b.** $^{18}_{9}F \longrightarrow \beta^+ + ^{18}_{8}O$
c. $^{133}_{56}Ba + ^{0}_{-1}e \longrightarrow ^{133}_{55}Cs$ **d.** $^{114}_{47}Ag \longrightarrow \beta^- + ^{114}_{48}Cd$
e. $^{96}_{42}Mo + p^+ \longrightarrow ^{96}_{43}Tc + n$ **20.24** 13 g **20.26** 54 d
20.28 36.8 min **20.30** 2.60 y **20.32** 36.0 mg **20.34**
5.4% **20.38** 4.79 MeV **20.40** 3.93 MeV **20.42**
1784.0 MeV **20.44** 7.4757 MeV/nucleon for ^{15}N,
7.4758 MeV for ^{14}N **20.46** 8.791 MeV/nucleon **20.56**
2.555 MeV **20.60** 3.7×10^4 y **20.62** **a.** $^{17}_{9}F \longrightarrow ^{17}_{8}O +$
β^+ **b.** $^{21}_{9}F \longrightarrow ^{21}_{10}Ne + \beta^-$ **c.** $^{216}_{84}Po \longrightarrow ^{212}_{82}Pb + \alpha$
d. $^{19}_{10}Ne \longrightarrow ^{19}_{9}F + \beta^+$ **e.** $^{15}_{6}C \longrightarrow ^{15}_{7}N + \beta^-$ **20.64**
$^{15}_{7}N$ **20.68** **a.** 238.0289 amu **b.** 1.87 times as abundant
c. 237.1 amu

Chapter 21

21.2 H; none; none; C; N, P; O, S, Se; F, Cl, Br, I, At **21.4**
Al, Ga, In, Tl; Sn, Pb; Bi; none; none **21.6** $+1$; $+2$; $+3$ (also

+1 at bottom of group); +4, +2, (−4); −3, +5, +3, +1; −2, +2, +4, +6; −1, +7, +5, +3, +1 **21.8** See Tables 21.1 through 21.8. **21.10** increases from the bottom to the top and from the left to the right **21.14** CaH_2 is ionic, GaH_3 covalent. **21.16** $NaH > MgH_2 > BeH_2 > CH_4 > SiH_4$. **21.18** They react with water and molecular oxygen. **21.26 a.** $CaCO_3(s) + 2HCl(aq) \longrightarrow$
$$CaCl_2(aq) + H_2O(l) + CO_2(g)$$
b. $CaSO_4(aq) + Na_2CO_3 \longrightarrow CaCO_3 + Na_2SO_4$
$CaCO_3(s) + 2HCl(aq) \longrightarrow CaCl_2(aq) + H_2O(l) + CO_2(g)$

$CaCl_2(l) \xrightarrow{\text{electrolysis}} Ca(s) + Cl_2(g)$
21.28 a. MgO **b.** $MgO(s) + H_2O(l) \longrightarrow Mg(OH)_2(aq)$
21.30 B_2H_6; see Figure 21.9A.
21.36 $2AlCl_3(aq) + 3H_2O(l) \longrightarrow Al_2O_3(s) + 6HCl(aq)$
21.40 $SrC_2 + 2H_2O \longrightarrow Sr(OH)_2 + C_2H_2$

21.42 $SiO_2(s) + C(s) \xrightarrow{\Delta} Si(l) + CO_2(g)$
21.44 See Figure 21.15. **21.46** $K_3N > TiN > BN > NH_3$.
21.48 $3Cu(s) + 8H^+(aq) + 2NO_3^-(aq) \longrightarrow$
$$2NO(g) + 4H_2O(l) + 3Cu^{2+}(aq)$$
$Cu(s) + 4H^+(aq) + 2NO_3^-(aq) \longrightarrow$
$$2NO_2(g) + 2H_2O(l) + Cu^{2+}(aq)$$
21.50 $Ca_3(PO_4)_2 + 5C + 3SiO_2 \longrightarrow$
$$3CaSiO_3 + 2P + 5CO$$
21.56 $H_2Te < H_2Se < H_2S < H_2O$. **21.58** Fluorine is the strongest oxidizing agent known.
21.60 $2F_2(g) + 2H_2O(l) \longrightarrow 4HF(aq) + O_2(g)$
$Cl_2(aq) + H_2O(l) \rightleftharpoons H^+(aq) + Cl^-(aq) + HOCl(aq)$
21.64 $:\!\ddot{O}\!:\!\ddot{C}l\!:\!\ddot{O}\!:$ paramagnetic **21.68** $(CN)_2 +$
$H_2O \longrightarrow HCN + HOCN$ **21.70** BeH_2 is covalent, BaH_2 ionic. **21.72** $0.103\ M$ **21.74** no **21.78** $0.0353\ M\ N_2O_4$, $0.0294\ M\ NO_2$ **21.80** 192 kJ, −98 kJ, −314.3 J/mol K, 2×10^{17} **21.82** $4Al + 3PbO_2 \longrightarrow 2Al_2O_3 + 3Pb$, exothermic, −2512 kJ, spontaneous

Chapter 22

22.2 Electrolysis of molten salts, chemical reduction, roasting and carbon reduction of sulfide ores, thermal decomposition, or found free in nature **22.6** $Ti > V > Cr > Mn$, $Cu > Co$
22.8 a. Mn +2, +4, +7 **b.** Fe +2, +3 **c.** Cr +2, +3, +6
d. Co +2, +3 **e.** V +3, +4, +5 **f.** Sc +3
22.16 Ti^{2+}, Ti^{3+}, TiO^{2+}
22.20 $TiO_2(s) + 2H^+(aq) \longrightarrow TiO^{2+}(aq) + H_2O(l)$
$TiO_2(s) + 2OH^-(aq) \longrightarrow TiO_3^{2-}(aq) + H_2O(l)$
22.24 +2, +3, +4, and +5

22.26 $3V_2O_5(s) + 10Al(s) \xrightarrow{\Delta} 6V(l) + 5Al_2O_3(s)$
22.28 a. yes **b.** possible but not observed **c.** no **d.** yes, slowly **e.** no **22.32** CrO, Cr_2O_3, CrO_2, CrO_3 **22.38** 2.67, which is a mixture of oxidation numbers +2 and +3

22.40 $Mn(s) + 2H_2O(l) \xrightarrow{\Delta} Mn(OH)_2(s) + H_2(g)$
22.42 a. $MnO_4^-(aq) + 5Fe^{2+}(aq) + 8H^+(aq) \longrightarrow$
$$Mn^{2+}(aq) + 5Fe^{3+}(aq) + 4H_2O(l)$$
b. $MnO_4^-(aq) + 3[Fe(CN)_6]^{4-}(aq) + 2H_2O(l) \longrightarrow$
$$MnO_2(s) + 3[Fe(CN)_6]^{3-}(aq) + 4OH^-(aq)$$

c. $2MnO_4^-(aq) + 3Mn^{2+}(aq) + 2H_2O(l) \longrightarrow$
$$5MnO_2(s) + 4H^+(aq)$$
d. $3MnO_4^{2-}(aq) + 4H^+(aq) \longrightarrow$
$$2MnO_4^-(aq) + MnO_2(s) + 2H_2O(l)$$
22.44 a. Mn^{2+} **b.** MnO_2 **c.** MnO_4^{2-}
22.48 a. $4Fe(NO_3)_2(aq) + O_2(g) + 4H^+(aq) +$
$$4NO_3^-(aq) \longrightarrow 4Fe(NO_3)_3(aq) + 2H_2O(l)$$
b. $Fe(s) + H_2SO_4(aq) \xrightarrow{\Delta} FeSO_4(aq) + H_2(g)$
c. $Fe(s) + Fe_2(SO_4)_3(aq) \longrightarrow 3FeSO_4(aq)$
d. $FeCl_3(aq) + 3NaOH(aq) \longrightarrow Fe(OH)_3(s) + 3NaCl(aq)$
e. $Fe(OH)_3(s) + OH^-(aq) \longrightarrow$ no reaction
22.50 coordination number 4 **22.54** $AgCl(s)$ (immediately), $AgBr(s)$ (slowly) **22.58** precipitation, complex-ion formation
22.62 Y, La, Ce, Pr, Nd, Pm, Sm, Eu, Gd, Tb, Dy, Ho, Er, Tm, Yb, Lu

22.70 $(NH_4)_2Cr_2O_7(s) \xrightarrow{\Delta} Cr_2O_3(s) + 4H_2O(g) + N_2(g)$
22.76 $Cr(OH)_3(s)$, $PbCrO_4(s)$
$FeCr_2O_4(s) + 8HNO_3(aq) \longrightarrow$
$$Fe(NO_3)_2(aq) + 2Cr(NO_3)_3(aq) + 4H_2O(l)$$
$Cr(NO_3)_3(aq) + 3NH_3(aq) + 3H_2O(l) \longrightarrow$
$$Cr(OH)_3(s) + 3NH_4NO_3(aq)$$
$4Cr(OH)_3 + 4K_2CO_3 + 3O_2(g) \xrightarrow{\Delta}$
$$4K_2CrO_4 + 4CO_2(g) + 6H_2O(l)$$
$Pb(NO_3)_2(aq) + K_2CrO_4(aq) \longrightarrow PbCrO_4(s) + 2KNO_3(aq)$

Chapter 23

23.4 a. $CH_3-CH_2-CH_3$ **b.**
c.
d.
23.6 $CH_3CH_2CH_2CH_2CH_3$, $CH_3CH(CH_3)CH_2CH_3$, $CH_3C(CH_3)_2CH_3$
23.8
cis-1,2-dibromoethene   trans-1,2-dibromoethene

23.10 a.
b. $HC\equiv C-CH_3$
c. $CH_2=CH-CH=CH-CH_2-CH_3$
d.

23.12 Butene will react with bromine in the dark; butane will not. **23.14 a.** 1,2-dibromobenzene **b.** 3-chloro-1-ethyl-benzene **c.** 3-methylphenol **d.** aminobenzene or aniline

23.16 a. $C_6H_6 + Cl_2 \xrightarrow{FeCl_3} C_6H_5Cl + HCl$

b. $C_6H_6 + CH_3Cl \xrightarrow{AlCl_3} C_6H_5CH_3 + HCl$

c. $C_6H_6 + HCl \longrightarrow$ no reaction

d. $C_6H_5CH_3 + MnO_4^- \xrightarrow{\Delta} C_6H_5COOH$

e. $C_6H_5CH_3 + Cl_2 \xrightarrow{light} C_6H_5CH_2Cl + HCl$

23.18 **a.** $C_6H_6 + CH_3Cl \xrightarrow{AlCl_3} C_6H_5CH_3 + HCl$

$C_6H_5CH_3 + MnO_4^- \xrightarrow{\Delta} C_6H_5COOH$

b. $C_6H_6 + Cl_2 \xrightarrow{FeCl_3} C_6H_5Cl + HCl$

c. $C_6H_6 + CH_3CH_2Cl \xrightarrow{AlCl_3} C_6H_5CH_2CH_3 + HCl$

23.20 **a.** 2-methyl-2-penten-5-ol **b.** 2-butanol **c.** 1,3-pentanediol **d.** 4-methyl-1-pentanol **23.22** 2-pentanone

23.24 $CH_3CH_2CH_2OH \xrightarrow{heat, Al_2O_3, H_2SO_4} CH_3CH=CH_2$
$CH_3CH=CH_2 + Cl_2 \longrightarrow CH_3CHClCH_2Cl$

23.28 **a.** primary **b.** primary **c.** primary **d.** secondary **e.** tertiary **f.** secondary **23.30** **a.** methyl 2-methyl-2-propyl ether **b.** ethyl methyl ether **c.** diethyl ether

23.32 $2CH_3OH \xrightarrow{heat, H_2SO_4} CH_3-O-CH_3 + H_2O$

23.36 **a.** butanal **b.** 3-pentanone or diethyl ketone **c.** propanone or acetone **d.** methanal or formaldehyde **e.** ethanal or acetaldehyde

23.38 **a.** $(CH_3)_2C=O + H_2 \xrightarrow{Pd} (CH_3)_2CHOH$

b. $(CH_3)_2C=O + LiAlH_4 \longrightarrow (CH_3)_2CHOH$

c. $CH_3CHO + KMnO_4 \xrightarrow{\Delta} CH_3COOH$

23.40 **a.** propanoic acid **b.** methyl propanoate **c.** methyl formate **d.** ethyl benzoate **e.** trichloroacetic acid **f.** 2-methyl propanoic acid **23.42** 2-propyl acetate **23.46** **a.** 3-chloro-1-propene **b.** 2,5-dibromo-1,1-dichloropentane **c.** 2-fluoro-2-methylpropane

23.48 **a.** $CH_3CH_2CHBrCH_2CH_3 + KOH \xrightarrow{\Delta}$
$CH_3CH_2CH=CHCH_3 + KBr + H_2O$

b. $CH_3CH_2Br + OH^- \longrightarrow CH_3CH_2OH + Br^-$

c. $CH_3CH_2Br + NaOCH_3 \longrightarrow$
$CH_2CH_2-O-CH_3 + NaBr$

d. $CH_3Br + KCN \longrightarrow CH_3CN + KBr$

23.52 **a.** $(CH_3)_2N(C_3H_7)$ **b.** $C_6H_5NH_2$ **c.** $(CH_3)_4N^+Br^-$ **d.** $(C_3H_7)_2NH$ **e.** $(CH_3)NH(C_2H_5)$

23.54 **a.**
$$CH_3-\overset{\overset{\displaystyle CH_3}{|}}{\underset{\underset{\displaystyle CH_3}{|}}{C}}-CH_2-\overset{\overset{\displaystyle CH_3}{|}}{CH}-CH_3$$

b.
$$CH_3-CH=\overset{\overset{\displaystyle C_2H_5}{|}}{C}-CH_2-CH_3$$

c. $CH_2=CH-CH=CH_2$

d.
$$\begin{array}{c} H_2C-CH_2 \\ | \quad\quad | \\ H_2C-CH_2 \end{array}$$

e.

f.

23.56 **a.** $(CH_3)_2C=O$ **b.** CH_3COOH **c.** CH_3-O-CH_3 **d.** CH_3CHO **e.** CH_3OH **f.** $CH_2=CH_2$ **g.** CH_3COOCH_3 **h.** C_6H_{12}

Chapter 24

24.6
$$\underset{\overset{||}{O}}{HOC}-C_6H_4-\underset{\overset{||}{O}}{COH} + HO-CH_2C_6H_{10}CH_2-OH + \underset{\overset{||}{O}}{HOC}-C_6H_4-\underset{\overset{||}{O}}{COH} + HO-CH_2C_6H_{10}CH_2-OH \longrightarrow$$
$$-\underset{\overset{||}{O}}{OC}-C_6H_4-\underset{\overset{||}{O}}{CO}-CH_2C_6H_{10}CH_2-\underset{\overset{||}{O}}{OC}-C_6H_4-\underset{\overset{||}{O}}{CO}-CH_2C_6H_{10}CH_2-O-+2nH_2O$$

24.12 $(n + 1)\ HOCH_2CH_2OH + (n + 1)\ \underset{\overset{||}{O}}{HOC}-C_6H_4-\underset{\overset{||}{O}}{COH} \longrightarrow$
$$HOCH_2CH_2O\left[\underset{\overset{||}{O}}{C}-C_6H_4-\underset{\overset{||}{O}}{C}-O-CH_2CH_2-O\right]_n\underset{\overset{||}{O}}{C}-C_6H_4\underset{\overset{||}{O}}{COH} + 2nH_2O$$

24.14 **a.** $nHOOC-CH_2-COOH + nHO-CH_2-CH_2-OH \longrightarrow$
$-[OC-CH_2-COO-CH_2CH_2-]_n + 2nH_2O$

b. polyester

24.16
$$\overset{\underset{\displaystyle |}{CN}}{} \quad \overset{\underset{\displaystyle |}{C_6H_5}}{} \quad \overset{\underset{\displaystyle |}{CN}}{} \quad \overset{\underset{\displaystyle |}{C_6H_5}}{}$$

$-CH-CH_2-CH-CH_2-CH-CH_2-CH-CH_2-$

24.18
$-\overset{\overset{\displaystyle O}{||}}{C}-(CH_2)_4-\overset{\overset{\displaystyle O}{||}}{C}-\overset{\overset{\displaystyle H}{|}}{N}-(CH_2)_{10}-\overset{\overset{\displaystyle H}{|}}{N}-\overset{\overset{\displaystyle O}{||}}{C}-(CH_2)_4-\overset{\overset{\displaystyle O}{||}}{C}-\overset{\overset{\displaystyle H}{|}}{N}-(CH_2)_{10}-\overset{\overset{\displaystyle H}{|}}{N}-$

24.20 **a.** $-CH_2-\overset{\overset{\displaystyle C\equiv N}{|}}{CH}-$ **b.** $-CH_2-\overset{\overset{\displaystyle C_6H_5}{|}}{CH}-$ **c.** $-CH_2-\overset{\overset{\displaystyle CH_3}{|}}{CH}-$

d. $-CH_2-CH=CH-CH_2-$

e. $-CH_2-CH=CH-CH_2-CH_2-\overset{\overset{\displaystyle C_6H_5}{|}}{CH}-$

f. $-\overset{\overset{\displaystyle F}{|}}{\underset{\underset{\displaystyle F}{|}}{C}}-\overset{\overset{\displaystyle F}{|}}{\underset{\underset{\displaystyle F}{|}}{C}}-$

24.24 $-CO-CH(CH_3)-NH-CO-CH(CH_3)-NH-CO-CH(CH_3)-NH-$

24.26 $H_2N-\overset{\overset{\displaystyle COOH}{|}}{\underset{\underset{\displaystyle R}{|}}{C}}-H$

24.30 $HOOC-CH(CH_3)-NH_2 + HOOC-CH_2-NH_2 \longrightarrow$
$HOOC-CH(CH_3)-NH-CO-CH_2-NH_2 + H_2O$

24.42   T-G-A-T-T-C-T-A   **24.46**   **a.** $C_6H_{12}O_6$, glucose
b. $C_{12}H_{22}O_{11}$,   sucrose   **c.** $(C_6H_{10}O_5)_n$,   starch   **24.48**
a. glucose and fructose **b.** glucose **c.** glucose **d.** glucose
24.58 $HOOC-CH(CH(CH_3)_2)-NH-CO-CH(CH_2SH)-NH-COCH(CH_2C_6H_4OH)-NH_2$

Glossary

absolute entropy *(S)* (12.4) a measure of the disorder or randomness of a system, measured relative to the entropy of a perfect crystal at 0 K, which is zero

absolute zero (1.3) theoretically, the lowest possible temperature; corresponds to $-273.15°C$ and to 0 K

absorption spectrum (6.2) a spectrum generated by the absorption of light by a sample; most wavelengths pass through, while some wavelengths are absorbed and appear as dark lines on the spectrum

accuracy (1.4) the difference between the value of a measured number and the correct or expected value

acid (3.5) a compound that releases hydrogen ions when dissolved in water

acid-ionization constant *(K_a)* (15.3) an equilibrium constant expressing the strength of a weak acid as $K_a = [H_3O^+][A^-]/[HA]$

acidic anhydride (4.3) a substance, usually a nonmetal oxide, that can combine with water to give an acid

actinide (3.1) any of the inner-transition elements following actinium in the periodic table; the elements thorium through lawrencium are actinides

activated complex (13.3) a high-energy arrangement through which molecules must pass on their way from reactants to products

activation energy *(E_a)* (13.3) the energy barrier that reactants must overcome before they can be converted to products; the difference between the energy of the activated complex and the average energy of the reactants

active site (24.4) the site in an enzyme at which catalytic activity occurs

activity (20.3) the number of nuclear disintegrations per unit time

activity series (4.2) a list of metals in order of decreasing activity

actual yield (4.4) the amount of product actually obtained in the laboratory from a reaction

acyclic hydrocarbon (23.2) a hydrocarbon that contains no ring structures

addition polymerization (24.3) formation of polymers by addition to a double bond

addition reaction (4.2) a reaction in which an element and a compound join to form a new compound

adduct (15.1) substance resulting from the formation of a coordinate covalent bond between two molecules

alcohol (8.4) a hydrocarbon derivative containing the —OH group

aldehyde (23.7) a hydrocarbon with the functional group —CHO

aliphatic (23.2) a hydrocarbon that contains localized single, double, or triple bonds

alkali metal (3.1) any Group IA (1) element except hydrogen; Li, Na, K, Rb, Cs, and Fr are alkali metals

alkaline-earth metal (3.1) any Group IIA (2) element; Be, Mg, Ca, Sr, Ba, and Ra

alkanes (8.4) hydrocarbons containing only carbon–carbon single bonds

alkenes (8.4) hydrocarbons containing carbon–carbon double bonds

alkyl group (23.3) the group that remains when one hydrogen is removed from an alkane

alkyl halide (23.7) a hydrocarbon that contains a halogen in place of a hydrogen atom

alkynes (8.4) hydrocarbons containing carbon–carbon triple bonds

allotropes (9.9) different forms of the same element in the same physical state

alloy (4.2) a homogeneous mixture formed by the interaction of two or more metals

alpha-amino acid (24.4) an organic molecule containing an amine group and a carboxylic acid group, both bonded to the same carbon at the end of a chain

alpha helix (24.4) the secondary structure adopted by many proteins, in which the protein chain is twisted into a helix and is parallel to itself within the helix

amine (23.7) ammonia containing one or more hydrocarbon substituents

amorphous solid (10.4) a solid that is made up of particles arranged in a random, nonrepeating fashion

amphoteric (4.3) capable of behaving either as an acid or as a base

amphoteric substance (15.1) a substance that can act as either an acid or a base

amplitude (6.1) the height of a wave at the crest (maximum); related to the intensity of radiation

anhydride (15.1) form of an oxoacid or a hydroxide base from which the components of water have been removed

anhydrous (4.2) free of water

anion (2.3) a negatively charged ion, such as the chloride ion, Cl^-

anode (18.1) the electrode that attracts anions and at which oxidation occurs

antibonding molecular orbital (8.5) a molecular orbital that concentrates electron density outside the nuclei, leading to repulsive forces between adjacent nuclei

aromatic hydrocarbon (23.2) a cyclic hydrocarbon with delocalized bonding, which imparts special stability

Arrhenius acid (15.1) any substance that generates an excess of hydrogen ions in aqueous solution

Arrhenius base (15.1) any substance that generates an excess of hydroxide ions in aqueous solution

Arrhenius equation (13.3) an equation that describes the temperature dependence of rate constants; $k = Ae^{-E_a/RT}$

aryl group (23.5) the group that remains when one hydrogen is removed from a cyclic aromatic hydrocarbon

atmosphere (atm) (9.2) a unit of pressure equal to the pressure of the atmosphere at sea level and 0°C during normal weather conditions; 760 torr

atom (2.1) the smallest particle of an element that retains the characteristic chemical properties of that element

atomic mass unit (amu) (2.4) the basic unit of mass of atoms and molecules; exactly 1/12 the mass of one ^{12}C atom. 1 amu = 1.6606×10^{-24} g

atomic number (Z) (2.2) the number of protons in the nucleus of a given atom

atomic weight (2.4) the average mass of an atom of an element, given in atomic mass units

Aufbau principle (6.5) an approach to writing electronic configurations of atoms by successively filling orbitals in order of increasing energy

autoionization (15.2) a reaction in which a solvent, such as water, ionizes to form both an acid and a base

average bond energy (7.8) the average energy required to break a given bond in any compound in which it occurs

average rate (13.1) reaction rate calculated over a finite time interval

Avogadro's hypothesis (9.3) the proposal that equal volumes of all gases contain equal numbers of moles or molecules (at constant temperature and pressure); $V = k_A n$

Avogadro's number (2.5) the number of fundamental particles (atoms, molecules, or ions) in one mole of any substance; 6.022×10^{23} particles/mole

axial position (8.1) a position above or below the triangular plane in a trigonal bipyramid

azimuthal quantum number (6.4) the quantum number that specifies the shape of the orbital occupied by an electron

balanced equation (4.1) a chemical equation in which the number of atoms of each element is the same on both sides of the equation

barometer (9.2) a closed tube filled with mercury and inverted into a pool of mercury, used to measure atmospheric pressure

base (4.2) a substance that is capable of neutralizing an acid

base-ionization constant (K_b) (15.4) an equilibrium constant expressing the strength of a weak base as $K_b = [B^+][OH^-]/[BOH]$

basic anhydride (4.3) a metal oxide that can combine with water to give a base

battery (18.2) a series of interconnected voltaic cells

bidentate ligand (19.2) a ligand that bonds to a metal ion through two atoms

big bang theory (20.6) a theory of the creation of the universe that holds that densely packed nuclear particles exploded outward to disperse matter throughout the universe

bimolecular reaction (13.4) an elementary reaction involving two reactant molecules

boiling point (10.6) the temperature at which the vapor pressure of a liquid equals the external pressure

boiling-point constant (K_b) (11.6) proportionality constant relating boiling-point elevation to the molal concentration of a solution

bombardment reaction (20.2) a nuclear reaction that is induced by the impact of a nuclide or a subatomic particle on another nuclide

bond angle (8.1) the angle between the two lines defined by a central atom attached to two surrounding atoms

bond energy (bond dissociation energy) (7.2) the energy required to break two bonded atoms into separate gaseous atoms

bond length (7.3) the distance between the centers of bonded atoms

bond order (8.6) the number of electron-pair bonds between two atoms; it is calculated as 1/2 (number of bonding electrons − number of antibonding electrons)

bonding molecular orbital (8.5) a molecular orbital that concentrates electron density between nuclei, leading to stable bonding

Born–Haber cycle (7.2) a cycle in which the formation of an ionic salt from its elements is represented by a sequence of processes whose energy changes are known or can be calculated and whose sum must equal the heat of formation

Boyle's law (9.3) a law stating that at constant temperature, the volume occupied by a fixed amount of a gas varies inversely with pressure; $V = k_a/P$

Brønsted-Lowry acid (15.1) any substance that can donate a proton to another substance

Brønsted-Lowry base (15.1) any substance that can accept a proton from another substance

buffer (16.3) a solution that resists changes in pH when small amounts of H^+ or OH^- are added. Buffers are formed from a weak acid and its conjugate base or from a weak base and its conjugate acid.

buffer range (16.3) the range of concentrations of a weak acid and its conjugate base (or of a weak base and its conjugate acid) in which effective buffering action can be observed

calorie (cal) (5.1) the amount of heat required to raise the temperature of 1 g of water by 1°C; 1 calorie = 4.18400 joules

Calorie (5.1) a kilocalorie (kcal); a unit used by nutritionists to measure the energy value of foods

calorimeter (5.2) a vessel used to measure the heat change accompanying a process; the system is insulated from the surroundings, causing heat changes from the process to change the temperature of the system

carbohydrate (24.6) an organic molecule with the general formula $C_n(H_2O)_m$

carboxylic acid (23.7) a hydrocarbon containing the functional group —COOH

catalyst (4.6) a substance that increases the rate of reacton but can be recovered unchanged from the reaction

catalytic cracking (23.6) a process for breaking larger alkanes into smaller alkanes and alkenes, in the presence of a catalyst

catalytic reforming (23.6) a process that changes straight-chain alkanes into branched-chain and cyclic alkanes, in the presence of a catalyst

cathode (18.1) the electrode that attracts cations and at which reduction occurs

cation (2.3) a positively charged ion, such as the sodium ion, Na^+

cell potential (18.1) the voltage difference between the electrodes in an electrochemical cell

Celsius scale (1.3) a temperature scale that defines 0°C as the freezing point of water and 100°C as the boiling point of water

chain reaction (20.5) a reaction in which the product of one step is a reactant in another step

chalcogens (21.22) the group VIA (16) elements

Charles's law (9.3) a law stating that at constant pressure, the volume occupied by a fixed amount of gas is directly proportional to its absolute temperature; $V = k_C T$

chelate (19.2) a complex ion in which the ligand is polydentate and wraps around the metal ion

chemical activity series (17.1) an arrangement of elements in order of decreasing reducing strength

chemical bond (p. 260) the force that holds atoms together in a molecule or compound

chemical change (1.1) a change in which substances are converted into new substances that have compositions and properties different from those of the original substances

chemical energy (5.1) the energy associated with a chemical reaction

chemical equation (p. 118) an abbreviated representation of a chemical reaction consisting of chemical symbols and formulas

chemical formula (2.3) a shorthand representation of the chemical composition of a substance using symbols of the elements and numerical subscripts to indicate how many atoms of each element are present

chemical property (1.1) a characteristic of a substance involving the possible transformations that the substance can undergo to produce a new substance

chemical reaction (1.1) a process in which at least one substance is changed in composition and identity

chemisorption (13.5) absorption of molecules onto a solid surface, followed by chemical activation of the molecules

coagulation (11.8) the aggregation of colloidal particles into larger particles

codon (24.5) series of three bases in a nucleic acid that defines the amino acid to be used in the synthesis of a protein

colligative properties (11.6) solution properties that depend only on the relative number of solute and solvent particles, not on the identity of the solute

collision theory (13.3) a theory that explains the temperature dependence of rate constants on the basis of collisions, molecular orientation, and kinetic energy

colloid (11.8) a mixture containing particles larger than most molecules but smaller than 200 nm

combined gas law (9.3) a law that describes the relationship among pressure, volume, and temperature for a fixed amount of gas; $P_1 V_1 / T_1 = P_2 V_2 / T_2$

common ion (16.2) an ion that is identical to one participating in an equilibrium reaction but is introduced from another source

common-ion effect (16.2) a change in the position of an ionic equilibrium caused by addition of a common ion

complex ion (16.4) charged coordination compound formed between a metal ion and other ions or molecules called ligands

compound (chemical compound) (1.1) a substance composed of two or more elements combined in definite proportions

compressibility factor (9.8) the ratio PV/nRT, used as a measure of the ideality of a gas by its conformance to the ideal-gas law

concentrated solution (4.5) a solution that contains a relatively high concentration of solute

concentration (4.5) the relative amounts of solute and solvent in a solution

condensation (10.6) the return of gas-phase molecules to the liquid state

condensation polymerization (24.3) a polymerization reaction in which small molecules are split out of the monomers during the polymerization process

conduction band (10.5) in the band theory of bonding, an energy band in which electrons are free to move

conjugate acid–base pair (15.1) two substances that can be formed from one another by the gain or loss of a proton

continuous spectrum (6.2) a spectrum in which all the component wavelengths of light are present

conversion factor (1.5) a ratio of equivalent quantities, accompanied by their units, that is mathematically equal to 1

coordinate covalent bond (7.6) a chemical bond formed when one atom donates both the electrons that are shared between two atoms

coordination number (7.2) the number of atoms or ions that immediately surround an atom or ion of opposite charge; the number of ligands that surround the central metal ion in a complex ion

coordination sphere (19.2) the central metal ion surrounded by its ligands in a coordination compound

copolymer (24.1) polymer formed between two or more different monomers

covalent bonding (7.1) the chemical bonding between atoms resulting from the sharing of electron pairs

covalent compound (3.2) a compound in which electrons are shared between atoms rather than transferred from one atom to another

covalent radius (6.8) one-half the distance between the centers of two atoms in a gaseous diatomic molecule

critical mass (20.5) the smallest amount of fissionable material necessary to support a continuing chain reaction

critical point (10.6) conditions of temperature and pressure above which the liquid state can no longer exist

critical pressure (10.6) the minimum pressure that causes a gas to liquefy at the critical temperature

critical temperature (9.8) the temperature above which a gas cannot be liquefied simply by the application of pressure

crystal (10.4) a solid that has a shape bounded by planes intersecting at characteristic angles

crystal field splitting energy (Δ) (19.6) the energy separation between the two sets of d orbitals that are split when placed in a crystal field

crystal field theory (19.6) bonding theory for complex ions that assumes a purely electrostatic attraction between the central metal cation and the ligands, which are assumed to be located at points in space

crystal lattice (7.2) the repeating pattern of atoms or ions in a crystal

crystalline solid (10.4) a substance in which the particles are arranged in a regular, repeating geometric structure

cubic closest-packed structure (10.4) arrangement of closest-packed layers of atoms in the repeating pattern ABCABCABC. . .

curie (20.3) a measurement of activity equal to 3.7×10^{10} disintegrations per second

cyclic hydrocarbon (23.2) a hydrocarbon that contains a ring structure

cycloalkane (23.3) an alkane that contains a ring structure

Dalton's law of partial pressures (9.5) a law stating that gases in a mixture behave independently and exert the same pressure they would if they were in the container alone; $P_{\text{total}} = P_A + P_B + \cdots$

decomposition reaction (4.2) a chemical reaction in which a substance breaks down into simpler compounds or elements

degenerate (6.5) having the same energy

density (1.3) mass per unit volume

deoxyribonucleic acid (DNA) (24.5) a high-molar-mass nucleic acid that stores genetic information

diamagnetic substance (8.5) a substance that contains no unpaired electrons and is weakly repelled by a magnetic field

diatomic molecule (2.3) a molecule containing two atoms, such as O_2 or CO

diffusion (9.6) the movement of gas molecules through space

dilute solution (4.5) a solution that contains a relatively low concentration of solute

dilution (4.5) the process of adding more solvent to give a solution of lower concentration

dipole moment (8.2) the degree of charge separation in a bond or molecule, defined as the product of the charge magnitude and the distance between the charges

diprotic acid (15.1) an acid that can donate two protons

direct synthesis reaction (combination reaction) (4.2) a reaction in which two elements, an element and a compound, or two compounds join to form a new compound

disproportionation (17.5) a reaction in which an element undergoes a change from one species to two others, one having a higher oxidation number and the other a lower oxidation number

doping (10.5) the addition of impurities to elements to create or enhance semiconductor behavior

double bond (7.3) a covalent bond that involves the sharing of two pairs of electrons

double-displacement reaction (4.2) a reaction in which two compounds exchange ions or elements to form new compounds

effective nuclear charge (6.8) the nuclear charge actually felt by an electron, which is reduced from the total nuclear charge because of shielding by inner electrons

effusion (9.6) the movement of gas molecules through a pinhole or porous barrier

elastomer (24.1) a polymer, such as rubber, that has elastic properties

electrical potential (17.5)   see *electromotive force*

electrode (18.1)   a solid material that conducts electrical charge and is used as part of an electrochemical cell; one electrode (the anode) acts as the site of oxidation, and the other (the cathode) acts as the site of reduction in electrochemical reactions

electrolysis (5.1)   a chemical reaction that occurs when electricity flows through a solution or molten solid

electrolysis cell (18.1)   an electrochemical cell through which electrical current is passed to cause a nonspontaneous oxidation-reduction reaction to occur

electrolyte (11.3)   a substance that dissociates into ions upon dissolution

electrolytic cell (18.1)   see *electrolysis cell*

electromagnetic radiation (6.1)   a form of energy that can be described as being propagated through space by waves that have electrical and magnetic components

electromotive force (emf) (17.5)   the chemical driving force in an oxidation–reduction reaction, measured in volts

electromotive force series (17.5)   a series of elements and other species arranged in order of increasing reduction potential

electron (2.2)   a negatively charged subatomic particle

electron affinity (6.8)   the enthalpy change that occurs when an electron is added to a gaseous atom

electron capture (20.2)   incorporation of an inner-shell electron into the nucleus, resulting in the conversion of a proton into a neutron

electron-dot formula (7.3)   see *Lewis formula*

electron-dot symbol (7.1)   see *Lewis symbol*

electronegativity (7.7)   the ability of an atom to attract electrons within a bond to itself

electronic configuration (6.5)   the distribution of electrons in atomic orbitals

electroplating (18.5)   the deposition of one metal onto another by the electrochemical reduction of a metal salt

element (1.1)   a pure substance that cannot be broken down into simpler stable substances in a chemical reaction

elementary reaction (13.4)   steps in a reaction mechanism that cannot be broken down into further sequences of steps

emission spectrum (6.2)   a spectrum generated by the emission of certain wavelengths of light when a sample is excited by heating or some other process

empirical formula (2.3)   a formula written with the simplest ratios of atoms or ions (the smallest whole-number subscripts)

emulsifying agent (11.8)   a substance that stabilizes an emulsion by simultaneous dissolution in the two substances at their interface

emulsion (11.8)   a colloid consisting of a liquid dispersed in a liquid

entantiomers (19.4)   a pair of optical isomers—mirror images of one another that cannot be superimposed

endothermic process (5.2)   a physical or chemical change that absorbs heat

energy (1.3)   the capacity to do work or to transfer heat

enthalpy (*H*) (5.3)   the energy stored within a substance that can be transferred as heat by a process occurring at constant pressure

entropy (11.2)   a measure of the tendency for matter to become randomly distributed

enzyme (13.5)   a protein molecule that acts as a catalyst specific to a single reaction or to a narrow range of reactions

equatorial position (8.1)   a position in the triangular plane in a trigonal bipyramid

equilibrium (10.6)   a state of balance between opposing processes

equilibrium constant (*K*) (12.7 and 14.1)   the reaction quotient for a state of equilibrium. It is a ratio of rate constants for forward and reverse reactions that is characteristic of the position of equilibrium for a given temperature.

equivalence point (15.5)   the point in a titration at which the two reagents are present in chemically equivalent amounts

equivalent weight (18.4)   the molecular weight of a substance divided by the number of electrons transferred to or from the substance; the mass of a substance that can be oxidized or reduced by the transfer of 1 mol of electrons

ester (23.7)   a condensation product of carboxylic acids and alcohols, having the general formula *RCOOR'*

ether (23.7)   an organic compound that contains the —O— functional group

evaporation (10.6)   the process by which molecules pass from the liquid state to the gaseous state

excited state (6.5)   a state of energy in an atom that is higher than the ground state

exothermic process (5.2)   a physical or chemical change that gives off heat

Fahrenheit scale (1.3)   a temperature scale that defines 32°F as the freezing point of water and 212°F as the boiling point of water

Faraday constant (18.3)   a constant equal to the charge on 1 mol of electrons (9.650×10^4 coul/mol e^-)

Faraday's law (18.4)   law stating that the quantity of a substance produced by electrolysis with a given amount of direct current is proportional to that substance's equivalent weight

fatty acid (24.7)   a long-chain carboxylic acid

ferromagnetism (22.1)   strong magnetism caused by permanent magnetic fields in a metal

fission (20.2)   the splitting of a heavy atom into two or more lighter atoms and some number of neutrons

fluid (10.2)   a substance that can flow (a gas or a liquid)

formal charge (7.4) the charge obtained for an atom in a molecule or ion by arbitrarily assigning one electron of a shared electron pair to each atom

formation constant (K_f) (16.4) the equilibrium constant for the formation of a complex ion

formula unit (2.3) the smallest repeating unit in an ionic compound

formula weight (2.4) the mass of one formula unit in atomic mass units

fractional distillation (23.6) separation of components by heating and condensation at different temperatures

free energy (G) (12.5) maximum amount of energy available to do useful work on the surroundings

freezing point (10.6) the temperature at which the liquid and solid states of a substance are in equilibrium (identical to the melting point)

freezing-point constant (K_f) (11.6) proportionality constant relating freezing-point depression to the molal concentration of a solution

frequency (6.1) the number of wave crests (or other recurring points on a wave that pass a stationary point in one second

frequency factor (13.3) a proportionality constant (A) in the Arrhenius equation: $k = Ae^{-E_d/RT}$

functional group (8.4) a group of atoms that is substituted for hydrogen on a hydrocarbon and that gives the compound its characteristic properties

fusion (20.2) the combination of light nuclei to form heavier nuclei

galvanic cell (18.1) see *voltaic cell*

gas (1.1) the physical state in which matter has no fixed shape or volume but expands to fill its container completely

Gay-Lussac's law of combining volumes (9.3) a law stating that in chemical reactions that occur at constant pressure and temperature, the volumes of gaseous reactants and products are in the ratio of small whole numbers

gel (11.8) a nonfluid colloid consisting of a solid dispersed in a liquid

geometrical isomers (19.4) stereoisomers that differ in the way the ligands are arranged around the central metal ion

Gibbs free energy (12.5) see *free energy*

Graham's law (9.6) a law stating that the relative rates of diffusion or effusion of two gases are inversely proportional to the square root of the molar masses (or densities) of the gases at constant temperature and pressure; $r_1/r_2 = \sqrt{MM_2}/\sqrt{MM_1}$

ground state (6.5) the state of lowest energy in an atom

group (3.1) a vertical column of related elements in the periodic table

Haber process (14.5) a process in which ammonia is formed by direct reaction of nitrogen and hydrogen gases

half-life (13.2) the time required for a given concentration of reactant to decrease to half its value

half-reaction (17.1) a reaction in which only oxidation or only reduction occurs

half-reaction method (17.4) a method for balancing oxidation–reduction equations by using half-reactions; also called the ion–electron method

half-reaction potential (18.1) the voltage developed by a half-reaction relative to the H_2/H^+ half-reaction, which is defined as having zero voltage

halogen (3.1) any Group VIIIA (17) element; the nonmetals F, Cl, Br, I, and At are halogens

halogenation reaction (23.8) replacement of a hydrogen atom by a halogen atom

hard water (11.5) water that contains metal ions, which cause precipitate formation when the water is boiled or when soap is added

heat (5.2) the energy that is transferred between two objects because of a difference in their temperatures

heat (enthalpy) of vaporization (10.6) the amount of heat that must be supplied to evaporate 1 mol of a liquid at constant temperature

heat capacity (5.2) the amount of heat that is required to raise the temperature of a substance by 1°C

heat (enthalpy) of formation (ΔH_f) (5.5) the enthalpy change associated with the formation of 1 mol of a substance from its elements in their stable form

heat of fusion (7.2) the energy required to melt 1 mol of a solid

heat (enthalpy) of reaction (5.4) the enthalpy change accompanying a chemical reaction

heat of solution (11.2) heat energy released or absorbed when a solute dissolves

heat of vaporization (7.2) the energy required to vaporize 1 mol of a liquid

Heisenberg uncertainty principle (6.3) a principle stating that it is not possible to know both the velocity and the position of an atomic particle with a high level of certainty

Henderson–Hasselbalch equation (16.3) an equation relating pH to the concentrations of a weak acid and its conjugate base: $pH = pK_a + \log [A^-]/[HA]$

Henry's law (11.2) law stating that the solubility of a gas in a liquid is directly proportional to the partial pressure of the gas above the liquid

hertz (Hz) (6.1) a unit of frequency equal to 1 s^{-1}, or cycle per second

Hess's law (5.4) a law stating that ΔH is independent of the number and nature of the intermediate steps that occur in an overall process; if a reaction can be regarded as the sum of two or more other reactions, ΔH for the overall reaction equals the sum of the enthalpy changes for the component reactions

heterogeneous catalyst (13.5) catalyst that is in a different physical state from the reaction system

heterogeneous mixture (1.1) a nonuniform mixture whose component substances can be distinguished visually

hexagonal closest-packed structure (10.4) arrangement of closest-packed layers of atoms in the repeating pattern ABABABAB. . .

high-spin complex (19.6) a complex ion that contains the maximum number of unpaired electrons and the minimum number of paired electrons

homogeneous catalyst (13.5) catalyst that is in the same physical state as the reaction system

homogeneous mixture (solution) (1.1) a mixture with uniform composition

Hund's rule (6.5) a rule stating that electrons are distributed in a set of orbitals of identical energy in such a way as to give the maximum number of unpaired electrons

hybrid orbitals (8.3) the equivalent orbitals formed by the process of hybridization

hybridization (8.3) the process of mixing two or more nonequivalent atomic orbitals from the same atom to form an equal number of equivalent orbitals suitable for bonding

hydrate (3.5) a chemical compound that contains molecules of water in a characteristic proportion as part of its solid structure

hydration reaction (4.3) a chemical reaction in which water is added to a compound to form a hydrate

hydrocarbon (8.4) a class of chemical compounds that contain only carbon and hydrogen

hydrogen bond (10.1) especially strong dipole–dipole forces between polar molecules that contain hydrogen attached to a highly electronegative element

hydrolysis (4.2) the reaction of water with a substance to form OH^- or H^+ or both

hydrolysis constant (K_h) (16.1) the equilibrium constant for a hydrolysis reaction

hydrometallurgy (17.6) extraction of metals from ores by means of processes occurring in aqueous solution

hydronium ion (15.1) the aqueous hydrogen ion

hygroscopic (4.3) capable of taking up water from the atmosphere

hypothesis (1.2) a tentative explanation for a set of observations that can be tested by further experimentation

ideal gas (9.3) a gas that follows predicted behavior as represented by the ideal gas law

ideal-gas law (9.3) a law stating the relationship among pressure, volume, temperature, and amount of an ideal gas; $PV = nRT$

indicator (4.5) a substance added to a solution being titrated that will change color at the endpoint

induced dipole (10.1) a temporary dipole formed by attraction of the electrons in the molecule to a nearby dipole or ion

initial rate (13.1) rate measured at the very beginning of a reaction

inner-transition element (3.1) an element that falls between lanthanum and hafnium or between actinium and element 104 in the periodic table; a lanthanide or actinide

instantaneous dipole (10.1) a temporary dipole formed by the movement of electrons within a molecule

instantaneous rate (13.1) rate calculated from the tangent to the concentration–time curve for a reaction

intermediate (13.4) a species produced in a reaction mechanism that is relatively unstable and reacts in a subsequent step of the mechanism

internal energy (5.1) the sum of the kinetic and potential energies of the fundamental particles that make up a substance or object

ion (2.3) an atom or molecule bearing an electrical charge, formed by addition or removal of electrons

ion–electron method (17.4) see *half-reaction method*

ion-product constant of water (K_w) (15.2) the product of hydrogen-ion and hydroxide-ion concentrations, which is constant in aqueous solutions at constant temperature; $K_w = [H_3O^+][OH^-] = 1.0 \times 10^{-14}$ at 25°C

ionic bonding (7.1) the chemical bonding between cations and anions that results from electrostatic attractions of opposite charges

ionic compound (2.3) a compound consisting of cations and anions in proportions that give electrical neutrality

ionic crystal (7.2) a solid structure in which the ions are arranged in a regular repeating pattern

ionic equation (11.4) a form of reaction equation in which ionic species are represented as the separated ions

ionic radius (6.8) the radius of an ion in a solid

ionization energy (ionization potential) (6.8) the minimum energy required to remove the outermost electron from a gaseous atom

isoelectronic series (6.6) a series of atoms or ions that have the same total number and configuration of electrons

isomers (19.4) two different compounds that have the same formula

isotopes (2.2) atoms that have the same number of protons but different numbers of neutrons (the same atomic number but different mass numbers)

joule (J) (5.1) the SI unit of energy

Kelvin scale (1.3) the SI temperature scale whose base unit is the kelvin and which is defined such that the lowest possible temperature is 0 K

ketone (23.7) a hydrocarbon containing an oxygen double-bonded to a carbon, which is also bonded to two other carbons: $R_2C{=}O$

kinetic energy (5.1) the energy possessed by an object because of its motion

kinetic-molecular theory of gases (9.7) a model for describing and predicting ideal-gas behavior; it assumes that gas molecules occupy no volume and do not attract one another

kinetics $(p.\ 529)$ the study of the rates of reactions and their mechanisms

lanthanide (3.1) any of the inner-transition elements following lanthanum in the periodic table; the elements cerium through lutetium are lanthanides

lanthanide contraction (22.1) a shrinking in the size of the lanthanides with atomic number, caused by poor nuclear shielding by the f orbitals

lattice energy (7.2) the energy change that occurs when gaseous ions combine to form a crystalline ionic solid

lattice plane (10.4) a plane that passes through equivalent points in successive unit cells in a crystal lattice

law (1.2) an explanation or relationship that appears to be universal

law of conservation of mass (2.1) a law stating that the mass of the substances produced in a chemical reaction equals the mass of the substances that reacted

law of conservation of energy (5.1) a law stating that energy can be converted from one form to another but cannot be created or destroyed

law of definite proportions (2.1) a law stating that all samples of the same pure substance will always contain the same proportions by mass of the component elements

law of mass action (14.1) a law stating that at a given temperature, a reaction will reach the same state of equilibrium, no matter what the starting conditions

Le Chatelier's principle (14.3) a principle stating that a system at equilibrium reacts to a stress such as a change in concentration, pressure, or temperature is such a way as to counteract the stress and establish a new equilibrium

leveling effect (15.1) the strongest acid that can exist in a solvent is the conjugate acid of that solvent; the strongest base that can exist in a solvent is the conjugate base of that solvent

Lewis acid (15.1) any species that can accept a share in a pair of electrons (an electron-pair acceptor)

Lewis base (15.1) any species that can share an electron pair with another substance (an electron-pair donor)

Lewis formula (electron-dot formula) (7.3) a formula of a compound consisting of the symbols of the component elements, each surrounded by dots that represent shared and unshared electrons

Lewis symbol (electron-dot symbol) (7.1) a symbol of an atom consisting of the letter symbol for the element surrounded by dots that represent valence electrons

ligand (16.4) a molecule or ion that binds to another species by donation of a pair of electrons to form a coordinate covalent bond

limiting reactant (limiting reagent) (4.4) the reactant that is used up because it is present in the smallest amount and that consequently determines the maximum amount of product that can be formed from that reaction

line spectrum (6.2) a spectrum in which only certain wavelengths of light are present

linear combination of atomic orbitals (8.5) a method of constructing molecular orbitals by adding or subtracting atomic orbitals

lipid (24.7) one of a general class of biological molecules that can be extracted from cells by nonpolar solvents

liquid (1.1) the physical state in which matter has no characteristic shape but takes the shape of the filled portion of its container

London dispersion forces (9.8) the attractive forces that result from polarization of adjacent nonpolar molecules or atoms

low-spin complex (19.6) complex ion that contains the maximum number of paired electrons and the minimum number of unpaired electrons

magnetic quantum number (6.4) the quantum number that specifies the orientation of an orbital in space

main-group element (representative element) (3.1) a member of one of the A groups of elements in the periodic table

mass (1.3) the quantity of matter

mass defect (20.4) the discrepancy between the mass of a nuclide and the sum of the masses of its component subatomic particles

mass number (A) (2.2) the sum of the number of protons and the number of neutrons in the nucleus of an atom

matter (1.1) anything that occupies space and is perceptible to the senses

mechanism (13.4) a detailed description of the way a reaction takes place; usually, a sequence of reaction steps that makes up the overall reaction

melting point (10.6) see *freezing point*

meniscus (10.2) curved upper surface of a liquid column

messenger RNA (24.5) the form of RNA that takes the genetic code from DNA and carries it to ribosomes

metal (1.1) an element that typically has a characteristic luster, is a good conductor of electricity and heat, is opaque, may be melted, and has a tendency to lose electrons to form cations when it forms compounds

metallic bonding (7.1) the chemical bonding between metal atoms that results from the sharing of a cloud of free and mobile valence electrons

metallic radius (6.8) one-half the distance between the centers of atoms in a solid metal

metalloid (semiconductor, semimetal) (1.1) an element that has properties intermediate between those of metals and nonmetals; metalloids resemble metals physically and nonmetals chemically

miscible liquids (11.1) liquids that can mix in all proportions

mixture (1.1) a combination of two or more substances that can be separated by physical means

molality (11.1) moles of solute per kilogram of solvent

molar heat capacity (5.2) the amount of heat required to raise the temperature of 1 mol of a substance by 1°C

molar mass (2.5) the mass of one mole of any substance, in units of grams per mole

molar volume (9.3) the volume occupied by 1 mol of a gas, which equals 22.414 L at STP for an ideal gas

molarity (M) (4.5) the number of moles of solute per liter of solution

mole (2.5) the amount of a substance that contains the same number of fundamental particles (atoms, molecules, or formula units) as there are atoms in exactly 12 g of ^{12}C (Avogadro's number)

mole fraction (11.1) moles of a solution component divided by total moles in the solution

molecular equation (11.4) a form of reaction equation in which species are represented as if they were in their molecular forms (using formula units), even though the substances may exist in solution as ions

molecular formula (2.3) a formula that indicates the actual number of atoms present in a molecule

molecular orbital (8.5) an area in space around a molecule in which it is highly probable that electrons will be found

molecular orbital theory (8.5) an approach to covalent bonding in which wave functions are used to describe how electrons are distributed in molecular orbitals

molecular weight (2.4) the mass of one molecule in units of atomic mass units

molecularity (13.4) the number of reactant molecules in an elementary reaction

molecule (2.3) the smallest particle of a pure substance that retains the composition and chemical properties of that substance and can exist independently

monatomic ion (2.3) an ion containing only one atom, such as Na^+

monodentate ligand (19.2) a ligand that bonds to a metal ion through only one atom

monomer (24.1) one of the individual units that combine to form a polymer

monoprotic acid (15.1) an acid that can donate only one proton

Nernst equation (18.2) an equation relating the cell potential (E) to the standard cell potential $(E°)$: $E = E° - (0.05916/n)\log Q$, where n is the number of electrons transferred in the reaction, and Q is the reaction quotient.

net ionic equation (11.4) a form of reaction equation in which ionic species are represented as the separated ions, and spectator ions are eliminated

neutralization reaction (4.2) a reaction of an acid and a base to form a salt and water; transfer of a hydrogen ion from an acid to a base

neutron (2.2) an electrically neutral subatomic particle

neutron number (N) (2.2) the number of neutrons in the nucleus of an atom

noble gas (3.1) any of the relatively nonreactive gases that make up Group VIIIA (18); He, Ne, Ar, Kr, Xe, and Rn are noble gases

nonbonding molecular orbital (8.5) a molecular orbital that is essentially an atomic orbital in that it concentrates electron density on a single atom

nonelectrolyte (11.3) a substance that retains its molecular identity upon dissolution

nonmetal (1.1) an element that typically has a dull appearance, is a poor conductor of electricity and heat, and has a tendency to form covalent compounds or to gain electrons to form anions when it forms compounds

nonpolar covalent bond (7.7) a chemical bond that results from the equal sharing of electron pairs

normal boiling point (10.6) the temperature at which the vapor pressure of a liquid equals the standard atmospheric pressure of 1 atm

normal freezing point (10.6) the temperature at which the liquid and solid states of a substance are in equilibrium at a pressure of 1 atm

normality (11.1) equivalents of solute per liter of solution; equal to moles per liter of the reactive species

nuclear binding energy (20.4) the amount of energy that would be released if appropriate numbers of subatomic particles were combined to give an atom

nuclear bombardment (20.2) a nuclear reaction that is induced by the impact of a nuclide or a subatomic particle on another nuclide

nucleic acid (24.5) one of a class of biomolecules that carry genetic information and regulate the growth of proteins

nucleon (20.1) a nuclear particle (neutron or proton)

nucleotide (24.5) the fundamental unit of a nucleic acid polymer, consisting of a sugar molecule bonded to both a phosphate group and a base

nucleus (2.2)   the central core of the atom, which contains the positive electrical charges and most of the mass of the atom

octahedral holes (10.4)   holes in a crystal lattice that are surrounded by six lattice points in an octahedral arrangement

octane rating (23.6)   a relative scale for gasoline that measures the tendency of the fuel to cause automobile engines to knock

octet rule (7.1)   a rule stating that atoms tend to gain, lose, or share electrons to achieve an electronic configuration with eight valence electrons

optical isomers (19.4)   isomers that are mirror images of one another and that rotate plane-polarized light in opposite directions

orbital (6.4)   the region in space around a nucleus where there is a high probability of finding an electron

ore (3.6)   a naturally occurring material from which metals can be economically obtained

organic chemistry (23.1)   the study of organic compounds

osmosis (11.6)   passage of solvent molecules, but not solute particles, through a membrane from a solution that is less concentrated in solute to a solution that is more concentrated

osmotic pressure (11.6)   the pressure that must be exerted on a solution to prevent water from passing through a semipermeable membrane and diluting the solution

overvoltage (18.4)   the difference between the potential actually required to carry out an electrolysis reaction and the potential calculated from standard electrode potentials

oxidation number (3.2)   a number assigned to an atom to indicate how many electrons it controls in a compound compared with the number of electrons it would have as an uncombined neutral atom

oxidation potential (17.5)   the voltage generated by an oxidation half-reaction

oxidation–reduction reaction (redox reaction) (4.2)   a reaction that involves a change in the oxidation number of one or more elements

oxidation-number-change method (17.4)   a method for balancing oxidation–reduction equations by equalizing the change in oxidation number for the species being oxidized and that for the species being reduced

oxidize (4.2)   to cause an element to increase its oxidation number

oxidizing agent (17.1)   an atom, ion, or molecule that brings about an increase in the oxidation number of another substance

oxoacid (3.5)   an acid that contains hydrogen, a nonmetal (or rarely, a metal), and oxygen

oxoanion (2.3)   an anion containing oxygen attached to some other element

pairing energy (6.8)   the energy required to place two electrons in a single orbital

paramagnetic substance (8.5)   a substance that contains unpaired electrons and is strongly attracted to a magnetic field

Pauli exclusion principle (6.5)   a principle stating that no two electrons in an atom may have the same set of four quantum numbers

percent by mass (11.1)   mass of solute divided by mass of solution and multiplied by 100

percent by volume (11.1)   volume of solute divided by volume of solution and multiplied by 100

percent yield (4.4)   the mass of product obtained divided by the theoretical yield and multiplied by 100%

period (3.1)   a horizontal row of elements in the periodic table

periodic law (3.1)   a law stating that the properties of the elements are periodic functions of their atomic numbers

periodic table (3.1)   an arrangement of the elements that illustrates the periodic law

pH (15.2)   a measure of the acidity of aqueous solutions; pH = $-\log [H_3O^+]$

phase (10.6)   a homogeneous part of a system in a given state

phase diagram (10.6)   a pressure–temperature plot that diagrams the conditions under which the various states of matter can exist

photon (6.3)   a ''particle'' of energy existing as electromagnetic radiation

physical change (1.1)   a process characterized by changes only in the physical properties of a substance, not in its composition

physical property (1.1)   a characteristic of a substance that can be observed without changing its composition

physical state (1.1)   the form of matter: solid, liquid, or gas

pi (π) bond (8.3)   a covalent bond characterized by a two-lobed cross section with greatest electron density on either side of the bond axis; formed by side-to-side overlap of two p orbitals

Planck's constant (6.3)   proportionality constant relating the energy and the frequency of light; 6.626×10^{-34} J s

plastics (24.1)   common name for synthetic polymers that can be molded and set into a hard mass

polar covalent bond (7.7)   a chemical bond that results from the unequal sharing of electron pairs

polarity (7.7)   a measure of the separation of electronic charge within a bond (or molecule)

polarizability (9.8)   the tendency of an atom's electron cloud to be deformed by nearby charges

polyatomic ion (2.3)   an ion containing two or more atoms, such as SO_4^{-2}

polyatomic molecule (2.3)   a molecule containing two or more atoms, such as S_8

polydentate ligand (19.2)   a ligand that bonds to a metal ion through more than one atom

polymer (p. 1000)   a giant molecule consisting of many connected, identical units

polyprotic acid (15.1)   an acid that can donate more than one proton

positron (20.1) a particle that has the same mass as an electron but the opposite electrical charge

potential energy (5.1) the energy possessed by an object because of its position

precipitate (4.2) an insoluble solid deposited from a solution

precipitation reaction (4.2) a reaction in which the reactants in solution exchange ions to form a solid

precision (1.4) the extent of agreement in a set of repeated measurements of the same quantity

pressure (9.2) the amount of force applied per unit area

primary structure (24.4) the arrangement of amino acid units in a protein

principal quantum number (6.4) the quantum number that describes the main energy level of an electron

product (1.1) substance formed from another substance during a chemical reaction

protein (24.4) one of a class of organic polymers making up most body tissue; characterized by the presence of peptide bonds and formed by the polymerization of amino acids

proton (2.2) a positively charged subatomic particle

pure substance (1.1) a substance that has the same composition throughout and from sample to sample and that cannot be separated into its components by physical means

pyrolysis (23.8) cracking reaction

quantum mechanical model (6.4) a mathematical model of the atom that describes the behavior of electrons in atoms in terms of wave functions that are characterized by quantum numbers

quantum number (6.3) an integer whose value specifies the energy of a system

quantum theory (6.3) the proposition that energy exists in tiny discrete units called quanta and can be transferred only in whole units

quaternary structure (24.4) the structure that results from the association of two or more protein chains

radioactivity (20.1) the emission of radiation by an atom

radiocarbon dating (20.3) a technique for establishing the age of a carbon-containing object by determining the amount of carbon-14 present in the object

Raoult's law (11.6) law stating that the partial pressure of a component of a solution equals the vapor pressure of the pure component multiplied by its mole fraction in the solution

rare-earth metals (22.9) yttrium, lanthanum, and the lanthanide elements

rate constant (13.2) the proportionality constant that relates the rate of a reaction to some dependence on concentrations of reactants and other species

rate-determining step (13.4) the slowest step in the sequence of steps that make up a reaction mechanism

rate equation (13.2) an equation that relates the rate of reaction to some function of concentrations

rate law (13.2) see *rate equation*

reactant (1.1) a substance that is converted to another substance during a chemical reaction

reaction order (13.2) the power to which the concentration of a substance is raised in the rate law

reaction quotient (Q) (12.7) a measure of the extent of reaction, equal to the product pressures or concentrations raised to a power equal to their coefficients in the balanced equation times each other, divided by the similar factor for the reactants. Pressure or concentration terms for pure solids and liquids are not included in the expression for Q.

reaction rate (13.1) the change in concentration of a reactant or product per unit of time

reactive substance (1.1) a substance that has a tendency to participate in chemical reactions

reduce (4.2) to cause an element to decrease its oxidation number

reducing agent (17.1) an atom, ion, or molecule that brings about a decrease in the oxidation number of another substance

reduction potential (17.5) the voltage generated by a reduction half-reaction

representative element (3.1) see *main-group element*

resonance hybrid (7.5) an average or composite Lewis structure that is derived from two or more valid Lewis formulas and that closely represents the bonding in a molecule

reverse osmosis (11.7) process by which application of pressure causes solvent to flow through a membrane from a solution that is more concentrated in solute to a solution that is less concentrated

ribonucleic acid (RNA) (24.5) low-molar-mass nucleic acid involved in the use of genetic information to regulate the synthesis of proteins

salt bridge (18.1) a device used in electrochemical cells that allows migration of ions to maintain electrical neutrality but prevents mixing of the anode and cathode solutions

saturated hydrocarbon (23.2) hydrocarbon that contains only carbon–carbon single bonds

saturated solution (11.1) a solution that is in equilibrium with excess solute

scientific method (1.2) a method of inquiry or investigation that involves cycles of observation and interpretation

scientific notation (1.3) a system in which numbers are expressed as products of a number between 1 and 10 multiplied by the appropriate power of 10

second law of thermodynamics (12.3) law stating that the entropy of the universe tends to approach a maximum

secondary structure (24.4) the structure adopted by chains of amino acid units in proteins, usually a sheet or helix

semiconductor (3.1) see *metalloid*

semimetal (3.1) see *metalloid*

semipermeable membrane (11.6)   a membrane that allows the passage of solvent but not of solute particles

shell (6.4)   an energy level containing a set of electrons that have the same principal quantum number

shielding effect (6.8)   the cancellation of the effect of nuclear charge on an electron by electrons located closer to the nucleus

SI (1.3)   *Système International,* the most modern version of the metric system, built on seven base units

sigma (σ) bond (8.3)   a covalent bond characterized by a circular cross section with greatest electron density along the bond axis; formed by overlap of two *s* orbitals or of an *s* orbital and a *p* orbital or by end-to-end overlap of two *p* orbitals

significant figures (1.4)   the digits in a number that are known for certain, plus one digit that is somewhat uncertain

single bond (7.3)   a covalent bond that involves the sharing of one pair of electrons

single-displacement reaction (4.2)   a reaction in which a free element displaces another element from a compound to produce a different compound and a different free element

smoke (11.8)   a colloid consisting of a solid dispersed in a gas

sol (11.8)   a fluid colloid consisting of a solid dispersed in a liquid

solid (1.1)   the physical state of matter characterized by a fixed shape and low compressibility

solubility (11.1)   a ratio that describes the maximum amount of a solute that dissolves in a particular solvent to form an equilibrium solution under specified conditions

solubility product constant (K_{sp}) (16.5)   the equilibrium constant for an insoluble ionic solid in equilibrium with its ions in solution. K_{sp} is equal to the product of the equilibrium concentrations of ions formed in the dissolution of the salt, each raised to the power of its stoichiometric coefficient.

solute (4.5)   the substance being dissolved; usually, that component of a solution that is present in the lesser amount

solution (1.1)   a homogeneous mixture of two or more substances dispersed at a molecular or ionic level that can be prepared in variable proportions

solvent (4.5)   the substance doing the dissolving; usually, that component of a solution that is present in the larger amount

specific heat (5.2)   the amount of heat required to raise the temperature of 1 g of a substance by 1°C

spectrochemical series (19.6)   arrangement of ligands in order of decreasing crystal field splitting energy

spectrum (6.2)   a resolution, or separation, of radiation into its component wavelengths

spin quantum number (6.4)   the quantum number that specifies the direction of spin of an electron

spontaneous reaction (p. 493)   a reaction that occurs by itself without external intervention

stable substance (1.1)   a substance that is not easily decomposed or otherwise chemically changed

standard enthalpy of formation (ΔH_f°) (5.5)   the enthalpy change accompanying the formation of 1 mol of a substance from its elements in their stable form at 25°C and 1 atm

standard entropy (S°) (12.4)   entropy for a substance in its standard state

standard free energy of formation (ΔG_f°) (12.6)   free-energy change for the process in which 1 mol of a substance is formed from its component elements, all substances being in their standard states

standard free-energy change (ΔG°) (12.6)   the change in free energy measured with all substances in their standard states

standard hydrogen electrode (18.1)   an electrode consisting of a platinum foil covered in platinum black, over which hydrogen gas at 1 atm pressure is bubbled; the potential of this electrode is the standard zero point against which other electrode potentials are measured

standard state (5.4)   the physical state in which a substance is stable at 1 atm pressure and a specified temperature, usually 25°C

standard temperature and pressure (STP) (9.3)   0°C and 1 atm

state function (5.4)   a quantity whose value is independent of the path taken to get from one state, or set of conditions, to another

stereoisomers (19.4)   isomers in which the bonding in the two compounds is identical

stoichiometry (4.4)   the study of quantitative relationships between substances involved in chemical reactions

structural isomers (19.4)   isomers in which there is a difference in the way ligands bond to the metal

sublimation (10.6)   vaporization of a solid followed by recondensation of the vapor to the solid state; also used to mean any vaporization of a solid

subshell (6.4)   an energy level containing a set of electrons that have the same principal quantum number and the same azimuthal quantum number

substance (1.1)   a kind of matter that has constant composition

supercooling (10.6)   cooling a liquid to a temperature below the freezing point without causing the formation of a solid

supersaturated solution (11.1)   an unstable solution that contains more dissolved solute than the maximum dictated by the solubility

surface tension (10.2)   the work necessary to expand the surface of a liquid; the property of a liquid surface that causes it to behave like a stretched membrane

surroundings (5.2)   all of the universe that is not considered to be part of a particular system

suspension (11.8)   a mixture containing particles larger than 200 nm; the mixture will undergo separation under the influence of gravity

system (5.2)   that part of the universe in which we are interested at a particular time

temperature (1.3)   the degree of hotness or coldness of an object, measured on a definite scale

termolecular reaction (13.4)   an elementary reaction involving three reactant molecules

tertiary structure (24.4)   the structure shown by helical chains of proteins; the folding of the cylinder defined by alpha-helical chains

tetrahedral holes (10.4)   holes in a crystal lattice that are surrounded by four lattice points in a tetrahedral arrangement

theoretical yield (4.4)   the maximum amount of product that can be produced in a chemical reaction from known amounts of reactants

theory (1.2)   an explanation of sets of observations, hypotheses, or laws that is applicable in a relatively wide variety of circumstances

thermite reaction (p. 118)   a reaction between iron oxide and aluminum powder that produces molten iron and aluminum oxide

thermochemical equation (5.4)   a balanced chemical equation that includes the value of the enthalpy change for the reaction

thermochemistry (5.4)   the measurement and manipulation of ΔH values associated with chemical processes

thermodynamics (p. 494)   study of energy changes in chemical systems

thermoplastic (24.2)   polymer that becomes soft and can be re-formed when heated

thermosetting polymer (24.2)   polymer that becomes hard when heated and that does not soften when reheated after it has set

third law of thermodynamics (12.4)   law stating that the absolute entropy of a perfect crystal of a pure substance at 0 K is zero

torr (9.2)   a unit of pressure equal to 1 mm Hg

transfer RNA (24.5)   the form of RNA that brings the appropriate amino acid to the site of protein synthesis

transition element (3.1)   a member of one of the B groups of elements in the periodic table

transition state (13.3)   the state of maximum energy corresponding to the arrangement (activated complex) through which molecules must pass on their way from reactants to products

transition-state theory (13.3)   a theory that explains the temperature dependence of rate constants on the basis of the formation of an activated complex, or transition state, to overcome an energy barrier

triple bond (7.3)   a covalent bond that involves the sharing of three pairs of electrons

triple point (10.6)   conditions of temperature and pressure at which three different phases coexist in a state of equilibrium

triprotic acid (15.1)   an acid that can donate three protons

Tyndall effect (11.8)   the scattering of light rays by colloidal particles

unimolecular reaction (13.4)   an elementary reaction involving a single reactant molecule

unit cell (10.4)   the smallest unit of a crystal lattice, which generates that lattice if repeated indefinitely in all three dimensions

unsaturated hydrocarbon (23.2)   hydrocarbon that contains carbon–carbon double or triple bonds

unsaturated solution (11.1)   solution that contains less than the maximum amount of solute possible in a stable system

valence band (10.5)   in the band theory of bonding, a filled energy band in which electrons are localized on the atoms or used in forming covalent bonds

valence bond theory (8.3)   a theory that explains the formation of covalent bonds in terms of the overlap of half-filled atomic orbitals in the valence shells of the bonded atoms

valence electrons (6.5)   the electrons in the highest or outermost principal energy level or shell

valence shell (6.5)   the outermost principal energy level or shell

valence-shell electron-pair repulsion (VSEPR) theory (8.1)   a theory stating that the shapes of molecules result from the tendency of electron pairs to minimize repulsions by maximizing the distance between pairs

van der Waals equation (9.8)   a correction to the ideal gas law that provides for nonideality by including terms for molecular volume and intermolecular forces; $(P + an^2/V^2)(V - nb) = nRT$

van't Hoff factor (11.3)   the average number of particles produced by one solute formula unit when it dissolves

vapor pressure (10.6)   the partial pressure of gas molecules above a liquid (or solid) when the two states are in equilibrium

viscosity (10.2)   resistance to flow

voltaic cell (18.1)   an electrochemical cell in which an oxidation–reduction reaction causes electrons to flow through an external circuit; also known as a galvanic cell

volume (1.3)   the space occupied by a sample of matter

wave mechanics (6.4)   the calculation of the allowed energies of atoms by solution of equations that describe the wave behavior of electrons in atoms

wavelength (6.1)   the distance between two corresponding points (such as the maxima or minima) on a wave

weight (1.3)   the force exerted on an object by gravity

work (5.1)   the result of a force acting over a distance

zone refining (21.4)   a purification process for crystals that involves melting small zones within the crystal

Index

Entropy (*S*), 452, 493, 495
  absolute, 503
  change, 500–502
    phase changes, 504, 507
    second law of thermodynamics, 502–503
    spontaneous processes, 508
    temperature and, 503–504
  environmental effects, 519–520
  standard, 505–507
  third law of thermodynamics and, 503–504
Environmental Protection Agency (EPA), 466
Enzymes
  active sites, 1019
  catalytic properties, 568
Epoxy, 311
Equations, *see* Chemical equations
Equatorial position, 306
Equilibrium
  Brønsted-Lowry acids and bases, 630
  catalyst effects, 602
  complex ion concentrations, 702–794
  concentration change effects, 595–597
  constant, *see* Equilibrium constant
  cooling curves, 428
  dissolved and undissolved solutes, 442
  expression, heterogeneous equilibria, 586–587
  free energy changes and, 513–518
  industrial gas-phase processes, 613–619
  Le Chatelier's principle, 594
  liquid-gas phase changes, 424
  phase changes, 428
  pressure change effects, 594, 597–600
  rate constants and, 579–582
  respiration and, 603
  temperature change effects, 600–602
  weak acid calculations, 650–656
Equilibrium constant, 516–518, 581
  acid-ionization constants, 647
  calculations with, 604–613
    addition at equilibria, 612–613
    ignoring small quantities, 608–609
    known initial concentrations, 605–611
    perfect squares, 607–608
    quadratic equations, 610–611
    unknown pressure or concentration, 604–605
  concentration units (K_c), 581
  determination of, 588–589
    partial pressures, 584–585
  free energy change and, 591–594
  hydrolysis reactions, 685
  K_p and K_c relationship, 590–591
  law of mass action, 582–583
  manipulation of, 587–589
  multiplication of, 589, 661
  partial pressure (K_p), 584–585, 590–591
  reaction spontaneity and, 593–594, 787
  solubility product constant, 705–706

thermodynamic analysis, 601
  units of, 585
  water self-ionization, 639
  weak base solutions, 659
Equivalence point, 153, 663
Equivalent, 449
Equivalent weight, 794
Erbium, 960, 962
Esters, 987
  chemical reactions, 994
  functional group, 324
  triglycerides, 1032
Ethane, 322, 322n, 970, 972
  petroleum fraction, 980
  reaction with hydrogen, 564
Ethanol (ethyl alcohol), 325, 325n, 983
  dehydration, 670
  freezing-point and boiling-point constants, 473
  polarity and miscibility, 453–454
  reaction with chromium(VI) oxide, 508
Ethanolamine, 615n, 659
Ethene, 323, 323n, 970. *See also* Ethylene
Ethers, 984, 985
Ethyl alcohol, *see* Ethanol
Ethylamine, 659
Ethylene, 11, 970, 976
  hybrid orbitals, 318–319
  platinum complex (Zeise's salt), 831
  polymerization, 1005–1007
  production and uses, 111, 112
Ethylenediamine, 818, 818n, 819, 822
Ethylenediaminetetraacetic acid (EDTA), 818, 820
Ethylene dichloride, 111
Ethylene glycol, 453n, 472, 983
  polarity and miscibility, 453–454
  terephthalic acid copolymerization, 1012
Ethylene-propylene rubber, 1010
Ethyne, 970. *See also* Acetylene
Europium, 960, 961, 962
Evaporation, 401, 420, 422–423, 469. *See also* Vaporization
  condensation equilibrium, 424
  flash evaporation, 481
  intermolecular forces and, 422
  molar heat of vaporization, 185
Excited state electrons, 225, 315
Exothermic reactions, 174
  explosives, 501
  light emissions, 206
  solubility, 455–456
  spontaneous reactions, 493, 495
  temperature vs. equilibrium constant, 601
Experimentation, 12
Explosives, 501
Extrinsic semiconductors, 419

Face-centered cubic structure, 404–405
Fahrenheit scale, 18
Fajans, Kazimir, 267n

Families of elements, 81
Faraday, Michael, 783n, 800n, 1001
Faraday constant, 783, 795
Faraday's law, 794
Fats, 896, 1032
Fatty acids, 1032
  soaps, 468
Feedstock, 194
Fermium, 960
Ferric ion, 106
Ferricyanide ion, 813
Ferromagnetism, 941, 958. *See also* Magnetism
Ferrous ion, 106
Fireworks, 223, 501
First coordination sphere, 815–816
  geometry of, 816–818
First law of thermodynamics, 496
  enthalpy changes and, 499–500
First-order kinetics, 538
First-order rate law, 541–543
First-order reaction, 538
  half-life, 545
Fischer-Tropsch synthesis, 196–197
Fission, 172, 175, 860–861, 873–877
  energy of, 872–873
  products, 877
  reactors, 875–877
Flash distillation, 480
Flash evaporation, 481
Fluids, 398. *See also* Liquid(s)
Fluoride, 107
  in toothpaste, 108
Fluorine, 53, 85, 253n, 735, 930
  electronic configuration, 227
    bonding, 262–263
  molecular structure, 274
  oxidation number, 95
  oxides of, 89–90, 932
  oxidizing properties, 735
  reactions, 254
    with boron, 905
    with sodium, 262
    with uranium, 367
  related elements, 79. *See also* Halogens
  storage, 930n
Fluoristan, 66
Fluorite, 411, 411n
Fluoroapatite, 922n
Fluoro ligand, 819
Fluorospar, 411n, 930, 932n
Fluosilicic acid, 914
Foam, 486
f-orbitals, 220–221
  lanthanide contraction, 942–943
Force, SI unit for, 344
Formal charge, 279–280
Formaldehyde, 617n, 985
  phenol copolymerization, 1013
Formalin, 985
Formation, heat of, *see* Heat of formation

Credits

All photographs by Sean Brady unless otherwise noted.

Figure 1.1 Jim Scherer **Figure 1.2 (left)** Dr. Jeremy Burgess/Science Photo Library/Photo Researchers **Figure 1.2 (right)** Science Photo Library/Photo Researchers **Figure 1.8** Nelson Morris/Photo Researchers **Figure 1.9** © Jennifer Waddell **Figure 1.14** Courtesy of NASA **Figure 2.1** Chemical Heritage Foundation **Figure 2.2** David, *Antoine Laurent Lavoisier and His Wife.* The Metropolitan Museum of Art, Purchase, Mr. and Mrs. Charles Wrightsman Gift, in honor of Everett Fahy, 1977. (1977.10). **Figure 2.6** Bruce Scharft Ph.D., courtesy of Digital Instruments **Figure 2.8** IBM Corporation, Research Division, Almaden Research Center **Figure 2.14 (left)** Tripos Associates, Inc. **Figure 2.14 (right)** Photo courtesy of Ward's Natural Science Establishment, Inc. **Figure 2.15** Tripos Associates, Inc. **Figure 2.17** Courtesy of NASA **Figure 2.22 (right)** Photo courtesy of Hewlett-Packard Company **Figure 3.3** Chemical Heritage Foundation **Figure 3.4** Chemical Heritage Foundation **Figure 3.5** Chemical Heritage Foundation **Figure 3.13** Cameramann International **Figure 3.22** Courtesy of NASA **Figure 3.28** James Scherer **Figure 4.1** Tripos Associates, Inc. **Figure 4.3** James Scherer **Figure 4.6** James Scherer **Figure 4.9** James Scherer **Figure 4.11 (right)** Courtesy of Brian Deedy, Beth Israel Hospital Radiologic Foundation **Figure 4.17** Lee Snyder/ Photo Researchers **Figure 4.18** Portland Cement Association **Figure 4.19** James Scherer **Figure 4.23 (right)** © Janice Rubin/Black Star **Figure 5.5 (left)** Alabama Power Company **Figure 6.8 (left)** © Tony Clark/The Image Works **Figure 6.19** © Bob Daemmrich/The Image Works **Figure 6.24 (right)** © Alan Carey/The Image Works **Figure 6.28** Argonne National Laboratory **Figure 6.29** Goodyear Tire and Rubber Company **Figure 6.30** © Michal Grecco/Stock Boston **Figure 7.1 (right)** Akzo Salt, Inc. **Figure 7.4** Tripos Associates, Inc. **Figure 7.16** James Scherer **Figure 7.17** Argonne National Laboratory **Figure 8.1** Tripos Associates, Inc. **Figure 8.10** Tom Walker/Stock Boston **Figure 8.23** James Scherer **Figure 8.24** Tripos Associates, Inc. **Figure 8.28** © Yoav Levy/Phototake **Chapter 9 Opener** Courtesy of NASA **Figure 9.7** James Scherer **Figure 9.14** Department of Energy **Figure 9.24** Johnson Matthey Plc **Figure 9.26** James Scherer **Figure 9.28** Photo courtesy of Ward's Natural Science Establishment, Inc. **Figure 9.29** © Fletcher & Baylis/Photo Researchers **Chapter 10 Opener** Sven-Olof Lindblad/Photo Researchers **Figure 10.7** © D. Wells/The Image Works **Figure 10.10 (right)** R.L. Armstrong. WDC/NSIDC **Figure 10.11** Photo courtesy of Ward's Natural Science Establishment, Inc. **Figure 10.17** Thomas L. Groy, X-ray Diffraction Facility, Arizona State University **Figure 10.25** Tripos Associates, Inc. **Figure 10.30** Tripos Associates, Inc. **Figure 10.35** A. Saupe, Liquid Crystal Institute **Figure 10.36** Photo provided courtesy of Hallcrest Products, Inc., Glenview, IL, USA **Figure 11.8** James Scherer **Figure 11.10** James Scherer **Figure 11.21** © David Phillips/Science Source/Photo Researchers **Figure 11.23 (bottom)** Weir Westgarth **Figure 11.27** R. Weldon/Gemological Institute of America **Figure 12.3** D. & J. Heaton/Stock Boston **Figure 12.5** Jim Zipp/Photo Researchers **Figure 12.10** Pacific Gas & Electric **Figure 12.12 (left)** © W. Hill/The Image Works **Figure 13.21** Kristen Brochmann/Fundamental Photographs, New York **Figure 13.25** Courtesy GMC, AC Rochester Division **Figure 13.26** Tripos Associates, Inc. **Figure 14.8** James Scherer **Figure 14.9** Tripos Associates, Inc. **Figure 14.12 (left)** Charlton Photos **Figure 14.12 (right)** Grant Heilman/Grant Heilman Photography **Figure 15.13** © Dr. E. R. Degginger **Figure 15.14** Richard Megna/Fundamental Photographs, New York **Figure 15.16** Courtesy Atotech USA Inc. **Figure 16.9** Tripos Associates, Inc. **Figure 16.12** © Dr. E. R. Degginger **Figure 17.7** Carol Palmer **Figure 17.14** Philip M. Hocker/Mineral Policy Center **Figure 18.9** Diana Gongora/Fundamental Photographs, New York **Figure 18.10** Tom & Pat Leeson/Photo Researchers **Figure 18.20** Courtesy Kennecott Corporation **Figure 19.5** Tripos Associates, Inc. **Figure 19.6** Tripos Associates, Inc. **Figure 19.7** Tripos Associates, Inc. **Figure 19.8** Tripos Associates, Inc. **Figure 19.11** Tripos Associates, Inc. **Figure 19.23** R. Weldon/Gemological Institute of America **Chapter 20 Opener** EME Corporation **Figure 20.6** Courtesy NITON **Figure 20.7** Courtesy of Bicron (Solon, Ohio) **Figure 20.13** Earl Roberge/Photo Researchers **Figure 20.18** Fermilab Visual Media Services **Figure 20.19** © CNRI/Phototake NYC **Figure 20.20** Image courtesy of Dr. William J. Powers, Washington University School of Medicine, St. Louis, Missouri **Figure 20.21** Courtesy Nordion International Inc. **Figure 21.3** James Scherer **Figure 21.5** Smithsonian Institution Photo No. 83-580. **Figure 21.6** Chemical Heritage Foundation **Figure 21.7 (right)** Sol Mednick **Figure 21.9** Tripos Associates, Inc. **Figure 21.10** James Scherer **Figure 21.12** Tripos Associates, Inc. **Figure 21.16** © Dr. E. R. Degginger **Figure 21.17** Paul Silverman/Fundamental Photographs, New York **Figure 21.20** © Dr. E. R. Degginger **Figure 21.25** The Corning Museum of Glass, Corning NY **Figure 22.8** James Scherer **Figure 23.4** James Scherer **Figure 23.7** Photo courtesy of Amoco Corporation **Figure 23.10** James Scherer **Figure 24.1** Tripos Associates, Inc. **Figure 24.8** James Scherer **Figure 24.27** James Holmes/Cellmark Diagnostics/Science Photo Library/Photo Researchers

Table of Atomic Numbers and Atomic Weights

Name	Symbol	Atomic number	Atomic weight	Name	Symbol	Atomic number	Atomic weight
Actinium	Ac	89	(227)	Neptunium	Np	93	(237)
Aluminum	Al	13	26.981539	Nickel	Ni	28	58.6934
Americium	Am	95	(243)	Niobium	Nb	41	92.90638
Antimony	Sb	51	121.757	Nitrogen	N	7	14.00674
Argon	Ar	18	39.948	Nobelium	No	102	(259)
Arsenic	As	33	74.92159	Osmium	Os	76	190.2
Astatine	At	85	(210)	Oxygen	O	8	15.9994
Barium	Ba	56	137.327	Palladium	Pd	46	106.42
Berkelium	Bk	97	(247)	Phosphorus	P	15	30.973762
Beryllium	Be	4	9.012182	Platinum	Pt	78	195.08
Bismuth	Bi	83	208.98037	Plutonium	Pu	94	(244)
Boron	B	5	10.811	Polonium	Po	84	(209)
Bromine	Br	35	79.904	Potassium	K	19	39.0983
Cadmium	Cd	48	112.411	Praseodymium	Pr	59	140.90765
Calcium	Ca	20	40.078	Promethium	Pm	61	(145)
Californium	Cf	98	(251)	Protactinium	Pa	91	(231)
Carbon	C	6	12.011	Radium	Ra	88	(226)
Cerium	Ce	58	140.115	Radon	Rn	86	(222)
Cesium	Cs	55	132.90543	Rhenium	Re	75	186.207
Chlorine	Cl	17	35.4527	Rhodium	Rh	45	102.90550
Chromium	Cr	24	51.9961	Rubidium	Rb	37	85.4678
Cobalt	Co	27	58.93320	Ruthenium	Ru	44	101.07
Copper	Cu	29	63.546	Samarium	Sm	62	150.36
Curium	Cm	96	(247)	Scandium	Sc	21	44.955910
Dysprosium	Dy	66	162.50	Selenium	Se	34	78.96
Einsteinium	Es	99	(252)	Silicon	Si	14	28.0855
Erbium	Er	68	167.26	Silver	Ag	47	107.8682
Europium	Eu	63	151.965	Sodium	Na	11	22.989768
Fermium	Fm	100	(257)	Strontium	Sr	38	87.62
Fluorine	F	9	18.9984032	Sulfur	S	16	32.064
Francium	Fr	87	(223)	Tantalum	Ta	73	180.9479
Gadolinium	Gd	64	157.25	Technetium	Tc	43	(98)
Gallium	Ga	31	69.723	Tellurium	Te	52	127.60
Germanium	Ge	32	72.61	Terbium	Tb	65	158.92534
Gold	Au	79	196.96654	Thallium	Tl	81	204.3833
Hafnium	Hf	72	178.49	Thorium	Th	90	232.0381
Helium	He	2	4.002602	Thulium	Tm	69	168.93421
Holmium	Ho	67	164.93032	Tin	Sn	50	118.710
Hydrogen	H	1	1.00794	Titanium	Ti	22	47.88
Indium	In	49	114.82	Tungsten	W	74	183.85
Iodine	I	53	126.90447	Unnilennium	Une	109	(267)
Iridium	Ir	77	192.22	Unnilhexium	Unh	106	(263)
Iron	Fe	26	55.847	Unniloctium	Uno	108	(265)
Krypton	Kr	36	83.80	Unnilpentium	Unp	105	(262)
Lanthanum	La	57	138.9055	Unnilquadium	Unq	104	(261)
Lawrencium	Lr	103	(262)	Unnilseptium	Uns	107	(262)
Lead	Pb	82	207.2	Uranium	U	92	238.0289
Lithium	Li	3	6.941	Vanadium	V	23	50.9415
Lutetium	Lu	71	174.967	Xenon	Xe	54	131.29
Magnesium	Mg	12	24.3050	Ytterbium	Yb	70	173.04
Manganese	Mn	25	54.93805	Yttrium	Y	39	88.90585
Mendelevium	Md	101	(258)	Zinc	Zn	30	65.39
Mercury	Hg	80	200.59	Zirconium	Zr	40	91.224
Molybdenum	Mo	42	95.94				
Neodymium	Nd	60	144.24				
Neon	Ne	10	20.1797				

Atomic weights in this table are from the IUPAC report "Atomic Weights of the Elements 1989," *Pure and Applied Chemistry*, Vol. 63, No. 7 (1991), pp. 975–1002. (©1991 IUPAC.)

A value in parentheses is the mass number of the isotope of longest half-life.